Pathogens and Toxins in Foods

CHALLENGES AND INTERVENTIONS

Pathogens and Toxins in Foods

CHALLENGES AND INTERVENTIONS

EDITORS

VIJAY K. JUNEJA

Microbial Food Safety Research Unit, Eastern Regional Research Center,
Agricultural Research Service, U.S. Department of Agriculture,
Wyndmoor, Pennsylvania

JOHN N. SOFOS

Center for Meat Safety & Quality, Food Safety Cluster of Infectious Diseases Supercluster,
Department of Animal Sciences, Colorado State University,
Fort Collins, Colorado

ASM PRESS *Washington, DC*

Address editorial correspondence to ASM Press, 1752 N St. NW, Washington, DC 20036-2904, USA

Send orders to ASM Press, P.O. Box 605, Herndon, VA 20172, USA
Phone: 800-546-2416; 703-661-1593
Fax: 703-661-1501
E-mail: books@asmusa.org
Online: estore.asm.org

Library of Congress Cataloging-in-Publication Data

Pathogens and toxins in foods : challenges and interventions / editors, Vijay K. Juneja, John N. Sofos.
      p. ; cm.
   Includes bibliographical references and index.
   ISBN 978-1-55581-459-5 (hardcover)
   1. Food—Microbiology.   2. Food—Toxicology.   I. Juneja, Vijay K., 1956-   II. Sofos, John Nikolaos.
   [DNLM: 1. Food Microbiology.   2. Food Contamination—prevention & control.
   3. Food Handling. QW 85 P297 2009]
   QR115.P383 2009
   664.001′579—dc22

                                                                                      2009017076

Current printing (last digit)
10 9 8 7 6 5 4 3 2 1

Cover illustration: Fluorescent microscopic image of green fluorescent protein
(GFP)-expressing *Escherichia coli* O157:H7 cells. Image has been converted to red.
(Provided by John Sofos and his collaborators.)

CONTENTS

Contributors • vii

Preface • xi

1. *Bacillus cereus* and Other *Bacillus* spp. • 1
Mansel W. Griffiths

2. *Campylobacter jejuni* and Other Campylobacters • 20
Nelson A. Cox, L. Jason Richardson, and Michael T. Musgrove

3. *Clostridium botulinum* • 31
Michael W. Peck

4. *Clostridium perfringens* • 53
Vijay K. Juneja, John S. Novak, and Ronald J. Labbe

5. Diarrheagenic *Escherichia coli* • 71
Catherine S. Beauchamp and John N. Sofos

6. *Listeria monocytogenes* • 95
Anna C. S. Porto-Fett, Jeffrey E. Call, Peter M. Muriana, Timothy A. Freier, and John B. Luchansky

7. *Salmonella* • 108
Stan Bailey, L. Jason Richardson, Nelson A. Cox, and Douglas E. Cosby

8. Staphylococcal Food Poisoning • 119
Keun Seok Seo and Gregory A. Bohach

9. *Shigella* • 131
Keith A. Lampel

10. Pathogenic Vibrios in Seafood • 146
Anita C. Wright and Keith R. Schneider

11. *Yersinia enterocolitica* and *Yersinia pseudotuberculosis* • 164
Maria Fredriksson-Ahomaa, Miia Lindström, and Hannu Korkeala

12. Other Bacterial Pathogens: *Aeromonas, Arcobacter, Helicobacter, Mycobacterium, Plesiomonas,* and *Streptococcus* • 181
Elaine M. D'Sa and Mark A. Harrison

13. Food-Borne Parasites • 195
Dolores E. Hill and J. P. Dubey

14. Human Pathogenic Viruses in Food • 218
Lee-Ann Jaykus and Blanca Escudero-Abarca

15. Seafood Toxins • 233
Sherwood Hall

16. Biogenic Amines in Foods • 248
K. Koutsoumanis, C. Tassou, and G.-J. E. Nychas

17. Fungal and Mushroom Toxins • 275
Charlene Wolf-Hall

18. Critical Evaluation of Uncertainties of Gluten Testing: Issues and Solutions for Food Allergen Detection • 286
Carmen Diaz-Amigo and Jupiter M. Yeung

19. Naturally Occurring Toxins in Plants • 301
Andrea R. Ottesen and Bernadene A. Magnuson

20.   Chemical Residues: Incidence in the
United States   •   314
Stanley E. Katz and Paula Marie L. Ward

21.   A European Food Safety Perspective
on Residues of Veterinary Drugs and
Growth-Promoting Agents   •   326
Martin Danaher and Deirdre M. Prendergast

22.   Prions and Prion Diseases   •   343
Dragan Momcilovic

23.   Interventions for Hazard Control in Foods
Preharvest   •   357
*Jarret D. Stopforth, Balasubrahmanyam Kottapalli,
and John N. Sofos*

24.   Interventions for Hazard Control in Foods
during Harvesting   •   379
*Mayra Márquez-González, Kerri B. Harris,
and Alejandro Castillo*

25.   Interventions for Hazard Control during
Food Processing   •   396
Ifigenia Geornaras and John N. Sofos

26.   Interventions for Hazard Control
in Retail-Handled Ready-To-Eat
Foods   •   411
Alexandra Lianou and John N. Sofos

27.   Interventions for Hazard Control
at Food Service   •   436
O. Peter Snyder, Jr.

28.   Recent Developments in Rapid
Detection Methods   •   450
*Lawrence D. Goodridge and Mansel W.
Griffiths*

29.   Molecular Subtyping and Tracking of
Food-Borne Bacterial Pathogens   •   460
*Brandon A. Carlson and Kendra K.
Nightingale*

30.   Food Safety Management
Systems   •   478
Virginia N. Scott and Yuhuan Chen

Index   •   493

CONTRIBUTORS

Stan Bailey
Biomerieux, Inc., Hazelwood, MI 63042

Catherine S. Beauchamp
Colorado Premium Foods, 2035 2nd Ave., Greeley, CO 80631

Gregory A. Bohach
Department of Microbiology, Molecular Biology, and Biochemistry, University of Idaho, Moscow, ID 83844

Jeffrey E. Call
Microbial Food Safety Research Unit, U.S. Department of Agriculture, Agricultural Research Service, Eastern Regional Research Center, Wyndmoor, PA 19038

Brandon A. Carlson
Department of Animal Sciences, Colorado State University, Fort Collins, CO 80523

Alejandro Castillo
Department of Animal Science, Texas A&M University, College Station, TX 77843-2471

Yuhuan Chen
Grocery Manufacturers Association, 1350 I St. N.W., Suite 300, Washington, DC 20005

Douglas E. Cosby
USDA-ARS-Bacterial Epidemiology and Antimicrobial Resistant Research Unit, Russell Research Center, Athens, GA 30605

Nelson A. Cox
USDA-ARS-Poultry Microbiological Safety Research Unit, Russell Research Center, Athens, GA 30605

Martin Danaher
Food Safety Department, Teagasc, Ashtown Food Research Centre, Ashtown, Dublin 15, Ireland

Carmen Diaz-Amigo
Center for Food Safety and Applied Nutrition, U.S. Food and Drug Administration, Laurel, MD 20708

Elaine M. D'Sa
Department of Foods and Nutrition, The University of Georgia, Athens, GA 30602

J. P. Dubey
U.S. Department of Agriculture, Agricultural Research Service, Animal and Natural Resources Institute, Animal Parasitic Diseases Laboratory, Building 1001, BARC-East, Beltsville, MD 20705-2350

Blanca Escudero-Abarca
Department of Food, Bioprocessing and Nutrition Sciences, North Carolina State University, Raleigh, NC 27695-7624

Maria Fredriksson-Ahomaa
Institute of Hygiene and Technologie of Food of Animal Origin, Ludwig-Maximilians University Munich, Schoenleutnerstrasse 8, D-85764 Oberschleissheim, Germany

Timothy A. Freier
Corporate Food Safety and Regulatory Affairs, Cargill, Inc., Minneapolis, MN 55440-9300

Ifigenia Geornaras
Center for Meat Safety & Quality, Food Safety Cluster of Infectious Diseases Supercluster, Department of Animal Sciences, Colorado State University, Fort Collins, CO 80523-1171

Lawrence D. Goodridge
Department of Animal Sciences, Colorado State University, Fort Collins, CO 80523-1171

Mansel W. Griffiths
Canadian Research Institute for Food Safety and Department of Food Science, University of Guelph, Guelph, Ontario, Canada N1G 2W1

Sherwood Hall
Chemical Contaminants Branch HFS-716, Division of Bioanalytical Chemistry, Office of Regulatory Science, Center for Food Safety and Applied Nutrition, U.S. FDA, College Park, MD 20740

Kerri B. Harris
Department of Animal Science, Texas A&M University, College Station, TX 77843-2471

Mark A. Harrison
Department of Food Science and Technology, The University of Georgia, Athens, GA 30602

Dolores E. Hill
U.S. Department of Agriculture, Agricultural Research Service, Animal and Natural Resources Institute, Animal Parasitic Diseases Laboratory, Building 1044, BARC-East, Beltsville, MD 20705-2350

Lee-Ann Jaykus
Department of Food, Bioprocessing and Nutrition Sciences, North Carolina State University, Raleigh, NC 27695-7624

Vijay K. Juneja
Microbial Food Safety Research Unit, Eastern Regional Research Center, Agricultural Research Service, U.S. Department of Agriculture, 600 E. Mermaid Lane, Wyndmoor, PA 19038

Stanley E. Katz
Department of Biochemistry and Microbiology, Rutgers, The State University of New Jersey, School of Environmental & Biological Sciences, 76 Lipman Drive, New Brunswick, NJ 08901-8525

Hannu Korkeala
Department of Food and Environmental Hygiene, University of Helsinki, P.O. Box 66, 00014 Helsinki University, Finland

Balasubrahmanyam Kottapalli
Kraft Foods, 200 Deforest Ave., East Hanover, NJ 07936

K. Koutsoumanis
Faculty of Agriculture, Department of Food Science and Technology, Laboratory of Food Microbiology and Hygiene, Aristotle University of Thessaloniki, P.O. Box 265, 54124, Thessaloniki, Greece

Ronald J. Labbe
Department of Food Science, University of Massachusetts, Amherst, MA 01003-1021

Keith A. Lampel
Food and Drug Administration, HFS-710, 5100 Paint Branch Parkway, College Park, MD 20740

Alexandra Lianou
Laboratory of Food Microbiology and Hygiene, Department of Food Science and Technology, Faculty of Agriculture, Aristotle University of Thessaloniki, Thessaloniki 54124, Greece

Miia Lindström
Department of Food and Environmental Hygiene, University of Helsinki, P.O. Box 66, 00014 Helsinki University, Finland

John B. Luchansky
Microbial Food Safety Research Unit, U.S. Department of Agriculture, Agricultural Research Service, Eastern Regional Research Center, Wyndmoor, PA 19038

Bernadene A. Magnuson
Cantox Health Sciences International, 2233 Argentia Road, Suite 308, Mississauga, Ontario, Canada L5N 2X7

Mayra Márquez-González
Department of Animal Science, Texas A&M University, College Station, TX 77843-2471

Dragan Momcilovic
Center for Veterinary Medicine, Food and Drug Administration, Rockville, MD 20855

Peter M. Muriana
Department of Animal Science, Food and Agricultural Products Research and Technology Center, Oklahoma State University, Stillwater, OK 74078

Michael T. Musgrove
Egg Safety and Quality Research Unit, Russell Research Center, Athens, GA 30605

Kendra K. Nightingale
Department of Animal Sciences, Colorado State University, Fort Collins, CO 80523

John S. Novak
American Air Liquide, Delaware Research & Technology Center, 200 GBC Drive, Newark, DE 19702-2462

G.-J. E. Nychas
Agricultural University of Athens, Department of Food Science and Technology, Laboratory of Microbiology & Biotechnology of Foods, Iera Odos 75, Athens 11855, Hellas

Andrea R. Ottesen
Plant Sciences and Landscape Architecture, University of Maryland, College Park, MD 20742

Michael W. Peck
Institute of Food Research, Norwich Research Park, Colney, Norwich NR4 7UA, United Kingdom

Anna C. S. Porto-Fett
Microbial Food Safety Research Unit, U.S. Department of Agriculture, Agricultural Research Service, Eastern Regional Research Center, Wyndmoor, PA 19038

Deirdre M. Prendergast
Food Safety Department, Teagasc, Ashtown Food Research Centre, Ashtown, Dublin 15, Ireland

L. Jason Richardson
USDA-ARS-Poultry Microbiological Safety Research Unit, Russell Research Center, Athens, GA 30605

Keith R. Schneider
University of Florida, Food Science and Human Nutrition Department, 359 FSHN Bldg., Newell Drive, Gainesville, FL 32611

Virginia N. Scott
Grocery Manufacturers Association, 1350 I St. N.W., Suite 300, Washington, DC 20005

Keun Seok Seo
Department of Microbiology, Molecular Biology, and Biochemistry, University of Idaho, Moscow, ID 83844

O. Peter Snyder, Jr.
Hospitality Institute of Technology and Management, 670 Transfer Road, Suite 21A, St. Paul, MN 55114

John N. Sofos
Center for Meat Safety & Quality, Food Safety Cluster of Infectious Diseases Supercluster, Department of Animal Sciences, Colorado State University, 1171 Campus Delivery, Fort Collins, CO 80523-1171

Jarret D. Stopforth
PURAC, Arkelsedijk 46, 4206 AC, Gorinchem, The Netherlands

C. Tassou
National Agricultural Research Foundation, Institute of Technology of Agricultural Products, S. Venizelou 1, Lycovrisi 14123, Athens, Hellas

Paula Marie L. Ward
Department of Biochemistry and Microbiology, Rutgers, The State University of New Jersey, School of Environmental & Biological Sciences, 76 Lipman Drive, New Brunswick, NJ 08901-8525

Charlene Wolf-Hall
Great Plains Institute of Food Safety, 1523 Centennial Blvd., 114A Van Es Hall, North Dakota State University, Fargo, ND 58105

Anita C. Wright
University of Florida, Food Science and Human Nutrition Department, 359 FSHN Bldg., Newell Drive, Gainesville, FL 32611

Jupiter M. Yeung
Grocery Manufacturers Association, Washington, DC 20005

PREFACE

Our understanding of the contamination of food matrices with human health hazards, including microbial pathogens and various toxic agents, has increased significantly in the last 2 decades. Microorganisms previously unknown or not considered to be causes of food-borne illnesses, as well as the reasons for their occurrence in foods, are continually being recognized and linked to documented outbreaks of illnesses. Foods previously thought not to be involved in food-borne illnesses or believed to be infrequent sources of specific food-borne hazards have been associated with major outbreaks or sporadic episodes of sometimes fatal illnesses. The complexity of the preharvest, harvest, and postharvest environments makes it challenging to trace or control all potential sources of microbial contamination in foods. These food safety concerns are magnified because of increased international travel and trade, globalization of the food system, and consumer preferences for natural, minimally processed foods that are prepared with minimum preservatives, that are safe and easily available, and that offer convenience and require minimum preparation before consumption. Hence, there has been a concerted effort by the industry and researchers to continually make advances in food processing and preservation and apply antimicrobial interventions at all stages of the food chain in order to control pathogens and toxic agents in foods and ensure their safety.

Applicable and effective preharvest controls to minimize microbial contamination sources of food products of animal and plant origin, and postharvest intervention technologies, in conjunction with the tracking of food-borne pathogens with appropriate and effective diagnostic methods, can assist in developing and implementing effective hazard analysis critical control point (HACCP) plans, with the primary objective of delivering safe foods to consumers. These developments and interest by regulatory and public health agencies, as well as our involvement through the years in food safety research, led us to the conclusion that a comprehensive book that provides the reader with the latest research advances and insights into the safety of foods is timely. Accordingly, this compilation, written by select experts who represent the best in the field of food safety, covers current, definitive, and factual material on topics of practical worldwide importance, such as food-borne hazard issues, concerns, and challenges, as well as food processing operations and interventions to control hazards. The emphasis of each chapter is on the type of illness and characteristics of the hazard, sources, and incidence of the hazard in the environment and foods; intrinsic and extrinsic factors that affect survival and growth in food products and contribute to illness; food processing operations that influence the level, spread, or characteristics of the hazard; recent advances in biological, chemical, and physical interventions to guard against food-borne hazards; and discriminative detection methods for confirmation and trace-back of contaminated products. Authors explore different intervention approaches to killing, removing, or reducing pathogens; controlling toxins; and offering quality, nutritious, safe, low-cost food products to consumers. Accordingly, recent developments in intervention strategies for control of food-borne hazards preharvest; during harvesting, food processing, and retail handling of ready-to-eat foods; and at food service have been addressed.

It is necessary for the food industry and regulatory agencies to have personnel who are knowledgeable about methodologies applied in the control or inactivation of microorganisms that may be present in foods. This knowledge contributes to the development of regulations and optimization of HACCP. Until now, such information has been presented in a variety of sources which are not always readily available.

Accordingly, this book brings together these latest advances and should be of special benefit to those looking for a resource along with or in place of additional classroom training. This book should be a valuable tool for those who are directly or indirectly involved in the production, handling, processing, distribution, and serving of food; control of hazards and spoilage of food products; inspection of food processing facilities; or research studies on microbial control or inactivation. Those in academic, industrial, and government institutions, including federal, state, private, and local agencies, as well as food consultants and lobbyists, should find the book helpful in their work.

The credit for making this book a reality goes to the authors and coauthors of various chapters. We commend the authors for their endeavor in compiling the necessary information in their chapters. We hope that the excellent work of the contributors will serve as a basis for new and innovative approaches to controlling food-borne hazards and significantly contributing to technologies that decrease the incidence of food-borne illnesses. Furthermore, we look forward to the breakthrough discoveries that will emerge in the future as a result of the information presented in this book.

Vijay K. Juneja
John N. Sofos

Pathogens and Toxins in Foods: Challenges and Interventions
Edited by V. K. Juneja and J. N. Sofos
© 2010 ASM Press, Washington, DC

Chapter 1

Bacillus cereus and Other *Bacillus* spp.

MANSEL W. GRIFFITHS

BACILLUS SPP.

Bacillus spp. are gram-positive, endospore-forming facultatively anaerobic bacteria. The resistance of their spores to adverse conditions has resulted in widespread distribution of the organism. They have been isolated from air, soil, and water, as well as animal and plant material. The ubiquitous nature of these organisms makes contamination of food materials a common occurrence.

In 1980, with the publication of the *Approved Lists of Bacterial Names*, taxonomic experts cooperated in establishing a comprehensive and agreed-upon list of accepted names of bacterial species. At this time, 38 species of aerobic, endospore-forming bacteria were listed, of which 31 were allocated to the genus *Bacillus* (Skerman et al., 1980). Since 1980, the taxonomy of these organisms has been restructured based on the application of better enrichment and isolation procedures that take into account their physiology and nutritional and cultural requirements and on the development of molecular methods, especially 16S rRNA/DNA sequence analysis. The latter has led to the separation of groups of species from the core genus *Bacillus* to form new genera (Fritze, 2004).

The genus *Bacillus* can be split into two groups: the *B. subtilis* group and the *B. cereus* group. Species of the *B. subtilis* group are closely related and include *B. subtilis, B. licheniformis,* and *B. pumilus.* These species are generally mesophilic, with cells less than 1 μm wide, sporangia not swollen, and ellipsoidal spores. All members of the group are placed in 16S rRNA/DNA group 1. Only some of the species in this group can be differentiated based on classical phenotypic tests. However, all species can be differentiated by molecular techniques, and further study may result in the identification of additional genospecies.

A second group of species of major interest is the *B. cereus* group, which is comprised of *B. cereus, B. thuringiensis, B. mycoides,* and *B. anthracis.* A psychrotrophic species, *B. weihenstephanensis,* has also been isolated (Lechner et al., 1998). While the members of this group are easily distinguished from other spore-formers by their inability to produce acid from mannitol and their production of lecithinase, they are difficult to distinguish from each other. Cells of these organisms are wider than 1 μm, sporangia are not swollen, and their spores are ellipsoidal. *B. cereus* and *B. thuringiensis* are usually motile, whereas *B. cereus, B. thuringiensis,* and *B. mycoides* are hemolytic and penicillin resistant. *B. anthracis* is exclusively lysed by the gamma phage. Many of the so-called typical reactions, such as the rhizoid growth of *B. mycoides,* the formation of crystalline parasporal inclusion bodies by *B. thuringiensis,* and the pathogenicity of *B. anthracis* are culture dependent and/or plasmid encoded. There is still significant controversy surrounding the taxonomy of the *B. cereus* group. Helgason et al. (2000) have proposed that *B. anthracis, B. cereus,* and *B. thuringiensis* should be regarded as one species with subspecies, while Jackson et al. (1999) found an extensive genotypic diversity among the species, suggesting the presence of even more species (or subspecies) within this group.

Multiple locus sequence typing of strains of *B. cereus* has revealed three lineages, with the majority of food-related isolates belonging to lineage I (Cardazzo et al., 2008). This work also demonstrated that there is widespread horizontal gene transfer among the strains, and this has contributed significantly to the evolution of toxin-encoding genes.

Mansel W. Griffiths • Canadian Research Institute for Food Safety and Department of Food Science, University of Guelph, Guelph, Ontario, Canada N1G 2W1.

Characteristics of *Bacillus cereus*

Arguably the most important characteristic of *Bacillus* spp. is their ability to form refractile endospores. These spores are more resistant than vegetative cells to heat, drying, food preservatives, and other environmental challenges. The bacteria of the genus *Bacillus* are usually free living, that is, not host adapted, and their spores are widely distributed throughout nature. The spores of *B. cereus* are ellipsoidal and central to subterminal and do not distend the sporangia (Griffiths and Phillips, 1990b).

The spore

Spores from strains commonly associated with food poisoning have a decimal reduction time at 95°C (D_{95}) of approximately 24 minutes, with other strains showing a wider range of heat resistance (D_{95} of 1.5 to 36 minutes) (Meer et al., 1991). The z value varies between about 6°C and 9°C (Byrne et al., 2006). However, there appears to be substantial heterogeneity among spore populations in their response to heat treatment (Cronin and Wilkinson, 2008). Models describing the effects of temperature and pH on thermal inactivation (Fernandez et al., 1999, 2002) and germination and outgrowth (Gaillard et al., 2005) of *B. cereus* spores have been generated. There was little difference in the heat resistances of spores in the presence and absence of nisin, which is known to prevent spore outgrowth, but spores were 10 times more sensitive to high-pressure treatments (517 MPa) when nisin was present (Cruz and Montville, 2008). The spore is also more resistant to irradiation, and the dose for a 90% reduction in count lies between 1.25 and 4 kGy. The corresponding dose for vegetative cells is 0.17 to 0.65 kGy (De Lara et al., 2002).

Spore germination can occur over the temperature range of 5°C to 50°C in cooked rice and between −1°C and 59°C in laboratory media (Griffiths and Schraft, 2002). *Bacillus cereus* spores germinate in response to particular nutrients, such as glycine or a neutral L-amino acid and purine ribosides, which they sense by receptors encoded by the *gerA* family of operons (Hornstra et al., 2005), or in response to physical treatments, such as temperature (Griffiths and Phillips, 1990b) and high pressures of 500 MPa (Black et al., 2007). It has been demonstrated that simulation of passage through the gastrointestinal tract at 37°C resulted in better germination of spores of mesophilic rather than psychrotrophic strains of *B. cereus* (Wijnands et al., 2006b). The ability of spores to survive transit through the intestinal tract also depends on the food matrix and stomach acidity (Clavel et al., 2004). Germination can be inhibited by a variety of compounds, and these can be assigned to either of two broad groups: those that inhibit the action of, or response to, one nutrient receptor on the spore; or those that prevent the action of, or response to, several or all of the spore's nutrient receptors (Cortezzo et al., 2004). Organic acids can be used to control germination and outgrowth of *B. cereus* spores in foods, but only when the storage temperature is ≤8°C (Del Torre et al., 2001).

Their hydrophobic nature, coupled with the presence of appendages on their surfaces, enables spores to adhere to several types of surfaces, including epithelial cells (Faille et al., 2002; Klavenes et al., 2002; Peng et al., 2001; Stalheim and Granum, 2001; Tauveron et al., 2006). However, damage to the exosporium, such as that which would occur during sporulation under unfavorable conditions or by shear stress that may occur during processes such as microfluidization (Feijoo et al., 1997b), results in a decrease in the ability of the spore to adhere to surfaces (Faille et al., 2007).

Vegetative growth

Early in the growth cycle vegetative cells are gram positive, but cells may become gram variable when in late log or stationary phase. Colonies on agar media have a dull or frosted appearance.

The emergence of psychrotrophic strains of the organism, which have been classified as *Bacillus weihenstephanensis*, now means that the temperature range for growth is between approximately 5°C and 50°C (Lechner et al., 1998). Generation times at 7°C lie between 9.4 and 75 hours, but in boiled rice at 30°C the generation times can range between 26 and 57 minutes (Dufrenne et al., 1995; Meer et al., 1991). The organism can produce toxin at low temperatures, but this phenomenon appears to be strain dependent (Fermanian et al., 1997). Several strains can grow slowly at sodium chloride concentrations of 10%, with a minimum water activity (a_w) for growth between 0.91 and 0.93. *B. cereus* has an absolute requirement for amino acids as growth factors, but vitamins are not required. The organism grows over a pH range of approximately 4.4 to 9.3. These limits for growth are not absolute and are dependent on several factors, including strain (Baker and Griffiths, 1993).

A surface structure of *B. cereus* cells, the S-layer, is involved in the adhesion of the organism to host cells and increased radiation resistance (Kotiranta et al., 2000). It has also been shown that the ability of *B. cereus* cells to tolerate bile depends on strain and the food in which the cells are present (Clavel et al., 2007).

CHARACTERISTICS OF *BACILLUS CEREUS* FOOD POISONING

Two distinct types of illness have been attributed to the consumption of foods contaminated with

B. cereus: (i) the diarrheal syndrome, which has an incubation time of 4 to 16 h and is manifested as abdominal pain and diarrhea that usually subsides within 12 to 24 h (the infective dose is reported as 10^5 to 10^7 cells ingested) and (ii) the emetic syndrome, which has an incubation time of 1 to 5 h, causing nausea and vomiting that last for 6 to 24 h. To cause emetic *B. cereus* food poisoning, the food involved will typically contain 10^5 to 10^8 cells/g. In some *B. cereus* outbreaks there appears to be a clear overlap of the diarrheal and emetic syndromes (Kramer and Gilbert, 1989).

As well as causing enteric illness, *B. cereus* has been responsible for postoperative infections, especially in immunocompromised patients. Although a rare event, such an infection can result in bacteremia, septicemia, endocarditis, meningitis, and pneumonia, among other symptoms (Drobniewski, 1993; Schoeni and Wong, 2005). The organism is also commonly associated with eye infections, such as posttraumatic endophthalmitis and metastatic endophthalmitis, and can very quickly result in irreversible tissue damage (Schoeni and Wong, 2005). There have been some reports of neonatal infections due to *B. cereus*, and it has been proposed that the systemic complications observed with these cases are associated with enterotoxins (Girisch et al., 2003).

Bacillus cereus Diarrheal Syndrome

The disease

The diarrheal illness caused by *B. cereus* is self-limiting, and no treatment is necessary. In severe cases, fluid replacement therapy may be required. This type of *B. cereus* food poisoning was first described in detail by Hauge (1955), who found that an outbreak of gastroenteritis in a hospital was associated with vanilla pudding contaminated with high numbers (up to 10^8 CFU/ml) of *B. cereus*. Since that first publication, reports on the incidence of *B. cereus* diarrheal disease have increased worldwide.

The incidence of food-borne illness caused by *B. cereus* varies from country to country. Although information on the extent of the problem is scant, it has been reported that outbreaks due to *B. cereus* can be the most common cause of food-borne illness in some countries, such as The Netherlands (Simone et al., 1997) and Taiwan (Chang and Chen, 2003). In France it has been estimated that there are 219 to 701 cases of *B. cereus*-associated food-borne illness annually, resulting in 26 to 84 hospitalizations (Vaillant et al., 2005). Data for the United States suggest that more than 27,000 cases of *B. cereus* gastrointestinal infections may occur annually (Mead et al., 1999). For the period between 1998 and 2006, the Centers for Disease Control and Prevention reported 90 outbreaks and 1,623 cases of infection caused by *Bacillus*

spp. This accounted for 3.4% of outbreaks and 1.6% of cases of bacterial food-borne illness reported during this period (Anonymous, 2008).

Pathogenesis of the diarrheal syndrome

Goepfert et al. (1972) were the first to relate the mechanism of pathogenicity of *B. cereus* to a possible enterotoxin responsible for fluid accumulation in the ligated rabbit ileal loop (LRIL) test measured after injection of *B. cereus* cultures or culture supernatants (Spira and Goepfert, 1972). Initial attempts to purify the enterotoxin of *B. cereus* revealed a protein complex, hemolysin BL (HBL), consisting of three proteins, termed B, L_1, and L_2, that exhibited positive results in the LRIL and vascular permeability assays (Beecher and Wong, 1994, 1997; Ryan et al., 1997). All three components were apparently required to produce maximal fluid accumulation in the LRIL assay. The HBL enterotoxin is very heterogeneous, and some *B. cereus* strains may produce more than one set of the three HBL components (Beecher and Wong, 2000; Schoeni and Wong, 1999). The nucleotide and deduced amino acid sequences have been reported for all three components of HBL (Heinrichs et al., 1993; Okstad et al., 1999; Ryan et al., 1997). The proteins exhibit 20 to 24% homology and have a structure that is almost entirely α-helix, indicating that they have evolved from the duplication of a single gene (Schoeni and Wong, 2005).

The HBL enterotoxin complex produces a unique discontinuous pattern of hemolysis on blood agar (Beecher and Macmillan, 1990; Beecher and Wong, 1994, 1997), which involves the B and L_1 components.

As well as causing fluid accumulation in the LRIL assay, HBL is dermonecrotic (Beecher and Wong, 1994) and cytotoxic (Beecher et al., 1995). The toxin acts by binding to the membrane of eukaryotes, where it oligomerizes to form pores (Beecher and Wong, 1997). A possible pathway of pore formation has been suggested based on the X-ray crystal structure of the B component (Madegowda et al., 2008).

In contrast, Granum's research group in Norway has purified and characterized a nonhemolytic enterotoxin (Nhe) (Granum et al., 1996; Lund and Granum, 1996). This toxin is also composed of three protein components (39, 45, and 105 kDa), which demonstrate homology with each other and the components of HBL (Schoeni and Wong, 2005). The 105-kDa protein is a protease that is not part of the Nhe complex (Lund and Granum, 1999). The genes encoding the Nhe complex are situated in an operon containing three open reading frames (ORF_s), *nheA*, *nheB*, and *nheC* (Granum et al., 1999). NheB is the only protein of the complex that binds to Vero cells, but all three proteins are needed for activity. The function of NheC is not yet understood, but it probably functions as a "catalyst," either by bringing NheA and NheB together following binding of NheB to the target cells or by enhancing conformational changes (Lindback et al., 2004). Whereas the exact mode of action is unclear, plasma membrane disruption was observed with epithelia exposed to Nhe from culture supernatants of *B. cereus,* but not with those exposed to supernatants from a mutant strain lacking NheB and NheC (Fagerlund et al., 2008). The purified Nhe components also combined to form large pores in lipid bilayers, which led to the hypothesis that, because of their common structural and functional properties, the HBL/Nhe and ClyA families of toxins belong to a superfamily of pore-forming cytotoxins. Sequencing and gene expression assays show that the genes responsible for both HBL and Nhe are transcribed from one operon, with maximum enterotoxin activity produced during late exponential or early stationary growth (Granum et al., 1999). Early work suggested that enterotoxigenic *B. cereus* isolates produce either both Nhe and HBL toxins or only one (Lund and Granum, 1997). A more extensive study, in which PCR primer pairs targeting *hblC, hblD, hblA, nheA, nheB, nheC, cytK,* and *entFM* were used to analyze 411 *B. cereus* strains and 205 *B. thuringiensis* strains, confirmed this finding and identified four groups, based on the presence of virulence genes. In group I, all eight genes were present; groups II and III contained isolates devoid of all *hbl* genes and *cytK* genes, respectively; and group IV strains lacked both the *hbl* and *cytK* genes. Both *nhe* and *entFM* genes were present in all isolates tested (Ngamwongsatit et al., 2008).

Agata et al. (1995) described yet another protein, *B. cereus* enterotoxin T (BcET), which showed positive reactions in vascular permeability reaction VPR and LRIL tests. The protein and the encoding DNA sequence are not related to HBL or Nhe. However, additional research indicated that this protein is not likely to cause diarrheal disease (Choma and Granum, 2002) and that the *bceT* gene was, in fact, not a single gene but four independent, ligated DNA fragments (Hansen et al., 2003). One of these fragments had 93% homology to an ORF (ORF 101) located within the pathogenic island of the *Bacillus anthracis* pXO1 virulence plasmid. The authors concluded that the enterotoxic activity of the original cloned *bceT* construct could be due to either this fusion gene or the fragment with homology to ORF 101.

A fourth possible *B. cereus* enterotoxin (EntFM), which was thought to be similar to the gene encoding the 45-kDa cytotoxic protein described by Shinagawa and colleagues (Shinagawa, 1990; Shinagawa et al., 1991), has been cloned and sequenced by Asano et al. (1997). It has been suggested that EntFM has similar properties to a phosphatase-associated protein with cell wall hydrolase activity from *B. subtilis* (Margot et al., 1998; Schoeni and Wong, 2005).

A cytotoxic protein (CytK) implicated in necrotic enteritis has been isolated from a *B. cereus* strain involved in a severe outbreak of *B. cereus* diarrheal disease, which resulted in three deaths. The protein was very similar to ß-barrel channel-forming toxins found in *Staphylococcus aureus* and *Clostridium perfringens* (Lund et al., 2000). Again this protein was capable of forming pores in lipid bilayers (Hardy et al., 2001). The *cytK* gene was detected in 73% of *B. cereus* isolates involved in diarrheal disease, but only in 37% of isolates from food not implicated in disease (Guinebretiere et al., 2002, 2006). Subsequently, two variants of CytK, CytK1 and CytK2, have been identified. The isolated CytK2 proteins were hemolytic, cytotoxic, and capable of forming membrane pores, but their toxicity was about 20% of that of CytK1 (Fagerlund et al., 2004; Guinebretiere et al., 2006). Brillard and Lereclus (2004) reported that *cytK* was transcribed more strongly in an isolate of *B. cereus* linked to human illness than in a reference strain and that *cytK* was controlled by the global regulator PlcR.

It was originally thought that the *B. cereus* diarrheal syndrome was the result of intoxication, due to the ingestion of toxin produced during growth of *B. cereus* in the food. However, it has been postulated that the disease results from ingestion of *B. cereus* cells that grow and produce enterotoxin within the intestinal tract of the patient (toxico-infection) (Granum and Lund, 1997; Griffiths and Schraft, 2002; Schraft and Griffiths, 2006). This claim is based on three observations. (i) Most strains produce enterotoxin in significant amounts only after reaching cell concentrations of 10^7/ml, while the infectious dose in many food-borne outbreaks has been determined to be 10^3 to 10^4/ml. (ii) The toxin is rapidly degraded at pH 3 (a pH similar to that found in the stomach) and hydrolyzed by the proteolytic enzymes of the digestive tract (Granum et al., 1993). However, these studies were conducted using enterotoxin produced in defined or semidefined media, and it should be noted that enterotoxin produced and ingested in a food matrix may be protected from damage through low pH and enzymes (Baker and Griffiths, 1995). (iii) Incubation times of more than 24 h have been observed for many outbreaks, and this is too long for simple enterotoxin action and may reflect the time required for spore germination. However, several *B. cereus* isolates are able to grow well under anaerobic conditions at 37°C and produce significant levels of enterotoxin within 6 h (Glatz and Goepfert, 1976; Granum et al., 1993). Other studies have shown that enterotoxin was not produced in defined medium during anaerobic growth (Beattie and Williams, 2002). Both HBL and Nhe production are influenced by carbon source and growth rate of the organism during anaerobic growth (Ouhib et al., 2006). At low oxidoreduction potentials, there is a strong stimulation of HBL production and, to a lesser extent, Nhe production. Nhe was produced first, early in the exponential growth phase, and HBL was synthesized later, during the early-stationary-growth phase (Zigha et al., 2006). This response is controlled by a two-component regulatory system, ResDE, and by a protein similar to the Fur redox regulator of *B. subtilis* (Duport et al., 2006; Zigha et al., 2007).

It may well be possible that both modes of pathogenesis may be possible, especially since several diarrhegenic enterotoxins have been described. For example, a toxico-infection might be the mode of action for a more severe form of *B. cereus* diarrhegenic syndrome described by Granum (1997). In one outbreak, 17 out of 24 people were affected after eating stews, and 3 of the patients were hospitalized, 1 for 3 weeks. The infective dose observed in this outbreak was 10^4 to 10^5 cells, which is lower than that usually associated with *B. cereus*. The time to onset of symptoms for these patients was more than 24 hours, also longer than commonly observed for *B. cereus*. In another outbreak, competitors at a skiing event in Norway became ill after drinking milk. The illness was restricted to young skiers between the ages of 16 and 19, while the older coaches and officials did not exhibit symptoms. Again, the incubation period was greater than 24 hours in some cases and symptoms persisted for between 2 and several days. Andersson et al. (1998a) postulate that this more severe form of *B. cereus* gastroenteritis results from adhesion of spores to the epithelial cells, followed by germination and enterotoxin production within the intestinal tract. The longer incubation time seen in these cases may be due to the time required for germination of the spores. It has been demonstrated that spores of *B. cereus* strains with high hydrophobicity are more adherent to CaCo-2 cells than spores with low hydrophobicity and that two of these highly hydrophobic strains were isolated during the outbreak involving meat stew (Anderson et al., 1998a). After adhesion, the spores were able to germinate and produce enterotoxins.

Only limited studies have been carried out on the mode of action of *Bacillus* enterotoxin. The enterotoxins identified for *B. cereus* are capable of forming pores in lipid bilayers, and a mechanism of cytotoxicity for Nhe, comprising osmotic lysis following pore formation in the plasma membrane, has been proposed by Fagerlund et al. (2008). *B. cereus* enterotoxins reverse absorption of fluid, Na^+, and Cl^- by epithelial cells and cause malabsorption of glucose and amino acids, as well as necrosis and mucosal damage. It has been suggested that the effects on fluid absorption are due to stimulation of adenylate cyclase (Kramer and Gilbert, 1989).

Bacillus cereus Emetic Syndrome

The disease

A second type of *B. cereus* gastroenteritis, the emetic syndrome, was first identified in the 1970s and was associated with the consumption of fried rice (Kramer and Gilbert, 1989). The emetic illness has an incubation time of 1 to 5 h and is manifested by nausea and vomiting that last for 6 to 24 h. Diarrhea is observed infrequently. The symptoms are usually self-limiting, and treatment is seldom necessary. To cause emesis, the food involved will typically contain 10^5 to 10^8 cells/g. On rare occasions, emetic *B. cereus* may also cause more severe disease. An outbreak was described in which a 17-year-old boy exhibited *B. cereus* emetic syndrome and died of liver failure within 2 days (Mahler et al., 1997). His father suffered hyperbilirubinemia and rhabdomyolysis, but he recovered. High levels of *B. cereus* emetic toxin were found in the pan used to reheat the food and in the boy's liver and bile. *B. cereus* was also cultured from both intestinal contents and pan residues. It was concluded that liver failure was due to the emetic toxin causing inhibition of hepatic mitochondrial fatty-acid oxidation. A subsequent outbreak linked to emetic *B. cereus* and involving five children from a family in Belgium who had eaten a contaminated pasta salad led to the death of the youngest child due to liver failure (Dierick et al., 2005).

Pathogenesis of the emetic syndrome

Early experiments relying on monkey feeding trials identified the cause of the emetic syndrome as a toxin because cell-free supernatants produced the same symptoms as cell cultures (Kramer and Gilbert, 1989). For many years the structure of the emetic toxin was elusive, partly because there was no convenient assay. It was thought to be a lipid and was known to be nonantigenic and extremely resistant to heat, alkaline and acid pH, and proteolysis. The emetic toxin has now been identified as a ring-structured peptide composed of three repeat sequences of four amino and/or oxy-acids. The dodecadepsipeptide, termed cereulide, has a molecular mass of 1.2 kDa and the following composition: [D-O-leu-D-ala-L-O-val-L-val]₃.

It closely resembles the potassium ionophore valinomycin (Agata et al., 1994). The toxin was stable at pH 2 to 11, withstood heating at 121°C for 90 min, and was not inactivated by treatment with trypsin and pepsin (Shinagawa et al., 1995, 1996). Oxygen is required for the production of the toxin (Jaaskelainen et al., 2004). Cereulide was formed when *B. cereus* emetic strains were present at cell numbers of $\geq 10^6$ CFU/g in products with a_w values of >0.953 and pH values of >5.6 (Jaaskelainen et al., 2003). High levels of toxin (0.3 to 5.5 μg/g) were produced during growth of *B. cereus* in rice-containing pastries at ambient temperatures (21 to 23°C). However, cereulide was not formed in products stored at 4 to 8°C (Jaaskelainen et al., 2003).

Cereulide stimulates the vagus afferent by binding to 5-HT$_3$ receptors and induces swelling of mitochondria, with toxic effects due to potassium-ionophoretic properties (Agata et al., 1994; Mikkola et al., 1999; Sakurai et al., 1994). Cytotoxicity is accompanied by inhibition of RNA synthesis and cell proliferation at cereulide concentrations of 2 nM (Andersson et al., 2007). At high doses, cereulide can cause a massive degeneration of hepatocytes (Yokoyama et al., 1999). The toxin also inhibits natural killer cells in vitro, which suggests that the toxin may possess immunomodulating properties (Paananen et al., 2002).

Cereulide is produced by a nonribosomal peptide synthetase (Ehling-Schulz et al., 2005b), which is encoded by a cereulide synthetase *(ces)* gene cluster (Ehling-Schulz et al., 2006). These genes are found only in emetic strains of *B. cereus* and are located on a 208-kb megaplasmid. The *ces* gene cluster (24 kb) comprises seven coding sequences (CDSs). These include the typical nonribosomal peptide synthetase genes, encoding a phosphopantetheinyl transferase and two enzymes for the activation and incorporation of monomers in the growing peptide chain, a CDS encoding a putative hydrolase (upstream region), and an ABC transporter (downstream). Regions flanking the *ces* gene locus have high homology to virulence plasmids of *B. cereus*, *B. thuringiensis*, and *B. anthracis*. This has led to the conclusion that, besides the insecticidal and anthrax toxins of *B. thuringiensis* and *B. anthracis*, respectively, a third type of *B. cereus* group toxins exists, and these toxins are encoded on megaplasmids.

The emetic toxin can be assayed by determining vacuolation induced in HEp-2 cells (Hughes et al., 1988), and this has been aided by the development of colorimetric modifications (Finlay et al., 1997; Mikami et al., 1994; Szabo et al., 1991). A rapid bioassay, which tests toxicity of cereulide to boar spermatozoa (Andersson et al., 1998b, 2004), and a chemical assay, which is based on high-performance liquid chromatography and mass spectroscopy (Haggblom et al., 2002), are now available. Kawamura-Sato et al. (2005) have described a quantitative assay for cereulide based on uncoupling of the respiratory activity of rat liver mitochondria.

Unlike the diarrheal strains of *B. cereus* that are extremely heterogeneous, emetic strains have very low diversity and consist of a single, distinct cluster of isolates that are unable to degrade starch and do not ferment salicin (Ehling-Schulz et al., 2005a).

Incidence and Transmission through Food

The mild and transient character of the *B. cereus* food-borne diseases likely results in the disease being vastly underreported. None of the reports on food-borne disease outbreaks caused by *B. cereus* distinguishes between emetic and diarrheal syndromes, but it is likely that the outbreaks reflect the diarrheal syndrome. The incidence of *B. cereus* food-borne diseases seems to be higher in countries of northern Europe (20 to 33%) and in Canada (14%). England and Wales, Japan, and the United States (1.2 to 4.4%) have a clearly lower incidence. The reasons for these differences are not known. Foods most often implicated as vehicles for transmission of the diarrheal syndrome include meat and meat dishes, vegetables, cream, spices, poultry, and eggs. The emetic syndrome is most frequently associated with starchy foods, especially rice dishes (Kramer and Gilbert, 1989).

The amount of emetic toxin in 13 of 14 food samples implicated in emetic-type illness ranged from 0.01 to 1.28 μg/g (Agata et al., 2002). When challenge studies were performed using an emetic toxin-producing strain of *B. cereus*, the organism was capable of rapid growth in boiled rice and produced emetic toxin at both 30 and 35°C. In farinaceous foods, the production of emetic toxin was as high as that in the foods implicated in illness. Low levels of emetic toxin were detectable in eggs and meat and their products, and a small quantity of toxin was detectable in milk and soy milk, when not aerated. Bacterial growth and toxin production were inhibited in foods cooked with vinegar, mayonnaise, and catsup, presumably due to the low pH generated by acetic acid (Agata et al., 2002). When potato puree and penne pasta were inoculated with cereulide-producing strains of *B. cereus* and stored without shaking, higher cereulide concentrations were achieved (4 μg/g in puree and 3.2 μg/g in penne) than in boiled rice (2 μg/g) (Rajkovic et al., 2006), despite counts in excess of 10^8 CFU/g being attained in all three products. With aeration, cereulide production in these products was more than 10-fold lower. Cereulide was not observed following growth of emetic strains of *B. cereus* in aerated milk, but levels reached 1.1 μg/ml in statically incubated milk. Thus, a number of factors, including food type, temperature, pH, aeration, and strain, determine the amount of cereulide that will be produced. Cereulide is also remarkably resistant to heat and alkaline pH (Rajkovic et al., 2008).

It has also been shown that emetic strains of *B. cereus* were not able to grow at temperatures below 10°C but were able to grow at 48°C. In general, spores of emetic strains also were more resistant to heat (90°C) (Carlin et al., 2006). Carlin et al. suggested that these properties make emetic strains a special risk in heat-processed foods or preheated foods that are kept warm, but they should not pose a risk in refrigerated foods.

An infant food containing cereal and dried infant milk formula was associated with illness in two infants, which presented as rapid projectile vomiting (Duc et al., 2005). Both *B. subtilis* and *B. cereus* were isolated from the product, but cereulide was not detected. This led the investigators to suggest that the cause of the illness was due either to a new cereulide-type toxin or to the high levels of *B. subtilis* present. It has been reported that infant food containing both cereal and milk powder are able to support greater levels of cereulide production than other infant formulas and that mishandling and temperature abuse of reconstituted formula were high-risk activities (Shaheen et al., 2006).

Altayar and Sutherland (2006) found that 45.8% of 271 samples of soils, animal feces, and raw and processed vegetables were positive for *B. cereus*. Of the 325 isolates obtained, only 3 were positive for emetic toxin production. All the positive isolates came from washed or unwashed potato skins, and one strain was psychrotrophic.

PRESENCE, GROWTH, AND SURVIVAL OF *B. CEREUS* IN FOODS

Presence of *B. cereus* in Foods

Members of the *B. cereus* group are ubiquitously distributed in the environment, mainly because of their spore-forming capabilities. Thus, *B. cereus* can easily contaminate various types of foods, especially products of plant origin. The organism is also frequently isolated from milk and dairy products, meat and meat products, pasteurized liquid egg, rice, ready-to-eat vegetables, and spices (te Giffel et al., 1996). The last can be a source of contamination of prepared meals (Banerjee and Sarkar, 2004). *B. cereus* was also isolated from vacuum-packaged and film-packaged soya- and cereal-based vegetarian foods (Fang et al., 1999).

Of seven strains isolated from these products, one produced diarrheal toxin and four caused lethality in mice. Because of the ubiquitous distribution of the organism, it is virtually impossible to obtain raw products that are free from *B. cereus* spores. Vegetable purees have been implicated in outbreaks of *B. cereus* illness (Lund et al., 2000), and the source of the contamination is the raw vegetable and the texturing agents, such as milk protein and starch, used in the formulation (Guinebretiere et al., 2003). No spores were detected on the processing equipment.

It was also postulated that *B. cereus* could be present in paper pulp and the products made from it; however, it has been determined that the level of cereulide retained in paper is too low to present a risk to human health (Hoornstra et al., 2006).

The dairy industry faces problems due to *B. cereus* because it is virtually impossible to exclude the organism from milk; their spores readily attach to processing equipment, and pasteurization does not eliminate them (Andersson et al., 1995). Indeed, in many cases, pasteurization may result in activation and germination of spores (Griffiths and Phillips, 1990b; Hanson et al., 2005), which has resulted in shelf-life issues with milks pasteurized at temperatures in excess of the recommended minimum for high-temperature, short holding time processing. Fortunately, it seems that the majority of *B. cereus* strains isolated from dairy products are not highly cytotoxic (Arnesen et al., 2007).

The appearance of psychrotrophic strains in the dairy industry has added a new dimension to *B. cereus* surveillance in food. Studies indicate that both raw and pasteurized milk will harbor psychrotrophic *B. cereus,* with a prevalence of 9 to 37% in raw milk and 2 to 35% in pasteurized milk (te Giffel et al., 1995), and that the source of the contamination of raw milk was primarily silage (te Giffel et al., 2002). Subsequent work has suggested that the presence of *B. cereus* in pasteurized milk may be the result of postpasteurization contamination at the filler (Eneroth et al., 2001). These psychrotrophic strains have an average generation time of 17 h at 6°C and produce enterotoxin during extended storage at slightly abusive temperatures (Christiansson et al., 1989; Griffiths, 1990; Griffiths and Phillips, 1990a). In coffee creamers stored at ambient temperatures, *B. cereus* can grow to levels of public health concern within 11 h (Feijoo et al., 1997a). For example, Odumeru et al. (1997) found that 38% of pasteurized milks stored at 10°C until their expiry date contained *B. cereus* enterotoxin. Similarly, te Giffel et al. (1997) reported that 40% of pasteurized milks sampled from household refrigerators in The Netherlands were contaminated with *B. cereus* and that about 75% of these were enterotoxigenic.

The psychrotrophic *B. cereus* isolates described in these earlier reports may be strains of *B. weihenstephanensis* (Lechner et al., 1998). However, of the 39 *Bacillus* isolates obtained from whipping cream, only 4 were identified as *B. weihenstephanensis* and none of these were cytotoxic, although all possessed at least one enterotoxin (Stenfors et al., 2002). This species has also been isolated from liquid egg (Baron et al., 2007).

Bacillus cereus has also been isolated from highly controlled foods, such as infant formula (Becker et al., 1994; Rowan and Anderson, 1998a), dietary supplements for hospitalized human immunodeficiency virus patients (Rowan and Anderson, 1998b), and airline meals (Hatakka, 1998a, 1998b). The ability of *B. cereus* to produce enterotoxin in infant milk formula as well as coffee creamer was confirmed, but no toxin production was observed with cocoa powder (Cooper and McKillip, 2006).

Wijnands et al. (2006a) categorized randomly selected foods into eight product groups: oils, fats, and oil and fat products; fish and fish products; meat and meat products; flavorings; milk and milk products; ready-to-eat foods; vegetables and vegetable products; and pastry. They determined the presence and number of *B. cereus* cells in each of the food groups by using mannitol-egg yolk-polymyxin agar and characterized the isolates for growth temperature range and the presence of toxin genes by PCR. The genes for Nhe, HBL, and CytK were present in >97%, approximately 66%, and 50% of the isolates, respectively. A relatively low percentage of isolates from the flavorings group contained genes encoding Nhe, and a higher-than-average number of isolates in the pastry group possessed the genes encoding both HBL and Nhe. Cereulide was produced by 8.2% of the isolates, but all those isolated also had genes for one or more of the other virulence factors. Perhaps surprisingly, although there was a relatively high incidence of *B. cereus* in pasta and rice products, the isolates obtained from these products did not exhibit an elevated prevalence of the cereulide gene. Most of the strains isolated during the study were mesophilic (89.9%), while only 4.4% of the isolates were psychrotrophic. The incidence of the latter was greatest in the milk-and-milk products' group. In the product groups' flavorings, milk and milk products, vegetable(s) and vegetable products, pastry, and ready-to-eat foods, 0.14%, 0.19%, 0.16%, 0.04%, and 0.29%, respectively, contained *B. cereus* counts of >10^5 CFU/g.

As well as the foods themselves, processing plant equipment can harbor *B. cereus* (Faille et al., 2000, 2001). For example, the organism was isolated from evaporators and drip trays of chilling units in catering plants and in plants processing cooked meats (Evans et al., 2004).

Growth and Survival

An excellent overview of the factors affecting growth and survival of *B. cereus* can be found in the monograph *Microorganisms in Foods 5* published by the International Committee on Microbiological Specifications of Foods (ICMSF) in 1996. The most important aspects are summarized below.

Under ideal conditions *B. cereus* has an optimum growth temperature of 30 to 40°C, with growth possible between 4 and 55°C. Strains that grow at 7°C or below are identified as psychrotrophic and will not grow above 43°C. Mesophilic *B. cereus* usually grows at temperatures of 15°C to 50° or 55°C. There is evidence that *B. cereus* can demonstrate a cold adaptation response following repeated exposure to low temperatures (Foegeding and Berry, 1997). Psychrotrophic and mesophilic strains of *Bacillus* can be distinguished based on PCR targeting of genes responsible for encoding the major cold shock protein (Francis et al., 1998). The optimum pH for *B. cereus* growth has been reported as 6.0 to 7.0, with generation times of approximately 23 minutes at 30°C. The minimum pH for growth is 5.0, and the maximum pH is 8.8. It has been shown that strains of *B. cereus* associated with food-borne illness are more resistant to adverse pH environments than those that did not produce gastrointestinal infections (Garcia-Arribas and Kramer, 1990). In the presence of NaCl as a humectant, *B. cereus* will not grow at an a_w of 0.93. However, when glycerol was used as a humectant, growth was possible at an a_w of 0.93, but not 0.92. Only a few reports are available on the effects of organic acids and chemical preservatives on *B. cereus*. In rice filling held at 23°C, growth was inhibited by 0.26% sorbic acid and by 0.39% potassium sorbate. Butylated hydroxyanisole at concentrations of 0.1 to 0.5% inhibited *B. cereus* growth at 32°C over 24 h.

The response of *B. cereus* to a number of stresses, such as heat, acid, and osmotic stress, has been described by Browne and Dowds (2001, 2002).

Heat Resistance and Germination of Spores

Effective destruction of *B. cereus* spores is the ultimate goal for safe foods with an extended shelf-life. A $D_{121.1}$ value of 0.03 to 2.35 minutes (z value = 7.9 to 9.9°C) has been reported for spores suspended in phosphate buffer at pH 7.0. For spores suspended in milk, the D_{95} value ranged from 1.8 to 19.1 min. A similar wide range for D_{95} (2.7 to 15.3 minutes) was observed for spores in milk-based infant formula. Dufrenne et al. (1995) reported that D_{90} values for spores of 12 psychrotrophic strains of *B. cereus* ranged from 2.2 to 9.2 min, but the value for one strain was >100 min. In general, spores of psychrotrophic *B. cereus* are less heat resistant than spores of mesophilic strains (Dufrenne et al., 1994; Notermans and Batt, 1998). In pork luncheon meat, *D* values of *B. cereus* spores ranged from about 32 min at 85°C to about 2 min at 95°C (*z* value = 6.6) (Byrne et al., 2006). Mathematical models to describe the heat resistance of *B. cereus* at different pH values and temperatures (Collado et al., 2003) and the effect of heat shock and germinant concentration on germination (Collado et al., 2006) have been published.

Germination of spores requires the presence of purine ribosides and glycine or a neutral L-amino acid. L-alanine is the most effective amino acid stimulating germination. Spores are able to sense the presence of these germinants in the environment by receptors encoded by the *gerA* family of operons (Hornstra et al., 2005). Germination temperatures vary greatly between strains and are strongly influenced by the substratum. In laboratory media, germination has been observed at −1°C to 59°C; in cooked rice, germination temperatures ranged from 5°C to 50°C (Kramer and Gilbert, 1989).

ISOLATION AND IDENTIFICATION

Cultural Methods

Both emetic and diarrhegenic *B. cereus* have a relatively high infective dose, and detection of the organism is achieved primarily by direct plating onto selective agar media. Two selective media are widely used for the isolation and presumptive identification of *B. cereus*: mannitol-egg yolk-polymyxin (MYP) agar and polymyxin-pyruvate-egg yolk-mannitol-bromothymol blue agar (PEMBA). Both rely on the presence of lecithinase (phospholipase C) and the absence of mannitol fermentation in *B. cereus*. Presumptive colonies are surrounded by an egg yolk precipitate and have a violet-red background on MYP and a turquoise or peacock blue color on PEMBA. Both media contain polymyxin to inhibit growth of competitive organisms (van Netten and Kramer, 1992). A medium that requires no polymyxin, VRM, has been described; Tris and resazurin are included in VRM to inhibit growth of *B. megaterium* and many other competitive bacteria present in foods (DeVasconellos and Rabinovitch, 1995). Peng et al. (2001) developed the BCM *B. cereus*/*B. thuringiensis* chromogenic agar plating medium, which is based on the detection of the phosphatidylinositol-specific phospholipase C enzyme present in *B. cereus*.

Fricker et al. (2008) evaluated two chromogenic media (chromogenic *B. cereus* agar and BCM) and two standard media (PEMBA and MYP) for the isolation and identification of *B. cereus*. They showed that the chromogenic media performed as well as the conventional standard media and were useful if staff were not highly trained in identification of the organism, because the use of conventional media could result in substantial misidentification and underestimation of *B. cereus*. However, the chromogenic media could not detect some *B. cereus* strains. Sequence analysis of the *plcR* gene, a pleiotropic regulator of various virulence factors and *B. cereus*-specific enzymes, revealed a significant correlation between atypical colony appearance and specific variances within the *plcR* gene of those strains.

If enrichment is required for detection of low numbers in foods, a nutrient-rich broth, such as brain heart infusion broth or trypticase soy broth, supplemented with polymyxin (1,000 IU/liter) is recommended (Van Netten and Kramer, 1992).

A method for the detection of *B. cereus* spores in milk has been described by Christiansson et al. (1997). The procedure includes a heat treatment at 72°C and incubation with trypsin and Triton X-100 at 55°C. The samples are filtered through a membrane filter to concentrate the spores. The filter is then placed onto solid agar media to allow for growth of the spores.

Presumptive colonies isolated from selective media are characterized by various confirmatory tests; *B. cereus* and closely related species will be gram-positive, catalase-positive rods with spores that do not swell the sporangium. In addition, they will produce acid from glucose anaerobically and show a positive reaction for nitrate reduction, Voges-Proskauer assays, L-tyrosine degradation, and growth in 0.001% lysozyme. A biochemical *Bacillus* identification system, the API 50 CHB test, is available from bioMérieux. The test determines the ability of an isolate to assimilate 49 carbohydrates and has become a reference method for the biochemical identification of *Bacillus* species.

Noncultural Detection Methods

With the advent of rapid microbiological methods for the detection of bacteria in foods, several culture-independent approaches have been proposed.

Molecular-based assays

PCR is now routinely used for the detection of food-borne pathogens, and methods for the detection of *B. cereus* and related organisms have been described. One group of these assays aims at detecting *B. cereus* per se, while other assays target specific subgroups of *B. cereus*, such as psychrotrophic or enterotoxigenic strains (Hsieh et al., 1999).

Schraft and Griffiths (1995) reported a sensitive test for the detection of *B. cereus* in milk, which is based on primers targeting the phospholipase gene of *B. cereus*. This gene encodes the enzyme responsible for the positive egg yolk reaction used for the presumptive identification of the organism on selective plating media. The authors used a DNA purification procedure that allowed the detection of less than 1 CFU per ml of milk without a preenrichment step. Damgaard et al. (1996a) combined magnetic capture hybridization with a nested PCR assay targeting the phospholipase gene of *B. thuringiensis* to detect *B. cereus* and *B. thuringiensis* in naturally contaminated rhizosphere samples. Although no lower detection limits were reported, this magnetic capture hybridization PCR assay might be useful for the detection of *B. cereus* in foods. A PCR-based assay for the detection of all members of the *B. cereus* group, which targets 16S rRNA genes, has been developed by Hansen et al. (2001).

Yamada et al. (1999) designed a primer set for a PCR assay based on the gyrase B gene *(gyrB)* of *B. cereus*. They evaluated the assay using experimentally inoculated rice homogenates and found that a detection limit of less than 1 CFU/g could be achieved when a 15-h preenrichment and two-step filtration were applied before the PCR amplification.

PCR protocols have been described that allow the selective detection of psychrotrophic *B. cereus*. The primers used in the assay target the psychrotrophic and mesophilic rRNA gene signature sequences (von Stetten et al., 1998) and gene sequences for the major *B. cereus* cold shock proteins CspF and CspA (Francis et al., 1998). However, some *B. cereus* strains that do not possess these rRNA gene signature sequences can grow at low temperatures (Stenfors and Granum, 2001).

Several PCR assays have been described for the detection of enterotoxigenic *B. cereus*. These tests are based on the gene sequence of the toxin to be detected. As described above, there are several different proteins or protein complexes that have been implicated in the *B. cereus* diarrheal syndrome; assays that simultaneously target all potential enterotoxin genes would seem to have the greatest potential. Protocols for the detection of the HBL and BcET toxin complex (Mantynen and Lindstrom, 1998) and for Nhe and CytE (Granum et al., 1999; Guinebretiere et al., 2002) have been described.

Real time PCR-based methods for the identification of emetic strains of *B. cereus* have been described

based on amplification of sequences of the cereulide synthetase genes (Fricker et al., 2007).

Other amplification technologies, such as nucleic acid sequence-based amplification, have been used to detect *hblC* expression in *Bacillus* spp. (Gore et al., 2003). More recently, DNA microarrays have been used to detect the presence of genes encoding the *B. cereus* toxins (Liu et al., 2007).

Antibodies against spores and vegetative cells

Koo et al. (1998b) have engineered a monoclonal single-chain antibody, which is reported to be specific for *B. cereus* T spores. The antibody was fused with a streptavidin molecule and the fusion protein expressed in *Escherichia coli* (Koo et al., 1998a). This antibody-streptavidin protein was then used in a magnetic bead-based immunoassay to concentrate *B. cereus* spores from liquid samples. When samples were inoculated with 5×10^4 *B. cereus* spores, the assay was capable of removing more than 90% of the spores from phosphate buffer and 37% from whole milk. Charni et al. (2000) reported the production of monoclonal antibodies which reacted only with *B. cereus* vegetative cells, but not with spores. More recently, real-time, antibody-based biosensors for *B. anthracis* spores using evanescent waves (Love et al., 2008) and a piezoelectric excited cantilever (Campbell and Mutharasan, 2007; McGovern et al., 2008) for detection have been described.

Other detection methods

Methods for the detection of *Bacillus* using bacteriophage in a variety of ways have been reported. High-affinity, cell wall-binding domains of bacteriophage endolysins have been used to capture cells of *B. cereus* (Kretzer et al., 2007). In addition, an amperometric, phage-based biosensor capable of detecting 10 cells/ml within 8 h has been developed and measures α-glucosidase activity released from cells upon phage lysis (Yemini et al., 2007).

Fourier transform infrared spectroscopy can provide a way to differentiate bacterial genera and species through unique Fourier transform infrared vibrational combination bands produced from active components of bacterial cells. Applying principal component analysis to these spectra reveals clear distinctions between different bacterial strains, even in food matrices (Al-Holy et al., 2006). The technique is able to differentiate between *B. cereus* and other species, when they are present both as vegetative cells (Lin et al., 1998; Winder and Goodacre, 2004) and as spores (Perkins et al., 2005). Matrix-assisted laser desorption ionization–time of flight (mass spectrometry) can also be used to detect spores and vegetative cells of *B. anthracis* (Lasch et al., 2008).

Typing Methods

Fast and accurate identification methods for *B. cereus* are also important in the food industry. Odumeru et al. (1999) evaluated two automated microbial identification systems for their ability to accurately and reproducibly identify 40 *B. cereus* isolates. The API 50 CHB biochemical test was used as a reference test. The fatty acid-based microbial identification system (MIDI Inc., Newark, DE) had a slightly better (55%) sensitivity, i.e., the proportion of reference positive strains correctly identified, than the biochemical Vitek system (42.5%; bioMérieux Vitek, Hazelwood, MO). When the definition of a correct identification was expanded to include closely related members of the *B. cereus* group, e.g., *B. mycoides* and *B. thuringiensis*, the sensitivity of the systems increased to 82.5% and 67.5%, respectively. The specificity, i.e., proportion of reference negative strains not identified as *B. cereus*, was very good (97.5%) for both systems.

When *B. cereus* is repeatedly isolated in finished products or suspected of causing a food-borne disease outbreak, methods for tracing possible contamination sources throughout the production chain are of crucial importance.

Traditionally, biochemical profiles, serology, phage typing, and fatty acid analysis have been used for typing. Mostly phage typing and fatty acid profiling have been applied in more recent studies. Väisänen et al. (1991) found phage typing could be used to exclude packaging materials as a potential source of *B. cereus* in dairy products, and Ahmed et al. (1995) described a 12-phage typing scheme which could be used in epidemiological investigations for the identification of sources of *B. cereus* food poisoning. Since phage typing requires continuous phage propagation, it has not found widespread application in the food industry.

Fatty acid analysis can now be performed with a fully automated system, the microbial identification system described above, for identification. Whole-cell fatty acid analysis has been used to trace and identify *B. cereus* strains colonizing the processing environment of dairy plants (Pirttijarvi et al., 1998; Schraft et al., 1996).

Pyrolysis mass spectrometry is capable of differentiating strains of *B. anthracis* but was of limited value for differentiation of *B. cereus* strains (Helyer et al., 1997).

In addition to phenotypic typing, genotypic methods are frequently used for epidemiological typing of *B. cereus*. For example, randomly amplified polymorphic DNA (RAPD) has been applied to type over 2,000 mesophilic and thermophilic *Bacillus* species isolated from an industrial setting (Ronimus et al., 1997, 2003). The authors were able to trace individual species from the feedstock to the finished product in a food processing plant. RAPD typing also allowed the separation of mesophilic from psychrotrophic *B. cereus* isolates (Lechner et al., 1998). Amplified fragment length polymorphism (AFLP) is a genotyping method with significantly higher reproducibility than that of RAPD, and it has been used to differentiate strains of the *B. cereus* group (Hill et al., 2004). Ripabelli et al. (2000) evaluated AFLP for epidemiological typing of *B. cereus* with 21 cultures from seven different outbreaks. All isolates could be typed with high reproducibility, and a unique AFLP pattern was generated for isolates of each outbreak. Helgason et al. (2004) have described a multilocus sequence typing scheme for the *B. cereus* group, which is based on the use of primers designed for conserved regions of housekeeping genes, and sequencing of 330- to 504-bp internal fragments of seven of these genes, *adk*, *ccpA*, *ftsA*, *glpT*, *pyrE*, *recF*, and *sucC*. The introduction of a rapid DNA preparation method for *Bacillus cereus* will facilitate large-scale typing of *B. cereus* with either RAPD or AFLP (Nilsson et al., 1998).

An automated ribotyping instrument, the RiboPrinter, has been developed by Qualicon Inc. Andersson et al. (1999) compared RAPD typing with automated ribotyping and found the latter to be a fast and reliable method, which was only slightly less discriminatory than the more labor-intensive RAPD technique. A molecular typing method based on sequencing of the *rpoB* gene has been applied for the differentiation of *Bacillus* spp. in milk (Durak et al., 2006; Huck et al., 2007).

Detection of Toxins

Two commercial kits are available for the detection of *B. cereus* diarrhegenic toxins. Both kits are immunoassays, but they detect different antigens (Day et al., 1994). The reverse passive latex agglutination enterotoxin assay produced by Oxoid reacts with a 43-kDa protein, the subunit L_2 of the tripartite enterotoxin complex (HBL) described by Beecher and Wong (1994). The second assay uses the enzyme-linked immunosorbent assay format. This kit is marketed under the name *Bacillus* diarrhoeal enterotoxin visual immunoassay kit (Tecra). It detects a 41-kDa protein (NheA), which is part of the nonhemolytic enterotoxin complex (Nhe) described by Lund and Granum (1996). Any enterotoxigenic *B. cereus* strain may produce either HBL or Nhe or both. Thus, it is not surprising that early comparisons of the two commercial test kits have triggered contradicting reports on their accuracy in detecting the enterotoxin. Based on today's knowledge about *B. cereus* enterotoxins, any isolate that reacts positively with one or both of the commercial kits should be considered enterotoxigenic (Granum and Lund, 1997). Final confirmation of enterotoxigenicity of an isolate can be obtained by cytotoxicity tests using Vero cells, Chinese hamster ovary (CHO) cells, or human embryonic lung cells (Fletcher and Logan, 1999).

A cell-based biosensor containing a B-cell hybridoma, Ped-2E9, encapsulated in a collagen matrix has been described for the rapid detection of enterotoxin from *Bacillus* species. The sensor measures the alkaline phosphatase released from infected Ped-2E9 cells using a colorimetric assay (Banerjee et al., 2008).

Dietrich et al. (1999) have described monoclonal antibodies that target each of the proteins in the HBL complex. The same group has developed polyclonal and monoclonal antibodies against the proteins of the Nhe complex (Dietrich et al., 2005).

The bioassay using boar spermatozoa and the high-pressure liquid chromatography–mass spectrometry procedure described in *Bacillus cereus* Emetic Syndrome: pathogenesis of the emetic syndrome are the only tests reported for the detection of the emetic toxin cereulide.

OUTBREAKS CAUSED BY *BACILLUS* SPP. OTHER THAN *B. CEREUS*

Bacillus species that do not belong to the *B. cereus* group have been reported to cause foodborne disease (Kramer and Gilbert, 1989). *B. subtilis* has caused vomiting within a few hours of ingesting the incriminated food (meat, seafood with rice, and pastry). High numbers of *B. subtilis* cells (10^7 CFU/g) were isolated from the patients' vomitus. *B. licheniformis* has caused diarrheal illness, with an incubation time of about 8 h. High numbers (10^7 CFU/g) of *B. licheniformis* cells were isolated from the implicated foods (meat, poultry or vegetable pies, and stews).

An outbreak involving three people who became ill after eating dinner in a Chinese restaurant was described by From et al. (2007a). Symptoms included dizziness, headache, chills, and back pain, which developed during the meal; these were followed, a few hours later, by stomach cramps and diarrhea which lasted for several days. High levels of *B. pumilus* were detected in cooked, reheated rice, and this organism was found to produce a complex of lipopeptides,

pumilacidins. The *B. pumilus* isolate produced a seven-component complex identical in molecular mass to pumilacidins A, B, C, D, E, F, and G. These lipopeptides have previously been regarded as harmless to humans (Pedersen et al., 2002), but a lipopeptide, lichenysin, produced by *B. licheniformis* isolated from an infant milk formula was thought to be responsible for the death of an infant (Salkinoja-Salonen et al., 1999). The pumilacidin complex isolated from the *B. pumilus* associated with the restaurant outbreak was active in the boar sperm assay and was cytotoxic (From et al., 2007a, 2007b). From and colleagues also demonstrated that *B. mojavensis*, *B. subtilis*, and *B. fusiformis* produce lipopeptides with cytotoxic activity, and they called for a reevaluation of the role of these compounds in food-borne illness (From et al., 2005, 2007b).

Production of toxins by non-*B. cereus* isolates has been confirmed. The *B. cereus* enterotoxin reverse passive latex agglutination assay (specific for the L_2 subunit of the HBL enterotoxin) was positive for isolates of *B. licheniformis*, *B. subtilis*, *B. circulans*, and *B. megaterium*, indicating that these *Bacillus* species may potentially cause food-borne diarrhegenic disease (Rowan et al., 2001). Inhibition of boar sperm motility, an indicator for *B. cereus* emetic toxin, was demonstrated for approximately 50% of *B. licheniformis* cells isolated from food (Salkinoja-Salonen et al., 1999).

Cell-free cultures of toxin-producing isolates of *B. licheniformis* obtained from foods involved in outbreaks, from raw milk, and from industrially produced baby food, inhibited sperm motility, damaged cell membrane integrity, depleted cellular ATP, and caused swelling of the acrosomes of spermatozoa, but no mitochondrial damage was observed (Salkinoja-Salonen et al., 1999). These effects were due to a molecule smaller than 10 kDa, with physicochemical properties resembling those of cereulide. Similar results have been observed with *B. megaterium*, *B. simplex*, and *B. firmus* (Taylor et al., 2005). The agent was nonproteinaceous, soluble in 50 and 100% methanol, and resistant to heat, protease, and acid or alkali. The isolates associated with illness were β-hemolytic and grew anaerobically and at temperatures of 55°C, but not at 10°C. Phenotypically, they were indistinguishable from the type strain of *B. licheniformis*.

Strains of the psychrotolerant *B. weihenstephanensis* have been shown to produce cereulide at temperatures as low as 8°C (Thorsen et al., 2006).

B. thuringiensis, closely related to *B. cereus* and still widely used as an insecticide, has been reported to contain *B. cereus* enterotoxins (Gaviria Rivera et al., 2000; Honda et al., 1991; Jackson et al., 1999;

Perani et al., 1998; Rivera et al., 2000; Rusul and Yaacob, 1995). Strains of *B. thuringiensis* isolated from food have been shown to be cytotoxic (Damgaard et al., 1996b), and enterotoxigenic *B. thuringiensis* strains used as commercial bioinsecticides have been isolated from fresh fruits and vegetables, including cabbage, tomatoes, cucumbers, and peppers, for sale for human consumption (Frederiksen et al., 2006; Hendriksen and Hansen, 2006). Enterotoxin genes have also been found in mosquito-larvicidal *B. sphaericus* (Yuan et al., 2002). Such findings should be of concern to regulators approving the use of *Bacillus* species as biological pesticides.

More recently, concern has been expressed regarding the use of *B. anthracis* as an agent of bioterror. Issues related to the contamination of food by *B. anthracis* have been reviewed (Erickson and Kornacki, 2003). The disease caused by *B. anthracis*, termed anthrax, is primarily a disease of herbivores. Humans can become infected through consumption of contaminated food or by contact with contaminated animal products. Anthrax is prevalent in many parts of the world, and cases have been reported in most countries. Three forms of the disease have been recognized: cutaneous, gastrointestinal, and pulmonary. The major virulence factors of *B. anthracis* are a poly-D-glutamic acid capsule and a three-component protein exotoxin consisting of a protective antigen (83 kDa), lethal factor (90 kDa), and edema factor (89 kDa). The genes encoding the toxin and the enzymes responsible for capsule production are carried on plasmids pXO1 and pXO2, respectively. The mechanisms of pathogenicity in *B. anthracis* have been reviewed (Bhatnagar and Batra, 2001).

TREATMENT AND PREVENTION

The symptoms of *B. cereus* infection are usually mild and self-limiting and do not normally require treatment.

Because *B. cereus* can almost invariably be isolated from foods and can survive extended storage in dried food products, it is not practical to eliminate low numbers of spores from foods. Control against food poisoning should be directed at preventing germination of spores and minimizing growth of vegetative cells. It has been established that peracetic acid-based sanitizers offer the best approach to reduce spore loads in food processing and catering establishments (Ernst et al., 2006). Arguably, the most effective control is to ensure that foods are rapidly and efficiently cooled to less than 7°C or maintained above 60°C, and they should be thoroughly reheated before serving (Hillers et al., 2003).

REFERENCES

Agata, N., M. Mori, M. Ohta, S. Suwan, I. Ohtani, and M. Isobe. 1994. A novel dodecadepsipeptide, cereulide, isolated from *Bacillus cereus* causes vacuole formation in Hep-2 cells. *FEMS Microbiol. Lett.* 121:31–34.

Agata, N., M. Ohta, Y. Arakawa, and M. Mori. 1995. The *bceT* gene of *Bacillus cereus* encodes an enterotoxic protein. *Microbiology* 141:983–988.

Agata, N., M. Ohta, and K. Yokoyama. 2002. Production of *Bacillus cereus* emetic toxin (cereulide) in various foods. *Int. J. Food Microbiol.* 73:23–27.

Ahmed, R., P. Sankarmistry, S. Jackson, H. W. Ackermann, and S. S. Kasatiya. 1995. *Bacillus cereus* phage typing as an epidemiologic tool in outbreaks of food poisoning. *J. Clin. Microbiol.* 33:636–640.

Al-Holy, M. A., M. Lin, A. G. Cavinato, and B. A. Rasco. 2006. The use of Fourier transform infrared spectroscopy to differentiate *Escherichia coli* 0157:H7 from other bacteria inoculated into apple juice. *Food Microbiol.* 23:162–168.

Altayar, M., and A. D. Sutherland. 2006. *Bacillus cereus* is common in the environment but emetic toxin producing isolates are rare. *J. Appl. Microbiol.* 100:7–14.

Andersson, A., P. E. Granum, and U. Ronner. 1998a. The adhesion of *Bacillus cereus* spores to epithelial cells might be an additional virulence mechanism. *Int. J. Food Microbiol.* 39:93–99.

Andersson, A., U. Ronner, and P. E. Granum. 1995. What problems does the food industry have with the spore-forming pathogens *Bacillus cereus* and *Clostridium perfringens*? *Int. J. Food Microbiol.* 28:145–155.

Andersson, A., B. Svensson, A. Christiansson, and U. Ronner. 1999. Comparison between automatic ribotyping and random amplified polymorphic DNA analysis of *Bacillus cereus* isolates from the dairy industry. *Int. J. Food Microbiol.* 47:147–151.

Andersson, M. A., R. Mikkola, J. Helin, M. C. Andersson, and M. Salkinoja-Salonen. 1998b. A novel sensitive bioassay for detection of *Bacillus cereus* emetic toxin and related depsipeptide ionophores. *Appl. Environ. Microbiol.* 64:1338–1343.

Andersson, M. A., P. Hakulinen, U. Honkalampi-Hamalainen, D. Hoornstra, J. C. Lhuguenot, J. Maki-Paakkanen, M. Savolainen, I. Severin, A. L. Stammati, L. Turco, A. Weber, A. von Wright, F. Zucco, and M. Salkinoja-Salonen. 2007. Toxicological profile of cereulide, the *Bacillus cereus* emetic toxin, in functional assays with human, animal and bacterial cells. *Toxicon* 49:351–367.

Andersson, M. A., E. L. Jaaskelainen, R. Shaheen, T. Pirhonen, L. M. Wijnands, and M. S. Salkinoja-Salonen. 2004. Sperm bioassay for rapid detection of cereulide-producing *Bacillus cereus* in food and related environments. *Int. J. Food Microbiol.* 94:175–183.

Anonymous. 3 March 2008, access date. Outbreak surveillance data. Centers for Disease Control and Prevention, Washington, DC. http://www.cdc.gov/foodborneoutbreaks/outbreak.data.htm.

Arnesen, L. P. S., K. O'Sullivan, and P. E. Granum. 2007. Food poisoning potential of *Bacillus cereus* strains from Norwegian dairies. *Int. J. Food Microbiol.* 116:292–296.

Asano, S. I., Y. Nukumizu, H. Bando, T. Iizuka, and T. Yamamoto. 1997. Cloning of novel enterotoxin genes from *Bacillus cereus* and *Bacillus thuringiensis*. *Appl. Environ. Microbiol.* 63:1054–1057.

Baker, J. M., and M. W. Griffiths. 1995. Evidence for increased thermostability of *Bacillus cereus* enterotoxin in milk. *J. Food Prot.* 58:443–445.

Baker, J. M., and M. W. Griffiths. 1993. Predictive modeling of psychrotrophic *Bacillus cereus*. *J. Food Prot.* 56:684–688.

Banerjee, M., and P. K. Sarkar. 2004. Growth and enterotoxin production by sporeforming bacterial pathogens from spices. *Food Control* **15**:491–496.

Banerjee, P., D. Lenz, J. P. Robinson, J. L. Rickus, and A. K. Bhunia. 2008. A novel and simple cell-based detection system with a collagen-encapsulated B-lymphocyte cell line as a biosensor for rapid detection of pathogens and toxins. *Lab. Invest.* **88**:196–206.

Baron, F., M. F. Cochet, N. Grosset, M. N. Madec, R. Briandet, S. Dessaigne, S. Chevalier, M. Gautier, and S. Jan. 2007. Isolation and characterization of a psychrotolerant toxin producer, *Bacillus weihenstephanensis*, in liquid egg products. *J. Food Prot.* **70**:2782–2791.

Beattie, S. H., and A. G. Williams. 2002. Growth and diarrhoeagenic enterotoxin formation by strains of *Bacillus cereus* in vitro in controlled fermentations and in situ in food products and a model food system. *Food Microbiol.* **19**:329–340.

Becker, H., G. Schaller, W. von Wiese, and G. Terplan. 1994. *Bacillus cereus* in infant foods and dried milk products. *Int. J. Food Microbiol.* **23**:1–15.

Beecher, D. J., and J. D. Macmillan. 1990. A novel bicomponent hemolysin from *Bacillus cereus*. *Infect. Immun.* **58**:2220–2227.

Beecher, D. J., J. S. Pulido, N. P. Barney, and A. C. L. Wong. 1995. Extracellular virulence factors in *Bacillus cereus* endophthalmitis: methods and implication of involvement of hemolysin BL. *Infect. Immun.* **63**:632–639.

Beecher, D. J., and A. C. L. Wong. 1994. Improved purification and characterization of hemolysin BL, a hemolytic dermonecrotic vascular permeability factor from *Bacillus cereus*. *Infect. Immun.* **62**:980–986.

Beecher, D. J., and A. C. L. Wong. 2000. Tripartite haemolysin BL: isolation and characterization of two distinct homologous sets of components from a single *Bacillus cereus* isolate. *Microbiology* **146**:1371–1380.

Beecher, D. J., and A. C. L. Wong. 1997. Tripartite hemolysin BL from *Bacillus cereus*. Hemolytic analysis of component interaction and model for its characteristic paradoxical zone phenomenon. *J. Biol. Chem.* **272**:233–239.

Bhatnagar, R., and S. Batra. 2001. Anthrax toxin. *Crit. Rev. Microbiol.* **27**:167–200.

Black, E. P., P. Setlow, A. D. Hocking, C. M. Stewart, A. L. Kelly, and D. G. Hoover. 2007. Response of spores to high-pressure processing. *Compr. Rev. Food Sci. Food Saf.* **6**:103–119.

Brillard, J., and D. Lereclus. 2004. Comparison of cytotoxin *cytK* promoters from *Bacillus cereus* strain ATCC 14579 and from a *B. cereus* food-poisoning strain. *Microbiology* **150**:2699–2705.

Browne, N., and B. C. A. Dowds. 2002. Acid stress in the food pathogen *Bacillus cereus*. *J. Appl. Microbiol.* **92**:404–414.

Browne, N., and B. C. A. Dowds. 2001. Heat and salt stress in the food pathogen *Bacillus cereus*. *J. Appl. Microbiol.* **91**:1085–1094.

Byrne, B., G. Dunne, and D. J. Bolton. 2006. Thermal inactivation of *Bacillus cereus* and *Clostridium perfringens* vegetative cells and spores in pork luncheon roll. *Food Microbiol.* **23**:803–808.

Campbell, G. A., and R. Mutharasan. 2007. Method of measuring *Bacillus anthracis* spores in the presence of copious amounts of *Bacillus thuringiensis* and *Bacillus cereus*. *Anal. Chem.* **79**:1145–1152.

Cardazzo, B., E. Negrisolo, L. Carraro, L. Alberghini, T. Patarnello, and V. Giaccone. 2008. Multiple-locus sequence typing and analysis of toxin genes in *Bacillus cereus* food-borne isolates. *Appl. Environ. Microbiol.* **74**:850–860.

Carlin, F., M. Fricker, A. Pielaat, S. Heisterkamp, R. Shaheen, M. S. Salonen, B. Svensson, C. Nguyen-The, and M. Ehling-Schulz. 2006. Emetic toxin-producing strains of *Bacillus cereus* show distinct characteristics within the *Bacillus cereus* group. *Int. J. Food Microbiol.* **109**:132–138.

Chang, J. M., and T. H. Chen. 2003. Bacterial foodborne outbreaks in central Taiwan, 1991–2000. *J. Food Drug Anal.* **11**:53–59.

Charni, N., C. Perissol, J. Le Petit, and N. Rugani. 2000. Production and characterization of monoclonal antibodies against vegetative cells of *Bacillus cereus*. *Appl. Environ. Microbiol.* **66**:2278–2281.

Choma, C., and P. E. Granum. 2002. The enterotoxin T (BcET) from *Bacillus cereus* can probably not contribute to food poisoning. *FEMS Microbiol. Lett.* **217**:115–119.

Christiansson, A., K. Ekelund, and H. Ogura. 1997. Membrane filtration method for enumeration and isolation of spores of *Bacillus cereus* from milk. *Int. Dairy J.* **7**:743–748.

Christiansson, A., A. S. Naidu, I. Nilsson, T. Wadstrom, and H. E. Pettersson. 1989. Toxin production by *Bacillus cereus* dairy isolates in milk at low temperatures. *Appl. Environ. Microbiol.* **55**:2595–2600.

Clavel, T., F. Carlin, C. Dargaignaratz, D. Lairon, C. Nguyen-The, and P. Schmitt. 2007. Effects of porcine bile on survival of *Bacillus cereus* vegetative cells and haemolysin BL enterotoxin production in reconstituted human small intestine media. *J. Appl. Microbiol.* **103**:1568–1575.

Clavel, T., F. Carlin, D. Lairon, C. Nguyen-The, and P. Schmitt. 2004. Survival of *Bacillus cereus* spores and vegetative cells in acid media simulating human stomach. *J. Appl. Microbiol.* **97**:214–219.

Collado, J., A. Fernandez, L. M. Cunha, M. J. Ocio, and A. Martinez. 2003. Improved model based on the Weibull distribution to describe the combined effect of pH and temperature on the heat resistance of *Bacillus cereus* in carrot juice. *J. Food Prot.* **66**:978–984.

Collado, J., A. Fernandez, M. Rodrigo, and A. Martinez. 2006. Modelling the effect of a heat shock and germinant concentration on spore germination of a wild strain of *Bacillus cereus*. *Int. J. Food Microbiol.* **106**:85–89.

Cooper, R. M., and J. L. McKillip. 2006. Enterotoxigenic Bacillus spp. DNA fingerprint revealed in naturally contaminated nonfat dry milk powder using rep-PCR. *J. Basic Microbiol.* **46**:358–364.

Cortezzo, D. E., B. Setlow, and P. Setlow. 2004. Analysis of the action of compounds that inhibit the germination of spores of Bacillus species. *J. Appl. Microbiol.* **96**:725–741.

Cronin, U. P., and M. G. Wilkinson. 2008. *Bacillus cereus* endospores exhibit a heterogeneous response to heat treatment and low-temperature storage. *Food Microbiol.* **25**:235–243.

Cruz, J., and T. J. Montville. 2008. Influence of nisin on the resistance of *Bacillus anthracis* sterne spores to heat and hydrostatic pressure. *J. Food Prot.* **71**:196–199.

Damgaard, P. H., C. S. Jacobsen, and J. Sorensen. 1996a. Development and application of a primer set for specific detection of *Bacillus thuringiensis* and *Bacillus cereus* in soil using magnetic capture hybridization and PCR amplification. *Syst. Appl. Microbiol.* **19**:436–441.

Damgaard, P. H., H. D. Larsen, B. W. Hansen, J. Bresciani, and K. Jorgensen. 1996b. Enterotoxin-producing strains of *Bacillus thuringiensis* isolated from food. *Lett. Appl. Microbiol.* **23**:146–150.

Day, T. L., S. R. Tatani, S. Notermans, and R. W. Bennett. 1994. A comparison of ELISA and RPLA for detection of *Bacillus cereus* diarrhoeal enterotoxin. *J. Appl. Bacteriol.* **77**:9–13.

De Lara, J., P. S. Fernandez, P. M. Periago, and A. Palop. 2002. Irradiation of spores of *Bacillus cereus* and *Bacillus subtilis* with electron beams. *Innovat. Food Sci. Emerg. Techn.* **3**:379–384.

Del Torre, M., M. Della Corte, and M. L. Stecchini. 2001. Prevalence and behaviour of *Bacillus cereus* in a REPFED of Italian origin. *Int. J. Food Microbiol.* **63**:199–207.

Devasconcellos, F. J. M., and L. Rabinovitch. 1995. A new formula for an alternative culture-medium, without antibiotics, for isolation and presumptive quantification of *Bacillus cereus* in foods. *J. Food Prot.* 58:235–238.

Dierick, K., E. Van Coillie, I. Swiecicka, G. Meyfroidt, H. Devlieger, A. Meulemans, G. Hoedemaekers, L. Fourie, M. Heyndrickx, and J. Mahillon. 2005. Fatal family outbreak of *Bacillus cereus*-associated food poisoning. *J. Clin. Microbiol.* 43:4277–4279.

Dietrich, R., C. Fella, S. Strich, and E. Martlbauer. 1999. Production and characterization of monoclonal antibodies against the hemolysin BL enterotoxin complex produced by *Bacillus cereus*. *Appl. Environ. Microbiol.* 65:4470–4474.

Dietrich, R., M. Moravek, C. Burk, P. E. Granum, and E. Martlbauer. 2005. Production and characterization of antibodies against each of the three subunits of the *Bacillus cereus* nonhemolytic enterotoxin complex. *Appl. Environ. Microbiol.* 71:8214–8220.

Drobniewski, F. A. 1993. *Bacillus cereus* and related species. *Clin. Microbiol. Rev.* 6:324–338.

Duc, L. H., T. C. Dong, N. A. Logan, A. D. Sutherland, J. Taylor, and S. M. Cutting. 2005. Cases of emesis associated with bacterial contamination of an infant breakfast cereal product. *Int. J. Food Microbiol.* 102:245–251.

Dufrenne, J., M. Bijwaard, M. Tegiffel, R. Beumer, and S. Notermans. 1995. Characteristics of some psychrotrophic *Bacillus cereus* isolates. *Int. J. Food Microbiol.* 27:175–183.

Dufrenne, J., P. Soentoro, S. Tatini, T. Day, and S. Notermans. 1994. Characteristics of *Bacillus cereus* related to safe food production. *Int. J. Food Microbiol.* 23:99–109.

Duport, C., A. Zigha, E. Rosenfeld, and P. Schmitt. 2006. Control of enterotoxin gene expression in *Bacillus cereus* F4430/73 involves the redox-sensitive ResDE signal transduction system. *J. Bacteriol.* 188:6640–6651.

Durak, M. Z., H. I. Fromm, J. R. Huck, R. N. Zadoks, and K. J. Boor. 2006. Development of molecular typing methods for Bacillus spp. and Paenibacillus spp. isolated from fluid milk products. *J. Food Sci.* 71:M50–M56.

Ehling-Schulz, M., M. Fricker, H. Grallert, P. Rieck, M. Wagner, and S. Scherer. 2006. Cereulide synthetase gene cluster from emetic *Bacillus cereus*: structure and location on a mega virulence plasmid related to *Bacillus anthracis* toxin plasmid pXO1. *BMC Microbiol.* 6:20.

Ehling-Schulz, M., B. Svensson, M. H. Guinebretiere, T. Lindback, M. Andersson, A. Schulz, M. Fricker, A. Christiansson, P. E. Granum, E. Martlbauer, C. Nguyen-The, M. Salkinoja-Salonen, and S. Scherer. 2005a. Emetic toxin formation of *Bacillus cereus* is restricted to a single evolutionary lineage of closely related strains. *Microbiology* 151:183–197.

Ehling-Schulz, M., N. Vukov, A. Schulz, R. Shaheen, M. Andersson, E. Martlbauer, and S. Scherer. 2005b. Identification and partial characterization of the nonribosomal peptide synthetase gene responsible for cereulide production in emetic *Bacillus cereus*. *Appl. Environ. Microbiol.* 71:105–113.

Eneroth, A., B. Svensson, G. Molin, and A. Christiansson. 2001. Contamination of pasteurized milk by *Bacillus cereus* in the filling machine. *J. Dairy Res.* 68:189–196.

Erickson, M. C., and J. L. Kornacki. 2003. Bacillus anthracis: current knowledge in relation to contamination of food. *J. Food Prot.* 66:691–699.

Ernst, C., J. Schulenburg, P. Jakob, S. Dahms, A. M. Lopez, G. Nychas, D. Werber, and G. Klein. 2006. Efficacy of amphoteric surfactant- and peracetic acid-based disinfectants on spores of *Bacillus cereus* in vitro and on food premises of the German armed forces. *J. Food Prot.* 69:1605–1610.

Evans, J. A., S. L. Russell, C. James, and J. E. L. Corry. 2004. Microbial contamination of food refrigeration equipment. *J. Food Eng.* 62:225–232.

Fagerlund, A., T. Lindback, A. K. Storset, P. E. Granum, and S. P. Hardy. 2008. *Bacillus cereus* Nhe is a pore-forming toxin with structural and functional properties similar to the ClyA (HlyE, SheA) family of haemolysins, able to induce osmotic lysis in epithelia. *Microbiology* 154:693–704.

Fagerlund, A., O. Ween, T. Lund, S. P. Hardy, and P. E. Granum. 2004. Genetic and functional analysis of the *cytK* family of genes in *Bacillus cereus*. *Microbiology* 150:2689–2697.

Faille, C., F. Fontaine, and T. Benezech. 2001. Potential occurrence of adhering living *Bacillus* spores in milk product processing lines. *J. Appl. Microbiol.* 90:892–900.

Faille, C., C. Jullien, F. Fontaine, M. N. Bellon-Fontaine, C. Slomianny, and T. Benezech. 2002. Adhesion of *Bacillus* spores and *Escherichia coli* cells to inert surfaces: role of surface hydrophobicity. *Can. J. Microbiol.* 48:728–738.

Faille, C., J. M. Membre, J. P. Tissier, M. N. Bellon-Fontaine, B. Carpentier, M. A. Laroche, and T. Benezech. 2000. Influence of physicochemical properties on the hygienic status of stainless steel with various finishes. *Biofouling* 15:261–274.

Faille, C., G. Tauveron, C. L. Gentil-Lelievre, and C. Slomianny. 2007. Occurrence of *Bacillus cereus* spores with a damaged exosporium: consequences on the spore adhesion on surfaces of food processing lines. *J. Food Prot.* 70:2346–2353.

Fang, T. J., C. Y. Chen, and W. Y. Kuo. 1999. Microbiological quality and incidence of *Staphylococcus aureus* and *Bacillus cereus* in vegetarian food products. *Food Microbiol.* 16:385–391.

Feijoo, S. C., L. N. Cotton, C. E. Watson, and J. H. Martin. 1997a. Effect of storage temperatures and ingredients on growth of *Bacillus cereus* in coffee creamers. *J. Dairy Sci.* 80:1546–1553.

Feijoo, S. C., W. W. Hayes, C. E. Watson, and J. H. Martin. 1997b. Effects of Microfluidizer technology on *Bacillus licheniformis* spores in ice cream mix. *J. Dairy Sci.* 80:2184–2187.

Fermanian, C., C. Lapeyre, J. M. Fremy, and M. Claisse. 1997. Diarrhoeal toxin production at low temperature by selected strains of *Bacillus cereus*. *J. Dairy Res.* 64:551–559.

Fernandez, A., J. Collado, L. M. Cunha, M. J. Ocio, and A. Martinez. 2002. Empirical model building based on Weibull distribution to describe the joint effect of pH and temperature on the thermal resistance of *Bacillus cereus* in vegetable substrate. *Int. J. Food Microbiol.* 77:147–153.

Fernandez, A., M. J. Ocio, P. S. Fernandez, M. Rodrigo, and A. Martinez. 1999. Application of nonlinear regression analysis to the estimation of kinetic parameters for two enterotoxigenic strains of *Bacillus cereus* spores. *Food Microbiol.* 16:607–613.

Finlay, W. J. J., N. A. Logan, and A. D. Sutherland. 1997. Semi-automated metabolic staining assay for *Bacillus cereus* emetic toxin. *Appl. Environ. Microbiol.* 65:1811–1812.

Fletcher, P., and N. A. Logan. 1999. Improved cytotoxicity assay for *Bacillus cereus* diarrhoeal enterotoxin. *Lett. Appl. Microbiol.* 28:394–400.

Foegeding, P. M., and E. D. Berry. 1997. Cold temperature growth of clinical and food isolates of *Bacillus cereus*. *J. Food Prot.* 60:1256–1258.

Francis, K. P., R. Mayr, F. von Stetten, G. S. A. B. Stewart, and S. Scherer. 1998. Discrimination of psychrotrophic and mesophilic strains of the *Bacillus cereus* group by PCR targeting of major cold shock protein genes. *Appl. Environ. Microbiol.* 64:3525–3529.

Frederiksen, K., H. Rosenquist, K. Jorgensen, and A. Wilcks. 2006. Occurrence of natural *Bacillus thuringiensis* contaminants and residues of *Bacillus thuringiensis*-based insecticides on fresh fruits and vegetables. *Appl. Environ. Microbiol.* 72:3435–3440.

Fricker, M., U. Messelhausser, U. Busch, S. Scherer, and M. Ehling-Schulz. 2007. Diagnostic real-time PCR assays for the detection of emetic *Bacillus cereus* strains in foods and recent food-borne outbreaks. *Appl. Environ. Microbiol.* 73:1892–1898.

Fricker, M., R. Reissbrodt, and M. Ehling-Schulz. 2008. Evaluation of standard and new chromogenic selective plating media for isolation and identification of *Bacillus cereus*. *Int. J. Food Microbiol.* **121**:27–34.

Fritze, D. 2004. Taxonomy of the genus *Bacillus* and related genera: the aerobic endospore-forming bacteria. *Phytopathology* **94**:1245–1248.

From, C., V. Hormazabal, and P. E. Granum. 2007a. Food poisoning associated with pumilacidin-producing *Bacillus pumilus* in rice. *Int. J. Food Microbiol.* **115**:319–324.

From, C., V. Hormazabal, S. P. Hardy, and P. E. Granum. 2007b. Cytotoxicity in *Bacillus mojavensis* is abolished following loss of surfactin synthesis: implications for assessment of toxicity and food poisoning potential. *Int. J. Food Microbiol.* **117**:43–49.

From, C., R. Pukall, P. Schumann, V. Hormazabal, and P. E. Granum. 2005. Toxin-producing ability among *Bacillus* spp. outside the *Bacillus cereus* group. *Appl. Environ. Microbiol.* **71**:1178–1183.

Gaillard, S., I. Leguerinel, N. Savy, and P. Mafart. 2005. Quantifying the combined effects of the heating time, the temperature and the recovery medium pH on the regrowth lag time of *Bacillus cereus* spores after a heat treatment. *Int. J. Food Microbiol.* **105**:53–58.

Garcia-Arribas, M. L., and J. M. Kramer. 1990. The effect of glucose, starch, and pH on growth, enterotoxin and haemolysin production by strains of *Bacillus cereus* associated with food poisoning and non-gastrointestinal infection. *Int. J. Food Microbiol.* **11**:21–33.

Gaviria Rivera, A. M., P. E. Granum, and F. G. Priest. 2000. Common occurrence of enterotoxin genes and enterotoxicity in *Bacillus thuringiensis*. *FEMS Microbiol. Lett.* **190**:151–155.

Girisch, M., M. Ries, M. Zenker, R. Carbon, R. Rauch, and M. Hofbeck. 2003. Intestinal perforations in a premature infant caused by *Bacillus cereus*. *Infection* **31**:192–193.

Glatz, B. A., and J. M. Goepfert. 1976. Defined conditions for synthesis of *Bacillus cereus* enterotoxin by fermenter-grown cultures. *Appl. Environ. Microbiol.* **32**:400–404.

Goepfert, J. M., W. M. Spira, and H. U. Kim. 1972. *Bacillus cereus*: food poisoning organism. A review. *J. Milk Food Technol.* **35**:213–227.

Gore, H. M., C. A. Wakeman, R. M. Hull, and J. L. McKillip. 2003. Real-time molecular beacon NASBA reveals *hblC* expression from *Bacillus* spp. in milk. *Biochem. Biophys. Res. Comm.* **311**:386–390.

Granum, P. E. 1997. *Bacillus cereus*, p. 327–336. *In* M. P. Doyle, L. R. Beuchat, and T. J. Montville (ed.), *Food Microbiology Fundamentals and Frontiers*, 1st ed. ASM Press, Washington, DC.

Granum, P. E., A. Andersson, C. Gayther, M. T. Giffel, H. Larsen, T. Lund, and K. O'Sullivan. 1996. Evidence for a further enterotoxin complex produced by *Bacillus cereus*. *FEMS Microbiol. Lett.* **141**:145–149.

Granum, P. E., S. Brynestad, K. O'Sullivan, and H. Nissen. 1993. Enterotoxin from *Bacillus cereus* - Production and biochemical characterization. *Neth. Milk Dairy J.* **47**:63–70.

Granum, P. E., and T. Lund. 1997. *Bacillus cereus* and its food poisoning toxins. *FEMS Microbiol. Lett.* **157**:223–228.

Granum, P. E., K. O'Sullivan, and T. Lund. 1999. The sequence of the non-haemolytic enterotoxin operon from *Bacillus cereus*. *FEMS Microbiol. Lett.* **177**:225–229.

Griffiths, M. W. 1990. Toxin production by psychrotrophic *Bacillus* spp. present in milk. *J. Food Prot.* **53**:790–792.

Griffiths, M. W., and J. D. Phillips. 1990a. Incidence, source and some properties of psychrotrophic *Bacillus* spp. found in raw and pasteurized milk. *J. Soc. Dairy Technol.* **43**:62–66.

Griffiths, M. W., and J. D. Phillips. 1990b. Strategies to control the outgrowth of spores of psychrotrophic *Bacillus* in dairy products.

II. Use of heat-treatments. *Milchwiss. Milk Sci. Int.* **45**:719–721.

Griffiths, M. W., and H. Schraft. 2002. *Bacillus cereus* food poisoning, p. 261–270. *In* D. Cliver (ed.), *Foodborne Diseases*, 2nd ed. Elsevier Science Ltd., London, England.

Guinebretiere, M. H., V. Broussolle, and C. Nguyen-The. 2002. Enterotoxigenic profiles of food-poisoning and food-borne *Bacillus cereus* strains. *J. Clin. Microbiol.* **40**:3053–3056.

Guinebretiere, M. H., A. Fagerlund, P. E. Granum, and C. Nguyen-The. 2006. Rapid discrimination of cytK-1 and cytK-2 genes in *Bacillus cereus* strains by a novel duplex PCR system. *FEMS Microbiol. Lett.* **259**:74–80.

Guinebretiere, M. H., H. Girardin, C. Dargaignaratz, F. Carlin, and C. Nguyen-The. 2003. Contamination flows of *Bacillus cereus* and spore-forming aerobic bacteria in a cooked, pasteurized and chilled zucchini puree processing line. *Int. J. Food Microbiol.* **82**:223–232.

Haggblom, M. M., C. Apetroaie, M. A. Andersson, and M. S. Salkinoja-Salonen. 2002. Quantitative analysis of cereulide, the emetic toxin of *Bacillus cereus*, produced under various conditions. *Appl. Environ. Microbiol.* **68**:2479–2483.

Hansen, B. M., P. E. Hoiby, G. B. Jensen, and N. B. Hendriksen. 2003. The *Bacillus cereus* bceT enterotoxin sequence reappraised. *FEMS Microbiol. Lett.* **223**:21–24.

Hansen, B. M., T. D. Leser, and N. B. Hendriksen. 2001. Polymerase chain reaction assay for the detection of *Bacillus cereus* group cells. *FEMS Microbiol. Lett.* **202**:209–213.

Hanson, M. L., W. L. Wendorff, and K. B. Houck. 2005. Effect of heat treatment of milk on activation of *Bacillus* spores. *J. Food Prot.* **68**:1484–1486.

Hardy, S. P., T. Lund, and P. E. Granum. 2001. CytK toxin of *Bacillus cereus* forms pores in planar lipid bilayers and is cytotoxic to intestinal epithelia. *FEMS Microbiol. Lett.* **197**:47–51.

Hatakka, M. 1998a. Microbiological quality of cold meals served by airlines. *J. Food Saf.* **18**:185–195.

Hatakka, M. 1998b. Microbiological quality of hot meals served by airlines. *J. Food Prot.* **61**:1052–1056.

Hauge, S. 1955. Food poisoning caused by aerobic spore forming bacilli. *J. Appl. Bacteriol.* **18**:591–595.

Heinrichs, J. H., D. J. Beecher, J. D. Macmillan, and B. A. Zilinskas. 1993. Molecular-cloning and characterization of the *hbla* gene encoding the B component of hemolysin BL from *Bacillus cereus*. *J. Bacteriol.* **175**:6760–6766.

Helgason, E., O. A. Okstad, D. A. Caugant, H. A. Johansen, A. Fouet, M. Mock, I. Hegna, and A. B. Kolsto. 2000. *Bacillus anthracis, Bacillus cereus,* and *Bacillus thuringiensis*—one species on the basis of genetic evidence. *Appl. Environ. Microbiol.* **66**:2627–2630.

Helgason, E., N. J. Tourasse, R. Meisal, D. A. Caugant, and A. B. Kolsto. 2004. Multilocus sequence typing scheme for bacteria of the *Bacillus cereus* group. *Appl. Environ. Microbiol.* **70**:191–201.

Helyer, R. J., T. Kelley, and R. C. W. Berkeley. 1997. Pyrolysis mass spectrometry studies on *Bacillus anthracis, Bacillus cereus* and their close relatives. *Zentralb. Bakteriol.* **285**:319–328.

Hendriksen, N. B., and B. M. Hansen. 2006. Detection of *Bacillus thuringiensis* kurstaki HD1 on cabbage for human consumption. *FEMS Microbiol. Lett.* **257**:106–111.

Hill, K. K., L. O. Ticknor, R. T. Okinaka, M. Asay, H. Blair, K. A. Bliss, M. Laker, P. E. Pardington, A. P. Richardson, M. Tonks, D. J. Beecher, J. D. Kemp, A. B. Kolsto, A. C. L. Wong, P. Keim, and P. J. Jackson. 2004. Fluorescent amplified fragment length polymorphism analysis of *Bacillus anthracis, Bacillus cereus,* and *Bacillus thuringiensis* isolates. *Appl. Environ. Microbiol.* **70**:1068–1080.

Hillers, V. N., L. Medeiros, P. Kendall, G. Chen, and S. DiMascola. 2003. Consumer food-handling behaviors associated with prevention of 13 foodborne illnesses. *J. Food Prot.* 66:1893–1899.

Honda, T., A. Shiba, S. Seo, J. Yamamoto, J. Matsuyama, and T. Miwatani. 1991. Identity of hemolysins produced by *Bacillus thuringiensis* and *Bacillus cereus*. *FEMS Microbiol. Lett.* 79:205–210.

Hoornstra, D., O. Dahlman, E. Jaaeskelainen, M. A. Andersson, A. Weber, B. Aurela, H. Lindell, and M. S. Salkinoja-Salonen. 2006. Retention of *Bacillus cereus* and its toxin, cereulide, in cellulosic fibres. *Holzforschung* 60:648–652.

Hornstra, L. M., Y. P. de Vries, W. M. de Vos, T. Abee, and M. H. Wells-Bennik. 2005. *gerR*, a novel *ger* operon involved in L-alanine- and inosine-initiated germination of *Bacillus cereus* ATCC 14579. *Appl. Environ. Microbiol.* 71:774–781.

Hsieh, Y. M., S. J. Sheu, Y. L. Chen, and H. Y. Tsen. 1999. Enterotoxigenic profiles and polymerase chain reaction detection of *Bacillus cereus* group cells and *B. cereus* strains from foods and food-borne outbreaks. *J. Appl. Microbiol.* 87:481–490.

Huck, J. R., N. H. Woodcock, R. D. Ralyea, and K. J. Boor. 2007. Molecular subtyping and characterization of psychrotolerant endospore-forming bacteria in two New York State fluid milk processing systems. *J. Food Prot.* 70:2354–2364.

Hughes, S., B. Bartholomew, J. C. Hardy, and J. M. Kramer. 1988. Potential application of a HEp-2 cell assay in the investigation of *Bacillus cereus* emetic syndrome food poisoning. *FEMS Microbiol. Lett.* 52:7–11.

Jaaskelainen, E. L., M. M. Haggblom, M. A. Andersson, and M. S. Salkinoja-Salonen. 2004. Atmospheric oxygen and other conditions affecting the production of cereulide by *Bacillus cereus* in food. *Int. J. Food Microbiol.* 96:75–83.

Jaaskelainen, E. L., M. M. Haggblom, M. A. Andersson, L. Vanne, and M. S. Salkinoja-Salonen. 2003. Potential of *Bacillus cereus* for producing an emetic toxin, cereulide, in bakery products: quantitative analysis by chemical and biological methods. *J. Food Prot.* 66:1047–1054.

Jackson, P. J., K. K. Hill, M. T. Laker, L. O. Ticknor, and P. Keim. 1999. Genetic comparison of *B. anthracis* and its close relatives using AFLP and PCR analysis. *J. Appl. Microbiol.* 87:263–269.

Kawamura-Sato, K., Y. Hirama, N. Agata, H. Ito, K. Torii, A. Takeno, T. Hasegawa, Y. Shimomura, and M. Ohta. 2005. Quantitative analysis of cereulide, an emetic toxin of *Bacillus cereus*, by using rat liver mitochondria. *Microbiol. Immunol.* 49:25–30.

Klavenes, A., T. Stalheim, O. Sjovold, K. Josefsen, and P. E. Granum. 2002. Attachment of *Bacillus cereus* spores with and without appendages to stainless steel surfaces. *Food Bioprod. Process.* 80:312–318.

Koo, K., P. M. Foegeding, and H. E. Swaisgood. 1998a. Construction and expression of a bifunctional single-chain antibody against *Bacillus cereus* spores. *Appl. Environ. Microbiol.* 64:2490–2496.

Koo, K., P. M. Foegeding, and H. E. Swaisgood. 1998b. Development of a streptavidin-conjugated single-chain antibody that binds *Bacillus cereus* spores. *Appl. Environ. Microbiol.* 64:2497–2502.

Kotiranta, A., K. Lounatmaa, and M. Haapasalo. 2000. Epidemiology and pathogenesis of *Bacillus cereus* infections. *Microbes Infect.* 2:189–198.

Kramer, J. M., and R. J. Gilbert. 1989. *Bacillus cereus* and other *Bacillus* species, p. 21–70. *In* M. P. Doyle (ed.), *Foodborne Bacterial Pathogens*, Marcel Dekker, New York, NY.

Kretzer, J. W., R. Lehmann, M. Schmelcher, M. Banz, K. P. Kim, C. Korn, and M. J. Loessner. 2007. Use of high-affinity cell wall-binding domains of bacteriophage endolysins for immobiliza-

tion and separation of bacterial cells. *Appl. Environ. Microbiol.* 73:1992–2000.

Lasch, P., H. Nattermann, M. Erhard, M. Stammler, R. Grunow, N. Bannert, B. Appel, and D. Naumann. 2008. MALDI-TOF mass spectrometry compatible inactivation method for highly pathogenic microbial cells and spores. *Anal. Chem.* 80:2026–2034.

Lechner, S., R. Mayr, K. P. Francis, B. M. Pruss, T. Kaplan, E. Wiessner-Gunkel, G. S. A. B. Stewart, and S. Scherer. 1998. *Bacillus weihenstephanensis* sp. nov. is a new psychrotolerant species of the *Bacillus cereus* group. *Int. J. Syst. Bacteriol.* 48:1373–1382.

Lin, S. F., H. Schraft, and M. W. Griffiths. 1998. Identification of *Bacillus cereus* by Fourier transform infrared spectroscopy (FTIR). *J. Food Prot.* 61:921–923.

Lindback, T., A. Fagerlund, M. S. Rodland, and P. E. Granum. 2004. Characterization of the *Bacillus cereus* Nhe enterotoxin. *Microbiology* 150:3959–3967.

Liu, Y. L., B. Elsholz, S. O. Enfors, and M. Gabig-Ciminska. 2007. Confirmative electric DNA array-based test for food poisoning *Bacillus cereus*. *J. Microbiol. Methods* 70:55–64.

Love, T. E., C. Redmond, and C. N. Mayers. 2008. Real time detection of anthrax spores using highly specific anti-EA1 recombinant antibodies produced by competitive panning. *J. Immunol. Methods* 334:1–10.

Lund, T., M. L. De Buyser, and P. E. Granum. 2000. A new cytotoxin from *Bacillus cereus* that may cause necrotic enteritis. *Mol. Microbiol.* 38:254–261.

Lund, T., and P. E. Granum. 1996. Characterisation of a non-haemolytic enterotoxin complex from *Bacillus cereus* isolated after a foodborne outbreak. *FEMS Microbiol. Lett.* 141:151–156.

Lund, T., and P. E. Granum. 1997. Comparison of biological effect of the two different enterotoxin complexes isolated from three different strains of *Bacillus cereus*. *Microbiology* 143:3329–3336.

Lund, T., and P. E. Granum. 1999. The 105-kDa protein component of *Bacillus cereus* non-haemolytic enterotoxin (Nhe) is a metalloprotease with gelatinolytic and collagenolytic activity. *FEMS Microbiol. Lett.* 178:355–361.

Madegowda, M., S. Eswaramoorthy, S. K. Burley, and S. Swaminathan. 2008. X-ray crystal structure of the B component of hemolysin BL from *Bacillus cereus*. *Proteins* 71:534–540.

Mahler, H., A. Pasi, J. M. Kramer, P. Schulte, A. C. Scoging, W. Bar, and S. Krahenbuhl. 1997. Fulminant liver failure in association with the emetic toxin of *Bacillus cereus*. *New Engl. J. Med.* 336:1142–1148.

Mantynen, V., and K. Lindstrom. 1998. A rapid PCR-based DNA test for enterotoxic *Bacillus cereus*. *Appl. Environ. Microbiol.* 64:1634–1639.

Margot, P., M. Whalen, A. Gholamhuseinian, P. Piggot, and D. Karmata. 1998. The *lytE* gene of *Bacillus subtilis* 168 encodes a cell wall hydrolase. *J. Bacteriol.* 180:749–752.

McGovern, J. P., W. Y. Shih, R. Rest, M. Purohit, Y. Pandya, and W. H. Shih. 2008. Label-free flow-enhanced specific detection of *Bacillus anthracis* using a piezoelectric microcantilever sensor. *Analyst* 133:649–654.

Mead, P. S., L. Slutsker, V. Dietz, L. F. McCaig, J. S. Bresee, C. Shapiro, P. M. Griffin, and R. V. Tauxe. 1999. Food related illness and death in the United States. *Emerg. Infect. Dis.* 5:607–625.

Meer, R. R., J. Baker, F. W. Bodyfelt, and M. W. Griffiths. 1991. Psychrotrophic *Bacillus* spp. in fluid milk products: a review. *J. Food Prot.* 54:969–979.

Mikami, T., T. Horikawa, T. Murakami, T. Matsumoto, A. Yamakawa, S. Murayama, S. Katagiri, K. Shinagawa, and M. Suzuki. 1994. An improved method for detecting cytostatic toxin (emetic toxin) of *Bacillus cereus* and its application to food samples. *FEMS Microbiol. Lett.* **119**:53–57.

Mikkola, R., N. E. L. Saris, P. A. Grigoriev, M. A. Andersson, and M. S. Salkinoja-Salonen. 1999. Ionophoretic properties and mitochondrial effects of cereulide: the emetic toxin of *Bacillus cereus*. *Eur. J. Biochem.* **263**:112–117.

Ngamwongsatit, P., W. Buasri, P. Pianariyanon, C. Pulsrikarn, M. Ohba, A. Assavanig, and W. Panbangred. 2008. Broad distribution of enterotoxin genes (hblCDA, nheABC, cytK, and entFM) among *Bacillus thuringiensis* and *Bacillus cereus* as shown by novel primers. *Int. J. Food Microbiol.* **121**:352–356.

Nilsson, J., B. Svensson, K. Ekelund, and A. Christiansson. 1998. A RAPD-PCR method for large-scale typing of *Bacillus cereus*. *Lett. Appl. Microbiol.* **27**:168–172.

Notermans, S., and C. A. Batt. 1998. A risk assessment approach for food-borne *Bacillus cereus* and its toxins. *J. Appl. Microbiol.* **84**:51S–61S.

Odumeru, J. A., M. Steele, L. Fruhner, C. Larkin, J. D. Jiang, E. Mann, and W. B. McNab. 1999. Evaluation of accuracy and repeatability of identification of food-borne pathogens by automated bacterial identification systems. *J. Clin. Microbiol.* **37**:944–949.

Odumeru, J. A., A. K. Toner, C. A. Muckle, M. W. Griffiths, and J. A. Lynch. 1997. Detection of *Bacillus cereus* diarrheal enterotoxin in raw and pasteurized milk. *J. Food Prot.* **60**:1391–1393.

Okstad, O. A., M. Gominet, B. Purnelle, M. Rose, D. Lereclus, and A. B. Kolsto. 1999. Sequence analysis of three *Bacillus cereus* loci carrying PlcR-regulated genes encoding degradative enzymes and enterotoxin. *Microbiology* **145**:3129–3138.

Ouhib, O., T. Clavel, and P. Schmitt. 2006. The production of *Bacillus cereus* enterotoxins is influenced by carbohydrate and growth rate. *Curr. Microbiol.* **53**:222–226.

Paananen, A., R. Mikkola, T. Sareneva, S. Matikainen, M. Hess, M. Andersson, I. Julkunen, M. S. Salkinoja-Salonen, and T. Timonen. 2002. Inhibition of human natural killer cell activity by cereulide, an emetic toxin from *Bacillus cereus*. *Clin. Exp. Immunol.* **129**:420–428.

Pedersen, P. B., M. E. Bjornvad, M. D. Rasmussen, and J. N. Petersen. 2002. Cytotoxic potential of industrial strains of *Bacillus* spp. *Regul. Toxicol. Pharmacol.* **36**:155–161.

Peng, J. S., W. C. Tsai, and C. C. Chou. 2001. Surface characteristics of *Bacillus cereus* and its adhesion to stainless steel. *Int. J. Food Microbiol.* **65**:105–111.

Perani, M., A. H. Bishop, and A. Vaid. 1998. Prevalence of beta-exotoxin, diarrhoeal toxin and specific delta-endotoxin in natural isolates of *Bacillus thuringiensis*. *FEMS Microbiol. Lett.* **160**:55–60.

Perkins, D. L., C. R. Lovell, B. V. Bronk, B. Setlow, P. Setlow, and M. L. Myrick. 2005. *In* (IMS 2005) Proceedings of the 2005 IEEE International Workshop on Measurement Systems for Homeland Security, Contraband Detection and Personal Safety. IEEE, Los Alamitos, CA.

Pirttijarvi, T. S., L. M. Ahonen, L. M. Maunuksela, and M. S. Salkinoja-Salonen. 1998. *Bacillus cereus* in a whey process. *Int. J. Food Microbiol.* **44**:31–41.

Rajkovic, A., M. Uyttendaele, S. A. Ombregt, E. Jaaskelainen, M. Salkinoja-Salonen, and J. Debevere. 2006. Influence of type of food on the kinetics and overall production of *Bacillus cereus* emetic toxin. *J. Food Prot.* **69**:847–852.

Rajkovic, A., M. Uyttendaele, A. Vermeulen, M. Andjelkovic, I. Fitz-James, P. in't Veld, Q. Denon, R. Verhe, and J. Debevere. 2008. Heat resistance of *Bacillus cereus* emetic toxin, cereulide. *Lett. Appl. Microbiol.* **46**:536–541.

Ripabelli, G., J. McLauchlin, V. Mithani, and E. J. Threlfall. 2000. Epidemiological typing of *Bacillus cereus* by amplified fragment length polymorphism. *Lett. Appl. Microbiol.* **30**:358–363.

Rivera, A. M. G., P. E. Granum, and F. G. Priest. 2000. Common occurrence of enterotoxin genes and enterotoxicity in *Bacillus thuringiensis*. *FEMS Microbiol. Lett.* **190**:151–155.

Ronimus, R. S., L. E. Parker, and H. W. Morgan. 1997. The utilization of RAPD-PCR for identifying thermophilic and mesophilic *Bacillus* species. *FEMS Microbiol. Lett.* **147**:75–79.

Ronimus, R. S., L. E. Parker, N. Turner, S. Poudel, A. Ruckert, and H. W. Morgan. 2003. A RAPD-based comparison of thermophilic bacilli from milk powders. *Int. J. Food Microbiol.* **85**:45–61.

Rowan, N. J., and J. G. Anderson. 1998a. Diarrhoeal enterotoxin production by psychrotrophic *Bacillus cereus* present in reconstituted milk-based infant formulae (MIF). *Lett. Appl. Microbiol.* **26**:161–165.

Rowan, N. J., and J. G. Anderson. 1998b. Growth and enterotoxin production by diarrhoeagenic *Bacillus cereus* in dietary supplements prepared for hospitalized HIV patients. *J. Hosp. Infect.* **38**:139–146.

Rowan, N. J., K. Deans, J. G. Anderson, C. G. Gemmell, I. S. Hunter, and T. Chaithong. 2001. Putative virulence factor expression by clinical and food isolates of *Bacillus* spp. after growth in reconstituted infant milk formulae. *Appl. Environ. Microbiol.* **67**:3873–3781.

Rusul, G., and N. H. Yaacob. 1995. Prevalence of *Bacillus cereus* in selected foods and detection of entertoxin using Tecra-Via and BCET-RPLA. *Int. J. Food Microbiol.* **25**:131–139.

Ryan, P. A., J. D. Macmillan, and B. A. Zilinskas. 1997. Molecular cloning and characterization of the genes encoding the L(1) and L(2) components of hemolysin BL from *Bacillus cereus*. *J. Bacteriol.* **179**:2551–2556.

Sakurai, N., K. Kolke, Y. Irie, and H. Hayashi. 1994. The rice culture filtrate of *Bacillus cereus* isolated from emetic type food poisoning causes mitochondrial swelling in a HEp-2 cell. *Microbiol. Immunol.* **38**:337–343.

Salkinoja-Salonen, M. S., R. Vuorio, M. A. Andersson, P. Kampfer, M. C. Andersson, T. Honkanen-Buzalski, and A. C. Scoging. 1999. Toxigenic strains of *Bacillus licheniformis* related to food poisoning. *Appl. Environ. Microbiol.* **65**:4637–4645.

Schoeni, J. L., and A. C. L. Wong. 2005. *Bacillus cereus* food poisoning and its toxins. *J. Food Prot.* **68**:636–648.

Schoeni, J. L., and A. C. L. Wong. 1999. Heterogeneity observed in the components of hemolysin BL, an enterotoxin produced by *Bacillus cereus*. *Int. J. Food Microbiol.* **53**:159–167.

Schraft, H., and M. W. Griffiths. 2006. *Bacillus cereus* gastroenteritis, p. 563–582. *In* H. P. Riemann and D. O. Cliver (ed.), *Foodborne Infections and Intoxications*, 3rd ed. Academic Press (Elsevier), Amsterdam, The Netherlands.

Schraft, H., and M. W. Griffiths. 1995. Specific oligonucleotide primers for detection of lecithinase-positive *Bacillus* spp. by PCR. *Appl. Environ. Microbiol.* **61**:98–102.

Schraft, H., M. Steele, B. McNab, J. Odumeru, and M. W. Griffiths. 1996. Epidemiological typing of *Bacillus* spp. isolated from food. *Appl. Environ. Microbiol.* **62**:4229–4232.

Shaheen, R., M. A. Andersson, C. Apetroaie, A. Schulz, M. Ehling-Schulz, V. M. Ollilainen, and M. S. Salkinoja-Salonen. 2006. Potential of selected infant food formulas for production of *Bacillus cereus* emetic toxin, cereulide. *Int. J. Food Microbiol.* **107**:287–294.

Shinagawa, K. 1990. Analytical methods for *Bacillus cereus* and other *Bacillus* species. *Int. J. Food Microbiol.* **10**:125–141.

Shinagawa, K., H. Konuma, H. Sekita, and S. Sugii. 1995. Emesis of rhesus-monkeys induced by intragastric administration with the Hep-2 vacuolation factor (cereulide) produced by *Bacillus cereus*. *FEMS Microbiol. Lett.* **130**:87–90.

Shinagawa, K., S. Ueno, H. Konuma, N. Matsusaka, and S. Sugii. 1991. Purification and characterization of the vascular permeability factor produced by *Bacillus cereus*. *J. Vet. Med. Sci.* **53**:281–286.

Shinagawa, K., Y. Ueno, D. L. Hu, S. Ueda, and S. Sugii. 1996. Mouse lethal activity of a HEp-2 vacuolation factor, cereulide, produced by *Bacillus cereus* isolated from vomiting-type food poisoning. *J. Vet. Med. Sci.* **58:**1027–1029.

Simone, E., M. Goosen, S. H. Notermans, and M. W. Borgdorff. 1997. Investigations of foodborne diseases by food inspection services in the Netherlands, 1991 to 1994. *J. Food Prot.* **60:**442–446.

Skerman, V. B. D., V. McGowan, and P. H. A. Sneath. 1980. *Approved Lists of Bacterial Names.* American Society for Microbiology, Washington, DC.

Spira, W. M., and J. M. Goepfert. 1972. *Bacillus cereus* induced fluid accumulation in rabbit ileal loops. *Appl. Microbiol.* **24:**341–348.

Stalheim, T., and P. E. Granum. 2001. Characterization of spore appendages from *Bacillus cereus* strains. *J. Appl. Microbiol.* **91:**839–845.

Stenfors, L. P., and P. E. Granum. 2001. Psychrotolerant species from the *Bacillus cereus* group are not necessarily *Bacillus weihenstephanensis.* *FEMS Microbiol. Lett.* **197:**223–228.

Stenfors, L. P., R. Mayr, S. Scherer, and P. E. Granum. 2002. Pathogenic potential of fifty *Bacillus weihenstephanensis* strains. *FEMS Microbiol. Lett.* **215:**47–51.

Szabo, R. A., J. L. Speirs, and M. Akhtar. 1991. Cell culture detection and conditions for production of a *Bacillus cereus* heat-stable toxin. *J. Food Prot.* **54:**272–276.

Tauveron, G., C. Slomianny, C. Henry, and C. Faille. 2006. Variability among *Bacillus cereus* strains in spore surface properties and influence on their ability to contaminate food surface equipment. *Int. J. Food Microbiol.* **110:**254–262.

Taylor, J. M. W., A. D. Sutherland, K. E. Aidoo, and N. A. Logan. 2005. Heat-stable toxin production by strains of *Bacillus cereus, Bacillus firmus, Bacillus megaterium, Bacillus simplex* and *Bacillus licheniformis.* *FEMS Microbiol. Lett.* **242:**313–317.

Te Giffel, M. C., R. R. Beumer, P. E. Granum, and F. M. Rombouts. 1997. Isolation and characterisation of *Bacillus cereus* from pasteurised milk in household refrigerators in the Netherlands. *Int. J. Food Microbiol.* **34:**307–318.

Te Giffel, M. C., R. R. Beumer, B. A. Slaghuis, and F. M. Rombouts. 1995. Occurrence and characterization of (psychrotrophic) *Bacillus cereus* on farms in the Netherlands. *Neth. Milk Dairy J.* **49:**125–138.

Te Giffel, M. C., A. Wagendorp, A. Herrewegh, and F. Driehuis. 2002. Bacterial spores in silage and raw milk. *Antonie van Leeuwenhoek* **81:**625–630.

Te Giffel, M. C., R. R. Beumer, S. Leijendekkers, and F. M. Rombouts. 1996. Incidence of *Bacillus cereus* and *Bacillus subtilis* in foods in the Netherlands. *Food Microbiol.* **13:**53–58.

Thorsen, L., B. M. Hansen, K. F. Nielsen, N. B. Hendriksen, R. K. Phipps, and B. B. Budde. 2006. Characterization of emetic *Bacillus weihenstephanensis,* a new cereulide-producing bacterium. *Appl. Environ. Microbiol.* **72:**5118–5121.

Vaillant, V., H. de Valk, E. Baron, T. Ancelle, P. Colin, M.-C. Delmas, B. Dufour, R. Pouillot, Y. le Strat, P. Weinbreck, E. Jougla, and J. C. Desenclos. 2005. Foodborne infections in France. *Foodborne Pathog. Dis.* **2:**221–232.

Vaisanen, O. M., N. J. Mwaisumo, and M. S. Salkinoja-Salonen. 1991. Differentiation of dairy strains of the *Bacillus cereus* group by phage typing, minimum growth temperature, and fatty acid analysis. *J. Appl. Bacteriol.* **70:**315–324.

Van Netten, P., and J. M. Kramer. 1992. Media for the detection and enumeration of *Bacillus cereus* in foods: a review. *Int. J. Food Microbiol.* **17:**85–99.

von Stetten, F., K. P. Francis, S. Lechner, K. Neuhaus, and S. Scherer. 1998. Rapid discrimination of psychrotolerant and mesophilic strains of the *Bacillus cereus* group by PCR targeting 16s rDNA. *J. Microbiol. Methods* **34:**99–106.

Wijnands, L. M., J. B. Dufrenne, F. M. Rombouts, P. H. In't Veld, and F. M. Van Leusden. 2006a. Prevalence of potentially pathogenic *Bacillus cereus* in food commodities in The Netherlands. *J. Food Prot.* **69:**2587–2594.

Wijnands, L. M., J. B. Dufrenne, M. H. Zwietering, and F. M. van Leusden. 2006b. Spores from mesophilic *Bacillus cereus* strains germinate better and grow faster in simulated gastro-intestinal conditions than spores from psychrotrophic strains. *Int. J. Food Microbiol.* **112:**120–128.

Winder, C. L., and R. Goodacre. 2004. Comparison of diffuse-reflectance absorbance and attenuated total reflectance FT-IR for the discrimination of bacteria. *Analyst* **129:**1118–1122.

Yamada, S., E. Ohashi, N. Agata, and K. Venkateswaran. 1999. Cloning and nucleotide sequence analysis of *gyrB* of *Bacillus cereus, B. thuringiensis, B. mycoides,* and *B. anthracis* and their application to the detection of *B. cereus* in rice. *Appl. Environ. Microbiol.* **65:**1483–1490.

Yemini, M., Y. Levi, E. Yagil, and J. Rishpon. 2007. Specific electrochemical phage sensing for *Bacillus cereus* and *Mycobacterium smegmatis. Bioelectrochemistry* **70:**180–184.

Yokoyama, K., M. Ito, N. Agata, M. Isobe, K. Shibayama, T. Horii, and M. Ohta. 1999. Pathological effect of synthetic cereulide, an emetic toxin of *Bacillus cereus,* is reversible in mice. *FEMS Immunol. Med. Microbiol.* **24:**115–120.

Yuan, Z. M., B. M. Hansen, L. Andrup, and J. Eilenberg. 2002. Detection of enterotoxin genes in mosquito-larvicidal *Bacillus* species. *Curr. Microbiol.* **45:**221–225.

Zigha, A., E. Rosenfeld, P. Schmitt, and C. Duport. 2006. Anaerobic cells of *Bacillus cereus* F4430/73 respond to low oxidoreduction potential by metabolic readjustments and activation of enterotoxin expression. *Arch. Microbiol.* **185:**222–233.

Zigha, A., E. Rosenfeld, P. Schmitt, and C. Duport. 2007. The redox regulator *fnr* is required for fermentative growth and enterotoxin synthesis in *Bacillus cereus* F4430/73. *J. Bacteriol.* **189:**2813–2824.

Pathogens and Toxins in Foods: Challenges and Interventions
Edited by V. K. Juneja and J. N. Sofos
© 2010 ASM Press, Washington, DC

Chapter 2

Campylobacter jejuni and Other Campylobacters

Nelson A. Cox, L. Jason Richardson, and Michael T. Musgrove

CHARACTERISTICS OF THE ORGANISM AND TYPES OF ILLNESS

The taxonomy has changed considerably over the years and could change in the future, but to date the family *Campylobacteraceae* includes the genera *Campylobacter, Arcobacter,* and *Sulfurospirillum* and the generically misclassified *Bacteroides ureolyticus* (Vandamme, 2000). In regard to the genus *Campylobacter,* there are 14 species, and of these species, several are considered pathogenic to humans, causing enteric and extraintestinal illnesses. *Campylobacter* species are gram-negative, microaerophilic, non-spore-forming organisms with curved or small spiral-shaped cells that have characteristic rapid, darting, reciprocating motility (corkscrew-like motion by means of a single polar flagellum at one or both ends of the cell) and can occur in short or long chains. They range in width from 0.2 to 0.9 μm, and in length from 0.5 to 5 μm, and most species have an optimum temperature range for growth of 30 to 37°C, except for the thermophilic *Campylobacter* spp., which grow optimally at 42°C. A few strains can grow aerobically or anaerobically. An atmosphere containing increased nitrogen may be needed for optimum growth of certain strains. The cells can form spherical or coccoid bodies as cultures age, and it has been postulated that certain species can have the characteristics of a viable, but not culturable, state (Rollins and Colwell, 1986; Moran and Upton, 1987).

Campylobacter species have a chemoorganotrophic metabolism, and energy is derived from amino acids or tricarboxylic acid cycle intermediates due to their inability to oxidize or ferment carbohydrates. The majority of *Campylobacter* spp. reduce nitrate and nitrite. *Campylobacter* spp. have typical biochemical characteristics, which include the reduction of fumarate to succinate; a negative reaction to methyl red, acetone, and indole production; negative hippurate hydrolysis (except for most *C. jejuni* strains); and positivity for oxidase activity (Vandamme, 2000; Sellars et al., 2002). *Campylobacter* spp. can be either catalase positive or negative. Broadly speaking, catalase-positive *Campylobacter* spp. are most often associated with human disease, but not in all cases.

The pathogenic mechanisms by which campylobacteriosis occurs are not totally understood, and information on species other than *C. jejuni* is scarce, but some important virulence factors include motility, ability to translocate, chemotaxis, and production of toxins (Walker et al., 1986; Ketley, 1997; Wassenaar, 1997). It appears that the virulence factors involved in the infection greatly influence the symptoms of the disease, and pathogenesis results from several different bacterial properties and host defenses. Motility, which is achieved by means of a single flagellum at one or both ends of the bacterium, has an extremely important role in virulence because it is required for the bacterium to reach the attachment sites and penetrate into the intestinal cells. If the bacterium loses its motility, its ability to colonize the gastrointestinal tract and cause infection is lost.

C. jejuni contains two flagellin genes, *flaA* and *flaB;* the wild-type bacterium expresses *flaA* only, but *flaB* can be expressed under certain conditions. The flagellum of *Campylobacter* spp. plays a much more important function than just motility. *C. jejuni* flagella may also play a role in the dissemination and internalization of the organism. In addition, flagellin has been proposed as an adhesin in the binding to culture cells. It has been shown that *C. jejuni* is able to become internalized within human intestinal epithelial cells and traverse monolayers of polarized human colonic carcinoma cells, allowing access to

Nelson A. Cox and L. Jason Richardson • USDA-ARS-Poultry Microbiological Safety Research Unit, Russell Research Center, Athens, GA 30605. Michael T. Musgrove • Egg Safety and Quality Research Unit, Russell Research Center, Athens, GA 30605.

submucosal tissue, which leads to tissue inflammation and damage (Grant et al., 1993; Ketley, 1997).

Toxin production also plays a role in pathogenicity. In regard to *C. jejuni*, the organism synthesizes several toxins, classified mainly as enterotoxins or cytotoxins (Wassenaar, 1997). Spikes in levels of all immunoglobulin classes have been shown with humans after infection. Even though the synthesis of several toxins has been reported, their mechanism of action and importance in regard to disease still remain unclear. The problem in determining the aforementioned is that researchers have not been able to detect toxin produced by *Campylobacter* spp. A more comprehensive review of the pathogenesis of *Campylobacter* spp. can be found in review articles by Walker et al. (1986), Ketley (1997), and Wassenaar (1997).

For approximately 3 decades, the genus *Campylobacter* has had increased focus as a threat to food safety, due to the rise in enteritis in humans caused by consumption or handling of foods contaminated with the organisms. Four species *(C. jejuni, C. coli, C. lari,* and *C. upsaliensis)* are known as "thermophilic campylobacters" and are clinically significant due to their roles as dominant causative agents of human campylobacteriosis (Blaser et al., 1982; Jacobs-Reitsma, 2000; Keener et al., 2004). *C. jejuni* is the predominant species that causes bacterial gastroenteritis in the United States and in many other developed countries. *C. coli* is responsible for the majority of other cases of illness. In the United States, campylobacteriosis and salmonellosis go back and forth as the leading cause of bacterial food-borne illness. Transmission of *Campylobacter* spp. to humans generally occurs by either ingestion of contaminated food or water or by direct contact with fecal material from infected animals or persons. In humans, there are two types of illnesses associated with *Campylobacter* infections, and they are intestinal and extraintestinal infections. Two types of diarrhea are usually observed with campylobacteriosis: (i) an inflammatory diarrhea, with slimy, bloody stools containing leukocytes and fever and (ii) noninflammatory diarrhea, with watery stools and the absence of blood and leukocytes (Wassenaar, 1997). In some cases, intense abdominal pain, headaches, cramping, and vomiting can occur. Serious complications, such as Reiter's syndrome, Guillain-Barré syndrome (GBS), osteomyelitis, pancreatitis, nephritis, myocarditis, cystitis, septic abortion, and bacteremia in certain cases, can arise (Altekruse et al., 1999; Winer, 2001; Keener et al., 2004). Although campylobacteriosis does not usually lead to death, it has been estimated that as many as 730 people in the United States with *Campylobacter* infections die annually, often due to secondary complications (Saleha et al., 1998).

In the vast majority of cases, campylobacteriosis is mainly a self-limiting bacterial gastroenteritis, and recovery is completed in approximately 8 days, either spontaneously or after appropriate antimicrobial therapy. However, in some instances symptoms can persist longer than 2 weeks. The population that is most susceptible to illness includes children less than 1 year of age, young adults aged 15 to 25 years, and immunosuppressed individuals (Keener et al., 2004). A concern for those suffering from campylobacteriosis is that they will suffer from neurological sequelae months or even years afterward. Two neuropathies associated with *C. jejuni* infections are GBS and Miller Fisher syndrome. Both of these syndromes are characterized by being acute or subacute, immune-mediated neuropathies.

In regard to GBS, this is characterized by alexia, motor paralysis, acellular increase in the total protein content in the cerebrospinal fluid, and inflammatory demyelinating polyradiculoneuropathy. GBS occurs in approximately 1 out of 1,000 cases (Winer, 2001). GBS cases are associated with nerve roots, causing mononuclear infiltration of peripheral nerves, and this eventually leads to primary axonal degeneration or demyelination. Molecular mimicry is believed to be the cause of GBS because a few peripheral nerves of the human neurological system share molecules similar to those of antigens on the surface of *C. jejuni* cells (Winer, 2001). Since *C. jejuni* contains a lipopolysaccharide structure (LPS) attached to the outer membrane, the core oligosaccharides of its LPS contain ganglioside-like structures, which are similar to certain human gangliosides (Ang et al., 2001). The LPS structure is very antigenic, and upon exposure to *C. jejuni,* the immune system produces antibodies against the LPS structure as an attempt to fight the infection. Due to the similarity of the core oligosaccharides of the LPS and the gangliosides, after the infection, antibodies attack the gangliosides on the neuromuscular junction, contributing to the appearance of neurological symptoms (Lindsay, 1997; Ang et al., 2001). A more detailed review of GBS can be found in review articles by Lindsay (1997), Willison and O'Hanlon (2000), and Winer (2001).

SOURCES AND INCIDENCE IN THE ENVIRONMENT AND FOODS

Unlike many other enteric pathogens, *Campylobacter* spp. have limited spread from host to host. *Campylobacter* spp. may not be recovered by conventional cultural methods outside of the host if exposed to dry conditions or atmospheric oxygen levels for extended periods of time. *Campylobacter* enteritis can be classified as a zoonosis, because animals are the main reservoir of these organisms. *Campylobacter* spp. exist as commensals in many wild and domestic animals (Keener et al., 2004). This presents a risk to food safety due to the contamination of carcasses at slaughter and other foodstuffs by cross-contamination when raw or undercooked meat is handled. Contamination with this pathogen can occur at numerous stages along the food chain. This includes, but is not limited to, production, processing, distribution, handling, and preparation. *Campylobacter* spp. are fastidious organisms that are capable of existing in a broad range of environments and have been sporadically recovered from rivers, coastal waters, shellfish, and vegetables but routinely recovered from sheep, cattle, swine, rodents, and avian species (Jacobs-Reitsma, 2000; Kemp et al., 2005). Certain species of *Campylobacter* are routinely associated with certain species of animals. In poultry and cattle, *C. jejuni* is the predominant species, and *C. coli* is the common species recovered from swine. The majority of *Campylobacter* infections are sporadic, and outbreaks are rare but have been traced back to contaminated water, raw milk, poultry, beef, eggs, fruits, and contact with farm animals and pets (Altekruse et al., 1999; Friedman et al., 2004). Generally speaking, the primary source of contamination of the environment and foods is believed to be from animal feces (Brown et al., 2004).

Avian species, particularly poultry, are the most common host for *Campylobacter* spp.; therefore, poultry is considered the main source of human illness. Studies have shown that as much as 70% of human illnesses due to *Campylobacter* spp. are caused by the consumption or handling of raw or undercooked poultry or poultry products. Increased attention has been given to reducing the level of *Campylobacter* spp. in poultry pre- and postharvest to reduce the level and incidence of raw product contamination (Allos, 2001; Friedman et al., 2004; Keener et al., 2004). The ecology of *Campylobacter* spp. in poultry is not fully elucidated. Numerous studies are being conducted to determine when and how *Campylobacter* spp. gain entry into poultry flocks so that more effective intervention strategies can be employed.

Campylobacter spp. colonize the mucus layer of the intestinal tract but have been recovered from numerous tissues and organs within the bird, suggesting it is not limited to the digestive tract. In addition to the digestive tract, *Campylobacter* spp. have been isolated from the circulating blood, thymus, spleen, liver, gallbladder, unabsorbed yolk sacs, ovarian follicles, and reproductive tracts of commercial poultry (Cox et al., 2005, 2006, 2007; Richardson et al., 2007a). In regard to the digestive tract, levels up to 10^9 CFU/g of fecal content have been shown (Altekruse et al., 1999). Two modes of transmission of *Campylobacter* into poultry flocks occur and they are horizontal and vertical transmission (Keener et al., 2004; Byrd et al., 2007). It has been shown that if a single bird in a flock is colonized, then the spread to adjacent rearing mates is rapid, and within a week *Campylobacter* prevalence in the flock can reach 100% (Beery et al., 1988; Gregory et al., 1997; Wallace et al., 1998).

The prevalence of *Campylobacter* contamination of carcasses and poultry products can vary greatly, depending on the sensitivity of the cultural procedures utilized and by the point along the process chain at which sampling is being conducted. The type of methodology employed significantly affects prevalence rates of *Campylobacter* spp. from carcasses at the final stages of processing. For example, if a survey is being conducted on the prevalence of *Campylobacter* on poultry carcasses postchill and enrichment of the sample is utilized, then as much as 70% to 100% of the samples can be positive for *Campylobacter*. However, if a less sensitive method is utilized, such as direct plating onto selective agar, which may exclude sublethally injured cells, then the number of samples detected as positive could be greatly reduced. Including both direct plating and enrichment often allows the best probability for recovery. Even though the enrichment used is designed to be selective for *Campylobacter* spp., the organisms can be culturally fragile, to the extent that they can be overgrown by organisms that were meant to be suppressed (Musgrove et al., 2001). A question often asked is whether the injured or stressed cells could have the ability to infect humans and cause illness. This is one reason why studies on the incidence of *Campylobacter* in poultry processing plants vary, and it is critical to consider the cultural procedures utilized and the impact those choices have on sensitivity to recover or detect the organism.

A significantly high prevalence rate of *Campylobacter* spp. contamination can be found in retail poultry and poultry products and is often directly related to the prevalence rate at the farm. The reported prevalence rate for the farm continuum varies between studies, but on average greater than 70% of the birds are *Campylobacter* positive (Maekin et al., 2008; Allen et al., 2007). In a study of supermarkets, *Campylobacter* spp. were isolated from 82%, 82%, and 71% of whole chickens, breast with skin attached, and pieces, respectively (Harrison et al., 2001). However, *Campylobacter* incidence within the farm continuum does not generally relate to an increase in retail positives in other types of animal products.

Campylobacter prevalence rates in one study reported 47% of cattle and 46% of swine harbored the organism within the farm continuum (Nielsen et al., 1997). However, comparatively low prevalence rates (less than 2%) of *Campylobacter* spp. have been found in these meat products at the retail level (Zhao et al., 2001). This could be due to, but is not limited to, a number of factors such as commensal level in poultry; skin removal from the carcasses of other animals during processing, unlike poultry; processing procedures utilized for poultry carcasses; and the sheer number of poultry carcasses being processed in the plant each day.

INTRINSIC AND EXTRINSIC FACTORS THAT AFFECT SURVIVAL AND GROWTH IN FOOD PRODUCTS AND CONTRIBUTE TO OUTBREAKS

Cross-contamination of food products is a major factor that contributes to human illness. *Campylobacter* spp. can be sensitive to environmental conditions outside of an animal's intestinal tract. Even though *Campylobacter* spp. are sensitive to drying, high oxygen concentration, and low pH (less than or equal to 4.7), they are still one of the biggest causes of gastroenteritis. Several studies have shown that strong acids, such as formic, acetic, ascorbic, and lactic acids, rapidly inhibit the growth and survival of *Campylobacter* spp. *Campylobacter* spp. are sensitive to sodium chloride, with an optimum growth concentration of 0.5% (Doyle and Roman, 1982a). Sensitivity to salt also shows a temperature-dependent effect. At 4°C, *C. jejuni* was sensitive to 1.0% or more NaCl, but the rate of death at this temperature was much less than that at 25°C. A 3 $\log_{10}$ decrease of cells occurred in 4.5% NaCl after 1.25 to 2.1 days at 25°C, but a similar reduction took about 2 weeks at the same salt concentration when a temperature of 4°C was maintained.

The decimal reduction time for *Campylobacter* spp. varies, depending on the type of food product, but survival kinetics generally follow a rapid decline in numbers, which is followed by a slower rate of inactivation. This may explain the high survival rate of *Campylobacter* spp. on poultry carcasses due to the high levels of the organisms in the bird's digestive tract at the time of processing. Studies have shown that the potential for survival decreases to a few hours at temperatures of 37°C and increases to a few days at temperatures of 4°C. However, *Campylobacter* spp. have been shown to survive for several weeks in groundwater (Buswell et al., 1998).

When environments become unfavorable for growth, *C. jejuni*, it is postulated, can enter into a viable but nonculturable (VBNC) state. The cells are metabolically active and show signs of respiratory activity but are unable to be cultured through conventional methodology procedures. The VBNC stage was first described by Rollins and Colwell (1986), who postulated that it could play a role in human infection and illness. The VBNC state arises from exposure to sublethal adverse environmental conditions, and recovery occurs by passage of the organism to a susceptible host. Several studies have explored the recovery of VBNC forms of *Campylobacter* cells (Jones et al., 1991; Saha and Sanyel, 1991; Chaveerach et al., 2003; Richardson et al., 2007b). Nonculturable *C. jejuni* and *C. coli* after subjection to acid stress were shown to be viable by injecting the cultures into the amniotic fluid and yolk sac of fertilized eggs (Chaveerach et al., 2003). In a separate study, *Campylobacter* cells were subjected to dry stress on filter and chick paper pads and after becoming nonculturable were determined to be viable utilizing a chick bioassay (Richardson et al., 2007b). In addition, freeze-thaw-injured *C. jejuni* cells that were nonculturable were converted back to culturable after passage through the rat gut (Saha et al., 1991). The significance of the VBNC state remains unclear and controversial, but as the understanding of this phenomenon unfolds, this could shed light on how *Campylobacter* spp. survive and persist in certain food commodities and go undetected in dry environments.

FOOD PROCESSING OPERATIONS THAT INFLUENCE THE NUMBERS, SPREAD, OR CHARACTERISTICS

While outbreaks of human campylobacteriosis have been associated with raw milk and untreated water, poultry meat, which is frequently contaminated with the organism, may be responsible for as much as 70% of sporadic campylobacteriosis (Skirrow, 1991). Contamination is thought to originate from the intestinal tract of primarily avian species (mainly poultry) and then spread to the meat during transport and processing, though it has also been demonstrated that broiler crops, particularly after a feed withdrawal prior to transport to the processing facility, may harbor large numbers of *Campylobacter* bacteria (Musgrove et al., 1997; Berrang et al., 2001). Crops burst more often than cecal pouches or other parts of the gut and can contaminate previously

Campylobacter-free carcasses. As birds enter the plant, levels of *Campylobacter* in the intestinal tract can be as high as 10^7 CFU/g cecal contents, and when whole carcasses with feathers are rinsed, 10^6 CFU/ml of rinse can be recovered (Berrang et al., 2004; Northcutt, 2005). External contamination often increases during transport from grow-out houses to the processing plant. Generally, *Campylobacter* counts decrease in the scalding tank, increase during removal of feathers (picking), and are at their highest immediately following evisceration (Berrang and Cason, 2000). In most commercial processing facilities, carcasses are chilled in a series of tanks using recirculated cool water. This part of the process has a tendency to dilute *Campylobacter* numbers, especially on those carcasses that are highly contaminated. Since passage of hazard analysis and critical control points, the amount of water used during processing has increased dramatically, and many researchers report a decreased incidence of *Campylobacter* contamination on many poultry products.

Campylobacter is a culturally fastidious organism. Recovering this organism from samples which are low in biologically available moisture can be especially challenging. This has contributed to the perception that *Campylobacter* contamination of egg may not be important. Doyle (1984) reported that only 8.1% of layer hens shed *Campylobacter* chronically and found ~1% of shell-contaminated table eggs. Izat and Gardner (1988) sampled two commercial egg processing facilities. Samples analyzed included eggs, egg products, egg wash water, and surfaces within the facilities. *Campylobacter* was never recovered from any of the sample types. However, cultural media for, expertise in, and knowledge of *Campylobacter* have seen many changes in the last 15 to 20 years. Cox sampled spent hens and a small number of eggs from a commercial shell egg washing facility. While *Campylobacter* spp. were not recovered from either egg contents or shells, most of the hens were positive for the organism, including the reproductive tract (Cox et al., 2006). Musgrove and Jones (2006) sampled packer head brushes at two commercial shell egg processing facilities. These brushes assist in gently transferring eggs from the weighing/grading equipment to a series of belts, some of the last surfaces to touch the eggs before they reach retail packaging. Though low rates of recovery were observed with packer head brushes (1.5%) and pooled crushed egg shells/membranes (4.2%), the presence of the bacteria bears consideration, particularly since an egg-borne outbreak of campylobacteriosis has been reported (Finch and Blake, 1985).

RECENT ADVANTAGES IN BIOLOGICAL, CHEMICAL, AND PHYSICAL INTERVENTIONS TO GUARD AGAINST THE PATHOGEN

In developed countries worldwide a great deal of effort is being expended on developing interventions in *Campylobacter* contamination of poultry and poultry products. Preharvest efforts include improved biosecurity, training of farm personnel, drinking water amendments, and even treatment with other microorganisms or their products (Humphrey et al., 1993; Gibbens et al., 2001). Competitive exclusion products can be supplied to hatching chicks to prevent enteric colonization by human pathogens. Generally, more success has been achieved in suppressing *Salmonella* than *Campylobacter*, but some degree of protection has been observed for various formulations. Commercial products, many of which are of a defined nature, are now available. However, undefined cultures made from mucosal scrapings have been the most effective. Truly effective vaccines and application strategies have yet to be developed for prevention of avian intestinal colonization. A formalin-inactivated *C. jejuni* whole-cell vaccine was shown to reduce colonization in immunized chickens by 16% to 93%, and purification of flagellin antigens by preservation of conformational epitopes may enhance their use as vaccines (Widders et al., 1998; Muir et al., 2000). Other interventions have included use of other microorganisms, such as *Pseudomonas aeruginosa* or excretory-secretory products of *Trichuris suis* to affect attachment to skin or intestinal epithelial cells (Abner et al., 2002). *Campylobacter* colonization of broilers has been affected by application of bacteriophages (Carillo et al., 2005; Wagenaar et al., 2005).

Postharvest efforts include making plants sanitary, ensuring clean water for processing, maintaining increased water during processing, adding new equipment, evaluating chill tank water sanitizers, providing carcasses frozen instead of fresh, irradiating, and antimicrobial packaging. In order to reduce the chances of cross-contamination, amendments such as sodium hypochlorite, acidified sodium chlorite, and trisodium phosphate may be added to the inside-outside bird washer or to the chill tank (Kemp et al., 2001; Chantarapagant et al., 2002). Prior to further processing or transport to a retail market, carcasses and other poultry products may be stored under refrigerated or frozen conditions. Survivability of the organism is favored at refrigerated temperatures but may also occur under frozen conditions. Oosterom et al. (1983) reported that freezing affected *C. jejuni* only during the first few hours. They detected an initial drop but determined that the organism could survive on chicken carcasses and chicken livers at $-20°C$ for more than 64 and 84 days, respectively. *Campylobacter* spp. die faster at 25°C than at either 4 or 30°C. Heat injury of *C. jejuni* occurs at 46°C, and death occurs at 48°C. Doyle (1982b) determined rates of thermal inactivation for five strains of *C. jejuni* in a skim-milk heat menstruum, and at 48°C, D values ranged from 7.2 to 12.8 min, while at 55°C they ranged from 0.74 to 1.00 min. These data indicate that ordinary cooking temperatures should be sufficient to destroy campylobacters contaminating poultry or other meat samples.

Campylobacter spp. are sensitive to certain spices and food ingredients, such as sodium chloride. Deibel and Banwart (1994) conducted a study on the effects of various concentrations of oregano, sage, and ground cloves on the growth of *C. jejuni* in a liquid growth medium incubated at 4°C, 25°C, and 42°C over a 48-h period. At 25°C, more than a 3 $\log_{10}$ decrease in cell numbers was observed with the suspensions containing sage or oregano. At 4°C, less than a 1 $\log_{10}$ reduction was observed for any of the three spices. Koidis and Doyle (1983) analyzed the inhibitory effects of garlic, onion, black pepper, and oregano on *C. jejuni* in an enrichment nutrient broth stored at 4°C. A 3.9 $\log_{10}$ decrease was noted for the broth containing onion, and 3.0 $\log_{10}$ reductions below the respective controls were noted for the maximum concentrations of garlic, pepper, and oregano. This may suggest that further processing, particularly marination, may create a more hostile environment for *Campylobacter* spp.

DISCRIMINATIVE DETECTION METHODS FOR CONFIRMATION AND TRACE-BACK OF CONTAMINATED PRODUCTS

Isolation, identification, and confirmation of *Campylobacter* isolates by traditional methods require plating enrichment broth samples after incubation onto selective agar plates. Typical colonies are smooth, convex, and glistening, with a distinct edge, or flat, shiny, translucent, and spreading, with an irregular edge. They are colorless or grayish or light cream and may range from pinpoint to 4 to 5 mm in diameter. Growth may be confluent without distinct colonies, particularly on wet agar. Presumptive *Campylobacter* isolates should be observed using dark-field or phase-contrast microscopy for characteristic morphology and motility; however, cells from cultures older than 24 h may appear coccoidal and nonmotile (Musgrove et al., 2001). An isolated

colony can be useful in further characterization, in terms of antimicrobial resistance, speciation, relatedness to other isolates, or pinpointing sources for epidemiological purposes, though some information can be obtained without obtaining a specific isolate (Lu et al., 2003). In addition, the method used for isolation can affect genotyping results (Newell et al., 2001).

Campylobacter spp. can be confirmed using latex agglutination assays, which are commercially available. A variety of other automated kits and systems based on either immunological or molecular criteria are now available, and some methods combine various identification approaches (Cloak et al., 2001; Padungtod et al., 2002). Immunomagnetic separation of *Campylobacter* spp. coupled with PCR-based detection systems are being developed for detection with food samples (Docherty et al., 1996).

Typing of *Campylobacter* strains is needed for epidemiological purposes. By 1982, the most widely used serological schemes were published and in use (Penner and Hennessy, 1980; Lior et al., 1982). Typing methods can be divided into categories, one of which is typing based on phenotypic characteristics, such as heat-stable or heat-labile antigens or antimicrobial susceptibility patterns. A second category consists of the various genotyping analyses. Included in this category are restriction fragment length polymorphism (RFLP) of selected genomes, such as done with ribotyping or other genomes; bacterial restriction endonuclease DNA analysis (BRENDA); and pulsed-field gel electrophoresis. The usefulness of any of these approaches depends upon availability of resources and personnel time and reproducibility and discriminatory capability of the method, as well as the openness of management to embrace new and sometimes daunting technologies. In the last 15 years, strain differentiation has shifted to highly specific genotyping analyses, including flagellin typing, randomly amplified polymorphic DNA, pulsed-field gel electrophoresis of chromosomal DNA, multiplex PCR-RFLP, and ribotyping. Genotyping techniques can be used to determine sources of infection and routes of transmission in humans and animals, though the variety of approaches published in studies worldwide make interlaboratory comparisons difficult (Wassenaar and Newell, 2000).

Grajewski et al. (1985) originally described a set of 14 phage types which discriminated about 90% of 255 human isolates tested. Therefore, to be of value, it appears that phage typing should be used together with another typing method to confidently discriminate among isolates. Another phenotypic assay used to discriminate between isolates involves the analysis of the outer membrane protein contents of *Campylobacter* by sodium dodecyl sulfate polyacrylamide gel electrophoresis. Although discrimination is possible, limitations of sodium dodecyl sulfate-polyacrylamide gel electrophoresis to discriminate between isolates have also been reported (Derclay et al., 1989). In addition, considerable technician time is required for such analyses, and interlaboratory reproducibility has been poor.

An early comparison of 10 methods for subgrouping *Campylobacter* strains was published in 1991 (Patton et al., 1991). The genotyping methods studied included multilocus enzyme electrophoresis (MEE), BRENDA, restriction digests of selected bacterial chromosomal DNA (ribotyping), and plasmid profile analysis. For the MEE approach, isolates were assayed for enzymatic activity and the patterns were then statistically analyzed to determine the genetic distances between strains. BRENDAs entailed electrophoresing DNA fragments, and observing patterns were revealed by UV illumination following ethidium bromide staining. Ribotyping analysis required characterizing the strains for restriction patterns resulting from the digest of the ribosomal DNA. The authors concluded that MEE, BRENDA, and ribotyping were the most sensitive methods that were capable of identifying nine types among 22 strains. The authors observed that MEE and ribotyping had several advantages over the other methods because they measured relatively stable and significant chromosomal differences and were applicable to other species and genera.

Hernandez et al. (1996) analyzed 39 strains of *Campylobacter* by ribotyping and confirmed the above favorable conclusions. Modification to the method by Patton et al. (1991) employed use of the HaeIII enzyme to digest the isolated chromosomal DNA. Upon electrophoresis of the digest product, they were able to distinguish 32 different band profiles that allowed discrimination between the strains analyzed. Each ribotyping pattern comprised between 3 and 11 fragments of 1 to 10 kb.

A typing scheme for *Campylobacter* spp. has been described (Nachamkin et al., 1996) and entails probing the flagellin gene, *flaA*, for differences or similarities between strains. The method uses a PCR and RFLP. Meinersmann et al. (1997) have described a system that allows differentiation of strains based on differences in *flaA* gene sequencing. Methods such as these are now the standard means for determining outbreak strains and for more in-depth analysis of epidemiological data. Techniques such as these allow for a greater level of discrimination than was previously possible. However, for many methods, genotype instability has also been demonstrated. Multilocus

sequence typing has shown that *C. jejuni* is partly nonclonal, having a natural ability to take up DNA. This genome plasticity, allowing for recombinations within loci used in typing, makes for complex population genetics and further complicates interpretations of typing data (Wassenaar, 2004).

CONCLUDING REMARKS

Commercial poultry have been shown to be a major source of *Campylobacter* gastroenteritis in humans. *C. jejuni* and *C. coli* are the species most often associated with human illness and isolated from poultry. *Campylobacter* spp. are widespread in poultry, and in order to reduce the contamination level of poultry products, reductions within the preharvest continuum will have to be accomplished in order to reduce the level of contamination entering processing facilities. In order for this to be accomplished, a large portion of microbiologists will have to stop arguing whether *Campylobacter* is transmitted via the egg from parent birds to progeny. Many continue to say that there is no evidence to suggest egg transmission and hatchery contamination when the evidence is not only present, but in recent years overwhelming (Clarke and Bueschkens, 1985; Lindblom et al., 1986; Maruyama and Katsube, 1990; Maruyama et al., 1995; Chuma et al., 1994, 1997; Pearson et al., 1996; Cox et al., 2002; Hiett et al., 2002, 2003; Acevedo, 2005; Byrd et al., 2007). It has never been said that vertical transmission is the only source of contamination of a flock, but it is definitely a source, and additional evidence continues to mount. If the research community continues to ignore published facts, then this source will always be present and the level of contamination of commercial poultry will never be eliminated in reference to *Campylobacter*.

There is conflicting evidence in the published literature as to whether *Campylobacter* can pass from breeder hen to progeny through the fertile egg, whether it is vertical or horizontal contamination of the egg and its developing embryo. It is easy to take issue with some of the studies which support the hypothesis that *Campylobacter* is not transmitted by the egg. The number of samples tested in some studies is very small, and then strong conclusions, such as "*Campylobacter* cannot be transmitted through the egg," are made. For example, Acuff et al. (1982) stated that no *Campylobacter* could be found in turkey eggs or young turkey poults; however, only 20 samples of each were tested, with sampling methods that were not very sensitive. Callicott et al. (2006) stated that they could not find any evidence for vertical transmission of *Campylobacter* when only 13 pooled fluff samples were tested out of 60,000 parent breeders. Studies such as these leave the impression that transmission via the egg is highly improbable; however, the majority of the published data support just the opposite conclusion.

For example, when fertile chicken eggs were inoculated with *Campylobacter* by pressure differential, 11% of these chicks at hatch had the inoculated microorganism in their intestinal tract (Clarke and Bueschkens, 1985). Also, *Campylobacter* could be transmitted to the offspring via the egg following oral inoculation of Japanese quails (Maruyama and Katsube, 1990). Chickens raised in a laboratory environment without exposure to any farm environment still became colonized by *Campylobacter* (Lindblom et al., 1986). Studies using a sensitive detection method (colony DNA hybridization) indicated the carrier rate of *Campylobacter* in the cecal contents of newly hatched chicks to be as much as 35%, suggesting that the chicks were already contaminated with *Campylobacter* before they were delivered to the farm (Chuma et al., 1994). More recently, *Campylobacter* isolates from two independent commercial broiler breeder flocks, as well as from their respective progeny, were shown to be clonal in origin using both ribotyping and DNA sequencing analysis (Cox et al., 2002). Through molecular testing, *Campylobacter* has been found in hatchery fluff, intestinal tracts of developing embryos, and newly hatched chicks (Chuma et al., 1994, 1997; Hiett et al., 2002, 2003). Also, epidemiological surveys have traced the source of broiler flock infections to hatcheries (Pearson et al., 1996).

However, even in light of these numerous peer-reviewed published studies demonstrating the transmission of *Campylobacter* from breeders and hatcheries to broiler flocks, obvious deficiencies in the standard cultural methodologies prevent this hypothesis from being universally accepted. The actual samples may be called negative utilizing traditional cultural detection methods, when *Campylobacter* actually is present, but in low numbers or in a VBNC state. Some recent studies support this hypothesis. Byrd et al. (2007), using a modified methodology procedure, were able to culture *Campylobacter* from three different commercial hatchery chick pads, and the breeders providing eggs to these hatcheries were also positive for *Campylobacter*. In addition, *Campylobacter* was cultured from a commercial incubator and from the interior egg content and egg surfaces of fertile commercial breeder eggs (Acevedo, 2005).

These studies conclusively prove that *Campylobacter* is present in the hatchery, breeder flocks, and chicks prior to exposure to any possible environmental source. As we continue to improve our laboratory

methods to detect *Campylobacter*, both viable and dry-stressed viable, but presently nonculturable, it will be accepted as fact that the fertile egg is a significant source of *Campylobacter* spp. Given that other microorganisms, such as *E. coli* and *Salmonella* spp., have been demonstrated to transmit from one generation of chickens to another via fertile eggs (Gordon and Tucker, 1965; Humphrey et al., 1991; Petersen et al., 2006), it seems strange that some scientists refuse to consider that *Campylobacter* is also doing the same. Perhaps, the main reason is the inability of routine cultural methods to consistently recover these organisms from many dry types of samples (e.g., egg shells, hatching debris, etc.). Modifications in standard laboratory procedures have led to recent discoveries. As methodology procedures improve and our understanding of the *Campylobacter* ecology improves, vertical transmission will not be debated, but will become part of a more focused and effective set of intervention strategies. Due to the array of environmental conditions in the poultry continuum, this provides an excellent means of studying the ecology of *Campylobacter* spp., and with knowledge, vision, and persistence, numerous advances and discoveries can be achieved.

REFERENCES

Abner, S. R., D. E. Hill, J. R. Turner, E. D. Black, P. Bartlett, J. F. Urban, and L. S. Mansfield. 2002. Response of intestinal epithelial cells to *Trichuris suis* excretory-secretory products on the influence on *Campylobacter jejuni* invasion under *in vitro* conditions. *J. Parasitol.* **88:**738–745.

Acevedo, Y. M. 2005. Deteccion de *Campylobacter jejuni* en las Incumadoras de una planta procesadora de pollos parrilleros de Puerto Rico. M.S. thesis. Universidad de Puerto Rico, San Juan, Puerto Rico.

Acuff, G. R., C. Vanderzandt, F. A. Gardner, and F. A. Golan. 1982. Examination of turkey eggs, poults, and brooder house facilities for *Campylobacter jejuni*. *J. Food Prot.* **45:**1279–1281.

Allen, A. V., S. A. Bull, J. E. L. Corry, G. Domingue, F. Jørgensen, J. A. Frost, R. Whyte, A. Gonzalez, N. Elviss, and T. J. Humphrey. 2007. *Campylobacter* spp. contamination of chicken carcasses during processing in relation to flock colonization. *Int. J. Food Microbiol.* **113:**45–61.

Allos, B. M. 2001. *Campylobacter jejuni* infections: update on emerging issues and trends. *Clin. Infect. Dis.* **32:**1201–1206.

Altekruse, S. F., N. J. Stern, P. I. Fields, and D. L. Swerdlow. 1999. *Campylobacter jejuni*—an emerging foodborne pathogen. *Emerg. Infect. Dis.* **5:**28–35.

Ang, C. W., M. A. de Klerk, H. P. Endtz, B. C. Jacobs, J. D. Laman, F. G. A. van der Meché, and P. A. van Doorn. 2001. Guillain-Barré syndrome and Miller Fisher syndrome-associated *Campylobacter jejuni* lipopolysaccharides induce anti-GM$_1$ and anti-GQ$_{1b}$ antibodies in rabbits. *Infect. Immun.* **69:**2462–2469.

Beery, J. T., M. B. Hugdahl, and M. P. Doyle. 1988. Colonization of gastrointestinal tracts of chicks by *Campylobacter jejuni*. *Appl. Environ. Microbiol.* **54:**2365–2370.

Berrang, M. E., R. J. Buhr, J. A. Cason, and J. A. Dickson. 2001. Broiler carcass contamination with *Campylobacter* from feces during defeathering. *J. Food Prot.* **64**:2063–2066.

Berrang, M. E., and J. A. Cason. 2000. Presence and level of *Campylobacter* spp. on broiler carcasses throughout the processing plant. *J. Appl. Poult. Res.* **9**:43–47.

Berrang, M. E., J. K. Northcutt, and J. A. Cason. 2004. Recovery of *Campylobacter* from broiler feces during extended storage of transport cages. *Poult. Sci.* **83**:1213–1217.

Blaser, M. J., J. L. Penner, and J. G. Wells. 1982. Diversity of serotypes involved in outbreaks of *Campylobacter* eneteritis. *J. Infect. Dis.* **146**:825–829.

Brown, P. E., O. F. Christensen, H. E. Clough, P. J. Diggle, C. A. Hart, S. Hazel, R. Kemp, A. J. Leatherbarrow, A. Moore, J. Sutherst, J. Turner, N. J. Williams, E. J. Wright, and N. P. French. 2004. Frequency and spatial distribution of environmental *Campylobacter* spp. *Appl. Environ. Microbiol.* **70**:6501–6511.

Buswell, C. M., Y. M. Herlithy, L. M. Lawrence, J. T. M. McGuiggan, P. D. Marsh, C. W. Keevil, and S. A. Leach. 1998. Extended survival and persistence of *Campylobacter* spp. in water and aquatic biofilms and their detection by immunofluorescent antibody and rRNA staining. *Appl. Environ. Microbiol.* **64**:733–741.

Byrd, J. A., R. H. Bailey, R. W. Wills, and D. J. Nisbet. 2007. Presence of *Campylobacter* in day-of-hatch broiler chicks. *Poult. Sci.* **86**:26–29.

Callicot, K. A., V. Frioriksdottir, J. Reierson, R. Lowman, J. R. Bisaillon, E. Gunnarsson, E. Berndtson, K. L. Hiett, D. S. Needleman, and N. J. Stern. 2006. Lack of evidence for vertical transmission of *Campylobacter* spp. in chickens. *Appl. Environ. Microbiol.* **72**:5794–5798.

Carillo, C. L., R. J. Atterbur, A. El-Shibiny, P. L. Connerton, E. Dillon, A. Scott, and I. F. Connerton. 2005. Bacteriophage therapy to reduce *Campylobacter jejuni* colonization of broiler chickens. *Appl. Environ. Microbiol.* **71**:6554–6563.

Chantarapagant, W., M. E. Berrang, and J. F. Frank. 2002. Numbers of viable cells observed on skin by direct microscopic counts compared to those obtained by plating. *J. Food Prot.* **65**:475–478.

Chaveerach, P., A. A. H. M. ter Huurne, L. J. A. Lipman, and F. van Knapen. 2003. Survival and resuscitation of ten strains of *Campylobacter jejuni* and *Campylobacter coli* under acid conditions. *Appl. Environ. Microbiol.* **69**:711–714.

Chuma, T., T. Yamada, K. Yano, K. Okamoto, and H. Yugi. 1994. A survey of *Campylobacter jejuni* in broilers from assignment to slaughter using DNA-DNA hybridization. *J. Vet. Med. Sci.* **56**:697–700.

Chuma, T., K. Yano, H. Omori, K. Okamoto, and H. Yugi. 1997. Direct detection of *Campylobacter jejuni* in chicken cecal contents by PCR. *J. Vet. Med. Sci.* **59**:85–87.

Clarke, A. G., and D. H. Bueschkens. 1985. Laboratory infection of chicken eggs with *Campylobacter jejuni* by using temperature or pressure differentials. *Appl. Environ. Microbiol.* **49**:1467–1471.

Cloak, O. M., G. Duffy, J. J. Sheridan, L. S. Blair, and D. A. McDowell. 2001. A survey on the incidence of *Campylobacter* spp. and the development of a surface adhesion polymerase chain reaction (SA-PCR) assay for the detection of *Campylobacter jejuni* in retail meat products. *Food Microbiol.* **18**:287–298.

Cox, N. A., N. J. Stern, K. L. Hiett, and M. E. Berrang. 2002. Identification of a new source of *Campylobacter* contamination in poultry: transmission from breeder hens to broiler chickens. *Avian Dis.* **46**:535–541.

Cox, N. A., J. S. Bailey, L. J. Richardson, R. J. Buhr, D. E. Cosby, J. L. Wilson, K. L. Hiett, G. R. Siragusa, and D. V. Bourassa.

2005. Presence of naturally occurring *Campylobacter* and *Salmonella* in the mature and immature ovarian follicles of late-life broiler breeder hens. *Avian Dis.* **49**:285–287.

Cox, N. A., L. J. Richardson, R. J. Buhr, P. J. Fedorka-Cray, J. S. Bailey, J. L. Wilson, and K. L. Hiett. 2006. Natural presence of *Campylobacter* spp. in various internal organs of commercial broiler breeder hens. *Avian Dis.* **50**:450–453.

Cox, N. A., L. J. Richardson, R. J. Buhr, J. K. Northcutt, J. S. Bailey and P. J. Fedorka-Cray. 2007. Recovery of *Campylobacter* and *Salmonella* serovars from the spleen, liver/gallbladder, and ceca of 6 and 8 week old commercial broilers. *J. Appl. Poult. Res.* **16**:477–480.

Deibel, K. E., and G. J. Banwart. 1984. Effect of spices on *Campylobacter jejuni* at three temperatures. *J. Food Saf.* **6**:241–251.

Derclay, I., E. Delor, M. Van Bouchaute, P. Moureau, G. Wauters, and G. R. Cornelis. 1989. Identification of *Campylobacter jejuni* and *C. coli* by gel electrophoresis of the outer membrane proteins. *J. Clin. Microbiol.* **27**:1072–1076.

Docherty, L., M. R. Adams, P. Patel, and J. McFadden. 1996. The magnetic immuno-polymerase chain reaction assay for the detection of *Campylobacter* in milk and poultry. *Lett. Appl. Microbiol.* **22**:288–292.

Doyle, M. P., and D. J. Roman. 1982a. Response of *Campylobacter jejuni* to sodium chloride. *Appl. Environ. Microbiol.* **43**:561–565.

Doyle, M. P., and D. J. Roman. 1982b. Sensitivity of *Campylobacter jejuni* to drying. *J. Food Prot.* **45**:507–510.

Doyle, M. P. 1984. Association of *Campylobacter jejuni* with laying hens and eggs. *Appl. Environ. Microbiol.* **47**:533–536.

Finch, M. J., and P. A. Blake. 1985. Food-borne outbreaks of campylobacteriosis: the United States experience, 1980–1982. *Am. J. Epidemiol.* **122**:262–268.

Friedman, C. R., R. M. Hoekstra, M. Samuel, R. Marcus, J. Bender, B. Shiferaw, S. Reddy, D. Ahuja, L. Helfrick, F. Hardnett, M. Carter, B. Anderson and R. Tauxe for the Emerging Infectious Program FoodNet Working Group. 2004. Risk factors for sporadic *Campylobacter* infection in the United States: a case-control study in FoodNet sites. *Comm. Infec. Dis.* **38**:285–296.

Gibbens, J. C., S. J. S. Pascoe, S. J. Evans, R. H. Davides, and A. R. Sayers. 2001. A trial of biosecurity as a means to control *Campylobacter* infection of broiler chickens. *Prev. Vet. Med.* **48**:85–99.

Gordon, R. F., and J. F. Tucker. 1965. The epizootiology of *Salmonella menstron* infection of fowls and the effect of feeding poultry food artificially infested with *Salmonella*. *Br. Poult. Sci.* **6**:251–253.

Grajewski, B. A., J. W. Kusek, and H. M. Gelfand. 1985. Development of a bacteriophage typing system for *Campylobacter jejuni* and *Campylobacter coli*. *J. Clin. Microbiol.* **22**:13–18.

Grant, C. C., M. E. Konkel, W. Cieplak, Jr., and L. S. Tompkins. 1993. Role of flagella in adherence, internalization, and translocation of *Campylobacter jejuni* in nonpolarized and polarized epithelial cell cultures. *Infect. Immun.* **61**:1764–1771.

Gregory, E., H. Barnhart, D. W. Dreesen, N. J. Stern, and J. L. Corn. 1997. Epidemiological study of *Campylobacter* spp. in broilers: source, time of colonization, and prevalence. *Avian Dis.* **41**:890–898.

Harrison, W. A., C. J. Griffith, D. Tennant, and A. C. Peters. 2001. Incidence of *Campylobacter* and *Salmonella* isolated from retail chicken and associated packaging in South Wales. *Lett. Appl. Microbiol.* **33**:450–454.

Hernandez, J., A. Fayos, J. L. Alonso, and R. J. Owen. 1996. Ribotypes and AP-PCR fingerprints of thermophilic campylobacters from marine recreational waters. *J. Appl. Bacteriol.* **80**:157–164.

Hiett, K. L., N. A. Cox, and N. J. Stern. 2002. Direct polymerase chain reaction detection of *Campylobacter* spp. in poultry hatchery samples. *Avian Dis.* **46:**219–223.

Hiett K. L., N. A. Cox, and K. M. Phillips. 2003. PCR detection of naturally occurring *Campylobacter spp.* in commercial chicken embryos, p. 137. *In Proceedings of the 11th International Workshop on Campylobacter, Helicobacter and Related Organisms.* University of Freiburg, Freiburg, Germany.

Humphrey, T. J., A. Whitehead, H. L. Gawler, A. Henley, and B. Rowe. 1991. Numbers of *Salmonella enteritidis* in the contents of naturally contaminated hens' eggs. *Epidemiol. Infect.* **106:**489–496.

Humphrey, T. J., A. Henley, and D. G. Lanning. 1993. The colonization of broiler chickens with *Campylobacter jejuni:* some epidemiological investigations. *Epidemiol. Infect.* **110:**601–687.

Izat, A. L., and F. A. Gardner. 1988. Incidence of *Campylobacter jejuni* in processed egg products. *Poult. Sci.* **67:**1431–1435.

Jacobs-Reitsma, W. 2000. *Campylobacter* in the food supply, p. 467–481. *In* I. Nachamkin and M. J. Blaser (ed.), *Campylobacter.* American Society for Microbiology, Washington, DC.

Jones, D. M., E. M. Sutcliffe, and A. Curry. 1991. Recovery of viable but non-culturable *Campylobacter jejuni. J. Gen. Microbiol.* **137:**2477–2482.

Keener, K. M., M. P. Bashor, P. A. Curtis, B. W. Sheldon, and S. Kathariou. 2004. Comprehensive review of *Campylobacter* and poultry processing. *Compr. Rev. Food Sci. Food Saf.* **3:**105–116.

Kemp, G. K., M. L. Aldrich, M. L. Guerra, and K. R. Schneider. 2001. Continuous online processing of fecal- and ingesta-contaminated poultry carcasses using an acidified sodium chlorite antimicrobial intervention. *J. Food Prot.* **64:**807–812.

Kemp, R., A. J. H. Leatherbarrow, N. J. Williams, C. A. Hart, H. E. Clough, J. Turner, E. J. Wright, and N. P. French. 2005. Prevalence and diversity of *Campylobacter* spp. in environmental water samples from a 100-square-kilometer predominantly dairy farming area. *Appl. Environ. Microbiol.* **71:**1876–1882.

Ketley, J. M. 1997. Pathogenesis of enteric infection by *Campylobacter. J. Microbiol.* **143:**5–21.

Koidis, P., and M. P. Doyle. 1983. Survival of *Campylobacter jejuni* in fresh and heated red meat. *J. Food Prot.* **46:**771–774.

Lindblom, G. B., E. Sjorgren, and B. Kaijser. 1986. Natural *Campylobacter* colonization in chickens raised under different environmental conditions. *J. Hyg.* **96:**385–391.

Lindsay, J. A. 1997. Chronic sequelae of foodborne disease. *Emerg. Infect. Dis.* **3:**443–452.

Lior, H., D. L. Woodward, J. A. Edgar, L. J. Laroche, and P. Gill. 1982. Sero-typing of *Campylobacter jejuni* by slide agglutination based on heat-labile antigenic factors. *J. Clin. Microbiol.* **15:**761–768.

Lu, J., S. Sanchez, C. Hofacre, J. J. Maurer, B. G. Harmon, and M. D. Lee. 2003. Evaluation of broiler litter with reference to the microbial composition as assessed by using 16S rRNA and functional gene markers. *Appl. Environ. Microbiol.* **69:**901–908.

Maekin, E., J. Popowski, and L. Szponar. 2008. Thermotolerant Campylobacter spp.–report on monitoring studies performed in 2004–2005 in Poland. *Food Control* **19:**219–222.

Maruyama, S., and Y. Katsube. 1990. Isolation of *Campylobacter jejuni* from the eggs and organs in experimentally infected laying Japanese quails (Coturnix coturnix japonica). *Jpn. J. Vet. Med. Sci.* **52:**671–674.

Maruyama, S., Y. Morita, and Y. Katsube. 1995. Invasion and viability of *Campylobacter jejuni* in experimentally contaminated Japanese quails' eggs. *Jpn. J. Vet. Med. Sci.* **57:**587–590.

Meinersmann, R. J., L. O. Helsel, P. I. Fields, and K. L. Hiett. 1997. Discrimination of *Campylobacter jejuni* strains by *fla* gene sequencing. *J. Clin. Microbiol.* **35:**2810–2814.

Moran, A. P., and M. E. Upton. 1987. Factors affecting production of coccoid forms by *Campylobacter jejuni* on solid media during incubation. *J. Appl. Bacteriol.* **62:**527–537.

Muir, W. I., W. L. Bryden, and A. J. Husband. 2000. Immunity, vaccination and the avian intestinal tract. *Dev. Comp. Immunol.* **24:**325–342.

Musgrove, M. T., M. E. Berrang, J. A. Byrd, N. J. Stern, and N. A. Cox. 2001. Detection of *Campylobacter* spp. in ceca and crops with and without enrichment. *Poult. Sci.* **80:**825–828.

Musgrove, M. T., J. A. Cason, D. L. Fletcher, N. J. Stern, N. A. Cox, and J. S. Bailey. 1997. Effect of cloacal plugging on microbial recovery from partially processed broilers. *Poult. Sci.* **76:**530–533.

Musgrove, M. T., and D. R. Jones. 2006. Sanitation in US shell egg processing, p. FS13–FS17. *In Proceedings of the United States-Japan Cooperative Program in Natural Resources (UJNR).* Sonoma, CA.

Nachamkin, I., H. Ung, and C. M. Patton. 1996. Analysis of HL and O serotypes of *Campylobacter* strains by the flagellin gene typing system. *J. Clin. Microbiol.* **34:**277–281.

Newell, D. G., J. E. Shreeve, M. Toszeghy, G. Domingue, S. Bull, T. Humphrey, and G. Mead. 2001. Changes in carriage of *Campylobacter* strains by poultry carcasses during processing in abattoirs. *Appl. Environ. Microbiol.* **67:**2636–2640.

Nielsen, E. M., J. Engberg, and M. Madsen. 1997. Distribution of serotypes of *Campylobacter jejuni* and *C. coli* from Danish patients, poultry, cattle, and swine. *Immun. Med. Microbiol.* **46:**199–205.

Northcutt, J. K. 2005. Effect of broiler age, feed withdrawal, and transportation on levels of coliforms, *Campylobacter, Escherichia coli* and *Salmonella* on carcasses before and after slaughter. *J. Food Prot.* **57:**1101–1121.

Oosterom, J., G. J. A. de Wilde, E. de Boer, L. H. de Blaauw, and H. Karman. 1983. Survival of *Campylobacter jejuni* during poultry processing and pig slaughtering. *J. Food Prot.* **46:**702–706.

Padungtod, P., R. Hanson, D. L. Wilson, J. Bell, J. E. Linz, and J. B. Kaneene. 2002. Identification of *Campylobacter jejuni* isolates from cloacal and carcass swabs of chickens in Thailand by a 5′ nuclease fluorogenic polymerase chain reaction assay. *J. Food Prot.* **65:**1712–1716.

Patton, C. M., I. K. Wachsmuth, G. M. Evans, J. A. Kiehlbauch, B. D. Plikaytis, N. Troup, L. Tompkins, and H. Lior. 1991. Evaluation of 10 methods to distinguish epidemic-associated *Campylobacter* strains. *J. Clin. Microbiol.* **29:**680–688.

Pearson, A. D., M. H. Greenwood, R. K. Feltham, T. D. Healing, J. Donaldson, D. M. Jones, and R. R. Colwell. 1996. Microbial ecology of *Campylobacter jejuni* in United Kingdom chicken supply chain: intermittent common source, vertical transmission, and amplification by flock propagation. *Appl. Environ. Microbiol.* **62:**4614–4620.

Penner, J. L., and J. N. Hennessy. 1980. Passive hemagglutination technique for sero-typing *Campylobacter fetus* subsp. *jejuni* on the basis of soluble heat-stable antigens. *J. Clin. Microbiol.* **29:**680–688.

Petersen, A., J. P. Christensen, P. Kuhnert, M. Bisgaard, and J. E. Olsen. 2006. Vertical transmission of a flouroquinolone-resistant Escherichia coli within an integrated broiler operation. *Vet. Microbiol.* **116:**120–128.

Richardson L. J., R. J. Buhr, N. A. Cox, K. L. Hiett, and M. A. Harrison. 2007a. Isolation of *Campylobacter* from circulating blood of commercial broilers via vena-puncture of exposed/unexposed brachial veins. *Zoonoses Public Health* **54**(S1):98.

Richardson L. J., N. A. Cox, R. J. Buhr, K. L. Hiett, J. S. Bailey, and M. A. Harrison, 2007b. Recovery of viable non-culturable dry stressed *Campylobacter jejuni* from inoculated samples utilizing a chick bioassay. *Zoonoses Public Health* 54(S1):119.

Rollins, D. M., and R. Colwell. 1986. Viable but non-culturable stage of *Campylobacter jejuni* and its role in survival in the natural aquatic environment. *Appl. Environ. Microbiol.* 45:531–538.

Saha, S. K., and S. C. Sanyel. 1991. Recovery of injured *Campylobacter jejuni* cells after animal passage. *Appl. Environ. Microbiol.* 57:3388–3389.

Saleha, A. A., G. C. Mead, A. L. Mead, and A. L. Ibrahim. 1998. *Campylobacter jejuni* in poultry production and processing in relation to public health. *J. World Poult. Sci.* 54:49–58.

Sellars, M. J., S. J. Hall, and D. J. Kelly. 2002. Growth of *Campylobacter jejuni* supported by respiration of fumarate, nitrate, nitrite, trimethylamine-*N*-oxide, or dimethyl sulfoxide requires oxygen. *J. Bacteriol.* 184:4187–4196.

Skirrow, M. B. 1991. Epidemiology of *Campylobacter* enteritis. *Int. J. Food Microbiol.* 12:9–16.

Vandamme, P. 2000. Taxonomy of the family *Campylobacteraceae*, p. 3–26. *In* I. Nachamkin and M. J. Blaser (ed.), *Campylobacter*. American Society for Microbiology, Washington, DC.

Wagenaar, J. A., M. A. P. Van Bergen, M. A. Mueller, T. M. Wassenaar, and R. M. Carlton. 2005. Phage therapy reduces *Campylobacter jejuni* colonization in broilers. *Vet. Microbiol.* 109:275–283.

Walker, R. I., M. B. Caldwell, E. C. Lee, P. Guerry, T. J. Trust, and G. M. Ruiz-Palacios. 1986. Pathophysiology of *Campylobacter* enteritis. *Microbiol. Rev.* 50:81–94.

Wallace, J. S., K. N. Stanley, and K. Jones. 1998. The colonization of turkeys by thermophilic campylobacters. *J. Appl. Microbiol.* 85:224–230.

Wassenaar, T. M. 1997. Toxin production by *Campylobacter* spp. *Clin. Microbiol. Rev.* 10:466–476.

Wassenaar, T. M., and D. G. Newell. 2000. Genotyping of *Campylobacter* spp. *Appl. Environ. Microbiol.* 66:1–9.

Wassenaar, T. M. 2004. Genetic differentiation of *Campylobacter jejuni*. *Int. J. Infect. Dis.* 6:S22–S25.

Widders, P. R., L. M. Thomas, K. A. Long, M. A. Tokhi, M. Panaccio, and E. Apos. 1998. The specificity of antibody in chickens immunized to reduce intestinal colonization with *Campylobacter jejuni*. *Vet. Microbiol.* 64:39–50.

Willison, H. J., and G. M. O'Hanlon. 2000. Antiglycosphingolipid antibodies and Guillain-Barré syndrome, p. 259–285. *In* I. Nachamkin and M. J. Blaser (ed.), *Campylobacter*. American Society for Microbiology, Washington, DC.

Winer, J. B. 2001. Guillain-Barre syndrome. *Mol. Pathol.* 54:381–385.

Zhao, C., G. Beilei, J. de Villena, R. Sudler, E. Yeh, S. Zhao, D. G. White, D. Wagner, and J. Meng. 2001. Prevalence of *Campylobacter* spp., *Escherichia coli*, and *Salmonella* serovars in retail chicken, turkey, pork, and beef from the greater Washington, DC area. *Appl. Environ. Microbiol.* 67:5431–5436.

Pathogens and Toxins in Foods: Challenges and Interventions
Edited by V. K. Juneja and J. N. Sofos
© 2010 ASM Press, Washington, DC

Chapter 3

Clostridium botulinum

Michael W. Peck

CHARACTERISTICS OF *CLOSTRIDIUM BOTULINUM* AND ITS NEUROTOXINS

Characteristics of *Clostridium botulinum*

The ability to form botulinum neurotoxin is restricted to strains of *Clostridium botulinum* and some strains of *C. baratii* and *C. butyricum*. *C. botulinum* is not a homogeneous species, but a collection of four physiologically and genetically distinct bacteria (Hatheway, 1993). All are gram-positive, endospore-forming anaerobes. The name *C. botulinum* is retained for this heterogeneous species in order to emphasize the importance of neurotoxin formation. For each of the six neurotoxigenic clostridia, a nonneurotoxigenic phylogenetically equivalent organism is described (Table 1). Thus, for example, *C. botulinum* group I (proteolytic *C. botulinum*) is actually more closely related to *C. sporogenes* than to *C. botulinum* group II, *C. botulinum* group III, or *C. botulinum* group IV. In general terms, four neurotoxigenic clostridia are associated with botulism in humans: *C. botulinum* group I (proteolytic *C. botulinum*), *C. botulinum* group II (nonproteolytic *C. botulinum*), and neurotoxigenic strains of *C. baratii* and *C. butyricum*. *C. botulinum* group III is associated with botulism in birds and animals, while strains of *C. botulinum* group IV *(C. argentinense)* have been isolated from the environment but not associated with botulism.

Strains of proteolytic *C. botulinum* form either type A, B, or F neurotoxin. Dual-neurotoxin-forming strains have been reported, while many single-neurotoxin-forming strains also carry an incomplete neurotoxin gene. This pathogen is a mesophile with growth not reported at 10°C or below. It forms spores with high heat resistance and is of particular concern in the production of canned foods. Sequencing of the genome of proteolytic *C. botulinum* type A strain Hall A (ATCC 3502, NCTC 13319) has revealed

new information on the biology of this organism (Sebaihia et al., 2007). The genome consists of a chromosome (3.9 Mb) and a plasmid (16.3 kb), which carry 3,650 and 19 predicted genes, respectively. The plasmid encodes a bacteriocin (called a boticin). Approximately 43% of the predicted genes present in the Hall A genome are absent from other sequenced clostridia *(C. acetobutylicum, C. difficile, C. perfringens,* and *C. tetani)*, while only 16% of the predicted genes are present in all five sequenced clostridia. There is a significant lack of recently acquired DNA, indicating a stable genomic content. This contrasts with the genome of *C. difficile*, which is highly mobile (Sebaihia et al., 2006). Although proteolytic *C. botulinum* is a highly dangerous pathogen, it is well adapted to a saprophytic lifestyle. In addition to the expected large number of genes encoding secreted proteases and enzymes involved in uptake and metabolism of amino acids, there are also genes encoding enzymes involved in chitin degradation (Sebaihia et al., 2007).

Nonproteolytic *C. botulinum* is a psychrotroph, with growth and neurotoxin formation reported at 3.0°C but not at 2.5°C. Energy is derived from the degradation of sugars. Strains form either type B, type E, or type F neurotoxin. There appears to be an absence of dual-neurotoxin-forming strains or strains carrying a second incomplete neurotoxin gene. Spores formed by nonproteolytic *C. botulinum* are of moderate heat resistance. The foods most commonly involved in outbreaks are fermented marine products, dried fish, and vacuum-packed fish. While there is a low genetic relatedness between the genomes of proteolytic *C. botulinum* and nonproteolytic *C. botulinum*, they share some common features. For example, both genomes are AT-rich (GC content, 26% to 28%), have a similar genome size (3.6 to 4.1 Mb), and possess plasmids (Strom et al., 1984; Lund and Peck, 2000; Lindström and Korkeala, 2006; Sebaihia

Michael W. Peck • Institute of Food Research, Norwich Research Park, Colney, Norwich NR4 7UA, United Kingdom.

Table 1. The six physiologically and phylogenetically distinct clostridia that form the botulinum neurotoxin

Neurotoxigenic clostridia	Neurotoxins formed	Equivalent nonneurotoxigenic clostridia
C. botulinum group I (proteolytic C. botulinum)	A, B, F	C. sporogenes
C. botulinum group II (nonproteolytic C. botulinum)	B, E, F	No name given
C. botulinum group III	C, D	C. novyi
C. botulinum group IV (C. argentinense)	G	C. subterminale
C. baratii	F	All typical strains
C. butyricum	E	All typical strains

et al., 2007). Evidence for the low genetic relatedness comes from studies of DNA homology (Hatheway, 1993) and tests with a DNA microarray based on the genome sequence of the proteolytic *C. botulinum* strain Hall A. When DNA from the nonproteolytic *C. botulinum* strain Eklund 17B was applied to the microarray, it was too genetically diverse to give a meaningful hybridization (Sebaihia et al., 2007). The genetic diversity of nonproteolytic *C. botulinum* strains appears greater than that of proteolytic *C. botulinum* strains (Lindström and Korkeala, 2006). Neurotoxigenic strains of *C. baratii* and *C. butyricum* form type F or type E neurotoxin, respectively. Both are mesophiles and have a minimum growth temperature within the range of 10°C to 15°C (S. C. Stringer and M. W. Peck, unpublished data). The physiological properties of these two organisms have not been studied extensively.

Characteristics of Botulinum Neurotoxins

The botulinum neurotoxins are heat-labile proteins. Heat inactivation is rapid at temperatures greater than 80°C but nonlinear, and some foods afford protection (Siegel, 1993). Heating at 85°C for 5 min reduced the concentration of active neurotoxin by a factor of 10,000. Freezing does not inactivate botulinum neurotoxins. There are seven major botulinum neurotoxins (types A to G). The subdivision into seven distinguishable serotypes was originally made on the basis that polyclonal antibodies raised against purified neurotoxins neutralized the toxicity of neurotoxins of the same serotype, but not those of different serotypes, in the mouse test. The robustness of this scheme has been confirmed by sequencing of neurotoxin genes and studies of the site of action of the neurotoxins. Recently, the presence of a number of botulinum neurotoxin subtypes has been reported; for example, four subtypes of type A neurotoxin (termed A1, A2, A3, and A4) have been distinguished (Smith et al., 2005; Arndt et al., 2006). Variability within the various subtypes is between 3% and 32% at the amino acid level, compared to up to 70% between different botulinum neurotoxins (Smith et al., 2005). The neurotoxins are commonly associated with other proteins, such as hemagglutinin and nontoxin-nonhemagglutinin, that serve to protect the neurotoxin and assist in its absorption into the body.

The botulinum neurotoxins are 150-kDa proteins and comprise a heavy chain (100 kDa) and a light chain (50 kDa) linked by a disulfide bond. The heavy chain has two functional domains, a C-terminal binding domain that facilitates binding to synaptic membrane vesicle proteins (synaptotagmins or SV2) at the nerve terminal and an N-terminal translocation domain that delivers the light chain into the cytosol of the nerve cell following receptor-mediated endocytosis (Simpson, 2004; Dong et al., 2006; Jahn, 2006; Mahrhold et al., 2006). The light chains possess zinc endopeptidase activity and cleave protein components of the acetylcholine-containing synaptic vesicle docking/fusion complex within the cytosol of the nerve. Each light chain cleaves a specific protein in the synaptic vesicle docking/fusion complex at a unique site. The light chains of type A and E neurotoxins target the synaptosome-associated 25-kDa protein (SNAP-25); type B, D, F, and G light chains cleave the vesicle-associated membrane protein (VAMP); and the type C light chain cleaves both SNAP-25 and syntaxin (Simpson, 2004; Montecucco and Molgo, 2005; Rossetto et al., 2006). Cleavage of any one of SNAP-25, VAMP, and syntaxin prevents binding of acetylcholine-containing synaptic vesicles and consequently neurotransmitter release, leading to flaccid paralysis of the muscle. Symptoms of botulism are primarily neurological and frequently commence with blurred vision, often leading to dysphagia (difficulty swallowing), dysphonia (difficulty speaking), paralysis of face muscles, generalized weakness, descending bilateral flaccid paralysis, nausea/vomiting, dizziness/vertigo, and muscle weakness. Flaccid paralysis of the respiratory or cardiac muscles can result in death if not treated. In many countries, equine antitoxin is administered to patients suffering from food-borne or wound botulism (but not infant botulism). Rapid treatment with equine antitoxin and supportive therapy has led to a reduction in the fatality rate to approximately 5 to 10% of cases, although complete recovery can be slow, may require mechanical ventilation, and can take many months or even longer.

Botulism — the Disease

Food-borne botulism is a severe but rare neuroparalytic intoxication resulting from consumption of preformed botulinum neurotoxin. Since some strains of *C. botulinum* form as much as 1,000 human lethal doses of neurotoxin per gram, consuming only a few micrograms of food in which *C. botulinum* or other neurotoxin-producing clostridia have grown could result in food-borne botulism. The botulinum neurotoxin is the most potent substance known, with as little as 30 ng of neurotoxin sufficient to cause illness and even death (Lund and Peck, 2000). This is more than a million-fold lower than the fatal oral dose for alkali cyanides, which is estimated to be in the region of 50 to 100 mg (CDC, 2007). Neurotoxins of types A, B, E, and occasionally F are associated with botulism in humans.

There is anecdotal evidence that food-borne botulism may have occurred in ancient cultures and that some dietary laws and taboos may have evolved to prevent the disease (Smith, 1977). The word "botulism" is derived from the Latin word "botulus," meaning sausage, and was given to a disease associated with consumption of blood sausage, characterized by muscle paralysis, breathing difficulties, and a high fatality rate, in central Europe in the 18th and 19th century. In 1897, van Ermengem first isolated a causative organism from raw homemade salted ham and the spleen of a man who later died of botulism. The isolate is now lost, but reports of its properties suggest that it would be classified as nonproteolytic *C. botulinum* type B. A great number of botulism outbreaks occurred in the early part of the 20th century and were often associated with commercial and home canning processes. Through the identification and implementation of appropriate effective control measures, the incidence of food-borne botulism is lower than a century ago. Today, outbreaks are often associated with home-prepared foods, for which known control measures have not been implemented. Botulism outbreaks involving commercial processing are less common but can be associated with significant medical and economic consequences (e.g., the cost per botulism case in the United States has been estimated at $30 million [Setlow and Johnson, 1997]).

Infection and then colonization of the gastrointestinal tracts of susceptible infants by proteolytic *C. botulinum* (or, more rarely, neurotoxigenic strains of *C. baratii* or *C. butyricum*) can lead to infant botulism. The first clinical cases were described in 1976, although subsequent investigations revealed earlier cases (Arnon, 2004). Infant botulism has been reported in many countries and has been suggested as one possible cause of sudden infant death syndrome (Fox et al., 2005; Nevas et al., 2005a). Honey and general environmental contamination (e.g., soil and dust) have been identified as sources of spores (Lund and Peck, 2000; Aureli et al., 2002; Arnon, 2004). Approximately half of the European cases of infant botulism have been associated with honey consumption (Aureli et al., 2002). This association has led to recommendations in several countries that honey jars should carry a warning label indicating that the product is not suitable for infants less than 12 months of age. It is estimated that between 10 and 100 spores are sufficient to bring about infection (Arnon, 2004). This is based on reports that honey samples that have been associated with infant botulism contain 5 to 25 spores per gram (Midura et al., 1979) and 5 to 70 spores per gram (Arnon et al., 1979). Once spores enter the body, an immature intestinal flora fails to prevent colonization. Spore germination is followed by cell multiplication and neurotoxin formation. Infants aged between 2 weeks and 6 months are most susceptible (Arnon, 2004). Typical symptoms include extended constipation and flaccid paralysis. The disease is rarely fatal. Equine antitoxin is not given in cases of infant botulism. As an alternative, the use of human botulism immune globulin has been developed (Arnon et al., 2006). This treatment has been shown to be safe and effective in bringing about a significant reduction in the length and cost of the hospital stay and the severity of illness. Infectious botulism is a similar but rarer disease that affects adults. It occurs when competing bacteria in the normal intestinal flora of adults have been suppressed (e.g., by antibiotic treatment).

Wound botulism is also an infection, in which growth and neurotoxin formation by proteolytic *C. botulinum* neurotoxin type A or B occurs in a wound in the body. This disease was first described in 1943, and initially most cases involved gross trauma (e.g., crush injury) to an extremity (Passaro et al., 1998; Lund and Peck, 2000). Since the early 1990s, however, there has been a significant increase in reports of wound botulism in many countries associated with intravenous drug users (Passaro et al., 1998; Werner et al., 2000; Sandrock and Murin, 2001; Brett et al., 2004). For example, prior to 2000, wound botulism had not been reported in the United Kingdom, but more than 100 suspected cases were reported between 2000 and 2005, all involving heroin injection (Akbulut et al., 2005; CDSC, 2006).

Work to develop the use of botulinum neurotoxin as a potential biological weapon began more than half a century ago. Although the 1972 Biological and Toxin Weapons Convention prohibited offensive research and production of biological weapons, in some countries (e.g., the Soviet Union and Iraq) further developments continued (Arnon et al., 2001). The use of botulinum neurotoxin as a weapon by bioterrorists remains a concern, and the implications of a bioterrorism attack in which botulinum neurotoxin is deliberately added to milk have been assessed (Wein and Liu, 2005). On at least three occasions between 1990 and 1995, aerosols of botulinum neurotoxin were dispersed by the Aum Shinrikyo cult in Japan, although these attacks apparently failed (Arnon et al., 2001). Inadvertent inhalation of aerosolized botulinum neurotoxin type A by veterinary personnel disposing of contaminated animal fur did, however, lead to three cases of botulism in Germany in 1962 (Arnon et al., 2001). A number of botulinum neurotoxin preparations (e.g., Botox, Dysport, and Myobloc/NeuroBloc) have been approved in various countries for therapeutic and cosmetic use. The use of an unlicensed, highly concentrated botulinum neurotoxin preparation, however, led to four cases of botulism in the United States in 2004. The injections were given for cosmetic purposes, and it is estimated that 2,857 times the human lethal dose may have been delivered (Chertow et al., 2006).

EPIDEMIOLOGY OF FOOD-BORNE BOTULISM

General Overview

Food-borne botulism is not a reportable disease in all countries; thus, the extent of reporting and investigation of cases varies from country to country (Therre, 1999). Reported data are therefore likely to be far from complete. Many recent food-borne botulism outbreaks have been associated with home-prepared foods, for which known control measures have not been implemented. In some countries with a particularly high incidence, this has often been associated with an increased reliance on home preservation/bottling/canning of foods, reflecting difficult economic conditions at the time. In recent years, between 20 and 40 cases of food-borne botulism have been recorded annually in France, Georgia, Germany, Italy, and the United States, with a higher incidence in China and Poland (Table 2). More than 2,500 cases of food-borne botulism were reported in Europe in 1999 and 2000 (Peck, 2006). Countries with a high reported incidence include Armenia, Azerbaijan, Belarus, Georgia, Poland, Russia, Turkey, and Uzbekistan. A total of 887 cases were reported in Russia in 1999 and 2000 and were often associated with consumption of smoked, salted, and dried fish and home-canned vegetables (Peck, 2006). Georgia has one of the highest nationally reported rates of food-borne botulism in the world, with most cases attributed to home-preserved vegetables and fish (Varma et al., 2004). One large outbreak at a wedding in 1994 affected 173 people and was associated with consumption of contaminated fish.

Most cases of food-borne botulism are associated with proteolytic *C. botulinum* or nonproteolytic *C. botulinum*. Occasionally, neurotoxigenic strains of *C. baratii* and *C. butyricum* are implicated (Anniballi et al., 2002; Harvey et al., 2002). In order to understand the causes of botulism outbreaks and to prevent any recurrence, it is highly desirable to

Table 2. Recorded food-borne botulism in different countries[a]

Country	Period	No. of cases	
		Total	Avg per yr
Belgium	1982–2000	32	2
Canada	1971–2005	439	13
China	1958–1983	4,377	168
Denmark	1984–2000	18	1
France	1971–2003	1,286	39
Georgia	1980–2002	879	40
Germany	1983–2000	376	22
Italy	1979–2000	750	34
Japan	1951–1987	479	13
Poland	1971–2000	9,219	307
Spain	1971–1998	277	10
Sweden	1969–2000	13	1
United Kingdom	1971–2005	38	1
United States	1971–2003	934	28

[a]Table compiled from information in works by Hauschild and Gauvreau (1985), Hauschild (1993), Galazka and Przbylska (1999), Therre (1999), Lund and Peck (2000), Varma et al. (2004), CDC (2005), Popoff and Carlier (2004), McLauchlin et al. (2006), J. W. Austin (personal communication), WHO (2000, 2003). Data may include some cases of human botulism that are not food borne.

identify which of the neurotoxigenic clostridia is responsible. In some food-borne botulism outbreaks, the type of neurotoxin formed is determined but not the organism. This has happened most frequently when type B neurotoxin has been involved, when it is not established whether proteolytic *C. botulinum* or nonproteolytic *C. botulinum* is responsible (Table 3). In some botulism outbreaks, the type of neurotoxin is also not reported (Table 3). Since proteolytic *C. botulinum* and nonproteolytic *C. botulinum* are physiologically distinct and present different hazard scenarios, it is appropriate to consider the epidemiology of food-borne botulism for each separately.

Food-Borne Botulism Outbreaks Associated with Proteolytic *C. botulinum*

Outbreaks of food-borne botulism have often been associated with the inadequate processing of canned or bottled foods, both on a commercial scale and at home (Table 4). One very large outbreak in rural Thailand in March 2006 was associated with inadequately home-canned bamboo shoots. Subsequent ambient storage led to growth and neurotoxin formation by proteolytic *C. botulinum* type A (CDC, 2006a). The bamboo shoots were consumed at a local religious rite, where 209 villagers contracted botulism and 134 were hospitalized, with 42 requiring mechanical ventilation (Ungchusak et al., 2007). It is likely that there would have been significant mortality if the Thai authorities had not rapidly mobilized 42 ventilators.

Proteolytic *C. botulinum* has also been associated with botulism outbreaks in which foods intended for refrigerated storage have been held at ambient temperature (Table 4). The only barrier to controlling neurotoxin formation by proteolytic *C. botulinum* in these foods appears to be refrigerated storage. Proteolytic *C. botulinum* type B grew and formed neurotoxin in a bottle of chopped garlic in soybean oil stored at ambient temperature (and not refrigerated, as indicated on the label) for 8 months in a restaurant in Canada. The garlic was subsequently used in sandwiches (St. Louis et al., 1988). Growth and neurotoxin

formation by strains of proteolytic *C. botulinum* type A led to botulism outbreaks associated with vacuum-packed clam chowder and with a black bean dip in the United States in 1994 (Anonymous, 1995). Both products were purchased from the refrigeration section of a supermarket, were labeled "keep refrigerated" but stored at ambient temperature by consumers, and were consumed despite seeming spoiled (Anonymous, 1995). One case in France in 1999 involved commercial chilled fish soup in a carton. This soup was temperature-abused in the home, and this allowed proteolytic *C. botulinum* type A to grow and form neurotoxin (Carlier et al., 2001). An outbreak in September 2006 in Canada (Toronto) and the United States (Florida and Georgia) was associated with growth and neurotoxin formation by proteolytic *C. botulinum* type A in commercial chilled carrot juice (Anonymous, 2006; CDC, 2006b). This outbreak was severe, with all six cases requiring mechanical ventilation (Anonymous, 2006; CDC 2006b).

Food-borne botulism has also occurred when ingredients containing preformed botulinum neurotoxin have been added to a correctly refrigerated product. Examples include the large outbreaks associated with hazelnut yogurt in the United Kingdom and skordalia in the United States (Table 4). The largest recorded botulism outbreak in the United Kingdom occurred in 1989 (27 cases, one fatal) and involved yogurt prepared with hazelnut conserve contaminated with type B neurotoxin (O'Mahony et al., 1990). Heat processing of the hazelnut conserve was insufficient to destroy spores of proteolytic *C. botulinum* type B. The spores subsequently germinated, leading to growth and neurotoxin formation in the hazelnut conserve. A large restaurant-associated outbreak in the United States in 1994 (30 cases) involved skordalia prepared with potato contaminated with type A neurotoxin (Angulo et al., 1998). Potatoes wrapped in foil were baked and then stored at ambient temperature, permitting growth and neurotoxin formation by proteolytic *C. botulinum* type A. The neurotoxin-containing potatoes were then added to yogurt to make skordalia.

Table 3. Examples of recent incidents of food-borne botulism in which the neurotoxigenic clostridia and/or neurotoxin have not been reported

Yr	Country	Product	Toxin type	No. of cases (deaths)	Reference
1998	Croatia	Ham	B	20	Pavic et al. (2001)
1999	Morocco	Commercial mortadella sausage	B	78 (20)	Ouagari et al. (2002)
2003	France	Commercial halal sausage	B	4	Espie and Popoff (2003)
2004	Italy	Restaurant-preserved green olives in saline	B	16	Cawthorne et al. (2005)
1994	Georgia	Fish consumed at wedding	?	173	Varma et al. (2004)
1998	Algeria	Kashir (meat dish)	?	1,400 (17)	Anonymous (1998a)
1999	Azerbaijan	Fish consumed in restaurant	?	90 (4)	Anonymous (1999)

Table 4. Examples of recent incidents of food-borne botulism involving proteolytic *C. botulinum*

Yr	Country	Product	Toxin type	No. of cases (deaths)	Factors contributing to botulism outbreak[a]	Reference(s)
1985	Canada	Commercial garlic in oil	B	36	Bottled; no preservatives; temperature abuse	St. Louis et al. (1988)
1987	Canada	Bottled mushrooms	A	11	Underprocessing and/or inadequate acidification	CDC (1987), McLean et al. (1987)
1989	United Kingdom	Commercial hazelnut yogurt	B	27 (1)	Hazelnut conserve underprocessed	O'Mahony et al. (1990)
1993	United States	Restaurant, commercial process cheese sauce	A	8 (1)	Contaminated after opening, then temperature abused	Townes et al. (1996)
1993	Italy	Commercial canned roasted aubergine in oil	B	7	Insufficient heat treatment; improper acidification	CDC (1995)
1994	United States	Restaurant; potato dip ("skordalia") and aubergine dip ("meligianoslata")	A	30	Baked potatoes held at room temperature	Angulo et al. (1998)
1994	United States	Commercial clam chowder	A	2	No secondary barrier; temperature abuse	Anonymous (1995)
1994	United States	Commercial black bean dip	A	1	No secondary barrier; temperature abuse	Anonymous (1995)
1996	Italy	Commercial mascarpone cheese	A	8 (1)	No competitive microflora; pH >6, temperature abuse	Franciosa et al. (1999), Aureli et al. (2000)
1997	Italy	Homemade pesto/oil	B	3	pH 5.8, a_w 0.97	Chiorboli et al. (1997)
1997	Germany	Home-prepared beans	A	1	Poor preparation	Gotz et al. (2002)
1997	Iran	Traditional cheese preserved in oil	A	27 (1)	Unsafe process	Pourshafie et al. (1998)
1998	Thailand	Home-canned bamboo shoots	A	13 (2)	Inadequate processing (?)	CDC (1999)
1998	Argentina	Meat roll ("matambre")	A	9	Cooked and heat-shrunk plastic wrap; temperature abuse	Villar et al. (1999)
1998	United Kingdom	Home-bottled mushrooms in oil (imported from Italy)	B	2 (1)	Unsafe process	CDSC (1998), Roberts et al. (1998)
1999	France	Commercial chilled fish soup	A	1	Temperature abuse at home	Carlier et al. (2001)
2000	France	Homemade asparagus soup	B	9	Inadequate processing (?)	Abgueguen et al. (2003)
2001	United States	Commercial frozen chilli sauce	A	16	Temperature abuse at salvage store	Kalluri et al. (2003)
2002	South Africa	Commercial tinned pilchards	A	2 (2)	Corrosion of tin; permitted secondary contamination	Frean et al. (2004)
2002	Canada	Restaurant-baked potato in aluminium foil	A	1	Baked potato held at room temperature (?)	Bhutani et al. (2005)
2005	United Kingdom	Not known (travel from Georgia)	A	1	Not known	McLauchlin et al. (2006)
2005	Turkey	Homemade suzme (condensed) yogurt	A	10 (2)	Dangerous process	Akdeniz et al. (2007)
2006	Thailand	Home-canned bamboo shoots	A	209	Inadequate processing	CDC (2006a), Ungchusak et al. (2007)
2006	United States	Home-fermented tofu	A	2	Dangerous process	Meyers et al. (2007)
2006	Canada/ United States	Commercial refrigerated carrot juice	A	6	Temperature abuse	Anonymous (2006), CDC (2006b)

[a] a_w, water activity.

Food-Borne Botulism Outbreaks Associated with Nonproteolytic *C. botulinum*

Recorded outbreaks of botulism involving nonproteolytic *C. botulinum* are generally associated with strains forming type E neurotoxin (Table 5).

However, it appears likely that many European outbreaks associated with type B neurotoxin are probably due to nonproteolytic *C. botulinum* type B (Lucke, 1984; Hauschild, 1993). Other recent botulism outbreaks involving smoked, dried, or salted fish, in

Table 5. Examples of recent incidents of food-borne botulism involving nonproteolytic *C. botulinum*

Yr	Country	Product	Toxin type	No. of cases (deaths)	Factors contributing to botulism outbreak	Reference(s)
1981	United States	Uneviscerated, salted, air-dried fish ("kapchunka")	B	1	Poorly controlled salting, lack of refrigeration	CSHD (1981)
1982	Madagascar	Commercial pork sausage	E	60 (30)	Inadequate preservation	Viscens et al. (1985)
1985	United States	Uneviscerated, salted, air-dried fish ("kapchunka")	E	2 (2)	Poorly controlled salting, lack of refrigeration	CDC (1985)
1987	United States/ Israel	Commercial uneviscerated, salted, air-dried fish ("kapchunka")	E	8 (1)	Poorly controlled salting, lack of refrigeration	Slater et al. (1989)
1991	Sweden	Vacuum-packed, hot-smoked rainbow trout	E	Not known	Not known	Korkeala et al. (1998)
1991	Egypt	Commercial uneviscerated, salted fish ("faseikh")	E	>91 (18)	Putrefaction of fish before salting	Weber et al. (1993)
1992	United States	Commercial uneviscerated, salted fish ("moloha")	E	8	Insufficient salt	CDC (1992)
1994	Sweden	Vacuum-packed, hot-smoked rainbow trout	E	Not known	Not known	Korkeala et al. (1998)
1995	Canada	"Fermented" seal or walrus (4 outbreaks)	E	9	Unsafe process	Proulx et al. (1997)
1997	France	Fish	E	1	Not known	Boyer et al. (2001)
1997	Germany	Commercial hot-smoked, vacuum-packed fish ("Raucherfisch")	E	2	Suspected temperature abuse	Jahkola and Korkeala (1997), Korkeala et al. (1998)
1997	Argentina	Home-cured ham	E	6	Not known	Rosetti et al. (1999)
1997	Germany	Home-smoked, vacuum-packed fish ("Lachsforellen")	E	4	Temperature abuse	Anonymous (1998b)
1998	Germany	Commercial smoked, vacuum-packed fish	E	4	Not known	Therre (1999)
1998	France	Commercial frozen, vacuum-packed scallops	E	1	Temperature abuse (?)	Boyer et al. (2001)
1998	France	Commercial frozen, vacuum-packed prawns	E	1	Temperature abuse (?)	Boyer et al. (2001)
1999	Finland	Whitefish eggs	E	1	Temperature abuse	Lindström et al. (2004)
1999	France	Salmon or fish soup	E	1	Not known	Boyer et al. (2001)
1999	France	Grey mullet	E	1	Temperature abuse (?)	Boyer et al. (2001)
2001	Australia	Reheated chicken	E	1	Poor temperature control	Mackle et al. (2001)
2001	United States	Homemade fermented beaver tail and paw	E	3	Temperature abuse	CDC (2001)
2001	Canada	Homemade fermented salmon roe (2 outbreaks)	E	4	Unsafe process	Anonymous (2002)
2002	United States	Homemade "muktuk" (from Beluga whale)	E	12	Unsafe process	McLaughlin et al. (2004)
2003	Germany	Home-salted, air-dried fish	E	3	Temperature abuse (?)	Eriksen et al. (2004)
2004	Germany	Commercial vacuum-packed smoked salmon	E	1	Consumed after "use-by date"	Dressler (2005)
2006	Finland	Commercial vacuum-packed smoked whitefish	E	1	Temperature abuse (?)	Lindström et al. (2006a)
2006	Iran	Traditional soup (Ashmast)	E	11	Not known	Vahdani et al. (2006)

which the responsible organism/toxin has not been identified, are also likely to be associated with non-proteolytic *C. botulinum*. Food-borne botulism outbreaks associated with nonproteolytic *C. botulinum* have been associated most frequently with

- fish (e.g., salted, vacuum, smoked, or dried)
- meat (e.g., sausage and home-cured ham)

- homemade foods prepared by the peoples of Alaska and north Canada (e.g., fermented beaver tail and paw, fermented seal or walrus, fermented salmon roe, and "muktuk")

A large outbreak involving consumption of commercially produced uneviscerated salted fish ("faseikh") was recorded in Egypt in 1991. Time and/

or temperature abuse of commercial refrigerated foods has led to food-borne botulism (Table 5). For example, outbreaks in Germany (1997) and Finland (2006) were associated with suspected temperature abuse of commercial hot-smoked vacuum-packed fish. An outbreak in Germany in 2004 was associated with consumption of commercial vacuum-packed salmon after the use-by date. Outbreaks of botulism have been associated with temperature abuse of commercial products intended for frozen storage in France in 1998 (Table 5). Home-prepared foods have also been associated with botulism (Table 5).

Minimally heat-processed refrigerated foods have an excellent safety record with respect to food-borne botulism. There are no reports of food-borne botulism that involved neurotoxin formation by nonproteolytic *C. botulinum* in commercial chilled foods for which the shelf-life and storage temperature were maintained as specified by manufacturers (Peck et al., 2008). It is essential, however, that safety is maintained as new types of food are developed. A change in the etiology of food-borne botulism in France has been reported over the last few years. There has been an increased association with type E neurotoxin compared to type B neurotoxin and with commercial foods compared to home-prepared foods. It has been suggested that these changes might be associated with an increased consumption of vacuum-packed foods (Boyer et al., 2001; Boyer and Salah, 2002; Popoff and Carlier, 2004).

INCIDENCE OF *CLOSTRIDIUM BOTULINUM* SPORES IN THE ENVIRONMENT AND FOODS

In order to assess the risk presented by *C. botulinum*, it is necessary to have quantitative data on its incidence in the environment and foods. In response, there have been many surveys of the incidence of *C. botulinum*. One difficulty with such work is the need to detect the four physiologically different clostridia that comprise the *C. botulinum* species. In view of this problem and other issues, it should be recognized that some of these surveys may be subject to various limitations (Lund and Peck, 2000). For example, an underestimate of the extent of contamination may arise because (i) the detection limit is not as low as assumed (for example, because of the presence of competing bacteria, bacteriophages or bacteriocins, or other inhibitors in the test samples [i.e., detection of one spore per bottle is not achieved]) and/or (ii) the use of heat treatments may inactivate spores of nonproteolytic *C. botulinum*. A further complicating factor is that many studies have ascertained only the type of neurotoxin formed; consequently, it can be difficult to establish whether proteolytic *C. botulinum* or nonproteolytic *C. botulinum* is present (e.g., if type B or type F neurotoxin is formed). This can be a severe limitation if attempting to evaluate the hazard presented by just one of the two (proteolytic *C. botulinum* and nonproteolytic *C. botulinum*). Nonetheless, these surveys present a consistent picture of widespread contamination of the environment and foods, albeit generally at a low concentration. This subject has been reviewed extensively previously (Dodds, 1993a, 1993b; Lund and Peck, 2000).

Incidence of *Clostridium botulinum* in the Environment

The natural habitat of proteolytic *C. botulinum* and nonproteolytic *C. botulinum* is soil and sediments. Food ingredients originating from soils or sediments in geographic regions of high spore incidence may themselves have high loadings (e.g., vegetables and fish). Conversely, ingredients taken from other sources may contain lower numbers of spores (e.g., meat and dairy products). In some circumstances, however, the position might be more complicated. For example, foods may consist of a blend from a variety of geographic regions (e.g., honey), spores may be inadvertently added to foods (e.g., from spices), or processing conditions and plant hygiene may influence the contamination of foods.

In many circumstances, however, the prevalence of spores and their concentration in the environment and hence in foods are related to the incidence of botulism. For example, proteolytic *C. botulinum* type A is reported to be prevalent in soils in the western continental United States, China, Brazil, and Argentina, and most cases of food-borne botulism in these locations are associated with this organism. While proteolytic *C. botulinum* type B appears more prevalent in soils in the eastern continental United States, nonproteolytic *C. botulinum* type B appears to predominate in much of Europe. Nonproteolytic *C. botulinum* type E is associated primarily with marine and freshwater sediments in temperate regions of the world and is the predominant cause of food-borne botulism in Scandinavia, Canada, Alaska, and Japan (Dodds, 1993a).

A most-probable-number approach was used by Dodds (1993a) to summarize data on the incidence of *C. botulinum* in the environment. The overall picture is that of a widespread but generally low contamination. In surveys of soil and sediment in Europe, there were 23 reports of less than 100 spores/kg, 7 reports of 100 to 1,000 spores/kg, and 3 reports of more than 1,000 spores/kg. In similar surveys in North America, there were 16 reports of less than 100 spores/kg,

7 reports of 100 to 1,000 spores/kg, and 1 report of more than 1,000 spores/kg. A similar picture has been presented in other parts of the world. One survey in China reported 25,000 spores/kg of proteolytic *C. botulinum* type A (Dodds, 1993a).

Incidence in Foods

Thirteen surveys of meat found a consistent but very low contamination, and the highest most-probable-number estimate was 7 spores/kg. Sixteen surveys of fruits and vegetables found a consistent but low contamination (Lund and Peck, 2000). The largest most-probable-number estimate was 410 spores/kg. Fish and seafood appear to be more heavily contaminated. While many studies have reported a low contamination (e.g., 38 out of 283 French fish/shellfish samples were positive for *C. botulinum*, with the most probable spore-loading estimated at <10 spores/kg [Fach et al., 2002; Carlin et al., 2004]), several studies reported more than 100 spores/kg and two reported more than 1,000 spores/kg, the highest being 5,300 spores/kg (Huss et al., 1974; Rouh-bakhsh-Khaleghdoust, 1975; Hauschild et al., 1975; Lindroth and Genigeorgis, 1986; Garcia and Genige-orgis, 1987; Baker et al., 1990; Lund and Peck, 2000). Extremely high loadings of honey with spores of *C. botulinum* have been reported in surveys carried out in Japan and North America, with 4 out of 14 reports indicating more than 1,000 spores/kg. In view of the ubiquitous nature of spores of *C. botulinum*, albeit often at low numbers, it is apparent that raw products cannot be guaranteed free of spores. Foods which are, or can become, anaerobic may allow growth of *C. botulinum* and must therefore be subjected to treatments that destroy spores or stored under conditions that prevent growth and neurotoxin formation.

CONTROL OF PROTEOLYTIC *CLOSTRIDIUM BOTULINUM* IN FOOD PROCESSING OPERATIONS

The minimum temperature at which growth and neurotoxin production occurs is within the range of 10°C to 12°C (Peck and Stringer, 2005). Growth and neurotoxin formation at 3 to 4 weeks at 12°C with a large inoculum have been reported. The optimum growth temperature is 37°C. At this temperature it is estimated that under good growth conditions, the time to a three-decimal (3-D) increase from a spore inoculum is 12.6 hours (Peck, 2006). Growth of proteolytic *C. botulinum* is prevented at a pH value of <4.6, or by 10% NaCl, and the minimum water activity permitting growth is 0.94 and 0.93 with NaCl and glycerol, respectively, as humectants (Table 6). The use of other factors and combinations of factors to control or prevent growth of proteolytic *C. botulinum* has been described, and predictive models have been developed (Dodds, 1993c; Lund and Peck, 2000). Some predictive models for proteolytic *C. botulinum* are freely available through software packages, such as ComBase Predictor and the Pathogen Modeling Program (www.combase.cc/predictor.html; www.ars.usda.gov/Services/docs.htm?docid=6796 [both accessed 22 May 2007]). Predictions generally compare well with observed growth and neurotoxin formation in independent tests, giving the user confidence that the models can be used to target, effectively, challenge tests. Published and unpublished original growth and death curves are available free of charge in ComBase (www.combase.cc [accessed 22 May 2007]).

Proteolytic *C. botulinum* produces spores with high heat resistance and is the principal concern for the safe production of low-acid canned foods (Table 6). A heat treatment at 121.1°C for 3 min (or

Table 6. Effect of environmental factors on the growth and survival of proteolytic *C. botulinum* and nonproteolytic *C. botulinum*[b]

Factor	Proteolytic *C. botulinum*	Nonproteolytic *C. botulinum*
Neurotoxins formed	A, B, F	B, E, F
Minimum growth temperature	10–12°C	2.5–3.0°C
Optimum growth temperature	37°C	25°C
Minimum pH for growth	4.6	5.0
NaCl concentration preventing growth (%)	10	5
Minimum water activity for growth		
NaCl as humectant	0.94	0.97
Glycerol as humectant	0.93	0.94
Spore heat resistance	$D_{121°C} = 0.21$ min	$D_{82.2°C} = 2.4/231$ min[a]
Spore radiation resistance	$D = 2.0–4.5$ kGy	$D = 1.0–2.0$ kGy
Foods involved in botulism outbreaks	Home-canned foods, faulty commercial processing	Fermented marine products, dried fish, vacuum-packed fish

[a] Heat resistance data without/with lysozyme during recovery.
[b] Modified from the work of Lund and Peck (2000).

equivalent heat treatment at another temperature, based on a z value of 10°C) has been adopted as the minimum standard for a "botulinum cook" for canned foods (Stumbo et al., 1975). The application of this heat treatment has ensured the safe production of low-acid canned foods, and botulism outbreaks have occurred only when the full heat treatment has not been appropriately delivered for home-prepared or commercial ambient-stored foods (Table 4). This heat treatment is intended to deliver a 12-D reduction in the number of spores of proteolytic *C. botulinum,* and although the safety of this heat treatment has been demonstrated over many decades, it has been questioned whether a 12-D reduction is actually delivered (Casolari, 1994; Anderson et al., 1996).

CONTROL OF NONPROTEOLYTIC *CLOSTRIDIUM BOTULINUM* IN FOOD PROCESSING OPERATIONS

Growth and neurotoxin have been reported at 3.0°C to 3.3°C at 5 to 7 weeks but not at 2.1°C to 2.5°C at 12 weeks (Ohye and Scott, 1957; Schmidt et al., 1961; Eklund et al., 1967a, 1967b; Graham et al., 1997). The optimum growth temperature is 25°C. At this temperature, when other environmental factors are also optimal, the time to a 3-D increase from a spore inoculum is estimated to be 10.4 hours (Peck, 2006). Growth and neurotoxin formation are not reported below a pH of 5.0 or at a NaCl concentration above 5%. The minimum water activity permitting growth is 0.97 and 0.94 with NaCl and glycerol, respectively, as the humectants (Hauschild, 1989; Lund and Peck, 2000). The effect of other preservative factors (alone and in combination) on growth and neurotoxin formation has been reported previously, and predictive models have been developed (Graham et al., 1996a; Whiting and Oriente, 1997; Lund and Peck, 2000; Fernandez et al., 2001; Peck and Stringer, 2005; Cawley et al., 2006). Published and unpublished original data and some of these predictive models are freely available through the Internet (as noted above for proteolytic *C. botulinum*).

Developments in predictive microbiology over the last few decades have enabled reliable and accurate predictions to be made of the effect of environmental conditions on the growth rate of nonproteolytic *C. botulinum* and other food-borne pathogens (Baranyi, 1998). Lag time, however, is variable and more difficult to predict reliably, especially for spore-forming bacteria. It is influenced by spore history (e.g., sublethal damage) as well as current growth conditions. Variability in the population lag time is dependent on variability in the lag times of individual spores. In the case of nonproteolytic *C. botulinum,* this variability is particularly important, as growth is likely to result from one or a small number of spores. In these circumstances, time to neurotoxin formation will be closely related to the distribution of individual lag times within the population. Recent studies have begun to extend understanding and prediction of lag time in nonproteolytic *C. botulinum* (Stringer et al., 2005; Webb et al., 2007). The large variability in lag time was found to result from variability in each stage of lag (germination, emergence, time for maturation of one cell, and time to two cells), and the duration of these various stages was not correlated. The lag time of individual spores could not be predicted from their germination time, and the first spore to germinate was not the first to two cells (Stringer et al., 2005; Webb et al., 2007). Spores formed in medium to which 3% NaCl was added were sublethally damaged. In growth media with and without added NaCl, they had a longer lag time and lower probability of reaching one cell than spores formed in medium without added NaCl (Webb et al., 2007). In these circumstances the duration of all stages of lag was extended.

Spores of nonproteolytic *C. botulinum* are of moderate heat resistance (Table 6). Heating for a few minutes at 80°C to 85°C inactivates the germination system, resulting in sublethal injury (Peck et al., 1992; Lund and Peck, 1994). Lysozyme, however, is able to diffuse through the coat of a fraction (typically 0.1 to 1%) of these sublethally heat-damaged spores inducing germination by hydrolyzing peptidoglycan in the cortex, giving growth and biphasic survival curves (Peck and Stringer, 2005). Lysozyme-permeable spores are approximately 100 times more heat resistant than those that are not permeable (Table 6). This effect of lysozyme may be important for food safety, as lysozyme is present in many foods and relatively heat stable (Stringer et al., 1999; Peck and Stringer, 2005). For example, while heat treatments of 85°C for 36 min, 90°C for 10 min, and 95°C for 15 min each prevented an inoculum of 10^6 spores of nonproteolytic *C. botulinum,* leading to growth and neurotoxin formation at 25°C in 60 days in a model food with no added lysozyme, when hen egg white lysozyme was added prior to heating at 85°C for 84 min, 90°C for 34 min, or 95°C for 15 min, growth was observed at 25°C after 13, 14, and 32 days, respectively (Peck et al., 1995; Peck and Fernandez, 1995; Graham et al., 1996b; Fernandez and Peck, 1997, 1999). Naturally occurring lysozyme in crabmeat (estimated at 200 μg/g prior to heating) (Lund and Peck, 1994) may facilitate recovery of sublethally damaged spores, since heat treatments at 90.6°C for 65 min, 92.2°C for 35 min, or 94.4°C for 15 min were required to prevent growth and neurotoxin formation from 10^6 spores of nonproteolytic *C. botulinum* at 27°C in 150 days (Peterson et al., 1997). The heat resistance of spores of nonproteolytic *C. botulinum* is not related to the type of neurotoxin formed (Table 7).

Alternative processing technologies may also be effective in controlling growth and neurotoxin formation by nonproteolytic *C. botulinum*. High-hydrostatic-pressure-treated high-acid foods (e.g., fruit juices) first became commercialized in Japan in the early 1980s. Bacterial spores are particularly pressure resistant, and although viable spores remained in these high-acid foods after pressure treatment, an acidic environment prevented subsequent growth, creating intrinsically safe products. The product range has now been extended to include products with a high pH and water activity, and it is essential that the risk of neurotoxin formation by nonproteolytic *C. botulinum* is controlled. Reddy et al. (1999) reported that a combination of a pressure of 827 MPa and 40°C/10 min or 50°C/5 min delivered a 5-D reduction in the number of spores of nonproteolytic *C. botulinum*. A proposed alternative approach is to apply a two-stage high-hydrostatic pressure treatment, the first treatment to germinate the spores and the second to kill germinated spores (Heinz and Knorr, 2001). Such an approach requires caution, however, as a small proportion of spores are likely to resist pressure-induced germination. Spores of nonproteolytic *C. botulinum* are more sensitive to irradiation than those of proteolytic *C. botulinum* (Table 6). Low irradiation doses (1 to 3 kGy), a process sometimes called "irradiation-pasteurization," had little effect on the control of neurotoxin formation by nonproteolytic *C. botulinum* in fish (Eklund, 1982). Indeed in some tests, irradiation-pasteurization led to more rapid neurotoxin formation, presumably

because the competing microflora were inactivated or damaged (Eklund, 1982).

Nonproteolytic *C. botulinum* has been identified as the principal microbiological safety hazard in minimally heated refrigerated foods (also known as refrigerated processed foods of extended durability, cook-chill foods, sous-vide foods, and ready meals) (Peck, 1997; Peck and Stringer, 2005; Peck, 2006). It is estimated that more than 1.5×10^{10} of these meals have been sold in Europe over the last 20 years, with sales increasing by about 10% per annum. These foods are not sterile and are often packed under a modified low-oxygen atmosphere or vacuum. Safety and quality are dependent on a combination of

- minimal heat treatment (typical maximum of 70°C to 95°C)
- refrigerated storage temperature (typically ≤8°C)
- restricted shelf-life (which can be up to 42 days)
- other intrinsic/extrinsic factors that may also be important (e.g., pH and water activity)

Recommendations from the United Kingdom Advisory Committee on the Microbiological Safety of Food (ACMSF) on the safe production of vacuum and modified atmosphere packed chilled foods, with respect to *C. botulinum* and the associated food-borne botulism hazard, are summarized in Table 8. It is recommended that the heat treatments or combination processes deliver a safety factor of 10^6 (a 6-D

Table 7. Effect of heat treatment at 85°C/60 min on the inactivation of spores of nonproteolytic *C. botulinum* recovered on PYGS medium with lysozyme[a]

Strain	Neurotoxin type	Log kill after 60 min/85°C
Foster B96	E	1.9
Eklund 17B	B	2.7
Eklund 2B	B	3.1
2129B	B	3.6
Hazen 36208	E	3.7
Sebald P34	E	3.7
Hobbs FT50	B	3.7
Dolman VH	E	3.8
Beluga	E	3.8
Eklund 202F	F	4.0
Craig 610	F	4.2
Colworth 151	B	5.0

[a]PYGS, peptone, yeast extract, glucose, starch medium (Stringer et al., 1999).

Table 8. Recommended procedures to ensure the safety of minimally heated refrigerated foods with respect to nonproteolytic *C. botulinum*[a]

Recommendation[b]
Storage at <3.0°C
Storage at ≤8°C and a shelf-life of ≤10 days
Storage at chill temperature[†] combined with heat treatment of 90°C for 10 min or equivalent lethality (e.g., 80°C for 129 min, 85°C for 36 min)[‡]
Storage at chill temperature combined with pH ≤ 5.0 throughout the food
Storage at chill temperature combined with a salt concentration of ≥3.5% throughout the food
Storage at chill temperature combined with an a_w ≤ 0.97 throughout the food
Storage at chill temperature combined with a combination of heat treatment and other preservative factors which can be shown consistently to prevent growth and neurotoxin production by *C. botulinum*

[a]Information derived from the work of the ACMSF (1992, 1995, 2006). a_w, water activity.
[b]†, chill temperature is specified as 8°C in England and Wales; ‡, alternative heat treatments of 80°C for 270 min and 85°C for 52 min are recommended by the European Chilled Food Federation (ECFF, 1996; CFA, 2006).

process), with regard to spores of nonproteolytic *C. botulinum* (ACMSF, 1992; ECFF, 1996; CFA, 2006).

Barker and colleagues have used a probabilistic modeling approach to assess the risk of food-borne botulism presented by nonproteolytic *C. botulinum* (Barker et al., 2002, 2005). This approach considers the entire food process, from production to consumption, as a series of linked operations and utilizes recent developments in mathematics and computing to quantify the magnitude of the risks and to identify hazard scenarios. A probability distribution is established for each variable (e.g., initial contamination load) and incorporates an intrinsic representation of uncertainties. All the distributions are then incorporated in a belief network to give a clear picture of information dependencies and system complexity. Outputs from these probabilistic risk models can be used to identify and prioritize steps that minimize the effects of detrimental events and that maximize awareness of process control options. A probabilistic risk model has been developed for the exposure of consumers to neurotoxin formed by nonproteolytic *C. botulinum* in gnocchi, a minimally heated refrigerated potato product with an extended shelf-life (Barker et al., 2005). The production process includes mixing of ingredients (e.g., raw potato flakes and starch), delivery of a minimal heat treatment, cooling, and packaging. Account is taken of consumer behavior, with respect to temperature of storage and time prior to consumption. By combining data and information from several sources, it was demonstrated that gnocchi is particularly safe with respect to nonproteolytic *C. botulinum* hazards (Barker et al., 2005). Output from the process risk model was consistent with an assessment of the prevalence and behavior of nonproteolytic *C. botulinum* in gnocchi (Del Torre et al., 2004).

METHODS FOR THE DETECTION OF *CLOSTRIDIUM BOTULINUM* AND ITS NEUROTOXINS

In view of the danger presented by *C. botulinum* and its potent neurotoxins, practical work is restricted to containment laboratories offering an appropriate degree of protection. It must be ensured that all appropriate safety guidelines are in place before work is started.

Cultural Methods for Proteolytic *C. botulinum* and Nonproteolytic *C. botulinum*

Culture of *C. botulinum* may be required for studies of the extent of contamination of the environment or foods and for detection of the organisms in foods or clinical samples suspected of being implicated in botulism outbreaks. As proteolytic *C. botulinum* and nonproteolytic *C. botulinum* are physiologically distinct, a single enrichment procedure cannot be relied on for both organisms. Conditions that are optimum for enrichment and isolation of one are often not the best for the other. Furthermore, in attempts to culture *C. botulinum* from the environment or foods, it is important to determine the efficiency of the method(s) used by demonstrating that strains of all relevant groups of *C. botulinum* can be isolated following addition of low numbers of bacterial spores to test samples (Sugiyama et al., 1970; Fach et al., 2002; Del Torre et al., 2004). The type of problems that can be encountered is illustrated by the work of Sugiyama et al. (1970). These authors found that sediment samples from the Fox River in the United States were so inhibitory to nonproteolytic *C. botulinum* type E that it was necessary to add 10^6 spores of this bacterium to one gram of sediment before type E neurotoxin could be detected in enrichment culture.

Competing vegetative bacteria can be eliminated, prior to enrichment, by the use of a heat treatment. While heating at 75°C to 80°C for 10 to 15 min is useful in culturing from spores of proteolytic *C. botulinum*, heating at 60°C is more appropriate for use during isolation of nonproteolytic *C. botulinum*, and the addition of lysozyme to the enrichment medium can increase recovery of heated spores (Peck et al., 1993; Sebald and Petit, 1994). The use of ethanol is an alternative to heat treatment (Smith and Sugiyama, 1988; Solomon and Lilly, 2001). Unheated samples may also be tested in order to isolate vegetative bacteria or spores that are not fully heat resistant. It is essential that the enrichment media is fully anaerobic (Hatheway, 1988; Kautter et al., 1992; Sebald and Petit, 1994; Solomon and Lilly, 2001), and the redox dye resazurin can be a useful indicator (Lund and Wyatt, 1984). To enrich for proteolytic *C. botulinum*, samples should be inoculated into cooked meat medium and incubated at 35°C, while for nonproteolytic *C. botulinum*, chopped meat glucose starch medium (Hauschild, 1989) or trypticase-peptone-glucose-yeast extract broth containing trypsin is suggested, with incubation at 26°C (Solomon and Lilly, 2001). Trypsin is added to inactivate bacteriocins that may be produced by closely related clostridia. An extended period of incubation (7 days or longer) should be allowed for growth, sporulation, and neurotoxin formation. A portion of the enriched culture can be plated onto a suitable solid medium, and another portion used to test for botulinum neurotoxin or for neurotoxin gene by PCR. Egg-yolk agar is a favored nonselective solid medium, as colonies of proteolytic *C. botulinum* and nonproteolytic *C. botulinum* have a typical appearance associated with their lipase activity. Selective plating media include *C. botulinum* isolation agar (Dezfulian et al., 1981) and botulinum selective medium (Mills et al., 1985). The trimethoprim in these media may inhibit nonproteolytic *C. botulinum* (Hatheway, 1988).

Detection and Quantification of *C. botulinum* Neurotoxins

It may be necessary to test for botulinum neurotoxin in samples of food, clinical samples, or enrichment media. The standard method for detection and identification of botulinum neurotoxin has been intraperitoneal injection into mice (Hatheway, 1988; Solomon and Lilly, 2001). Specificity is achieved by the use of specific antisera and by observations of typical symptoms of botulism in the mice prior to death. It is necessary to treat samples with trypsin to detect neurotoxin formed by nonproteolytic *C. botulinum*. Trypsin is required to convert the single-chain neurotoxin to the more toxic dichain form. The mouse test is extremely sensitive (5 to 10 pg of neurotoxin), measures the biological activity of the neurotoxin, and is both repeatable and reproducible (Kautter and Solomon, 1977; Hatheway, 1988). Additionally, there is the potential to detect previously undescribed neurotoxins, atypical neurotoxins, and antigenic variants. Disadvantages of the mouse test are the ethical issues concerning the use of animals, the need to wait several days before a sample can be judged negative, the expense, and the need for skilled personnel. A number of alternative animal methods have been described, some of which are nonlethal (Sesardic et al., 1996; Pearce et al., 1997).

Immunochemical methods for the detection of botulinum neurotoxins include enzyme-linked immunosorbent assays (ELISA), lateral flow immunoassays, and various other tests (Scarlatos et al., 2005; Sharma and Whiting, 2005; Lindström and Korkeala, 2006). These methods are generally cheaper and much easier to use than the mouse test, and some have the same sensitivity and specificity. The various immunochemical methods may be subject to various limitations, however; for example, (i) some may react differently with neurotoxins of a specific type produced by different strains, since these neurotoxins may differ in antigenicity (Gibson et al., 1987, 1988; Huhtanen et al., 1992; Doellgast et al., 1993; Ekong et al., 1995; Smith et al., 2005); (ii) some may react with biologically inactive neurotoxin; (iii) some use antibodies that were raised to preparations containing a mixture of antigens, so that the tests

may not be specific for neurotoxins (Sakaguchi, 1979; Huhtanen et al., 1992; Potter et al., 1993); and (iv) many tests require complex and expensive amplification systems to achieve the sensitivity of the mouse test (e.g., Shone et al., 1985; Modi et al., 1986; Doellgast et al., 1993). Despite these limitations, ELISAs and other immunochemical methods have been used widely for detection of *C. botulinum* neurotoxins (Table 9). An ELISA developed by Ferreira and colleagues (Ferreira and Crawford, 1998; Ferreira, 2001; Ferreira et al., 2001, 2004) offers independent detection of types A, B, E, and F neurotoxins and has been used to detect neurotoxin in food associated with botulism outbreaks in the United States (Ferreira et al., 2001, 2004). The test is not as sensitive as the mouse test. Sharma et al. (2006) reported on an amplified ELISA that detected <5 50% minimum lethal doses (MLD_{50})/ml of type A, B, E, and F neurotoxins in casein buffer and <60 MLD_{50}/ml in various foods. A simple, rapid, cheap, sensitive, chemiluminescent slot blot immunoassay has been used for the quantification of type E neurotoxin (Cadieux et al., 2005). Lateral flow immunoassays are extremely easy to use and provide a very rapid detection of the target, typically in 15 to 30 min (Aldus et al., 2003; Capps et al., 2004; Sharma et al., 2005). These assays are qualitative and appear to show promise only as an initial screening tool for botulinum neurotoxin, due to their relatively high detection limit (1,000 to 2,000 MLD_{50}/ml for type A, B, and F neurotoxins [Sharma et al., 2005; Gessler et al., 2007]). A further method uses paramagnetic bead-based electrochemiluminescence to detect botulinum neurotoxins. In this test the signal is generated after capture, by a magnet, of complexes containing paramagnetic beads complexed with capture and detection antibodies and botulinum neurotoxin. The test permits rapid throughput of samples, requiring 2 hours to complete. Detection limits for type A, B, E, and F neurotoxins ranged from 2 to 70 MLD_{50}/ml in clinically relevant matrices and <1 to 70 MLD_{50}/ml in selected food matrices (Rivera et al., 2006). A number of other immunochemical methods have been developed for quantification of botulinum neurotoxins, such as fiber-optic evanescent-wave immunosensors (Scarlatos et al., 2005; Sharma and Whiting, 2005; Lindström and Korkeala, 2006). Colony immunoblot techniques have been used to identify colonies of proteolytic *C. botulinum* types A and B (Goodnough et al., 1993) and nonproteolytic *C. botulinum* type E (Dezfulian, 1993; Goodnough et al., 1993).

A number of in vitro assays have been developed that quantify the highly specific endopeptidase activity of the botulinum neurotoxin light chains. The first assays to be developed used an immunochemical

Table 9. Examples of sensitive quantitative immunochemical methods for detection of *C. botulinum* neurotoxins

Neurotoxins detected (MLD_{50}/ml)	Comments	Reference(s)
A (5–10)	ELISA. Failed to detect neurotoxin produced by one type A strain. No cross-reaction with other clostridia, denatured neurotoxin, or other neurotoxin types. Complex amplification system. Used with foods.	Shone et al. (1985), Gibson et al. (1987)
B (20)	ELISA. Failed to detect neurotoxin produced by one type B strain. No cross-reaction with other clostridia or other neurotoxin. Used with foods. Complex amplification system.	Modi et al. (1986), Gibson et al. (1988)
A (1–32), B (<1–16)	ELISA. May respond to antigens with no neurotoxicity. Correlation between the response from ELISA and mouse test not always consistent. Used with foods.	Huhtanen et al. (1992)
A (9), B (<1), E (<1)	ELISA. Cross-reaction with other clostridia. Used extensively to measure formation of type A, B, and E neurotoxin in meat and in vegetable preparations. Also reacted with type F neurotoxin.	Potter et al. (1993), Carlin and Peck (1995), Fernandez and Peck (1999), Stringer et al. (1999)
A (1), B (1), E (1)	ELISA. Weak reaction with neurotoxin from some strains. Used to measure neurotoxin production by nonproteolytic *C. botulinum* in fish fillets. Complex amplification system.	Doellgast et al. (1993, 1994), Roman et al. (1994)
A (1–20)	ELISA. Developed for therapeutic preparations.	Ekong et al. (1995)
E (1–10)	ELISA. No cross-reaction with other neurotoxins or other clostridia. Used with foods.	Wong (1996)
A (10), B (10), E (10), F (10)	ELISA. Tested in a ring trial. Used to quantify neurotoxin present in food samples associated with botulism outbreaks.	Ferreira and Crawford (1998), Ferreira (2001), Ferreira et al. (2001, 2004)
E (4)	Chemiluminescent slot blot immunoassay. Used with bacterial cultures, naturally contaminated soil, inoculated fish. Some cross-reaction.	Cadieux et al. (2005)
A (2), B (2), E (5), F (<1)	ELISA. Low detection limit in casein buffer, higher in food samples (60 MLD_{50}/ml). High specificity.	Sharma et al. (2006)
A (50), B (70), E (5), F (<1)	Paramagnetic bead-based electrochemiluminescence. Sensitive and specific detection in food and clinically relevant matrices.	Rivera et al. (2006)

approach to detect cleavage of SNAP-25 or VAMP by botulinum neurotoxins and were often as sensitive as the mouse test. In these assays, a fragment of the target protein (SNAP-25 or VAMP) is attached to a microtiter plate and serves as the substrate. The sample containing neurotoxin is then added, and after a period of incubation, specific antibodies are added that bind to the cleaved target protein and then secondary antibodies and a detection system are added. Advantages over ELISA procedures follow: (i) the tests measure the biological activity of the neurotoxin light chain (but not the heavy chain); (ii) variations in the antigenicity of neurotoxins of a specific type do not influence the response; and (iii) the problem resulting from antibodies having been raised to a mixture of antigens is eliminated. Endopeptidase assays for type A neurotoxin and type B neurotoxin were specific, did not cross-react with each other or other neurotoxins, and after signal amplification, had detection limits of 30 to 40 MLD_{50}/ml (Hallis et al., 1996). A highly sensitive endopeptidase assay developed for therapeutic preparations of type A neurotoxin correlated well with the mouse test and had a detection limit of 0.2 to 1.0 MLD_{50}/ml (Ekong et al., 1997). In a slightly different format, a highly sensitive assay for type B neurotoxin has been developed that captures the neurotoxin on an immunoaffinity column or microtiter plate, prior to an endopeptidase assay (Wictome and Shone, 1998; Wictome et al., 1999a, 1999b). This method worked well with foods including meat, fish, and cheese and was able to distinguish between type B neurotoxin formed by proteolytic *C. botulinum* and nonproteolytic *C. botulinum*. Alternative assays employ a synthetic peptide (to mimic SNAP-25 or VAMP) labeled with quenched fluorophores as the substrate. The endopeptidase activity of the botulinum neurotoxins leads to cleavage of these peptides and a release of fluorescence (Schmidt et al., 2001; Schmidt and Stafford, 2003). The endopeptidase activity of botulinum neurotoxins has also been quantified using mass spectrometry. Synthetic substrates are incubated with the test sample, and the product peptides identified on the basis of their mass. The concentrations of all seven neurotoxin types were measured simultaneously in a single test sample, and with samples in buffer, this assay was more sensitive than the mouse test (Barr et al., 2005; Boyer et al., 2005). To avoid problems encountered with nonspecific proteases, this method has been expanded to include an antibody capture method to partially purify and concentrate neurotoxin (e.g., from serum or stool), prior to quantification of endopeptidase activity (Kalb et al., 2006). Detection limits for samples spiked with type A, B, E, and F neurotoxins were <1 to 20 MLD_{50}/ml for human serum samples and <1 to 200 MLD_{50}/ml for stool samples. The entire method could be performed in 4 hours, but limitations are the high equipment costs and the need for highly trained personnel. Potential limitations of all endopeptidase activity tests include (i) interference by other proteases (although in some tests this has been addressed by capture of the neurotoxin prior to measurement of endopeptidase activity) and (ii) positive reaction with neurotoxin that is inactive in vivo, (because the endopeptidase assays relate to the biological activity of the light chain and would not be affected by inactivation of the heavy chain).

Molecular Methods for the Detection and Characterization of *C. botulinum*

A number of tests have been developed that use PCR to detect neurotoxin genes. While these tests do not detect the neurotoxin, they have been shown to be of considerable utility in the investigation of botulism outbreaks and in the testing of enrichment media during surveys for the presence of *C. botulinum* in food, clinical, and environmental samples. Following a cultural enrichment, these methods correlated well with tests for neurotoxin using mice and, with the inclusion of a most-probable-number series of dilutions in the cultural enrichment, have been used for quantitative detection of bacteria containing these genes in relevant samples (e.g., Hielm et al., 1996, 1998; Aranda et al., 1997; Lindström et al., 2001; Carlin et al., 2004; Nevas et al., 2005b; Myllykoski et al., 2006; Lindström and Korkeala, 2006; Merivirta et al., 2006). The use of a cultural enrichment ensures good sensitivity, while minimizing possible problems due to the presence of extracellular DNA or dead bacteria. Probes have been constructed that enable PCR tests for the nonspecific detection of all botulinum neurotoxin genes (Campbell et al., 1993) and for the specific detection of genes encoding each neurotoxin (e.g., Szabo et al., 1993; Franciosa et al., 1994; Fach et al., 1995; Ferreira and Hamdy, 1995; Alsallami and Kotlowski, 2001; Wu et al., 2001; Fach et al., 2002; Carlin et al., 2004; Lindström and Korkeala, 2006). An important step forward has been the development of multiplex PCR methods for simultaneous detection of type A, B, E, and F neurotoxin genes in a single reaction (Lindström et al., 2001, 2006b; Gauthier et al., 2005). Detection in these tests was achieved by gel electrophoresis confirmation of product size or by

further hybridization onto a polyester cloth membrane coated with cDNA probes to the PCR products. In some of these tests, retargeting of the PCR primers may be required in order to ensure the detection of all recently described neurotoxin subtypes (Smith et al., 2005). There is also merit in confirming the specificity of a positive PCR result, for example, by sequencing of the PCR product. The use of molecular methods for the detection of *C. botulinum* has been reviewed recently (Lindström and Korkeala, 2006).

A number of typing tools have been used for the molecular characterization of strains of *C. botulinum*. These include ribotyping, pulsed-field gel electrophoresis (PFGE), DNA sequencing, DNA microarrays, PCR-based methods (e.g., repetitive element sequence-based PCR, amplified fragment length polymorphism [AFLP], and randomly amplified polymorphic DNA analysis), and focal plane array-Fourier transform infrared spectroscopy. Ribotyping is based on analysis of conservative ribosomal genes and is widely used for molecular typing of unknown bacteria and distinguished strains of proteolytic *C. botulinum* and nonproteolytic *C. botulinum* (Lindström and Korkeala, 2006). For PFGE, genomic DNA that has been digested with rare-cutting restriction enzymes is separated by electrophoresis. In a recent study, this technique was used to investigate the diversity of 55 strains of proteolytic *C. botulinum* (Nevas et al., 2005c). PFGE has a higher discriminatory power than ribotyping (Lindström and Korkeala, 2006). The genome sequence of proteolytic *C. botulinum* strain Hall A (ATCC 3502, NCTC 13319) has now been published (Sebaihia et al., 2007), and the sequencing of genomes of other strains of proteolytic *C. botulinum* and nonproteolytic *C. botulinum* is in progress. Comparative genomic indexing using a DNA microarray based on the Hall A genome sequence was an effective tool to discriminate strains of proteolytic *C. botulinum* but not nonproteolytic *C. botulinum* (Sebaihia et al., 2007). Unlike other typing methods, information is also provided on the genome content of the tested strains. It was found that 87% to 96% of the Hall A presumptive genes were possessed by nine other strains of proteolytic *C. botulinum* and that 84% to 87% of the presumptive genes were shared with two strains of *C. sporogenes*. Two prophages present in the Hall A strain were absent in the 11 test strains. For repetitive element sequence-based PCR, PCR is targeted at conservative repetitive extragenic elements. A species-specific fingerprint can be derived from the size and number of amplification products, while strain-specific differentiation is limited. Hyytiä et al. (1999) used this method to characterize strains of proteolytic *C. botulinum* and nonproteolytic *C. botulinum*. For AFLP, genomic

DNA that has been digested with a pair of restriction enzymes is ligated with restriction site-specific adaptors, and a subset of the fragments amplified by PCR. AFLP has been used to characterize 33 strains of proteolytic *C. botulinum* and 37 strains of nonproteolytic *C. botulinum* (Keto-Timonen et al., 2005). This method clearly differentiated between proteolytic *C. botulinum* and nonproteolytic *C. botulinum* and was suitable for typing at the strain level. Randomly amplified polymorphic DNA analysis involves PCR amplification with randomly annealing universal primers under conditions of low stringency. This method and ribotyping, however, were not as effective as PFGE in discriminating strains of nonproteolytic *C. botulinum* type E from the Canadian arctic (Leclair et al., 2006). Focal plane array-Fourier transform infrared spectroscopy has been used for whole-organism fingerprinting of 44 strains of proteolytic *C. botulinum* and nonproteolytic *C. botulinum* (Kirkwood et al., 2006). This method provided rapid discrimination of the two *C. botulinum* groups.

CONCLUDING REMARKS

C. botulinum organisms are a heterogeneous group of bacteria that forms the botulinum neurotoxin, the most potent substance known. Since spores of *C. botulinum* are ubiquitous in food, albeit often at a low concentration, it is necessary to apply treatments that destroy spores or store the food under conditions that prevent growth and neurotoxin formation. In view of the severity of food-borne botulism, vigilance is needed to ensure that *C. botulinum* does not become an emerging pathogen. Nonproteolytic *C. botulinum* has been identified as the principal safety hazard in minimally heated refrigerated foods, and it is essential that research continues to underpin the safe development of these novel foods. It is important that as new food processes are applied (e.g., high hydrostatic pressure), the principle of equivalence with existing safe processes is adopted. The increased movement of raw materials and foods, through globalization of trade, may also bring an increased risk. For example, three of the four food-borne botulism outbreaks reported in the United Kingdom between 2002 and 2005 were associated with food from eastern European countries with a higher botulism incidence than that of the United Kingdom. In comparison, previously there were only four United Kingdom outbreaks between 1956 and 2001 (McLauchlin et al., 2006). There is also the need to be alert to the potential transfer of botulinum neurotoxin genes to other bacteria. A more recent concern is the deliberate introduction of *C. botulinum* or its neurotoxin into the food chain through a bioterrorism act.

Acknowledgments. I am grateful for funding from the Competitive Strategic Grant of the BBSRC and other funders of the Institute of Food Research.

REFERENCES

Abgueguen, P., V. Delbos, J. M. Chennebault, S. Fanello, O. Brenet, P. Alquier, J. C. Granry, and E. Pichard. 2003. Nine cases of foodborne botulism type B in France and literature review. *Eur. J. Clin. Microbiol. Infect. Dis.* **22:**749–752.

ACMSF. 1992. *Report on Vacuum Packaging and Associated Processes.* Her Majesty's Stationery Office, London, United Kingdom.

ACMSF. 1995. *Annual Report 1995.* Her Majesty's Stationery Office, London, United Kingdom.

ACMSF. 2006. ACMSF minutes 8 June 2006. ACMSF Secretariat, London, United Kindgom. www.food.gov.uk/science/ouradvisors/microbiogsafety/acmsfmeets/acmsf2006/acmsfmeet080606/acmsfmin080606. Accessed 1 June 2007).

Akbulut, D., J. Dennis, M. Gent, K. A. Grant, V. Hope, C. Ohai, J. McLauchlin, V. Vithani, O. Mpamugo, F. Ncube, and L. de Souza-Thomas. 2005. Wound botulism in injectors of drugs: upsurge in cases in England during 2004. *Eurosurveillance* **10(9):**1–6.

Akdeniz, H., T. Buzgan, M. Tekin, H. Karsen, and M. K. Karahocagil. 2007. An outbreak of botulism in a family in Eastern Anatolia associated with eating süzme yoghurt buried under soil. *Scan. J. Infect. Dis.* **39:**108–114.

Aldus, C. F., A. van Amerogen, R. M. C. Ariens, M. W. Peck, J. H. Wichers, and G. M. Wyatt. 2003. Principles of some novel rapid dipstick methods for detection and characterisation of verotoxigenic *Escherichia coli. J. Appl. Microbiol.* **95:**380–389.

Alsallami, A. A., and R. Kotlowski. 2001. Selection of primers for specific detection of *Clostridium botulinum* types B and E neurotoxin genes using a PCR method. *Int. J. Food Microbiol.* **69:**247–253

Anderson, W. A., P. J. McClure, A. C. Baird-Parker, and M. B. Cole. 1996. The application of a log logistic model to describe the thermal inactivation of *Clostridium botulinum* 213B at temperatures below 121.1°C. *J. Appl. Bacteriol.* **80:**283–290.

Angulo, F. J., J. Getz, J. P. Taylor, K. A. Hendricks, C. L. Hatheway, S. S. Barth, H. M. Solomon, A. E. Larson, E. A. Johnson, L. N. Nickey, and A. A. Ries. 1998. A large outbreak of botulism: the hazardous baked potato. *J. Infect. Dis.* **178:**172–177.

Annibali, F., L. Fenicia, G. Franciosa, and P. Aureli. 2002. Influence of pH and temperature on the growth of and toxin production by neurotoxingenic strains of *Clostridium butyricum* type E. *J. Food Prot.* **65:**1267–1270.

Anonymous. 1995. Foodborne outbreaks in California, 1993–1994. *Dairy Food Environ. Sanit.* **15:**611–615.

Anonymous. 1998a. Botulism, human – Algeria. ProMED archive no. 19980723.1393. www.promedmail.org.

Anonymous. 1998b. Fallbericht: botulismus nach dem Verzehr von geraucherten Lachsforellen. *Epidemiol. Bull.* **4:**20.

Anonymous. 1999. Botulism - Azerbaijan (Baju). ProMED archive no.19991222.2193. www.promedmail.org.

Anonymous. 2002. Two outbreaks of botulism associated with fermented salmon roe – British Columbia – August 2001. *Can. Commun. Dis. Rep.* **28/06:**1–4.

Anonymous. 2006. Botulism: Ontario. *Infect. Dis. News Brief* **13/10:**1.

Aranda, E., M. M. Rodriguez, M. A. Asensio, and J. J. Córdoba. 1997. Detection of *Clostridium botulinum* types A, B, E and F in food by PCR and DNA probe. *Lett. Appl. Microbiol.* **25:**186–190.

Arndt, J. W., M. J. Jacobson, E. E. Abola, C. M. Forsyth, W. H. Tepp, J. D. Marks, E. A. Johnson, and R. C. Stevens. 2006. A structural perspective of the sequence variability within botulinum neurotoxin subtypes A1–A4. *J. Mol. Biol.* **362:**733–742.

Arnon, S. S. 2004. Infant botulism, p. 1758–1766. *In* R. D. Feigin, and J. D. Cherry (ed.), *Textbook of Pediatric Infectious Disease*, 5th ed. Saunders, Philadelphia, PA.

Arnon, S. S., T. F. Midura, K. Damus, B. Thompson, R. M. Wood, and J. Chin. 1979. Honey and other environmental risk factors for infant botulism. *J. Pediatr.* **94:**331–336.

Arnon, S. S., R. Schechter, T. V. Inglesby, D. A. Henderson, J. G. Bartlett, M. S. Ascher, E. Eitzen, A. D. Fine, J. Hauer, M. Layton, S. Lillibridge, M. T. Osterholm, T. O'Toole, G. Parker, T. M. Perl, P. K. Russell, D. L. Swerdlow, and K. Tonat. 2001. Botulinum toxin as a biological weapon. *JAMA* **285:**1059–1070.

Arnon, S. S., R. Schechter, S. E. Maslanka, N. P. Jewell, and C. L. Hatheway. 2006. Human botulism immune globulin for the treatment of infant botulism. *N. Engl. J. Med.* **354:**462–471.

Aureli, P., M. Di Cunto, A. Maffei, G. De Chiara, G. Franciosa, L. Accorinti, A. M. Gambardella, and D. Greco. 2000. An outbreak in Italy of botulism associated with a dessert made with mascarpone cream cheese. *Eur. J. Epidemiol.* **16:**913–918.

Aureli, P., G. Franciosa, and L. Fenicia. 2002. Infant botulism and honey in Europe: a commentary. *Pediatr. Infect. Dis. J.* **21:**866–888.

Baker, D. A., C. Genigeorgis, and G. Garcia. 1990. Prevalence of *Clostridium botulinum* in seafood and significance of multiple incubation temperatures for determination of its presence and type in fresh retail fish. *J. Food Prot.* **53:**668–673.

Baranyi, J. 1998. Comparison of stochastic and deterministic concepts of bacterial lag. *J. Theor. Biol.* **192:**403–408.

Barker, G. C., P. K. Malakar, M. Del Torre, M. L. Stecchini, and M. W. Peck. 2005. Probabilistic representation of the exposure of consumers to *Clostridium botulinum* neurotoxin in a minimally processed potato product. *Int. J. Food Microbiol.* **100:**345–357.

Barker, G. C., N. L. C. Talbot, and M. W. Peck. 2002. Risk assessment for *Clostridium botulinum*: a network approach. *Int. Biodeterioration Biodegradation* **50:**167–175.

Barr, J. R., H. Moura, A. E. Boyer, A. R. Woolfitt, S. R. Kalb, A. Pavlopoulos, L. G. McWilliams, J. G. Schmidt, R. A. Martinez, and D. L. Ashley. 2005. Botulinum neurotoxin detection and differentiation by mass spectrometry. *Emerg. Infect. Dis.* **11:**1578–1583.

Bhutani, M., E. Ralph, and M. D. Sharpe. 2005. Acute paralysis following a "bad potato": a case of botulism. *Can. J. Anesth.* **52:**433–436.

Boyer, A., C. Girault, F. Bauer, J. M. Korach, J. Salomon, E. Moirot, J. Leroy, and G. Bonmarchand. 2001. Two cases of foodborne botulism type E and review of epidemiology in France. *Eur. J. Clin. Microbiol. Infect. Dis.* **20:**192–195.

Boyer, A., and A. Salah. 2002. Le botulisme en France: épidémiologie et clinique. *Ann. Med. Intern.* **153:**300–310.

Boyer A. E., H. Moura, A. R. Woolfitt, S. R. Kalb, L. G. McWilliams, A. Pavlopoulos, J. G. Schmidt, D. L. Ashley, and J. G. Barr. 2005. From the mouse to the mass spectrometer: detection and differentiation of the endoproteinase activities of botulinum neurotoxins A-G by mass spectrometry. *Anal. Chem.* **77:**3916–3924.

Brett, M. M., G. Hallas, and O. Mpamugo. 2004. Wound botulism in the UK and Ireland. *J. Med. Microbiol.* **53:**555–561.

Cadieux, B., B. Blanchfield, J. P. Smith, and J. W. Austin. 2005. A rapid chemiluminescent slot blot immunoassay for the detection and quantification of *Clostridium botulinum* neurotoxin type E, in cultures. *Int. J. Food Microbiol.* **101:**9–16.

Campbell, K. D., M. D. Collins, and A. K. East. 1993. Gene probes for identification of the botulinal neurotoxin gene and specific identification of neurotoxin types B, E and F. *J. Clin. Microbiol.* **31:**2255–2262.

Capps, K. L., E. M. McLaughlin, A. W. A. Murray, C. F. Aldus, G. M. Wyatt, M. W. Peck, A. van Amerongen, R. M. C. Ariëns, J. H. Wichers, C. Baylis, D. R. Wareing, and F. J. Bolton. 2004. Validation of three rapid screening methods for the detection of verotoxin-producing *Escherichia coli* in foods: an interlaboratory study. *J. AOAC Int.* **87:**68–77.

Carlier, J. P., C. Henry, V. Lorin, and M. R. Popoff. 2001. Le botulisme en France a la fin du deuxieme millenaire (1998–2000). *Bull. Epidemiol. Hebd.* 2001(9):37–39.

Carlin, F., V. Broussolle, S. Perelle, S. Litman, and P. Fach. 2004. Prevalence of *Clostridium botulinum* in food raw materials used in REPFEDs manufactured in France. *Int. J. Food Microbiol.* **91:**141–145.

Carlin, F., and M. W. Peck. 1995. Growth and toxin production by non-proteolytic and proteolytic *Clostridium botulinum* in cooked vegetables. *Lett. Appl. Microbiol.* **20:**152–156.

Casolari, A. 1994. About basic parameters of food sterilization technology. *Food Microbiol.* **11:**75–84.

Cawley, G. C., N. L. C. Talbot, G. Janacek, and M. W. Peck. 2006. Sparse Bayesian survival analysis for modelling the growth domain of microbial pathogens. *IEEE Trans. Neural. Netw.* **17:**471–481.

Cawthorne, A., L. P. Celentano, F. D'Ancona, A. Bella, M. Massari, F. Anniballi, L. Fenicia, P. Aureli, and S. Salmaso. 2005. Botulism and preserved green olives. *Emerg. Infect. Dis.* **11:**781–782.

CDC. 1985. Botulism associated with commercially distributed Kapchunka—New York City. *MMWR Morb. Mortal. Wkly. Rep.* **34:**546–547.

CDC. 1987. Restaurant-associated botulism from mushrooms bottled in-house—Vancouver, British Columbia, Canada. *MMWR Morb. Mortal. Wkly. Rep.* **36:**103.

CDC. 1992. Outbreak of type E botulism associated with an uneviscerated, salt-cured fish product—New Jersey, 1992. *MMWR Morb. Mortal. Wkly. Rep.* **41:**521–522.

CDC. 1995. Type B botulism associated with roasted eggplant in oil—Italy, 1993. *MMWR Morb. Mortal. Wkly. Rep.* **44:**33–36.

CDC. 1999. Foodborne botulism associated with home-canned bamboo shoots—Thailand, 1998. *MMWR Morb. Mortal. Wkly. Rep.* **48:**437–439.

CDC. 2001. Botulism outbreak associated with eating fermented food—Alaska 2001. *MMWR Morb. Mortal. Wkly. Rep.* **50:**680–682.

CDC. 2005. Summary of notifiable diseases—United States 2003. *MMWR Morb. Mortal. Wkly. Rep.* **52:**1–85.

CDC. 2006a. Botulism from home-canned bamboo shoots—Nan providence, Thailand, March 2006. *MMWR Morb. Mortal. Wkly. Rep.* **55:**389–392.

CDC. 2006b. Botulism associated with commercial carrot juice—Georgia and Florida, September 2006. *MMWR Morb. Mortal. Wkly. Rep.* **55:**1098–1099.

CDC. 2007. Cyanides. www.cdc.gov/Niosh/idlh/cyanides.html. Accessed 1 June 2007.

CDSC. 1998. Botulism associated with home-preserved mushrooms. *CDR Wkly.* **8:**159–162.

CDSC. 2006. Wound botulism in injecting drug users in the United Kingdom. *CDR Wkly.* **16:**2–3.

CFA. 2006. *Best Practice Guidelines for the Production of Chilled Foods*, 4th ed. The Stationery Office, Norwich, United Kingdom.

Chertow, D. S., E. T. Tan, S. E. Maslanka, J. Schulte, E. A. Bresnitz, R. S. Weisman, J. Bernstein, S. M. Marcus, S. Kumar, J. Malecki, J. Sobel, and C. R. Braden. 2006. Botulism in 4 adults following cosmetic injections with an unlicensed, highly concentrated botulinum preparation. *JAMA* **296:**2476–2479.

Chiorboli, E., G. Fortina, and G. Bona. 1997. Flaccid paralysis caused by botulinum toxin type B after pesto ingestion. *Pediatr. Infect. Dis. J.* **16:**725–726.

CSHD. 1981. Botulism Alert. California morbidity (November 6, 1981).

Del Torre, M., M. L. Stecchini, A. Braconnier, and M. W. Peck. 2004. Prevalence of *Clostridium* species and behaviour of *Clostridium botulinum* in gnocchi, a REPFED of Italian origin. *Int. J. Food Microbiol.* **96:**115–131.

Dezfulian, M. 1993. A simple procedure for identification of *Clostridium botulinum* colonies. *World J. Microbiol. Biotechnol.* **9:**125–127.

Dezfulian, M., L. M. McCroskey, C. L. Hatheway, and V. R. Dowell. 1981. Selective medium for isolation of *Clostridium botulinum* from human faeces. *J. Clin. Microbiol.* **13:**526–531.

Dodds, K. L. 1993a. *Clostridium botulinum* in the environment, p. 21–51. *In* A. H. W. Hauschild and K. L. Dodds (ed.), Clostridium botulinum. *Ecology and Control in Foods.* Marcel Dekker, New York, NY.

Dodds, K. L. 1993b. *Clostridium botulinum* in foods, p. 53–68. *In* A. H. W. Hauschild and K. L. Dodds (ed.), Clostridium botulinum. *Ecology and Control in Foods.* Marcel Dekker, New York, NY.

Dodds, K. L. 1993c. An introduction to predictive microbiology and the development of probability models with *Clostridium botulinum. J. Ind. Microbiol.* **12:**139–143.

Doellgast, G. J., G. A. Beard, J. D. Bottoms, T. Cheng, B. H. Roh, M. G. Roman, P. A. Hall, and M. X. Triscott. 1994. Enzyme-linked immunosorbent assay and enzyme-linked coagulation assay for detection of *Clostridium botulinum* neurotoxins A, B, and E and solution-phase complexes with dual-label antibodies. *J. Clin. Microbiol.* **32:**105–111.

Doellgast, G. J., M. X. Triscott, G. A. Beard, J. D. Bottoms, T. Cheng, B. H. Roh, M. G. Roman, P. A. Hall, and J. E. Brown. 1993. Sensitive enzyme-linked immunosorbent assay for detection of *Clostridium botulinum* neurotoxins A, B, and E using signal amplification via enzyme-linked coagulation assay. *J. Clin. Microbiol.* **31:**2402–2409.

Dong, M., F. Yeh, W. H. Tepp, C. Dean, E. A. Johnson, R. Janz, and E. R. Chapman. 2006. SV2 is the protein receptor for botulinum neurotoxin A. *Science* **312:**592–596.

Dressler, D. 2005. Botulismus durch Raucherlachsverzehr. *Der Nervenartz* **76:**763–766.

ECFF. 1996. *Guidelines for the Hygienic Manufacture of Chilled Foods.* The European Chilled Food Federation, Helsinki, Finland.

Eklund, M. W. 1982. Significance of *Clostridium botulinum* in fishery products preserved short of sterilization. *Food Technol.* **36:**107–112.

Eklund, M. W., F. T. Poysky, and D. I. Wieler. 1967a. Characteristics of *Clostridium botulinum* type F isolated from the Pacific coast of the United States. *Appl. Microbiol.* **15:**1316–1323.

Eklund, M. W., D. I Wieler, and F. T. Poysky. 1967b. Outgrowth and toxin production of nonproteolytic type B *Clostridium botulinum* at 3.3°C to 5.6°C. *J. Bacteriol.* **93:**1461–1462.

Ekong, T. A. N., I. M. Feavers, and D. Sesardic. 1997. Recombinant SNAP-25 is an effective substrate for *Clostridium botulinum* type A toxin endopeptidase activity *in vitro. Microbiology* **143:**3337–3347.

Ekong, T. A. N., K. McLellan, and D. Sesardic. 1995. Immunological detection of *Clostridium botulinum* toxin type A in therapeutic preparations. *J. Immunol. Methods* **180:**181–191.

Eriksen, T., A. B. Brantsaeter, W. Kiehl, and I. Steffens. 2004. Botulism infection after eating fish in Norway and Germany: two outbreak report. *Eurosurveillance Wkly.* **8**(3):1–2.

Espie, E., and M. R. Popoff. 2003. France recalls internationally distributed halal meat products from the plant implicated as the source of a type B botulism outbreak. *Eurosurveillance Wkly.* **7**(38):1–2.

Fach, P., M. Gibert, R. Griffais, J. P. Guillou, and M. R. Popoff. 1995. PCR and gene probe identification of botulinum neurotoxin A-, B-, E-, F-, and G-producing *Clostridium* spp. and evaluation of food samples. *Appl. Environ. Microbiol.* **61:**389–392.

Fach, P., S. Perelle, F. Dilasser, J. Grout, C. Dargaignaratz, L. Botella, J. M. Gourreau, F. Carlin, M. R. Popoff, and V. Broussolle. 2002. Detection by PCR-enzyme-linked immunosorbent assay of *Clostridium botulinum* in fish and environmental samples from a coastal area in northern France. *Appl. Environ. Microbiol.* **68:**5870–5876.

Fernandez, P. S., J. Baranyi, and M. W. Peck. 2001. A predictive model of growth from spores of nonproteolytic *Clostridium botulinum* in the presence of different CO_2 concentrations as influenced by chill temperature, pH and NaCl. *Food Microbiol.* **18:**453–462.

Fernandez, P. S., and M. W. Peck. 1997. A predictive model that describes the effect of prolonged heating at 70-80°C and incubation at refrigeration temperatures on growth and toxigenesis by non-proteolytic *Clostridium botulinum. J. Food Prot.* **60:**1064–1071.

Fernandez, P. S., and M. W. Peck. 1999. Predictive model that describes the effect of prolonged heating at 70-90°C and subsequent incubation at refrigeration temperatures on growth and toxigenesis by nonproteolytic *Clostridium botulinum* in the presence of lysozyme. *Appl. Environ. Microbiol.* **65:**3449–3457.

Ferreira, J. L. 2001. Comparison of amplified ELISA and mouse bioassay procedures for determination of botulinal toxins A, B, E, and F. *J. AOAC Int.* **84:**85–88.

Ferreira, J. L., and R. G. Crawford. 1998. Detection of type a botulinal toxin-producing organisms subcultured from cheese using an amplified ELISA system. *J. Rapid Methods Automation Microbiol.* **6:**289–296.

Ferreira, J. L., S. J. Eliasberg, P. Edmonds, and M. A. Harrison. 2004. Comparison of the mouse bioassay and enzyme-linked immunosorbent assay procedures for the detection of type A botulinal toxin in food. *J. Food Prot.* **67:**203–206.

Ferreira, J. L., S. J. Eliasberg, M. A. Harrison, and P. Edmonds. 2001. Detection of preformed type A botulinal toxin in hash brown potatoes by using the mouse bioasssay and a modified ELISA test. *J. AOAC Int.* **84:**1460–1464.

Ferreira, J. L., and M. K. Hamdy. 1995. Detection of botulinal toxin genes: types A and E or B and F using the muliplex polymerase chain reaction. *J. Rapid Methods Automation Microbiol.* **3:**177–183.

Fox, C. K., C. A. Keet, and J. B. Strober. 2005. Recent advances in infant botulism. *Pediatr. Neurol.* **32:**149–154.

Franciosa, G., J. L. Ferreira, and C. L. Hatheway. 1994. Detection of type A, B, and E genes in *Clostridium botulinum* and other

Clostridium species by PCR: evidence of unexpressed type B toxin genes in type A toxigenic organisms. *J. Clin. Microbiol.* 32:1911–1917.

Franciosa, G., M. Pourshaban, M. Gianfranceschi, A. Gattuso, L. Fenicia, A. M. Ferrini, V. Mannoni, G. De Luca, and P. Aureli. 1999. *Clostridium botulinum* spores and toxin in marscapone cheese and other milk products. *J. Food Prot.* 62:867–871.

Frean, J., L. Arntzen, and J. van den Heever. 2004. Fatal type A botulism in South Africa, 2002. *Trans. R. Soc. Trop. Med. Hyg.* 98:290–295.

Galazka, A., and A. Przbylska. 1999. Surveillance of foodborne botulism in Poland. *Eurosurveillance* 4:69–72.

Garcia, G., and C. Genigeorgis. 1987. Quantitative evaluation of *Clostridium botulinum* nonproteolytic types B, E and F growth risk in fresh salmon homogenates stored under modified atmospheres. *J. Food Prot.* 50:390–397, 400.

Gauthier, M., B. Cadieux, J. W. Austin, and B. W. Blais. 2005. Cloth-based hybridization array system for the detection of *Clostridium botulinum* type A, B, E, and F neurotoxin genes. *J. Food Prot.* 68:1477–1483.

Gessler, F., S. Pagel-Wieder, M. A. Avondet, and H. Bohnel. 2007. Evaluation of lateral flow assays for the detection of botulinum neurotoxin type A and their application in laboratory diagnosis of botulism. *Diagn. Microbiol. Infect. Dis.* 57:243–249.

Gibson, A. M., N. K. Modi, T. A. Roberts, C. C. Shone, P. Hambleton, and J. Melling. 1987. Evaluation of a monoclonal antibody-based immunoassay for detecting type A *Clostridium botulinum* toxin produced in pure culture and an inoculated model cured meat system. *J. Appl. Bacteriol.* 63:217–226.

Gibson, A. M., N. K. Modi, T. A. Roberts, P. Hambleton, and J. Melling. 1988. Evaluation of a monoclonal antibody-based immunoassay for detecting type B *Clostridium botulinum* toxin produced in pure culture and an inoculated model cured meat system. *J. Appl. Bacteriol.* 64:285–291.

Goodnough, M. C., B. Hammer, H. Sugiyama, and E. A. Johnson. 1993. Colony immunoblot assay of botulinal toxin. *Appl. Environ. Microbiol.* 59:2339–2342.

Gotz, J., J. Carlsson, R. Schipmann, M. Rohde, E. Sorges, F. Manz, and U. Tebbe. 2002. 37-jahrige patientin mit augenmuskel-lahmungen und rasch progredienter ateminsuffizienz. *Internist* 43:548–553.

Graham, A. F., D. R. Mason, F. J. Maxwell, and M. W. Peck. 1997. Effect of pH and NaCl on growth from spores of non-proteolytic *Clostridium botulinum* at chill temperatures. *Lett. Appl. Microbiol.* 24:95–100.

Graham, A. F., D. R. Mason, and M. W. Peck. 1996a. A predictive model of the effect of temperature, pH and sodium chloride on growth from spores of non-proteolytic *Clostridium botulinum*. *Int. J. Food Microbiol.* 31:69–85.

Graham, A. F., D. R. Mason, and M. W. Peck. 1996b. Inhibitory effect of combinations of heat treatment, pH and sodium chloride on growth from spores of non-proteolytic *Clostridium botulinum* at refrigeration temperatures. *Appl. Environ. Microbiol.* 62:2664–2668.

Hallis, B., B. A. F. James, and C. C. Shone. 1996. Development of novel assays for botulinum type A and B neurotoxins based on their endopeptidase activities. *J. Clin. Microbiol.* 34:1934–1938.

Harvey, S. M., J. Sturgeon, and D. E. Dassey. 2002. Botulism due to *Clostridium baratii* type F toxin. *J. Clin. Microbiol.* 40:2260–2262.

Hatheway, C. L. 1988. Botulism, p. 111–133. *In* A. Balows, W. J. Hausler, M. Ohashi, and A. Turano (ed.), *Laboratory Diagnosis of Infectious Diseases: Principles and Practice.* Springer-Verlag, London, United Kingdom.

Hatheway, C. L. 1993. *Clostridium botulinum* and other clostridia that produce botulinum neurotoxin, p. 3–20. *In* A. H. W. Hauschild and K. L. Dodds (ed.), Clostridium botulinum. *Ecology and Control in Foods.* Marcel Dekker, New York, NY.

Hauschild, A. H. W. 1989. *Clostridium botulinum*, p. 112–189. *In* M. P. Doyle (ed.), *Foodborne Bacterial Pathogens.* Marcel Dekker, New York, NY.

Hauschild, A. H. W. 1993. Epidemiology of foodborne botulism, p. 69–104. *In* A. H. W. Hauschild and K. L. Dodds (ed.), Clostridium botulinum. *Ecology and Control in Foods.* Marcel Dekker, New York, NY.

Hauschild, A. H. W., B. J. Aris, and R. Hilsheimer. 1975. *Clostridium botulinum* in marinated products. *Can. Inst. Food Sci. Technol. J.* 8:84–87.

Hauschild, A. H. W., and L. Gauvreau. 1985. Food-borne botulism in Canada, 1971–84. *Can. Med. Assoc. J.* 133:1141–1146.

Heinz, V., and D. Knorr. 2001. Effects of high pressure on spores, p. 77–113. *In* M. E. G. Hendrickx and D. Knorr (ed.), *Ultra High Pressure Treatments of Foods.* Kluwer Academic/Plenum, New York, NY.

Hielm, S., E. Hyytiä, A. B. Andersin, and H. Korkeala. 1998. A high prevalence of *Clostridium botulinum* type E in Finnish freshwater and Baltic Sea sediment samples. *J. Appl. Microbiol.* 84:133–137.

Hielm, S., E. Hyytiä, J. Ridell, and H. Korkeala. 1996. Detection of *Clostridium botulinum* in fish and environmental samples using polymerase chain reaction. *Int. J. Food Microbiol.* 31:357–365.

Huhtanen, C. N., R. C. Whiting, A. J. Miller, and J. E. Call. 1992. Qualitative correlation of the mouse neurotoxin and enzyme-linked immunoassay for determining *Clostridium botulinum* types A and B toxins. *J. Food Saf.* 12:119–127.

Huss, H. H., A. Pedersen, and D. C. Cann. 1974. The incidence of *Clostridium botulinum* in Danish trout farms. Distribution in fish and their environment. *J. Food Technol.* 9:445–450.

Hyytiä, E., J. Bjorkroth, S. Hielm, and H. Korkeala. 1999. Characterisation of *Clostridium botulinum* Groups I and II by randomly amplified polymorphic DNA analysis and repetitive element sequence-based PCR. *Int. J. Food Microbiol.* 48:179–189.

Jahkola, M., and H. Korkeala. 1997. Botulismi saksassa suomessa pakatusta savustista. *Kansanterveys* 3:8–9.

Jahn, R. 2006. A neuronal receptor for botulinum toxin. *Science* 312:540–541.

Kalb, S. R., H. Moura, A. E. Boyer, L. G. McWilliams, J. L. Pirkle, and J. R. Barr. 2006. The use of Endopep-MS for the detection of botulinum toxins A, B, E, and F in serum and stool samples. *Anal. Biochem.* 351:84–92.

Kalluri, P., C. Crowe, M. Reller, L. Gaul, J. Hayslett, S. Barth, S. Eliasberg, J. Ferreira, K. Holt, S. Bengston, K. Hendricks, and J. Sobel. 2003. An outbreak of foodborne botulism associated with food sold at a salvage store in Texas. *Clin. Infect. Dis.* 37:1490–1495.

Kautter, D. A., and H. M. Solomon. 1977. Collaborative study of a method for the detection of *Clostridium botulinum* and its toxins in foods. *J. AOAC Int.* 60:541–545.

Kautter, D. A., H. M. Solomon, D. E. Lake, D. T. Bernard, and D. C. Mills. 1992. *Clostridium botulinum* and its toxins, p. 605–621. *In* C. Vanderzant and D. F. Splittstoesser (ed.), *Compendium of Methods for the Microbiological Examination of Foods*, 3rd ed. American Public Health Association, Washington, DC.

Keto-Timonen, R., M. Nevas, and H. Korkeala. 2005. Efficient DNA fingerprinting of *Clostridium botulinum* types A, B, E, and F by amplified fragment length polymorphism analysis. *Appl. Environ. Microbiol.* 71:1148–1154.

Kirkwood, J., A. Ghetler, J. Sedman, D. Leclair, F. Pagotto, J. W. Austin, and A. A. Ismail. 2006. Differentiation of Group I and Group II strains of *Clostridium botulinum* by focal plane array Fourier transform infrared spectroscopy. *J. Food Prot.* **69:**2377–2383.

Korkeala, H., G. Stengel, E. Hyytiä, B. Vogelsang, A. Bohl, H. Wihlman, P. Pakkala, and S. Hielm. 1998. Type E botulism associated with vacuum-packaged hot-smoked whitefish. *Int. J. Food Microbiol.* **43:**1–5.

Leclair, D., F. Pagotto, J. M. Farber, B. Cadieux, and J. W. Austin. 2006. Comparison of DNA fingerprinting methods for use in investigation of type E botulism outbreaks in the Canadian arctic. *J. Clin. Microbiol.* **44:**1635–1644.

Lindroth, S. E., and C. A. Genigeorgis. 1986. Probability of growth and toxin production by nonproteolytic *Clostridium botulinum* in rockfish stored under modified atmospheres. *Int. J. Food Microbiol.* **3:**167–181.

Lindström, M., S. Hielm, M. Nevas, S. Tuisku, and H. Korkeala. 2004. Proteolytic *Clostridium botulinum* type B in the gastric content of a patient with type E botulism due to whitefish eggs. *Foodborne Pathog. Dis.* **1:**53–58.

Lindström, M., R. Keto, A. Markkula, M. Nevas, S. Hielm, and H. Korkeala. 2001. Multiplex PCR assay for detection and identification of *Clostridium botulinum* types A, B, E, and F in food and fecal material. *Appl. Environ. Microbiol.* **67:**5694–5699.

Lindström, M., and H. Korkeala. 2006. Laboratory diagnostics of botulism. *Clin. Microbiol. Rev.* **19:**298–314.

Lindström, M., M. Vuorela, K. Hinderink, H. Korkeala, E. Dahlsten, M. Raahenmaa, and M. Kuusi. 2006a. Botulism associated with vacuum-packed smoked whitefish in Finland, June-July 2006. *Eurosurveillance* **11**(7):1–2.

Lindström, M., M. Nevas, and H. Korkeala. 2006b. Detection of *Clostridium botulinum* by multiplex PCR in foods and feces. *Foodborne Pathog. Dis.* **21:**37–45.

Lucke, F. K. 1984. Psychrotrophic *Clostridium botulinum* strains from raw hams. *Syst. Appl. Microbiol.* **5:**274–279.

Lund, B. M., and M. W. Peck. 1994. Heat resistance and recovery of non-proteolytic *Clostridium botulinum* in relation to refrigerated processed foods with an extended shelf-life. *J. Appl. Bacteriol.* **76:**115–128.

Lund, B. M., and M. W. Peck. 2000. *Clostridium botulinum*, p.1057–1109. *In* B. M. Lund, T. C. Baird-Parker, and G. W. Gould (ed.), *The Microbiological Safety and Quality of Food.* Aspen, Gaithersburg, MD.

Lund, B. M., and G. M. Wyatt. 1984. The effect of redox potential, and its interaction with sodium chloride concentration, on the probability of growth of *Clostridium botulinum* type E from spore inocula. *Food Microbiol.* **1:**49–65.

Mackle, I. J., E. Halcomb, and M. J. Parr. 2001. Severe adult botulism. *Anaesth. Intensive Care* **29:**297–300.

Mahrhold, S., A. Rummel, H. Bigalke, B. Davletov, and T. Binz. 2006. The synaptic vesicle protein 2C mediates the uptake of botulinum neurotoxin A into phrenic nerves. *FEBS Lett.* **580:**2011–2014.

McLauchlin, J., K. A. Grant, and C. A. Little. 2006. Foodborne botulism in the United Kingdom. *J. Public Health* **28:**337–342.

McLaughlin, J. B., J. Sobel, T. Lynn, E. Funk, and T. Middaugh. 2004. Botulism type E outbreak associated with eating a beached whale, Alaska. *Emerg. Infect. Dis.* **10:**1685–1687.

McLean, H. E., S. Peck, F. J. Blatherwick, W. A. Black, J. Fung, R. G. Mathias, M. E. Milling, K. A. Catherwood, and L. J. Lukey. 1987. Restaurant-associated botulism from in-house bottled mushrooms - British Columbia. *Can. Dis. Wkly. Rep.* **13:**35–36.

Merivirta, L. O., M. Lindström, K. J. Bjorkroth, and H. J. Korkeala. 2006. The prevalence of *Clostridium botulinum* in

European river lamprey (*Lampetra fluviatilis*) in Finland. *Int. J. Food Microbiol.* **109:**234–237.

Meyers, H, G. Inami, J. Rosenberg, J. Mohle-Boetani, D. Vugia, and J. Yuan. 2007. Foodborne botulism from home-prepared fermented tofu—California, 2006. *JAMA* **297:**1311–1312.

Midura, T. F., S. Snowden, R. M. Wood, and S. S. Arnon. 1979. Isolation of *Clostridium botulinum* from honey. *J. Clin. Microbiol.* **9:**282–283.

Mills, D. C., T. F. Midura, and S. S. Arnon. 1985. Improved selective medium for the isolation of lipase-positive *Clostridium botulinum* from feces of human infants. *J. Clin. Microbiol.* **21:**947–950.

Modi, N. K., C. C. Shone, P. Hambleton, and J. Melling. 1986. Monoclonal antibody based enzyme-linked-immunosorbent-assays for *Clostridium botulinum* toxin types A and B, p. 1184–1188. *In Proceedings of the 2nd World Congress Foodborne Infections and Intoxications*, vol. 2. Berlin (West), Germany.

Montecucco, C., and J. Molgo. 2005. Botulinal neurotoxins: revival of an old killer. *Curr. Opin. Pharmacol.* **5:**274–279.

Myllykoski, J., M. Nevas, M. Lindström, and H. Korkeala. 2006. The detection and prevalence of *Clostridium botulinum* in pig intestinal samples. *Int. J. Food Microbiol.* **110:**172–177.

Nevas, M., M. Lindström, A. Virtanen, S. Hielm, M. Kuusi, S. S. Arnon, E. Vuori, and H. Korkeala. 2005a. Infant botulism acquired from household dust presenting as sudden infant death syndrome. *J. Clin. Microbiol.* **43:**511–513.

Nevas, M., M. Lindström, K. Hautamaki, S. Puoskari, and H. Korkeala. 2005b. Prevalence and diversity of *Clostridium botulinum* types A, B, E and F in honey produced in the Nordic countries. *Int. J. Food Microbiol.* **105:**145–151.

Nevas, M., M. Lindström, S. Hielm, J. Bjorkroth, M. W. Peck, and H. Korkeala. 2005c. Diversity of proteolytic *Clostridium botulinum* strains, determined by a pulsed-field gel electrophoresis approach. *Appl. Environ. Microbiol.* **71:**1311–1317.

Ohye, D. F., and W. J. Scott. 1957. Studies in the physiology of *Clostridium botulinum* type E. *Aust. J. Biol. Sci.* **10:**85–94.

O'Mahony, M. O., E. Mitchell, R. J. Gilbert, D. N. Hutchinson, N. T. Begg, J. C. Rodehouse, and J. E. Morris. 1990. An outbreak of foodborne botulism with contaminated hazelnut yoghurt. *Epidemiol. Infect.* **104:**389–395.

Ouagari, Z., A. Chakib, M. Sodqi, L. Marih, K. M. Filali, A. Benslama, L. Idrissi, S. Moutawakkil, and H. Himmich. 2002. Le botulisme á Casablanca. *Bull. Soc. Pathol. Exot.* **95:**272–275.

Passaro, D. J., S. B. Werner, J. McGee, W. R. MacKenzie, and D J. Vugia. 1998. Wound botulism associated with black tar heroin among injecting drug users. *JAMA* **279:**859–863.

Pavic, S., D. Lastre, S. Bukovski, M. Hadziosmanovic, B. Miokovic, and L. Kozacinski. 2001. Convenience of *Clostridium botulinum* agar for selective isolation of type B *Clostridium botulinum* from dry cured Dalmatian ham during a family outbreak. *Archi. Lebensmit.* **52:**43–45.

Pearce, L. B., E. R. First, R. D. MacCallum, and A. Gupta. 1997. Pharmacologic characterization of botulinum toxin for basic science and medicine. *Toxicon* **35:**1373–1412.

Peck, M. W. 1997. *Clostridium botulinum* and the safety of refrigerated processed foods of extended durability. *Trends Food Sci. Technol.* **8:**186–192.

Peck, M. W. 2006. *Clostridium botulinum* and the safety of minimally heated chilled foods: an emerging issue? *J. Appl. Microbiol.* **101:**556–570.

Peck, M. W., D. A. Fairbairn, and B. M. Lund. 1992. Factors affecting growth from heat-treated spores of non-proteolytic *Clostridium botulinum*. *Lett. Appl. Microbiol.* **15:**152–155.

Peck, M. W., D. A. Fairbairn, and B. M. Lund. 1993. Heat-resistance of spores of non-proteolytic *Clostridium botulinum* estimated on medium containing lysozyme. *Lett. Appl. Microbiol.* **16:**126–131.

Peck, M. W., and P. S. Fernandez. 1995. Effect of lysozyme concentration, heating at 90°C, and then incubation at chilled temperatures on growth from spores of non-proteolytic *Clostridium botulinum. Lett. Appl. Microbiol.* **21:**50–54.

Peck, M. W., K. E. Goodburn, R. P. Betts, and S. C. Stringer. 2008. Assessment of the potential for growth and neurotoxin formation by non-proteolytic *Clostridium botulinum* in short shelf-life commercial foods designed to be stored chilled. *Trends Food Sci. Technol.* **19:**207–216.

Peck, M. W., B. M. Lund, D. A. Fairbairn, A. S. Kaspersson, and P. C. Undeland. 1995. Effect of heat treatment on survival of, and growth from, spores of nonproteolytic *Clostridium botulinum* at refrigeration temperatures. *Appl. Environ. Microbiol.* **61:**1780–1785.

Peck, M. W., and S. C. Stringer. 2005. The safety of pasteurised in-pack chilled meat products with respect to the foodborne botulism hazard. *Meat Sci.* **70:**461–475.

Peterson, M. E., G. A. Pelroy, F. T. Poysky, R. N. Paranjpye, F. M. Dong, G. M. Pigott, and M. W. Eklund. 1997. Heat-pasteurization process for inactivation of nonproteolytic types of *Clostridium botulinum* in picked Dungeness crabmeat. *J. Food Prot.* **60:**928–934.

Popoff, M. R., and J. P. Carlier. 2004. Botulism in France, p. 83–87. *In* C. Duchesnes, J. Mainil, M. W. Peck, and P. E. Granum (ed.), *Food Microbiology and Sporulation of the Genus Clostridium.* University Press, Liege, Belgium.

Potter, M. D., J. Meng, and P. Kimsey. 1993. An ELISA for detection of botulinal toxin types A, B and E in inoculated food samples. *J. Food Prot.* **56:**856–861.

Pourshafie, M. R., M. Saifie, A. Shafiee, P. Vahdani, M. Aslani, and J. Salemian. 1998. An outbreak of foodborne botulism associated with contaminated locally made cheese in Iran. *Scand. J. Infect. Dis.* **30:**92–94.

Proulx, J. F., V. Milot-Roy, and J. Austin. 1997. Four outbreaks of botulism in Ungava Bay, Nunavik, Quebec. *Can. Commun. Dis. Rep.* **23**(4):30–32.

Reddy, N. R., H. M. Solomon, G. A. Fingerhut, E. J. Rhodehamel, V. M. Balasubramaniam, and S. Palaniappan. 1999. Inactivation of *Clostridium botulinum* type E spores by high pressure processing. *J. Food Saf.* **19:**277–288.

Rivera, V. R., J. J. Gamez, W. K. Keener, J. A. White, and M. A. Poli. 2006. Rapid detection of *Clostridium botulinum* toxins A, B, E, and F in clinical samples, selected food matrices, and buffer using paramagnetic bead-based electrochemiluminescence detection. *Anal. Biochem.* **353:**248–256.

Roberts, E., J. M. Wales, M. M. Brett, and P. Bradding. 1998. Cranial-nerve palsies and vomiting. *Lancet* **352:**1674.

Roman, M. G., J. Y. Humber, P. A. Hall, N. R. Reddy, H. M. Solomon, M. X. Triscott, G. A. Beard, J. D. Bottoms, T. Cheng, and G. J. Doellgast. 1994. Amplified immunoassay ELISA-ELCA for measuring *Clostridium botulinum* type E neurotoxin in fish fillets. *J. Food Prot.* **57:**985–990.

Rosetti, F., E. Castelli, J. Labbe, and R. Funes. 1999. Outbreak of type E botulism associated with home-cured ham consumption. *Anaerobe* **5:**171–172.

Rossetto, O., L. Morbiato, P. Caccin, M. Rigoni, and C. Montecucco. 2006. Presynaptic enzymatic neurotoxins. *J. Neurochem.* **97:**1534–1545.

Rouhbakhsh-Khaleghdoust, A. 1975. The incidence of *Clostridium botulinum* type E in fish and bottom deposits in the Caspian sea coastal waters. *Pahlavi Med. J.* **6:**550–556.

Sakaguchi, G. 1979. Botulism, p. 389–442. *In* H. Riemann and F. L. Bryan (ed.), *Foodborne Infections and Intoxications,* 2nd ed. Academic Press, New York, NY.

Sandrock, C. E., and S. Murin. 2001. Clinical predictors of respiratory failure and long-term outcome in black tar heroin-associated wound botulism. *Chest* **120:**562–566.

Scarlatos, A., B. A. Welt, B. Y. Cooper, D. Archer, T. DeMarse, and K. V. Chau. 2005. Methods for detecting botulinum toxin with applicability to screening foods against biological terrorist attacks. *J. Food Sci.* **70:**R121–R130.

Schmidt, C. F., R. V. Lechowich, and J. F. Folinazzo. 1961. Growth and toxin production by type E *Clostridium botulinum* below 40°F. *J. Food Sci.* **26:**626–630.

Schmidt, J. J., and R. G. Stafford. 2003. Fluorigenic substrates for the protease activities of botulinum neurotoxins, serotypes A, B, and F. *Appl. Environ. Microbiol.* **69:**297–303.

Schmidt, J. J., R. G. Stafford, and C. B. Millard. 2001. High-throughput assays for botulinum neurotoxin proteolytic activity: serotypes A, B, D and F. *Anal. Chem.* **296:**130–137.

Sebaihia, M., M. W. Peck, N. P. Minton, N. R. Thomson, M. T. G. Holden, W. J. Mitchell, A. T. Carter, S. D. Bentley, D. R. Mason, L. Crossman, C. J. Paul, A. Ivens, M. H. J. Wells-Bennik, I. J. Davis, A. M. Cerdeno-Tarraga, C. Churcher, M. A. Guail, T. Chillingworth, T. Feltwell, A. Fraser, I. Goodhead, Z. Hance, K. Jagels, N. Larke, M. Maddison, S. Moule, K. Mungall, H. Norbertczak, E. Rabbinowitsch, M. Sanders, M. Simmonds, B. White, S. Whithead, and J. Parkhill. 2007. Genome sequence of proteolytic (Group I) *Clostridium botulinum* strain Hall A and comparative analysis of the clostridial genomes. *Genome Res.* **17:**1082–1092.

Sebaihia, M., B. W. Wren, P. Mullany, N. F. Fairweather, N. Minton, R. Stabler, N. R. Thomson, A. P. Roberts, A. M. Cerdeno-Tarraga, H. Wang, M. T. G. Holden, A. Wright, C. Churcher. M. A. Quail, S. Baker, N. Bason, K. Brooks, T. Chillingworth, A. Cronin, P. Davis, L. Dowd, A. Fraser, T. Feltwell, Z. Hance, S. Holroyd, K. Jagels, S. Moule, K. Mungall, C. Price, E. Rabbinowitsch, S. Sharp, M. Simmonds, K. Stevens, L. Unwin, S. Whithead, B. Dupuy, G. Dougan, B. Barrell, and J. Parkhill. 2006. The multidrug-resistant human pathogen *Clostridium difficile* has a highly mobile, mosaic genome. *Nat. Genet.* **38:**779–786.

Sebald, M., and J. C. Petit. 1994. *Méthodes de laboratoire bacteréries anaérobes et leur identification.* Institut Pasteur, Paris, France.

Sesardic, D., K. McLellan, T. A. N. Ekong, and R. Das Gaines. 1996. Refinement and validation of an alternative bioassay for potency testing of therapeutic botulinum type A toxin. *Pharmacol. Toxicol.* **78:**283–288.

Setlow, P., and E. A. Johnson. 1997. Spores and their significance, p. 30–65. *In* M. P. Doyle, L. R. Beuchat, and T. J. Montville (ed.), *Food Microbiology: Fundamentals and Frontiers.* ASM Press, Washington DC.

Sharma, S. K., B. S. Eblen, R. L. Bull, D. H. Burr, and R. C. Whiting. 2005. Evaluation of lateral-flow *Clostridium botulinum* neurotoxin detection kits for food analysis. *Appl. Environ. Microbiol.* **71:**3935–3941.

Sharma, S. K., J. L. Ferreira, B. S. Eblen, and R. C. Whiting. 2006. Detection of type A, B, E, and F *Clostridium botulinum* neurotoxins in foods by using an amplified enzyme-linked immunosorbent assay with digoxigenin-labeled antibodies. *Appl. Environ. Microbiol.* **72:**1231–1238.

Sharma, S. K., and R. C. Whiting. 2005. Methods for detection of *Clostridium botulinum* toxin in foods. *J. Food Prot.* **68:**1256–1263.

Shone, C., P. Wilton-Smith, N. Appleton, P. Hambleton, N. Modi, S. Gatley, and J. Melling. 1985. Monoclonal antibody-based immunoassay for type A *Clostridium botulinum* toxin is comparable to the mouse bioassay. *Appl. Environ. Microbiol.* **50:**63–67.

Siegel, L. S. 1993. Destruction of botulinum toxins in food and water, p. 323–341. *In* A. H. W. Hauschild and K. L. Dodds (ed.), Clostridium botulinum. *Ecology and Control in Foods.* Marcel Dekker, New York, NY.

Simpson, L. L. 2004. Identification of the major steps in botulinum toxin action. *Ann. Rev. Pharmacol. Toxicol.* **44:**167–193.

Slater, P. E., D. G. Addiss, A. Cohen, A. Leventhal, G. Chassis, H. Zehavi, A. Bashari, and C. Costin. 1989. Foodborne botulism: an international outbreak. *Int. J. Epidemiol.* **18:**693–696.

Smith, L. D. S. 1977. *Botulism*, 1st ed. Charles C. Thomas, Springfield, IL.

Smith, L.D.S., and H. Sugiyama. 1988. *Botulism*, 2nd ed. Charles C. Thomas, Springfield, IL.

Smith, T. J., J. Lou, I. N. Geren, C. M. Forsyth, R. Tsai, S. L. LaPorte, W. H. Tepp, M. Bradshaw, E. A. Johnson, L. A. Smith, and J. D. Marks. 2005. Sequence variation within botulinum neurotoxin serotypes impacts antibody binding and neutralization. *Infect. Immun.* **73:**5450–5457.

Solomon, H. M., and T. Lilly. 2001. *Clostridium botulinum. In Bacteriological Analytical Manual Online.* Food and Drug Administration, Washington, DC. www.cfsan.fda.gov/~ebam/bam-toc.html. Accessed 6 January 2007.

St. Louis, M. E., S. H. S. Peck, D. Bowering, G. B. Morgan, J. Blatherwick, S. Banerjee, G. D. M. Kettyls, W. A. Black, M. E. Milling, A. H. W. Hauschild, R. V. Tauxe, and P. A. Blake. 1988. Botulism from chopped garlic: delayed recognition of a major outbreak. *Ann. Intern. Med.* **108:** 363–368.

Stringer, S. C., N. Haque, and M. W. Peck. 1999. Growth from spores of non-proteolytic *Clostridium botulinum* in heat-treated vegetable juice. *Appl. Environ. Microbiol.* **65:**2136–2142.

Stringer, S. C., M. D. Webb., S. M. George, C. Pin, and M. W. Peck. 2005. Heterogeneity of times for germination and outgrowth from single spores of nonproteolytic *Clostridium botulinum. Appl. Environ. Microbiol.* **71:**4998–5003.

Strom, M. S., M. W. Eklund, and F. T. Poysky. 1984. Plasmids in *Clostridium botulinum* and related *Clostridium* species. *Appl. Environ. Microbiol.* **48:**956–963.

Stumbo, C. R., K. S. Purohit, and T. V. Ramakrishna. 1975. Thermal process lethality guide for low-acid foods in metal containers. *J. Food Sci.* **40:**1316–1323.

Sugiyama, H., T. L. Bott, and E. M. Foster. 1970. *Clostridium botulinum* type E in an inland bay (Green Bay of Lake Michigan), p. 287–291. *In* M. Herzberg (ed.), *Proceedings 1st U.S.-Japan Conference on Toxic Micro-organisms.* U.S. Department of the Interior, Washington, DC.

Szabo, E. A., J. M. Pemberton, and P. M. Desmarchelier. 1993. Detection of the genes encoding botulinum neurotoxin types A to E by the polymerase chain reaction. *Appl. Environ. Microbiol.* **59:**3011–3020.

Therre, H. 1999. Botulism in the European Union. *Eurosurveillance* **4:**2–7.

Townes, J. M., P. R. Cieslak, C. L. Hatheway, H. M. Solomon, J. T. Holloway, M. P. Baker, C. F. Keller, L. M. McCroskey, and P. M. Griffin. 1996. An outbreak of type A botulism associated with a commercial cheese sauce. *Ann. Intern. Med.* **125:**558–563.

Ungchusak, K., S. Chunsuttiwat, C. R. Braden, W. Aldis, K. Ueno, S. J. Olsen, and S. Wiboolpolprasert. 2007. The need for global planned mobilization of essential medicine: lessons from a massive Thai outbreak. *Bull. W. H. O.* **85:**238–240.

Vahdani, P., D. Yadegarinia, Z. Aminzadeh, M. Z. Dehabadi, and O. Eilami. 2006. Outbreak of botulism type E associated with eating traditional soup in a family group, Loghman Hakim Hospital, Tehran, Iran. *Iranian J. Clin. Infect. Dis.* **1:**43–46.

Varma, J. K., G. Katsitadze, M. Moiscrafishvili, T. Zardiashvili, M. Chokheli, N. Tarkhashvili, E. Jhorjholiani, M. Chubinidze, T. Kukhalashvili, I. Khmaladze, N. Chakvetadze, P. Imnadez, and J. Sobel. 2004. Foodborne botulism in the Republic of Georgia. *Emerg. Infect. Dis.* **10:**1601–1605.

Villar, R. G., R. L. Shapiro, S. Busto, C. Riva-Posse, G. Verdejo, M. I. Farace, F. Rosetti, J. A. San Juan, C. M. Julia, J. Becher, S. E. Maslanka, and D. L. Swerdlow. 1999. Outbreak of type A botulism and development of a botulism surveillance and antitoxin release system in Argentina. *JAMA* **281:**1334–1338.

Viscens, R., N. Rasolofonirina, and P. Coulanges. 1985. Premiers cas humains de botulisme alimentaire a Madagascar. *Arch. Inst. Pasteur Madagascar* **52:**11–22.

Webb, M. D., C. Pin, M. W. Peck, and S. C. Stringer. 2007. Historical and contemporary NaCl concentrations affect the duration and distribution of lag times from individual spores of nonproteolytic *Clostridium botulinum. Appl. Environ. Microbiol.* **73:**2118–2127.

Weber, J. T., R. G. Hibbs, A. Darwish, B. Mishu, A. L. Corwin, M. Rakha, C. L. Hatheway, S. Elsharkawy, S. A. Elrahim, M. F. S. Alhamd, P. A. Blake, and R. V. Tauxe. 1993. A massive outbreak of type E botulism associated with traditional salted fish in Cairo. *J. Infect. Dis.* **167:**51–54.

Wein, L. M., and Y. Liu. 2005. Analyzing a bioterror attack on the food supply: the case of botulinum toxin in milk. *Proc. Natl. Acad. Sci. USA* **102:**9984–9989.

Werner, S. B., D. Passaro, J. McGee, R. Schechter, and D. C. Vugia. 2000. Wound botulism in California, 1951-1998: recent epidemic in heroin injectors. *Clin. Infect. Dis.* **31:**1018–1024.

Whiting, R. C., and J. C. Oriente. 1997. Time-to-turbidity model for non-proteolytic type B *Clostridium botulinum. Int. J. Food Microbiol.* **35:**49–60.

WHO. 2000. WHO surveillance programme for control of foodborne infections and intoxications in Europe. 7th report, 1993–1998. www.bfr.bund.de/internet/7threport/7threp_fr.htm. Accessed 8 March 2006.

WHO. 2003. WHO surveillance programme for control of foodborne infections and intoxications in Europe. 8th report, 1999–2000. www.bfr.bund.de/internet/8threport/8threp_fr.htm. Accessed 12 October 2005.

Wictome, M., and C. C. Shone. 1998. Botulinum neurotoxins: mode of action and detection. *J. Appl. Microbiol.* **84:**87s–97s.

Wictome, M., K. A. Newton, K. Jameson, P. Dunnigan, S. Clarke, J. Gaze, A. Tauk, K. A. Foster, and C. C. Shone. 1999a. Development of in vitro assays for the detection of botulinum toxins in foods. *FEMS Immunol. Med. Microbiol.* **24:**319–323.

Wictome, M., K. Newton, K. Jameson, B. Hallis, P. Dunnigan, E. Mackay, S. Clarke, R. Taylor, J. Gaze, K. A. Foster, and C. C. Shone. 1999b. Development of an in vitro bioassay for *Clostridium botulinum* type B neurotoxin in foods that is more sensitive than the mouse bioassay. *Appl. Environ. Microbiol.* **65:**3787–3792.

Wong, P. C. K. 1996. Detection of *Clostridium botulinum* type E toxin by monoclonal antibody enzyme immunoassay. *J. Rapid Methods Automation Microbiol.* **4:**191–206.

Wu, H. C., Y. L. Huang, S. C. Lai, Y. Y. Huang, and M. F. Shalo. 2001. Detection of *Clostridium botulinum* neurotoxin type A using immuno-PCR. *Lett. Appl. Microbiol.* **32:**321.

Pathogens and Toxins in Foods: Challenges and Interventions
Edited by V. K. Juneja and J. N. Sofos
© 2010 ASM Press, Washington, DC

Chapter 4

Clostridium perfringens

VIJAY K. JUNEJA, JOHN S. NOVAK, AND RONALD J. LABBE

CHARACTERISTICS OF THE ORGANISM AND TYPE OF ILLNESS

Clostridium perfringens is an anaerobic (microaerophilic), gram-positive, nonmotile, spore-forming, rod-shaped bacterium. The length of the rods depends upon the growth environment. In glucose-rich medium, the rods are short, whereas in starch-based sporulation medium, the rods are long. Spores are oval and subterminal. Microbiological testing for confirmation is based upon the ability of the pathogen to hydrolyze gelatin, reduce nitrate to nitrite, and ferment lactose (stormy fermentation of lactose in milk). Lecithinase (alpha-toxin) activity also characterizes the isolates and is demonstrated by a zone of hemolysis (often a double zone) on blood agar or a pearly opalescence around colonies on egg yolk agar. This activity can be inhibited by type A antitoxin (Nagler reaction). On sulfite-containing agars, the colonies appear black due to sulfite reduction. *C. perfringens* also produces a wide variety of extracellular toxins (soluble antigens) (Table 1). The types (A, B, C, D, and E) are based upon the production of four extracellular toxins: alpha, beta, epsilon, and iota (Petit et al., 1999). These toxins are not associated with food poisoning. Types A, C, and D are pathogenic to humans. Types B, C, D, E, and maybe A affect animals. The enterotoxin produced by type A and C is distinct from the exotoxins and is responsible for the typical symptoms of food poisoning.

C. perfringens is ubiquitous in the environment and is found in soil, dust, raw ingredients, such as spices used in food processing, and in the intestines of humans and animals (ICMSF, 1996; Juneja et al., 2006a). Thus, raw protein foods of animal origin are frequently contaminated with *C. perfringens*. However, a large proportion of *C. perfringens* found in raw foods is enterotoxin negative, lacking *cpe* (Saito, 1990). In a survey conducted by Lin and Labbe (2003), none of 40 *C. perfringens* isolates from 131 retail food samples in the United States were enterotoxin-positive *C. perfringens*, containing *cpe*, whereas Wen and McClane (2004) found a prevalence rate of 1.7% in U.S. retail food. Foods that are associated with food poisoning outbreaks contain large numbers of enterotoxin-positive *C. perfringens* organisms.

The incidence of *C. perfringens* gastrointestinal illnesses in the United States has been estimated by the Centers for Disease Control and Prevention to be 248,000 cases per year, leading to 41 hospitalizations and 7 deaths each year, with 100% of these cases being due to food-borne transmission of the pathogen (Mead et al., 1999) and with an estimated cost of $200 per case (Todd, 1989). In another report from 1993 to 1997, *C. perfringens* accounted for 2.1% of the outbreaks and 3.2% of the cases of food-borne illnesses (CDC, 2000). Most cases of *C. perfringens* food poisoning are mild and are not reported. In 1994, the total cost of illnesses due to *C. perfringens* was estimated at $123 million in the United States (Anonymous, 1995). The estimated large number of illnesses due to *C. perfringens* clearly stresses the importance of cooling foods quickly after cooking, with proper refrigeration during shelf-life storage.

C. perfringens outbreaks usually result from improper handling and preparation of foods, such as inadequate cooling at the home, retail, or food service level, rarely involving commercial meat processors (Bryan, 1988; Taormina et al., 2003; CDC, 2000; Bean et al., 1997; Bean and Griffin, 1990). Major

Vijay K. Juneja • Microbial Food Safety Research Unit, Eastern Regional Research Center, Agricultural Research Service, U.S. Department of Agriculture, 600 E. Mermaid Lane, Wyndmoor, PA 19038. **John S. Novak** • American Air Liquide, Delaware Research & Technology Center, 200 GBC Drive, Newark, DE 19702-2462. **Ronald J. Labbe** • Department of Food Science, University of Massachusetts, Amherst, MA 01003-1021.

Table 1. Toxins of *Clostridium perfringens*

Lethal toxin (types)	Activity of toxins
Alpha-toxin (A, B, C, D, E)	Lethal, necrotizing, hemolytic lecithinase C
Beta-toxin (B, C)	Lethal, necrotizing
Gamma-toxin (B, C)	Lethal
Delta-toxin (B, C)	Hemolytic, lethal
Epsilon-toxin (B, D)	Lethal, necrotizing (activated by trypsin)
Eta-toxin (A)	Lethal (validity questionable)
Theta-toxin (A, B, C, D, E)	Hemolytic (oxygen labile, lethal)
Iota-toxin (E)	Necrotizing, lethal (activated by trypsin)
Kappa-toxin (A, C, D, E)	Collagenase (lethal, necrotizing, gelatinase)
Lambda-toxin (B, D, E)	Proteinase (disintegrates Azocoll and hide powder but not collagen; gelatinase)
Mu-toxin (A, B, D)	Hyaluronidase
Nu-toxin (A, B, C, D, E)	Deoxyribonuclease
Enterotoxin (A, C, D)	Complex with plasma membrane

contributing factors leading to food poisoning associated with *C. perfringens* include its ability to form heat-resistant spores that can survive commercial cooking operations, as well as the ability to germinate, outgrow, and multiply at a very rapid rate during postcook handling, primarily under conditions conducive to germination. Such conditions occur when the cooling of large batches of cooked foods is not fast enough to inhibit bacterial growth or when the foods are held at room temperature for an extended period or are temperature abused. Germination and outgrowth of *C. perfringens* spores during cooling of thermally processed meat products have been reported extensively (Juneja et al., 1994c, 1999). Accordingly, improper cooling (40.9%) of food products has been cited as the most common cause of *C. perfringens* outbreaks (Angulo et al., 1998).

C. perfringens food poisoning is one of the most common types of food-borne illness (Labbe, 1989) and occurs typically from the ingestion of >10^6 viable vegetative cells of the organism in temperature-abused foods (Labbe and Juneja, 2002). Acidic conditions encountered in the stomach may actually trigger the initial stages of sporulation of the vegetative cells of *C. perfringens*. Once in the small intestine, the vegetative cells sporulate, releasing an enterotoxin upon sporangial autolysis. The enterotoxin is responsible for the pathological effects in humans, as well as the typical symptoms of acute diarrhea with severe abdominal cramps and pain (Duncan and Strong, 1969; Duncan et al., 1972). Pyrexia and vomiting are usually not encountered in affected individuals. The typical incubation period before onset of symptoms is 8 to 24 hours, and acute symptoms usually last less than 24 hours. Full recovery within 24 to 48 hours is normal. Fatalities are rare in healthy individuals. Food-borne outbreaks of *C. perfringens* can be confirmed if >10^5 CFU/g of

the organism or >10^6 spores/g are detected in the implicated food or feces, respectively (Labbe and Juneja, 2002).

Meat and poultry products were associated with the vast majority of *C. perfringens* outbreaks in the United States, probably due to the fastidious requirement of more than a dozen amino acids and several vitamins for the organism to grow in these products (Labbe and Juneja, 2002; Brynestad and Granum, 2002). Roast beef, turkey, meat-containing Mexican foods, and other meat dishes have been associated with *C. perfringens* food poisoning outbreaks (Bryan, 1969, 1988). Roast beef and other types of cooked beef were implicated primarily as vehicles of transmission for 26.8% of 190 *C. perfringens* enteritis outbreaks in the United States from 1973 to 1987 and 33.9% of 115 outbreaks from 1977 to 1984, although poultry products were also commonly implicated (Bean and Griffin, 1990; Bryan, 1988).

In retrospect, most outbreaks of *C. perfringens* food poisoning can be avoided by adequate cooking of meat products followed by holding at hot temperatures or rapid cooling.

SOURCES AND INCIDENCE IN THE ENVIRONMENT AND FOODS

Clostridium perfringens is commonly found on vegetable products and in other raw and processed foods. The organism is found frequently in meats, generally through fecal contamination of carcasses; contamination from other ingredients, such as spices; or postprocessing contamination. *C. perfringens* was detected in 36%, 80%, and 2% of fecal samples from cattle, poultry, and pigs, respectively (Tschirdewahn et al., 1991). The organism was found on 29, 66, and 35% of beef, pork, and lamb carcasses, respectively

(Smart et al., 1979). *C. perfringens* was isolated from 43.1% of processed and unprocessed meat samples tested in one study, including beef, veal, lamb, pork, and chicken products (Hall and Angelotti, 1965). Many areas within broiler chicken processing plants are contaminated with the organism (Craven, 2001), and the incidence of *C. perfringens* on raw poultry ranges from 10 to 80% (Waldroup, 1996). *C. perfringens* was detected in 47.4% of raw ground beef samples (Ladiges et al., 1974), and a mean level of 45.1 *C. perfringens* CFU per cm^2 was detected on raw beef carcass surface samples (Sheridan et al., 1996). *C. perfringens* was detected on 38.9% of commercial pork sausage samples (Bauer et al., 1981) and on raw beef, equipment, and cooked beef in food service establishments (Bryan and McKinley, 1979). In a survey conducted in the United Kingdom by Hobbs et al. (1965), 67% of the vacuum-packaged fish product samples were positive for clostridia, predominantly *C. perfringens*. About 50% of raw or frozen meat and poultry contains *C. perfringens* (Labbe, 1989). *C. perfringens* spores were isolated from 80% of 54 different spices and herbs (Deboer et al., 1985). This presents a public health hazard since spices and herbs are commonly used in meat and meat products. However, early surveys of foods did not determine the enterotoxin-producing ability of isolates. In recent surveys on the incidence of *C. perfringens* in raw and processed foods, an incidence level of 30 to 80% has been found (Lin and Labbe, 2003). The mild symptoms of the illness result in underreporting, as was previously explained in this chapter. Underreporting, combined with an absence of active surveillance for this microorganism, leads to only a select number of cases that are voluntarily reported to the Centers for Disease Control and Prevention in any given year. It is currently estimated that the incidence of food-borne illness from *C. perfringens* is a factor of 10 to 350 times the number of cases actually reported (Mead et al., 1999). In the United States from 1983 to 1997, *C. perfringens* was the third most common cause of confirmed outbreaks and cases of food-borne illness (CDC, 2000). There is no evidence to suggest that the contrary exists today.

Determining the level of contamination, Kalinowski et al. (2003) reported that out of 197 raw comminuted meat samples analyzed, all but 2 samples had undetectable levels (<3 spores per g) and 2 ground pork samples contained 3.3 and 66 spores per g. In another survey, Taormina et al. (2003) examined a total of 445 whole-muscle and ground or emulsified raw pork, beef, and chicken product mixtures acquired from industry sources for *C. perfringens* vegetative cells and spores. Out of 194 cured whole-muscle samples examined, 1.6% were

positive for vegetative cells, and spores were not detected. Out of 152 cured ground or emulsified samples, 48.7% and 5.3% were positive for vegetative cells and spores, respectively. Populations of vegetative cells and spores did not exceed 2.72 and 2.00 log$_{10}$ CFU/g. These studies suggest a low incidence of spores in raw, cured, whole-muscle ham, as well as low levels of spores in cured, ground, emulsified meats and in raw comminuted meat samples.

Li et al. (2007) surveyed soils and home kitchen surfaces in the Pittsburgh, PA (United States) area to determine the prevalence of *cpe*-positive *C. perfringens* isolates in these two environments and reported that neither soil nor home kitchen surfaces represent major reservoirs for type A isolates with chromosomal *cpe* that cause food poisoning, although soil does appear to be a reservoir for *cpe*-positive isolates causing non-food-borne gastrointestinal diseases. Rahmati and Labbe (2008) reported 17 samples positive for *C. perfringens*, one possessing the enterotoxin gene, out of 347 fresh and processed seafood samples examined. In another study (Wen and McClane, 2004), a survey of American retail foods did report that approximately 1.7% of raw meat, fish, and poultry items sold in retail food stores contain type A isolates carrying a chromosomal *cpe* gene, and no plasmid *cpe* gene was found in any of those surveyed retail foods. In a national survey of the retail meats conducted in Australia, *C. perfringens* was not recovered from any of the 94 ground beef samples and was isolated from 1 of 92 samples of diced lamb (Phillips et al., 2008). These surveys indicate a low incidence of *C. perfringens* in retail meats.

INTRINSIC AND EXTRINSIC FACTORS THAT AFFECT SURVIVAL AND GROWTH IN FOOD PRODUCTS AND CONTRIBUTE TO OUTBREAKS

Temperature

Although technically an anaerobe, *C. perfringens* is quite aerotolerant. The optimal growth temperature range for the organism is 43 to 45°C, although growth can occur between 15 and 50°C (Labbe and Juneja, 2002). Growth outside this range of temperatures may be characterized by extended lag and longer generation times and might require strict anaerobic conditions. Growth at 52.3°C was observed only under strict anaerobic conditions (Shoemaker and Pierson, 1976). Populations of *C. perfringens* at temperatures below 10°C have been reported to decline or remain stable, suggesting that the pathogen does not grow at proper refrigeration temperatures (Traci and Duncan, 1974; Strong et al., 1966). Thus, refrigeration of foods contaminated with spores and vegetative cells of *C. perfringens* will not provide favorable conditions for growth.

C. perfringens is capable of extremely rapid growth in meat systems, which makes the organism a particular concern to meat processors as well as the food service industry. Hall and Angelotti (1965) found that *C. perfringens* vegetative cells inoculated into meat began multiplying without any lag phase, at 45°C. *C. perfringens* grew faster at 45°C in autoclaved ground beef than in broth media at the same temperature (Willardsen et al., 1978, 1979). One strain of *C. perfringens* had a generation time of 7.1 minutes in autoclaved ground beef held at 41°C, although the mean generation time for an eight-strain mixture ranged from 19.5 minutes at 33°C to 8.8 minutes at 45°C (Willardsen et al., 1978). Because of their rapid growth, numbers of *C. perfringens* cells sufficient to cause illness, that being $>10^6$ cells, can rapidly be reached under optimal conditions in meats and meat products. Geopfert and Kim (1975) reported that *C. perfringens* growth does not initiate in raw ground beef stored at 15°C or below, even after extended storage. However, Solberg and Elkind (1970) reported that *C. perfringens* vegetative cells in frankfurters increased by 3 $\log_{10}$ cycles in 3 days at 15°C and in 5 days at 12°C, but the growth was restricted at 10 and 5°C, respectively.

Oxygen

C. perfringens does not grow very well in the presence of oxygen. Juneja et al. (1994a, 1994b) examined growth of vegetative *C. perfringens* cells in cooked ground beef and cooked turkey at various temperatures and determined that *C. perfringens* was able to grow in both products under both anaerobic and aerobic conditions, although growth was faster in both products held under anaerobic conditions. *C. perfringens* grew in ground beef to >7 logs within 12 h at 28, 37, and 42°C under anaerobic atmosphere and at 37 and 42°C under aerobic conditions. At 28°C under aerobic conditions, growth was relatively slow and total viable count increased to >6 logs within 36 h. Similarly, growth at 15°C in air was both slower and less than that under vacuum. This suggests that aerobic conditions can delay growth of *C. perfringens*, even during temperature abuse. Regardless of packaging, the organism either declined or did not grow at 4, 8, and 12°C. Temperature abuse (28°C storage) of refrigerated products for 6 h will not permit *C. perfringens* growth (Juneja et al., 1994a). Nevertheless, temperature abuse of a product

stored under aerobic conditions is unlikely to inhibit *C. perfringens* growth completely. *C. perfringens* grew in cooked turkey to about 7 logs within 9 h anaerobically and by 24 h aerobically at 28°C. While aerobic growth was slow at 15°C, mean $\log_{10}$ CFU/g increased anaerobically by 4 to 4.5 logs by day 8 for both strains (Juneja et al., 1994b).

a_w

Growth of *C. perfringens* is inhibited by water activity (a_w) ranging from 0.93 to 0.97 (Kang et al., 1969; Strong et al., 1970; Labbe and Juneja, 2006). Within this range, inhibition is dependent on the solute used to adjust the a_w and on factors such as inoculum size, pH, temperature, oxidation reduction potential, and the presence of various nutrients (Craven, 1980). While germination of spores was inhibited on laboratory medium at an a_w value of 0.93, a *C. perfringens* population at an a_w of 0.95 showed an increase of 1.8 log CFU/ml by 5 h and 4.9 $\log_{10}$ CFU/ml by 12 h at 37°C (Kang et al., 1969). Therefore, foods contaminated with low levels of *C. perfringens* would not result in substantial growth within 5 h of temperature abuse at 37°C. Growth of *C. perfringens* vegetative cells at 37°C was not observed until 12 and 20 h at a_w values of 0.965 and 0.96, respectively (Strong et al., 1970). These data suggest that low-a_w foods are unlikely to support *C. perfringens* growth sufficiently to cause illness, though growth can occur at an a_w of 0.93. Some sausages, salamis, hams, pepperonis, soups, and chipped and dried beef products have a_w values below this level (Holley et al., 1988).

pH

Optimum pH for *C. perfringens* growth is between 6.0 and 7.0, and the range is between pH 5.0 and 9.0. Some growth may be expected at pH values of ≤ 5.0 and ≥ 8.3 (Hobbs, 1979; Labbe 1989). In general, acidic foods (pH ≤ 5.0) would not foster growth of *C. perfringens*. However, certain *C. perfringens* strains have been found to be relatively more acid tolerant than others (de Jong, 1989). Labbe (1989) suggested that an acidic pH in foods can act synergistically with other hurdles, such as curing salts, to restrict *C. perfringens* growth.

Curing Salts

The sodium chloride level in a food affects the ability of *C. perfringens* to grow. Published research suggests that *C. perfringens* can grow in media supplemented with various levels of curing salts. While growth was not inhibited by 4% (wt/vol) NaCl, some strains do not grow in 5 to 6% NaCl, and most strains failed to grow in 7 to 8% NaCl (Roberts and Derrick, 1978; Mead et al., 1999; Gibson and Roberts, 1986). These levels are higher than those typically found in ready-to-eat (RTE) foods.

Studies have been conducted with laboratory media on the antimicrobial effects of nitrites, with respect to the growth of *C. perfringens* vegetative cells and germination of spores (Gough and Alford, 1965; Perigo and Roberts, 1968; Labbe and Duncan, 1970; Riha and Solberg, 1975; Roberts and Derrick, 1978; Gibson and Roberts, 1986). Traditionally, sodium nitrite has been used as a preservative in cured meats to inhibit *Clostridium botulinum* growth and subsequent deadly neurotoxin production. *C. perfringens* appears to be more resistant to nitrites. Gough and Alford (1965) reported that *C. perfringens* growth was not inhibited at 8,000 ppm of sodium nitrite but was inhibited when the concentration was increased to 12,000 ppm. Further, it has been demonstrated that 0.02% and 0.01% sodium nitrite slows germination of both heat-resistant and heat-sensitive *C. perfringens* spores (Labbe and Duncan, 1970; Sauter et al., 1977).

It is worth mentioning that the inhibitory effect of sodium nitrite is enhanced at elevated temperatures (Davidson and Juneja, 1990), since heat-injured spores are more sensitive to the effects of nitrite (Chumney and Adams, 1980; Labbe and Duncan, 1970). Anaerobic conditions can also increase the antimicrobial effectiveness of sodium nitrite. Further, the inhibitory effect of nitrite can be enhanced under acidic conditions and by prior heating of the medium containing nitrite (Labbe, 1989; Riha and Solberg, 1975).

The ability of *C. perfringens* to grow in many cured meats is well documented. For instance, *C. perfringens* spores grew rapidly in frankfurters at 37°C and 23°C, and although growth was slower at lower temperatures, populations increased 2 log CFU/g after 2 days at 15°C and 3 days at 12°C (Solberg and Elkind, 1970). Most processed meat products contain about 2.75 and 3.25% salt (Maurer, 1983), and sodium nitrite is allowed in cured meats at ingoing (initial) levels of 156 ppm. Vareltzis et al. (1984) found that chicken frankfurters containing 2.6% salt and 150 ppm sodium nitrite did not allow *C. perfringens* growth from spores (starting inoculum of 4.7 log) over 9 days at 20°C. Hallerbach and Potter (1981) reported that frankfurters containing 2.2% salt and 140 ppm nitrite did not support *C. perfringens* vegetative cell growth for 74 h at 20°C. Further, Amezquita et al. (2004) reported no growth of *C. perfringens* at 21 and 17°C after 10 and 21 days, respectively, in cured ham (initial sodium nitrite concentration of 156 ppm). Gibson and Roberts (1986)

reported that the inhibitory concentrations of sodium nitrite can be lowered if combined with other curing salts. In their study, *C. perfringens* growth at 20°C was inhibited by 200 μg of nitrite/ml and 3% NaCl or 50 μg of nitrite/ml and 4% NaCl at pH 6.2 in a laboratory medium. In another study, the levels of sodium nitrite necessary to inhibit the strains tested dropped from 300 ppm to 25 ppm when the concentration of NaCl was increased from 3 to 6% (Roberts and Derrick, 1978). Thus, sodium nitrite acts synergistically with NaCl to inhibit *C. perfringens* growth. The differences in the minimum growth temperatures could be a function of product characteristics, such as meat species, pH, and concentrations of functional ingredients, such as NaCl, phosphates, and other antimicrobial ingredients, such as nitrates/nitrites or salts of organic acids. These ingredients have been shown to affect the growth of *C. perfringens* from spore inocula from specific meat products (Juneja and Thippareddi, 2004a, 2004b; Thippareddi et al., 2003; Zaika 2003).

FOOD PROCESSING OPERATIONS THAT INFLUENCE THE NUMBERS, SPREAD, OR CHARACTERISTICS

Slow cooking associated with low-temperature, long-time cooking will not eliminate *C. perfringens* spores. Cooking temperatures, if designed to inactivate *C. perfringens* spores, may negatively impact the product quality and desirable organoleptic attributes of foods. Therefore, spores are likely to survive the normal pasteurization/cooking temperatures applied to these foods. Cooking usually increases the anaerobic environment in food and reduces the numbers of competing spoilage organisms, which is ecologically important because *C. perfringens* competes poorly with the spoilage flora of many foods. Mean generation times in autoclaved ground beef during slow heating from 35 to 52°C ranged from 13 to 30 minutes, with temperature increases of 6 to 12.5°C/hour (Willardsen et al., 1978). Another study also demonstrated growth of the organism in autoclaved ground beef during linear temperature increases (4.1°C to 7.5°C/hour) from 25°C to 50°C. (Roy et al., 1981).

Cooking of foods can also heat shock *C. perfringens* spores, since germination activation of *C. perfringens* spores can occur at temperatures between 60 and 80°C (Walker, 1975). Similar to spores of other bacterial species, spores of *C. perfringens* germinate at a higher rate after heat shock. For instance, while only 3% of inoculated *C. perfringens* spores germinated in raw beef without prior heat shock, almost all spores germinated after the beef received a heat treatment

(Barnes et al., 1963). As a result of such heat shock conditions, a physiological response is triggered in organisms, leading to the synthesis of a specific set of proteins known as heat shock proteins (HSPs). Synthesis of HSPs has been observed in bacterial as well as mammalian cells (Lindquist and Craig, 1988). After heat shock, germination and outgrowth of spores and *C. perfringens* vegetative growth are likely to occur in cooked foods if the rate and extent of cooling are not sufficient or if the processed foods are temperature abused. The abuse may occur during transportation, distribution, storage, or handling in supermarkets, or during preparation of foods by consumers which includes low-temperature–long-time cooking of foods as well as scenarios in which foods are kept on warming trays before final heating or reheating. Studies have described growth of *C. perfringens* during cooling of cooked, uncured meat products. In a study by Tuomi et al. (1974), when cooked ground beef gravy inoculated with a mixture of vegetative cells and spores of *C. perfringens* NCTC 8239 was cooled in a refrigerator, rapid growth of the organism was reported to occur during the first 6 h of cooling, when the gravy temperature was in an ideal growth temperature range. While the study by Tuomi et al. (1974) identified the cooling stage as most critical in ensuring the safety of such products, the experimental design included both the vegetative cells and spores of *C. perfringens*; the vegetative cells would not likely be present in a cooked product. Shigehisa et al. (1985) reported on germination and growth characteristics of *C. perfringens* spores inoculated into ground beef at 60°C and cooled to 15°C at a linear cooling rate of 5 to 25°C/h. They observed that the organism did not grow during exposure to falling temperature rates of 25 to 15°C/h. However, multiplication of the organism was observed when the rate was less than 15°C/h. Interestingly, the total population did not change for the first 150 min, regardless of the cooling rate. This study is not totally applicable to typical retail food operations because cooling is not linear; it is exponential.

Steele and Wright (2001) evaluated growth of *C. perfringens* spores in turkey roasts cooked to an internal temperature of 72°C and then cooled in a walk-in cooler from 48.9°C to 12.8°C in 6, 8, or 10 h. Results of that study indicated that an 8.9-hour cooling period was adequate to prevent growth of *C. perfringens* with a 95% tolerance interval. To simulate commercial chili cooling procedures, Blankenship et al. (1988) conducted exponential cooling experiments in which the cooling time was 4 and 6 h for a temperature decline from 50 to 25°C. This is approximately equivalent to a cooling rate of 12 h and 18 h for temperatures of 54.4 to 7.2°C. They

observed a declining growth rate in the case of 4-h and 6-h cooling times. Juneja et al. (1994c) reported that no appreciable growth (<1.0 $\log_{10}$ CFU/g) occurred if cooling took 15 h or less when cooked ground beef inoculated with heat-activated *C. perfringens* spores was cooled from 54.4 to 7.2°C at an exponential rate, that being more rapid cooling at the beginning followed by a slower cooling rate later. However, *C. perfringens* grew by 4 to 5 $\log_{10}$ CFU/g if the cooling time was greater than 18 h. This implies that *C. perfringens* is capable of rapid growth in meat systems, making this organism a particular concern to meat processors, as well as to the food service industry.

A limited amount of published research is available regarding growth of the pathogen in cooked cured meats during cooling. Taormina et al. (2003) inoculated bologna and ham batter with *C. perfringens* spores and then subjected them to cooking and either cooling procedures typically used in industry or extended chilling. In that study, growth of the organism was not detected in any of the products tested during chilling from 54.4 to 7.2°C. Zaika (2003) reported complete inhibition of *C. perfringens* germination and growth in cured hams with NaCl concentrations of 3.1%, when cooled exponentially from 54.4 to 7.2°C within 15, 18, or 21 h. Cooked cured turkey cooled from 48.9 to 12.8°C did not support *C. perfringens* growth in 6 h; however, a 3.07 log increase was observed following a 24-h cooling time (Kalinowski et al., 2003).

RECENT ADVANCES IN BIOLOGICAL, CHEMICAL, AND PHYSICAL INTERVENTIONS TO GUARD AGAINST THE PATHOGEN

The presence of inhibitory agents in the products can affect germination of *C. perfringens* spores and may also affect the minimum growth temperatures for the germinated spores. Recent studies have shown the efficacy of certain antimicrobial agents against the growth of *C. perfringens* during cooling of meat products. For instance, Sabah et al. (2003) found that 0.5 to 4.8% sodium citrate inhibited growth of *C. perfringens* in cooked, vacuum-packaged, restructured beef cooled from 54.4°C to 7.2°C within 18 h. The same researchers also demonstrated growth inhibition of the microorganism by oregano in combination with organic acids during cooling of sous-vide-cooked ground beef products (Sabah et al., 2004). Organic acid salts, such as 1 to 1.5% sodium lactate, 1% sodium acetate, 0.75 to 1.3% buffered sodium citrate (with or without sodium diacetate), and 1.5% sodium lactate supplemented with sodium diacetate, inhibited germination and outgrowth of *C. perfringens* spores during the chilling of marinated ground turkey breast (Juneja and Thippareddi, 2004a, 2004b). In another study, Thippareddi et al. (2003) reported complete inhibition of *C. perfringens* spore germination and outgrowth by sodium salts of lactic and citric acids (2.5 and 1.3%, respectively) in roast beef, pork ham, and injected turkey products. Incorporation of plant-derived natural antimicrobials, such as thymol (1 to 2%), cinnamaldehyde (0.5 to 2%), oregano oil (2%), and carvacrol (1 to 2%), as well as the biopolymer chitosan (0.5 to 3%) derived from shellfish and green tea catechins (0.5 to 3%), individually inhibited *C. perfringens* germination and outgrowth during exponential cooling of ground beef and turkey (Juneja et al., 2006a, 2006b, 2007; Juneja and Friedman, 2007). Therefore, natural compounds can be used as ingredients in processed meat products to provide an additional measure of safety to address the *C. perfringens* hazard during chilling and subsequent refrigeration of meat products, thus further minimizing risk to the consumer.

Numerous studies have examined the heat resistance of *C. perfringens* spores and/or vegetative cells. Heat resistance varies among strains of *C. perfringens*, although both heat-resistant and heat-sensitive strains can cause food poisoning (Labbe and Juneja, 2006). Environmental stresses, such as storage and holding temperatures and low-temperature–long-time cooking, expose the contaminating vegetative and spore-forming food-borne pathogens to conditions similar to heat shock, thereby rendering the heat-shocked organisms more resistant to subsequent lethal heat treatments. Researchers have reported on the quantitative assessment of heat resistance, the heat shock response, and the induced thermotolerance to assist food processors in designing thermal processes for the inactivation of *C. perfringens*, thereby ensuring the microbiological safety of cooked foods (Juneja et al., 2001a, 2003). Heat shocking vegetative cells suspended in beef gravy at 48°C for 10 min allowed the microorganisms to survive longer and increased the heat resistance by as high as 1.5-fold (Juneja et al., 2001a). The thermal resistance (*D* values in min) of *C. perfringens* cells heated in beef gravy at 58°C ranged from 1.21 min (C 1841 isolate) to 1.60 min (F 4969 isolate). Compared to the control (no heat shock), the increase in heat resistance after heat shocking at 58°C ranged from 1.2-fold (B 40 isolate) to 1.5-fold (NCTC 8239 isolate). The *D* values of *C. perfringens* spores heated in beef gravy at 100°C ranged from 15.50 min (NCTC 8239 isolate) to 21.40 min (NB 16 isolate) (Juneja et al., 2003). Compared to the control (no heat shock), the increase in heat resistance of *C. perfringens* spores

at 100°C after heat shocking ranged from 1.1-fold (NCTC 8238 isolate) to 1.5-fold (F 4969 isolate). Similar results were obtained by Heredia et al. (1997), who found that a sublethal heat shock at 55°C for 30 minutes increased tolerance of both spores and vegetative cells to a subsequent heat treatment. In their study, when the heat resistance of *C. perfringens* vegetative cells, grown at 43°C in fluid thioglycolate medium to an A_{600} of 0.4 to 0.6, was determined, *D* values at 55°C of 9 and 5 min for the FD-1 and FD1041 strains, respectively, were reported. The *D* values of the heat-shocked cells were 85 and 10 min, respectively. Heredia et al. (1997) heat shocked sporulating cells of *C. perfringens* at 50°C for 30 min and then determined the *D* values at 85 or 90°C. The authors reported that a sublethal heat shock increased the thermotolerance of *C. perfringens* spores by at least 1.7- to 1.9-fold; the *D* values at 85°C increased from 24 to 46 min and at 95°C from 46 to 92 min, respectively. Bradshaw et al. (1982) reported that *D* values at 99°C for *C. perfringens* spores suspended in commercial beef gravy ranged from 26 to 31.4 min. Miwa et al. (2002) found that spores of enterotoxin-positive *C. perfringens* strains were more heat resistant than enterotoxin-negative strains. Similarly, food poisoning isolates of the organism are generally more heat resistant than *C. perfringens* spores from other sources (Labbe, 1989). Sarker et al. (2000) reported *D* values at 100°C for 12 isolates of *C. perfringens* spores, carrying either the chromosomal *cpe* gene or the plasmid *cpe* gene, in DS medium ranged from 0.5 min to 124 min. Sarker et al. (2000) reported *D* values at 55°C for *C. perfringens* cells grown at 37°C in FTG of 12.1 min and 5.6 min for the E13 and F5603 strains, respectively. These authors also reported a strong association of the food poisoning isolates and increased heat resistance; the *D* values at 55 or 57°C for the *C. perfringens* chromosomal *cpe* isolates were significantly higher ($P < 0.05$) than the *D* values of the *C. perfringens* isolates carrying a plasmid *cpe* gene; however, differences in heat resistance levels were not observed at higher temperatures. Nevertheless, understanding these variations in heat resistance is certainly necessary in order to design adequate cooking regimes to eliminate *C. perfringens* vegetative cells in RTE foods.

Studies have been conducted to assess the effects and interactions of multiple food formulation factors on the heat resistance of spores and vegetative cells of *C. perfringens*. In a study by Juneja and Marmer (1996), when the thermal resistance of *C. perfringens* spores (expressed as *D* values in min) in turkey slurries that included 0.3% sodium pyrophosphate at pH 6.0 and salt levels of 0, 1, 2, or 3% was assessed, the *D* values at 99°C decreased from 23.2 min (no salt) to 17.7 min (3% salt). In a beef slurry, the D values significantly decreased ($P < 0.05$) from 23.3 min (pH 7.0, 3% salt) to 14.0 min (pH 5.5, 3% salt) at 99°C (Juneja and Majka 1995). While addition of increasing levels (1 to 3%) of salt in turkey (Juneja and Marmer, 1996) or a combination of 3% salt and pH 5.5 in beef (Juneja and Majka, 1995) can result in a parallel increase in the sensitivity of *C. perfringens* spores at 99°C, mild heat treatments given to minimally processed foods will not eliminate *C. perfringens* spores. Juneja and Marmer (1998) examined the heat resistance of vegetative *C. perfringens* cells in ground beef and turkey containing sodium pyrophosphate (SPP). The D values for beef that included no SPP were 21.6, 10.2, 5.3, and 1.6 min at 55, 57.5, 60, and 62.5°C, respectively; the values for turkey ranged from 17.5 min at 55°C to 1.3 min at 62.5°C. Addition of 0.15% SPP resulted in a concomitant decrease in heat resistance, as evidenced by reduced bacterial D values. The D values for beef that included 0.15% SPP were 17.9, 9.4, 3.5, and 1.2 min at 55, 57.5, 60, and 62.5°C, respectively; the values for turkey ranged from 16.2 min at 55°C to 1.1 at 62.5°C. The heat resistance was further decreased when the SPP levels in beef and turkey were increased to 0.3%. Heating such products to an internal temperature of 65°C for 1 min killed >8 $\log_{10}$ CFU/g. The z values for beef and turkey for all treatments were similar, ranging from 6.22 to 6.77°C. Thermal death time values from this study should assist institutional food service settings in designing thermal processes that ensure safety against *C. perfringens* in cooked beef and turkey.

Researchers have assessed the efficacy of added preservatives on inhibiting or delaying the growth of *C. perfringens* in extended shelf-life, refrigerated, processed meat products. When ground turkey containing 0.3% SPP and 0, 1, 2, or 3% salt was sous vide processed (71.1°C) and held at 28°C, lag times of 7.3, 10.6, 11.6, and 8.0 h were observed for salt levels of 0, 1, 2, and 3%, respectively (Juneja and Marmer 1996). Growth of *C. perfringens* spores in cooked ground turkey with added 0.3% SPP was inhibited for 12 h at 3% salt, pH 6.0, and 28°C. After 16 h, spores germinated and grew at 28°C from 2.25 to >5 $\log_{10}$ CFU/g in sous-vide-processed (71.1°C) turkey samples, regardless of the presence or absence of salt (Juneja and Marmer, 1996). While *C. perfringens* spores germinated and grew at 15°C to >5 $\log_{10}$ CFU/g in turkey with no salt by day 4, the presence of 3% salt in samples at 15°C completely inhibited germination and subsequent multiplication of vegetative cells, even after 7 days of storage (Juneja and Marmer, 1996). Thus, the addition of 3% salt in sous-vide-processed ground turkey containing 0.3% SPP delayed growth for 12 h at 28°C and completely

inhibited the outgrowth of spores at 15°C (Juneja and Marmer, 1996). However, 3% salt in RTE products will not inhibit germination and growth of *C. perfringens* spores if refrigerated products are temperature abused to 28°C for an extended period. In another study (Juneja and Majka, 1995), the combination of 3% salt and pH 5.5 inhibited *C. perfringens* growth from spores in sous-vide-processed ground beef supplemented with 0.3% SPP at 15 and 28°C. Growth from *C. perfringens* spores occurred within 6 days in sous-vide-processed (71.1°C and pH 7.0) ground beef samples but was delayed until day 8 in the presence of 3% salt at pH 5.5 at 15°C (Juneja and Majka, 1995). *C. perfringens* growth from a spore inoculum at 4°C was not observed with sous-vide-cooked turkey or beef samples (Juneja and Marmer, 1996; Juneja and Majka, 1995). In a related study, Juneja et al. (1996) showed that *C. perfringens* growth in cooked turkey can be effectively inhibited in an atmosphere containing 25 to 75% CO_2, 20% O_2, and a balance of N_2 in conjunction with good refrigeration; however, the atmosphere cannot be relied upon to eliminate the risk of *C. perfringens* food poisoning in the absence of proper refrigeration (Juneja et al., 1996). Kalinowski et al. (2003) investigated the fate of *C. perfringens* in cooked-cured and uncured turkey at refrigeration temperatures. In their study, *C. perfringens* decreased by 2.52, 2.54, and 2.75 $\log_{10}$ CFU/g in cured turkey held at 0.6, 4.4, and 10°C, respectively, and the reductions in levels were similar in uncured turkey.

The efficacy of sodium lactate (NaL) in inhibiting the growth from spores of *C. perfringens* in a sous-vide-processed food has been assessed. Inclusion of 3% NaL in sous vide beef goulash inhibited *C. perfringens* growth at 15°C, delayed growth for a week at 20°C, and had little inhibitory effect at 25°C (Aran, 2001). While addition of 4.8% NaL restricted *C. perfringens* growth from spores for 480 h at 25°C in sous-vide-processed (71.1°C) marinated chicken breast, it delayed growth for 648 h at 19°C. *C. perfringens* growth was not observed at 4°C, regardless of NaL concentration (Juneja, 2006). These studies suggest that NaL can have significant bacteriostatic activity against *C. perfringens* and may provide sous-vide-processed foods with a degree of protection against this microorganism, particularly if employed in conjunction with adequate refrigeration.

Predictive bacterial growth models that describe *C. perfringens* spore germination and outgrowth during cooling of food systems have been generated by researchers using constant temperature data. Juneja et al. (1999) presented a model for predicting the relative growth of *C. perfringens* from spores, through lag, exponential, and stationary phases of growth, at temperatures spanning the entire growth temperature range of about 10 to 50°C. Huang (2003a, 2003b, 2003c) used different mathematical methods to estimate the growth kinetics of *C. perfringens* in ground beef during isothermal, square-waved, linear, exponential, and fluctuating cooling temperature profiles. Juneja et al. (2001b) developed a predictive cooling model for cooked, cured beef based on growth rates of the organism at different temperatures, which estimated that exponential cooling from 51 to 11°C in 6, 8, or 10 h would result in an increase of 1.43, 3.17, and 11.8 $\log_{10}$ CFU/g, respectively, when assuming that the ratio of the mathematical lag time to the generation time for cells in the exponential phase of growth was equal to 8.068, the estimated geometric mean. A similar model was later developed for cooked, cured chicken (Juneja and Marks, 2002). Juneja et al. (2006c) developed a model for predicting growth of *C. perfringens* from spore inocula in cured pork ham. In their study, isothermal growth of *C. perfringens* at various temperatures from 10 to 48.9°C was evaluated using a methodology that employed a numerical technique to solve a set of differential equations, simulating the dynamics of bacterial growth; the authors described the effect of temperature on the kinetic parameter K (dissociation constant) by the modified Ratkowski model. According to the coefficient of the model, the estimated theoretical minimum and maximum growth temperatures of *C. perfringens* in cooked cured pork were 13.5 and 50.6°C, respectively. The kinetic and growth parameters obtained from these studies can be used in evaluating growth of *C. perfringens* from spore populations during dynamically changing temperature conditions, such as those encountered in meat processing.

In a model for growth of *C. perfringens* during cooling of cooked uncured beef (Juneja et al., 2008), for a temperature decline from 54.4°C to 27°C in 1.5 h, the models predicted a $\log_{10}$ relative growth of about 1.1, with a standard error of about 0.08 $\log_{10}$, while observed results for two replicates were 0.43 and 0.90 $\log_{10}$; for the same temperature decline in 3 h, the predicted $\log_{10}$ relative growth was about 3.6 $\log_{10}$ (with a standard error of about 0.07), and the observed $\log_{10}$ relative growths were 2.4 and 2.5 $\log_{10}$. When the cooling scenarios extended to lower temperatures, the predictions were somewhat better, taking into account the larger relative growth. For a cooling scenario of 54.4°C to 27°C in 1.5 h and 27°C to 4°C in 12.5 h, the average observed and predicted $\log_{10}$ relative growths were 2.7 $\log_{10}$ and 3.2 $\log_{10}$, respectively; when cooling was extended from 27°C to 4°C in 15 h, the average observed and predicted $\log_{10}$ relative growths were 3.6 $\log_{10}$ and 3.7 $\log_{10}$,

respectively. For the latter cooling scenario the levels were greater than 6 $\log_{10}$, still less than stationary levels of about 7 or 8 $\log_{10}$. The differences of the estimates obtained for the models were insignificant. Smith-Simpson and Schaffner (2005) collected data under changing temperature conditions and developed a model to predict growth of *C. perfringens* in cooked beef during cooling. It was suggested that the accuracy of the germination, outgrowth, and lag time model has a profound influence upon the overall prediction, with small differences in the germination, outgrowth, and lag time prediction (~1 h) having a very large effect on the predicted final concentration of *C. perfringens*. Amezquita et al. (2004) developed an integrated model for heat transfer and dynamic growth of *C. perfringens* during cooling of cured ham and demonstrated that the effective integration of engineering and microbial modeling is a useful quantitative tool for ensuring microbiological safety. The above models can be successfully used to design microbiologically "safe" cooling regimes for cooked meat and poultry products.

Recent research has focused on combining traditional inactivation, survival, and growth-limiting factors at subinhibitory levels with emerging novel nonthermal intervention food preservation techniques using ionizing radiation, high hydrostatic pressure, or exposure to ozone. For example, the efficacy of high pressure is considerably enhanced when combined with heat, antimicrobials, or ionizing radiation. The effect of the combined intervention strategies is either additive or synergistic, in which the interaction leads to a combined effect of greater magnitude than the sum of the constraints applied individually. For example, the lethal effect of heat on spores can be enhanced after exposure to ozone. Novak and Yuan (2004) reported that the spores were more sensitive to heat at 55 or 75°C following 5 ppm of aqueous ozone for 5 min. Shelf-life extension of meat processed with 5 ppm O_3 for 5 min and containing *C. perfringens* spores combined with modified atmosphere packaging as a "hurdle" technology was proven to be effective in inhibiting spore germination and outgrowth over 10 days of storage at CO_2 concentrations above 30% and 4°C (Novak and Yuan, 2004). Likewise, *C. perfringens* cells exposed to 3 ppm O_3 for 5 min following mild heat exposure (55°C for 30 min) were more susceptible to ozone treatment.

When *C. perfringens* spores were suspended in peptone solution and exposed to combination treatments of hydrostatic pressure (138 to 483 MPa), time (5 min), and temperature (25 to 50°C), inactivation of spores ranged between 0.1 and 0.2 log cycles (Kalchayanand et al., 2004). When suspended spores were pressurized at 50°C for 5 min and stored at 4°C or 25°C for 24 h, 12 to 52% spores germinated, indicating that germination increased both at 4°C and 25°C during 24 h. Log_{10} reductions of spores were 6.1 log/ml when bacteriocins were supplemented in the recovery medium. These results show that germinated spores at high levels could be killed by using a bacteriocin-based preservative in foods.

Regulatory Requirements

Due to its ubiquitous nature and rapid growth in meat products, *C. perfringens* can be a potential hazard in processed meat and poultry products. The FDA (2001) Food Code dictates that cooked potentially hazardous foods, such as meats, should be cooled from 60 to 21°C within 2 h, and from 60 to 5°C within 6 h. In the United Kingdom, it is recommended that uncured cooked meats be cooled from 50 to 12°C within 6 h and from 12 to 5°C within 1 h (Gaze et al., 1998). Safe cooling times for cured meats may be up to 25% longer (Gaze et al., 1998). The USDA/FSIS compliance guidelines (USDA/FSIS, 2001) for chilling of thermally processed meat and poultry products state that these products should be chilled according to the prescribed chilling rates or require that process authorities validate the safety of customized chilling rates to control spore-forming bacteria. Specifically, the guidelines state that cooling from 54.4 to 26.7°C should take no longer than 1.5 h and that cooling from 26.7 to 4.4°C should take no longer than 5 h (USDA/FSIS, 2001). Additional guidelines allow for the cooling of certain cured cooked meats from 54.4 to 26.7°C in 5 h and from 26.7 to 7.2°C in 10 h (USDA/FSIS, 2001). If meat processors are unable to meet the prescribed time-temperature cooling schedule, they must be able to document that the customized or alternative cooling regimen used will result in a less than $1-log_{10}$-CFU increase in *C. perfringens* in the finished product. If the cooling guidelines cannot be achieved, computer modeling and/or product sampling can be used to evaluate the severity and microbiological risk of the process deviation, and additional challenge studies may be necessary to determine whether performance standards have been met.

Detection of *C. perfringens* in Foods

The ability of *C. perfringens* to cause food-borne illness and occasional associated outbreaks necessitates effective discriminatory detection methods for this pathogen in order to ensure reliable and confirmatory epidemiological screening of suspected foods. The many available methods of detection can be categorized as (i) metabolite-based biochemical (phenotyping/biotyping) assays, (ii) toxin, antigen-based immunological methods, and (iii) nucleic acid-based molecular techniques. Each characterization scheme also contains various advantages and disadvantages, with respect to sensitivity and discrimination against similar related clostridia, ease of analysis, time to results, and expense of reagents and/or equipment and materials (Table 2).

According to the U.S. Food and Drug Administration Center for Food Safety and Applied Nutrition (FDA/CFSAN), *Bacteriological Analytical Manual*, methods are prescribed for the enumeration and identification of *C. perfringens* in foods (Rhodehamel and Harmon, 1998). These methods are based upon the microorganism's innate ability to reduce sulfite to sulfide, producing a distinctive black ferrous sulfide precipitate on tryptose-sulfite-cycloserine (TSC) agar (Harmon et al., 1971). With added egg yolk emulsion, *C. perfringens* colonies produce a 2- to 4-mm opaque white zone surrounding the colonies, as a result of lecithinase activity (Rhodehamel and Harmon, 1998). Further presumptive tests include rapid coagulation of an iron-milk medium, liquefaction of gelatin in lactose-gelatin medium, fermentation of lactose-producing acid and gas, lack of motility in motility-nitrate medium, and an ability to reduce nitrates to nitrites (Rhodehamel and Harmon, 1998).

Oxoid Ltd. (Basingstoke, Hampshire, United Kingdom) has recently developed a chromogenic medium for identification and enumeration of *C. perfringens* colonies in water samples and claims to provide increased sensitivity and specificity compared to TSC agar (Oxoid Ltd., 2002). The chromogenic m-CP medium differentiates yellow colonies of *C. perfringens* based upon the microorganism's ability to ferment sucrose, whereas non-*C. perfringens* colonies that hydrolyze indoxyl-β-D-glucoside turn purple. Additionally, exposure to ammonium hydroxide causes a dark pink color change in phosphatase-producing colonies (Oxoid Ltd., 2002). A recent European patent, EP1816209, describes a new medium for selective differentiation of *C. perfringens* that utilizes sodium bisulfite and ammonium ferric citrate to distinguish sulfite reduction capability combined with a fluorogenic substrate, 4-methylumbelliferyl phosphate disodium salt, which is used for acid phosphatase detection (Araujo, 2007).

Different media are continually being compared for their use in the enumeration of *C. perfringens* bacteria from foods. A recent report found that TSC was still the medium of choice when compared with iron sulfite agar, Shahidi-Ferguson perfringens agar, sulfite cycloserine azide (SCA), differential clostridial agar (DCA), and oleandomycin polymyxin sulfadiazine perfringens agar (de Jong et al., 2003). Although all

Table 2. Comparison of commonly used discriminatory methods for detection of *C. perfringens*[a]

Method	Detection limits	Considerations	Reference
Mouse neutralization	6 MLD_{50}/ml	Inhumane and costly	Uzal et al. (2003)
ELISA	5 ng CPE/g feces	Sporulation variability in foods	Bartholomew et al. (1985)
4-Layer ELISA	6.25 ng/ml toxin	Proteolysis	El Idrissi (1989)
PC-ELISA	0.075 MLD_{50}/ml	False positives	Uzal et al. (2003)
MC-ELISA	0.75 MLD_{50}/ml	Lower sensitivity	Uzal et al. (2003)
CIEP	200 MLD_{50}/ml	False negatives	Uzal et al. (2003)
Vero cell assay	40 ng CPE/g feces	Sensitivity and reproducibility	Berry et al. (1988)
Reversed passive latex agglutination	50–100 ng CPE/g feces	Nonspecific interference	Brett (1998)
Colony hybridization	10 CFU/g food	48-h detection	Baez and Juneja (1995)
Hydrophobic grid membrane filter	10 CFU/g food	Isolation of *cpe*-positives	Heikinheimo et al. (2004)
Real-time PCR	50 fg genomic DNA 20 cells (pure culture)	Unidentified PCR inhibitors	Wise and Siragusa (2005)
Multiplex PCR	Enrichment culture	False positives	Baums et al. (2004)
PFGE	DNA preparations from enriched samples	Problematic nucleases	Maslanka et al. (1999)
AFLP	DNA preparations from enriched samples	Multiple genetic elements	Keto-Timonen et al. (2006)
Ribotyping	DNA preparations from enriched samples	Costly equipment and existing databases	Schalch et al. (1997)
Oligonucleotide microarrays	DNA preparations from enriched samples	Fluorescently labeled DNA probes hybridized to single-stranded DNA on a chip	Al-Khaldi et al. (2004)

[a] This list is representative of the many types of methods available, and although it may not be totally inclusive, many other methods that are not listed here may represent only small variations or improvements over the techniques represented here.

of the media tested were effective for enumeration studies, SCA and DCA were less effective at low concentrations of *C. perfringens* in food (de Jong et al., 2003). Shahidi Ferguson perfringens and oleandomycin polymyxin sulfadiazine perfringens agars produced low counts at times, and DCA and SCA were very laborious to prepare (de Jong et al., 2003). Iron sulfite agar lacked selectivity (de Jong et al., 2003).

In another study, fluorocult-supplemented TSC agar was compared with TSC agar, m-CP agar, tryptose-sulfite-neomycin agar, sulfite polymyxin sulfadiazine agar, and Wilson-Blair agar for detection and enumeration of *C. perfringens* spores in water (Araujo et al., 2004). It was determined that membrane filtration with fluorogenic TSC agar performed better than any of the other tested enumeration media for *C. perfringens* (Araujo et al., 2004). In addition, a low confirmation rate for *C. perfringens* on m-CP medium was explained by the very subjective color differentiation of pink colonies on the agar following exposure to ammonia fumes (Araujo et al., 2004). Lactose-sulfite broth was suggested to have greater sensitivity than that of TSC for enumerating *C. perfringens* bacteria but was not exclusively selective for *C. perfringens*, whereas cooked meat medium was less favorable for observing *C. perfringens* growth (Xylouri et al., 1997). Therefore, in terms of differential selective media and after more than 30 years of medium

innovations, TSC still appears to be best for detection and enumeration of *C. perfringens* bacteria from foods (Emswiler et al., 1977).

The many toxins produced by *C. perfringens* are listed in Table 1. Such a range of products specific to *C. perfringens* provides an ample library of antigenic determinants for subsequent immunological detection schemes. As a result, the toxinogenic typing (type A, B, C, D, or E) of *C. perfringens* is not based on the serologic specificity of *C. perfringens* enterotoxin (CPE)-related food-borne illness but on the many other exotoxins produced by the microorganism and is designated alpha, beta, epsilon, and iota (Petit et al., 1999; Brown 2000). The slide agglutination serotyping technique was developed by Stringer et al. (1980). Whole-cell preparations of isolated *C. perfringens* cultures were used as antigens against antisera raised using confirmed type strains (Stringer et al., 1980). Agglutination to specific antisera resulted in the designation of specific serotypes (Stringer et al., 1980). The mouse neutralization protection test was used initially to type *C. perfringens* strains (Sterne and Batty, 1975). Neutralization of toxin-type specific pathological effects was accomplished using appropriate antisera in laboratory mouse models (Sterne and Batty, 1975). Unfortunately, toxin-type positive results were at the expense of the animal and may be considered inhumane. Typically intestinal fluid samples are typed by combined inoculation of known antitoxins into mice, which allows the mice to survive if protected through the neutralization of only those toxins in the intestinal fluid samples that match the known antitoxins. Although both gas-liquid chromatography and nuclear magnetic resonance spectroscopy have shown the ability to discriminate among strains of *C. perfringens* due to capsular polysaccharide characteristics, their use is of limited practical routine practice due to being laborious, expensive, and time consuming (Paine and Cherniak 1975; Sheng and Cherniak 1998).

A number of immunological tests have been evaluated for detection of *C. perfringens* type D epsilon-toxin, including a polyclonal capture enzyme-linked immunosorbent assay (PC-ELISA), a monoclonal capture ELISA (MC-ELISA), and counterimmunoelectrophoresis (CIEP) (Naik and Duncan, 1977; Uzal et al., 2003). With respect to ability to detect epsilon-toxin in the intestinal contents and body fluids of sheep and goats, the PC-ELISA was most sensitive in enabling the detection of a 0.075 50% mouse lethal dose (MLD_{50})/ml compared to the MC-ELISA (25 MLD_{50}/ml) and CIEP (50 MLD_{50}/ml) (Uzal et al., 2003). The mouse neutralization test was also used as a standard (6 MLD_{50}/ml) (Uzal et al., 2003). The sensitivity results were in marked contrast to specificity results for the PC-ELISA (31.57%), MC-ELISA (57.89%), and CIEP (84.21%) (Uzal et al., 2003). Due to the inconsistencies among the diagnostic tests for enterotoxemia from *C. perfringens* epsilon-toxin, the authors recommended that absolute confirmation should include combined clinical and pathological data as well.

Another group of researchers has developed a four-layer sandwich ELISA for the detection of type D epsilon-toxin with a sensitivity limit of 6.25 ng/ml for purified toxin preparations (El Idrissi, 1989). Double antibody sandwich techniques have been used previously (Weddell and Worthington, 1984; Naylor et al., 1987). Serological tests that have been used for epsilon-toxin detection include radial immunodiffusion and reverse passive hemagglutination (Beh and Buttery 1978). A sensitive two-site enzyme-linked immunoassay utilizing electrochemical detection has been developed for *C. perfringens* phospholipase C (alpha-toxin) at detection limits of 67.1 and 13.0 ng/ml based on use of microtiter plates and polydispersed polymeric microbeads, respectively (Cardosi et al., 2005). Western immunoblotting for CPE typically suffers from the disadvantage of isolates poorly sporulating in vitro in lab media, preventing the detection of the sporulation-associated enterotoxin (Kokai-Kun et al., 1994).

Type A food-borne CPE production is considered a relatively rare characteristic, present in approximately only 6% of tested *C. perfringens* strains (Van Damme-Jongsten et al., 1989). It has been reported that all of the food poisoning *C. perfringens* isolates carried a chromosomal *cpe* gene (Collie and McClane, 1998). There are a number of methods used for the detection of CPE in feces. The tissue culture assay using Vero cells has been determined to have relatively low sensitivity, having a limit of detection of 40 ng enterotoxin/g feces compared to an ELISA (Tech Lab, Inc.) that can detect 5 ng/g (Bartholomew et al., 1985; Berry et al., 1988). A reversed passive latex agglutination assay for CPE quantitation is also available from Oxoid. In the case of isolates (as opposed to feces), these assays depend on the in vitro sporulation of such isolates. Many strains of *C. perfringens* sporulate only reluctantly (Hsieh and Labbe, 2007).

Current innovations in epidemiological detection of food-borne CPE include a multitude of DNA-based technologies. Baez and Juneja (1995) used a filter colony hybridization assay with digoxigenin-labeled DNA specific for the *C. perfringens* type A enterotoxin gene to enumerate enterotoxigenic *C. perfringens* bacteria from raw beef. The PCR was used to amplify and incorporate digoxigenin dUTP into a 364-bp internal region of the *cpe* gene (Baez and Juneja, 1995). Following hybridization of filter membranes containing fixed DNA from *C. perfringens* cells from meat samples with anti-digoxigenin antibody conjugated to alkaline phosphatase, visualization was accomplished with a blue precipitate using the substrates nitroblue tetrazolium chloride and 5-bromo-4-chloro-3-indolyl-phosphate (Baez and Juneja, 1995). The 2-day assay was sensitive enough to detect ≤10 CFU/g in the presence of a heterogeneous bacterial flora containing 10^6 CFU/g (Baez and Juneja, 1995). In a similar manner, 380 samples of spices and herbs were analyzed for *C. perfringens* and enterotoxigenicity with a dot-blot technique incorporating a *cpe*-gene digoxigenin-labeled DNA probe, which detected *cpe* in 4.25% of 188 confirmed isolates of *C. perfringens* (Rodriguez-Romo et al., 1998). Another similar method has been described for the enumeration and facilitated isolation of *cpe*-positive *C. perfringens* from fecal samples using a hydrophobic grid membrane filter colony hybridization method (Heikinheimo et al., 2004).

There are numerous PCR methods for detection of *C. perfringens* from foods and stool samples. DNA samples are isolated from suspected isolates of *C. perfringens* on selective agar and then amplified using specific primer pairs for a gene encoding a product such as alpha-toxin (Kalender and Ertas, 2005).

Specific PCR assays have been developed for the detection of the alpha-, beta-, and epsilon-toxin genes that offer a reliable and quicker alternative to the biological mouse assay for toxin typing of *C. perfringens* strains (Miserez et al., 1998). The presence of alpha-, beta-, beta-2-, epsilon-, iota-, and enterotoxin toxin genes in *C. perfringens* isolates from poultry across Sweden has been analyzed using PCR (Engstrom et al., 2003). Certainly with respect to specific components of foods, such as collagen, there may be inhibitory barriers to PCR use (Kim et al., 2000). Other limitations to PCR use include accurate product size determinations and nonspecific target amplifications. In addition to single gene PCR assays, multiplex PCR assays have been developed to simultaneously detect more than one gene target, such as the alpha-toxin gene present in all *C. perfringens* strains, as well as the *cpe* gene present only in food poisoning isolates of *C. perfringens* (Kanakaraj et al., 1998). Another application advantage for multiplex PCR would be the comprehensive detection of several toxin genes, such as *cpa*, *cpb*, *etx*, *iap*, *cpe*, and *cpb*2, providing very useful *C. perfringens* genotyping diagnoses (Baums et al., 2004). A quantitative real-time PCR assay, through the combination of PCR with fluorescently incorporated nucleotide base labels in the DNA amplification process for specific gene targets, enables an even more rapid detection (within a few hours) and enumeration scheme for *C. perfringens* isolates in a food or fecal sample (Wise and Siragusa, 2005). The analytical sensitivity limit of the real-time PCR assay was reported to be 100 CFU of *C. perfringens*/g in ileal samples, but 10,000 *C. perfringens* CFU/g in cecal samples as a result of unidentified PCR inhibitors (Wise and Siragusa, 2005).

Pulsed-field gel electrophoresis (PFGE) has been successfully used to type *C. perfringens* isolates (Maslanka et al., 1999). The change in the direction of the electric field and current during a preprogrammed electrophoretic migration of genomic DNA allows the separation of very large (>10-Mb) DNA fragments. Unlike standard restriction enzyme patterns for DNA agarose gel electrophoresis, restriction enzymes that cut DNA less frequently (6- to 8-bp cutters) are used in order to enhance discriminatory differences in DNA banding patterns. The inability to subtype 8% of the *C. perfringens* isolates analyzed using PFGE could be explained by the presence of strong nucleases from *C. perfringens* strains that degrade the genomic DNA preparations (Maslanka et al., 1999). PFGE has been shown to be an effective means to analyze the genetic diversity among *C. perfringens* isolates (Lin and Labbe, 2003; Nauerby et al., 2003).

With respect to extracellular DNase production limiting *C. perfringens* PFGE results, there are reports that amplified fragment length polymorphism (AFLP) analysis can provide an alternative solution to those problems (Keto-Timonen et al., 2006). In AFLP analysis the total DNA from an isolate is digested with restriction enzymes, and then adapters that allow amplification of subsets of fragments by PCR are ligated, producing a fingerprint that can be used with fluorescently labeled primers to allow computer automated data collection and database searches for comparisons (Vos et al., 1995). The AFLP procedure does not need to involve a complex mixture of complementary restriction enzymes, as a recent assay using a single restriction enzyme, Hind III, was used to provide a quick, economic method with good discrimination and reproducibility for *C. perfringens* strains of animal origin (Shinya et al., 2006). The AFLP method allowed for PCR amplification of the alpha-, beta-, epsilon-, iota-, enterotoxin, and beta-2-toxin genes of *C. perfringens* (Shinya et al., 2006). A drawback of the technique was the presence of multiple *C. perfringens* biotypes in the same animal, as well as different biotypes in the same AFLP profile that was attributed to mobile genetic elements, such as transposons, insertion sequences, plasmids, and bacteriophages (Shinya et al., 2006).

Ribotyping is a method that utilizes the highly conserved nature of ribosomal DNA sequences as a means of identifying discriminatory strain differences. Ribotyping was used to determine the genetic relationship of *C. perfringens* isolates from foodborne poisoning cases (Schalch et al., 1997). Twelve distinct ribotype patterns were discovered among 34 *C. perfringens* isolates analyzed (Schalch et al., 1997). EcoRI-digested DNA fragments were probed with digoxigenin-labeled 16S and 23S rRNA from *Escherichia coli*, resulting in ribopatterns with greater discernibility than that of the profiling of plasmid DNA (Schalch et al., 1997). Dupont Qualicon (Wilmington, DE) offers an automated RiboPrint system. Using repetitive-element PCR and EcoRI ribotyping, molecular subtyping of *C. perfringens* isolates was dramatically improved over PFGE (Siragusa et al., 2006). At the 90% correlation level, repetitive-element PCR with Dt primers demonstrated greater discrimination (0.938) than that with ribotyping (0.873) (Siragusa et al., 2006).

A further advancement of multiplex PCR amplification of DNA sequences involves the use of a microarray-based method for the characterization of six *C. perfringens* toxin genes, including the genes for iota-toxin *(iA)*, alpha-toxin *(cpa)*, enterotoxin E *(cpe)*, epsilon-toxin *(etxD)*, beta-1-toxin *(cpb1)*, and beta-2-toxin *(cpb2)* (Al-Khaldi et al., 2004). Hybridization of fluorescently labeled isolate DNA to complementary single-strand DNA attached to computer chips or microscope slides enables the simultaneous screening of large numbers of *C. perfringens* genes (Al-Khaldi et al., 2004). Using this technique, 16 out of 17 *C. perfringens* strains were correctly isotyped (Al-Khaldi et al., 2004). Although recent advances in molecular typing of strain isolates have improved the identification of many *C. perfringens* specimens for epidemiological purposes, none are infallible alone. The best scheme for discriminatory trace-back of suspected food-borne contamination from *C. perfringens* strains will continue to involve a combination of multiple methods, time permitting, to allow greater confidence and confirmation of resulting strain identifications.

CONCLUDING REMARKS

The incidence of *C. perfringens* food poisoning is quite common and costly. Although somewhat fastidious in growth characteristics with the use of synthetic laboratory media, the microorganism is very prolific when found in food products. Despite the pathogen's ubiquity in the natural environment, food-borne illnesses arise from the improper handling and preparation of foods. Complete eradication of the microorganism from foods is not possible, largely due to the ability to form highly resistant spores. Control measures in place take advantage of the microorganism's limitations of growth with respect to oxygen, a_w, pH, curing salts, organic acids, and natural inhibitors. Many predictive growth models have been developed to accurately estimate *C. perfringens* survival following various types of food processing scenarios. The best strategy to control *C. perfringens* appears to be a hurdle approach combined with careful handling of foods to avoid temperature abuse. Regulatory requirements for *C. perfringens* in foods in the United States follow the USDA/FSIS compliance guidelines. Discriminatory methods for evaluation and trace-back of *C. perfringens* food-borne illness have evolved over the years from culture-based methods to serological typing to more sophisticated and rapid molecular-based technologies. Awareness, preventive measures for control, and multiple hurdles appear to provide the greatest opportunities for success.

REFERENCES

Al-Khaldi, S. F., K. M. Myers, A. Rasooly, and V. Chizhikov. 2004. Genotyping of *Clostridium perfringens* toxins using multiple oligonucleotide microarray hybridization. *Mol. Cell. Probes* 18:359–367.

Amezquita, A., C. L. Weller, L. Wang, H. Thippareddi, and D. E. Burson. 2004. Development of an integrated model for heat transfer and dynamic growth of *Clostridium perfringens* during the cooling of cooked boneless ham. *Int. J. Food Microbiol.* 101:123–144.

Angulo, F. L., A. C. Voetsch, D. Vugia, J. L. Hadler, M. Farley, C. Hedberg, P. Cieslak, D. Morse, D. Dwyer, and D. L. Swerdlow. 1998. Determining the burden of human illness from foodborne diseases—CDC's emerging infectious disease program Foodborne Diseases Active Surveillance Network (FoodNet). *Vet. Clin. N. Amer. Food Anim. Pr.* 14:165–172.

Anonymous. 1995. Food poisoning—an overview. *Int. Poul. Prod.* 4:20–21.

Aran, N. 2001. The effect of calcium and sodium lactates on growth from spores of *Bacillus cereus* and *Clostridium perfringens* in a 'sous-vide' beef goulash under temperature abuse. *Int. J. Food Microbiol.* 63:117–123.

Araujo, P. M. August 2007. Culture medium for the detection of *Clostridium perfringens*. European Patent EP1816209 A2.

Araujo, M., R. A. Sueiro, M. J. Gomez, and M. J. Garrido. 2004. Enumeration of *Clostridium perfringens* spores in groundwater samples: comparison of six culture media. *J. Microbiol. Methods.* 57:175–180.

Baez, L. A., and V. K. Juneja. 1995. Nonradioactive colony hybridization assay for detection and enumeration of enterotoxigenic *Clostridium perfringens* in raw beef. *Appl. Environ. Microbiol.* 61:807–810.

Barnes, E. M., J. E. Despaul, and M. Ingram. 1963. The behavior of a food poisoning strain of *Clostridium welchii* in beef. *J. Appl. Bacteriol.* 26:415–427.

Bartholomew, B. A., M. F. Stringer, G. N. Watson, and R. J. Gilbert. 1985. Development and application of an enzyme linked immunosorbent assay for *Clostridium perfringens* type A enterotoxin. *J. Clin. Pathol.* 38:222–228.

Bauer, F. T., J. A. Carpenter, and J. O. Reagan. 1981. Prevalence of *Clostridium perfringens* in pork during processing. *J. Food Prot.* 44:279–283.

Baums, C. G., U. Schotte, G. Amtsberg, and R. Goethe. 2004. Diagnostic multiplex PCR for toxin genotyping of *Clostridium perfringens* isolates. *Vet. Microbiol.* 100:11–16.

Bean, N. H., and P. M. Griffin. 1990. Foodborne disease outbreaks in the United States, 1973–1987: pathogens, vehicles, and trends. *J. Food Prot.* 9:804–817.

Bean, N. H., J. S. Goulding, M. T. Daniels, and F. J. Angulo. 1997. Surveillance for foodborne disease outbreaks—United States, 1988–1992. *J. Food Prot.* 60:1265–1286.

Beh, K. J., and S. H. Buttery. 1978. Reverse phase passive haemagglutination and single radial immunodiffusion to detect epsilon antigen of *Clostridium perfringens* type D. *Aus. Vet. J.* 54:541–544.

Berry, P. R., J. C. Rodhouse, S. Hughes, B. A. Bartholomew, and R. J. Gilbert. 1988. Evaluation of ELISA, RPLA, and Vero cell assays for detecting *Clostridium perfringens* enterotoxin in faecal specimens. *J. Clin. Pathol.* 41:458–461.

Blankenship, L. C., S. E. Craven, R. G. Leffler, and C. Custer. 1988. Growth of *Clostridium perfringens* in cooked chili during cooling. *Appl. Environ. Microbiol.* 53:1104–1108.

Bradshaw, J., G. Stelma, V. Jones, J. Peeler, J. Wimsatt, J. Corwin, and R. Twedt. 1982. Thermal inactivation of *Clostridium perfringens* enterotoxin in buffer and in chicken gravy. *J. Food Sci.* 47:914–916.

Brett, M. M. 1998. Kits for the detection of some bacterial food poisoning toxins: problems, pitfalls, and benefits. *J. Appl. Microbiol.* 84:110S–118S.

Brown, K. L. 2000. Control of bacterial spores. *Bri. Med. Bull.* 56:158–171.

Bryan, F. L. 1969. What the sanitarian should know about *Clostridium perfringens* foodborne illness. *J. Milk Food Technol.* 32:381–389.

Bryan, F. L. 1988. Risks associated with vehicles of foodborne pathogens and toxins. *J. Food Prot.* 51:498–508.

Bryan, F. L., and T. W. McKinley. 1979. Hazard analysis and control of roast beef preparation in foodservice establishments. *J. Food Prot.* 42:4–18.

Brynestad, S., and P. E. Granum. 2002. *Clostridium perfringens* and foodborne infections. *Int. J. Food Microbiol.* 74:195–202.

Cardosi, M., S. Birch, J. Talbot, and A. Phillips. 2005. An electrochemical immunoassay for *Clostridium perfringens* phospholipase C. *Electroanalysis* 3:169–176.

CDC. 2000. Surveillance for foodborne-disease outbreaks—United States, 1993–1997. *MMWR Morb. Mortal. Wkly. Rep.* 49:1–51.

Chumney, R. K., and D. M. Adams. 1980. Relationship between the increased sensitivity of heat injured *Clostridium perfringens* spores to surface-active antibiotics and to sodium chloride and sodium nitrite. *J. Appl. Bacteriol.* 49:55–63.

Collie, R. E., and B. A. McClane. 1998. Evidence that the enterotoxin gene can be episomal in *Clostridium perfringens* isolates associated with non-food-borne human gastrointestinal diseases. *J. Clin. Microbiol.* 36:30–36.

Craven, S. E. 2001. Occurrence of *Clostridium perfringens* in the broiler chicken processing plant as determined by recovery in iron milk medium. *J. Food Prot.* 64:1956–1960.

Craven, S. E. 1980. Growth and sporulation of *Clostridium perfringens* in foods. *Food Technol.* 34:80–87, 95.

Davidson, P. M., and V. K. Juneja. 1990. Antimicrobial agents, p. 83. *In* A. L. Branen, P. M. Davidson, and S. Salminen (ed.), *Food Additives.* Marcel Dekker, Inc., New York, NY.

Deboer, E., W. Spiegelenberg, and F. Jenssen. 1985. Microbiology of spices and herbs. *Antonie van Leeuwenhoek* 51:435–438.

de Jong, A. E. I., G. P. Eijhusen, E. J. F. Brouwer-Post, M. Grand, T. Johansson, T. Karkkainen, J. Marugg, P. H. in't Veld, F. H. M. Warmerdam, G. Worner, A. Zicavo, F. M. Rombouts, and R. R. Beumer. 2003. Comparison of media for enumeration of *Clostridium perfringens* from foods. *J. Microbiol. Meth.* 54:359–366.

de Jong, J. 1989. Spoilage of an acid food product by *Clostridium perfringens*, *C. barati*, and *C. butyricum. Int. J. Food Microbiol.* 8:121–132.

Duncan, C. L., and D. H. Strong. 1969. Ileal loop fluid accumulation and production of diarrhea in rabbits by cell-free products of *Clostridium perfringens. J. Bacteriol.* 100:86–94.

Duncan, C. L., D. H. Strong, and M. Sebald. 1972. Sporulation and enterotoxin production by mutants of *Clostridium perfringens. J. Bacteriol.* 110:378–391.

El Idrissi, A. H. 1989. A 4-layer sandwich ELISA for detection of *Clostridium perfringens* epsilon toxin. FAO Corporate Document Repository. http://www.fao.org/Wairdocs/ILRI/x5489B/x5489b14.htm. Accessed 25 January 2008.

Engstrom, B. E., C. Fermer, A. Lindberg, E. Saarinen, V. Baverud, and A. Gunnarsson. 2003. Molecular typing of isolates of *Clostridium perfringens* from healthy and diseased poultry. *Vet. Microbiol.* 9:225–235.

Emswiler, B. S., C. J. Pierson, and A. W. Kotula. 1977. Comparative study of two methods for detection of *Clostridium perfringens* in ground beef. *Appl. Environ. Microbiol.* 33:735–737.

FDA. 2001. Limitation of growth of organisms of public health concern. 2001 Food Code. Part 3-5.

Gaze, J. E. R. Shaw, and J. Archer. 1998. *Identification and Prevention of Hazards Associated with Slow Cooling of Hams and Other Large Cooked Meats and Meat Products,* review no. 8. Campden and Chorleywood Food Research Association, Gloucestershire, United Kingdom.

Gibson, A. M., and T. A. Roberts. 1986. The effect of pH, sodium chloride, sodium nitrite, and storage temperature on the growth of *Clostridium perfringens* and fecal streptococci in laboratory media. *Int. J. Food Microbiol.* 3:195–210.

Goepfert, J. M., and H. K. Kim. 1975. Behavior of selected foodborne pathogens in raw ground beef. *J. Milk Food Technol.* 38:449–456.

Gough, B. J., and J. A. Alford. 1965. Effect of curing agents on the growth and survival of food-poisoning strains of *Clostridium perfringens. J. Food Sci.* 30:1025–1028.

Hall, H. E., and R. Angelotti. 1965. *Clostridium perfringens* in meat and meat products. *Appl. Microbiol.* 13:352–357.

Hallerbach, C. M., and N. N. Potter. 1981. Effects of nitrite and sorbate on bacterial populations in frankfurters and thuringer cervelat. *J. Food Prot.* 44:341–346.

Harmon, S. M., D. A. Kautter, and J. T. Peeler. 1971. Improved medium for enumeration of *Clostridium perfringens. Appl. Microbiol.* 22:688–692.

Heikinheimo, A., M. Lindstrom, and H. Korkeala. 2004. Enumeration and isolation of *cpe*-positive *Clostridium perfringens* spores from feces. *J. Clin. Microbiol.* 42:3992–3997.

Heredia, N. L., G. A. Garcia, R. Luevanos, R. G. Labbe, and J. S. Garcia-Alvarado. 1997. Elevation of the heat resistance of vegetative cells and spores of *Clostridium perfringens* type A by sublethal heat shock. *J. Food Prot.* 60:998–1000.

Hobbs, G., D. Cann, B. Wilson, and J. Shewan. 1965. The incidence of organisms of the genus *Clostridium* in vacuum-packed fish in the United Kingdom. *J. Appl. Bacteriol.* 28:265–270.

Hobbs, B. C. 1979. *Clostridium perfringens* gastroenteritis, p. 131–167. *In* H. Riemann and F. L. Bryan (ed.), *Food-Borne Infections and Intoxications*, 2nd ed. Academic Press, Inc., New York, NY.

Holley R. A., A. M. Lammerding, and F. Tittiger. 1988. Microbiological safety of traditional and starter-mediated processes for the manufacture of Italian dry sausage. *Int. J. Food Microbiol.* 7:49–62.

Hsieh, P., and R. Labbe. 2007. Influence of petone source on sporulation of *Clostridium perfringens* type A. *J. Food Protect.* 70:1730–1734.

Huang, L. 2003a. Growth kinetics of *Clostridium perfringens* in cooked beef. *J. Food Saf.* 23:91–105.

Huang, L. 2003b. Estimation of growth of *Clostridium perfringens* in cooked beef under fluctuating temperature conditions. *Food Microbiol.* 20:549–559.

Huang, L. 2003c. Dynamic computer simulation of *Clostridium perfringens* growth in cooked ground beef. *Int. J. Food Microbiol.* 87:217–227.

ICMSF. 1996. *Clostridium perfringens*, p. 112–115. *Microorganisms in Foods 5: Characteristics of Microbial Pathogens.* Blackie Academic and Professional, London, United Kingdom.

Juneja, V. K., H. Marks, and H. Thippareddi. 2008. Predictive model for growth of *Clostridium perfringens* during cooling of cooked uncured beef. *Food Microbiol.* 25:42–55.

Juneja, V. K., and M. Friedman. 2007. Carvacrol, cinnamaldehyde, oregano oil, and thymol inhibit *Clostridium perfringens* spore germination and outgrowth in ground turkey during chilling. *J. Food Prot.* 70: 218–222.

Juneja, V. K., M. L. Bari, Y. Inatsu, S. Kawamoto, and M. Friedman. 2007. Control of *Clostridium perfringens* by green tea leaf extracts during cooling of cooked ground beef, chicken, and pork. *J. Food Prot.* 70:1429–1433.

Juneja, V. K., H. Thippareddi, and M. Friedman. 2006a. Control of *Clostridium perfringens* in cooked ground beef by carvacrol, cinnamaldehyde, thymol or oregano oil during chilling. *J. Food Prot.* **69**:1546–1551.

Juneja, V. K., H. Thippareddi, M. L. Bari, Y. Inatsu, S. Kawamoto, and M. Friedman. 2006b. Chitosan protects cooked ground beef and turkey against *Clostridium perfringens* spores during chilling. *J. Food Sci.* **71**:M236–M240.

Juneja, V. K., L. Huang, and H. Thippareddi. 2006c. Predictive model for growth of *Clostridium perfringens* in cooked cured pork. *Int. J. Food Microbiol.* **110**:85–92.

Juneja, V. K. 2006. Delayed *Clostridium perfringens* growth from a spore inocula by sodium lactate in sous-vide chicken products. *Food Microbiol.* **23**:105–111.

Juneja, V. K., and H. Thippareddi. 2004a. Inhibitory effects of organic acid salts on growth of *Clostridium perfringens* from spore inocula during chilling of marinated ground turkey breast. *Int. J. Food Microbiol.* **93**:155–163.

Juneja, V. K., and H. Thippareddi. 2004b. Control of *Clostridium perfringens* in a model roast beef by salts of organic acids during chilling. *J. Food Saf.* **24**:95–108.

Juneja, V. K., J. S. Novak, L. Huang, and B. S. Eblen. 2003. Increased thermotolerance of *Clostridium perfringens* spores following sublethal heat shock. *Food Control* **14**:163–168.

Juneja, V. K., and H. M. Marks. 2002. Predictive model for growth of *Clostridium perfringens* during cooling of cooked cured chicken. *Food Microbiol.* **19**:313–327.

Juneja, V. K., J. S. Novak, B. S. Eblen, and B. A. McClane. 2001a. Heat resistance of *Clostridium perfringens* vegetative cells as affected by prior heat shock. *J. Food Saf.* **21**:127–139.

Juneja, V. K., J. S. Novak, H. M. Marks, and D. E. Gombas. 2001b. Growth of *Clostridium perfringens* from spore inocula in cooked cured beef: development of a predictive model. *Innov. Food Sci. Emer. Technol.* **2**:289–301.

Juneja, V. K., R. C. Whiting, H. M. Marks, and O. P. Snyder. 1999. Predictive model for growth of *Clostridium perfringens* at temperatures applicable to cooling of cooked meat. *Food Microbiol.* **16**:335–349.

Juneja, V. K., and B. S. Marmer. 1998. Thermal inactivation of *Clostridium perfringens* vegetative cells in ground beef and turkey as affected by sodium pyrophosphate. *Food Microbiol.* **15**:281–287.

Juneja, V. K., and B. S. Marmer. 1996. Growth of *Clostridium perfringens* from spore inocula in sous-vide turkey products. *Int. J. Food Microbiol.* **21**:115–123.

Juneja, V. K., B. S. Marmer, and J. E. Call. 1996. Influence of modified atmosphere packaging on growth of *Clostridium perfringens* in cooked turkey. *J. Food Saf.* **16**:141–150.

Juneja, V. K., and W. M. Majka. 1995. Outgrowth of *Clostridium perfringens* spores in cook-in-bag beef products. *J. Food Saf.* **15**:21–34.

Juneja, V. K., J. E. Call, B. S. Marmer, and A. J. Miller. 1994a. The effect of temperature abuse on *Clostridium perfringens* in cooked turkey stored under air and vacuum. *Food Microbiol.* **11**:187–193.

Juneja, V. K., B. S. Marmer, and A. J. Miller. 1994b. Growth and sporulation potential of *Clostridium perfringens* in aerobic and vacuum-packaged cooked beef. *J. Food Prot.* **57**:393–398.

Juneja, V. K., O. P. Snyder, and M. Cygnarowicz-Provost. 1994c. Influence of cooling rate on outgrowth of *Clostridium perfringens* spores and cooked ground beef. *J. Food Prot.* **57**:1063–1067.

Kalchayanand, N., C. P. Dunne, A. Sikes, and B. Ray. 2004. Germination induced and inactivation of *Clostridium* spores at medium-range hydrostatic pressure treatment. *Innov. Food. Sci. Emerg. Tech.* **5**:277–283.

Kalender, H., and H. B. Ertas. 2005. Isolation of *Clostridium perfringens* from chickens and detection of the alpha toxin gene by polymerase chain reaction (PCR). *Turk. J. Vet. Anim. Sci.* **29**:847–851.

Kalinowski, R. M., R. B. Tompkin, P. W. Bodnaruk, and W. P. Pruett. 2003. Impact of cooking, cooling, and subsequent refrigeration on the growth or survival of *Clostridium perfringens* in cooked meat and poultry products. *J. Food Prot.* **66**:1227–1232.

Kanakaraj, R., D. L. Harris, J. G. Songer, and B. Bosworth. 1998. Multiplex PCR assay for detection of *Clostridium perfringens* in feces and intestinal contents of pigs and in swine feed. *Vet. Microbiol.* **63**:29–38.

Kang C. K., M. Woodburn, A. Pagenkopf, and R. Cheney. 1969. Growth, sporulation, and germination of *Clostridium perfringens* in media of controlled water activity. *Appl. Microbiol.* **18**:798–805.

Keto-Timonen, R., A. Heikinheimo, E. Eerola, and H. Korkeala. 2006. Identification of *Clostridium* species and DNA fingerprinting of *Clostridium perfringens* by amplified fragment length polymorphism analysis. *J. Clin. Microbiol.* **44**:4057–4065.

Kim, S., R. G. Labbe, and S. Ryu. 2000. Inhibitory effects of collagen on the PCR for detection of *Clostridium perfringens*. *Appl. Environ. Microbiol.* **66**:1213–1215.

Kokai-Kun, J. F., J. G. Songer, J. R. Czeczulin, F. Chen, and B. A. McClane. 1994. Comparison of Western immunoblots and gene detection assays for identification of potentially enterotoxigenic isolates of *Clostridium perfringens*. *J. Clin. Microbiol.* **32**:2533–2539.

Labbe, R. J., and V. K. Juneja. 2006. *Clostridium perfringens* gastroenteritis. *In* H. Riemann and D. O. Cliver (ed.), *Foodborne Infections and Intoxications*. Academic Press, Inc., San Diego, CA.

Labbe, R. J., and V. K. Juneja. 2002. *Clostridium perfringens*, p. 192–126. *In* D. O. Cliver and H. Riemann (ed.), *Foodborne Diseases*. Academic Press, Inc., San Diego, CA.

Labbe, R. G. 1989. *Clostridium perfringens*, p. 192–234. *In* M. P. Doyle (ed.), *Foodborne Bacterial Pathogens*. Marcel Dekker, Inc., New York, NY.

Labbe, R. G., and C. L. Duncan. 1970. Growth from spores of *Clostridium perfringens* in the presence of sodium nitrite. *Appl. Microbiol.* **19**:353–359.

Ladiges, W. C., J. F. Foster, and W. M. Ganz. 1974. Incidence and viability of *Clostridium perfringens* in ground beef. *J. Milk Food Technol.* **37**:622–626.

Li, J., S. Sayeed, and B. A. McClane. 2007. Prevalence of enterotoxigenic *Clostridium perfringens* isolates in Pittsburgh (Pennsylvania) area soils and home kitchens. *Appl. Environ. Microbiol.* **73**:7218–7224.

Lin, Y. T., and R. Labbe. 2003. Enterotoxigenicity and genetic relatedness of *Clostridium perfringens* isolates from retail food. *Appl. Environ. Microbiol.* **69**:1642–1646.

Lindquist, S., and E. A. Craig. 1988. The heat-shock proteins. *Annu. Rev. Genet.* **22**: 631–677.

Maslanka, S. E., J. G. Kerr, G. Williams, J. M. Barbaree, L. A. Carson, J. M. Miller, and B. Swaminathan. 1999. Molecular subtyping of *Clostridium perfringens* by pulsed-field gel electrophoresis to facilitate food-borne-disease outbreak investigations. *J. Clin. Microbiol.* **37**:2209–2214.

Maurer, A. J. 1983. Reduced sodium usage in poultry muscle foods. *Food Technol.* **37**:60–65.

Mead, P. S., L. Slutsker, V. Dietz, L. F. McCaig, J. S. Bresee, C. Shapiro, P. M. Griffin, and R. V. Tauxe. 1999. Food-related illness and death in the United States. *Emerg. Infect. Dis.* **5**:607–625.

Miserez, R., J. Frey, C. Buogo, S. Capaul, A. Tontis, A. Burnens, and J. Nicolet. 1998. Detection of α- and ε-toxigenic *Clostridium perfringens* type D in sheep and goats using a DNA amplification technique (PCR). *Lett. Appl. Microbiol.* **26:**382–386.

Miwa, N., T. Masuda, A. Kwamura, K. Terai, and M. Akiyama. 2002. Survival and growth of enterotoxin-positive and enterotoxin-negative *Clostridium perfringens* in laboratory media. *Int. J. Food Microbiol.* **72:**233–238.

Naik, H., and C. Duncan. 1977. Rapid detection and quantitation of *Clostridium perfringens* enterotoxin by counterimmunoelectrophoresis. *Appl. Environ. Microbiol.* **34:**125–128.

Nauerby, B., K. Pedersen, and M. Madsen. 2003. Analysis by pulsed-field gel electrophoresis of the genetic diversity among *Clostridium perfringens* isolates from chickens. *Vet. Microbiol.* **94:**257–266.

Naylor, R. D., P. K. Martin, and R. T. Sharpe. 1987. Detection of *Clostridium perfringens* epsilon toxin by ELISA. *Res. Vet. Sci.* **42:**255–285.

Novak, J. S., and Y. T. C. Yuan. 2004. The fate of *Clostridium perfringens* spores exposed to ozone and/or mild heat pretreatment on beef surfaces followed by modified atmosphere packaging. *Food Microbiol.* **21:**667–673.

Oxoid Ltd. 2002. Improved detection of *Clostridium perfringens* in water samples. http://www.rapidmicrobiology.com/news/603h15.php. Accessed 25 January 2008.

Paine, C. M., and R. Cherniak. 1975. Composition of the capsular polysaccharides of *Clostridium perfringens* as a basis for their classification by chemotypes. *Can. J. Microbiol.* **21:**181–185.

Perigo, J. A., and T. A. Roberts. 1968. Inhibition of clostridia by nitrate. *J. Food. Technol.* **39:**91.

Petit, L., M. Gilbert, and M. R. Popoff. 1999. *Clostridium perfringens*: toxinotype and genotype. *Trends Microbiol.* **7:**104–110.

Phillips, D., D. Jordan, S. Morris, I. Jenson, and J. Sumner. 2008. A national survey of the microbiological quality of retail raw meats in Australia. *J. Food Prot.* **71:**1232–1236.

Rahmati, T., and R. Labbe. 2008. Levels and toxigenicity of Bacillus cereus and Clostridium perfringens from retail seafood. *J. Food Prot.* **71:**1178–1185.

Rhodehamel, J., and S. Harmon. 1998. *Clostridium perfringens*, p. 16.01–16.06. *In* R. W. Bennett (ed.), *FDA Bacteriological Analytical Manual*, 8th ed. AOAC International, Gaithersburg, MD.

Riha, W. E., and M. Solberg. 1975. *Clostridium perfringens* inhibition by sodium nitrite as a function of pH, inoculum size, and heat. *J. Food Sci.* **40:**439–442.

Roberts, T. A., and C. M. Derrick. 1978. The effect of curing salts on the growth of *Clostridium perfringens* (welchii) in a laboratory medium. *J. Food Technol.* **13:**349–356.

Rodriguez-Romo, L. A., N. L. Heredia, R. G. Labbe, and J. S. Garcia-Alvarado. 1998. Detection of enterotoxigenic *Clostridium perfringens* in spices used in Mexico by dot blotting using a DNA probe. *J. Food Prot.* **61:**201–204.

Roy, R. J., F. F. Busta, and D. R. Thompson. 1981. Thermal inactivation of *Clostridium perfringens* after growth at several constant and linearly rising temperatures. *J. Food Sci.* **46:**1586–1591.

Sabah, J. R., H. Thippareddi, J. L. Marsden, and D. Y. C. Fung. 2003. Use of organic acids for the control of *Clostridium perfringens* in cooked vacuum-packaged restructured roast beef during an alternative cooling procedure. *J. Food Prot.* **66:**1408–1412.

Sabah, J. R., V. K. Juneja, and D. Y. C. Fung. 2004. Effect of spices and organic acids on the growth of *Clostridium perfringens* during cooling of cooked ground beef. *J. Food Prot.* **67:**1840–1847.

Saito, M. 1990. Production of enterotoxin by *Clostridium perfringens* derived from humans, animals, foods, and the natural environment in Japan. *J. Food Prot.* **53:**115–118.

Sarker, M. R., R. P. Shivers, S. G. Sparks, V. K. Juneja, and B. A. McClane. 2000. Comparative experiments to examine the effects of heating on vegetative cells and spores of *Clostridium perfringens* isolates carrying plasmid genes versus chromosomal enterotoxin genes. *Appl. Environ. Microbiol.* **66:**3234–3240.

Sauter, E. A., J. D. Kemp, and B. E. Langlois. 1977. Effect of nitrite and erythorbate on recovery of *Clostridium perfringens* spores in cured pork. *J. Food Sci.* **42:**1678–1679.

Schalch, B., J. Bjorkroth, H. Eisgruber, H. Korkeala, and A. Stolle. 1997. Ribotyping for strain characterization of *Clostridium perfringens* isolates from food poisoning cases and outbreaks. *Appl. Environ. Microbiol.* **63:**3992–3994.

Sheng, S., and R. Cherniak. 1998. Structure of the capsular polysaccharide of *Clostridium perfringens* Hobbs 10 determined by NMR spectroscopy. *Carbohydr. Res.* **305:**65–72.

Sheridan, J. J., R. L. Buchanan, and T. J. Montville (ed.). 1996. *HACCP: an Integrated Approach to Assuring the Microbiological Safety of Meat and Poultry.* Food & Nutrition Press, Trumbull, CT.

Shigehisa, T., T. Nakagami, and S. Taji. 1985. Influence of heating and cooling rates on spore germination and growth of *Clostridium perfringens* in media and in roast beef. *Jpn. J. Vet. Sci.* **47:**259–267.

Shinya, L. T., M. R. Baccaro, and A. M. Moreno. 2006. Use of single-enzyme amplified fragment length polymorphism for typing *Clostridium perfringens* isolated from diarrheic piglets. *Braz. J. Microbiol.* **37:**385–389.

Shoemaker, S. P., and M. D. Pierson. 1976. "Phoenix phenomenon" in the growth of *Clostridium perfringens*. *Appl. Environ. Microbiol.* **32:**803–807.

Siragusa, G. R., M. D. Danyluk, K. L. Hiett, M. G. Wise, and S. E. Craven. 2006. Molecular subtyping of poultry-associated type A *Clostridium perfringens* isolates by repetitive-element PCR. *J. Clin. Microbiol.* **44:**1065–1073.

Smart, J. L., T. A. Roberts, M. F. Stringer, and N. Shah. 1979. The incidence and serotypes of *Clostridium perfringens* on beef, pork and lamb carcasses. *J. Appl. Bacteriol.* **46:**377–383.

Smith-Simpson, S., and D. W. Schaffner. 2005. Development of a model to predict growth of *Clostridium perfringens* in cooked beef during cooling. *J. Food Prot.* **68:**336–341.

Solberg, M., and B. Elkind. 1970. Effect of processing and storage conditions on the microflora of *Clostridium perfringens*-inoculated frankfurters. *J. Food Sci.* **35:**126–131.

Steele, F. M., and K. H. Wright. 2001. Cooling rate effect on outgrowth of *Clostridium perfringens* in cooked, ready-to-eat turkey breast roasts. *Poult. Sci.* **80:**813–816.

Sterne, M., and I. Batty. 1975. *Pathogenic Clostridia*, p. 79–122. Butterworths, London, England.

Stringer, M. F., P. C. B. Turnbull, and R. J. Gilbert. 1980. Application of serological typing to the investigation of outbreaks of *Clostridium perfringens* food poisoning, 1970–1978. *J. Hyg.* **84:**443–456.

Strong, D. H., E. F. Foster, and C. L. Duncan. 1970. Influence of water activity on the growth of *Clostridium perfringens*. *Appl. Microbiol.* **19:**980–987.

Strong, D. H., K. F. Weiss, and L. W. Higgins. 1966. Survival of *Clostridium perfringens* in starch pastes. *J. Am. Diet. Assoc.* **49:**191–195.

Taormina, P. J., G. W. Bartholomew, and W. J. Dorsa. 2003. Incidence of *Clostridium perfringens* in commercially produced cured raw meat product mixtures and behavior in cooked products during chilling and refrigerated storage. *J. Food Prot.* **66:**72–81.

Thippareddi, H., V. K. Juneja, R. K. Phebus, J. L. Marsden, and C. L. Kastner. 2003. Control of *Clostridium perfringens* germination and outgrowth by buffered sodium citrate during chilling of ground roast beef and injected pork. *J. Food Prot.* **66:**376–381.

Todd, E. C. D. 1989. Costs of acute bacterial foodborne disease in Canada and the United States. *Int. J. Food Microbiol.* **9:**313–326.

Traci, P. A., and C. L. Duncan. 1974. Cold shock lethality and injury in *Clostridium perfringens. Appl. Microbiol.* **28:**815–821.

Tschirdewahn, B., S. Notermans, K. Wernars, and F. Untermann. 1991. The presence of enterotoxigenic *Clostridium perfringens* strains in feces of various animals. *Int. J. Food Microbiol.* **14:**175–178.

Tuomi, S., M. E. Matthews, and E. H. Marth. 1974. Behavior of *Clostridium perfringens* in precooked chilled ground beef gravy during cooling, holding, and reheating. *J. Milk Food Technol.* **37:**494–498.

USDA/FSIS. 2001. Performance standards for the production of certain meat and poultry products, final rule. *Fed. Regist.* **64:**732–749.

Uzal, F. A., W. R. Kelly, R. Thomas, M. Hornitzky, and F. Galea. 2003. Comparison of four techniques for the detection of *Clostridium perfringens* type D epsilon toxin in intestinal contents and other body fluids of sheep and goats. *J. Vet. Diagn. Invest.* **15:**94–99.

Van Damme-Jongsten, M., M. K. Wernars, and S. Notermans. 1989. Cloning and sequencing of the *Clostridium perfringens* enterotoxin gene. *Antonie van Leeuwenhoek* **56:**181–190.

Vareltzis, K., E. M. Buck, and R. G. Labbe. 1984. Effectiveness of a Betalains/Potassium Sorbate system versus sodium nitrite for color development and control of total aerobes, *Clostridium perfringens* and *Clostridium sporogenes* in chicken frankfurters. *J. Food Prot.* **47:**532–536.

Vos, P., R. Hogers, M. Bleeker, M. Reijans, T. van de Lee, M. Hornes, A. Frijters, J. Pot, J. Peleman, M. Kulper, and M. Zabeau. 1995. AFLP: a new technique for DNA fingerprinting. *Nucleic Acid Res.* **23:**4407–4414.

Waldroup, A. L. 1996. Contamination of raw poultry with pathogens. *World Poult. Sci. J.* **52:**7–25.

Walker, H. W. 1975. Foodborne illness from *Clostridium perfringens. CRC Crit. Rev. Food Sci. Nutr.* **7:**71–104.

Weddell, W., and R. W. Worthington. 1984. An enzyme labeled immunosorbent assay for measuring *Clostridium perfringens* epsilon toxin gut contents. *N. Z. Vet. J.* **33:**36–37.

Wen, Q., and B. McClane. 2004. Detection of enterotoxigenic *Clostridium perfringens* type A isolates in American retail foods. *Appl. Environ. Microbiol.* **70:**2685–2691.

Willardsen, R. R., F. F. Busta, and C. E. Allen. 1979. Growth of *Clostridium perfringens* in three different beef media and fluid thioglycollate medium at static and constantly rising temperatures. *J. Food Prot.* **42:**144–148.

Willardsen, R. R., F. F. Busta, C. E. Allen, and L. B. Smith. 1978. Growth and survival of *Clostridium perfringens* during constantly rising temperatures. *J. Food Sci.* **43:**470–475.

Wise, M., and G. R. Siragusa. 2005. Semi-quantitative detection of *Clostridium perfringens* in the broiler fowl gastrointestinal tract by real-time PCR. *Appl. Environ. Microbiol.* **71:**3911–3916.

Xylouri, E., C. Papadopoulou, G. Antoniadis, and E. Stoforos. 1997. Rapid identification of *Clostridium perfringens* in animal feedstuffs. *Anaerobe* **3:**191–193.

Zaika, L. L. 2003. Influence of NaCl content and cooling rate on outgrowth of *Clostridium perfringens* spores in cooked ham and beef. *J. Food Prot.* **66:**1599–1603.

Pathogens and Toxins in Foods: Challenges and Interventions
Edited by V. K. Juneja and J. N. Sofos
© 2010 ASM Press, Washington, DC

Chapter 5

Diarrheagenic *Escherichia coli*

CATHERINE S. BEAUCHAMP AND JOHN N. SOFOS

Existing estimates attribute 76 million domestic illnesses and 5,000 deaths to food-borne diseases each year (Mead et al., 1999). While this information may be outdated, the incidence of documented food-borne diarrheagenic *Escherichia coli* infections has remained fairly constant over the past 30 years, even though surveillance programs have been improved and include more extensive geographical areas; there has, however, been a shift in the vehicles of pathogen transmission (CDC, 2006; Rangel et al., 2005). Traditionally, outbreaks of food-borne diarrheagenic *E. coli* infections were primarily associated with raw, undercooked, and ready-to-eat beef products (Bell et al., 1994; CDC, 1995a, 1995b, 2000; Griffin and Tauxe, 1991; Riley et al., 1983; Tilden et al., 1996; Tuttle et al., 1999). While potentially contaminated beef products continue to be recalled, other types of food products have also been implicated in outbreaks, including unpasteurized (raw) milk, butter and soft cheeses made from raw milk, cheese curds, yogurt and ice cream, unpasteurized apple juice and cider, melons, grapes, radish and alfalfa sprouts, carrots, coleslaw, and bagged lettuce and spinach (Besser et al., 1993; CDC, 1997; Karmali, 1989; Morgan et al., 1993; NIIDIDCD/MHWJ, 1997; Rangel et al., 2005; http://www.fda.gov/ola/2006/foodsafety1115.html; http://www.cfsan.fda.gov/~comm/ift3-4a.html). The ability of diarrheagenic *E. coli* to make its way into and persist in food products, in conjunction with an increased incidence and severity of diarrheagenic *E. coli* infections in the young, the elderly, and immunocompromised individuals, necessitates the routine assessment and validation of established and novel pathogen mitigation strategies throughout the food production chain. This chapter presents an overview of diarrheagenic *E. coli*, its incidence in the environment and in food, factors associated with its association, survival and growth in food products, and recent advances in control measures used in food processing, as well as discriminative detection methods for confirmation of these pathogens and trace-back of contaminated food products.

THE ORGANISM

Characteristics

Escherichia coli is a member of the *Enterobacteriaceae* family, along with other enteric gram-negative bacteria, such as *Salmonella* and *Shigella*, and is known for its ability to adapt and colonize a diverse array of reservoirs, including open-air environments and the gastrointestinal (GI) tracts of mammals and birds (Falkow, 1996; Souza et al., 2002). Such flexibility is permitted through facultative respiration, during which citrate, NO_2, and NO_3 take the place of oxygen in the electron transport chain (Stewart, 1988). All compounds required for the metabolic function of *E. coli* can be synthesized from glucose (Sussman, 1997). *E. coli* bacteria are mesophilic organisms which replicate at temperatures of 7 to 45°C. Under optimal temperature conditions (35 to 42°C and 95 to 104°F) the pathogen can replicate at pH values of 4 to 10 and in the presence of up to 8% sodium chloride. A minimum water activity of 0.95 is required for growth (ICMSF, 1980, 1996). It is important to note, however, that the manipulation of one or more growth factors can influence the minimum or maximum values of other factors involved in microbial survival and/or growth (Bacon and Sofos, 2003).

While most strains of *E. coli* are not human pathogens, certain serotypes are responsible for three types of human illness: (i) neonatal meningitis, (ii) chronic urinary tract infections, and (iii) gastroenteritis (Buchanan and Doyle, 1997; Johnson et al., 2002). The six classes of diarrheagenic *E. coli*

Catherine S. Beauchamp • Colorado Premium Foods, 2035 2nd Ave., Greeley, CO 80631. John N. Sofos • Center for Meat Safety & Quality, Food Safety Cluster of Infectious Diseases Supercluster, Department of Animal Sciences, Colorado State University, 1171 Campus Delivery, Fort Collins, CO 80523-1171.

associated with human gasteroenteritis are entero-pathogenic *E. coli* (EPEC), enteroaggregative *E. coli* (EAEC or EAggEC), enteroinvasive *E. coli* (EIEC), enterotoxigenic *E. coli* (ETEC), enterohemorrhagic *E. coli* (EHEC), and diffuse adherent *E. coli* (DAEC) (Begue et al., 1998; Bacon and Sofos, 2003; Buchanan and Doyle, 1997; Guion et al., 2008; Nataro and Kaper, 1998).

Diarrheagenic Illness

The characteristics and symptoms associated with each class of diarrheagenic *E. coli* infection are found in Table 1. Common symptoms associated with human diarrheagenic *E. coli* infections (Table 1) generally include loose, watery stools (diarrhea) and mild to severe abdominal pain and/or cramping. Other symptoms may include mild fever, headache, nausea, or vomiting. Illness is uncommon in healthy adults, and symptoms generally approximate a mild bout with the flu. The incidence of certain classes of diarrheagenic *E. coli* infection is elevated among healthy adults traveling to foreign destinations, specifically developing countries (http://www.cfsan.fda.gov/~ebam/bam-4a.html). Some individuals may be genetically predisposed to infection due to their blood type, and susceptibility may also be heightened when using oral contraceptives or antibiotics (Begue et al., 1998; Berger et al., 1989; Besser et al., 1999; Wittels and Lichtman, 1986).

Persons with immature or impaired immune systems are more prone to both mild and severe illnesses and are also more likely to sustain secondary infections and/or other health complications (Table 1) (Besser et al., 1999). Symptoms associated with severe infections of certain serotypes include bloody diarrhea (hemorrhagic colitis), as well as renal (kidney) malfunction and failure, thrombocytopenia (inadequate platelet count), microangiopathic hemolytic anemia (lysis of red blood cells), or hemolytic uremic syndrome (HUS), in which patients exhibit all of these symptoms (Bacon and Sofos, 2003; Begue et al., 1998). Traditional HUS is most commonly observed in children less than 5 years of age, while adults are more likely to develop an atypical form of HUS characterized by a lack of prodromal diarrhea (D-HUS) (McCrae and Cines, 2000). D-HUS is commonly misdiagnosed as thrombotic thrombocytopenic purpura, a strikingly similar type of thrombotic microangiopathy, also characterized by extensive and irreversible neurological damage (Boyce et al., 1995). Seizure, stroke, herniated bowels, and/or chronic renal malfunction may also accompany severe infections (Buchanan and Doyle, 1997). Prolonged or chronic cases of diarrhea may result in severe dehydration and/or malnutrition due to decreased nutrient uptake by damaged intestinal lining (http://www.cfsan.fda.gov/~ebam/bam-4a.html).

Mechanisms of Virulence

E. coli bacteria are particularly promiscuous microorganisms and continually exchange genetic material with other related and unrelated species of

Table 1. Most common mode of transmission, host, symptoms, and characteristics of illness associated with the different classes of diarrheagenic *Escherichia coli*[a]

Class	Classic host	Symptoms	Incubation; duration (days)	Acute or chronic presentation	Infectious dose
EPEC	Infants (<6 mo); more prevalent in developing countries	Severe diarrhea, fever, vomiting	Variable	Chronic diarrhea, malnutrition	High; low in infants
EAEC	Children; more prevalent in developing countries	Watery or bloody diarrhea, fever	Variable	Chronic watery diarrhea, severe dehydration	High
EIEC	Children; more prevalent in developing countries	Watery diarrhea, abdominal cramping, fever	1–3; self-limiting	Dysentery syndrome	Low
ETEC	Travelers and infants native to developing countries	Watery diarrhea, abdominal cramping, mild fever, nausea	1–3; 3–7	Cholera-like	High
EHEC	Children, elderly	Diarrhea, bloody diarrhea, abdominal pain, vomiting	1–8; 4–10	Bloody diarrhea (hemorrhagic colitis), HUS, kidney failure	Low

[a]Data derived from Bacon and Sofos (2003), Begue et al. (1998), Nataro and Kaper (1998), Nguyen et al. (2004, 2005), http://www.cfsan.fda.gov/~mow/chap14.html, http://www.cfsan.fda.gov/~mow/chap16.html, and http://www.cidrap.umn.edu/cidrap/content/fs/food-disease/causes/ecolioview.html.

Table 2. Mechanisms of pathogenicity which differentiate the classes of diarrheagenic *Escherichia coli*[a]

Class	Adhesion site	Adhesion mediator	Invasion potential	Toxins	Other virulence factors
EPEC	Small intestine	Intimin	Moderate	Possible enterotoxin (EAST1)	EAF plasmid, LEE island, flagellin, CDT
EAEC	Small and large intestine	AAF	None	EAST1, Pet, Pic	Flagellin
EIEC	Large intestine (colon)	Unclear	High	Enterotoxin	Cell-to-cell spread (IcsA), serine-protease (SepA)
ETEC	Small intestine	Fimbrial CFs	None	LT, ST	CDT
EHEC	Large intestine (colon)	Intimin	Moderate	Stx, enterohemolysin, α-hemolysin, EAST1	LEE island, pO157, flagellin, CDT, CNF

[a]AAF, set of three fimbrial adhesions expressed by EAEC and required for formation of distinctive stacked-brick AA to host cells (Nataro et al., 1994); CDT, cytolethal distending toxin; CNF, cytotoxic necrotizing factor; EAST1, enteroaggregative ST toxin; Stx, Shiga toxins, Shiga-like toxins, or verotoxins. Data derived from Aragon et al. (1997), Bacon and Sofos (2003), Begue et al. (1998), Beutin et al. (1994), Berin et al. (2002), Blanco et al. (1992), Friedrich et al. (2006), Hicks et al. (1996), Kaper et al. (2004), Nataro and Kaper (1998), O'Brien and Holmes (1996), Robinson et al. (2006), Savarino et al. (1996), Schmidt et al. (1996), Torres et al. (2005), Tzipori et al. (1988), Zychlinsky et al. (1994), http://www.cfsan.fda.gov/~ebam/bam-4a.html, and http://www.cidrap.umn.edu/cidrap/content/fs/food-disease/causes/ecolioview.html.

bacteria. Correspondingly, the virulence genes associated with each class of diarrheagenic *E. coli* (Table 2) were acquired through extensive lateral transfers of genetic material (Souza et al., 2002). Colonization of host GI tissue is rather methodical, and evasion of host defenses, colonization, replication, and damage to the host are observed with all diarrheagenic *E. coli* infections (Nataro and Kaper, 1998). However, the sites of pathogen adhesion to host GI tissue and mechanism of attachment to host cells, as well as the organization of the actual adhesive structure, differentiate each of the six classes (Table 2).

Colonization of Host Cells

The formation of attaching and effacing (A/E) lesions on host intestinal cells is characteristic of EPEC and EHEC (Torres et al., 2005). The factors involved in the formation of A/E lesions are genetically encoded on the locus of enterocyte effacement (LEE), also known as the pathogenic LEE island. EPEC and EHEC LEE islands are similar, although not identical in their genetic homology or subsequent host interaction (Blattner et al., 1997; DeVinney et al., 2001; Elliott et al., 1998, 1999; Frankel et al., 1998; Nataro and Kaper, 1998; Torres et al., 2005; www.blackwell-synergy.com/doi/abs/10.1111/j.1462-5822.2006.00794.x). The type III protein secretion system is a critical component of the LEE island and encodes a number of factors involved in the formation of A/E lesions, including the type III translocon (comprised of EspA, EspB, and EspD) and Tir (or EspE). The type III translocon locates target host cells and inserts Tir into the plasma membrane. Tir then functions as a protein receptor, or "beacon," for the adhesin protein, intimin (EaeA) (DeVinney et al., 1999; Hartland et al., 1999; Isberg

et al., 1987; Kenny et al., 1997; Taylor et al., 1998). The coupling of Tir and intimin initiates the degradation of host epithelial cell microvilli, which provides a surface for bacterial effacement, the reorganization of the host cell cytoskeletal, and pedestal formation (DeVinney et al., 1999; Knutton et al., 1998; Moon et al., 1983; Singleton, 2004). Host colonization by Tir⁻ cells is unlikely, although some forms of intimin can attach to host cells in the absence of Tir (Hartland et al., 1999; Kenny et al., 1997). Once the lesion is formed and the pedestal protrudes into the lumen, other Type III proteins and toxins enter the cell, initiating signal transduction pathways and activating protein kinase C, tumor necrosis factor, and interleukin-8 (Moon et al., 1983; Knutton et al., 1998; Singleton, 2004). In humans, tumor necrosis factor induces increased (10- to 100-fold) expression of Shiga toxin receptors on the surface of vascular endothelial cells (van de Kar et al., 1992). Interleukin-8 is involved in the translocation of host neutrophils (Zhou et al., 2003; http://www.cidrap.umn.edu/cidrap/content/fs/food-disease/causes/ecolioview.html). The resulting migration of neutrophils across the epithelium damages host mucosal lining and results in fluid secretion (Zhou et al., 2003; http://www.cidrap.umn.edu/cidrap/content/fs/food-disease/causes/ecolioview.html).

The attachment of EPEC, ETEC, and EAEC to host cells is mediated by long, thin filaments called adhesive fimbriae (Table 2) (Collinson et al., 1992). EPEC expresses bundle-forming fimbriae (Bfp), which are encoded on the EPEC adherence factor plasmid (Girón et al., 1991; Nataro and Kaper, 1998), while ETEC expresses fimbrial colonization factors (Torres et al., 2005). Aggregative adhesion (AA) fimbriae (AAFs) I, II, and III are involved in the binding of most EAEC bacteria to host cells, although afimbrial modes of attachment have been suggested (Garcia et al., 1996, 2000; Torres et al., 2005). Following fimbrial/afimbrial attachment, EAEC forms thick, dense biofilms at the intestinal epithelium and begins diffusing cytotoxins into host epithelial cells; in vivo biofilms exhibit an increased resistance to host defense systems and antibiotics (Okeke and Nataro, 2001; http://www.cidrap.umn.edu/cidrap/content/fs/food-disease/causes/ecolioview.html).

The mechanism by which EIEC adheres to host cells is not well defined. Once associated with host cells, the EIEC Ipa complex is inserted into and forms pores in the plasma membrane, allowing effector proteins to enter the cell and trigger signal transduction pathways involved in the symptoms of human infection (Small and Falkow, 1988; Sansonetti et al., 1999; Torres et al., 2005). While initial host cell invasion occurs through the lumen of the intestine, EIEC also expresses another factor (IcsA), which facilitates the spread of the pathogen from cell to cell (Nataro and Kaper, 1998).

Toxins

The toxins involved in diarrheagenic *E. coli* infections are also unique to each class (Table 2). While many types of *E. coli* toxins exist, Shiga toxins (also known as Shiga-like toxins or verotoxins) are similar to those produced by *Shigella dysenteriae* (Sandvig, 2001). These toxins are the primary virulence factor of Shiga toxin-producing *E. coli* (STEC), including *E. coli* O157:NM and O157:H7, and the principal cause of hemorrhagic colitis and HUS in humans (Dean-Nystrom et al., 2003; O'Brien and Holmes, 1987). Upon interaction with host cell receptors (Gb_3 and Gb_4), Shiga toxins interrupt protein synthesis and trigger cell death (Sandvig, 2001). In the early stages of childhood development, renal cells express a greater number of toxin receptors, which may help explain their predisposal to STEC/EHEC infections, hemorrhagic colitis, and HUS (Lingwood, 1994). Currently available data indicate that strains that produce only Stx2 are more commonly associated with HUS infections than strains that produce both Stx1 and Stx2, or only Stx1 (Louise and Obrig, 1995). This trend may be due, in part, to the cumulative effects of (i) the higher level of Stx2 production in species which only produce the one toxin; (ii) a higher affinity of human host cells for Stx2 compared to Stx1; (iii) a greater cytotoxic activity of Stx2 on human target cells; and (iv) multiple Stx2 toxin variants which induce mild to severe diarrheic symptoms (Beutin et al., 1989; Friedrich et al., 2002; Louise and Obrig, 1995; Obrig et al., 1988; Schmidt and Karch, 1996).

Hemolysins lyse red blood cells and erythrocytes, impede cytokine release, and interfere with neutrophile function (Bauer and Welsh, 1996; König et al., 1994; Russo et al., 2005). Enterohemolysins can be distinguished from α- and β-hemolysins by biochemical characteristics, extracellular location, and specificities between different species of hosts (O'Brien and Holmes, 1996). While it is not an obligatory factor in the development of HUS, α-hemolysin is involved in the majority of HUS infections. A subsequent increase in systemic iron, a metabolic requirement of *E. coli*, due to the hemolytic activity of the toxin, may help explain this phenomenon (Schmidt and Karch, 1996). Enterotoxins include heat-labile (LT) and heat-stable (ST) derivatives and the ST EAEC enterotoxin (EAST1). LT toxin and the cholera toxin are similar in genotype, structure, function, and action upon host cells (Dallas and Falkow, 1980), and signal transduction pathways initiated by LT toxin eventually trigger the onset of osmotic diarrhea. There are two forms of ST toxin: STa and STb. While both induce diarrheic symptoms, STa (or STI) appears to antagonize the intestinal regulation of sodium chloride and STb (or STII) damages the intestinal mucosa and interferes with fluid absorption (Nataro and Kaper, 1998).

Two serine protease inhibitors, Pet and Pic, also appear to exhibit toxigenic activity during EAEC infections (Table 2) (Bellini et al., 2005). Other significant yet less common toxins associated with diarrheagenic *E. coli* include cytolethal distending toxin and cytotoxic necrotizing factor (Table 2) (Aragon et al., 1997; Marques et al., 2003).

Other Virulence Factors

All *E. coli* O157:NM and O157:H7 strains possess a distinguishing virulence plasmid, pO157, which encodes genes for adherence factors, toxins, a catalase-peroxidase, and other factors of unknown importance (Brunder et al., 1996; Nataro and Kaper, 1998). The *toxB* gene, which may indirectly influence pathogenicity, can also be found on the plasmid (Tatsuno et al., 2001). The presence of the pO157 plasmid may be correlated to the development of HUS in STEC patients (Schmidt and Karch, 1996).

Motility factors, defined by the presence of flagellar (H) antigens or the *fliC* gene, may also be involved in pathogenicity. For example, researchers have described a possible relationship between flagellin proteins (Table 2) and the degree of host intestinal mucosa inflammation (Berin et al., 2002; Steiner et al., 2000; Zhou et al., 2003). Flagellin aggravates the release of interleukin-8, which is involved in the translocation of neutrophils across the host epithelium (Zhou et al.,

2003; http://www.cidrap.umn.edu/cidrap/content/fs/food-disease/causes/ecolioview.html).

Classification

Specific characteristics, symptoms of illness, and virulence mechanisms associated with each class of diarrheagenic *E. coli* are found in Tables 1 and 2. Relevant information pertaining to DAEC was insufficient for its inclusion in the aforementioned tables. Although human illnesses have been reported (Girón et al., 1991; Gunzberg et al., 1993; Levine et al., 1993), little is known about the characteristics, route of transmission, or clinical presentation of DAEC infections. DAEC appears to initiate host colonization via fimbrial attachment mechanisms (Bilge et al., 1989, 1993) and also induces finger-like protrusions from host HEp-2 cells, which engulf the pathogen and serve as a shield against antimicrobial treatments (Cookson and Nataro, 1996; Yamamoto et al., 1994). A classic human host is not well defined, although DAEC infections are observed generally in young children, but not in infants (<1 yr) (Gunzberg et al., 1993; Scaletsky et al., 2002; Spano et al., 2008). While not commonly recovered from clinical cases in most countries, DAEC is the primary class of diarrheagenic *E. coli* associated with acute diarrhea in Brazilian children (Spano et al., 2008).

While all classes of diarrheagenic *E. coli* can cause serious illness in humans, STEC and specifically *E. coli* O157:H7 cause the most severe types of infection, including hemorrhagic colitis, HUS, and D-HUS. According to older estimates, STEC causes approximately 110,000 domestic food-borne illnesses each year, 73,000 of which are caused by *E. coli* O157:H7 (Mead et al., 1999). Persons with O157 STEC infections are also more likely to experience bloody diarrhea and develop HUS than non-O157 STEC-infected individuals (CDC, 2007). As with DAEC, regional disparities are also observed for different STEC serotypes recovered from clinical cases; while O157:NM and O157:H7 are most commonly recovered from U.S. patients, non-O157 STEC infections are more common in Australia, Canada, South America, and Europe (Blanco et al., 2004; Boerlin et al., 1999; Goldwater and Bettelheim, 1995; Lopez et al., 1998). Predominant non-O157 STEC serotypes associated with food-borne human infections include O26, O103, and O111 (Table 3).

SOURCES OF DIARRHEAGENIC *E. COLI*

Environment

E. coli organisms are usually nonpathogenic and account for the majority of facultative flora found in the GI tracts of most vertebrates, including humans (Falkow, 1996; Nataro and Kaper, 1998); colonized hosts then shed these organisms into the environment. A review of existing information indicates that diarrheagenic *E. coli* bacteria have been recovered from a plethora of environmental locations that include, but are not limited to, ranches and livestock feeding operations, livestock harvesting facilities, natural and human-made bodies of water, municipal water sources, compost, urban and rural soils and landscapes, sewage treatment effluent, and agricultural fairs, as well as from hospitals and day care centers (Barham et al., 2002; Belongia et al., 2003; Callaway et al., 2004; CDC, 2001a; Childs et al., 2006; Keene et al., 1994; Keen et al., 2003; Mulvey et al., 2005; Sofos et al., 1999a, 1999b; Swerdlow et al., 1992; WHO, 1998; http://www.oznet.k-state.edu/library/fntr2/mf2269.pdf). Animals shown to shed the organism include beef and dairy cattle, sheep, swine, horses, rodents, dogs, deer, and wild and domesticated birds (Buchanan and Doyle, 1997; Keene et al., 1997; Kudva et al., 1996; Moss and Frost, 1984; Rice et al., 1995; Travena et al., 1996; Vernozy-Rosand et al., 2002; Wallace et al., 1997; http://www.fda.gov/ola/2006/foodsafety1115.html; http://www.oznet.k-state.edu/library/fntr2/mf2269.pdf).

Foods

The ubiquitous nature of diarrheagenic *E. coli* organisms facilitates their ease of entry into the food

Table 3. Virulence factors and phenotypic indicators associated with common Shiga toxin-producing *E. coli* (STEC) serotypes[a]

O group	Flagellar antigens	Hyl	Intimin	Sor	LDC	Suc	Rha	Raf	Dul	Ara
O22	HNM, H11	+	+	+	+	+	−	+	−	+
O103	HNM, H2, H7, H21, H25	+	+	+	+	+		−	−	
O111	HNM, H2, H7, H8, H11	+/−	+	+	−	+/−	+	+	+	+
O157	HNM, H7	+	+	−	+	+	+	+	+	+

[a]Hyl, enterohemolysin; Sor, sorbitol; LDC, lysine decarboxylase; Suc, sucrose; Rha, rhamnose; Raf, raffinose; Dul, dulcitol; Ara, arabinose. Data derived from Bielaszewska and Karch (2000), Guth et al. (2002), Hiramatsu et al. (2002), Hussein and Bollinger (2005), Irino et al. (2007), Possé et al. (2008), Vaz et al. (2004), and http://nzfsa.govt.nz/science/data-sheets/non-o157-stec.pdf.

supply. Animal feces are the source of contamination for animal hides, water, soil, and inanimate objects, leading to transfer of contamination to raw food ingredients, processing water, equipment, and workers. During harvest processes, microorganisms from the gastrointestinal tract and feces of animals may contaminate meat and poultry carcasses, as well as the processing environment, which may then become a secondary source of contamination or cross-contamination (Scanga, 2005). Raw produce may become contaminated through intimate contact with contaminated water and/or soil and animal manure during cultivation or harvest, leading to contamination of associated products, such as juices, salads, and coleslaw (Lee et al., 2004). In addition, the innate characteristics of produce of neutral or slightly acidic pH (i.e., high water activity and high nutrient content) may support survival and growth of diarrheagenic *E. coli* (Lee et al., 2004), especially acid-tolerant strains in acidic products such as fruit juices. Shelled and chopped nuts may also become contaminated in a manner similar to produce, and such contamination may then be transferred to the nutmeat during shell removal (Meyer and Vaughn, 1969). Other food products derived from cattle, including raw milk, have also been associated with food-borne diarrheagenic *E. coli* infections (Karmali, 1989; Quinto and Cepeda, 1997). In reality, however, nearly all food products (raw and processed) are subject to cross-contamination, which may occur at nearly all processing or handling steps leading up to consumption.

INTRINSIC AND EXTRINSIC FACTORS INVOLVED IN OUTBREAKS AND RECALLS

E. coli can withstand nutrient starvation, as well as exposure to low pH (≤ 4.5) and/or thermal and osmotic stresses, and has been shown to persist in both predictable and unusual locations for remarkable lengths of time (LeJeune et al, 2004; Nychas et al., 2007; Stopforth et al., 2002, 2003a, 2003b). Thus, concerns regarding the development and maintenance of biofilms within the food industry are justified, and increasing evidence of and repercussions associated with such niches are accumulating. The likelihood of their presence and the lack of a clear understanding of the fate of biofilms when exposed to common intervention technologies, in conjunction with the apparent resistance of cells in biofilm to various control measures, have undoubtedly contributed to the persistence of environmental and food-borne contamination and subsequent outbreaks of food-borne illness.

While early prevalence data were based on sub-optimal detection methodologies and tended to overlook season and geographical information, such details are generally present in more recent publications. As mesophiles, diarrheagenic *E. coli* logically should flourish during warmer months and be of less concern during colder seasons (http://www.fsis.usda.gov/OPPDE/rdad/FSISNotices/63-06.pdf); however, not all prevalence studies have observed such trends (Alam and Zurek, 2004). Nevertheless, much of the available data indicate a higher prevalence during summer and/or fall months, with peaks in prevalence being determined by regional temperature and climate (Barkocy-Gallagher et al., 2003; Childs et al., 2006; Michel et al., 1999). While risk of exposure to diarrheagenic *E. coli* is determined by numerous factors and can vary at any given time, more food-borne diarrheagenic *E. coli* outbreaks are reported to the Centers for Disease Control and Prevention (CDC) generally during the summer and fall months (Hedberg et al., 1997; Rangel et al., 2005). Higher levels of environmental contamination, prevalence in contaminated food products, and frequency of catered/picnic/outdoor grilling events may contribute to this increase. The considerable numbers of persons being fed during such events may also be a contributing factor to the apparent upsurge in outbreaks during warmer months. In addition, according to Allos et al. (2004), modernization within the food industry has led to increased availability and decreased cost of countless foodstuffs. The nationwide distribution of contaminated products may lead to an upsurge in food-borne illnesses that emerge as an outbreak or appear to be sporadic (Killalea et al., 1996).

FOOD PROCESSING CONDITIONS ASSOCIATED WITH OUTBREAKS AND RECALLS

Although outbreaks do not account for all or even the majority of food-borne diarrheagenic *E. coli* infections, the food processing factors implicated in the development of an outbreak may also lead to isolated cases of illness (Rangel et al., 2005). The five most significant "foodborne illness risk factors" include (i) the acquisition of products from unsafe sources, (ii) poor personal hygiene, (iii) contaminated processing equipment, (iv) inadequate heat treatments, and (v) improper holding temperatures (http://www.cfsan.fda.gov/~dms/retrsk2.html). According to the CDC, 46% of the food-borne outbreaks from 1998 to 2002 involved at least one of the aforementioned risk factors (CDC, 2006). Since 1982, multiple types of outbreaks have been attributed to diarrheagenic

E. coli, including home-level, person-to-person, animal-to-person, laboratory-related, multistate, and water-borne outbreaks, as well as those with unknown routes of transmission (Rangel et al., 2005). For all intents and purposes, only those outbreaks associated with contamination introduced at or before the commercial food processing level are discussed in detail.

Meat Products

In 1982, *E. coli* O157:H7 was first recognized and classified as a human pathogen after being linked to illnesses stemming from consumption of improperly cooked hamburgers served at fast-food restaurants in Oregon and Michigan, United States (Riley et al., 1983). Another landmark outbreak was documented 10 years later, during a 4-month span of time in 1992 and 1993, when improperly cooked hamburgers contaminated with *E. coli* O157:H7 were again linked to 501 illnesses, 45 cases of HUS, and three deaths (Bell et al., 1994). From 1995 to 2004, ground and mechanically tenderized beef products were associated with multiple outbreaks of *E. coli* O157:H7 infections (CDC, 1996, 2002; FSIS, 2005). Drastic measures have been taken to control the association of *E. coli* O157:H7 with beef products, and the Food Safety and Inspection Service (FSIS) reported that the number of ground beef samples positive for *E. coli* O157:H7 decreased by 80% between 2001 and 2006 (<1% of samples were positive in 2001) (http://www.beefusa.org/uDocs/ecolinumbersdecline. pdf). Despite such achievements, the incidence of beef samples which were positive for *E. coli* O157:H7 increased in 2007, and multistate outbreaks and large recalls continue (CDC, 2002; USDA/FSIS, 2007). For these reasons, additional sources of contamination, such as lymph nodes and food contact surfaces, should be investigated and more effective controls should be sought.

In response to international outbreaks (1994 to 1998) of *E. coli* O157:H7 infections associated with dry fermented sausages (Alexander et al., 1995; Tilden et al., 1996; Williams et al., 2000) and published data demonstrating the inability of existing fermentation, drying, and/or storage methods to inactivate 4 log cycles of *E. coli* O157:H7 in dry fermented sausages (Glass et al., 1992), the FSIS mandated that all processors implement a combination of treatments/processes capable of inactivating ≥5 log cycles of *E. coli* O157:H7 in dry fermented sausage before retail distribution (FSIS, 1996). Unfortunately, fermented sausages continue to be associated with outbreaks of food-borne *E. coli* O157:H7 infections (MacDonald et al., 2004). In response, it has been suggested that cells which survive the fermentation process may not be as susceptible to ensuing heat (56°C) treatments during processing (Ryu and Beuchat, 1998; Buchanan and Edelson, 1999). However, most of these products do not support extended survival or permit the growth of *E. coli* O157:H7 (Hammes et al., 1990; http://www.pork. org/PorkScience/documents/SEMIDRYSAUSAGE. pdf). Even so, *E. coli* O157:H7 is tolerant of acidic environments, and its inactivation on such products may not be immediate (Glass et al., 1992). Therefore, the pathogen may survive long enough to cause food-borne illness when contaminated sausages are consumed shortly after production (Naim et al., 2004). Outbreak-related contamination events may also occur during postprocessing handling (e.g., slicing and repackaging) at the retail level.

Unpasteurized Dairy Products

Ordinarily, pasteurized and fermented dairy products are not a threat to human health, and in developed countries consumption of raw milk and dairy products is limited to an unconventional (or unsuspecting) minority. Raw milk and cheese derived from raw milk have been implicated in repeated outbreaks of human *E. coli* O157:NM and O157:H7, *Salmonella, Campylobacter, Yersinia, Listeria monocytogenes,* and *Cryptosporidium* infections; *Staphylococcus* intoxication; terminal HUS; tuberculosis; and brucellosis (Bielaszewska et al., 1997; Blaser et al., 1979; Doyle and Roman, 1982; Keene et al., 1997; Potter et al., 1984; Taylor et al., 1982; Taylor and Perdue, 1989). Following many of these outbreaks, organisms genetically homologous to those involved in such illnesses, or antibodies indicating previous exposure, were recovered from the feces and/or milk of lactating animals on implicated farms. One outbreak of *E. coli* O157:H7 was associated with consumption of fresh cheddar cheese curds made from pasteurized milk (CDC, 2000). It was eventually determined that the curds were stored in a non-sanitized vat which had previously held cheddar cheese curds made with unpasteurized milk (CDC, 2000). Aside from educating/reminding milk producers and consumers of the substantial risks associated with these products, little can be done to dissuade the small communities that continue to prefer raw milk/ dairy foods.

Unpasteurized Juices

Documented outbreaks of human STEC infection have been attributed to unpasteurized apple juice and cider (Besser et al., 1993; CDC, 1997; Millard et al., 1994; Steele et al., 1982). As specific examples, two outbreaks of *E. coli* O157:H7 infections were caused by the consumption of unpasteurized apple cider made from both hanging and fallen (drop) apples (CDC, 1997). All apples had been brushed and washed in potable water before pressing, and 0.1% potassium sorbate was added to the cider before bottling (CDC, 1997). The low infectious dose associated with clinical cases, in combination with the addition of preservatives and the acidic pH (3 to 4) of apple juice/cider, generated considerable alarm among consumers and health officials in response to the outbreaks (CDC, 1997). Also in 1996, an unrelated outbreak of illnesses stemming from consumption of commercial unpasteurized apple juice was implicated in the death of a young girl in Colorado, United States (http://www. cfsan.fda.gov/~comm/ift3-4a.html). While apple juice/ cider has been implicated in the majority of juice-related outbreaks, orange juice produced and sold by a small roadside vendor also caused a small outbreak of ETEC infections in 1992 (http://www. cfsan.fda.gov/~comm/ift3-4a.html). These outbreaks further reinforce the acid-tolerant nature of *E. coli* O157:H7 and demonstrate the inability of preservatives used within the commercial juice/cider industry to consistently control these pathogens (CDC, 1997).

Produce

In 1996, radish sprouts fed to Japanese school children caused over 6,000 *E. coli* O157:H7 infections and three deaths (Michino et al., 1999; NIIDIDCD/MHWJ, 1997). Over the next 2 years, smaller outbreaks in Japan and the United States were also traced back to radish, alfalfa, and clover sprouts contaminated with *E. coli* O157:H7 (http://www. cfsan.fda.gov/~comm/ift3-4a.html). Seeds which are contaminated before germination or exposed to contaminated soil/water may develop into sprouts with pathogens attached to the root system or internalized into the inner tissue of the young plant (Solomon et al., 2002a, 2002b; Warriner et al., 2003). Temperature (25 to 30°C) and humidity conditions during sprouting may then facilitate the growth of existing STEC contamination (Hara-Kudo et al., 1997). Processes/control measures designed to address microbial contamination associated with seeds or sprouts were not in place prior to the series of international sprout-associated outbreaks of *E. coli* O157:H7 infections (Brooks et al., 2001; CDC, 2001b). Although control measures have since been implemented, STEC, including *E. coli* O157:H7, continues to be recovered from retail-level sprout samples (Samadpour et al., 2006), and thus, children (≤5 years old), the elderly, and the immunocompromised are encouraged not to eat sprouts (CDC, 2001b).

In recent years, fresh-cut produce sales have generated about $12 billion in annual revenue, and the market share for these products continues to grow (http://www.cfsan.fda.gov/~dms/prodgui4. html). Even so, fresh-cut leafy green produce has fallen under intense scrutiny due to the size of recent outbreaks and recalls associated with these products. Between 1995 and 2005, lettuce and mixed leafy green salads were associated with 14 domestic outbreaks of *E. coli* O157:H7 infections (http:// www.cspinet.org/foodsafety/outbreak/pathogen. php). In late 2006, shredded lettuce, served at multiple restaurants from a single fast-food chain, was again implicated in a multistate outbreak of *E. coli* O157:H7 infections (http://www.fda.gov/ola/2006/ foodsafety1115.html). Also in 2006, 204 illnesses, 31 cases of HUS, and three deaths were associated with a highly publicized outbreak of *E. coli* O157:H7 infections eventually traced back to fresh-cut bagged spinach (http://www.fda.gov/ola/2006/ foodsafety1115.html). Initial contamination was most likely introduced at the farm level, as a genetically homologous strain of *E. coli* O157:H7 was recovered from water sources used to irrigate the spinach and from the feces of hogs which had reportedly broken into the spinach fields (http:// www.fda.gov/ola/2006/foodsafety1115.html). Similar findings were reported for pears and apples irrigated with *E. coli*-contaminated water (Riordan et al., 2001). As with sprouts, it has been shown that *E. coli* can be incorporated into the root systems (at the root junctions or rhizosphere) of traditionally and hydroponically grown spinach seedlings in the early stages of cultivation and recovered from the leaves, stems, and root system of plants at harvest (Warriner et al., 2003). According to Solomon et al. (2002a, 2000b), *E. coli* O157:H7 may be internalized into the edible portion of lettuce plants. While Wachtel et al. (2002) and Warriner et al. (2005) did not recover internalized *E. coli* O157:H7 from the edible portion of spinach or cabbage plants which were inoculated during germination, any degree of pathogen internalization is alarming.

ETEC O169:H41 has emerged as the primary serotype involved in outbreaks of food-borne ETEC infections, with raw herbs and vegetables being implicated in the majority of such outbreaks (Beatty et al., 2004). Cut fruit and fruit salads have also been implicated in outbreaks of food-borne *E. coli* O157:H7 infections, although the pathogen was most likely introduced during cross-contamination events at the food preparation level (http://www.cfsan.fda. gov/~comm/ift3-4a.html). In 1993, a total of 168 hotel and airline patrons in Rhode Island and New Hampshire became ill after eating shredded carrots contaminated with ETEC serotype O169:H41 (http://www.cfsan.fda.gov/~comm/ift3-4a.html). The carrots served at both venues were grown in the same state, although a single source was never identified. In 1998, two outbreaks stemming from food served in Minnesota restaurants which contained fresh chopped parsley resulted in a total of 77 illnesses (Naimi et al., 2003). It was later determined that the parsley was grown at a single farm in Baja, Mexico, which used inadequately chlorinated water during parsley processing and packing operations (Naimi et al., 2003). The majority of fresh produce is washed in 100- to 200-ppm sodium hypochlorite (bleach) solutions during processing (Warriner et al., 2003). Unfortunately, when used under typical processing conditions, sodium hypochlorite is only moderately effective in reducing pathogen counts (Beuchat, 1998; Beuchat and Ryu, 1997) because of its rapid inactivation in the presence of organic material (Lillard, 1980; Russell and Axtell, 2005). Thus, successive rinsing and sanitizing steps may be more effective against surface contamination. Surface contamination can also be transferred into previously sterile tissue through cuts/punctures in the skin of fruits and vegetables (Warriner et al., 2003). This route of contamination warrants significant attention, as it was also implicated in multiple outbreaks of *E. coli* O157:H7 infections associated with unpasteurized apple juice and cider (CDC, 1997).

ADVANCES IN DIARRHEAGENIC *E. COLI* CONTROL MEASURES DURING PRODUCTION AND PROCESSING

Safer Supply Chains

The volume and international breadth of human illnesses associated with alfalfa and radish sprouts justified extensive modifications to pathogen control programs within the sprout industry. In 1998, the FDA issued the *Guide to Minimize Microbial Contamination of Fresh Fruits and Vegetables*. As a result, vast improvements have been made to the conventional germination, cultivation, harvesting, and processing of sprouts. Among the most significant of the improvements was the screening of water used during hydroponic cultivation for *E. coli* (http://www.cfsan.fda.gov/~dms/prodguid.html). Seed decontamination technologies include the application of acidified sodium chlorite, quaternary ammonium compounds, ozone gas, and irradiation, all of which are capable of reducing levels of *E. coli* O157:H7 contamination

(Warriner et al., 2005; Bari et al., 2004). In addition, the lot numbers on individual packages/bags of certified seed can now be traced back to the original premises at which the seeds were grown. Other efforts include increased pest control and sanitation programs at the processing level, the use of backlighting equipment and air curtains, adoption of hazard analysis and critical control point (HACCP) concepts, and final product testing for *E. coli* O157 and *Salmonella* (http://www.cfsan.fda.gov/~dms/prodguid.html). In February of 2008, most likely in response to the massive outbreak associated with fresh-cut bagged spinach in 2006, the FDA released another guidance document entitled *Guide to Minimize Microbial Food Safety Hazards of Fresh-Cut Fruits and Vegetables*, which outlines similar microbiological control measures, including recommendations on worker health and hygiene and sanitation operations and for the preparation and maintenance of processing water (http://www.cfsan.fda.gov/~dms/prodgui4.html).

The level of hide contamination has been correlated to the subsequent proportion of contaminated carcasses (Woerner et al., 2006), and technologies which minimize hide contamination are under extensive investigation. By reducing the number of animals shedding the pathogen, hide contamination and/or carcass contamination events during hide removal and evisceration may be minimized. Proposed strategies include transport and lairage management (cleaning holding areas and separating "clean" versus heavily soiled lots of cattle), hide decontamination, and improved hide removal technologies (Gill, 2005; Hadley et al., 1997; Sofos et al., 1999c). Measures to control the colonization of cattle by *E. coli* O157:H7 include the feeding of probiotics, prebiotics or other feed additives/ingredients, water chlorination, vaccination programs, behavioral management, and feedlot pen maintenance (Brashears et al., 2005; Callaway et al., 2005; Diez-Gonzalez, 2005; Stopforth and Sofos, 2006). Although their application is unlikely to be counterproductive, such programs have yet to provide repeatable, lasting results when applied under realistic parameters.

Minimizing Contamination Events during Processing

Contamination may be introduced by many different routes (e.g., animal to environment, human to food product, food product to food product, and processing equipment to food product) (Bell, 1997; Sofos, 1994, 2004). However, the most significant advances in process control almost certainly include those which address worker education and hygiene (9 CFR 416.5, Code of Federal Regulations, 2001), as workers can serve as original vectors of enterically derived contamination or transfer existing contamination to unsoiled areas or products. In-plant contamination issues may be reduced by simply reviewing the basic fundamentals related to personal hygiene and hand washing, product and equipment handling, and appropriate behavior during illness (FSIS, 1999). Employee education and training courses are typically included in the prerequisite program portion of a HACCP system and can be customized to fit the needs of an individual operation. Once proper behaviors are introduced and improper practices have been corrected, supervisors must provide access to and/or enforce the use of appropriate restroom and eating facilities and good hygiene practices. The advent of mechanical devices as processing aids, which limit worker contact with food products, has also reduced cross-contamination events during processing (Scanga, 2005).

When present, microorganisms can contaminate previously sterile inner tissue through cuts or tears in outer skin, rinds, or hides (CDC, 1997; Sofos, 1994, 2005; Sofos et al., 1999c). Therefore, it is imperative that injurious events to protective outer surfaces are minimized, raw produce with surface defects are rejected, and the levels of contamination on outer surfaces are minimized before inner tissues are exposed. The outer surfaces/trimmings from carcasses are generally intended for ground meat production, even though the outer surface typically encounters a higher level of contamination than any other carcass region (Gill, 2005). Although cooking (72°C or 160°F) raw ground beef before consumption should inactivate pathogens of concern, improper cooking is very likely to occur and should be anticipated. Thus, it is not surprising that the majority of fresh-meat-related outbreaks and recalls involve ground meat (Rangel et al., 2005). Carcasses are sterile before hide removal, at which time contamination may be transferred from the hide to the carcass (Sofos, 1994; Gill, 2005). Advances in mechanical hide removal should reduce the incidence of hide-to-carcass transfer of pathogenic *E. coli* and also reduce the number of carcasses which fall from overhead rails and onto the ground and must then be diverted to areas reserved for trimming and additional FSIS inspection, as well as the incidence of hide and carcass defects (Scanga, 2005). In addition, the number of potential harbors for pathogenic contamination may be reduced by covering tears in the subcutaneous fat layer with plastic film prior to washing carcasses (Simpson et al., 2006). The evisceration process also has the potential to generate high levels/incidence of contamination on carcasses (Gill, 2005).

Knife trimming has been implemented to address the "zero tolerance" policy, which requires the removal of all visible soil, feces, ingesta, and milk contaminants from the surface of nonintact beef (FSIS, 1993). Knives are invaluable food processing tools but may also be contaminated with pathogens (Bell, 1997). According to the section of the Code of Federal Regulations (9 CFR 415.4, Code of Federal Regulations, 2001) describing sanitary operations, "all [direct and indirect] food-contact surfaces including...utensils and equipment, must be sanitized as frequently as necessary to prevent the creation of insanitary conditions and the adulteration of product." The dual-knife system has been a great asset to the meat packing industry. In an effort to reduce the incidence of cross-contamination events between carcasses, knives are rotated between consecutive carcasses (one knife is applied to a carcass, while the other knife is submerged in sanitizer) (Scanga, 2005). Other carcass-spot decontamination technologies include steam pasteurization or steam vacuum units, which are used to remove visible contamination or to systematically treat the area most likely to become contaminated during hide removal (Sofos and Smith, 1998; Sofos et al., 1999a, 1999b). While zero tolerance implies that all such matter is removed, attempts to remove contamination from the entire surface of every carcass using spot decontamination strategies is not feasible, especially with the chain speeds used in many domestic commercial facilities. Inappropriate terminology aside, spot decontamination efforts effectively reduce microbial counts and limit the spread of particulate matter over carcass surfaces during subsequent carcass washing steps (Sofos et al., 1999a, 1999b).

Inactivation

Decontamination with scientifically validated decontamination fluids may be valuable in minimizing the incidence of diarrheagenic *E. coli*, specifically STEC contamination, on surfaces of seeds, produce, eggs, nuts, and fresh or processed meat products (Beuchat, 1998; Sapers et al., 1999; Sofos et al., 2006; Warriner et al., 2005). In lieu of pasteurization, such treatments may be used to decontaminate the outer surfaces of fruits destined for juice production and only when accompanied by rigid microbiological "hold and test" programs; unpasteurized juice must be labeled as such (21 CFR 120.24, Code of Federal Regulations, 2002). Available decontamination fluids include chlorine and chlorine derivatives, organic acids, iodophors, trisodium phosphate, peroxides and peroxyacid solutions, ozone and electrolyzed oxidizing water, and protein compounds (Acuff, 2005; Beuchat, 1998; Cords et al, 2005; Sapers et al., 1999; Sofos, 2005; Sofos et al., 1999a, 1999b;

Warriner et al., 2005). It is extremely important that these solutions be applied under appropriate conditions, as sublethal exposures may only enhance the stress tolerance of diarrheagenic *E. coli* (Davidson and Harrison, 2002; Duffy et al., 2000; Samelis and Sofos, 2003; Sofos and Smith, 1998). Lipids, proteins, other organic compounds, water hardness, contact time, and pressure during application influence the killing power of many decontamination fluids (Arrit et al., 2002; Cutter et al., 1997; Pordesimo et al., 2002; Russell and Axtell, 2005; Sofos et al., 1999, Sofos and Smith, 1998; Sofos et al., 2006). In general, many decontamination treatments tend to be more effective when used sequentially or when combined with other antimicrobial factors, such as heat (Bacon et al., 2003; Hardin et al., 1995; Leistner and Gould, 2002; Sofos et al., 1999a, 1999b, 2006). Spray applications may be more convenient than immersion treatments (Beuchat, 1998), although all surfaces of a product may not be exposed or adequately decontaminated (Warriner et al., 2005).

Appropriate cooking and handling recommendations are available for many raw or partially cooked food products, and adherence to such recommendations should eliminate the risk associated with raw foods. However, possible pathogen contamination must be addressed during processing of products intended for immediate consumption (i.e., ready to eat). In 1998, it became mandatory for all noncitrus juices entering interstate commerce to be pasteurized or be clearly labeled as an unpasteurized product, or as an alternative to pasteurization, fruit juice manufacturers may adopt HACCP programs which include alternative scientifically validated processes which are capable of reducing ≥5 log cycles of *E. coli* O157:H7 (21 CFR 120.24). Pasteurization is recommended (21 CFR 120.24, Code of Federal Regulations, 2002), and a thermal pasteurization guide for milk, *Grade "A" Pasteurized Milk Ordinance,* is available at http://www.cfsan.fda.gov/~ear/pmo01toc.html. The successful pasteurization of milk and other dairy products can be determined by screening for the presence of postpasteurization phosphatase activity (http://www.cfsan.fda.gov/~ebam/bam-27.html). Alkaline phosphatase is an LT enzyme naturally found in raw milk and is inactivated at temperatures just above those required to inactivate most milk-borne microorganisms, including diarrheagenic *E. coli* (http://www.hc-sc.gc.ca/fn-an/alt_formats/hpfb-dgpsa/pdf/res-rech/mfo25-eng.pdf). For these reasons, alkaline phosphatase inactivation is used to verify the successful application of pasteurization treatments (15 min at 71.7 or 30 min at 62.8°C) (http://www.cfsan.fda.gov/~ebam/bam-27.html). In general, the efficacy of pasteurization treatments depends on the

process, as well the components (e.g., pulp and fat content), pH, and level of target microorganisms associated with a product (NACMCF, 2006).

Low doses of irradiation are highly lethal for most microorganisms, specifically *E. coli*, and many improvements to irradiation technologies have been developed in recent years (http://www.wisc.edu/fri/briefs/foodirrd.htm). Irradiation programs have been scientifically validated and are available to inactivate *E. coli* O157:H7 associated with juices, seeds, sprouts, spices, and meat products (Bari et al., 2004; Foley et al., 2004; Murano et al. 1998; NACMCF, 2006). With adherence to the guidelines for optimization of temperature, packaging conditions, dose, and depth of penetration, the administration of ≥3 kGy successfully inactivates the pathogen, with little impact on the flavor, texture, or color of treated ground beef samples (Arthur et al., 2005; Murano et al., 1998; Vickers and Wang, 2002). Nonmeat ingredients also appear to increase overall efficacy of irradiation treatments (Bricher, 2003; Kamat et al., 1997; Lacroix et al., 2004; Smith and Pillai, 2004; http://www.wisc.edu/fri/briefs/foodirrd.htm). The generation of free radicals is also responsible for the generation of hydrogen peroxide and quality defects in irradiated produce, which include mushiness, alterations in flavor profiles, and rapid oxidation (Kuby, 1997; Lewis et al., 2002). Other nonthermal processes include high hydrostatic pressure, shock waves, ultrasonication, and pulsed ultraviolet light or electric field treatments (Guan and Hoover, 2005; NACMCF, 2006).

Inhibition

Postlethality contamination events pose a significant risk to human health, and tools are available to inhibit pathogen growth in both processed and unprocessed products. Antimicrobial agents within processed food product formulations and sprays applied to final products or in packaging material may be used to control postlethal pathogen growth. However, outbreaks of *E. coli* infections are still associated with fresh, minimally processed, and/or ready-to-eat food products, and the use of antimicrobial ingredients to control diarrheagenic *E. coli* is limited. Dry fermented sausages are the exception; the use of various lactic acid bacteria cultures, which produce inhibitory compounds (e.g., acids, bacteriocins, diacetyl, and ethanol) are added to raw sausage formulations to ensure proper fermentation and products of acceptable quality and safety (Lahti et al., 2001; Kang and Fung, 1999; Jay, 1982).

The rate and extent of product chilling, as affected by packaging, stocking density, air flow, relative humidity, temperature, and time, can influence the fate of microorganisms during storage. As oxygen and other gases in the environment affect microbial growth, modified atmosphere packaging is used for microbial inhibition. Such technologies manipulate individual levels of the main components of air (O_2, N_2, and CO_2) and include common vacuum packaging (Phillips, 1996). The performance of *E. coli* O157:H7 under modified atmospheres is not well understood and may be similar to its performance under aerobic conditions (Hao and Brackett, 1993).

Product quality and safety are almost certainly compromised after encounters of abusive temperatures for sporadic/extended periods of time during holding, transport, or distribution (Koutsoumanis and Taoukis, 2005). Diarrheagenic *E. coli* can survive but will not grow at recommended refrigeration temperatures (4°C or 40°F), but it is capable of growth at temperatures that barely exceed (7 to 9°C or 45°F) the recommendations (ICMSF, 1996). Thus, to effectively inhibit the proliferation of these pathogens during cold storage of produce and meat products, temperature must be lowered as quickly as possible and maintained for the duration of product holding (Hao and Brackett, 1993; Simpson et al., 2006). Chilling food products too rapidly or storing them at >4°C may result in quality defects (e.g., cold-shortening in rapidly chilled beef, lamb, and pork or freeze-damaged fresh produce) (Lee et al., 2004). Temperature monitoring/recording systems, equipped with audible default alarms, can be used to effectively maintain the correct temperature during each phase of cold storage.

METHODS FOR ISOLATION, DETECTION, IDENTIFICATION, AND CONFIRMATION OF DIARRHEAGENIC *E. COLI*

The FDA and USDA conduct routine sampling and testing of food products, and processors are also encouraged to adopt scientifically validated microbiological testing programs, although the need for such programs remains a topic of debate. The low incidence of *E. coli* O157:H7 in food products further compromises the integrity of microbiological testing programs (Buchanan and Doyle, 1997). In any respect, microbiological analysis of raw materials and finished food products should not be used in place of a comprehensive pathogen control program or to "confirm" product safety. Instead, routine microbiological testing should be incorporated as part of a multifaceted program to validate/verify the efficacy of antimicrobial interventions. Microbiological tests can be used to detect, identify, or quantify the presence of specific microorganisms. While

many tests can be used for more than one of these objectives, it is important to employ the most appropriate method regarding the type of information needed, sample composition, and the expected level of contamination.

Enrichment and Isolation

Enrichment protocols usually precede most conventional and rapid detection methods. Although results of microbiological analyses may be delayed as a result, enrichment protocols are critical when attempting to isolate very low levels of pathogenic *E. coli* from samples with or without excessive levels of background flora (fresh produce and fermented foods), as well as in the recovery of injured cells (Hepburn et al., 2002; Sanderson et al., 1995). Selective *E. coli* enrichment broths include brain heart infusion, tryptic soy broth, and gram-negative or *E. coli* broth supplemented with novobiocin (Hajna, 1955; Hussein et al., 2008). Additional compounds added to selective media to suppress the growth of background flora include antimicrobials to which *E. coli* are particularly resistant, such as cefixime, cefsulodin, potassium tellurite, and vancomycin (Hussein et al., 2008). Hussein et al. (2008) found that brain heart infusion supplemented with potassium tellurite, novobiocin, vancomycin, and cefixime was more successful than other selective enrichment broths in the suppression of background flora during enrichment of STEC recovered from cattle feces.

In general, detection of microorganisms in food product samples can be problematic due to low levels of contamination, a nonhomogeneous distribution of cells throughout the product/sample, or interactions between food components and reagents used in detection protocols (Gill, 2005; Smith et al., 2000). Practical detection methods must be sensitive, specific, repeatable, and economical and also generate results expressed in practical units in a timely manner. To date, conventional culture methods remain the gold standard in pathogen detection, and numerous selective and differential media, which require as little as 18 h for the detection of diarrheagenic *E. coli*, are available (Zadik et al., 1993; http://www.cfsan.fda.gov/~ebam/bam-a1.html). Popular solid media used to isolate generic biotype I *E. coli*, EPEC, ETEC, and EAEC include violet red bile agar, Levine's eosine methylene blue agar, MacConkey agar (MAC), and 3M PetriFilm rehydratable films (http://www.cfsan.fda.gov/~ebam/bam-a1.html). MAC and Hektoen enteric agar are most useful when isolating EIEC (Doyle and Padhye, 1989).

Sorbitol is added to MAC (SMAC) to differentiate *E. coli* O157 from non-O157 strains, which readily ferment sorbitol, while *E. coli* O157 strains typically do not (Schmidt et al., 1999). It should be noted, however, that some strains of *E. coli* O157 can ferment sorbitol, and some species other than *E. coli* are sorbitol negative and can induce false-positive results (Borczyk et al., 1990). In addition to SMAC, other media used to differentiate *E. coli* O157:NM and O157:H7 from non-O157 *E. coli* include SMAC supplemented with cefixime and potassium tellurite, rainbow agar, hemorrhagic colitis agar, and other proprietary chromogenic media (Huang et al., 1997; March and Ratnam, 1986; Zadik et al., 1993; http://www.cfsan.fda.gov/~ebam/bam-4a.html). Other differential substrates include rhamnose, 5-bromo-5-chloro-3-indoxyl-β-D-glucoronide (β-galactosidase), and 4-methylum-belliferyl-β-D-glucoronide (MUG) (Chapman et al., 1991; Okrend et al., 1992; Thompson et al., 1990).

Immunomagnetic separation is commonly used to concentrate target cells which are present in low numbers within large samples and can select both sorbitol-positive and -negative *E. coli* O157 and O157:H7 strains (Karch et al., 1996). To do so, antibody-coated magnetic beads are added to samples during (pre-) enrichment steps and interact with the antigens present on the surfaces of target cells; beads are gathered together using a magnet, which then allows for the removal of unbound sample components (Karch et al., 1996).

Rapid detection methods have become increasingly popular in the food industry for the detection of specific pathogens and toxins (Lazcka et al., 2007). Available methods are not as specific or sensitive as conventional methods, and therefore, presumptive positive samples identified by rapid technologies must be confirmed by conventional culture methods (http://www.cfsan.fda.gov/~ebam/bam-a1.html; http://www.cfsan.fda.gov/~ebam/bam-4a.html). Moreover, the majority of rapid detection methods, specific to STEC and EHEC, are not designed to indicate the presence of other classes of diarrheagenic *E. coli* (Meng et al., 2001a). The optimal sensitivity and specificity of a test are observed when detecting pathogens in broth systems, and food components tend to antagonize the results of some types of rapid tests (http://www.cfsan.fda.gov/~ebam/bam-a1.html). Currently, the most popular combinations of detection and confirmation methods are those which employ conventional culturing and enumeration and PCR technologies (Lazcka et al., 2007). The most commonly used DNA-based detection technologies include genetic probes, PCR, and bacteriophages (http://www.cfsan.fda.gov/~ebam/bam-a1.html), and many commercial kits are available and approved for use to detect *E. coli* O157:H7 and O157 or

customized combinations of *E. coli* virulence genes or target sequences.

PCR is the most extensively used rapid technology for detection of STEC and involves the extraction of genetic material; DNA hybridization, or the coupling of specific probes with target cell DNA; and the amplification of the target sequence (Lazcka et al., 2007; Paton et al., 1993; http://www.cfsan.fda.gov/~ebam/bam-a1.html). In general, PCR products are visualized via electrophoresis on agarose gels and enzyme-linked hybridizations, or Southern blot assays using labeled probes can be used to confirm the identity of a product (Fratamico et al., 2005). Hybridized probes labeled with fluorescent substrates; enzyme-binding proteins, like biotin-streptavidin; or radioactive isotopes (less common) are adsorbed onto a thin membrane (solid-phase hybridization) or captured from liquid-phase assays by a solid support, and then visualized (Entis et al., 2001). Multiplex-PCR protocols are capable of detecting multiple target sequences and can be used to screen samples for multiple pathogens, or more specifically identify a microorganism. The target sequences in many multiplex-PCR protocols used to detect STEC strains, specifically *E. coli* O157:H7, include $rfbE_{O157}$, $hylA$, $fliC_{H7}$, $eaeA$, stx_1, and stx_2 (Gannon et al., 1997; Wang et al., 2002; Fratamico et al., 2000). Real-time PCR utilizes the fluorescent signal of a reporter dye which increases in intensity as the target sequence is amplified (Fratamico et al., 2005; Higuchi et al., 1992, 1993; Livak et al., 1995; http://www.cfsan.fda.gov/~comm/ift3-4a.html). Software is available which generates a curve of intensity based on initial fluorescence and level of fluorescence at each ensuing cycle; levels of the target organism can be quantified by comparing these curves to standardized models (Fratamico et al., 2005). Popular automated real-time multiplex PCR systems used to detect *E. coli* O157 include BAX (DuPont Qualicon; http://www2.dupont.com/Qualicon/en_US/products/BAX_System/bax_ecolimp.html) and TaqMan (Applied Biosytems; http://www3.appliedbiosystems.com/cms/groups/applied_markets_marketing/documents/generaldocuments/cms_039421.pdf) systems. Guion et al. (2008) have developed a real-time fluorescence-based multiplex PCR protocol capable of detecting all six classes of diarrheagenic *E. coli*. In brief, specifically designed primer sets are used to amplify eight different virulence genes in a single PCR reaction, including *aggR* (transcriptional activator of AAF I expression) for EAEC (Nataro et al., 1994), *stIa/stIb* and *lt* (ST and LT enterotoxins) for ETEC, *eaeA* (intimin) for EPEC and STEC/EHEC, stx_1 and stx_2 (Shiga toxins) for STEC/EHEC, *ipaH* for EIEC, and *daaD* for DAEC. A similar single-reaction protocol was developed by Moyo et al. (2007) for use in the detection of EAEC, EPEC,

ETEC, EIEC, and EHEC but not DAEC. In general, the efficacy of any PCR protocol directly depends on the proper enrichment and purification of food samples, as many substances can antagonize PCR reagents (Entis et al., 2001).

Enzyme-linked immunosorbent assays and immunoprecipitation are both "sandwich" assays, during which target antigens interact with two types of antibodies; the first type is isolated by precipitation because it is covalently bound or adsorbed to a solid surface, and particles that do not interact with the antibodies are washed away; and the second type is attached to an enzyme-labeled "indicator" compound (Fratamico et al., 2005; http://www.cfsan.fda.gov/~ebam/bam-a1.html). These assays are unable to differentiate viable from nonviable cells (Fratamico et al., 2005). The use of enzyme immunoassays in clinical diagnostic laboratories to screen for STEC is becoming increasingly common. Correspondingly, while O157:H7 has historically been the most common serotype (80%) associated with STEC infections (http://www.cdc.gov/foodborneoutbreaks/ecoli/STEC2002.pdf), the incidence of non-O157 STEC infections among health districts which use an enzyme immunoassay for Shiga toxin to screen clinical stool samples is disproportionately higher than that among districts which use culture-based screening methods (CDC, 2007).

Confirmation or Identification

Phenotypic and genotypic characteristics can be used to confirm the identity of or classify a microorganism. Phenotyping methods include the screening of known biochemical characteristics, serotyping, and phage typing (Hiett, 2005). Manual, rapid, and fully automated versions of standardized biochemical tests which identify the standard biochemical characteristics of E. coli and E. coli O157:H7 are available (Table 4). Serotyping of diarrheagenic E. coli by agglutination with O- and H-group antisera is also used to confirm the identity of pure cultures (Meng et al., 2001a). Agglutination assays utilize antibody-coated particles which interact with target antigens on cell surfaces or with soluble toxin antigens (Hajra et al., 2007).

Pulsed-field gel electrophoresis (PFGE) is the DNA-based molecular typing assay used by PulseNet and FoodNet, two branches of the CDC which investigate, compare, and track human clinical cases of food-borne illness and human pathogens recovered from food products (Swaminathan et al., 2001). Briefly, the genetic material of a pure cell culture is cut apart by enzymes which target very specific DNA sequences; the resulting fragments are then separated by electrophoresis on an agarose gel. The resulting pattern of DNA fragments can then be compared to the DNA patterns of other isolates, and genetic homology can be measured (Farber, 1996). The use of multiple enzymes increases the specificity of digestive enzyme-based typing assays (Hiett, 2005).

Recent Advances in Analytical Methods

Possé et al. (2008) recently described a series of plating media used to differentiate and confirm the identity of STEC serotypes O26, O103, O111, and O145, as well as sorbitol-positive and -negative O157. Briefly, non-O157 serotypes were differentiated by general appearance and color of colonies following spread plating and incubation on MAC

Table 4. Phenotypic indicators of E. coli and E. coli O157/O157:H7[a]

Phenotypic indicator	Generic biotype I E. coli	E. coli O157, O157:H7
MUG	+	−
ONPG	+	+
Lysine decarboxylase	+	+
Ornithine decarboxylase	+	+
H$_2$S	−	−
Citrate utilization	−	−
Methyl red	+	+
Voges-Proskauer	−	−
Indole	+	+
Urease	−	−
Arabinose	+	+
Sucrose	+	−
Sorbitol	+	−
Rhamnose	+	−
Cellobiose	−	−

[a]Data derived from Meng et al. (2001b) and http://www.cfsan.fda.gov/~ebam/bam-4a.html.

supplemented with sucrose, sorbose, bile salts no. 3, 5-bromo-4-chloro-3-indolyl-β-D-galactopyranoside (X-Gal), isopropyl-β-D-thiogalactopyranoside (IPTG), novobiocin, and potassium tellurite. A second medium was used to confirm the identity of suspect colonies and was composed of a phenol red base supplemented with dulcitol, L-rhamnose, D-raffinose, or D-arabinose. Sorbitol-positive and -negative O157 strains were plated onto differential media consisting of MAC supplemented with sorbitol, bile salts no. 3, X-Gal, IPTG, novobiocin, and potassium tellurite. Following incubation, the identity of suspect sorbitol-positive colonies (purple) was confirmed using a phenol red base plus L-rhamnose (Possé et al., 2008).

Bacteriophages are species-specific viruses that, when labeled with an indicator compound (i.e., fluorescent protein), can be used to detect target organisms among a diverse background of unrelated microflora. Labeled *E. coli*-specific bacteriophages infect and replicate within viable cells and can indicate the presence of extremely low levels of *E. coli* within 1 to 2 h (Goodridge et al., 1999; Tanji et al., 2004). In one study, the ability of green fluorescent protein-labeled PP01 bacteriophage to detect *E. coli* O157:H7 was examined, and using an epifluorescence microscope, adsorption of green fluorescent protein-PP01 to cells was observed after only 10 min (4°C); maximum fluorescence was observed after 3 h (Oda et al., 2004). Brigati et al. (2007) described the use of a recombinant phage specific for *E. coli* O157:H7, which initiates the production of N-(3-oxohexanoyl)-L-homosterine lactone (OHHL) in infected cells. Bioreporter cells carrying the *Vibrio fischeri lux* operon were added to the reaction and emitted a bioluminescent signal in response to OHHL production by infected cells after as few as 4 h (Brigati et al., 2007). While these and other phage assays demonstrate potential for use in the rapid detection of *E. coli* O157:H7, overall sensitivity and specificity in the presence of food components and background flora remain unproven. Extended enrichment protocols may also be necessary when using bacteriophages to detect *E. coli* in food samples (Goodridge et al., 1999). Additional drawbacks include bacteriophages with a very broad/narrow target host range and the eventual development of resistance by target host cells (Greer and Dilts, 1990; Loessner and Busse, 1990; McGrath et al., 2002; Summers, 2001).

Biosensors are devices associated or integrated with cellular transduction systems and include cell components and biologically or synthetically engineered compounds (Lazcka et al., 2007). A multitude of enzyme-, nucleic acid-, or antibody-based biosensor assays have been developed for use in the detection of pathogens, although antibody-based assays are the most popular (Nakamura and Karube, 2003). Briefly, pathogen-specific antibodies, including those for *E. coli* O157:H7 and/or Shiga toxins, are stabilized by avidin-biotin systems, by adsorption to gold, or by forming self-assembled monolayers, and deviations in optical, fluorescent, refractive, or electrochemical impedance indices are then used as indicators (Bergwerff and Knapen, 2006; Lazcka et al., 2007; Radke and Alocilja, 2003; Ruan et al., 2002). Although most require at least 1 to 3 h, some biosensor assays may detect *E. coli* O157:H7 in as little as 5 min (Radke and Alocilja, 2003). While the sensitivity and specificity of some biosensor assays rival or exceed those of conventional detection methods, a much more thorough examination of each method is required before it will be accepted as a viable option for the detection of STEC/EHEC in foods (Lazcka et al., 2007; http://www.cfsan.fda.gov/~ebam/bam-4a.html).

SURVEILLANCE AND TRACE-BACK PROGRAMS

Locating the food product responsible for human illness can be rather difficult for a number of reasons: (i) only a small proportion of the human population is at risk for severe infection which requires medical attention, and thus, many infections go unreported; (ii) an accurate recollection of all activities and foods eaten before the onset of symptoms is nearly impossible; and (iii) the documentation of epidemiologically and genetically related illnesses and/or contaminated food products may not be available and/or correct (CDC, 2006; Cieslack et al., 1997; Mead et al., 1999; Rangel et al., 2005; Swaminathan et al., 2001). When attribution is possible, the CDC estimates a 2- to 3-week interval between consumption of contaminated food and being classified as part of an *E. coli* outbreak (http://www.cdc.gov/ecoli/reportingtimeline.htm). This interval includes incubation time, onset and progression of symptoms which require medical attention, diagnostic testing by state health authorities, and completion of the analyses required to link a patient to other isolated cases of illness, to a contaminated food product, or to an established outbreak (http://www.cdc.gov/ecoli/reportingtimeline.htm).

PFGE, the molecular typing assay used to link human clinical and/or food isolates, is conducted by trained laboratory technicians with standardized equipment and protocols and generates distinctive DNA fragment patterns or genetic "fingerprints." Once generated, PFGE profiles are made available to other regional laboratories with access to the national

CDC database (http://www.cdc.gov/pulsenet/whatis/htm). PulseNet and FoodNet generate and manage the PFGE profiles of human clinical and food-derived isolates, respectively, and the databases are routinely compared (http://www.cdc.gov/pulsenet/whatis/htm). The admitted limitations of PFGE analysis include extensive time requirements, variation between technicians and/or human error, the generation of measurements that do not relate to actual phylogenetic homology, the genetic instability of bacteria, and an inability to generate profiles for all strains (Davis et al., 2003; http://www.cdc.gov/pulsenet/whatis. htm). The limitations of PulseNet include the dissimilar prioritization, and a lack of financial and informational resources within local, state, and federal agencies (http://www.cdc.gov/pulsenet/whatis.htm). Future improvements may include the update and streamlining of current PFGE protocols, the adoption of new subtyping technologies, and an increased number of participants and contributors with an enhanced degree of communication (http://www.cdc.gov/pulsenet/whatis.htm).

Despite the inability of routine microbiological testing to confirm the safety of food products and the limitations of current surveillance programs, Rangel et al. (2005) evaluated the epidemiology and incidence of *E. coli* O157:H7 outbreaks between 1982 and 2002 and found that the annual incidence of outbreaks remained fairly constant, while the median size of each outbreak has decreased over time. The documentation of smaller and smaller outbreaks is indicative of successful surveillance efforts that include the refinement and standardization of laboratory techniques, improved reporting and sharing of information, and induction of more regional testing facilities which cater to a larger proportion of the national population. In addition to the surveillance of human clinical cases, the routine testing of environmental and food samples has led to the recall of hundreds of millions of pounds of potentially contaminated food products, which may have led to many illnesses, hospitalizations, and deaths.

REFERENCES

Acuff, G. R. 2005. Chemical decontamination strategies for meat, p. 350–363. *In* J. N. Sofos, *Improving the Safety of Fresh Meat.* CRC Press, Boca Raton, FL.

Alam, M. J., and L. Zurek. 2004. Association of *Escherichia coli* O157:H7 with houseflies on a cattle farm. *Appl. Environ. Microbiol.* 70:7578–7580.

Alexander, E. R., J. Boase, M. Davis, L. Kirchner, C. Osaki, T. Tanino, P. Samadpour, M. Goldoft, P. Bradley, B. Hinton, P. Tighe, B. Pearson, and G. R. Flores. 1995. *Escherichia coli* O157:H7 outbreak linked to commercially distributed dry-cured salami. *JAMA* 273:985–986.

Allos, B. M., M. R. Moore, P. M. Griffin, and R. V. Tauxe. 2004. Surveillance for sporadic foodborne disease in the 21st century: the FoodNet perspective. *Clin. Infect. Dis.* 38:S115–S120.

Aragon, V., K. L. Chao, and L. A. Dreyfus. 1997. Effect of cytolethal distending toxin on F-actin assembly and cell division in Chinese hamster ovary cells. *Infect. Immun.* 65:3774–3780.

Arrit, F. M., J. D. Eifert, M. D. Pierson, and S. S. Sumner. 2002. Efficacy of antimicrobials against *Campylobacter jejuni* on chicken breast skin. *J. Appl. Poultry Res.* 11:358–366.

Arthur, T. M., T. L. Wheeler, S. D. Shackelford, J. M. Bosilevac, X. W. Nou, and M. Koohmaraie. 2005. Effects of low-dose, low-penetration electron beam irradiation of chilled beef carcass surface cuts on *Escherichia coli* O157:H7 and meat quality. *J. Food Prot.* 68:666–672.

Bacon, R. T., and J. N. Sofos. 2003. Characteristics of biological hazards in foods, p. 157–195. *In* R. H. Schmidt and G. E. Rodrick, *Food Safety Handbook.* John Wiley & Sons, Hoboken, NJ.

Bacon, R. T., J. R. Ransom, J. N. Sofos, P. A. Kendall, K. E. Belk, and G. C. Smith. 2003. Thermal inactivation of susceptible and multiantimicrobial-resistant *Salmonella* strains grown in the absence or presence of glucose. *Appl. Environ. Microbiol.* 69:4123–4128.

Barham, A. R., B. L. Barham, A. K. Johnson, D. M. Allen, J. R. Blanton, and M. F. Miller. 2002. Effects of the transportation of beef cattle form feedyard to the packing plant on prevalence levels of *Escherichia coli* O157 and *Salmonella* spp. *J. Food Prot.* 65:280–283.

Bari, M. L., M. I. Al-Haq, T. Kawasaki, M. Nakauma, S. Todoriki, S. Kawamoto, and K. Isshiki. 2004. Irradiation to kill *Escherichia coli* O157:H7 and *Salmonella* on ready-to-eat radish and mung bean sprouts. *J. Food Prot.* 67:2263–2268.

Barkocy-Gallagher, G. A., T. M. Arthur, M. Rivera-Betancourt, X. W. Nou, S. D. Shackelford, T. L. Wheeler, and M. Koohmaraie. 2003. Seasonal prevalence of Shiga toxin-producing *Escherichia coli*, including O157:H7 and non-O157 serotypes, and *Salmonella* in commercial beef processing plants. *J. Food Prot.* 66:1978–1986.

Bauer, M. E., and R. A. Welch. 1996. Characterization of an RTX toxin from enterohemorrhagic *Escherichia coli* O157:H7. *Infect. Immun.* 64:167–175.

Beatty, M. E., C. A. Bopp, J. G. Wells, K. D. Greene, N. D. Puhr, and E. D. Mintx. 2004. Enterotoxin-producing *Escherichia coli* O169:H41, United States. *Emerg. Infect. Dis.* 10:518–521.

Begue, R. E., D. I. Mehta, and U. Bleckler. 1998. *Escherichia coli* and the hemolytic-uremic syndrome. *South. Med. J.* 91:798–804.

Bell, B. P., M. Goldoft, P. M. Griffin, M. A. Davis, D. C. Gordon, P. I. Tarr, C. A. Bartleson, J. H. Lewis, T. J. Barrett, J. G. Wells, R. Baron, and J. Kobayashi. 1994. A multistate outbreak of *Escherichia coli* O157:H7 associated with bloody diarrhea and hemolytic-uremic syndrome from hamburgers—the Washington experience. *JAMA* 272:1349–1353.

Bell, R. G. 1997. Distribution and sources of microbial contamination on beef carcasses. *J. Appl. Microbiol.* 82:292–300.

Bellini, E. M., W. P. Elias, T. A. T. Gomes, T. L. Tanaka, C. R. Taddei, R. Huerta, F. Navarro-Garcia, and M. B. Martinez. 2005. Antibody response against plasmid-encoded toxin (Pet) and the protein involved in intestinal colonization (Pic) in children with diarrhea produced by enteroaggregative *Escherichia coli*. *FEMS Immunol. Med. Microbiol.* 43:259–264.

Belongia, E. A., P.-H. Chyou, R. T. Greenlee, G. Perez-Perez, W. F. Bibb, and E. O. DeVries. 2003. Diarrhea incidence and farm-related risk factors for *Escherichia coli* O157:H7 and *Campylobacter jejuni* antibodies among rural children. *J. Infect. Dis.* 187:1460–1468.

Berger, S. A., N. A. Young, and S. C. Edberg. 1989. Relationship between infectious diseases and human blood type. *Eur. J. Clin. Microbiol. Infect. Dis.* 8:681–689.

Bergwerff, A. A., and F. van Knapen. 2006. Surface plasmon resonance biosensors for detection of pathogenic microorganisms: strategies to secure food and environmental safety. *J. AOAC Int.* 89:826–831.

Berin, M. C., A. Darfeuille-Michaud, L. J. Egan, Y. Miyamoto, and M. F. Kagnoff. 2002. Role of EHEC O157:H7 virulence factors in the activation of intestinal epithelial cell NF-β and MAP kinase pathways and the upregulated expression of interleukin 8. *Cell. Microbiol.* 4:635–647.

Besser, R. E., P. M. Griffin, and L. Slutsker. 1999. *Escherichia coli* O157:H7 gastroenteritis and the hemolytic uremic syndrome: an emerging infectious disease. *Annu. Rev. Med.* 50:355–367.

Besser, R. E., S. M. Lett, J. T. Weber, M. P. Doyle, T. J. Barrett, J. G. Wells, and P. M. Griffin. 1993. An outbreak of diarrhea and hemolytic uremic syndrome from *Escherichia coli* O157:H7 in fresh-pressed apple cider. *JAMA* 269:2217–2220.

Beuchat, L. R. 1998. Surface decontamination of fruits and vegetables eaten raw: a review. WHO/FSF/FOS/98.2. World Health Organization, Food Safety Unit.

Beuchat, L. R., and J. H. Ryu. 1997. Produce handling and processing practices. *Emerg. Infect. Dis.* 3:459–465.

Beutin, L., M. A. Montenegro, I. Orskov, F. Orskov, J. Prada, S. Zimmermann, and R. Stephan. 1989. Close association of verotoxin Shiga-like toxin production with enterohemolysin production in strains of *Escherichia coli*. *J. Clin. Microbiol.* 27:2559–2564.

Beutin, L., S. Aleksic, S. Zimmermann, and K. Gleier. 1994. Virulence factors and phenotypical traits of verotoxigenic strains of *Escherichia coli* isolated from human patients in Germany. *Med. Microbiol. Immun.* 183:13–21.

Bielaszewska, M., and H. Karch. 2000. Non-O157 Shiga toxin verocytotoxin-producing *Escherichia coli* strains: epidemiological significance and microbiological diagnosis. *World J. Microbiol. Biotech.* 16:711–718.

Bielaszewska, M., J. Janda, K. Blahova, H. Minarikova, E. Jikova, M. A. Karmali, J. Laubova, J. Sikulova, M. A. Preston, R. Khakhria, H. Karch, H. Klazarova, and O. Nyc. 1997. Human *Escherichia coli* O157:H7 infection associated with the consumption of unpasteurized goat's milk. *Epidemiol. Infect.* 119:299–305.

Bilge, S. S., C. R. Clausen, W. Lau, and S. L. Moseley. 1989. Molecular characterization of a fimbrial adhesin, F1845, mediating diffuse adherence of diarrhea–associated *Escherichia coli* to HEp-2 cells. *J. Bacteriol.* 171:4281–4289.

Bilge, S. S., J. M. Apostol, Jr., K. J. Fullner, and S. L. Moseley. 1993. Transcriptional organization of the F1845 fimbrial adhesin determinant of *Escherichia coli*. *Mol. Microbiol.* 7:993–1006.

Blanco, J. E., M. Blanco, M. P. Alonso, A. Mora, G. Dahbi, M. A. Coira, and J. Blanco. 2004. Serotypes, virulence genes, and intimin types of Shiga toxin verotoxin-producing *Escherichia coli* isolates from human patients: prevalence in Lugo, Spain, from 1992 through 1999. *J. Clin. Microbiol.* 43:311–319.

Blanco, J., M. Blanco, M. P. Alonso, J. E. Blanco, E. A. González, and J. I. Garabal. 1992. Characteristics of haemolytic *Escherichia coli* with particular reference to production of cytotoxic necrotizing factor type 1 (CNF1). *Res. Microbiol.* 143:869–878.

Blaser, M. J., J. Cravens, B. W. Powers, F. M. Laforce, and W. L. L. Wang. 1979. *Campylobacter enteritis* associated with unpasteurized milk. *Am. J. Med.* 67:715–718.

Blattner, F. R., F. Plunkett III, C. A. Bloch, N. T. Perna, V. Burland, M. Riley, J. Collado-Vides, J. D. Glasner, C. K. Rode, G. F. Mayhew, J. Gregor, N. W. Davis, H. A. Kirkpatrick, M. A. Goeden, D. J. Rose, B. Mau, and Y. Shao. 1997. The complete genome sequence of *Escherichia coli* K-12. *Science* 227:1453–1462.

Boerlin, P., S. A. McEwen, F. Boerlin-Petzold, J. B. Wilson, R. P. Johnson, and C. L. Gyles. 1999. Associations between virulence factors of Shiga toxin-producing *Escherichia coli* and disease in humans. *J. Clin. Microbiol.* 37:497–503.

Borczyk, A. A., N. Harnett, M. Lombos, and H. Lior. 1990. False positive identification of *Escherichia coli* O157 by commercial latex agglutination tests. *Lancet* 336:946–947.

Boyce, T. G., A. G. Pemberton, J. G. Wells, and P. M. Griffin. 1995. Screening for *Escherichia coli* O157:H7–a nationwide survey of clinical laboratories. *J. Clin. Microbiol.* 33:3275–3277.

Brashears, M., G. Loneragan, and S. Younts–Dahl. 2005. Controlling microbial contamination on the farm: an overview, p. 156–174. In J. N. Sofos (ed.), *Improving the Safety of Fresh Meat*. CRC Press, Boca Raton FL.

Bricher, J. L. 2003. Technology round-up: innovations in microbial interventions. *Food Saf. Mag.* 11:29–33.

Brigati, J. R., S. A. Ripp, C. M. Johnson, P. A. Iakova, P. Jegier, and G. S. Sayler. 2007. Bacteriophage-based bioluminescent bioreporter for the detection of *Escherichia coli* O157:H7. *J. Food Prot.* 70:1386–1392.

Brooks, J. T., S. Y. Rowe, P. Shillam, D. M. Heltzel, S. B. Hunter, L. Slutsker, R. M. Hoekstra, and S. P. Luby. 2001. *Salmonella* Typhimurium infections transmitted by chlorine–pretreated clover sprout seeds. *Am. J. Epidemiol.* 154:1020–1028.

Brunder, W., H. Schmidt, and H. Karch. 1996. KatP, a novel catalase-peroxidase encoded by the large plasmid of enterohaemorrhagic *Escherichia coli* O157:H7. *Microbiology* 142:3305–3315.

Buchanan, R. L., and M. P. Doyle. 1997. Foodborne disease significance of *Escherichia coli* O157:H7 and other enterohemorrhagic *E. coli*. *Food Technol.* 51:69–76.

Buchanan, R. L., and S. G. Edelson. 1999. Effect of pH–dependent, stationary phase acid resistance on the thermal tolerance of *Escherichia coli* O157:H7. *Food Microbiol.* 16:447–458.

Callaway, T. R., R. C. Anderson, T. S. Edrington, K. J. Genovese, T. L. Poole, R. B. Harvey, D. J. Nisbet, and K. D. Dunkley. 2005. Probiotics, vaccines, and other interventions for pathogen control in animals, p. 192–213. In J. N. Sofos (ed.), *Improving the Safety of Fresh Meat*. CRC Press, Boca Raton, FL.

Callaway, T. R., R. C. Anderson, T. S. Edrington, K. J. Genovese, K. M. Bischoff, T. L. Poole, Y. S. Jung, R. B. Harvey, and D. J. Nisbet. 2004. What are we doing about *Escherichia coli* O157:H7 in cattle? *J. Anim. Sci.* 82:93–99.

CDC. 1995a. *Escherichia coli* O157:H7 outbreak linked to commercially distributed dry-cured salami—Washington and California, 1994. *MMWR Morb. Mortal. Wkly. Rep.* 44:157–159.

CDC. 1995b. Community outbreak of hemolytic uremic syndrome attributable to *Escherichia coli* O111:NM—South Australia, 1995. *MMWR Morb. Mortal. Wkly. Rep.* 44:550–551.

CDC. 1996. Outbreak of *Escherichia coli* O157:H7 infection: Georgia and Tennessee, June 1995. *MMWR Morb. Mortal. Wkly. Rep.* 45:249–251.

CDC. 1997. Outbreaks of *Escherichia coli* O157:H7 infection and cryptosporidiosis associated with drinking unpasteurized apple cider—Connecticut and New York, October 1996. *MMWR Morb. Mortal. Wkly. Rep.* 46:4–8.

CDC. 2000. Outbreak of *Escherichia coli* O157:H7 infection associated with eating fresh cheese curds. *MMWR Morbid. Mortal. Wkly. Rep.* 49:911–913.

CDC. 2001a. Outbreaks of *Escherichia coli* O157:H7 infections among children associated with farm visits—Pennsylvania and Washington, 2000. *MMWR Morb. Mortal. Wkly. Rep.* 50:293–297.

CDC. 2001b. Outbreaks of *Escherichia coli* O157:H7 infection associated with eating alfalfa sprouts—Michigan and Virginia. *MMWR Morb. Mortal. Wkly. Rep.* 46:741–744.

CDC. 2002, Multistate outbreak of *Escherichia coli* O157:H7 infections associated with ground beef—United States, June–July, 2002. *MMWR Morb. Mortal. Wkly. Rep.* 51:637–639.

CDC. 2006. Surveillance for foodborne-disease outbreaks—United States, 1998–2002. *MMWR Morb. Mortal. Wkly. Rep.* 55:1–34.

CDC. 2007. Laboratory-confirmed non-O157 Shiga toxin—producing *Escherichia coli*—Connecticut, 2000–2005. *MMWR Morb. Mortal. Wkly. Rep.* 56:29–31.

Chapman, P. A., C. A. Siddon, P. M. Zadik, and L. Jewes. 1991. An improved selection medium for the isolation of *Escherichia coli* O157. *J. Med. Microbiol.* 35:107–110.

Childs, K. D., C. A. Simpson, W. Warren–Serna, G. Bellenger, B. Centrella, R. A. Bowling, J. Ruby, J. Stefanek, D. J. Vote, T. Choat, J. A. Scanga, J. N. Sofos, G. C. Smith, and K. E. Belk. 2006. Molecular characterization of *Escherichia coli* O157:H7 hide contamination routes: feedlot to harvest. *J. Food Prot.* 69:1240–1247.

Cieslak, P. R., S. J. Noble, D. J. Maxson, L. C. Empey, O. Ravenholt, G. Legarza, J. Tuttle, M. P. Doyle, T. J. Barrett, J. G. Wells, A. M. McNamara, and P. M. Griffin. 1997. Hamburger-associated *Escherichia coli* O157:H7 infection in Las Vegas: a hidden epidemic. *Am. J. Public Health* 87:176–180.

Code of Federal Regulations. 2001. Title 9, chapter III, Food Safety and Inspection Service, Department of Agriculture, parts 300–592. Document 9CFR300–592. U.S. Government Printing Office, Washington, DC.

Code of Federal Regulations. 2002. Title 21, chapter I, Food and Drug Administration, Department of Health and Human Services, parts 1–1299. Document 21CFR1–1299. U.S. Government Printing Office, Washington, DC.

Collinson, S. K., L. Emody, T. J. Trust, and W. W. Kay. 1992. Thin aggregative fimbriae from diarrheagenic *Escherichia coli*. *J. Bacteriol.* 174:4490–4495.

Cookson, S. T., and J. P. Nataro. 1996. Characterization of HEp-2 cell projection formation induced by diffusely adherent *Escherichia coli*. *Microb. Pathog.* 21:421–434.

Cords, B. R., S. L. Burnett, J. Hilgren, M. Finley, and J. Magnuson. 2005. Sanitizers: halogens, surface-active agents, and peroxides, p. 507–572. *In* P. M. Davidson, J. N. Sofos, and A. L. Branen (ed.), *Antimicrobials in Food*, 3rd ed. Taylor and Francis, Boca Raton, FL.

Cutter, C. N., W. J. Dorsa, and G. R. Siragusa. 1997. Parameters affecting the efficacy of spray washes against *Escherichia coli* O157:H7 and fecal contamination on beef. *J. Food Prot.* 60:614–618.

Dallas, W. S., and S. Falkow. 1980. Amino acid sequence homology between cholera toxin and *Escherichia coli* heat–labile toxin. *Nature* 288:499–501.

Davidson, P. M., and M. A. Harrison. 2002. Resistance and adaptation to food antimicrobials, sanitizers, and other process controls. *Food Tech.* 56:69–78.

Davis, M. A., D. D. Hancock, T. E. Besser, and D. R. Call. 2003. Evaluation of pulsed-field gel electrophoresis as a tool for determining the degree of genetic relatedness between strains of *Escherichia coli* O157:H7. *J. Clin. Microbiol.* 41:1843–1849.

Dean-Nystrom, E. A., A. R. Melton-Celsa, J. F. L. Pohlanz, H. W. Moon, and A. D. O'Brien. 2003. Comparative pathogenicity of *Escherichia coli* O157 and intimin-negative non-O157 shiga toxin–producing *E. coli* strains in neonatal pigs. *Infect. Immun.* 71:6526–6533.

DeVinney, R., M. Stein, D. Reinscheid, A. Abe, S. Ruschkowski, and B. B. Finlay. 1999. Enterohemorrhagic *Escherichia coli* O157:H7 produces Tir, which is translocated to the host cell membrane but is not tyrosine phosphorylated. *Infect. Immun.* 67:2389–2398.

DeVinney, R., J. L. Puente, A. Gauthier, D. Goosney, and B. B. Finlay. 2001. Enterohaemorrhagic and enteropathogenic *Escherichia coli* use a different Tir-based mechanism for pedestal formation. *Mol. Microbiol.* 41:1445–1458.

Diez-Gonzalez, F. 2005. The use of diet to control pathogens in animals, p. 175–191. *In* J. N. Sofos (ed.), *Improving the Safety of Fresh Meat*. CRC Press, Boca Raton, FL.

Doyle M., and V. Padhye. 1989. *Escherichia coli*, p. 236–282. *In* M. P. Doyle (ed.), *Foodborne Bacterial Pathogens*. Marcel Dekker Inc., New York, NY.

Doyle, M. P., and D. J. Roman. 1982. Prevalence and survival of *Campylobacter jejuni* in unpasteurized milk. *Appl. Environ. Microbiol.* 44:1154–1158.

Duffy, L. L., F. H. Grau, and P. B. Vanderlinde. 2000. Acid resistance of enterohaemorrhagic and generic *Escherichia coli* associated with foodborne disease and meat. *Int. J. Food Microbiol.* 60:83–89.

Elliott, S. J., L. A. Wainwright, T. K. McDaniel, K. G. Jarvis, Y. Deng, L. Lai, B. P. McNamara, M. S. Donnenberg, and J. B. Kaper. 1998. The complete sequence of the locus of enterocyte effacement LEE from enteropathogenic *Escherichia coli* E2348/69. *Mol. Microbiol.* 28:1–4.

Elliott, S. J., S. W. Hutcheson, M. S. Dubois, J. L. Mellies, L. A. Wainwright, M. Batchelor, G. Frankel, S. Knutton, and J. B. Kaper. 1999. Identification of CesT, a chaperone for the type III secretion of Tir in enteropathogenic *Escherichia coli*. *Mol. Microbiol.* 33:1176–1189.

Entis, P., D. Y. C. Fung, M. W. Griffiths, L. McIntre, S. Russell, and A. N. Sharpe. 2001. Rapid methods for detection, identification, and enumeration, p. 89–126. *In* F. Pouch and K. Ito (ed.), *Compendium of Methods for the Microbial Examination of Foods*, 4th ed. American Public Health Association, Washington, DC.

Falkow, S. 1996. The evolution of pathogenicity in *Escherichia coli*, *Shigella*, and *Salmonella*, p. 2723–2769. *In* F. C. Neidhardt (ed.), *Escherichia coli* and *Salmonella: Cellular and Molecular Biology*, 2nd ed. American Society for Microbiology. Washington, DC.

Farber, J. M. 1996. An introduction to the hows and whys of molecular typing. *J. Food Prot.* 59:1091–1101.

Foley, D., M. Euper, F. Caporaso, and A. Prakash. 2004. Irradiation and chlorination effectively reduces *Escherichia coli* O157:H7 inoculated on cilantro (Coriandrum sativum) without negatively affecting quality. *J. Food Prot.* 67:2092–2098.

Frankel, G., A. D. Phillips, I. Rosenshine, G. Dougan, J. B. Kaper, and S. Knutton. 1998. Enteropathogenic and enterohemorrhagic *Escherichia coli*: more subversive elements. *Mol. Microbiol.* 30:911–921.

Fratamico, P. N., L. K. Bagi, and T. Pepe. 2000. A multiplex polymerase chain reaction assay for rapid detection and identification of *Escherichia coli* O157:H7 in foods and bovine feces. *J. Food Prot.* 63:1032–1037.

Fratamico, P. M., A. Gehring, J. Karns, and J. van Kessel. 2005. Detecting pathogens in cattle and meat, p. 24–55. *In* J. N. Sofos (ed.), *Improving the Safety of Fresh Meat*. CRC Press, Boca Raton, FL.

Friedrich, A. W., M. Bielaszewska, W. L. Zhang, M. Pulz, T. Kuczius, A. Ammon, and H. Karch. 2002. *Escherichia coli* harboring Shiga toxin 2 gene variants: frequency and association with clinical symptoms. *J. Infect. Dis.* **185:**74–84.

Friedrich, A. W., S. Lu, M. Bielaszewska, R. Prager, P. Bruns, J-G. Xu, H, Tschäpe, and H. Karch. 2006. Cytolethal distending toxin in *Escherichia coli* O157:H7: spectrum of conservation, structure, and endothelial toxicity. *J. Clin. Microbiol.* **44:**1844–1846.

FSIS. 1993. FSIS correlation packet, interim guidelines for inspectors. United States Department of Agriculture, Washington, DC.

FSIS. 1996. Pathogen reduction; hazard analysis and critical control point (HACCP) systems; final rule. 9 CFR Part 304. *Fed. Regist.* **61:**38805–38989.

FSIS. 1999. Rules and regulations: sanitation requirements for official meat and poultry establishments. *Fed. Regist.* **64:**56400–56418.

FSIS. 2005. Federal Register notice: HACCP plan reassessment for mechanically tenderized beef products. *Fed. Regist.* **70:**30331–30334.

Gannon, V. P. J., S. Dsouza, T. Graham, and R. K. King. 1997. Specific identification of *Escherichia coli* O157:H7 using a multiplex PCR assay, p. 81–82. *In Mechanisms in the Pathogenesis of Enteric Diseases*, vol. 412. Plenum Press, New York, NY.

Garcia, M. I., P. Gounon, P. Courcoux, A. Labigne, and C. Le Bouguenec. 1996. The afimbrial adhesive sheath encoded by the *afa-3* gene cluster of pathogenic *Escherichia coli* is composed on two adhesins. *Mol. Microbiol.* **19:**683–693.

Garcia, M. I., M. Jouve, J. P. Nataro, P. Gounon, and C. Le Bouguenec. 2000. Characterization of the AfaD-like family of invasins encoded by pathogenic *Escherichia coli* associated with intestinal and extra–intestinal infections. *FEBS Lett.* **479:**111–117.

Gill, C. O. 2005. Sources of microbial contamination at slaughtering plants, p. 231–243. *In* J. N. Sofos (ed.), *Improving the Safety of Fresh Meat*. CRC Press, Boca Raton, FL.

Girón, J. A., A. S. Ho, and G. K. Schoolnik. 1991. An inducible bundle-forming pilus of enteropathogenic *Escherichia coli*. *Science* **254:**710–713.

Glass, K. A., J. M. Loeffelholz, J. P. Ford, and M. P. Doyle. 1992. Fate of *Escherichia coli* O157:H7 as affected by pH or sodium chloride and in fermented, dry sausage. *Appl. Environ. Microbiol.* **58:**2513–2516.

Goldwater, P. N., and K. A. Bettelheim. 1995. The role of enterohemorrhagic *Escherichia coli* serotypes other than O157:H7 as causes of disease in Australia. *Commun. Dis. Intell.* **19:**2–4.

Goodridge, L., J. R. Chen, and M. Griffiths. 1999. The use of a fluorescent bacteriophage assay for detection of *Escherichia coli* O157:H7 in inoculated ground beef and raw milk. *Int. J. Food Microbiol.* **47:**43–50.

Greer, G. G., and B. D. Dilts. 1990. Inability of a bacteriophage pool to control beef spoilage. *Int. J. Food Microbiol.* **10:**331–342.

Griffin, P. M., and R. V. Tauxe. 1991. The epidemiology of infections caused by *Escherichia coli* O157:H7, other enterohemorrhagic *E. coli*, and the associated hemolytic uremic syndrome. *Epidemiol. Rev.* **13:**60–98.

Guan, D., and D. G. Hoover. 2005. Emerging decontamination techniques for meat, p. 388–417. *In* J. N. Sofos (ed.), *Improving the Safety of Fresh Meat*. CRC Press, New York, NY.

Guion, C. E., T. J. Ochoa, C. M. Walker, F. Barletta, and T. G. Cleary. 2008. Detection of diarrheagenic *Escherichia coli* by use of melting-curve analysis and real-time multiplex PCR. *J. Clin. Microbiol.* **46:**1752–1757.

Gunzberg, S. T., J. Chang, S. J. Elliott, V. Burke, and M. Gracey. 1993. Diffuse and enteroaggregative patterns of adherence of enteric *Escherichia coli* isolated from aboriginal children from the Kimberly Region of Western Australia. *J. Infect. Dis.* **167:**755–758.

Guth, B. E. C., S. Ramos, A. M. F. Cerqueira, J. R. C. Andrade, and T. A. T. Gomes. 2002. Phenotypic and genotypic characteristics of Shiga toxin-producing *Escherichia coli* strains isolated from children in Sao Paulo, Brazil. *Mem. Inst. Oswaldo Cruz* **97:**1085–1089.

Hadley, P. J., J. S. Holder, and M. H. Hinton. 1997. Effects of fleece soiling and skinning method on the microbiology of sheep carcasses. *Vet. Rec.* **140:**570–574.

Hajna, A. A. 1955. A new enrichment broth medium for gram-negative organisms of the intestinal group. *Public Health Lab.* **13:**83–89.

Hajra, T. K., P. K. Bag, S. C. Das, S. Mukherjee, A. Khan, and T. Ramamurthy. 2007. Development of a simple latex agglutination assay for detection of Shiga toxin-producing *Escherichia coli* (STEC) by using polyclonal antibody against STEC. *Clin. Vacc. Immun.* **14:**600–604.

Hammes, W. P., A. Bantleon, and S. Min. 1990. Lactic acid bacteria in meat fermentation. *FEMS Microbiol. Rev.* **87:**165–174.

Hao, Y. Y., and R. E. Brackett. 1993. Growth of *Escherichia coli* O157:H7 in modified atmosphere. *J. Food Prot.* **56:**330–332.

Hara-Kudo, Y., H. Konuma, M. Iwaki, F. Kasuga, Y. Sugita-Konishi, Y. Itoh, and S. Kumagai. 1997. Potential hazard of radish sprouts as a vehicle of *Escherichia coli* O157:H7. *J. Food Prot.* **60:**1125–1127.

Hardin, M. D., G. R. Acuff, L. M. Lucia, J. S. Oman, and J. W. Savell. 1995. Comparison of methods for decontamination of beef carcass surfaces. *J. Food Prot.* **58:**368–374.

Hartland, E. L., M. Batchelor, R. M. Delahay, C. Hale, S. Matthews, G. Dougan, S. Knutton, I. Connerton, and G. Frankel. 1999. Binding of intimin from enteropathogenic *Escherichia coli* to Tir and to host cells. *Mol. Microbiol.* **32:**151–158.

Hedberg, C. W., S. J. Savarino, J. M. Besser, C. J. Paulus, V. M. Thelen, L. J. Myers, D. N. Cameron, T. J. Barrett, J. B. Kaper, M. T. Osterholm, W. Boyer, F. Kairis, L. Gabriel, J. Soler, L. Gyswyt, S. Bray, R. Carlson, C. Hooker, A. Fasano, K. Jarvis, T. McDaniel, and N. Tornieporth. 1997. An outbreak of foodborne illness caused by *Escherichia coli* O39:NM, an agent not fitting into the existing scheme for classifying diarrheogenic *E. coli*. *J. Infect. Dis.* **176:**1625–1628.

Hepburn, N. F., M. MacRae, and L. D. Ogden. 2002. Survival of *Escherichia coli* O157 in abattior waste products. *Lett. Appl. Microbiol.* **35:**223–236.

Hicks, S., D. C. A. Candy, and A. D. Phillips. 1996. Adhesion of enteroaggregative *Escherichia coli* to pediatric intestinal mucosa in vitro. *Infect. Immun.* **64:**4751–4760.

Hiett, K. L. 2005. Molecular typing methods for tracking pathogens, p. 591–605. *In* J. N. Sofos (ed.), *Improving the Safety of Fresh Meat*. CRC Press, Boca Raton, FL.

Higuchi, R., G. Dollinger, P. S. Walsh, and R. Griffith. 1992. Simultaneous amplification and detection of specific DNA sequences. *BioTechniques* **10:**413–417.

Higuchi, R., C. Fockler, G. Dollinger, and R. Watson. 1993. Kinetic PCR analysis—real-time monitoring of DNA amplification reactions. *BioTechniques* **11:**1026–1030.

Hiramatsu, R., M. Matsumoto, Y. Miwa, Y. Suzuki, M. Saito, and Y. Miyazaki. 2002. Characterization of Shiga toxin-producing *Escherichia coli* O26 strains and establishment of selective isolation media for these strains. *J. Clin. Microbiol.* **40:**922–925.

Huang, S. W., C. H. Chang, T. F. Tai, and T. C. Chang. 1997. Comparison of the beta-glucuronidase assay and the conventional method for identification of *Escherichia coli* on eosin–methylene blue agar. *J. Food Prot.* **60:**6–9.

Hussein, H. S., and L. M. Bollinger. 2005. Prevalence of Shiga toxin-producing *Escherichia coli* in beef cattle. *J. Food Prot.* **68:**2224–2241.

Hussein, H. S., L. M. Bollinger, and M. K. Hall. 2008. Growth and enrichment medium for detection and isolation of Shiga toxin-producing *Escherichia coli* in cattle feces. *J. Food Prot.* **71:**927–933.

ICMSF. 1980. *Microbial Ecology of Foods*, vol. 1. *Factors Affecting Life and Death of Microorganisms.* Academic Press, New York, NY.

ICMSF. 1996. *Microorganisms in Foods*, vol. 5. *Characteristics of Microbial Pathogens.* Blackie Academic & Professional, New York, NY.

Irino, K., T. M. I. Vaz, M. I. C. Medeiros, M. Kato, T. A. T. Gomes, M. A. M. Vieira, and B. E. C. Guth. 2007. Serotype diversity as a drawback in the surveillance of Shiga toxin-producing *Escherichia coli* infections in Brazil. *J. Med. Microbiol.* **56:**565–567.

Isberg, R. R., D. L. Voorhis, and S. Falkow. 1987. Identification of invasion: a protein that allows enteric bacteria to penetrate cultured mammalian cells. *Cell* **50:**769–776.

Jay, J. M. 1982. Antimicrobial properties of diacetyl. *Appl. Environ. Microbiol.* **44:**525–532.

Johnson, J. R., E. Oswald, T. T. O'Bryan, M. A. Kuskowski, and L. Spanjaard. 2002. Phylogenetic distribution of virulence–associated genes among *Escherichia coli* isolates associated with neonatal bacterial meningitis in The Netherlands. *J. Infect. Dis.* **185:**774–784.

Kamat, A. S., S. Khare, T. Doctor, and P. M. Nair. 1997. Control of *Yersinia enterocolitica* in raw pork and pork products by gamma-irradiation. *Int. J. Food Microbiol.* **36:**69–76.

Kang, D.-H., and D. Y. C. Fung. 1999. Effect of diacetyl on controlling *Escherichia coli* O157:H7 and *Salmonella* Typhimurium in the presence of starter culture in a laboratory medium and during meat fermentation. *J. Food Prot.* **62:**975–979.

Kaper, J. B., J. P. Nataro, and H. L. T. Mobley. 2004. Pathogenic *Escherichia coli*. *Nat. Rev. Microbiol.* **2:**123–140.

Karch, H., C. Janetzki-Mittmann, S. Aleksic, and M. Datz. 1996. Isolation of enterohemorrhagic *Escherichia coli* O157 strains from patients with hemolytic-uremic syndrome by using immunomagnetic separation, DNA-based methods, and direct culture. *J. Clin. Microbiol.* **34:**516-519.

Karmali, M. A. 1989. Infection by verocytotoxin-producing *Escherichia coli*. *Clin. Microbiol. Rev.* **2:**15–38.

Keen, J. E., T. E. Wittum, J. R. Dunn, J. L. Bono, and M. E. Fontenot. 2003. Occurrence of STEC O157, O111, and O26 in livestock at agricultural fairs in the United States, p. 22. *Proceedings of the 5th International Symposium on Shiga Toxin-Producing* Escherichia coli *Infection.* Edinburgh, United Kingdom.

Keene, W. E., J. M. McAnulty, F. C. Hoesley, L. P. Williams, K. Hedberg, G. L. Oxman, T. J. Barrett, M. A. Pfaller, and D. W. Fleming. 1994. A swimming-associated outbreak of hemorrhagic colitis caused by *Escherichia coli* O157:H7 and Shigella sonnei. *N. Engl. J. Med.* **331:**579–584.

Keene, W. E., K. Hedberg, D. E. Herriott, D. D. Hancock, R. W. McKay, T. J. Barrett, and D. W. Fleming. 1997. Prolonged outbreak of *Escherichia coli* O157:H7 infections caused by commercially distributed raw milk. *J. Infect. Dis.* **176:**815–818.

Kenny, B., R. DeVinney, M. Stein, D. J., Reinscheid, E. A. Frey, and B. B. Finlay. 1997. Enteropathogenic *Escherichia coli* (EPEC) transfers its receptors for intimate adherence into mammalian cells. *Cell* **28:**511–520.

Killalea, D., L. R. Ward, D. Roberts, J. deLouvois, F. Sufi, J. M. Stuart, P. G. Wall, M. Susman, M. Schwieger, P. J. Sanderson, I. S. T. Fisher, P. S. Mead, O. N. Gill, C. L. R. Bartlett, and B. Rowe. 1996. International epidemiological and microbiological study of outbreak of *Salmonella* agona infection from a

ready to eat savoury snack—England and Wales and the United States. *Brit. Med. J.* **313:**1105–1107.

Knutton, S., I. Rosenshine, M. J. Pallen, I. Nisan, B. C. Neves, C. Bain, C. Wolff, G. Dougan, and G. Frankel. 1998. A novel EspA–associated surface organelle of enteropathogenic *Escherichia coli* involved in protein translocation into epithelial cells. *EMBO J.* **17:**2166–2176.

König, B., A. Ludwig, and W. König. 1994. Pore formation by the *Escherichia coli* α–hemolysin: role for mediator releases from human inflammatory cells. *Infect. Immun.* **62:**4611–4617.

Koutsoumanis, K., and P. S. Taoukis. 2005. Meat safety, refrigerated transport and storage: modeling and management, p. 503–561. *In* J. N. Sofos (ed.), *Improving the Safety of Fresh Meat.* CRC Press, Boca Raton, FL.

Kuby, J. 1997. Overview of the immune system, p. 3–24. *In* D. Allen (ed.), *Immunology*, 3rd ed. W. H. Freeman & Co., New York, NY.

Kudva, I. T., P. G. Hatfield, and C. J. Hovde. 1996. *Escherichia coli* O157:H7 in microbial flora of sheep. *J. Clin. Microbiol.* **34:**431–433.

Lacroix, M., F. Chiasson, J. Borsa, and B. Ouattara. 2004. Radio sensitization of *Escherichia coli* and *Salmonella* typhi in presence of active compounds. *Radiat. Phys. Chem.***71:**65–68.

Lahti, E., T. Johansson, T. Honkanen–Buzalski, P. Hill, and E. Nurmi. 2001. Survival and detection of *Escherichia coli* O157:H7 and *Listeria monocytogenes* during the manufacture of dry sausage using two different starter cultures. *Food Microbiol.* **18:**75–85.

Lazcka, O., F. J. del Campo, and F. X. Munoz. 2007. Pathogen detection: a perspective of traditional methods and biosensors. *Biosens. Bioelectron.* **22:**1205–1217.

Lee, S. Y., M. Costello, and D. H. Kang. 2004. Efficacy of chlorine dioxide gas as a sanitizer of lettuce leaves. *J. Food Prot.* **67:**1371–1376.

Leistner, L., and G. W. Gould. 2002. *Multiple Hurdle Technologies.* Kluwer Academic Publishers, New York, NY.

LeJeune, J. T., T. E. Besser, D. H. Rice, J. L. Berg, R. P. Stilborn, and D. D. Hancock. 2004. Longitudinal study of fecal shedding of *Escherichia coli* O157:H7 in feedlot cattle: predominance and persistence of specific clonal types despite massive cattle population turnover. *Appl. Environ. Microbiol.* **70:**377–384.

Levine, M. M., C. Ferreccio, V. Prado, M. Cayazzo, P. Abrego, J. Martinez, L. Maggi, M. M. Baldini, W. Martin, D. Maneval, B. Kay, L. Guers, H. Lior, S. S. Wasserman, and J. P. Nataro. 1993. Epidemiologic studies of Escherichia coli diarrheal infections in a low socioeconomic level peri-urban community in Santiago, Chile. *Am. J. Epidemiol.* **138:**849–869.

Lewis, S. J., A. Velasquez, S. L. Cuppett, and S. R. McKee. 2002. Effect of electron beam irradiation on poultry meat safety and quality. *Poult. Sci.* **81:**896–903.

Lillard, H. S. 1980. Effect on broiler carcasses and water of treating chiller water with chlorine or chlorine dioxide. *Poult. Sci.* **59:**1761–1766.

Lingwood, C. A. 1994. Verotoxin-binding in human renal sections. *Nephron* **66:**21–28.

Livak, K. J., S. J. A. Flood, J. Marmaro, W. Giusti, and K. Deetz. 1995. Oligonucleotides with fluorescent dyes at opposite ends provide a quenched probe system useful for detecting PCR product and nucleic-acid hybridization. *PCR Methods Appl.* **4:**357–362.

Loessner, M. J., and M. Busse. 1990. Bacteriophage-typing of *Listeria* species. *Appl. Environ. Microbiol.* **56:**1912–1918.

Lopez, E. L., M. M. Contrini, and M. F. de Rosa. 1998. Epidemiology of Shiga toxin-producing *Escherichia coli* in South America, p. 15–22. *In* J. B. Kaper and A. D. O'Brien (ed.), Escherichia coli

O157:H7 and Other Shiga Toxin-Producing E. coli *Strains.* ASM Press, Washington, DC.

Louise, C. B., and T. G. Obrig. 1995. Specific interaction of *Escherichia coli* O157:H7-derived Shiga-like toxin II with human renal endothelial cells. *J. Infect. Dis.* **172:**1397–1401.

MacDonald D. M., M. Fyfe, A. Paccagnella, A. Trinidad, K. Louie, and D. Patrick. 2004. *Escherichia coli* O157:H7 outbreak linked to salami, British Columbia, Canada, 1999. *Epidemiol. Infect.* **132:**283–289.

March, S. B., and S. Ratnam. 1986. Sorbitol MacConkey medium for detection of *Escherichia coli* O157–H7 associated with hemorrhagic colitis. *J. Clin. Microbiol.* **23:**869–872.

Marques, L. R. M., A. T. Tavechio, C. M. Abe, and T. A. T. Gomes. 2003. Search for cytolethal distending toxin production among fecal *Escherichia coli* isolates from Brazilian children with diarrhea and without diarrhea. *J. Clin. Microbiol.* **41:**2206–2208.

McCrae, K. R., and D. Cines. 2000. Throbocytopenic purpura and hemolytic uremic syndrome, p. 2126–2138. *In* R. Hoffman, E. J. Benz, Jr., S. J. Shattil, H. J. Cohen, L. E. Silberstein, and P. McGlave (ed.), *Hematology: Basic Principles and Practice*, 3rd ed. Churchill Livingston, Philadelphia, PA.

McGrath, S., G. F. Fitzgerald, and D. van Sinderen. 2002. Identification and characterization of phage-resistance genes in temperate lactococcal bacteriophages. *Mol. Microbiol.* **43:**509–520.

Mead, P. S., L. Slutsker, V. Dietz, L. F. McCaig, J. S. Bresee, and C. Shapiro. 1999. Food-related illness and death in the United States. *Emerg. Infect. Dis.* **5:**607–625.

Meng J., M. P. Doyle, T. Zhao, and S. Zhao. 2001a. Enterohemorrhagic *Escherichia coli*, p. 193–213. *In* M. P. Doyle, L. R. Beuchat, and T. J. Montville, *Food Microbiology: Fundamentals and Frontiers*, 2nd ed. ASM Press, Washington, DC.

Meng, J., P. Feng, and M. P. Doyle. 2001b. Pathogenic *Escherichia coli*, p. 331–342. *In* C. Vanderzant and D. Splittstoesser (ed.), *Compendium of Methods for the Microbiological Examination of Foods*, 4th ed. American Public Health Association, Washington, DC.

Meyer, M. T., and R. H. Vaughn. 1969. Incidence of *Escherichia coli* in black walnut meats. *Appl. Microbiol.* **18:**925–931.

Michel, P., J. B. Wilson, S. W. Martin, R. C. Clark, S. A. McEwen, and C. L. Gyles. 1999. Temporal and geographical distributions of reported cases of *Escherichia coli* O157:H7 infection in Ontario. *Epidemiol. Infect.* **122:**193–200.

Michino, H., K. Araki, S. Minami, S. Takaya, N. Sakai, M. Miyazaki, A. Ono, and H. Yanagawa. 1999. Massive outbreak of *Escherichia coli* O157:H7 infection in schoolchildren in Sakai City, Japan, associated with consumption of white radish sprouts. *Am. J. Epidemiol.* **150:**787–796.

Millard, P. S., K. F. Gensheimer, D. G. Addiss, D. M. Sosin, G. A. Beckett, A. Houckjankoski, and A. Hudson. 1994. An outbreak of cryptosporidiosis from fresh-pressed apple cider. *JAMA* **272:**1592–1596.

Moon, H. W., S. C. Whipp, R. A. Argenzio, M. M. Levine, and R. A. Giannella. 1983. Attaching and effacing activities of rabbit and human enteropathogenic *Escherichia coli* in pig and rabbit intestines. *Infect. Immun.* **41:**1340–1351.

Morgan, D., C. P. Newman, D. N. Hutchinson, A. M. Walker, B. Rowe, and F. Majid. 1993. Verotoxin producing *Escherichia coli* O157 infections associated with the consumption of yoghurt. *Epidemiol. Infect.* **111:**181–187.

Moss, S., and A. J. Frost. 1984. The resistance to chemotherapeutic agents of *Escherichia coli* from domestic dogs and cats. *Aust. Vet. J.* **61:**82–84.

Moyo, S. J., S. Y. Maselle, M. I. Matee, N. Langeland, and H. Mylvaganam. 2007. Identification of diarrheagenic

Escherichia coli isolated from infants and children in Dar es Salaam, Tanzania. *BMC Infect. Dis.* **7:**92.

Mulvey, M. R., E. Bryce, D. A. Boyd, M. Ofner-Agostini, A. M. Land, A. E. Simor, and S. Paton. 2005. Molecular characterization of cefoxitin-resistant *Escherichia coli* from Canadian hospitals. *Antimicrob. Agents Chemother.* **49:**358–365.

Murano, P. S., E. A. Murano, and D. G. Olson. 1998 Irradiated ground beef: sensory and quality changes during storage under various packaging conditions. *J. Food Sci.* **63:**548–551.

NACMCF. 2006. Requisite scientific parameters for establishing the equivalence for alternative methods of pasteurization. *J. Food Prot.* **69:**1190–1216.

Naim, F., S. Messier, L. Saucier, and G. Piette. 2004. Postprocessing in vitro digestion challenge to evaluate survival of *Escherichia coli* O157:H7 in fermented dry sausages. *Appl. Environ. Microbiol.* **70:**6637–6642.

Naimi, T. S., J. H. Wicklund, S. J. Olsen, G. Krause, J. G. Wells, J. M. Bartkus, D. J. Boxrud, M. Sullivan, H. Kassenborg, J. M. Besser, E. D. Mintz, M. T. Osterholm, and C. W. Hedberg. 2003. Concurrent outbreaks of *Shigella sonnei* and enterotoxigenic *Escherichia coli* infections associated with parsley: Implications for surveillance and control of foodborne illness. *J. Food Prot.* **66:**535–541.

Nakamura, H., and I. Karube. 2003. Current research activity in biosensors. *Anal. Bioanal. Chem.* **377:**446–468.

Nataro, J. P., and J. B. Kaper. 1998. Diarrheagenic *Escherichia coli. Clin. Microbiol Rev.* **11:**142–201.

Nataro, J. P., D. Yikang, D. Yingkang, and K. Walker. 1994. Aggr, a transcriptional activator of aggregative adherence fimbria I expression in enteroaggregative *Escherichia coli. J. Bacteriol.* **176:**4691–4699.

Nguyen, T. V., P. L. Van, C. L Huy, K. N. Gia, and A. Weintraub. 2005. Detection and characterization of diarrheagenic *Escherichia coli* from young children in Hanoi, Vietnam. *J. Clin. Microbiol.* **43:**755–760.

Nguyen, T. V., P. Le Van, C. Le Huy, and A. Weintraub. 2004. Diarrhea caused by rotavirus in children less than 5 years of age in Hanoi, Vietnam. *J. Clin. Microbiol.* **42:**5745–5750.

NIIDIDCD/MHWJ. 1997. Verocytotoxin-producing *Escherichia coli* enterohemorrhagic *E. coli* infection, Japan, 1996–June 1997. *Infect. Agents Surveill. Rep.* **18:**153–154.

Nychas, G. -J. E., D. L. Marshall, and J. N. Sofos. 2007. Meat, poultry and seafood spoilage, p. 105–140. *In* M. P. Doyle and L. R. Beuchat, *Food Microbiology: Fundamentals and Frontiers*, 3rd ed. ASM Press, Washington, DC.

O'Brien, A. D., and R. K. Holmes. 1987. Shiga and Shiga-like toxins. *Microbiol. Rev.* **51:**206–220.

O'Brien, A. D., and R. K. Holmes. 1996. Protein toxins of *Escherichia coli* and *Salmonella*, p. 2788–2802. *In* F. C. Neidhardt (ed.), Escherichia coli *and* Salmonella. ASM Press, Washington, DC.

Obrig, T. G., P. J. Del Vecchio, J. E. Brown, T. P. Moran, B. M. Rowland, T. K. Judge, and S. W. Rothman. 1988. Direct action of cytotoxic Shiga toxin on human vascular endothelial cells. *Infect. Immun.* **56:**2373–2378.

Oda, M., M. Morita, H. Unno, and Y. Tanji. 2004. Rapid detection of *Escherichia coli* O157:H7 by using green fluorescent protein-labeled PP01 bacteriophage. *Appl. Environ. Microbiol.* **70:**527–534.

Okeke, I. N., and J. P. Nataro. 2001. Enteroaggregative *Escherichia coli. Lancet* **i:**304–313.

Okrend, A. J. G., B. E. Rose, and C. P. Lattuada. 1992. Isolation of *Escherichia coli* O157:H7 using O157 specific antibody coated magnetic beads. *J. Food Prot.* **55:**214–217.

Paton, A. W., J. C. Paton, P. N. Goldwater, and P. A. Manning. 1993. Direct detection of *Escherichia coli* Shiga-like toxin genes in primary fecal cultures by polymerase chain reaction. *J. Clin. Microbiol.* **31:**3063–3067.

Phillips, C. A. 1996. Modified atmosphere packaging and its effects on the microbiological quality and safety of produce. *Int. J. Food Sci. Tech.* **31:**463–479.

Pordesimo, L. O., E. G. Wilkerson, A. R. Womac, and C. N. Cutter. 2002. Process engineering variables in the spray washing of meat and produce. *J. Food Prot.* **65:**222–237.

Possé, B., L. De Zutter, M. Heyndrickx, and L. Herman. 2008. Novel differential and confirmation plating media for Shiga toxin-producing *Escherichia coli* serotypes O26, O103, O111, O145 and sorbitol-positive and -negative O157. *FEMS Microbiol. Lett.* **282:**124–131.

Potter, M. E., A. F. Kaufmann, P. A. Blake, and R. A. Feldman. 1984. Unpasteurized milk—the hazards of a health fetish. *JAMA* **252:**2048–2052.

Quinto, E. J., and A. Cepeda. 1997. Incidence of toxigenic *Escherichia coli* in soft cheese made with raw or pasteurized milk. *Lett. Appl. Microbiol.* **24:**291–295.

Radke, S., and E. Alocilja. 2003. Impedimetric biosensor for the rapid detection of *Escherichia coli* O157:H7, abstr. 037060. *2003 Ann. Int. Meet. Soc. Eng. Agric. Food Biol. Sys.* Las Vegas, NV, 27 to 30 July 2003.

Rangel, J. M., P. H. Sparling, C. Crowe, P. M. Griffin, and D. L. Swerdlow. 2005. Epidemiology of *Escherichia coli* O157:H7 outbreaks, United States, 1982–2002. *Emerg. Infect. Dis.* **11:**603–609.

Rice, D. H., D. D. Hancock, and T. E. Besser. 1995. Verotoxigenic *Escherichia coli* O157:H7 colonization of wild deer and range cattle. *Vet. Rec.* **137:**524.

Riley, L. W., R. S. Remis, S. D. Helgerson, H. B. McGee, J. B. Wells, and B. R. Davis. 1983. Hemorrhagic colitis associated with a rare *Escherichia coli* subtype. *N. Engl. J. Med.* **308:**681–685.

Riordan, D. C., G. M. Sapers, T. R. Hankinson, M. Magee, A. M. Mattrazzo, and B. A. Annous. 2001. A study of U.S. orchards to identify potential sources of *Escherichia coli* O157:H7. *J. Food Prot.* **64:**1320–1327.

Robinson, C. M., J. F. Sinclair, M. J. Smith, and A. D. O'Brien. 2006. Shiga toxin of enterohemorrhagic *Escherichia coli* type O157:H7 promotes intestinal colonization. *Proc. Natl. Acad. Sci. USA* **103:**9667–9672.

Ruan, C., H. Wang, and Y. Li. 2002. A bienzyme electrochemical biosensor coupled with immunomagnetic separation for rapid detection of *Escherichia coli* O15:H7 in food samples. *Trans. ASAE* **45:**249–255.

Russell, S. M., and S. P. Axtell. 2005. Monochloramine versus sodium hypochlorite as antimicrobial agents for reducing populations of bacteria on broiler chicken carcasses. *J. Food Prot.* **68:**758–763.

Russo, T. A., B. A. Davidson, S. A. Genagon, N. M. Warholic, U. MacDonald, P. D. Pawlicki, J. M. Beanan, R. Olson, B. A. Holm, and P. R. Knight III. 2005. *E. coli* virulence factor hemolysin induces neutrophil apoptosis and necrosis/lysis in vitro and necrosis/lysis and lung injury in a rat pneumonia model. *Am. J. Physiol. Lung Cell. Mol. Physiol.* **289:**L207–L216.

Ryu, J. H., and L. R. Beuchat. 1998. Influence of acid tolerance responses on survival, growth, and thermal cross-protection of *Escherichia coli* O157:H7 in acidified media and fruit juices. *Int. J. Food Microbiol.* **45:**185–193.

Samadpour, M., M. W. Barbour, T. Nguyen, T. M. Cao, F. Buck, G. A. Depavia, E. Mazengia, P. Yang, D. Alfi, M. Lopes, and J. D. Stopforth. 2006. Incidence of enterohemorrhagic *Escherichia coli*, *Escherichia coli* O157, *Salmonella*, and *Listeria monocytogenes* in retail fresh ground beef, sprouts, and mushrooms. *J. Food Prot.* **69:**441–443.

Samelis, J., and J. N. Sofos. 2003. Strategies to control stress-adapted pathogens and provide safe food, p. 303–351. *In* A. E. Yousef and V. K. Juneja (ed.), *Microbial Adaptation to Stress and Safety of New-Generation Foods.* CRC Press, Boca Raton, FL.

Sanderson, M. W., J. M. Gay, D. D. Hancock, C. C. Gay, L. K. Fox, and T. E. Besser. 1995. Sensitivity of bacteriological culture for detection of *Escherichia coli* O157:H7 in bovine feces. *J. Clin. Microbiol.* **33:**2616–2619.

Sandvig, K. 2001. Shiga toxins. *Toxicon* **39:**1629–1635.

Sansonetti, P. J., G. T. Van Nhieu, and C. Egile. 1999. Rupture of the intestinal epithelial barrier and mucosal invasion by *Shigella flexneri*. *Clin. Infect. Dis.* **28:**466–475.

Sapers, G. M., R. L. Miller, and A. M. Mattrazzo. 1999. Effectiveness of sanitizing agents in inactivating *Escherichia coli* in Golden Delicious apples. *J. Food Sci.* **64:**734–737.

Savarino, S. J., A. McVeigh, J. Watson, A. Cravioto, J. Molina, P. Echeverria, M. K. Bhan, M. M. Levine, and A. Fasano. 1996. Enteroaggregative *Escherichia coli* heat-stable enterotoxin is not restricted to enteroaggregative *E. coli*. *J. Infect. Dis.* **173:**1019–1022.

Scaletsky, I. C. A., S. H. Fabbricotti, R. L. B. Carvalho, C. R. Nunes, H. S. Maranhão, M. B. Marais, and U. Fagundes-Neto. 2002. Diffusely adherent *Escherichia coli* as a cause of acute diarrhea in young children in northeast Brazil: a case-control study. *J. Clin. Microbiol.* **40:**645–648.

Scanga, J. A. 2005. Slaughter and fabrication/boning processes and procedures, p. 259–272. *In* J. N. Sofos (ed.), *Improving the Safety of Fresh Meat.* CRC Press, Boca Raton, FL.

Schmidt, H., and H. Karch. 1996. Enterohemolytic phenotypes and genotypes of Shiga toxin-producing *Escherichia coli* O111 strains from patients with diarrhea and hemolytic-uremic syndrome. *J. Clin. Microbiol.* **34:**2364–2367.

Schmidt, H., C. Kernbach, and H. Karch. 1996. Analysis of the EHEC *hly* operon and its location in the physical map of the large plasmid of enterohaemorrhagic *Escherichia coli* O157:H7. *Microbiology* **142:**907–914.

Schmidt, H., J. Scheef, H. I. Huppertz, M. Frosch, and H. Karch. 1999. *Escherichia coli* O157:H7 and O157:H⁻ strains that do not produce Shiga toxin: phenotype and genetic characteristics of isolates associated with diarrhea and hemolytic-uremic syndrome. *J. Clin. Microbiol.* **37:**3491–3496.

Simpson, C. A., J. R. Ransom, J. A. Scanga, K. E. Belk, J. N. Sofos, and G. C. Smith. 2006. Changes in microbiological populations on beef carcass surfaces exposed to air- or spray-chilling and characterization of hot box practices. *Food Prot. Trends* **26:**226–235.

Singleton, P. 2004. *Bacteria in Biology, Biotechnology and Medicine*, 6th ed. John Wiley & Sons, Ltd., Chichester, United Kingdom.

Small, P. L., and S. Falkow. 1988. Identification of regions on a 230-kilobase plasmid from enteroinvasive *Escherichia coli* that are required for entry into HEp-2 cells. *Infect. Immun.* **56:**225–229.

Smith, J. S., and S. Pillai. 2004. Irradiation and food safety. *Food Tech.* **58:**48–55.

Smith, T. J., L. O'Connor, M. Glennon, and M. Maher. 2000. Molecular diagnostics in food safety: rapid detection of food-borne pathogens. *Irish J. Agr. Food Res.* **39:**309–319.

Sofos, J. N. 1994. Antimicrobial agents, p. 501–529. *In* A. T. Tu and J. A. Maga (ed.), *Handbook of Toxicology: vol. 1., Food Additive Toxicology.* Marcel Dekker, New York, NY.

Sofos, J. N. 2004. Pathogens in animal products: sources and control, p. 701–703. *In* W. Pond and A. Bell (ed.), *Encyclopedia of Animal Science.* Marcel Dekker, New York, NY.

Sofos, J. N. 2005. *Improving the Safety of Fresh Meat.* CRC Press, Boca Raton, FL.

Sofos, J. N., and G. C. Smith. 1998. Evaluation of various treatments to reduce contamination on carcass tissue, p. 316–317. In *Proceedings of the 44th International Congress of Meat Science and Technology*. Barcelona, Spain.

Sofos, J. N., C. A. Simpson, K. E. Belk, J. A. Scanga, and G. C. Smith. 2006. *Salmonella* interventions for beef. In *Proceedings of the 59th Reciprocal Meat Conference*. American Meat Science Association, Savoy, IL.

Sofos, J. N., K. E. Belk, and G. C. Smith. 1999c. Processes to reduce contamination with pathogenic microorganisms in meat, p. 596–605. In *Proceedings of the 45th International Congress of Meat Science and Technology*. Yokohama, Japan.

Sofos, J. N., S. L. Kochevar, G. R. Bellinger, D. R. Buege, D. D. Hancock, S.C. Ingram, J. B. Morgan, J. O. Reagan, and G. C. Smith. 1999a. Sources and extent of microbiological contamination of beef carcasses in seven United States slaughtering plants. *J. Food Prot.* 62:140–145.

Sofos, J. N., S. L. Kochevar, J. O. Reagan, and G. C. Smith. 1999b. Extent of beef carcass contamination with *Escherichia coli* and probabilities of passing US regulatory criteria. *J. Food Prot.* 62:234–238.

Solomon, E. B., C. J. Potensky, and K. R. Matthew. 2002a. Effect of irrigation method on transmission to and persistence of *Escherichia coli* O157:H7 on lettuce. *J. Food Prot.* 65:673–676.

Solomon, E. B., S. Yaron, and K. R. Matthews. 2002b. Transmission of *Escherichia coli* O157:H7 from contaminated manure and irrigation water to lettuce plant tissue and its subsequent internalization. *Appl. Environ. Microbiol.* 68:397–400.

Souza, V., A. Castillo, and L. E. Equiarte. 2002. The evolutionary ecology of *Escherichia coli*. *Am. Sci.* 90:332–341.

Spano, L. C., A. D. I. Sadovsky, P. A. Segui, K. W. Saick, S. M. S. Kitagawa, F. E. L. Pereira, U. Fagundes-Neto, and I. C. A. Scaletsky. 2008. Age-specific prevalence of diffusely adherent *Escherichia coli* in Brazilian children and diarrhea. *J. Med. Microbiol.* 57:359–363.

Steele, B. T., N. Murphy, G. S. Arbus, and C. P. Rance. 1982. An outbreak of hemolytic uremic syndrome associated with ingestion of fresh apple juice. *J. Pediatr.* 101:963–965.

Steiner, T. S., J. P. Nataro, C. E. Poteet–Smith, J. A. Smith, and R. L. Guerrant. 2000. Enteroaggregative *Escherichia coli* expresses a novel flagellin that causes IL-8 release from intestinal epithelial cells. *J. Clin. Investig.* 105:1769–1777.

Stewart, V. 1988. Nitrate respiration in relation to facultative metabolism in enterobacteria. *Microbiol. Rev.* 52:190–232.

Stopforth, J. D., and J. N. Sofos. 2006. Recent advances in pre- and post-slaughter intervention strategies for control of meat contamination, p. 66–68. *In* V. J. Juneja, J. P. Cherry, and M. H. Tunick (ed.), *Advances in Microbial Food Safety*. Oxford University Press, Washington, DC.

Stopforth, J. D., J. Samelis, J. N. Sofos, P. A. Kendall, and G. C. Smith. 2002. Biofilm formation by *Listeria monocytogenes* in fresh beef decontamination washings, p. 202–203. In *Proceedings of the 48th International Congress of Meat Science and Technology*, Rome, Italy.

Stopforth, J. D., J. Samelis, J. N. Sofos, P. A. Kendall, and G. C. Smith. 2003a. Influence of extended acid stressing in fresh beef decontamination runoff fluids on sanitizer resistance of acid-adapted *Escherichia coli* O157:H7 in biofilms. *J. Food Prot.* 66:2258–2266.

Stopforth, J. D., J. Samelis, J. N. Sofos, P. A. Kendall, and G. C. Smith. 2003b. Influence of organic acid concentration on survival of *Listeria monocytogenes* and *Escherichia coli* O157:H7 in beef carcass wash water and on model equipment surfaces. *Food Microbiol.* 20:651–660.

Summers, W. C. 2001. Bacteriophage therapy. *Ann. Rev. Microbiol.* 55:437–451.

Sussman, M. 1997. *Escherichia coli: Mechanisms of Virulence*. Cambridge University Press, Cambridge, United Kingdom.

Swaminathan, B., and G. M. Matar. 1993. Molecular typing methods, p. 26–50. *In* T. F. S. D. H. Persing, F. C. Tenover, and T. J. White (ed.), *Diagnostic Molecular Microbiology: Principles and Applications*. ASM Press, Washington, DC.

Swaminathan, B., T. J. Barrett, S. B. Hunter, and R. V. Tauxe. 2001. PulseNet: the molecular subtyping network for foodborne bacterial disease surveillance, United States. *Emerg. Infect. Dis.* 7:382–389.

Swerdlow, D. L., B. A. Woodruff, R. C. Brady, P. M. Griffin, S. Tippen, H. D. Donnel, E. Geldreich, B. J. Payne, A. Meyer, and J. G. Wells 1992. A waterborne outbreak in Missouri of *Escherichia coli* O157:H7 associated with bloody diarrhea and death. *Ann. Intern. Med.* 117:812–819.

Tanji, Y., C. Furukawa, S. H. Na, T. Hijikata, K. Miyanaga, and H. Unno. 2004. *Escherichia coli* detection by GFP-labeled lysozyme-inactivated T4 bacteriophage. *J. Biotech.* 114: 11–20.

Tatsuno, I., M. Horie, H. Abe, T. Miki, K. Makino, H. Shinagawa, H, Taguchi, S. Kamiya, T. Hayashi, and C. Sasakawa. 2001. *toxB* gene on pO157 of enterohemorrhagic *Escherichia coli* O157:H7 is required for full epithelial cell adherence pathways. *Infect. Immun.* 69:6660–6669.

Taylor, D. N., J. M. Bied, J. S. Munro, and R. A. Feldman. 1982. *Salmonella* Dublin infections in the United States, 1979–1980. *J. Infect. Dis.* 146:322–327.

Taylor, J. P., and J. N. Perdue. 1989. The changing epidemiology of human brucellosis in Texas, 1977–1986. *Am. J. Epidemiol.* 130:160–165.

Taylor, K. A., C. B. O'Connell, P. W. Luther, and M. S. Donnenberg. 1998. The EspB protein of enteropathogenic *Escherichia coli* is targeted to the cytoplasm of infected HeLa cells. *Infect. Immun.* 66:5501–5507.

Thompson, J. S., D. S. Hodge, and A. A. Borczyk. 1990. Rapid biochemical test to identify verocytotoxin-positive strains of *Escherichia coli* serotype O157. *J. Clin. Microbiol.* 28:2165–2168.

Tilden, J., W. Young, A. M. McNamara, C. Custer, B. Boesel, M. A. Lambert–Fair, J. Majkowski, D. Vugia, S. B. Werner, J. Hollingsworth, and J. G. Morris, Jr. 1996. A new route of transmission for *Escherichia coli* O157:H7: infection from dry fermented salami. *Public. Health Briefs* 86:1142–1145.

Torres, A. G., X. Zhou, and J. B. Kaper. 2005. Adherence of diarrheagenic *Escherichia coli* strains to epithelial cells. *Infect. Immun.* 73:18–29.

Travena, W. B., G. A. Willshaw, T. Cheasty, C. Wray, and J. Gallagher. 1996. Verocytotoxin-producing *Escherichia coli* O157 infection associated with farms. *Lancet* 347:60–61.

Tuttle, J., T. Gomez, M. P. Doyle, J. G. Wells, T. Zhao, R. V. Tauxe, and P. M. Griffin. 1999. Lessons from a large outbreak of *Escherichia coli* O157:H7 infections: insights into the infectious dose and method of widespread contamination of hamburger patties. *Epidemiol. Infect.* 122:185–192.

Tzipori, S., I. K. Wachsmuth, J. Smithers, and C. Jackson. 1988. Studies in gnotobiotic piglets on non-O157:H7 *Escherichia coli* serotypes isolated from patients with hemorrhagic colitis. *Gastroenterology* 94:590–597.

USDA/FSIS. 2007. California firm expands recall of ground beef for possible *E. coli* O157:H7 contamination. Recall Release FSIS–RC–025–2007. Food Safety and Inspection Service, U.S. Department of Agriculture, Washington, DC.

van de Kar, N., L. A. H. Monnens, M. A. Karmali, and V. W. M. Vanhinsbergh. 1992. Tumor necrosis factor and interleukin-1 induce expression of the verocytotoxin receptor globotriaosylceramide on human endothelial cells—implications for the pathogenesis of the hemolytic uremic syndrome. *Blood* 80:2755–2764.

Vaz, T. M. I., K. Irino, M. A. M. F. Kato, A. M. G. Dias, T. A. T. Gomes, M. I. C. Medeiros, M. M. M. Rocha, and B. E. C. Guth. 2004. Virulence properties and characteristics of Shiga toxin-producing *Escherichia coli* in Sao Paulo, Brazil, from 1976 through 1999. *J. Clin. Microbiol.* 42:903–905.

Vernozy-Rosand, C., M. P. Montet, F. Lequerrec, E. Serillon, B. Tilly, C. Bavai, S. Ray-Gueniot, J. Bouvet, C. Mazuy-Cruchaudet, and Y. Richard. 2002. Prevalence of verotoxin-producing *Escherichia coli* (VTEC) in slurry, farmyard manure and sewage sludge in France. *J. Appl. Microbiol.* 93:473–478.

Vickers, Z. M., and J. Wang. 2002. Liking of ground beef patties is not affected by irradiation. *J. Food Sci.* 67:380–383.

Wachtel, M. R., L. C. Whitehand, and R. E. Mandrell. 2002. Prevalence of *Escherichia coli* associated with a cabbage crop inadvertently irrigated with partially treated sewage wastewater. *J. Food Prot.* 65:471–475.

Wallace, J. S., T. Cheasty, and K. Jones. 1997. Isolation of verocytotoxin-producing *Escherichia coli* from wild birds. *J. Appl. Microbiol.* 82:399–404.

Wang, G. -H., C. G. Clark, and F. G. Rodgers. 2002. Detection in *Escherichia coli* of the genes encoding the major virulence factors, the genes defining the O157:H7 serotype, and components of the type 2 Shiga toxin family by multiplex PCR. *J. Clin. Microbiol.* 40:3613–3619.

Warriner, K., F. Ibrahim, M. Dickinson, C. Wright, and W. M. Waites. 2003. Isolation of *Escherichia coli* with growing salad spinach sprouts. *J. Food Prot.* 66:1790–1797.

Warriner, K., F. Ibrahim, M. Dickinson, C. Wright, and W. M. Waites. 2005. Seed decontamination as an intervention step for eliminating *Escherichia coli* on salad vegetables and herbs. *J. Sci. Food Agric.* 85:2307–2313.

WHO. 1998. *Zoonotic Non-O157 Shiga Toxin-Producing* Escherichia coli *(STEC). Report of a WHO Scientific Working Group Meeting.* WHO/CSR/APH/98.8. Geneva, Switzerland.

Williams, R. C., S. Isaacs, M. L. Decou, E. A. Richardson, M. C. Buffett, R. W. Slinger, M. H. Brodsky, B. W. Ciebin, A. Ellis, and A. Hockin. 2000. Illness outbreak associated with *Escherichia coli* O157:H7 in Genoa salami. *CMAJ* 162:1409–1413.

Wittles, E. G., and H. C. Lichtman. 1986. Blood group incidence and *Escherichia coli* bacterial sepsis. *Transfusion* 26:533–535.

Woerner, D. R., J. R. Ransom, J. N. Sofos, G. A. Dewell, G. C. Smith, M. D. Salman, and K. E. Belk. 2006. Determining the prevalence of *Escherichia coli* O157 in cattle and beef from the feedlot to the cooler. *J. Food Prot.* 69:2824–2827.

Yamamoto, T., M. Kaneko, S. Changchawalit, O. Serichantalergs, S. Ijuin, and P. Echeverria. 1994. Actin accumulation associated with clustered and localized adherence in *Escherichia coli* isolated from patients with diarrhea. *Infect. Immun.* 62:2917–2929.

Zadik, P. M., P. A. Chapman, and C. A. Siddons. 1993. Use of tellurite for the selection of verocytotoxigenic *Escherichia coli* O157. *J. Med. Microbiol.* 39:155–158.

Zhou, X., J. A. Girón, A. G. Torres, J. A. Crawford, E. Negrete, S. N. Vogel, and J. B. Kaper. 2003. Flagellin of enteropathogenic *Escherichia coli* stimulates interleukin-8 production from T84 cells. *Infect. Immun.* 71:2120–2129.

Zychlinsky, A., B. Kenny, R. Menard, M. C. Prevost, I. B. Holland, and P. J. Sansonetti. 1994. IpaB mediates macrophage apoptosis induced by *Shigella flexneri*. *Mol. Microbiol.* 11:619–627.

Pathogens and Toxins in Foods: Challenges and Interventions
Edited by V. K. Juneja and J. N. Sofos
© 2010 ASM Press, Washington, DC

Chapter 6

Listeria monocytogenes

Anna C. S. Porto-Fett, Jeffrey E. Call, Peter M. Muriana,
Timothy A. Freier, and John B. Luchansky

TYPE OF ILLNESS AND CHARACTERISTICS OF THE ORGANISM

In 1924 in Cambridge, England, what is now known as *Listeria monocytogenes* was first documented by E. G. D. Murray and colleagues (1926) as the causative agent of a septic illness, peripheral monocytosis, in laboratory rabbits. A few years later Nyfeldt (1929) reported the first case of illness in humans that was attributed to this bacterium. *L. monocytogenes* is a gram-positive, non-spore-forming rod that exhibits tumbling end-over-end motility at 22°C. Biochemically, it is catalase positive and oxidase negative, and it hydrolyzes esculin and exhibits slight β-hemolysis on blood agar (Ryser and Marth, 1999). There are currently 13 serotypes (1/2a, 1/2b, 1/2c, 3a, 3b, 3c, 4a, 4ab, 4b, 4c, 4d, 4e, and 7) (Khelef et al., 2006). The optimum growth temperature is 30 to 37°C, but it can grow at −0.4 and 45°C and can survive freezing for prolonged periods (Rowan and Anderson, 1998). The optimum pH range for growth is pH 6 to 8, but growth has been observed between pH 4.4 and 9.6 (Pearson and Marth, 1989; Ryser and Marth, 1999). In general, the pH that will support growth of *L. monocytogenes* increases as the temperature decreases. *L. monocytogenes* can also survive and/or grow at water activity (a_w) levels that are usually too low for other bacteria to survive, that being a_w values of ≤0.911 (Nolan et al., 1992). The ability to survive and/or grow over a wide range of temperatures, pH, and a_w values is of great concern to the food industry, because this provides an opportunity for the pathogen to survive in a variety of niches at various points within food processing operations and ultimately contaminate product contact surfaces and finished products (Tompkin, 2002).

LISTERIOSIS

There is still much that is not known about how this disease is manifested in the human body (McLauchlin, 1996). Because the pathogen is opportunistic and is not necessarily adapted to humans, it probably has multiple routes of infection and types of symptoms, depending on how it enters the body; however, the principal route of infection for food-borne listeriosis begins with the ingestion of contaminated food. Although healthy adults and children can also suffer from listeriosis, young, old, pregnant, and immune-compromised individuals ("YOPIs") are particularly susceptible (CDC, 2000). Of the estimated 2,500 food-borne cases per year in the United States, of which about 500 are fatal, about one-third involve pregnant women (Mead et al., 1999). Symptoms range from an asymptomatic infection to flu-like symptoms, such as fever and muscle aches in healthy individuals, to septicemia and meningitis in immune-compromised patients, intrauterine infections in pregnant women, and severe systemic infections in the unborn or neonates. Until recently, it was thought that pregnant women were more susceptible because their immune system was changed and/or compromised during pregnancy. New information indicates that once the pathogen has been ingested and disseminated to the maternal organs, it is able to migrate to the placenta, wherein it is protected from the maternal immune system (Bakardjiev et al., 2006). Between 1976 and 2002, there were 27 outbreaks of food-borne listeriosis reported worldwide, with about 2,900 cases and about 260 deaths (mortality rate of ca. 9.0%) (McLauchlin et al., 2004). Between 1998 and 2002, although listeriosis represented only 0.7% of bacterial food-borne illnesses over this 5-year

Anna C. S. Porto-Fett, Jeffrey E. Call, and John B. Luchansky • Microbial Food Safety Research Unit, U.S. Department of Agriculture, Agricultural Research Service, Eastern Regional Research Center, Wyndmoor, PA 19038. **Peter M. Muriana** • Department of Animal Science, Food and Agricultural Products Research and Technology Center, Oklahoma State University, Stillwater, OK 74078. **Timothy A. Freier** • Corporate Food Safety and Regulatory Affairs, Cargill, Inc., Minneapolis, MN 55440-9300.

period, it was responsible for 54% of all deaths (CDC, 2006). The use of genomic and proteomic tools has generated appreciable insight into the genes and proteins that make *L. monocytogenes* virulent. Such progress has also been accelerated by the availability and comparison of the whole genomes of select serotype 4b and 1/2a strains (Nelson et al., 2004; Glaser et al., 2001). More detailed information on molecular characterization of the determinants involved in invasion of host cells, intracellular motility, and cell-to-cell spread is beyond the scope of this chapter but can be found elsewhere (Kathariou, 2002; Paoli et al., 2005). Further expansion of our knowledge of the mechanism(s) of pathogenesis at the molecular level will ultimately result in more direct and more effective interventions to better manage the occurrence, persistence, and numbers of this pathogen in our food supply.

SOURCES AND INCIDENCE OF *L. MONOCYTOGENES* IN THE ENVIRONMENT AND FOODS

L. monocytogenes is widespread in nature, being associated with plants, soil, water, sewage, feed, and animals raised as food (CFSAN, 2003). It has also been recovered from stool samples of an estimated 1 to 10% of healthy humans (Farber and Peterkin, 1991). The pathogen has also been recovered from a variety of raw and ready-to-eat (RTE) foods (Meng and Doyle, 1997). The majority of cases linked to foods are associated with refrigerated, RTE foods that were consumed without reheating. Refrigerated, RTE foods provide an ideal environment for *L. monocytogenes,* primarily because the pathogen can grow at low temperatures, whereas competitors cannot.

The bacterium is found in numerous foods with a prevalence ranging from 13% in raw meats to 3% in dairy products, 11% in fresh vegetables, and 3% in seafoods. When found in RTE foods, levels are typically low. For example, Gombas et al. (2003) reported a prevalence of 1.82% (577 of 31,705 samples) and levels of <0.3 most probable numbers (MPN)/g to 1.5 × 10⁵ CFU/g in six categories of RTE foods. Likewise, Wallace et al. (2003) reported a prevalence of 1.6% (532 of 32,800 samples) among 12 brands of commercial frankfurters and levels of 1.0 to 5.0 $\log_{10}$ CFU per package/pound. As summarized by Farber and Peterkin (1991), when found in various meat and dairy products, the pathogen is usually present at levels of ca. 1.0 to 3.0 $\log_{10}$ CFU/g but sometimes can be found at levels as high as 7.0 $\log_{10}$ CFU/g. A Joint Agency risk assessment prioritizing the relative risk of listeriosis among 20 food categories identified deli meats,

undercooked chicken, and Latin-style cheese as foods of higher risk (CFSAN, 2003). Efforts to fill research voids needed to conduct risk assessments are aided by the ability to predict the fate of the pathogen on various foods using the tools of predictive microbiology. In this regard, ComBase, the Pathogen Modeling Program (PMP), and the Predictive Microbiology Informational Portal are worth mentioning. ComBase is a relational database that contains over 40,000 data sets/records of the growth, survival, and/or inactivation of *L. monocytogenes* and other food-borne pathogens under diverse environments relevant to food processing operations. The PMP is a stand-alone online application used by ca. 30% of the food industry for hazard analysis of critical control point validation. It is accessed by ca. 5,000 online users in some 35 countries worldwide each year to predict the fate of pathogens and spoilage microbes in a variety of foods. The PMP contains some 40 models, including 23 models for growth, 3 for heat inactivation, 4 for survival, 4 for cooling, and 6 for irradiation. The Predictive Microbiology Informational Portal is a comprehensive website that presently is comprised of predictive models, research data, relevant regulatory policies, and guidelines pertaining to *L. monocytogenes* and RTE meat and poultry products (http://portal.arserrc.gov).

INTRINSIC AND EXTRINSIC FACTORS THAT AFFECT SURVIVAL AND GROWTH IN FOOD PRODUCTS AND CONTRIBUTE TO OUTBREAKS

The psychrotrophic nature of *L. monocytogenes* is of particular concern, given that refrigeration is one of the most common interventions used to ensure safety and extend shelf-life (Gandhi and Chikindas, 2007) and primarily because RTE foods, such as Latin-style cheese and deli meats, may be consumed without any reheating and/or further preparation. *L. monocytogenes* growth/survival at low temperatures requires maintenance of membrane fluidity for appropriate enzymatic activity and transport of solutes across the membrane, as well as for structure stabilization of macromolecules, such as ribosomes, and/or the uptake or synthesis of compatible cryoprotectant solutes, such as glycine betaine and carnitine (Chattopadhyay, 2006; Gandhi and Chikindas, 2007). *L. monocytogenes* may exhibit a thermotolerance response when exposed to temperatures above its optimum growth temperature, given that these temperatures may trigger physiological responses that induce synthesis of polypeptides, namely, heat shock proteins (Rowan and Anderson, 1998). Many factors affect thermotolerance of *L. monocytogenes* in foods, such as the proximate composition of the food; differences in the strains used or their physiological state; presence of deleterious chemicals in the growth media; environmental stresses, such as osmotic and acidic shocks; history/adaptation of the cells prior to inoculation; and the microbiological media and/or incubation conditions used to recover injured cells (Doyle et al., 2001).

Acidic stress may also increase the ability of the pathogen to survive in foods and/or may facilitate expression of its virulence genes (Ryser and Marth, 1999). The minimum pH requirement reported for *L. monocytogenes* growth in foods is pH 4.4, when at near-optimum temperatures. However, its growth at low pH is dependent on the medium or food composition, the strain and its physiological state, and the preincubation temperature (Phan-Thanh et al., 2000). For example, Parish and Higgins (1989) reported that *L. monocytogenes* survived for 21 and 5 days at 4 and 30°C, respectively, in orange serum samples adjusted to pH 3.6 with HCl. Its ability to adapt to acidic environments is of particular concern because the pathogen encounters these environments in low-pH foods, such as fermented meat and dairy products; in the acid conditions within the gastrointestinal tract; and in the phagosomes of macrophages (O'Driscoll et al., 1996; Cotter and Hill, 2003). The adaptation of *L. monocytogenes* to low-pH environments is the result of the activation of stress adaptation mechanisms, such as induction of proteins, pH homeostasis, the glutamate decarboxylase system, sigma factor (σ^B), and a two-component regulatory system, comprised of *lis*R and *lis*K (Shen et al., 2006; Gandhi and Chikindas, 2007). Tolerance of acidic environments can be induced in *L. monocytogenes* by exposure to sublethal pH. O'Driscoll et al. (1996) reported that, regardless of the acid used to adjust the pH, acid-adapted (pH 5.5) cells of *L. monocytogenes* strain LO28 showed greater tolerance toward lethal pH (pH 3.5) compared to those that were not acid adapted (pH 7.0). As another example, Greenacre et al. (2003) reported that cells of *L. monocytogenes* strain EGD-e also developed an acid tolerance response at 20°C when cells were acid adapted (pH 5.0 to 5.5) with either acetic acid or lactic acid prior to exposure to a lethal pH (pH 3.0).

The lowering of a_w in foods is another strategy widely used for controlling food-borne pathogens. Ingham et al. (2006) evaluated the survival of *L. monocytogenes* on vacuum-packaged beef jerky and related products stored at 21°C for up 28 days. The authors reported that jerky products with an a_w of ≤0.87 did not support growth of *L. monocytogenes,* but the organism survived in jerky at a_w values ≤0.47. The a_w of foods can be altered by addition of osmolitic compounds, by removal of water using physical methods, and/or by binding of water to a range of macromolecular components (Stecchini et al., 2004). The stress response of *L. monocytogenes* to a low-a_w environment is induced, at least in part, by accumulation of solutes, such as glycine betaine and carnitine (Bayles and Wilkinson, 2000).

Studies have shown that glycine betaine and carnitine are the most effective solutes for osmotic adaptation of *L. monocytogenes* (Bayles and Wilkinson, 2000; Angelidis and Smith, 2003). The ability of *L. monocytogenes* to accumulate these intracellular solutes allows it to grow at NaCl concentrations of ≤12% (Razavilar and Genigeorgis, 1998), thus challenging control of this pathogen in salted RTE foods, such as soft cheese and deli meats. The intracellular solutes that are synthesized or that are taken up from the food or natural environment compensate for the external osmotic strength without affecting the macromolecular structure of the cell (Duché et al., 2002; Smith, 1996). As one example, Smith (1996) evaluated the adaptation of osmotically stressed cells of *L. monocytogenes* on processed meat surfaces. Their results showed that *L. monocytogenes* accumulated higher levels of glycine betaine and carnitine when inoculated onto the surfaces of bologna, frankfurters, wieners, ham, and bratwurst at refrigerated temperatures than when inoculated in liquid media supplemented with either of the osmolytes. According to Garner et al. (2006), exposure of *L. monocytogenes* to environmental conditions present in RTE meats, such as the addition of NaCl and organic acid salts, may enhance its pathogenicity and consequently the association of listeriosis with these foods.

FOOD PROCESSING OPERATIONS THAT INFLUENCE THE NUMBERS, SPREAD, OR CHARACTERISTICS

Control of *L. monocytogenes* is one of the most difficult challenges faced by manufacturers and handlers of RTE meat, poultry, seafood, and dairy products. Overcoming this challenge requires the flawless execution of numerous food safety-related systems and operations, including proper sourcing and storage of ingredients, segregation of raw product areas from cooked/pasteurized product areas, heat treatment or other listericidal steps, sanitary design of equipment and facilities, sanitation, environmental monitoring, and cold chain management. Other important control considerations, such as the intrinsic and extrinsic factors of the food and specific *L. monocytogenes* control interventions, are addressed elsewhere in this chapter.

While most other microbial food-borne pathogens can be controlled by a well-validated cooking or pasteurization step, *L. monocytogenes* thrives in the cool, damp environment typical of the postcook or postpasteurization environments of RTE manufacturing facilities, retail slicing areas, and even home refrigerators. Thus, protection of cooked or pasteurized RTE products from recontamination by *L. monocytogenes* and from exposure to conditions conducive to *L. monocytogenes* growth is key to protecting public health.

Sourcing and Storage of Ingredients

A consideration of all inputs to the process or product is a key step in controlling *L. monocytogenes*. Ingredients in which *L. monocytogenes* is capable of multiplication should be stored at temperatures and for times that will not allow for an increase in numbers of the pathogen that might subsequently overwhelm the final kill step. Ingredients added after cooking, such as sauces, dry spice rubs, and heat-sensitive nutrient supplements, are even more critical. If there is a regulatory or public health-determined "zero tolerance" for *L. monocytogenes* in the final product, ingredients added postcook must also be free of any viable *L. monocytogenes*. Processing aids, packaging materials, and brine chill solutions, as well as water and air, should also be carefully considered if they will be contacting the post-*L. monocytogenes*-kill-step product.

Segregation of Raw and Treated Areas

Because *L. monocytogenes* is ubiquitous in the environment and is often present on raw ingredients, it is important to minimize the presence and levels of *L. monocytogenes* in the posttreatment environment, especially when the product is directly exposed to that environment. Complete exclusion of *L. monocytogenes* is not attainable, but limiting the occurrence of *L. monocytogenes* in the posttreatment environment to the minimum level possible will greatly protect the treated/finished product (Tompkin, 2002). The RTE industry has used numerous practices to attain this goal. For example, segregation of employees that work in raw versus treated areas can best be attained by having dedicated employees for each area. Many facilities even provide separate employee welfare areas (locker rooms, cafeterias, and restrooms) to avoid cross-contamination. For many operations, employing two complete sets of workers is impractical. Carefully planned procedures then become critical for limiting *L. monocytogenes* cross-contamination when employees transition from raw to treated areas. One of the most important considerations is provision of

footwear and outer clothing that are worn only in the high-care areas and that are adequately cleaned and sanitized on a routine basis. A well-designed hand wash station incorporating such things as touchless water, soap, paper towel, and final hand sanitizer dispenser controls and with comfortable water temperature and adequate sink space (so employees are not rushed) is just one of the recommended requisites. Separate sets of tools, utensils, and equipment for production, maintenance, and sanitation should be maintained for the pretreatment and posttreatment areas. Manufacturers should also consider how ingredients, packaging materials, sanitation equipment and supplies, garbage/inedible containers, pallets, quality assurance supplies, reworked product, fork lifts/hand trucks, lubricants, and other processing aids will enter the treated product area. Air pressure should be highest in the most critical high-care area, becoming more negative in the raw areas. Air balances should be conducted to ensure that these pressures are maintained during all phases of operation, especially including the sanitation shift, when the generation of microaerosols could spread *L. monocytogenes*. Air balances should be repeated whenever new equipment is added or whenever the facility is modified in ways that could impact the air balance. Facility air chilling systems should be designed so that *L. monocytogenes* is not able to multiply within the system or be disseminated in the air stream. Modern systems that treat the air supply with heat, HEPA filtration, and/or UV radiation are available. These systems must also be designed to quickly reduce the humidity generated during sanitation and maintain low humidity levels during operation.

Kill Step

The capacity to kill *L. monocytogenes* during the cooking or pasteurization step has been determined for many products and processes (CDC, 1988; Juneja and Eblen, 1999). This kill capacity is not unlimited, so worst-case estimations of the levels potentially present before the kill step should be conducted to assess the adequacy of this control point. Baseline studies conducted by the USDA/FSIS indicated the highest levels of *L. monocytogenes* detected in several types of raw meat and poultry (Table 1). Unfortunately, public data do not exist for many meat and poultry ingredients, nor do they exist for most other ingredients, such as spice blends. If any ingredients added prior to cooking are suspected of having occasional high levels of *L. monocytogenes*, quantitative testing should be conducted to determine maximum levels. A critical limit for the level of kill necessary to produce a safe product needs to be determined based

Table 1. Levels of *L. monocytogenes* found in USDA baseline sampling[a]

Type of animal product (sample type)	No. sampled	Highest level (MPN/cm^2)
Market hogs (carcass swab)	2,112	46
Broiler chickens (carcass rinse)	1,297	51
Cows and bulls (carcass swab)	2,112	43
Young turkeys (carcass rinse)	1,221	0.3

[a]Source: http://www.fsis.usda.gov/Science/Baseline_Data/index.asp. Accessed 15 April 2008. MPN, most probable numbers.

on a worst-case estimation of the load of *L. monocytogenes* that could potentially be in the product before the kill step. These data can be generated by conducting quantitative testing of the uncooked product or ingredients, or adapted from existing published data (Table 1). It is generally accepted that the heating times and temperatures necessary to eliminate *Salmonella* in cooked meat and poultry products are also adequate to eliminate *L. monocytogenes*. Traditional milk pasteurization times/temperatures are also considered generally adequate to eliminate the typically low levels of *L. monocytogenes* that may be present in raw milk. Once validated, however, the kill step must undergo routine verifications to ensure its continuing efficacy.

Sanitary Design of Equipment and Facilities

The most typical cause of the contamination of RTE product with *L. monocytogenes* is from a growth niche that exists in or near product contact surfaces on equipment in the exposed posttreatment product area. A growth niche is an area that contains the essential ingredients of food, water, and time within the *L. monocytogenes* growth temperature range. The food source is typically residue from the product being manufactured, although it can sometimes come from other sources, such as dust from a cardboard box. Water can be supplied from numerous sources, including juices from the product, residual water left from the cleaning/sanitation process, discharged chilling water from packaging equipment, leaking steam and water valves, condensation, and even poorly managed antimicrobial foot baths or floor foaming systems. The time necessary for *L. monocytogenes* growth depends on temperature and other extrinsic conditions. The most problematic areas of equipment that can become growth niches for *L. monocytogenes* are the areas that do not receive physical disruption (scrubbing, turbulent flow of clean-in-place systems, and high-pressure spray) and/or do not receive adequate exposure to cleaning and sanitizing chemicals. Examples of growth niches are hollow equipment elements that are not hermetically sealed (support structures, conveyor rollers, motors, etc.), gaskets, pumps, rubber floor mats, bearings, drive

chains, cracks, bolt/screw/rivet penetrations, electrical panels, equipment controls, compressed air lines, and cooling coils. Anytime two solid pieces of material are combined, if moisture can infiltrate the junction, the potential exists for an "*L. monocytogenes* sandwich." The most common harborage sites for *L. monocytogenes* are typically the floor and drain systems. The most effective way to limit the levels of *L. monocytogenes* in the overall facility environment is to strive for clean, sealed, unbroken floors that remain dry during food production shifts. Excellent information on best practices for the sanitary design of equipment and facilities has been compiled by the American Meat Institute and is available at www.meatami.org (accessed 15 April 2008).

Sanitation

Sanitation, especially in critical high-care areas, is one of the most important *L. monocytogenes* control points. Cleaning is the critical step that removes food residues and allows the chemical sanitizer to function optimally. A four-step cleaning process is recommended. The first step is dry cleaning, removing as much product residue as possible by scraping, sweeping, and shoveling. Care should be taken to pick up and dispose of this residue and not rinse it down the drainage system, where clogs and backups can occur, leading to a high-risk *L. monocytogenes* situation. The second step is the rough rinse with hot water (approximately 140°F). This temperature is important, because if it is too cool, then fats and oils will not be melted and removed, and if too hot, then protein residues can be baked onto the equipment. The third step is applying the appropriate detergent. This detergent should be matched to the application by a cleaning chemical expert, as it needs to match the cleaning needs (clean in place, foaming, scrubbing, etc.), water softness, residues to be removed, etc. Detergent is commonly applied as a foam solution. It is important that the foam not be allowed to dry on the equipment, as it will not rinse properly if allowed to dry. The final cleaning step is the final rinse, where it is critical that all remaining food and detergent residue be completely rinsed from the equipment. When the equipment and room are clean, the appropriate sanitizer can be applied. The most important consideration for sanitizer application is coverage. Flood sanitizing (low-pressure and high-volume) application of the sanitizer is most effective. Other methods (fogging and garden sprayer) may not give 100% coverage and should be avoided, except in special circumstances. Many facilities double and even triple sanitize to ensure for the most complete kill of *L. monocytogenes* and other microorganisms. Often, a fast-acting oxidizing sanitizer, such as a mixed peracid, is used first and is followed by a quaternary ammonium solution that is slower acting but has residual killing activity. Most sanitizers should be applied as a mixture with cool water. If regulations allow, the final sanitizer should be applied at a level that can be left on equipment without a final rinse.

For very difficult-to-sanitize equipment, heat can be effectively used to kill *L. monocytogenes*. Equipment can be covered with a tarp and steamed, placed in an oven, or immersed in hot water. The sanitizing method should be carefully evaluated for employee safety and degradation of heat-sensitive machine components. Clean-out-of-place tanks can be very useful for cleaning and heat sanitizing smaller parts. Heat sanitizing processes should be validated for the ability to kill *L. monocytogenes* in the central, most heat-protected areas of the equipment. When using wet heat, the times and temperatures necessary to kill *L. monocytogenes* in food products will also kill *L. monocytogenes* in or on equipment. Dry heat has much less killing capacity, so higher temperatures and longer application times are typically required.

Environmental Monitoring

Areas capable of harboring *L. monocytogenes* are not always visually apparent. Environmental monitoring gives manufacturers the ability to find and eliminate *L. monocytogenes* harborage sites and growth niches. The two biggest keys to success are aggressive searching and diligent corrective action. One recommendation for aggressive searching is to reward employees for finding positive environmental samples. Others include using the fastest, most sensitive methods possible; sampling large areas; and sampling areas that are the most difficult to reach and clean. Using a broader indicator group, such as *Listeria* spp. or *Listeria*-like organisms, enhances the ability to find growth niches, in contrast to looking specifically for *L. monocytogenes*. The RTE meat and poultry industry has found the greatest success in finding *L. monocytogenes* growth niches by doing routine monitoring several hours after the start of processing operations or at the end of production but before sanitation. Sampling devices should be optimized to pick up the target group from the equipment

so that it can be transferred into the detection system. Microcellulose sponges, single-ply tissues, and gauze all work well and pick up the target group more efficiently than cotton-tipped swabs or direct-contact plates (Vorst et al., 2004). A neutralizing buffer should be used to moisten the sampling device. Samples should be kept refrigerated and be processed in a timely manner to prevent overgrowth by competitors or die-off of the target group.

Cold Chain Management

Protection from *L. monocytogenes* growth conditions does not end when the product leaves the manufacturing facility. If the product is temperature abused at any step of manufacture, storage, distribution, retail, or home use, the potential for *L. monocytogenes* to grow to infectious numbers increases; thus, the key to minimizing listeriosis is to prevent *L. monocytogenes* multiplication (Chen et al., 2003). Many manufacturers occasionally send temperature monitoring devices with the product during transit and storage and then review the monitoring information to ensure that the cold chain is routinely being maintained. In addition, retailers and display case manufacturers have made improvements in maintaining cold temperatures during retail display. A joint risk assessment conducted by the FDA and the USDA indicated that if all home refrigerators could be maintained at or below the recommended 41°F, cases of listeriosis could be cut by >98%, highlighting the importance of the cold chain in minimizing this disease (CFSAN, 2003).

RECENT ADVANCES IN BIOLOGICAL, CHEMICAL, AND PHYSICAL INTERVENTIONS TO GUARD AGAINST THE PATHOGEN

L. monocytogenes is arguably the food-borne pathogen that has consistently garnered the most regulatory attention over the past 25 years. The severity and mortality of the disease, the nature and number of the most susceptible members of our society, and the frequency and magnitude of product recalls have resulted in the "zero tolerance" policy in effect at present in the United States, that being an allowable limit of ≤1 cell/25 g food (0.04 cell/g) (Shank et al., 1996). However, as pointed out by Buchanan and colleagues (1997), it is noteworthy that the relative frequency of listeriosis in countries that practice zero tolerance, such as the United States, is about the same as that in countries that do not, such as Germany (about 4 or 5 cases per million inhabitants). Despite advances that reduced by about 40% both the prevalence of the pathogen in certain foods and the occurrence of listeriosis (CDC, 2006), additional guidelines are in place for producers of red meat and poultry products (Anonymous, 2003). This ruling provides manufacturers with three options for the plant/product which ultimately determine the frequency of regulatory testing: use of both a post-process lethality step and an antimicrobial to control outgrowth (lowest testing frequency) (alternative 1); use of either a postprocessing lethality step or an antimicrobial to control outgrowth (moderate testing frequency) (alternative 2); or use of appropriate sanitation alone (most testing) (alternative 3). Due to this "incentivized" offering by USDA/FSIS, efforts have intensified to develop and implement biological, chemical, and/or physical interventions to control the pathogen both pre- and postprocessing. In addition, USDA/FSIS also provides a venue for companies to implement antimicrobial approaches via plant trials to provide efficacy data for USDA acceptance that can be listed on the USDA's New Technologies web page, along with a description of the antimicrobial process (http://www.fsis.usda.gov/Regulations_&_Policies/New_Technologies/index.asp).

BIOLOGICAL INTERVENTIONS

Bacteriophage

Several companies have proposed utilizing bacteriophage as a biological anti-listerial intervention for food and other applications (Intralytics, Inc., Baltimore, MD; OmniLytics, Salt Lake City, UT; EBI Food Safety, Wageningen, The Netherlands). One virulent lytic bacteriophage, phage P100, displayed a host range against 95% of the 250 *Listeria* isolates tested (Carlton et al., 2005). When the bacteriophage was applied to soft cheese, Carlton et al. (2005) observed a 2.0- to 3.0-$\log_{10}$ decrease in viable counts of inoculated *L. monocytogenes* (1.5×10^8 PFU/ml phage titer), which improved to a 3.5-$\log_{10}$ reduction with higher phage titers (3×10^9 PFU/ml). This has been developed into a product, Listex-P100 (EBI), that has been granted GRAS (generally recognized as safe) status by the FDA on a petition submitted by Intralytics and that has been granted approval as a food additive for use on RTE meat and poultry products (FDA, 2006).

Bacteriocins

Bacteriocins are inhibitory peptides and proteins produced by bacteria. Those produced by lactic acid bacteria have long been proposed as food "biopreservatives" because of the historical use of these organisms in food fermentations, their generalized GRAS status for use in foods, and their inhibitory activity against various food-borne pathogens (Muriana, 1996). Use of purified bacteriocins would constitute a direct food additive and require FDA approval. The only bacteriocin that has been granted such approval is nisin, which is allowed for use in low-moisture/low-salt pasteurized processed cheese. Other bacteriocins have been used in this regard by employing bacteriocin-producing lactic acid bacteria for production of cultured milk or whey preparations that are used directly as ingredients. In efforts to expand the application of nisin as an antimicrobial, researchers have been examining its use in a variety of different food applications, for which it has not yet been approved. Using nisin-coated plastic films for vacuum-packaged cold-smoked salmon, Neetoo et al. (2008) have shown a 3.9-$\log_{10}$-CFU/cm^2 reduction of *L. monocytogenes* relative to growth in controls at the highest level of nisin used (2,000 IU/cm^2). Concern has also surfaced for the resistance of *L. monocytogenes* to nisin and other bacteriocins that may occur during repetitive selective pressure in food applications.

PHYSICAL INTERVENTIONS

High-Pressure Processing

High-pressure processing (HPP) represents a promising nonthermal process for the preservation of sensitive products, such as sliced deli meats, that are prone to quality changes using typical thermal processes. Pressures of 100 to 600 MPa are being used to control microbial growth at low or moderate temperatures, without affecting organoleptic properties. The level of inactivation by HPP depends on the type of microorganism, pressure, treatment time, temperature, pH, water activity, and the composition of foodstuffs (Hugas et al., 2002). Cellular functions sensitive to pressure include modification of membrane permeability, fatty acid composition, cell and membrane morphology, protein denaturation, and inhibition of enzyme activity; however, a change in membrane structure is believed to be the main cause of inactivation (Lado and Yousef, 2002). Several disadvantages of HPP are batch-wise processing, the cost of scale-up to handle commercial-sized quantities of product, and the fact that not all organisms are equally affected.

Irradiation

Irradiation, also known as cold pasteurization, is an effective control measure in maintaining the quality of raw, cooked, and minimally processed meat products (Molins et al., 2001). The FDA/WHO Codex

Alimentarious Commission considers irradiation a safe technology for controlling *L. monocytogenes* in raw and uncooked meat. Among the different forms of irradiation, UV, gamma radiation, and electron beam (generated by electricity) are considered to be bactericidal. Because of their poor penetration power, UV rays are restricted to the treatment of food or equipment surfaces and eradication of airborne contaminants and, thus, can be of some practical use in reducing *L. monocytogenes* in food production and storage areas. In 2001, 40 countries permitted the use of irradiation in different types of foods, including 12 countries that allow irradiation for control of pathogens in poultry, 8 countries that permit its use in meat, and 13 that allow its use for fish and seafood (Molins et al., 2001). Zhu et al. (2005) demonstrated a 2.0- to 5.0-$\log_{10}$ reduction of *L. monocytogenes* on hams at 1.0 to 2.5 kGy, respectively, while examining electron beam irradiation of RTE products in the presence of various acidulants. Although flavor changes of raw meats induced by irradiation may be masked during cooking, flavor and quality challenges still exist for its application to RTE meat products.

Thermal Surface Pasteurization

Listeria contamination on RTE products often represents postprocess surface contamination, for which treatment of the entire mass of product may not be necessary. Submersed-water, postpackage pasteurization has been used commercially as a postprocess lethality step for surface pasteurization of RTE deli products capable of providing a 2.0- to 4.0-$\log_{10}$ decrease after 2 to 10 min of exposure at 91 to 96°C (Muriana et al., 2002). Radiant heat prepackage surface pasteurization has also achieved similar lethalities of *L. monocytogenes* (2.0- to 2.8-$\log_{10}$ reduction) in less time (60 to 75 s), with reduced purge loss or need for special pasteurization bags and reduced chilling requirements (Gande and Muriana, 2003). Prepackage pasteurization requires the product to be packaged immediately upon exit from the oven to eliminate the possibility of recontamination; however, a combination of the two processes eliminates this aspect while providing a 3.0- to 4.0-$\log_{10}$ reduction of *L. monocytogenes* (Muriana et al., 2004). Either method alone provides an alternative 2 process or in combination with chemical antimicrobials could provide for an alternative 1 process.

Other Physical Methods

As reviewed elsewhere in detail (NACMCF, 2006), additional methods that could be used as microbial interventions may also include microwave processing, ohmic heating, pulsed electric fields, nonthermal plasma, oscillating magnetic fields, ultrasound, and even filtration.

CHEMICAL INTERVENTIONS

Organic Acids

Many organic acids have GRAS status as food ingredients, and studies have shown their effectiveness under various conditions and in different foodstuffs. Such chemicals are more effective when used at a pH below the pK_a of the acid. Under such conditions, more of the undissociated form of an organic acid can enter cells and dissociate to generate the toxic anion, thus inhibiting the organism. The problem with many foods in which this form of inhibition takes place is that direct application of the acidulant may create a temporary acidified film around the product, which may be absorbed and neutralized with time by the food itself. The combination of the sodium/potassium salts of lactate and diacetate has become the standard ingredient for the processed meat industry ever since the USDA/FSIS increased the allowable level of lactate to 4.8% and of diacetate to 0.25% of the total formulation (Anonymous, 2000). Additional research demonstrated that this combination of acidulants had appreciable inhibitory activity for suppressing the growth of *L. monocytogenes* in formulated/processed RTE meats (Barmpalia et al., 2004; Bedie et al., 2001; Stekelenburg, 2003).

Oxidizing Solutions

Ozone is a strong oxidizing agent approved with GRAS status for the sanitization of bottled water in 1982. More recently, it was approved as an antimicrobial agent on food, including meat and poultry (Anonymous, 2001). As a general oxidant, it attacks chemical/organic constituents of the cell wall/membrane of bacteria and other microbes, including *L. monocytogenes*. The use of ozone is limited by its short half-life, which requires that it be used quickly upon generation. When foods are treated directly, its effectiveness is also limited by the degree of organic material, which reduces the oxidizing capacity of the solution, and therefore, it shows more consistency in application on surfaces with minimal organic background, such as fruits, vegetables, and environmental surface sanitation. Wade et al. (2003) found a greater efficacy of the reduction of *L. monocytogenes* (1.48 $\log_{10}$ CFU/g) on alfalfa sprouts and seeds that were continuously sparged for 5 to 20 min with ozonated water than that on those treated with water alone.

Smoke-Derived Flavorings and Extracts

Traditional smoking of commercial food products has been replaced largely by "liquid smoke" treatment because of the quick and easy application; however, liquid smoke is used mostly for flavor attributes and does not adequately provide the preservation characteristics of traditional smoked products. Liquid smoke is known to possess a variety of phenolic compounds that have demonstrated inhibitory properties toward pathogens and spoilage organisms. Faith et al. (1992) demonstrated the ability of liquid smoke to inactivate *L. monocytogenes* in hot dog exudates. In further testing with 11 different phenols, isoeugenol, a major phenolic component of liquid smoke, was the only phenol shown to appreciably inhibit *L. monocytogenes* (Faith et al., 1992). Other components derived from liquid smoke, including carbonyl compounds, have also demonstrated effective antilisterial properties on RTE meats and have suppressed growth of *L. monocytogenes* when inoculated at low levels or provided as much as a 5.3-$\log_{10}$ reduction when combined with heat pasteurization (Gedela et al., 2007a, 2007b).

Lauric Arginate

Current practices, such as the use of postprocess thermal treatments and/or application of biological and food-grade chemicals, have met with varying levels of success at reducing the prevalence of this pathogen. Regarding the latter, most studies evaluated the delivery of food-grade chemicals as an ingredient and/or a bath, dip, or spray for the finished product, and/or that were applied into or onto the packaging material. Instead, the sprayed lethality in container (SLIC) method was developed to introduce an antimicrobial solution into the packaging container just prior to when the finished meat or poultry product is placed within the package and then rely on the vacuum-packaging step to evenly distribute the antimicrobial purge, such that total coverage of both product and package is achieved (Luchansky et al., 2005). Depending on the product, this system was effective at delivering about a 2.0- to 5.0-$\log_{10}$ reduction of *L. monocytogenes* within 24 hours at refrigeration temperatures. Relative to other interventions that use food-grade chemicals, food treated via SLIC tastes better because there is less impact on flavor. Moreover, production costs are appreciably reduced to about $0.002 per pound, resulting in cost savings of $0.5 to $2 million per year for a small- or medium-sized frankfurter processing plant. Additionally, using this system would result in significantly shorter processing times (2 or 3 s), and the compact SLIC system fits directly on existing production lines at a nominal, one-time cost of $5,000 to $10,000 for the equipment.

DISCRIMINATIVE DETECTION METHODS FOR CONFIRMATION AND TRACE-BACK OF CONTAMINATED PRODUCTS

Subtyping offers an approach for investigating the relatedness of isolates and identifying and tracing the sources of epidemics (Lyytikäinen et al., 2000; de Valk et al., 2000; Sim et al., 2002). Subtyping has also been of great value in identifying sources of contamination in food processing plants (Berrang et al., 2000; Rørvik et al., 2003). As summarized elsewhere (Ryser and Marth, 1999), there are several practical (cost and ease-of-use) and scientific (typeability, reproducibility, and discriminatory power) considerations for selecting a suitable method for subtyping. The more often a particular method is practiced, the easier and less costly it becomes. Regarding scientific criteria, all strains should be unambiguously typeable by the method(s) used, the same results should be obtained each time by all persons using a standardized method, and strains that are indistinguishable should be shown as such, whereas strains that are nonidentical should be shown as different. Regardless, the results obtained with a given typing method must also be epidemiologically relevant (Tenover et al., 1997). The literature is replete with various methods for typing and/or tracking listeriae (Sauders et al., 2006). Although all methods provide useful information, some methods are better than others, relative to typeability, reproducibility, and/or discriminatory capability (Arbeit, 1995). For example, despite a general usefulness in studies of food-borne illness, it is often difficult to differentiate isolates of *L. monocytogenes* using only phenotypic methods: between 20 to 40% of strains are non-phage-typeable, the same or similar O antigens are found in several species/strains, serotype 4b strains are responsible for all major outbreaks and about 50% of sporadic cases, and the majority of the strains involved in food-borne outbreaks comprise two distinct genomic divisions that correlate with flagellar antigens (Rocourt, 1994). Thus, research on the implementation and optimization of typing strategies, used alone or in combination, has intensified to establish unequivocal relationships among strains for molecular subtyping studies.

It became quite apparent by the late 1980s that the overwhelming number of food-borne listeriosis cases were caused by only three serotypes, namely, serotypes 1/2a, 1/2b, and 4b. Thus, serotyping has limited value for epidemiologic applications. Another approach that was used to type *L. monocytogenes* was multilocus enzyme electrophoresis; isolates were grouped into one of two types (based on their enzyme locus profiles), I (4b, 1/2b, and 3b) and II (1/2a and 1/2c), which also coincided with groupings based on flagellar antigens a and c and on b (Piffaretti et al., 1989). Interestingly, isolates from sporadic cases of listeriosis and animal isolates could not be grouped into distinctive subsets based on multilocus enzyme electrophoresis types. In the mid-1990s the techniques of pulsed-field gel electrophoresis and ribotyping confirmed that *L. monocytogenes* grouped into these same two distinct groups (Brosh et al., 1994). Over the past decade, nucleic acid sequence-based typing schemes have flourished, given the advancements in PCR amplification and sequencing technologies. Automated DNA sequencers with laser fluorescence detection using a single gel lane can provide upwards of 500 bp of DNA sequence for as little as US$10.00 (core facility charge). One common method is multilocus sequence typing, whereby specific genetic loci are identified, amplified, sequenced, placed together as artificially contiguous DNA sequences, and analyzed in comparison to similar loci from other strains by multiple sequence alignment analyses (Maiden et al., 1998). One of the main advantages of multilocus sequence typing over DNA fragment-based typing schemes is the elimination of the ambiguity in fragment migration, especially in databases containing gels of DNA fragments produced by scientists in different labs. Automated DNA sequencing is more accurate than ever before, and DNA sequences are readily analyzed by portable computer software for simultaneous comparisons of various clones.

There are numerous examples already published describing the use of molecular methods to subtype and source *L. monocytogenes* on farms, in plants, and/or with foods. Due to space constraints, we will illustrate the approach using a study that we recently completed (Brito et al., 2008). Based on the association of *L. monocytogenes* with raw and pasteurized milk, dairy farms, cheese processing plants, and Latin-style cheese and, in turn, on the potential threat of listeriosis, we used pulsed-field gel electrophoresis to identify possible sources of contamination in a dairy processing plant producing Minas frescal cheese (MFC). Samples from 9 of 10 MFC brands from retail sites located in Juiz de Fora, Minas Gerais, Brazil, tested negative for *L. monocytogenes;* however, 6 of 10 from "brand F" of MFC tested positive. Thus, the farm/plant that produced brand F MFC was sampled; several sites within the processing plant and MFC samples tested positive. All 344 isolates recovered from retail MFC, plant F

MFC, and plant F environmental samples were serotype 1/2a and displayed the same AscI or ApaI fingerprints. These results helped establish that the storage coolers served as the contamination source of the MFC. Following renovation, samples from sites that previously tested positive for the pathogen were collected from the processing environment and MFC on multiple visits; all tested negative for *L. monocytogenes.* Our results validated that systematic culturing and analyses of products and processing facilities can identify areas that harbor *L. monocytogenes.* In addition, we also demonstrated that it is possible, with appropriate interventions, to eliminate harborage points within a food processing facility.

CONCLUDING REMARKS

Although in theory and from a policy perspective one can talk about eliminating *L. monocytogenes* from our food supply, given its ubiquity, persistence, and pathogenicity, in practice it may be more a question of better managing efforts to identify harborage points and applying interventions to lessen the prevalence and levels of this pathogen and, in turn, the threat to public health. To this end, we must be ever vigilant in following appropriate good manufacturing practices, good agricultural practices, standard operating protocols, and hazard analysis and critical control point programs, so as to minimize the load and occurrence of the pathogen, and concomitantly continue efforts to develop and implement effective interventions to ensure that an infectious dose of *L. monocytogenes* will not reach the consumer's plate. We must also be pragmatic with regard to when, where, and how much of our resources should be directed toward developing policies and hiring inspectors versus directed toward filling research voids. A wealth of information collected from the numerous studies conducted thus far confirms that some foods present a greater risk for listeriosis due to their ability to support survival/growth of the pathogen (e.g., $\geq$pH 4.4, $\geq a_w$ 0.92, $\geq 4°$C, absence of inhibitory substances, and/or combinations thereof) and that some people (i.e., "YOPIs") are at greater risk due to their age and/or health status. As such, it may be time to revisit the zero tolerance policy and consider allowing minimal levels of *L. monocytogenes* (e.g., $\leq$100 cells/g) in lower risk foods that do not support its growth. This would allow food safety professionals to focus available resources on foods that present a greater threat to human health and to ostensibly reduce the occurrence and severity of listeriosis worldwide.

REFERENCES

Angelidis, A. S., and G. M. Smith. 2003. Role of glycine betaine and carnitine transporters in adaptation of *Listeria monocytogenes* to chill stress in defined medium. *Appl. Environ. Microbiol.* **69:**7492–7498.

Anonymous. 2000. Food additives for use in meat and poultry products: sodium diacetate, sodium acetate, sodium lactate and potassium lactate. *Fed. Reg.* **65:**3121–3123.

Anonymous. 2001. Secondary direct food additives permitted in food for human consumption. *Fed. Reg.* **66:**33829–33830.

Anonymous. 2003. Control of *L. monocytogenes* in ready-to-eat meat and poultry products; final rule. *Fed. Reg.* **68:**34207–34254.

Arbeit, R. D. 1995. Laboratory procedure for the epidemiologic analysis of microorganisms, p. 190–208. *In* P. R. Murray, E. J. Baron, M. A. Phaller, F. C. Tenover, and R. H. Yolken (ed.), *Manual of Clinical Microbiology*, 7th ed. ASM Press, Washington, DC.

Bakardjiev, I., J. A. Theriot, and D. A. Portnoy. 2006. *Listeria monocytogenes* traffics from maternal organs to the placenta and back. *PloS Pathog.* **2:**0623–0631.

Barmpalia, I. M., I. Geornaras, K. E. Belk, J. A. Scanga, P. A. Kendall, G. C. Smith, and J. N. Sofos. 2004. Control of *Listeria monocytogenes* on frankfurters with antimicrobials in the formulation and by dipping in organic acid solutions. *J. Food Prot.* **67:**2456–2464.

Bayles, D. O., and B. J. Wilkinson. 2000. Osmoprotectants and cryoprotectants for *Listeria monocytogenes*. *Lett. Appl. Microbiol.* **30:**23–27.

Bedie, G. K., J. Samelis, J. N. Sofos, K. E. Belk, J. A. Scanga, and G. C. Smith. 2001. Antimicrobials in the formulation to control *Listeria monocytogenes*. Postprocessing contamination on frankfurters stored at 4°C in vacuum packages. *J. Food Prot.* **64:**1949–1955.

Berrang, M. E., R. J. Meinersmann, J. K. Northcutt, and D. P. Smith. 2000. Molecular characterization of *Listeria monocytogenes* isolated from a poultry further processing facility and from fully cooked product. *J. Food Prot.* **65:**1574–1579.

Brito, J. R., E. M. Santos, M. M. P. O. Cerqueira, E. F. Arcuri, C. C. Lange, M. A. Brito, G. N. Souza, J. M. Beltran, J. E. Call, Y. Liu, A. C. S. Porto-Fett, and J. B. Luchansky. 2008. Retail survey of Brazilian milk and Minas frescal cheese and a contaminated dairy plant to establish the prevalence, relatedness, and sources of *Listeria monocytogenes* isolates. *Appl. Environ. Microbiol.* **74:**4954–4961.

Brosch, R., J. Chen, and J. B. Luchansky. 1994. Pulsed-field fingerprinting of listeriae: identification of genomic divisions for *Listeria monocytogenes* and their correlation with serovar. *Appl. Environ. Microbiol.* **60:**2584–2592.

Buchanan, R. L., W. G. Damert, R. C. Whiting, and M. van Schothorst. 1997. Use of epidemiologic food survey data to estimate a purposefully conservative dose-response relationship for *Listeria monocytogenes* levels and incidence. *J. Food Prot.* **60:**918–922.

Carlton, R. M., W. H. Noordman, B. Biswas, E. D. de Meester, and M. J. Loessner. 2005. Bacteriophage P100 for control of *Listeria monocytogenes* in foods: genome sequence, bioinformatic analyses, oral toxicity study, and application. *Reg. Toxicol. Pharmacol.* **43:**301–312.

CDC. 1988. Epidemiologic notes and reports update—listeriosis and pasteurized milk. *MMWR Morbid. Mortal. Wkly. Rpt.* **37:**764–766.

CDC. 2000. Multi-state outbreak of listeriosis—United States, 2000. *MMWR Morbid. Mortal. Wkly. Rpt.* **49:**1129–1130.

CDC. 2006. Surveillance for foodborne-disease outbreaks—United States, 1998–2002. *MMWR Morbid. Mortal. Wkly. Rpt.* **55:**1–42.

CFSAN. 2003. Interpretive summary: quantitative assessment of the relative risk to public health from foodborne *Listeria monocytogenes* among selected categories of ready-to-eat foods. http://www.foodsafety.gov/~dms/lmr2-su.html. Accessed 15 April 2008.

Chattopadhyay, M. K. 2006. Mechanism of bacterial adaptation to low temperature. *J. Biosci.* **31:**101–109.

Chen, Y., W. H. Ross, V. N. Scott, and D. E. Gombas. 2003. *Listeria monocytogenes:* low levels equal low risk. *J. Food Prot.* **66:**570–577.

Cotter, P. D., and C. Hill. 2003. Surviving acid test: responses of gram-positive bacteria to low pH. *Microbiol. Mol. Biol. Rev.* **67:**429–453.

de Valk, H., V. Vaillant, and V. Goulet. 2000. Epidemiology of human *Listeria* infections in France. *Bull. Acad. Natl. Med.* **184:**267–274.

Doyle, M. E., A. S. Mazzotta, T. Wang, D. W. Wiseman, and V. N. Scott. 2001. Heat resistance of *Listeria monocytogenes*. *J. Food Prot.* **64:**410–429.

Duché, O., F. Trémoulet, P. Glaser, and J. Labadie. 2002. Salt stress proteins induced in *Listeria monocytogenes*. *Appl. Environ. Microbiol.* **68:**1491–1498.

Faith, N. G., A. E. Yousef, and J. B. Luchansky. 1992. Inhibition of *Listeria monocytogenes* by liquid smoke and isoeugenol, a phenolic component found in smoke. *J. Food Saf.* **12:**303–314.

Farber, J. M. 1991. *Listeria monocytogenes* in fish products. *J. Food Prot.* **54:**922–924, 934.

Farber, J. M., and P. I. Peterkin. 1991. *Listeria monocytogenes*, a food-borne pathogen. *Microbiol. Mol. Biol. Rev.* **55:**476–511.

FDA. 2006. Food additives permitted for direct addition to food for human consumption; bacteriophage preparation. *Fed. Reg.* **71:**47729–47732.

Gande, N., and P. M. Muriana. 2003. Pre-package surface pasteurization of ready-to-eat deli meats for reduction of *Listeria monocytogenes*. *J. Food Prot.* **66:**1623–1630.

Gandhi, M., and M. L. Chikindas. 2007. *Listeria*: a foodborne pathogen that knows how to survive. *Int. J. Food Microbiol.* **113:**1–15.

Garner, M. R., K. E. James, M. C. Callahan, M. Wiedmann, and K. J. Boor. 2006. Exposure to salt and organic acids increases the ability of *Listeria monocytogenes* to invade Caco-2 cells but decreases its ability to survive gastric stress. *Appl. Environ. Microbiol.* **72:**5384–5395.

Gedela, S., J. R. Escoubas, and P. M. Muriana. 2007a. Effect of inhibitory liquid smoke fractions on *Listeria monocytogenes* during long-term storage of frankfurters. *J. Food Prot.* **70:**386–391.

Gedela, S., R. Wright, S. Macwana, J. R. Escoubas, and P. M. Muriana. 2007b. Effect of inhibitory extracts derived from liquid smoke combined with postprocess pasteurization for control of *Listeria monocytogenes* on ready-to-eat meats. *J. Food Prot.* **70:**2749–2756.

Glaser, P., L. Frangeul, C. Buchrieser, C. Rusniok, A. Amend, F. Baquero, P. Berche, H. Bloecker, P. Brandt, T. Chakraborty, A. Charbit, F. Chetouani, E. Couvé, A. de Daruvar, P. Dehoux, E. Domann, G. Domínguez-Bernal, E. Duchaud, L. Durant, O. Dussurget, K.-D. Entian, H. Fsihi, F. Garcia-Del Portillo, P. Garrido, L. Gautier, W. Goebel, N. Gómez-López, T. Hain, J. Hauf, D. Jackson, L.-M. Jones, U. Kaerst, J. Kreft, M. Kuhn, F. Kunst, G. Kurapkat, E. Madueño, A. Maitournam, J. Mata Vicente, E. Ng, H. Nedjari, G. Nordsiek, S. Novella, B. de Pablos, J.-C. Pérez-Diaz, R. Purcell, B. Remmel, M. Rose, T. Schlueter, N. Simoes, A. Tierrez, J.-A. Vázquez-Boland, H. Voss, J. Wehland, and P. Cossart. 2001. Comparative genomics of *Listeria* species. *Science* **294:**849–852.

Gombas, D. E., Y. Chen, R. S. Clavero, and V. N. Scott. 2003. Survey of *Listeria monocytogenes* in ready-to-eat foods. *J. Food Prot.* **66**:559–569.

Greenacre, E. J., T. F. Brocklehurst, C. R. Waspe, D. R. Wilson, and P. D. G. Wilson. 2003. *Salmonella enterica* serovar Typhimurium and *Listeria monocytogenes* acid tolerance response induced by organic acids at 20°C: optimization and modeling. *Appl. Environ. Microbiol.* **69**:3945–3951.

Hugas, M., M. Garriga, and J. M. Monfort. 2002. New mild technologies in meat processing: high pressure as a model technology. *Meat Sci.* **62**:359–371.

Ingham, S. C., G. Searls, S. Mohanan, and D. R. Buege. 2006. Survival of *Staphylococcus aureus* and *Listeria monocytogenes* on vacuum-packaged beef jerky and related products stored at 21°C. *J. Food Prot.* **69**:2263–2267.

Juneja, V. K., and B. S. Eblen. 1999. Predictive thermal inactivation model for *Listeria monocytogenes* with temperature, pH, NaCl, and sodium pyrophosphate as controlling factors. *J. Food Prot.* **62**:986–993.

Kathariou, S. 2002. Listeria monocytogenes virulence and pathogenicity, a food safety perspective. *J. Food Prot.* **65**:1811–1829.

Khelef, N., M. Lecuit, C. Buchrieser, D. Cabanes, O. Dussurget, and P. Cossart. 2006. *Listeria monocytogenes* and the genus *Listeria*. *Prokaryotes* **4**:404–476.

Lado, B. H., and A. E. Yousef. 2002. Alternative food-preservation technologies: efficacy and mechanisms. *Microbes Infect.* **4**:433–440.

Luchansky, J. B., J. E. Call, B. Hristova, L. Rumery, L. Yoder, and A. Oser. 2005. Viability of *Listeria monocytogenes* on commercially-prepared hams surface treated with acidified calcium sulfate and lauric arginate and stored at 4°C. *Meat Sci.* **71**:92–99.

Lyytikäinen, O., R. Maijala, A. Siitonen, T. Autio, T. Honkanen-Buzalski, M. Hatakka, J. Mikkola, V. Anttila, and P. Ruutu. 2000. An outbreak of *Listeria monocytogenes* serotype 3a from butter in Finland. *J. Infec. Dis.* **181**:1838–1841.

Maiden, M. C. J., J. A. Bygraves, E. Feil, G. Morelli, J. E. Russell, R. Urwin, Q. Zhang, J. Zhou, K. Zurth, D. A. Caugant, I. M. Feavers, M. Achtman, and B. G. Spratt. 1998. Multilocus sequence typing: a portable approach to the identification of clones within populations of pathogenic microorganisms. *Proc. Natl. Acad. Sci. USA* **95**:3140–3145.

McLauchlin, J. 1996. The relationship between *Listeria* and listeriosis. *Food Control* **7**:187–193.

McLauchlin, J., R. T. Mitchell, W. J. Smerdon, and K. Jewell. 2004. *Listeria monocytogenes* and listeriosis: a review of hazard characterisation for use in microbiological risk assessment of foods. *Int. J. Food Microbiol.* **92**:15–33.

Mead, P. S., L. Slutsker, V. Dietz, L. F. McCaig, J. S. Bresee, C. Shapiro, P. M. Griffin, and R. V. Tauxe. 1999. Food-related illness and death in the United States. *Emerg. Infect. Dis.* **5**:607–625.

Meng, J., and M. P. Doyle. 1997. Emerging issues in microbiological food safety. *Ann. Rev. Nutrit.* **17**:255–275.

Molins, R. A., Y. Motarjemi, and F. K. Kaferstein. 2001. Irradiation: a critical control point in ensuring the microbiological safety of raw foods. *Food Control* **12**:347–356.

Muriana, P. M. 1996. Bacteriocins for control of *Listeria*. *J. Food Prot.* **59** (Suppl.):54–63.

Muriana, P. M., N. Gande, W. Robertson, B. Jordan, and S. Mitra. 2004. Effect of prepackage and postpackage pasteurization on postprocess elimination of *Listeria monocytogenes* on deli turkey products. *J. Food Prot.* **67**:2472–2479.

Muriana, P. M., W. Quimby, C. Davidson, and J. Grooms. 2002. Post-package pasteurization of RTE deli meats by submersion heating for reduction of *Listeria monocytogenes*. *J. Food Prot.* **65**:963–969.

Murray, E. G. D., R. A. Webb, and M. B. R. Swann. 1926. A disease of rabbit characterised by a large mononuclear leucocytosis, caused by a hitherto undescribed bacillus *Bacterium monocytogenes* (n.sp.). *J. Pathol. Bacteriol.* **29**:407–439.

NACMCF. 2006. Requisite scientific parameters for establishing the equivalence of alternative methods of pasteurization. *J. Food Prot.* **69**(Suppl.):1190–1216.

Neetoo, H., M. Ye, H. Chen, R. D. Yoerger, D. T. Hicks, and D. G. Hoover. 2008. Use of nisin-coated plastic films to control *Listeria monocytogenes* on vacuum-packaged cold-smoked salmon. *Int. J. Food Microbiol.* **122**:8–15.

Nelson, K. E., D. E. Fouts, E. F. Mongodin, J. Ravel, R. T. DeBoy, J. F. Kolonay, D. A. Rasko, S. V. Angiuoli, S. R. Gill, I. T. Paulsen, J. Peterson, O. White, W. C. Nelson, W. Nierman, M. J. Beanan, L. M. Brinkac, S. C. Daugherty, R. J. Dodson, A. S. Durkin, R. Madupu, D. H. Haft, J. Selengut, S. Van Aken, H. Khouri, N. Fedorova, H. Forberger, B. Tran, S. Kathariou, L. D. Wonderling, G. A. Uhlich, D. O. Bayles, J. B. Luchansky, and C. M. Fraser. 2004. Whole genome comparisons of serotype 4b and 1/2a strains of the food-borne pathogen *Listeria monocytogenes* reveal new insights into the core genome components of this species. *Nucleic Acids Res.* **32**:2386–2395.

Nolan, D. A., D. C. Chamblin, and J. A. Troller. 1992. Minimal water activity levels for growth and survival of *Listeria monocytogenes* and *Listeria inocua*. *Int. J. Food Microbiol.* **16**:323–335.

Nyfeldt, A. 1929. Etiologie de la mononucleose infectieuse. *Soc. Biol.* **101**:590–592.

O'Driscoll, B., C. G. M, Gahan, and C. Hill. 1996. Adaptive acid tolerance response in *Listeria monocytogenes*: isolation of an acid tolerant mutant which displays increased virulence. *Appl. Environ. Microbiol.* **62**:1693–1698.

Paoli, G. C., A. K. Bhunia, and D. O Bayles. 2005. *Listeria monocytogenes*, p. 295–340. *In* P. M. Fratamico, A. K. Bhunia, and J. L. Smith (ed.), *Foodborne Pathogens: Microbiology and Molecular Biology*. Caister Academic Press, Norfolk, United Kingdom.

Parish, M. E., and D. P. Higgins. 1989. Survival of *Listeria monocytogenes* in low pH model broth systems. *J. Food Prot.* **52**:144–147.

Pearson, L. J., and E. H. Marth. 1989. *Listeria monocytogenes*—threat to a safe food supply: a review. *J. Dairy Sci.* **73**:912–928.

Phan-Thanh, L., F. Mahouin, and S. Aligé. 2000. Acid responses of *Listeria monocytogenes*. *Int. J. Food Microbiol.* **55**:121–126.

Piffaretti, J. C., H. Kressebuch, M. Aeschbacher, J. Bille, E. Bannerman, J. M. Musser, R. K. Selander, and J. Rocourt. 1989. Genetic characterization of clones of the bacterium *Listeria monocytogenes* causing epidemic disease. *Proc. Natl. Acad. Sci.* **86**:3818–3822.

Razavilar, V., and C. Genigeorgis. 1998. Prediction of *Listeria* spp. growth as affected by various levels of chemicals, pH, temperature and storage time in a model broth. *Int. J. Food Microbiol.* **40**:149–157.

Rocourt, J. 1994. *Listeria monocytogenes*: the state of the science. *Dairy Food Environ. Sanit.* **14**:70–82.

Rørvik, L. M., B. Aase, T. Alvestad, and D. A. Caugant. 2003. Molecular epidemiological survey of *Listeria monocytogenes* in broilers and poultry products. *J. Appl. Microbiol.* **94**:633–640.

Rowan, N. J., and J. G. Anderson. 1998. Effects of above-optimum growth temperature and cell morphology on thermotolerance of *Listeria monocytogenes* cells suspended in bovine milk. *Appl. Environ. Microbiol.* **64**:2065–2071.

Ryser, E. T., and E. H. Marth. 1999. *Listeria, Listeriosis, and Food Safety*, 2nd ed. Marcel Dekker, New York, NY.

Sauders, B. D., Y Schukken, L. Kornstein, V. Reddy, T. Bannerman, E. Salehi, N. Dummas, B. J. Anderson, J. P. Massey, and M. Wiedmenn. 2006. Molecular epidemiology and cluster analysis of human listeriosis cases in three U.S. states. *J. Food Prot.* 69:1680–1689.

Shank, F. R., E. L. Elliot, I. K. Wachsmuth, and M. E. Losikoff. 1996. US position on *Listeria monocytogenes* in foods. *Food Control* 7:229–234.

Shen, Y., Y. Liu, Y. Zhang, J. Cripe, W. Conway, J. Meng, G. Hall, and A. A. Bhaguat. 2006. Isolation and characterization of *Listeria monocytogenes* from ready-to-eat foods in Florida. *Appl. Environ. Microbiol.* 72:5073–5076.

Sim, J., D. Hood, L. Finnie, M. Wilson, C. Graham, M. Brett, and J. A. Hudson. 2002. Series of incidents of *Listeria monocytogenes* non-invasive febrile gastroenteritis involving ready-to-eat meats. *Lett. Appl. Microbiol.* 35:409–413.

Smith, L. T. 1996. Role of osmolytes in adaptation of osmotically stressed and chill-stressed *Listeria monocytogenes* grown in liquid media and on processed meat surfaces. *Appl. Environ. Microbiol.* 62:3088–3093.

Stecchini, M. L., M. Del Torre, and E. Venir. 2004. Growth of *Listeria monocytogenes* as influenced by viscosity and water activity. *Int. J. Food Microbiol.* 96:181–187.

Stekelenburg, F. K. 2003. Enhanced inhibition of *Listeria monocytogenes* in frankfurter sausage by the addition of potassium lactate and sodium diacetate mixtures. *Food Microbiol.* 20:133–137.

Tenover, F. C., R. D. Arbeit, and R. V. Goering. 1997. How to select and interpret molecular strain typing methods for epidemiological studies of bacterial infections: a review for healthcare epidemiologists. *Infect. Control Hosp. Epidemiol.* 18:426–439.

Tompkin, R. B. 2002. Control of *Listeria monocytogenes* in the food-processing environment. *J. Food Prot.* 65:709–725.

Vorst, K. L., E. C. D. Todd, and E. T. Ryser. 2004. Improved quantitative recovery of *Listeria monocytogenes* from stainless steel surfaces using a one-ply composite tissue. *J. Food Prot.* 67:2212–2217.

Wade, W. N., A. J. Scouten, K. H. McWatters, R. L. Wick, A. Demerci, W. F. Fett, and L.R. Beuchat. 2003. Efficacy of ozone in killing *Listeria monocytogenes* on alfalfa seeds and sprouts and effects on sensory quality of sprouts. *J. Food Prot.* 66:44–51.

Wallace, F. M., J. E. Call, A. C. S. Porto, G. J. Cocoma, ERRC Special Projects Team, and J. B. Luchansky. 2003. Recovery rate of *Listeria monocytogenes* from commercially prepared frankfurters during extended refrigerated storage. *J. Food Prot.* 66:584–591.

Zhu, M. J., A. Mendonca, H. A. Ismail, M. Du, E. J. Lee, and D. U. Ahn. 2005. Impact of antimicrobial ingredients and irradiation on the survival of *Listeria monocytogenes* and the quality of ready-to-eat turkey ham. *Poult. Sci.* 84:613–620.

Pathogens and Toxins in Foods: Challenges and Interventions
Edited by V. K. Juneja and J. N. Sofos
© 2010 ASM Press, Washington, DC

Chapter 7

Salmonella

STAN BAILEY, L. JASON RICHARDSON, NELSON A. COX, AND DOUGLAS E. COSBY

Clinical pathologists in France in the early 19th century first identified a typhoid bacillus that was the causative agent for typhoid fever. This agent was eventually identified as *Salmonella*. In the United States, Salmon and Smith (1885) first isolated *Bacillus cholerae-suis*, now known as *Salmonella cholerasuis*, from swine with "hog cholera" (Le Minor, 1981). Theobald Smith, a researcher under Daniel E. Salmon in the USDA's Bureau of Animal Industry, was the first American to identify *Salmonella* as a separate strain or genus. Although Smith actually identified the bacteria, Salmon's name as administrator was listed first on the research paper, so the new bacterium was named for Salmon. In this chapter, the sections are intended to give the reader a basic understanding of the characteristics and illness, along with the sources and incidence of *Salmonella* in the environment and food commodities. Factors affecting survival in food processing operations, use of interventions, and discriminative detection methods are also discussed.

TYPE OF ILLNESS AND CHARACTERISTICS OF THE ORGANISM

Salmonellosis is a human infection caused by *Salmonella* bacteria. The establishment of a human *Salmonella* infection rests on the ability of the organism to attach (colonization) and enter (invasion) intestinal columnar epithelial cells (enterocytes) and specialized M cells overlying Peyer's patches (Ponka et al., 1995). The majority of persons infected with *Salmonella* develop diarrhea, fever, and abdominal cramps 12 to 72 hours after infection. The illness usually lasts 4 to 7 days, and most persons recover without treatment. However, in some persons the diarrhea may be so severe that the patient needs to be hospitalized. In these patients, the *Salmonella* infection may spread from the intestines to the blood stream and then to other body sites and can cause death, unless the person is treated promptly with antibiotics. The elderly, infants, and those with impaired immune systems are more likely to have a severe illness. In rare instances, chronic conditions, such as reactive arthritis, Reiter's syndrome, and ankylosing spondylitis, have resulted from *Salmonella* infections in patients. Bacterial prerequisites for the onset of these chronic diseases include the ability of the bacterial strain to infect mucosal surfaces, the presence of outer membrane lipopolysaccharides, and a propensity to invade host cells (Hakalehto et al., 2007; Burslem et al., 1990; Sockett et al., 1993). Voetsch et al. (2004) determined that the average confirmed annual human illness in the United States caused by nontyphoidal *Salmonella* infections was approximately 15,000 hospitalizations and 400 deaths. However, approximately 1.4 million cases of illness have been estimated to occur in the United States annually. The annual cost associated with salmonellosis has been estimated to be several billion dollars (Voetsch et al., 2004).

Salmonella spp. are facultatively anaerobic gram-negative non-spore-forming rods belonging to the family *Enterobacteriaceae*. *Salmonella* spp. are capable of adapting to extreme environmental conditions. The majority of *Salmonella* spp. are motile by peritrichous flagella. However, nonflagellated variants, such as *S. enterica* serovar Pullorum and *S. enterica* serovar Gallinarum, and nonmotile *Salmonella* strains resulting from dysfunctional flagella do occur. *Salmonella* spp. have the ability to metabolize nutrients by

Stan Bailey • Biomerieux, Inc., Hazelwood, MI 63042. L. Jason Richardson and Nelson A. Cox • USDA-ARS-Poultry Microbiological Safety Research Unit, Russell Research Center, Athens, GA 30605. Douglas E. Cosby • USDA-ARS-Bacterial Epidemiology and Antimicrobial Resistant Research Unit, Russell Research Center, Athens, GA 30605.

the respiratory and fermentative pathways. The optimal temperature of growth is between 35 and 40°C. However, dependent on the *Salmonella* strain and the type of food matrix, the range of growth can occur between 2 and 54°C. Furthermore, *Salmonella* strains have an optimum pH for sustained growth between 6.5 and 7.5.

The nomenclature of *Salmonella* has changed significantly over time and has progressed through a succession of taxonomical schemes based on biochemical and serological characteristics, principles of numerical taxonomy, and DNA homology. The Centers for Disease Control and Prevention (CDC) has adopted as its official nomenclature the scheme in which the genus *Salmonella* contains two species, each of which contains multiple serotypes (Brenner et al., 2000). The two species are *S. enterica* and *S. bongori*. *S. enterica* is divided into six subspecies, each referred to by a Roman numeral and a name (I, *S. enterica* subsp. *enterica*; II, *S. enterica* subsp. *salamae*; IIIa, *S. enterica* subsp. *arizonae*; IIIb, *S. enterica* subsp. *diarizonae*; IV, *S. enterica* subsp. *houtenae*; and VI, *S. enterica* subsp. *indica*). *S. enterica* subspecies are differentiated biochemically and through genomic relatedness. A total of approximately 2,450 serotypes have been identified, with the largest number of these in the *S. enterica* subsp. *enterica* group (D'Aoust, 2000).

Salmonella strains vary in their pathogenic abilities. *S. enterica* subspecies are considered significant etiological agents for human food-borne-related illnesses. However, it is important to remember that public health agencies in the United States consider all species of *Salmonella* important to public health. Therefore, isolation procedures should recover all serotypes of *Salmonella*.

SOURCES AND INCIDENCE IN THE ENVIRONMENT AND FOODS

The natural habitat of *Salmonella* is the gastrointestinal tract of animals. However, it is ubiquitous and has been isolated from numerous other sources, such as seafood, fruits, and vegetables. These sources usually become exposed to *Salmonella* by either direct or indirect contact with contaminated fecal matter. Historically though, *Salmonella* has been thought of as primarily a pathogen of poultry meat, eggs, swine, and other farm animals. USDA, Food Safety and Inspection Service (FSIS) hazard analysis critical control point (HACCP) samplings confirm this association of *Salmonella* with raw meat and egg products (Table 1). Broiler chickens (11.4%), turkeys (20.3%), and ground chicken (45.0%) had the highest

Table 1. Percent positive *Salmonella* tests in FSIS HACCP verification samples—2006[a]

Product	% Positive
Broilers	11.4
Market hogs	4.0
Cows/bulls	0.8
Steers/heifers	0.3
Ground beef	2.0
Ground chicken	45.0
Ground turkey	20.3
Turkeys	7.1

[a]Source: www.fsis.usda.gov/Science/Progress_Report_Salmonella_Testing_Tables/index.asp.

incidence among the FSIS-regulated meat products in 2006. Because chicken is the meat product with the highest frequency of *Salmonella*-positive samples, many people assume that the serotypes most frequently associated with human illness would be the serotypes most commonly found on poultry. The *Salmonella* incidence data from the CDC for humans and from the National Antimicrobial Resistance Monitoring System for farm animals are relatively consistent from year to year. As seen in the 2004 data (Table 2), *S. enterica* serovar Typhimurium, *S. enterica* serovar Enteritidis, and *S. enterica* serovar Newport are the most frequently isolated *Salmonella* serovars from human salmonellosis cases. However, *S. enterica* serovar Kentucky, *S. serovar* Typhimurium, and *S. enterica* serovar Heidelberg are the serotypes most frequently associated with poultry. The biggest change in poultry isolations in recent years is the rise in the incidence of *S. serovar* Kentucky and the increase in the incidence of *S. serovar* Enteritidis in broiler chickens.

Of particular concern since the mid-1990s has been a pandemic associated with *S. serovar* Enteritidis associated with raw or lightly cooked shell eggs.

Table 2. *Salmonella enterica* serotypes in humans versus chickens (slaughter), 2004[a]

Rank	Human Serovar	%	Chicken (slaughter) Serovar	%
1	Typhimurium	19.2	Kentucky	44.4
2	Enteritidis	14.1	Typhimurium	13.4
3	Newport	9.3	Heidelberg	13.1
4	Javiana	5.0	Enteritidis	6.6
5	Heidelberg	4.9	Schwarzengrund	2.8
6	Montevideo	2.4	Montevideo	2.3
7	1,4,[5],12:i:-	2.1	Thompson	1.7
8	Muenchen	2.1	Mbandaka	1.5
9	Saintpaul	1.9	Infantis	1.5
10	Braenderup	1.9	1,4,[5],12:i:-	1.4

[a]Values are percentages of total isolates positive for the serotypes.

The overall incidence of serovar Enteritidis contamination of eggs from commercial flocks in the United States has been estimated at around 0.005% (Ebel and Schlosser, 2000). Like most other paratyphoid (non-host-adapted) *Salmonella* serotypes, serovar Enteritidis is usually introduced to chickens via the gastrointestinal tract. Of particular concern is that after invasion through mucosal epithelial cells serovar Enteritidis can be disseminated systemically to a wide array of internal organs, including reproductive tissues (Gast and Beard, 1990; Humphrey et al., 1993). However, studies have also shown that several other serotypes can also be recovered from internal organs and tissues (Bailey et al., 2005; Cox et al., 2006, 2007a). By colonizing the ovary (the site of yolk maturation and release) and the oviduct (the site of albumen secretion around the descending yolk), serovar Enteritidis appears to gain access to the contents of eggs (Miyamoto et al., 1997; De Buck et al., 2004). The prevalence of serovar Enteritidis in egg layers and the incidence of serovar Enteritidis in human illnesses peaked in the late 1990s in the United States. Egg quality assurance programs have been sponsored by the industry, states, and the federal government, and their effectiveness in influencing the epidemiology of serovar Enteritidis in the United States is documented by Mumma et al. (2004).

Despite the widespread perception of meats and eggs as the predominate source of *Salmonella,* an analysis of the 23 *Salmonella* outbreaks in which the etiology of the causative food was known, reveals that only 5 outbreaks were definitively linked to poultry meat or eggs. The remaining 18 outbreaks were associated with nonmeat products, including multiple outbreaks associated with tomatoes, cantaloupe, and raw milk (http://www.about-salmonella.com, 2007). The largest outbreaks in 2006 were associated with tomatoes and peanut butter and in 2007 with pot pies (http://cdc.gov). While meat and poultry must always be considered potential sources of *Salmonella,* the recent and numerous outbreaks associated with other vectors, primarily produce, suggest that public health, regulatory, and food production industries must be aware of the almost ubiquitous presence of *Salmonella.*

INTRINSIC AND EXTRINSIC FACTORS THAT AFFECT SURVIVAL AND GROWTH IN FOOD PRODUCTS AND CONTRIBUTE TO OUTBREAKS

Although the potential growth of food-borne *Salmonella* is of primary importance in safety assessments, the propensity of these pathogens to persist in hostile environments further heightens public health

concerns and the need to control this pathogen in the food chain (Humphrey, 2004). This significantly contributes to outbreaks occurring from contaminated food products, whether the products are from animal origin or another source. Intrinsic factors that affect survival and growth of *Salmonella* in food products include, but are not limited to, acidity, pH, moisture content, nutrient content, competitive microflora, and natural/added antimicrobials in the food.

The growth of *Salmonella* in acidic environments has been demonstrated to be significantly more robust if the bacterium has been preconditioned by previous exposure to lower pH conditions (Huhtanen, 1975). Acid stress can also trigger enhanced bacterial resistance to other adverse environmental conditions, as demonstrated by the growth of *S.* serovar Typhimurium at pH 5.8, leading to an increased thermal resistance at 50°C, an enhanced tolerance to high osmotic stress (2.5 M NaCl) ascribed to the induced synthesis of the Omp C outer membrane proteins, a greater surface hydrophobicity, and an increased resistance to the antibacterial lactoperoxidase system and surface-active agents, such as crystal violet and polymyxin B (Leyer and Johnson, 1993). The presence of *Salmonella* spp. that have increased acid tolerances in foods heightens the level of public health hazard because this could minimize the antimicrobial action of gastric acidity (pH 2.5) and promote the survival of *Salmonella* within the digestive system of humans (D'Aoust, 1991a).

High salt concentrations have long been recognized for their ability to extend the shelf-life of foods by inhibiting the growth of endogenous microflora (Pohl et al., 1993). This bacteriostatic effect results from a dramatic decrease in water activity (a_w). *Salmonella* growth is inhibited in the presence of 3 to 4% NaCl; however, salt tolerance increases with increasing temperature in the range of 10 to 30°C (D'Aoust, 1991b). *Salmonella* will not grow in foods with an a_w value of <0.93. The a_w in fresh meat, poultry, and fish ranges between 0.99 and 1.00 and in eggs, natural cheeses, fresh fruits, and vegetables from 0.95 to 1.00. *Salmonella* has been shown to proliferate at pH values ranging from 4.5 to 9.5, even though the optimum pH for growth is 6.5 to 7.5. Meat and poultry have a pH range of 5.1 to 6.4. Fish and shellfish have a pH range of 5.5 to 7.0. Most fresh fruits have a pH range of 1.8 to 6.7, and vegetables have a pH range of 3.8 to 7.3.

Extrinsic factors that affect survival and growth of *Salmonella* in food products include, but are not limited to, temperature, storage conditions, and packaging/atmosphere. *Salmonella* can readily adapt to extreme environmental conditions. Heat is used widely in food manufacturing processes to control the bacterial quality and safety of end products. *Salmonella* has the ability to acquire greater heat resistance following exposure to sublethal temperatures. The rapid adaptation of the organism to rising temperatures in the microenvironment with a level of enhanced thermotolerance is quite distinct from that described in conventional time-temperature curves of thermal lethality. This adaptive response uncovers potentially serious implications for the safety of thermal processes that expose or maintain food products at marginally lethal temperatures.

Salmonella can actively grow within a wide temperature range and has also exhibited psychrotrophic properties, as reflected in the ability to grow in foods stored at 2°C to 4°C (D'Aoust, 1994). In addition, exposing cells to low temperatures can increase the ability of some *Salmonella* spp. to grow and survive at refrigeration temperatures (Alpuche-Aranda et al., 1994; D'Aoust, 1991b). It has been shown that the composition of the freezing menstruum, the kinetics of the freezing process, the physiological state of *Salmonella*, and the serovar-specific responses to extreme temperatures determine the fate of *Salmonella* during freezer storage of foods (Corry, 1976). This has raised concerns for the efficacy of chill temperatures in ensuring food safety. Widespread refrigerated storage of foods packaged under vacuum or modified atmosphere to prolong shelf-life has also been utilized. Gaseous mixtures consisting of 60 to 80% (vol/vol) CO_2 with varying proportions of N_2 and/or O_2 have been found to inhibit the growth of aerobic spoilage microorganisms, such as *Pseudomonas* spp., without promoting the growth of *Salmonella* spp. (D'Aoust, 1994). A more in-depth review of stress responses and the ability of *Salmonella* to survive in foods can be found in the work of Humphrey (2004).

FOOD PROCESSING OPERATIONS THAT INFLUENCE THE NUMBERS, SPREAD, OR CHARACTERISTICS

The ecology and behavior of *Salmonella* in foods of animal origin have been studied extensively. However, the behavior of *Salmonella* within other types of commodities, such as fruits and vegetables, is not fully elucidated (Beuchet, 2002). The occurrence of *Salmonella* in food products which are capable of harboring the organism, but traditionally not considered to harbor the organism, presents serious problems to food safety.

With regard to meat products, specifically poultry, the problem with eliminating *Salmonella* is complicated by the fact that these organisms are natural commensal microflora of the digestive tract and usually present in very high numbers. The prevalence rate of *Salmonella* in preharvest poultry is higher than any of the other meat commodities; therefore, reducing/eliminating the organism at the processing plant becomes a challenge. Once the birds reach market age, prior to catching, the feed is withdrawn to allow clearance of the gastrointestinal tract. This process may increase *Salmonella* incidence in the crop due to consumption of litter and fecal matter. The increase of the pH and decrease in lactic acid within the crop may actually be the cause of *Salmonella* increase within the crop or a combination of the two (Corrier et al., 1999; Hinton et al., 2000; Smith and Berrang, 2006). During transportation to the processing plant, cross-contamination from transport coops can occur. Once at the processing facility, the birds enter the killing stage, and cross-contamination from the neck-cutting knife can occur (Mead et al., 1994). Scalding, defeathering, and evisceration also cause cross-contamination between carcasses. Immersion chilling can be a source of cross-contamination if no antimicrobials are used; but with the addition of chlorine or other antimicrobials, the chiller has been shown to greatly reduce *Salmonella* and other microorganisms on the carcass. Air chilling is also utilized and is effective in reducing *Salmonella* cross-contamination but is less effective in reducing total bacterial load because there is no washing effect, as is seen in immersion chilling. However, it is important to point out that, as previously discussed, rapid exposure to decreased temperatures can increase the ability of some *Salmonella* spp. to grow and survive at refrigeration temperatures. This could be of particular concern if the *Salmonella* spp. become entrapped and/or attached within feather follicles, which could provide protection from interventions utilized in the chill tanks.

The emergence of antimicrobial resistance in *Salmonella* which can be traced to food-producing animals has generated much controversy and attention. The use of antimicrobial agents in production and processing facilities has contributed to the emergence and persistence of resistant strains. The meat animal and aquaculture industries are believed to contribute to increased *Salmonella*-resistant bacteria (Threlfall, 2002; Angulo et al., 2000). Furthermore, utilization of antibiotics at subtherapeutic levels as growth promoters could also be contributing to resistance (Antunes et al., 2003).

Fruit and vegetable contamination can come from the use of noncomposted animal manure, untreated sewage or irrigation water, or animals directly or indirectly in the fields. Increases in geographic distribution have also caused changes in the handling and storage practices for these commodities. This in turn could possibly influence the ability of *Salmonella* to survive in these products. Cross-contamination of the products could occur during the collection, washing, or packaging steps within the processing chain. Sanitizers utilized to decontaminate the surface of these products lack the ability to infiltrate the cracks and intercellular spaces; therefore, *Salmonella* can survive and proliferate on the product (Mahmoud and Linton, 2008; Takeuchi and Frank, 2000; Zhuang and Beuchat, 1996).

Salmonella has been recovered from numerous other types of products. Interventions are usually not aimed against *Salmonella* in these types of processing facilities. Therefore, the *Salmonella* presence increases the likelihood of the resulting product being contaminated and a possible illness or outbreak occurring from consumption of the product. An example is *Salmonella* in dry-roasted cocoa beans which are utilized to produce chocolate products. Thermal inactivation of *Salmonella* in molten chocolate is difficult because the time-temperature conditions required to effectively eliminate the pathogen in this product would likely result in an organoleptically unacceptable product. The more troubling concern is that *Salmonella* has the ability to survive for many years in the finished product when stored at ambient temperature (D'Aoust, 1984). In 2007, *Salmonella* was isolated from peanut butter, which has never been considered a product in which *Salmonella* would be of concern. It was found that the contamination of the product was due to cross-contamination from environmental sources, not from the ingredients themselves. Therefore, from a processing view, no matter the product being processed, it is important to remember that *Salmonella* and other pathogens could potentially contaminate the products; therefore, adequate standard operating procedures, sanitation standard operating procedures, good agricultural practices, HACCP, or other practices should be in place to prevent, it is hoped, pathogenic organisms from adulterating the food product.

RECENT ADVANCES IN BIOLOGICAL, CHEMICAL, AND PHYSICAL INTERVENTIONS TO GUARD AGAINST THE PATHOGEN

Probably more work has been done to control *Salmonella* in chicken production and processing than in any other food. A comprehensive review of these efforts is discussed below.

Evidence of the role that on-farm interventions can play in effectively helping to control *Salmonella* can be seen in Sweden and Denmark, where on-farm control programs have significantly suppressed *Salmonella* in broiler chicken production. Sweden's and

Denmark's programs were initiated in about 1990 and 1995, respectively. In both programs, extensive testing of feed, the environment, and processed chickens was conducted. No *Salmonella*-positive feed was permitted, and all breeder birds that tested positive for *Salmonella* were eradicated. In Sweden, if any *Salmonella*-positive flocks are identified, they must be killed and properly disposed of. In Denmark, the initial phases of the program are as in Sweden, but *Salmonella*-positive grow-out broilers are processed separately and can be sold to the consumer. Initially the cost of implementing the programs in both countries was paid for by the government, but now the programs are paid for through industry checkoffs. Final economic analysis for a similar program in the United States is ongoing, but it will likely not be economically feasible to implement this same program in the United States. However, alternative methods of achieving similar results may be possible.

The European experience has confirmed that the best way to control *Salmonella* in food systems is to control the pathogen on the farm and to prevent it from ever entering the processing plant. The size and competitive nature of the U.S. poultry industry make implementation of new *Salmonella* intervention technologies that would significantly increase costs of production a challenge, unless there is a concomitant decision by the entire industry to implement the technology. There are several technologies that are currently being researched and in some cases used by the U.S. poultry industry which offer the opportunity to significantly reduce *Salmonella*. Research and anecdotal evidence suggest that the use of live and killed-cell vaccines in breeders, hatch cabinet disinfection, competitive exclusion treatments in breeders and broilers, and extensive biosecurity in breeder and broiler operations should yield beneficial results, without the extensive costs of eradication programs.

Interviews with U.S. poultry producers suggest that as much as 50% of the industry is now using killed-cell vaccination or a combination of killed-cell and live vaccines to help reduce *Salmonella* in broiler breeders and their progeny. Killed vaccines can be more universal in nature, consisting of strains of *S.* serovar Typhimurium and *S.* serovar Enteritidis and some other strain, such as *S.* serovar Heidelberg or *S.* serovar Kentucky. In other cases, the killed vaccines are a collection of autogenous strains associated with a particular company or production complex. Live vaccines are often deletion mutant strains of *S.* serovar Typhiumurium and are used to boost immune response and will not persist in the birds or the environment for long periods of time.

There is a paucity of peer-reviewed research concerning the long-term effectiveness of these programs when implemented in large-scale commercial operations; however, anecdotal responses from many in the poultry industry suggest that after 1 or 2 years, the level of *Salmonella* can be reduced by 20 to 50% from prevaccination levels.

Outside of preventing *Salmonella* from ever entering the breeder or grow-out farms and being an effective method for preventing *Salmonella* in processed broilers, the development and use of undefined competitive exclusion (CE) cultures probably have been the most effective interventions (Bailey et al., 2000). In 1972, Esko Nurmi and coworkers were the first to publish on the use of CE cultures to help control *Salmonella* in broiler chickens. Nurmi took intestinal material from healthy adult birds and fed the material back to newly hatched chicks, making them, in a very short time, much more resistant to becoming colonized with *Salmonella* than chicks of the same age with no added microflora. Several commercial products were eventually made using the concepts initiated by Nurmi. These products are used in many countries around the world today, but not in the United States (at least not legally with pathogen claims) because they were comprised of many strains of bacteria (undefined culture), some of which were not or could not be identified. The FDA has taken jurisdiction over these types of products and handles them as they do all "drugs," which has made getting approval more problematic. Many people have made "probiotic" or defined CE products (in which all the component bacteria are known), but to this point no products have been demonstrated to be as effective as the undefined CE products. There was one product, Preempt, which was a continuous-flow defined culture comprised of 21 specific strains of bacteria, which gained FDA approval and was used briefly by some companies in the United States in the late 1990s or early 2000s. Preempt never proved to be particularly effective in reducing *Salmonella*, and there were production problems in maintaining the balance of organisms in the continuous flow. Therefore, Preempt is no longer commercially available.

The other area of "on-farm" pathogen control falls under the general category of biosecurity. For the purposes of this discussion, cleaning and disinfection, rodent control, and control of movement of people and their vehicles (anything entering the property or facilities) are linked under this category. Cleaning and disinfection programs vary widely. In Scandinavia, where all of the houses are built on concrete floors and out of materials designed to handle the harsh winters, the houses are cleaned down to the concrete and disinfected with chemical washes by a contract cleaner between every flock. In the United States, some houses have concrete floors, but many have dirt floors, and most are not built to be completely washed and cleaned between flocks. Growers within companies will either clean out the houses between flocks or rear broilers on what is called "built-up" litter, a process in which wood pine (or some other product) shavings are put down on top of the old bedding material between every flock. This process can often continue for 1 or 2 years (5 to 10 flocks) before cleanout. Several factors, such as company policy, flock health, availability of bedding material, and effective disposal of the litter, play into this decision. Rodent control has been shown to be critical for control of *S.* serovar Enteritidis in egg-laying flocks and likely plays a role in the control of all types of *Salmonella* in broiler grow-out. Most U.S. companies have a policy on rodent control, but implementation of the program is left up to the individual grower. Effective programs are available for those that choose to follow them.

Biosecurity varies widely in the poultry industry. Most biosecurity programs are implemented primarily to control and prevent movement of viruses and other potential disease-causing organisms which will affect the health and production of the animals. Help in controlling the movement of bacterial pathogens associated with food safety is a side benefit of these biosecurity programs. At the broiler breeder level, entry into and exit from the facilities always involve foot baths and can also include shower-in for entry into the facility. At the broiler grow-out level, biosecurity generally involves restricting access to the facilities to those authorized and the use of dedicated or disposable footwear and footbaths with disinfectants, such as quaternary ammonium compounds or iodine compounds. A few companies also have installed disinfection stations that will spray the tires and undercarriages of trucks and other vehicles entering and leaving the grow-out facilities.

In the integrated poultry production operations, the hatchery may be the most critical area for *Salmonella* control for two reasons. First, the newly hatched chick is more susceptible to colonization by *Salmonella* than older birds. As few as 1 to 10 cells can colonize a day-of-hatch chick compared to the more than 1,000 cells needed to colonize an older chicken (Cox et al., 1990a). Second, because there are often between 10,000 and 40,000 hatching eggs in a single hatch cabinet, even if only one or a few eggs are contaminated with *Salmonella,* when these eggs hatch they can spread *Salmonella* to many other chicks in the hatch cabinet. These chicks can then become colonized with *Salmonella* and subsequently spread the *Salmonella* to other chicks during transport or in the grow-out house. An association with *Salmonella* serotypes recovered from the hatchery can be found on the final processed carcass (Bailey et al., 2002). Many studies have shown the need to properly clean eggs and to disinfect between hatch groups (Bailey et al., 1994; Cox et al., 1990b, 1991, 1997, 2007b). Several chemicals, including hydrogen peroxide or properly controlled formalin, can effectively prevent the spread of *Salmonella* during the actual hatch period (last 2 or 3 days in the hatch cabinet).

Since the implementation of the HACCP pathogen reduction plan in 1996 (http://www.fsis.usda.gov/oa/background/finalrul.htm), the primary means of attempting to control *Salmonella* in poultry has been the use of chemical disinfection in the processing plant. The primary reason for this approach is to try to emulate the pasteurization of milk, in which a terminal treatment for pathogens is applied at the end of processing, just before the product goes to the consumer.

On 23 December 1999, the Food Safety and Inspection Service (FSIS) published in the *Federal Register* a final rule, "Food ingredients and sources of radiation listed or approved for use in the production of meat and poultry products." In January 2000, FDA and FSIS entered into a memorandum of understanding that outlines the procedures that are followed by FDA and FSIS regarding the joint review of requests for the use of food ingredients and sources of radiation in meat and poultry products. Except in certain circumstances, FDA will now list in its regulations (21 CFR) food additives and sources of radiation that are safe and suitable for use in the production of meat or poultry products. Approved chemicals with application for addition to meat and poultry production are listed (http://www.fsis.usda.gov/OPPDE/rdad/FSISDirectives/7120.1.htm). Chemicals used in poultry processing to eliminate or reduce *Salmonella* are listed in Table 3.

In 2006, a survey of intervention treatments at five processing plant locations (prescalder brushes, online reprocessing [OLR], chiller treatments, chiller pH control [acidified], and postchill treatments) was responded to by eight integrated companies with a total of 100 processing plants. Chemical applications could be applied by spray with or without scrubber brushes or as dips. More intervention efforts were

Table 3. Commercial antimicrobials commonly used during poultry processing

Antimicrobial	Use	Suggested mode of action
Acidified sodium chlorite	Spray of dip solution: 500–1,200 ppm sodium chlorite and any GRAS acid to achieve pH 2.3–2.9; prechiller or chiller solution: 50–150 ppm sodium chlorite and any GRAS acid to achieve pH 2.8–3.2	Broad-spectrum germicides: oxchlorous compounds act by breaking bonds on cell membrane surfaces
Cetylpyridinium chloride	Not to exceed 0.3 g/lb of poultry, propylene glycol concentration 1.5 times that of cetylpyridium chloride	Hydrophilic portion reacts with the cell membrane, resulting in the leakage of the cellular components, leading to cell death
Chlorine-sodium hypochlorite	20–50 ppm in chill water; also used in processing water at 10–30 ppm	Oxidation of cell components, resulting in cell death
Chlorine dioxide	Not to exceed 3 ppm in poultry process water contacting whole fresh carcass	Oxidation of the cellular membrane and cellular constituents at high concentrations; it breaks the cell wall
Ozone	Antimicrobial agent as stated in 21 CFR 170.3(o)(2), used in gaseous or aqueous phase	Direct molecular reaction and indirect reactions involving free radicals, oxidation of cell membrane
Peroxyacetic acid	Maximum concentration of 220 ppm peroxyacetic acid, 110 ppm hydrogen peroxide, and 13 ppm 1-hydroxy-ethylidene-1, 1-diphosphoric acid	Strong oxidation of cell membrane and other cell components, resulting in cell death
Trisodium phosphate	8–12% solution in conjunction with a water spray containing 20 ppm chlorine; solution can be applied by spraying or dipping chilled or prechilled carcasses for up to 15 s	Disruption of cell membrane causing leakage of intracellular fluid; details of the antimicrobial mechanism have not been completely elucidated

made in the chiller treatments (93%) and in OLR (86%) than with the scalder (18%) or postchill dips (12%). Chiller treatments were overwhelmingly chlorine (hypochlorous acid) followed by peracetic acid and chlorine dioxide. The key to effective chlorine treatments in the chiller is to maintain chiller water at or near pH 7.0. Depending on the area of the country where the plant is located or whether trisodium phosphate is being used in the plant, the chill water may need to be acidified. Ninety plants used carbon dioxide to control pH in the chill tank compared to three plants that used citric acid. Chemical treatments used for OLR were far more diversified. Sodium chlorite (33%) and trisodium phosphate (24%) were used most frequently. Only 12 plants used a postchill treatment, and 8 of these used sodium chlorite. Food safety managers at the processing plants felt strongly that multiple hurdles will have to be used for successful reduction of *Salmonella* and that none of the interventions will work without attention to the entire process.

As seen in this recent industry survey, chemical treatments in the processing plant are the principal method that companies are currently using to control *Salmonella* and indirectly *Campylobacter* in chicken production and processing. There are potentially some concerns that may arise from this extensive use of chemicals during processing. The first would be direct health concerns that some of the chemicals may complex with the proteins in the meat and potentially be a human health concern. The second is the perception of many consumers that the use of chemicals is bad for them (note the rapid interest in all things organic), and finally, there are export issues with many countries that do not allow the use of any chemicals during processing.

DISCRIMINATIVE DETECTION METHODS FOR CONFIRMATION AND TRACE-BACK OF CONTAMINATED PRODUCTS

Isolation and identification of *Salmonella* by traditional cultural methods requires a series of steps for isolation and identification, which takes 4 to 6 days to complete. To reduce the screening time, rapid methods (i.e., miniaturized biochemical kits, antibody/DNA-based tests, and modifications to conventional tests) are utilized to detect the presence of *Salmonella* in samples (McMeekin, 2003). With most of the rapid methods commercially available, screening time for a *Salmonella* presence in the sample can be performed in as few as 2 days. The 2-day limitation is due mainly to the samples having to be enriched prior to testing to increase sensitivity and specificity (Feng, 1997). However, a positive result through utilization of rapid

methods still needs to be confirmed by traditional cultural methods (Feng, 1996). A more detailed review of rapid methods can be found in the work of Feng (1996, 1997) and McMeekin (2003).

Traditional cultural isolation from samples involves an overnight preenrichment step utilizing a nonselective enrichment, such as buffered peptone water or lactose broth. After preenrichment, the sample is then transferred to selective enrichments, such as tetrathionate, Rappaport-Vassiliadis, and selenite cystine broths, and incubated for an additional 24 hours. After enrichment, the samples are then plated onto selective agar plates, such as brilliant green sulfa, bismuth sulfite, xylose lysine desoxycholate, XLT4, modified lysine iron, and Hektoen enteric agars, and incubated for 24 h. In addition, chromogenic plating media can be utilized (Schonenbrucher et al., 2008). After incubation, presumptive positive samples are then subjected to biochemical screening on triple sugar iron (TSI) and lysine iron agars, which involves another day of incubation. Briefly, *Salmonella* spp. are oxidase and catalase negative, generally produce hydrogen sulfide and decarboxylate lysine, and do not hydrolyze urea. Many of the traits listed above traditionally have formed the basis for the presumptive biochemical identification of *Salmonella* isolates (Cox and Mercuri, 1976). Typical *Salmonella* isolates will produce acid and gas from glucose in TSI agar medium and will not utilize lactose or sucrose in TSI or in differential plating media, such as brilliant green, xylose lysine deoxycholate, and Hektoen enteric agars. *Salmonella* spp. produce an alkaline reaction from the decarboxylation of lysine to cadaverine in lysine ion (LI) agar, generate hydrogen sulfide gas in TSI and LI media, and fail to hydrolyze urea. The biochemical classification of food-borne and clinical *Salmonella* isolates is generally coupled with serological confirmation evaluating the somatic (O) lipopolysaccharides on the bacterial outer membrane, flagellum (H) antigens on the peritrichous flagella, and the capsular (Vi) antigen.

The significant number of illnesses due to *Salmonella* has made epidemiological trace-back to the cause of infection a very valuable tool to correct problems and reduce the incidence of food-borne outbreaks. Phenotyping and genotyping methods exist in order to evaluate *Salmonella* strains in research and from a public health standpoint. Serotyping, phage typing, biotyping, and R typing are examples of phenotyping methods. The enormous number of *Salmonella* serotypes, the various sources from which *Salmonella* can be isolated, and genetic differences that occur within a single serotype mean that it is not sufficient to carry out only serotyping in order to accurately determine the source of infection from an epidemiological standpoint (Reeves, 1993), due mainly to the fact that the method does not have the discriminatory power needed to accurately access problems from an epidemiological standpoint. Phage typing and biotyping have been utilized in epidemiological and surveillance studies, but they also lack some of the discriminative power to identify strain differences (Rankin and Platt, 1995). Antibiotic-resistant typing has been beneficial in evaluating the evolving increase in antibiotic-restraint strains of *Salmonella*; however, the method should be combined with other epidemiology tracking methods (De Oliveira et al., 2007).

Genotyping is usually performed by extrachromosomal analysis (plasmid profiling and restriction digestion of plasmids) or through chromosomal analysis. A few examples of DNA-based molecular methods are PCR, ribotyping, and pulsed-field gel electrophoresis (PFGE). Multilocus sequence typing is an example of an RNA-based method. Examples of PCR-based methods include random amplified polymorphic DNA and amplified fragment length polymorphism (AFLP). Amplified fragment length polymorphism has been shown to be more discriminatory than other DNA-based methods, such as ribotyping and PFGE (Nair et al., 2000; Ross and Heuzenroeder, 2005). It was reported that multilocus sequence typing was more discriminatory than PFGE (Kotetishvili et al., 2002). While a few of the molecular methods are gaining attention due to their discrimination power, PFGE is still considered the "gold standard" for the subtyping of *Salmonella* strains and has been highly effective in epidemiological investigations tracing back sources of outbreaks (Swaminathan et al., 2001). PFGE is utilized by PulseNet, which is the national surveillance network for bacterial foodborne disease in the United States. PFGE has gained much recognition since being implemented around 1998, due to the reproducibility between laboratories, 1-day standardized protocols, and the simplicity of analysis of strain relatedness.

All the methods discussed in this section have been developed and utilized for a number of reasons. These include but are not limited to conventional isolation and monitoring, evaluation of evolutionary changes, amplification of the discriminatory ability to detect differences between strains, and the monitoring and accessing of changes in gene expressions. It is imperative to remember that when considering a method, certain circumstances may call for one method, and another situation for a totally different method. On the practical side, a combination of the discussed methods should increase obtainable information, whether for control programs or in trace-back studies.

CONCLUDING REMARKS

Historically, *Salmonella* has been thought of as a problem associated almost exclusively with meat and poultry products. Advances in methods to genetically characterize *Salmonella* have greatly assisted epidemiologists in their ability to accurately determine the source and spread of *Salmonella*. Recent epidemiological studies have clearly demonstrated that if it were ever true that meat and poultry were the primary sources of outbreaks, that may no longer be the case. Outbreaks associated with tomatoes, cantaloupes, peanut butter, and other produce suggest that vigilance and interventions from the farm all the way through food processing must be maintained at all times in order to ensure the production of safe and wholesome foods without *Salmonella*.

It is clear that the problem of *Salmonella* in the global food chain and its current and projected repercussions for human health are causes for concern. Numerous studies have suggested that antimicrobial resistance among bacteria is on the rise, and this has led to changes in control and treatment strategies. Increased understanding of *Salmonella* at the molecular level will possibly lead to better intervention strategies, real-time screening methods, and a dramatic reduction of *Salmonella* in certain food products. However, with the increase in globalization, efforts to control *Salmonella* will continue to be a significant issue well into the future as new products are continually introduced into the food arena.

REFERENCES

Alpuche-Aranda, C. M., E. L. Racoussin, J. A. Swanson, and S. I. Miller. 1994. *Salmonella* stimulate macrophage macropinocytosis and persist within spacious phagosomes. *J. Exp. Med.* **179**:601–608.

Angulo, F. J., K. Johnson, R. V. Tauxe, and M. L. Cohen. 2000. Significance and sources of antimicrobial-resistant nontyphoidal *Salmonella* infections in the United States. *Microb. Drug Res.* **6**:77–83.

Antunes, P., C. Reu, J. C. Sousa, L. Peixe, and N. Pestana. 2003. Incidence of *Salmonella* from poultry products and their susceptibility to antimicrobial agents. *Int. J. Food Microbiol.* **82**:97–103.

Bailey J. S., N. A. Cox, D. E. Cosby, and L. J. Richardson. 2005. Movement and persistence of *Salmonella* in broiler chickens following oral or intracloacal inoculation. *J. Food Prot.* **68**:2698–2701.

Bailey, J. S., N. A. Cox, S. E. Craven, and D. E. Cosby. 2002. Serotype tracking of *Salmonella* through integrated broiler chicken operations. *J. Food Prot.* **65**:742–745.

Bailey J. S., N. J. Stern, and N. A. Cox. 2000. Commercial field trial evaluation of mucosal starter culture to reduce *Salmonella* incidence in processed broiler carcasses. *J. Food Prot.* **63**:867–870.

Bailey J. S., N. A. Cox, and M. E. Berrang. 1994. Hatchery-acquired salmonellae in broiler chicks. *Poult. Sci.* **73**:1153–1157.

Beuchat, L. R. 2002. Ecological factors influencing survival and growth of human pathogens on raw fruit and vegetables. *Microb. Infect.* 4:413–423.

Brenner, F. W., R. G. Villar, F. J. Angulo, R. Tauxe, and B. Swaminathan. 2000. *Salmonella* nomenclature. *J. Clin. Microbiol.* 38:2465–2467.

Burslem, C. D., M. J. Kelly, and F. S. Preston. 1990. Food poisoning–a major threat to airline operations. *J. Soc. Occup. Med.* 40:97–100.

Corrier, D. E., J. A. Byrd, B. M. Hargis, M. E. Hurne, R. H. Bailey, and L. H. Stanker. 1999. Survival of *Salmonella* in the crop contents of market-age broilers during feed withdrawal. *Avian Dis.* 43:453–460.

Corry, J. E. L. 1976. The safety of intermediate moisture foods with respect to *Salmonella*, p. 215–238. *In* R. Davies, G. G. Birch, and K. J. Parker (ed.), *Intermediate Moisture Foods*. Applied Science Publishers Ltd, London, England.

Cox, N. A., L. J. Richardson, R. J. Buhr, J. K. Northcutt, J. S. Bailey, and P. J. Fedorka-Cray. 2007a. Recovery of *Campylobacter* and *Salmonella* serovars from the spleen, liver/gallbladder, and ceca of six and eight week old commercial broilers. *J. Appl. Poult. Res.* 16:477–480.

Cox, N. A., L. J. Richardson, R. J. Buhr, M. T. Musgrove, M. E. Berrang, and W. Bright. 2007b. Bactericidal effect of several chemicals on hatching eggs inoculated with *Salmonella* serovar Typhimurium. *J Appl. Poult. Res.* 16:623–627.

Cox, N. A., L. J. Richardson, R. J. Buhr, J. K. Northcutt, P. J. Fedorka-Cray, J. S. Bailey, B. D. Fairchild, and J. M. Mauldin. 2006. Natural occurrence of *Campylobacter* spp., *Salmonella* serovars, and other bacteria in unabsorbed yolks of market-age commercial broilers. *J. Appl. Poult. Res.* 15:551–557.

Cox, N. A., J. S. Bailey, M. E. Berrang, and J. M. Mauldin. 1997. Diminishing incidence and level of salmonellae in commercial broiler hatcheries. *J. Appl. Poult. Res.* 6:90–93.

Cox, N. A., J. S. Bailey, J. M. Mauldin, L. C. Blankenship, and J. L. Wilson. 1991. Extent of salmonellae contamination in breeder hatcheries. *Poult. Sci.* 70:416–418.

Cox, N. A., J. S. Bailey, L. C. Blankenship, R. J. Meinersmann, N. J. Stern, and F. McHan. 1990a. Fifty percent colonization dose for *Salmonella typhimurium* administered orally and intracloacally to young broiler chicks. *Poult. Sci.* 69:1809–1812.

Cox, N. A., J. S. Bailey, J. M. Mauldin, and L. C. Blankenship. 1990b. Presence and impact of *Salmonella* contamination in commercial broiler hatcheries. *Poult. Sci.* 69:1606–1609.

Cox, N. A., and A. J. Mercuri. 1976. Rapid confirmation of suspect *Salmonella* colonies by use of the Minitek system in conjunction with serological tests. *J. Appl. Bacteriol.* 41:389–394.

D'Aoust, J.-Y. 1984. Effective enrichment-plating conditions for detection of Salmonella in foods. *J. Food Prot.* 47:588–590.

D'Aoust, J. Y. 1991a. Psychrotrophy and foodborne *Salmonella*. *Int. J. Food Microbiol.* 13:207–216.

D'Aoust, J. Y. 1991b. Pathogenicity of foodborne *Salmonella*. *Int. J. Food Microbiol.* 12:17–40.

D'Aoust, J. Y. 1994. *Salmonella* and the international food trade. *Int. J. Food Microbiol.* 24:11–31.

D'Aoust, J. Y. 2000. *Salmonella*, p. 1234. *In* B. M. Lund, A. C. Baird-Parker, and G. W. Gould (ed.), *The Microbiological Safety and Quality of Food*. Aspen Publishers, Gaithersburg, MD.

De Buck J. F. Pasmans, F. Van Immerseel, F. Hasebrouck, and R. Ducatelle. 2004. Tubular glands of the isthmus are the predominant colonization site of *Salmonella* Enteritidis in the upper oviduct of laying hens. *Poult. Sci.* 83:352–358.

De Oliveira, F. A., A. P. G. Frazzon, A. Brandelli, and E. C. Tondo. 2007. Use of PCR-ribotyping, RAPD, and antimicrobial resistance for typing of *Salmonella* enteritidis involved in food-borne

outbreaks in southern Brazil. *J. Infect. Dev. Countries* 1:170–176.

Ebel, E., and W. Schlosser. 2000. Estimating the annual fraction of eggs contaminated with *Salmonella* enteritidis in the United States. *Int. J. Food Microbiol.* 61:51–62.

Feng, P. 1997. Impact of molecular biology on the detection of foodborne pathogens. *Mol. Biotech.* 7:267–278.

Feng, P. 1996. Emergence of rapid methods for identifying microbial pathogens in foods. *J. Assoc. Off. Anal. Chem. Int.* 79:809–812.

Gast, R. K., and C. W. Beard. 1990. Isolation of *Salmonella* enteritidis from internal organs of experimentally infected hens. *Avian Dis.* 34:991–993.

Hakalehto, E., J. Pesola, L. Heitto, A. Narvanen, and A. Heitto. 2007. Aerobic and anaerobic growth modes and expression of type 1 fimbriae in *Salmonella*. *Pathophysiology* 14:61–69.

Hinton, A., R. J. Buhr, and K. D. Ingram. 2000. Reduction of *Salmonella* in the crop of broiler chickens subjected to feed withdrawal. *Poult. Sci.* 79:1566–1570.

Huhtanen, C. N. 1975. Use of pH gradient plates for increasing the acid tolerance of salmonellae. *Appl. Microbiol.* 29:309–312.

Humphrey, T. J. 2004. *Salmonella*, stress responses and food safety. *Nat. Rev.* 2:504–509.

Humphrey T. J., A. Baskerville, A. Whitehead, B. Rowe, and A. Henley. 1993. Influence of feeding patterns on the artificial infection of laying hens with *Salmonella* enteritidis phage type 4. *Vet. Rec.* 132:407–409.

Kotetishvili, M., O. C. Stine, A. Kregar, J. G. Morris, and A. Sulakvelidize. 2002. Multilocus sequence typing for characteristics of clinical and environmental *Salmonella* strains. *J. Clin. Microbiol.* 40:1626–1635.

Le Minor, L. 1981. The genus *Salmonella*, p. 1148–1159. *In* M. P. Starr, H. Stolp, H. G. Truper, A. Balows, and H. G. Schlegel (ed.), *The Prokaryotes*. Springer-Verlag, New York, NY.

Leyer, G. J., and E. A. Johnson. 1993. Acid adaptation induces cross-protection against environmental stresses in *Salmonella typhimurium*. *Appl. Environ. Microbiol.* 59:1842–1847.

Mahmoud, B. S. M., and R. H. Linton. 2008. Inactivation kinetics of inoculated *Escherichia coli* O157:H7 and *Salmonella enterica* on lettuce by chlorine dioxide gas. *Food Microbiol.* 25:244–252.

McMeekin, T. A. 2003. *Detecting Pathogens in Food*. Woodhead Publishing, Cambridge, England.

Mead, G. C., W. R. Hudson, and M. H. Hinton. 1994. Use of a marker organism in poultry processing to identify sites of cross-contamination and evaluate possible control measures. *Br. Poult. Sci.* 35:345–354.

Miyamoto, T, E. Baba, T. Tanaka, K. Sasai, T. Fukata, and A. Arakawa. 1997. *Salmonella enteritidis* contamination of eggs from hens inoculated by vaginal, cloacal, and intravenous routes. *Avian Dis.* 41:296–303.

Mumma, G. A., P. M. Griffin, M. I. Meltzer, C. R. Braden, and R. V. Tauxe. 2004. Egg quality assurance programs and egg-associated *Salmonella* Enteritidis infections, United States. *Emerg. Infect. Dis.* 10:1782–1789.

Nair, S., E. Schreiber, K. Thong, T. Pang, and M. Altwegg. 2000. Genotypic characterization of *Salmonella typhi* by amplified fragment length polymorphism fingerprinting provides increased discrimination as compared to pulsed-field gel electrophoresis and ribotyping. *J. Microbiol. Methods* 41:35–43.

Pohl, P., Y. Glupczynskj, M. Marin, G. van Robaeys, P. Lintermans, and M. Couturier. 1993. Replicon typing characterization of plasmids encoding resistance to gentamicin and apramycin in *Escherichia coli* and *Salmonella typhimurium* isolated from human and animal sources in Belgium. *Epidemiol. Infect.* 3:229–238.

Ponka, A., Y. Andersson, A. Sütonen, B. de Jong, M. Jahkola, O. Haikala, A. Kuhmonen, and P. Pakkala. 1995. *Salmonella* in alfalfa sprouts. *Lancet* 345:462–463.

Rankin, A. M., and D. J. Platt. 1995. Phage conversion in *Salmonella enterica* serotype Enteritidis: implication for epidemiology. *Epidemiol. Infect.* 114:227–236.

Reeves, P. 1993. Evolution of *Salmonella* O antigen variation by interspecific gene transfer on a large scale. *Trends Genet.* 9:17–22.

Ross, I. L., and M. W. Heuzenroeder. 2005. Use of AFLP and PFGE to discriminate between *Salmonella enterica* serovar Typhimurium DT126 isolates from separate food-related outbreaks in Australia. *Epidemiol. Infect.* 133:635–644.

Salmon, D. E., and T. Smith. 1885. Report on swine plague. USDA Bureau of Animal Ind. 2nd Annual Report. USDA, Washington, DC.

Schonenbrucher, V., E. T. Mallinson, and M. Bulte. 2008. A comparison of standard cultural methods for the detection of foodborne *Salmonella* species including three new chromogenic plating media. *Int. J. Food Microbiol.* 123:61–66.

Smith, D. P., and M. E. Berrang. 2006. Prevalence and numbers of bacteria in broiler crop and gizzard contents. *Poult. Sci.* 85:144–147.

Sockett, P. N., J. M. Cowden, S. LeBaigne, D. Ross, G. K. Adak, and H. Evans. 1993. Foodborne disease surveillance in England and Wales: 1989–1991. *Comm. Dis. Rep.* 3:159–173.

Swaminathan, B., T. Barrett, S. Hunter, R. Tauxe, and the CDC PulseNet Taskforce. 2001. PulseNet: the molecular subtyping network for foodborne bacterial disease surveillance, United States. *Emerg. Infect. Dis.* 7:382–389.

Takeuchi, K., and J. F. Frank. 2000. Penetration of *Escherichia coli* O157:H7 into lettuce tissues as affected by inoculum size and temperature, and the effect of chlorine treatment on cell viability. *J. Food Prot.* 63:434–440.

Threlfall, E. J. 2002. Antimicrobial drug resistance in *Salmonella*: problems and perspectives in food and water-borne infection. *FEMS Microbiol. Rev.* 26:141–148.

Voetsch, A. C., T. J. Van Gilder, F. J. Angulo, M. M. Farley, S. Shallow, R. Marcus, P. R. Cieslak, V. C. Deneen, and R. T. Tauxe. 2004. Foodnet estimates of the burden of illness caused by nontyphoidal *Salmonella* infections in the United States. *Clin. Infect. Dis.* 38:127–134.

Zhuang, R. Y., and L. R. Beuchat. 1996. Effectiveness of trisodium phosphate for killing *Salmonella* Montevideo on tomatoes. *Lett. Appl. Microbiol.* 22:97–100.

Pathogens and Toxins in Foods: Challenges and Interventions
Edited by V. K. Juneja and J. N. Sofos
© 2010 ASM Press, Washington, DC

Chapter 8

Staphylococcal Food Poisoning

KEUN SEOK SEO AND GREGORY A. BOHACH

Staphylococcus aureus, the etiological agent of staphylococcal food poisoning (SFP), has been extensively studied and is known to produce a cadre of extracellular pathogenic factors. In addition to SFP, *S. aureus* is implicated in a variety of illnesses, including skin infections, endocarditis, and osteomyelitis, plus several toxin-mediated diseases, such as toxic shock syndrome. This chapter focuses predominantly on SFP, another toxin-mediated illness. SFP, also known as staphylococcal gastroenteritis, is not an infection but is a foodborne intoxication caused by one or more staphylococcal enterotoxins (SEs). It appears to have been initially reported in 1894 as a family outbreak due to undercooked meat from a sick cow that became contaminated with *S. aureus* (Denys, 1894). In 1914, Barber purposely drank stored milk from a cow showing signs of mastitis and provided the first evidence that a soluble toxin was responsible (Barber, 1914). Although the true morbidity and incidence of SFP are unknown, an estimate is that the number of U.S. cases annually exceeds 185,000, with 1,750 hospitalizations, at a cost of more than $1.5 billion. SFP also constitutes an estimated 4.5% of food-borne illnesses in the European Union (WHO Surveillance Programme, 2000).

CHARACTERISTICS OF THE ORGANISM

Taxonomy

According to *Bergey's Manual of Systematic Bacteriology* (Schleifer, 1986), members of the genus *Staphylococcus,* also known informally as staphylococci, are located within the family *Micrococcaceae,* which includes the genera *Micrococcus, Staphylococcus, Stomatococcus,* and *Planococcus. Staphylococcus* members are gram-positive, catalase-positive and nonmotile cocci occurring singly, in pairs, or in irregular clusters. The name *Staphylococcus* is derived from the

Greek words "staphyle" and "kokkos," which mean grapes and cocci, respectively, describing their Gram-stained appearance under the microscope (Schleifer, 1986). Currently, 53 *Staphylococcus* species and subspecies are recognized (Kloos et al., 1991).

Staphylococci are often divided into coagulase-positive staphylococci and coagulase-negative staphylococci (CNS), on the basis of coagulase production or the presence of a coagulase gene. Coagulase-positive staphylococci include *S. aureus, S. intermedius, S. delphini, S. hyicus,* and *S. schleiferi.* CNS comprise a heterogeneous group of many species. Although several species, including some CNS, have been implicated in SFP (Breckinridge and Bergdoll, 1971; Crass and Bergdoll, 1986), clearly most cases are attributable to *S. aureus.*

Methods used for phenotypic identification of staphylococci include a conventional scheme (Kloos et al., 1991) and commercial kits. The International Organization for Standardization suggests Baird-Parker agar (ISO6888-1) and rabbit plasma fibrinogen agar (ISO6888-2) to enumerate coagulase-positive staphylococci from food and animal feeding stuffs (De Buyser et al., 2003). Other media, such as mannitol-salt agar, are useful. Due to the complexity and prolonged time of conventional methods, clinical laboratories may use key tests for presumptive identification of *S. aureus,* including the coagulase tube test alone or in combination with Gram-staining, catalase, thermostable nuclease production, and colony morphology. Additional biochemical tests can provide further speciation and confirmation, including novobiocin susceptibility, urease, pyrrolidonyl arylamidase, ornithine decarboxylase, and acid from aerobic mannose tests (De Paulis et al., 2003). Speciation by phenotypic properties can lead to misidentification. For example, a coagulase-deficient phenotype, caused by a transposon or transcriptional and post-transcriptional defects of the coagulase gene, has been

Keun Seok Seo and Gregory A. Bohach • Department of Microbiology, Molecular Biology, and Biochemistry, University of Idaho, Moscow, ID 83844.

reported (Fox et al., 1996; Vandenesch et al., 1994). One study reported that 54 of 110 *S. aureus* human clinical isolates (49.1%) had a coagulase-deficient phenotype (Luczak-Kadlubowska et al., 2006).

Test kits that distinguish species by biochemical reactions are frequently used. Currently, several manufacturers supply kits, such as API (bioMérieux), Vitek-2 (bioMérieux), Walk/Away (Dade MicroScan), BD Phoenix (Becton Dickinson), and Replianalyzer (Oxoid). The accuracy of these systems is usually adequate to identify *S. aureus* and most common CNS clinical isolates. However, most commercial kits are not designed to identify veterinary or food-borne CNS. For example, *S. chromogenes* was misidentified as *S. hyicus, S. epidermidis,* or *S. simulans* with the systems above (Matthews et al., 1990).

In recent decades, advances in taxonomy have been accomplished through techniques based on genotypic analysis. Genetic methods to identify *Staphylococcus* spp. offer increased speed and accuracy and include 16S rDNA sequencing (Becker et al., 2004), PCR amplification of the 16S to 23S rRNA gene intergenic spacer (Martin et al., 2004), real-time PCR of 16S rRNA genes (Skow et al., 2005), and the development of species-specific PCR amplification primers targeting 16S rRNA genes (Yamashita et al., 2005) or the other genes, such as the *sodA* (Poyart et al., 2001) and *tuf* genes (Heikens et al., 2005). In particular, 16S rRNA gene sequence analysis has emerged as one preferred technique (Clarridge, 2004) because the 16S rRNA gene contains regions well conserved in all organisms that are ideal for PCR and sequencing. It also contains species-specific variable regions. Since the 5' end of the 16S rRNA gene contains enough information to identify most *Staphylococcus* spp., partial gene sequencing is widely used. In a study by Becker et al. (2004), 49 species and subspecies of staphylococci were identified by partial 16S rRNA gene sequencing. At the subspecies level, sequencing of the 5' 16S rRNA gene fragment could discriminate among subspecies of *S. carnosus, S. cohnii, S. hominis, S. schleiferi,* and *S. succinus.* In another study, a similar technique unambiguously identified 97.4% of staphylococci (Jousson et al., 2007).

Some drawbacks of molecular genetics-based species-level identification are associated with the developing databases and include faulty, incomplete, and/or redundant sequence entries; outdated nomenclature; and poor quality assurance. Additionally, sequencing is relatively expensive for routine identification in many clinical diagnostic settings. Thus, other relatively rapid and inexpensive molecular methods, such as PCR-based DNA fingerprinting, have been developed and used successfully for *Staphylococcus* sp. identification at the species level. In general, rapid bacterial identification by PCR-restriction fragment length polymorphism (RFLP) uses species-specific and ubiquitous DNA as a target. Universal pathway and function genes, whose nucleotide sequences are fairly homologous among bacteria, are being used more frequently as PCR targets. The *gap* gene, encoding glyceraldehyde-3-phosphate dehydrogenase, is a well-conserved gene in *Staphylococcus* spp., and it has been suggested that a *gap* gene PCR-RFLP assay may be a useful tool for differentiating clinical isolates of 24 staphylococcal species (Yugueros et al., 2000).

Physiology

Growth

The optimum growth temperature for *S. aureus* is 37°C (range of 6 to 48°C). It has been reported that, under some conditions, *S. aureus* may grow at <10°C and that conventional refrigeration may not always ensure the safety of foods (Angelotti et al., 1961). The upper limit of growth can be extended to >44°C by addition of NaCl and monosodium glutamate (Jay et al., 2005). *S. aureus* is facultative. Its growth can be delayed in the presence of 80% CO_2 (Genigeorgis et al., 1969). Although the precise pH extremes tolerated are dependent on several factors, the optimum growth pH is 7.0 to 7.5 (range of 4.2 to 9.3). Growth is inhibited in the presence of 0.1% acetic acid (pH 5.1) or at a pH of 4.8 with 5% NaCl, and acid tolerance is dramatically decreased in the presence of potassium sorbate, glycerol, sucrose, and NaCl (Stewart et al., 2002). *S. aureus* is able to grow at a lower water activity (a_W) than other commonly encountered mesophilic bacteria. It has been reported to be capable of growing at an a_W as low as 0.83 in the presence of NaCl, sucrose, or glycerol humectants, although 0.86 is generally accepted as the minimum (Marshall et al., 1971).

Survival and resistance

Although *S. aureus* is known for acquiring genetic elements conferring resistance to heavy metals and antimicrobial agents, except for osmotolerance, its resistance to common food preservative materials is generally unremarkable. *S. aureus* has an a_W as low as 0.83, which is equivalent to 3.5 M NaCl (Townsend and Wilkinson, 1992). Staphylococcal osmotolerance is problematic for food safety since blocked growth of competing bacteria can result in a more likely overgrowth of *S. aureus.* The molecular mechanisms of staphylococcal osmotolerance are based on intracellular levels of osmoprotectants, proline, and glycine betaine which increase corresponding with the level of NaCl in the environment (Townsend and Wilkinson,

1992). High- and low-affinity transport systems operate in *S. aureus*. Both depend on Na^+ for activity; the high-affinity system is stimulated by a lower concentration of Na^+ and is proline specific, whereas the low-affinity system has broader substrate specificity. Osmoprotection systems in other bacteria require activation of K^+ transportation systems and a subsequent synthesis of osmoprotectant transporters under osmotic stress. In contrast, intracellular K^+ levels are high in unstressed *S. aureus*, suggesting that the transport systems are constitutively active (Townsend and Wilkinson, 1992).

Survival during food processing is an important area of applied research. Jerky is a dried stable product made from lightly salted and spiced meat. When drying to reduce the a_w below 0.86 is conducted within 3 hours, problems with food-borne pathogens are minimized. However, *S. aureus* may survive when drying is extended and at temperatures over 60°C (Holley, 1985). Cavett (1962) and Tonge et al. (1964) found that when high-salt vacuum-packed bacon was held at 20°C for 22 days, catalase-positive cocci dominated, whereas, at 30°C, CNS became dominant. In low-salt bacon (5 to 7% NaCl) held at 20°C, the micrococci became dominant. At 30°C, CNS, *Enterococcus faecalis*, and micrococci were dominant. Curing is a traditional means of meat preservation using NaCl, nitrate, and sugar to develop flavor and color. In a study of Iberian drycured hams, over 97% of the isolates were CNS, including *S. equorum*, *S. xylosus*, *S. saprophyticus*, and *S. cohnii* (Rodriguez et al., 1996). The survival of *S. aureus* in country-cured hams was investigated by spraying fresh hams with *S. aureus* and then curing, cold smoking, and aging them. After 4 months of aging, no detectable *S. aureus* was found (Portocarrero et al., 2002). In a study of

Italian dry fermented sausages, the most frequently isolated staphylococcus was *S. xylosus*, followed by *S. saprophyticus*, *S. aureus*, and *S. sciuri*. *S. xylosus* appears to be the most frequently isolated from several Italian dry sausages. In Iberian dry cured hams, the two predominant organisms in the ripening process are *S. equorum* and *S. xylosus*, and both are believed to contribute to product flavor (Gardini et al., 2003).

Reservoirs

Staphylococci are ubiquitous in air, dust, sewage, water, milk, and many foods and on food equipment, environmental surfaces, humans, and animals. Humans and animals are the primary reservoirs. *S. aureus* is efficiently carried in the nares of at least 20 to 30% of the general population. While colonized food handlers are the main source of dissemination of staphylococci to others and in outbreaks, equipment and environmental surfaces can also be implicated. Foods that require considerable handling during preparation and that are left at room temperature without subsequent cooking or adequate heating are usually involved. Food products with high protein contents are particularly good growth substrates for *S. aureus*. Frequently implicated foods include raw and processed meat; poultry and eggs; bakery products, such as custard- or cream-filled pastries, cream pies, and chocolate éclairs; shellfish in a mayonnaise-like dressing; egg, tuna, chicken, potato, and pasta salads; sandwich fillings; milk; milk products; ice cream; and soft cheeses. Some commonly contaminated foods are summarized in Table 1.

Many animal species are often heavily colonized and can directly contaminate food during processing of domestic animals and their products. This high

Table 1. Prevalence of *S. aureus* in several foods

Product	No. of samples tested	% Positive for *S. aureus*	No. of *S. aureus* CFU/g	% Positive for SE	Reference
Raw pork	135	57.7	NA[a]	34.6	Atanassova et al. (2001)
Raw meat	139	2.8	NA	7.8	Moon et al. (2007)
Raw milk	714	7.9	NA	31.8	Moon et al. (2007)
Ground beef	1,090	30	<100	NA	USDA (1996)
Ground turkey	50	6	>10	NA	Guthertz et al. (1977)
Frozen prawns	46	23.9	>3	26	Sanjeev et al. (1987)
Shrimp	1,468	27	>3	NA	Swartzentruber et al. (1980)
Beef carcasses	1,155	28.7	<10	NA	Phillips et al. (2006)
Boneless beef	1,082	20.3	<10	NA	Phillips et al. (2006)
Pork carcasses	1,650	4.8	<10	NA	Yeh et al. (2005)
Ready-to-eat fast food	3,332	8.6	NA	47	Oh et al. (2007)
Vegetables	616	7.9	NA	73.1	Moon et al. (2007)
Bakery products	214	9.8	NA	30	Summer et al. (1993)

[a]NA, not available.

prevalence of staphylococci in raw foods of animal origin emphasizes the need for effective processing to ensure safety (Genigeorgis, 1989). For example, one serious problem for the dairy industry is mastitis. Mastitis caused by *S. aureus* is a potential public health concern due to the high prevalence of SE genes in mastitis isolates (Fitzgerald et al., 2000).

Staphylococci, other than *S. aureus*, are often considered potential pathogens and are common inhabitants of the skin; for example *S. epidermidis* and *S. haemolyticus* are associated with infections in hospitalized patients (Crass and Bergdoll, 1986). *S. chromogenes*, *S. sciuri*, *S. xylosus*, and *S. simulans* are commonly isolated from dairy samples (Taponen et al., 2006) and *S. intermedius* is commonly isolated from canine samples (Khambaty et al., 1994). Recently, evidence has been mounting to indicate that SEs are expressed by other species and that their risk for SFP is underestimated (see below).

PROPERTIES OF SEs AND RELATED PROTEIN TOXINS

SE Classification

For many years, five classical and antigenically distinct SEs (SEA, SEB, SEC, SED, and SEE) were recognized. In addition, minor differences in the antigenicity of SEC resulted in its differentiation into subtypes, SEC1, SEC2, and SEC3. A related toxin, toxic shock syndrome toxin 1 (TSST-1), was originally designated SEF but was subsequently shown to lack emetic properties, and therefore, the SEF designation is no longer appropriate. Since the improvement of genomic analysis, which has facilitated sequencing capabilities, 13 novel SEs and related proteins (types G through R and U), three novel variants of SEC (SEC-bovine, SEC-ovine, and SEC-caprine), and four additional toxins (types Gv, Iv, Nv, and Uv) have been identified. Although all proteins in this family share significant homology, some have been shown to lack emetic properties or have not yet been tested. Therefore, the International Nomenclature Committee for Staphylococcal Superantigens (INCSS) recommends that proteins related to the SEs, but not confirmed to exhibit emetic activity in the monkey feeding assay, be designated SE-like (SEl) until their enterotoxigenic activity can be confirmed (Lina et al., 2004). Major features of SEs and SEls are summarized in Table 2.

Basic Biophysical Characteristics

SEs and SEls are moderately sized (~25 KDa) structurally stable proteins. SEs are partially resistant to gastrointestinal and other proteases, such as trypsin, chymotrypsin, rennin, and papain, but sensitive to pepsin at pH 2 (Balaban and Rasooly, 2000; Bergdoll, 1967; Dinges et al., 2000). Cleavage of SEs by some proteases does not always result in the loss of biological activity. They may be denatured in high concentrations of urea or guanidine hydrochloride. However, once the denaturing conditions are removed, SEs may regain biological activity (Spero et al., 1976).

The heat resistance of SEs has been well documented and complicates approaches to prevent SFP since heating may kill staphylococci but not necessarily inactivate the toxins already expressed. Pasteurization has minimal effect on SE biological activity. SEB activity is not altered by heating for 16 h at 60°C in pH 7.3 (Schantz et al., 1972). SEA is thermally inactivated by heating at 121°C for 8 min (Denny et al., 1966). SEC retains its serological activity after heating for 30 min at 60°C (Borja and Bergdoll, 1967). Another study showed that thermal processing of canned infant formula alters SEA and SED serological activity, but not their biological activity (Bennett and Berry, 1987), suggesting that their antigenic determinant confirmation is not directly related to their biological properties. Crude toxin preparations are more resistant than purified toxins, suggesting that other organic molecules help stabilize the SE structure.

SE Environmental Regulation

SE production is affected by temperature, pH, and osmotic conditions and is also regulated at the molecular level. SEA, SEB, SEC, and SED production was detected in ground beef, ham, and bologna at as low as 10°C and as high as 46°C (Vandenbosch et al., 1973). The optimum temperature for SEB and SEC production was 40°C in protein hydrolysate medium. SEB production in brain heart infusion medium is inversely related to the NaCl concentration (Pereira et al., 1982). Production occurs maximally in the absence of NaCl and is nearly abrogated in >6% NaCl.

Except for SEA, which is produced relatively early, SEs are expressed minimally during early growth but rapidly increase during the late-exponential or stationary growth phases. This effect is largely due to the staphylococcal quorum sensing Agr system which regulates staphylococcal virulence factors in response to the density of an autoinducing peptide (AIP). Inactivation of Agr results in decreased expression of SEs, except SEA (Tremaine et al., 1993). SEC expression is also linked to the level of pH in the culture, which regulates AIP levels. AIP levels and SE expression occur maximally at neutral pH. In the presence of excessive glucose, a reduced pH decreases

Table 2. Biochemical and functional properties of SEs and SEls (grouped according to sequence and functional similarities)

SE or SEl[a]	MW	Emesis[b]	Genetic location	Human TCR interaction(s)[c]	Reference(s)
SEA	27.1	Yes	φSa3mu,[d] φSa3mw[d]	1.1, 5.3, 6.3, 6.4, 6.9, 7.3, 7.4, 9, 16, 18, 21.3, 22.1, 23	Betley and Mekalanos (1988)
SED	26.9	Yes	pIB485[e]	1, 5.3, 6.9, 7.4, 8, 12	Bayles and Iandolo (1989), Kappler et al. (1989)
SEE	26.8	Yes		5.1, 6.3, 6.4, 8.1, 8.2, 13.1, 18, 21.6	Couch et al. (1988)
SElJ	31.2	ND[f]	pIB485, pF5[e]	ND	Zhang et al. (1998)
SElP	27.1	ND	φSa3n[d]	5.1, 6, 8, 16, 18, 21.3	Omoe et al. (2005b)
SElN	26.1	ND	Type I υSaβ,[d] egc[g]	9	Jarraud et al. (2001)
SElO	26.8	ND	Type I υSaβ, egc	5, 7, 22	Jarraud et al. (2001)
SElH	25.1	ND		Vα10	Petersson et al. (2003)
SEB	28.4	Yes	SaPI3[h]	1, 3.2, 6.4, 12, 14, 15, 17, 20	Ranelli et al. (1985)
SEC1	27.5	Yes	Type II υSa3,[d] SaPI$_{bov}$[h]	3.2, 5, 6.4, 6.9, 12, 13.2,15, 17, 20	Deringer et al. (1997)
SEC2	26.6	Yes	Type II υSa3, SaPI$_{bov}$	5, 12, 13.1, 13.2, 14, 15, 17, 20	Deringer et al. (1997)
SEC3	26.6	Yes	Type II υSa3, SaPI$_{bov}$	2, 3, 5, 12, 13.1, 13.2, 14, 17, 20	Deringer et al. (1997)
SElU	27.2	ND	Type I υSaβ, egc	ND	Letertre et al. (2003b)
SEG	27.0	Yes	Type I υSaβ, egc	3, 12, 13.6, 14	Munson et al. (1998)
SElR	27.0	ND	pIB485, pF5	3, 11, 12, 13.2, 14	Omoe et al. (2004)
SEI	24.9	Weak	Type I υSaβ, egc	1, 5, 5.3, 6, 23	Munson et al. (1998)
SElK	30.0	ND	φSa3mw, SaPI1,[h] SaPI3	5.1, 5.2, 6.7	Orwin et al. (2001)
SElL	26.8	No	Type II υSa3, SaPI$_{bov}$	5.1, 5.2, 6.7, 16, 22	Orwin et al. (2003)
SElM	24.8	ND	Type I υSaβ, egc	6, 8, 9, 18, 21.3	Jarraud et al. (2001)
SElQ	25.0	No	φSa3mw, SaPI3	2, 5.1, 21.3	Orwin et al. (2002)

[a]Nomenclature in this table follows INCSS guidelines (Lina, 2004).
[b]Emesis as determined by primate feeding assay.
[c]Numbers indicate human TCR Vβ elements, except for SElH.
[d]Genomic island, defined as gene cluster encoding pathogenic or virulence factors, such as drug resistance or toxins. Some genomic islands are similar to SaPI but have key features missing.
[e]Plasmids encoding SEs.
[f]ND, not determined.
[g]egc, enterotoxin gene cluster harboring the *sei*, *sem*, *sen*, and *seo* genes with or without *seg*.
[h]SaPI (designation of *S. aureus* pathogenicity island). SaPI is a mobile pathogenicity island encoding SE genes and other virulence factors.

AIP levels and concurrently SEC expression (Regassa et al., 1991). Another global regulator, the Sar locus, also affects expression of SEB by regulating Agr and Rot regulation (McNamara et al., 2000).

The alternative sigma factor (σ^B) responds to an environmental stimulus, such as high temperature, alkaline pH, high levels of NaCl, or catabolites (glucose, galactose, sucrose, glycerol, and maltose), by recognizing its target promoter, GTTT(N_{14-17})GGG-TAT (Kullik and Giachino, 1997). SEB expression is negatively regulated by σ^B. However, the promoter of *seb* differs from the target promoter for σ^B, suggesting that expression of SEB is not directly regulated by σ^B (Schmidt et al., 2004).

SE Structural Characteristics

The three-dimensional topologies of the SEs and SEls are highly conserved; the SEC3 crystal structure is shown in Figure 1 as a representative. The proteins are ellipsoidal and folded into two domains, connected by a central α-helix. The oligonucleotide-oligosaccharide-binding fold containing a β-barrel structure capped with an α-helix comprises much of one domain and is present in other toxins including

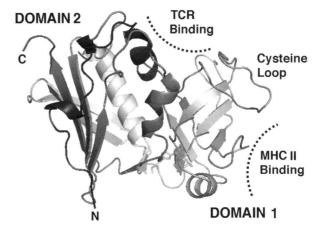

Figure 1. Ribbon diagram of SEC3. The figure is oriented with domain 1 containing the oligonucleotide-oligosaccharide-binding fold on the right and the N and C termini on the left. Major features including the TCR-cell receptor binding site, MHC II binding site, and cysteine loop are indicated.

cholera toxin, pertussis toxin, and Shiga toxin. Most SEs contain two cysteines forming a variable length flexible cysteine loop. Toxins lacking one or both cysteine residues are generally nonemetic or weakly emetic (Bergdoll, 1988). The cysteine loop is not directly associated with emesis and is believed to provide proper orientation of critical residues located near the disulfide linkage (Warren et al., 1974). SEA, SEC1, SEC2, SEC3, SED, SEE, and SElH bind zinc which is required by some toxins to bind major histocompatibility complex class II (MHC II) (Baker and Acharya, 2004) (see below). Depending on the toxin, zinc binding can occur at the outer surface or in a classical motif (H-E-X-X-H) in the groove between two domains (Baker and Acharya, 2004).

SAg Activity

SEs and SEls belong to a family of toxins produced by *S. aureus* and *Streptococcus pyogenes,* which share several properties, including superantigen (SAg) activity. SAgs stimulate a high percentage of T cells. Most bind to the outside of the peptide binding groove of MHC II molecules on antigen-presenting cells (APCs) and to T-cell receptors (TCRs) bearing specific beta chain Vβ sequences on T cells. Binding activates APCs and extensive proliferation of T cells, resulting in an acute uncontrolled release of proinflammatory cytokines and immunomodulatory cytokines in a chronic response (McCormick et al., 2001). Recent studies demonstrated that long-term stimulation by SAgs induces development of $CD4^+$ $CD25^+$ regulatory T cells in humans, rodents, and bovines, capable of suppressing the responses of responder T-cell proliferation (Seo et al., 2007). This suggests that immunomodulation induced by SAgs has an important role in staphylococcal diseases.

Emesis

Emesis involves coordinated gastrointestinal activity, including contraction of abdominal muscles, forceful contractions of the stomach pylorus, and relaxation of the fundus, cardiac sphincter, and esophagus. Studies with primates demonstrated that SEs stimulate neural receptors in the abdomen, which transmit impulses through the vagus and sympathetic nerves, ultimately stimulating the central vomiting center in the fourth ventricle (Sugiyama and Hayama, 1965). However, little is known about the molecular and cellular mechanisms by which SEs induce emesis. This gap in knowledge is due to a lack of a convenient and inexpensive animal model confirmed to mimic human SFP following oral administration. The monkey feeding assay described by Bergdoll (1988) is considered to be the standard method. Other animal models, including kittens, piglets, house musk shrews, and ferrets, have been used (Hu et al., 2007). Results with these models, particularly those that require systemic administration rather than oral feeding, must

be interpreted with caution, since they may mimic toxic shock syndrome rather than SFP. A recent study using the house musk shrew demonstrated that SEA induces a release of 5-hydroxytrptamine (5-HT) in the intestine and that the 5-HT_3 receptors on vagal afferent neurons are essential for SEA-induced emesis (Hu et al., 2007).

SE Potency In Vivo

Several studies assessed the amount of SE required to initiate SFP symptoms. An early study in which human volunteers ingested partially purified toxin demonstrated that 20 to 25 μg of SEB (~0.4 μg/ kg of body weight) is required to cause emesis (Raj and Bergdoll, 1969). However, in an outbreak in the United States in which students consumed 2% chocolate milk contaminated with SEA, the average dose of SEA in the carton was 144 $\pm$ 50 ng (Evenson et al., 1988). A massive outbreak in Japan caused by low-fat milk contaminated with SEA showed that the total intake of SEA per individual was estimated at approximately 20 to 100 ng (Asao et al., 2003).

The minimal number of *S. aureus* required to produce detectable and toxic levels of SE in food has been investigated. Considering the variability among strains, variable environmental conditions in food, and the possible interference by food in SE detection, the relevance of this information for SFP in general must be cautiously considered. A strain producing SEA, SEB, and SED produced detectable levels of SEB and SED (1 ng/ml) at a density of 6×10^6 CFU/ml, and then SEA (4 ng/ml) was detected at a cell density of 3×10^7 CFU/ml (Noleto and Bergdoll, 1982). In milk, SEA and SED were detected as low as 10^4 and 10^7 CFU/ml, respectively.

FOOD-BORNE OUTBREAKS

SFP Characteristics

SFP is caused by ingestion of food containing SE preformed by metabolically active staphylococci. SFP is usually a self-limiting illness, often abrupt and severe in onset, with a short incubation period (1 to 8 h). Victims experience severe nausea and characteristically projectile vomiting (Balaban and Rasooly, 2000; Dinges et al., 2000; Le Loir et al., 2003). The severity depends on an individual's susceptibility to the SE, the amount of contaminated food eaten, the amount of toxin in the food ingested, and the general health of the victim. Other symptoms include convulsive retching, diarrhea, abdominal cramps, pain, headache, and muscle ache. The illness may be accompanied by subnormal temperature and transient changes in blood pressure and pulse rate. Typically, the duration of SFP varies from several hours to 1 day. Dehydration, prostration, and shock can occur in severe cases which may require hospitalization. Deaths are rare, with fatality rates ranging from 0.03% for the general public to 4.4% for the very young, the elderly, or chronically ill patients. Fatal cases are usually associated with fluid and metabolic abnormalities, despite intravenous electrolyte and fluid replacement. It is estimated that ~1,750 hospitalizations and two deaths typically occur each year from SFP in the United States (Mead et al., 1999).

SFP Incidence

SFP occurs as isolated cases or outbreaks. Although it has likely afflicted humankind for centuries, the exact incidence is unknown for several reasons. Although there is a national surveillance system for food-borne disease in the United States through FoodNet (CDC, 1998), SFP is not officially reportable. In addition, since it is not usually life-threatening, definitive diagnoses of isolated cases are not usually accomplished, or affected individuals may not seek treatment. Confirmatory diagnosis is not always practical since it usually requires identification of SE-producing S. aureus in food which may have been discarded. Since SFP is not an infection, attempts to isolate the causative staphylococci from patients are not warranted.

Outbreaks typically receive considerable publicity. Some examples of recent outbreaks are summarized in Table 3. In the United States, 47 outbreaks of SFP involving 3,181 cases were reported between 1983 and 1987 and 42 reported outbreaks involving 1,413 cases occurred between 1993 and 1997, accounting for 1.6% of total food-borne disease cases (Olsen et al., 2000). It has been estimated that the mean annual number of cases from 1982 through 1997 could have been 185,060 (Mead et al., 1999), and the cost of SFP in the United States is approximately $1.5 billion annually (Harvey and Gilmour,

2000). There has been one report of a U.S. outbreak by shredded pork and coleslaw contaminated with a community-acquired methicillin-resistant S. aureus, which is known to frequently express SEs (Jones et al., 2002). Due to the lack of a unified surveillance system for the European Union, the reported prevalence of SFP varies greatly from country to country. SFP accounted for an average of 4.5% of food-borne disease cases (926 outbreaks) between 1993 and 1998 in 15 European Union countries (WHO Surveillance Programme, 2000). In Japan, 2,525 outbreaks were reported, involving 59,964 persons between 1980 and 1999 (Shimizu et al., 2000).

Distribution of SEs among Staphylococci

Many SE and SEl genes are associated with genomic islands and mobile genetic elements such as the staphylococcal pathogenicity island (designated SaPI), prophages, or plasmids (Baba et al., 2002). The known locations of SE and SEl genes are summarized in Table 2. Recent studies demonstrated that SaPI is mobilized by staphylococcal bacteriophages and endogenous prophages (Lindsay et al., 1998), suggesting that SE genes might be transferred horizontally. Numerous studies have assessed SE and SEl profiles of S. aureus from various sources. Although there is considerable variability among strains from different geographic regions, the most commonly observed SEs are those harbored on the enterotoxin gene cluster (egc), namely, those encoded by the sei, sem, sen, and seo genes with or without the seg gene, followed by those on the genetic element named bovine staphylococcal pathogenicity island (SaPI bov) (sec, tst, and sel) (Fitzgerald et al., 2000; Kwon et al., 2004; Lawrynowicz-Paciorek et al., 2007; Srinivasan et al., 2006). In contrast, the SE most frequently produced by S. aureus implicated in SFP is SEA, followed by SED, SEI, and SEC (Chiang et al., 2008; Kerouanton et al., 2007). Another coagulase-positive species, S. intermedius, is a well-known SE producer that possesses a variant SEC gene, sec_{canine}, as well as genes for

Table 3. Large outbreaks of SFP

Yr	Site or country	Product	No. of cases	Reference
1975	Commercial airline	Omelets and ham for airplane meal in Anchorage, AL	197	State of Alaska (1975)
1983	Caribbean cruise ship	Cream-filled pastries	215	CDC (1983)
1986	United States	Low-fat chocolate milk	ca. 1,000	Evenson et al. (1988)
1990	Thailand	Éclairs	485	Thaikruea et al. (1995)
1992	United States	Chicken salad	1,364	USFDA (1992)
2000	Japan	Low-fat milk	14,870	Asao et al. (2003)
2004	Brazil	Ordination meal	ca. 8,000	Do Carmo et al. (2004)

other SEs (SEA, SEB, SEC, SED, and SEE) and a novel SE-related gene, *se-int* (Futagawa-Saito et al., 2004). A 1991 outbreak of SFP in the United States involving 265 persons was conclusively linked to an SEA-producing strain of *S. intermedius* (Khambaty et al., 1994). Therefore, staphylococci other than *S. aureus* should not be neglected in regard to risk of SFP.

In fact, CNS also harbor SE genes, albeit less frequently than *S. aureus*. *S. xylosus* and *S. cohnii* from dry cured ham produce SEC and SED (Rodriguez, 1996). *S. equorum*, *S. capitis*, *S. lentus*, *S. gallinarum*, and *S. cohnii* from goat milk and cheese produce SEE (Vernozy-Rozand et al., 1996). *S. capitis*, *S. cohnii*, *S. haemolyticus*, *S. hominis*, *S. hyicus* subspecies *hyicus*, *S. saprophyticus*, *S. schleiferi*, *S. warneri*, and *S. xylosus* from various sources reportedly produce SEs (Bautista et al., 1988; Breckinridge and Bergdoll, 1971; Hoover et al., 1983; Olsvik et al., 1982; Udo et al., 1999; Valle et al., 1990). It is unclear whether these species have the capacity to frequently cause SFP since one study reported that CNS produce very small amounts of SE (<10 ng/ml) (Bergdoll, 1995).

SE Detection Methods

There is considerable interest in detecting SEs or their genes for epidemiological purposes. Currently, commercial kits are available to detect the classical SEs (SEA to SEE) in food. These are usually based on passive agglutination or the enzyme-linked immunosorbent assay and include the SET-RPLA kit (Denka Seiken), the visual immunoassay (Tecra), BioDetect Test Strip (Alexeter Technologies), TransiaTube (Diffchamb AB), and Ridascreen (rBiopharma). Detection of more recently identified SEs is problematic, since fewer reagents are available, particularly antisera. T-cell mitogenicity has been used to detect novel SEs based on their SAg activity (Holtfreter et al., 2004). Some laboratories have raised antibodies for the detection of certain SEs, such as SEG, SElH, and SEI (Omoe et al., 2002).

PCR is an extremely powerful tool to rapidly screen *S. aureus* strains for SE-encoding genes. Several multiplex PCR methods have been developed. Monday and Bohach (1999) developed an assay which, in a single reaction, detects nine SE genes. Others extended this method and developed PCR techniques for larger numbers of SE genes in multiple reactions (Omoe et al., 2005b). Multiplex PCR methods that subtype the *sec* gene into its variants *sec1*, *sec2*, and *sec3* have been described (Chen et al., 2001). Letertre et al. (2003b) developed a real-time PCR to detect the *sea* to *sej* genes. Sergeev et al. (2004) described an assay which involves PCR amplification of SE genes using degenerate primers, followed by characterization of the amplicons by microchip hybridization with oligonucleotide probes specific for SE genes. This assay has the advantage that it can detect previously unidentified SE genes.

Despite the usefulness of methods described above, their performance time and compromised accuracy due to food matrices limit their effectiveness. Recently the ability of biosensors to overcome these shortcomings has been investigated. These systems typically link a biological component, such as an antibody, with a physicochemical transducer to convert and amplify minute signals from the biological component into a measurable signal. Enzymatic bio-nanotransduction systems use specific antibody conjugated to nano-signal-producing DNA templates. Signals may be amplified by in vitro transcription of DNA templates bound to target molecules producing RNA nano-signals specific for multiple targets in the sample. Recent application of this method detected SEB at a level of 0.11 ng/ml (Branen et al., 2007). Biomolecular interaction analysis mass spectrometry was developed and applied to detect SEB and TSST-1. Biomolecular interaction analysis-mass spectometry uses surface plasmon resonance to detect SEB binding to antibody immobilized on a gold electrode, followed by matrix-assisted laser desorption ionization–time-of-flight to identify and differentiate the bound toxins. This method detected and differentiated SEB and TSST-1 at levels of 1 ng/ml in milk and mushrooms (Nedelkov et al., 2000).

CONCLUDING REMARKS

SFP has been a major concern of the food industry and medical profession for many decades, and great advances have been made in regard to its prevention, treatment, diagnosis, and etiology. Although it is not usually life-threatening, SFP produces considerable morbidity and economic impact. An understanding of the exact impact in the United States would require more stringent and uniform reporting requirements. In recent years, major advances were made in understanding the complicated epidemiology of SFP, and several past dogmas are now considered obsolete. For example, we have only recently learned that, in addition to the five classical SEs, many additional molecular variants and recently identified SEs and SEIs exist. Some of these related proteins are confirmed to lack emetic activity, suggesting that emesis is not necessarily a feature of the toxins that provides a selective advantage to staphylococci. It is more likely that enterotoxicity is a secondary effect of SE-induced immunomodulation and activity as SAgs.

Presumably, both biological properties are endowed by the unique and conserved molecular structures of SEs, which have been revealed in the past decade. The identification of novel toxins and improved methods of detection have also indicated that SE production is much more widespread than originally thought and includes species of staphylococci other than *S. aureus*. One key area of understanding that is still lacking includes the molecular and cellular mechanisms by which SEs induce SFP. It is becoming clear that induction of emesis and SAg activity are not directly related, but the lack of a convenient and inexpensive animal model has hindered progress in this area.

Acknowledgments. Our efforts in preparation of the manuscript were supported by grants from the PHS (U54AI57141, RR15587, and RR00166) and the Idaho Agricultural Experiment Station.

REFERENCES

Angelotti, R., M. J. Foter, and K. H. Lewis. 1961. Time-temperature effects on salmonellae and staphylococci in foods. I. Behavior in refrigerated foods. II. Behavior at warm holding temperatures. *Am. J. Public Health Nations Health* 51:76–88.

Asao, T., Y. Kumeda, T. Kawai, T. Shibata, H. Oda, K. Haruki, H. Nakazawa, and S. Kozaki. 2003. An extensive outbreak of staphylococcal food poisoning due to low-fat milk in Japan: estimation of enterotoxin A in the incriminated milk and powdered skim milk. *Epidemiol. Infect.* 130:33–40.

Atanassova, V., A. Meindl, and C. Ring. 2001. Prevalence of *Staphylococcus aureus* and staphylococcal enterotoxins in raw pork and uncooked smoked ham—a comparison of classical culturing detection and RFLP-PCR. *Int. J. Food Microbiol.* 68:105–113.

Baba, T., F. Takeuchi, M. Kuroda, H. Yuzawa, K. Aoki, A. Oguchi, Y. Nagai, N. Iwama, K. Asano, T. Naimi, H. Kuroda, L. Cui, K. Yamamoto, and K. Hiramatsu. 2002. Genome and virulence determinants of high virulence community-acquired MRSA. *Lancet* 359:1819–1827.

Baker, M. D., and K. R. Acharya. 2004. Superantigens: structure-function relationships. *Int. J. Med. Microbiol.* 293:529–537.

Balaban, N., and A. Rasooly. 2000. Staphylococcal enterotoxins. *Int. J. Food Microbiol.* 61:1–10.

Barber, M. A. 1914. Milk poisoning due to a type of *Staphylococcus albus* occurring in the udder of a healthy cow. *Philipp. J. Sci.* 9:515–519.

Bautista, L., P. Gaya, M. Medina, and M. Nunez. 1988. A quantitative study of enterotoxin production by sheep milk staphylococci. *Appl. Environ. Microbiol.* 54:566–569.

Bayles, K. W., and J. J. Iandolo. 1989. Genetic and molecular analyses of the gene encoding staphylococcal enterotoxin D. *J. Bacteriol.* 171:4799–4806.

Becker, K., D. Harmsen, A. Mellmann, C. Meier, P. Schumann, G. Peters, and C. von Eiff. 2004. Development and evaluation of a quality-controlled ribosomal sequence database for 16S ribosomal DNA-based identification of *Staphylococcus* species. *J. Clin. Microbiol.* 42:4988–4995.

Bennett, R. W., and M. R. Berry, Jr. 1987. Serological reactivity and in vivo toxicity of *Staphylococcus aureus* enterotoxin A and D in selected canned foods. *J. Food Sci.* 52:416–418.

Bergdoll, M. S. 1967. The staphylococcal enterotoxin, p. 1–25. *In* R. I. Mateles and G. N. Wogan (ed.), *Biochemistry of Some Foodborne Microbial Toxins*, MIT Press, Cambridge, MA.

Bergdoll, M. S. 1988. Monkey feeding test for staphylococcal enterotoxin. *Methods Enzymol.* 165:324–333.

Bergdoll, M. S. 1995. Importance of staphylococci that produce nanogram quantities of enterotoxin. *Zentralbl. Bakteriol.* 282:1–6.

Betley, M. J., and J. J. Mekalanos. 1988. Nucleotide sequence of the type A staphylococcal enterotoxin gene. *J. Bacteriol.* 170:34–41.

Borja, C. R., and M. S. Bergdoll. 1967. Purification and partial characterization of enterotoxin C produced by *Staphylococcus aureus* strain 137. *Biochemistry* 6:1467–1473.

Branen, J. R., M. J. Hass, E. R. Douthit, W. C. Maki, and A. L. Branen. 2007. Detection of *Escherichia coli* O157, *Salmonella enterica* serovar *Typhimurium*, and staphylococcal enterotoxin B in a single sample using enzymatic bio-nanotransduction. *J. Food Prot.* 70:841–850.

Breckinridge, J. C., and M. S. Bergdoll. 1971. Outbreak of foodborne gastroenteritis due to a coagulase-negative enterotoxin-producing staphylococcus. *N. Engl. J. Med.* 284:541–543.

Cavett, J. J. 1962. The microbiology of vacuum packed sliced bacon. *J. Appl. Bacteriol.* 25:282–289.

CDC. 1983. Staphylococcal food poisoning on a cruise ship. *MMWR Morb. Mortal. Wkly. Rep.* 32:294–295.

CDC. 1998. FoodNet:What We Are, Where We Are, and What We're Doing in 1998. http://www.cdc.gov/ncidod/dbmd/foodnet/foodnet.htm.

Chen, T. R., M. H. Hsiao, C. S. Chiou, and H. Y. Tsen. 2001. Development and use of PCR primers for the investigation of C1, C2 and C3 enterotoxin types of *Staphylococcus aureus* strains isolated from food-borne outbreaks. *Int. J. Food Microbiol.* 71:63–70.

Chiang, Y. C., W. W. Liao, C. M. Fan, W. Y. Pai, C. S. Chiou, and H. Y. Tsen. 2008. PCR detection of Staphylococcal enterotoxins (SEs) N, O, P, Q, R, U, and survey of SE types in *Staphylococcus aureus* isolates from food-poisoning cases in Taiwan. *Int. J. Food Microbiol.* 121:66–73.

Clarridge, J. E., III. 2004. Impact of 16S rRNA gene sequence analysis for identification of bacteria on clinical microbiology and infectious diseases. *Clin. Microbiol. Rev.* 17:840–862.

Couch, J. L., M. T. Soltis, and M. J. Betley. 1988. Cloning and nucleotide sequence of the type E staphylococcal enterotoxin gene. *J. Bacteriol.* 170:2954–2960.

Crass, B. A., and M. S. Bergdoll. 1986. Involvement of coagulase-negative staphylococci in toxic shock syndrome. *J. Clin. Microbiol.* 23:43–45.

De Buyser, M. L., B. Lombard, S. M. Schulten, P. H. In't Veld, S. L. Scotter, P. Rollier, and C. Lahellec. 2003. Validation of EN ISO standard methods 6888 part 1 and part 2: 1999—enumeration of coagulase-positive staphylococci in foods. *Int. J. Food Microbiol.* 83:185–194.

Denny, C. B., P. L. Tan, and C. W. Bohrer. 1996. Heat inactivation of staphylococcal enterotoxin. *J. Food Sci.* 31:762–767.

Denys, J. 1894. Empoisonnement par la viande contenant le *Staphylocoque pyogène. Sem. Méd.* 14:441.

De Paulis, A. N., S. C. Predari, C. D. Chazarreta, and J. E. Santoianni. 2003. Five-test simple scheme for species-level identification of clinically significant coagulase-negative staphylococci. *J. Clin. Microbiol.* 41:1219–1224.

Deringer, J. R., R. J. Ely, S. R. Monday, C. V. Stauffacher, and G. A. Bohach. 1997. Vβ-dependent stimulation of bovine and human T cells by host-specific staphylococcal enterotoxins. *Infect. Immun.* 65:4048–4054.

Dinges, M. M., P. M. Orwin, and P. M. Schlievert. 2000. Exotoxins of *Staphylococcus aureus. Clin. Microbiol. Rev.* 13:16–34.

Do Carmo, L. S., C. Cummings, V. R. Linardi, R. S. Dias, J. M. De Souza, M. J. De Sena, D. A. Dos Santos, J. W. Shupp,

R. K. Pereira, and M. Jett. 2004. A case study of a massive staphylococcal food poisoning incident. *Foodborne Pathog. Dis.* 1:241–246.

Evenson, M. L., M. W. Hinds, R. S. Bernstein, and M. S. Bergdoll. 1988. Estimation of human dose of staphylococcal enterotoxin A from a large outbreak of staphylococcal food poisoning involving chocolate milk. *Int. J. Food Microbiol.* 7:311–316.

Fitzgerald, J. R., P. J. Hartigan, W. J. Meaney, and C. J. Smyth. 2000. Molecular population and virulence factor analysis of *Staphylococcus aureus* from bovine intramammary infection. *J. Appl. Microbiol.* 88:1028–1037.

Fox, L. K., T. E. Besser, and S. M. Jackson. 1996. Evaluation of a coagulase-negative variant of *Staphylococcus aureus* as a cause of intramammary infections in a herd of dairy cattle. *J. Am. Vet. Med. Assoc.* 209:1143–1146.

Futagawa-Saito, K., M. Suzuki, M. Ohsawa, S. Ohshima, N. Sakurai, W. Ba-Thein, and T. Fukuyasu. 2004. Identification and prevalence of an enterotoxin-related gene, *se-int*, in *Staphylococcus intermedius* isolates from dogs and pigeons. *J. Appl. Microbiol.* 96:1361–1366.

Gardini, F., R. Tofalo, and G. Suzzi. 2003. A survey of antibiotic resistance in *Micrococcaceae* isolated from Italian dry fermented sausages. *J. Food Prot.* 66:937–945.

Genigeorgis, C., H. Riemann, and W. W. Sadler. 1969. Production of enterotoxin B in cured meats. *J. Food Sci.* 32:62–68.

Genigeorgis, C. A. 1989. Present state of knowledge on staphylococcal intoxication. *Int. J. Food Microbiol.* 9:327–360.

Guthertz, L. S., J. T. Fruin, R. L. Okoluk, and J. L. Fowler. 1977. Microbial quality of frozen comminuted turkey meat. *J. Food Sci.* 42:1344–1347.

Harvey, J., and A. Gilmour. 2000. *Staphylococcus aureus*, p. 2066–2071. *In* C. A. Batt and R. D. Patel (ed.), *Encyclopedia of Food Microbiology*, 1st ed. Academic Press, London, United Kingdom.

Heikens, E., A. Fleer, A. Paauw, A. Florijn, and A. C. Fluit. 2005. Comparison of genotypic and phenotypic methods for species-level identification of clinical isolates of coagulase-negative staphylococci. *J. Clin. Microbiol.* 43:2286–2290.

Holley, R. A. 1985. Beef jerky: fate of *Staphylococcus aureus* in marinated and corned beef during jerky manufacture and 2.5°C storage. *J. Food Prot.* 48:107–111.

Holtfreter, S., K. Bauer, D. Thomas, C. Feig, V. Lorenz, K. Roschack, E. Friebe, K. Selleng, S. Lovenich, T. Greve, A. Greinacher, B. Panzig, S. Engelmann, G. Lina, and B. M. Broker. 2004. *egc*-encoded superantigens from *Staphylococcus aureus* are neutralized by human sera much less efficiently than are classical staphylococcal enterotoxins or toxic shock syndrome toxin. *Infect. Immun.* 72:4061–4071.

Hoover, D. G., S. R. Tatini, and J. B. Maltais. 1983. Characterization of staphylococci. *Appl. Environ. Microbiol.* 46:649–660.

Hu, D. L., G. Zhu, F. Mori, K. Omoe, M. Okada, K. Wakabayashi, S. Kaneko, K. Shinagawa, and A. Nakane. 2007. Staphylococcal enterotoxin induces emesis through increasing serotonin release in intestine and it is downregulated by cannabinoid receptor 1. *Cell. Microbiol.* 9:2267–2277.

Jarraud, S., M. A. Peyrat, A. Lim, A. Tristan, M. Bes, C. Mougel, J. Etienne, F. Vandenesch, M. Bonneville, and G. Lina. 2001. *egc*, a highly prevalent operon of enterotoxin gene, forms a putative nursery of superantigens in *Staphylococcus aureus*. *J. Immunol.* 166:669–677.

Jay, J. M., M. J. Loessner, and D. A. Golden. 2005. Staphylococcal gastroenteritis, p. 544–566. *In* D. R. Heldman (ed.), *Modern Food Microbiology*, 7th ed. Springer Science and Business Inc., New York, NY.

Jones, T. F., M. E. Kellum, S. S. Porter, M. Bell, and W. Schaffner. 2002. An outbreak of community-acquired foodborne illness caused by methicillin-resistant *Staphylococcus aureus*. *Emerg. Infect. Dis.* 8:82–84.

Jousson, O., D. Di Bello, M. Vanni, G. Cardini, G. Soldani, C. Pretti, and L. Intorre. 2007. Genotypic versus phenotypic identification of staphylococcal species of canine origin with special reference to *Staphylococcus schleiferi* subsp. *coagulans*. *Vet. Microbiol.* 123:238–244.

Kappler, J., B. Kotzin, L. Herron, E. W. Gelfand, R. D. Bigler, A. Boylston, S. Carrel, D. N. Posnett, Y. Choi, and P. Marrack. 1989. V beta-specific stimulation of human T cells by staphylococcal toxins. *Science* 244:811–813.

Kerouanton, A., J. A. Hennekinne, C. Letertre, L. Petit, O. Chesneau, A. Brisabois, and M. L. De Buyser. 2007. Characterization of *Staphylococcus aureus* strains associated with food poisoning outbreaks in France. *Int. J. Food Microbiol.* 115:369–375.

Khambaty, F. M., R. W. Bennett, and D. B. Shah. 1994. Application of pulsed-field gel electrophoresis to the epidemiological characterization of *Staphylococcus intermedius* implicated in a food-related outbreak. *Epidemiol. Infect.* 113:75–81.

Kloos, W. E., K. H. Schleifer, and R. Götz. 1991. The genus *Staphylococcus*, p. 1369–1420. *In* A. Balows, H. G. Trüper, M. Dworkin, W. Harder, and K. H. Schleifer (ed.), *The Prokaryotes. A Handbook on the Biology of Bacteria: Ecophysiology, Isolation, Identification, Applications*, 2nd ed., vol. 2. Springer, New York, NY.

Kullik, I. I., and P. Giachino. 1997. The alternative sigma factor sigmaB in *Staphylococcus aureus*: regulation of the *sigB* operon in response to growth phase and heat shock. *Arch. Microbiol.* 167:151–159.

Kwon, N. H., S. H. Kim, K. T. Park, W. K. Bae, J. Y. Kim, J. Y. Lim, J. S. Ahn, K. S. Lyoo, J. M. Kim, W. K. Jung, K. M. Noh, G. A. Bohach, and Y. H. Park. 2004. Application of extended single-reaction multiplex polymerase chain reaction for toxin typing of *Staphylococcus aureus* isolates in South Korea. *Int. J. Food Microbiol.* 97:137–145.

Lawrynowicz-Paciorek, M., M. Kochman, K. Piekarska, A. Grochowska, and B. Windyga. 2007. The distribution of enterotoxin and enterotoxin-like genes in *Staphylococcus aureus* strains isolated from nasal carriers and food samples. *Int. J. Food Microbiol.* 117:319–323.

Le Loir, Y., F. Baron, and M. Gautier. 2003. *Staphylococcus aureus* and food poisoning. *Genet. Mol. Res.* 2:63–76.

Letertre, C., S. Perelle, F. Dilasser, and P. Fach. 2003a. Detection and genotyping by real-time PCR of the staphylococcal enterotoxin genes *sea* to *sej*. *Mol. Cell. Probes* 17:139–147.

Letertre, C., S. Perelle, F. Dilasser, and P. Fach. 2003b. Identification of a new putative enterotoxin SEU encoded by the *egc* cluster of *Staphylococcus aureus*. *J. Appl. Microbiol.* 95:38–43.

Lina, G., G. A. Bohach, S. P. Nair, K. Hiramatsu, E. Jouvin-Marche, and R. Mariuzza. 2004. Standard nomenclature for the superantigens expressed by *Staphylococcus*. *J. Infect. Dis.* 189:2334–2336.

Lindsay, J. A., A. Ruzin, H. F. Ross, N. Kurepina, and R. P. Novick. 1998. The gene for toxic shock toxin is carried by a family of mobile pathogenicity islands in *Staphylococcus aureus*. *Mol. Microbiol.* 29:527–543.

Luczak-Kadlubowska, A., J. Krzyszton-Russjan, and W. Hryniewicz. 2006. Characteristics of *Staphylococcus aureus* strains isolated in Poland in 1996 to 2004 that were deficient in species-specific proteins. *J. Clin. Microbiol.* 44:4018–4024.

Marshall, B. J., D. F. Ohye, and J. H. Christian. 1971. Tolerance of bacteria to high concentrations of NaCl and glycerol in the growth medium. *Appl. Microbiol.* 21:363–364.

Martin, M. C., J. M. Fueyo, M. A. Gonzalez-Hevia, and M. C. Mendoza. 2004. Genetic procedures for identification of enterotoxigenic strains of *Staphylococcus aureus* from three food poisoning outbreaks. *Int. J. Food Microbiol.* **94:**279–286.

Matthews, K. R., S. P. Oliver, and S. H. King. 1990. Comparison of Vitek gram-positive identification system with API Staph-Trac system for species identification of staphylococci of bovine origin. *J. Clin. Microbiol.* **28:**1649–1651.

McCormick, J. K., J. M. Yarwood, and P. M. Schlievert. 2001. Toxic shock syndrome and bacterial superantigens: an update. *Annu. Rev. Microbiol.* **55:**77–104.

McNamara, P. J., K. C. Milligan-Monroe, S. Khalili, and R. A. Proctor. 2000. Identification, cloning, and initial characterization of *rot*, a locus encoding a regulator of virulence factor expression in *Staphylococcus aureus*. *J. Bacteriol.* **182:**3197–3203.

Mead, P. S., L. Slutsker, V. Dietz, L. F. McCaig, J. S. Bresee, C. Shapiro, P. M. Griffin, and R. V. Tauxe. 1999. Food-related illness and death in the United States. *Emerg. Infect. Dis.* **5:**607–625.

Monday, S. R., and G. A. Bohach. 1999. Use of multiplex PCR to detect classical and newly described pyrogenic toxin genes in staphylococcal isolates. *J. Clin. Microbiol.* **37:**3411–3414.

Moon, J. S., A. R. Lee, S. H. Jaw, H. M. Kang, Y. S. Joo, Y. H. Park, M. N. Kim, and H. C. Koo. 2007. Comparison of antibiogram, staphylococcal enterotoxin productivity, and coagulase genotypes among *Staphylococcus aureus* isolated from animal and vegetable sources in Korea. *J. Food Prot.* **70:**2541–2548.

Munson, S. H., M. T. Tremaine, M. J. Betley, and R. A. Welch. 1998. Identification and characterization of staphylococcal enterotoxin types G and I from *Staphylococcus aureus*. *Infect. Immun.* **66:**3337–3348.

Nedelkov, D., A. Rasooly, and R. W. Nelson. 2000. Multitoxin biosensor-mass spectrometry analysis: a new approach for rapid, real-time, sensitive analysis of staphylococcal toxins in food. *Int. J. Food Microbiol.* **60:**1–13.

Noleto, A. L., and M. S. Bergdoll. 1982. Production of enterotoxin by a *Staphylococcus aureus* strain that produces three identifiable enterotoxins. *J. Food Prot.* **45:**1096–1097.

Oh, S. K., N. Lee, Y. S. Cho, D. B. Shin, S. Y. Choi, and M. Koo. 2007. Occurrence of toxigenic *Staphylococcus aureus* in ready-to-eat food in Korea. *J. Food Prot.* **70:**1153–1158.

Olsen, S. J., L. C. MacKinnon, J. S. Goulding, N. H. Bean, and L. Slutsker. 2000. Surveillance for foodborne-disease outbreaks—United States, 1993–1997. *MMWR Surveill. Summ.* **49:**1–62.

Olsvik, O., K. Fossum, and B. P. Berdal. 1982. Staphylococcal enterotoxin A, B, and C produced by coagulase-negative strains within the family *Micrococcaceae*. *Acta Pathol. Microbiol. Immunol. Scand. Sect. B* **90:**441–444.

Omoe, K., D. L. Hu, H. Takahashi-Omoe, A. Nakane, and K. Shinagawa. 2005a. Comprehensive analysis of classical and newly described staphylococcal superantigenic toxin genes in *Staphylococcus aureus* isolates. *FEMS Microbiol. Lett.* **246:**191–198.

Omoe, K., K. Imanishi, D. L. Hu, H. Kato, Y. Fugane, Y. Abe, S. Hamaoka, Y. Watanabe, A. Nakane, T. Uchiyama, and K. Shinagawa. 2005b. Characterization of novel staphylococcal enterotoxin-like toxin type P. *Infect. Immun.* **73:**5540–5546.

Omoe, K., K. Imanishi, D. L. Hu, H. Kato, H. Takahashi-Omoe, A. Nakane, T. Uchiyama, and K. Shinagawa. 2004. Biological properties of staphylococcal enterotoxin-like toxin type R. *Infect. Immun.* **72:**3664–3667.

Omoe, K., M. Ishikawa, Y. Shimoda, D. L. Hu, S. Ueda, and K. Shinagawa. 2002. Detection of *seg*, *seh*, and *sei* genes in *Staphylococcus aureus* isolates and determination of the enterotoxin productivities of *S. aureus* isolates harboring *seg*, *seh*, or *sei* genes. *J. Clin. Microbiol.* **40:**857–862.

Orwin, P. M., J. R. Fitzgerald, D. Y. Leung, J. A. Gutierrez, G. A. Bohach, and P. M. Schlievert. 2003. Characterization of *Staphylococcus aureus* enterotoxin L. *Infect. Immun.* **71:**2916–2919.

Orwin, P. M., D. Y. Leung, H. L. Donahue, R. P. Novick, and P. M. Schlievert. 2001. Biochemical and biological properties of staphylococcal enterotoxin K. *Infect. Immun.* **69:**360–366.

Orwin, P. M., D. Y. Leung, T. J. Tripp, G. A. Bohach, C. A. Earhart, D. H. Ohlendorf, and P. M. Schlievert. 2002. Characterization of a novel staphylococcal enterotoxin-like superantigen, a member of the group V subfamily of pyrogenic toxins. *Biochemistry* **41:**14033–14040.

Pereira, J. L., S. P. Salzberg, and M. S. Bergdoll. 1982. Effect of temperature, pH and sodium chloride concentrations on production of staphylococcal enterotoxin A and B. *J. Food Prot.* **45:**1306–1309.

Petersson, K., H. Pettersson, N. J. Skartved, B. Walse, and G. Forsberg. 2003. Staphylococcal enterotoxin H induces V alpha-specific expansion of T cells. *J. Immunol.* **170:**4148–4154.

Phillips, D., D. Jordan, S. Morris, I. Jenson, and J. Sumner. 2006. A national survey of the microbiological quality of beef carcasses and frozen boneless beef in Australia. *J. Food Prot.* **69:**1113–1117.

Portocarrero, S. M., M. Newman, and B. Mikel. 2002. *Staphylococcus aureus* survival, staphylococcal enterotoxin production and shelf stability of country-cured hams manufactured under different processing procedures. *Meat Sci.* **62:**267–273.

Poyart, C., G. Quesne, C. Boumaila, and P. Trieu-Cuot. 2001. Rapid and accurate species-level identification of coagulase-negative staphylococci by using the *sodA* gene as a target. *J. Clin. Microbiol.* **39:**4296–4301.

Raj, H. D., and M. S. Bergdoll. 1969. Effect of enterotoxin B on human volunteers. *J. Bacteriol.* **98:**833–834.

Ranelli, D. M., C. L. Jones, M. B. Johns, G. J. Mussey, and S. A. Khan. 1985. Molecular cloning of staphylococcal enterotoxin B gene in *Escherichia coli* and *Staphylococcus aureus*. *Proc. Natl. Acad. Sci. USA* **82:**5850–5854.

Regassa, L. B., J. L. Couch, and M. J. Betley. 1991. Steady-state staphylococcal enterotoxin type C mRNA is affected by a product of the accessory gene regulator *(agr)* and by glucose. *Infect. Immun.* **59:**955–962.

Rodriguez, M., F. Nunez, J. J. Cordoba, E. Bermudez, and M. A. Asensio. 1996. Gram-positive, catalase-positive cocci from dry cured Iberian ham and their enterotoxigenic potential. *Appl. Environ. Microbiol.* **62:**1897–1902.

Sanjeev, S., T. S. Gopalakrishna Iyer, P. R. Varma, C. C. Panduranga Rao, and K. Mahadeva Iyer. 1987. Carriage of enterotoxigenic staphylococci in workers of fish processing factories. *Indian J. Med. Res.* **85:**262–265.

Schantz, E. J., W. G. Roessler, M. J. Woodburn, J. M. Lynch, H. M. Jacoby, S. J. Silverman, J. C. Gorman, and L. Spero. 1972. Purification and some chemical and physical properties of staphylococcal enterotoxin A. *Biochemistry* **11:**360–366.

Schleifer, K. H. 1986. Gram positive cocci, p. 999–1100. *In* P. A. Sneath (ed.), *Bergey's Manual of Systematic Bacteriology*, 1st ed., vol. 2. Williams & Wilkins Co., Baltimore, MD.

Schmidt, K. A., N. P. Donegan, W. A. Kwan, Jr., and A. Cheung. 2004. Influences of sigmaB and *agr* on expression of staphylococcal enterotoxin B *(seb)* in *Staphylococcus aureus*. *Can. J. Microbiol.* **50:**351–360.

Seo, K. S., S. U. Lee, Y. H. Park, W. C. Davis, L. K. Fox, and G. A. Bohach. 2007. Long-term staphylococcal enterotoxin C1 exposure induces soluble factor-mediated immunosuppression by bovine CD4+ and CD8+ T cells. *Infect. Immun.* **75:**260–269.

Sergeev, N., D. Volokhov, V. Chizhikov, and A. Rasooly. 2004. Simultaneous analysis of multiple staphylococcal enterotoxin genes by an oligonucleotide microarray assay. *J. Clin. Microbiol.* **42:**2134–2143.

Shimizu, A., M. Fujita, H. Igarashi, M. Takagi, N. Nagase, A. Sasaki, and J. Kawano. 2000. Characterization of *Staphylococcus aureus* coagulase type VII isolates from staphylococcal food poisoning outbreaks (1980–1995) in Tokyo, Japan, by pulsed-field gel electrophoresis. *J. Clin. Microbiol.* **38:**3746–3749.

Skow, A., K. A. Mangold, M. Tajuddin, A. Huntington, B. Fritz, R. B. Thomson, Jr., and K. L. Kaul. 2005. Species-level identification of staphylococcal isolates by real-time PCR and melt curve analysis. *J. Clin. Microbiol.* **43:**2876–2880.

Spero, L., B. Y. Griffin, J. L. Middlebrook, and J. F. Metzger. 1976. Effect of single and double peptide bond scission by trypsin on the structure and activity of staphylococcal enterotoxin C. *J. Biol. Chem.* **251:**5580–5588.

Srinivasan, V., A. A. Sawant, B. E. Gillespie, S. J. Headrick, L. Ceasaris, and S. P. Oliver. 2006. Prevalence of enterotoxin and toxic shock syndrome toxin genes in *Staphylococcus aureus* isolated from milk of cows with mastitis. *Foodborne Pathog. Dis.* **3:**274–283.

State of Alaska. 1975. Staphylococcal foodborne outbreak. *State Alaska Epidemiol. Bull.* **4:**1–2.

Stewart, C. M., M. B. Cole, J. D. Legan, L. Slade, M. H. Vandeven, and D. W. Schaffner. 2002. *Staphylococcus aureus* growth boundaries: moving towards mechanistic predictive models based on solute-specific effects. *Appl. Environ. Microbiol.* **68:**1864–1871.

Sugiyama, H., and T. Hayama. 1965. Abdominal viscera as site of emetic action for staphylococcal enterotoxin in the monkey. *J. Infect. Dis.* **115:**330–336.

Summer, S. S., J. A. Albrecht, and D. L. Peters. 1993. Occurrence of enterotoxigenic strains of *Staphylococcus aureus* and enterotoxin production in bakery products. *J. Food Prot.* **56:**722–724.

Swartzentruber, A., A. H. Schwab, A. P. Duran, B. A. Wentz, and R. B. Read, Jr. 1980. Microbiological quality of frozen shrimp and lobster tail in the retail market. *Appl. Environ. Microbiol.* **40:**765–769.

Taponen, S., H. Simojoki, M. Haveri, H. D. Larsen, and S. Pyorala. 2006. Clinical characteristics and persistence of bovine mastitis caused by different species of coagulase-negative staphylococci identified with API or AFLP. *Vet. Microbiol.* **115:**199–207.

Thaikruea, L., J. Pataraarechachai, P. Savanpunyalert, and U. Naluponjiragul. 1995. An unusual outbreak of food poisoning. *Southeast Asian J. Trop. Med. Public Health* **26:**78–85.

Tonge, R. J., A. C. Baird-Parker, and J. J. Cavett. 1964. Chemical and microbiological changes during storage of vacuum packed sliced bacon. *J. Appl. Bacteriol.* **27:**252–264.

Townsend, D. E., and B. J. Wilkinson. 1992. Proline transport in *Staphylococcus aureus:* a high-affinity system and a low-affinity system involved in osmoregulation. *J. Bacteriol.* **174:**2702–2710.

Tremaine, M. T., D. K. Brockman, and M. J. Betley. 1993. Staphylococcal enterotoxin A gene *(sea)* expression is not affected by the accessory gene regulator *(agr)*. *Infect. Immun.* **61:**356–359.

Udo, E. E., M. A. Al-Bustan, L. E. Jacob, and T. D. Chugh. 1999. Enterotoxin production by coagulase-negative staphylococci in restaurant workers from Kuwait City may be a potential cause of food poisoning. *J. Med. Microbiol.* **48:**819–823.

USDA. 1996. Nationwide federal plant raw ground beef microbiological survey. U.S. Department of Agriculture, Washington, DC.

USFDA. 1992. *Staphylococcus aureus. Bad Bug Book, Foodborne Pathogenic Microorganisms and Natural Toxins Handbook.* U.S. Food and Drug Administration, Washington, DC. http://vm.cfsan.fda.gov/~mow/chap3.html.

Valle, J., E. Gomez-Lucia, S. Piriz, J. Goyache, J. A. Orden, and S. Vadillo. 1990. Enterotoxin production by staphylococci isolated from healthy goats. *Appl. Environ. Microbiol.* **56:**1323–1326.

Vandenbosch, L. L., D. Y. Fung, and M. Widomski. 1973. Optimum temperature for enterotoxin production by *Staphylococcus aureus* S-6 and 137 in liquid medium. *Appl. Microbiol.* **25:**498–500.

Vandenesch, F., C. Lebeau, M. Bes, D. McDevitt, T. Greenland, R. P. Novick, and J. Etienne. 1994. Coagulase deficiency in clinical isolates of *Staphylococcus aureus* involves both transcriptional and post-transcriptional defects. *J. Med. Microbiol.* **40:**344–349.

Vernozy-Rozand, C., C. Mazuy, G. Prevost, C. Lapeyre, M. Bes, Y. Brun, and J. Fleurette. 1996. Enterotoxin production by coagulase-negative staphylococci isolated from goats' milk and cheese. *Int. J. Food Microbiol.* **30:**271–280.

Warren, J. R., L. Spero, and J. F. Metzger. 1974. Stabilization of native structure by the closed disulfide loop of staphylococcal enterotoxin B. *Biochim. Biophys. Acta* **359:**351–363.

WHO Surveillance Programme. 2000. 7th Report (1993–1998), p. 13–393. *In* K. Schmidt and C. Tirado (ed.), *WHO Surveillance Programme 7th Report.* WHO, Berlin, Germany.

Yamashita, K., A. Shimizu, J. Kawano, E. Uchida, A. Haruna, and S. Igimi. 2005. Isolation and characterization of staphylococci from external auditory meatus of dogs with or without otitis externa with special reference to *Staphylococcus schleiferi* subsp. *coagulans* isolates. *J. Vet. Med. Sci.* **67:**263–268.

Yeh, K. S., S. P. Chen, and J. H. Lin. 2005. One-year (2003) nationwide pork carcass microbiological baseline data survey in Taiwan. *J. Food Prot.* **68:**458–461.

Yugueros, J., A. Temprano, B. Berzal, M. Sanchez, C. Hernanz, J. M. Luengo, and G. Naharro. 2000. Glyceraldehyde-3-phosphate dehydrogenase-encoding gene as a useful taxonomic tool for *Staphylococcus* spp. *J. Clin. Microbiol.* **38:**4351–4355.

Zhang, S., J. J. Iandolo, and G. C. Stewart. 1998. The enterotoxin D plasmid of *Staphylococcus aureus* encodes a second enterotoxin determinant *(sej)*. *FEMS Microbiol. Lett.* **168:**227–233.

Pathogens and Toxins in Foods: Challenges and Interventions
Edited by V. K. Juneja and J. N. Sofos
© 2010 ASM Press, Washington, DC

Chapter 9

Shigella

KEITH A. LAMPEL

Diarrheal diseases afflict a significant number of the world's population each year. The number of deaths associated with this type of illness is estimated to be in the millions and is considered the fourth leading cause of mortality in the world (Murray and Lopez, 1997). Etiological agents comprise a diverse group of pathogens, including bacterial, viral, and parasitic agents. Although *Shigella* spp. have not drawn as much attention as other human microbial pathogens, particularly in developed countries where the mortality rate due to these pathogens is quite low, they remain a scourge throughout the developing world and continue to be a significant health concern.

Estimates of disease caused by *Shigella* spp. on a yearly basis worldwide range from 164.7 to 200 million people infected, with nearly 1.1 million deaths attributed to this pathogen (Kotloff et al., 1999). Most of the 163.2 million people affected live in developing countries. Children under the age of 5 years, particularly those who are malnourished, are the most susceptible and have the highest mortality rates. Several factors are responsible for differences in the morbidity and mortality rates between developing countries and developed (industrialized) countries and include socioeconomic trends and nutrition. Also, the severity of illness is dependent upon which *Shigella* spp. is the causative agent of disease.

The genus *Shigella* comprises four serotypes based on the antigenicity of each lipopolysaccharide (somatic O antigens): *Shigella dysenteriae, Shigella flexneri, Shigella boydii,* and *Shigella sonnei.* In developing countries, *S. flexneri* is the most common *Shigella* spp. isolated, followed by *S. sonnei.* The numbers of isolated *Shigella* varieties are nearly reversed in industrialized countries where 16% and 77% of infections are caused by *S. flexneri* and *S. sonnei* isolates, respectively. *S. dysenteriae*, notably type 1, which is the only *Shigella* sp. to carry the

genetic information for the Shiga toxins, is a concern primarily in developing countries. *S. boydii* is commonly isolated in the Indian subcontinent, and some cases have been reported recently from Europe and the United States.

As with other enteric pathogens, *Shigella* spp. owe their pathogenesis to genetic determinants that reside on pathogenicity islands, regions of either chromosomal or plasmid DNA that encode for most of the genes responsible for virulence. Briefly, the pathogenesis of *Shigella* spp. follows a cascade of events modulated by intrinsic factors provided by the host. After consumption of contaminated foods or water, or simply the transfer of *Shigella* via the fecal-oral route, this pathogen is able to survive and transit the stomach milieu to the intestine.

As with other gram-negative pathogens, *Shigella* spp. invade selected target host cells, in this case colonic epithelial cells, and, by a series of sophisticated interactions between the host immune system and their own genetic factory, establish an infection by intracellular multiplication and intercellular spread. Common to some bacterial pathogens, the host-pathogen interactions are mediated by bacterial pathogen effector molecules that utilize the type III secretion system to effectively deliver these proteins to specific sites within the host cell. Destruction of the intestinal epithelial cells and mucosal inflammation is a consequence of the host's polymorphonuclear leukocytes and a subsequent recruitment influx of chemokines and cytokines at the sites of *Shigella* invasion. Consequently, the infected host is unable to retain fluids, leading to diarrheal episodes (Sansonetti et al., 1999). A short, excellent review of the key components of the "cross-talk" between the host and bacterial pathogen and the role of host cell proinflammatory responses is provided elsewhere (Sansonetti, 2006).

Keith A. Lampel • Food and Drug Administration, HFS-710, 5100 Paint Branch Parkway, College Park, MD 20740.

TYPE OF ILLNESS AND CHARACTERISTICS OF THE ORGANISM

Bacillary dysentery or shigellosis is caused by all four members of the *Shigella* genus. These bacteria are members of the family *Enterobacteriaceae*, tribe *Escherichiaeae*. The four species of *Shigella* are divided into 55 serotypes and subserotypes. *S. dysenteriae* (group A) contains 15 serotypes, *S. flexneri* (group B) 8 serotypes and 11 subserotypes, *S. boydii* (group C) 20 serotypes, and *S. sonnei* (group D) one serotype (Table 1). Enteroinvasive *Escherichia coli* (EIEC) is capable of causing dysentery, distinguishing itself from other pathogenic *E. coli* strains. EIEC has pathogenic factors (i.e., contains a homologous virulence plasmid of shigellae) and biochemical properties similar to those of *Shigella*. Other attributes that distinguish EIEC from other *E. coli* bacteria are that they are nonmotile and are nonlactose fermenters. Some serotypes of EIEC, such as EIEC 0124, also share identical O antigens with *Shigella* (Sansonetti et al., 1985). The evolutionary relatedness of EIEC to *Shigella* has been discussed by Lan et al. (2004), and their conclusion is that these two pathogens comprise a pathovar of *E. coli*.

Shigellae are short (1 to 3 μm), gram-negative, nonmotile, non-spore-forming, nonencapsulated, rod-shaped cells. Important physiological/biochemical traits follow. They are facultative anaerobes, lactose negative (some biotypes of *S. flexneri* 6 are positive; positive strains of *S. boydii* 13 and 14 have been noted [NB: *S. boydii* 13 is the exception to many genetic biochemical characteristics; it has been noted (Strockbine and Maurelli, 2005) that this organism may not be *Shigella*]), lysine decarboxylase negative (*S. boydii* serotype 13 is positive), do not produce gas from glucose utilization (*S. flexneri* and *S. boydii* serotypes 13 and 14 produce gas), do not produce hydrogen sulfide, and are ornithine decarboxylase negative (*S. sonnei* is positive [Edwards and Ewing, 1972]).

Genetically, *Shigella* spp. are closely related evolutionarily to *E. coli* and are considered to be clones of this species. One of the major differences between *Shigella* spp. and pathogenic *E. coli* strains relates to the physical mechanism of entering host cells; *Shigella* spp. do not have any adherence factors such as those that enable *E. coli* pathogens to attach to the host cell surface prior to invasion. In addition, the identity of "black holes," deletions of genetic regions of the *Shigella* chromosome compared to the genetic map of *E. coli* (Maurelli et al., 1998), suggests that these two bacteria are genetically drifting apart from each other. In addition, the more selective ecological niche that shigellae occupy–for instance, they are host adapted to humans and higher primates in contrast to *E. coli*–may also support the notion that *Shigella* spp. are independent from *E. coli*.

Proponents of reclassifying *Shigella* spp. within *E. coli* posit that shigellae are clones of *Escherichia* spp. This is based on DNA-DNA reassociation studies and transconjugants between *Shigella* and *E. coli* (Luria and Burrous, 1957; Falkow et al., 1963; Brenner et al., 1969). Recent analysis using multilocus enzyme electrophoresis, ribotyping, and DNA sequencing of select genetic loci, such as the malate dehydrogenase *(mdh)* gene, provided additional support that *Shigella* spp. are, in essence, strains of *E. coli* (Lan and Reeves, 2002). Further corroborative data to place shigellae and *E. coli* in the same phylogenetic group come from sequence analysis of the 16S ribosomal RNA genes (Cilia et al., 1996). Accordingly, *Shigella* species are grouped as *E. coli* strains and are placed within ECOR groups A or B1 (Johnson, 1999). Further evidence of the relatedness of the *E. coli* and *Shigella* chromosome was provided by data generated from the sequence of the *S. flexneri* 2a genome (Jin et al., 2002; Wei et al., 2003).

Table 1. Characteristics of *Shigella* spp.

Species	Serogroup	Serotypes and subtypes	Geographic distribution	Distinguishing characteristics
S. dysenteriae	A	15 serotypes	Indian subcontinent, Africa, Asia, Central America	Type I produces Shiga toxin, causes most severe dysentery; high mortality rate if untreated
S. flexneri	B	8 serotypes, 11 subtypes	Most common isolate in developing countries	Elicits less severe dysentery than *S. dysenteriae*
S. boydii	C	20 serotypes	Indian subcontinent, rarely isolated in developed countries	Biochemically identical to S. *flexneri*; distinguished by serology
S. sonnei	D	1[a] serotype	Most common isolate in developed countries	Produces mildest form of shigellosis

[a]Forms I and II are serotypically distinguishable.

Shigellosis is a highly communicable disease due in part to the rapid spread of the pathogen within certain populations, particularly in crowded communities and/or in environments with poor sanitary conditions. Another contributing factor is the low infectious dose of *Shigella* that is required to cause disease. Studies with volunteers have shown that ingestion of only 100 virulent cells is sufficient to elicit clinical symptoms (DuPont et al., 1989).

Clinical presentations range from mild diarrhea to severe dysenteric syndrome, the latter consisting of abdominal pain, tenesmus, and bloody, mucoid stools. Fever is common, and copious amounts of watery diarrhea are accompanied by large numbers of leukocytes. This may be a reflection of the anatomical events of bacterial destruction of the colonic mucosa and the inability of the host to reabsorb fluids, manifesting in diarrhea (Rout et al., 1975; Kinsey et al., 1976). The presence of high numbers of leukocytes and shigellae in stool specimens occurs in the early stages of the disease. Later in infection, the bacteria spread from cell to cell, producing more destruction and sloughing of the colonic mucosal cells, as evident by the presence of blood, pus, and mucus in stools.

The onset of shigellosis usually occurs within 2 days after ingestion and is usually a self-limiting disease that resolves itself within 5 to 7 days. The illness can persist in a patient for 1 to 2 weeks. The disease can be fatal in immunocompromised individuals and malnourished people, particularly children under the age of 5 and the elderly. HIV (human immunodeficiency virus)-infected individuals can experience both chronic diarrhea and dysentery from a *Shigella* infection and usually manifest more severe clinical symptoms (http://www.who.int/vaccine_research/diseases/diarrhoeal/en/index6.html). Preferred treatment is hydration, and for patients with severe symptoms, antibiotic therapy with ampicillin, trimethoprim/sulfamethoxazole, nalidixic acid, or ciprofloxacin may be implemented.

The mildest form of the disease is associated with *S. sonnei*, which is endemic in developed countries. Infections caused by *S. flexneri* and *S. boydii* can range from mild to severe. *S. dysenteriae* type 1 carries the genetic information for a potent enterotoxin, Shiga toxin, and causes the most severe form of the disease. It accounts for deadly epidemics of bacillary dysentery and frequent pandemics in Central America, parts of south Asia, sub-Saharan Africa, and particularly in human-made and natural disasters. *S. dysenteriae* is isolated from 6% of patients with bacillary dysentery in developing countries, where poverty and crowded, unsanitary conditions exist (Kotloff et al., 1999). Of these, 30% of the isolates are *S. dysenteriae* type 1.

This may account for its ability to cause the most severe and prolonged form of dysentery with the highest frequency of fatality.

Further complications of shigellosis have been identified as toxic megacolon, dehydration, intestinal perforation, seizures, reactive arthritis, and hemolytic uremic syndrome (HUS) (Bennish, 1991). As with other gram-negative bacteria, reactive arthritis is associated with individuals of the HLA-B27 histocompatibility tissue type (Brewerton et al., 1973; Simon et al., 1981; Bunning et al., 1988). HUS, more commonly associated with *E. coli* O157:H7, is a rare postinfection sequela of *Shigella*, and the sole etiological species is *S. dysenteriae* 1 (Raghupathy et al., 1978). Shiga toxin, as with Shiga-like toxin of *E. coli* O157:H7, may be involved with HUS by damaging the vascular endothelial cells of the kidney (Karmali et al., 1985; Lopez et al., 1989). Septicemia is not a usual complication from shigellosis, although cases have been reported.

Overall, due to the low infectious dose of shigellae, the dissemination of the bacteria from person to person can be extremely swift and can be responsible for the high secondary attack rate when introduced within environs such as crowded and institutionalized populations. Children ages 1 to 5 are most susceptible to infections due to shigellae. This phenomenon is compounded by pediatric malnutrition in developing nations. These children face an increase in attack, relapse, and mortality rates.

Many other human bacterial pathogens may have multiple animal or environmental niches. For *Shigella* species, the principal natural reservoir is humans. In many geographical areas, shigellosis outbreaks are seasonal, with summer months showing the highest number of incidences. In warmer months, people's interactions increase; this augments the likelihood of symptomatic and asymptomatic carriers coming into contact with uninfected and susceptible populations. Asymptomatic carriers of *Shigella* may exacerbate the spread and maintenance of this pathogen in developing countries. Two studies, one in Bangladesh (Hossain et al., 1994) and another in Mexico (Guerrero et al., 1994), showed that *Shigella* was isolated from stool samples collected from asymptomatic children under the age of 5 years. *Shigella* spp. are rarely found in infants under the age of 6 months. Higher primates have also been known to harbor *Shigella*, particularly at zoos and primate centers (Line et al., 1992; Banish et al., 1993). In one instance (Kennedy et al., 1993), three animal caretakers at a monkey house presented with diarrheal symptoms. Further examination showed that *S. flexneri* 1b was isolated from stool samples of these employees and that four monkeys were shedding the identical serotype. The disease was spread by direct contact of the caretakers with excrement from the infected monkeys. In a recent study of free-ranging mountain gorillas in two national parks in Uganda, *S. flexneri*, *S. sonnei*, and *S. boydii* were isolated from preadult to adult gorillas. An increase in human exposure may have contributed to zoonotic transmission of these *Shigella* spp. (Nizeyi et al., 2001).

SOURCES AND INCIDENCES IN THE ENVIRONMENT AND FOODS

Foods

The primary means of human-to-human transmission of *Shigella* is by the fecal-oral route. Most cases of shigellosis are caused by the ingestion of fecal-contaminated food or water. In the case of foods, the major contributing factor for contamination is the poor personal hygiene of food handlers. From carriers, this pathogen is spread by "the four F's": food, fingers, feces, and flies; of late, a relatively new category, fomites, which are inanimate objects (e.g., utensils and cutting surfaces), have been added (Islam et al., 2001). Flies can transmit the bacteria from fecal matter to foods. Improper storage of contaminated foods is the second most common factor accounting for food-borne outbreaks due to *Shigella* (Smith, 1987).

Other contributing factors are inadequate cooking, contaminated equipment, and food obtained from unsafe sources (Bean and Griffin, 1990). To reduce the spread of shigellosis, infected patients, particularly in day care centers, are monitored until stool samples are negative for *Shigella*. In the United

States, outbreaks of shigellosis and other diarrheal diseases in day care centers are increasing, as more single- and two-parent working families turn to these facilities to care for their children (Pickering et al., 1986; Levine and Levine, 1994). Transmission of *Shigella* in this population is very efficient, and the low infectious dose for causing disease increases the risk for shigellosis. Increased risk also extends to family contacts of day care attendees (Weissman et al., 1974).

Shigellosis can be endemic in other institutional settings, such as prisons, mental hospitals, and nursing homes, where crowding and/or insufficient hygiene create an environment for direct fecal-oral contamination. In other cases, such as when natural or human-made disasters destroy the sanitary waste treatment and water purification infrastructure, developed countries assume the conditions of developing countries. These conditions place a population at risk for diarrheal diseases, such as cholera and dysentery. Examples include the war in Bosnia-Herzegovina, famine and political upheaval in Somalia, and massive population displacement, (e.g., refugees fleeing from Rwanda into Zaire in 1994), which led to explosive epidemics of diarrheal disease caused by *Vibrio cholerae* and *S. dysenteriae* 1 (Goma Epidemiology Group, 1995).

As evident in Table 2, sources of contaminated foods implicated in *Shigella* outbreaks varied extensively. Since humans are basically the sole source of *Shigella*, foods are considered factors only when contaminated by human waste. As such, food preparers are a common means for transmitting *Shigella* from an infected host to susceptible populations. Many of the outbreaks listed in Table 2 represent this type of

Table 2. Outbreaks of shigellosis

Yr	Location; source of contamination	Isolate	Reference(s)
1987	Rainbow Family gathering; food handlers	*S. sonnei*	Wharton et al. (1990)
1988	Outdoor music festival, Michigan; food handlers	*S. sonnei*	Lee et al. (1991)
1988	Commercial airline; cold sandwiches	*S. sonnei*	Hedberg et al. (1992)
1991	Alaska; moose soup, food preparer	*S. sonnei*	Gessner and Beller (1994)
1992–1993	Operation Restore Hope, Somalia, U.S. troops	*Shigella* spp.	Sharp et al. (1995)
1994	Europe; shredded lettuce from Spain	*S. sonnei*	Kapperud et al. (1995)
1994	Cruise ship	*S. flexneri*	CDC (1994)
1998	Multistate; fresh parsley	*S. sonnei*	CDC (1999), Naimi et al. (2003)
2000	Multistate; bean dip	*S. sonnei*	CDC (2000), Wetherington et al. (2000), Kimura et al. (2004)
2001	New York; tomato	*S. flexneri*	Reller et al. (2006)
2001	Japan; oysters	*S. sonnei*	Terajima et al. (2004)

passage, as exemplified by moose soup prepared by a recovering individual in Galena, AL, who contaminated the food and infected 25 other people. In some cases, rapid dissemination of the pathogen occurred, causing additional cases, not only within the immediate population but also within a wider population, thus elevating the secondary attack rate. Furthermore, with increasing globalization of commerce, the amount of foods being transported from one country to another has increased. In the context of food-borne illnesses, this raises the issue of how to ensure that imported foods are safe for the consumers.

Although a wide range of food sources have been implicated as the source of *Shigella*, produce commodities have been a major route of dissemination in developed countries. This has become a much more acute issue as food has become an integral part of the global economy. Standards for food safety seem to vary greatly around the globe, and unless each shipment of food imported into or exported from one country to another is sampled, the prospect of microbial food contamination still persists.

One of the striking features about food-borne outbreaks caused by shigellae is that often contamination of foods may not have originated at the processing plant but rather from either a food handler or feces-containing water used for irrigation or used at other points in food processing. Foods are not routinely tested for the presence of *Shigella*, and testing is done usually if a food is suspected of being contaminated.

Surveillance of food-borne illnesses caused by *Shigella* spp. continues in many countries, with many reporting to a central repository at the World Health Organization. In the United States, regulatory agencies, such as the Food and Drug Administration, monitor food supplies through several different mechanisms. One effort was the initiation of a surveillance program to monitor imported and domestically grown produce. For instance, the U.S. FDA issued a Domestic and Imported Produce Assignment for sampling specified produce commodities starting in 1999 (FDA survey of imported fresh produce, FY 1999 field assignment; http://www.cfsan.fda.gov/~dms/prodsur6.html) and has continued this program to the present. In the first survey, samples were collected specifically from high-volume imported fresh produce from 21 countries, as well as produce grown in the United States.

Produce that were of interest to the U.S. FDA included broccoli, cantaloupe, celery, cilantro, looseleaf lettuce, parsley, scallions (green onions), strawberries, and tomatoes. Most of the commodities collected were analyzed for *Salmonella, Shigella,* and *E. coli* O157:H7. Out of the 1,003 imported samples

that were collected, 44 samples (4.5%) were found to be contaminated with either *Shigella* (1.0%; 9/1,003) or *Salmonella* (3.5%; 35/1,003); no *E. coli* O157:H7 was isolated from any sample. In one case, an investigation of a farm in which *Shigella* was found in parsley revealed that the well water used for cooling and washing the produce was contaminated with this pathogen. Contamination most likely occurred by an influx of water from a river in close proximity to the wells. Not uncommonly, *Shigella* spp. are found in water polluted with human feces. Subsequent findings of the U.S. FDA domestic and imported surveys can be found at http://www.cfsan.fda.gov/~dms/prodsur7.html (2001 assignment) and http://www.cfsan.fda.gov/~dms/prodsur9.html (2000 domestic interim results). Data from the 2005 import assignment, in which 327 samples were analyzed, showed that 8 samples were found to be positive with either *Salmonella* (5 samples) or *Shigella* (3 samples); none were found for *E. coli* O157:H7 (http://intranet.cfsan.fda.gov/OC/pages/programs/impstatR.pdf). In the latest posted results for 2006 (http://intranet.cfsan.fda.gov/OC/pages/programs/impsumry.pdf), of the 486 samples that were tested, 7 were positive for *Salmonella* and 1 was positive for *Shigella*.

Environment

In addition to analyzing food directly for microbial pathogens, the United States does monitor recreational water, whether treated or nontreated. As reported in 2006 from a CDC surveillance report, from 1995 to 2004 there were 136 outbreaks attributed to *Shigella*, representing 9.6% of the total cases of gastrointestinal illnesses. From 2003 to 2004, shigellosis was reported for 56 cases from treated recreational (fountain) water, whereas 23 cases were reported for untreated lake water; all cases were due to *S. sonnei* (CDC, 2006a; http://www.cdc.gov/mmwr/preview/mmwrhtml/ss5512a1.htm?s_cid=ss5512a1_e).

In addition, since 1971, CDC, the U.S. Environmental Protection Agency (EPA), and the Council of State and Territorial Epidemiologists have collaborated to collect data in reference to the sources of waterborne disease outbreaks (http://www.epa.gov/nheerl/articles/2006/waterborne_disease/waterborne_outbreaks.pdf; Craun et al., 2006). In this report, shigellosis accounted for 16% of all illnesses associated with waterborne human pathogens from 1961 to 1970. During this time span, *Shigella* was one of the two most common etiological agents for waterborne disease outbreaks. In later reporting periods, the numbers dropped to 6% for the years 1971 to 1990 and subsequently to 4% for 1991 to 2002.

Waterborne outbreaks due to *Shigella* spp. are obviously not confined to the United States. Epidemiological data collected from several outbreaks in Greece indicate the widespread nature of this pathogen. In regions of Greece and other Mediterranean countries, the availability of chlorinated water may be scarce. In two reports covering the recent outbreaks over the past 3 decades, *S. sonnei* was responsible for all of the outbreaks reported (Alamanos et al., 2000; Koutsotoli et al., 2006).

Other countries have followed suit and recognized the importance of identifying environmental conditions that may influence the spread of pathogens such as *Shigella*. A report from New Zealand indicates that 18,000 to 34,000 cases occur each year; five outbreaks were reported in 2004. *Shigella* was listed as one of the principal microbial pathogens responsible for most of the cases of waterborne gastroenteritis in New Zealand (http://www.moh.govt.nz/moh.nsf/pagesmh/5821/$File/water-borne-disease-burden-prelim-report-feb07.doc).

Water- and food-borne diseases due to *Shigella* spp. have also been reported on cruise ships (Rooney et al., 2004). In two incidents, one in October 1989 and the other in August 1994, passengers and crew members reported having gastrointestinal symptoms; 14% of the passengers and 3% of the crew members from the 1989 outbreak reported having diarrhea (Lew et al., 1991), and 37% of the passengers and 4% of the crew from the 1994 outbreak reported having diarrhea, with one death occurring (CDC, 1994). A multiple-antibiotic-resistant strain of *S. flexneri* 2a was isolated from several ill passengers and crew members in 1989, and the source of this outbreak was identified as German potato salad. Contamination was introduced by infected food handlers, initially in the country where the food was originally prepared and second, only by a member of the galley crew on the cruise ship. In August 1994, *S. flexneri* 2a was isolated from a patient, and the suspected source of contamination was spring onions (http://www.publichealthreports.org/userfiles/119_4/119435.pdf).

One of the lessons learned from a recent *E. coli* O157:H7 outbreak related to consumption of spinach (CDC, 2006b) was to identify potential sources of contamination for food commodities that are presented to the consumer in their raw state. Besides good manufacturing practices, good agricultural practices (GAP), and hazard analytical critical control point (HACCP), attention to details in all the growing and processing steps is required for reducing the advent of a food-related outbreak. This includes following GAP and good manufacturing practices (GMP) during the growth, harvesting,

transporting, and processing stages. In addition, in the processing plants, HACCP principles can be applied.

Data from the CDC on reported food-borne cases in the United States show a decline in the number of illnesses due to *Shigella* spp. over the past few years (CDC, 2007b). Partially, this can be attributed to better vigilance on the part of the regulatory agencies, as well as an increased scrutiny by the produce growers and the food industry overall.

INTRINSIC AND EXTRINSIC FACTORS THAT AFFECT SURVIVAL AND GROWTH IN FOOD PRODUCTS AND CONTRIBUTE TO OUTBREAKS

Laboratory Conditions

The ability of *Shigella* spp. to either grow or survive under a myriad of laboratory conditions in broth cultures is obviously influenced by temperature and pH, the latter also dependent on the type of acid supplemented in the medium. Various studies have addressed the impact of these environmental concerns on how storage conditions of foods can affect the growth or survival of the pathogen. *Shigella* spp. are considered more acid tolerant than some other enteric bacteria, and this characteristic may enable these pathogens to transit through the acidic conditions of the human digestive system. Yet, under some laboratory conditions and in fecal samples, shigellae are unable to grow and do not survive well in an acidic environment.

Growth of *Shigella* spp. in broth medium is observed over a range of temperatures. *S. sonnei* can grow at temperatures as low as 6°C and *S. flexneri* at 7°C and up to 48°C. Under laboratory conditions, *S. sonnei* and *S. flexneri* grow in culture media with nearly the same pH values, between 4.5 and 9.3. Under certain culture conditions with media supplemented with organic compounds, such as formic or acetic acid, salts (3.8 to 5.2% NaCl), or nitrite (300 to 700 ppm), no growth of shigellae is observed. Although no growth is observed, survival of shigellae can be extended, depending on the temperature; lower temperature appears to support the survival time of *S. flexneri* in broth cultures at different pH values with different organic acids (Zaika, 2002a). Salt (NaCl) concentration and its effect on the growth and survival of *S. flexneri* and *S. sonnei* are greatly influenced by temperature and pH. In media with a pH of 4, no growth is observed and survival of the bacteria declines. With brain heart infusion medium, growth for *S. sonnei* and *S. flexneri* was observed at a minimum pH of 4.50 and 4.75, respectively (Bagamboula et al., 2002a).

Survival and Growth in Foods

Smith (1987) summarized which foods have commonly been associated with food-borne outbreaks caused by *Shigella* spp. Recently, efforts have been directed to determine how well this pathogen grows and survives in foods, with specific attention to produce/vegetables. Growth can occur in foods under the right environmental conditions, including temperatures above suggested proper storage settings, nondeleterious pH values, and lack of chemical and biological inhibitors. *Shigella* spp., as will be exemplified below, can survive in a broad milieu of food matrices. For instance, *S. sonnei* and *S. flexneri* can survive at 4°C for 21 days in foods commonly implicated in outbreaks, such as cheese, potato salad, and mayonnaise. In foods, survival time is quite different at −20°C, 4°C, and room temperature and with a brief exposure to 80°C; *S. flexneri* and *S. sonnei* can survive for the longest period of time at room temperature. In acidic foods, such as citrus juices and carbonated soft drinks, the survival time for *S. flexneri* and *S. sonnei* is from 4 h to 10 days (ICMSF, 1996). In a more detailed analysis, *S. sonnei* and *S. flexneri* did not grow at 22°C in either apple juice, pH 3.3 to 3.4, or tomato juice, pH 3.9 to 4.1 (Bagamboula et al., 2002a). Other examples include citric juices (orange, grape, and lemon), carbonated beverages, and wine; variable recovery of shigellae was obtained after 1 to 6 days (ICMSF, 1996). In neutral pH foods, such as butter or margarine, shigellae can be recovered after 100 days when stored frozen or at 6°C. *S. dysenteriae*, tested in orange juice at 4°C and grape juice at 20°C, survived up to 170 hours and 2 to 28 hours, respectively (ICMSF, 1996).

Depending upon the in vitro conditions, *Shigella* sp. can survive in media with a pH range of 2 to 3 for several hours. Acid resistance seems to be modulated by a sigma factor encoded by the *rpoS (katF)* gene, while at least two other genes, *hdeA* and *gadC*, are involved (Waterman and Small, 1996). Few studies have examined the survival and growth of *Shigella* in foods commonly associated with food-borne outbreaks. In order to address these concerns, a variety of produce was tested with regard to their effect on the growth and survival of *Shigella* (Rafii et al., 1995; ICMSF, 1996; Zaika and Scullen, 1996; Rafii and Lunsford, 1997; Bagamboula et al., 2002b). Although different strains, inoculation numbers, incubation conditions, and recovery/enumeration parameters were reported in these studies, some data were reported that indicate how these pathogens may survive or grow in foods. In most vegetables and other types of foods tested, such as carrot, cauliflower, radish, celery, broccoli, green pepper, onion, cabbage, strawberry, fresh fruit salad, pea, and broths (beef, chicken, and vegetable), *Shigella* survived for more than 10 days at refrigerated temperatures (4 to 10°C).

Composition of the food matrix and resident microbial flora had a pronounced effect on survival. For example, data on the recovery of *S. flexneri* or *S. sonnei*, seeded onto strawberries (pH 3.47) and fresh fruit salad (pH 3.73) and incubated at 4°C for 4 to 48 h varied. *S. sonnei* was recovered from both commodities, whereas *S. flexneri* was isolated only from fruit salad and not strawberries. In some cases, lactic acid bacteria were identified as part of the microbial flora. Under conditions favoring growth, it was speculated that these bacteria lowered the pH of the environmental milieu and may have produced conditions unfavorable for both growth and survival of *Shigella*. As noted above, *Shigella* spp. do not grow in broth medium when the pH is less than 4.5. Growth of *Shigella* at elevated temperatures (12 to 37°C) was mixed. In most cases, growth was observed, and in some foods (e.g., carrots), initial growth did occur and was followed by a significant decline in the number of recovered cells (Bagamboula et al., 2002b). A recent paper (Warren et al., 2007) examined survival of *S. sonnei* on three different foods: tomato (smooth surface), potato salad, and raw ground beef. Under normal storage conditions for tomatoes and refrigerated temperatures (2 to 8°C) for potato salad and ground beef, no *S. sonnei* was recovered from tomatoes after 2 days, whereas with the other two commodities, no significant decline in the number of *S. sonnei* colony-forming units recovered was observed compared to the initial inoculation level.

Survival of shigellae in foods under acidic conditions depends on temperature and type of acid. *Shigella* can survive a temperature range of −20°C to room temperature. Survival of shigellae was longer in foods stored frozen or at refrigeration temperatures than at room temperature. In foods, such as salads with mayonnaise and some cheese products, *Shigella* survived for 13 to 92 days. These organisms can survive on dried surfaces for an extended period of time and in foods (e.g., shrimp, ice cream, and minced pork meat) stored frozen. Growth and survival rates of *Shigella* are impeded in the presence of 3.8 to 5.2% NaCl at pH 4.8 to 5.0, in 300 to 700 mg/liter $NaNO_2$ and in 0.5 to 1.5 mg/liter of sodium hypochlorite (NaClO) in water at 4°C. Data from Zaika (2002b) indicated that *S. flexneri* survived well in broth with pH values higher than 4, with NaCl concentrations from 0.5% to 7%, and at temperatures of 12°C, 19°C, 28°C, and 37°C. They are sensitive to ionizing radiation, and a reduction of 10% at 3 kGy is observed.

Factors that affect growth and survival of *Shigella* spp. in foods include not only pH, water activity, salt concentration, and temperature, but also the chemical components of the food matrix and the indigenous microbial flora. Understanding the dynamics of the aforementioned factors, either as individual or collective components, can provide more in-depth insight into the effect that these environmental parameters have on the number of *Shigella* spp. present in food during short or prolonged storage. Whether these conditions lead to either an increase in the total count of the pathogen or the maintenance of sufficient numbers of infectious agents, knowing the potential for contamination is important to prevent contaminated foods from reaching the consumer. For example, in some cases of shigellosis caused by the ingestion of contaminated produce, foods are rendered unsanitary at the site of production/harvest by fecal-laden irrigated water or poor sanitary hygienic practices by those responsible for harvest or by a food handler postharvest. This has been noted both with a broad stroke with other microbial pathogens (Tyrrel et al., 2006) and with specific bacteria, such as *E. coli* O157:H7 (Wachtel et al., 2002).

From several studies, it is apparent that *Shigella* spp. survive well in most foods; the ability to increase in numbers may not be a significant point, considering that the infectious dose is 10 to 200 organisms and the number of shigellae on the foods may greatly exceed this figure. From epidemiological data, it becomes obvious how one contaminated commodity can transverse the commercial food process maze and affect a great number of unsuspecting consumers. Therefore, can methods be developed that would eradicate pathogens from foods yet retain the quality of that particular commodity? As simple an idea as this is, the concept of actually implementing such a process becomes a complex issue. The food industry and regulatory agencies must grapple with whatever system is developed to reach a satisfactory approach that meets the demands of all concerned parties, including the health and safety of the consumer.

Recovery of shigellae is a daunting process, as indicated by the CDC data detailing the low number of confirmed sources of food-borne outbreaks (CDC, 2007a). For *Shigella*, the physiological state of the pathogen may be a significant factor in why shigellae are not likely to be isolated from foods. In addition, as *Shigella* may be present in low numbers and/or stressed/injured in the suspected food samples, recovery may require special enrichment procedures for successful isolation of *Shigella* (Andrews, 1989).

FOOD PROCESSING OPERATIONS THAT INFLUENCE THE NUMBERS, SPREAD, OR CHARACTERISTICS

Over the past few years, the number of laboratory-confirmed cases of food-borne illnesses caused by *Shigella* spp. has consistently marked this pathogen as the third leading cause of bacterial food-borne diseases in the United States (http://www.cdc.gov/mmwr/preview/mmwrhtml/mm5614a4.htm; CDC, 2007b). Of the 17,252 cases identified through FoodNet surveillance (CDC Foodborne Diseases Active Surveillance Network), 2,736 were attributed to *Shigella*, representing 1.91 incidences per 100,000 population; *Salmonella* and *Campylobacter* were the leading causes of food-borne illnesses reported by FoodNet (CDC, 2007a). Produce commodities have been a common vehicle for food-borne diseases caused by *Shigella* spp. worldwide (Smith, 1987; Kapperud et al., 1995; CDC, 1999).

One of the traditional approaches to address the problem of microbial contaminated foods in the processing plant is to inspect the final product. Several drawbacks to this approach have been elucidated (Hall, 1994), but the primary reason was the lack of effectiveness in sampling and analysis of the final product. In addition, current bacteriological methods are oftentimes consuming and laborious. An alternative to end product testing is the HACCP (http://vm.cfsan.fda.gov/~lrd/haccp.html) system, comprised of seven principles, including the responsibility of the food industry to identify certain points of the processing system that may be most vulnerable to microbial, chemical, and physical hazards. This approach would be instituted as a preventive program, with less reliance on end product testing. A monitoring system such as HACCP may be well suited for pathogenic bacteria, such as *E. coli* O157:H7 and *Salmonella* spp., that are known to be associated with specific foods (e.g., meats and egg products, respectively).

In contrast, establishing specific critical control points (CCP) for preventing *Shigella* contamination of foods is not always suitable under the HACCP concept. This pathogen is usually introduced into the food supply by an infected person, such as a food handler with poor personal hygiene. In some cases, this may occur at the manufacturing site but more likely happens at a point between the processing plant and the consumer. However, a CCP could be directed at preventing ill workers from entering the workplace or instituting proper sanitation.

Another factor is that foods, such as vegetables (e.g. lettuce and parsley), can be contaminated at the site of collection and shipped directly to market.

Although HACCP is a method for controlling food safety and preventing food-borne outbreaks, pathogens such as *Shigella*, which are not indigenous to, but rather introduced into, foods, are most likely to be undetected. An example of this problem was recently exemplified in a survey conducted in Guadalajara, Mexico, in which freshly squeezed, unpasteurized orange juice, sold by either street vendors or small grocery stores, and other samples, such as wiping cloths, were analyzed for the presence of *Shigella* (Castillo et al., 2006). Briefly, all four *Shigella* spp., including *S. dysenteriae,* were isolated from all types of samples (6 to 17%). Total aerobic and *E. coli* counts indicated that the samples were contaminated with fecal matter. Overall, HACCP would work at the processing plant but would be more difficult to implement at the site of production, where control over all CCPs may be more problematic.

RECENT ADVANCES IN BIOLOGICAL, CHEMICAL, AND PHYSICAL INTERVENTIONS TO GUARD AGAINST THE PATHOGEN

Introduction of shigellae into foods, particularly raw vegetables/produce, most likely occurs during processing, including irrigation, harvesting, and hand packaging. Means to control microbial pathogen contamination may involve chemical treatment or irradiation. Although public attention has been drawn to the control of other microbial pathogens, such as *Salmonella* and *E. coli* O157:H7, foods potentially contaminated with *Shigella* could also be successfully treated with ionizing radiation. Debate over whether there is an adverse effect on the "taste" of the raw commodity after irradiation or over the cost of implementing irradiation on a larger scale may lessen the appeal of this process to reduce the number of shigellae in foods. Irradiation of foods is supported by U.S. government agencies, such as the FDA (FDA's Center for Food Safety and Applied Nutrition; "Food Irradiation: a Safe Measure," http://vm.cfsan.fda.gov/~dms/opa-bckg.html) and the CDC (http://www.cdc.gov/ncidod/dbmd/diseaseinfo/foodirradiation.htm), and at the international level by the Codex Alimentarius Commission (Codex, 2003a, 2003b; www.Codexalimentarius.net).

In the United States and other countries (e.g., Canada [http://www.foodproductiondaily.com/news/ng.asp?id=28788-canada-extends-food]), the application of food irradiation is slowly being accepted by consumers. In contrast, in Europe, acceptance is moving more slowly. Irradiating foods is an accepted process in 40 countries, but only 10 European Union members have approved facilities for food processing (http://www.foodproductiondaily.com/news/printNewsBis.asp?id=77092). The U.S. FDA has approved the use of irradiation for meats, poultry, fresh fruits, vegetables, and spices (21 CFR 179.26; http://www.fda.gov/opacom/catalog/irradbro.html). A recent consumer report on the application of irradiation of fresh produce covers many pertinent topics ranging from food safety to the question of whether there are any adverse effects of irradiation of foods (Groth, 2007). Another approach is chemical applications with acetic acid or chlorinated water. These have been shown to be effective, simple methods to reduce or eliminate the pathogen load in raw vegetables, as demonstrated with cut parsley artificially inoculated with *S. sonnei* (Wu et al., 2000).

Although the CDC has reported that food-borne outbreaks of shigellosis in the United States have declined recently, shigellosis continues to be a major public health concern. In compiling data from 1982 to 1997, which included reported cases (from 1983 to 1992) and the number of cases from passive (1992 to 1997) and active surveillance via FoodNet (http://www.cdc.gov/ncidod/dbmd/foodnet), the CDC estimated that there were approximately 448,000 cases of shigellosis in the United States, the third leading cause of food-borne outbreaks by bacterial pathogens (CDC, 2005).

The FDA published a GAP document and the *FDA's Guide to Minimize Microbial Food Safety Hazards for Fresh Fruits and Vegetables* in 2007 (http://www.foodsafety.gov/~dms/prodguid.html). In June 2004, the FDA further committed to addressing the problem by releasing an action plan to reduce produce-related illness, *Produce Safety from Production to Consumption: 2004 Action Plan to Minimize Food-borne Illness Associated with Fresh Produce Consumption* (http://www.cfsan.fda.gov/~dms/prodpla2.html). The plan had several objectives, including (i) preventing contamination of fresh produce with pathogens; (ii) minimizing the public health impact when contamination of fresh produce occurs; (iii) improving communication with producers, preparers, and consumers about fresh produce; and (iv) facilitating and supporting research relevant to fresh produce. For each objective, FDA's action plan identifies steps that could contribute to the achievement of the objective (http://www.cfsan.fda.gov/~dms/prodpla2.html).

A point of emphasis is that foods are not routinely examined for the presence of *Shigella* unless epidemiological data suggest that it may be the source of an outbreak. Vegetables and salads are commonly contaminated sources (Smith, 1987). U.S. government regulatory departments and agencies have implemented programs to test imported and domestic produce for the presence of select bacterial agents as a means to obtain baseline data and to take regulatory action when warranted. A survey of how extensive this problem is and how the U.S. FDA is addressing this issue has been published (Beru and Salsbury, 2002; www.cfsan.fda.gov/~dms/ prodact.html).

Discriminative Detection Methods for Confirmation and Trace-Back of Contaminated Produce

Methods used to isolate, detect, and identify *Shigella* spp. from foods can be divided into three major categories: conventional bacteriological, nucleic acid based, and biosensors. Presently, there is not one definitive method that has the robustness, rapidity, and efficacy to be effective in isolating *Shigella* from foods. Clinical samples, which reflect a much smaller number of matrices than foods, pose a challenge to laboratory personnel isolating *Shigella* (e.g., from feces). Currently, no enrichment medium exists to selectively grow *Shigella* from either food or fecal samples. This severely hampers the ability of laboratories to isolate *Shigella* from foods, particularly when this pathogen competes with the indigenous flora present in foods. Most methods entail growth in broth medium which is followed by plating on selective agars. In the *Bacteriological Analytical Manual* (Andrews and Jacobsen, 2001; www.cfsan.fda. gov/~ebam/bam-6.html), the authors detail one method to isolate shigellae from foods. Health Canada (http://www.hc-sc.gc.ca/fn-an/res-rech/analy-meth/ microbio/index_e.html) uses a very similar scheme, with slight modifications. A protocol from the International Organization of Standards (ISO 21567:2004, Microbiology of food and animal feeding stuffs— Horizontal method for the detection of Shigella spp.) offers another method.

A range of selective agar media is recommended to plate cultures after growth overnight in broths. Two or three different selective media should be used to increase the chance of recovering *Shigella*. Growth of *Shigella* on MacConkey agar, a low-selectivity medium, is used to screen for lactose-negative colonies (*Shigella* are lactose negative). Eosin methylene blue and Tergitol-7 agar are alternative low-selectivity agars. Desoxycholate and xylose-lysine-desoxycholate agars are intermediate selective media and are preferred media to isolate *Shigella* spp. Although most *Shigella* spp. do not ferment xylose, some species, e.g., *S. boydii*, have variable reactions and may be missed. Highly selective media include *Salmonella-Shigella* and Hektoen agars. Some *Shigella* spp., such

as *S. dysenteriae* type I, are unable to grow on highly selective *Salmonella-Shigella* medium.

Media containing chromogenic or fluorogenic indicators have been applied to isolating and detection regimens for a number of pathogens, notably *E. coli* O157:H7 and *Salmonella* spp. As for *Shigella*, one medium containing a chromogenic indicator has been tested and should be commercially available in the near future. In that study chromogenic agar was used with tomatoes artificially contaminated with *S. boydii* and *S. sonnei* (Warren et al., 2005). However, this medium fares poorly with foods having a high number of indigenous microbial populations.

Confirmation is usually performed using biochemical tests, either automated or commercially available biochemical strips. Suspected colonies that are gram-negative, nonmotile rods are inoculated onto lysine iron or Kliger iron agar and incubated 18 to 24 hours at 37°C. *Shigella* spp. produce alkaline (red) slants, acid (yellow) butt, no H_2S, and no gas, with the exception of *S. flexneri* serotype 6, on these agars. Similar to other enteric bacteria, *Shigella* spp. are oxidase negative, ferment glucose, and, except for *S. dysenteriae* type 1, are catalase positive. Further biochemical characterizations show that *Shigella* spp. are negative for H_2S production, phenylalanine deaminase, and sucrose and lactose fermentation (although *S. sonnei* may after a long period of incubation); do not utilize citrate, acetate, potassium cyanide, malonate, inositol, adonitol, and salicin; and lack lysine decarboxylase. Shigellae are negative for the Voges-Proskauer test (*S. sonnei* and *S. boydii* serotype 13 are positive); however, all shigellae are methyl red positive and are unable to produce acid from glucose and other carbohydrates (acid and gas production occurs with *S. flexneri* serotype 6, *S. boydii* serotypes 13 and 14, and *S. dysenteriae* 3). One *Shigella* sp., *S. dysenteriae*, is catalase negative and has ornithine decarboxylase activity.

These biochemical tests and those presented in several laboratory protocols aim to differentiate *Shigella* spp. from *E. coli* and also to distinguish *Shigella* spp. from each other. Growth on Christensen citrate, sodium mucate, or acetate agar is one characteristic that discriminates between *E. coli* and *Shigella* spp.; shigellae are unable to utilize citrate, acetate, or mucate as a sole carbon source. Other biochemical tests are used to identify the serotypes of *Shigella*. The ability to utilize mannitol, dulcitol, xylose, rhamnose, raffinose, glycerol, and indole and the presence of ornithine decarboxylase and o-nitrophenyl-β-D-galactopyranosidase have been used to physiologically discriminate between *Shigella* spp.

Serological testing using polyvalent antiserum is used to identify the *Shigella* groups A to D. A note of caution should be taken. EIEC causes the same disease, bacillary dysentery, as do the shigellae. Some O-antigen structures of EIEC strains share homology with O-antigen structures of some *Shigella* serotypes (Tulloch et al., 1973). Several serotypes of *S. dysenteriae*, *S. flexneri*, and *S. boydii* have reciprocal cross-reactivity with *E. coli* O antigens of the Alkalescens-Dispar bioserogroup or EIEC.

Although *S. sonnei* has only one serotype, two forms of this pathogen exist on agar plates. Form (or phase) I is virulent and has a smooth colony texture, whereas form II is irreversibly avirulent and has a rough colony phenotype on agar surfaces. Both forms carry distinct antigens; therefore, antiserum targeting both forms should be applied to identify immunologically *S. sonnei* isolates.

DNA-based assays

Several types of DNA-based assays have been applied to detect *Shigella* spp. in foods. Initially, DNA probes and PCR were developed to detect *Shigella* from clinical samples. PCR, an in vitro amplification system, is a much more sensitive means of detecting the presence of microbial pathogens than conventional bacteriology and the use of DNA probes. As for using a PCR-based assay to detect *Shigella* in foods, primers are selected that target one specific virulence gene, *ipaH* (Wang et al., 1997). The advantage of these primers is that the *ipaH* gene is present in multiple copies in the virulence plasmid (five copies) and chromosome (seven genes/homologues; cognates) of *Shigella* (Jin et al., 2002; Wei et al., 2003). Hence, the 5 to 12 target sequences increase the sensitivity of the assay compared to primers targeting just one copy of a gene. Also, in instances when the large virulence plasmid is lost, these particular primers do target copies of the *ipaH* gene in the chromosome, ensuring an opportunity for a successful reaction. Unlike other methods that may take several days to draw a conclusion regarding the presence of *Shigella* in foods, PCR assays can yield a result in less than 1 day. One potential problem for PCR-based assays is the presence of inhibitors of the reaction deriving from the food matrix. To ensure that negative results are truly that, control reactions using washes or homogenates with target DNA added, are essential for accuracy. Therefore, template preparation is one of the key steps in ensuring the correct result in analyzing foods by PCR.

Two approaches are being entertained, namely, whether or not to include an enrichment step prior to PCR. Several manufacturers provide kits, and most, if not all, require enrichment in broth. This added step may be sufficient for some pathogens, but shigellae pose a more difficult task. Injured, weak, viable but nonculturable cells may persist in the population and may not be resuscitated by enrichment. Therefore, false-negative reactions may not be truly indicative of contamination. One alternative to enrichment uses a filter that lyses the bacterial cell and sequesters the DNA in the membrane and can be used directly as a template in the reaction (Lampel et al., 2000). Real-time PCR has the advantage of reducing the time for analysis. There are several different platforms that are available, each with their own strengths. In each system, the amplification of PCR products can be monitored in real time, and in some cases, an extra step, such as using a probe within the reaction, can increase the specificity of the reaction in a very short period of time. Also, much of this type of analysis is automated, meaning less chance of mistakes and reduced labor. As with conventional PCR, the same problems exist; templates in sufficient quantity and free of PCR inhibitors prepared from all types of food matrices are critical. Most of these protocols require enrichment in broth for 6 to 18 h.

Another aspect of identifying causative agents of food-borne outbreaks of diarrheal disease is the ability to link the people affected in one region to other geographical locales and ultimately to the source of contamination. In this regard, pulse-field gel electrophoresis surpasses most other nucleic acid-based typing methods. The CDC has established several surveillance programs to monitor food-borne outbreaks in the United States. These include PulseNet (http://www.cdc.gov/pulsenet/) and Food-Net (http://www.cdc.gov/foodnet/); others can be found at http://www.cdc.gov/ and, specifically for *Shigella*, at http://www.cdc.gov/ncidod/dbmd/phlis data/shigella.htm. The World Health Organization (WHO; http://www.who.int/emc/) also monitors outbreaks and publishes some of their findings in the *Weekly Epidemiological Record* (http://www.who.int/wer/index.html). Ribotyping, another means of subtyping microbial pathogens that uses gel electrophoresis patterns of PCR-amplified RNA genes, subsequently digested with a specific endonuclease, has been used to establish a database to type and identify *Shigella* strains (Coimbra et al., 2001).

Other technologies

An alternative to agarose gel electrophoresis is the application of an enzyme-linked immunosorbent assay at the conclusion of a PCR run to detect the *ipaH* gene of *Shigella* spp. from stool specimens (Sethabutr et al., 2000). Immunomagnetic separation technology, in which beads coated with antibodies specific to *Shigella*, has been tested using clinical samples. Its application for foods has been limited, although more commercially available instruments that use immunomagnetic separation are now marketed. One limiting step is the incorporation of an antibody or antibodies that target each *Shigella* serotype and subserotypes.

Technological improvements in the field of biosensors have been significant. The sensitivity of these instruments is approaching the level of molecular-based methods, such as PCR. In one report, *Shigella* was artificially inoculated into buffer and several food and beverage samples. The limit of detection for *S. dysenteriae* in both buffer and chicken carcass wash was 4.9×10^4 CFU/ml. This figure was obtained from direct wash; no preenrichment was performed (Sapsford, et al., 2004).

CONCLUDING REMARKS

Diseases caused by the four species of *Shigella* remain a global health issue, whether in developed or developing countries. Epidemic and pandemic outbreaks are attributed to these pathogens and are particularly acute in human environments that are crowded, have poor sanitation, and present an appropriate milieu for facilitating rapid transmission of shigellae via the fecal-oral route. Children under the age of 5 years often have the highest mortality rate in developing countries, usually for the aforementioned conditions. The emergence of multiple antibiotic resistance strains poses another formidable challenge to health care workers treating bacillary dysentery. Although considered a self-limiting disease, in susceptible populations *Shigella* spp. are responsible for nearly 1.1 million deaths worldwide each year. WHO and public health researchers around the globe have recognized the necessity for a vaccine against *Shigella* (WHO, 1997; http://www.who.int/docstore/wer/pdf/1997/wer7211.pdf; http://www.who.int/vaccine_research/diseases/shigella/en/). Yet an effective vaccine that could substantially influence the mortality rate in developing countries is not available.

Although *Shigella* spp. can be routinely grown in laboratory conditions, the ability to isolate this pathogen from foods may be challenging. Even in fecal samples in which ample numbers of shigellae are shed in infected individuals, the fact remains that these bacteria are fastidious and do not survive well in fecal material; hence, repeated calls are made for development of better transport medium in clinical

laboratories. Foods present a different scenario. Food matrices have diverse effects on the ability of the pathogen to either grow or survive. In addition, the matrices pose a difficult challenge to the food analyst attempting to isolate and even detect the presence of *Shigella* in foods. An effective means of isolating and detecting the presence of *Shigella* in foods remains elusive. However, recent advances in instrument technology are now at the stage of assessing these instruments for analysis of different environmental matrices. Technology of today can be the basis of instruments in the near future that will result in analysis being completed in real time and being automated and portable, two assets that will definitely impact food safety and food defense.

REFERENCES

Alamanos, Y., V. Maipa, S. Levidiotou, and E. Gessouli. 2000. A community waterborne outbreak of gastro-enteritis attributed to *Shigella sonnei. Epidemiol. Infect.* **125**:499–503.

Andrews, W. H. 1989. Methods for recovering injured classical enteric pathogenic bacteria (*Salmonella, Shigella*, and enteropathogenic *Escherichia coli*) from foods, p. 55–113. *In* B. Ray (ed.), *Injured Index and Pathogenic Bacteria: Occurrence and Detection in Foods, Water, and Feeds*, CRC Press, Boca Raton, FL.

Andrews, W. H., and A. Jacobsen. 2001. Chapter 6, *Shigella*, p. 6.01–6.06. *In Bacteriological Analytical Manual*, 8th ed. International AOAC, Gaithersburg, MD.

Bagamboula, C. F., M. Uyttendaele, and J. Debevere, J. 2002a. Acid tolerance of *Shigella sonnei* and *Shigella flexneri. J. Appl. Microbiol.* **93**:479–486.

Bagamboula, C. F., M. Uyttendaele, and J. Debevere. 2002b. Growth and survival of *Shigella sonnei* and *S. flexneri* in minimal processed vegetables packed under equilibrium modified atmosphere and stored at 7°C and 12°C. *Food Microbiol.* **19**:529–536.

Banish, L. S., R. Sims, M. Bush, D. Sack, and R. J. Montali. 1993. Clearance of *Shigella flexneri* carriers in a zoologic collection of primates. *J. Am. Vet. Med. Assoc.* **203**:133–136.

Bean, N. H., and P. M. Griffin. 1990. Foodborne disease outbreaks in the United States, 1973–1987: pathogens, vehicles, and trends. *J. Food Protect.* **53**:804–817.

Bennish, M. L. 1991. Potentially lethal complications of shigellosis. *Rev. Infect. Dis.* **13**(Suppl 4):S319–S324.

Beru, N., and P. Salsbury. 2002. FDA's produce safety activities. *Food Saf. Mag.* **8**:14–19.

Brenner, D. J., G. E. Fanning, K. E. Johnson, R. V. Citarella, and S. Falkow. 1969. Polynucleotide sequence relationships among members of the *Enterobacteriaceae. J. Bacteriol.* **98**:637–650.

Brewerton, D. A., M. Caffrey, A. Nicholls, D. Walters, J. K. Oates, and D. C. O. James. 1973. Reiter's disease and HLA-B27. *Lancet* **i**:996–998.

Bunning, V. K., R. B. Raybourne, and D. L. Archer. 1988. Foodborne enterobacterial pathogens and rheumatoid disease. *J. Appl. Bacteriol. Symp.* **65**(Suppl.):87S–107S.

Castillo, A., A. Villarruel-Lopez, V. Navarro-Hidalgo, N. E. Martinez-Gonzalez, and M. R. Torres-Vitela. 2006. *Salmonella* and *Shigella* in freshly squeezed orange juice, fresh oranges, and wiping cloths collected from public markets and street booths in Guadalajara, Mexico: incidence and comparison of analytical routes. *J. Food Prot.* 69:2595–2599.

CDC. 1994. Outbreak of *Shigella flexneri* 2a infections on a cruise ship. *MMWR Morb. Mortal. Wkly. Rep.* 43:657.

CDC. 1999. Outbreaks of Shigella sonnei infection associated with eating fresh parsley—United States and Canada. July-August, 1998. *MMWR Morbid. Mortal. Wkly Rep.* 48:285–289.

CDC. 2000. Public health dispatch: outbreak of *Shigella sonnei* infections associated with eating a nationally distributed dip—California, Oregon, and Washington, January 2000. *MMWR Morb. Mortal.Wkly. Rep.* 49:60–61.

CDC. 2005. Preliminary FoodNet data on the incidence of infections with pathogens transmitted commonly through food—10 sites, United States. *MMWR Morb. Mortal.Wkly. Rep.* 54:352–356.

CDC. 2006a. Surveillance for waterborne disease and outbreaks associated with recreational water—United States, 2003–2004. *MMWR Morb. Mortal. Wkly. Rep.* 55:1–24.

CDC. 2006b. Ongoing multistate outbreak of *Escherichia coli* serotype O157:H7 infections associated with consumption of fresh spinach—United States, September 2006. *MMWR Morbid. Mortal. Wkly Rep.* 55:1045–1046.

CDC. 2007a. Preliminary FoodNet data on the incidence of infection with pathogens transmitted commonly through food—10 States, 2006. *MMWR Morb. Mortal. Wkly. Rep.* 56:336–339.

CDC. 2007b. Summary of notifiable diseases—United States, 2005. *MMWR Morb. Mortal. Wkly. Rep.* 54:2–92.

Cilia, V., B. Lafay, and R. Christen. 1996. Sequence heterogeneities among 16S ribosomal RNA sequences, and their effect on phylogenetic analyses at the species level. *Mol. Biol. Evol.* 13:451–461.

Codex. 2003a. Codex general standard for irradiated foods. CODEX STAN 106-1983, Rev.1–2003. Codex Alimentarius Commission, Food and Agriculture Organization and World Health Organization, Rome, Italy.

Codex. 2003b. Recommended international code of practice for the radiation processing of food. CAC/RCP 19–1979, Rev.1–2003. Codex Alimentarius Commission, Food and Agriculture Organization and World Health Organization, Rome, Italy.

Coimbra, R. S., G. Nicastro, P. A. D. Grimont, and F. Grimont. 2001. Computer identification of *Shigella* species by rRNA gene restriction patterns. *Res. Microbiol.* 152:47–55.

Craun, M. F., G. F. Craun, R. L. Calderon, and M. J. Beach. 2006. Waterborne outbreaks reported in the United States. *J. Water Health* 4(Suppl. 2):19–30.

DuPont, H. L., M. M. Levine, R. B. Hornick, and S. B. Formal. 1989. Inoculum size in shigellosis and implications for expected mode of transmission. *J. Infect. Dis.* 159:1126–1128.

Edwards, P. R. and W. H. Ewing. 1972. Identification of *Enterobacteriaceae*. Burgess Publishing Co., Minneapolis, MN.

Falkow, S., H. Schneider, T. J. Magnani, and S. B. Formal. 1963. Virulence of *Escherichia-Shigella* genetic hybrids for the guinea pig. *J. Bacteriol.* 86:1251–1258.

Gessner, B. D., and M. Beller. 1994. Moose soup shigellosis in Alaska. *West. J. Med.* 160:430–433.

Goma Epidemiology Group. 1995. Public health impact of Rwandan refugee crisis: what happened in Goma, Zaire, in July, 1994? *Lancet* 345:339–344.

Groth, E. 2007. Food irradiation for fresh produce. The Organic Center critical issue report. www.organic-center.org.

Guerrero, L., J. J. Calva, A. L. Morrow, F. R. Velazquez, F. Tuz-Dzib, Y. Lopez-Vidal, H. Ortega, H. Arroyo, T. G. Cleary, L. K. Pickering, and G. M. Ruiz-Palacios. 1994. Asymptomatic *Shigella* infections in a cohort of Mexican children younger than two years of age. *Pediatr. Infect. Dis. J.* 13:597–602.

Hall, P. A. 1994. Scope for rapid microbiological methods in modern food production, p. 255–267. *In* P. Patel (ed.), *Rapid Analysis Techniques in Food Microbiology.* Blackie Academic & Professional, New York, NY.

Hedberg, C. W., W. C. Levine, K. E. White, R. H. Carlson, D. K. Winsor, D. N. Cameron, K. L. MacDonald, and M. T. Osterholm. 1992. An international food borne outbreak of shigellosis associated with a commercial airline. *JAMA* 268:3208–3212.

Hossain, M. A., K. Z. Hasan, and M. J. Albert. 1994. *Shigella* carriers among non-diarrhoeal children in an endemic area of shigellosis in Bangladesh. *Trop. Geogr. Med.* 46:40–42.

ICMSF. 1996. *Shigella*, p. 280–298. *Microorganisms in Foods* 5. Blackie Academic & Professional, London, United Kingdom.

Islam, M. S., M. A. Hossain, S. I. Khan, M. N. H. Khan, R. B. Sack, M. J. Albert, A. Hug, and R. R. Colwell. 2001. Survival of *Shigella dysenteriae* type 1 on fomites. *J. Health Popul. Nutr.* 19:177–182.

Jin, Q., Z. Yuan, J. Xu, Y. Wang, Y. Shen, W. Lu, J. Wang, H. Liu, J. Yang, F. Yang, X. Zhang, J. Zhang, G. Yang, H. Wu, D. Qu, J. Dong, L. Sun, Y. Xue, A. Zhao, Y. Gao, J. Zhu, B. Kan, K. Ding, S. Chen, H. Cheng, Z. Yao, B. He, R. Chen, D. Ma, B. Qiang, Y. Wen, Y. Hou, and J. Yu. 2002. Genome sequence of *Shigella flexneri* 2a: insights into pathogenicity through comparison with genomes of *Escherichia coli* K12 and O157. *Nucleic Acids Res.* 30:4432–4441.

Johnson, J. R. 1999. *Shigella* and *E. coli. ASM News* 65:460–461.

Kapperud, G., L. M. Rorvik, V. Hasseltvedt, E. A. Hoiby, B. G. Iversen, K. Staveland, G. Johnsen, J. Leitao, H. Herikstad, Y. Andersson, G. Langeland, B. Gondrosen, and J. Lassen. 1995. Outbreak of *Shigella sonnei* infection traced to imported iceberg lettuce. *J. Clin. Microbiol.* 33:609–614.

Karmali, M. A., M. Petric, C. Lim, P. C. Fleming, G. S. Arbus, and H. Lior. 1985. The association between idiopathic hemolytic syndrome and infection by verotoxin-producing *Escherichia coli. J. Infect. Dis.*151:775–782.

Kennedy, F. M., J. Astbury, J. R. Needham, and T. Cheasty. 1993. Shigellosis due to occupational contact with non-human primates. *Epidemiol. Infect.* 110:247–257.

Kimura, A. C., K. Johnson, M. S. Palumbo, J. Hopkins, J. C. Boase, R. Reporter, M. Goldoft, K. R. Stefonek, J. A. Farrar, T. J. Van Gilder, and D. J. Vugia. 2004. Multistate shigellosis outbreak and commercially prepared food, United States. *Emerg. Infect. Dis.* 10:1147–1149.

Kinsey, M. D., S. B. Formal, G. J. Dammin, and R. A. Giannella. 1976. Fluid and electrolyte transport in rhesus monkeys challenged intracecally with *Shigella flexneri* 2a. *Infect. Immun.* 14:368–371.

Kotloff, K. L., J. P. Winickoff, B. Ivanoff, J. D. Clemens, D. L. Swerdlow, P. J. Sansonetti, G. K. Adak and M. M. Levine. 1999. Global burden of *Shigella* infections: implications for vaccine development and implementation of control strategies. *Bull. W.H.O.* 77:651–666.

Koutsotoli A. D., M. E. Papassava, V. E. Maipa, and Y. P. Alamanos. 2006. Comparing *Shigella* waterborne outbreaks in four different areas in Greece: common features and differences. *Epidemiol. Infect.* 134:157–162.

Lampel, K. A., L. Kornegay, and P. A. Orlandi. 2000. Improved template preparation for PCR-based assays for the detection of food-borne bacterial pathogens. *Appl. Environ. Microbiol.* 66:4539–4542.

Lan, R., M. C. Alles, K. Donohoe, M. B. Martinez, and P. R. Reeves. 2004 Molecular evolutionary relationships of enteroinvasive *Escherichia coli* and *Shigella* spp. *Infect. Immun.* **72:**5080–5088.

Lan, R., and P. R. Reeves. 2002. *Escherichia coli* in disguise: molecular origins of *Shigella. Microbes Infect.* **4:**1125–1132.

Lee, L. A., S. M. Ostroff, H. B. McGee, D. R. Johnson, F. P. Downes, D. N. Cameron, N. H. Bean, and P. M. Griffin. 1991. An outbreak of shigellosis at an outdoor music festival. *Am. J. Epidemiol.* **133:**608–615.

Levine, M. M., and O. S. Levine. 1994. Changes in human ecology and behavior in relation to the emergence of diarrheal diseases, including cholera. *Proc. Natl. Acad. Sci. USA* **91:**2390–2394.

Lew, J. F., D. L. Swerdlow, M. E. Dance, P. M. Griffin, C. A. Bopp, M. J. Gillenwater, T. Mercatante, and R. I. Glass. 1991. An outbreak of shigellosis aboard a cruise ship caused by a multiple-antibiotic-resistant strain of *Shigella flexneri. Am. J. Epidemiol.* **134:**413–420.

Line, A. S., J. Paul-Murphy, D. P. Aucoin, and D. C. Hirsh. 1992. Enrofloxacin treatment of long-tailed macaques with acute bacillary dysentery due to multiresistant *Shigella flexneri* IV. *Lab. Anim. Sci.* **42:**240–244.

Lopez, E. L., M. Diaz, S. Grinstein, S. Dovoto, F. Mendila-harzu, B. E. Murray, S. Ashkenazi, E. Rubeglio, M. Woloj, M. Vasquez, M. Turco, L. K. Pickering, and T. G. Cleary. 1989. Hemolytic uremic syndrome and diarrhea in Argentine children: the role of Shiga-like toxins. *J. Infect. Dis.* **160:**469–475.

Luria, S. E., and J. W. Burrous. 1957. Hybridization between *Escherichia coli* and *Shigella. J. Bacteriol.* **74:**461–476.

Maurelli A. T., R. E. Fernandez, C. A. Bloch, C. K. Rode, and A. Fasano. 1998. "Black holes" and bacterial pathogenicity: a large genomic deletion that enhances the virulence of *Shigella* spp. and enteroinvasive *Escherichia coli. Proc. Natl. Acad. Sci. USA* **95:**3943–3948.

Murray, C. J., and A. D. Lopez. 1997. Mortality by cause for eight regions of the world: global burden of disease study. *Lancet* **349:**1436–1442.

Naimi, T. S. J. H. Wicklund, S. J. Olsen, G. Krause, J. G. Wells, J. M. Bartkus, D. J. Boxrud, M. Sullivan, H. Kassenborg, J. M. Besser, E. D. Mintz, M. T. Osterholm, and C. W. Hedberg. 2003. Outbreaks of *Shigella sonnei* and enterotoxigenic *Escherichia coli* infections associated with parsley: implications for surveillance and control of food borne illness. *J. Food Prot.* **66:**535–541.

Nizeyi, J. B., R. B. Innocent, J. Erume, G. R. Kalema, M. R. Cranfield, and T. K. Gracyzk. 2001. Campylobacteriosis, salmonellosis, and shigellosis in free-ranging human-habituated mountain gorillas of Uganda. *J. Wildl. Dis.* **37:**239–244.

Pickering, L. K., A. V. Bartlett, and W. E. Woodward. 1986. Acute infectious diarrhea among children in day-care: epidemiology and control. *Rev. Infect. Dis.* **8:**539–547.

Rafii, F., M. A. Holland, W. E. Hill, and C. E. Cerniglia. 1995. Survival of *Shigella flexneri* on vegetables and detection by polymerase chain reaction. *J. Food Prot.* **58:**727–732.

Rafii, F., and P. Lunsford. 1997. Survival and detection of *Shigella flexneri* in vegetables and commercially prepared salads. *J. AOAC Int.* **80:**1191–1197.

Raghupathy, P., A. Date, J. C. M. Shastry, A. Sudarsanam, and M. Jadhav. 1978. Haemolytic-uraemic syndrome complicating shigella dysentery in south Indian children. *Br. Med. J.* **1:**1518–1521.

Reller, M. E., J. M. Nelson, K. Mølbak, D. M. Ackman, D. J. Schoonmaker-Bopp, T. P. Root, and E. D. Mintz. 2006. A large, multiple-restaurant outbreak of infection with *Shigella flexneri* serotype 2a traced to tomatoes. *Clin. Infect. Dis.* **42:**163–169.

Rooney, R. M., J. K. Bartram, E. H. Cramer, S. Mantha, G. Nichols, R. Suraj, and E. C. D. Todd. 2004. A review

of outbreaks of waterborne disease associated with ships: evidence for risk management. *Public Health Rep.* **119:**435–442.

Rout, W. R., S. B. Formal, R. A. Giannella, and G. J. Dammin. 1975. Pathophysiology of *Shigella* diarrhea in the rhesus monkey: intestinal transport, morphological, and bacteriological studies. *Gastroenterology* **68:**270–278.

Sansonetti, P. J. 2006. The bacterial weaponry. Lessons from *Shigella. Ann. N. Y. Acad. Sci.* **1072:**307–312.

Sansonetti, P. J., T. L. Hale, and E. V. Oaks. 1985. Genetics of virulence in enteroinvasive *Escherichia coli*, p. 74–77. *In* L. Leive, P. F. Bonventre, J. A. Morello, S. Schlesinger, S. D. Silver, and H. C. Wu (ed.), *Microbiology—1985.* American Society for Microbiology, Washington, DC.

Sansonetti, P. J., J. Arondel, M. Huerre, A. Harada, and K. Matsushima. 1999. Interleukin-8 controls bacterial transepithelial translocation at the cost of epithelial destruction in experimental shigellosis. *Infect. Immun.* **67:**1471–1480.

Sapsford K. E., A. Rasooly, C. R. Taitt, and F. S. Ligler. 2004. Detection of *Campylobacter* and *Shigella* species in food samples using an array biosensor. *Anal. Chem.* **76:**433–440.

Sethabutr, O., M. Venkatesan, S. Yam, L. W. Pang, B. L. Smoak, W. K. Sang, P. Echeverria, D. N. Taylor, and D. W. Isenbarger. 2000. Detection of PCR products of the *ipaH* gene from *Shigella* and enteroinvasive *Escherichia coli* by enzyme linked immunosorbent assay. *Diagn. Microbiol. Infect. Dis.* **37:**11–16.

Sharp, T. W., S. A. Thornton, M. R. Wallace, R. F. Defraites, J. L. Sanchez, R. A. Batchelor, P. J. Rozmajzl, R. K. Hanson, P. Echeverria, A. Z. Kapikian, X. J. Xiang, M. K. Estes, and J. P. Burans. 1995. Diarrheal disease among military personnel during Operation Restore Hope, Somalia, 1992–1993. *Am. J. Trop. Med. Hyg.* **52:**188–193.

Simon, D. G., R. A. Kaslow, J. Rosenbaum, R. L. Kaye, and A. Calin. 1981. Reiter's syndrome following epidemic shigellosis. *J. Rheumatol.* **8:**969–973.

Smith, J. L. 1987. *Shigella* as a foodborne pathogen. *J. Food Prot.* **50:**788–801.

Strockbine, N., and A. T. Maurelli. 2005. *Shigella*, p. 811–823. *In* D. J. Brenner, N. R. Krieg, and T. E. Staley (ed.), *Bergey's Manual of Systematic Bacteriology.* Springer, New York, NY.

Terajima, J., K. Tamura, K. Hirose, H. Izumiya, M. Miyahara, H. Konuma, and H. Watanabe. 2004. A multi-prefectural outbreak of *Shigella sonnei* infections associated with eating oysters in Japan. *Microbiol. Immunol.* **48:**49–52.

Tulloch, E. F., K. J. Ryan, S. B. Formal, and F. A. Franklin. 1973. Invasive enteropathic *Escherichia coli* dysentery. *Ann. Intern. Med.* **79:**13–17.

Tyrrel, S. F., J. W. Knox and E. K. Weatherhead. 2006. Microbiological water quality requirements for salad irrigation in the United Kingdom. *J. Food Prot.* **69:**2029–2035.

Wachtel, M. R., L. C. Whitehead, and R. E. Mandrell. 2002. Association of *Escherichia coli* O157:H7 with preharvest leaf lettuce upon exposure to contaminated irrigation water. *J. Food Prot.* **65:**18–25.

Wang, R.-F., W. -W. Cao, and C. E. Cerniglia. 1997. A universal protocol for PCR detection of 13 species of food borne pathogens in foods. *J. Appl. Microbiol.* **83:**727–736.

Warren, B. R., M. E. Parish, and K. R. Schneider. 2005. Comparison of chromogenic *Shigella* spp. plating medium with standard media for the recovery of *Shigella boydii* and *Shigella sonnei* from tomato surfaces. *J. Food Prot.* **68:**621–624.

Warren, B. R., H.-G. Yuk, and K. R. Schneider. 2007. Survival of *Shigella sonnei* on smooth tomato surface, in potato salad and in raw ground beef. *Int. J. Food Microbiol.* **116:**400–404.

Waterman, S. R., and P. L. Small. 1996. Identification of sigma S-dependent genes associated with the stationary-phase acid-resistance phenotype of *Shigella flexneri*. *Mol. Microbiol.* 21:925–940.

Wei, J., M. B. Goldberg, V. Burland, M. Venkatesan, W. Deng, G. Fournier, G. F. Mayhew, G. Plunkett, D. J. Rose, A. Darling, B. Mau, N. T. Perna, S. M. Payne, L. J. Runyer-Janecky, S. Zhou, D. C. Schwartz, and F. R. Blattner. 2003. Complete genome sequence and comparative genomics of *Shigella flexneri* serotype 2a strain 2457T. *Infect. Immun.* 71:1919–1928.

Weissman, J. B., A. Schmerler, P. Weiler, G. Filice, N. Godby, and I. Hansen. 1974. The role of preschool children and day-care centers in the spread of shigellosis in urban communities. *J. Pediatr.* 84:797–802.

Wetherington, J., J. L. Bryant, K. A. Lampel, and J. M. Johnson. 2000. PCR screening and isolation of *Shigella sonnei* from layered party dip. *Lab. Inform. Bull.* 16:1–8.

Wharton, M., R. A. Spiegel, J. M. Horan, R. V. Tauxe, J. G. Wells, N. Barg, J. Herndon, R. A. Meriwether, J. N. MacCormack, and R. H. Levine. 1990. A large outbreak of antibiotic resistant shigellosis at a mass gathering. *J. Infect. Dis.* 162:1324–1328.

WHO. 1997. Vaccine research and development. New strategies for accelerating *Shigella* vaccine development. *Wkly. Epidemiol. Rec.* 72:73–80.

Wu, F. M., M. P. Doyle, L. R. Beuchat, J. G. Wells, E. D. Mintz, and B. Swaminathan. 2000. Fate of *Shigella sonnei* on parsley and methods of disinfection. *J. Food Prot.* 63:568–572.

Zaika, L. L. 2002a. Effect of organic acids and temperature on survival of *Shigella flexneri* in broth at pH 4. *J. Food Prot.* 65:1417–1421.

Zaika, L. L. 2002b. The effect of NaCl on survival of *Shigella flexneri* in broth as affected by temperature and pH. *J. Food Prot.* 65:774–779.

Zaika, L. L., and O. J. Scullen. 1996. Growth of *Shigella flexneri* in foods: comparison of observed and predicted growth kinetics parameters. *Int. J. Food Prot.* 32:91–102.

Pathogens and Toxins in Foods: Challenges and Interventions
Edited by V. K. Juneja and J. N. Sofos
© 2010 ASM Press, Washington, DC

Chapter 10

Pathogenic Vibrios in Seafood

ANITA C. WRIGHT AND KEITH R. SCHNEIDER

The primary pathogens in the genus *Vibrio* are *Vibrio parahaemolyticus*, *V. cholerae*, and *V. vulnificus*. Other *Vibrio* species that can cause human disease include *V. hollisae*, *V. alginolyticus*, *V. damsela*, *V. mimicus*, *V. fluvialis*, *V. metschnikovii*, *V. furnissii*, *V. cincinnatiensis*, and *V. carchariae* (Hlady and Klontz, 1996), but reported cases are either relatively rare or do not involve food-borne transmission. Vibrios are gram-negative, aquatic bacteria that commonly inhabit coastal ecosystems throughout the world, and several species are also pathogens of fish and shellfish. All species appear as curve-shaped rods with one or two single-polar flagella in standard culture (Baumann and Schubert, 1984). Metabolically they are moderate halophiles and facultative anaerobes that use glucose as a sole carbon source. The various species demonstrate a range of fermented sugars and enzymatic profiles, but they all secrete chitinase. Kaneko and Colwell (1975a) postulated that the distribution of vibrios in marine environments is determined by their adsorption to and utilization of the chitinous exoskeletons of planktonic organisms. Figure 1 shows attachment of *V. vulnificus* cells to a planktonic diatom in culture, demonstrating the close association of these two marine species. Furthermore, the initiation of vibrio microcolonies on these animals is thought to be critical for carbon cycling in the world's oceans (Keyhani and Roseman, 1999).

Because vibrios are indigenous to aquatic environments, related diseases are generally transmitted to humans via consumption of contaminated seafood or drinking water. Vibrios account for about 75% of seafood-borne bacterial illness in the United States (Feldhusen, 2000). Skin infections from *Vibrio* species are also common and are associated with exposure of wounds to seawater or by handling of seafood during processing. Although most bacterial food-borne infections have declined in recent years, the estimated annual incidence of vibrio disease in the United States actually increased by 78% in 2006 compared to 1996 to 1998 (CDC, 2007). *Vibrio* species are responsible for both sporadic and epidemic disease, and symptoms vary greatly among the different pathogenic species. As summarized in Table 1, gastrointestinal symptoms range from mild watery diarrhea and vomiting to the severe, purging diarrhea associated with cholera epidemics. Many species produce rapidly lethal septicemia; however, systemic disease seldom occurs in healthy individuals, and susceptible persons have predisposing conditions, usually related to hepatic or immune systems.

PATHOGENS: TYPE OF ILLNESS AND CHARACTERISTICS OF THE ORGANISMS

V. cholerae

This species was first described as the causative agent of epidemic cholera disease by Pacini in 1854 and later by Koch in 1883, and it remains one of the deadliest bacterial disease agents worldwide (Lipp et al., 2002). Although cholera epidemics have been essentially eliminated in the United States, the disease is still endemic in parts of Asia and Africa and more recently South and Central America. The global incidence of this disease remains devastatingly high: approximately 100,000 to 300,000 cases are reported annually worldwide to the WHO, which estimates that these figures probably reflect only 1% of the actual disease incidence (Zuckerman et al., 2007). Nonepidemic *V. cholerae* is also responsible for occasional seafood-borne disease in the United States and can be distinguished from epidemic disease by the severity of symptoms, the serogroups of associated strains, and the capacity for global spread.

Anita C. Wright and Keith R. Schneider • University of Florida, Food Science and Human Nutrition Department, 359 FSHN Bldg., Newell Drive, Gainesville, FL 32611.

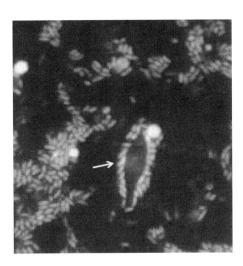

Figure 1. *V. vulnificus* bacteria that are attached to diatom are indicated by arrow. (Figure provided by Maria Chatzidaki-Livanis.)

Epidemic cholera is characterized by massive watery diarrhea, which produces rapid fluid loss that, if untreated, leads to frequently fatal dehydration. The disease requires a relatively large infectious dose, estimated to be about 1 million bacteria, and susceptibility to the disease varies widely among healthy adults (Sack et al., 1972). Following ingestion, the bacterium multiplies in the lumen of the small intestine and uses pili to anchor to enterocytes in the gut (Taylor et al., 1987b). The primary virulence factor is the cholera toxin (CT) that is secreted from the bacterium during infection and is responsible for the characteristic diarrhea (Finkelstein and LoSpalluto, 1969). The activity of CT has been studied extensively, and it serves as a model for the study of other bacterial toxins with similar structures (Sandvig and van Deurs, 2002; Lyerly et al., 1982).

As shown in Figure 2, CT is comprised of two protein subunits designated A and B, and these proteins combine to make the active holotoxin (Kaper et al., 1995). The single A subunit has the enzymatic activity responsible for symptoms and is surrounded by five B subunits that provide the binding capacity of the toxin. After docking to the intestinal cells, the toxin enters the cell and transverses across the cytoplasm, releasing the A subunit into the cytoplasm. Activity of the A subunit blocks the host machinery for recycling the intracellular signaling molecule, cyclic AMP. Specifically, the toxin transfers ADP ribose to G proteins that are involved in this process. The net result is the permanent activation of host adenylate cyclase, which causes significant increases in cyclic AMP levels, leading to secretion of massive amounts of chloride ions into the lumen of the intestine. Subsequent osmotic imbalance induces dramatic water loss from the cells and tissues, resulting in characteristic voluminous diarrhea. This rapid loss of large fluid volumes accounts for the severity and high mortality associated with cholera. *V. cholerae* bacteria lacking CT are not able to cause epidemic disease (Morris, 2003).

Other virulence factors include the toxin coregulated pilus (TCP) that is expressed during colonization of the small intestine (Taylor et al., 1987a). Interestingly, this pilus also serves as the receptor for a lysogenic filamentous bacteriophage (CTXφ) that carries the genetic material for the production of CT (Waldor and Mekalanos, 1996). Nontoxigenic *V. cholerae* strains are converted to toxigenic, potentially epidemic, strains following infection by this phage. Genes for TCP and CTX are both regulated by the same transcriptional activator, ToxT (DiRita et al., 1991). Motility is also required for virulence (Richardson, 1991), and chemotaxis is important for colonization and induction of CT expression in vivo (Lee et al., 2001). Other possible virulence factors include outer membrane proteins or porins (Provenzano and Klose, 2000), and an inner membrane protein ToxR that is produced by all *Vibrio* species and coordinately regulates the other virulence factors (Miller and Mekalanos, 1984).

Treatment of cholera includes oral or intravenous rehydration and antibiotic therapy. *V. cholerae* is sensitive to a variety of antibiotics, especially tetracyclines, but multiply drug-resistant strains have been reported (Morris, 2003). Vaccines offer protection against disease but are expensive to produce and deliver, and the immunity provided may be limited (Taylor et al., 1988; Zuckerman et al., 2007; Ford et al., 2007). Genetically

Table 1. Typical symptoms associated with different *Vibrio* species

Vibrio species (type)	Symptoms			
	Gastroenteritis	Severe diarrhea	Septicemia	Wounds
V. cholerae (epidemic)	Less common	Common	No	Rare
V. cholerae (nonepidemic)	Common	No	Rare	Rare
V. parahaemolyticus	Common	No	Rare	Less common
V. vulnificus	Less common	No	Common	Common

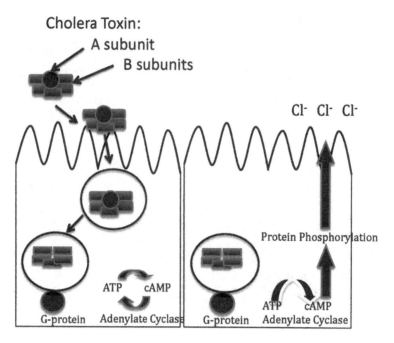

Figure 2. Activity of CT. CT B subunits bind host intestinal cells, followed by endocytosis and release of the A subunit (black circle), which ribosylates G proteins to activate adenylate cyclase. The subsequent increase in cyclic AMP (cAMP) results in protein phosphorylation, extrusion of chloride ions, and massive diarrhea. (Adapted from Kaper et al., 1995.)

engineered, live attenuated oral vaccines hold great promise and generally involve deletion of the gene (*ctxA*) encoding the enzymatic A subunit of CT (Kaper et al., 1995). Unfortunately, these vaccine strains still retain some reactogenicity, and emerging strains can be antigenically different from vaccine strains, necessitating the construction of new vaccines.

V. parahaemolyticus

V. parahaemolyticus was first described as the causative agent of outbreaks of gastrointestinal illness in Japan (Fujino, 1953). Symptoms associated with these infections contrast greatly with epidemic cholera and generally present as mild gastroenteritis with watery diarrhea that resolves without treatment. Occasional dysentery, wound infections, and septicemia are also caused by *V. parahaemolyticus*, but fatalities are rare and generally follow septicemia (Su and Liu, 2007). *V. parahaemolyticus* produces several hemolytic toxins. The first described toxin was the thermal stable direct hemolysin or TDH, whose activity is characterized by the Kanagawa phenomenon that is based on lysis of human red blood cells on Wagatsuma agar (Miyamoto et al., 1969). Some clinical isolates produce a genetically similar toxin called the TDH-related hemolysin (TRH), and other strains produce both TDH and TRH (Honda et al., 1988). Experimental genetic

deletion of the gene (*tdh*) encoding TDH corresponds to loss of symptoms in animal models (Nishibuchi et al., 1992). Diarrhea may result from toxin-mediated induction of Ca^{2+}-activated chloride channels, leading to fluid accumulation (Baffone et al., 2005). Toxic molecules related to type three secretion systems have also been implicated in disease and are under investigation (Liverman et al., 2007).

Treatment for *V. parahaemolyticus* involves rehydration, especially for persistent diarrhea, and infected persons may also benefit from antibiotic therapy with tetracyclines or quinolones (Morris, 2003). A vaccine is not available for *V. parahaemolyticus*, and vaccine development seems unlikely due to the typically mild nature of symptoms. However, recent increases in the global spread of epidemic *V. parahaemolyticus* disease and better understanding of virulence mechanisms may lead to improved prevention strategies.

V. vulnificus

V. vulnificus was first described by Hollis et al. (1976), and typical symptoms are dramatically different from those of other pathogenic vibrios. Disease can result either as a consequence of seafood consumption or from exposure of wounds to seawater or through the handling of seafood (Blake et al., 1979). Disease incidence is relatively rare compared to the other vibrio pathogens, and only 30 to 60 cases are reported annually in the United States (Feldhusen, 2000). However, this single species is responsible for nearly all of the deaths related to bacterial contamination of seafood in the United States. The mortality rate is exceptionally high (>50%), and risk of disease may be greater in countries outside the United States, where increased exposure of consumers to raw seafood and lower immune status is more common (Park et al., 1991). A recent survey in Japan detected six cases of *V. vulnificus* septicemia during a 20-day period (Ono, 2001).

Unlike the other vibrio pathogens, illnesses caused by *V. vulnificus* are rarely seen in healthy adults or children. This species is very much an opportunistic pathogen, and persons who are at risk for this disease generally exhibit some type of underlying condition that includes alcoholic cirrhosis, hepatitis C, diabetes, hemochromatosis (iron overload), and immune system dysfunction (Hlady and Klontz, 1996; CDC, 2005). In these persons, a fulminating, primary septicemia can occur within hours of exposure and produce symptoms that are similar to other types of gram-negative toxic shock, including rapid drop in blood pressure, intravascular coagulation, and multiple organ failure. A distinguishing characteristic of *V. vulnificus* sepsis is the appearance of bollous lesions on the extremities, as shown in Figure 3. *V. vulnificus* also causes gastrointestinal illness with mild diarrhea and vomiting. This organism is the most common cause of serious wound infections associated with *Vibrio* species, and these infections may result from exposure of breached skin surface to seawater or contaminated seafood handling (Shapiro et al., 1998; Howard and Lieb, 1988) (reviewed by Oliver [2005]). In compromised hosts, wound infections may also lead to septic shock and death.

Experimental evidence supports epidemiological data indicating that host iron status contributes to

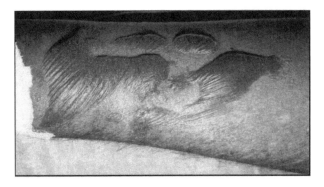

Figure 3. Bollous lesions associated with *V. vulnificus* septicemia.

V. vulnificus disease. For example, excess physiological iron as a consequence of hereditary hemochromatosis increases susceptibility to disease (Blake et al., 1979). Wright et al. (1981) found that exogenously administered iron in mice dramatically reduced the 50% lethal dose from about 10,000 to <10 bacteria. Production of bacterial iron-sequestering siderophores also contributes significantly to virulence (Litwin et al., 1996). Other reported virulence factors include flagella (Lee et al., 2006; Kim and Rhee, 2003), pili (Paranjpye and Strom, 2005; Paranjpye et al., 2007), and quorum sensing (Kim et al., 2003), as well as numerous secreted molecules (proteases, phospholipases, DNases, etc.) whose relevance to virulence is unclear, as reviewed by Gulig et al. (2005). *V. vulnificus* hemolysin is unrelated to the hemolysins of *V. parahaemolyticus* but shows genetic similarity to those found in *V. cholerae* and other *Vibrio* species (Yamamoto et al., 1990). Furthermore, the *V. vulnificus* hemolysin is unlikely to be a virulence factor because inactivation of the *vvhA* gene encoding the hemolysin does not reduce virulence in mice (Wright and Morris, 1991; Fan et al., 2001). Another hemolysin, referred to as repeats in toxin or RTX, has been implicated in disease in animal models (Lee et al., 2007).

Encapsulation by polysaccharide is a primary virulence factor and contributes to the ability of *V. vulnificus* to cause systemic disease by providing protection from the lytic activity of serum and from phagocytosis during systemic disease (Tamplin et al., 1985; Yoshida et al., 1985; Simpson et al., 1987). Encapsulation is marked by formation of opaque colonies; however, this phenotype may undergo spontaneous phase variation to an alternate colony type, which appears more translucent and has reduced capsular polysaccharide (CPS) expression and thus decreased virulence in animal models (Yoshida et al., 1985; Amako et al., 1984). The different colony types are shown in Figure 4 and have been observed with other *Vibrio* species. A CPS operon was recently described for *V. vulnificus* and showed genetic similarity to the *Esherichia coli* group 1 capsule (Chatzidaki-Livanis et al., 2006b; Chatzidaki-Livanis et al., 2006a; Wright et al., 2001). Multiple types of translucent-phase variants have been identified with differing degrees of encapsulation and variation in the genetic organization of the CPS operon. Acapsular variants show deletion mutations in this operon and are locked in the translucent phase, while strains with partial CPS expression retain an intact operon and reversible phase variation.

V. vulnificus septicemia is rapidly fatal, and the early administration of antibiotics is critical to the outcome of disease (Klontz et al., 1988). Combinations of antimicrobials may be more therapeutic and

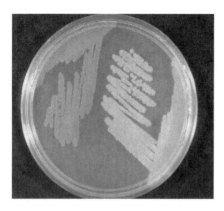

Figure 4. Colony morphology of *V. vulnificus* showing opaque (right) and translucent (left) phenotypes associated with virulence and capsule expression.

increase survival, and minocycline combined with cefotzxime or fluoroquinolone is recommended (Chuang et al., 1998). A vaccine against *V. vulnificus* is not available.

ENVIRONMENTAL SOURCES: INCIDENCE IN THE ENVIRONMENT AND FOODS

Vibrio species are found naturally in aquatic environments and are particularly evident in estuaries with temperate climates that support their preference for moderate salinity, warmer water temperature, and neutral to alkaline pH (Tantillo et al., 2004). They also form intimate associations with marine plankton and mollusks through adaptations that enhance their environmental survival. For example, larvae of bobtail squid *(Euprymna scolopes)* are colonized by *V. fischeri,* and the development of a light organ in this squid is dependent on colonization by these bioluminescent bacteria, thereby providing camouflage for the adult squid (Ruby, 1996; Ruby and Lee, 1998). Presumably, the relationship of bacteria and squid is beneficial for both organisms. Furthermore, some pathogenic *Vibrio* species are recovered from oysters in high numbers (>1,000 bacteria/g) that generally exceed their distribution in the water column (Tamplin et al., 1982; Wright et al., 1996; Motes et al., 1998), and even higher levels are recovered from the intestinal tracts of fish that feed upon shellfish (DePaola et al., 1994). Although *Vibrio* species can be pathogens of fish, mollusks, and crustaceans, they commonly appear to coexist without damage to the host. Thus, the distribution of vibrios in the environment is likely to be determined by adaptations that evolved to allow coexistence with these hosts in marine ecosystems.

Specific physical parameters greatly influence the environmental distribution of the different pathogenic *Vibrio* species; however, responses to these parameters vary somewhat among species. These bacteria grow optimally at temperatures between 22 and 42°C within a pH range of 5 to 11 for salinities ranging from 1 to 8% (Kaspar and Tamplin, 1993; Motes et al., 1998; Kelly and Stroh, 1988; Kelly, 1982). In general, the numbers of all *Vibrio* species peak in warmer months and decline after extended exposure to cold temperatures, but tolerances to extremes in temperature and salinity vary with species. *V. parahaemolyticus* can be recovered from colder, more saline environments that generally do not support survival of *V. vulnificus* or *V. cholerae* (Kaneko and Colwell, 1973; Kelly and Stroh, 1988; Kelly, 1982). Thus, extremes in temperature and salinity may limit distribution of a particular *Vibrio* species in shellfish. It should also be noted that combined parameters of temperature, pH, and salinity have interactive effects on growth and survival and vary with depth. Additional biological factors that are likely to influence the prevalence of these species in the environment include fluctuations in nutrient availability, algal blooms, competition with other bacterial flora, ultraviolet irradiation, and availability of invertebrate or vertebrate hosts.

V. cholerae

Since 1817 there have been at least seven pandemics of *V. cholerae* that have spread throughout the world (Lipp et al., 2002). In 1849 John Snow famously traced the source of a cholera outbreak in London to a tainted well, laying the foundations for the field of epidemiology and establishing principles of modern public health. In cholera-endemic regions of the world, *V. cholerae* is transmitted primarily through contaminated drinking water. Both natural and human-made disasters exacerbate outbreaks of cholera due to a breakdown of infrastructure for sewage disposal and water treatment and to delay of medical treatment. Although cultural taboos often prohibit the consumption of shellfish in regions where cholera has been endemic historically, more recent epidemics in South America clearly implicate raw or undercooked seafood and shellfish. Sporadic cases reported in New Jersey and Florida were attributed to imported blue crab (CDC, 1991; Tauxe et al., 1995). Improved understanding of the biology of *Vibrio* species supports the essential link of this bacterium to the aquatic environment as the primary reservoir of disease (Colwell et al., 1977). *V. cholerae* has a relatively low salinity requirement compared to other pathogenic species, and therefore, its distribution includes freshwater ponds and river basins. Epidemics generally follow the monsoon cycles that facilitate disease by extensive flooding that alters the local estuarine environment.

Most environmental isolates of *V. cholerae* commonly lack the toxin genes *(ctxA* and *ctxB)* associated with epidemic disease and therefore are relatively avirulent. Genes encoding CT are carried by bacteriophage viruses that infect the bacterium and transfer the toxin genes to the bacterial genome to convert nontoxigenic strains to toxigenic variants (Waldor and Mekalanos, 1996). Gene transfer requires that the bacteria express TCP as the receptor for phage, and most environmental strains also lack genes for this receptor. Transformation to toxigenic *V. cholerae* may be facilitated during human passage as a consequence of the close proximity of bacteria and phage in the human intestinal tract, and environmental conditions that enhance uptake of DNA are also likely to increase the prevalence of pathogenic *V. cholerae* (Lipp et al., 2002).

Prior to 1992, all epidemic cholera was caused by a single O1 *V. cholerae* serotype; environmental isolates were serologically heterogenic and were collectively referred to as non-O1 or nonagglutinating strains. In 1992 a newly emergent epidemic *V. cholerae* serotype, O139, originated in the Bay of Bengal and spread to the Western Hemisphere (Nair et al., 1994). This strain is now pandemic in India, Bangladesh, and Indonesia, as well as in South and Central American countries. This strain also produces a unique polysaccharide capsule (Waldor et al., 1994). Genetic analyses indicated that this strain evolved from existing seventh pandemic O1 strains via acquisition of genes for encapsulation. The acquisition of the capsule probably facilitated environmental distribution and survival against host immune defenses. Strains continue to evolve, and epidemic strains that are non-O1, non-O139 are now observed.

V. parahaemolyticus

Similar to the epidemiology of *V. cholerae*, most environmental isolates of *V. parahaemolyticus* also do not cause disease, because they lack the genes for TDH and TRH toxins. However, the epidemiology and environmental distribution of *V. parahaemolyticus* have been rapidly changing. Prior to 1992, most outbreaks occurred only in Japan and were generally associated with uncooked fish products (Fujino et al., 1953). U.S. cases were infrequent and often transmitted to humans via contaminated blue crabs (Molenda et al., 1972). Strains isolated during these earlier outbreaks showed very diverse serotypes. However, this profile changed as a "pandemic" strain of *V. parahaemolyticus* emerged in India and Indonesia in 1992. This emergent type was genetically and biologically distinct from strains recovered in prior epidemics, and it appeared to result from clonal expansion of a single *V. parahaemolyticus* strain designated serotype O3:K6 (Nair et al., 2007). As shown in Figure 5, this

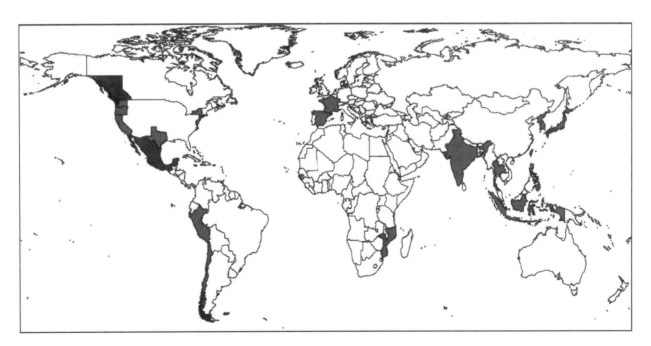

Figure 5. Global distribution of "pandemic" *V. parahaemolyticus* disease. (Reprinted from Nair et al., 2007.)

disease now has much wider distribution in the Western Hemisphere and is largely associated with raw oyster consumption. Outbreaks have now occurred along the Pacific coast from Alaska to Chile, with large outbreaks reported in the United States beginning in 1998 in Texas, New York, New Jersey, and Connecticut. After 2004, outbreaks were also reported in California, Oregon, and Alabama, as well as in Canada. As the "pandemic" strain spread throughout the world, the O3:K6 serotype subsequently diverged, and strains now exhibit multiple serotypes due to genetic mutations affecting lipopolysaccharide and CPS genes, but these strains still appear to be genetically related.

The emergence of epidemic *V. parahaemolyticus* in regions previously unaffected by this disease has followed global warming trends; however, it is unclear why this strain is so successful in its dissemination. Seasonal distribution of *V. parahaemolyticus* was first observed in the Chesapeake Bay by Kaneko and Colwell (1973), and the number of total *V. parahaemolyticus* generally showed linear increase with temperature. *V. parahaemolyticus* was isolated in samples from the Pacific Northwest only when temperatures exceeded 15°C (Kaysner et al., 1990). Recent outbreaks of *V. parahaemolyticus* in Alaska also correlated with water temperatures exceeding 15°C (McLaughlin et al., 2005). Coincidental changes in the genetic background of clinical isolates may also contribute to greater capacity for environmental survival and/or enhanced ability to cause disease. Investigation of the environmental distribution of this species is hampered by the relative scarcity of virulent phenotypes compared to avirulent phenotypes. A survey of Mobile Bay, AL, showed that only 0.08% of total *V. parahaemolyticus* isolates were TDH positive, and their prevalence was inversely correlated with water temperature (DePaola et al., 2003a). Risk assessment analyses estimated that the ratio of virulent to avirulent phenotypes was about 1:10,000, and these numbers are used to determine regulatory policy as to the acceptable levels of *V. parahaemolyticus* in commercially processed oysters.

V. vulnificus

As with other vibrios, distribution of *V. vulnificus* is seasonal and related to water temperature. Seafood-borne infections related to *V. vulnificus* are almost always associated with raw oyster consumption and particularly oysters harvested from the Gulf Coast (Blake, 1979). Clams, mussels, and fish are rarely implicated in disease, as these products are more likely to be cooked prior to consumption. Most (90%) oyster samples from the U.S. West Coast were negative for this bacterium, while 95% of Gulf Coast and 51% of mid-Atlantic samples were positive (Cook et al., 2002). This differential geographic distribution is likely to reflect lower water temperatures in Pacific and North Atlantic shellfish harvesting waters relative to the Gulf of Mexico. *V. vulnificus* is generally not recovered from shellfish at temperatures below 18°C (Tamplin et al., 1982; Motes et al., 1998; Tilton and Ryan, 1987; Cook, 1994). However, the reason for the greater disease incidence of *V. vulnificus* in the Gulf Coast compared to that of Atlantic oysters is not entirely clear, as levels in the mid-Atlantic oysters can be similar to those from the Gulf Coast (Wright et al., 1996).

Unlike other pathogenic vibrios, *V. vulnificus* does not express a toxin marker equivalent to CT or TDH that is indicative of virulence, and most isolates recovered from oysters are virulent in animal models (DePaola et al., 2003b). However, genetic analyses showed clinical strains are associated with specific genotypes that have relatively infrequent distribution in most environment isolates (Chatzidaki-Livanis et al., 2006a; Rosche et al., 2005; Nilsson et al., 2003). These data support the hypothesis that this "clinical" genotype may have increased virulence, and geographic disparity in distribution of more virulent strains could account for the predilection of disease coming from Gulf Coast oysters. A survey of Galveston Bay found that "clinical" genotypes were seasonal and had wider distribution in late summer months, when the number of cases is generally higher (Lin et al., 2003). Recent examination with a mouse model also showed that strains with the "clinical" genotype were significantly more associated with systemic disease compared to those with "environmental" types (P. A. Gulig and A. C. Wright, unpublished data).

V. vulnificus is also a pathogen in marine fish and invertebrates and is a significant problem for aquacultured fish. Preventative measures, including probiotics and vaccines, are being developed to ensure the survival of these animals and prevent transmission of disease to humans (Ravi et al., 2007; Esteve-Gassent et al., 2004; Novotny et al., 2004). Fish pathogens are generally distinct from most human disease isolates and can be differentiated by biotype (Tison et al., 1982). Genetic studies confirm biotype 1 was more likely to be derived from human infection via oyster consumption, while biotype 2 was specific for fish infections. Furthermore, biotype 3 was pathogenic in humans and transmitted by fish (Bisharat et al., 1999). Genetic analysis of biotype 3 indicated that these strains were a genetic hybrid derived from recombination of the fish and oyster

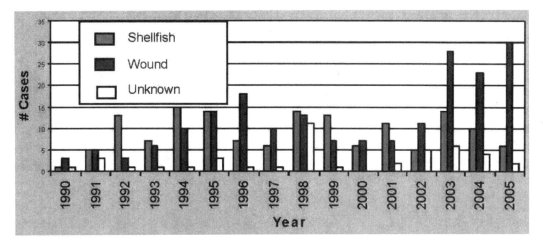

Figure 6. Reported cases of *V. vulnificus* in Florida from 1990 to 2005. Recent reports show increases in the number of cases associated with wounds compared to other sources. (Data provided by Roberta Hammond.)

genotypes (Bisharat et al., 2005). Recent trends suggest increased incidence of wound infections in Texas and Florida (Oliver, 2005) and during exposures resulting from Hurricane Katrina (CDC, 2005; Oliver, 2005). A survey of Florida cases over recent years supports this trend toward increases in wound infections (Figure 6).

INTRINSIC AND EXTRINSIC FACTORS: ROLE IN SURVIVAL AND GROWTH IN FOOD PRODUCTS

Intrinsic Factors

Most food-borne vibrio infections are caused by consumption of raw oysters. Unlike fish, in which intact muscle tissue is sterile, molluscan shellfish are filter-feeding organisms, and their bacterial flora reflects the content of their environment. Furthermore, oysters and clams are consumed frequently without processing or cooking. Therefore, the most important intrinsic factor for vibrio-associated diseases is the initial load of bacteria in the shellfish. Disease incidence dramatically increases when shellfish are harvested from water at higher temperatures, and 86% of vibrio-induced septicemia occurs from April until October (Hlady and Klontz, 1996). Epidemiological data indicate that *V. parahaemolyticus* disease incidence increases as temperatures approach 15°C. Increased incidence of disease also corresponds historically to increased summer harvesting of oysters, which rose from 8% in 1970 to 30% in 1994 (Altekruse et al., 1997). *Vibrio* species are ubiquitous in shellfish harvesting waters in summer months, and standard monitoring methods for fecal pollution

do not prevent or predict vibrio contamination of seafood (Tamplin et al., 1982). Seasonal fluctuations in the environmental distribution of vibrios are likely to be aggravated by the progression of global warming, which not only increases water temperatures but also decreases salinities along coastlines. Both factors may work synergistically to increase vibrio distribution (Lipp et al., 2002).

Other factors related to vibrio contamination of shellfish are algae or zooplankton that serve as the primary diet for mollusks. Vibrios absorb to these chitinaceous organisms, and algal blooms may influence the density of the vibrio load in oysters, as they are ingested with the plankton. Finally, oyster health may contribute to the vibrio burden of oysters. Parasitic disease may debilitate oyster immunity, resulting in increased bacterial populations. For example, the oyster parasite *Perkinsus marinus* produces an immunosuppressant protease that was shown to decrease in vitro clearance of *V. vulnificus* in oyster hemocytes (Tall et al., 1999). It should be noted that although vibrios are maintained at levels ranging from 100 to 10,000 CFU/g in summer months, they generally do not appear to have deleterious effects on oyster health.

Extrinsic Factors

Temperature is the primary parameter for controlling growth of *Vibrio* species in seafood products during storage and processing. Optimum growth for pathogenic vibrios in oysters is between 30 and 37°C (Kaspar and Tamplin, 1993; Kelly, 1982). Limiting growth is particularly critical for *V. parahaemolyticus*, as populations may double within 8 to 9 minutes under optimum conditions (Joseph et al., 1982).

Consequently, strict time and temperature regulations have been established to limit exposure of seafood to elevated temperatures that will promote the growth of these species (NSSP, 2005). Vibrio pathogens die or become nonculturable at refrigeration temperatures. Experimental evidence indicates that *Vibrio* species grown at refrigeration temperatures remain viable but cannot be recovered by standard culture methods. Entry into a dormant "viable but nonculturable" state has been postulated (Xu et al., 1982; Oliver et al., 1991). Furthermore, viable but nonculturable bacteria retain virulence in animal models but with increased infectious dose (Oliver and Bockian, 1995). Vibrios are also sensitive to elevated temperatures, with upper limits for growth temperature in *V. parahaemolyticus* and *V. vulnificus* at 44 and 42°C, respectively. However, these bacteria die at even mildly elevated temperatures (45 to 55°C), and infections are very rarely associated with properly cooked seafood. Beuchat (1973) reported thermal resistance at 47°C (D_{47} values) ranging from 0.8 to 65.1 minutes, but heat resistance increased at higher inocula and with addition of 7% NaCl. Comparison of heat resistances for pathogenic vibrios showed that D_{55} values varied among species (Johnston and Brown, 2002). *V. parahaemolyticus* (D_{55} of 1.75 min) was considerably more resistant compared to *V. vulnificus* and *V. cholerae* (12 and 22.5 s, respectively). However, standard pasteurization conditions (70°C for 2 min) produced a >7-log reduction in all three species. Some food processing, such as mild steaming, may not thoroughly heat the entire oyster and result in survival of vibrios postcooking.

Vibrios are fairly tolerant of high pH but will not grow below pH 6 (Beuchat, 1973). Decimal reduction time was reduced significantly at 53°C for pH 5 to 6 compared to pH 7. The primary enrichment medium (alkaline peptone water) uses pH 8 to achieve selective growth of these species. Survival and growth of *Vibrio* species in seafood are also related to salinity. All vibrios require some salt for growth, but species differ greatly in their responses to salinity. For example, *V. parahaemolyticus* and *V. vulnificus* are both moderate halophiles and will not grow below 0.1% NaCl. However, *V. parahaemolyticus* grows well at 8% NaCl, while *V. vulnificus* does not survive at 8.5% and growth is inhibited at 6.5% NaCl (Kelly, 1982). In contrast, *V. cholerae* will grow in culture medium without additional salt, and *V. alginolyticus* is differentiated from the other pathogens by the ability to grow in 10% NaCl. Survival of *V. parahaemolyticus* at lower temperatures can increase with addition of NaCl or by adjusting the pH of growth medium (Beuchat, 1973).

FOOD PROCESSING AND RECENT ADVANCES IN BIOLOGICAL, CHEMICAL, AND PHYSICAL INTERVENTIONS

Historically, one of the oldest processing practices for molluscan shellfish sanitation was depuration, which has been practiced throughout the world for approximately 100 years (Canzonier, 1991). This application was driven by a series of typhoid fever outbreaks during the 1920s that were associated with the consumption of raw oysters (Canzonier, 1991) and led to the establishment of the National Shellfish Sanitation Program (NSSP). The process involves the removal of potential pathogens by placement of shellfish in sanitized seawater that is usually treated either by ozonation (Schneider et al., 1991) or by UV light (Tamplin and Capers, 1992) during recirculation into wet storage tanks (Furfari, 1991). UV light irradiation is currently the most common depuration method in the United States and the United Kingdom. UV light allows continuous dosage, and numerous studies have shown that it is effective in inactivating bacteria (Cabelli and Haffernau, 1971; Cortelyou et al., 1954; Kelly, 1961; Kawabata and Harada, 1959) and viruses (Hill et al., 1969, 1970). The bactericidal activity disrupts unsaturated bonds, particularly purine and pyrimidines found in DNA, and lethal changes are directly related to the depth of penetration and amount of particulate matter (Huff et al., 1965). Depuration is the preferred method for removal of fecal coliforms, enteric bacteria, and viral pathogens in many countries and has a distinct advantage over other treatments in that it does not kill mollusks (Richards, 1990). Unfortunately, the depuration process does not effectively remove *Vibrio* species from shellfish (Vasconcelos and Lee, 1972; Tripp, 1960; Colwell and Liston, 1960), as vibrios persisted in shellfish for up to 96 hours of treatment at concentrations that were similar to those in untreated shellfish (Rodrick and Schneider, 1991). Thus, vibrios appear to be firmly attached to the shellfish tissues and are not effectively removed by depuration.

In 2003 the FDA and the NSSP set forth guidelines intended to minimize consumer risk from naturally occurring *V. vulnificus* in oysters. The guidance document focused on consumer education, harvest restrictions, and postharvest treatments (PHP) to ensure product safety. The oyster industry responded to these guidelines by implementing pre- and postharvest controls aimed at reducing levels of *V. vulnificus* and *V. parahaemolyticus*. Control methods have been evaluated and adjusted in an effort to reduce annual vibrio infections. Implementation of these methods requires validation and verification that these treatments meet standards for reduction of vibrios in oysters. Approved and validated treatments include high hydrostatic pressure, pasteurization (heat shock), and individual quick freezing. These treatments are generally used in combination with approved transport and storage practices involving icing, refrigeration, and/or freezing.

Refrigeration

Seafood held at ambient temperatures can support rapid growth of *Vibrio* species (Gooch et al., 2002; Cook, 1994); therefore, both warm and cold temperature processing have been used in controlling and killing pathogenic *Vibrio* species in molluscan shellstock. The simplest and most common processing practice for all fish and shellfish handling is the immediate transfer of the product to ice or refrigerated storage. Refrigeration of oysters is required within 10 hours of harvest in summer months; however, the time-temperature matrix varies with the ambient air temperature and infection risk. In shellfish areas with confirmed association with a vibrio outbreak, the allowable time between harvest and refrigeration is reduced. Bacterial levels decline somewhat with refrigeration, but vibrios are not eliminated in oysters, even with an extended exposure of 10 to 15 days (Kaspar and Tamplin, 1993; Cook, 1994; Murphy and Oliver, 1992). The rate of decline is inoculum dependent, and survival at low temperature is also enhanced by more gradual rates of cooling (Bryan et al., 1999). Furthermore, growth was observed with food products at 9.5 to 10°C. Temperature controls are even more relevant for intertidal oysters, as these oysters are exposed to ambient air temperatures and may warm to levels that promote growth of *Vibrio* species. Therefore, harvesting during intertidal exposure is avoided, or oysters are transferred to deeper, cooler seawater prior to harvest. Maintenance of refrigerated or freezing temperatures during transport and storage is also regulated, and lack of compliance with these recommendations may account for a significant proportion of disease.

Unfortunately, refrigerated storage results in only moderate bacterial reductions and is not appropriate for PHP that is aimed at risk reduction. Immersion of oysters in ice prior to refrigeration results in greater reduction compared to simple refrigeration but also does not completely eliminate *V. vulnificus* in oysters (Quevedo et al., 2005). Furthermore, this study also found that ice immersion could increase the number of fecal coliforms in oysters, suggesting that this method should be used with caution. Thus, product with a significant bacterial load at harvest will still present increased risk to the consumer, even following rapid decreases in temperature. After prolonged exposure to low temperature storage, vibrios can also become nondetectable on standard media but still be viable and in virulence (Baffone et al., 2003).

Ultra-Low Temperature Treatment

Recently, ultra-low temperature treatment ($<-70°C$) has been shown to effectively reduce vibrios when it was followed by extended frozen storage at $-20°C$ for 1 to 2 weeks, depending on the process. Ultra-low freezing can be achieved by immersion of shellstock in liquid nitrogen or CO_2, and liquid nitrogen treatment was recently validated as an oyster PHP for reduction of *V. vulnificus*. Vibrio content in oysters following nitrogen treatment and frozen storage was significantly reduced to levels that were below those specified by the NSSP (Wright et al., 2007). This treatment has the advantage of producing a high-quality product with extended shelf life.

Heat Treatment

Heating is also an effective treatment for elimination of *Vibrio* species, as evidenced by the lack of disease association with cooked seafood products. All *Vibrio* species die rapidly at temperatures exceeding 55°C (Johnston and Brown, 2002). Heat shock treatment involves immersion in hot water (50°C for 5 to 10 minutes) combined with frozen storage and achieves the desired reduction within 2 weeks (Andrews et al., 2000). Traditional pasteurization is also done at 75°C for 8 min and can reduce vibrio counts to safe levels for the consumer but produces some biochemical and sensory changes in the meats (Cruz-Romero et al., 2007). Heat treatment of oysters also improves shucking yields compared to untreated oysters and benefits both the consumer and the distributor.

HPP

Recent progress in the high-pressure processing (HPP) of foods in general, and oysters in particular, suggests that this method can be applied to shellfish. Application of elevated pressure dramatically reduces the numbers of *Vibrio* species, and produces a "self-shucking" product. Treatment of oysters at 400 MPa for 10 minutes resulted in an about 5-log-unit reduction of total viable count, H_2S-producing organisms, lactic acid bacteria, and coliforms (Lopez-Caballero et al., 2000). The use of HPP at 310 MPa caused the opening of the shell with 100% efficiency as the adductor muscle of the oyster was detached in the process (He et al., 2002). Recent studies have shown the ability of freezing coupled with frozen storage as a means of reducing the number of recoverable

V. vulnificus bacteria in Gulf Coast oysters (Parker et al., 1994).

Irradiation

The United States has lagged behind other countries in the application of food irradiation technologies partially due to the U.S. Food, Drug, and Cosmetic Act of 1958, which classified irradiation as a food additive. In the 1980s, the FDA approved irradiation as a safe and effective method of decreasing or eliminating harmful bacteria in herbs, spices and vegetable seasoning, fruits, vegetables, and grains (Morrison, 1986). Meat, poultry, and other foods, including oysters, were not approved by the FDA until 2006. Oysters appear tolerant to irradiation processing at levels of a <2.5-kGy absorbed dose, as normal shelf-life is maintained with no increase in mortality compared with untreated control oysters (Mallett et al., 1987). However, higher levels of irradiation increased mortality and resulted in a yellow exudate and an unpalatable product. Furthermore, *V. parahaemolyticus* cultures were less sensitive to irradiation compared to *V. vulnificus.*

Relaying

The basic concept of relaying involves transferring shellfish from restricted, polluted waters to areas approved for harvesting and permitting the process of natural biological purification. Filter-feeding shellfish can purge themselves of certain abiotic and biotic substances when transferred or "relayed" to relatively clean seawater. The rate of purging depends on the type of contaminant, species of shellfish, water quality, length and method of relay, and various environmental factors. Erdman and Tennant (1956) showed an average 85% reduction in the number of coliform bacteria in the soft-shell clam *(Mya arenaria)*. Relaying also includes transporting oysters to areas with higher salinity in order to reduce levels of *V. vulnificus* (Motes et al., 1998). Problems associated with relaying are that it is a labor-intensive process and that up to 50% of the original harvest can be lost (Furfari, 1979). Unfortunately, depuration and relaying to higher salinities present logistical and regulatory problems and could increase contamination by other more halophilic *Vibrio* species.

Food Additives and Biological Controls

Although consumer markets for the treated seafood have been growing, a substantial number of shellfish consumers still prefer live, fresh, raw oysters. Various food additives have also been investigated but were not effective in live oysters (Sun and Oliver, 1994; Birkenhauer and Oliver, 2003). Probiotics or biological controls use microbiological products or microorganisms as food additives to reduce pathogen load. These "beneficial" bacteria can outcompete pathogens, limit their growth, or be lethal to potential pathogens. Antibiotics are generally not added to foods, but an exception to this rule is a class of antimicrobial molecules, termed bacteriocins, that are produced by bacteria and are specifically lethal to other bacteria. Applications are primarily against gram-positive bacteria (Lipinska, 1977), and the bacteriocin nisin is now permitted in over 40 countries worldwide, including the United Kingdom, the European Union, Russia, and India (http://www.foodnavigator.com). Acceptance of these molecules as biological controls is a consequence of their natural presence in milk products, lack of activity against higher organisms, and demonstration of effectiveness toward *Listeria monocytogenes* (Farber, 1993). Unfortunately, currently accepted bacteriocins are not effective against gram-negative bacteria and thus will not clear pathogenic vibrios in oysters.

Government Regulation

Standard mitigation and monitoring strategies for shellfish were originally based on fecal coliform content, but this type of analysis is ineffective in preventing naturally occurring contamination of vibrios in shellfish. Warning labels on oyster products and extensive educational programs targeted at-risk individuals with underlying diseases, such as cirrhosis, hemochromatosis (iron overload), diabetes, or immune system dysfunction. Unfortunately, these strategies did not completely eliminate disease mortality from *V. vulnificus* diseases (NSSP, 2003), and a small number of cases are still reported annually, presumably because refrigerated temperature guidelines limit growth but do not eliminate vibrios in oysters. Furthermore, problems with seafood safety have been exacerbated by recent outbreaks of *V. parahaemolyticus*. PHP of oysters to reduce the numbers of vibrios has been mandated for Gulf Coast oysters. PHP is now implemented for >25% of oyster shellstock from the Gulf Coast and requires a minimum reduction of 3.52 log most probable numbers (MPN)/g reduction in vibrio content in order to validate a treatment for commercial processing. California also requires all oysters harvested from Gulf Coast states to be treated during certain months. Increased incidence of *V. parahaemolyticus* disease necessitated expanding PHP programs to include Pacific Coast oysters, and new guidelines provide for monitoring oysters with recommended closures when *V. parahaemolyticus* levels exceeded 10,000 bacteria/g.

DETECTION METHODS FOR CONFIRMATION AND TRACE-BACK OF CONTAMINATED PRODUCTS

Selective Differential Agar Media

Several media are available to isolate *Vibrio* species; however, in practice, species identification cannot be determined upon primary isolation and requires additional testing for confirmation. Thiosulphate citrate bile salts sucrose (TCBS) agar was originally developed for isolation of pathogenic *V. cholerae* from stool samples and utilizes alkaline pH (8.6), ox bile, and NaCl (1%) to suppress the growth of nontarget bacteria (Kobayashi et al., 1963). This medium differentiates *V. cholerae* from *V. parahaemolyticus* and *V. vulnificus* on the basis of sucrose fermentation (Kelly, 1982; Karunasagar et al., 1986). Unfortunately other *Vibrio* species, such as *V. alginolyticus*, also ferment sucrose and are common isolates on this agar. *V. parahaemolyticus* and *V. vulnificus* do not ferment sucrose and therefore cannot be discriminated from each other on this medium. In one study, 49% of the bacteria isolated on TCBS were reported as not vibrio, and aeoromaonads are common isolates on TCBS (Cleland, 1985). Additionally, TCBS can also inhibit recovery of *Vibrio* species from the environment compared to recovery on nonselective agars (Wright et al., 1993). Therefore, growth in enrichment broth, usually alkaline peptone water at pH 8.0 with 1% peptone and 1 to 2% NaCl, is often employed prior to plating (Kaysner and DePaola, 2004).

Media recommended by the FDA to differentiate *V. vulnificus* and *V. parahaemolyticus* include modified cellobiose polymyxin B colistin agar (mCPC) (Tamplin et al., 1991) and cellobiose colistin agar (Hoi et al., 1998). Peptide antibiotics, colistin, and/or polymyxin are used as selective agents, and cellobiose fermentation is used for differentiation of *V. vulnificus*, which yields typical yellow colonies surrounded by a yellow halo on this medium. Incubation at 40°C further inhibits nontarget bacteria and is important for maintaining the selectivity of mCPC. *V. parahaemolyticus* generally is not recovered on these media. Therefore, TCBS is still used to isolate this species. Recently, a new chromogenic agar medium (CHROMagar Vibrio CV; CHROMagar Microbiology, Paris, France) was described for more effective recovery of *V. parahaemolyticus* and appears to be more sensitive and accurate than TCBS for primary isolation of this species (Hara-Kudo et al., 2001). Unfortunately, all of these isolation media lack adequate specificity to reliably identify species, and confirmation by additional assays using antibodies or molecular probes is required.

Species Identification

Commercial formats, such as the API 20E and Biolog systems, are available for biochemical identification of vibrios. However, these assays are costly, and the phenotypic plasticity of these species frequently complicates their application. Antibodies for species-, serotype-, and toxin-specific identification are also available but rely upon expressed antigen and may require cell processing for detection. Lipopolysaccharide antibody or "O" antigen has been used traditionally to identify and classify epidemic *V. cholerae*. Clinical strains *V. vulnificus* and *V. parahaemolyticus* show more heterogeneous serotypes, which are not useful for species identification. Increasingly, molecular methods are used for species identification because they improve specificity and sensitivity of detection. A nonradioactive DNA probe format, using alkaline phosphatase-labeled oligonucleotides, is the FDA-specified method for vibrio identification (Kaysner and DePaola, 2004). DNA probes for *V. cholerae* recognize genes for CT (Wright et al., 1992), and species-specific identification of *V. parahaemolyticus* targets the gene for a thermal-labile hemolysin, while potentially pathogenic isolates are discriminated by probes detecting genes for TDH and TRH virulence-associated hemolysins, described above (Nordstrom et al., 2006). Probes for *V. vulnificus* are based upon a cytolysin gene, *vvhA* (Wright et al., 1993).

A number of PCR methods are also used to amplify species- or virulence-specific DNA (Brauns et al., 1991), and more recent methods include real-time detection and quantitative PCR (Panicker et al., 2004; Panicker and Bej, 2005; Campbell and Wright, 2003; Tarr et al., 2007; Blackstone et al., 2003; Lyon, 2001). Multiplex PCR analyses for simultaneous identification of multiple food-borne pathogens, including vibrios, have been reported (Wang et al., 1997; Lee et al., 2003; Brasher et al., 1998). Unfortunately, PCR reactions are often inhibited by food matrices and generally require strain isolation and some purification of the DNA template for optimum sensitivity. Improved separation technologies may function to concentrate and increase sample size, while removing inhibitors (Stevens and Jaykus, 2004).

Enumeration of Pathogenic Vibrios

The Food and Drug Administration (FDA) *Bacteriological Analytical Manual* (http://www.cfsan.fda.gov/~ebam/bam-9.html) specifies two basic approaches for enumeration of vibrios: (i) MPN and (ii) colony counts determined by direct spread plating using DNA probe identification of colonies (Kaysner and DePaola, 2004). Enumeration of vibrios generally begins with homogenization of fish or shellfish tissue samples in phosphate-buffered saline. MPN is based on titration of homogenates into multiple enrichment cultures, usually in alkaline peptone water, described above. Species-specific growth is indicated by isolation of selective agars (usually TCBS, colistin agar, or mCPC), with confirmation by biochemical, antibody, or molecular identification of species. The number of tubes with positive growth for each dilution is used to calculate a statistically valid assessment of the MPN in the sample. Alternatively, colony counts are derived from dilutions of homogenate that are spread plated on nonselective media, and numbers are determined by DNA probe identification of colonies. Both assays can be complicated by the overgrowth of competing nontarget organisms.

Ideally, shellfish monitoring should discriminate pathogenic vibrios in hours rather than days. Molecular techniques, such as PCR, offer more rapid analysis and increase assay sensitivity and specificity necessary to accurately detect low numbers of *V. vulnificus* organisms. Quantitative or real-time PCR (QPCR) assays provide quantitative analysis while reducing the need for postassay processing. In these assays, the formation rate of the PCR products is detected by increased fluorescence signal that corresponds directly to the DNA target concentration. Sensitivity of QPCR assays is limited by the small sample volume size, but molecular technologies can be combined with standard enrichment protocols to increase both sensitivity and sample size. Several studies have shown that PCR, when used in combination with MPN methodology, can reduce the limit of detection to 1 bacterium/g (Panicker and Bej, 2005; Blackstone et al., 2003; Wright et al., 2007). Recent side-by-side comparison of the standard FDA MPN analysis versus direct QPCR confirmation of positive MPN enrichment samples demonstrated that the QPCR-MPN assay had excellent correlation with the standard method and enhanced the sensitivity of the assay (Wright et al., 2007). Evaluation of postharvest processed oysters disputed the possibility that dead cells may be enumerated by QPCR-MPN, as only viable cells were detected.

Emerging technologies for detection of *Vibrio* species include DNA microarrays, biosensors, and spectroscopic methods. DNA microarrays have been utilized to explore chitin utilization (Meibom et al., 2004) and the genetic basis of endemic versus pandemic disease (Dziejman et al., 2002) in *V. cholerae*. Biosensors that recognize specific cellular targets or products, such as cell wall proteins, nucleic acids, and toxins, have been employed to detect pathogens such as *E. coli* O157:H7 (DeMarco et al., 1999) and *V. cholerae* (Carter et al., 1995) without culturing the organisms. However, recovery of cells and subsequent culture are often preferable for confirmation or additional assessment (Tims and Lim, 2003).

CONCLUSIONS

Vibrios are potentially lethal food-borne pathogens that naturally inhabit estuaries and river basins. The perceived threat of diseases associated with these organisms has impacted tremendously the seafood industry. Concerted efforts by government agencies and the seafood industry have successfully implemented postharvest processing in strategic states on the Gulf Coast. CDC (2005) surveillance spanning from 1997 to 2004 documented a 78% increase in disease related to *Vibrio* species. These recent increases of vibrio infections in the United States are primarily a consequence of the number of cases related to outbreaks of *V. parahaemolyticus* (CDC, 2007). Monitoring of this organism in shellfish is now common practice on the Pacific Coast. However, monitoring is complicated by the facts that their presence does not correlate with conventional measurements of seafood safety, such as fecal coliform analysis, and that virulent strains are difficult to detect in the environment. *Vibrio* species show intimate associations with marine fauna, and the contributions of these associations to the evolution of human diseases are unknown. Greater understanding of the role of these bacteria in estuarine ecosystems and the risks associated with environmental, bacterial, and host factors is crucial for control and safety of seafood products.

REFERENCES

Altekruse, S. F., M. L. Cohen, and D. L. Swerdlow. 1997. Emerging foodborne diseases. *Emerg. Infect. Dis.* 3:285–293.

Amako, K., K. Okada, and S. Miake. 1984. Evidence for the presence of a capsule in *Vibrio vulnificus. J. Gen. Microbiol.* 130:2741–2743.

Andrews, L. S., D. L. Parks, and Y. P. Chen. 2000. Low temperature pasteurization to reduce the risk of vibrio infections from raw shell-stock oysters. *Food Addit. Contam.* 17:787–791.

Baffone, W., A. Casaroli, R. Campana, B. Citterio, E. Vittoria, L. Pierfelici, and G. Donelli. 2005. 'In vivo' studies on the pathophysiological mechanism of *Vibrio parahaemolyticus* TDH+-induced secretion. *Microb. Pathog.* 38:133–137.

Baffone, W., B. Citterio, E. Vittoria, A. Casaroli, R. Campana, L. Falzano, and G. Donelli. 2003. Retention of virulence in viable but non-culturable halophilic Vibrio spp. *Int. J. of Food Microbiol.* 89:31–39.

Baumann, P., and R. H. W. Schubert. 1984. Family II. Vibrionaceae, p. 516–550. In N. R. Krieg and J. G. Holt (ed.), *Bergey's Manual of Systematic Bacteriology.* Williams & Wilkins, Baltimore, MD.

Beuchat, L. R. 1973. Interacting effects of pH, temperature, and salt concentration on growth and survival of Vibrio parahaemolyticus. *Appl. Microbiol.* 25:844–846.

Birkenhauer, J. M., and J. D. Oliver. 2003. Use of diacetyl to reduce the load of Vibrio vulnificus in the Eastern oyster, Crassostrea virginica. *J. Food Prot.* **66:**38–43.

Bisharat, N., V. Agmon, R. Finkelstein, R. Raz, G. Ben-Dror, L. Lerner, S. Soboh, R. Colodner, D. N. Cameron, D. L. Wykstra, D. L. Swerdlow, and J. J. Farmer III. 1999. Clinical, epidemiological, and microbiological features of Vibrio vulnificus biogroup 3 causing outbreaks of wound infection and bacteraemia in Israel. Israel Vibrio Study Group. *Lancet* **354:**1421–1424.

Bisharat, N., D. I. Cohen, R. M. Harding, D. Falush, D. W. Crook, T. Peto, and M. C. Maiden. 2005. Hybrid Vibrio vulnificus. *Emerg. Infect. Dis.* **11:**30–35.

Blackstone, G. M., J. L. Nordstrom, M. C. Vickery, M. D. Bowen, R. F. Meyer, and A. DePaola. 2003. Detection of pathogenic Vibrio parahaemolyticus in oyster enrichments by real time PCR. *J. Microbiol. Methods* **53:**149–155.

Blake, P. A., M. H. Merson, R. E. Weaver, D. G. Hollis, and P. C. Heublein. 1979. Disease caused by a marine Vibrio. Clinical characteristics and epidemiology. *N. Engl. J. Med.* **300:**1–5.

Brasher, C. W., A. DePaola, D. D. Jones, and A. K. Bej. 1998. Detection of microbial pathogens in shellfish with multiplex PCR. *Curr. Microbiol.* **37:**101–107.

Brauns, L. A., M. C. Hudson, and J. D. Oliver. 1991. Use of the polymerase chain reaction in detection of culturable and noncul-turable Vibrio vulnificus cells. *Appl. Environ. Microbiol.* **57:**2651–2655.

Bryan, P. J., R. J. Steffan, A. DePaola, J. W. Foster, and A. K. Bej. 1999. Adaptive response to cold temperatures in Vibrio vulnifi-cus. *Curr. Microbiol.* **38:**168–175.

Cabelli, V. J., and. W. P. Heffernau. 1971. Seasonal factors relevant to coliform levels in northern quahaugs. *Proc. Natl. Shellfish Assoc.* **61:**95–101.

Campbell, M. S., and A. C. Wright. 2003. Real-time PCR analysis of Vibrio vulnificus from oysters. *Appl. Environ. Microbiol.* **69:**7137–7144.

Canzonier, W. J. 1991. *Historical Perspective on Commercial Depuration of Shellfish.* CRC Press, Inc., Boca Raton, FL.

Carter, R. M., J. J. Mekalanos, M. B. Jacobs, G. J. Lubrano, and G. G. Guilbault. 1995. Quartz crystal microbalance detection of Vibrio cholerae O139 serotype. *J. Immunol. Methods* **187:**121–125.

CDC. 1991. Epidemiologic notes and reports: cholera—New Jersey and Florida. *MMWR Morb. Mortal. Wkly. Rep.* **40:**287–289.

CDC. 2005. Vibrio illnesses after Hurricane Katrina—multiple states, August–September 2005. *MMWR Morb. Mortal. Wkly. Rep.* **54:**928–931.

CDC. 2007. Preliminary FoodNet data on the incidence of infec-tion with pathogens transmitted commonly through food—10 States, 2006. *MMWR Morb. Mortal. Wkly. Rep.* **56:**336–339.

Chatzidaki-Livanis, M., M. A. Hubbard, K. Gordon, V. J. Harwood, and A. C. Wright 2006a. Genetic distinctions among clinical and environmental strains of Vibrio vulnificus. *Appl. Environ. Microbiol.* **72:**6136–6141.

Chatzidaki-Livanis, M., M. K. Jones, and A. C. Wright. 2006b. Genetic variation in the Vibrio vulnificus group 1 capsular poly-saccharide operon. *J. Bacteriol.* **188:**1987–1998.

Chuang, Y.-C., W.-C. Ko, S.-T. Wang, J.-W. Liu, C.-F. Kuo, J.-J. Wu, and K.-Y. Huang. 1998. Minocycline and cefotaxime in the treatment of experimental murine Vibrio vulnificus infec-tion. *Antimicrob. Agents Chemother.* **42:**1319–1322.

Cleland, D., M. B. Thomas, D. Strickland, and J. Oliver. 1985. A comparison of media for the isolation of Vibrio spp. from envi-ronmental sources, abstr. N-16, p. 220. *Abstr. Annu. Meet. Am. Soc. Microbiol.*

Colwell, R. R., J. Kaper, and S. W. Joseph. 1977. Vibrio cholerae, Vibrio parahaemolyticus, and other vibrios: occurrence and dis-tribution in Chesapeake Bay. *Science* **198:**394–396.

Colwell, R. R., and J. Liston. 1960. Microbiology of shellfish. Bacteriological study of the natural flora of Pacific oysters (*Crassostrea gigas*). *Appl. Environ. Microbiol.* **8:**104–109.

Cook, D. W. 1994. Effect of time and temperature on multiplica-tion of Vibrio vulnificus in postharvest Gulf Coast shellstock oysters. *Appl. Environ. Microbiol.* **60:**3483–3484.

Cook, D. W., P. Oleary, J. C. Hunsucker, E. M. Sloan, J. C. Bowers, R. J. Blodgett, and A. Depaola. 2002. Vibrio vulnificus and Vibrio parahaemolyticus in U.S. retail shell oysters: a national survey from June 1998 to July 1999. *J. Food Prot.* **65:**79–87.

Cortelyou, J. R., M. A. McWhinnie, M. S. Riddiford, and J. E. Semrad. 1954. The effects of ultraviolet irradiation on large populations of certain water-borne bacteria in motion. II. Some physical factors affecting the effectiveness of germicidal ultravio-let irradiation. *Appl. Microbiol.* **2:**269–273.

Cruz-Romero, M., A. L. Kelly, and J. P. Kerry. 2007. Effects of high-pressure and heat treatments on physical and biochemical characteristics of oysters (Crassostrea gigas). *Innov. Food Sci. Emerg. Technol.* **8:**30–38.

DeMarco, D. R., E. W. Saaski, D. A. McCrae, and D. V. Lim. 1999. Rapid detection of Escherichia coli O157:H7 in ground beef using a fiber-optic biosensor. *J. Food Prot.* **62:**711–716.

DePaola, A., G. M. Capers, and D. Alexander. 1994. Densities of Vibrio vulnificus in the intestines of fish from the U.S. Gulf Coast. *Appl. Environ. Microbiol.* **60:**984–988.

DePaola, A., J. L. Nordstrom, J. C. Bowers, J. G. Wells, and D. W. Cook. 2003a. Seasonal abundance of total and patho-genic Vibrio parahaemolyticus in Alabama oysters. *Appl. Environ. Microbiol.* **69:**1521–1526.

DePaola, A., J. L. Nordstrom, A. Dalsgaard, A. Forslund, J. Oliver, T. Bates, K. L. Bourdage, and P. A. Gulig. 2003b. Analysis of Vibrio vulnificus from market oysters and septice-mia cases for virulence markers. *Appl. Environ. Microbiol.* **69:**4006–4011.

DiRita, V. J., C. Parsot, G. Jander, and J. J. Mekalanos. 1991. Regulatory cascade controls virulence in Vibrio cholerae. *Proc. Nat. Acad. Sci. USA* **88:**5403–5407.

Dziejman, M., E. Balon, D. Boyd, C. M. Fraser, J. F. Heidelberg, and J. J. Mekalanos. 2002. Comparative genomic analysis of Vibrio cholerae: genes that correlate with cholera endemic and pandemic disease. *Proc. Nat. Acad. Sci. USA* **99:**1556–1561.

Erdman, I. E., and A. D. Tennant. 1956. The self-cleansing of soft-shell clams: bacteriological and public health aspects. *Can. J. Public Health* **47:**196–202.

Esteve-Gassent, M. D., R. Barrera, and C. Amaro. 2004. Vaccination of market-size eels against vibriosis due to Vibrio vulnificus sero-var E. *Aquaculture* **241:**9–19.

Fan, J. J., C. P. Shao, Y. C. Ho, C. K. Yu, and L. I. Hor. 2001. Isolation and characterization of a Vibrio vulnificus mutant defi-cient in both extracellular metalloprotease and cytolysin. *Infect. Immun.* **69:**5943–5948.

Farber, J. M. 1993. Current research on Listeria monocytogenes in foods: an overview. *J. Food Prot.* **56:**640–643.

Feldhusen, F. 2000. The role of seafood in bacterial foodborne dis-eases. *Microbes Infect.* **2:**1651–1660.

Finkelstein, R. A., and J. J. LoSpalluto. 1969. Pathogenesis of experimental cholera. Preparation and isolation of choleragen and choleragenoid. *J. Exp. Med.* **130:**185–202.

Ford, L., D. G. Lalloo, and D. R. Hill. 2007. Cholera vaccines—authors' reply. *Lancet Infect. Dis.* **7:**178.

Fujino, T. Y., Y. Okuno, D. Nakada, A. Aoyama, K. Fukai, T. Mukai, and T. Ueho. 1953. On the bacteriological examina-tion of Shirasu food poisoning. *Med. J. Osaka Univ.* **4:**299–304.

Furfari, S. A. 1979. Monitoring shellfish purification, p. 1–13. *In Source Book on Shellfish Purification.* USPHS, FDA, Northeast Tech, Davisville, RI.

Furfari, S. A. 1991. *Design of Depuration Systems.* CRC Press, Inc., Boca Raton, FL.

Gooch, J. A., A. DePaola, J. Bowers, and D. L. Marshall. 2002. Growth and survival of Vibrio parahaemolyticus in postharvest American oysters. *J. Food Prot.* 65:970–974.

Gulig, P. A., K. L. Bourdage, and A. M. Starks. 2005. Molecular pathogenesis of Vibrio vulnificus. *J. Microbiol.* 43:118–131.

Hara-Kudo, Y., T. Nishina, H. Nakagawa, H. Konuma, J. Hasegawa, and S. Kumagai. 2001. Improved method for detection of Vibrio parahaemolyticus in seafood. *Appl. Environ. Microbiol.* 67:5819–5823.

He, H., R. M. Adams, D. F. Farkas, and M. T. Morrissey. 2002. Use of high-pressure processing for oyster shucking and shelf-life extension. *J. Food Sci.* 67:640–645.

Hill, W. F., Jr., F. E. Hamblet, and W. H. Benton. 1969. Inactivation of poliovirus type 1 by the Kelly-Purdy ultraviolet seawater treatment unit. *Appl. Microbiol.* 17:1–6.

Hill, W. F., Jr., F. E. Hamblet, W. H. Benton and E. W. Akin. 1970. Ultraviolet devitalization of eight selected enteric viruses in estuarine water. *Appl. Microbiol.* 19:805–812.

Hlady, W. G., and K. C. Klontz. 1996. The epidemiology of Vibrio infections in Florida, 1981–1993. *J. Infect. Dis.* 173:1176–1183.

Hoi, L., I. Dalsgaard, and A. Dalsgaard. 1998. Improved isolation of Vibrio vulnificus from seawater and sediment with cellobiose-colistin agar. *Appl. Environ. Microbiol.* 64:1721–1724.

Hollis, D. G., R. E. Weaver, C. N. Baker, and C. Thornsberry. 1976. Halophilic *Vibrio* spp. isolated from blood cultures. *J. Clin. Microbiol.* 3:425.

Honda, T., Y. X. Ni, and T. Miwatani. 1988. Purification and characterization of a hemolysin produced by a clinical isolate of Kanagawa phenomenon-negative Vibrio parahaemolyticus and related to the thermostable direct hemolysin. *Infect. Immun.* 56:961–965.

Howard, R. J., and S. Lieb. 1988. Soft-tissue infections caused by halophilic marine vibrios. *Arch. Surg.* 123:245–249.

Huff, C. B., H. F. Smith, W. D. Boring, and N. A. Clarke. 1965. Study of ultraviolet disinfection of water and factors in treatment efficiency. *Public Health Rep.* 80:695–705.

Johnston, M. D., and M. H. Brown. 2002. An investigation into the changed physiological state of Vibrio bacteria as a survival mechanism in response to cold temperatures and studies on their sensitivity to heating and freezing. *J. Appl. Microbiol.* 92:1066–1077.

Joseph, S. W., R. R. Colwell, and J. B. Kaper. 1982. Vibrio parahaemolyticus and related halophilic vibrios. *Crit. Rev. Microbiol.* 10:77–124.

Kaneko, T., and R. R. Colwell. 1973. Ecology of Vibrio parahaemolyticus in Chesapeake Bay. *J. Bacteriol.* 113:24–32.

Kaneko, T., and R. R. Colwell. 1975a. Adsorption of Vibrio parahaemolyticus onto chitin and copepods. *Appl. Microbiol.* 29:269–274.

Kaneko, T., and R. R. Colwell. 1975b. Incidence of Vibrio parahaemolyticus in Chesapeake Bay. *Appl. Microbiol.* 30:251–257.

Kaper, J. B., J. G. Morris, Jr., and M. M. Levine. 1995. Cholera (Erratum, 8:316.) *Clin. Microbiol. Rev.* 8:48–86.

Karunasagar, I., M. N. Venugopal, and K. Segar. 1986. Evaluation of methods for enumeration of Vibrio parahaemolyticus from seafood. *Appl. Environ. Microbiol.* 52:583–585.

Kaspar, C. W., and M. L. Tamplin. 1993. Effects of temperature and salinity on the survival of Vibrio vulnificus in seawater and shellfish. *Appl. Environ. Microbiol.* 59:2425–2429.

Kawabata, T., and T. Harada. 1959. The disinfection of water by germicidal lamp. *J. Illumination Soc.* 36:89.

Kaysner, C. A., and A. DePaola, Jr. 2004. Chapter 9. *Vibrio cholerae, V. parahaemolyticus, V. vulnificus,* and other *Vibrio* spp. *In Bacteriological Analytical Manual,* 8th ed. U.S. FDA, Washington, DC.

Kaysner, C. A., M. M. Wekell, and C. Abeyta, Jr. 1990. Enhancement of virulence of two environmental strains of Vibrio vulnificus after passage through mice. *Diagn. Microbiol. Infect. Dis.* 13:285–288.

Kelly, C. B. 1961. Disinfection of sea water by ultraviolet radiation. *Am. J. Public Health Nations Health* 51:1670–1680.

Kelly, M., and E. Stroh. 1988. Occurrence of Vibrionaceae in natural and cultivated oyster populations in the Pacific Northwest. *Diagn. Microbiol. Infect. Dis.* 9:1–5.

Kelly, M. T. 1982. Effects of temperature and salinity on Vibrio (Beneckea) vulnificus occurrence in a Gulf Coast environment. *Appl. Environ. Microbiol.* 44:820–824.

Keyhani, N. O., and S. Roseman. 1999. Physiological aspects of chitin catabolism in marine bacteria. *Biochim. Biophys. Acta* 1473:108–122.

Kim, S. Y., S. E. Lee, Y. R. Kim, C. M. Kim, P. Y. Ryu, H. E. Choy, S. S. Chung, and J. H. Rhee. 2003. Regulation of Vibrio vulnificus virulence by the LuxS quorum-sensing system. *Mol. Microbiol.* 48:1647–1664.

Kim, Y. R., and J. H. Rhee. 2003. Flagellar basal body flg operon as a virulence determinant of Vibrio vulnificus. *Biochem. Biophys. Res. Commun.* 304:405–410.

Klontz, K. C., S. Lieb, M. Schreiber, H. T. Janowski, L. M. Baldy, and R. A. Gunn. 1988. Syndromes of Vibrio vulnificus infections. Clinical and epidemiologic features in Florida cases, 1981–1987. *Ann. Intern. Med.* 109:318–323.

Kobayashi, T., S. Enomoto, R. Sakazaki, and S. Kuwahara. 1963. A new selective isolation medium for the Vibrio group; on a modified Nakanishi's medium (TCBS agar medium). *Nippon Saikingaku Zasshi* 18:387–392.

Lee, C.-Y., G. Panicker, and A. K. Bej. 2003. Detection of pathogenic bacteria in shellfish using multiplex PCR followed by CovaLink NH microwell plate sandwich hybridization. *J. Microbiol. Methods* 53:199–209.

Lee, J. H., M. W. Kim, B. S. Kim, S. M. Kim, B. C. Lee, T. S. Kim, and S. H. Choi. 2007. Identification and characterization of the Vibrio vulnificus rtxA essential for cytotoxicity in vitro and virulence in mice. *J. Microbiol.* 45:146–152.

Lee, S. E., S. Y. Kim, B. C. Jeong, Y. R. Kim, S. J. Bae, O. S. Ahn, J. J. Lee, H. C. Song, J. M. Kim, H. E. Choy, S. S. Chung, M. N. Kweon, and J. H. Rhee. 2006. A bacterial flagellin, Vibrio vulnificus FlaB, has a strong mucosal adjuvant activity to induce protective immunity. *Infect. Immun.* 74: 694–702.

Lee, S. H., S. M. Butler, and A. Camilli. 2001. Selection for in vivo regulators of bacterial virulence. *Proc. Natl. Acad. Sci. USA* 98:6889–6894.

Lin, M., D. A. Payne, and J. R. Schwarz. 2003. Intraspecific diversity of Vibrio vulnificus in Galveston Bay water and oysters as determined by randomly amplified polymorphic DNA PCR. *Appl. Environ. Microbiol.* 69:3170–3175.

Lipinska, E. 1977. Nisin and its applications, p. 103–130. *In* M. Woodbine (ed.), *Antimicrobials and Antibiosis in Agriculture,* Butterworths, London, United Kingdom.

Lipp, E. K., A. Huq, and R. R. Colwell. 2002. Effects of global climate on infectious disease: the cholera model. *Clin. Microbiol. Rev.* 15:757–770.

Litwin, C. M., T. W. Rayback, and J. Skinner. 1996. Role of catechol siderophore synthesis in Vibrio vulnificus virulence. *Infect. Immun.* 64:2834–2838.

Liverman, A. D. B., H.-C. Cheng, J. E. Trosky, D. W. Leung, M. L. Yarbrough, D. L. Burdette, M. K. Rosen and K. Orth.

2007. Arp2/3-independent assembly of actin by Vibrio type III effector VopL. *Proc. Nat. Acad. Sci. USA* **104:**17117–17122.

Lopez-Caballero, M. E., M. Perez-Mateos, P. Montero, and A. J. Borderias. 2000. Oyster preservation by high-pressure treatment. *J. Food Prot.* **63:**196–201.

Lyerly, D. M., D. E. Lockwood, S. H. Richardson, and T. D. Wilkins. 1982. Biological activities of toxins A and B of *Clostridium difficile. Infect. Immun.* **35:**1147–1150.

Lyon, W. J. 2001. TaqMan PCR for detection of *Vibrio cholerae* O1, O139, non-O1, and non-O139 in pure cultures, raw oysters, and synthetic seawater. *Appl. Environ. Microbiol.* **67:**4685–4693.

Mallett, J. C., L. E. Beghian, and T. Metcalf. 1987. Potential of irradiation technology for improved shellfish sanitation, p. 247–258. *In* G. E. Rodrick, R. E. Martin, and W. S. Otwell (ed.), *Molluscan Shellfish Depuration*, CRC Press, Inc., Boca Raton, FL.

McLaughlin, J. B., A. DePaola, C. A. Bopp, K. A. Martinek, N. P. Napolilli, C. G. Allison, S. L. Murray, E. C. Thompson, M. M. Bird, and J. P. Middaugh. 2005. Outbreak of Vibrio parahaemolyticus gastroenteritis associated with Alaskan oysters. *N. Engl. J. Med.* **353:**1463–1470.

Meibom, K. L., X. B. Li, A. T. Nielsen, C. Y. Wu, S. Roseman, and G. K. Schoolnik. 2004. The Vibrio cholerae chitin utilization program. *Proc. Natl. Acad. Sci. USA* **101:**2524–2529.

Miller, V. L., and J. J. Mekalanos. 1984. Synthesis of cholera toxin is positively regulated at the transcriptional level by toxR. *Proc. Natl. Acad. Sci. USA* **81:**3471–3475.

Miyamoto, Y., T. Kato, Y. Obara, S. Akiyama, K. Takizawa, and S. Yamai. 1969. In vitro hemolytic characteristic of *Vibrio parahaemolyticus:* its close correlation with human pathogenicity. *J. Bacteriol.* **100:**1147–1149.

Molenda, J. R., W. G. Johnson, M. Fishbein, B. W. I. J. Mehlman, J. Thoburn, and A. Dadisman. 1972. *Vibrio parahaemolyticus* gastroenteritis in Maryland: laboratory aspects. *Appl. Microbiol.* **24:**444–448.

Morris, J. G. J. 2003. Cholera and other types of vibriosis: a story of human pandemics and oysters on the half shell. *Clin. Infect. Dis.* **37:**272–280.

Morrison, R. M. 1986. Food irradiation: a look at regulatory status, consumer acceptance, and economies of scale. *J. Food Distrib. Res.* **29:**38–84.

Motes, M. L., A. DePaola, D. W. Cook, J. E. Veazey, J. C. Hunsucker, W. E. Garthright, R. J. Blodgett, and S. J. Chirtel. 1998. Influence of water temperature and salinity on *Vibrio vulnificus* in Northern Gulf and Atlantic Coast oysters (*Crassostrea virginica*). *Appl. Environ. Microbiol.* **64:**1459–1465.

Murphy, S. K., and J. D. Oliver. 1992. Effects of temperature abuse on survival of *Vibrio vulnificus* in oysters. *Appl. Environ. Microbiol.* **58:**2771–2775.

Nair, G. B., T. Ramamurthy, S. K. Bhattacharya, B. Dutta, Y. Takeda, and D. A. Sack. 2007. Global dissemination of *Vibrio parahaemolyticus* serotype O3:K6 and its serovariants. *Clin. Microbiol. Rev.* **20:**39–48.

Nair, G. B., T. Ramamurthy, S. K. Bhattacharya, A. K. Mukhopadhyay, S. Garg, M. K. Bhattacharya, T. Takeda, T. Shimada, Y. Takeda, and B. C. Deb. 1994. Spread of Vibrio cholerae O139 Bengal in India. *J. Infect. Dis.* **169:**1029–1034.

Nilsson, W. B., R. N. Paranjype, A. DePaola, and M. S. Strom. 2003. Sequence polymorphism of the 16S rRNA gene of *Vibrio vulnificus* is a possible indicator of strain virulence. *J. Clin. Microbiol.* **41:**442–446.

Nishibuchi, M., A. Fasano, R. G. Russell, and J. B. Kaper. 1992. Enterotoxigenicity of *Vibrio parahaemolyticus* with and without genes encoding thermostable direct hemolysin. *Infect. Immun.* **60:**3539–3545.

Nordstrom, J. L., R. Rangdale, M. C. Vickery, A. M. Phillips, S. L. Murray, S. Wagley, and A. DePaola. 2006. Evaluation of an alkaline phosphatase-labeled oligonucleotide probe for the detection and enumeration of the thermostable-related hemolysin (trh) gene of Vibrio parahaemolyticus. *J. Food Prot.* **69:**2770–2772.

Novotny, L., L. Dvorska, A. Lorencova, V. Beran, and I. Pavlik. 2004. Fish: a potential source of bacterial pathogens for human beings. *Vet. Med.* **49:**343–358.

NSSP. 2003. Control of shellfish harvesting. *In Guide for the Control of Molluscan Shellfish.* US FDA, Washington, DC.

NSSP. 2005. Florida Vibrio vulnificus risk reduction plan for oysters. Department of Agriculture and Consumer Services Report No. DACS-P-0/496.

Oliver, J. D. 2005. Wound infections caused by Vibrio vulnificus and other marine bacteria. *Epidemiol. Infect.* **133:**383–391.

Oliver, J. D., and R. Bockian. 1995. In vivo resuscitation, and virulence towards mice, of viable but nonculturable cells of *Vibrio vulnificus. Appl. Environ. Microbiol.* **61:**2620–2623.

Oliver, J. D., L. Nilsson, and S. Kjelleberg. 1991. Formation of nonculturable *Vibrio vulnificus* cells and its relationship to the starvation state. *Appl. Environ. Microbiol.* **57:**2640–2644.

Ono, M., Y. Inoue, and M. Yokoyama. 2001. A cluster of *Vibrio vulnificus* infection in Kumamoto. *Infect. Agents Surveill. Rep.* No. 22.

Panicker, G., and A. K. Bej. 2005. Real-time PCR detection of *Vibrio vulnificus* in oysters: comparison of oligonucleotide primers and probes targeting *vvhA. Appl. Environ. Microbiol.* **71:**5702–5709.

Panicker, G., M. L. Myers, and A. K. Bej. 2004. Rapid detection of *Vibrio vulnificus* in shellfish and Gulf of Mexico water by real-time PCR. *Appl. Environ. Microbiol.* **70:**498–507.

Paranjpye, R. N., A. B. Johnson, A. E. Baxter, and M. S. Strom. 2007. Role of type IV pilins in persistence of *Vibrio vulnificus* in *Crassostrea virginica* oysters. *Appl. Environ. Microbiol.* **73:**5041–5044.

Paranjpye, R. N., and M. S. Strom. 2005. A *Vibrio vulnificus* type IV pilin contributes to biofilm formation, adherence to epithelial cells, and virulence. *Infect. Immun.* **73:**1411–1422.

Park, S. D., H. S. Shon, and N. J. Joh. 1991. Vibrio vulnificus septicemia in Korea: clinical and epidemiologic findings in seventy patients. *J. Am. Acad. Dermatol.* **24:**397–403.

Parker, R. W., E. M. Maurer, A. B. Childers, and D. H. Lewis. 1994. Effect of frozen storage and vacuum-packaging on survival of Vibrio vulnificus in Gulf Coast oysters (Crassostrea virginica). *J. Food. Prot.* **57:**604–606.

Provenzano, D., and K. E. Klose. 2000. Altered expression of the ToxR-regulated porins OmpU and OmpT diminishes Vibrio cholerae bile resistance, virulence factor expression, and intestinal colonization. *Proc. Natl. Acad. Sci. USA* **97:**10220–10224.

Quevedo, A. C., J. G. Smith, G. E. Rodrick, and A. C. Wright. 2005. Ice immersion as a postharvest treatment of oysters for the reduction of Vibrio vulnificus. *J. Food Prot.* **68:**1192–1197.

Ravi, A. V., K. S. Musthafa, G. Jegathammbal, K. Kathiresan, and S. K. Pandian. 2007. Screening and evaluation of probiotics as a biocontrol agent against pathogenic Vibrios in marine aquaculture. *Lett. Appl. Microbiol.* **45:**219–223.

Richards, G. P. S. D. 1990. Shellfish depuration, p. 395–428. *In* C. R. Hackney and D. R. Ward (ed.), *Microbiology of Marine Food Products.* Van Nostrand Reinhold, New York, NY.

Richardson, K. 1991. Roles of motility and flagellar structure in pathogenicity of *Vibrio cholerae*: analysis of motility mutants in three animal models. *Infect. Immun.* **59:**2727–2736.

Rodrick, G. E., and K. R. Schneider. 1991. Vibrios in depuration, p. 115–125. *In* W. Otwell, G. E. Roderick, and R. E. Martin (ed.), *Molluscan Shellfish Depuration*. CRC Press, Inc., Boca Raton, FL.

Rosche, T. M., Y. Yano, and J. D. Oliver. 2005. A rapid and simple PCR analysis indicates there are two subgroups of Vibrio vulnificus which correlate with clinical or environmental isolation. *Microbiol. Immunol.* **49:**381–389.

Ruby, E. G. 1996. Lessons from a cooperative, bacterial-animal association: the Vibrio fischeri-Euprymna scolopes light organ symbiosis. *Annu. Rev. Microbiol.* **50:**591–624.

Ruby, E. G., and K.-H. Lee. 1998. The *Vibrio fischeri-Euprymna* scolopes light organ association: current ecological paradigms. *Appl. Environ. Microbiol.* **64:**805–812.

Sack, G. H., Jr., N. F. Pierce, K. N. Hennessey, R. C. Mitra, R. B. Sack, and D. N. Mazumder. 1972. Gastric acidity in cholera and noncholera diarrhoea. *Bull. W. H. O.* **47:**31–36.

Sandvig, K., and B. van Deurs. 2002. Transport of protein toxins into cells: pathways used by ricin, cholera toxin and Shiga toxin. *FEBS Lett.* **529:**49–53.

Schneider, K. R., F. S. Steslow, F. S. Sierra, G. E. Rodrick, and C. I. Noss. 1991. Ozone depuration of Vibrio vulnificus from the southern quahog clam, Mercenaria campechiensis. *J. Invertebr. Pathol.* **57:**184–190.

Shapiro, R. L., S. Altekruse, L. Hutwagner, R. Bishop, R. Hammond, S. Wilson, B. Ray, S. Thompson, R. V. Tauxe, P. M. Griffin, and the Vibrio Working Group. 1998. The role of Gulf Coast oysters harvested in warmer months in Vibrio vulnificus infections in the United States, 1988–1996. *J. Infect. Dis.* **178:**752–759.

Simpson, L. M., V. K. White, S. F. Zane, and J. D. Oliver. 1987. Correlation between virulence and colony morphology in *Vibrio vulnificus*. *Infect. Immun.* **55:**269–272.

Stevens, K. A., and L. A. Jaykus. 2004. Bacterial separation and concentration from complex sample matrices: a review. *Crit. Rev. Microbiol.* **30:**7–24.

Su, Y.-C., and C. Liu. 2007. Vibrio parahaemolyticus: a concern of seafood safety. *Food Microbiol.* **24:**549–558.

Sun, Y., and J. D. Oliver. 1994. Antimicrobial action of some GRAS compounds against Vibrio vulnificus. *Food Addit. Contam.* **11:**549–558.

Tall, B. D., J. F. La Peyre, J. W. Bier, M. D. Miliotis, D. E. Hanes, M. H. Kothary, D. B. Shah, and M. Faisal. 1999. *Perkinsus marinus* extracellular protease modulates survival of *Vibrio vulnificus* in eastern oyster (*Crassostrea virginica*) hemocytes. *Appl. Environ. Microbiol.* **65:**4261–4263.

Tamplin, M., G. E. Rodrick, N. J. Blake, and T. Cuba. 1982. Isolation and characterization of *Vibrio vulnificus* from two Florida estuaries. *Appl. Environ. Microbiol.* **44:** 1466–1470.

Tamplin, M. L., and G. M. Capers. 1992. Persistence of *Vibrio vulnificus* in tissues of Gulf Coast oysters, *Crassostrea virginica*, exposed to seawater disinfected with UV light. *Appl. Environ. Microbiol.* **58:**1506–1510.

Tamplin, M. L., A. L. Martin, A. D. Ruple, D. W. Cook, and C. W. Kaspar. 1991. Enzyme immunoassay for identification of *Vibrio vulnificus* in seawater, sediment, and oysters. *Appl. Environ. Microbiol.* **57:**1235–1240.

Tamplin, M. L., S. Specter, G. E. Rodrick, and H. Friedman. 1985. *Vibrio vulnificus* resists phagocytosis in the absence of serum opsonins. *Infect. Immun.* **49:**715–718.

Tantillo, G. M., M. Fontanarosa, A. Di Pinto, and M. Musti. 2004. Updated perspectives on emerging vibrios associated with human infections. *Lett. Appl. Microbiol.* **39:**117–126.

Tarr, C. L., J. S. Patel, N. D. Puhr, E. G. Sowers, C. A. Bopp, and N. A. Strockbine. 2007. Identification of *Vibrio* isolates by a multiplex PCR assay and *rpoB* sequence determination. *J. Clin. Microbiol.* **45:**134–140.

Tauxe, R. V., E. D. Mintz, and R. E. Quick. 1995. Epidemic cholera in the new world: translating field epidemiology into new prevention strategies. *Emerg. Infect. Dis.* **1:**141–146.

Taylor R, S. C., K. Peterson, P. Spears, and J. Mekalanos. 1988. Safe, live Vibrio cholerae vaccines? *Vaccine* **6:**151–154.

Taylor, R. K., V. L. Miller, D. B. Furlong, and J. J. Mekalanos. 1987a. Use of *phoA* gene fusions to identify a pilus colonization factor coordinately regulated with cholera toxin. *Proc. Natl. Acad. Sci. USA* **84:**2833–2837.

Taylor, R. K., V. L. Miller, D. B. Furlong, and J. J. Mekalanos. 1987b. Use of *phoA* gene fusions to identify a pilus colonization factor coordinately regulated with cholera toxin. *Proc. Natl. Acad. Sci.* **84:**2833–2837.

Tilton, R. C., and R. W. Ryan. 1987. Clinical and ecological characteristics of Vibrio vulnificus in the northeastern United States. *Diagn. Microbiol. Infect. Dis.* **6:**109–117.

Tims, T. B., and D. V. Lim. 2003. Confirmation of viable E. coli O157:H7 by enrichment and PCR after rapid biosensor detection. *J. Microbiol. Methods* **55:**141–147.

Tison, D. L., M. Nishibuchi, J. D. Greenwood, and R. J. Seidler. 1982. *Vibrio vulnificus* biogroup 2: new biogroup pathogenic for eels. *Appl. Environ. Microbiol.* **44:**640–646.

Tripp, M. R. 1960. Mechanisms of removal of injected microorganisms from the American oyster, Crassostrea virginica (Gmelin). *Biol. Bull.* **119:**273–282.

Vasconcelos, G. J., and J. S. Lee. 1972. Microbial flora of Pacific oysters (*Crassostrea gigas*) subjected to ultraviolet-irradiated seawater. *Appl. Environ. Microbiol.* **23:**11–16.

Waldor, M. K., R. Colwell, and J. J. Mekalanos. 1994. The Vibrio cholerae O139 serogroup antigen includes an O-antigen capsule and lipopolysaccharide virulence determinants. *Proc. Natl. Acad. Sci. USA* **91:**11388–11392.

Waldor, M. K., and J. J. Mekalanos. 1996. Lysogenic conversion by a filamentous phage encoding cholera toxin. *Science* **272:**1910–1914.

Wang, R. F., W. W. Cao, and C. E. Cerniglia. 1997. A universal protocol for PCR detection of 13 species of foodborne pathogens in foods. *J. Appl. Microbiol.* **83:**727–736.

Wright, A. C., V. Garrido, G. Debuex, M. Farrell-Evans, A. A. Mudbidri, and W. S. Otwell. 2007. Evaluation of postharvest processed oysters using PCR-based most-probable-number for *Vibrio vulnificus*. *Appl. Environ. Microbiol.* **73:**7477–7481.

Wright, A. C., Y. Guo, J. A. Johnson, J. P. Nataro, and J. G. Morris, Jr. 1992. Development and testing of a nonradioactive DNA oligonucleotide probe that is specific for *Vibrio cholerae* cholera toxin. *J. Clin. Microbiol.* **30:**2302–2306.

Wright, A. C., R. T. Hill, J. A. Johnson, M. C. Roghman, R. R. Colwell, and J. G. Morris, Jr. 1996. Distribution of *Vibrio vulnificus* in the Chesapeake Bay. *Appl. Environ. Microbiol.* **62:**717–724.

Wright, A. C., G. A. Miceli, W. L. Landry, J. B. Christy, W. D. Watkins, and J. G. Morris, Jr. 1993. Rapid identification of *Vibrio vulnificus* on nonselective media with an alkaline phosphatase-labeled oligonucleotide probe. *Appl. Environ. Microbiol.* **59:**541–546.

Wright, A. C., and J. G. Morris, Jr. 1991. The extracellular cytolysin of *Vibrio vulnificus:* inactivation and relationship to virulence in mice. *Infect. Immun.* **59:**192–197.

Wright, A. C., J. L. Powell, J. B. Kaper, and J. G. Morris, Jr. 2001. Identification of a group 1-like capsular polysaccharide operon for *Vibrio vulnificus*. *Infect. Immun.* **69:**6893–6901.

Wright, A. C., L. M. Simpson, and J. D. Oliver. 1981. Role of iron in the pathogenesis of *Vibrio vulnificus* infections. *Infect. Immun.* **34:**503–507.

Xu, H. S., N. Robert, F. L. Singleton, R. W. Attwel, D. J. Grimes, and R. R. Colwell. 1982. Survival and viability of non-culturable

Escherichia coli and *Vibrio cholerae* in the estuarine and marine environment. *Microb. Ecol.* 8:313–323.

Yamamoto, K., A. C. Wright, J. B. Kaper, and J. G. Morris, Jr. 1990. The cytolysin gene of *Vibrio vulnificus:* sequence and relationship to the *Vibrio cholerae* El Tor hemolysin gene. *Infect. Immun.* 58:2706–2709.

Yoshida, S., M. Ogawa, and Y. Mizuguchi. 1985. Relation of capsular materials and colony opacity to virulence of *Vibrio vulnificus. Infect. Immun.* 47:446–451.

Zuckerman, J. N., L. Rombo, and A. Fisch. 2007. The true burden and risk of cholera: implications for prevention and control. *Lancet Infect. Dis.* 7:521–530.

Pathogens and Toxins in Foods: Challenges and Interventions
Edited by V. K. Juneja and J. N. Sofos
© 2010 ASM Press, Washington, DC

Chapter 11

Yersinia enterocolitica and *Yersinia pseudotuberculosis*

Maria Fredriksson-Ahomaa, Miia Lindström, and Hannu Korkeala

TYPE OF ILLNESS AND CHARACTERISTICS OF THE ORGANISM

Yersiniosis

Yersiniosis is an infectious disease caused by *Yersinia*, food-borne yersiniosis being due to *Yersinia enterocolitica* or *Yersinia pseudotuberculosis*. Most human illness worldwide is caused by *Y. enterocolitica*. Most reported cases of yersiniosis are sporadic, and outbreaks are uncommon. In recent years, however, *Y. pseudotuberculosis* has emerged as an outbreak-associated organism (Table 1). Outbreaks of *Y. pseudotuberculosis* infection, often in schoolchildren, have occurred mainly in Finland and Japan (Tsubokura et al., 1989; Nuorti et al., 2004; Jalava et al., 2004, 2006).

In 2004, mandatory notifications of yersiniosis were reported to have taken place in most of the member states of the European Union; however, only a few countries conduct national surveillance for *Y. pseudotuberculosis* (EFSA, 2006). The highest incidences of 14.6, 12.2, 7.6, and 6.8 cases per 100,000 inhabitants were observed in Lithuania, Finland, Sweden, and Germany, respectively. In Austria, the incidence was only 1.7; however, seroprevalence of anti-Yop antibodies among 750 healthy Austrians was shown to be 29.7%, which indicates that the majority of infections are either subclinical or mild (Tomaso et al., 2006). Of the 9,533 confirmed cases in the European Union, the majority was found to be *Y. enterocolitica*. Due to several outbreaks, the incidence of *Y. pseudotuberculosis* is high in Finland (Figure 1).

Yersiniosis caused by *Y. enterocolitica* and *Y. pseudotuberculosis* is transmitted by the fecal-oral route. Infections are acquired primarily through the gastrointestinal tract as a result of contaminated foods or water (Bottone, 1999; Smego et al., 1999). After ingestion, *Y. enterocolitica* and *Y. pseudotuberculosis*

initially travel to the small intestine and bind to the intestinal epithelium of the terminal ileum. These bacteria penetrate the intestinal mucosa through M cells, the specialized cells of the follicle-associated epithelia that function in antigen uptake. After penetration of the intestinal epithelium, enteropathogenic *Yersinia* replicates within the local lymphoid follicles known as Peyer's patches. The bacteria can then spread to mesenteric lymph nodes, resulting in mesenteric lymphadenitis, which is more typical for *Y. pseudotuberculosis* than for *Y. enterocolitica* (Wren, 2003). Enteropathogenic *Yersinia* can also disseminate from the intestinal tract to the major lymphatic organs, like the spleen, liver, and lungs. However, most infections are localized and self-limiting because the host's inflammatory response is usually able to eliminate the invaders (Bottone, 1999; Smego et al., 1999).

Infection with *Y. enterocolitica* and *Y. pseudotuberculosis* can cause a variety of symptoms, depending on the age of the person infected. Common symptoms are fever, abdominal pain, and diarrhea (Bottone, 1999). Diarrhea is a more dominant symptom of *Y. enterocolitica* infections, while abdominal pain and fever are more frequent in *Y. pseudotuberculosis* infection (Wren, 2003). The minimal infective dose for humans has not been determined. Symptoms typically develop 4 to 7 days after exposure and may last 5 to 14 days, although diarrhea may persist for several weeks. Sometimes diarrhea may be bloody and fever may be high, especially in infants (Jones et al., 2003). In older children, right-sided abdominal pain and fever can be the predominant symptoms and may be confused with appendicitis (Smego et al., 1999). Unnecessary appendectomies have also been observed with yersiniosis patients. The linking of unnecessary appendectomies can help to identify a *Yersinia* outbreak (Nuorti et al., 2004). Occasionally,

Maria Fredriksson-Ahomaa • Institute of Hygiene and Technologie of Food of Animal Origin, Ludwig-Maximilians University Munich, Schoenleutnerstrasse 8, D-85764 Oberschleissheim, Germany. Miia Lindström and Hannu Korkeala • Department of Food and Environmental Hygiene, University of Helsinki, P.O. Box 66, 00014 Helsinki University, Finland.

Table 1. Reported *Yersinia pseudotuberculosis* outbreaks

Yr	Country	Location	Serotype	No. of cases	Source	Reference(s)
1977	Japan	Kindergarten	O:1b	82	Water?	Tsubokura et al. (1989)
1981	Finland	School	O:3	19	Unknown	Tertti et al. (1984)
1981	Japan	School	O:5a	188	Vegetable?	Tsubokura et al. (1989)
1982	Japan	Mountain area	O:4b	140	Water	Tsubokura et al. (1989)
1984	Japan	School	O:3	63	Unknown	Tsubokura et al. (1989)
1984	Japan	Mountain area	O:4b	11	Water	Tsubokura et al. (1989)
1984	Japan	Restaurant	O:5a	39	Barbecue?	Nakano et al. (1989)
1985–86	Japan	School	O:4b	609	Unknown	Tsubokura et al. (1989)
1987	Finland	School	O:1a	34	Unknown	Tertti et al. (1989)
1997	Finland	School	O:3	36	Vegetables?	Hallanvuo et al. (2003)
1998	Canada	Household	O:1b	74	Milk	Nowgesic et al. (1999)
1998–2000	Finland	School, cafeteria	O:3	122	Iceberg lettuce	Hallanvuo et al. (2003), Nuorti et al. (2004)
2001	Finland	School	O:1, O:3	89	Iceberg lettuce	Jalava et al. (2004)
2003	Finland	School	O:1	76	Carrot	Jalava et al. (2006)

typically among adults, complications such as joint pain (reactive arthritis) and skin rash (erythema nodosus) may occur. The reactive arthritis is associated with the presence of the HLA-B27 antigen. The joint pains, most commonly in the knees, ankles, or wrists, usually develop within 1 week to 1 month after the initial infection and generally resolve after 1 to 6 months (Granfors et al., 1989; Hannu et al., 2003).

Uncomplicated cases of diarrhea usually resolve on their own without antimicrobial treatment. In more severe infections, like systemic infection and bacteremia, antimicrobials may be useful. A number of antimicrobial agents are active in vitro against enteropathogenic *Yersinia* strains isolated from human and nonhuman sources. These include aminoglycosides, the third-generation cephalosporins, co-trimoxazole, tetracyclines, chloramphenicol, and fluoroquinolones (Stock and Wiedemann, 1999; Baumgartner et al., 2006).

Characteristics of *Y. enterocolitica* and *Y. pseudotuberculosis*

The genus *Yersinia* is a member of the family *Enterobacteriaceae* and is composed of at least

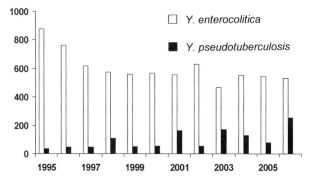

Figure 1. Number of yersiniosis cases in Finland from 1995 to 2006 (http://www3.ktl.fi/stat/).

12 species: *Y. aldovae, Y. aleksiciae, Y. bercovieri, Y. enterocolitica, Y. frederiksenii, Y. intermedia, Y. kristensenii, Y. mollaretii, Y. pestis, Y. pseudotuberculosis, Y. rohdei,* and *Y. ruckeri* (Bottone, 1999; Sprague and Neubauer, 2005). The newest species, *Y. aleksiciae,* has recently been separated from *Y. kristensenii* (Sprague and Neubauer, 2005). Three species, *Y. enterocolitica, Y. pseudotuberculosis,* and *Y. pestis,* are pathogenic for humans. The genome of these pathogens is very similar in size (4.6 to 4.7 Mb) (Thomson et al., 2006). The G+C content of *Y. enterocolitica* and of *Y. pseudotuberculosis* and *Y. pestis* is 47.3 and 47.6, respectively, which is lower than that of most members of the family *Enterobacteriaceae.* Multilocus sequence analysis and DNA-DNA hybridization studies suggest that *Y. enterocolitica* and *Y. pseudotuberculosis* diverged within the last 200 million years and that *Y. pestis* is a clone of *Y. pseudotuberculosis* that has emerged within the last 1,500 to 20,000 years (Achtman et al., 1999; Wren, 2003). Genome sequence data confirm that *Y. pestis* and *Y. pseudotuberculosis* are closely related, with DNA sequence homology of nearly 97%, whereas both share only about 50% homology with *Y. enterocolitica* (Wren, 2003). However, like *Y. enterocolitica, Y. pseudotuberculosis* is an enteric pathogen that causes gastroenteritis through oral infection, while *Y. pestis* is transmitted by fleas or in aerosols.

The genus *Yersinia* is a group of gram-negative, oxidase-negative, catalase-positive, non-spore-forming rods or coccobacilli, which are facultative anaerobes and have the ability to grow at refrigeration temperatures. *Y. enterocolitica* is heterogeneous in its biochemical and antigenic properties. The vast majority of clinical isolates belong to bioserotypes 1B/O:8, 2-3/O:5,27, 2-3/O:9, and 3-4/O:3. These bioserotypes have been shown to have different geographical distributions (Table 2). Strains of bioserotype 1B/O:8 have been found mostly in the United States and

Table 2. Ecological and geographical distribution of most common serotypes of *Yersinia enterocolitica* and *Yersinia pseudotuberculosis*

Species	Biotype(s)	Serotype(s)	Hosts	Geographical distribution
Y. enterocolitica	1B	O:8	Human, pig, rodent	North America, France, Germany, Italy, Japan
	2 and 3	O:5,27	Human, pig, dog, sheep	Australia, Europe, Japan, North America
	2 and 3	O:9	Human, pig, cattle, goat, sheep, dog, cat, rat	Europe, Australia, Canada, Japan
	3	O:1,2,3	Chinchilla	Europe, United States
	3 and 4	O:3	Human, pig, dog, cat, rat, monkey	All continents
	5	O:2,3	Hare, goat, sheep, rabbit	Europe, Australia
Y. pseudotuberculosis		O:1–O:3	Human, pig, pets, wild animals	Canada, Europe, Far East
		O:4 and O:5	Human, pig, pets, wild animals	Far East

Canada and only seldom in Japan and Europe. Strains that are largely responsible for human yersiniosis in Europe, Japan, Canada, and the United States belong to bioserotype 4/O:3 (Bottone, 1999). Strains of bioserotype 3/O:3 have been recovered mostly in Japan and China (Fukushima et al., 1997). Strains belonging to biotype 1A are considered to be nonpathogenic; however, strains of this biotype have constituted quite a large fraction of strains from patients with gastroenteritis and may act as opportunistic pathogens (Pujol and Bliska, 2005). Among *Y. pseudotuberculosis* strains there is little variation in biochemical reactions, except with the sugars melibiose and raffinose, and citrate, which can be used to divide the species into four biotypes (Tsubokura and Aleksic, 1995). *Y. pseudotuberculosis* strains can be classified into 21 serotypes (Tsubokura and Aleksic, 1995). The most common serotypes of *Y. pseudotuberculosis* are serotypes O:1 to O:5, which differ in their geographical distribution (Table 2). All *Y. pseudotuberculosis* strains are considered to be pathogenic.

Virulence Factors of *Y. enterocolitica* and *Y. pseudotuberculosis*

Several virulence factors have been identified that are common to the pathogenic *Y. enterocolitica* and *Y. pseudotuberculosis* (Table 3). All pathogenic *Yersinia* species harbor an approximately 70-kb plasmid, termed pYV (plasmid for *Yersinia* virulence), that is essential for bacterial replication in host tissue (Cornelis et al., 1998). This pYV codes for an outer membrane protein, YadA (*Yersinia* adhesin A); a set of secreted proteins called Yops (*Yersinia* outer membrane proteins); and their secretion apparatus called Ysc (Yop secretion). With Ysc, extracellularly located *Yersinia* cells that are in close contact with the eukaryotic cell deliver Yops into the cytosol of the target cell. Yops protect *Yersinia* from the macrophage by destroying the phagocytic and signaling capacities of the macrophage and finally by inducing apoptosis. The *virF* gene is an important activator of the production of Yops.

The plasmid alone is not sufficient for the full expression of virulence in *Yersinia*. At least two chromosomal genes, *inv* and *ail*, are needed for mammalian cell invasion. The *inv* gene, which is homologous for *Y. enterocolitica* and *Y. pseudotuberculosis*, codes for Inv, an outer membrane protein, which appears to play a vital role in promoting entry into epithelial cells of the ileum during the initial stage of infection (Isberg and Barnes, 2001). The *ail* gene, which is highly correlated with virulence, codes for a surface protein, Ail, which promotes serum resistance in *Y. enterocolitica* and *Y. pseudotuberculosis* (Miller et al., 2001). The *yst* gene in the chromosome of *Y. enterocolitica*, but not of *Y. pseudotuberculosis*, encodes a heat-stable enterotoxin, Yst (*Yersinia* stable toxin) (Delor et al., 1990). Yersiniabactin is a siderophore, which is synthesized by *Y. enterocolitica* belonging to biotype 1B and

Table 3. Virulence factors of *Yersinia enterocolitica* and *Yersinia pseudotuberculosis*

Genomic origin	Gene	Determinant	Function	*Y. enterocolitica*	*Y. pseudotuberculosis*
Plasmid	*ysc-yop*	Yops	Resistance to phagocytosis	+	+
	yadA	YadA	Attachment, invasion, serum resistance	+	+
Chromosome	*inv*	Invasin	Attachment, invasion	+	+
	ail	Ail	Attachment, invasion	+	−
			Serum resistance	+	+
	ystA	YstA	Fluid secretion in intestine	+	−
	ypm	YPM	Stimulation of T cells	−	+
	irp	HPI	Iron uptake	+[a]	+

[a]Strains belonging to biotype 1B.

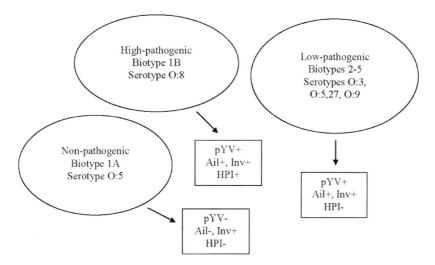

Figure 2. Distribution of different virulence determinants among high-, low-, and nonpathogenic *Yersinia enterocolitica* strains belonging to the most common bioserotypes.

Y. pseudotuberculosis only (Figure 2). The yersinia-bactin genes are clustered within a region of the chromosome referred to as a high-pathogenicity island (Carniel, 1999). High-pathogenicity-island-positive *Y. enterocolitica* strains are also called high-pathogenic strains because they can cause lethal infections in mice. *Y. pseudotuberculosis* can also synthesize a superantigenic toxin, YPM (*Y. pseudo-tuberculosis*-derived mitogen). The frequency of YPM-producing strains has been shown to be higher in the Far East than in Europe, where systemic symptoms in infected patients were less common than those in the Far East (Fukushima et al., 2001).

SOURCES AND PREVALENCE IN THE ENVIRONMENT AND FOODS

Animal Reservoirs of *Y. enterocolitica* and *Y. pseudotuberculosis*

Animals have long been suspected of being reservoirs for pathogenic *Y. enterocolitica* and *Y. pseudo-tuberculosis* and, hence, sources of human infections. These bacteria have been recovered from diverse animal sources, ranging from farm animals, domestic pets, and experimental animals to wild and captive animals (Table 4). Pathogenic *Y. enterocolitica* bacteria have frequently been isolated only from slaughtered pigs. However, the seroprevalence of anti-Yop antibodies has been shown to be high in cattle (76%) and goats (70%) in Germany (Tomaso et al., 2006). The highest prevalence of pathogenic *Y. enterocolitica* has been observed with pig tonsils, with serotype O:3 being the most common (Fredriksson-Ahomaa and Korkeala, 2003a). This serotype has a worldwide distribution in the pig population and also has been shown to be the most common serotype in the

United States (Bhaduri and Wesley, 2006). In the United Kingdom, serotypes O:5,27 and O:9 are common in pigs, and serotype O:5,27 in cattle and sheep (McNally et al., 2004). *Y. enterocolitica* O:9 has frequently been isolated from stools of cattle and goats in France (Gourdon et al., 1999). Serotypes O:3 and O:8 have sporadically been isolated from captive monkeys (Iwata et al., 2005; Fredriksson-Ahomaa et al., 2007b), and serotypes O:3 and O:5,27 from dogs and cats (Fukushima et al., 1985; Fredriksson-Ahomaa et al., 2001c). In Japan, serotypes O:3, O:8, and O:9 have been isolated from wild rodents, especially from field mice (Hayashidani et al., 1995).

Several studies have been conducted to isolate *Y. pseudotuberculosis* from farm animals, pets, and wild animals (Table 5). The host range is broad, but the principal reservoir hosts are believed to be rodents, wild birds, and domestic animals (especially pigs and ruminants). Most animals are healthy carriers, but they may become ill and excrete the bacteria under stress, such as cold and humid weather or starvation. Jerret et al. (1990) reported that *Y. pseudotuberculosis* is one of the most common infectious causes of death among farmed deer in Australia. Several outbreaks of *Y. pseudotuberculosis* infection among captive monkeys have been reported in Japan (Kageyama et al., 2002).

Some studies have been conducted to identify the primary reservoir of *Y. enterocolitica* and *Y. pseudo-tuberculosis* at the farm level for pigs (Gürtler et al., 2005; Bowman et al., 2007; Nesbakken et al., 2007; Niskanen et al., 2008). In all studies, there was a trend of increasing prevalence as pigs mature. The highest prevalence was observed among fattening pigs. In contrast, the lowest prevalence was found mostly among suckling pigs, weaned pigs, and sows;

Table 4. Animal reservoirs for *Yersinia enterocolitica* and *Yersinia pseudotuberculosis*

Reservoir type	Animal source	Y. enterocolitica	Y. pseudotuberculosis
Farm animals	Pig	O:3, O:5,27, O:8, O:9	O:1–O:4
	Cattle, sheep, goat	O:5,27, O:9	O:1–O:3
Pets	Dog, cat	O:3, O:5,27	O:1–O:5
	Chinchilla	O:1,2,3	O:3
Wild animals	Bird		O:1, O:2
	Deer		O:1–O:3
	Hare, rabbit	O:2,3, O:8	O.1–O:5
	Rodent	O:3, O:8, O:9	O:1–O:5
Zoo animals	Monkey	O:3, O:8	O:1–O:3

however, Bowman et al. (2007) found 9% of the gestating sows positive but none of the farrowing sows. Nesbakken et al. (2007) have shown that piglets under 2 months of age are *Y. enterocolitica* negative in both the tonsils and feces, and no antibodies against *Y. enterocolitica* can be measured. The young pigs became carriers in the tonsils and feces when they were about 60 to 80 days old and seropositive shortly thereafter. The tonsils remained positive for *Y. enterocolitica* during the last weeks of the lives of fattening pigs, which explains the high prevalence of this pathogen in the tonsils of slaughter pigs.

PCR has been applied in only a few studies investigating the prevalence of pathogenic *Y. enterocolitica* in pigs (Fredriksson-Ahomaa et al., 2000a, 2007a; Boyapalle et al., 2001; Korte et al., 2004; Bhaduri et al., 2005). In these studies, the detection rate was shown to be clearly higher with PCR than with culture methods. Most (88%) of the pig tonsils were shown to be positive with PCR when pathogenic *Y. enterocolitica* could be isolated from only 35% of the tonsils (Fredriksson-Ahomaa et al., 2007a). Bhaduri et al. (2005) detected pathogenic *Y. enterocolitica* in 12% of fecal samples in the United States using PCR

compared to 4% using a culture method. Boyapalle et al. (2001) found 40% of the mesenteric lymph nodes of pigs *Y. enterocolitica* positive using PCR but none with culturing.

The Environment Contaminated with *Y. enterocolitica* and *Y. pseudotuberculosis*

Y. enterocolitica strains belonging to pathogenic bioserotypes have rarely been isolated from the environment. However, strains of bioserotype 4/O:3 have occasionally been isolated from the environment in slaughterhouses and butcher shops (Nesbakken, 1988; Fredriksson-Ahomaa et al., 2000b, 2004). Pathogenic *Y. enterocolitica* has been detected by PCR on a variety of environmental sources in a Finnish slaughterhouse (Fredriksson-Ahomaa et al., 2000b). This pathogen was detected on the brisket saw, the hook from which the pluck set (tongue, esophagus, trachea, lungs, heart, diaphragm, liver, and kidneys) hangs, evisceration knife, aprons used by trimming workers, the computer keyboard used in the meat inspection area, and the handle of the coffee maker used by slaughterhouse workers. Additionally, pathogenic *Y. enterocolitica* was isolated from the air in the

Table 5. Prevalence of *Yersinia pseudotuberculosis* in animal sources

Source	No. of samples	No. of positive samples	% Positive	Country	Reference
Pig	1,200	52	4	Japan	Fukushima et al. (1989)
	210	8	4	Finland	Niskanen et al. (2002)
Cattle	2,639	185	7	Australia	Slee et al. (1988)
Sheep	449	21	5	Australia	Slee and Skilbeck (1992)
Dog	252	16	6	Japan	Fukushima et al. (1985)
Cat	318	4	1	Japan	Fukushima et al. (1985)
Bird	259	2	1	Japan	Fukushima and Gomyoda (1991)
	468	3	1	Sweden	Niskanen et al. (2003)
Deer	153	29	18	Australia	Jerret et al. (1990)
	215	8	4	Japan	Fukushima and Gomyoda (1991)
Hare	139	2	1	Japan	Fukushima and Gomyoda (1991)
Rabbit	148	4	3	China	Zheng et al. (1995)
Mouse	107	9	8	China	Zheng et al. (1995)
Rat	148	4	3	China	Zheng et al. (1995)
	237	4	2	Japan	Kageyama et al. (2002)

bleeding area. PCR-positive samples were also obtained from the floor in the eviscerating and weighing areas and from the table in the meat-cutting area. Pathogenic *Y. enterocolitica* 4/O:3 has also been isolated occasionally from water (Thompson and Gravel, 1986). In Brazil, natural water and sewage have shown to be sporadically contaminated with pathogenic *Y. enterocolitica* belonging to serotype O:5,27 (Falcao et al., 2004). In Japan, *Y. enterocolitica* O:8 strains have been isolated from stream water (Iwata et al., 2005). Sandery et al. (1996) have shown with PCR that pathogenic strains of *Y. enterocolitica* can frequently be detected in environmental waters. In a case-control study, untreated drinking water has been reported to be a risk factor for sporadic *Y. enterocolitica* infections in Norway (Ostroff et al., 1994).

Y. pseudotuberculosis is widely spread in the environment (soil, water, vegetables, etc.), where it can survive for a long time. The environment can be contaminated by the feces of infected animals, mainly wild animals, like deer, rodents, and birds (Fukushima et al., 1998). *Y. pseudotuberculosis* has been isolated from fresh water, such as river, well, and mountain stream water, at a considerably high rate (Tsubokura et al., 1989). In a large point source outbreak caused by raw carrots contaminated by *Y. pseudotuberculosis*, the epidemic strain was also found on the soil and on the production line (washing and peeling equipment) at the farm of origin (Jalava et al., 2006).

Foods Contaminated with *Y. enterocolitica* and *Y. pseudotuberculosis*

Food has often been suggested to be the main source of *Y. enterocolitica* and *Y. pseudotuberculosis,* although pathogenic isolates have seldom been recovered from food samples. In epidemiological studies, *Y. enterocolitica* O:3 infections have been associated with eating raw or undercooked pork within 2 weeks before onset (Tauxe et al., 1987; Ostroff et al., 1994). In the United States, *Y. enterocolitica* O:3 infections have been associated with the household preparation of chitterlings (intestines of pigs, which are a traditional dish in the southern United States) (Lee et al., 1990; Jones et al., 2003). Raw pork products have been investigated widely due to the association between *Y. enterocolitica* and pigs. The only pathogenic bioserotype found in northern Europe and Germany is 4/O:3 (Johannessen et al., 2000; Fredriksson-Ahomaa et al., 2001a; Fredriksson-Ahomaa and Korkeala, 2003b; Thisted Lambertz and Danielsson-Tham, 2005). Bioserotypes 2/O:5,27 and 2/O:9 have sporadically been found in the United Kingdom, and 3/O:3 and 3/O:5,27 in Japan in pork (Logue et al., 1996; Fukushima et al., 1997).

Y. pseudotuberculosis has very rarely been isolated from foods. This pathogen has sporadically been isolated from pork in Japan (Tsubokura et al., 1989; Fukushima et al., 1997). *Y. pseudotuberculosis* has been isolated from iceberg lettuce and raw carrots implicated in food-borne outbreaks in Finland (Nuorti et al., 2004; Jalava et al., 2006). The detection rate of pathogenic *Y. enterocolitica* in foods has been shown to be clearly higher by PCR than by culturing (Fredriksson-Ahomaa and Korkeala, 2003a). The highest detection rate has been obtained from pig offal, including pig tongues, livers, hearts, and kidneys, and from chitterlings (Table 6). Lambertz et al. (2007) reported recently a relatively high prevalence (11%) of pathogenic *Y. enterocolitica* in fermented pork sausages by use of PCR. Vishnubhatla et al. (2001) detected a high prevalence of *yst*-positive *Y. enterocolitica* in ground beef; however, pathogenic strains have only sporadically been isolated from beef. No pathogenic *Y. enterocolitica* has been detected in fish and chicken, but this pathogen has been detected in 3% of lettuce samples with PCR (Fredriksson-Ahomaa and Korkeala, 2003b). In Korea, Lee et al. (2004) isolated one *ail*-positive *Y. enterocolitica* strain of bioserotype 3/O:3 from 673 ready-to-eat vegetables.

Transmission of *Y. enterocolitica* and *Y. pseudotuberculosis* to Humans

The primary transmission route of human yersiniosis is proposed to be fecal-oral via contaminated food (Figure 3). In particular, pork and pork products have been implicated as the major source of human *Y. enterocolitica* infection, with some epidemiological studies linking consumption of uncooked or undercooked pork (Tauxe et al., 1987; Ostroff et al., 1994; Satterthwaite et al., 1999; Fredriksson-Ahomaa et al., 2001b, 2006b). Consumption of raw pork may play an important role in countries like Belgium, Germany, and The Netherlands, where raw minced pork with pepper and onion is a delicacy that can be purchased in ready-to-eat form from butcher shops. Transmission is likely to occur more often via cross-contamination of cooked pork or foods not normally harboring *Y. enterocolitica*. Another transmission route may be from person to person, which can be direct or indirect. Direct person-to-person contact has not been demonstrated, but Lee et al. (1990) reported *Y. enterocolitica* O:3 infections in infants who were probably exposed to infection by their caretakers. This may happen when basic hygiene and hand-washing habits are inadequate. Indirect person-to-person transmission has apparently occurred in several instances by transfusion

Table 6. Prevalence of pathogenic *Yersinia enterocolitica* in foods with PCR and culture methods

Sample	No. of samples	No. of culture-positive samples (%)[a]	No. of PCR-positive samples (%)	Reference
Pig tongues	15	7 (47)	10 (67)	Vishnubhatla et al. (2001)
	99	79 (80)	82 (83)	Fredriksson-Ahomaa and Korkeala (2003b)
Pig offal[b]	110	38 (35)	77 (70)	Fredriksson-Ahomaa and Korkeala (2003b)
Chitterling	350	8 (2)	278 (79)	Boyapalle et al. (2001)
Ground pork	350	0	133 (38)	Boyapalle et al. (2001)
	100	32 (32)	47 (47)	Vishnubhatla et al. (2001)
Ground beef	100	23 (23)	31 (31)	Vishnubhatla et al. (2001)
Minced pork	255	4 (2)	63 (25)	Fredriksson-Ahomaa and Korkeala (2003b)
	100	5 (5)	35 (35)	Lambertz et al. (2007)
Pork[c]	300	6 (2)	50 (17)	Johannessen et al. (2000)
	91	6 (7)	9 (10)	Thisted Lambertz and Danielsson-Tham (2005)
	150	0	9 (6)	Fredriksson-Ahomaa et al. (2007a)
	97	0	11 (11)	Lambertz et al. (2007)
Chicken	43	0	0	Fredriksson-Ahomaa and Korkeala (2003b)
Fish	200	0	0	Fredriksson-Ahomaa and Korkeala (2003b)
Lettuce	101	0	3 (3)	Fredriksson-Ahomaa and Korkeala (2003b)
Tofu	50	0	6 (12)	Vishnubhatla et al. (2001)

[a] Pathogenicity of isolates confirmed.
[b] Liver, heart, and kidney.
[c] Except pig tongues and offal.

of contaminated blood products (Bottone, 1999). In these cases, the most likely source of *Yersinia* has been blood donors with subclinical bacteremia. Direct contact with pigs, a common risk for pig farmers and slaughterhouse workers, may also be a transmission route. Elevated serum antibody concentrations have been found among people involved in swine breeding or pork production (Seuri and Granfors, 1992). Pet animals have also been suspected as being sources of human yersiniosis through close contact with humans, especially young children. Pathogenic *Y. enterocolitica* may be transmitted to humans indirectly from pork and offal via dogs and cats (Fredriksson-Ahomaa et al., 2001c).

Y. pseudotuberculosis is also a food-borne pathogen, but this pathogen has rarely been isolated from foods implicated in the illness. In the reported outbreaks, fresh produce and untreated surface water have been possible infection sources. Some recent epidemiologic investigations in Finland have linked outbreaks of *Y. pseudotuberculosis* to domestically grown iceberg lettuce and carrots (Nuorti et al., 2004; Jalava et al., 2006). *Y. pseudotuberculosis* occurs in water and in the environment and has been isolated from various animals; however, the transmission routes are unclear. In the outbreak linked to carrots, washing and peeling equipment was shown to be contaminated with the outbreak strain; however, the

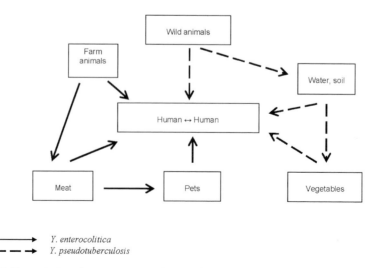

Figure 3. Transmission of *Yersinia enterocolitica* and *Yersinia pseudotuberculosis* to humans.

exact mechanism of contamination of carrots on the farm could not be clarified (Jalava et al., 2006). A combination of direct contact with wildlife feces during storage and cross-contamination of the equipment was the most likely contributing factor. In another outbreak, iceberg lettuce was probably contaminated with irrigation water. *Y. pseudotuberculosis* could have been transmitted from infected animals to the lettuce and irrigation water by feces at the farm before distribution (Nuorti et al., 2004). Unchlorinated drinking water from wells, springs, and streams contaminated with feces of wild animals is considered an important transmission route in mountainous areas in Japan (Fukushima et al., 1998). In Korea, untreated mountain spring water could be linked to *Y. pseudotuberculosis* infection (Han et al., 2003). An epidemiological study has shown evidence that *Y. pseudotuberculosis* was transmitted from infected pets to humans in a familial outbreak in Japan (Fukushima et al., 1994).

INTRINSIC AND EXTRINSIC FACTORS THAT AFFECT SURVIVAL AND GROWTH IN FOOD PRODUCTS AND CONTRIBUTE TO OUTBREAKS

Intrinsic Factors

The most important intrinsic factors are nutrition, pH, and water activity. Nutritionally, *Yersinia* is not a fastidious organism. This bacterium is also able to multiply over a wide pH range from approximately pH 4 to 10, with an optimum pH of around 7.6. However, as only few foods have an alkaline pH, the tolerance of a high pH is relatively unimportant. *Yersinia* is generally considered to be sensitive to low pH conditions. Citric acid is less inhibitory than acetic acid. With citric acid, growth was detected even at a pH of 3.81 obtained with a concentration of 0.31% (vol/vol), while acetic acid inhibited the growth at pH 5.58 obtained with a concentration greater than 0.16% (vol/vol) (Karapinar and Gonul, 1992).

Salt tolerance of *Yersinia* is moderate and strongly dependent on storage temperature. While 7% or more of NaCl is bacteriostatic (Robins-Browne, 1997) for *Yersinia*, the bacterium is able to grow in 5% NaCl. Thus, at the levels most commonly present in foods, salt alone will not prevent growth, and other preservative hurdles are required. *Y. enterocolitica* can tolerate both sodium nitrate and nitrite of up to 20 mg/ml for 48 h in vitro (De Giusti and De Vito, 1992). However, a nitrite concentration of only 80 mg/kg has been reported to inhibit the growth of *Y. enterocolitica* in fermented sausages (Asplund et al., 1993).

Extrinsic Factors

The most important extrinsic factors include temperature and gas atmosphere. *Yersinia* is unusual among pathogenic enterobacteria in being psychrotrophic, thus having the ability to grow at refrigerator temperatures. The doubling time at the optimum growth temperature (approximately 28 to 30°C) is around 34 min (Schiemann, 1989). Although *Yersinia* can grow at temperatures as low as 0°C, the organism grows much more slowly as temperatures drop below 5°C (Goverde et al., 1994; Harrison et al., 2000). It has been shown that the number of *Y. enterocolitica* on pork can reach 10^9 CFU per cm^2 after 5 days at 10°C (Nissen et al., 2001). In some Finnish *Y. pseudotuberculosis* outbreaks detected in spring, the vehicles have been vegetables of the crop stored at chilled temperature from a previous year. *Yersinia* can withstand freezing and can survive in frozen foods for extended periods, even after repeated freezing and thawing (Toora et al., 1992).

At the cellular level, the acute cold shock response, induced by any temperature downshift of 10°C or more, and longer-term adaptation to low temperature involve different mechanisms (Bresolin et al., 2006). As in most bacteria, the cold shock response in Yersiniae is featured by the expression of cold shock proteins (Csp). Csp bind single-stranded nucleic acids and might thus play a role as RNA chaperones and transcription antiterminators (Jones and Inouye, 1994; Bae et al., 2000). Csp also bind mRNA and regulate ribosomal translation, mRNA decay, and termination of transcription (Ermolenko and Makhatadze, 2002; Chattopadhyay, 2006). At the acclimation phase, the production of Csp declines. The early and mid-exponential growth of *Y. enterocolitica* at low temperature involves the activation of genes encoding environmental sensors and regulators related to signal transduction (Bresolin et al., 2006), catabolic events, utilization of endogenous resources, and defense against oxidative stress. Also, flagellar synthesis and chemotaxis (Bresolin et al., 2006) are induced, a well-known phenomenon in many pathogens exposed to temperatures outside of the mammalian host. The late exponential and early stationary phase of adaptation of *Y. enterocolitica* to low temperature is dominated by mechanisms involved in biodegradative metabolism (Bresolin et al., 2006). Membrane rigidity and fluidity are maintained by changing the membrane composition. At low temperature, unsaturated fatty acids and phospholipids predominate, while saturated and cyclic fatty acid contents become a majority upon a rising temperature (Nagamachi et al., 1991; Goverde et al., 1998).

Bacterial virulence functions are often temperature-regulated due to the high energy cost of unnecessary expression of virulence genes (Hurme and Rhen, 1998). Thus, many of the *Yersinia* virulence factors are expressed at 26°C but not at 37°C (Table 7).

The maximum growth temperature of *Yersinia* is between 42 and 44°C. *Yersinia* is relatively sensitive to heat, with thermal treatments employed in food processing destroying it readily. It is also destroyed by pasteurization at 72°C for 15 to 20 s. As a facultatively anaerobic bacterium, *Yersinia* can grow in anaerobic conditions. It can also grow well in modified atmospheres (Harrison et al., 2000), but with higher levels of CO_2 the length of the lag phase will increase and growth will be slower (Pin et al., 2000). *Y. enterocolitica* has been shown to grow well on meat when packaged by vacuum or in modified atmosphere and stored at 5°C (Doherty et al., 1995; Bodnaruk and Draughon, 1998), even in the presence of high numbers of other bacteria (Barakat and Harris, 1999; Bredholt et al., 1999). *D* values, or thermal resistance, at 50°C (D_{50}), D_{55}, and D_{60} of 1 min, 1 min, and 0.55 min, respectively, have been reported for vacuum-packaged minced beef (Bolton et al., 2000).

FOOD PROCESSING OPERATIONS THAT INFLUENCE THE NUMBER, SPREAD, OR CHARACTERISTICS OF ENTEROPATHOGENIC *YERSINIA*

Slaughter pigs often carry pathogenic *Y. enterocolitica* and *Y. pseudotuberculosis* in the oral cavity, particularly in tonsils and feces (Fredriksson-Ahomaa and Korkeala, 2003b; Niskanen et al., 2002, 2008). The presence of symptomless carriers together with the widespread occurrence of pathogenic *Yersinia* in herds renders the control of this bacterium at farm level difficult. Furthermore, it is impossible to reject asymptomatic pigs contaminated with pathogenic *Yersinia* at postmortem meat inspection. There are several meat inspection procedures, like evisceration and incision of the submaxillary lymph nodes, in which contamination of carcass and offal can occur. During evisceration, the spread of pathogenic *Yersinia* from tonsils to pluck set and other parts of the carcass by knives and hands is unavoidable because the tonsils are usually removed together with the pluck set (tongue, esophagus, trachea, lungs, heart, diaphragm, liver, and kidneys) (Fredriksson-Ahomaa et al., 2000b, 2001a). By removing the head together with the tonsils and tongue prior to evisceration, the contamination of pathogenic *Yersinia* could probably be reduced. Additionally, the use of a mechanized bung cutter in connection with enclosing the anus and rectum in a plastic bag to minimize fecal contamination has been shown to reduce feces contamination of carcasses (Nesbakken et al., 1994). The inspection of head, tonsils, and tongue should occur in a separate room with separate equipment. Furthermore, the head meat and tongue should be pasteurized before delivery. However, strict slaughter hygiene remains important in reducing contamination in slaughterhouses. *Yersinia* contamination in the later stages of the food chain has not been studied sufficiently.

Fresh produce may become contaminated with pathogenic *Yersinia*, especially with *Y. pseudotuberculosis*, during irrigation, harvesting, packing, shipping, and processing. Preventing the access of wild animals to irrigation water and fields could reduce the risk of contamination of surface water and soil. Furthermore, the access by rodents and birds to storage facilities should be prevented, and the cleaning of processing equipment should be adequate (Nuorti et al., 2004; Jalava et al., 2006).

Table 7. Effect of temperature on activity of virulence-related genes in *Y. enterocolitica*[b]

Genes/function	Effect of temp on gene expression[a]	
	26°C	37°C
inv/invasion	↑	↓
irp/iron uptake	↑	↓
rfb/lipopolysaccharide	↑	↓
yst/enterotoxin production	↑	↓
virF/transcriptional activator	↓	↑
yadA/adhesion to animal cells, serum resistance	↓	↑
ysc, yop/secretion of Yops	↓	↑
ail/attachment, serum resistance	↓	↑

[a]↑, upregulation; ↓, downregulation.
[b]Modified from the work of Straley and Perry (1995).

DISCRIMINATIVE DETECTION METHODS FOR CONFIRMATION AND TRACE-BACK OF CONTAMINATED PRODUCTS

Several difficulties have been associated with isolating pathogenic *Yersinia* from food and environmental samples. Conventional culture-dependent methods have several limitations, such as long incubation steps taking up to 4 weeks, lack of identification between species, and lack of discrimination between pathogenic and nonpathogenic strains (Fredriksson-Ahomaa and Korkeala, 2003a). In addition, the ability of bacteria, including *Y. enterocolitica*, to persist in samples in a viable but noncultivable state can be a problem (Alexandrino et al., 2004). Thus, investigations have been undertaken to develop rapid and reliable methods, such as PCR, for detecting pathogenic *Yersinia*, especially *Y. enterocolitica*, in food samples.

Detection of *Y. enterocolitica* and *Y. pseudotuberculosis* Using PCR

Several methods have been developed to detect *Y. enterocolitica* in food and environmental samples using PCR (Fredriksson-Ahomaa and Korkeala, 2003a). Many of these methods use primers targeted to the *virF* or *yadA* gene located on pYV (Table 8). Because of possible plasmid loss during culturing, PCR methods targeted to chromosomal virulence genes have also been established. The *ail* gene located in the chromosome of pathogenic *Y. enterocolitica* strains is the most frequently used target. Some of the PCR methods have been created to detect more than one gene at the same time. The method most frequently used to detect PCR products has been electrophoresis in an agarose gel; however, the use of real-time monitoring by fluorescence techniques has substantially increased. There are already some real-time PCR methods designed for qualitative detection of pathogenic *Y. enterocolitica* in food samples (Table 8). The *Yersinia* methods are based on dual-labeled probes (Jourdan et al., 2000; Boyapalle et al., 2001; Fredriksson-Ahomaa et al., 2007a) or SYBR green dye (Fukushima et al., 2003). There are, so far, two standardized conventional PCR methods for detection of pathogenic *Y. enterocolitica* in foods. Method A is based on a one-step PCR with primers targeted to the *ail* gene in the chromosome, while method B uses a two-step PCR with primers targeted to *yadA* on the pYV. These two methods have been published by the Nordic Committee on Food Analysis (NCFA, 1998).

Numerous commercial DNA purification kits are available, and some of these kits have also been used in PCR assays designed for *Y. enterocolitica*. DNA extraction procedures using silica particles or chelex resin have commonly been used, as they are rapid and simple to use; however, they are not necessarily the most effective methods to remove inhibitors from complex matrices (Kaneko et al., 1995; Sandery et al., 1996; Jourdan et al., 2000; Boyapalle et al., 2001; Vishnubhatla et al., 2001; Fredriksson-Ahomaa et al., 2007a). Buoyant density centrifugation, with some variations, has been used to separate *Y. enterocolitica* from food samples to remove inhibitors and to prevent false-positive results due to DNA originating from dead cells (Thisted Lambertz et al., 2000; Wolffs et al., 2004; Fukushima et al., 2007). Knutsson et al. (2002) have designed a selective enrichment medium, *Yersinia* PCR-compatible enrichment medium, for *Y. enterocolitica*, which can be used prior to PCR to increase the sensitivity and decrease the risk of false-positive results due to detection of dead cells.

Detection of *Y. enterocolitica* and *Y. pseudotuberculosis* Using Culture Methods

It is generally easier to find *Y. enterocolitica* and *Y. pseudotuberculosis* in clinical specimens of infected individuals than to find them in asymptomatic carriers or in foods and environmental samples. During acute gastroenteritis or with organ abscesses, pathogenic *Yersinia* organisms are often dominant bacteria and

Table 8. Detection of *Yersinia enterocolitica* and *Yersinia pseudotuberculosis* in food samples using PCR

Gene region(s) or gene(s)	Type of PCR	Detection[a]	Y. enterocolitica	Y. pseudotuberculosis	Reference(s)
Plasmid					
virF	Nested	AGE	×		NCFA (1998)
yadA	Nested	AGE	×		NCFA (1998)
	Single	Real time	×		Wolffs et al. (2005)
Chromosome					
ail	Semi-nested	AGE	×		Sandery et al. (1996)
	Nested	AGE	×		NCFA (1998)
	Single	Real time	×		Jourdan et al. (2000), Fredriksson-Ahomaa et al. (2007a)
yst	Single	AGE	×		Özbas et al. (2000)
	Single	Real time	×		Vishnubhatla et al. (2001)
Plasmid and chromosome					
virF, ail, inv	Multiplex	AGE	×	×	Kaneko et al. (1995)
virF, ail	Multiplex	AGE	×		Nilsson et al. (1998)
yadA, 16S rRNA	Multiplex	AGE	×		Lantz et al. (1999)
yadA, ail	Multiplex	AGE	×		Boyapalle et al. (2001)

[a]AGE, agarose gel electrophoresis; Real time, real-time monitoring.

can easily be isolated by direct plating on conventional enteric media. Because of the high number of other bacteria and the low number of pathogenic strains of *Yersinia* in food and environmental samples, direct isolation, even on selective media, is seldom successful (Fredriksson-Ahomaa and Korkeala, 2003a). To increase the number of *Yersinia* strains in these samples, enrichment in liquid media prior to isolation on solid media is required. Several different methods are available for isolation of *Y. enterocolitica* from food and environmental samples (Table 9). Far less information is available on reliable methods for recovery of *Y. pseudotuberculosis* from foods. The most widely used selective enrichment broth is irgasan-ticarcillin-potassium chlorate broth, which has been shown to be efficient for recovery of strains of bioserotype 4/O:3. Niskanen et al. (2002) have shown that cold enrichment for 2 weeks in phosphate-buffered saline broth, supplemented with 1% mannitol and 0.15% bile salts, is productive for isolation of *Y. pseudotuberculosis*, especially followed by alkali treatment. Alkali treatment with 0.25 to 0.50% KOH for 20 s has widely been used for isolation of *Yersinia* in foods (FDA, 2001; ISO, 2003; NCFA, 2003).

Many different selective plating media developed for enteropathogens have been used for isolation of *Yersinia* (Fredriksson-Ahomaa and Korkeala, 2003a). Of the traditional enteric media, the most widely used is MacConkey (MAC) agar. Several entirely new media for isolation of *Yersinia* have been designed to improve selectivity. Most of the media have been designed for *Y. enterocolitica*; however, *Y. pseudotuberculosis* is usually also able to grow on them, but its growth is mostly poor and delayed. Cefsulodin-irgasan-novobiocin (CIN) agar, according to Schiemann (1979), and *Salmonella-Shigella* deoxycholate calcium chloride agar, according to Wauters et al. (1988), are widely used for bioserotype 4/O:3 due to their high selectivity and commercial availability (ISO, 2003; NCFA, 2003).

However, differentiation of *Yersinia* from competing organisms, such as *Aeromonas, Citrobacter, Enterobacter, Klebsiella, Morganella, Proteus,* and *Serratia,* can be difficult. The CIN agar is most commonly used for both *Y. enterocolitica* and *Y. pseudotuberculosis*. MAC agar has been used alongside CIN and is generally considered to be useful for the isolation of *Y. pseudotuberculosis*, despite its low selectivity.

Y. enterocolitica and *Y. pseudotuberculosis* can be identified with commercial rapid identification kits, like API 20E, API Rapid 32 IDE, and Micronaute E. The API 20E system, widely used for identification of presumptive *Yersinia* isolates, has been shown to be accurate in identifying *Y. enterocolitica* and *Y. pseudotuberculosis*. This kit has high positive identification rates of 93% and 90% for *Y. enterocolitica* and *Y. pseudotuberculosis*, respectively, when incubated at a temperature between 25°C and 30°C instead of 37°C for 18 to 24 h (Neubauer et al., 1998).

The majority of *Y. enterocolitica* isolates recovered from nonhuman sources are considered nonpathogenic; thus, it is important to assess the pathogenicity of isolates (Fredriksson-Ahomaa and Korkeala, 2003a). This can be done rapidly and with high specificity with DNA-based methods. Several PCR and DNA colony hybridization assays have been designed to verify the pathogenicity of *Y. enterocolitica* isolates. The methods are based on specific segments of the virulence plasmid or the chromosomal DNA with known virulence functions. Currently, two oligonucleotide probes are available from the FDA's (U.S. Food and Drug Administration) Center for Food Safety and Applied Nutrition for identification of pathogenic *Y. enterocolitica* (FDA, 2001). These probes are specific for the *ail* gene in the chromosome and the *virF* on the pYV. In addition, microarray chip technology has shown increased importance and potential in rapidly determining a complete gene profile in pathogenic microorganisms. Myers et al. (2006) have recently developed a microarray chip combined

Table 9. Isolation methods of *Yersinia enterocolitica* and *Yersinia pseudotuberculosis* most commonly used for food samples[a]

Enrichment medium (conditions)	Isolation medium/media (conditions)	*Y. enterocolitica*	*Y. pseudotuberculosis*	Reference
Not selective				
TSB (22–25°C, 1 day)	CIN (32°C, 18 h); MAC (22–25°C, 48 h)	×	×	APHA (1992)
PBS (4°C, 14–21 days)	CIN (32°C, 18 h); MAC (22–25°C, 48 h)	×	×	APHA (1992)
Slightly selective				
PSB (10°C, 10 days)	CIN and MAC (30°C, 1 day)	×		FDA (2001)
PSB (25°C, 1–3 days)	CIN (30°C, 24 h)	×	×	ISO (2003)
Selective				
ITC (25°C, 2 days)	SSDC (30°C, 24 h)	×		ISO (2003)
MRB (22–25°C, 2–3 days)	CIN (32°C, 18 h)	×		APHA (1992)

[a] PBS, phosphate-buffered saline; PSB, phosphate-buffered saline broth with sorbitol and bile salts; ITC, irgasan-ticarcillin-potassium chlorate broth; SSDC, *Salmonella-Shigella*-sodium deoxycholate-calcium chloride agar plate; MRB, modified Rappaport broth; TSB, tripticase soy broth.

with multiplex PCR for detection and characterization of four virulence genes in *Y. enterocolitica*. Multiplex PCR was used to amplify the *virF, ail,* and *yst* genes. Using this technique, important virulence genes of *Y. enterocolitica* could simultaneously be detected directly from pasteurized whole milk.

Molecular Characterization

In epidemiological studies, differentiation of *Y. enterocolitica* and *Y. pseudotuberculosis* into subtypes is necessary to identify contamination sources and routes. Molecular techniques represent valuable alternatives for subtyping of *Yersinia*. Several DNA-based methods have been used in molecular typing of *Y. enterocolitica* and *Y. pseudotuberculosis* (Table 10). Pulsed-field gel electrophoresis (PFGE) has proved to be highly discriminatory for molecular typing of *Yersinia*; thus, this method is a commonly used technique in epidemiological studies of *Y. enterocolitica* and *Y. pseudotuberculosis*. PFGE allows subtyping of *Yersinia* strains belonging to the same serotype (Niskanen et al., 2002; Fredriksson-Ahomaa et al., 2006a). Amplified fragment length polymorphism, a recently adopted PCR-based typing method, also allows differentiation of *Y. enterocolitica* strains within serotype-related clusters (Fearnley et al., 2005). Both techniques have been used to investigate the relationship between genotypes and host sources (Fredriksson-Ahomaa et al., 2006b; Kuehni-Boghenbor et al., 2006). Ribotyping, random amplification of polymorphic DNA (RAPD) assay, and restriction endonuclease analysis of the plasmid (REAP) have also frequently been applied to characterize enteropathogenic *Yersinia* strains (Table 10). Ribotyping has been shown to be a useful tool for molecular typing, especially to characterize *Y. pseudotuberculosis* strains (Voskressenskaya et al., 2005). Fukushima et al. (1994, 1998) have demonstrated differences in the geographical distribution among *Y. pseudotuberculosis* strains belonging to the same serotype using REAP.

The RAPD assay, which is simple and quick to perform but may possess a low reproducibility, has been used in several studies to characterize *Y. enterocolitica* strains (Fredriksson-Ahomaa et al., 2006a). It allows discrimination between strains belonging to different bioserotypes and also, in some cases, between strains belonging to the same bioserotype. RAPD is also able to distinguish *Y. pseudotuberculosis* strains at the subserotype level; thus, it has been used in two outbreak studies by Makino et al. (1994) and Kageyama et al. (2002).

CONCLUDING REMARKS

Yersiniosis is an infectious disease caused by *Yersinia*, food-borne yersiniosis being due to *Yersinia enterocolitica* or *Yersinia pseudotuberculosis*. Most human illness worldwide is caused by *Y. enterocolitica*. The reported cases of yersiniosis are mainly sporadic and outbreaks are uncommon. However, outbreaks of *Y. pseudotuberculosis* infection, often in schoolchildren, have occurred in Finland and Japan. The primary transmission route of human yersiniosis is proposed to be fecal-oral via contaminated food. Some epidemiological studies have linked *Y. enterocolitica* infection with consumption of uncooked or undercooked pork. In the United States, household preparation of chitterlings (traditional dish made of intestines of pigs in the southern United States) has often been associated with *Y. enterocolitica* O:3 infections. Pathogenic *Y. enterocolitica* has frequently been recovered only from slaughter pigs. By removing the head together with tonsils and tongue prior to evisceration and by using a mechanized bung cutter in connection with enclosing the anus and rectum in a plastic bag, the contamination of pathogenic *Yersinia* could probably be reduced. Despite the fact that *Y. enterocolitica* and *Y. pseudotuberculosis* are important food-borne pathogens, these bacteria have seldom been isolated from foods. The low isolation rate

Table 10. Frequently used methods for molecular subtyping of *Yersinia enterocolitica* and *Yersinia pseudotuberculosis*

Typing method[a]	*Y. enterocolitica*	*Y. pseudotuberculosis*	Reference(s)
AFLP	×		Fearnley et al. (2005), Kuehni-Boghenbor et al. (2006)
RAPD	×		Fredriksson-Ahomaa et al. (2006a)
		×	Makino et al. (1994), Kageyama et al. (2002)
REAP	×		Fredriksson-Ahomaa et al. (2006a)
		×	Fukushima et al. (1994, 1998), Han et al. (2003)
Ribotyping	×		Fredriksson-Ahomaa et al. (2006a)
		×	Voskressenskaya et al. (2005)
PFGE	×		Favier et al. (2005), Fredriksson-Ahomaa et al. (1999, 2006a, 2006b, 2007b), Falcao et al. (2006), Lambertz et al. (2007)
		×	Niskanen et al. (2002, 2003, 2008), Nuorti et al. (2004), Jalava et al. (2004, 2006)

[a]AFLP, amplified fragment length polymorphism.

is probably due to the low sensitivity of the culture methods available. Using PCR, pathogenic *Y. enterocolitica* has been detected with a high frequency in raw pork and pork products. The host range of *Y. pseudotuberculosis* is broad, but the principal reservoir hosts are believed to be rodents, wild birds, and domestic animals (especially pigs and ruminants). In the reported outbreaks of *Y. pseudotuberculosis*, fresh produce and untreated surface water have been implicated in the illness. A combination of direct contact with wildlife feces during the storage and cross-contamination of the equipment are the most likely contributing factors. Although several studies on the epidemiology of enteropathogenic *Yersinia* have been conducted, a lot of questions remain to be solved using DNA-based detection and characterization methods in the future.

REFERENCES

Achtman, M., K. Zurth, G. Morelli, G. Torrea, A. Guiyoule, and E. Carniel. 1999. *Yersinia pestis*, the cause of plague, is a recently emerged clone of *Yersinia pseudotuberculosis*. *Proc. Natl. Acad. Sci. USA* **96:**14043–14048.

Alexandrino, M., E. Grohmann, and U. Szewzyk. 2004. Optimization of PCR-based methods for rapid detection of *Campylobacter jejuni*, *Campylobacter coli* and *Yersinia enterocolitica* serovar O:3 in waste water samples. *Water Res.* **38:**1340–1346.

APHA. 1992. *Yersinia*, p. 433–450. *In* C. Vanderzant and D. F. Splittstoesser (ed.), *Compendium of Methods for the Microbiological Examination of Foods*, 3rd ed. American Public Health Association, Washington, DC.

Asplund, K., E. Nurmi, J. Hirn, T. Hirvi, and P. Hill. 1993. Survival of *Yersinia enterocolitica* in fermented sausages manufactured with different levels of nitrite and different starter cultures. *J. Food Prot.* **56:**710–712.

Bae, W., B. Xia,, M. Inouye, and K. Severinov. 2000. *Escherichia coli* CspA-family RNA chaperones are transcription antiterminators. *Proc. Natl. Acad. Sci. USA* **97:**7784–7789.

Barakat, R. K., and L. J. Harris. 1999. Growth of *Listeria monocytogenes* and *Yersinia enterocolitica* on cooked modified-atmosphere-packaged poultry in the presence and absence of natural occurring microbiota. *Appl. Environ. Microbiol.* **65:**342–345.

Baumgartner, A., M. Küffer, D. Suter, T. Jemmi, and P. Rohner. 2006. Antimicrobial resistance of *Yersinia enterocolitica* strains from human patients, pigs and retail pork in Switzerland. *Int. J. Food Microbiol.* [Epub ahead of print.] doi: 10.1016/j. i; food micro. 2006.10.008.

Bhaduri, S., and I. Wesley. 2006. Isolation and characterization of *Yersinia enterocolitica* from swine feces recovered during the National Animal Health Monitoring System Swine 2000 Study. *J. Food Prot.* **69:**2107–2112.

Bhaduri, S., I. Wesley, and E. J. Bush. 2005. Prevalence of pathogenic *Yersinia enterocolitica* strains in pigs in the United States. *Appl. Environ. Microbiol.* **71:**7117–7121.

Bodnaruk, P. W., and F. A. Draughon. 1998. Effect of packaging atmosphere and pH on the virulence and growth of *Yersinia enterocolitica* on pork stored at 4°C. *Food Microbiol.* **15:**129–136.

Bolton, D. J., C. M. McMaghon, A. M. Doherty, J. J. Sheridan, D. A. McDowell, I. S. Blair, and D. Harrington. 2000. Thermal inactivation of *Listeria monocytogenes* and *Yersinia enteroco-*

litica in minced beef under laboratory conditions and in sous-vide prepared minced and solid beef cooked in a commercial retort. *J. Appl. Microbiol.* **88:**626–632.

Bottone, E. J. 1999. *Yersinia enterocolitica*: overview and epidemiologic correlates. *Microb. Infect.* **1:**323–333.

Bowman, A. S., C. Glendeing, T. E. Wittum, J. T. LeJeune, R. W. Stich, and J. A. Funk. 2007. Prevalence of *Yersinia enterocolitica* in different phases of production on swine farms. *J. Food Prot.* **70:**11–16.

Boyapalle, S., I. V. Wesley, H. S. Hurd, and P. G. Reddy. 2001. Comparison of culture, multiplex, and 5′ nuclease polymerase chain reaction assay for the rapid detection of *Yersinia enterocolitica* in swine and pork products. *J. Food Prot.* **64:**1352–1361.

Bredholt, S., T. Nesbakken, and A. Holck. 1999. Protective cultures inhibit growth of *Listeria monocytogenes* and *Escherichia coli* O157:H7 in cooked, sliced, vacuum- and gas-packaged meat. *Int. J. Food Microbiol.* **53:**43–52.

Bresolin, G., K. Neuhaus, S. Scherer, and T. M. Fuchs. 2006. Transcriptional analysis of long-term adaptation of *Yersinia enterocolitica* to low-temperature growth. *J. Bacteriol.* **188:**2945–2958.

Carniel, E. 1999. The Yersinia high-pathogenicity island. *Int. Microbiol.* **2:**161–167.

Chattopadhyay, M. K. 2006. Mechanism of bacterial adaptation to low temperature. *J. Biosci.* **31:**157–165.

Cornelis, G. R., A. Boland, A. P. Boyd, C. Geuijen, M. Iriarte, C. Neyt, M. P. Sory, and I. Stainier. 1998. The virulence plasmid of *Yersinia*, an antihost genome. *Microbiol. Mol. Biol. Rev.* **62:**1315–1352.

De Giusti, M., and E. De Vito. 1992. Inactivation of *Yersinia enterocolitica* by nitrite and nitrate in food. *Food Addit. Contam.* **9:**405–408.

Delor, I., A. Kaeckenbeek, G. Wauters, and G. R. Cornelis. 1990. Nucleotide sequence of *yst*, the *Yersinia enterocolitica* gene encoding the heat-stable enterotoxin, and prevalence of the gene among pathogenic and nonpathogenic yersiniae. *Infect. Immun.* **58:**2983–2988.

Doherty, A., J. J. Sherida, P. Allen, D. A. McDowell, I. S. Blair, and D. Harrington. 1995. Growth of *Yersinia enterocolitica* O:3 on modified atmosphere packaged lamb. *Food Microbiol.* **12:**251–257.

EFSA. 2006. The community summary report on trends and sources of zoonoses, zoonotic agents, antimicrobial resistance and foodborne outbreaks in the European Union in 2005. *EFSA J.* **94:**159–165.

Ermolenko, D. N., and G. I. Makhatadze. 2002. Bacterial cold-shock proteins. *Cell. Mol. Life Sci.* **59:**1902–1913.

Falcao, J. P., M. Brocchi, J. L. Proenca-Modena, G. O. Acrani, E. F. Correa, and D. F. Falcao. 2004. Virulence characteristics and epidemiology of *Yersinia enterocolitica* and yersiniae other than *Y. pseudotuberculosis* and *Y. pestis* isolated from water and sewage. *J. Appl. Microbiol.* **96:**1230–1235.

Falcao, J. P., D. P. Falcao, A. Pitondo-Silva, A. C. Malaspina, and M. Brocchi. 2006. Molecular typing and virulence markers of *Yersinia enterocolitica* strains from human, animal and food origins isolated between 1969 and 2000 in Brazil. *J. Med. Microbiol.* **55:**1539–1548.

Favier, G. I., M. E. Escudero, and M. S. de Guzman. 2005. Genotypic and phenotypic characterisation of *Yersinia enterocolitica* isolated from the surface of chicken eggshells obtained in Argentina. *J. Food Prot.* **68:**1812–1815.

FDA. 2001. *Yersinia enterocolitica* and *Yersinia pseudotuberculosis*. *In* George J. Jackson, Robert I. Merker, and Ruth Bander (ed.), *Bacteriological Analytical Manual* online. Food and Drug Administration, Washington, DC. http://www.cfsan.fda.gov/-ebam/bam-8.html.

Fearnley, C., S. L. W. On, B. Kokotovic, G. Manning, T. Cheasty, and D. G. Newell. 2005. Application of fluorescent amplified length polymorphism for comparison of human and animal isolates of *Yersinia enterocolitica*. *Appl. Environ. Microbiol.* 71:4960–4965.

Fredriksson-Ahomaa, M., and H. Korkeala. 2003a. Low occurrence of pathogenic *Yersinia enterocolitica* in clinical, food and environmental samples: a methodological problem. *Clin. Microbiol. Rev.* 16:220–229.

Fredriksson-Ahomaa, M., and H. Korkeala. 2003b. Molecular epidemiology of *Yersinia enterocolitica* 4/O:3. *Adv. Exp. Med. Biol.* 529:295–369.

Fredriksson-Ahomaa, M., T. Autio, and H. Korkeala. 1999. Efficient subtyping of *Yersinia enterocolitica* bioserotype 4/O:3 with pulsed-field gel electrophoresis. *Lett. Appl. Microbiol.* 29:308–312.

Fredriksson-Ahomaa, M., J. Björkroth, S. Hielm, and H. Korkeala. 2000a. Prevalence and characterization of pathogenic *Yersinia enterocolitica* in pig tonsils from different slaughterhouses. *Food Microbiol.* 17:93–101.

Fredriksson-Ahomaa, M., M. Bucher, C. Hank, A. Stolle, and H. Korkeala. 2001a. High prevalence of *Yersinia enterocolitica* 4:O3 on pig offal: a slaughtering technique problem. *System. Appl. Microbiol.* 24:457–463.

Fredriksson-Ahomaa, M., S. Hallanvuo, T. Korte, A. Siitonen, and H. Korkeala. 2001b. Correspondence of genotypes of sporadic *Yersinia enterocolitica* 4/O:3 strains from human and porcine origin. *Epidemiol. Infect.* 127:37–47.

Fredriksson-Ahomaa, M., B. Hartmann, P. Scheu, and A. Stolle. 2007a. Detection of pathogenic *Yersinia enterocolitica* in meat using real-time PCR. *J. Verbr. Lebensm.* 2:202–208

Fredriksson-Ahomaa, M., U. Koch, C. Klemm, M. Bucher, and A. Stolle. 2004. Different genotypes of *Yersinia enterocolitica* 4/O:3 strains widely distributed in butcher shops in the Munich area. *Int. J. Food Microbiol.* 15:89–94.

Fredriksson-Ahomaa, M., T. Korte, and H. Korkeala. 2000b. Contamination of carcasses, offals and the environment with *yadA*-positive *Yersinia enterocolitica* in a pig slaughterhouse. *J. Food Prot.* 63:31–35.

Fredriksson-Ahomaa, M., T. Korte, and H. Korkeala. 2001c. Transmission of *Yersinia enterocolitica* 4/O:3 to pets via contaminated pork. *Lett. Appl. Microbiol.* 32:308–312.

Fredriksson-Ahomaa M., T. Naglic, N. Turk, B. Šeol, Ž. Grabarevic´, I. Bata, D. Perkovic, and A. Stolle. 2007b. Yersiniosis in zoo marmosets *(Callitrix jacchuss)* caused by *Yersinia enterocolitica* 4/O:3. *Vet. Microbiol.* 121:363–367.

Fredriksson-Ahomaa, M., A. Stolle, and H. Korkeala. 2006a. Molecular epidemiology of *Yersinia enterocolitica* infections. *FEMS Immunol. Med. Microbiol.* 47:315–329.

Fredriksson-Ahomaa, M., A. Stolle, A. Siitonen, and H. Korkeala. 2006b. Sporadic human *Yersinia enterocolitica* infections caused by bioserotype 4/O:3 originate mainly from pigs. *J. Med. Microbiol.* 55:747–749.

Fukushima, H., and M. Gomyoda. 1991. Intestinal carriage of *Yersinia pseudotuberculosis* by wild birds and mammals in Japan. *Appl. Environ. Microbiol.* 57:1152–1155.

Fukushima, H., M. Gomyoda, N. Hashimoto, I. Takashima, F. N. Shubin, L. M. Isachikova, I. K. Paik, and X. B. Zheng. 1998. Putative origin of *Yersinia pseudotuberculosis* in western and eastern countries. A comparison of restriction endonuclease analysis of virulence plasmid. *Zentralbl. Bakteriol.* 288:93–102.

Fukushima, H., M. Gomyoda, S. Kaneko, M. Tsubokura, N. Takeda, T. Hongo, and F. N. Shubin. 1994. Restriction endonuclease analysis of virulence plasmid for molecular epidemiology of *Yersinia pseudotuberculosis* infection. *J. Clin. Microbiol.* 32:1410–1413.

Fukushima, H., K. Hoshina, H. Itowa, and M. Gomyoda. 1997. Introduction into Japan of pathogenic *Yersinia* through imported pork, beef and fowl. *Int. J. Food Microbiol.* 35:205–212.

Fukushima, H., K. Katsube, Y. Hata, R. Kishi, and S. Fujiwara. 2007. Rapid separation and concentration of food-borne pathogens in food samples prior to quantification by viable-cell counting and real-time PCR. *Appl. Environ. Microbiol.* 73:92–100.

Fukushima, H., K. Maruyama, I. Omori, K. Ito, and M. Iorihara. 1989. Role of the contaminated skin of pigs in faecal *Yersinia* contamination of pig carcasses at slaughter. *Fleischwirtschaft* 69:409–413.

Fukushima, H., Y. Matsuda, R. Seki, M. Tsubokura, N. Takeda, F. N. Shubin, I. K. Paik, and X. B. Zheng. 2001. Geographical heterogeneity between Far Eastern and western countries in prevalence of the virulence plasmid, the superantigen *Yersinia pseudotuberculosis*-derived mitogen, and the high-pathogenicity island among *Yersinia pseudotuberculosis* strains. *J. Clin. Microbiol.* 39:3541–3547.

Fukushima, H., R. Nakamura, S. Iitsuka, Y. Ito, and K. Saito. 1985. Presence of zoonotic pathogens (*Yersinia* spp., *Campylobacter jejuni*, *Salmonella* spp., and *Leptospira* spp.) simultaneously in dogs and cats. *Zentrabl. Bakteriol. Hyg. Abt. 1 Orig. B* 181:430–440.

Fukushima, H., Y. Tsunomori, and R. Seki. 2003. Duplex real-time SYBR green PCR assay for detection of 17 species of food- or waterborne pathogens in stools. *J. Clin. Microbiol.* 41:5134–5146.

Gourdon, F., J. Beytout, A. Reynaud, J. P. Romaszko, D. Perre, P. Theodore, H. Soubelet, and J. Sirt. 1999. Human and animal epidemic of *Yersinia enterocolitica* O:9, 1989–1997, Auvergne, France. *Emerg. Infect. Dis.* 5:719–721.

Goverde, R. L. J., J. H. J. Huis in't Veld, J. G. Kusters, and F. R. Mooi. 1998. The psychrotrophic bacterium *Yersinia enterocolitica* requires expression of *pnp*, the gene for polynucleotide phosporylase, for growth at low temperature (5°C). *Mol. Microbiol.* 28:555–569.

Goverde, R. L., J. G. Kusters, and J. H. Huis in't Veld. 1994. Growth rate and physiology of *Yersinia enterocolitica*; influence of temperature and presence of the virulence plasmid. *J. Appl. Bacteriol.* 77:96–104.

Granfors, K., S. Jalkanen, R. von Essen, R. Lahesmaa-Rantala, O. Isomäki, K. Pekkola-Heikkola, R. Merilahti-Palo, R. Saario, H. Isomäki, and A. Toivanen. 1989. Yersinia antigens in synovial-fluid cells from patients with reactive arthritis. *N. Engl. J. Med.* 320:216–221.

Gürtler, M., T. Alter, S. Kasimir, M. Linnebur, and K. Fehlhaber. 2005. Prevalence of *Yersinia enterocolitica* in fattening pigs. *J. Food Prot.* 115:640–644.

Hallanvuo, S., P. Nuorti, U. M. Nakari, and A. Siitonen. 2003. Molecular epidemiology of the five recent outbreaks of Yersinia pseudotuberculosis in Finland. *Adv. Exp. Med. Biol.* 529:309–312.

Han, T. H., I. K. Paik, and S. J. Kim. 2003. Molecular relatedness between isolates of *Yersinia pseudotuberculosis* from a patient and an isolate from mountain spring water. *J. Korean Med. Sci.* 18:425–428.

Hannu, T., L. Mattila, J. P. Nuorti, P. Ruutu, J. Mikkola, A. Siitonen, and M. Leirisalo-Repo. 2003. Reactive arthritis after an outbreak of *Yersinia pseudotuberculosis* serotype O:3 infection. *Ann. Rheum. Dis.* 62:866–869.

Harrison, W. A., A. C. Peters, and L. M. Fielding. 2000. Growth of *Listeria monocytogenes* and *Yersinia enterocolitica* colonies under modified atmospheres at 4 and 8°C using a model food system. *J. Appl. Microbiol.* 88:38–44.

Hayashidani, H., Y. Ohtomo, Y. Toyokawa, M. Saito, K. I. Kaneko, J. Kosuge, M. Kato, M. Ogawa, and G. Kapperud. 1995. Potential sources of sporadic human infections with *Yersinia enterocolitica* serovar O:8 in Aomori Prefecture, Japan. *J. Clin. Microbiol.* **33**:1253–1257.

Hurme, R., and M. Rhen. 1998. Temperature sensing in bacterial gene regulation—what it all boils down to. *Mol. Microbiol.* **30**:1–6.

Isberg, R. R., and P. Barnes. 2001. Subversion of integrins by enteropathogenic *Yersinia*. *J. Cell Sci.* **114**:21–28.

ISO. 2003. Horizontal method for the detection of presumptive pathogenic *Yersinia enterocolitica*. ISO 10273. International Organisation for Standardisation, Geneva, Switzerland.

Iwata, T., Y. Une, T. A. Okatani, S. Kaneko, S. Namai, S. Yoshida, T. Horisaka, T. Horikita, A. Nakadai, and H. Hayashidani. 2005. *Yersinia enterocolitica* serovar O:8 infection in breeding monkeys in Japan. *Microbiol. Immunol.* **49**:1–7.

Jalava, K., M. Hakkinen, M. Valkonen, U. M. Nakari, T. Palo, S. Hallanvuo, J. Ollgren, A. Siitonen, and P. Nuorti. 2006. An outbreak of gastrointestinal illness and erythema nodosum from grated carrots contaminated with *Y. pseudotuberculosis*. *J. Infect. Dis.* **194**:1209–1216.

Jalava, K., S. Halanvuo, U. M. Nakari, P. Ruutu, E. Kela, T. Heinäsmäki, A. Siitonen, and J. P. Nuorti. 2004. Multiple outbreaks of *Yersinia pseudotuberculosis* infections in Finland. *J. Clin. Microbiol.* **42**:2789–2791.

Jerret, I. V., K. J. Slee, and B. I. Robertson. 1990. Yersiosis in farmed deer. *Aust. Vet. J.* **67**:212–214.

Johannessen, G. S., G. Kapperud, and H. Kruse. 2000. Occurrence of pathogenic *Yersinia enterocolitica* in Norwegian pork products determined by a PCR method and a traditional culturing method. *Int. J. Food Microbiol.* **54**:75–80.

Jones, P. G., and M. Inouye. 1994. The cold-shock response—a hot topic. *Mol. Microbiol.* **11**:811–818.

Jones, T. F., S. C. Buckingham, C. A. Bopp, E. Ribot, and W. Schaffner. 2003. From pig to pacifier: chitterling-associated yersiniosis outbreak among black infants. *Emerg. Infect. Dis.* **9**:1007–1009.

Jourdan, A. D., S. C. J. Johnson, and I. V. Wesley. 2000. Development of a fluorogenic 5′ nuclease PCR assay for detection of the *ail* gene of pathogenic *Yersinia enterocolitica*. *Appl. Environ. Microbiol.* **66**:3750–3755.

Kageyama, T., A. Ogasawara, and R. Fukuhara. 2002. *Yersinia pseudotuberculosis* infection in breeding monkeys: detection and analysis of strain diversity by PCR. *J. Med. Primatol.* **31**:129–135.

Kaneko, S., N. Ishizaki, and Y. Kokubo. 1995. Detection of pathogenic *Yersinia enterocolitica* and *Yersinia pseudotuberculosis* from pork using polymerase chain reaction. *Contr. Microbiol. Immunol.* **13**:153–155.

Karapinar, M., and S. A Gonul. 1992. Effects of sodium bicarbonate, vinegar, acetic and citric acids on growth and survival of *Y. enterocolitica*. *Int. J. Food Microbiol.* **16**:343–347.

Knutsson, R., M. Fontanesi, H. Grage, and P. Rådström. 2002. Development of a PCR-compatible enrichment medium for *Yersinia enterocolitica*: amplification precision and dynamic detection range during cultivation. *Int. J. Food Microbiol.* **72**:185–201.

Korte, T., M . Fredriksson-Ahomaa, T. Niskanen, and H. Korkeala. 2004. Low prevalence of pathogenic *Yersinia enterocolitica* in sow tonsils. *Foodborne Pathog. Dis.* **1**:45–52.

Kuehni-Boghenbor, K., S. L. W. On, B. Kokotovic, A. Baumgartnr, T. M. Wassenaar, M. Wittwer, B. Bissig-Choisat, and J. Frey. 2006. Genotyping of human and porcine *Yersinia enterocolitica*, *Yersinia intermedia*, and *Yersinia bercovieri* strains from

Switzerland by amplified fragment length polymorphism analysis. *Appl. Environ. Microbiol.* **72**:4061–4066.

Lambertz, S. T., K. Granath, M. Fredriksson-Ahomaa, K. E. Johansson, amd M. L. Danielsson-Tham. 2007. Evaluation of a combined culture and PCR method (NMKL–163) for detection of presumptive pathogenic *Yersinia enterocolitica* in pork products. *J. Food Prot.* **70**:335–340.

Lantz, P. G., R. Knutsson, Y. Blixt, W. Abu Al-Soud, E. Borch, and P. Rådström. 1999. Detection of pathogenic *Yersinia enterocolitica* in enrichment media and pork by a multiplex-PCR: a study of sample preparation and PCR-inhibitory components. *Int. J. Food Microbiol.* **45**:93–105.

Lee, L. A., A. R. Gerber, M. S. Lonsway, J. D. Smith, G. P. Carter, D. P. Nancy, C. M. Parrish, R. K. Sikes, R. J. Finton, and R.V. Tauxe. 1990. *Yersinia enterocolitica* O:3 infections in infants and children, associated with the household preparation of chitterlings. *N. Engl. J. Med.* **14**:984–987.

Lee, T. S., S. W. Lee, W. S. Seok, M. Y. Yoo, J. W. Yoon, B. K. Park, K. D. Moon, and D. H. Oh. 2004. Prevalence, antibiotic susceptibility, and virulence factors of *Yersinia enterocolitica* and related species from ready-to-eat vegetables available in Korea. *J. Food Prot.* **67**:1123–1127.

Logue, C. M., J. J Sheridan, G. Wauters, D. A. McDowell, and I. S. Blair. 1996. *Yersinia* spp. and numbers, with particular reference to *Y. enterocolitica* bio/serotypes, occurring on Irish meat and meat products, and the influence of alkali treatment on their isolation. *Int. J. Food Microbiol.* **33**:257–274.

Makino, S. I., Y. Okada, T. Maruyama, S. Kaneko, and C. Sasakawa. 1994. PCR-based random amplified polymorphic DNA fingerprinting of *Yersinia pseudotuberculosis* and its practical applications. *J. Clin. Microbiol.* **32**:65–69.

McNally, A., C. Fearnley, R. W. Dalzier, G. A. Paiba, G. Manning, and D. G. Newell. 2004. Comparison of the biotypes of *Yersinia enterocolitica* isolated from pigs, cattle and sheep at slaughter and from humans with yersiniosis in Great Britain during 1999–2000. *Lett. Appl. Microbiol.* **39**:103–108.

Miller, V. L., K. B. Beer, G. Heusipp, B. M. Young, and M. R. Wachtel. 2001. Identification of regions of Ail required for the invasion and serum resistance phenotypes. *Mol. Microbiol.* **41**:1053–1062.

Myers, K. M., J. Gaba, and S. F. Al-Khaldi. 2006. Molecular identification of *Yersinia enterocolitica* isolated from pasteurized whole milk using DNA microarray chip hybridization. *Mol. Cell. Probes* **20**:71–80.

Nagamachi, E., S. Shibuya, Y. Hirai, O. Matsushita, K. Tomochika, and Y. Kanemasa. 1991. Adaptational changes of fatty acid composition and the physical state of membrane lipids following the change of growth temperature in *Yersinia enterocolitica*. *Microbiol. Immunol.* **35**:1085–1093.

Nakano, T., H. Kawaguchi, K. Nakao, T. Maruyama, H. Kamiya, and M. Sakurai. 1989. Two outbreaks of *Yersinia pseudotuberculosis* 5a infection in Japan. *Scand. J. Infect. Dis.* **21**: 175–179.

NCFA. 1998. Pathogenic *Yersinia enterocolitica*. PCR methods for detection in foods. NMKL method No. 163. Nordic Committee on Food Analysis, Oslo, Norway.

NCFA. 2003. *Yersinia enterocolitica*. Detection in food. NMKL method No. 117, 4th ed. Nordic Committee on Food Analysis, Oslo, Norway.

Nesbakken, T. 1988. Enumeration of *Yersinia enterocolitica* O:3 from the porcine oral cavity, and its occurrence on cut surface of pig carcasses and the environment in a slaughterhouse. *Int. J. Food Microbiol.* **6**:287–293.

Nesbakken, T., T. Iversen, K. Eckner, and B. Lium. 2007. Testing of pathogenic *Yersinia enterocolitica* in pig herds based on the natural dynamic of infection. *Int. J. Food Microbiol.* **111:** 99–104.

Nesbakken, T., E. Nerbrink, O. J. Røtterud, and E. Borch. 1994. Reduction of *Yersinia enterocolitica* and *Listeria* spp. on pig carcasses by enclosure of the rectum during slaughter. *Int. J. Food Microbiol.* **23:**197–208.

Neubauer, H., T. Sauer, H. Becker, S. Aleksic, and H. Meyer. 1998. Comparison of systems for identification and differentiation of species within the genus *Yersinia. J. Clin. Microbiol.* **36:**3366–3368.

Nilsson, A., S. T. Lambertz, P. Stålhandske, P. Norberg, and M. L. Danielsson-Tham. 1998. Detection of *Yersinia enterocolitica* in food by PCR amplification. *Lett. Appl. Microbiol.* **26:**140–141.

Niskanen, T., M. Fredriksson-Ahomaa, and H. Korkeala. 2002. *Yersinia pseudotuberculosis* with limited genetic diversity is a common finding in tonsils of fattening pigs. *J. Food Prot.* **65:** 540–545.

Niskanen, T., R. Laukkanen, M. Fredriksson-Ahomaa, and H. Korkeala. 2008. Distribution of *virF/lcrF*-positive *Yersinia pseudotuberculosis* O:3 at farm level. *Zoonoses Public Health* **55:**214–221.

Niskanen, T., J. Waldenström, M. Fredriksson-Ahomaa, B. Olsen, and H. Korkeala. 2003. *virF*-positive *Yersinia pseudotuberculosis* and *Yersinia enterocolitica* found in migratory birds in Sweden. *Appl. Environ. Microbiol.* **69:**4670–4675.

Nissen, H., T. Maugesten, and P. Lea. 2001. Survival and growth of *Escherichia coli* O157:H7, *Yersinia enterocolitica* and *Salmonella enteritidis* on decontaminated and untreated meat. *Meat Sci.* **57:**291–298.

Nowgesic, E., M. Fyfe, J. Hockin, A. King, A. Ng, A. Paccagnella, A. Trinidad, L. Wilcott, R. Smith, A. Denney, L. Sruck, G. Embree, K. Higo, J. I. Chan, P. Markey, S. Martin, and D. Bush. 1999. Outbreak of *Yersinia pseudotuberculosis* in British Columbia—November 1998. *Can. Commun. Disp. Rep.* **25:**97–100.

Nuorti, J. P., T. Niskanen, S. Hallanvuo, J. Mikkola, E. Kela, M. Hatakka, M. Fredriksson-Ahomaa, O. Lyytikäinen, A. Siitonen, H. Korkeala, and P. Ruutu. 2004. A widespread outbreak of *Y. pseudotuberculosis* O:3 infections from iceberg lettuce. *J. Infect. Dis.* **189:**766–774.

Ostroff, S. M., G. Kapperud, L. C. Huteagner, T. Nesbakken, N. H. Bean, J. Lassen, and R. V. Tauxe. 1994. Sources of sporadic *Yersinia enterocolitica* infections in Norway: a prospective case-control study. *Epidemiol. Infect.* **112:**133–141.

Özbas, Z. Y., A. Lehner, and M. Wagner. 2000. Development of a multiplex and semi-nested PCR assay for detection of *Yersinia enterocolitica* and *Aeromonas hydrophila* in raw milk. *Food Microbiol.* **17:**197–203.

Pin, C., J. Baranyi, and G. Garcia de Fernando. 2000. Predictive model for the growth of *Yersinia enterocolitica* under modified atmosphere. *J. Appl. Microbiol.* **88:**521–530.

Pujol, C., and J. B. Bliska. 2005. Turning *Yersinia* pathogenesis outside in: subversion of macrophage function by intracellular yersiniae. *Clin. Immunol.* **114:**216–226.

Robins-Browne, R. M. 1997. *Yersinia enterocolitica,* p. 192–215. *In* M. P. Doyle, L. R. Beuchat, and T. J. Montville (ed.), *Food Microbiology: Fundamentals and Frontiers.* ASM Press, Washington, DC.

Sandery, M., T. Stinear, and C. Kaucner. 1996. Detection of pathogenic *Yersinia enterocolitica* in environmental water by PCR. *J. Appl. Bacteriol.* **80:**327–332.

Satterthwaite, P., K. Pritchard, D. Floyd, and B. Law. 1999. A case-control study of *Yersinia enterocolitica* infections in Auckland. *Aust. N. Z. J. Public Health* **23:**482–485.

Schiemann, D. A. 1979. Synthesis of a selective agar medium for *Yersinia enterocolitica. Can. J. Microbiol.* **25:**1298–1304.

Schiemann, D. A. 1989. *Yersinia enterocolitica* and *Yersinia pseudotuberculosis,* p. 601–672. *In* M. P. Doyle (ed.), *Foodborne Bacterial Pathogens,* Marcel Dekker, New York. NY.

Seuri, M., and K. Granfors. 1992. Possible confounders of the relationship between occupational swine contact and *Yersinia enterocolitica* O:3 and O:9 antibodies. *Eur. J. Epidemiol.* **8:** 532–538.

Slee, K. J., and N. W. Skilbeck. 1992. Epidemiology of *Yersinia pseudotuberculosis* and *Yersinia enterocolitica* infections in sheep in Australia. *J. Clin. Microbiol.* **30:**712–715.

Slee, K. J., P. Brightling, and R. J. Seiler. 1988. Enteritis in cattle due to *Yersinia pseudotuberculosis* infection. *Aust. Vet. J.* **65:**271–275.

Smego, R. A., J. Frean, and H. J. Koornhof. 1999. Yersiniosis I: microbiological and clinicoepidemiological aspects of plague and non-plague *Yersinia* infections. *Eur. J. Clin. Microbiol. Infect. Dis.* **18:**1–15.

Sprague, L. D., and H. Neubauer. 2005. *Yersinia aleksiciae* sp. nov. *Int. J. Syst. Evol. Microbiol.* **55:**831–835.

Stock, I., and B. Wiedemann. 1999. An in-vitro study of the antimicrobial susceptibility of *Yersinia enterocolitica* and the definition of a database. *J. Antimicrob. Chemother.* **43:** 37–45.

Straley, S. C., and R. D. Perry. 1995. Environmental modulation of gene expression and pathogenesis in *Yersinia. Trends Microbiol.* **3:**310–317.

Tauxe, R. V., J. Vandepitte, G. Wauters, S. M. Martin, V. Goossens, P. De Moll, R. van Noyen, and G. Theirs. 1987. *Yersinia enterocolitica* infections and pork: the missing link. *Lancet* i:1129–1132.

Tertti, R., K. Granfors, O. P. Lehtonen, J. Mertsola, A. I. Mäkelä, L. Välimäki, M. Hänninen, and A. Toivanen. 1984. An outbreak of *Yersinia pseudotuberculosis* infection. *J. Infect. Dis.* **149:**245–250.

Tertti, R., R. Vuento, P. Mikkola, K. Granfors, A. I. Mäkelä, and A. Toivanen. 1989. Clinical manifestation of *Yersinia pseudotuberculosis* infection in children. *Eur. J. Clin. Microbiol. Infect. Dis.* **8:**587–591.

Thisted Lambertz, S., and M. L. Danielsson-Tham. 2005. Identification and characterization of pathogenic *Yersinia enterocolitica* isolates by PCR and pulsed-field gel electrophoresis. *Appl. Environ. Microbiol.* **71:**3674–3681.

Thisted Lambertz, S., R. Lindqvist, A. Ballagi-Pordany, and M.-L. Danielsson-Tham. 2000. A combined culture and PCR method for detection of pathogenic *Yersinia enterocolitica* in food. *Int. J. Food Microbiol.* **57:**63–73

Thompson, J. S., and M. Gravel. 1986. Family outbreak of gastroenteritis due to *Yersinia enterocolitica* serotype O:3 from well water. *Can. J. Microbiol.* **32:**700–701.

Thomson, N. R., S. Howard, and B. W. Wren. 2006. The complete genome sequence and comparative genome analysis of the high pathogenicity *Yersinia enterocolitica* strain 8081. *PLoS Genet.* **2:**2039–2051.

Tomaso, H., G. Mooseder, S. Al Dahouk, C. Bartling, H. C. Scholz, R. Strass, T. M. Treu, and H. Neubauer. 2006. Seroprevalence of anti-*Yersinia* antibodies in healthy Austrians. *Eur. J. Epidemiol.* **21:**77–81.

Toora, S., E. Budu-Amoako, R. F. Ablett, and J. Smith. 1992. Effect of high-temperature short-time pasteurisation, freezing and thawing and constant freezing, on survival of *Yersinia enterocolitica* in milk. *J. Food Prot.* **55:**803–805.

Tsubokura, M., and S. Aleksic. 1995. A simplified antigenic schema for serotyping of *Yersinia pseudotuberculosis:* phenotypic characterisation of reference strains and preparation of O and H factor sera. *Contrib. Microbiol. Immunol.* **13:**99–105.

Tsubokura, M., K. Otsuki, K. Sato, M. Tanaka, T. Hongo, H. Fukushima, T. Maruyama, and M. Inoue. 1989. Special features of distributions of *Yersinia pseudotuberculosis* in Japan. *J. Clin. Microbiol.* **27:**790–791.

Vishnubhatla, A., R. D. Oberst, D. Y. C. Fung, W. Wonglumsom, M. P. Hays, and T. G. Nagaraja. 2001. Evaluation of a 5′-nuclease (TaqMan) assay for the detection of virulent strains of *Yersinia enterocolitica* in raw meat and tofu samples. *J. Food Prot.* **64:**355–360.

Voskressenskaya, E., A. Leclercq, G. Tseneva, and E. Carniel. 2005. Evaluation of ribotyping as a tool for molecular typing of *Yersinia pseudotuberculosis* strains of worldwide origin. *J. Clin. Microbiol.* **43:**6155–6160.

Wauters, G., V. Goossens, M. Janssens, and J. Vandepitte. 1988. New enrichment method for isolation of pathogenic *Yersinia enterocolitica* serogroup O:3 from pork. *Appl. Environ. Microbiol.* **54:**851–854.

Wolffs P., B. Norling, and P. Rådström. 2005. Risk assessment of false-positive quantitative real-time PCR results in food, due to detection of DNA originating from dead cells. *J. Microbiol. Methods* **60:**315–323.

Wolffs, P., R. Knutsson, B. Norling, and P. Rådström. 2004. Rapid quantification of *Yersinia enterocolitica* in pork samples by a novel sample preparation method, flotation, prior to real-time PCR. *J. Clin. Microbiol.* **42:**1042–1047.

Wren, B. W. 2003. The yersiniae—a model genus to study the rapid evolution of bacterial pathogens. *Nat. Rev. Microbiol.* **1:** 55–64.

Zheng, X. B., M. Tsubokura, Y. Wang, C. Xie, T. Nagano, K. Someya, T. Kiyohara, K. Suzuki, and T. Sanekata. 1995. *Yersinia pseudotuberculosis* in China. *Microbiol. Immunol.* **39:**821–824.

Pathogens and Toxins in Foods: Challenges and Interventions
Edited by V. K. Juneja and J. N. Sofos
© 2010 ASM Press, Washington, DC

Chapter 12

Other Bacterial Pathogens: *Aeromonas, Arcobacter, Helicobacter, Mycobacterium, Plesiomonas,* and *Streptococcus*

ELAINE M. D'SA AND MARK A. HARRISON

The occurrence of food-borne illness may be dominated in the media and public awareness by outbreaks related to a small group of microorganisms, but the role of lesser-known food-borne pathogens in the causation of illness cannot be discounted. This chapter covers some of these bacterial pathogens and will attempt to give a concise, thorough overview of the significance, characteristics, and food safety concerns involving each pathogen.

AEROMONAS SPP.

Characteristics of Organisms and Type of Illness

These gram-negative bacteria are of the order *Aeromonadales* and family *Aeromonadaceae,* and *Aeromonas hydrophila* is the type species. They are coccobacilli or bacilli, occurring singly or in pairs, with rare short chains. Strains may be motile with single, polar flagella or lateral or peritrichous flagella. Carbohydrates are utilized both oxidatively and fermentatively.

In accordance with the Safe Drinking Water Act, *A. hydrophila* is listed in the Environmental Protection Agency's (EPA) first and second Contaminant Candidate List as a "potential waterborne pathogen." An EPA method (method 1605) exists for the detection and enumeration of *A. hydrophila* bacteria in drinking water. *Aeromonas* species (aeromonads) are ubiquitous, globally occurring aquatic, oxidase-positive facultative anaerobes. Environments in which they have been detected include drinking and wastewater, the ground, surfaces, and marine water bodies. There is evidence that some species are pathogenic for humans and warm-blooded and cold-blooded animals, including domestic animals and birds. *A. hydrophila, A. veronii* biovar *sobria, A. trota, A. caviae, A. jandaei, A. veronii* biovar *veronii,* and *A. schubertii* have been associated with human infections. *A. salmonicida* is a significant pathogen of fish (USEPA, 2006).

Bergey's Manual of Systematic Bacteriology (Brenner et al., 2005) lists 17 DNA-DNA hybridization groups and 14 phenospecies of this genus. Additional species have been described more recently. To a lesser extent, 16S ribosomal DNA and *gyrB* sequence studies have been used to establish taxonomic species relatedness. Serotyping studies determined the existence of 44 serogroups based on O antigen studies, with an additional 52 provisional groups. Serotypes O:11, O:16, and O:34 were found to predominate in clinical species. The serological heterogeneity of several strains precludes its usefulness for taxonomic studies, though serotyping may be useful as a rapid method for identification of aeromonads to the genus level.

Aeromonas species are indicated as important human pathogens causing gastrointestinal and other infections in healthy and immunocompromised hosts. Chronic diarrheal disease caused by this genus occurs most commonly in children, the elderly, and the immunocompromised (von Graevenitz and Mensch, 1968). Aeromonads have been implicated as a cause of traveler's diarrhea, nosocomial infections, and community-acquired infections. Conditions caused by *Aeromonas* in adults include diarrhea, wound infections, endocarditis, meningitis, septicemia, and bile tract infections. The most implicated clinical species is *A. hydrophila* (48%), followed by *A. sobria* and *A. caviae* (about 25% each). Aeromonads also cause disease in fish, especially in crowded, aquaculture

Elaine M. D'Sa • Department of Foods and Nutrition, The University of Georgia, Athens, GA 30602. • Mark A. Harrison
• Department of Food Science and Technology, The University of Georgia, Athens, GA 30602.

environments, and thus have resulted in notable losses to the aquaculture industry (Moyer, 2006).

Virulence in *Aeromonas* species may be due to cell-associated or extracellular factors. Enzymes produced include toxins, hemolysins, lipases, adhesins, proteases, and agglutinins. Other enzymes like amylase, chitinase, elastase, lecithinase, and nuclease have also been identified, in addition to cell-associated flagella, outer membrane proteins, pili, capsules, and lipopolysaccharide (LPS). Serine proteases and metalloproteases function in toxin activation, and a lipase that digests host cell plasma membranes is also produced.

Presence and adherence of pili are essential in virulence. Two important families of type IV pili produced in gastroenteritis-associated *Aeromonas* species are the Tap (short, rigid) pili, associated with the Tap gene cluster (Tap A), and Bfp (bundle-forming) long, wavy pili, associated with the *tapABCD* gene cluster (Barnett et al., 1997). Tap pili attach to enterocytes, while Bfp pili are thought to be important in attachment to human erythrocytes and in intestinal colonization. Both the polar flagella and multiple lateral flagella facilitate adhesion and thus are thought to be important virulence factors and significant in the formation of biofilms that are resistant to disinfection (Kirov et al., 2004).

"S-layers" are cell surface proteins that are associated with LPS. Strains that produce these are believed to be more pathogenic for fish, but their role in human infections has not been established (USEPA, 2006).

"Aerolysins" are toxins produced by some *Aeromonas* species that have hemolytic, enterotoxic, and cytolytic activity. One type is produced from cells as "proaerolysin"; proteolytic cleavage of a C-terminal peptide from this produces an active fragment that is approximately 40 amino acids long. Another aerolysin is a dimer which produces a pore in the outer membrane protein, facilitating altered cell permeability, leakage of cell contents, and cell destruction (USEPA, 2006).

Sources and Incidence in the Environment and Foods

Aeromonas species have been detected in a variety of foods, including raw produce, raw meats, seafood, ready-to-eat meats and dairy products, and treated drinking water. They are part of the normal microflora of healthy animals and humans and are found at high levels in sewage (USEPA, 2006). *Aeromonas* may be ingested from water and food through the environment, but most such host-microbe interactions do not result in disease. A high infectious dose ($>10^{10}$) has been reported, which may explain why no outbreaks associated with treated drinking water

have been described (Rusin et al., 1997). Lack of personal hygiene, water reuse, and consumption of untreated water containing large numbers of aeromonads are risk factors for the spread of disease. Increased transmission rates occur in day care environments, nursing homes, and intensive care units.

Intrinsic and Extrinsic Factors That Affect Survival and Growth in Food Products and Contribute to Outbreaks

Aeromonas species occur in biofilms that resist disinfection, though no outbreaks due to the presence of these organisms in water biofilms have been reported. A seasonal association is observed between aeromonads present in drinking water supplies and those identified in human cases of gastrointestinal disease. Numbers of aeromonads in water are reported to be higher in summer months (Janda and Abbott, 1998). The optimum growth temperature for *Aeromonas* spp. is 22 to 35°C, with an extended temperature range of 0 to 45°C for some. Some species (including most *A. salmonicida*) do not grow at 35°C or above (Mateos et al., 1993). The optimum pH range is from 5.5 to 9.0; a wider range of 4.5 to 9.0 is tolerated. Optimum NaCl concentration is from 0 to 4%.

Most motile aeromonads are resistant to penicillin, ampicillin, carbenicillin, and ticarcillin (with the exception of *A. trota* and 30% of *A. caviae*, which are susceptible to ampicillin). They are susceptible to second- and third-generation cephalosporins, chloramphenicols, tetracyclines, quinolones, aminoglycosides, carbapenems, and trimethoprim-sulfamethoxazole.

Aeromonas spp. are differentiated from *Vibrio* and *Plesiomonas* spp. by their resistance to 150 µg O/129 and variable reaction to ornithine decarboxylase. Aeromonads are unable to grow in 6.5% NaCl, liquefy gelatin, do not ferment i-inositol, and have a negative string test. Phenotypically, they are unable to grow on thiosulfate-citrate-bile salts-sucrose agar; most are able to ferment D-mannitol and sucrose (USEPA, 2006).

Food specimens are transported and handled according to the procedures in the FDA's *Bacteriological Analytical Manual* (Andrews and Hammack, 2005), while environmental samples are handled according to methods in APHA's *Standard Methods for Examination of Water and Wastewater,* 21st edition (APHA, 2005). One or more broth enrichment procedures (with or without antibiotics) may be needed to recover aeromonads where low numbers of organisms are expected. No single culture medium is considered optimal for recovery of all aeromonads, with combinations of plating, membrane filtration, and multiple tube tests commonly used. The EPA method 1605, which is a membrane filtration method using ampicillin dextrin agar (ADA) with vancomycin has been validated for isolation of *A. hydrophila* from drinking water (USEPA, 2006). Modified bile salts irgasan brilliant green agar has been used for recovery of 2 to 7 $\log_{10}$ aeromonads from foods, even in the presence of 5 to 6 $\log_{10}$ background flora. Starch ampicillin agar and commercially available *Aeromonas* medium (Ryan's medium) have also been used. Screening of colonies may be done by the oxidase test, and identification by commercial kits or biochemical methods. Phenotypic diversity makes identification based on biochemical tests challenging. Within the group, colistin resistance can be used as a phenotypic marker. Antigen-antibody methods are not commonly used for identification of aeromonads.

For recovery of aeromonads from clinical specimens, MacConkey agar, cefsulodin irgasan novobiocin or blood ampicillin agar may be used, followed by screening and identification, as mentioned above. Commercial identification systems have been found to give inaccurate results. The characterization of 193 strains using carbohydrate, tube, and plate tests (Abbott et al., 2003) was carried out and is recommended for identification.

Molecular methods for detection of aeromonads have been studied. Microarray technology is useful for the study of populations in biofilms (USEPA, 2006). Typing methods are useful in epidemiological studies, due to the heterogeneity observed. Pulsed-field gel electrophoresis has been found to be increasingly useful and is replacing ribotyping as a typing method.

Food Processing Operations That Influence the Numbers, Spread, or Characteristics

Aeromonads are inactivated by commonly used disinfectants used in water treatment and by routinely used food processing and preparation methods.

Recent Advances in Biological, Chemical, and Physical Interventions To Guard against the Pathogen

With little evidence of *Aeromonas* causing food-borne illness, typical food-borne pathogen intervention methods are effective in inactivating *Aeromonas* spp. in foods.

Discriminative Detection Methods for Confirmation and Trace-Back of Contaminated Products

Several methods have been used for the detection, enumeration, and isolation of aeromonads.

Among these, the membrane filtration method using ADA has been included in the APHA's *Standard Methods for Examination of Water and Wastewater* (APHA, 2005) and EPA method 1605 (ADA with vancomycin). Concentrations of 10 to 30 mg/liter ampicillin have been used in ADA. Molecular detection methods used for clinical or environmental samples have a detection limit greater than 3 to 4 $\log_{10}$ CFU/ml and generally need an enrichment step. Some of these methods target virulence genes, with a view to identifying pathogenic strains.

Detection of aeromonads in environmental samples is influenced by temperature (should be greater than 15°C) and residual chlorine levels (should be less than 0.2 mg/liter). The tendency of the cells to form biofilms makes accurate determination of their numbers in water systems a difficult task (USEPA, 2006). This is not considered a significant problem though, since the level of aeromonads in water is not close to the infectious dose for humans. Hence, the genus is not considered to be a public health threat in water. Additionally, *Aeromonas* levels do not correlate well with commonly used indicator organisms for fecal pollution of drinking water.

Concluding Remarks

The importance of aeromonads as pathogens of food-borne origin dates back to the 1950s, following their isolation from humans. Their presence in water sources and raw and ready-to-eat foods lends authenticity to their capacity to cause outbreaks of food-borne disease under the right circumstances. Maintenance of sanitary food handling and distribution practices lowers the threat from this genus in the causation of food-borne illness.

ARCOBACTER SPP.

Characteristics of Organisms and Type of Illness

These "arc-shaped" bacteria, first observed in veterinary specimens, were designated species of a distinct genus in 1991 (Vandamme et al., 1991), officially separating them from the closely related species of the genus *Campylobacter*, with which they share a relatively common background of sources, characteristics, and illness manifestations in humans and animals. Together with *Campylobacter* spp., they are placed in the family *Campylobacteraceae*. Seven species have been identified: *A. butzleri* (thought to be the primary human pathogen), *A. cryaerophilus* hybridization groups 1A and 1B, *A. skirrowii, A. nitrofigilis* (Houf et al., 2000), *A. cibarius* (Houf et al., 2005), *A. halophilus* (Donachie et al., 2005), and *A. sulfidicus* (Wirsen et al., 2002). *Campylobacter, Helicobacter, Wolinella*, and *Arcobacter* spp. make up the rRNA superfamily VI of the epsilon class of *Proteobacteria. Arcobacter* spp. are gram-negative, mesophilic, aerotolerant, nonsporing rods and have been known to cause gastritis (rarely bacteremia or appendicitis) in humans and abortions, mastitis, and gastritis in livestock (Wesley, 1994). Infections with these organisms cause economic losses in livestock herds and compromise human health.

Since these organisms are relatively metabolically inert, tests used in their identification include those used to identify *Campylobacter* spp. Definitive biochemical tests used to identify *Arcobacter* and distinguish between the species are limited and somewhat unreliable. They include growth temperature, catalase activity, nitrate reduction, fatty acid analysis, hydrolysis of indoxyl acetate, growth in the presence of glycine or sodium chloride, growth on MacConkey agar, the inability to hydrolyze hippurate, and cadmium chloride susceptibility (Wesley, 1994). Other tests suggested include arylsulfatase and pyrazinamidase activities and susceptibility to polymyxin B, since in contrast to *Campylobacter* strains, *Arcobacter* strains are negative for all three tests (Burnens and Nicolet, 1993). Susceptibility to antimicrobial agents has also been used to characterize this genus (Kiehlbauch et al., 1992). Molecular and nucleic acid-based techniques, using genus- and species-specific DNA probes complementary to the highly conserved 16S rRNA molecule or the *glyA* gene, have proven to be more reliable means of identification and distinction for *Arcobacter* (and between *Campylobacter* and *Arcobacter*) species. Plasmid analysis (with a plasmid detection rate of 33% for *Arcobacter* spp.) has also been used, along with biochemical tests and antimicrobial resistance patterns, for epidemiological typing (Harras et al., 1998).

Sources and Incidence in the Environment and Foods

Limited studies have been conducted globally on the prevalence of these organisms. This, together with the fact that classical laboratory testing is geared toward the detection of thermophilic *Campylobacter* spp. (Bolton et al., 1992), allowing *Arcobacter* spp. to possibly go undetected, makes the determination of the accurate prevalence of the genus a study in statistical extrapolation.

Lammerding et al. (1996) postulated that the genus has a significant reservoir in poultry products. Its presence in red meat animals has also been documented. Consequently, containment of intestinal contents and fecal material will limit the spread of this emerging pathogen in farm animal operations.

With the exception of *A. nitrofigilis*, the organisms have been isolated from water (surface and drinking water), sewage, raw foods (including poultry [chickens, ducks, and turkeys], beef, and pork), ready-to-eat poultry, chicken flocks, humans and animals with enteritis, and aborted animal fetuses (Wesley, 1996). Isolation rates of up to 43% from porcine abortions have been reported. One study found the distribution of *Arcobacter* spp. in porcine abortions in the United States to be *A. butzleri* at 8%, *A. cryaerophilus* 1A at 16%, and *A. cryaerophilus* 1B at 60% (Harmon and Wesley, 1997). They are also a part of the indigenous flora of healthy animals. Antigenically identical strains have been recovered from diseased and healthy animals, underlining the possibly opportunistic role of *Arcobacter* spp. in disease under suitable conditions. Their role in the causation of disease is underlined by higher isolation rates from aborted fetuses and cases of porcine infertility than from healthy animals (Ellis et al., 1977). *A. cryaerophilus* has been associated with cases of infectious abortion in cattle herds in Germany (Wesley, 1994).

Son et al. (2007) sampled 325 broiler carcasses during a three-month period in 2004, at the prescalding, prechilling, and postchilling steps of a commercial poultry operation. *Arcobacter* recovery rates were highest in prescalded carcasses (96.8%) compared to prechilled (61.3%) and postchilled carcasses (9.6%). *A. butzleri* was found to be the most prevalent species in 79.1% of the samples, followed by *A. cryaerophilus* 1B (18.6%) and *A. cryaerophilus* 1A (2.3%). *A skirrowii* was not isolated in this study. A total of 71.8% of *Arcobacter* isolates showed multiple antimicrobial resistance. A total of 125 (89.9%) of *A. butzleri* isolates showed resistance to clindamycin, 114 (82%) to azithromycin, and 33 (23.7%) to nalidixic acid.

The disease can be acquired from travel in developing areas/countries, unhygienic food preparation practices and environmental conditions, cross-contamination, and person-to-person contact. In vivo studies have suggested the potential virulence of *A. butzleri* (Wesley et al., 1996; Musmanno et al., 1997). The presence of *Arcobacter* spp. in seafood and raw milk remains unknown. Tertiary treatment of sewage effluent with 2 ppm chlorine dioxide removes 99.9% of *Arcobacter* spp. (Stampi et al., 1993).

Surveys of cattle have detected *Arcobacter* spp. in 10.52% of dairy cattle fecal samples (Wesley, 1997). Wesley et al. (2000) studied the shedding of *Campylobacter* and *Arcobacter* spp. in the feces of dairy cattle in various U.S. states and found 14.3% to be positive for *Arcobacter*. Analysis of farm management practices revealed that feeding of alfalfa and use of individual waterers protected the cattle from infection with

Arcobacter. Thus, feed and dietary supplements may alter gut homeostasis, influencing microbial colonization.

Intrinsic and Extrinsic Factors That Affect Survival and Growth in Food Products and Contribute to Outbreaks

Campylobacter spp. have been identified as the leading cause of bacterial food-borne diarrheal illness in humans (CDC, 2001). It has been postulated that some of these campylobacters may be misidentified *Arcobacter* spp. Given the high incidence of *Campylobacter* spp. in raw poultry and in the environment and the ability of arcobacters to grow at mesophilic temperatures (as low as 15°C), it is possible that *Arcobacter* spp. are also significant contributors to the global incidence of diarrheal illness. In 1983, *A. butzleri* was the only pathogen isolated from an outbreak of diarrheal illness in Italian schoolchildren (Vandamme et al., 1993). A few other cases of *A. butzleri* in hospitalized patients who responded to antibiotics and in which *A. butzleri* was isolated as the sole pathogen, point to the pathogenic and invasive capacity of the organism (D'Sa, 2002). Lior serogroup 1 predominates in samples of food and from patients with illness.

In a study carried out by D'Sa (2002) using four human isolates of *A. butzleri* and two human isolates of *A. cryaerophilus*, it was determined that the pH growth range for these isolates was 5.5 to 8.0. The optimum pH range for *A. butzleri* strains was 6.0 to 7.0 and for *A. cryaerophilus* strains was 7.0 to 7.5. The effect of NaCl on growth and survival of the genus was variable, with upper growth limits of 3.5 and 3.0% for *A. butzleri* and *A. cryaerophilus*, respectively. Survival at NaCl concentrations up to 5% was noted at 25°C for some *Arcobacter* strains. Calculated z values ranged from 5.20 to 6.28°C. The species studied were found to be cold sensitive, elucidated by the fact that a mild heat treatment followed by cold shock acted synergistically to produce a large reduction in cell numbers.

Food Processing Operations That Influence the Numbers, Spread, or Characteristics

The same food processing and handling practices that are recommended to reduce *Campylobacter* problems should control problems that could be associated with *Arcobacter* species. There have been a few studies focused on the inhibition or elimination of *Arcobacter* in foods.

Low doses of irradiation (1.5 to 4.5 kGy) were found to be sufficient to eliminate the presence of *A. butzleri* in ground pork, as it was found to possess a slightly higher radiation resistance than that of *Campylobacter jejuni*. A z value of 8.1 was reported for exponential phase *A. butzleri* cells (Hilton et al., 2001).

Lactic and citric acids at concentrations of 0.5, 1, and 2% were found to be inhibitory to the growth in culture of *A. butzleri*, with citric acid being more effective (Phillips, 1999). Nisin (500 IU/ml) was inhibitory, while sodium citrate was more effective than sodium lactate. Low levels of sodium tripolyphosphate can prevent growth (0.016%) and survival (0.02%) of *A. butzleri* (D'Sa and Harrison, 2005).

Hancock (2000) studied the effects of spices on *A. butzleri* in vitro and in spice marinades on fresh pork loins. Cinnamon aquaresin (1.56%) and cinnamon oleoresin (3.13%) were the most inhibitory to the two strains of *A. butzleri* studied. Pimento leaf essential oil (3.13%), barbeque spice aquaresin, and clove aquaresin also inhibited growth at 6.25%. In the marination study, pimento leaf essential oil (6%) reduced *A. butzleri* populations by 1.37 and 2.11 logs over 2 and 5 days, respectively. Cinnamon aquaresin (3%) reduced *A. butzleri* levels by 1.03 and 1.34 logs over the same time span.

Hancock (2000) also studied the efficacy of organic acids, both in vitro and sprayed onto fresh pork loins inoculated with *A. butzleri*. Acetic acid, citric acid monohydrate, and lactic acid were inhibitory to *A. butzleri* strains studied in vitro, at less than 0.5%. On pork loins, *A. butzleri* populations were reduced by 0.87 $\log_{10}$ using 4% lactic acid. Organic acid concentration and type significantly affected the effectiveness of antimicrobial activity. Inclusion of both spices and organic acids as "hurdle" components, in combination with other means of reducing these bacterial species on food, was suggested.

Due to the lower rate of recovery of *Arcobacter* from red meats that have a reduced surface moisture level, it has been postulated that the genus is sensitive to the effects of ambient drying (Nachamkin and Blaser, 2000). The organism is also cold sensitive and may be sublethally injured at lower temperatures. Heat resistance studies have shown that recommended temperatures for safe cooking of foods should be sufficient to kill *Arcobacter* spp. (D'Sa and Harrison, 2005; USDA/FSIS, 2007).

Recent Advances in Biological, Chemical, and Physical Interventions To Guard against the Pathogens

Cervenka et al. (2008) found that cinnamaldehyde, followed by thymol, carvacrol, caffeic acid, tannic acid, and eugenol, was inhibitory in vitro against three strains of *A. butzleri*, two strains of *A. cryaerophilus*, and one strain of *A. skirrowii*.

Discriminative Detection Methods for Confirmation and Trace-Back of Contaminated Products

Multiplex PCR assays, PCR-restriction fragment length polymorphism assays, and combined PCR-enzyme-linked immunosorbent assays targeting both the 16s and 23s rRNA regions have been developed and tested as a means of identification and distinction of *Arcobacter* spp. (Wesley et al., 1995; Cardarelli-Leite et al., 1996; Harmon and Wesley, 1997; Marshall et al., 1999; Winters and Slavik, 2000; Antolin et al., 2001). Use of selective and enrichment media is necessary to increase the rate of isolation of *Arcobacter* spp. from food, environmental, and patient samples (D'Sa, 2002).

Concluding Remarks

There is still uncertainty as to the significance of the threat posed by members of the genus *Arcobacter*. While there have been food-borne illness cases linked to foods contaminated with species of *Arcobacter*, their frequency has been sporadic to date. Whether this is related to the clinical and laboratory methods used, which might overlook species of *Arcobacter* or misidentify them as *Campylobacter*, is unknown. Considering the similarities of many of the characteristics of the genera *Arcobacter* and *Campylobacter*, it is reasonable to consider species of *Arcobacter* potential food-borne hazards. While the two genera are frequently found on similar foods and share many growth and survival traits, they also are controlled by the same food handling and processing practices.

HELICOBACTER

Characteristics of Organism and Type of Illness

Helicobacter pylori, first isolated in 1983, is the type species and most well-known species of this genus. It has been the target of intensive study because of its link to human illness, especially peptic ulcers, chronic gastritis, and stomach cancer, including mucosa-associated lymphoid tissue lymphoma (Hardin and Wright, 2002). It colonizes stomach walls, is present on the gastric mucosal layer of carriers, is recognized as the major cause of peptic ulcer disease globally, and is one of the most common bacterial infections in humans (Suerbaum and Michetti, 2002). The genus *Helicobacter* was created in 1989 and includes about 23 recognized species. Taxonomically, these are placed in the epsilon class of *Proteobacteria*.

H. pylori infections require an array of bacterial and host factors; significant among these are *flaA* and *flaB* (genes coding for flagellin), urease (which alters the gastric microenvironment), specific adhesins that facilitate adherence, cecropins that limit the growth of competing microflora, and a P-type adenosine-triphosphatase. The organism attaches to gastric mucosa and using various factors, including its own LPS, causes mucosal injury. Following gastric colonization, an inflammatory reaction is induced that initiates a cascade of reactions, resulting in the recognizable symptoms of infection. *H. pylori* also alters normal gastric secretion. Patients with duodenal ulcers have an elevated serum gastrin level that results in increased gastric acid levels. Asymptomatic *H. pylori* infections are frequent. The relatively inaccessible microenvironment of the organism during infection makes treatment challenging. Metronidazole-clarithromycin antibiotic combinations have been used commonly to treat infections, along with proton pump inhibitors like ranitidine bismuth citrate. Under adverse conditions, the species has been known to produce viable but noncultivable forms that retain infective capacity (Bode et al., 1993). It has been suggested that invasion of gastric epithelial cells could be a mechanism of infection and persistence in the occurrence of *H. pylori* disease (Petersen and Krogfelt, 2003).

The organisms are gram-negative, fastidious, microaerophilic, nonsporing, spiral, motile (with polar flagella), curved rods (Hill, 1997) that grow optimally at 35 to 37°C but not at 25°C. The optimal NaCl concentration for growth is 0.5 to 1% (Jiang and Doyle, 1998). Addition of 0.05% ferrous sulfate and 0.05% sodium pyruvate optimized the growth of *H. pylori* strains in broth (Jiang and Doyle, 2000). They are sensitive to acid but are protected from the effects of gastric acidity in their mucosal microenvironment. Noncultivable cells have been detected in the oral cavity and in feces and may be significant in the assessment of the routes of transmission of the organism.

Sources and Incidence in the Environment and Foods

A definitive source of the organisms is unknown (Hill, 1997), and there is no established link to zoonotic transmission (Martin and Penn, 2001). Food-borne transmission has been postulated by several authors but not definitively proven. Water (both surface and municipal) may also be linked to transmission (Gomes and De Martinis, 2004a). Studies have demonstrated the ability of *H. pylori* to survive in water globally (Bellack et al., 2006).

The incidence of *H. pylori* infection in developed countries (0.3 to 0.5%) was found to be on the decline. The organism is acquired mainly during childhood and primarily by the fecal-oral, oral-oral, and iatrogenic routes. Higher infection rates are linked to lower socioeconomic conditions and higher density of living (Gomes and De Martinis, 2004b). In other studies, it was determined that while the infection rate is 80% among middle-aged adults in developing countries (Suerbaum and Michetti, 2002), it may be carried by as much as 70 to 90% of the population in developing countries (Gomes and De Martinis, 2004b) and by 25 to 50% in developed countries (Gomes and De Martinis, 2004b).

Studies have revealed the capacity of the organism to survive in foods. Gomes and De Martinis (2004a) inoculated *H. pylori* on lettuce and carrot samples stored at 8°C under normal or modified atmosphere and observed that the species remained viable for up to 72 to 120 h, depending on sample preparation and recovery media. Jiang and Doyle (2002) studied the optimization of culture conditions aimed at detecting *H. pylori* in foods such as ground beef, sterile skim milk, and pasteurized fruit juices stored at 4°C. The organism was detected in inoculated ground beef (7 days), milk (up to 11 days), and apple juice (24 h). It was noted that survival and detection of the species were difficult in the presence of competing microflora and that the organism does not grow in foods or in samples treated with enrichment media. However, foods with a pH range of 4.9 to 6.0 and a water activity greater than 0.97 could allow survival of *H. pylori*, especially under refrigerated conditions (Beuchat, 2002; Gomes and De Martinis, 2004a; Jiang and Doyle, 1998).

Intrinsic and Extrinsic Factors That Affect Survival and Growth in Food Products and Contribute to Outbreaks

The capacity of the organism to invade and persist in gastric mucosal cells, together with the formation of viable but noncultivable forms, points to its capacity to survive in foods and tissues under adverse environmental conditions.

O'Gara et al. (2008) observed the rapid anti-*H. pylori* activity of garlic oil (16 and 32 micrograms/ml) in simulated gastric environments. It was suggested that the garlic components are acid stable and thus able to exert their inhibitory effects at the low gastric pH of 2.0 or below.

Food Processing Operations That Influence the Numbers, Spread, or Characteristics

It is not certain whether the organism can be transmitted through food, but some authors believe that this is possible. Quaglia et al. (2007) studied the survival of four strains in pasteurized and ultrahigh-temperature processed milk stored at 4°C and determined that *H. pylori* had a median survival time of 9 days in pasteurized milk and 12 days in ultra-heat-treated milk. Thus,

consumption of raw, contaminated food (fruits and vegetables) and postprocess contamination of foods are believed to be possibilities in the food-borne transmission of the organism.

Recent Advances in Biological, Chemical, and Physical Interventions To Guard against the Pathogen

Gancz et al. (2008) showed that increasing NaCl concentrations altered cell growth, morphology, survival, and virulence factor expression of *H. pylori*, with some strain differences observed. Tabak et al. (1999) tested the effects of cinnamon extracts (ethanol based and methylene chloride based) on growth and urease activity of *H. pylori*. It was determined that the methylene chloride-based extract inhibited *H. pylori* growth at the concentration of commonly used antibiotics (15 to 50 µg/ml), while the ethanol extract inhibited urease activity.

Discriminative Detection Methods for Confirmation and Trace-Back of Contaminated Products

The organism is difficult to grow in culture (Suerbaum and Michetti, 2002). Methods used to diagnose infection include histological examination of tissues obtained during endoscopy, culture, PCR-based techniques, breath-testing for urease during active infection, serological tests, stool antigen assays, restriction length polymorphism analysis, and identification of *H. pylori* in feces, saliva, and dental plaque (Gomes and De Martinis, 2004b). Neubauer and Hess (2006) developed a multiplex PCR procedure that would detect and distinguish *Helicobacter pullorum* in poultry and poultry products.

Concluding Remarks

Transmission of *H. pylori* through foods leading to incidences of food-borne illness has been speculated but not demonstrated to date. There is evidence that the bacterium can survive on some types of foods, but the lack of evidence linking illness due to the organism to food products would indicate that the typical route of exposure to *H. pylori* is not food related. *Helicobacter* also appears to possess no outstanding survival characteristics that would be of concern when applying typical food processing and handling practices.

MYCOBACTERIUM SPP.

Characteristics of Organisms and Type of Illness

The genus *Mycobacterium*, most well known for its historically important species *M. tuberculosis* and

M. leprae, causative agents of tuberculosis and leprosy, respectively, has more than 120 species currently recognized. Several are thought to be pathogenic for humans and/or animals (Greenstein, 2003). Organisms belonging to the family *Mycobacteriaceae* are gram-positive, aerobic, nonmotile, acid-fast, rod-shaped bacteria. *M. tuberculosis* is the type species. The organisms are extremely fastidious in their growth requirements and have a temperature growth range of 25 to 45°C, with an optimum of 39°C. NaCl concentrations below 5% allow growth, as does a pH of 5.5 and higher (Collins et al., 2001). These characteristics make mycobacteria very resistant to destruction in various environments.

Mycobacterium avium subsp. *paratuberculosis* (Johne's bacillus) has been linked to paratuberculosis (Johne's disease) in domestic and wild ruminants and some other animals. This chronic, wasting gastrointestinal disease is observed globally with dairy herds, with animals (including those that are nonsymptomatic) being vectors of the disease. Consequently, the organism is found in raw milk, with up to 10^4 CFU/ml being reported (Skovgaard, 2007). Levels of 2 to 8 CFU/50 ml have also been reported for asymptomatic cows (Sweeney et al., 1992). It is thought that by correlation it may also be present in raw meat. *M. avium* subsp. *paratuberculosis* has also been linked to Crohn's disease in humans, with milk being the potential vehicle of zoonotic transmission. Crohn's disease is a chronic inflammatory disease that affects the gastrointestinal tract, especially the distal ileum and colon. Since there is no known cure, disease management practices are emphasized with Crohn's disease patients (Griffiths, 2006). *M. avium* subsp. *paratuberculosis* is extremely slow growing, a characteristic that makes detection and study of the species difficult.

Sources and Incidence in the Environment and Foods

Sources of the nontuberculous mycobacteria (i.e., mycobacteria other than *M. tuberculosis*) include the environment, water, soil, dust, and aerosols. These have low virulence and lack a person-to-person transmission route (Tortoli, 2006). However, their significance may be higher in immunocompromised hosts.

M. avium subsp. *paratuberculosis* has been shown to survive low-temperature pasteurization in milk (Grant et al., 2002) and ripening in Swiss cheese (Spahr and Schafroth, 2001). It has been isolated from potable water and appears to be resistant to utility levels of chlorination. At high inoculum levels (for example,

10^6 CFU/ml, *M. avium* subsp. *paratuberculosis* strains were not completely destroyed (reductions of 1.32 to 2.82 $\log_{10}$s were observed) at up to 2 µg/ml chlorine for up to 30 minutes (Whan et al., 2001; Greenstein, 2003). Zoonotic routes of transmission, while still uncertain, are thought to most likely occur through milk, meat, water, and contact with infected animals (Grant, 2005).

Intrinsic and Extrinsic Factors That Affect Survival and Growth in Food Products and Contribute to Outbreaks/Food Processing Operations That Influence the Numbers, Spread, or Characteristics

Studies in cheese have shown that the organism is inactivated faster at a lower pH, and NaCl concentrations of 2 to 6% had little effect on the capacity of the organism to survive at any pH (Sung and Collins, 2000).

There have been several studies focusing on the potential for *M. avium* subsp. *paratuberculosis* to survive the typical pasteurization temperatures applied to fluid milk (Grant, 2005). These have ranged from laboratory studies to those using commercial-scale pasteurization procedures. It has been noted in several of these studies that *M. avium* subsp. *paratuberculosis* has a greater degree of heat resistance than those of other mycobacteria of concern in milk. In some studies, low numbers of viable *M. avium* subsp. *paratuberculosis* were recovered after pasteurization. One study in England and Wales detected *M. avium* subsp. *paratuberculosis* in postpasteurized milk by PCR, and another study of 241 dairies in England, Wales, Northern Ireland, and Scotland documented the occurrence of *M. avium* subsp. *paratuberculosis* in milk samples. Another concern with dairy products is whether or not *M. avium* subsp. *paratuberculosis* could be a factor that determines postpasteurization survival of cells (Slana et al., 2008).

M. avium subsp. *paratuberculosis* has also been detected in 10 samples of dried-milk infant food products in Europe and in cheese at 50, 12, and 5% rates in samples from Greece, Czechoslovakia, and the United States, respectively (Slana et al., 2008). An additional concern is the possibility that *M. avium* subsp. *paratuberculosis* could survive in and be transmitted through cheeses made from raw milk (Donaghy et al., 2004). Since Johne's disease affects cattle, it is logical to expect *M. avium* subsp. *paratuberculosis* to be a possible contaminant of beef. However, surveillance for *M. avium* subsp. *paratuberculosis* in ground beef has not revealed a problem to date (Grant, 2005).

Recent Advances in Biological, Chemical, and Physical Interventions To Guard against the Pathogen

Effective farm management (screening, paratuberculosis control, and disease-free certification of herds) and disease control practices that eliminate the occurrence of Johne's disease in cattle, including vaccination of animals, will lower the occurrence of the organism and will minimize the potential for contamination of food. Much of the concern with this potential food-borne pathogen has focused on possible survival of the organism during milk pasteurization. Coupled with the lack of evidence for the transmission of this pathogen in milk, current pasteurization methods appear to be sufficient for the safe processing of milk.

Discriminative Detection Methods for Confirmation and Trace-Back of Contaminated Products

In both humans and animals, M. avium subsp. paratuberculosis is thought to exist in protoplast form, which makes identification by acid-fast staining (Ziehl-Neelsen method) untenable.

In the past, enzyme-linked immunosorbent assay methods based on detection of antibodies to M. avium subsp. paratuberculosis, have been studied. These have been used both to detect M. avium subsp. paratuberculosis in milk and to establish the prevalence of disease within a herd, as antibodies to M. avium subsp. paratuberculosis are usually not produced during the early stages of the disease. A disadvantage is the occurrence of false-positive reactions due to animals being exposed to other atypical mycobacteria. The use of solid phase cytometry for detection of fluorescently labeled M. avium subsp. paratuberculosis in milk and bioluminescence to detect M. avium subsp. paratuberculosis in a variety of substrates has also been reported.

Several methods for the rapid detection of M. avium subsp. paratuberculosis in milk are being studied; most are molecular methods that focus on the detection of M. avium subsp. paratuberculosis DNA (Slana et al., 2008). Identification of the IS900 DNA sequence unique to M. avium subsp. paratuberculosis, although not conclusive in establishing M. avium subsp. paratuberculosis as the cause of Crohn's disease, has been used to establish the presence of the organism in the host system. The method, though, has shown variability in reproducibility (Greenstein, 2003). These newer approaches adopt the application of variations of immunomagnetic separation-PCR techniques (Metzger-Boddien et al., 2006; Slana et al., 2008). A disadvantage of these is their inability to differentiate between viable and nonviable M. avium subsp. paratuberculosis cells.

Concluding Remarks

The association between Crohn's disease and M. avium subsp. paratuberculosis is possible, although conflicting views on direct evidence linking the two are still lacking. The issue continues to be reviewed by expert groups seeking to determine whether a link exists and if food processing and handling methods need to be modified to avoid possible food-borne transmission of the organism.

PLESIOMONAS SPP.

Characteristics of Organisms and Type of Illness

Plesiomonas spp. belong to the gamma subclass of Proteobacteria. Plesiomonas shigelloides, the most significant species of interest, has been isolated by several researchers, from cases of human diarrheal illness, and is also suspected to be the cause of other generalized human infections. It was the sole pathogen isolated from two outbreaks of acute diarrheal disease in Japan in 1973 (978 cases) and 1974 (24 cases). Symptoms included diarrhea, abdominal pain, fever, and headache. The predominant fecal isolate from the 1973 outbreak, P. shigelloides O17:H2, was also isolated from tap water and mud samples, while P. shigelloides O24:H5 was the fecal isolate recovered from the 1974 outbreak (Tsukamoto et al., 1978).

P. shigelloides produces a heat-stable enterotoxin, but based on a lack of consistent in vitro and in vivo evidence, the species is thought to possess low pathogenicity (Abbott et al., 1991). Some researchers believe that the organism is an opportunistic pathogen, posing a significant risk in immunocompromised patients or those with preexisting illness. Food-borne transmission has not been documented.

Sources and Incidence in the Environment and Foods

The organism has been isolated from surface waters (especially in warm weather), possibly unchlorinated tap water, soil, fish, shellfish, aquatic species, and meat animals (Abbott et al., 1991).

Intrinsic and Extrinsic Factors That Affect Survival and Growth in Food Products and Contribute to Outbreaks

P. shigelloides grows optimally at 30 to 35°C, as low as 10°C, and at a pH as low as 4.5. It has been detected in association with river fish (Jay, 2005).

Food Processing Operations That Influence the Numbers, Spread, or Characteristics/Recent Advances in Biological, Chemical, and Physical Interventions To Guard against the Pathogen

With little evidence of the pathogen causing food-borne illness, typical food-borne pathogen intervention methods are effective in inactivating *P. shigelloides* in foods.

Concluding Remarks

Since the 1970s, *P. shigelloides* has received some attention as a possible food-borne pathogen. However, the lack of evidence linking contamination of food and subsequent illness in humans indicates at this time that the organism probably poses little risk.

STREPTOCOCCUS SPP.

Characteristics of Organisms and Type of Illness

The etiological role of *Streptococcus* species in both human and animal disease has been well documented over the past few centuries. The organisms are widely distributed, found in soil and water and on plants, associated with humans and other animals (including the gastrointestinal tract), and found in foods.

Microorganisms belonging to the genus *Streptococcus* are gram-positive, nonmotile, catalase-negative, microaerophilic cocci occurring in chains or pairs. Their growth range is 25 to 45°C, with an optimum of 37°C. For some strains, growth occurs in the presence of 6.5% NaCl and up to a pH of 9.6. Several species are pathogenic for humans and animals (Hardie and Whiley, 1997). Historically, based on hemolytic, antigenic, and physiological characteristics, the streptococci have been divided into Lancefield groups. Current classification of the genus is based on 16S rRNA gene sequencing. Accordingly, the genus is primarily divided into the beta-hemolytic streptococci and the non-beta-hemolytic streptococci. Ongoing taxonomic research may result in additions or transfers of strains between taxonomic groups.

The beta-hemolytic streptococci include *Streptococcus pyogenes* (Lancefield's group A streptococcus). *S. pyogenes* is the type species and causes scarlet fever, bacterial pharyngitis, and several invasive infectious conditions. The M-protein antigen is the major virulence factor for this organism. Non-*S. pyogenes* group A streptococci exist but are not common (Facklam, 2002). Other beta-hemolytic streptococci include *S. agalactiae* (Lancefield's group B streptococcus), *S. dysgalactiae* subsp. *equisimilis* (formerly Lancefield's group C streptococcus), *S. equi* subsp. *equi* (that causes infections in horses), *S. equi* subsp. *zooepidemicus* (has caused human infections, mainly through contaminated dairy products), *S. canis* (identified as having the Lancefield group G antigen), *S. anginosus* group (beta-hemolytic strains of *S. anginosus*, *S. constellatus*, *S. intermedius*), *S. constellatus* subsp. *pharyngis* (Lancefield group C antigen; causes pharyngitis in humans), *S. porcinus* (strains have Lancefield group E, P, U, and V antigens), *S. iniae* (zoonotic to humans through fish), *S. phocae*, and *S. didelphis*.

Non-beta-hemolytic streptococci include *S. pneumoniae* (major cause of pneumonia in communities), *S. bovis* group (*S. bovis*, *S. equinus*, *S. gallolyticus*, *S. infantarius*, *S. pasteurianus*, and *S. lutetiensis*), *S. suis* (causes meningitis or septicemia in humans; zoonotic through pigs), viridans streptococci (include 26 species), unusual *Streptococcus* species, and other gram-positive cocci in chains, and new species being tested (Facklam, 2002).

Sources and Incidence in the Environment and Foods

Various foods have been associated with outbreaks of disease caused by *S. pyogenes* strains, including dairy products (milk, ice cream, cream, custard, and rice pudding), eggs, cooked seafood, and potato/egg/shrimp salads. Unpasteurized milk, poor food handler hygiene, and unsafe food handling practices (including temperature abuse) have been implicated as causes in disease outbreaks.

Other species of streptococci (*S. faecalis*, *S. faecium*, *S. durans*, *S. avium*, and *S. bovis*) are thought to cause staphylococcal-type food-borne intoxications. Foods associated with outbreaks caused by these species include raw and pasteurized milk, evaporated milk, cheese, sausage, and cooked meat products. Disease is believed to occur through introduction of the microorganisms into the food through underprocessing of foods and unsafe food handling practices. The diarrhea is acute but self-limiting (USFDA-CFSAN, 2007).

S. zooepidemicus (*S. equi* subsp. *zooepidemicus*) was identified as the causative agent of streptococcal infection in 16 cases in New Mexico. Infection was associated with eating homemade queso blanco, a soft, unaged Mexican-style cheese made from raw cow's milk. This strain of *Streptococcus* is one hitherto associated with animal disease, usually rare in humans (CDC, 1983).

Another member of this genus, *Streptococcus iniae*, has been recovered from aquatic sources, including water, marine animals (dolphins), fish, and aquaculture farms worldwide, and is linked to cases of human infection in several countries. It is considered to be an emerging pathogen in aquaculture, infecting mainly adult fish. It is responsible for up to 50% of mortality in trout and about 10% of losses in sea bass and sea bream in the Eastern Mediterranean area (Ghittino et al., 2003). In humans, these infections are acquired mainly through open lesions and result in soft tissue infections (Fuller et al., 2001). They appear to have been acquired by handling and consumption of infected fish, mainly farm-raised tilapia. Fish species like tilapia and barramundi may carry *S. iniae* asymptomatically, but the infection takes the form of skin lesions and necrotizing myositis in red drum fish and is seen as meningoencephalitis and panophthalmitis in trout and tilapia (Lau et al., 2003).

More recently, Romalde et al. (2008) described *Streptococcus phocae* outbreaks in Atlantic salmon cage farmed in Chile. These outbreaks have occurred in the summer months, since 1999, and have affected up to 25% of the fish population. *S. phocae* is now considered an emerging pathogen for salmonid aquaculture in Chile.

Though the incidence of human infection is low, major manifestations of disease have been observed, including bacteremia, cellulitis (most common), and osteomyelitis. Skin is thus thought to be a common primary infection site. Increased age and underlying illness are risk factors for *S. iniae* infections.

Intrinsic and Extrinsic Factors That Affect Survival and Growth in Food Products and Contribute to Outbreaks/Food Processing Operations That Influence the Numbers, Spread, or Characteristics

Salad bars have been postulated as sources of outbreaks *(S. pyogenes)*, with contamination of food by an infected handler and subsequent spread to a large number of consumers. Spread through milk was more frequent before pasteurization became routine. Outbreaks involving other streptococci are thought to be less common (USFDA-CFSAN, 2007).

Use of good manufacturing practices and vigilant sanitation, including the use of vaccines, and when appropriate, medicated feed, is essential to control *S. iniae* infections in aquaculture (Ghittino et al., 2003).

Recent Advances in Biological, Chemical, and Physical Interventions To Guard against the Pathogens/Concluding Remarks

Earlier in the 20th century, there were efforts to reduce the incidence of food-borne illness due to *Streptococcus* species. While there are still occasional cases of illness reported, the intervention methods put in place have greatly reduced the likelihood of food-borne illness due to members of this genus.

REFERENCES

Abbott, S. L., W. K. W. Cheung, and J. M. Janda. 2003. The genus *Aeromonas*: biochemical characteristics, atypical reactions, and phenotypic identification schemes. *J. Clin. Microbiol.* **41:**2348–2357.

Abbott, S. L., R. P. Kokka, and J. M. Janda. 1991. Laboratory investigations on the low pathogenic potential of *Plesiomonas shigelloides. J. Clin. Microbiol.* **29:**148–153.

Andrews, W. H., and T. S. Hammack. 2005. Food sampling and preparation of sample homogenate. *In Bacteriological Analytical Manual,* 8th ed. U.S. FDA-CFSAN, Department of Health and Human Services, Washington, DC. http://www.cfsan.fda.gov/~ebam/bam-toc.html. Accessed 9 August 2008.

Antolin, A., I. Gonzalez, T. Garcia, P. Hernandez, and R. Martin. 2001. *Arcobacter* spp. enumeration in poultry meat using a combined PCR-ELISA assay. *Meat Sci.* **59:**169–174.

APHA. 2005. *Standard Methods for Examination of Water and Wastewater,* 21st ed. American Public Health Association Press, Washington, DC.

Barnett, T. C., S. M. Kirov, M. S. Strom, and K. Sanderson. 1997. *Aeromonas* spp. possess at least two distinct type IV pilus families. *Microb. Pathog.* **23:**241–247.

Bellack, N. R., M. W. Koehoorn, Y. C. MacNab, and M. G. Morshed. 2006. A conceptual model of water's role as a reservoir in *Helicobacter pylori* transmission: a review of the evidence. *Epidemiol. Infect.* **134:**439–449.

Beuchat, L. R. 2002. Ecological factors influencing survival and growth of human pathogens on raw fruits and vegetables. *Microbes Infect.* **4:**413–423.

Bode, G., F. Mauch, and P. Malfertheiner. 1993. The coccoid forms of *Helicobacter pylori*. Criteria for their viability. *Epidemiol. Infect.* **111:**483–490.

Bolton, F. J., D. R. A. Wareing, M. B. Skirrow, and D. N. Hutchinson. 1992. Identification and biotyping of *Campylobacters. In* R. G. Board, D. Jones, and F. A. Skinner (ed.), *Identification Methods in Applied and Environmental Microbiology.* Blackwell Scientific Publications, London, United Kingdom.

Brenner, D. J., N. R. Krieg, and J. T. Staley (ed.). 2005. *Bergey's Manual of Systematic Bacteriology,* 2nd ed., vol. 2, part B. Springer-Verlag, New York, NY.

Burnens, A. P., and J. Nicolet. 1993. Three supplementary diagnostic tests for *Campylobacter* species and related organisms. *J. Clin. Microbiol.* **31:**708–710.

Cardarelli-Leite, P., K. Blom, C. M. Patton, M. A. Nicholson, A. G. Steigerwalt, S. B. Hunter, D. J. Brenner, T. J. Barrett, and B. Swaminathan. 1996. Rapid identification of *Campylobacter* species by restriction fragment length polymorphism analysis of a PCR-amplified fragment of the gene coding for 16S rRNA. *J. Clin. Microbiol.* **34:**62–67.

CDC. 1983. Group C streptococcal infections associated with eating homemade cheese—New Mexico. *MMWR Morb. Mortal. Wkly. Rep.* **32:**510, 515–516.

CDC. 2001. Preliminary FoodNet data on the incidence of food-borne illnesses—selected sites, United States, 2000. *MMWR Morb. Mortal. Wkly. Rep.* **50:**241–246. http://www.cdc.gov/mmwr/preview/mmwrhtml/mm5013a1.htm. Accessed 9 August 2008.

Cervenka, L., I. Peskova, M. Pejchalova, and J. Vytrasova. 2008. Inhibition of *Arcobacter butzleri, Arcobacter cryaerophilus,* and

Arcobacter skirrowii by plant oil aromatics. *J. Food Prot.* **71:**165–169.

Collins, M. T., U. Spahr, and P. M. Murphy. 2001. Ecological characteristics of *M. paratuberculosis*. *Bull. Int. Dairy Fed.* **362:**32–40.

Donachie, S. P., J. P. Bowman, S. L. On, and M. Alam. 2005. *Arcobacter halophilus* sp. nov., the first obligate halophile in the genus *Arcobacter*. *Int. J. Syst. Evol. Microbiol.* **55:**1271–1277.

Donaghy, J. A., N. L. Totton, and M. T. Rowe. 2004. Persistence of *Mycobacterium paratuberculosis* during manufacture and ripening of cheddar cheese. *Appl Environ. Microb.* **70:**4899–4905.

D'Sa, E. M. 2002. Fate of *Arcobacter* spp. upon exposure to environmental stresses and predictive model development. Ph.D. dissertation. The University of Georgia, Athens, GA.

D'Sa, E. M., and M. A. Harrison. 2005. Effect of pH, NaCl content and temperature on growth and survival of *Arcobacter* spp. *J. Food Prot.* **68:**18–25.

Ellis, W. A., S. D. Neill, J. J. O'Brien, H. W. Ferguson, and J. Hanna. 1977. Isolation of *Spirillum/Vibrio*-like organisms from bovine fetuses. *Vet. Rec.* **100:**451–452.

Facklam, R. 2002. What happened to the streptococci: overview of taxonomic and nomenclature changes. *Clin. Microbiol. Rev.* **15:**613–630.

Fuller, J. D., D. J. Bast, V. Nizet, D. E. Low, and J. C. S. de Azavedo. 2001. *Streptococcus iniae* virulence is associated with a distinct genetic profile. *Infect. Immun.* **69:**1994–2000.

Gancz, H., K. R. Jones, and D. S. Merrell. 2008. Sodium chloride affects *Helicobacter pylori* growth and gene expression. *J. Bacteriol.* **190:**4100–4105.

Ghittino, C., M. Latini, F. Agnetti, C. Panzieri, L. Lauro, and R. Ciappelloni. 2003. Emerging pathogens in aquaculture: effects on production and food safety. *Vet. Res. Comm.* **27**(Suppl. 1):471–479.

Gomes, B. C., and E. C. P. De Martinis. 2004a. Fate of *Helicobacter pylori* artificially inoculated in lettuce and carrot samples. *Braz. J. Microbiol.* **35:**145–150.

Gomes, B. C., and E. C. P. De Martinis. 2004b. The significance of *Helicobacter pylori* in water, food and environmental samples. *Food Control* **15:**397–403.

Grant, I. R. 2005. Zoonotic potential of *Mycobacterium avium* ssp. *paratuberculosis*: the current position. *J. Appl. Microbiol.* **98:**1282–1293.

Grant, I. R., H. J. Ball, and M. T. Rowe. 2002. Incidence of *Mycobacterium paratuberculosis* in bulk raw and commercially pasteurized cow's milk from approved dairy processing establishments in the United Kingdom. *Appl. Environ. Microbiol.* **68:**2428–2435.

Greenstein, R. J. 2003. Is Crohn's disease caused by a *Mycobacterium*? Comparisons with leprosy, tuberculosis and Johne's disease. *Lancet Infect. Dis.* **3:**507–514.

Griffiths, M. W. 2006. *Mycobacterium paratuberculosis*. *In* Y. Motarjemi and M. Adams (ed.), *Emerging Foodborne Pathogens*. CRC Press, Woodhead Publishing Limited, Cambridge, England.

Hancock, R. T. 2000. Antimicrobial activity of selected spices and organic acids against *Arcobacter butzleri* in laboratory media and on fresh pork. Master's thesis. The University of Georgia, Athens, GA.

Hardie, J. M., and R. A. Whiley. 1997. Classification and overview of the genera *Streptococcus* and *Enterococcus*. *J. Appl. Microbiol.* **83**(Suppl.):1S–11S.

Hardin, F. J., and R. A. Wright. 2002. *Helicobacter pylori*: review and update. *Hosp. Physician* **38:**23–31.

Harmon, K. M., and I. V. Wesley. 1997. Multiplex PCR for the identification of *Arcobacter* and differentiation of *Arcobacter butzleri* from other arcobacters. *Vet. Microbiol.* **58:**215–227.

Harrass, B., S. Schwarz, and S. Wenzel. 1998. Identification and characterization of *Arcobacter* isolates from broilers by biochemical tests, antimicrobial resistance patterns and plasmid analysis. *J. Vet. Med.* **45:**87–94.

Hill, M. 1997. The microbiology of *Helicobacter pylori*. *Biomed. Pharmacother.* **51:**161–163.

Hilton, C. L., B. M. Mackey, A. J. Hargreaves, and S. J. Forsythe. 2001. The recovery of *Arcobacter butzleri* NCTC 12481 from various temperature treatments. *J. Appl. Microbiol.* **91:**929–932.

Houf, K., S. On, T. Coenye, J. Van Hoof, and P. Vandamme. 2005. *Arcobacter cibarius* sp. nov., isolated from broiler carcasses. *Int. J. Syst. Evol. Microbiol.* **55:**713–717.

Houf, K., A. Tutenel, L. De Zutter, J. Van Hoof, and P. Vandamme. 2000. Development of a multiplex PCR assay for the simultaneous detection and identification of *Arcobacter butzleri*, *Arcobacter cryaerophilus*, and *Arcobacter skirrowii*. *FEMS Microbiol. Lett.* **193:**89–94.

Janda, J. M., and S. L. Abbott. 1998. Evolving concepts regarding the genus *Aeromonas*: an expanding panorama of species. Disease presentations, and unanswered questions. *Clin. Infect. Dis.* **27:**332–344.

Jay, J. M., M. J. Loessner, and D. A. Golden. 2005. Viruses and some other proven and suspected foodborne biohazards. *In* *Modern Food Microbiology*, 7th ed. Springer Science and Business Inc., New York, NY.

Jiang, X. P., and M. P. Doyle. 1998. Effect of environmental and substrate factors on survival and growth of *Helicobacter pylori*. *J. Food Prot.* **61:**929–933.

Jiang, X. P., and M. P. Doyle. 2000. Growth supplements for *Helicobacter pylori*. *J. Clin. Microbiol.* **38:**1984–1987.

Jiang, X. P., and M. P. Doyle. 2002. Optimizing enrichment culture conditions for detecting *Helicobacter pylori* in foods. *J. Food Prot.* **65:**1949–1954.

Kiehlbauch, J. A., C. N. Baker, and I. K. Wachsmuth. 1992. In vitro susceptibilities of aerotolerant *Campylobacter* isolates to 22 antimicrobial agents. *Antimicrob. Agents Chemother.* **36:**717–722.

Kirov, S. M., M. Castrisios, and J. G. Shaw. 2004. *Aeromonas* flagella (polar and lateral) are enterocyte adhesins that contribute to biofilm formation on surfaces. *Infect. Immun.* **72:**1939–1945.

Lammerding, A., J. E. Harris, H. Lior, D. E. Woodward, L. Cole, and C. A. Muckle. 1997. Isolation methods for recovery of *Arcobacter butzleri* from fresh poultry and poultry products. *In* D. G. Newell, J. Ketley, and R. A. Feldman (ed.), *Campylobacters, Helicobacters, and Related Organisms*. Springer Publishing Co., New York, NY.

Lau, S. K. P., P. C. Y. Woo, H. Tse, K-W. Leung, S. S. Y. Wong, and K-Y. Yuen. 2003. Invasive *Streptococcus iniae* infections outside North America. *J. Clin. Microbiol.* **41:**1004–1009.

Marshall, S. M., P. L. Melito, D. L. Woodward, W. M. Johnson, F. G. Rodgers, and M. R. Mulvey. 1999. Rapid identification of *Campylobacter*, *Arcobacter*, and *Helicobacter* isolates by PCR-restriction length polymorphism analysis of the 16S rRNA gene. *J. Clin. Microbiol.* **37:**4158–4160.

Martin, A. C., and C. W. Penn. 2001. *Helicobacter pylori*, p. 1131–1354. *In* M. Sussman (ed.), *Molecular Medical Microbiology*. Academic Press, San Diego, CA.

Mateos, D., J. Anguita, G. Naharro, and C. Paniagua. 1993. Influence of growth temperature on the production of extracellular virulence factors and pathogenicity of environmental and human strains of *Aeromonas hydrophila*. *J. Appl. Bacteriol.* **74:**111–118.

Metzger-Boddien, C., D. Khaschabi, M. Schoenbauer, S. Boddien, T. Schleder, and J. Kehle. 2006. Automated high-throughput immunomagnetic separation-PCR for detection of *Mycobacterium avium* subsp. p*aratuberculosis* in bovine milk. *Int. J. Food Microbiol.* **110:**201–208.

Moyer, N. P. (revised by J. Standridge). 2006. *Aeromonas*. *In Waterborne Pathogens*, 2nd ed. American Water Works Association, Denver, CO.

Musmanno, R. A., M. Russi, H. Lior, and N. Figura. 1997. *In vitro* virulence factors of *Arcobacter butzleri* strains isolated from superficial water samples. *Microbiologica* **20:**63–68.

Nachamkin, I., and M. J. Blaser. 2000. *Campylobacter*, 2nd ed. ASM Press, Washington, DC.

Neubauer, C., and M. Hess. 2006. Detection and identification of foodborne pathogens of the genera *Campylobacter*, *Arcobacter* and *Helicobacter* by multiplex PCR in poultry and poultry products. *J. Vet. Med.* **53:**376–381.

O'Gara, E. A., D. J. Maslin, A. M. Nevill, and D. J. Hill. 2008. The effect of simulated gastric environments on the anti-*Helicobacter* activity of garlic oil. *J. Appl. Microbiol.* **104:**1324–1331.

Petersen, A. M., and K. A. Krogfelt. 2003. *Helicobacter pylori*: an invading microorganism? A review. *FEMS Immunol. Med. Microbiol.* **36:**117–126.

Phillips, C. A. 1999. The effect of citric acid, lactic acid, sodium citrate, and sodium lactate, alone, and in combination with nisin, on the growth of *Arcobacter butzleri*. *Lett. Appl. Microbiol.* **29:**424–428.

Quaglia, N. C., A. Dambrosio, G. Normanno, A. Parisi, A. Firinu, V. Lorusso, and G. V. Celano. 2007. Survival of *Helicobacter pylori* in artificially contaminated ultrahigh temperature and pasteurized milk. *Food Microbiol.* **24:**296–300.

Romalde, J. L., C. Ravelo, I. Valdes, B. Magarinos, E. de la Fuente, C. San Martin, R. Avendano-Herrera, and A. Toranzo. 2008. *Streptococcus phocae*, an emerging pathogen for salmonid culture. *Vet. Microbiol.* **130:**198–207.

Rusin, P. A., J. B. Rose, C. N. Haas, and C. P. Gerba. 1997. Risk assessment of opportunistic bacterial pathogens in drinking water. *Rev. Environ. Contam. Toxicol.* **152:**57–83.

Skovgaard, N. 2007. New trends in emerging pathogens. *Int. J. Food Microbiol.* **120:**217–224.

Slana, I., F. Paolicchi, B. Janstova, P. Navratilova, and I. Pavlik. 2008. Detection methods for Mycobacterium avium subsp. paratuberculosis in milk and milk products: a review. *Vet. Med. Czech.* **53:**283–306.

Son, I., M. D. Englen, M. E. Berrang, P. J. Fedorka-Cray, and M. A. Harrison. 2007. Prevalence of *Arcobacter* and *Campylobacter* on broiler carcasses during processing. *Int. J. Food Microbiol.* **113:**16–22.

Spahr, U., and K. Schafroth. 2001. Fate of *Mycobacterium avium* subsp. p*aratuberculosis* in Swiss hard and semihard cheese manufactured from raw milk. *Appl. Environ. Microbiol.* **67:**4199–4205.

Stampi, S., O. Varoli, F. Zanetti, and G. De Luca. 1993. *Arcobacter cryaerophilus* and thermophilic campylobacters in a sewage treatment plant in Italy: two secondary treatments compared. *Epidemiol. Infect.* **110:**633–639.

Suerbaum, S., and P. Michetti. 2002. *Helicobacter pylori* infection. *N. Engl. J. Med.* **347:**1175–1186.

Sung, N., and M. T. Collins. 2000. Effect of three factors in cheese production (pH, salt and heat) on *Mycobacterium avium* subsp. *paratuberculosis* viability. *Appl. Environ. Microb.* **66:**1334–1339.

Sweeney, R. W., R. H. Whitlock, and A. E. Rosenberger. 1992. *Mycobacterium paratuberculosis* cultured from milk and supramammary lymph nodes of infected asymptomatic cows. *J. Clin. Microbiol.* **30:**166–171.

Tabak, M., R. Armon, and I. Neeman. 1999. Cinnamon extracts' inhibitory effect on *Helicobacter pylori*. *J. Ethnopharmacol.* **67:**269–277.

Tortoli, E. 2006. The new mycobacteria: an update. *FEMS Immunol. Med. Microbiol.* **48:**159–178.

Tsukamoto, T., Y. Konoshita, T. Shimada, and R. Sakazaki. 1978. Two epidemics of diarrhoeal disease possibly caused by *Plesiomonas shigelloides*. *J. Hyg.* (Cambridge) **80:**275–280.

USDA/FSIS. 2007. Keep food safe! Food safety basics. Safe food handling fact sheet. Food Safety and Inspection Service, U.S. Department of Agriculture, Washington, DC. http://www.fsis. usda.gov/Factsheets/Keep_Food_Safe_Food_Safety_Basics/ index.asp. Accessed 9 August 2008.

USEPA. 2006. *Aeromonas*: human health criteria document. Office of Science and Technology, Office of Water, U.S. Environmental Protection Agency, Washington, DC.

USFDA-CFSAN. 2007. *Streptococcus*. *In Bad Bug Book*. Center for Food Safety and Applied Nutrition, U.S. Food and Drug Administration. http://vm.cfsan.fda.gov/~mow/chap21.html. Accessed 9 August 2008.

Vandamme, P., E. Falsen, R. Rossau, B. Hoste, P. Segers, R. Tytgat, and J. D Ley. 1991. Revision of *Campylobacter*, *Helicobacter* and *Wolinella* taxonomy: emendation of generic descriptions and proposal of *Arcobacter* gen. nov. *Int. J. Syst. Bacteriol.* **41:**88–103.

Vandamme, P., B. A. J. Giesendorf, A. Van Belkum, D. Pierard, S. Lauwers, K. Kersters, J. Butler, H. Goossens, and W. G. V. Quint. 1993. Discrimination of epidemic and sporadic isolates of *Arcobacter butzleri* by polymerase chain reaction-mediated DNA fingerprinting. *J. Clin. Microbiol.* **31:**3317–3319.

von Graevenitz, A., and A. H. Mensch. 1968. The genus *Aeromonas* in human bacteriology—report of 30 cases and a review of literature. *N. Engl. J. Med.* **278:**245–249.

Wesley, I. V. 1994. *Arcobacter* infections, p. 181–190. *In* G. W. Beran and J. H. Steele (ed.), *Handbook of Zoonoses. Section A: Bacterial, Rickettsial, Chlamydial and Mycotic*, 2nd ed. CRC Press, Boca Raton, FL.

Wesley, I. V., L. Schroeder-Tucker, A. I. Baetz, F. E. Dewhirst, and B. J. Paster. 1995. *Arcobacter*-specific and *Arcobacter butzleri*-specific 16s rRNA-based DNA probes. *J. Clin. Microbiol.* **33:**1691–1698.

Wesley, I. V. 1996. *Helicobacter* and *Arcobacter* species: risks for foods and beverages. *J. Food Prot.* **59:**1127–1132.

Wesley, I. V. 1997. *Helicobacter* and *Arcobacter*: potential human foodborne pathogens? *Trends Food Sci. Tech.* **8:**293–299.

Wesley, I. V., A. L. Baetz, and D. J. Larson. 1996. Infection of cesarean-derived, colostrum-deprived 1-day-old piglets with *Arcobacter butzleri*, *Arcobacter cryaerophilus* and *Arcobacter skirrowii*. *Infect. Immun.* **64:**2295–2299.

Wesley, I. V., S. J. Wells, K. M. Harmon, A. Green, L. Schroeder-Tucker, M. Glover, and I. Siddique. 2000. Fecal shedding of *Campylobacter* and *Arcobacter* spp. in dairy cattle. *Appl. Environ. Microbiol.* **66:**1994–2000.

Whan, L. B., I. R. Grant, H. J. Ball, R. Scott, and M. T. Rowe. 2001. Bactericidal effect of chlorine on *Mycobacterium paratuberculosis* in drinking water. *Lett. Appl. Microbiol.* **33:** 227–231.

Winters, D. K., and M. F. Slavik. 2000. Multiplex PCR detection of *Campylobacter jejuni* and *Arcobacter butzleri* in food products. *Mol. Cell. Probes* **14:**95–99.

Wirsen, C. O., S. M. Sievert, C. M. Cavanaugh, S. J. Molyneaux, A. Ahmed, L. T. Taylor, E. F. DeLong, and C. D. Taylor. 2002. Characterization of an autotrophic sulfide-oxidizing marine *Arcobacter* sp. that produces filamentous sulfur. *Appl. Environ. Microbiol.* **68:**316–325.

Pathogens and Toxins in Foods: Challenges and Interventions
Edited by V. K. Juneja and J. N. Sofos
© 2010 ASM Press, Washington, DC

Chapter 13

Food-Borne Parasites

Dolores E. Hill and J. P. Dubey

Food-borne infections are a significant cause of morbidity and mortality worldwide, and food-borne parasitic diseases, though not as widespread as bacterial and viral infections, are common on all continents and in most ecosystems, including arctic, temperate, and tropical regions (Eckert, 1996; Anantaphruti, 2001; Macpherson, 2005). Certain food-borne parasitic organisms, such as *Toxoplasma gondii*, are extremely widespread, affecting 40 to 50% of the population worldwide (Dubey and Beattie, 1988), while others, such as *Fasciolopsis buski*, are restricted to specific geographic regions as a result of the limited range of the required definitive or intermediate host (Mas-Coma et al., 2005). However, as international travel, large-scale movement of human populations, climate change, and global trade in agricultural products increase, so does the risk of introduction of food-borne parasitic organisms into the human food chain in new geographic areas (Slifko et al., 2000; Polley, 2005; Macpherson, 2005). Three phyla containing parasitic organisms (Protozoa, Platyhelminthes, and Nemata) include genera whose transmission is primarily food borne.

PLATYHELMINTHS

Digenetic Trematodes

Food-borne digenetic (two or more hosts required for development) trematodiasis caused by intestinal *(Fasciolopsis buski, Echinostoma* spp., *Heterophyes heterophyes,* and *Metagonimus yokogawai),* lung *(Paragonimus westermani),* or liver flukes *(Fasciola hepatica, Fasciola gigantica, Opisthorchis felineus, Opisthorchis viverrini, Clonorchis sinensis)* infects more than 30 million people worldwide (Kaewkes, 2003; Chai et al., 2005).

The lung and liver flukes have wide geographic distributions, largely as a result of the colonization success of the *Lymnaeidae, Thiaridae,* and *Bithyniidae* snail intermediate hosts, while *F. buski* requires *Planorbidae* snail hosts belonging to the genera *Segmentina, Hippeutis,* or *Gyraulus* and is restricted to Asia (Krejci and Fried, 1994; Graczyk and Fried, 1998; Procop et al., 2000; Pokora, 2001; Velez et al., 2002; Sithithaworn and Haswell-Elkins, 2003; Keiser and Utzinger, 2005; Mahajan, 2005; Mas-Coma, 2005). The life cycles of these large (approximately 1.0 to 7.5 cm), leaf-shaped flukes are similar, though major differences do occur (Table 1; Figures 1 to 3)(Kim, 1984; Graczyk et al., 2001; Chai et al., 2005). Operculated eggs are passed in host feces. Miracidia hatch from the operculate eggs in fresh water, penetrate an appropriate snail as the first intermediate host, and develop into sporocysts (germinal sac). Rediae develop within sporocysts, emerge, and produce cercariae. Cercariae emerge from the snail host into fresh water, swimming or crawling to water plants and encysting as metacercariae on them, or penetrating freshwater fishes, mollusks, and crustaceans and encysting in muscle tissues and viscera as metacercariae (Schmidt and Roberts, 2005a). Humans and other animals become infected by consuming uncooked aquatic vegetation containing encysted metacercariae or by consuming raw or undercooked freshwater fishes, mollusks, crabs, and crayfish. Morbidity associated with infection is high (Keiser and Utzinger, 2004; Gulsen et al., 2006). The common practice of using human and animal waste to fertilize plants in aquaculture ponds where farmed fish are produced virtually ensures continued transmission in areas where infection is endemic.

Dolores E. Hill • U.S. Department of Agriculture, Agricultural Research Service, Animal and Natural Resources Institute, Animal Parasitic Diseases Laboratory, Building 1044, BARC-East, Beltsville, MD 20705-2350. **J. P. Dubey** • U.S. Department of Agriculture, Agricultural Research Service, Animal and Natural Resources Institute, Animal Parasitic Diseases Laboratory, Building 1001, BARC-East, Beltsville, MD 20705-2350.

Table 1. Major food-borne trematodiasis in humans

Characteristic of trematodiasis	Details for flukes in:				
	Intestines		Liver		Lung
Digenetic trematode species	*Fasciolopsis buski*	*Echinostoma* spp., *Heterophyes heterophyes*, *Metagonimus yokogawai*	*Opisthorchis felineus*, *O. viverrini*, *Clonorchis sinensis*	*Fasciola hepatica*, *F. gigantica*	*Paragonimus westermani*, *Paragonimus* spp.
Eggs in feces released into fresh water	Miricidia hatch from egg in 3–7 wk at optimum temperature (27–30°C) and penetrate planorbid snail host *Segmentina*, *Gyraulus*, or *Hippeutis* snail first intermediate host	Miricidia hatch from egg after 2–5 wk (*E.* spp.). Miricidia hatch from egg when egg is eaten by snail (*H. heterophyes* and *M. yokogawai*) *Lymnaeidae*, *Physa*, and *Bithyniidae* snail first intermediate (*E.* spp.) *Pirenella*, *Certhidia*, *Tarebia* (*H. heterophyes*) *Semisulcospira* (*M. yokogawai*)	Miricidia hatch from egg when egg is eaten by snail, emerge in rectum of snail, penetrate gut wall *Bithynia* snail first intermediate host (*O. felineus* and *O. viverrini*) *Parafossarulus*, *Bulimus*, *Semisulcospira*, *Alocinma*, *Melanoides* snail first intermediate host (*C. sinensis*)	Miricidia hatch from egg after 9–15 days at optimum temperature (22°C) *Lymnaeidae* and *Bithyniidae* snail first intermediate	Miricidia hatch from egg in 2–4 wk, must find snail host within 24 hours *Thieridae* snail first intermediate host
Sporocyst generation	Sporocysts give rise to two redial generations	Single sporocyst generation gives rise to two redial generations	One generation of sporocysts; sporocyst forms within 4 hours of snail infection	Sporocysts give rise to two redial generations	Sporocysts give rise to two redial generations
Rediae generation	Develop within 10 days of snail penetration by miricidium; mother and daughter rediae produced	Mother and daughter rediae produced (*E.* spp.)	One generation of rediae produced within 17 days after infection	Mother and daughter rediae produced	Mother and daughter rediae produced
Cercarial generation	Rediae exit after 1 mo, emerge from snail after 50 days, can survive in water for 70 days	Swimming cercariae exit snail Cercariae penetrate second intermediate host (mollusks) Cercariae penetrate second intermediate host (freshwater fish)	Cercariae produced 3–4 wk postinfection Cercariae must penetrate second intermediate host (freshwater fish) within 24–48 hours	Cercariae exit snail 4–7 wk after miracidial penetration	Cercariae exit snail 11 wk after miracidial penetration Cercariae penetrate second intermediate host (crabs or crayfish)
Metacercarial generation	Encyst on aquatic vegetation or float on surface of water; excyst in duodenum, attach to intestinal wall of definitive host	Metacercariae encyst in mollusks and on aquatic vegetation (*E.* spp.); metacercariae encyst in tissues of freshwater fishes (*H. heterophyes*, *M. yokogawai*); excyst in small intestine of mammalian or avian definitive host (*E.* spp., *H. heterophyes*, *M. yokogawai*)	Encyst in muscles of freshwater cyprinid fishes or crayfish (*C. sinensis*); infective after 3 wk; excyst in duodenum of definitive host, migrates into the bile ducts within 4–7 hours	Encyst on vegetation or on surface of water; excyst in small intestine of definitive host, penetrate intestinal wall, penetrate liver, feed on the liver and grow for 2 mo before entering bile ducts	Encyst in tissues of crabs or crayfish; excyst in duodenum of definitive host, penetrate intestinal wall and enter abdominal wall for several days, reenter the coelom and penetrate the diaphragm and lung, form pairs, and encapsulate in lung bronchioles

Definitive hosts	Humans, pigs infected by consumption of uncooked aquatic vegetation or drinking water contaminated with floating metacercarial cysts	Humans, dogs, pigs, rats, cats, other fish-eating mammals infected by consumption of raw fish	Humans, many species of mammals and birds (E. spp.) Fish-eating mammals (H. heterophyes and M. yokogawai) infected by consumption of uncooked aquatic vegetation or raw mollusks (E. spp.) or consuming uncooked fish (H. heterophyes and M. yokogawai)	Humans, many species of mammals infected by consumption of uncooked aquatic vegetation or drinking water contaminated with floating metacercarial cysts	Humans, crab-eating felids, canids, rodents, pigs, viverrids, mustelids infected by consumption of uncooked crabs or crayfish
Disease symptoms/ diagnosis	Diarrhea, pain, nausea, intestinal edema, ulceration, obstruction, erosion of duodenal and jejunal mucosa; eggs in feces	Dependent upon worm burden; diarrhea, irritation and erosion of the bile duct, hepato- and splenomegaly, bile duct cancer (O. felineus and O. viverrini); eggs in feces Pipe-stem fibrosis, necrotic changes; rarely fatal, except in cases with large worm burdens (C. sinensis); eggs in feces	Diarrhea, pain, ulceration, intestinal perforations, myocarditis, and neurological disorders resulting from eggs or flukes lodging in extraintestinal organs; eggs in feces	Immature flukes cause extensive destruction of liver tissue, resulting in diarrhea, pain, and nausea; adult flukes reside in bile ducts, causing hemorrhage, gallstones, obstruction, and vessel atrophy; eggs in feces, serology	Granuloma formation in lungs around worms and eggs, breathing problems, cough, blood in sputum, ulceration, extrapulmonary invasion of spinal cord, heart, brain; sometimes fatal/eggs in sputum or feces
Prepatent period/ life span of adult fluke	3 Mo	3–4 Wk/10 yr (O. felineus and O. viverrini); 30 days/8–25 yr (C. sinensis)	7 Days/<1 yr in exptl animals	25–30 Days/10–12 yr	8–12 Wk/10–20 yr
Treatment	Oral triclabendazol, oxyclozanide, rafoxanide	Praziquantel	Praziquantel	Oral triclabendazol, bithionol	Praziquantel; bithionol
Reservoirs	Pigs	Felines (O. felineus and O. viverrini); wide variety of mammals, especially dogs and cats (C. sinensis)	Birds and mammals (E. spp.) Fish-eating mammals and possibly birds (H. heterophyes, M. yokogawai)	Sheep, cattle, and rabbits	Crab-eating felids, canids, rodents, pigs, viverrids, mustelids, and humans
Geographic locations	Central and southern China, Taiwan, throughout Southeast Asia	Eastern Europe, Turkey, southern Russia, Vietnam, India, Japan, Caribbean (O. felineus and O. viverrini) Southeast Asia (O. felineus and O. viverrini and C. sinensis)	Worldwide (E. spp.) North Africa, Eastern Europe, Asia, Hawaii (H. heterophyes) Far East, former Soviet Union, Spain, Balkans (M. yokogawai)	Worldwide, especially in cattle-raising areas; F. gigantica predominates in Asia	Asia, Oceana, sub-Saharan Africa, South and Central America

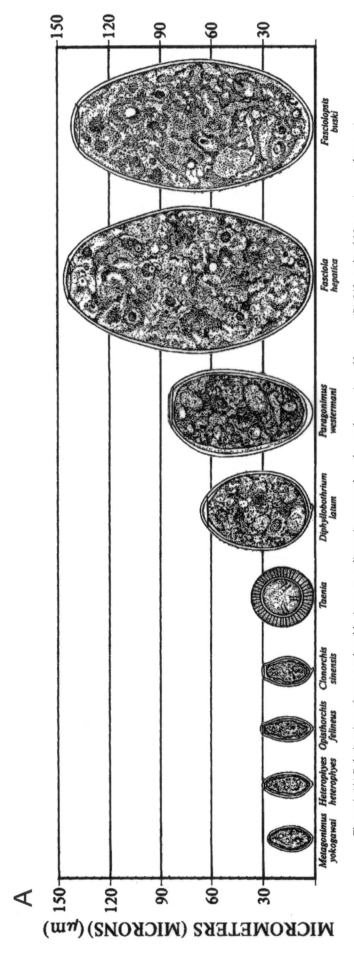

Figure 1. (A) Relative sizes of eggs produced by important digenetic trematode and cestode parasites of humans. (B) Life cycle of *Metagonimus yokogawai*. i, infective stage; d, diagnostic stage. 1d, embryonated eggs, each with a fully developed miracidium, are passed in feces. 2, snail host ingests eggs; miracidia emerge from eggs and penetrate snail intestine; and sporocysts, rediae, and then cercariae develop in snail tissues. 3, cercariae released from snail. 4, cercariae penetrate skin of fresh/brackish water fish and encyst as metacercariae in fish tissues. 5i, humans become infected by ingesting undercooked fish containing metacercariae. 6, metacercariae excyst in small intestine and develop to adult stage. 7, fish-eating mammals and birds can also be infected.

B

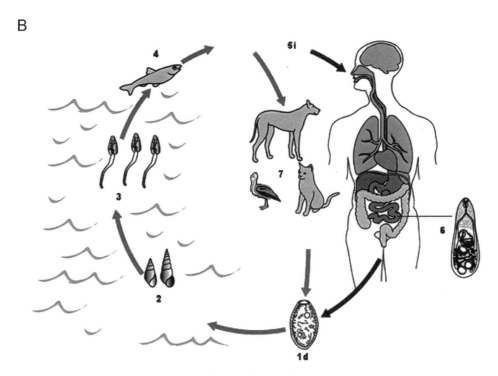

Figure 1. *Continued*

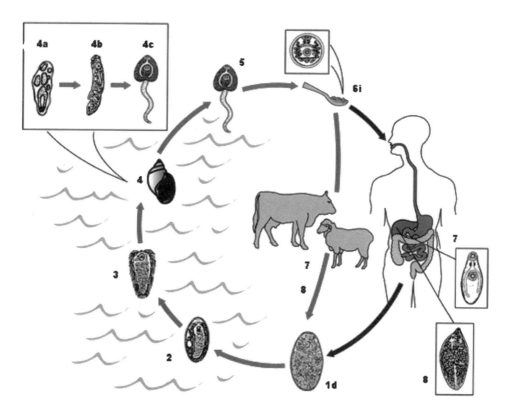

Figure 2. Life cycle of *Fasciola hepatica*. i, infective stage; d, diagnostic stage. 1d, unembryonated eggs passed in feces. 2, embryonated eggs contaminate a body of water. 3, miracidia hatch from embryonated eggs and penetrate snail. 4, miracidia develop in snail tissue into sporocysts (4a), rediae (4b), and then cercariae (4c). 5, free-swimming cercariae encyst on water plants. 6i, metacercariae on water plant ingested by humans, sheep, or cattle. 7, metacercariae excyst in duodenum, penetrate the intestinal wall, and enter the liver to feed. 8, adults reside in hepatic biliary ducts.

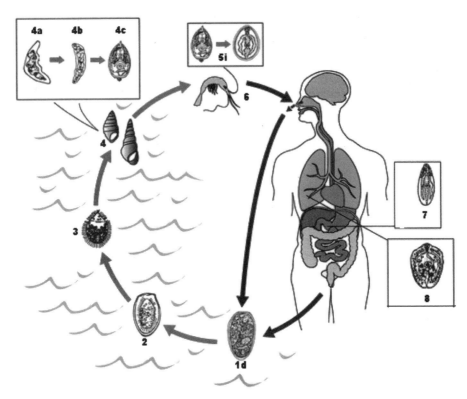

Figure 3. Life cycle of *Paragonimus westermani*. i, infective stage; d, diagnostic stage. 1d, unembryonated eggs are released in sputum or in feces. 2, eggs embryonate in water. 3, miracidia hatch from eggs in water and penetrate snails. 4, miracidia develop in snail tissues into sporocysts (4a), rediae (4b), and cercariae (4c). 5i, cercariae leave snails, enter a crustacean host, and encyst as metacercariae. 6, humans ingest inadequately cooked or pickled crustaceans containing metacercariae. 7, metacercariae excyst in the duodenum, penetrate the intestinal wall, and enter the lungs. 8, adults in cystic cavities in lungs lay eggs which are excreted in sputum or are swallowed and passed in stool.

CESTODES

Taenia Species

Taenia saginata, the beef tapeworm, and *T. asiatica* and *T. solium*, the pork tapeworms, cause taeniasis in humans. *Taenia saginata* and *T. solium* occur worldwide, endemically in underdeveloped regions where pork and beef are consumed and sporadically in developed nations, largely due to importation by immigrants and travelers. *Taenia asiatica*, a recently described parasite (species status currently undetermined) similar in morphology to *T. saginata* but epidemiologically similar to *T. solium* (Hoberg et al., 2000; McManus, 2006), is common in many Asian countries (Eom, 2006). Taeniasis results from the consumption of larval tapeworms (known as cysticerci, metacestodes, or bladderworms), which encyst in the muscles and viscera of infected cattle and pigs (Figure 4). Humans are the only definitive hosts for these large (3 to 15 m) cyclophyllidean tapeworm species (life cycle generally involves terrestrial hosts; larvae remain in eggs until eaten). Humans harbor the adult tapeworms, which are anchored in the small intestines by four powerful suckers *(T. saginata)*, four suckers and an unarmed rostellum (or reduced hooklets) *(T. asiatica)*, or four suckers and a rostellum armed with two rows of hooks *(T. solium)* (Figure 5A and B)(Eom and Rim, 1993; Fan et al., 1995; Eom, 2006). Gravid proglottids (segments), each containing thousands of eggs, detach from the end of the adult worm and exit the body in feces or crawl out of the anus independently (Bogtish et al., 2005). As the segment dries, it ruptures, releasing eggs which are immediately infective to intermediate hosts. Poor sanitary conditions and the use of night soil (human waste) as fertilizer result in egg contamination of the home, the surrounding environment, animal pasturage, and livestock feed. Eggs are ingested by cattle or swine and hatch in the small intestines, and the larvae (known as the oncosphere or hexacanth embryo) penetrate the intestinal wall, enter mesenteric circulation, and are distributed throughout the body. The larvae then exit the circulatory system and enter a muscle cell or organ, and each develops into a single cysticercus, a bladder-like structure, containing an invaginated scolex which is infective in about 8 weeks. The metacestode structures of *T. saginata, T. asiatica,* and *T. solium*

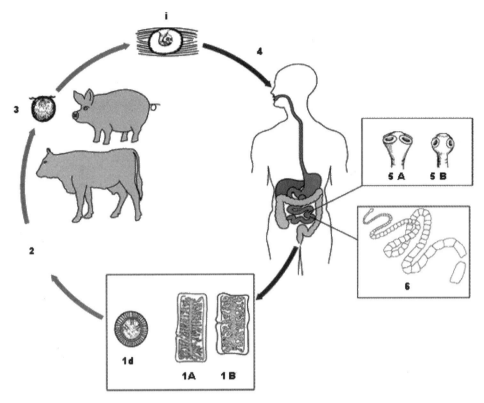

Figure 4. Life cycle of *Taenia solium* and *T. saginata*. i, infective stage; d, diagnostic stage. 1d, eggs or gravid proglottids of *T. saginata* (1A) or *T. solium* (1B) in feces are passed into the environment. 2, cattle *(T. saginata)* and pigs *(T. solium)* become infected by ingesting vegetation contaminated by eggs or proglottids. 3, oncospheres hatch from eggs in gut, penetrate intestinal wall, and circulate to musculature. i, oncospheres develop into cysticerci in muscle. 4, humans are infected by ingesting raw or undercooked infected meat. 5, the *T. saginata* (5A) or *T. solium* (5B) scolex evaginates and attaches to intestine using holdfast organs (hooks and suckers). 6, adult tapeworms reside in small intestine.

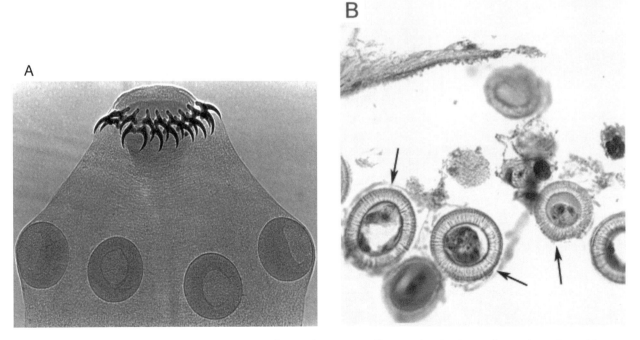

Figure 5. Scolex and egg of *T. solium*. (A) *Taenia solium* scolex, 1 mm in diameter, showing four suckers and rostellum with double row of hooks. Magnification, ×80. (B) *Taenia solium* egg (32 μm, with a range of 31 to 43 μm), indistinguishable from other taeniid eggs in feces; thick, striated shell enclosing hexacanth embryo (arrows). Magnification, ×400.

are easily large enough (up to 2 cm) to be seen with the naked eye in infected meat, which is known as measly beef or measly pork. People who consume these infected meats without thorough cooking become hosts to adult tapeworms. The scolex of the cysticercus evaginates, and the worm becomes anchored in the small intestine, while the bladder tissues are digested away. The neck of the tapeworm then begins the process of strobilation, the production of proglottids. The strobila will contain immature proglottids close to the scolex; as the strobila increases in length, distal proglottids mature and become gravid with eggs within 2 to 3 months. Gravid proglottids break off from the end of the strobila and exit the anus; *T. saginata* proglottids usually exit as singlets, whereas *T. solium* proglottids generally exit in groups of five or six (Bogtish et al., 2005). Adult tapeworms are generally benign, though symptoms such as abdominal pain, nausea, inappetance, weight loss, dizziness, and diarrhea are common (Flisser et al., 2004). Diagnosis is accomplished by identification of characteristic taeniid eggs in feces. Since it is impossible to distinguish *T. saginata* and *T. solium* eggs, it is preferable to collect the entire scolex from the expelled adult worm to determine the presence of an armed rostellum *(T. solium* and *T. asiatica)* versus an absent rostellum *(T. saginata)* to make an accurate diagnosis. The adult worms may live 20 to 25 years; elimination of adult tapeworms is accomplished by treatment with praziquantel or other taeniicides. Although direct pathogenesis by adult tapeworms is minimal, complications associated with the eggs released by the adults can be fatal (Flisser, 1994). Cysticercosis is a serious disease which occurs when humans act as the intermediate host of *T. solium;* humans do not appear to be suitable intermediate hosts for the cysticerci of *T. saginata* or *T. asiatica* (Schmidt and Roberts, 2005b; Eom, 2006). Eggs released from the gravid proglottids of *T. solium* tapeworms frequently contaminate clothing, bedding, and the household environment. When these eggs are ingested by humans, the larvae hatch from the egg, penetrate the intestinal mucosa, and are distributed throughout the body. Though cysticerci can form in virtually every organ, frequently they localize in the brain (Flisser et al., 2003). Severe headache, pressure necrosis, intracranial hypertension, paralysis, blindness, dementia, sudden onset of epilepsy, and hydrocephalus are common symptoms of neurocysticercosis. The course of treatment is dependent upon the number and location of the cysticerci in the brain, and the host response to the presence of the parasite (Flisser, 2001). Patients with active or calcified cysts in the brain parenchyma without hydrocephalus have a favorable prognosis. Anthelmintics (albendazole and praziquantel), antiseizure medications, and anti-inflammatory drugs are frequently prescribed, and outcomes are generally good. Cysticerci can also take on the much more dangerous racemous form, involving a large, lobulated bladder with no scolex and usually located in the cerebral ventricles or other extraparenchymal areas; these usually require surgical treatment, and the prognosis in these cases is poor (Garcia et al., 2002). Prevention of transmission and infection requires strict adherence to personal and environmental hygiene standards. Human defecation in areas not accessible to pigs, hand washing after defecation, elimination of night soils as fertilizer on crops, thorough washing of fruits and vegetables, freezing ($-10°C$ for 5 days) or thorough cooking of pork ($60°C$, no pink left) before eating, and anthelminthic treatment of adult tapeworm carriers are necessary to break the transmission cycle (Sarti et al., 2000).

Diphyllobothrium Species

The pseudophyllidean tapeworm (life cycle usually encompassing waterborne crustacean and piscine hosts; larvae escape from egg in water) *Diphyllobothrium latum* and its complex of associated species in the genus (Curtis and Bylund, 1991; Bogtish et al., 2005) is the largest tapeworm parasitizing humans, attaining a length of 9 m and a width of up to 2 cm. The parasite is found in freshwater lakes and rivers across northern Europe, Scandinavia, the old Soviet Union, and less commonly in North and South America (Dupouy-Camet and Peduzzi, 2004). Humans and other piscivorous mammalian hosts harbor the adult tapeworm, which is attached to the small intestinal surface by the scoop-shaped scolex and bothria (or grooves), which are characteristic holdfast organs of pseudophyllidean tapeworms (Figure 6). Eggs are released in feces and mature in water in 8 days to a few weeks, depending on environmental temperature. The mature coracidium emerges from the operculum of the egg and swims with its cilia until it is ingested by a copepod. Within the copepod, the coracidium enters the hemocoel and develops into a procercoid larva within 3 weeks. When the parasitized copepod is eaten by a fish, the procercoid larva burrows through the intestinal wall, enters the musculature or viscera of the fish, and develops into a plerocercoid larva, which may grow to several centimeters long and is visible to the naked eye (Guttowa and Moskwa, 2005). Many species of freshwater fishes serve as suitable second intermediate hosts, including arctic

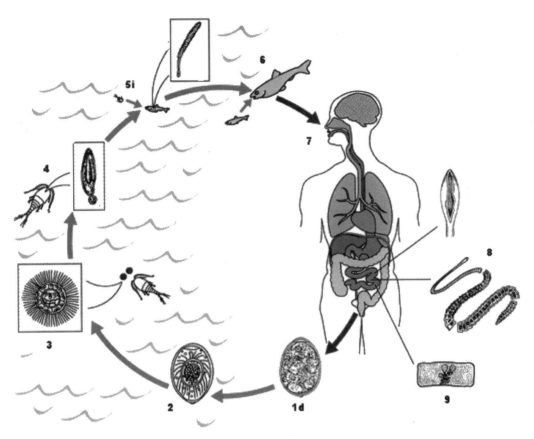

Figure 6. Life cycle of *Diphyllobothrium latum*. i, infective stage; d, diagnostic stage. 1d, unembryonated eggs are passed in feces. 2, eggs embryonate in water. 3, coracidia hatch from eggs and are ingested by crustaceans. 4, procercoid larvae form in the body cavity of crustaceans. 5i, infected crustaceans are eaten by small freshwater fish. Procercoid larvae are released from crustaceans and develop into plerocercoid larvae. 6, predator fish eats the small infected fish. 7, humans ingest raw or under-cooked infected fish. 8, adults live in small intestine. 9, proglottids release immature eggs.

char, trout, whitefish, pike, burbot, salmon, and perch (Lee et al., 2001). Dupouy-Camet and Peduzzi (2004) reported infestation rates with infectious plerocercoid larvae in perch from European lakes ranging from 3.7% in Lake Bienne in Switzerland to 33.3% in Lake Orta in Italy. If the infected fish is eaten in a raw or undercooked state, the plerocercoid becomes free in the small intestine of the consuming host and develops rapidly, beginning egg production within 2 weeks. Adult tapeworms are generally benign, but abdominal pain, nausea, and diarrhea may occur, and *D. latum* may absorb enough vitamin B_{12} to cause deficiency in malnourished individuals. Diagnosis is made by demonstration of proglottids or eggs in feces (Figure 7A to C) (Jacob et al., 2006). Removal of the adult worm is accomplished with praziquantel or niclosamide. Interruption of transmission requires proper sewage disposal, such that untreated sewage containing viable tapeworm eggs is not discharged into rivers and lakes, and proper cooking ($56°C$ for 5 minutes) or freezing

($-18°C$ for 24 hours) of fish before it is consumed (Jacob et al., 2006). Smoking, pickling, and marinating alone do not kill the parasite in fish. Humans can also become hosts to plerocercoid larvae by inadvertently swallowing infected copepods, amphibians, snakes, reptiles, and other procercoid hosts. The procercoid leaves the gut, develops into a plerocercoid larva, and migrates to other organs and tissues. The resulting phenomenon, known as sparganosis, can be life-threatening, depending on which tissue or organ is invaded and whether the sparganum proliferates by budding. Treatment is usually by surgical removal.

NEMATODES

Trichinella Species

Nematodes in the genus *Trichinella* are some of the most commonly recognized agents of food-borne parasitic disease. Human trichinellosis has historically been linked to the consumption of raw or undercooked

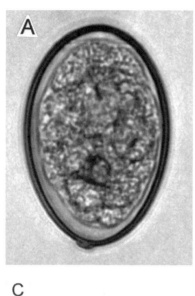

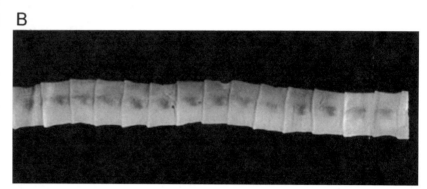

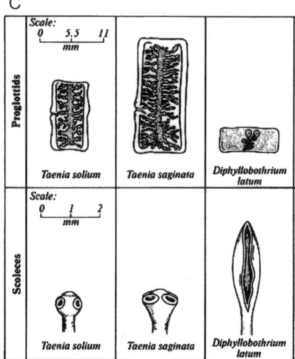

Figure 7. *Diphyllobothrium latum* egg and strobila morphology. (A) Operculated egg of *D. latum* in feces; egg size of 61 μm (range of 58 to 76 μm). Magnification, ×400. (B) Proglottids of *D. latum* showing the craspedote strobilar arrangement. (C) Proglottid and scolex structures of *T. solium*, *T. saginata*, and *D. latum*; acetabular (rounded sucker) structure of holdfast organs in *Taenia* species compared to bothridial (grooved sucker) holdfast structure in *D. latum*.

pork or certain game meats (e.g., bear and wild boar) (Murrell and Pozio, 2000). The horse is considered an aberrant host for *Trichinella*; however, numerous outbreaks linked to the consumption of raw horsemeat have been documented (Ancelle, 1998). *Trichinella* is transmitted from host to host by obligate carnivorism. Currently, eight sibling species, *T. spiralis* (T1), *T. nativa* (T2), *T. britovi* (T3), *T. pseudospiralis* (T4), *T. murrelli* (T5), *T. nelsoni* (T7), *T. papuae* (T10), and *T. zimbabwensis* (T11), and three genotypes of undetermined taxonomic status (T6, T8, and T9) have been identified in the genus (Pozio and Zarlenga, 2005; Zarlenga et al., 2006). All of these taxa (other than *T. zimbabwensis*) have been shown to infect humans after consumption of improperly prepared meats. The sylvatic genotypes (T2 through T11) of *Trichinella* are widespread in the environment due to an expansive geographic and host range (Murrell et al., 2000). Worldwide geographic distribution of these isolates has been described (Pozio et al., 1992; Zarlenga et al., 2006). Most cases of clinical disease

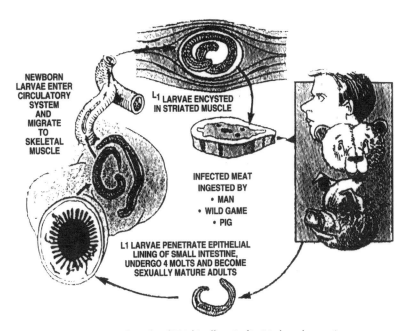

Figure 8. Life cycle of *Trichinella spiralis*. L1, larval stage 1.

and death from *Trichinella* infection, however, result from infection with *T. spiralis*, which is well adapted to the synanthropic cycle involving pigs, rats, and humans. Five of the sibling species, *T. spiralis*, *T. murrelli*, *T. pseudospiralis*, *T. nativa*, and *Trichinella* T6, occur in the continental United States and have been identified in a variety of mammalian species, including domestic swine and wild boar, black bear, opossums, raccoons, coyotes, wolves, dogs, foxes, skunks, rats, bobcats, cougars, and other carnivores. Although *T. spiralis* is virtually absent from the U.S. pig population (Gamble and Bush, 1999; Gamble et al., 1999b; USDA, 2001, 2007; 2000 and 2006 National Animal Health Monitoring System Survey, unpublished), some sylvatic isolates pose a risk for zoonotic transmission when pigs are exposed to *Trichinella*-infected wildlife in nonbiosecure pig barns or when pigs are managed in nonconfinement systems (van Knapen, 2000). Transmission occurs when undercooked meat containing infective muscle larvae is eaten (Figure 8). Digestion of infected meat releases the encysted first-stage muscle larvae, which penetrate epithelial cells lining the small intestinal mucosa. Larvae undergo four molts within 48 hours to the adult stage. Mating of adult male and female worms results within 1 week in the production of newborn larvae (eggs are not produced by the female), which enter the circulatory system and are disbursed throughout the body. The larvae penetrate individual striated muscle cells, and within 1 month encapsulate in an intracellular cystic collagenous structure known as a nurse cell (three

genotypes, *T. pseudospiralis*, *T. zimbabwensis*, and *T. papuae*, do not encapsulate) (Figures 9 and 10). Larvae can survive for years encysted in muscle. *Trichinella* remains a significant problem in commercially raised pork in some eastern European countries, Asia, South America, and Africa (Pozio et al., 1992; Pozio and Murrell, 2006). As a result, trading partners require testing for *Trichinella* in pig and horse carcasses destined for human consumption via export markets (Webster et al., 2006). A muscle burden of >1 larva per gram in meat is considered the infection level of public health concern which may cause clinical disease in humans. Many biological factors influence the final worm burden in a host, including species or strain of the parasite, infecting dose, previous exposure, and host immunocompetence; however, host animals commonly exhibit worm burdens from tens to several hundred to several thousand larvae per gram. Most human infections are subclinical; ingestion of larger numbers of infective larvae can result in mild to severe clinical symptoms requiring hospitalization, and even death from involvement of the heart and brain (neurotrichinellosis). Signs of infection are vague but include headache, nausea, sweating, rash, and diarrhea during the initial stages of infection, as worms initially penetrate the intestinal mucosa. Facial edema and fever may follow, and migration of muscle larvae may result in flu-like symptoms, muscle pain, facial edema, and heart, brain, or kidney damage (Dupouy-Camet et al., 2002). Penetration of striated muscle by juvenile worms may cause severe

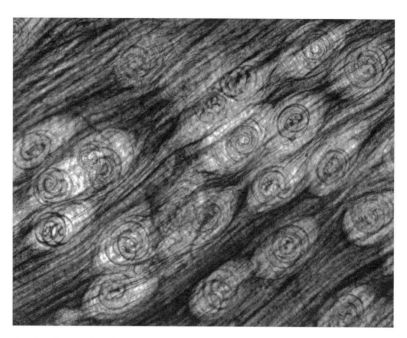

Figure 9. *Trichinella spiralis* infective muscle larvae in rat diaphragm muscle cells. Magnification, ×30.

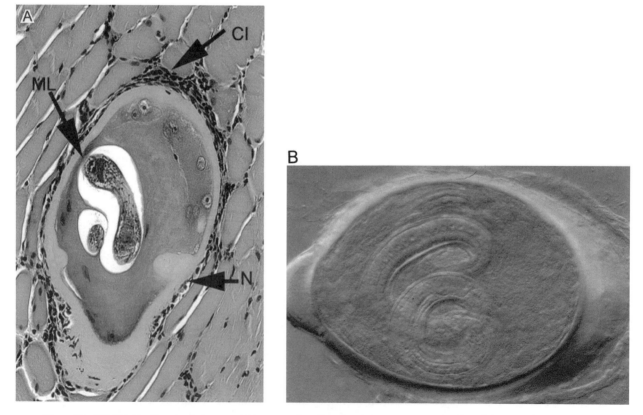

Figure 10. (A) *Trichinella spiralis* larvae within a muscle cell in horse tongue tissue. Hematoxylin-eosin-stained paraffin-embedded sections containing cellular infiltrates and encapsulated larvae in a nurse cell. CI, cellular infiltrates; ML, muscle larvae; N, nurse cell capsule. Magnification, ×160. (B) *T. spiralis* larvae within a nurse cell. Magnification, ×200.

fatigue, muscle pain, heart damage, and eosino-philia. Death may result from heart, kidney, or respiratory failure, or, rarely, neural invasion. Diagnosis is made by detection of antibodies in serum or by muscle biopsy. Treatment is targeted at worms in the intestine to reduce the number of female worms available to release muscle larvae. Benzimidazoles are frequently used for this purpose early in recognized infection, and supportive therapy for pain, nausea, and inflammation is administered in later stages of infection.

Currently available serological tests cannot be used as reliable indicators of infection in meat animals since seroconversion may be delayed for several weeks after infection (Kapel and Gamble, 2000). In addition, horses mount an antibody response to *Trichinella* infection that does not persist beyond 6 months, so infected horses cannot be diagnosed serologically (Hill et al., 2007). Consumers should be alerted to the risk of acquiring trichinellosis in undercooked meat products from any source and the importance of thoroughly cooking meat or obtaining assurance that adequate inspection has been performed using a validated method of artificial digestion before meat is consumed in an undercooked state.

Anisakid Species

Anisakis simplex, along with several other nematodes in the family *Anisakidae (Pseudoterranova, Contracaecum,* and *Hysterothylacium),* are large (5 to 15 cm) roundworms which utilize marine mammals (pinnipeds and ceteans), fish-eating birds, and large predatory fishes as the definitive hosts (Koie and Fagerholm, 1995; Chai et al., 2005). Adult worms are localized in the stomach; eggs are shed in feces into seawater, where they embryonate and hatch; and the free-living third-stage larvae are eaten by the first intermediate host, a marine crustacean (Figure 11). Pelagic and benthic fishes act as second intermediate hosts; larvae encyst in the musculature, body cavity, stomach wall, and liver of fishes, depending on the parasite species, the fish species, and on the geographic area that the fish inhabit, as liver infection is more common in Antarctic than in sub-Antarctic species (Palm, 1999; Szostakowska et al., 2005). Squid are also commonly utilized as second intermediate hosts. Humans become infected by consuming raw or undercooked fish or shellfish containing the infective third-stage larvae. Herring, cod, flounder, mackerel, Pacific salmon, haddock, monkfish, and squid are frequently infected. Larval attempts to penetrate the stomach or esophageal lining in order to complete development are usually the

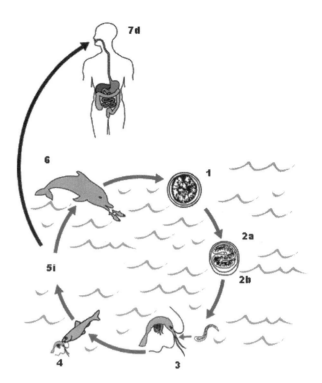

Figure 11. Life cycle of *Anisakis simplex*. i, infective stage; d, diagnostic stage. 1, marine mammals excrete unembryonated eggs. 2a, eggs are embryonated in water and second-stage larvae (L2) form. 2b, larvae hatch from eggs and become free swimming. 3, larvae are ingested by crustaceans and mature to L3. 4, infected crustaceans are eaten by fish and squid. 5i, larvae can be maintained as L3 in paratenic fish hosts by predation, or 6, fish or squid are consumed by marine mammals, and the larvae molt twice and develop into adult worms that produce eggs that are shed in feces. 7d, humans become infected by eating raw or undercooked seafood. Diagnosis can be made by gastroscopic examination, and the larvae can be removed with forceps.

first indication of human infection, as these burrowing attempts can result in severe epigastric pain and nausea within 2 hours of consuming the infected fish; symptoms may take as long as 10 days to become evident (Bouree et al., 1995). Perforation of the stomach or intestinal tract can occur, and, rarely, infection can be fatal. Larvae are usually removed by coughing or vomiting by the infected individual, or by a physician using a gastric probe fitted with forceps which can be used to grasp and remove the larvae. Removal of the larvae usually results in complete resolution of symptoms (Anonymous, 2006). Anisakids usually do not complete their life cycle in humans, and larvae are either expelled from the gut or die and are surrounded by eosinophils and other inflammatory cells, forming a granuloma, and destroyed. Diagnosis is made during acute infection by endoscopy; however, serological tests such as enzyme-linked immunosorbent assays (ELISA) and Western blot analyses have shown some utility in chronic cases in which granulomatous lesions have formed around dead larvae; histological examination of the lesion may confirm the diagnosis (Rodero et al., 2006; Ugenti et al., 2007). In cases in which the larvae are not surgically removed or spontaneously expelled by the host, symptoms usually resolve 8 to 10 weeks after infection as the larvae die and are destroyed by the inflammatory response.

Capillaria philippinensis

Capillaria philippinensis was first recognized as a serious human pathogen in the early 1960s in the Philippines and has been responsible for more than 2,000 confirmed human infections since its initial description (Cross and Basaca-Sevilla, 1991; Cross, 1992; Moravec, 2001). Most infections have been diagnosed in the Philippine islands; however, cases have occurred in Southeast Asia, Japan, Egypt, India, the United Arab Emirates, Italy, Spain, and Iran, with numerous outbreaks resulting in dozens of fatalities (Hwang, 1998; Lu et al., 2006). The complete life cycle of the parasite has not yet been elucidated fully, as reservoir hosts of the parasite have so far not been identified (Figure 12A). Piscivorous migratory birds are suspected both as reservoir hosts and as disseminators of the parasite outside of the Philippine islands (Cross and Basaca-Sevilla, 1983). Adult worms live in the small intestine, where female worms release both eggs and larval worms. Autoinfection by released larvae can result in an enormous worm burden, significant pathology, and enteropathy in the definitive host and probably accounts for the relatively high acute fatality rate associated with this parasite. Eggs pass out of the host with feces and embryonate in water in 5 to 10 days (Figure 12B). Embryonated eggs which are consumed by fish hatch in the fish intestine, where they complete development to the infective stage. Humans become infected by consuming raw or undercooked whole fishes containing the viscera.

Angiostrongylus Species

Angiostrongylus cantonensis is a nematode whose normal definitive hosts are rats and possibly other mammals found in Southeast Asia, but the parasite has now spread to the Pacific Islands, Australia, India, Africa, the Caribbean, and the United States. (Louisiana). A second species, *Angiostrongylus costaricensis,* has been implicated in human disease in Central America (Rodriguez et al., 2002). This global dispersion out of Southeast Asia is thought to have resulted from movement of ship rats in cargo after the Second World War (Kliks and Palumbo, 1992). Adult worms are found in the pulmonary arteries of the definitive host where they mate and release eggs; the eggs break through capillary walls and enter the alveoli (Figure 13). First-stage juvenile worms hatch in the alveoli and migrate up the trachea to the mouth, then are swallowed and pass out of the body in feces. Larvae are ingested by intermediate hosts, which include both aquatic and terrestrial gastropods, crustaceans, reptiles, and amphibians (Chau et al., 2003; Wan and Weng, 2004). Larvae develop to the third stage in the intermediate host and are infective to rats at this stage. When a rat consumes infected tissues from one of the intermediate hosts, the larvae burrow through the intestinal wall, migrate to the brain, and molt twice. The larvae then enter blood vessels in the brain, which carry them to the pulmonary arteries where they become sexually mature. People are infected by consumption of uncooked tissues from the intermediate hosts and can also become infected by consumption of fruit or vegetable products contaminated with third-stage larvae in the slime trails secreted by infected slugs (Pien and Pien, 1999; Alto, 2001). The parasite was thought to be of little consequence to human health until the link was made between eosinophilic meningoencephalitis and infection with *A. cantonensis*, which is now recognized as the most common cause of eosinophilic meningoencephalitis worldwide (Jindrak, 1975; Hughes and Biggs, 2002; Panackel et al., 2006; Kittimongkolma et al., 2007). After ingestion of third-stage larvae by a human, the larvae penetrate the intestinal mucosa and migrate to the brain and spinal cord. Humans are dead-end hosts for the parasite; larvae do not mature to the adult stage and are eventually killed by the immune response. Symptoms result from direct mechanical damage in the central nervous system caused by the migrating larvae and from the intense inflammatory response elicited by migrating and dying worms (Mojon, 1994). Worms have also been recovered from both chambers of the eye. Clinical manifestations of infection may include severe headache which progresses from intermittent to persistent, nausea, seizure (especially in children), stiff neck, involvement of the optical and facial nerves, and tingling or numbness of the trunk and limbs. The cerebrospinal fluid is abnormal in appearance and contains large numbers of white cells, eosinophils, and occasionally larvae. Treatment of infected individuals is largely supportive, and treatment with anthelminthics has been attempted; however, the susceptibility of the worms to anthelminthics is unknown. Pain relievers and corticosteroids are frequently administered to reduce pain and inflammation and as treatment to relieve cerebral pressure. Most infected individuals recover with treatment; however, heavy infections may leave some chronically disabled, and, rarely, death can occur.

A

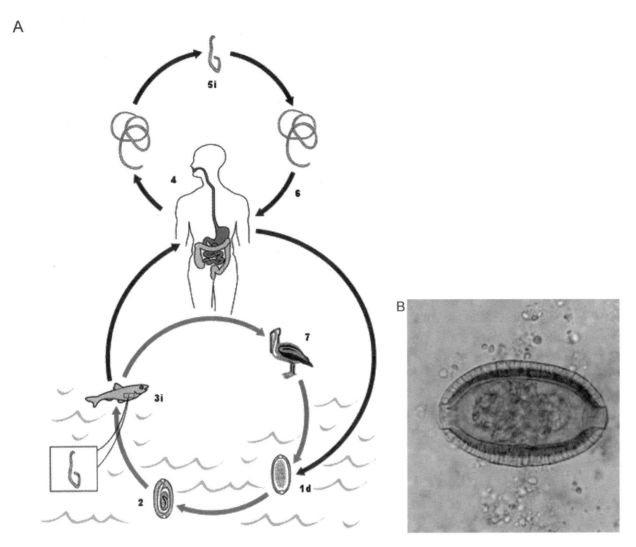

Figure 12. (A) Life cycle of *Capillaria philippinensis*. i, infective stage; d, diagnostic stage. 1d, unembryonated eggs are passed in feces of infected humans or birds. 2, eggs embryonate in water and are eaten by fish. 3i, larvae hatch from eggs and juveniles develop to the infective stage in the fish intestine. 4, humans and birds become infected by consuming undercooked or raw, whole fishes. 5i, adult worms in the human intestine can release eggs, or 6, juvenile worms can invade the intestinal mucosa, leading to massive autoinfections and severe pathology. (B) Egg of *Capillaria philippinensis* in feces; range of 36 to 45 μm in length. Characteristic thick, striated shell and polar plugs visible.

Gnathostoma

Gnathostomes are species belonging to a geographically widespread genus but are most commonly found as adult worms in carnivorous mammals in Asia (Ligon, 2005). Several species are known; however, most human cases have resulted from infection with *G. spinigerum*. Adult worms, whose head and stout body are covered in spines, are found attached to nodules in the stomach lining of carnivorous mammals; cats and dogs act as important definitive hosts for the parasite. Eggs are released in feces and embryonate in water in about 7 days (Figure 14). First-stage juveniles hatch from the eggs and are eaten by copepods; larvae develop to second-stage larvae in the

hemocoel. Copepods are eaten by a myriad of fish, amphibians, and crustacean species, and the freed second-stage larvae penetrate the intestine and molt to the third stage. A number of species of fish, chickens, ducks, shrimp, crabs, and frogs can also act as paratenic hosts for the third-stage juvenile parasite and represent an additional source of human infection (Rojekittikhun et al., 2002). Consumption of third-stage larvae in the tissues of undercooked intermediate or paratenic hosts results in human infection (Moore et al., 2003). The nematode migrates through a variety of organ systems and tissues, invoking an intense inflammatory response and resulting in clinical symptoms characteristic of gnathostomiasis (localized,

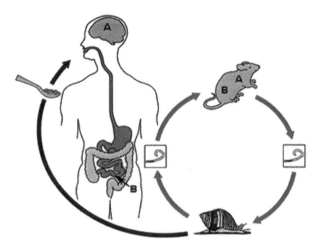

Figure 13. Life cycle of *Angiostrongylus cantonensis* and *A. costaricensis*. (A) *Angiostrongylus cantonensis* causes eosinophilic meningoencephalitis characterized by eosinophils in the cerebrospinal fluid. The infection is common in parts of Southeast Asia and the Pacific islands, Africa, and the Caribbean. Adult worms are found in pulmonary arteries of rodents; eggs are laid and hatch in the lungs. First-stage larvae migrate up the trachea, are coughed up, swallowed, and then are released in feces. Several types of aquatic and terrestrial mollusks serve as intermediate hosts; land crabs, shrimp, and frogs can serve as paratenic hosts. Infective third-stage larvae develop in intermediate hosts. Humans (and rats) become infected by eating undercooked and raw infected snails, slugs, crabs, or raw vegetables contaminated by snails or slugs. Humans are accidental hosts; completion of the life cycle has not been documented in humans. (B) *A. costaricensis* causes eosinophilic enteritis, an eosinophilic inflammation of the mesenteric arterioles of the ileocecal region of the gastrointestinal tract. Common in parts of Central and South America. Adult worms are found in mesenteric arteries of rodents; eggs are released in feces. Other aspects of the life cycle are similar to that of *A. cantonensis*, except that infective larvae invade the gastrointestinal tract in humans.

intermittent, migratory swellings; eosinophilia; and a positive serological response to a 24-kDa gnathostome antigen, linked with a history of travel to Southeast Asia) (Rusnak and Lucey, 1993). Subcutaneous migrations (cutaneous larva migrans) are common in humans and may appear as localized swellings which disappear spontaneously or as tracks in the skin caused by the meanderings of the larvae. Abscesses may also occur, as do eruptions of the larvae out of the skin. Larvae may also penetrate organ systems, the eye, or the central nervous system (visceral larva migrans), with serious and potentially fatal consequences (Hughes and Biggs, 2002). Human infection represents a dead end for the parasite, since the nematodes do not mature; gnathostomiasis is normally self-limiting in its manifestations, as worms are destroyed by the immune system after extended migration in humans. Cases of gnathostomiasis, though most common in Japan and Thailand, have recently been seen in Mexico and Central and South

America, due to the increasing habit of raw fish consumption (Clement-Rigolet et al., 2004). The infection is also seen in travelers to these areas who adopt local culinary habits. Treatment is effected most commonly with albendazole or ivermectin; a second course of treatment may be necessary.

PROTOZOANS

Toxoplasma gondii

Perhaps the most widespread protozoan parasite affecting humans is *Toxoplasma gondii*. *Toxoplasma* infects virtually all warm-blooded animals, including humans, livestock, birds, and marine mammals (Dubey and Beattie, 1988; Hill et al., 2005). There is only one species of *Toxoplasma*, *T. gondii*. Infection in humans occurs worldwide; however, prevalence varies widely from place to place. In the United States and the United Kingdom, it is estimated that 10 to 40% of people are infected, whereas in Central and South America and continental Europe, estimates of infection range from 50 to 80% (Dubey and Beattie, 1988; Jones et al., 2007). Most infections in humans are asymptomatic, but at times the parasite can produce devastating disease. Transmission of *T. gondii* occurs via the fecal-oral route (Figure 15), as well as through consumption of infected meat and by transplacental transfer from mother to fetus (Frenkel et al., 1970; Dubey and Beattie, 1988).

Oocysts, the environmentally resistant stage of the parasite, are shed in feces of the only definitive host, the cat. *Toxoplasma* oocysts are formed only in felids, probably in all members of the *Felidae*. Cats shed oocysts after ingesting any of the three infectious stages of *T. gondii*, i.e., rapidly multiplying tachyzoites (Figure 16A) in tissues of infected animals, slowly multiplying bradyzoites in tissue cysts (Figure 16B and C), and sporozoites contained within oocysts (Figure 16G) (Dubey and Frenkel, 1972, 1976; Dubey, 1996, 2002). Prepatent periods (time to the shedding of oocysts after initial infection) and frequency of oocyst shedding vary according to the stage of *T. gondii* ingested. Prepatent periods are 3 to 10 days after ingesting bradyzoites and 19 days or more after ingesting tachyzoites or oocysts (Dubey and Frenkel, 1972, 1976; Dubey, 1996). After the ingestion of tissue cysts (intracellular bradyzoites surrounded by a cyst wall) by cats, the tissue cyst wall is dissolved by proteolytic enzymes in the stomach and small intestine. The released bradyzoites penetrate the epithelial cells of the feline small intestine and initiate development (enteroepithelial cycle) of numerous generations of asexual and sexual cycles of *T. gondii* (Dubey and Frenkel, 1972); these stages are known as schizonts (Figure 16D). As the

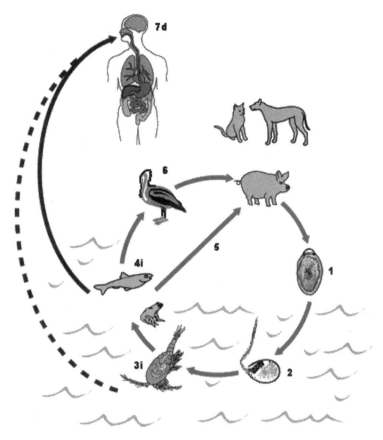

Figure 14. Life cycle of *Gnathostoma spinigerum* and *G. hispidum*. i, infective stage; d, diagnostic stage. 1, unembryonated eggs released in feces and embryonate in water. 2, first-stage larvae (L1) hatch from eggs. 3i, larvae enter a copepod and develop to the second stage. 4i, when the copepod is eaten by a vertebrate second intermediate host, the L2 penetrate the intestine, enter muscle, and develop to L3. 5, if the second intermediate host is eaten by an appropriate definitive host, adult worms develop in the stomach. 6, numerous paratenic hosts (including humans) have been identified in which the L3 may wander without development if the second intermediate host or a paratenic host is ingested. 7d, in humans, L3 may wander in the skin, eye, viscera, spinal cord, or brain, with devastating results. Diagnosis is usually made by recovery of the migrating larvae.

enteroepithelial cycle progresses, bradyzoites also penetrate the lamina propria of the feline intestine and multiply as tachyzoites. Within a few hours after infection of cats, *T. gondii* may disseminate to extraintestinal tissues. *Toxoplasma gondii* persists in intestinal and extraintestinal tissues of cats for at least several months and possibly for the life of the cat. In the feline intestine, organisms (merozoites) released from schizonts form male and female gametes. The male gamete has two flagella (Figure 16E), and it swims to and enters the female gamete. After the female gamete is fertilized by the male gamete, oocyst wall formation begins around the fertilized gamete. When oocysts are mature, they are discharged into the intestinal lumen by the rupture of intestinal epithelial cells and released into the environment in feces. In freshly passed feces, the spherical oocysts are unsporulated (noninfective) (Figure 16F). They sporulate (become infectious) outside the cat within 1 to 5 days, depending upon aeration and temperature.

Sporulated oocysts contain two ellipsoidal sporocysts (Figure 16G). Each sporocyst contains four sporozoites.

The tachyzoite, the rapidly dividing, extraintestinal stage of the parasite, enters the host cell by active penetration of the host cell membrane and can tilt, extend, and retract as it searches for a host cell. After entering the host cell, the tachyzoite becomes ovoid in shape and becomes surrounded by a parasitophorous vacuole. *Toxoplasma gondii* in a parasitophorous vacuole is protected from host defense mechanisms. The tachyzoite multiplies asexually within the host cell by repeated divisions in which two progeny form within the parent parasite, consuming it. Tachyzoites continue to divide until the host cell is filled with parasites. Cells rupture and release tachyzoites, which infect new cells. After a few such divisions, *T. gondii* forms tissue cysts, which remain intracellular. The tissue cyst wall is elastic and thin and may enclose thousands of the crescent-shaped, slender organisms in the *T. gondii* bradyzoite

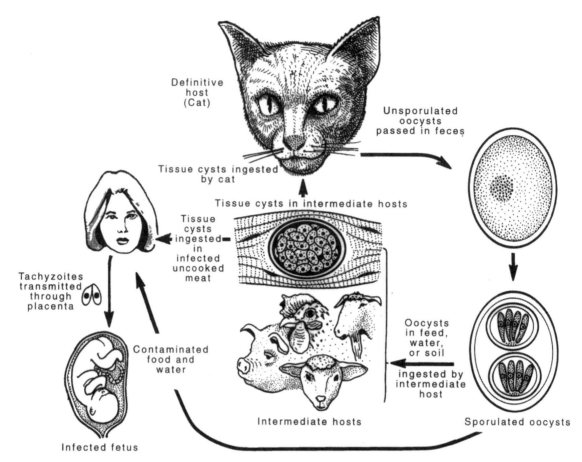

Figure 15. Life cycle of *Toxoplasma gondii.*

stage. Bradyzoites differ structurally only slightly from tachyzoites. Although tissue cysts containing bradyzoites may develop in visceral organs, including lungs, liver, and kidneys, they are more prevalent in muscular and neural tissues, including the brain, eye, skeletal, and cardiac muscle. Intact tissue cysts probably do not cause any harm and can persist for the life of the host.

Toxoplasmosis may be acquired by ingestion of parasites in the tissue-inhabiting stages. In the United States, infection in humans sometimes results from the ingestion of tissue cysts contained in undercooked meat (Dubey and Beattie, 1988; Roghmann et al., 1999; Lopez et al., 2000), though the exact contribution of food-borne toxoplasmosis versus oocyst-induced toxoplasmosis to human infection is currently unknown. In France, the prevalence of antibody to *T. gondii* is very high in humans; 84% of pregnant women in Paris have antibodies to *T. gondii,* while only 32% in New York City and 22% in London have such antibodies (Dubey and Beattie, 1988). The high incidence of *T. gondii* infection in France appears to be related in part to the French habit of eating

some meat products undercooked or uncooked. *Toxoplasma gondii* infection is common in many animals used for food, including sheep, pigs, goats, and rabbits (Smith et al., 1992; Dubey et al., 1995; Gamble et al., 1999a; Lehmann et al., 2003). The relative risk to U.S. consumers of acquiring *T. gondii* infection from undercooked meat, including pork, was recently determined in a nationwide retail meat survey. The survey of 698 retail outlets determined the prevalence of viable *T. gondii* tissue cysts in commercially available fresh pork products to be 0.38%; viable *T. gondii* tissue cysts were not found in chicken or beef (Dubey et al., 2005). Recent surveys in several European countries using serology and PCR technology to detect parasite DNA in meat and in meat animals have shown that infection rates in pigs and horses are negligible, while infection in sheep and cattle ranges from 1 to 6% (Warnekulasuriya et al., 1998; Wyss et al., 2000; Tenter et al., 2000; Aspinall et al., 2002). Serological surveys in eastern Poland revealed that 53% of cattle, 15% of pigs, and 0 to 6% of chickens, ducks, and turkeys were positive for *Toxoplasma* infection; nearly 50% of the people in the region were

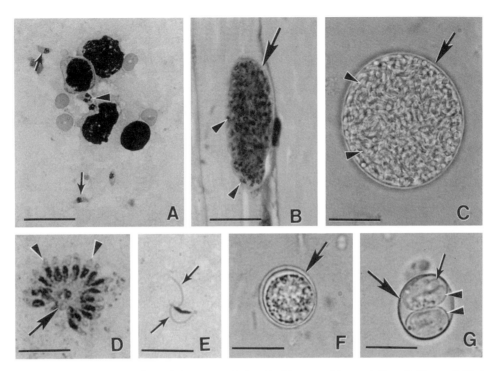

Figure 16. Stages of *Toxoplasma gondii*. Scale bar in panels A to D, 20 μm, and in panels E to G, 10 μm. A. Tachyzoites in impression smear of lung. Note sizes of crescent-shaped individual tachyzoites (arrows) and dividing tachyzoites (arrowhead) compared with sizes of host red blood cells and leukocytes. Giemsa stain. B. Tissue cysts in section of muscle. The tissue cyst wall is very thin (arrow) and encloses many tiny bradyzoites (arrowheads). Hematoxylin and eosin stain. C. Tissue cyst separated from host tissue by homogenization of infected brain. Note tissue cyst wall (arrow) and hundreds of bradyzoites (arrowheads). Unstained. D. Schizont (arrow) with several merozoites (arrowheads) separating from the main mass. Impression smear of infected cat intestine. Giemsa stain. E. A male gamete with two flagella (arrows). Impression smear of infected cat intestine. Giemsa stain. F. Unsporulated oocyst in fecal float of cat feces. Unstained. Note double-layered oocyst wall (arrow) enclosing a central undivided mass. G. Sporulated oocyst with a thin oocyst wall (large arrow) and two sporocysts (arrowheads). Each sporocyst has four sporozoites (small arrow) which are not in complete focus. Unstained.

also serologically positive for *Toxoplasma* infection (Sroka, 2001). Virtually all edible portions of an animal can harbor viable *T. gondii* tissue cysts, and tissue cysts can survive in food animals for years.

In humans, enlarged lymph nodes are the most frequently observed clinical form of toxoplasmosis (Table 2). Lymphadenopathy may be associated with fever, fatigue, muscle pain, sore throat, and headache. Encephalitis is the most important manifestation of toxoplasmosis in immunosuppressed patients (Dubey and Beattie, 1988; Luft and Remington, 1992). Patients may have headache, disorientation, drowsiness, hemiparesis, reflex changes, and convulsions, and many become comatose. Encephalitis caused by *T. gondii* is now recognized with great frequency with patients treated with immunosuppressive agents. In addition, toxoplasmosis ranks high on the list of diseases which lead to the deaths of patients with AIDS; approximately 10% of AIDS patients in the United States and up to 30% in Europe are estimated to die from toxoplasmosis (Luft and Remington, 1992). With AIDS patients, although any organ may be involved, including the testis, dermis, and the spinal cord, infection of the brain is most frequently reported. Most AIDS patients suffering from toxoplasmosis have bilateral, severe, and persistent headache which responds poorly to analgesics. As the disease progresses, the headache may give way to a condition characterized by confusion, lethargy, ataxia, and coma. The predominant lesion in the brain is necrosis, especially of the thalamus (Renold et al., 1992).

Diagnosis is made by biologic, serologic, or histologic methods or by some combination of the above. Clinical signs of toxoplasmosis are nonspecific and are not sufficiently characteristic for a definite diagnosis. Toxoplasmosis, in fact, mimics several other infectious diseases. Detection of *T. gondii* antibody in patients may aid diagnosis. There are numerous serologic procedures available for detection of humoral antibodies; these include the Sabin-Feldman dye test, the modified agglutination test, the indirect hemagglutination assay, the indirect fluorescent antibody assay, the direct agglutination test, the latex agglutination test, the ELISA, and the immunosorbent agglutination assay test. The indirect fluorescent antibody assay, immunosorbent agglutination assay

Table 2. Frequency of symptoms in people with postnatally acquired toxoplasmosis

Symptom	Patients with symptoms (%)	
	Atlanta outbreak (35 patients)[a]	Panama outbreak (35 patients)[b]
Fever	94	90
Lymphadenopathy	88	77
Headache	88	77
Myalgia	63	68
Stiff neck	57	55
Anorexia	57	NR
Sore throat	46	NR
Arthralgia	26	29
Rash	23	0
Confusion	20	NR
Earache	17	NR
Nausea	17	36
Eye pain	14	26
Abdominal pain	11	55

[a]From Teutsch et al. (1979).
[b]From Benenson et al. (1982). NR, not reported.

test, and ELISA have been modified to detect immunoglobulin M antibodies (Remington et al., 1995), which appear earlier during infection than immunoglobulin G and are indicative of a recent acute infection (Remingtonet al., 1995). Infection has been diagnosed by PCR using nested, stage-specific primers and cerebrospinal fluid with AIDS patients with suspected toxoplasmic encephalitis (Contini et al., 2002; Joseph et al., 2002), with immunocompromised patients undergoing hematopoietic stem cell transplantation (Lewis et al., 2002), and in suspected cases of fetal toxoplasmosis using amniotic fluid (Antsaklis et al., 2002). Sulfadiazine and pyrimethamine (Daraprim) are two drugs widely used for treatment of toxoplasmosis (Guerina et al., 1994; Chirgwin, et al., 2002). While these drugs have a beneficial action when given in the acute stage of the disease process when there is active multiplication of the parasite, they will not usually eradicate infection. It is believed that these drugs have little effect on subclinical infections, but the growth of tissue cysts in mice has been restrained with sulfonamides. Certain other drugs, like diaminodiphenylsulfone, atovaquone, spiramycin, and clindamycin, are also used to treat toxoplasmosis in difficult cases.

To prevent food-borne infection by *T. gondii*, the hands of people handling meat should be washed thoroughly with soap and water before they go to other tasks (Dubey and Beattie, 1988; Lopez et al., 2000). All cutting boards, sink tops, knives, and other materials coming in contact with uncooked meat should be washed with soap and water. Washing is effective because the stages of *T. gondii* in meat are killed by contact with soap and water (Dubey and

Beattie, 1988). *Toxoplasma gondii* in meat can be killed by freezing or cooking. Tissue cysts in meat are killed by heating the meat throughout to 67°C (Dubey et al., 1990) or by cooling to −13°C (Kotula et al., 1991). Meat of any animal should be cooked to 67°C before consumption, and tasting meat while cooking or while seasoning should be avoided.

REFERENCES

Alto, W. 2001. Human infections with *Angiostrongylus cantonensis. Pac. Health Dialog* **8**:176–182.

Anantaphruti, M. T. 2001. Parasitic contaminants in food. *Southeast Asian J. Trop. Med. Public Health* **32**(Suppl. 2):218–228.

Ancelle, T. 1998. History of trichinellosis outbreaks linked to horse meat consumption 1975-1998. *Euro. Surveill.* **3**:86–89.

Anonymous. 2006. *Anisakis simplex* and related worms. *In Bad Bug Book: Foodborne Pathogenic Microorganisms and Natural Toxins Handbook.* Center for Food Safety and Applied Nutrition, U.S. Food and Drug Administration. http://www.cfsan.fda.gov/~mow/chap25.html.

Antsaklis, A., G. Daskalakis, N. Papantoniou, A. Mentis, and S. Michalas. 2002. Prenatal diagnosis of congenital toxoplasmosis. *Prenatal Diag.* **22**:1107–1111.

Aspinall, T. V., D. Marlee, J. E. Hyde, and P. F. Sims. 2002. Prevalence of *Toxoplasma gondii* in commercial meat products as monitored by polymerase chain reaction—food for thought? *Int. J. Parasitol.* **32**:1193–1199.

Benenson, M. W., E. T. Takafuji, S. M. Lemon, R. L. Greenup, and A. J. Sulzer. 1982. Oocyst-transmitted toxoplasmosis associated with ingestion of contaminated water. *N. Engl. J. Med.* **307**:666–669.

Bogtish, B. J., C. E. Carter, and T. N. Oeltmann. 2005. Intestinal tapeworms, p. 266–274. *In* B. J. Bogtish, C. E. Carter, and T. N. Oeltmann (ed.), *Human Parasitology*, 3rd ed. Elsevier, Boston, MA.

Bouree, P., A. Paugam, and J C. Petithory. 1995. Anisakidosis: report of 25 cases and review of the literature. *Comp. Immunol. Microbiol. Infect. Dis.* **18**:75–84.

Chai, J. Y., K. D. Murrell, and A. J. Lymbery. 2005. Fish-borne parasitic zoonoses: status and issues. *Int. J. Parasitol.* **35**:1233–1254.

Chau, T. T., G. E. Thwaites, L. V. Chuong, D. X. Sinh, and J. J. Farrar. 2003. Headache and confusion: the dangers of a raw snail supper. *Lancet* **361**:1866.

Chirgwin, K., R. Hafner, C. Leport, J. Remington, J. Andersen, E. M. Bosler, C. Roque, N. Rajicic, V. McAuliffe, P. Morlat, D. T. Jayaweera, J. L. Vilde, and B. J. Luft, 2002. Randomized phase II trial of atovaquone with pyrimethamine or sulfadiazine for treatment of toxoplasmic encephalitis in patients with acquired immunodeficiency syndrome: ACTG 237/ANRS 039 study. *Clin. Infect. Dis.* **34**:1243–1250.

Clement-Rigolet, M. C., M. Danis, and E. Caumes. 2004. Gnathostomosis, an exotic disease increasingly imported into Western countries. *Presse Med.* **33**:1527–1532.

Contini, C., R. Cultrera, S. Seraceni, D. Segala, R. Romani, E. Fainardi, P. Cinque, A. Lazzarin, and S. Delia. 2002. The role of stage specific oligonucleotide primers in providing effective laboratory support for the molecular diagnosis of reactivated *Toxoplasma gondii* encephalitis in patients with AIDS. *J. Med. Microbiol.* **51**:879–890.

Cross, J. H., and V. Basaca-Sevilla. 1983. Experimental transmission of *Capillaria philippinensis* to birds. *Trans. R Soc. Trop. Med. Hyg.* **77**:511–514.

Cross, J. H., and V. Basaca-Sevilla. 1991. *Capillariasis philippinensis:* a fish-borne parasitic zoonosis. *Southeast Asian J. Trop. Med. Public Health* 22(Suppl.):153–157.

Cross, J. H. 1992. Intestinal capillariasis. *Clin. Microbiol. Rev.* 5: 120–129.

Curtis, M. A., and G. Bylund. 1991. Diphyllobothriasis: fish tapeworm disease in the circumpolar north. *Arctic Med. Res.* 50:18–24.

Dubey, J. P., A. W. Kotula, A. Sharar, C. D. Andrews, and D. S. Lindsay. 1990. Effect of high temperature on infectivity of *Toxoplasma gondii* tissue cysts in pork. *J. Parasitol.* 76:201–204.

Dubey, J. P., 1996. Infectivity and pathogenicity of *Toxoplasma gondii* oocysts for cats. *J. Parasitol.* 82:957–961.

Dubey, J. P., 2002. Tachyzoite-induced life cycle of *Toxoplasma gondii* in cats. *J. Parasitol.* 88:713–717.

Dubey, J. P., and J. K. Frenkel. 1972. Cyst-induced toxoplasmosis in cats. *J. Protozool.* 19:155–177.

Dubey, J. P., and J. K. Frenkel. 1976. Feline toxoplasmosis from acutely infected mice and the development of *Toxoplasma* cysts. *J. Protozool.* 23:537–546.

Dubey, J. P., and C. P. Beattie. 1988. *Toxoplasmosis of Animals and Man.* CRC Press, Boca Raton, FL.

Dubey, J. P., R. M. Weigel, A. M. Siegel, P. Thulliez, U. D. Kitron, M. A. Mitchell, A. Mannelli, N. E. Mateus-Pinilla, S. K. Shen, O. C. H. Kwok, and K. S. Todd. 1995. Sources and reservoirs of *Toxoplasma gondii* infection on 47 swine farms in Illinois. *J. Parasitol.* 81:723–729.

Dubey, J. P., D. E. Hill, J. L. Jones, A. W. Hightower, E. Kirkland, J. M. Roberts, P. L. Marcet, T. Lehmann, M. C. Vianna, K. Miska, C. Sreekumar, O. C. Kwok, S. K. Shen, and H. R. Gamble. 2005. Prevalence of viable *Toxoplasma gondii* in beef, chicken, and pork from retail meat stores in the United States: risk assessment to consumers. *J. Parasitol.* 91:1082–1093.

Dupouy-Camet, J., and R. Peduzzi. 2004. Current situation of human diphyllobothriasis in Europe. *Euro. Surveill.* 9:31–35.

Dupouy-Camet, J., W. Kociecka, F. Bruschi, F. Bolas-Fernandez, and E. Pozio. 2002. Opinion on the diagnosis and treatment of human trichinellosis. *Expert Opin. Pharmacother.* 3:1117–1130.

Eckert, J. 1996. Workshop summary: food safety: meat- and fish-borne zoonoses. *Vet. Parasitol.* 64:143–147.

Eom, K. S., and H. J. Rim. 1993. Morphologic descriptions of *Taenia asiatica* sp.n. *Korean J. Parasitol.* 31:1–6.

Eom, K. S. 2006. What is Asian Taenia? *Parasitol. Int.* 55(Suppl.): S137–S141.

Fan, P. C., C. Y. Lin, C. C. Chen, and W. C. Chung. 1995. Morphological description of *Taenia saginata asiatica* (Cyclophyllidea: Taeniidae) from man in Asia. *J. Helminthol.* 69:299–303.

Flisser, A. 1994. Taeniasis and cysticercosis due to *Taenia solium. Prog. Clin. Parasitol.* 4:77–116.

Flisser, A. 2001. Neurocysticercosis and epilepsy in developing countries. *J. Neurol. Neurosurg. Psychiatry* 70:707–708.

Flisser, A., E. Sarti, M. Lightowlers, and P. Schantz. 2003. Neurocysticercosis: regional status, epidemiology, impact and control measures in the Americas. *Acta Trop.* 87:43–51.

Flisser, A., A. E. Viniegra, L. Aguilar-Vega, A. Garza-Rodriguez, P. Maravilla, and G. Avila. 2004. Portrait of human tapeworms. *J. Parasitol.* 90:914–916.

Frenkel, J. K., J. P. Dubey, and N. L. Miller. 1970. *Toxoplasma gondii* in cats: fecal stages identified as coccidian oocysts. *Science* 167:893–896.

Gamble, H. R., and E. Bush. 1999. Seroprevalence of *Trichinella* infection in domestic swine based on the National Animal Health Monitoring System's 1990 and 1995 swine surveys. *Vet. Parasitol.* 80:303–310.

Gamble, H. R., R. C. Brady, and J. P. Dubey. 1999a. Prevalence of *Toxoplasma gondii* infection in domestic pigs in the New England states. *Vet. Parasitol.* 82:129–136.

Gamble, H. R., R. C. Brady, L. L. Bulaga, C. L. Berthoud, W. G. Smith, L. A. Detweiler, L. E. Miller, and E. A. Lautner. 1999b. Prevalence and risk association for *Trichinella* infection in domestic pigs in the northeastern United States. *Vet. Parasitol.* 82:59–69.

Garcia, H. H., C. A. Evans, T. E. Nash, O. M. Takayanagui, A. C. White, Jr., D. Botero, V. Rajshekhar, V. C. Tsang, P. M. Schantz, J. C. Allan, A. Flisser, D. Correa, E. Sarti, J. S. Friedland, S. M. Martinez, A. E. Gonzalez, R. H. Gilman, and O. H. Del Brutto. 2002. Current consensus guidelines for treatment of neurocysticercosis. *Clin. Microbiol. Rev.* 15:747–756.

Graczyk, T. K., and B. Fried. 1998. Echinostomiasis: a common but forgotten food-borne disease. *Am. J. Trop. Med. Hyg.* 58:501–504.

Graczyk, T. K., R. H. Gilman, and B. Fried. 2001. Fasciolopsiasis: is it a controllable food-borne disease? *Parasitol. Res.* 87:80–83.

Guerina, N. G., H. W. Hsu, H. C. Meissner, J. H. Maguire, R. Lynfield, B. Stechenberg, I. Abroms, M. S. Pasternack, R. Hoff, and R. B. Eaton for New England Regional Toxoplasma Working Group. 1994. Neonatal serologic screening and early treatment for congenital *Toxoplasma gondii* infection. *N. Engl. J. Med.* 330:1858–1863.

Gulsen, M. T., M. C. Savas, M. Koruk, A. Kadayifci, and F. Demirci. 2006. Fascioliasis: a report of five cases presenting with common bile duct obstruction. *Neth. J. Med.* 64:17–19.

Guttowa, A., and B. Moskwa. 2005. The history of the exploration of the *Diphyllobothrium latum* life cycle. *Wiad. Parazytol.* 51:359–364.

Hill, D. E., S. Chirukandoth, and J. P. Dubey. 2005. Biology and epidemiology of *Toxoplasma gondii* in man and animals. *Anim. Health Res. Rev.* 6:41–61.

Hill, D. E., L. Forbes, M. Kramer, A. Gajadhar, and H. R. Gamble. 2007. Larval viability and serological response in horses with long-term *Trichinella spiralis* infection. *Vet. Parasitol.* 146:107–116.

Hoberg, E. P., A. Jones, R. L. Rausch, K. S. Eom, and S. L. Gardner. 2000. A phylogenetic hypothesis for species of the genus *Taenia* (Eucestoda: Taeniidae). *J. Parasitol.* 86:89–98.

Hughes, A. J., and B. A. Biggs. 2002. Parasitic worms of the central nervous system: an Australian perspective. *Intern. Med. J.* 32:541–553.

Hwang, K. P. 1998. Human intestinal capillariasis *(Capillaria philippinensis)* in Taiwan. *Zhonghua Min Guo Xiao Er Ke Yi Xue Hui Za Zhi* 39:82–85.

Jacob, A. C., S. S. Jacob, and S. Jacob. 2006. Fecal fettuccine: a silent epidemic? *Am. J. Med.* 119:284–285.

Jindrak, K. 1975. *Angiostrongyliasis cantonensis* (eosinophilic meningitis, Alicata's disease). *Contemp. Neurol. Ser.* 12:133–164.

Jones, J. L., D. Kruszon-Moran, and M. Wilson. 2007. *Toxoplasma gondii* prevalence, United States. *Emerg. Infect. Dis.* 13:656. (Letter.)

Joseph, P., M. M. Calderon, R. H. Gilman, M. L. Quispe, J. Cok, E. Ticona, V. Chavez, J. A. Jimenez, M. C. Chang, M. J. Lopez, and C. A. Evans. 2002. Optimization and evaluation of a PCR assay for detecting toxoplasmic encephalitis in patients with AIDS. *J. Clin. Microbiol.* 40:4499–4503.

Kaewkes, S. 2003. Taxonomy and biology of liver flukes. *Acta Trop.* 88:177–186.

Kapel, C. M., and H. R. Gamble. 2000. Infectivity, persistence, and antibody response to domestic and sylvatic *Trichinella* spp. in experimentally infected pigs. *Int. J. Parasitol.* 30:215–221.

Keiser, J., and J. Utzinger. 2004. Chemotherapy for major food-borne trematodes: a review. *Expert Opin. Pharmacother.* 5:1711–1726.

Keiser, J., and J. Utzinger. 2005. Emerging foodborne trematodiasis. *Emerg. Infect. Dis.* **11:**1507–1514.

Kim, D. C. 1984. *Paragonimus westermani:* life cycle, intermediate hosts, transmission to man and geographical distribution in Korea. *Arzneimittelforschung* **34:**1180–1183.

Kittimongkolma, S., P. M. Intapan, K. Laemviteevanich, J. Kanpittaya, K. Sawanyawisuth, and W. Maleewong. 2007. Eosinophilic meningitis associated with angiostrongyliasis: clinical features, laboratory investigations and specific diagnostic IgG and IgG subclass antibodies in cerebrospinal fluid. *Southeast Asian J. Trop. Med. Public Health* **38:**24–31.

Kliks, M. M., and N. E. Palumbo. 1992. Eosinophilic meningitis beyond the Pacific Basin: the global dispersal of a peridomestic zoonosis caused by *Angiostrongylus cantonensis,* the nematode lungworm of rats. *Soc. Sci. Med.* **34:**199–212.

Koie, M., and H. P. Fagerholm. 1995. The life cycle of *Contracaecum osculatum* (Rudolphi, 1802) sensu stricto (Nematoda, Ascaridoidea, Anisakidae) in view of experimental infections. *Parasitol. Res.* **81:**481–489.

Kotula, A. W., J. P. Dubey, A. K. Sharar, C. D. Andrew, S. K. Shen, and D. S. Lindsay. 1991. Effect of freezing on infectivity of *Toxoplasma gondii* tissue cysts in pork. *J. Food Prot.* **54:**687–690.

Krejci, K. G., and B. Fried. 1994. Light and scanning electron microscopic observations of the eggs, daughter rediae, cercariae, and encysted metacercariae of *Echinostoma trivolvis* and *E. caproni. Parasitol. Res.* **80:**42–47.

Lee, K. W., H. C. Suhk, K. S. Pai, H. J. Shin, S. Y. Jung, E. T. Han, and J. Y. Chai. 2001. *Diphyllobothrium latum* infection after eating domestic salmon flesh. *Korean J. Parasitol.* **39:**319–321.

Lehmann, T., D. H. Graham, E. Dahl, C. Sreekumar, F. Launer, J. L. Corn, H. R. Gamble, and J. P. Dubey. 2003. Transmission dynamics of *Toxoplasma gondii* on a pig farm. *Infect. Genet. Evol.* **3:**135–141.

Lewis, J. S., H. Khoury, G. A. Storch, and J. DiPersio. 2002. PCR for the diagnosis of toxoplasmosis after hematopoietic stem cell transplantation. *Exp. Rev. Mol. Diag.* **2:**616–624.

Ligon, B. L. 2005. Gnathostomiasis: a review of a previously localized zoonosis now crossing numerous geographical boundaries. *Semin. Pediatr. Infect. Dis.* **16:**137–143.

Lopez, A., V. J. Dietz, M. Wilson, T. R. Navin, and J. L. Jones. 2000. Preventing congenital toxoplasmosis. *MMWR Recommend. Rep.* **49:**59–75.

Lu, L. H., M. R. Lin, W. M. Choi, K. P. Hwang, Y. H. Hsu, M. J. Bair, J. D. Liu, T. E. Wang, T. P. Liu, and W. C. Chung. 2006. Human intestinal capillariasis *(Capillaria philippinensis)* in Taiwan. *Am. J. Trop. Med. Hyg.* **74:**810–813.

Luft, B. J., and J. S. Remington. 1992. Toxoplasmic encephalitis in AIDS. *Clin. Infect. Dis.* **15:**211–222.

Macpherson, C. N. 2005. Human behaviour and the epidemiology of parasitic zoonoses. *Int. J. Parasitol.* **35:**1319–1331.

Mahajan, R. C. 2005. Paragonimiasis: an emerging public health problem in India. *Indian J. Med. Res.* **121:**716–718.

Mas-Coma, S. 2005. Epidemiology of fascioliasis in human endemic areas. *J. Helminthol.* **79:**207–216.

Mas-Coma, S., M. D. Bargues, and M. A. Valero. 2005. Fascioliasis and other plant-borne trematode zoonoses. *Int. J. Parasitol.* **35:**1255–1278.

McManus, D. P. 2006. Molecular discrimination of taeniid cestodes. *Parasitol. Int.* **55**(Suppl.): S31–S37.

Mojon, M. 1994. Human angiostrongyliasis caused by *Angiostrongylus costaricensis. Bull. Acad. Natl. Med.* **178:**625–631.

Moore, D. A., J. McCroddan, P. Dekumyoy, and P. L. Chiodini. 2003. Gnathostomiasis: an emerging imported disease. *Emerg. Infect. Dis.* **9:**647–650.

Moravec, F. 2001. Redescription and systematic status of *Capillaria philippinensis,* an intestinal parasite of human beings. *J. Parasitol.* **87:**161–164.

Murrell, K. D., and E. Pozio. 2000. Trichinellosis: the zoonosis that won't go quietly. *Int. J. Parasitol.* **30:**1339–1349.

Murrell, K. D., R. J. Lichtenfels, D. S. Zarlenga, and E. Pozio. 2000. The systematics of the genus *Trichinella* with a key to species. *Vet. Parasitol.* **93:**293–307.

Palm, H. W. 1999. Ecology of *Pseudoterranova decipiens* (Krabbe, 1878) (Nematoda: Anisakidae) from Antarctic waters. *Parasitol. Res.* **85:**638–646.

Panackel, C., Vishad, G. Cherian, K. Vijayakumar, and R. N. Sharma. 2006. Eosinophilic meningitis due to *Angiostrongylus cantonensis. Indian J. Med. Microbiol.* **24:**220–221.

Pien, F. D., and B. C. Pien. 1999. *Angiostrongylus cantonensis* eosinophilic meningitis. *Int. J. Infect. Dis.* **3:**161–163.

Pokora, Z. 2001. Role of gastropods in epidemiology of human parasitic diseases. *Wiad. Parazytol.* **47:**3–24.

Polley, L. 2005. Navigating parasite webs and parasite flow: emerging and re-emerging parasitic zoonoses of wildlife origin. *Int. J. Parasitol.* **35:**1279–1294.

Pozio, E., and K. D. Murrell. 2006. Systematics and epidemiology of *Trichinella. Adv. Parasitol.* **63:**367–439.

Pozio, E., and D. S. Zarlenga. 2005. Recent advances on the taxonomy, systematics and epidemiology of *Trichinella. Int. J. Parasitol.* **35:**1191–1204.

Pozio, E., G. La Rosa, P. Rossi, and K. D. Murrell. 1992. Biological characterization of *Trichinella* isolates from various host species and geographical regions. *J. Parasit.* **78:**647–653.

Procop, G. W., A. M. Marty, D. N. Scheck, D. R. Mease, and G. M. Maw. 2000. North American paragonimiasis. A case report. *Acta Cytol.* **44:**75–80.

Remington, J. S., R. McLeod, and G. Desmonts. 1995. Toxoplasmosis. *In* J. Remington and J. O. Klein (ed.), *Infectious Diseases of the Fetus and Newborn Infant,* 4th ed. W. B. Saunders Company, Philadelphia, PA.

Renold, C., A. Sugar, J. P. Chave, L. Perrin, J. Delavelle, G. Pizzolato, P. Burkhard, V. Gabriel, and B. Hirschel. 1992. *Toxoplasma* encephalitis in patients with the acquired immunodeficiency syndrome. *Medicine* **71:**224–239.

Rodero, M., C. Cuellar, S. Fenoy, C. del Aguila, T. Chivato, J. M. Mateos, and R. Laguna. 2006. ELISA antibody determination in patients with anisakiosis or toxocariosis using affinity chromatography purified antigen. *Allergy Asthma Proc.* **27:**422–428.

Rodriguez, R., A. A. Agostini, S. M. Porto, A. J. Olivaes, S. L. Branco, J. P. Genro, A. C. Laitano, R. L. Maurer, and C. Graeff-Teixeira. 2002. Dogs may be a reservoir host for *Angiostrongylus costaricensis. Rev. Inst. Med. Trop. Sao Paulo* **44:**55–56.

Roghmann, M. C., C. T. Faulkner, A. Lefkowitz, S. Patton, J. Zimmerman, and J. G. Morris, Jr. 1999. Decreased seroprevalence for *Toxoplasma gondii* in Seventh Day Adventists in Maryland. *Am. J. Trop. Med. Hyg.* **60:**790–792.

Rojekittikhun, W., J. Waikagul, and T. Chaiyasith. 2002. Fish as the natural second intermediate host of *Gnathostoma spinigerum. Southeast Asian J. Trop. Med. Public Health* **33**(Suppl. 3):63–69.

Rusnak, J. M., and D. R. Lucey. 1993. Clinical gnathostomiasis: case report and review of the English-language literature. *Clin. Infect. Dis.* **16:**33–50.

Sarti, E., P. M. Schantz, G. Avila, J. Ambrosio, R. Medina-Santillan, and A. Flisser. 2000. Mass treatment against human taeniasis for the control of cysticercosis: a population-based intervention study. *Trans. R. Soc. Trop. Med. Hyg.* **94:**85–89.

Schmidt, G. D., and L. S. Roberts. 2005a. Digeneans, p. 263–291. *In* L. S. Roberts and J. Janovy, Jr. (ed.), *Foundations of Parasitology,* 7th ed. McGraw-Hill, Boston, MA.

Schmidt, G. D., and L. S. Roberts. 2005b. Order Cyclophyllidea, p. 345–350. In L. S. Roberts, and J. Janovy, Jr. (ed.), Foundations of Parasitology, 7th ed. McGraw-Hill, Boston, MA.

Sithithaworn, P., and M. Haswell-Elkins. 2003. Epidemiology of Opisthorchis viverrini. Acta Trop. 88:187–194.

Slifko, T. R., H. V. Smith, and J. B. Rose. 2000. Emerging parasite zoonoses associated with water and food. Int. J. Parasitol. 30:1379–1393.

Smith, K. E., J. J. Zimmerman, S. Patton, G. W. Beran, and H. T. Hill. 1992. The epidemiology of toxoplasmosis on Iowa swine farms with an emphasis on the roles of free-living mammals. Vet. Parasitol. 42:199–211.

Sroka, J. 2001. Seroepidemiology of toxoplasmosis in the Lublin region. Ann. Agric. Environ. Med. 8:25–31.

Szostakowska, B., P. Myjak, M. Wyszynski, H. Pietkiewicz, and J. Rokicki. 2005. Prevalence of anisakin nematodes in fish from Southern Baltic Sea. Pol. J. Microbiol. 54(Suppl.):41–45.

Tenter, A. M., A. R. Heckeroth, and L. M. Weiss. 2000. Toxoplasma gondii: from animals to humans. Int. J. Parasit. 30:1217–1258.

Teutsch, S. M., D. D. Juranek, A. Sulzer, J. P. Dubey, and R. K. Sikes. 1979. Epidemic toxoplasmosis associated with infected cats. N. Engl. J. Med. 300:695–699.

Ugenti, I., S. Lattarulo, F. Ferrarese, A. De Ceglie, R. Manta, and O. Brandonisio. 2007. Acute gastric anisakiasis: an Italian experience. Minerva Chir. 62:51–60.

USDA. 2007. Part I. Reference of swine health and management. Practices in the United States, 2006, National Health Monitoring System. Publication no. N475.1007. USDA, APHIS, VS, Centers for Epidemiology and Animal Health, Fort Collins, CO.

USDA. 2001. Part I. Reference of swine health and management in the United States, 2000, National Health Monitoring System. Publication no. 338.0801. USDA, APHIS, VS, Centers for Epidemiology and Animal Health, Fort Collins, CO.

van Knapen, F. 2000. Control of trichinellosis by inspection and farm management practices. Vet. Parasitol. 93:385–392.

Velez, I. D., J. E. Ortega, and L. E. Velásquez. 2002. Paragonimiasis: a view from Columbia. Clin. Chest Med. 23:421–431, ix-x.

Wan, K. S., and W. C. Weng. 2004. Eosinophilic meningitis in a child raising snails as pets. Acta Trop. 90:51–53.

Warnekulasuriya, M. R., J. D. Johnson, and R. E. Holliman. 1998. Detection of Toxoplasma gondii in cured meats. Int. J. Food Microbiol. 45:211–215.

Webster, P., C. Maddox-Hyttel, K. Nöckler, A. Malakauskas, J. van der Giessen, E. Pozio, P. Boireau, and C. M. O. Kapel. 2006. Meat inspection for Trichinella in pork, horsemeat and game within the EU: available technology and its present implementation. Euro. Surveill. 11:50–55.

Wyss, R., H. Sager, N. Muller, F. Inderbitzin, M. Konig, L. Audige, and B. Gottstein. 2000. The occurrence of Toxoplasma gondii and Neospora caninum as regards meat hygiene. Schweiz. Arch. Tierheilkd. 142:95–108.

Zarlenga, D. S., B. M. Rosenthal, G. La Rosa, E. Pozio, and E. P. Hoberg. 2006. Post-Miocene expansion, colonization, and host switching drove speciation among extant nematodes of the archaic genus Trichinella. Proc. Natl. Acad. Sci. USA 103:7354–7359.

Pathogens and Toxins in Foods: Challenges and Interventions
Edited by V. K. Juneja and J. N. Sofos
© 2010 ASM Press, Washington, DC

Chapter 14

Human Pathogenic Viruses in Food

LEE-ANN JAYKUS AND BLANCA ESCUDERO-ABARCA

There are many viruses which have the potential to be transmitted by contaminated foods and subsequently cause disease in humans. Included among these are members of the following virus families: *Picornaviridae* (genera *Enterovirus, Parechovirus,* and *Kobuvirus* and hepatitis A virus [HAV]), *Caliciviridae* (genera *Norovirus* and *Sapovirus*), *Hepeviridae* (genus *Hepevirus*), *Reoviridae* (genus *Rotavirus*), *Astroviridae* (genus *Mamstrovirus*), and *Adenoviridae* (genus *Mastadenovirus*). Based on overall burden of disease, the human rotaviruses are probably the most significant of these agents. These viruses are an important cause of infant morbidity and mortality throughout the world, but they are usually spread by contaminated water and person-to-person contact, with food-borne transmission occurring infrequently. On the other hand, food-borne transmission of HAV and the noroviruses (NoV) is well documented. In fact, the NoV are recognized as a leading cause of food-borne disease in most of the developed world. The other viruses cited above probably cause food-borne disease, but the significance of foods in their transmission is poorly understood and likely to be less important than other routes of exposure. There are also some viruses for which food-borne transmission may be plausible (but not proven), including the Nipah virus (family *Paramyxoviridae*, genus *Henipavirus*), highly pathogenic avian influenza virus strain H5N1 (family *Orthomyxoviridae*, genus *Influenzavirus A*), and the virus causing severe acute respiratory syndrome (SARS) (family *Coronaviridae*, genus *Coronavirus*). This chapter will focus primarily on those viruses for which food-borne transmission is well documented (HAV and NoV), and therefore, control in food production, processing, and preparation is considered relevant. The other agents will be discussed when pertinent. Additional comprehensive reviews are provided by D'Souza et al. (2007), Koopmans and Duizer (2004), and Papafragkou et al. (2006).

CHARACTERISTICS OF THE FOOD-BORNE VIRUSES

Most of the viruses that can be transmitted by contaminated foods are small (20 to 40 nm in diameter) and of simple structure, containing a protein capsid and one (or more) nucleic acid molecules, usually in the form of single-stranded RNA. Because they lack a lipid envelope, which is an important reason for their environmental persistence and resistance, these are referred to as nonenveloped viruses. General information about structural, molecular, and environmental properties of those viruses specifically implicated in food-borne illness is detailed in Table 1.

Types of Illnesses

The viruses which can be transmitted by contaminated foods may be subdivided based on the target tissue for infection, i.e., those viruses infecting the gastrointestinal tract (enterotropic), the liver (hepatotropic), the nervous system (neurotropic), or the respiratory system (pneumotropic). The enterotropic viruses include the NoV, rotaviruses, sapoviruses, enteric adenoviruses, and others. HAV and hepatitis E virus (HEV) are hepatotropic, while the human enteroviruses and some of the important emerging viruses are neurotropic. The viruses that cause SARS and avian influenza cause pneumotropic disease.

Enterotropic viruses

NoV are an extremely common cause of acute viral gastroenteritis occurring both sporadically and

Lee-Ann Jaykus and Blanca Escudero-Abarca • Department of Food, Bioprocessing and Nutrition Sciences, North Carolina State University, Raleigh, NC 27695-7624.

Table 1. Human enteric viruses of epidemiological significance[a]

Virus family	Genus (type species)	Physical characteristics[b]	Disease syndrome(s)	Role of food-borne transmission
Picornaviridae	Enterovirus (poliovirus), Hepatovirus (HAV), Parechovirus (human parechovirus), Kabuvirus (Aichi virus)	Nonenveloped, icosahedral, featureless capsid; SS (+) sense RNA; 22–30 nm in diameter; Genome size of 7–8.8 kb; Members transmitted by food-borne routes; Usually resistant to extremes of pH, ionic conditions, organic solvents, and nonionic detergents; Most species have one or more cell culture-adapted strains	Human Enterovirus strains frequently cause asymptomatic or mild forms of gasteroenteritis, meningitis, encephalitis, myelitis, myocarditis, and conjunctivitis; HAV causes relatively mild form of acute hepatitis, usually self-limiting; symptoms and severity increase with age; Parechovirus respiratory and gastrointestinal symptoms in young children, with occasional infection of the central nervous system; Kabuvirus causes gastroenteritis in humans	Credible but not well documented for Enterovirus; Well documented for HAV; Credible but not well documented for Parechovirus; Recently documented for Kabuvirus (Aichi virus) via consumption of molluscan shellfish
Caliciviridae	Norovirus (Norwalk virus), Sapovirus (Sapporo virus)	Nonenveloped, icosahedral with cup-like depressions on capsid surface; SS (+) sense RNA; 27–40 nm in diameter; Genome size of 7.4–8.3 kb; Many strains stable in presence of organic solvents and at extremes of pH and ionic conditions; Human strains are nonculturable; cultivable surrogates include murine norovirus (MNV-1) and FCV	Norovirus causes a common form of acute viral gastroenteritis which is usually self-limiting; Sapovirus causes sporadic outbreaks and cases of gastroenteritis, also self-limiting	Well documented for Norovirus (genus forms a phylogenetic clade of five genogroups; genogroups I and II cause disease in humans, with genogroup II being more prevalent); Documented for Sapovirus; epidemiological importance of food-borne transmission unknown
Astroviridae	Mamastrovirus (human astrovirus)	Nonenveloped, icosahedral, spherical with star-like capsid appearance; SS (+) sense RNA; 28–30 nm in diameter; Genome size of 6.4–7.4 kb; Resistant to low pH, organic solvents, and nonionic, anionic, and zwitterionic detergents	Human astroviruses cause acute gastroeneritis in children and the immunocompromised	Credible but not well documented
Hepeviridae	Hepevirus (HEV)	Nonenveloped, icosahedral capsid; SS (+) sense RNA; 27–34 nm in diameter; Genome size of 7.2 kb; No reliable cell culture-adapted strain available	Causes outbreaks and cases of enterically transmitted acute hepatitis which are usually self-limiting, but pregnant women are at increased risk for mortality	Transmitted primarily by waterborne routes; food-borne transmission; credible but not yet documented
Reoviridae	Rotavirus (rotavirus A)	Nonenveloped, three-layered icosohedral capsid with wheel-like appearance; DS RNA; 80–100 nm in diameter; 11 Segments ranging from 11–125 kb each; Cell culture-adapted model strain available; Stable between pH 3 and 9, resistant to organic solvents and nonionic detergents; Infectivity lost by treatment with chlorine and 95% ethanol	Five species (types A–E); type A causes gastroenteritis and dehydration in infants <24 mo, milder disease in older children, life-threatening diarrhea in the malnourished; Type B most often associated with large epidemics involving human children and adults in China and causes severe gastroenteritis	Food-borne outbreaks documented but rare
Adenoviridae	Mastadenovirus (human enteric adenovirus, specifically types 40 and 41)	Nonenveloped, icosohedral capsid with protruding fibers; DS DNA, segmented; 70–90 nm in diameter; 30–37 kb, ranging from 11–125 kb each; Cell culture-adapted model strain available	Causes acute infantile gastroenteritis, second in prevalence to that caused by rotaviruses	Credible but not documented

[a] Information derived from the work of Fauquet et al. (2005).
[b] SS, single-stranded; DS, double-stranded.

as outbreaks. The disease caused by this genetically and antigenically diverse virus genus is characterized by nausea, vomiting (in >50% of patients), diarrhea, and abdominal pain with occasional headache and low-grade fever. Incubation periods range from 12 to 48 hours, with a mean of 24 hours, and illness usually lasts no more than 48 to 72 hours. The virus is spread rapidly between individuals in close contact, and outbreaks under such circumstances frequently have attack rates ranging from 30% to 80%. In most instances, severe illness or hospitalization is uncommon, except in children, the elderly, or the immunocompromised, where rehydration therapy may be necessary. Deaths have occurred, but rarely. Asymptomatic NoV infections are documented (Gallimore et al., 2004; Lindesmith et al., 2003), and virus can be shed in the stool both before symptoms occur and for at least 3 weeks after symptoms have abated (Rockx et al., 2002). The interested reader is referred to Green et al. (2001) and Atmar and Estes (2006) for a more comprehensive review of the human caliciviruses.

Aichi virus (genus *Kobuvirus*) was first reported in 1989 in association with an outbreak of classic nonbacterial gastroenteritis (diarrhea, nausea, and vomiting) caused by the consumption of contaminated oysters, and instances of shellfish-borne disease continue to occur. Prior to 2006, Aichi virus was isolated exclusively from individuals residing or traveling in Southeast Asia (Yamashita et al., 2000). More recently, the occurrence of this virus has been documented in clinical samples originating from Brazil and Germany (Oh et al., 2006), and there appears to be high seroprevalence (50 to 75%) in human adult populations around the world (Yamashita, 1993), although the significance of this is unknown.

Rotavirus is a common cause of childhood diarrhea, with worldwide distribution, albeit it is more commonly transmitted by contaminated water rather than food. The incubation period for rotavirus infection is less than 48 hours, and virus replication appears to be restricted to mature enterocytes on the tips of the small intestinal villi. The clinical presentation is acute gastroenteritis, which usually begins with the sudden onset of vomiting and fever lasting 1 or 2 days, which is followed by watery diarrhea which persists for several additional days. The disease occurs most commonly in children less than 1 year of age in the developing world and in those between 1 and 2 years of age in the developed world. While generally self-limiting, more severe disease has been documented with malnourished or immunocompromised individuals. Because of limited access to rehydration therapy in the developing world, rotavirus may be responsible for up to 5% of deaths among children

less than 5 years of age; in developed countries, it is a major cause of hospitalization in young children. Effective second-generation vaccines are available. The interested reader is referred to Franco and Greenburg (2002) for a more comprehensive review of the rotaviruses.

Astroviruses cause a diarrheal illness of short duration, sometimes with vomiting and fever. Longer-lasting illness can occur, particularly in association with serotype 3 strains. The disease appears to have a worldwide distribution and affects mostly young children and the aged (Guix et al., 2005). Enteric adenoviruses (adenovirus types 40 and 41) cause a similar disease and are well recognized as a cause of gastroenteritis in children (Klein et al., 2006). In recent studies, enteric adenoviruses have been isolated from naturally contaminated molluscan shellfish, as reviewed by Jaykus (2007). Nonetheless, the role of food in the transmission of these enterotropic viruses is currently unknown.

Hepatotropic viruses

The incubation period for HAV infection is 14 to 50 days, with an average of 28 days. The disease presents itself in four phases: (i) viral replication in the absence of symptoms; (ii) a preicteric phase during which patients may experience flu-like symptoms of nausea, vomiting, and fatigue; (iii) the icteric phase, which presents as darkening of the urine, pale-colored stools, jaundice, and right upper-quadrant pain with hepatomegaly; and (iv) the convalescent phase, in which symptoms resolve and liver enzymes return to normal. The frequency of icteric disease increases with age; fewer than 10% of children younger than 6 years of age show symptoms of hepatitis, while symptoms occur in 40 to 50% of older children and in 70 to 80% of adults. In most cases, infection is mild and self-limiting; however, in older patients more severe disease is possible. Chronic sequelae are rare. The overall mortality rate for HAV infection is <1%, immunity is lifelong, and effective vaccines are available. For more details on HAV, see Hollinger and Emerson (2001) and Cuthbert (2001).

HEV causes a moderately severe illness usually occurring in young adults and very similar to that caused by HAV. Like rotavirus, waterborne transmission is common, and food-borne transmission has limited documentation. The incubation period ranges from 15 to 60 days (mean of 40 days), and the disease usually resolves on its own, i.e., with no chronic sequelae. Mortality rates range from 0.5 to 3% for most individuals, although the development of cholestatic jaundice with obstruction of bile ducts can occur, which can be fatal if adequate medical care is not accessible. Pregnant women appear more prone to infection with HEV, and in a proportion of these cases, the disease progresses to fulminant hepatic failure, which is frequently accompanied by disturbance of neurological function. As a result, a mortality rate of 15 to 25% is estimated for this population subgroup. As is the case for HAV, HEV infection occurs at a much higher frequency than does icteric disease, and these subclinically infected individuals may shed virus in the stool, which serves as a source of contamination to food, water, or inanimate objects (fomites). Further details of HEV infection are provided by Smith (2001).

Neurotropic and pneumotropic viruses

The evidence for food-borne transmission of viruses causing neurological or respiratory disease is limited. Members of the *Picornaviridae* family (poliovirus and human enteroviruses) may be transmitted by this route, although poliovirus has been all but eradicated in much of the world, while the human enteroviruses usually cause somewhat mild and self-limiting infections, which are not reportable in the United States and hence difficult to track epidemiologically (Khetsuriani et al., 2006). Zoonotic transmission is a concern for some of the neurotropic and pneumotropic viruses (i.e., highly pathogenic avian influenza H5N1, SARS virus, and Nipah virus), and this will be discussed in greater detail below.

SOURCES OF CONTAMINATION AND INCIDENCE IN THE ENVIRONMENT AND FOODS

Sources of Contamination and At-Risk Foods

Enteric viruses are transmitted almost exclusively by the fecal-oral route through contact with human fecal material. Up to 10^8 virions/g (10^{11} virions/g for rotavirus) can be shed in the feces of clinically infected individuals, with lower numbers shed during pre- and postsymptomatic periods and in subclinical infection (Koopmans and Duizer, 2004). An important exception occurs in the case of NoV, which can also be shed in vomitus. In fact, as many as 30 million virus particles can be liberated from a single vomiting episode and aerosolization and deposition of these viruses serve as a source of contamination and subsequent transmission (Caul, 1994).

Although human fecal matter and vomitus (for NoV) are the definitive sources of contamination, specific transmission routes can be identified as direct (i.e., between individuals) or indirect (via consumption of contaminated food or water or contact with fomites) (Figure 1). Three types of food commodities are usually associated with viral disease outbreaks,

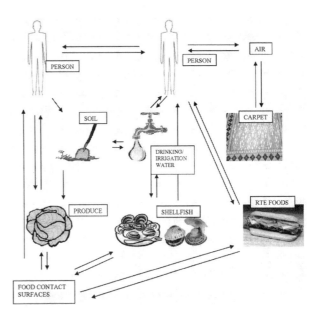

Figure 1. Transmission routes of food-borne viruses (reprinted with permission [D'Souza et al., 2007]).

those being (i) molluscan shellfish contaminated during production; (ii) fresh produce items contaminated during production, harvesting, or packing; and (iii) prepared foods contaminated during preparation. Food items specifically associated with recent outbreaks include raw molluscan shellfish, ice, fresh produce (green onions and raspberries), frozen produce (strawberries), and various prepared products, including salads, sandwiches, deli ham, and bakery items, as reviewed by Papafragkou et al. (2006).

Contamination of marine waters with human sewage is the critical factor causing virus contamination of molluscan shellfish (e.g., mussels, clams, cockles, and oysters). Sources of contamination in shellfish harvest waters include the illegal dumping of human waste, failing septic systems along shorelines, sewage treatment plants overloaded with storm water, and discharges of treated and untreated municipal wastewater and sludge (Jaykus, 2007). Molluscan shellfish are of particular concern because they are filter feeders and can concentrate microorganisms in their gut during the process of feeding. This is further complicated by the fact that enteric viruses are environmentally persistent and that shellfish are frequently consumed raw or only lightly cooked (Jaykus, 2007).

Most virus contamination of fresh produce occurs before the product reaches food service establishments (Koopmans and Duizer, 2004). Produce can become contaminated with viruses through the use of contaminated irrigation waters or when using contaminated water to reconstitute or dilute pesticides and herbicides. The use of improperly decontaminated sewage effluent for fertilization or conditioning can also be a source of virus contamination. Many of the produce items associated with viral food-borne disease outbreaks receive extensive human handling (berries and green onions) during harvest, and infected food handlers are probably a major source of contamination in these cases. Production in regions of the world in which HAV is endemic (Central and South America) may also be important, as small children accompanying parents working in fields can provide a source of fecal contamination which eventually reaches the product. Unfortunately, trace-back of produce items to the point at which contamination occurred is complicated and oftentimes impossible.

Poor personal hygiene practices of infected food handlers provide the source of contamination for prepared foods. This appears to be the most important factor influencing the food-borne transmission of enteric viruses, as U.S. Centers for Disease Control and Prevention (CDC) data indicate that 50 to 95% of viral food-borne disease outbreaks are attributable to poor personal hygiene of infected food handlers (Bean et al., 1997). Hands may become grossly contaminated by direct contact with fecal matter or can indirectly transfer viruses between surfaces by contact. The risk posed by infected food handlers depends on many factors, including phase of clinical infection, degree of virus shedding, hand hygiene habits, efficacy of virus removal and transfer, efficiency of virus transfer, and persistence.

Zoonotic Transmission

There are several viruses for which zoonotic transmission is recognized and for which food-borne transmission, although not yet proven, may be possible. Nipah virus, which presents as severe encephalitis with high mortality, serves as an excellent example. Bats are the natural reservoir for this virus, and they can transmit to pigs. From 1998 to 1999, clusters of human disease in Southeast Asia were noted with pig farmers and with persons working in abattoirs, leading to the conclusion that close contact with pigs is the primary source of human exposure to Nipah virus (Bellini et al., 2002). The recent emergence of the SARS coronavirus and the highly pathogenic avian influenza A H5N1 subtype in humans exposed to live poultry has also caused concern. Although the potential role for food-borne transmission cannot be ignored, it currently appears that on the rare occasions when humans are infected by these viruses, transmission occurs by direct contact with infected animals. Nonetheless, there remains significant concern that these viruses may eventually adapt, becoming more readily disseminated between individuals within human populations (Heeney, 2006).

Of greater concern may be the mounting evidence of animal reservoirs for both HEV and the

NoV. For example, HEV strains have been isolated from swine throughout the world, and there appears to be a high degree of genetic relatedness between swine and human isolates (Papafragkou et al., 2006). The presence of HEV RNA in commercial pig livers in the United States and the demonstration that positive liver homogenates could cause infection in intravenously inoculated pigs support a potential role for food-borne transmission of this virus, although further investigation is warranted (Feagins et al., 2007).

The swine enteric caliciviruses are genetically related to human NoV and actually form a distinct cluster in genogroup II of the *Norovirus* genus. In addition, a recent study demonstrated that swine NoV are antigenically related to the human genogroup II NoV, and there is a relatively high prevalence of antibodies to both swine and human NoV in pigs originating from the United States and Japan (Farkas et al., 2005). Further, the RNA corresponding to swine NoV has been identified with oysters harvested from various U.S. estuarine sites (Costani et al., 2006). The genetic diversity of the NoV is likely due to the high error rates and lack of proofreading mechanisms for the RNA-dependent RNA polymerase (Okamoto, 2007). As such, investigators are concerned about the potential for the emergence of animal strains, which can more easily cross the species barrier into human populations. There is also a concern that recombination events may result in virus strains of increased virulence. This is supported by the recent finding of recombinant NoV in clinical specimens derived from pediatric patients in Japan, with evidence of both intergenogroup and intersubgenotype recombination (Gia et al., 2007; Phan et al., 2006, 2007).

Incidence and Burden of Disease

NoV remain an extremely important cause of enteric disease. Nearly a decade ago, Mead et al. (1999) suggested that enteric viruses, particularly the NoV, are the leading cause of food-borne disease in the United States. Later, Widdowson et al. (2005) estimated that 50% of all food-borne disease outbreaks in the United States would be associated with the NoV if all specimens implicated in disease were screened for viruses. Unfortunately, there are a limited number of studies estimating the incidence of NoV infection, in part because these viruses cause a self-limiting disease for which medical treatment is frequently not sought, there is a lack of clinical diagnostic methods and/or thorough investigation of outbreaks, and very few population-based studies have been undertaken to date.

Another complicating factor is that, unlike many of the causes of bacterial food-borne illness, attributing the proportion of NoV disease caused specifically by food-borne transmission is difficult because these viruses are transmitted by a variety of routes. In fact, it is becoming evident that there are two major NoV transmission patterns, i.e., outbreaks occurring among individuals in close proximity (elder care facilities, health care settings, day cares, college campuses, and cruise ships) and food-borne transmission (Lopman et al., 2003a). There are only limited data on the proportion of disease that is actually attributable to food borne transmission. Fankhauser et al. (2002) reviewed 233 outbreaks of nonbacterial gastroenteritis reported to the CDC from 1997 to 2000 and identified that for 57% of these, the mode of transmission was food borne; for an additional 24%, the route of transmission was undetermined. In Europe, Lopman et al. (2003b) compiled data from 10 surveillance systems in the European Foodborne Virus Network, finding NoV to be the cause of >85% of all acute nonbacterial outbreaks of gastroenteritis reported between 1995 and 2000. Both the numbers and incidence of viral gastroenteritis differed by country, probably due to differences in surveillance systems. In general, European estimates of the proportion of viral gastroenteritis cases attributable to food-borne transmission range from 7% to 24%, although higher numbers have been reported (Lopman et al., 2003b).

Over the last decade or so, there has been an unprecedented increase in the annual incidence of NoV infection in the United States and the European Union. This appears to be associated with the emergence of genogroup II.4 strains, with cases of disease predominating in older individuals under institutional care (CDC, 2007; Lopman et al., 2004). It is not known whether these strains are more easily transmissible, have greater virulence, or are infecting a population with increased susceptibility, but deaths have been reported.

Although HAV infection occurs globally, it is considered highly endemic in the poorest areas of the world, where it is transmitted because of a general lack of adequate sanitary conditions. In much of Asia, Africa, and Central and South America, up to 95% of the population is infected before the age of 5 years, and most of these infections are subclinical. By the time adulthood is reached, a large proportion of the population is immune, and hence outbreaks rarely happen in these regions. In the developed world, endemicity is low and disease is relatively rare, so, much of the adult population is susceptible. In these areas, food-borne outbreaks can occur with relative ease. Before access to vaccination, it was estimated that 250,000 to 300,000 cases of HAV infection occurred annually in the United States. The greatest

risk factors for HAV infection in the United States and Europe are male homosexual behavior, international travel (most often to Central and South America), membership in certain ethnic groups (Hispanic, American Indian, and native Alaskan), and household contact with other infected individuals (Cuthbert, 2001; Wasley et al., 2006). Only a small percentage (<5%) of cases appear to be caused by food or waterborne exposures (Fiore, 2004; Wasley et al., 2006). Nonetheless, some very large outbreaks have occurred in the last decade (Hutin et al., 1999; Wheeler et al., 2005).

INTRINSIC AND EXTRINSIC FACTORS THAT AFFECT SURVIVAL IN FOOD PRODUCTS AND CONTRIBUTE TO OUTBREAKS

Intrinsic and extrinsic parameters commonly used by food processors include manipulation of temperature (cooking and heating, freezing, and refrigeration), water activity, pH, gaseous environment, natural and intentionally added inhibitors, and the presence of competitive microflora. Since enteric viruses cannot grow in foods, manipulation of the gaseous environment or competitive microflora is of little value. Currently, no "generally recognized as safe" substances with notable antiviral properties have been identified. Further, there is an inverse relationship between maintenance of virus infectivity and reduced temperature. For example, Pirtle and Beran (1991) demonstrated that enteroviruses artificially inoculated onto the surface of uncooked vegetables retained infectivity for 1 week and 1 month at room temperature and 10°C, respectively. Kingsley and Richards (2003) were able to detect infectious HAV for up to 3 weeks after artificial contamination of oysters. Freezing (at temperatures of <-20°C and preferably -80°C) is used to preserve enteric virus stock solutions, which can remain infectious for decades under these conditions. In fact, multiple freeze-thaw cycles may result in some loss of virus infectivity, but it is usually minimal.

Of particular note is the stability of enteric viruses at extremes of pH. For example, in simple cell culture studies, it has been shown that HAV can retain its infectivity at pH 1 for up to 5 hours; the human enteroviruses are stable at pH values as low as 3 (Pallansch and Roos, 2001). In foods, Hewitt and Greening (2004) demonstrated that HAV remained infectious in acidic marinade (pH ~3.75) when kept at 4°C for 4 weeks. Cannon et al. (2006) demonstrated the stability of murine NoV (a human NoV surrogate) at pH values ranging from 2 to 4, with infectivity reductions ranging from 0.5 to 1.0 $\log_{10}$, even after exposure to reduced pH for 2 hours. Epidemiological evidence also supports prolonged survival of NoV at low pH. In particular, early human challenge studies demonstrated infection in volunteers fed Norwalk virus which had been exposed to a pH of 2.7 for 3 hours (Dolin et al., 1972). A viral gastroenteritis outbreak associated with the consumption of contaminated orange juice has been documented (Fleet et al., 2000).

While control of water activity and relative humidity may have some impact on enteric virus persistence, epidemiological evidence suggests that most of these viruses are environmentally stable. In recent years, it has become more routine to process environmental swabs for detection of viral RNA by nucleic acid amplification when evaluating infection control strategies undertaken after NoV outbreaks in institutional settings (Liu et al., 2003). In these cases, viral RNA can be detected for weeks after removal of the source of contamination. Besides relative humidity, many other factors can influence the stability of the enteric viruses on surfaces and hands, including temperature, degree of inoculum drying, type of suspending medium (fecal or otherwise), virus type, and the type of surface contaminated (Abad et al., 1994; Mbithi et al., 1991). For HAV, Mbithi et al. (1991) observed a half-life of 7 days at relatively low humidity and 5°C, but a half-life of 2 hours at high humidity and 35°C. Doultree et al. (1999) reported that the NoV surrogate feline calicivirus (FCV) remained infectious on glass surfaces for over 60 days at 4°C, for 14 to 28 days at 25°C, and for 1 to 10 days at 37°C. HAV appears to be less stable on human skin than it is on surfaces; however, as much as one-third of the initial virus inoculum could be detected on finger pads 4 hours postinoculation (Mbithi et al., 1992), and similar results have been reported for rotavirus (Ansari et al., 1988). Transfer of viruses between surfaces occurs with relative ease, although its efficiency is affected by virus type, surface type, friction, pressure, and degree of drying of the virus inoculum. At its highest, investigators have documented as much as 10 to 50% of the initial virus inoculum transferred in a single tactile action, although transfer is likely to be somewhat less efficient in natural situations (Bidawid et al., 2000a, 2004; Mbithi et al., 1992). Further, the degree of virus transfer is quite variable, even when evaluated under highly controlled conditions. Nonetheless, the combined effect of environmental persistence and ease of transfer makes for a potentially dangerous situation if viruses are present in food production, processing, or preparation environments in which humans are handling food.

FOOD PROCESSING OPERATIONS THAT INFLUENCE VIRUS CONTAMINATION

Molluscan Shellfish

Molluscan shellfish may be consumed raw or cooked and do not usually receive any type of processing treatment prior to reaching the retail sector of the farm-to-fork continuum. Control of viral contamination in this product has been complicated by the fact that the numbers of fecal coliforms, a widely used indicator of the sanitary quality of shellfish and their harvesting waters, bear little relationship to the presence (or absence) of enteric viruses. There are several reasons for this, the most important of which is the extended persistence of viruses relative to the gram-negative indicator bacteria. Alternative indicators, including bacteriophages (F-specific coliphages, somatic coliphages, and phages of *Bacteroides fragilis*), enteroviruses, and adenoviruses, have been proposed (Lees, 2000; Muniain-Mujika et al., 2002). While bacteriophages tend to display persistence similar to that of the enteric viruses, the results of studies evaluating their usefulness as indicators for naturally occurring viral contamination of shellfish or their harvesting waters are inconclusive (Papafragkou et al., 2006). From the standpoint of other viral indicators, Beuret et al. (2003) demonstrated no significant correlation between the presence of human enteroviruses (as indicators) and NoV in estuarine waters in Switzerland. On the other hand, the levels of the human adenoviruses, which are environmentally stable, easy to detect, and appear to be more abundant than other viruses (Pina et al., 1998), have been found to correlate with the presence of other human enteric viruses in European shellfish harvesting waters (Formiga-Cruz et al., 2002). Until additional studies are completed, the fecal coliforms will remain the indicator of choice, meaning that the only definitive way to ensure that estuarine environments used for molluscan shellfish harvesting are free of enteric virus contamination is to prevent contamination with untreated human sewage.

Candidate postharvest controls include controlled purification (depuration and relaying), irradiation, heat, and high-pressure processing. Depuration and relaying are used extensively around the world and have been quite successful in reducing enteric bacterial illness associated with shellfish consumption. However, controlled purification cannot be relied upon to eliminate enteric viruses, particularly HAV, as well as the environmental *Vibrio* species (Papafragkou et al., 2006). Likewise, the use of ionizing radiation at the low doses (<3 kGy) necessary to protect the organoleptic properties of the product results in minimal (1 to 2 $\log_{10}$) inactivation of HAV (Mallet et al., 1991).

Of the enteric viruses most often epidemiologically implicated in shellfish-associated food-borne disease, HAV is the most resistant to heat. Even so, the failure of several cooking methods (grilling, stewing, and frying) to prevent a large oyster-associated (NoV) gastroenteritis outbreak has been reported (McDonnell et al., 1997). A number of studies have been undertaken in an effort to provide recommendations for the cooking of molluscan shellfish to maintain a "safe" product (NACMCF, 2007). These have culminated in United Kingdom standards for commercial molluscan shellfish cooking operations, which are based on demonstrating a 4-$\log_{10}$ inactivation of HAV and stipulate holding at an internal temperature of at least 90°C for 1.5 minutes (Lees, 2000). It is likely that this time-temperature combination also will inactivate the human NoV (Cannon et al., 2006), and it was recently documented that HEV is more heat labile than is HAV (Emerson et al., 2005). However, many consumers find that this stringent time-temperature combination results in a product that is perceived as overcooked.

Produce

Viral contamination most likely occurs on the surface of produce items, but there are a few studies that suggest the potential for uptake and translocation of viruses by plant tissue (Chancellor et al., 2006; Seymour and Appleton, 2001). Currently, the best approach to preventing contamination of produce items is reliance on the principles of good agricultural practices, with emphasis on ensuring that waters used in production (for irrigation and pesticide application) are not contaminated with human sewage, that farm workers have adequate education about hygiene and ready access to on-site toilet and hand-washing facilities, and that young children are excluded from production, picking, and packing operations.

Many produce items are washed at multiple steps along the postharvest continuum. A general rule of thumb is that water washing alone can remove approximately 1 to 2 $\log_{10}$ of microbiological contaminants, including viruses, from the surface of produce items. However, this estimate depends on a variety of factors, like produce type and condition (including surface characteristics and presence of cuts or abrasions), washing conditions (water temperature and rubbing), virus type, and degree of initial contamination (Papafragkou et al., 2006). To date, no promising produce washes with specific virucidal activity are on the horizon. Although many novel produce disinfection methods (ionizing radiation and

ozone) have been proposed and even investigated, their use directly on fruits and vegetables is frequently prohibitive due to unacceptable organoleptic changes to the product and/or lack of federal approval for application to produce (Bidawid et al., 2000b).

Prepared and RTE Foods

Ready-to-eat (RTE) or prepared foods are defined as those products that are edible without washing, cooking, or additional preparation by the consumer or by the food service establishment. In general, this means that such foods are not subjected to a terminal heating step prior to consumption. Consequently, control strategies must rely on prevention of contamination. Virus-contaminated food contact surfaces and hands serve as the source of the contaminants, the latter appearing to be the more important. Therefore, stressing that food handlers practice adequate personal hygiene coupled with the availability of virucidal agents for treatment of hands and surfaces are critical to controlling the food-borne transmission of these agents. Ensuring that appropriate hygiene is being practiced by all food handlers is complicated, and even the best programs suffer from some degree of noncompliance. Food handlers should have adequate training about proper hygiene and the risks of food-borne pathogen transmission in association with poor personal hygiene, be provided with a "no bare hand contact" policy for the handling of prepared foods, be provided with a sick-leave policy that encourages them to remain at home while symptomatic for gastrointestinal illness, and be provided adequate training about decontamination practices in the event of a vomiting episode (USFDA, 2005). While transmission via contaminated food or water may sometimes be prevented or contained by strict adherence to these measures, the potential for person-to-person spread cannot be eliminated.

For hand and surface decontamination to be successful in controlling contamination with the food-borne viruses, three elements must be in place: (i) an effective disinfecting agent, (ii) adequate use instructions, and (iii) regular compliance with cleaning and disinfection regimens (Sattar et al., 2002). There is limited information relative to the antiviral efficacy of most commercial disinfectants when used for surface disinfection. As a general rule of thumb, the majority of chemical disinfectants used in both institutional and domestic environments do not effectively inactivate enteric viruses on surfaces when used at manufacturer-recommended exposures. Only products containing gluteraldehyde (2%), sodium hypochlorite (>1,000 ppm), and a quaternary ammonium compound containing 23% HCl appeared to be effective for the inactivation of HAV and NoV on surfaces, but only when used at high concentrations and for long exposure times, frequently exceeding those recommended by the manufacturer (Jean et al., 2003; Mbithi et al., 1990). The data for hand disinfection agents are similar. The widely used alcohol-based hand sanitizers rarely achieve even a 1-$\log_{10}$ reduction in virus titers and usually much less than that (Papafragkou et al., 2006). In fact, ethanol-based hand rubs are not as effective as water and soap, supporting the continued need for education about and compliance with hand-washing policies in this sector of the industry (Bidawid et al., 2004). For more information, see Papafragkou et al. (2006).

Other Products

Recent laboratory evidence supports the commonly held belief that highly pathogenic avian influenza H5N1 is effectively destroyed by heat. More specifically, the use of USDA Food Safety and Inspection Service (FSIS) time-temperature recommendations for the cooking of chicken meat should completely inactivate the virus, even if the product is highly contaminated (Thomas and Swayne, 2007). A 5-$\log_{10}$ inactivation of HAV was demonstrated for artificially inoculated fluid dairy products at 85°C in less than 0.5 minute, with increased fat content shown to be protective (Bidawid et al., 2000c). This same study demonstrated that normal high-temperature–short-time pasteurization conditions inactivated less than 1 $\log_{10}$ of HAV in fluid milk, but the relevance of this finding is limited due to the unlikely occurrence of high levels of HAV in raw milk. While no specific work has been reported on the degree of inactivation of enteric viruses after cooking vegetables (e.g., boiling, steaming, microwaving, and broiling), it is likely that complete inactivation would occur, as viral contamination should be located on or near the surface of the product and treatment time and temperature combinations are usually quite rigorous.

RECENT ADVANCES IN BIOLOGICAL, CHEMICAL, AND PHYSICAL INTERVENTIONS FOR VIRUS CONTROL

High-Pressure Processing

Over the last decade, there has been great interest in the use of high pressure to inactivate enteric viruses. Most of the studies done to date have used viruses suspended in tissue culture medium, with critical process parameters identified as treatment temperature, pressure, time, and matrix effects (Grove et al., 2006). There are no virus family-specific trends in terms of sensitivity to high pressure. For example,

four members of the *Picornavirideae* family (Aichi virus A846/88, cocksackieviruses B5 and A9, and human parechovirus–1) demonstrated widely differing degrees of inactivation when exposed to pressures between 400 and 600 MPa. Specifically, coxsackievirus B5 and Aichi virus A846/88 were completely resistant to pressures as high as 600 MPa for 5 minutes, as evaluated by an infectivity assay. On the other hand, reductions exceeding 3 and 7 $\log_{10}$ were observed for coxsackievirus A9 after treatment at 400 and 600 MPa, respectively; close to 5 $\log_{10}$ of human parechovirus–1 was inactivated after 5 minutes of exposure to 600 MPa of pressure (Kingsley et al., 2004).

In similar studies, it was demonstrated that HAV and FCV (a NoV surrogate) suspended in tissue culture medium were eliminated after exposure to 450 MPa for 5 minutes and 275 MPa for 5 minutes, respectively (Kingsley et al., 2002). Later, Kingsley et al. (2006) reported that for HAV, treatment at elevated temperatures enhanced pressure efficacy, particularly at pressures of 350 MPa and above. Specifically, at 350 MPa and 450 MPa, the efficacy of pressure treatment improved by approximately 1.5 $\log_{10}$ and almost 4 $\log_{10}$, respectively, as temperature was increased from 5°C to 50°C. An improved efficacy of high pressure at both low (below 0°C) and high (above 50°C) temperature was also observed for FCV (Chen et al., 2005). When applied to murine NoV, probably the most relevant surrogate for human NoV strains, 4-$\log_{10}$ reductions were obtained at 400 MPa and 6.85-$\log_{10}$ reductions were achieved at 450 MPa when both treatments were done for 5 minutes at 30°C (Kingsley et al., 2007). In most instances, high-pressure inactivation kinetics cannot be described using conventional log-linear functions but are better expressed using models such as the Weibull, because of a significant tailing effect (Kingsley et al., 2006, 2007).

The exact mechanism of high-pressure inactivation of viruses has not been elucidated. The most compelling hypothesis is that inactivation is a consequence of pressure-induced changes to the integrity of the protein capsid. In early work with HAV, RNase protection experiments suggested that capsids may remain intact after pressure treatment and that loss of infectivity was due to more subtle changes to the capsid which might prevent virus attachment to the host cell or blockage of penetration of the viral nucleic acid (Kingsley et al., 2002). The enhanced inactivation of some viruses at low temperature may occur due to the phenomenon of cold denaturation (Grove et al., 2006). In addition, Chen et al. (2005) pointed out that temperature denaturation curves for proteins under pressure display second-order kinetics, consistent with the temperature effect observed in many high-pressure inactivation studies done with enteric viruses. Finally, when reverse transcription (RT)-PCR was done on pressure-inactivated HAV, amplicons could still be detected, suggesting that the integrity of the viral RNA was maintained even after pressure treatment (Kingsley et al., 2002).

A few studies have confirmed that high hydrostatic processing does indeed inactivate enteric viruses in the oyster matrix. Specifically, Calci et al. (2005) demonstrated that bioconcentrated HAV in shucked oysters was reduced by >1, 2, and 3 $\log_{10}$ when treated at 350, 375, and 400 MPa for 1 minute at a temperature of approximately 10°C. The log-linear inactivation curves were slightly less steep than those obtained for HAV suspended in cell culture media treated under the same pressure conditions, but for 5 minutes. In similar experiments, oysters that were allowed to bioaccumulate murine NoV and were then shucked and exposed to 350, 375, and 400 MPa for 5 minutes at 5°C demonstrated infectivity reductions of approximately 2 (350 and 375 MPa) and 4 (400 MPa) $\log_{10}$ (Kingsley et al., 2007). Although encouraging, particularly because the process produces a "raw-like" product, much more work will be necessary to validate high-pressure processing in whole-shell oysters, at various levels of salinity (which can be protective [Kingsley et al., 2002]) and against a panel of different human NoV strains. Interestingly, high pressure was recently applied to the inactivation of HAV in fruit puree and sliced green onions, with a >4-$\log_{10}$ inactivation documented at a pressure of 375 MPa for 5 minutes (Kingsley et al., 2005). However, the practical significance of using high pressure to control viral contamination of fresh produce remains to be seen.

Vaccination

In 1998, the first rotavirus vaccine was licensed for use in the United States, and it was quickly pulled from the market due to a reported increased risk of intussusception, or bowel obstruction (1/12,000 vaccinated infants) (Cunliffe et al., 2002). In 2006, two live attenuated rotavirus vaccines became available with no documented increase in the risk of bowel obstruction. Rotatrix (GlaxoSmithKline) is a monovalent live attenuated two-dose oral vaccine and RotaTeq (Merck and Co., Inc.) is a pentavalent reassortant live and attenuated three-dose oral vaccine. Both are licensed for use in the United States, United Kingdom, and many countries in Latin America, Asia, and Africa, and both appear to be efficacious, particularly in preventing severe disease. Studies are ongoing to evaluate the efficacy of these vaccines in the developing world,

where the burden of disease due to rotavirus infection is extensive (Girard et al., 2006).

Four inactivated HAV vaccines are currently available internationally; all four are given parenterally as a two-dose series. The vaccines have equal efficacy (94 to 100%), with provision of immunity for 20 years or more (Fiore, 2004). In the United States, HAV vaccination has been limited to high-risk groups and young children. Routine vaccination of food handlers is not currently recommended, although there is substantial debate about the cost-benefit relationship associated with such a policy. There is no NoV vaccine available, and the unique features of these viruses provide substantial technical difficulties in developing effective vaccines. Specifically, a high degree of antigenic and genetic variability, in conjunction with a lack of culturability and short-lived immunity, presents challenges that may limit NoV vaccine development efforts. For further information about vaccines for the enteric viruses, see Girard et al. (2006).

Surveillance

Recently, the U.S. Council for State and Territorial Epidemiologists passed a resolution making all acute gastroenteritis outbreaks reportable on a national basis, effective in 2008. The U.S. CaliciNet centralized database, which is managed by the CDC, provides a means of collecting and comparing NoV sequences, which should aid in epidemiological tracking of existing and emerging strains, as well as in determining the proportion of disease attributable to food-borne transmission (CDC, 2007). In Europe, DIVINE-Net provides a transnational surveillance system for monitoring NoV outbreaks, particularly those associated with food, water, and environmental sources. The system relies on rapid strain characterization by nucleic acid sequencing and provides early warning for "emerging" strains. DIVINE-Net is also seeking to harmonize laboratory diagnostics and strain typing and establish a web-based reporting system. Similar efforts are underway in Australia and Canada.

Discriminative Detection Methods

Despite extensive efforts to develop robust methods to detect viral contamination in foods, this remains challenging (D'Souza et al., 2006). Because the numbers of virus particles present in contaminated foods are usually low and no universal or rapid culture-based methods are available, cultural enrichment is not an option. Consequently, the viruses must be concentrated and purified from the food matrix prior to the application of detection methodology, with molecular amplification the detection method of choice. Sequence-based determination of amplicon identity is used for further confirmation, particularly in the case of outbreaks.

Virus concentration and purification schemes are designed to reduce sample volume and remove at least some of the matrix, while simultaneously recovering most of the contaminating viruses. In order to achieve these goals, sample manipulations are undertaken that capitalize on the behavior of enteric viruses to act as proteins in solutions, to cosediment by simple centrifugation when adsorbed to larger particles, and to remain infectious at extremes of pH or in the presence of organic solvents. Many virus concentration and purification methods use manipulation of pH and ionic conditions to favor virus adsorption to, or elution from, the food matrix. This is then followed by relatively low-speed centrifugation, after which the virus-containing phase (either precipitate or eluate) is recovered for further purification. Other steps in the virus concentration and purification process may include various forms of filtration (crude filtration and ultrafil-

tration), ultracentrifugation, precipitation (achieved through the addition of polyethylene glycol, organic flocculants, or by manipulation of pH), organic solvent extraction (to remove matrix-associated lipids), ligand-bound magnetic separation (using immunobeads or cationic particles), and/or enzyme pretreatment (to break down matrix-associated organic matter, particularly complex carbohydrates). In almost all instances, virus extraction is done by combining two or more of these steps in a series. A good virus concentration and purification method will be relatively simple and result in sample volume reductions of 10- to 1,000-fold and recovery of >90% of the target virus. However, recovery efficiency is both virus- and matrix-specific, and such a high degree of recovery is rarely achieved. Recently, scientists have begun using extraction process controls, which allow them to monitor the efficiency of the virus extraction process (Costafreda et al., 2006; Dreier et al., 2006). The European Union is presently engaged in multinational validation of methods to

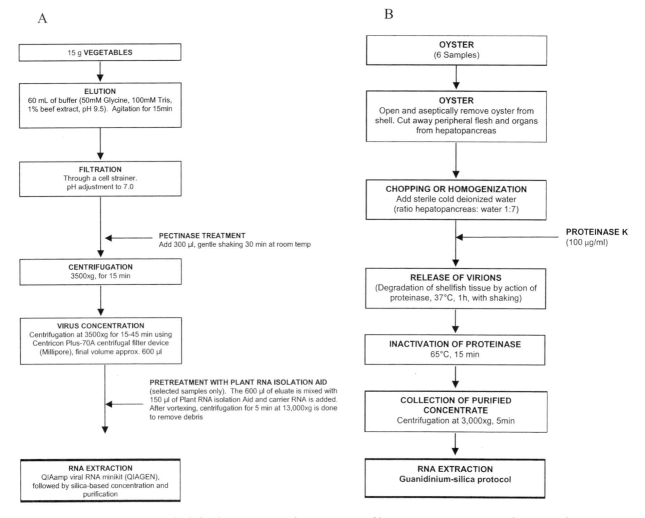

Figure 2. Representative methods for the extraction and concentration of human enteric viruses in produce (A) and oysters (B) based on protocols reported by Butot et al. (2007) and Jothikumar et al. (2005b), respectively.

concentrate viruses from fresh produce and molluscan shellfish (Figure 2). For all other commodities, there are no universally accepted virus extraction methods.

Although viral RNA can be released from capsids and made available for amplification using a simple heat treatment (99°C for 5 minutes), this is rarely done because, even with the best virus concentration method, residual matrix-associated components persist and oftentimes interfere with nucleic acid amplification. Therefore, an efficient nucleic acid extraction step is critical. The science of RNA extraction has developed rapidly over the last decade, and what was once a complex procedure is now much more simple and reliable. Nonetheless, whether using guanidinium isothiocyanate or column-based purification methods, matrix-associated inhibitors are difficult to eliminate, and it is often advisable to employ multiple RNA purification methods in tandem.

RT-PCR and nucleic acid sequence-based amplification are the two methods most often used for detection of viral RNA. The sensitivity and specificity of molecular amplification depend upon the efficiency of the upstream processing methods (virus concentration and purification and nucleic acid extraction), primer choice, and amplification conditions. The use of primers with low degeneracy and high melting temperature is recommended because this reduces the likelihood of nonspecific amplification, which is common when nucleic acids are extracted directly from a complex sample matrix. For HAV detection, broadly reactive primers with high annealing temperatures are available, usually corresponding to the VP1/2A junction or 5′ untranslated region of the viral genome (Jothikumar et al., 2005a). The high degree of genetic diversity for the NoV has made the development of broadly reactive primers difficult. Four "regions" of the NoV genome have been used for primer design (designated regions A, B, C, and D), but the ORF1-ORF2 junction (just downstream of region B) seems to be the most conserved and is frequently used for genogroup-specific detection (Jothikumar et al., 2005b; Kageyama et al., 2003). For strain comparison (as might be appropriate in outbreak investigation), primers corresponding to the NoV capsid region can be used, although these sometimes have a higher degree of degeneracy. It is recommended that multiple primer pairs be used when screening food samples for NoV contamination (Boxman et al., 2006, 2007), as one pair may perform better than another, depending on virus load and the matrix from which the sample was derived. In some instances, genotype-specific amplifications, such as for GII.4 strains, may be appropriate (Bull et al., 2006).

Since most groups screening for viral contamination of foods have moved to real-time amplification formats, confirmation and detection can theoretically be done at the same time. It is said that real-time assays can be made quantitative, but residual matrix-associated inhibitors frequently impact the reliable quantification of viral load in naturally contaminated food samples. Further, there is some debate as to whether an RNA or DNA standard is best when designing the standard curve for real-time quantification. While using a DNA standard provides a defined assessment of the efficiency of DNA amplification, it does not take into account an inefficient RT step, which may result in an underestimate of the number of viral genome copies in the sample.

When RT-PCR products are obtained from naturally contaminated foods, the amplicons must be sequenced for further confirmation and strain typing. Sequences derived from foods can be compared to those obtained from clinical specimens in an effort to make causal associations. Although this approach has been used successfully to link contaminated foods to outbreaks of viral gastroenteritis, more often than not it is impossible to achieve amplification of viral RNA from implicated foods (Rutjes et al., 2006). Historically, nested reactions have been used to increase the sensitivity of detection, particularly for naturally contaminated food products (Boxman et al., 2007).

Despite tremendous strides in virus detection, these protocols are applied infrequently and usually only in response to known or suspected food-borne disease outbreaks. The most important reasons for their limited use include the absence of universally applicable, collaboratively tested methods; high cost; and the need for highly trained personnel. In addition, care should be taken when interpreting both negative and positive test results. For example, while a sequence-confirmed positive test ensures the presence of the viral genome in the product, it does not necessarily confirm that the contaminant was infectious. Negative test results are fraught with interpretive challenges related to sampling, the potential for virus inactivation (particularly if extensive time occurs between sampling and testing), low levels of virus contamination, and the impact of the matrix on assay detection limits. In short, a negative test result cannot be relied upon to ensure the absence of virus contamination (Rutjes et al., 2006).

CONCLUDING REMARKS

Our understanding of the importance of enteric viruses in the overall burden of food-borne disease has increased dramatically over the past 20 years. Nonetheless, many unanswered questions remain. Specifically, more work is needed in efforts to develop effective virus inactivation methods, as applied to foods, surfaces, and hands; improved disease surveillance, including the collection of food attribution data; improved methods for the sensitive and specific detection of viruses in foods; and better characterization of viral food-borne disease risks. The field has been limited by the small number of scientists working with food-borne viruses, as well as the absence of effective in vitro culture methods, limiting accessibility to infectious virus strains. It does appear that this is changing, as more food microbiologists begin to tackle this fascinating and challenging subject. It is likely that the next decade will bring exciting new developments as we endeavor to understand and control these important causes of human food-borne disease.

REFERENCES

Abad, F. X., R. M. Pinto, and A. Bosch. 1994. Survival of enteric viruses on environmental fomites. *Appl. Environ. Microbiol.* 60:3704–3710.

Ansari, S. A., S. A. Sattar, V. S. Springthorpe, G. A. Wells, and W. Tostowaryk. 1988. Rotavirus survival on human hands and transfer of infectious virus to animate and nonporous inanimate surfaces. *J. Clin. Microbiol.* 26:1513–1518.

Atmar, R. L., and M. K. Estes. 2006. The epidemiologic and clinical importance of norovirus infection. *Gastroenterol. Clin. North Am.* 35:275–290.

Bean, N. H., J. S. Goulding, M. T. Daniels, and F. J. Angulo. 1997. Surveillance for foodborne disease outbreaks—United States, 1988–1992. *J. Food Prot.* 60:1265–1286.

Bellini, Q. J., P. A. Rota, and U. Parashar. 2002. Zoonotic paramyxoviruses, p. 845–855. *In* D. D. Richman, R. J. Whitley, and F. G. Hayden (ed.), *Clinical Virology*, 2nd ed. ASM Press, Washington, DC.

Beuret, C., A. Baumgartner, and J. Schluep. 2003. Virus-contaminated oysters: a three-month monitoring of oysters imported to Switzerland. *Appl. Environ. Microbiol.* 69:2292–2297.

Bidawid, S., N. Malik, O. Adegbunrin, S. A. Sattar, and J. M. Farber. 2004. Norovirus cross-contamination during food handling and interruption of virus transfer by hand antisepsis: experiments with feline calicivirus as a surrogate. *J. Food Prot.* 67:103–109.

Bidawid, S., J. M. Farber, and S. A. Sattar. 2000a. Contamination of foods by food handlers: experiments on hepatitis A virus transfer to food and its interruption. *Appl. Environ. Microbiol.* 66:2759–2763.

Bidawid, S., J. M. Farber, and S. Sattar. 2000b. Inactivation of hepatitis A virus (HAV) in fruits and vegetables by gamma irradiation. *Int. J. Food Microbiol.* 57:91–97.

Bidawid, S., J. M. Farber, S. A. Sattar, and S. Hayward. 2000c. Heat inactivation of hepatitis A virus in dairy foods. *J. Food Prot.* 63:522–528.

Boxman, I. L. A., J. J. H. C. Tilburg, N. A. J. M. te Loeke, H. Vennema, E. de Boer, and M. Koopmans. 2007. An efficient and rapid method for recovery of norovirus from food associated with outbreaks of gastroenteritis. *J. Food Prot.* 70:504–508.

Boxman, I. L. A., J. J. H. C. Tilburg, N. A. J. M. te Loeke, H. Vennema, K. Jonker, E. de Boer, and M. Koopmans. 2006. Detection of noroviruses in shellfish in the Netherlands. *Int. J. Food Microbiol.* 108:391–396.

Bull, R. A., E. T. V. Tu, C. J. McIver, W. D. Rawlinson, and P. A. Shite. 2006. Emergence of a new norovirus genotype II.4 variant associated with global outbreaks of gastroenteritis. *J. Clin. Microbiol.* 44:327–333.

Butot, S., T. Putallaz, and G. Sanchez. 2007. Procedure for rapid concentration and detection of enteric viruses from berries and vegetables. *Appl. Environ. Microbiol.* 73:186–192.

Calci, K. R., G. K. Meade, R. C. Tezloff, and D. H. Kingsley. 2005. High-pressure inactivation of hepatitis A virus within oysters. *Appl. Environ. Microbiol.* 71:339–343.

Cannon, J. L., E. Papafragkou, G. Park, J. Osborne, L. Jaykus, and J. Vinje. 2006. Surrogates for the study of norovirus stability and inactivation in the environment: a comparison of murine norovirus and feline calicivirus. *J. Food Prot.* 69:2761–2765.

Caul, E. O. 1994. Small round structured viruses: airborne transmission and hospital control. *Lancet* 343:1240–1241.

CDC. 2007. Norovirus activity—United States, 2006–2007. *MMWR Mor. Mortal. Wkly. Rep.* 56:842–846.

Chancellor, D. D., S. Tyagi, M. C. Bazaco, S. Bacvinskas, M. B. Chancellor, V. M. Dato, and F. DeMiguel. 2006. Green onions: potential mechanism for hepatitis A contamination. *J. Food Prot.* 69:1468–1472.

Chen, H., D. G. Hoover, and D. H. Kingsley. 2005. Temperature and treatment time influence high hydrostatic pressure inactivation of feline calicivirus, a norovirus surrogate. *J. Food Prot.* 68:2389–2394.

Costafreda, M. I., A. Bosch, and R. M. Pinto. 2006. Development, evaluation, and standardization of a real-time Taqman reverse transcription-PCR assay for quantification of hepatitis A in clinical and shellfish samples. *Appl. Environ. Microbiol.* 72:3846–3855.

Costani, V., F. Loisy, L. Joens, F. S. LeGuyader, and L. J. Saif. 2006. Human and animal enteric caliciviruses in oysters from different coastal areas of the United States. *Appl. Environ. Microbiol.* 72:1800–1809.

Cunliffe, N. A., J. S. Bresee, and C. A. Hart. 2002. Rotavirus vaccines: development, current issues and future prospects. *J. Infect.* 45:1–9.

Cuthbert, J. A. 2001. Hepatitis A: old and new. *Clin. Microbiol. Rev.* 14:38–58.

Dolin, R., N. R. Blacklow, H. DuPont, R. F. Buscho, R. G. Wyatt, J. A. Kasel, R. Hornick, and R. M. Chanock. 1972. Biological properties of Norwalk agent of acute infectious nonbacterial gastroenteritis. *Proc. Soc. Exp. Biol. Med.* 140:578–583.

Doultree, J. C., J. D. Druce, C. J. Birch, D. S. Bowden, and J. A. Marshall. 1999. Inactivation of feline calicivirus, a Norwalk virus surrogate. *J. Hosp. Infect.* 41:51–57.

Dreier, J., M. Stormer, D. Made, S. Burkhardt, and K. Kleesiak. 2006. Enhanced reverse transcription PCR assay for detection of norovirus genogroup I. *J. Clin. Microbiol.* 44:2714–2720.

D'Souza, D. H., C. L. Moe, and L. Jaykus. 2007. Foodborne viral pathogens, p. 581–607. *In* M. P. Doyle and L. R. Beuchat (ed.), *Food Microbiology: Fundamentals and Frontiers*, 3rd ed. ASM Press, Washington, DC.

D'Souza, D. H., J. Jean, and L. Jaykus. 2006. Methods for the detection of viral and parasitic protozoan pathogens in foods, p. 1–23. *In* Y. H. Hui (ed.), *Handbook of Food Science, Technology, and Engineering*, vol. 4. CRC-Taylor and Francis, Boca Raton, FL.

Emerson, S. U., V. A. Arankalle, and R. H. Purcell. 2005. Thermal stability of hepatitis E virus. *J. Infect. Dis.* 192:930–933.

Fankhauser, R. L., S. S. Monroe, J. S. Noel, C. D. Humphrey, J. S. Bresee, U. D. Parashar, T. Ando, and R. I. Glass. 2002. Epidemiologic and molecular trends of the "Norwalk-like viruses" associated with outbreaks of gastroenteritis in the United States. *J. Infect. Dis.* 186:1–7.

Farkas, T., S. Nakajima, M. Sugieda, X. Deng, W. Zhong, and X. Jiang. 2005. Seroprevalence of noroviruses in swine. *J. Clin. Microbiol.* **43**:657–661.

Fauquet, C. M., M. A. Mayo, J. Maniloff, U. Desselberger, and L. A. Ball. 2005. *Virus Taxonomy, Eighth Report of the International Committee on Taxonomy of Viruses.* Elsevier Academic Press, San Diego, CA.

Feagins, A. R., T. Opriessnig, D. K. Guenete, P. G. Halbur, and X.-J. Meng. 2007. Detection and characterization of infectious hepatitis E virus from commercial pig livers sold in local grocery stores in the USA. *J. Gen. Virol.* **88**:912–917.

Fiore, A. E. 2004. Hepatitis A transmitted by food. *Clin. Infect. Dis.* **38**:705–715.

Fleet, G. H., P. Heiskanen, I. Reid, and K. A. Buckle. 2000. Foodborne viral illness—status in Australia. *Int. J. Food Microbiol.* **59**:127–136.

Formiga-Cruz, M., G. Tofino-Quesada, S. Bofill-Mas, D. N. Lees, K. Henshilwood, A. K. Allard, A.-C. Conden-Hansson, B. E. Hernroth, A. Vantarakis, A. Tsibouxi, A. Vantarakis, and R. Girones. 2002. Distribution of human virus in contamination in shellfish from different growing areas in Greece, Spain, Sweden, and the United Kingdom. *Appl. Environ. Microbiol.* **68**:5990–5998.

Franco, M. A., and H. B. Greenberg. 2002. Rotaviruses, p. 743–762. *In* D. D. Richman, R. J. Whitley, and F. G. Hayden (ed.), *Clinical Virology,* 2nd ed. ASM Press, Washington, DC.

Gallimore, C. I., D. Cubitt, N. du Plessis, and J. J. Gray. 2004. Asymptomatic and symptomatic excretion of noroviruses during a hospital outbreak of gastroenteritis. *J. Clin. Microbiol.* **42**:2271–2274.

Gia, T., K. Kaneshi, Y. Ueda, S. Nakaya, S. Nishimura, A. Yamamoto, K. Sugita, S. Takanashi, S. Okitsu, and H. Ushijuma. 2007. Genetic heterogeneity, evolution, and recombination in noroviruses. *J. Med. Virol.* **76**:1388–1400.

Girard, M. P., D. Steele, C.-L. Chaignat, and M. P. Kieny. 2006. A review of vaccine research and development: human enteric infections. *Vaccine* **24**:2732–2750.

Green, K. Y., R. M. Chanock, and A. Z. Kapikian. 2001. Human caliciviruses, p. 841–874. *In* D. M. Knipe, P. M. Howley, D. E. Griffin, R. A. Lamb, M. A. Martin, B. Roizman, and S. E. Straus, (ed.), *Fields Virology,* 4th ed. Lippincott, Williams, and Wilkins, Philadelphia, PA.

Grove, S. F., A. Lee, T. Lewis, C. M. Stewart, H. Chen, and D. G. Hoover. 2006. Inactivation of foodborne viruses of significance by high pressure and other processes. *J. Food Prot.* **69**:957–968.

Guix, S., A. Bosch, and R. M. Pinto. 2005. Human astrovirus diagnosis and typing: current and future prospects. *Lett. Appl. Microbiol.* **41**:103–105.

Heeney, J. L. 2006. Zoonotic viral diseases and the frontier of early diagnosis, control and prevention. *J. Intern. Med.* **260**:399–408.

Hewitt, J., and G. E. Greening. 2004. Survival and persistence of norovirus, hepatitis A virus, and feline calicivirus in marinated mussels. *J. Food Prot.* **67**:1743–1750.

Hollinger, F. B., and S. U. Emerson. 2001. Hepatitis A virus, p. 799–840. *In* D. M. Knipe, P. M. Howley, D. E. Griffin, R. A. Lamb, M. A. Martin, B. Roizman, and S. E. Straus (ed.), *Fields Virology,* 4th ed. Lippincott, Williams, and Wilkins, Philadelphia, PA.

Hutin, Y. J., V. Pool, E. H. Cramer, O. V. Nainan, J. Weth, I. T. Williams, S. T. Goldstein, K. F. Gensheimer, B. P. Bell, C. N. Shapiro, M. J. Alter, and H. S. Margolis. 1999. A multistate foodborne outbreak of hepatitis A. *N. Engl. J. Med.* **340**:595–601.

Jaykus, L. 2007. Detection of the presence of bacteria, viruses, and parasitic protozoa in shellfish, p. 311–324. *In* C. J. Hurst (ed.), *Manual of Environmental Microbiology,* 3rd ed. ASM Press, Washington, DC.

Jean, J., J.-F. Vachon, O. Moroni, A. Darveau, I. Kukavica-Ibrulj and I. Fliss. 2003. Effectiveness of commercial disinfectants for inactivating hepatitis A virus on agri-food surfaces. *J. Food Prot.* **66**:115–119.

Jothikumar, N., T. L. Cromeans, M. D. Sobsey, and B. H. Robertson. 2005a. Development and evaluation of a broadly reactive TaqMan assay for rapid detection of hepatitis A virus. *Appl. Environ. Microbiol.* **71**:3359–3363.

Jothikumar, N., J. A. Lowther, K. Henshilwood, D. N. Lees, V. R. Hill, and J. Vinje. 2005b. Rapid and sensitive detection of noroviruses by using TaqMan-based one-step reverse transcription-PCR assays and application to naturally contaminated shellfish samples. *Appl. Environ. Microbiol.* **71**:1870–1875.

Kageyama, T., S. Kojima, M. Shinohara, K. Uchida, S. Fukushi, F. B. Hoshino, N. Takeda, and K. Katayama. 2003. Broadly reactive and highly sensitive assay for Norwalk-like viruses on real-time quantitative reverse transcription-PCR. *J. Clin. Microbiol.* **41**:1548–1557.

Khetsuriani, N., A. LaMonte-Fowlkes, M. S. Oberste, and M. A. Pallansch. 2006. Enterovirus surveillance—United States, 1970–2005. *MMWR Surveill. Summ.* **55**(SS–08):1–20.

Kingsley, D. H., D. R. Holliman, K. R. Calci, H. Chen, and G. J. Flick. 2007. Inactivation of a norovirus by high pressure processing. *Appl. Environ. Microbiol.* **73**:581–585.

Kingsley, D. H., D. Guan, D. G. Hoover, and H. Chen. 2006. Inactivation of hepatitis A virus by high-pressure processing: the role of temperature and pressure oscillation. *J. Food Prot.* **69**:2454–2459.

Kingsley, D. H., D. Guan, and D. G. Hoover. 2005. Pressure inactivation of hepatitis A virus in strawberry puree and sliced green onions. *J. Food Prot.* **68**:1748–1751.

Kingsley, D. H., H. Chen, and D. G. Hoover. 2004. Inactivation of selected picornaviruses by high hydrostatic pressure. *Virus Res.* **102**:221–224.

Kingsley, D. H., and G. P. Richards. 2003. Persistence of hepatitis A virus in oysters. *J. Food Prot.* **66**:331–334.

Kingsley, D. H., D. G. Hoover, E. Papafragkou, and G. P. Richards. 2002. Inactivation of hepatitis A virus and a calicivirus by high hydrostatic pressure. *J. Food Prot.* **65**:1605–1609.

Klein, E. J., D. R. Boster, J. R. Strapp, J. G. Wells, X. Qin, C. R. Clausen, D. L. Swerdlow, C. R. Braden, and P. I. Tarr. 2006. Diarrhea etiology in a children's hospital emergency department: a prospective cohort study. *Clin. Infect. Dis.* **43**:807–813.

Koopmans, M., and E. Duizer. 2004. Foodborne viruses: an emerging problem. *Int. J. Food Microbiol.* **90**:23–41.

Lees, D. 2000. Viruses and bivalve shellfish. *Int. J. Food Microbiol.* **59**:81–116.

Lindesmith, L., C. Moe, S. Marionneau, N. Ruvoen, X. Jiang, J. Lindblad, P. Stewart, J. LePendu, and R. Baric. 2003. Human susceptibility and resistance to Norwalk virus infection. *Nat. Med.* **9**:548–553.

Liu, B., P. Maywood, L. Gupta, and B. Campbell. 2003. An outbreak of Norwalk-like virus gastroenteritis in an aged-care residential hostel. *N. S. W. Public. Health Bull.* **14**:105–109.

Lopman, B., H. Vennema, E. Kohli, P. Pothier, A. Sanchez, A. Negredo, J. Buesa, E, Schreier, M. Reacher, D. Brown, J. Gray, M. Iturriza, C. Gallimore, B. Bottiger, K.-O. Hedlund, M. Torvén, C.-H. von Bonsdorff, L. Maunula, M. Poljsak-Prijatelj, J. Zimsek, G. Reuter, G. Szücs, B. Melegh, L. Svennson, Y. van Duijnhoven, and M. Koopmans. 2004. Increase in viral gastroenteritis outbreaks in Europe and epidemic spread of new norovirus variant. *Lancet* **363**:671–672.

Lopman, B. A., G. K. Adak, M. H. Reader, and D. W. Brown. 2003a. Two epidemiologic patterns of norovirus outbreaks: surveillance in England and Wales, 1992–2000. *Emerg. Infect. Dis.* 9:71–77.

Lopman, B. A., M. H. Reacher, Y. van Duijnhoven, F. X. Hanon, D. Brown, and M. Koopmans. 2003b. Viral gastroenteritis outbreaks in Europe, 1995–2000. *Emerg. Infect. Dis.* 9:90–96.

Mallet, J. C., L. E. Beghian, T. G. Metcalf, and J. D. Kaylor. 1991. Potential of irradiation technology for improving shellfish sanitation. *J. Food Saf.* 11:231–245.

Mbithi, J. N., V. S. Springthorpe, J. R. Boulet, and S. A. Sattar. 1992. Survival of hepatitis A virus on human hands and its transfer on contact with animate and inanimate surfaces. *J. Clin. Microbiol.* 30:757–763.

Mbithi, J. N., V. S. Springthorpe, and S. A. Sattar. 1991. Effect of relative humidity and air temperature on survival of hepatitis A virus on environmental surfaces. *Appl. Environ. Microbiol.* 57:1394–1399.

Mbithi, J. N., V. S. Springthorpe, and S. A. Sattar. 1990. Chemical disinfection of hepatitis A virus on environmental surfaces. *Appl. Environ. Microbiol.* 56:3601–3604.

McDonnell, S., K. B. Kirkland, W. G. Hlady, C. Aristeguieta, R. S. Hopcins, S. S. Monroe, and R. I. Glass. 1997. Failure of cooking to prevent shellfish-associated viral gastroenteritis. *Arch. Intern. Med.* 157:111–116.

Mead, P. S., L. Slutsker, V. Dietz, L. F. McCaig, J. S. Bresee, C. Shapiro, P. M. Griffin, and R.V. Tauxe. 1999. Food-related illness and death in the United States. *Emerg. Infect. Dis.* 5: 607–625.

Muniain-Mujika, I., R. Girones, G.Tofino-Quesada, M. Calvo, and F. Lucena. 2002. Depuration dynamics of viruses in shellfish. *Int. J. Food Microbiol.* 77:125–133.

NACMCF. 2008. Response to the questions posed by the U.S. Food and Drug Administration and National Marine Fisheries Service regarding determination of cooking parameters for safe seafood for consumers. *J. Food Prot.* 71:1287–1308.

Oh, D. Y., P. A. Silva, B. Hauroeder, S. Diedrich, D. D. Cardosos, and E. Schreier. 2006. Molecular characterization of the first Aichi viruses isolated in Europe and in South America. *Arch. Virol.* 151:1199–1206.

Okamoto, H. 2007. Genetic variability and evolution of hepatitis E virus. *Virus Res.* 127:216–228.

Pallansch, M. A., and R. P. Roos. 2001. Enteroviruses: polioviruses, coxsackieviruses, echoviruses, and newer enteroviruses, p. 723–775. *In* D. M. Knipe and P. M. Howley (ed). *Fields Virology,* 4th ed. Lippincott Williams and Wilkins, Philadelphia, PA.

Papafragkou, E., D. H. D'Souza, and L. Jaykus. 2006. Foodborne viruses: prevention and control, p. 289–330. *In* S. M. Goyal (ed.), *Viruses in Foods.* Springer Science + Business Media, LLC, New York, NY.

Phan, T. G., K. Kaneshi, Y. Ueda, S. Nakaya, S. Nishimura, A. Yamamotot, K. Sugita, S. Takanashi, S. Okitsu, and H. Ushijuma. 2007. Genetic heterogeneity, evolution, and recombination of noroviruses. *J. Med. Virol.* 79:1388–1400.

Phan, T. G., T. Kuroiwa, K. Kaneshi, Y. Ueda, S. Nakaya, S. Nishimura, A. Yamamoto, K. Sugita, T. Nishimura, F. Yagyu, S. Okitsu, W. E. Muller, N. Maneekarn, and H. Ushijuma. 2006. Changing distribution of norovirus genotypes and genetic analysis or recombinant GIIb among infants and children with diarrhea in Japan. *J. Med. Virol.* 78: 971–978.

Pina, S., M. Puig, F. Lucena, J. Jofre, and R. Girones. 1998. Viral pollution in the environment and in shellfish: human adenovirus detection by PCR as an index of human viruses. *Appl. Environ. Microbiol.* 64:3376–3382.

Pirtle, E. C., and G. W. Beran. 1991. Virus survival in the environment. *Rev. Sci. Technol.* 10:733–748.

Rockx, B., M. deWit, H. Vennema, J. Vinje, E. de Bruin, Y. van Duynhoven, and M. Koopmans. 2002. Natural history of human calicivirus infection: a prospective cohort study. *Clin. Infect. Dis.* 35:246–253.

Rutjes, S. A., F. Lodder-Verschoor, W. H. M. van der Poel, Y. T. H. P. van Duijnhoven, and A. M. de Roda Husman. 2006. Detection of noroviruses in foods: a study on virus extraction procedures in foods implicated in outbreaks of human gastroenteritis. *J. Food Prot.* 69:1949–1956.

Sattar, S. A., V. S. Springthorpe, J. Tetro, R. Vashon, and B. Keswick. 2002. Hygienic hand antiseptics: should they not have activity and label claims against viruses? *Am. J. Infect. Control* 30:355–372.

Seymour, I. J., and H. Appleton. 2001. Foodborne viruses and fresh produce. *J. Appl. Microbiol.* 91:759–773.

Smith, J. L. 2001. A review of hepatitis E virus. *J. Food Prot.* 64:572–586.

Thomas, C., and D. E. Swayne. 2007. Thermal inactivation of H5N1 high pathogenicity avian influenza virus in naturally infected chicken meat. *J. Food Prot.* 70:674–680.

USFDA. 2005. 2005 Food Code. U.S. Food and Drug Administration, U.S. Department of Health and Human Services, College Park, MD. http://www.cfsan.fda.gov/~dms/fc05-toc.html.

Wasley, A., A. Fiore, and B. P. Bell. 2006. Hepatitis A in the era of vaccination. *Epidemiol. Rev.* 28:101–111.

Wheeler, C., T. M. Vogt, G. L. Armstrong, G. Vaughan, A. Weltman, O. V. Nainan, V. Dato, G. Xia, K. Waller, J. Amon, T. M. Lee, A. Highbaugh-Battle, C. Hembree, S. Evenson, M. A. Ruta, I. T. Williams, A. E. Fiore, and B. P. Bell. 2005. An outbreak of hepatitis A associated with green onions. *N. Engl. J. Med.* 353:890–897.

Widdowson, M.-A., A. Sulka, S. N. Bulens, R. S. Beard, S. S. Chaves, R. Hammond, E. D. P. Salehi, E. Swanson, J. Totaro, R. Woron, P. S. Mead, J. S. Bresee, S. S. Monroe, and R. I. Glass. 2005. Norovirus and foodborne disease, United States, 1991–2000. *Emerg. Infect. Dis.* 11:95–102.

Yamashita, T., M. Sugiyama, H. Tsuzuki, K. Sakae, Y. Suzuki, and Y. Miyazaki. 2000. Application of a reverse transcription-PCR for identification and differentiation of Aichi virus, a new member of the Picornavirus family associated with gastroenteritis in humans. *J. Clin. Microbiol.* 38:2955–2961.

Yamashita, T., K. Sakae, Y. Ishihara, S. Isomura, and E. Utagawa. 1993. Prevalence of a newly isolated, cytopathic small round virus (Aichi strain) in Japan. *J. Clin. Microbiol.* 31:2938–2943.

Pathogens and Toxins in Foods: Challenges and Interventions
Edited by V. K. Juneja and J. N. Sofos
© 2010 ASM Press, Washington, DC

Chapter 15

Seafood Toxins

Sherwood Hall

In the sea, unicellular algae (phytoplankton) produce oxygen and convert solar energy into useful biomass, an important service to life on the planet. Unfortunately, some few species of phytoplankton produce potent toxins that can accumulate in seafood. In many locations the term "red tide" is used to describe such phenomena. The phrase is misleading since most toxicity is not associated with visible blooms and most visible blooms are not toxic, but convenience and popularity sustain its use, particularly in the media. The following examples are offered to give some idea of the nature of the issue.

Outbreaks of toxicity can have severe impacts. In 1987, on the Pacific coast of Guatemala, 26 people died in 3 days from paralytic shellfish poisoning (PSP) after consuming clams contaminated with the saxitoxins. There was no history of PSP in the region. People in the affected area had been consuming clams of the same species, from the same beach, for years without incident (Rodrigue et al., 1990). The phytoplankton source, *Pyrodinium bahamense*, was well known as a source of toxicity in Papua New Guinea, Indonesia, and the Philippines (Maclean, 1989). The same organism occurs in the tropical Atlantic but had been thought not to be toxic. In the wake of the Guatemala outbreak, one question was whether such toxicity could occur along the tropical southeastern coast of the United States. In 2002, several consumer illnesses were linked to otherwise nontoxic puffer fish from the Indian River Lagoon on the east coast of Florida. It was found that, rather than tetrodotoxin (TTX), the neurotoxin more commonly associated with puffer fish, these fish had accumulated saxitoxins produced by *P. bahamense* growing in the lagoon (Quilliam et al., 2004; Landsberg et al., 2006). As a result of this event, Florida implemented routine monitoring for the saxitoxins in shellfish.

New Zealand has a long tradition of producing high-quality shellfish but until 1993 had no history of biotoxin contamination and maintained only a minimal biotoxin monitoring program. Then, during the holiday season at the end of 1992 (austral summer), more than 180 people fell ill after consuming shellfish. These illnesses were not recognized as an outbreak until investigation of hind limb paralysis in two cats revealed that shellfish near Aukland were toxic, alerting authorities to the possibility of human cases. Although the primary cause was found to be neurotoxic shellfish poisoning (NSP), caused by brevetoxins, other toxins were detected, including domoic acid (DOM; responsible for amnesic shellfish poisoning) and diarrhetic shellfish poisoning (DSP) toxins related to okadaic acid. As a consequence of this episode, New Zealand implemented one of the most effective and technically sophisticated marine biotoxin management programs in the world.

In Prince Edward Island (PEI), Canada, a mussel culture industry developed through the 1980s, bringing significant economic benefits to the region. In late 1987, a large number of people fell ill after eating PEI mussels, 107, of whom eventually met a strict case definition. Three victims died (Perl et al., 1990; Teitelbaum et al., 1990). The causative agent was found to be DOM, a previously known compound not recognized to be a potent neurotoxin in humans (Todd, 1993). Unlike other seafood toxins, which are produced by dinoflagellates, DOM is produced by diatoms. Also in contrast to other seafood toxins, the neural damage caused by DOM is permanent. Survivors of PSP and most other seafood toxin syndromes tend to recover without permanent effects. In the wake of this outbreak, Canada implemented a monitoring program for DOM that, despite the severity of the syndrome, quickly restored market confidence in

Sherwood Hall • Chemical Contaminants Branch HFS-716, Division of Bioanalytical Chemistry, Office of Regulatory Science, Center for Food Safety and Applied Nutrition, U.S. FDA, College Park, MD 20740.

the safety of PEI shellfish. Extensive and costly surveys of seafood in the United States during the following 2 years revealed no DOM contamination. Then, in 1991, the investigation of deaths and erratic behavior of seabirds around Monterey Bay, CA (Work et al., 1993) revealed that anchovies in the region had become heavily contaminated with DOM. It was subsequently recognized that DOM contamination extended along the entire U.S. Pacific Coast and included crustaceans and bivalves as well as planktivorous fish. The extent of the problem and the drain on resources required to manage it prompted consideration of volunteer observer networks as an asset in the management of marine biotoxins. In some regions, particularly California, this concept has proven valuable.

The New England Red Tide of 2005 will serve as a final example. In the spring of 2005, an ongoing federally funded research program on harmful algal blooms detected a large bloom of *Alexandrium* sp. in the Gulf of Maine. Various species of *Alexandrium* are the primary sources of the saxitoxins along the Pacific and New England coasts. Marine biotoxin management programs in the New England states were alerted; concern was expressed in the media about a "disaster" due to the "red tide." PSP toxicity in shellfish reached very high levels, but through the coordinated efforts of all parties, there were no consumer illnesses and safe shellfish were available on the market and in restaurants throughout the event. Due to detailed, high-resolution toxicity monitoring, it was possible for states in the affected region to identify locations where the shellfish were safe and to keep them open for harvest.

In summary, marine biotoxins can have severe impacts on health and prosperity, but effective management can minimize the risk of such impacts.

GENERAL CHARACTERISTICS OF SEAFOOD TOXINS

The toxins under discussion here are generally low-molecular-weight, nonprotein compounds produced by smaller organisms (generally phytoplankton) and accumulated by larger organisms that eat them. The toxicity is notable in filter feeders (bivalve mollusks and planktivorous fish) but may accumulate at higher trophic levels (crabs and carnivorous fish). Toxic seafood tends to appear normal, not distinguishable by sight or taste from nontoxic. The vectoring of toxins to humans by herbivores is not unique to the marine environment. For instance, "milk sickness," due to tremetone passed through the milk of dairy cows that forage on white snakeroot, has had a significant impact on consumer health and

the viability of agriculture in some regions of the United States (Beier et al., 1993).

The distribution of phytoplankton populations is patchy and ephemeral; so, therefore, is the distribution of toxicity. Plankton populations move with the water masses of which they are a part, so a bloom can develop off the coast and be advected rapidly toward shore and shellfish beds or along the shore. If a bloom happens to be a toxic organism, this can result in a very rapid rise in shellfish toxicity. While the occurrence of toxic blooms may show some historical trends, episodes without historical precedent, as some noted above, can occur with severe consequences. The toxins tend to be very potent; thus, concentrations that are very low can present an acute risk to consumers. The detection of such low concentrations can be challenging.

SOME KNOWN FAMILIES OF SEAFOOD TOXINS AND ASSOCIATED SYNDROMES

Many toxic substances have been isolated from marine organisms. Discussion in this chapter is limited to those known to pose a significant risk to seafood consumers. The majority of these are accumulated from toxigenic plankton.

A recent consultation sponsored by the WHO and FAO, in part to provide guidance to the Codex Alimentarius, has summarized much of the current understanding of the various seafood toxins and may be downloaded from http://www.fao.org/ag/agn/agns/chemicals_biotoxins_en.asp. The FAO has prepared a food and nutrition paper on marine biotoxins, available on the web at http://www.fao.org/docrep/007/y5486e/y5486e01.htm. Two recent books on marine toxins edited by Botana (2007, 2008) will also be helpful.

Saxitoxins and PSP

In the summer of 1929, along the California coast near San Francisco, an outbreak of PSP caused six deaths and over 100 reported illnesses. Investigations during the years following this outbreak revealed that the source of the toxicity was a newly recognized dinoflagellate named *Gonyaulax catenella* (Sommer and Meyer, 1937). This was the first work to recognize plankton as a source of seafood toxins. Subsequent taxonomic study has assigned *G. catenella* and related species to the new genus *Alexandrium*. Other known sources of the saxitoxins include the marine dinoflagellates *Pyrodinium bahamense* (Steidinger et al., 1980; Harada et al., 1982) and *Gymnodinium catenatum* and various freshwater cyanophytes. While these taxa account for the majority of occurrences of the saxitoxins, the sources in some cases are less clear.

Significant concentrations of saxitoxins and novel saxitoxin derivatives have been found in various tropical crabs (Arakawa et al., 1994, 1995). *Alexandrium* from some locations has also been reported to contain both saxitoxins and TTX (Kodama et al., 1996).

Saxitoxin, the conceptual parent of the group, is a tricyclic diguanidinium compound, strongly polar and water soluble. More than 20 other saxitoxins are now known, consisting of various hydroxyl, sulfate, sulfo, and other modifications to the core saxitoxin skeleton (Hall et al., 1990). Most are also strongly polar and water soluble, but some recently discovered saxitoxins have substituents that make them sufficiently lipophilic that they can be lost in analytical cleanup procedures intended to remove lipophilic interferences. The saxitoxins tend to be vulnerable to oxidation at elevated pH, yielding fluorescent degradation products, a transformation that has proven useful for some detection methods.

The saxitoxins function by binding reversibly to site 1 of the voltage-activated sodium channel (Catterall, 1985). While a toxin molecule is bound, the normal passive influx of sodium ions through the channel cannot occur and the propagation of action potentials along nerve and muscle membranes is blocked. Victims of PSP experience symptoms related to such blockage in peripheral nerves and skeletal muscles, ranging from peripheral tingling in mild cases to muscular paralysis and, in severe cases without treatment, death from suffocation. Respiratory support will generally suffice to ensure survival. Onset is rapid (less than 2 hours). The toxins wash out of the system rapidly (Gessner et al., 1997; Garcia et al., 2004). Victims most often recover after a day or so, with no lasting effects.

Saxitoxins bearing a 21-sulfo substituent have greatly reduced potency due to reductions in both their on rates and dwell times on the binding site (Moczydlowski et al., 1984; Hall et al., 1990). These toxins can be transformed, either by hydrolysis of the 21-sulfo group or by loss of the entire sulfamate side chain, to other saxitoxins with higher on rates and longer dwell times, with increases in potency ranging from 5-fold to 70-fold. The 21-sulfo saxitoxins, which often predominate in natural mixtures, thus constitute a reservoir of potential toxicity that may be dangerously underestimated by some detection methods (Hall et al., 1980).

The particular saxitoxins present in a sample will vary significantly, depending on a number of factors. The composition present in *Alexandrium* varies with location (Hall, 1982; Oshima et al., 1982; Hall et al., 1990); other taxa have different compositions (Harada et al., 1982; Oshima et al., 1993; Negri et al., 2003, 2007). Toxin composition then undergoes diagenesis in accumulators due to metabolism, spontaneous transformations, and selective retention (Bricelj and Shumway, 1998).

TTX and Puffer Fish Poisoning

TTX and related compounds (Yasumoto and Yotsu-Yamashita, 1996) differ from the other seafood toxins discussed here, in that they appear to be endogenous or at least produced by symbionts, rather than accumulated from environmental sources (Noguchi and Hashimoto, 1984; Noguchi et al., 1986; Yasumoto et al., 1986b; Noguchi et al., 1987; Yotsu et al., 1987). Toxicity, though not invariant, is thus a relatively constant property of many puffer fish species. The pharmacology of TTX differs little from that of saxitoxin, both compounds binding to site 1 of the voltage-activated sodium channel (Catterall, 1985).

Puffer fish poisoning has been recognized since antiquity and occurs at a relatively high rate due to either accidental or intentional consumption of puffer fish. Under normal circumstances, TTX is localized in the skin and viscera of puffer fish, so careful cleaning of the flesh renders the product safe to consume. The toxicity of the toxic parts tends to be high enough that the cleaning must be carefully performed to ensure safety (Kao, 1966). The illnesses that occurred following consumption of normally nontoxic puffer fish in Florida in 2002 were in part due to a very different distribution of toxicity. It was found that the concentration of the saxitoxins in the flesh of Florida puffers was very high, so cleaning offered no protection to consumers (Quilliam et al., 2004; Landsberg et al., 2006).

DOM and Amnesic Shellfish Poisoning

DOM was originally discovered as the active ingredient in certain species of red seaweed ("domoi") traditionally used in Japan as a vermifuge (Daigo, 1958; Takemoto et al., 1966). It is structurally related to kainic acid, which has a similar history and has been used extensively in pharmacological studies of the central nervous system (CNS). Both are structurally related to the amino acid proline, and both act as glutamate analogues in the CNS, binding to one class of glutamate receptors. DOM binds strongly to glutamate receptors in the hippocampus, leading to sustained activation of the associated neurons. The toxicity of DOM is in part because of this strong binding and in part because the mechanisms that scavenge glutamate, keeping its concentration low and its stimulation brief, do not act on DOM. DOM in the CNS thus persists, and neural tissue is destroyed as a consequence of unremitting stimulation (Hynie and Todd, 1990; Iverson and Truelove, 1994).

While the majority of seafood toxins are produced by dinoflagellates, the primary sources of DOM in seafood are diatoms assigned to the genus *Pseudonitzschia* (Bates et al., 1989; Fritz et al., 1992; Garrison et al., 1992). DOM is distinct among the seafood toxins, in that it has a strong UV chromophore (due to conjugated double bonds on one side chain) that greatly simplifies high-performance liquid chromatography (HPLC) analysis. Furthermore, while isomers of DOM occur, they tend to have little or no toxicity and seldom predominate in natural mixtures (Wright et al., 1989, 1990; Wright, 1998). Detection methods to estimate consumer risk can therefore focus primarily on DOM itself. DOM acid has been found in many different types of organisms, including bivalves, crustaceans, and fish (Wekell et al., 1994).

Brevetoxins and NSP

Neurotoxic shellfish poisoning (NSP) was originally described as a "ciguatera-like" illness affecting several people who consumed oysters from the Gulf of Mexico (McFarren et al., 1965). The source organism was found to be a naked dinoflagellate now known as *Karenia brevis* (originally as *Gymnodinium breve* and for several years as *Ptychodiscus brevis*). Aside from the very large outbreak in New Zealand in 1992 (Jasperse, 1993), NSP has been recognized primarily in the Gulf of Mexico.

The brevetoxins bind reversibly to site 5 of the voltage-activated sodium channel (Catterall and Gainer, 1985), sustaining activation of the channel. The molecular mechanism is the opposite of the saxitoxins, sustaining rather than blocking the passive influx of sodium ions, so the symptoms of low doses are of sensory disruption rather than numbness. However, the eventual effect of prolonged sodium influx can be depolarization of the membrane, also leading to blockage of action potentials. There have been no reported deaths due to NSP, although two young victims in a recent outbreak might not have survived without rapid treatment (Poli et al., 2000).

The structures of the brevetoxins fall into two groups, based on two similar polycyclic core structures (Lin et al., 1981; Chou and Shimizu, 1982; Chou et al., 1985; Chou, 1986; Shimizu et al., 1986; Van Duyne, 1990; Murata et al., 1998). The brevetoxins produced by *K. brevis* tend to be strongly lipophilic but, following metabolism by accumulators, may become less so (Ishida et al., 1995, 1996; Morohashi et al., 1995, 1999).

Ciguatoxins and Ciguatera

Ciguatera results primarily from the consumption of tropical fish that have accumulated toxins similar to but critically different from the brevetoxins (Murata et al., 1989, 1990; Lewis et al., 1993; Satake et al., 1993; Yasumoto and Satake, 1996; Vernoux and Lewis, 1997). Ciguatera toxins appear to have a much stronger affinity than brevetoxins for site 5 on the voltage-activated sodium channel. While it is possible that other lipophilic toxins play a role in some illnesses diagnosed as ciguatera, these toxins appear to account for the bulk of the symptoms in the majority of cases. The symptoms of ciguatera and the structures of ciguatoxins differ slightly among regions. Ciguatera, like NSP, is characterized by sensory disruption, including hypersensitivity and reversal of hot and cold sensations. The neurological symptoms of ciguatera are more persistent and may reoccur weeks or months after intoxication (Withers, 1982; Friedman et al., 2008). Ciguatera is widespread in many tropical regions and, due to the persistence of its debilitating symptoms, has the greatest socioeconomic impact and causes the most suffering of the various seafood toxin syndromes (Lewis and Ruff, 1993; Friedman et al., 2008). Treatment with intravenous mannitol, pioneered by Palafox et al. (1988), has appeared to offer relief in several cases (Pearn et al., 1990; Mitchell, 2005), though a double-blind trial (Schnorf et al., 2002) questioned its efficacy.

In contrast to other seafood toxin syndromes in which the toxins are produced by plankton and concentrated by filter feeders, the ciguatoxins are produced by dinoflagellates, such as *Gambierdiscus toxicus*, that are largely benthic in habit and tend to encrust tropical reef surfaces and seaweed (Yasumoto et al., 1977, 1980a). The toxins are accumulated and metabolized by reef herbivores that feed on seaweed and scour coral surfaces. Being lipophilic, the ciguatoxins are bioconcentrated in higher trophic levels, the greatest risk to consumers tending to result from consumption of carnivores such as barracuda (Lewis and Endean, 1984). The risk is most severe with consumption of larger, older fish (Lewis and Holmes, 1993).

Okadaic Acid, Its Derivatives, and DSP

Yasumoto et al. (1978) described a large outbreak of diarrhea and other gastrointestinal symptoms in shellfish consumers in Japan. The syndrome, DSP, and the dinoflagellates that cause it are widespread, but tracking the extent of DSP illness is complicated by the similarity of the symptoms to the gastrointestinal distress caused by microbial contamination. Okadaic acid and compounds related to it are the primary cause of DSP (Murata et al., 1982; Yasumoto et al., 1985). Okadaic acid was originally isolated from two species of the sponge *Halichondria*—*H. okadai* from the coast of Japan and *H. melanodocia* from the coast of Florida (Tachibana et al., 1981), but the primary sources of okadaic acid and related compounds in seafood are dinoflagellates of the genera *Dinophysis* (Yasumoto et al., 1980b) and *Prorocentrum* (Murakami et al., 1982; Lee et al., 1989; Dickey, et al., 1990; Hu et al., 1993). While the DSP toxins, okadaic acid and the dinophysistoxins (Yasumoto et al., 1985), are mostly lipophilic, some are sulfated and thus water soluble (Hu et al., 1995a, 1995b). This can complicate reliable recovery for detection. Others occur as relatively inactive fatty acid esters that, when hydrolyzed, increase substantially in toxicity. This is analogous to the potential toxicity encountered with the 21-sulfo saxitoxins. Okadaic acid is a potent inhibitor of protein phosphatase. To explore this interaction, the crystal structure of a complex of okadaic acid and a protein phosphatase was determined by Maynes et al. (2001).

Azaspiracids and AZA Poisoning

In 1995, consumers of mussels from Ireland became ill with gastrointestinal symptoms resembling DSP. However, the absence of the known DSP toxins (okadaic acid and derivatives) and the planktonic organisms thought to be the source of DSP suggested that another family of toxins was responsible. Subsequent investigations identified a novel toxin, azaspiracid (AZA), and its derivatives as the cause (Satake et al., 1998; Ofuji et al., 1999, 2001; James et al., 2003b). The proximate source of AZA may be a predatory dinoflagellate (*Protoperidinium* sp.), which may prove to be the actual source or merely a vector for toxins accumulated from its prey (James et al., 2003a). The toxicology of AZA is not fully understood, although AZA may have an effect on the cytoskeleton. Although data are mixed, there are some indications from laboratory studies that the effects of AZA can be grave and permanent (Ito et al., 2000). AZA has been detected with both shellfish and crabs and has been reported from several countries in western Europe, and from Scandinavia to Morocco (James et al., 2001; Magdalena et al., 2003; Vale et al., 2008).

Palytoxins

Palytoxin, notorious as one of the most potent toxins known (Vick and Wiles, 1975), was originally isolated from zoanthids of the genus *Palythoa* (Moore and Scheuer, 1971) rather than from organisms used as food. Palytoxin and its derivatives have subsequently been detected in a wide range of accumulators and in dinoflagellates of the genus *Ostreopsis* (Usami et al., 1995; Ukena et al., 2001) and are now recognized as being responsible for some forms of seafood toxicity in fish (Melton et al., 1984; Fukui et al., 1987; Noguchi et al., 1988; Kodama et al., 1989) and crabs (Yasumoto et al., 1986a; Alcala et al., 1988; Fukui et al., 1988). The term "clupeotoxin" has been applied to toxicity from certain planktivorous fish, which now appears to be due to palytoxins (Onuma et al., 1999).

Palytoxin binds to the sodium pump, causing it to remain open and become a membrane pore (Hirsch and Wu, 1997). The cell is then unable to maintain a transmembrane potential.

Lipophilic Toxins of Uncertain Risk to Consumers

During investigations of seafood and plankton for DSP toxicity, many toxins have been found that are toxic to mice when injected intraperitoneally (i.p.), but they have not proven to have significant oral toxicity. While some of these co-occur with the known DSP toxins (okadaic acid and its derivatives), it has recently been recognized that, despite their i.p. toxicity to mice, there is no clear evidence that they have been responsible for human illnesses from seafood or that they have significant oral toxicity to test animals (Aune et al., 1998). These toxins include the yessotoxins (Murata et al., 1987), pectenotoxins (Murata et al., 1986), and several families of cyclic imine toxins, including gymnodimines (Seki et al., 1995), pinnatoxins, prorocentrolides (Torigoe et al., 1988), and spirolides (Hu et al., 1995c). These present a challenge when a mouse i.p. assay of lipophilic extracts is used to manage the risk of true lipophilic toxins. The best and most comprehensive option is to use liquid chromatography–mass spectrometry (LC/MS) for detection.

MANAGEMENT OF SEAFOOD TOXINS: OPTIONS AND CONSIDERATIONS

What are the options for protecting consumers from natural toxins in seafood? Can we take active measures to eliminate the problem? If blooms of toxic algae were caused by human activity, then it would, in principle, be possible to deal with the human activities that cause them. There is strong evidence that anthropogenic influences, particularly the eutrophication of coastal waters, may increase the frequency and severity of harmful algal blooms. However, we know from history that severe biotoxin outbreaks can occur in the absence of such inputs: the most severe outbreak of PSP on record occurred more than 200 years ago, in Alaska. While control of nutrient inputs from sewage and agriculture is necessary for the health of coastal ecosystems, it cannot be relied upon to eliminate the risk of seafood toxins.

Can growing areas be treated to exterminate toxic algae? Any treatment would have to be selective for harmful species and, with the exception of very small, intensive aquaculture systems, would have to treat vast amounts of water. There is little hope of eliminating toxic algal species from the waters where most seafood is produced.

Can processing seafood destroy or remove the toxins? Compared to pathogenic bacteria and proteinaceous food poisons, seafood toxins tend to be relatively heat stable. In processed seafood it is possible in some cases to reduce though not eliminate toxicity (Waskiewicz et al., 1951; Burdespahl et al., 1998). Careful monitoring of feedstock and product toxicity is necessary to ensure safety.

In some cases, if the toxicity is sufficiently localized within an animal, protection can be provided by removing the toxic parts. Sea scallop adductor muscles tend to accumulate little or no toxicity, even when the viscera are highly toxic. Therefore, scallops harvested along the northeast coast of the United States tend to be shucked at sea, and only the adductor muscles are brought to shore, avoiding concerns about toxicity. Geoducks are large clams harvested along the Pacific coast of North America. The geoduck visceral ball can accumulate significant toxicity, while the toxicity of the rest of the animal remains fairly low. Because of the size of the animal, it is practical to butcher and remove the visceral ball, leaving a product safe for consumption. Crabs and lobsters in North America have been found to accumulate both PSP and DOM, but the toxins tend to be localized in the viscera, principally in the hepatopancreas or "tomale." The risk has been managed either by advising consumers not to consume the tomale or by requiring processors to remove the viscera prior to sale. As noted above, puffer fish may contain high levels of TTX, but these are normally located in the skin and viscera such that careful cleaning can render the flesh safe for consumption. However, puffer fish accumulating saxitoxins from dietary sources had dangerous levels in the flesh.

But in the majority of cases, it is impossible or at least impractical to remove toxicity from seafood, leaving us with the option of simply avoiding the harvest of toxic product. This is often quite practical, since most seafood is safe in most locations most of the time. The challenge is to identify which seafood is toxic—on scales of space and time appropriate to the possible occurrence of toxicity.

It would be helpful to be able to predict when and where toxicity will occur. In the case of ciguatera, for which the source organism is benthic and thus associated with specific locations, it is possible to identify areas of relatively low or high risk.

The challenge then becomes distinguishing between fish species that tend to remain in a single location from those that move about and tracing fish harvested from known safe areas through market channels. In the case of toxins from planktonic organisms, prediction is not entirely impossible, in the sense that historical trends may continue for a given location. In temperate regions, toxicity tends to be less likely in winter. However, some of the more severe outbreaks have occurred during seasons or in locations where there was no historical precedent. Predicting the growth of toxic plankton on the basis of observable physical factors would be helpful, but remains a distant goal, widely pursued. However, prediction of toxicity on the basis of direct observations of plankton is practical and is discussed below.

Due to the intrinsic limitations of these other strategies, monitoring to detect toxicity before toxic seafood is consumed tends to be the core strategy for protecting consumers. However, as emphasized above, toxicity tends to be patchy and ephemeral and may increase rapidly compared to the time intervals between sampling, detection, and management decision and the time interval between harvest and table. Furthermore, because of the variability with space (vertical as well as horizontal) and time, sampling needs to be at a relatively high temporal and spatial density to offer a significant level of protection. The toxicity of a given sample has relatively little significance; the sample will not be consumed. The significance of samples lies in their ability to predict the mean and range of toxicity of what may be harvested from the population from which they are drawn. The reliability with which consumer risk can be estimated therefore depends very significantly on the density of the sampling in both space and time and the time interval from sampling to management decision versus the time interval from harvest to consumption. Density of sampling, both temporal and spatial, is constrained by cost. The time interval from sampling to management decision is largely limited by the needs for sample transport and preparation.

Unfortunately, no practical system can ensure absolute safety. The need is to find practical strategies that offer the best assurance of safety at an acceptable cost. Government resources for management tend to be limited; seafood safety competes with other priorities. Unfortunately, and quite naturally, biotoxin management tends to rise in priority in the wake of an outbreak of consumer illness and then subside relative to other issues of government concern. In the extreme case, biotoxin management programs may not be maintained as other needs compete for resources, thus again putting consumers at risk. The need is therefore to find affordable strategies for marine biotoxin management that are more likely to be sustained in the face of such competition.

The cost-effectiveness of a management program can be enhanced relative to that of toxicity testing alone by using other information to focus the toxicity testing on the times, places, and toxins of greatest concern. Plankton observations from networks of volunteer observers can be particularly useful, since they indicate the type of toxicity to be expected, may offer advanced warning, and can be conducted at relatively high temporal and spatial density at low cost. Other information includes observations of death or unusual behavior among birds, fish, and other animals along the coast. A key to the usefulness of such information is that it be communicated rapidly from the field to managers who need to decide where and when to sample for toxicity. Several states in the United States employ such systems, the program in California being particularly effective (G. Langlois, California Department of Public Health, personal communication).

From the standpoint of consumer health, the primary concern is to prevent the consumption of seafood with levels of toxins sufficient to cause illness. The principal threat is a steep increase for location or time, and the challenge is to implement measures that have a good chance of detecting high levels. The precision with which the toxicity of a population can be estimated is less of a concern than the reliability with which dangerously high levels can be detected.

DETECTION METHODS FOR SEAFOOD TOXINS

The following is intended to provide an overview of the principles of toxin detection and to present a few examples from the large number of detection methods that have been developed for research and regulatory use.

It is useful to consider detection methods as two distinct types, referred to here as assays and analyses (Hall and Reichardt, 1984). Assays are those detection methods that provide a single value for the content of a sample, while analyses are those that provide separate values for each constituent of a sample. For the response of any detection system, bioassay, instrumental, or other, to be useful for assessing consumer risk, it needs to be multiplied by a response factor which correlates system response to consumer risk for the substance or substances detected. In practice, this response factor may be broken down into concentration per system response and consumer risk per concentration, but the net result is that there is a

certain consumer risk (human oral potency [HOP]) per system response. In an analysis such as HPLC, which resolves a sample into separate peaks, it is possible to choose the appropriate response factor for each peak. The net risk indicated by a sample is the sum of those for the components. By definition, however, an assay gives only a single response, and one can thus use only one response factor. If composition (the relative amounts of the components to which a system responds) were constant, then a response factor could be determined that would correlate the assay system response to consumer risk. However, compositions tend to vary, and the ratio between assay response and consumer risk will vary, except to the extent that the response factors of the assay to each component are the same. This applies within toxin families, among toxin families, and to nontoxin interferences. The response factors of an assay system for each of the components to which it responds can be termed the response spectrum of the assay. An ideal response spectrum would be flat, the ratio between assay response and consumer risk for each component being the same.

ANALYSES

Most practical analyses for seafood toxins are based on HPLC, the compounds tending not to be amenable to gas chromatography. However, with the exception of DOM, for which HPLC with UV detection is straightforward (Lawrence et al., 1989; Quilliam et al., 1989), the toxins generally lack chromophores or fluorophores. Several methods employ pre- or postcolumn modification of the analytes to make them accessible to absorbance or fluorescence detection, but the most sweeping and useful development has been the coupling of HPLC with mass spectrometric detection techniques of increasing sophistication.

Okadaic acid, a component of DSP, can be transformed into a fluorescent derivative for HPLC with fluorescence detection (Lee et al., 1987; Marr et al., 1994; González et al., 1998). Fluorescent derivitization of DOM (Pocklington et al., 1990) allows HPLC with greater sensitivity than direct UV detection.

The saxitoxins can be oxidized to fluorescent products before HPLC (Janecek et al., 1993; Lawrence et al., 2005) or after. The earliest postcolumn reaction system (Buckley et al., 1978) employed hydrogen peroxide as an oxidant, following separation at low pressure on a relatively soft gel. It offered limited resolution and was relatively insensitive to the N-1-hydroxy saxitoxins. Sullivan (Sullivan and Iwaoka, 1983) introduced the use of paired ion chromatography reagents in the mobile phase, high-performance chromatography, and postcolumn oxidation using periodate rather than peroxide. Subsequent systems for HPLC of the saxitoxins have tended to use similar systems. Oshima (1995) employed isocratic rather than gradient elution. Boyer and Goddard (2000) used a postcolumn electrochemical oxidation system instead of adding oxidizing reagents.

TTX can be converted to fluorescent products with strong alkali, allowing HPLC with postcolumn reaction and fluorescence detection (Yasumoto and Michishita, 1985).

The coupling of a mass spectrometer to a gas chromatograph was natural and revolutionary—for analytes amenable to gas chromatography. Unfortunately, marine biotoxins tend not to be amenable. The interfaces that allow a mass spectrometer to be coupled to an HPLC were eagerly awaited and welcomed as they became available. Quilliam was one of the first to explore the utility of LC/MS as a research tool for marine biotoxins (for instance, Quilliam, 1995). LC/MS has become a key tool for marine biotoxin research (Lewis and Jones, 1997; James et al., 1999; Lewis et al., 1999; Lehane et al., 2002; James et al., 2003b; Suzuki et al., 2005; Turrell and Stobo, 2007; Vale et al., 2008). Despite the cost and complexity of LC/MS systems, the ability to concurrently analyze for several different toxins coupled with high throughput can make LC/MS a viable and attractive option for marine biotoxin monitoring, particularly for many lipophilic toxins for which good detection alternatives are not available. A laboratory in New Zealand (McNabb et al., 2005) has taken a leadership role in obtaining acceptance of LC/MS methods for regulatory monitoring.

ASSAYS

While it is desirable for an assay to have a flat response spectrum, many assays do not, and some of these are useful nevertheless. An assay for saxitoxin (Bates et al., 1978) in which the sample is oxidized with hydrogen peroxide in strong alkali has a very poor response to many of the saxitoxins but is still useful in special cases for quantifying a few of the individual saxitoxins.

Many assays can be considered as two steps: a selective interaction at the molecular level and transduction of that molecular event into something measurable.

The toxicity of the saxitoxins is due to their binding at site 1 on the exterior surface of the voltage-activated sodium channel. The limited data available (Hall et al., 1990) suggest that there is a good correspondence between this binding event and mouse i.p. potency. How the mouse i.p. potency of individual saxitoxins correlates with HOP remains to be established, but until data are available, it is reasonable to assume that they do correlate and, therefore, that binding is a reliable measure of consumer risk. Various options are available for the transduction of these molecular events into something observable.

One option is the traditional mouse bioassay for PSP (AOAC, 2005), in which occupancy of binding sites results in death of the mouse. Time from i.p. injection to death is measured and used to determine the toxin concentration. Data tend to be more reproducible when the sample dilution is adjusted to give a median 6-minute death time. With careful technique, the method is reproducible to within less than 20%. However, the performance of various laboratories may not be as favorable (LeDoux and Hall, 2000), and the method is subject to interference from zinc, which may accumulate to levels that, while innocuous to consumers, are lethal to mice i.p. (McCulloch et al., 1989). Nevertheless, the greatest problem with the mouse bioassay for PSP is that it works so well for consumer protection, providing a very rapid and reliable indication of dangerous levels of toxicity. It therefore establishes relatively high performance standards for alternative detection methods.

Cells cultured to express sodium channels can be tested using standard electrophysiological techniques and exposed to samples containing saxitoxins (Velez et al., 2001), providing a measure of toxicity similar to that provided by the mouse bioassay, but without killing animals. However, the assay system requires specialized skills and equipment.

Cultured cells can also be used to assay for saxitoxins and other sodium channel toxins by treating with drugs that perturb normal membrane functions (Jellett et al., 1992; Manger et al., 1993, 1995). Saxitoxins, brevetoxins (Dickey et al., 1999), or ciguatoxins in added sample then alter the metabolism of the cells, resulting in a visible color change. In a slightly different approach (Louzao et al., 2001; Manger et al., 2007), changes in membrane potential are indicated directly, rather than needing to wait until the metabolism of the cell responds.

Finally, binding of saxitoxins or TTX in a sample can be detected directly by their dilution of radiolabeled saxitoxin (Davio and Fontelo, 1984; Poli, 1996; Negri and Llewellyn, 1998). This is the most direct way to transduce the binding events into a measurable signal, but it requires dealing with radioisotope restrictions. The receptor binding assay has been put into a very efficient, high-throughput format by Van Dolah et al. (1994) (Powell and Doucette, 1999; Ruberu et al., 2003). The sensitivity of this assay, its efficiency, and its favorable response spectrum make it the current method of choice for routine monitoring of PSP. With different reagents, the same equipment can also be used to monitor for brevetoxins and DOM (Van Dolah et al., 1997).

There are receptor-based assays for other families of seafood toxins. The utility of the mouse i.p. assay for the lipophilic toxins (brevetoxins, DSP toxins, etc.) depends on the substances present, since there are many lipid-soluble substances that may be present in extracts for which the ratio of HOP to mouse i.p. toxicity (the response factor of the assay system) is relatively low, while the ratio of HOP to mouse i.p. toxicity for the actual toxins is quite high. The response spectrum of the assay is not at all flat and is very unfavorable. The interfering substances include free fatty acids and various lipophilic toxins with high acute toxicity to mice i.p. but are not known to be a risk to human consumers. The mouse i.p. assay for brevetoxins (McFarren, 1971) has been relatively useful for the management of NSP along the U.S. Gulf Coast but problematic in New Zealand, where various lipophilic toxins occur and interfere, having high potency to mice i.p. but little apparent significance to consumer health. With respect to DSP, there is the fundamental question of the relevance of mouse i.p. lethality to diarrhea resulting from consumption. Ciguatoxin can also be detected by mouse i.p. assay (Hoffman et al., 1983), although, as with assays for NSP and DSP, the detection limit by the assay is not far below the level of concern to human consumers.

Ciguatoxins and brevetoxins bind to the same receptor site on the voltage-activated sodium channel (Lombet et al., 1987) and thus may be determined with the same receptor binding assay, employing tritiated brevetoxin, and with the same cytotoxicity assay.

Okadaic acid and other DSP toxins can be detected by their inhibition of protein phosphatase, using chromogenic or fluorogenic substrates (Mountfort et al., 1999, 2000).

However, native receptors also have limitations. In the case of the site 1 toxins, attachment of reporter groups to reagent toxins tends to significantly perturb the binding interaction. While it is tempting to equate binding affinity with oral toxicity, this may not be the case. And to the extent that it is, it measures actual rather than potential toxicity. Alternative receptors may therefore offer benefits as well as weaknesses.

The saxiphillins (Mahar et al., 1991) are proteins of unknown function, found in various taxa (frogs, snakes, centipedes, etc.) that have high affinities for some of the saxitoxins and can be employed in assays for PSP (Negri and Llewellyn, 1998; Llewellyn and Doyle, 2001).

The binding interaction of antibodies with analytes can be more rugged than that of native receptors and more tolerant of reporter groups attached to reagent toxins.

Seafood toxins for which immunoassays have been developed include saxitoxins (Usleber et al., 1991; Jellett et al., 2002), brevetoxins (Trainer and Baden, 1991; Poli et al., 1995), DOM (Garthwaite et al., 1998), okadaic acids (Chin et al., 1995), and palytoxins (Levine et al., 1988a, 1988b; Bignami, 1993). Most are enzyme immunoassays, allowing the binding event to be read as a color change.

There are also instrumental approaches to transduction, including direct detection of binding through changes in surface plasmon resonance (Fonfría et al., 2007) and labeling with electrochemiluminescent groups (Poli et al., 2007). Gawley et al. (1999) have synthesized crown ethers as artificial receptors for the saxitoxins.

SUMMARY

Seafood is generally wholesome but may be contaminated with potent natural toxins which can cause illness and death in consumers. Such risks can be greatly reduced by the implementation of appropriate, sustainable monitoring strategies. While toxicity monitoring is essential, monitoring for toxicity alone is relatively expensive. The cost-effectiveness of a monitoring program can be enhanced by integrating various sources of environmental information to focus toxicity monitoring on the times, locations, and toxins of greatest concern.

REFERENCES

Alcala, A. C., L. C. Alcala, J. S. Garth, D. Yasumura, and T. Yasumoto. 1988. Human fatality due to ingestion of the crab *Demania reynaudii* that contained a palytoxin-like toxin. *Toxicon* **26**:105–107.

AOAC. 2005. Paralytic shellfish poison. Biological method. *In* W. Horwitz (ed.), *Official Methods of Analysis of AOAC International*, 18th ed. AOAC International, Gaithersburg, MD. Official method 959.08.

Arakawa, O., S. Nishio, T. Noguchi, Y. Shida, and Y. Onoue. 1995. A new saxitoxin analog from a xanthid crab *Atergatis floridus*. *Toxicon* **33**:1577–1584.

Arakawa, O., T. Noguchi, Y. Shida, and Y. Onoue. 1994. Occurrence of carbamoyl-N-hydroxy derivatives of saxitoxin and neosaxitoxin in a xanthid crab *Zosimus aeneus*. *Toxicon* **32**:175–183.

Aune, T., O. B. Stabell, K. Nordstoga, and K. Tjotta. 1998. Oral toxicity in mice of algal toxins from the diarrheic shellfish toxin (DST) complex and associated toxins. *J. Nat. Toxins* **7**:141–158.

Bates, H. A., R. Kostriken, and H. Rapoport. 1978. A chemical assay for saxitoxin. Improvements and modifications. *J. Agric. Food Chem.* **26**:252–254.

Bates, S. S., C. J. Bird, A. S. W. de Freitas, R. Foxhall, M. W. Gilgan, L. A. Hanic, G. E. Johnson, A. W. McColloch, P. Odense,

R. Pocklington, M. A. Quilliam, J. C. Sim, J. C. Smith, D. V. Subba Rao, E. C. D. Todd, J. A. Walter, and J. L. C. Wright. 1989. Pennate diatom *Nitschia pungens* as the primary source of domoic acid, a toxin in shellfish from eastern Prince Edward Island. *Can. J. Fish. Aquat. Sci.* 46:1203–1215.

Beier, R. C., J. O. Norman, J. C. Reagor, M. S. Rees, and B. P. Mundy. 1993. Isolation of the major component in white snakeroot that is toxic after microsomal activation: possible explanation of sporadic toxicity of white snakeroot plants and extracts. *Nat. Toxins* 1:286–293.

Bignami, G. S. 1993. A rapid and sensitive hemolysis neutralization assay for palytoxin. *Toxicon* 31:817–820.

Botana, L. M. (ed.). 2007. *Phycotoxins: Chemistry and Biochemistry.* Blackwell, Ames, IA.

Botana, L. M. (ed.). 2008. *Seafood and Freshwater Toxins: Pharmacology, Physiology, and Detection,* 2nd ed. CRC Press, Boca Raton, FL.

Boyer, G. L., and G. D. Goddard. 2000. High performance liquid chromatography coupled with post-column electrochemical oxidation for the detection of PSP toxins. *Nat. Toxins* 7:353–359.

Bricelj, V. M., and S. E. Shumway. 1998. Paralytic shellfish toxins in bivalve molluscs: occurrence, transfer kinetics, and biotransformation. *Rev. Fish. Sci.* 6:315–383.

Buckley, L. J., Y. Oshima, and Y. Shimizu. 1978. Construction of a paralytic shellfish toxin analyzer and its application. *Anal. Biochem.* 85:157–164.

Burdaspal, P. A., J. Bustos, T. M. Legarda, J. Olmeda, M. Vigo, L. Gonzalez, and J. A. Berenguer. 1998. Commercial processing of Acanthocardia tuberculatum L. naturally contaminated with PSP. Evaluation after one year industrial experience. Xunta de Galicia and Intergovernmental Oceanographic Commission, UNESCO, Santiago de Compostela, Spain.

Catterall, W. A. 1985. The voltage-sensitive sodium channel: a receptor for multiple toxins, p. 329–342. *In* D. M. Anderson, A. W. White, and D. G. Baden (ed.), *Toxic Dinoflagellates.* Elsevier, New York, NY.

Catterall, W. A., and M. Gainer 1985. Interaction of brevetoxin a with a new receptor site on the sodium channel. *Toxicon* 23:497–504.

Chin, J. D., M. A. Quilliam, J. M. Fremy, S. K. Mohapatra, and H. M. Sikorska. 1995. Screening for okadaic acid by immunoassay. *J. AOAC Int.* 78:508–513.

Chou, H. N. 1986. Isolation and characterization of toxins in the "Red Tide" dinoflagellate *Gymnodinium breve* Davis (= *Ptychodiscus brevis*), Ph.D. dissertation. University of Rhode Island, Kingston, RI.

Chou, H. N., and Y. Shimizu. 1982. A new polyether toxin from *Gymnodinium breve* Davis. *Tetrahedron Lett.* 23:5521–5524.

Chou, H. N., Y. Shimizu, G. Van Duyne, and J. Clardy. 1985. Isolation and structures of two new polycyclic ethers from *Gymnodinium breve* Davis (= *Ptychodiscus brevis*). *Tetrahedron Lett.* 26:2865–2868.

Daigo, K. 1958. Studies on the constituents of *Chondria armata*. I. Detection of the anthelmintical constituents. *J. Pharm. Soc. Jpn.* 79:350–353.

Davio, S. R., and P. A. Fontelo. 1984. A competitive displacement assay to detect saxitoxin and tetrodotoxin. *Anal. Biochem.* 141:199–204.

Dickey, R., E. Jester, R. Granade, D. Mowdy, C. Moncreiff, D. Rebarchik, M. Robl, S. Musser, and M. Poli. 1999. Monitoring brevetoxins during a *Gymnodinium breve* red tide: comparison of sodium channel specific cytotoxicity assay and mouse bioassay for determination of neurotoxic shellfish toxins in shellfish extracts. *Nat. Toxins* 7:157–165.

Dickey, R. W., S. C. Bobzin, D. J. Faulkner, F. A. Bencsath, and D. Andrzejewski. 1990. Identification of okadaic acid from a Caribbean dinoflagellate, *Prorocentrum concavum*. *Toxicon* 28:371–377.

Fonfría, E. S., N. Vilariño, K. Campbell, C. Elliott, S. A. Haughey, B. Ben-Gigirey, J. M. Vieites, K. Kawatsu, and L. M. Botana. 2007. Paralytic shellfish poisoning detection by surface plasmon resonance-based biosensors in shellfish matrixes. *Anal. Chem.* 79:6303–6311.

Friedman, M. A., L. E. Fleming, M. Fernandez, P. Bienfang, K. Schrank, R. Dickey, M.-Y. Bottein, L. Backer, R. Ayyar, R. Weisman, S. Watkins, R. Granade, and A. Reich. 2008. Ciguatera fish poisoning: treatment, prevention and management. *Mar. Drugs* 6:456–479.

Fritz, L., M. A. Quilliam, J. L. C. Wright, A. M. Beale, and T. M. Work. 1992. An outbreak of domoic acid poisoning attributed to the pennate diatom *Pseudonitzschia australis*. *J. Phycol.* 28:439–442.

Fukui, M., M. Murata, A. Inoue, M. Gawel, and T. Yasumoto. 1987. Occurrence of palytoxin in the trigger fish *Melichtys vidua*. *Toxicon* 25:1121–1124.

Fukui, M., D. Yasumura, M. Murata, A. C. Alcala, and T. Yasumoto. 1988. The occurrence of palytoxin in crabs and fish. *Toxicon* 26:20–21.

Garcia, C., M. del Carmen Bravo, M. Lagos, and N. Lagos. 2004. Paralytic shellfish poisoning: post-mortem analysis of tissue and body fluid samples from human victims in the patagonia fjords. *Toxicon* 43:149–158.

Garrison, D. L., S. M. Conrad, P. P. Eilers, and E. M. Waldron. 1992. Confirmation of domoic acid production by *Pseudonitzschia australis* (Bacillariophyceae) cultures. *J. Phycol.* 28:604–607.

Garthwaite, I., K. M. Ross, C. O. Miles, R. P. Hansen, D. Foster, A. L. Wilkins, and N. R. Towers. 1998. Polyclonal antibodies to domoic acid, and their use in immunoassays for domoic acid in sea water and shellfish. *Nat. Toxins* 6:93–104.

Gawley, R. E., Q. Zhang, P. I. Higgs, S. Wang, and R. M. Leblanc. 1999. Anthracylmethyl crown ethers as fluorescence sensors of saxitoxin. *Tetrahedron Lett.* 40:5461–5465.

Gessner, B. D., P. Bell, G. J. Doucette, E. Moczydlowski, M. A. Poli, F. Van Dolah, and S. Hall. 1997. Hypertension and identification of toxin in human urine and serum following a cluster of mussel-associated paralytic shellfish poisoning outbreaks. *Toxicon* 35:711–722.

González, J. C., J. M. Vieites, A. M. Botana, M. R. Vieytes, and L. M. Botana. 1998. Improved sample clean-up in the HPLC/fluorimetric determination of okadaic acid using 1-bromoacetylpyrene as derivatizing reagent. *J. Chromatogr. A* 793:63–70.

Hall, S. 1982. Toxins and toxicity of *Protogonyaulax* from the northeast Pacific. Ph.D. dissertation. University of Alaska, Fairbanks, AK.

Hall, S., and P. B. Reichardt. 1984. Cryptic paralytic shellfish toxins. *ACS Symp. Ser.* 262:113–123.

Hall, S., P. B. Reichardt, and R. A. Neve. 1980. Toxins extracted from an Alaskan isolate of *Protogonyaulax* sp. *Biochem. Biophys. Res. Commun.* 97:649–653.

Hall, S., G. Strichartz, E. Moczydlowski, A. Ravindran, and P. B. Reichardt. 1990. The saxitoxins. sources, chemistry, and pharmacology. *ACS Symp. Ser.* 418:29–65.

Harada, T., Y. Oshima, H. Kamiya, and T. Yasumoto. 1982. Confirmation of paralytic shellfish toxins in the dinoflagellate Pyrodinium bahamense var. compressa and bivalves in Palau. *Bull. Jpn. Soc. Sci. Fish* 48:821–825.

Hirsh, J. K., and C. H. Wu. 1997. Palytoxin-induced single-channel currents from the sodium pump synthesized by in vitro expression. *Toxicon* 35:169–176.

Hoffman, P. A., H. R. Granade, and J. P. McMillan. 1983. The mouse ciguatoxin bioassay: a dose-response curve and symptomatology analysis. *Toxicon* 21:363–369.

Hu, T., J. M. Curtis, Y. Oshima, M. A. Quilliam, J. A. Walter, W. M. Watson-Wright, and J. L. C. Wright. 1995a. Spirolides B and D, two novel macrocycles isolated from the digestive glands of shellfish. *J. Chem. Soc. Chem. Commun.* **20:**2159–2161.

Hu, T., J. M. Curtis, J. A. Walter, J. L. McLachlan, and J. L. C. Wright. 1995b. Two new water-soluble dsp toxin derivatives from the dinoflagellate prorocentrum maculosum: possible storage and excretion products. *Tetrahedron Lett.* **36:**9273–9276.

Hu, T., J. M. Curtis, J. A. Walter, and J. L. C. Wright. 1995c. Identification of DTX-4, a new water-soluble phosphatase inhibitor from the toxic dinoflagellate *Prorocentrum lima.* *J. Chem. Soc. Chem. Commun.* **5:**597–599.

Hu, T., A. S. W. deFreitas, D. J. Jackson, J. Marr, E. Nixon, S. Pleasance, M. A. Quilliam, J. A. Walter, and J. L. C. Wright. 1993. New DSP toxin derivates isolated from toxic mussels and the dinoflagellates, Prorocentrum lima and Prorocentrum concavum, p. 507–512. *In* T. J. Smayda and Y. Shimizu (ed.), *Toxic Phytoplankton Blooms in the Sea,* Elsevier, New York, NY.

Hynie, I., and E. C. D. Todd. 1990. Proceedings of a symposium. Domoic acid toxicity. Ottawa, Ontario. 11–12 April 1989. *Can. Dis. Wkly. Rep.* **16**(Suppl. 1E):1–123.

Ishida, H., N. Muramatsu, H. Nukaya, T. Kosuge, and K. Tsuji. 1996. Study on neurotoxic shellfish poisoning involving the oyster, *Crassostrea gigas,* in New Zealand. *Toxicon* **34:**1050–1053.

Ishida, H., A. Nozawa, K. Totoribe, N. Muramatsu, H. Nukaya, K. Tsuji, K. Yamaguchi, T. Yasumoto, H. Kaspar, N. Berkett, and T. Kosuge. 1995. Brevetoxin B$_1$, a new polyether marine toxin from the New-Zealand shellfish, *Austrovenus stutchburyi.* *Tetrahedron Lett.* **36:**725–728.

Ito, E., M. Satake, K. Ofuji, N. Kurita, T. McMahon, K. James, and T. Yasumoto. 2000. Multiple organ damage caused by a new toxin azaspiracid, isolated from mussels produced in Ireland. *Toxicon* **38:**917–930.

Iverson, F., and J. Truelove. 1994. Toxicology and seafood toxins: domoic acid. *Nat. Toxins* **2:**334–339.

James, K. J., A. G. Bishop, R. Draisci, L. Palleschi, C. Marchiafava, E. Ferretti, M. Satake, and T. Yasumoto. 1999. Liquid chromatographic methods for the isolation and identification of new pectenotoxin-2 analogues from marine phytoplankton and shellfish. *J. Chromatogr. A* **844:**53–65.

James, K. J., A. Furey, M. Satake, and T. Yasumoto. 2001. Azaspiracid poisoning (azp): a new shellfish toxic syndrome in Europe, p. 250–253. *In* G. M. Hallegraeff, S. I. Blackburn, C. J. Bolch, and R. J. Lewis (ed.), *Harmful Algal Blooms.* Intergovernmental Oceanographic Commission, UNESCO, Paris, France.

James, K. J., C. Moroney, C. Roden, M. Satake, T. Yasumoto, M. Lehane, and A. Furey. 2003a. A ubiquitous benign algae emerges as the cause of shellfish contamination responsible for the human toxin syndrome, azaspiracid poisoning. *Toxicon* **41:**145–151.

James, K. J., M. D. Sierra, M. Lehane, A. Brana Magdalena, and A. Furey. 2003b. Detection of five new hydroxyl analogues of azaspiracids in shellfish using multiple tandem mass spectrometry. *Toxicon* **41:**277–283.

Janecek, M., M. A. Quilliam, and J. F. Lawrence. 1993. Analysis of paralytic shellfish poisoning toxins by automated precolumn oxidation and microcolumn liquid chromatography with fluorescence detection. *J. Chromatogr. A* **644:**321–331.

Jasperse, J. A. (ed.). 1993. Marine toxins and New Zealand shellfish: proceedings of a workshop on research issues, 10–11 June 1993. Royal Society of New Zealand, Wellington.

Jellett, J. F., L. J. Marks, L. J. Marks, J. E. Stewart, M. L. Dorey, W. Watson Wright, and J. F. Lawrence. 1992. Paralytic shellfish poison (saxitoxin family) bioassays: automated endpoint determination and standardization of the *in vitro* tissue culture bioassay, and comparison with the standard mouse bioassay. *Toxicon* **30:**1143–1156.

Jellett, J. F., R. L. Roberts, M. V. Laycock, M. A. Quilliam, and R. E. Barrett. 2002. Detection of paralytic shellfish poisoning (PSP) toxins in shellfish tissue using MIST Alert, a new rapid test, in parallel with the regulatory AOAC mouse bioassay. *Toxicon* **40:**1407–1425.

Kao, C. Y. 1966. Tetrodotoxin, saxitoxin and their significance in the study of excitation phenomena. *Pharmacol. Rev.* **18:**997–1049.

Kodama, A. M., Y. Hokama, T. Yasumoto, M. Fukui, S. J. Manea, and N. Sutherland. 1989. Clinical and laboratory findings implicating palytoxin as cause of ciguatera poisoning due to Decapterus macrosoma (mackerel). *Toxicon* **27:**1051–1053.

Kodama, M., S. Sato, S. Sakamoto, and T. Ogata. 1996. Occurrence of tetrodotoxin in *Alexandrium tamarense,* a causative dinoflagellate of paralytic shellfish poisoning. *Toxicon* **34:**1101–1105.

Landsberg, J. H., S. Hall, J. N. Johannessen, K. D. White, S. M. Conrad, J. P. Abbott, L. J. Flewelling, R. W. Richardson, R. W. Dickey, E. L. E. Jester, S. M. Etheridge, J. R. Deeds, F. M. Van Dolah, T. A. Leighfield, Y. Zou, C. G. Beaudry, R. A. Benner, P. L. Rogers, P. S. Scott, K. Kawabata, J. L. Wolny, and K. A. Steidinger. 2006. Saxitoxin puffer fish poisoning in the United States, with the first report of pyrodinium bahamense as the putative toxin source. *Environ. Health Perspect.* **114:**1502–1507.

Lawrence, J. F., C. F. Charbonneau, C. Menard, M. A. Quilliam, and P. G. Sim. 1989. Liquid chromatographic determination of domoic acid in shellfish products using the paralytic shellfish poison extraction procedure of the Association of Official Analytical Chemists. *J. Chromatogr. A* **462:**349–356.

Lawrence, J. F., B. Niedzwiadek, and M. Cathie. 2005. Quantitative determination of paralytic shellfish poisoning toxins in shellfish using prechromatographic oxidation and liquid chromatography with fluorescence detection: collaborative study. *J. AOAC Int.* **88:**1714–1732.

LeDoux, M., and S. Hall. 2000. Proficiency testing of eight French laboratories in using the AOAC mouse bioassay for paralytic shellfish poisoning: interlaboratory collaborative study. *J. AOAC Int.* **83:**305–310.

Lee, J. S., T. Igarashi, S. Fraga, E. Dahl, P. Hovgaard, and T. Yasumoto. 1989. Determination of diarrhetic shellfish toxins in various dinoflagellate species. *J. Appl. Phycol.* **1:**147–152.

Lee, J. S., Y. Yanagi, R. Kenma, and T. Yasumoto. 1987. Fluorometric determination of diarrhetic shellfish toxins by high-performance liquid chromatography. *Agric. Biol. Chem.* **51:**877–881.

Lehane, M., A. Braña-Magdalena, C. Moroney, A. Furey, and K. J. James. 2002. Liquid chromatography with electrospray ion trap mass spectrometry for the determination of five azaspiracids in shellfish. *J. Chromatogr. A* **950:**139–147.

Levine, L., H. Fujiki, H. B. Gjika, and H. Van Vunakis 1988a. A radioimmunoassay for palytoxin. *Toxicon* **26:**1115–1121.

Levine, L., H. Fujiki, K. Yamada, M. Ojika, H. B. Gjika, and H. Van Vunakis. 1988b. Production of antibodies and development of a radioimmunoassay for okadaic acid. *Toxicon* **26:**1123–1128.

Lewis, R. J., and R. Endean 1984. Ciguatoxin from the flesh and viscera of the barracuda, *Sphyraena jello.* *Toxicon* **22:**805–810.

Lewis, R. J., and M. J. Holmes. 1993. Origin and transfer of toxins involved in ciguatera. *Comp. Biochem. Physiol. C* **106:**615–628.

Lewis, R. J., and A. Jones. 1997. Characterization of ciguatoxins and ciguatoxin congeners present in ciguateric fish by gradient reverse-phase high-performance liquid-chromatography mass-spectrometry. *Toxicon* 35:159–168.

Lewis, R. J., A. Jones, and J. P. Vernoux. 1999. HPLC/tandem electrospray mass spectrometry for the determination of sub-ppb levels of Pacific and Caribbean ciguatoxins in crude extracts of fish. *Anal. Chem.* 71:247–250.

Lewis, R. J., R. S. Norton, I. M. Brereton, and C. D. Eccles. 1993. Ciguatoxin-2 is a diastereomer of ciguatoxin-3. *Toxicon* 31:637–643.

Lewis, R. J., and T. A. Ruff. 1993. Ciguatera: ecological, clinical, and socioeconomic perspectives. *Crit. Rev. Environ. Sci. Tech.* 23:137–156.

Lin, Y. Y., M. Risk, S. M. Ray, D. Van Engen, J. Clardy, J. Golik, J. C. James, and K. Nakanishi. 1981. Isolation and structure of brevetoxin B from the "red tide" dinoflagellate *Ptychodiscus brevis (Gymnodinium breve)*. *J. Am. Chem. Soc.* 103:6773–6775.

Llewellyn, L. E., and J. Doyle. 2001. Microtitre plate assay for paralytic shellfish toxins using saxiphilin: gauging the effects of shellfish extract matrices, salts and pH upon assay performance. *Toxicon* 39:217–224.

Lombet, A., J. N. Bidard, and M. Lazdunski. 1987. Ciguatoxin and brevetoxins share a common receptor site on the neuronal voltage-dependent sodium channel. *FEBS Lett.* 219:355–359.

Louzao, M. C., M. R. Vieytes, J. M. V. Baptista de Sousa, F. Leira, and L. M. Botana. 2001. A fluorimetric method based on changes in membrane potential for screening paralytic shellfish toxins in mussels. *Anal. Biochem.* 289:246–250.

Maclean, J. L. 1989. Indo-Pacific red tides, 1985-1988. *Mar. Pollut. Bull.* 20:304–310.

Magdalena, A., M. Lehane, S. Krys, M. L. Fernandez, A. Furey, and K. James. 2003. The first identification of azaspiracids in shellfish from France and Spain. *Toxicon* 42:105–108.

Mahar, J., G. L. Lukacs, Y. Li, S. Hall, and E. Moczydlowski. 1991. Pharmacological and biochemical-properties of saxiphilin, a soluble saxitoxin-binding protein from the bullfrog *(Rana catesbeiana)*. *Toxicon* 29:51–71.

Manger, R. L., L. S. Leja, S. Y. Lee, J. M. Hungerford, Y. Hokama, R. W. Dickey, H. R. Granade, R. Lewis, T. Yasumoto, and M. M. Wekell. 1995. Detection of sodium-channel toxins: directed cytotoxicity assays of purified ciguatoxins, brevetoxins, saxitoxins, and seafood extracts. *J. AOAC Int.* 78:521–527.

Manger, R. L., L. S. Leja, S. Y. Lee, J. M. Hungerford, and M. M. Wekell. 1993. Tetrazolium-based cell bioassay for neurotoxins active on voltage-sensitive sodium-channels: semiautomated assay for saxitoxins, brevetoxins, and ciguatoxins. *Anal. Biochem.* 214:190–194.

Manger, R., D. Woodle, A. Berger, and J. Hungerford. 2007. Flow cytometric detection of saxitoxins using fluorescent voltage-sensitive dyes. *Anal. Biochem.* 366:149–155.

Marr, J. C., L. McDowell, and M. A. Quilliam. 1994. Investigation of derivatization reagents for the analysis of diarrhetic shellfish poisoning toxins by liquid chromatography with fluorescence detection. *Nat. Toxins* 2:302–311.

Maynes, J. T., K. S. Bateman, M. M. Cherney, A. K. Das, H. A. Luu, C. F. Holmes, and M. N. James. 2001. Crystal structure of the tumor-promoter okadaic acid bound to protein phosphatase-1. *J. Biol. Chem.* 276:44078–44082.

McCulloch, A. W., R. K. Boyd, A. S. de Freitas, R. A. Foxall, W. D. Jamieson, M. V. Laycock, M. A. Quilliam, J. L. Wright, V. J. Boyko, and J. W. McLaren. 1989. Zinc from oyster tissue as causative factor in mouse deaths in official bioassay for paralytic shellfish poison. *J. Assoc. Off. Anal. Chem.* 72:384–386.

McFarren, E. F. 1971. Assay and control of marine biotoxins. *Food Technol.* 25:38–46.

McFarren, E. F., F. J. Silva, H. Tanabe, W. B. Wilson, J. E. Campbell, and K. H. Lewis. 1965. The occurrence of a ciguatera-like poison in oysters, clams, and *Gymnodinium breve* cultures. *Toxicon* 3:111–123.

McNabb, P., A. I. Selwood, P. T. Holland, J. Aasen, T. Aune, G. Easlesham, P. Hess, M. Igarishi, M. Quilliam, D. Slattery, J. Van de Riet, H. Van Egmond, H. Van den Top, and T. Yasumoto. 2005. Multiresidue method for determination of algal toxins in shellfish: single-laboratory validation and inter-laboratory study. *J. AOAC Int.* 88:761–772.

Melton, R. J., J. E. Randall, N. Fusetani, R. S. Weiner, R. D. Couch, and J. K. Sims. 1984. Fatal sardine poisoning—a fatal case of fish poisoning in Hawaii associated with the Marquesan sardine. *Hawaii Med. J.* 43:114–124.

Mitchell, G. 2005. Treatment of a mild chronic case of ciguatera fish poisoning with intravenous mannitol, a case study. *Pac. Health Dialog.* 12:155–157.

Moczydlowski, E., S. Hall, S. S. Garber, G. S. Strichartz, and C. Miller. 1984. Voltage-dependent blockade of muscle Na+ channels by guanidinium toxins. *J. Gen. Physiol.* 84:687–704.

Moore, R. E., and P. J. Scheuer. 1971. Palytoxin: a new marine toxin from a coelenterate. *Science* 172:495–498.

Morohashi, A., M. Satake, K. Murata, H. Naoki, H. F. Kaspar, and T. Yasumoto. 1995. Brevetoxin B3, a new brevetoxin analog isolated from the greenshell mussel *Perna canaliculus* involved in neurotoxic shellfish poisoning in New Zealand. *Tetrahedron Lett.* 36:8995–8998.

Morohashi, A., M. Satake, H. Naoki, H. F. Kaspar, Y. Oshima, and T. Yasumoto. 1999. Brevetoxin B4 isolated from Greenshell mussels *Perna canaliculus*, the major toxin involved in neurotoxic shellfish poisoning in New Zealand. *Nat. Toxins* 7:45–48.

Mountfort, D. O., G. Kennedy, I. Garthwaite, M. Quilliam, P. Truman, and D. J. Hannah. 1999. Evaluation of the fluorometric protein phosphatase inhibition assay in the determination of okadaic acid in mussels. *Toxicon* 37:909–922.

Mountfort, D. O., T. Suzuki, and P. Truman. 2000. Protein phosphatase inhibition assay adapted for determination of total DSP in contaminated mussels. *Toxicon* 39:383–390.

Murakami, Y., Y. Oshima, and T. Yasumoto. 1982. Identification of okadaic acid as a toxic component of a marine dinoflagellate *Prorocentrum lima*. *Bull. Jap. Soc. Sci. Fish.* 48:69–72.

Murata, K., M. Satake, H. Naoki, H. F. Kaspar, and T. Yasumoto. 1998. Isolation and structure of a new brevetoxin analog, brevetoxin B2, from greenshell mussels from New Zealand. *Tetrahedron* 54:735–742.

Murata, M., M. Kumagai, J. S. Lee, and T. Yasumoto. 1987. Isolation and structure of yessotoxin, a novel polyether compound implicated in diarrhetic shellfish poisoning. *Tetrahedron Lett.* 28:5869–5872.

Murata, M., A. M. Legrand, Y. Ishibashi, M. Fukui, and T. Yasumoto. 1989. Structures of ciguatoxin and its congener. *J. Am. Chem. Soc.* 111:8929–8931.

Murata, M., A. M. Legrand, Y. Ishibashi, M. Fukui, and T. Yasumoto. 1990. Structures and configurations of ciguatoxin from the moray eel *Gymnothorax javanicus* and its likely precursor from the dinoflagellate *Gambierdiscus toxicus*. *J. Am. Chem. Soc.* 112:4380–4386.

Murata, M., M. Sano, T. Iwashita, H. Naoki, and T. Yasumoto. 1986. The structure of pectenotoxin-3, a new constituent of diarrhetic shellfish toxins. *Agric. Biol. Chem.* 50:2693–2695.

Murata, M., M. Shimatani, H. Sugitani, Y. Oshima, and T. Yasumoto. 1982. Isolation and structural elucidation of the causative toxin of the diarrhetic shellfish poisoning. *Bull. Jpn. Soc. Sci. Fish.* 48:549–552.

Negri, A. P., C. J. S. Bolch, S. Geiera, D. H. Green, T.-G. Park, and S. I. Blackburn. 2007. Widespread presence of hydrophobic paralytic shellfish toxins in Gymnodinium catenatum. *Harmful Algae* **6:**774–780.

Negri, A., and L. Llewellyn. 1998. Comparative analyses by HPLC and the sodium channel and saxiphilin 3H-saxitoxin receptor assays for paralytic shellfish toxins in crustaceans and molluscs from tropical North West Australia. *Toxicon* **36:**283–298.

Negri, A., D. Stirling, M. Quilliam, S. Blackburn, C. Bolch, I. Burton, G. Eaglesham, K. Thomas, R. H. Willis, and J. Walter. 2003. Three novel hydroxybenzoate saxitoxin analogues isolated from the dinoflagellate Gymnodinium catenatum. *Chem. Res. Toxicol.* **16:**1029–1033.

Noguchi, T., and K. Hashimoto. 1984. Current topics on swellfish poisons. Cultured swellfishes possess little or no poison. *Kagaku* **39:**155–160.

Noguchi, T., D. F. Hwang, O. Arakawa, K. Daigo, S. Sato, H. Ozaki, N. Kawai, M. Ito, and K. Hashimoto. 1988. Palytoxin as the causative agent in parrotfish poisoning. *Toxicon* **26:**34.

Noguchi, T., D. F. Hwang, O. Arakawa, H. Sugita, Y. Deguchi, Y. Shida, and K. Hashimoto. 1987. *Vibrio alginolyticus*, a tetrodotoxin-producing bacterium, in the intestines of the fish *Fugu vermicularis vermicularis*. *Mar. Biol.* **94:**625–630.

Noguchi, T., J. K. Jeon, O. Arakawa, H. Sugita, Y. Deguchi, Y. Shida, and K. Hashimoto. 1986. Occurrence of tetrodotoxin and anhydrotetrodotoxin in *Vibrio* sp. isolated from the intestines of a xanthid crab, *Atergatis floridus*. *J. Biochem.* (Tokyo) **99:**311–314.

Ofuji, K., M. Satake, T. McMahon, K. J. James, H. Naoki, Y. Oshima, and T. Yasumoto. 2001. Structures of azaspiracid analogs, azaspiracid-4 and azaspiracid-5, causative toxins of azaspiracid poisoning in Europe. *Biosci. Biotechnol. Biochem.* **65:**740–742.

Ofuji, K., M. Satake, T. McMahon, J. Silke, K. J. James, H. Naoki, Y. Oshima, and T. Yasumoto. 1999. Two analogs of azaspiracid isolated from mussels, Mytilus edulis, involved in human intoxication in Ireland. *Nat. Toxins* **7:**99–102.

Onuma, Y., M. Satake, T. Ukena, J. Roux, S. Chanteau, N. Rasolofonirina, M. Ratsimaloto, H. Naoki, and T. Yasumoto. 1999. Identification of putative palytoxin as the cause of clupeotoxism. *Toxicon* **37:**55–65.

Oshima, Y. 1995. Postcolumn derivatization liquid chromatographic method for paralytic shellfish toxins. *J. AOAC Int.* **78:**528–532.

Oshima, Y., S. I. Blackburn, and G. M. Hallegraeff. 1993. Comparative study on paralytic shellfish toxin profiles of the dinoflagellate *Gymnodinium catenatum* from three different countries. *Mar. Biol.* **116:**471–476.

Oshima, Y., T. Hayakawa, M. Hashimoto, Y. Kotaki, and T. Yasumoto. 1982. Classification of *Protogonyaulax tamarensis* from northern Japan into three strains by toxin composition. *Bull. Jpn. Soc. Sci. Fish.* **48:**851–854.

Palafox, N. A., L. G. Jain, A. Z. Pinano, T. M. Gulick, R. K. Williams, and I. J. Schatz. 1988. Successful treatment of ciguatera fish poisoning with intravenous mannitol. *JAMA* **259:**2740–2742.

Pearn, J. H., R. J. Lewis, T. Ruff, M. Tait, J. Quinn, W. Murtha, G. King, A. Mallett, and N. C. Gillespie. 1990. Ciguatera and mannitol: experience with a new treatment regimen. *Med. J. Aust.* **153:**306–307.

Perl, T. M., L. Bedard, T. Kosatsky, J. C. Hockin, E. C. Todd, and R. S. Remis. 1990. An outbreak of toxic encephalopathy caused by eating mussels contaminated with domoic acid. *N. Engl. J. Med.* **322:**1775–1780.

Pocklington, R., J. E. Milley, S. S. Bates, C. J. Bird, A. S. W. de Freitas, and M. A. Quilliam. 1990. Trace determination of domoic acid in seawater and phytoplankton by high-performance liquid chromatography of the fluorenylmethoxycarbonyl (FMOC) derivative. *Int. J. Environ. Anal. Chem.* **38:**351–368.

Poli, M. A. 1996. Three dimensional binding assays for the detection of marine toxins. *In* D. G. Baden (ed.), *Proceedings of the Workshop Conference on Seafood Intoxications: Pan American Implications of Natural Toxins in Seafood.* University of Miami, Miami, FL.

Poli, M. A., S. M. Musser, R. W. Dickey, P. P. Eilers, and S. Hall. 2000. Neurotoxic shellfish poisoning and brevetoxin metabolites: a case study from Florida. *Toxicon* **38:**981–993.

Poli, M. A., K. S. Rein, and D. G. Baden. 1995. Radioimmunoassay for PbTx-2-type brevetoxins: epitope specificity of two anti-PbTx sera. *J. AOAC Int.* **78:**538–542.

Poli, M. A., V. R. Rivera, D. D. Neal, D. G. Baden, S. A. Messer, S. M. Plakas, R. W. Dickey, K. E. Said, L. Flewelling, D. Green, and J. White. 2007. An electrochemiluminescence-based competitive displacement immunoassay for the type-2 brevetoxins in oyster extracts. *J. AOAC Int.* **90:**173–178.

Powell, C. L., and G. J. Doucette. 1999. A receptor binding assay for paralytic shellfish poisoning toxins: recent advances and applications. *Nat. Toxins.* **7:**393–400.

Quilliam, M. A. 1995. Analysis of diarrhetic shellfish poisoning toxins in shellfish tissue by liquid-chromatography with fluorometric and mass-spectrometric detection. *J. AOAC Int.* **78:**555–570.

Quilliam, M. A., P. G. Sim, A. W. McCulloch, and A. G. McInnes. 1989. High-performance liquid chromatography of domoic acid, a marine neurotoxin, with application to shellfish and plankton. *Int. J. Environ. Anal. Chem.* **36:**139–154.

Quilliam, M., D. Wechsler, S. Marcus, B. Ruck, M. Wekell, and T. Hawryluk. 2004. Detection and identification of paralytic shellfish poisoning toxins in Florida pufferfish responsible for incidents of neurologic illness, p. 116–118. *In* K. A. Steidinger, J. H. Landsberg, C. R. Tomas, and G. A. Vargo (ed.), *Harmful Algae 2002.* Florida Fish and Wildlife Conservation Commission, Florida Institute of Oceanography, and Inter-governmental Oceanographic Commission of United Nations Educational, Scientific and Cultural Organization, St. Petersburg, FL.

Rodrigue, D. C., R. A. Etzel, S. Hall, E. de Porras, O. H. Velasquez, R. V. Tauxe, E. M. Kilbourne, and P. A. Blake. 1990. Lethal paralytic shellfish poisoning in Guatemala. *Am. Soc. Trop. Med. Hyg.* **42:**267–271.

Ruberu, S. R., Y. G. Liu, C. T. Wong, S. K. Perera, G. W. Langlois, G. J. Doucette, and C. L. Powell. 2003. Receptor binding assay for paralytic shellfish poisoning toxins: optimization and inter-laboratory comparison. *J. AOAC Int.* **86:**737–745.

Satake, M., M. Murata, and T. Yasumoto. 1993. The structure of CTX3C, a ciguatoxin congener isolated from cultured *Gambierdiscus toxicus*. *Tetrahedron Lett.* **34:**1975–1978.

Satake, M., K. Ofuji, H. Naoki, K. J. James, A. Furey, T. McMahon, J. Silke, and T. Yasumoto. 1998. Azaspiracid, a new marine toxin having unique spiro ring assemblies, isolated from Irish mussels, Mytilus edulis. *J. Am. Chem. Soc.* **120:**9967–9968.

Schnorf, H., M. Taurarii, and T. Cundy. 2002. Ciguatera fish poisoning: a double-blind randomized trial of mannitol therapy. *Neurology* **58:**873–880.

Seki, T., M. Satake, L. Mackenzie, H. F. Kaspar, and T. Yasumoto. 1995. Gymnodimine, a new marine toxin of unprecedented structure isolated from New Zealand oysters and the dinoflagellate, *Gymnodinium* sp. *Tetrahedron Lett.* **36:**7093–7096.

Shimizu, Y., H. N. Chou, H. Bando, G. Van Duyne, and J. Clardy. 1986. Structure of brevetoxin A (GB-1 toxin), the most potent

246 HALL

toxin in the Florida red tide organism *Gymnodinium breve*
(*Ptychodiscus brevis*). *J. Am. Chem. Soc.* **108**:514–515.

Sommer, H., and K. F. Meyer 1937. Paralytic shellfish poisoning.
Arch. Pathol. **24**:560–598.

Steidinger, K. A., L. S. Tester, and F. J. R. Taylor. 1980. A redescription of Pyrodinium bahamense var. compressa (Böhm) stat. nov. from Pacific red tides. *Phycologia* **19**:329–337.

Sullivan, J. J., and W. T. Iwaoka. 1983. High pressure liquid chromatography determination of toxins associated with paralytic shellfish poisoning. *J. Assoc. Off. Anal. Chem.* **66**:297–303.

Suzuki, T., T. Jin, Y. Shirota, T. Mitsuya, Y. Okumura, and T. Kamiyama. 2005. Quantification of lipophilic toxins associated with diarrhetic shellfish poisoning in Japanese bivalves by liquid chromatography–mass spectrometry and comparison with mouse bioassay. *Fish. Sci.* **71**:1370–1378.

Tachibana, K., P. J. Scheuer, Y. Tsukitani, H. Kikuchi, D. Van Engen, J. Clardy, Y. Gopichand, and F. J. Schmitz. 1981. Okadaic acid, a cytotoxic polyether from two marine sponges of the genus *Halichondria*. *J. Am. Chem. Soc.* **103**:2469–2471.

Takemoto, T., K. Daigo, Y. Kondo, and K. Kondo. 1966. Studies on the constituents of *Chondria armata*. VIII. On the structure of domoic acid. I. *J. Pharm. Soc. Jpn.* **86**:874–877. (In Japanese.)

Teitelbaum, J. S., R. J. Zatorre, S. Carpenter, D. Gendron, A. C. Evans, A. Gjedde, and N. R. Cachman. 1990. Neurologic sequelae of domoic acid intoxication due to the ingestion of contaminated mussels. *N. Engl. J. Med.* **322**:1781–1787.

Todd, E. C. D. 1993. Domoic acid and amnesic shellfish poisoning: a review. *J. Food Prot.* **56**:69–83.

Torigoe, K., M. Murata, T. Yasumoto, and T. Iwashita. 1988. Prorocentrilide, a toxic nitrogenous macrocycle from a marine dinoflagellate, *Prorocentrum lima*. *J. Am. Chem. Soc.* **110**:7876–7877.

Trainer, V. L., and D. G. Baden. 1991. An enzyme immunoassay for the detection of Florida red tide brevetoxins. *Toxicon* **29**:1387–1394.

Turrell, E. A., and L. Stobo. 2007. A comparison of the mouse bioassay with liquid chromatography-mass spectrometry for the detection of lipophilic toxins in shellfish from Scottish waters. *Toxicon* **50**:442–447.

Ukena, T., M. Satake, M. Usami, Y. Oshima, H. Naoki, T. Fujita, Y. Kan, and T. Yasumoto. 2001. Structure elucidation of ostreocin D, a palytoxin analog isolated from the Dinoflagellate Ostreopsis siamensis. *Biosci. Biotechnol. Biochem.* **65**:2585–2588.

Usami, M., M. Satake, S. Ishida, A. Inoue, Y. Kan, and T. Yasumoto. 1995. Palytoxin analogs from the dinoflagellate Ostreopsis siamensis. *J. Am. Chem. Soc.* **117**:5389–5390.

Usleber, E., E. Schneider, and G. Terplan. 1991. Direct enzyme immunoassay in microtitration plate and test strip format for the detection of saxitoxin in shellfish. *Lett. Appl. Microbiol.* **13**:275–277.

Vale, P., R. Bireb, and P. Hess. 2008. Confirmation by LC–MS/MS of azaspiracids in shellfish from the Portuguese north-western coast. *Toxicon* **51**:1449–1456.

Van Dolah, F. M., E. L. Finley, B. L. Haynes, G. J. Doucette, P. D. Moeller, and J. S. Ramsdell. 1994. Development of rapid and sensitive high throughput pharmacologic assays for marine phycotoxins. *Nat. Toxins* **2**:189–196.

Van Dolah, F. M., T. A. Leighfield, B. L. Haynes, D. R. Hampson, and J. S. Ramsdell. 1997. A microplate receptor assay for the amnesic shellfish poisoning toxin, domoic acid, utilizing a cloned glutamate-receptor. *Anal. Biochem.* **245**:102–105.

Van Duyne, G. D. 1990. X-ray crystallographic studies of marine toxins. *ACS Symp. Ser.* **418**:144–165.

Velez, P., J. Sierralta, C. Alcayaga, M. Fonseca, H. Loyola, D. C. Johns, G. F. Tomaselli, E. Marban, and B. A. Suarez-Isla. 2001. A functional assay for paralytic shellfish toxins that uses recombinant sodium channels. *Toxicon* **39**:929–935.

Vernoux, J. P., and R. J. Lewis. 1997. Isolation and characterisation of Caribbean ciguatoxins from the horse-eye jack (*Caranx latus*). *Toxicon* **35**:889–900.

Vick, J. A., and J. S. Wiles. 1975. The mechanism of action and treatment of palytoxin poisoning. *Toxicol. Appl. Pharmacol.* **34**:214–223.

Waskiewicz, S., C. J. Carlson, H. W. Magnussen, E. J. Craven, and D. M. Galerman. 1951. *In* H. W. Magnussen and C. J. Carlson (ed.), *A Study of Processing Methods for Toxic Butter Clams*. Fisheries Experimental Commission of Alaska, Ketchikan, AK.

Wekell, J. C., E. J. Gauglitz, H. J. Barnett, C. L. Hatfield, and M. Eklund. 1994. The occurrence of domoic acid in razor clams (*Siliqua patula*), Dungeness crab (*Cancer magister*), and anchovies (*Engraulis mordax*). *J. Shellfish Res.* **13**:587–593.

Withers, N. W. 1982. Ciguatera fish poisoning. *Ann. R. Med.* **33**:97–111.

Work, T. M., B. Barr, A. M. Beale, L. Fritz, M. A. Quilliam, and J. L. C. Wright. 1993. Epidemiology of domoic acid poisoning in brown pelicans (*Pelecanus occidentalis*) and Brandt cormorants (*Phalacrocorax penicillatus*) in California. *J. Zoo Wild. Med.* **24**:54–62.

Wright, J. L. 1998. Domoic acid—ten years after. *Nat. Toxins* **6**:91–92.

Wright, J. L., C. J. Bird, A. S. de Freitas, D. Hampson, J. McDonald, and M. A. Quilliam. 1990. Chemistry, biology, and toxicology of domoic acid and its isomers. *Can. Dis. Wkly. Rep.* **16**(Suppl. 1E):21–26.

Wright, J. L. C., R. K. Boyd, A. S. W. de Freitas, M. Falk, R. A. Foxhall, W. D. Jamieson, M. V. Laycock, A. W. McCulloch, A. G. McInnes, P. Odense, V. P. Pathak, M. A. Quilliam, M. A. Ragan, P. G. Sim, P. Thibault, J. A. Walter, M. Gilgan, D. J. A. Richard, and D. Dewar. 1989. Identification of domoic acid, a neuroexcitatory amino acid, in toxic mussels from eastern Prince Edward Island. *Can. J. Chem.* **67**:481–490.

Yasumoto, T., A. Inoue, W. Sugawara, Y. Fukuyo, H. Oguri, T. Igarashi, and N. Fujita. 1980a. Environmental studies on a toxic dinoflagellate responsible for ciguatera. *Bull. Jpn. Soc. Sci. Fish.* **46**:1397–1404.

Yasumoto, T., and T. Michishita. 1985. Fluorometric determination of tetrodotoxin by high-performance liquid chromatography. *Agric. Biol. Chem.* **49**:3077–3080.

Yasumoto, T., M. Murata, Y. Oshima, M. Sano, G. K. Matsumoto, and J. Clardy. 1985. Diarrhetic shellfish toxins. *Tetrahedron* **41**:1019–1025.

Yasumoto, T., I. Nakajima, R. Bagnis, and R. Adachi. 1977. Finding of a dinoflagellate as a likely culprit of ciguatera. *Bull. Jpn. Soc. Sci. Fish.* **43**:1021–1026.

Yasumoto, T., Y. Oshima, W. Sugawara, Y. Fukuyo, H. Oguri, T. Igarashi, and N. Fujita. 1980b. Identification of *Dinophysis fortii* as the causative organism of diarrhetic shellfish poisoning. *Bull. Jpn. Soc. Sci. Fish.* **46**:1405–1412.

Yasumoto, T., Y. Oshima, and M. Yamaguchi. 1978. Occurrence of a new type of shellfish poisoning in the Tohoku District. *Bull. Jpn. Soc. Sci. Fish.* **44**:1249–1255.

Yasumoto, T., and M. Satake. 1996. Chemistry, etiology and determination methods of ciguatera toxins. *J. Toxicol. Toxin Rev.* **15**:91–107.

Yasumoto, T., D. Yasumura, Y. Ohizumi, M. Takahashi, A. C. Alcala, and L. C. Alcala. 1986a. Palytoxin in two species of xanthid crab from the Philippines. *Agric. Biol. Chem.* **50**:163–167.

Yasumoto, T., D. Yasumura, M. Yotsu, T. Michishita, A. Endo, and Y. Kotaki. 1986b. Bacterial production of tetrodotoxin and anhydrotetrodotoxin. *Agric. Biol. Chem.* **50**:793–795.

Yasumoto, T., and M. Yotsu-Yamashita. 1996. Chemical and etiologic studies on tetrodotoxin and its analogs. *J. Toxicol. Toxin Rev.* **15**:81–90.

Yotsu, M., T. Yamazaki, Y. Meguro, A. Endo, M. Murata, H. Naoki, and T. Yasumoto. 1987. Production of tetradotoxin and its derivatives by *Pseudomonas* sp. isolated from the skin of a pufferfish. *Toxicon* **25**:225–228.

Pathogens and Toxins in Foods: Challenges and Interventions
Edited by V. K. Juneja and J. N. Sofos
© 2010 ASM Press, Washington, DC

Chapter 16

Biogenic Amines in Foods

K. Koutsoumanis, C. Tassou, and G.-J. E. Nychas

Biogenic amines are basic nitrogenous compounds with low molecular weight and aliphatic, aromatic, or heterocyclic structures (Table 1). Histamine, putrescine, cadaverine, tyramine, tryptamine, phenylethylamine, spermine, spermidine, and agmatine are designated biogenic because they are formed by the action of living organisms. Biogenic amines are indispensable components of living cells with important physiological and biological activities. Their formation in food and beverages is a result of enzymic activity on food components or microbial decarboxylation of amino acids, but it has been found that some of the aliphatic amines can be formed by amination and transamination of aldehydes and ketones. The significance of biogenic amines in food science is based mainly on their implication as the causative agents in a number of food poisoning episodes and on the fact that they are able to initiate various physiological reactions. Histamine has been implicated in several outbreaks of food poisoning associated mainly with fish consumption, while tyramine has been proposed as the initiator of hypertensive crisis. Other amines like cadaverine, putrescine, and tyramine appear to enhance the toxicity of histamine. Moreover, biogenic amines are considered potential carcinogens because of their ability to react with nitrites and form carcinogenic nitrosamines. In addition to the above safety issues, the significance of biogenic amines is also related to their use as quality indicators. Indeed, the fact that most biogenic amines are absent or present in low levels in fresh foods, while their concentrations increase during storage as a result of microbial metabolic activity, has led many researchers to examine their use as potential indicators of food spoilage. The above characteristics have made

biogenic amines an important research field for many food scientists for the last 50 years. A lot of information on the significance of the biogenic amines in foods is given in a large number of reviews (Rice at al., 1976; ten Brink et al., 1990; Stratton et al., 1991; Halasz et al., 1994; Izquierdo-Pulido et al., 1994; Bardocz, 1995; Silla Santos, 1996; Shalaby, 1996; Hernandez-Jover et al., 1997; Lehane and Olley, 2000; Medina et al., 2003; Novella-Rodriguez et al., 2003; Suzzi and Gardini, 2003; Ruiz-Capillas and Jiménez-Colmenero, 2004; Kalac and Krizek, 2003; Garai et al., 2006; Onal, 2007). In the present chapter, the formation and degradation of biogenic amines, their occurrence in foods, their significance in food safety, their potential use as quality indicators, and the available methods for their determination are discussed.

FORMATION OF BIOGENIC AMINES

The main mechanism of biogenic formation is the decarboxylation of free amino acids by specific enzymes of microbial origin, which leads to the production of amines. The names of most biogenic amines correspond to the names of their originating amino acids. Thus, histamine, tyramine, tryptamine, and cadaverine are formed from histidine, tyrosine, tryptophan, and lysine, respectively. Arginine is converted to agmatine or, as a result of bacterial activity, can be degraded to ornithine, from which putrescine is formed by decarboxylation. Putrescine is also formed from agmatine. Spermidine and spermine can be formed from putrescine or from each other. In all cases, the amino acid decarboxylation takes place by removal of the α-carboxyl group to give the

K. Koutsoumanis • Faculty of Agriculture, Department of Food Science and Technology, Laboratory of Food Microbiology and Hygiene, Aristotle University of Thessaloniki, P.O. Box 265, 54124, Thessaloniki, Greece. **C. Tassou** • National Agricultural Research Foundation, Institute of Technology of Agricultural Products, S. Venizelou 1, Lycovrisi 14123, Athens, Hellas. **G.-J. E. Nychas** • Agricultural University of Athens, Department of Food Science and Technology, Laboratory of Microbiology & Biotechnology of Foods, Iera Odos 75, Athens 11855, Hellas.

Table 1. Most important biogenic amines in foods

Name	Classification	Precursors	Chemical structure
Histamine	Heterocyclic, monoamine	Histidine	$CH_2CH_2NH_2$ (imidazole ring)
Tyramine	Aromatic, monoamine	Tyrosine	$CH_2CH_2NH_2$ (phenol ring, OH)
Phenylethylamine	Aromatic, monoamine	Phenylalanine	$CH_2CH_2NH_2$ (benzene ring)
Tryptamine	Heterocyclic, monoamine	Tryptophan	$CH_2CH_2NH_2$ (indole ring)
Putrescine	Aliphatic, diamine	Ornithine, arginine, agmatine	H_2N——NH_2
Cadaverine	Aliphatic, diamine	Lysine	H_2N——NH_2
Spermidine	Aliphatic, polyamine	Putrescine, spermine	H_2N—N(H)—NH_2
Spermine	Aliphatic, polyamine	Putrescine, spermidine	H_2N—N(H)—N(H)—NH_2
Agmatine	Aliphatic, polyamine	Arginine	H_2N—(NH)(NH)—NH_2

corresponding amine. The key decarboxylation enzymes are histidine decarboxylase for histamine, arginase for the formation of ornithine from arginine, ornithine decarboxylase for putrescine, lysine decarboxylase for cadaverine, arginine decarboxylase for agmatine, agmatinase for the formation of putrescine from agmatine and spermidine, and spermine synthase for the formation of spermidine and spermine from putrescine (Bardocz, 1995).

For the formation of biogenic amines by amino acid decarboxylation, two specific mechanisms have been identified. The first is based on pyridoxal-5-phosphate, which can be considered to be part of the enzyme that catalyzes the decarboxylation reaction (Eitenmiller and De Souza, 1984). The active site of the decarboxylation enzyme is formed by pyridoxal phosphate, which is joined in a Schiff base linkage to the amino group of a lysyl residue. The carbonyl group of pyridoxal phosphate reacts with amino acids to form Schiff base intermediates, which are then decarboxylated to yield the corresponding amines (Eitenmiller and De Souza, 1984).

The second mechanism involves a pyruvoyl residue instead of pyridoxal phosphate (Snell et al., 1975). The pyruvoyl group is covalently bound to the amino group on the enzyme and acts in a similar way to pyridoxal phosphate in the decarboxylation reaction (Recsie and Snell, 1982; Eitenmiller and De Souza, 1984).

Based on the above mechanisms, the formation of biogenic amines in foods depends on (i) the availability of precursor free amino acids, (ii) the presence of decarboxylase-positive microorganisms, and (iii) the environmental conditions that allow bacterial growth, decarboxylase synthesis, and decarboxylase activity.

Availability of Precursor Amino Acids

The concentration of available free amino acids plays a fundamental role in the formation of amines in foods, in that they are their precursors and, moreover, constitute a substrate for microbial growth (Smith, 1980; Arnold and Brown, 1978). Free amino acids either occur as such in foods or can be formed through proteolysis. The presence of bacteria with high proteolytic enzyme activity potentially stimulates the formation of amines by increasing the availability of free amino acids. However, it has not been possible to establish a clear relationship between free amino acid concentration and formation of the corresponding biogenic amines (Eerola et al., 1996; Ruiz-Capillas and Moral, 2001). The rate of amino acid decarboxylation also depends on the concentration of the produced amine or the presence of other biogenic amines. For example, it has been reported that histidine-decarboxylase is repressed when the amount of histamine is accumulated in the medium (Omure et al., 1978; Vidal and Marine, 1984). Kurihara et al. (1993) reported that the presence of histamine had an inhibiting effect on the histidine decarboxylation activity of *Photobacterium histaminum* C–8, while histamine, agmatine, and putrescine inhibited the histidine decarboxylation activity of *Photobacterium phosphoreum*.

Microorganisms Responsible for Amine Formation

The microbial enzymes (decarboxylases) reported in the previous section are present in several bacterial species that are commonly found in foods, and as a consequence these species are able to produce biogenic amines. It should be noted, however, that not all the strains of these species are amine positive. Some strains have a rather wide spectrum and are able to decarboxylate many amino acids, whereas others have only strictly substrate-specific decarboxylases. In general, decarboxylases have been found in species of the genera *Bacillus* (Rodriguez-Jerez et al., 1994a), *Pseudomonas* (Tiecco et al., 1986), and *Photobacterium* (Morii et al., 1986, 1988; Jørgensen et al., 2000a), as well as in genera of the families *Enterobacteriaceae*, such as *Citrobacter, Klebsiella, Escherichia, Proteus, Salmonella* and *Shigella* (Edwards et al., 1987; Butturini et al., 1995; Roig-Sagues et al., 1996; Marino et al., 2000), and *Micrococcaceae*, such as *Staphylococcus, Micrococcus,* and *Kocuria* (Rodriguez-Jerez et al., 1994b; Martuscelli et al., 2000). Furthermore, many lactic acid bacteria (LAB), including *Lactobacillus, Enterococcus, Carnobacterium, Pediococcus, Lactococcus,* and *Leuconostoc,* are able to decarboxylate amino acids (Maijala et al., 1993; Edwards and Sandine, 1981; de Llano et al., 1998; Bover-Cid and Holzapfel, 1999; Lonvaud-Funel, 2001).

Histamine-producing microorganisms

Histamine-producing microorganisms include a large number of strains from various species and families, isolated from a variety of food products.

Histamine-forming bacteria isolated from fish include *Morganella (Proteus) morganii, Klebsiella pneumonia, Hafnia alvei, Proteus vulgaris, Proteus mirabilis, Clostridium perfringens, Enterobacter aerogenes, Acinetobacter lwoffi, Pseudomonas putrefaciens, Aeromonas hydrophila, Plesiomonas shigelloides,* and *Vibrio* spp. (Ferencik, 1970; Havelka, 1967; Lerke et al., 1978; Taylor et al., 1979; Kimata, 1961; Arnold and Brown, 1978; Arnold et al., 1980; Yoshinaga and Frank, 1982; Frank et al., 1985; Middlebrooks et al., 1988; Lepez-Sabater et al., 1994). Okuzumi et al. (1981) studied histamine-forming bacteria in addition to N-group (psychrophilic, halophilic, histamine-forming) bacteria in fresh fish. The histamine-forming bacteria were N-group bacteria, *P. morganii, P. vulgaris, H. alvei, Citrobacter* spp., *Vibrio* spp., and *Aeromonas* spp. Halotolerant histamine-forming strains have been identified for *Staphylococcus, Vibrio,* and *Pseudomonas* isolated from fish (Yatsunami and Echigo, 1991, 1992). Rodriguez-Jerez et al. (1994c) reported that among strains isolated from salted semipreserved anchovies, *Morganella morganii, Bacillus* spp., and *Staphylococcus xylosus* showed the highest histamine-forming activity, with capacities to form 2,123, 11, and 110 ppm, respectively, after 24 h of incubation at 37°C. Histidine decarboxylase activity has also been reported for many *Lactobacillus* spp. isolated from fish, such as *Lactobacillus sakei* (Dapkevicius et al., 2000).

Behling and Taylor (1982) divided histamine-producing bacteria into those species capable of producing large quantities of histamine (>100 mg/100 ml) in tuna fish infusion broth during a short incubation period (<24 h) at temperatures of >15°C and those capable of producing smaller quantities (<25 mg/100 ml) after prolonged incubation (>48 h) at 30°C or above. *M. morganii, Klebsiella pneumoniae,* and *Enterobacter aerogenes* were prolific histamine producers, and tested strains of *H. alvei, Citrobacter freundii,* and *Escherichia coli* were slow producers (Taylor et al., 1978; Behling and Taylor, 1982).

In meat products, many *Enterobacteriaceae* can also produce considerable levels of histamine (Halasz et al., 1994), particularly *Enterobacter cloacae, E. aerogenes,* and *Klebsiella oxytoca* (Roig-Sagues et al., 1996), as well as *Escherichia coli* (Silla Santos, 1998) and *Morganella (Proteus) morganii* (Bover-Cid et al., 2001b). Among LAB isolated from meat products, *Lactobacillus bulgaricus* (Chander et al., 1989; Bover-Cid and Holzapfel, 1999), *Lactobacillus acidophilus* (Bover-Cid and Holzapfel, 1999), and some leuconostocs (Dapkevicius et al., 2000) have been reported as histamine producing. Histidine decarboxylase activity was observed with some species belonging to the genera *Micrococcus* and *Staphylococcus* (Tiecco et al., 1986). Histamine production was observed with 76% of *Staphylococcus xylosus* strains isolated from Spanish sausages (Silla Santos, 1998) and in some strains of *Kocuria* spp. (Straub et al., 1995). *Staphylococcus carnosus* and *Staphylococcus piscifermentans* can have a high amino acid decarboxylase activity (Straub et al., 1995; Montel et al., 1999) and can produce histamine among other amines. Among yeasts, the genera *Debaryomyces* and *Candida* isolated from fermented meat were found to present histidine decarboxylase activity which was higher than that observed for LAB and staphylococci (Montel et al., 1999).

Histamine-producing bacteria isolated from dairy products include *Streptococcus lactis*, *Lactobacillus helveticus*, *Streptococcus faecium*, *Streptococcus mitis*, *Lactobacillus bulgaricus*, *Lactobacillus plantarum*, and propionibacteria (Stratton et al., 1991; Edwards and Sandine, 1981). *Lactobacillus casei*, *Lactobacillus acidophilus*, and *Lactobacillus arabinose* have also been shown to possess histidine decarboxylase activity (Stratton et al., 1991). Sumner et al. (1985) isolated from Swiss cheese a histamine-producing strain of *Lactobacillus buchneri* that is capable of producing 42 mg/100 ml of medium. Five strains that were very similar to the histamine-producing strain of *Lactobacillus buchneri* were also isolated from Gouda cheese (Joosten and Northold, 1989). Rodriguez-Jerez et al. (1994c) described the histamine activity of *Bacillus macerans*, isolated from Italian cheese samole during seasoning. This organism was capable of forming histamine at various temperatures from 4 to 43°C, but the maximum histamine formation was detected at 30°C with 4,285 μg of histamine per ml.

Lactobacillus spp. and *Pediococcus* spp. have been reported to produce histamine in beer, with the former being more effective amine producers (Kalac et al., 2002). In wine, *Pediococcus* spp. and *Oenococcus oeni* are considered to be responsible for histamine formation (Aerny, 1985; Lonvaud-Funel and Joyeux, 1994).

Tyramine-producing microorganisms

Edwards et al. (1987) reported that tyramine formation in meat products is restricted to lactobacilli, particularly *L. divergens* and *L. carnis*. These results have been confirmed by other studies which showed that strains of lactobacilli belonging to the species *L. buchneri*, *L. alimentarius*, *L. plantarum*, *L. curvatus*, *L. farciminis*, *L. bavaricus*, *L. homohiochii*, *L. reuteri*, and *L. sakei* were the most important tyramine-producing bacteria (Masson et al., 1996; Montel et al., 1999; Bover-Cid et al., 2001b;

Pereira et al., 2001). Masson et al. (1996) found that tyramine-producing strains in dry sausage belonged to *Carnobacterium*, *L. curvatus*, and *L. plantarum*, whereas *Micrococcaceae* and *L. sakei* did not produce tyramine. In the last study the authors reported considerable tyramine production by *Carnobacterium divergens*, *Carnobacterium piscicola*, and *Carnobacterium gallinarum*. Tyramine can also be formed by various strains of enterococci isolated from meat (Bover-Cid et al., 2001b; Masson et al., 1996; Gardini et al., 2001). Maijala et al. (1993) reported that some *Enterococcus faecalis* strains and coliform-related strains can also produce tyramine in meat products. In miso, tyrosine decarboxylase bacteria have been identified as *Enterococcus faecium* and *Lactobacillus bulgaricus* (Ibe et al., 1992a, 1992b).

Tyramine-producing bacteria isolated during beer fermentation were identified as *Pediococcus* spp., mainly *P. damnosus*. Tyramine formation was negligible at *Pediococcus* counts below 4×10^3 CFU per ml, while at counts over 1×10^5 CFU/ml, tyramine levels ranged between 15 and 25 mg/liter (Izquierdo-Pulido et al., 1996). In addition, considerable increases of tyramine were found in adequately pasteurized beers that were inoculated postpasteurization with mixed *Lactobacillus* spp. or *Pediococcus* spp. isolated from spoiled beer, and stored at 28°C until haze formation commenced (Kalac et al., 2002).

In a laboratory medium, tyramine was produced by *Enterococcus*, *Lactococcus*, *Proteus*, and *Pseudomonas* but not by *Acinetobacter*, *Bacillus*, *Escherichia*, *Hafnia*, *Salmonella arizonae*, *Serratia marcescens*, *Shigella*, or *Yersinia enterocolitica*, nor by the yeast species examined, with the exception of *Candida krusei* (von Beutling, 1993).

Putrescine- and cadaverine-producing microorganisms

Pseudomonads are the most commonly reported bacterial group for putrescine and cadaverine formation. Koutsoumanis et al. (1999) studied the changes in the concentration of nine biogenic amines in parallel with the development of the microbial population and sensory scores during the storage of Mediterranean gilt-head sea bream and reported a relationship between the pseudomonad population and putrescine and cadaverine levels. Similar results were reported by Okozumi et al. (1990), who investigated the relationship between microflora on horse mackerel and dominant spoilage bacteria. The results of their study showed that pseudomonads were dominant when high levels of putrescine and cadaverine were detected. Frank et al. (1985) found that decarboxylation of ornithine and lysine occurred in 38 and 92%, respectively, of the mesophyllic isolates from decomposed mahimahi, while 13 and 15% of psychrotrophic isolates were capable of decarboxylating ornithine and lysine, respectively.

Edwards et al. (1983) found that putrescine concentrations in intact beef, pork, lamb, and minced beef increased consistently with total aerobic viable count, but cadaverine concentrations increased only when high numbers of presumptive *Enterobacteriaceae* were present. Pure culture experiments have shown that during growth on aerobically stored pork many strains of *Enterobacteriaceae* produce appreciable amounts of cadaverine, while *Pseudomonas* produces mainly putrescine (Slemr, 1981). In fermented meat products, *Staphylococcus carnosus* and *Staphylococcus piscifermentans* have high amino acid decarboxylase activity (Straub et al., 1995; Montel et al., 1999) and can produce putrescine and cadaverine.

Environmental Factors Affecting Amine Formation

Since amines are formed mainly by bacterial decarboxylase activity, the environmental factors affecting biogenic amine formation are expected to be similar to those affecting growth and enzymatic activity of microorganisms. These factors include temperature, pH, water activity (a_w), packaging atmosphere, and addition of preservatives.

Temperature

Temperature has a marked effect on the formation of biogenic amines in foods. Several studies report that amine content depends on temperature and increases with time and storage temperature (Klausen and Lund, 1986; Diaz-Cinco et al., 1992; Halasz et al., 1994). Temperature-abusive conditions have the following twofold effect on amine formation: (i) an effect on proteolysis due to increased microbial growth promoting penetration in the muscle and increasing availability of precursor free amino acids and (ii) an effect on amino decarboxylase activity (Maijala et al., 1995).

Sayem-El-Daher et al. (1984) reported that the levels of biogenic amines in meat were positively correlated with both storage time and temperature. These results were confirmed by Klausen and Lund (1986), who reported that amine contents were temperature dependent and at 10°C were 2 to 20 times higher than at 2°C in both mackerel and herring. Ritchie and Mackie (1979) monitored the formation of histamine, putrescine, cadaverine, spermine, and spermidine in freshly caught mackerel and herring held ungutted on ice for 28 days at 1°C, in an incubator at 10°C, or in an insulated box at ambient temperature (25°C). In both mackerel and herring, the

concentrations of histamine and other amines were quite low after prolonged storage at 1°C, even when the fish were putrid. As expected, the amines were produced in relatively larger amounts at the two increased temperatures. Similar results were reported by Koutsoumanis et al. (1999), who found a strong correlation between temperature, microbial count, and amine level with fish. In the latter study, the authors reported that histamine was not produced at storage temperatures below 10°C. These results are in accordance with other studies that reported that histamine production is slowed at 10°C and nearly terminated at 5°C due to the slow growth of histamine-producing bacteria at low temperatures (Halasz et al., 1994). On the contrary, Ababouch et al. (1991) indicated that low storage temperatures are not sufficient to inhibit the formation of toxic amines such as histamine. While low-temperature storage effectively controls the growth of most histamine-producing bacteria, some strains that grow at refrigeration temperatures can produce smaller amounts of histamine in foods stored at temperatures between 0 and 10°C (Ritchie and Mackie, 1979; Klausen and Huss, 1987; Stratton and Taylor, 1991).

Under freezing conditions, most decarboxylases are unstable. For example, histidine decarboxylase becomes inactive after 8 to 15 days at –20°C (Halasz et al., 1994; Silla-Santos, 1996; Chen et al., 1994). In general, frozen storage does not favor the formation of biogenic amines. Hernandez-Jover et al. (1996a) found that the proportion of biogenic amines remained stable in pork stored for 12 days at –18°C. On the other hand, other authors have reported that biogenic amines may increase during storage at freezing conditions. Ben-Gigirey et al. (1998) investigated the changes in biogenic amines in white tuna during frozen storage at –18°C or –25°C. Putrescine decreased during storage after 6 months at either temperature but increased after 9 months. Similarly, histamine decreased after 3 months storage, but at –18°C it increased to a final concentration of 103% of the initial level after 9 months of storage. Cadaverine contents tripled or doubled after 3 months of storage.

Biogenic amine levels in foods are in general unaffected by thermal processing (cooking), with the exception of spermine, which decreased during heat treatment of cooked ground beef at 200°C for 2 h. Luten et al. (1992) and Wendakoon and Sakaguchi (1993) indicate that histamine is thermally stable during the cooking process. Given this fact, an amine presence in cooked products would appear to be closely linked to the quality of the raw material used.

Because of the importance of temperature in the production of histamine, several attempts have been made to predict histamine formation in spoiling fish at different temperatures. For example, Frank (1985) constructed normographs to predict histamine production in skipjack tuna, and Pan (1985) used Arrhenius plots for such estimations for mackerel and bonito. However, the methods involved too many assumptions and were inaccurate in their predictions. A further development, temperature function integrators, which take temperature fluctuations into account, was also considered unsuitable for predicting the complicated process of histamine production in spoiling fish (Olley and McMeekin, 1985).

pH

pH is a key factor influencing the amino acid decarboxylase activity and formation of biogenic amines in foods. It has been shown that amine formation by bacteria is a physiological mechanism to counteract an acid environment (Koessler et al., 1928; Buncic et al., 1993; Halasz et al., 1994; Teodorovic et al., 1994), and thus, amino acid decarboxylase activity is usually stronger in acid conditions, with optimum pH between 4.0 and 5.5 (Sinell, 1978; Teodorovic et al., 1994).

The optimum pH for tyramine synthesis in cheese was found to be 5.0 by Diaz-Cinco et al. (1992). Baranowski et al. (1985) found that the conversion of histidine to histamine by *Klebsiella pneumoniae* isolated from tuna has an optimum pH of 4.0, with 30% of this activity lost at pH 6.0. A correlation between amine production and the decrease of pH during fermentation of sausages has been reported (Eitenmiller et al., 1978; Santos-Buelga et al., 1986). However, this correlation has been attributed mainly to the effect of pH on the extent of growth of decarboxylating bacteria during fermentation (Yoshinaga and Frank, 1982).

a_w

High concentrations of solutes (e.g., sodium chloride) or processes that can decrease a_w (e.g., fermentation, smoking, etc.) have been reported to inhibit amine formation in foods, probably because low a_w retards bacterial growth. For example, NaCl at 3.5 ± 5.5% has been reported to inhibit histamine production by *K. pneumoniae* and *M. morganii* (gram-negative bacteria), whereas lower levels were not effective. Higher yields of histamine, tyramine, and tryptamine were produced by *Lactobacillus bulgaricus* in the absence of added sodium chloride in nutrient broth, and the rate of amine production was considerably reduced when the NaCl concentration of the broth was 2.0%. The rate of amine production of an *L. delbrueckii* subsp. *bulgaricus* strain was considerably reduced when salt concentration in the medium increased from 0% to 6% (Chander et al., 1989). Henry Chin and Koehler (1986) demonstrated that NaCl concentration ranging from 3.5% to 5.5% could inhibit histamine production. This influence can be attributed to reduced cell yields obtained in the presence of high NaCl concentrations and to a progressive disturbance of the membrane-located microbial decarboxylase enzymes (Sumner et al., 1990). A similar NaCl effect on cell yield and amine production has been reported for *E. faecalis* (Gardini et al., 2001). A study by Leroi and others (2000) showed that the inhibition of bacteria in cold-smoked salmon stored for 5 weeks at 5°C and salt (5%) and smoke was linearly proportional to the salt and smoke content (the higher the concentration, the greater the inhibition).

Packaging atmosphere

The packaging atmosphere also appears to have a significant effect on the formation of amines in foods. Halasz et al. (1994) reported that *Enterobacter cloacae* produces about half the quantity of putrescine and *Klebsiella pneumoniae* synthesizes significantly less cadaverine under anaerobic conditions compared with production under aerobic conditions. Watts and Brown (1982) and Vidal and Marine (1984) found that histidine decarboxylase activity of *Proteus morganii* is inhibited in atmospheres of 80% CO_2. On the other hand, it has been reported that reducing the redox potential of the medium stimulates histamine production, and histidine decarboxylase activity seems to be inactivated or destroyed in the presence of oxygen (Arnold and Brown, 1978; Halasz et al., 1994).

Addition of preservatives

The addition of preservatives to foods influences the microbial population dynamics and, consequently, the production of biogenic amines. Cantoni et al. (1994) reported that the addition of sodium nitrite (150 mg/kg) in Italian dry sausages was able to consistently reduce putrescine and cadaverine formation, but the presence of the curing agent caused a threefold increase of histamine concentration. Other authors (Santos et al., 1985; Straub et al., 1994) have reported that neither potassium nitrate nor potassium nitrite affected the production of biogenic amines in ripened meat products, but that the presence of these additives in soy hamburger texturizers increased production of tyramine.

The addition of sodium sulfide to slightly fermented sausages at concentrations of 500 and 1,000 mg/kg inhibited cadaverine accumulation but stimulated the production of tyramine (Bover-Cid et al., 2001a). Shalaby (1996) reported that potassium sorbate, frequently used as a microbial inhibitor, can limit the formation of biogenic amines in foods. Taylor and Speckhard (1984) observed that potassium sorbate at a concentration of 0.5% inhibited growth and histamine production at 10°C for up to 216 h and 32°C for up to 120 h.

DEGRADATION OF BIOGENIC AMINES

Biogenic amines in food can be physiologically inactivated by amine oxidases. These enzymes are present in bacterium, fungus, plant, and animal cells and are able to catalyze the oxidative deamination of amines with the production of aldehydes, hydrogen peroxide, and ammonia (Cooper, 1997). The potential role of microorganisms with amine oxidase activity has been investigated with the aim to prevent or reduce the formation of biogenic amines in foods. Leuschner et al. (1998) reported amine degradation by many strains belonging to the genera *Lactobacillus*, *Pediococcus*, and *Micrococcus*, as well as to the species *Staphylococcus carnosus* and *Brevibacterium linens*. Many *S. xylosus* strains isolated from fermented sausages in southern Italy also showed the ability to degrade biogenic amines in vitro (Martuscelli et al., 2000). Gardini et al. (2002) found a significant reduction in the concentration of tyramine and putrescine in the presence of *S. xylosus*. Histaminase activity has been detected with several types of bacteria, including *M. morganii*, *Vibrio* spp., and *Klebsiella* spp. (Gale, 1942; Satake et al., 1953; Ienistea, 1973). Ferencik (1970) reported that a strain of *M. morganii*, after being inoculated into a sterile tuna flesh homogenate, produced large amounts of histamine. However, a significant amount of histamine was soon decomposed by the organism. A subsequent addition of histidine to the inoculated homogenate again resulted in histamine formation, followed by histamine destruction. He concluded that the histamine production and destruction by the *M. morganii* strain was determined by the concentration of free histidine in the fish.

The environmental factors affecting amine degradation in foods are similar to those affecting the formation of amines. Leuschner et al. (1998) observed that tyramine degradation is strictly dependent on pH (with an optimum at 7.0), temperature, and NaCl, as well as glucose, oxygen availability, and hydralazine concentration. Gale (1942) found that bacterial histaminase is best produced under somewhat alkaline conditions (pH 7.5 to 8) but that moderate activity also occurs under slightly acidic conditions. Dapkevicius et al. (2000) reported that the highest degradation rate of this amine was observed at 37°C, but at 22 and 15°C, degradation was still considerable.

Similarly, Schomburg and Stephan (1993) showed that histaminase activity has its optimum temperature at 37°C and retains about 50% of its maximum activity at 20°C.

Based on the above observations, it can be concluded that the levels of biogenic amines in foods depend on the balance between amine production by amino acid decarboxylation and amine destruction by the amine oxidases of the contaminating microorganisms.

OCCURRENCE OF BIOGENIC AMINES IN FOODS

Theoretically, in all foods that contain free amino acids or proteins and are subject to conditions enabling microbial or enzymatic activity, biogenic amines can be expected. As mentioned above, the level of biogenic amines in foods depends on the type of food, quality of the raw material, type and extent of contamination, processing, distribution, and storage conditions. Biogenic amines are present in a wide range of food products, including fish products, meat products, dairy products, wine, beer, vegetables, fruits, nuts, and chocolate (Askar and Treptow, 1986; ten Brink et al., 1990). In general, scombroid fishes and some nonscombroid fishes (e.g., sardine, pilchards, anchovies, herring, and marline), as well as other foods, such as meat products, cheeses, some fermented products, and some beverages, have increased levels of biogenic amines and certainly deserve concerned attention. A review of the levels of biogenic amines in various food products is shown in Table 2.

SIGNIFICANCE OF BIOGENIC AMINES IN FOOD SAFETY

Toxicity of Biogenic Amines Present in Foods

The biogenic amines have important physiological effects in humans, generally either psychoactive or vasoactive (Lovenberg, 1973). Psychoactive amines affect the nervous system by acting on neural transmitters, while vasoactive amines act on the vascular system. As a result of their important physiological functions, exogenous amines absorbed from food may lead to toxic effects when dietary levels are high. Under normal circumstances, when low amounts of biogenic amines are ingested by a healthy person, no toxic effects are expected. In the latter case amines cannot gain access to the bloodstream due to an efficient detoxification system in the intestinal tract of mammals (Joosten, 1988; Huis in't Veld et al., 1990). The detoxification system consists of the enzymes monoamine oxidase (MAO), diamine oxidase (DAO), and polyamine oxidase (PAO), which metabolize

Table 2. Biogenic amine content in various food products and beverages

Products[c]	Biogenic amine[b]								No.[a]	Reference(s)
	HI	TY	TR	CA	PU	SD	SM	AG		
Fish products										
Canned tuna (mg/100 g)	4.5 (ND–18.7)	0.2 (ND–0.4)	0.2 (ND–0.3)	0.4 (ND–2.8)	0.8 (ND–3.5)	0.3 (ND–0.4)	4.1 (ND–20.2)	0.1 (ND–0.3)	21	Tsai et al. (2005a)
Canned mackerel (mg/100 g)	1.1 (ND–2.4)	ND	0.2 (ND–0.6)	0.6 (ND–2.6)	0.6 (ND–2.1)	0.2 (ND–0.5)	4.6 (ND–25.1)	ND	10	Tsai et al. (2005a)
Canned bonito (mg/100 g)	1.4 (ND–3.5)	0.3 (ND–0.5)	ND	0.8 (ND–1.8)	0.3 (ND–1.5)	0.4 (ND–0.6)	2.4 (ND–5.3)	ND	6	Tsai et al. (2005a)
Canned anchovy (mg/100 g)	3.2 (ND–7.5)	0.1 (ND–0.2)	0.8 (ND–1.6)	1.3 (ND–2.8)	1.4 (ND–3.4)	0.4 (ND–0.8)	2.1 (ND–4.7)	ND	5	Tsai et al. (2005a)
Canned Pacific saury (mg/100 g)	1.2 (ND–1.2)	0.2 (ND–0.5)	ND	1.0 (ND–2.2)	0.4 (ND–01.0)	ND	1.5 (ND–2.7)	ND	3	Tsai et al. (2005a)
Canned milk fish (mg/100 g)	0.2 (ND–0.5)	ND	ND	0.1 (ND–0.4)	0.2 (ND–0.5)	0.1 (ND–0.2)	ND	ND	3	Tsai et al. (2005a)
Northern Taiwan salted mackerel (retail market)	3.4 (ND–14.1)	ND	ND	2.1 (ND–26.0)	1.9 (ND–5.9)	3.9 (ND–42.0)	21.4 (ND–66.5)	1.2 (ND–8.0)	7	Tsai et al. (2005b)
Northern Taiwan salted mackerel (supermarket)	4.5 (ND–15.0)	ND	ND	5.0 (ND–21.0)	1.4 (ND–9.8)	ND	28.9 (ND–84.1)	1.4 (ND–3.5)	8	Tsai et al. (2005b)
Southern Taiwan salted mackerel (retail market)	26.0 (ND–120.2)	ND	ND	5.7 (ND–18.3)	2.3 (ND–2.4)	5.3 (ND–52.1)	26.9 (ND–54.5)	1.5 (ND–9.2)	10	Tsai et al. (2005b)
Southern Taiwan salted mackerel (supermarket)	21.4 (ND–70.1)	ND	ND	1.6 (ND–1.2)	ND	1.3 (ND–14.1)	21.2 (ND–46.0)	1.7 (ND–2.6)	8	Tsai et al. (2005b)
Vacuum-packed cold-smoked tuna (55 d, 5°C)	<5	170 ± 37e	<5	<5	<5	5.0 ± 1.1	20.0 ± 1.9	<5	3	Emborg and Dalgaard (2006)
Vacuum-packed cold-smoked tuna with starter 1 (42 d, 5°C)	12 ± 16	115 ± 5	13 ± 1	<5	<5	6 ± 0.5	41 ± 4	<5	3	Emborg and Dalgaard (2006)
Vacuum-packed cold-smoked tuna with starter 2 (26 d, 5°C)	256 ± 339	140 ± 64	13 ± 4	<5	<5	11 ± 4	52 ± 6	<5	3	Emborg and Dalgaard (2006)
Salt ripened anchovies	9.8 (4.5–19.4)	71.5 (40.2–89.5)	—	20.7 (7.1–47.8)	10.7 (6.5–17.2)	—	—	27.7 (12.7–57.5)	3	Pons-Sánchez-Cascado et al. (2005)
Fresh sardines (mg/100 g)	—	0.0	ND	0.39 ± 0.2	1.34 ± 0.6	0.12 ± 0.1	0.0	0.0	3	Özogul and Özogul (2006)
Sardines (air storage, 15 d) (mg/100 g)	—	1.6 ± 1.7	ND	10.0 ± 4.9	11.4 ± 2.6	0.76 ± 0.2	0.29 ± 0.2	2.7 ± 2.3	3	Özogul and Özogul (2006)
Sardines (MAP storage, 15 d) (mg/100 g)	—	0.0	ND	1.8 ± 1.0	2.1 ± 1.4	0.47 ± 0.3	0.26 ± 0.1	1.5 ± 1.3	3	Özogul and Özogul (2006)
Sardines (vacuum packaging, 15 d) (mg/100 g)	—	1.5 ± 1.0	ND	3.6 ± 2.7	4.7 ± 2.7	0.52 ± 0.4	0.31 ± 0.2	2.3 ± 2.0	3	Özogul and Özogul (2006)
Fresh rainbow trout	ND	ND	—	0.17 ± 0.16	0.42 ± 0.17	—	—	—	3	Rezaei et al. (2007)
Rainbow trout (storage in ice, 18 d)	1.6 ± 0.3	ND	—	6.8 ± 0.2	9.0 ± 0.2	—	—	—	3	Rezaei et al. (2007)
Meat products										
Belgian sausages	4.1 (0–19.7)	36.8 (10.2–150.6)	—	2.5 (0–5.6)	15.1 (3.1–40)	—	—	—	24	Vandekerckhove (1977)
Finnish sausages	54 (0–180)	88 (4–200)	14 (0–43)	50 (0–270)	79 (0–230)	4 (2–7)	31 (19–46)	—	11	Eerola et al. (1998a, 1998b)
Danish sausages	9 (1–56)	54 (5–110)	27 (0–91)	180 (0–790)	130 (0–450)	7 (3–9)	37 (23–47)	—	8	Eerola et al. (1998a, 1998b)
Russian sausages	89 (0–200)	110 (6–240)	22 (0–43)	10 (3–18)	93 (3–310)	5 (2–8)	33 (23–40)	—	4	Eerola et al. (1998a, 1998b)

255

Continued on following page

Table 2. *Continued*

Products[c]	Biogenic amine[b]								No.[a]	Reference(s)
	HI	TY	TR	CA	PU	SD	SM	AG		
Egyptian sausages	5.3 (7.5–41)	14 (9.5–53)	13 (2.5–34)	19 (5.6–39)	39 (12–100)	2.3 (5.3–12)	1.8 (1.5–5.3)	—	50	Shalaby (1993)
Norwegian sausages	1	12	9	1	1	4	24	—	5	Ansorena et al. (2002)
Italian sausages	0	187	19	1	1	6	18	—	5	Ansorena et al. (2002)
North Belgian sausages	1	76	17	0	33	3	12	—	5	Ansorena et al. (2002)
South Belgian sausages	18	176	39	5	125	4	15	—	5	Ansorena et al. (2002)
Turkish sucuk	12.5 ± 4.0[c]	167.6 ± 107.6	—	33.2 ± 41.85	82.7 ± 57.8	14.6 ± 2.2	54.6 ± 6.0	—	2	Genccelep et al. (2007)
Turkish sucuk with starters	10.6 (9.1–12.1)	19.3 (19.2–19.4)	—	13.8 (12.5–15.1)	32.0 (30.8–33.1)	13.9 (13.3–14.5)	57.6 (56.1–59.1)	—	2	Genccelep et al. (2007)
Turkish sucuk without nitrate	11.2 ± 5.1	97.5 ± 120.1	—	33.5 ± 42.0	63.0 ± 58.7	14.3 ± 2.9	56.6 ± 12.7	—	2	Genccelep et al. (2007)
Turkish sucuk with 150 ppm nitrate	12.1 ± 3.7	54.5 ± 84.5	—	13.1 ± 10.6	42.4 ± 27.9	14.0 ± 2.6	58.3 ± 7.7	—	2	Genccelep et al. (2007)
Fermented sausages	11 (1–3)	110 (40–10)	—	63 (1–150)	52 (1–190)	—	—	—	14	ten Brick et al. (1990)
Meetwurst	21 (0–170)	72 (5–320)	18 (0–54)	6 (0–16)	77 (2–580)	6 (3–11)	29 (22–38)	—	12	Eerola et al. (1998a, 1998b)
Saucisson (industrial)	71 (16–151)	220 (172–268)	3.9 (0–9)	103 (31–192)	279 (195–410)	5.1 (4–6.4)	91 (59–119)	—	5	Montel et al. (1999)
Saucisson (traditional)	15.3 (15–16)	164.3 (84–217)	0	71.3 (39–110)	223 (61–317)	4.3 (2–6)	83.7 (82–86)	—	3	Montel et al. (1999)
Soppressata	21.9 (0–100.9)	178 (0–556.9)	—	60.8 (0–271.4)	98.8 (0–416.1)	40 (0–91.3)	35.5 (0–97.9)	—	9	Parente et al. (2001)
Salsiccia	0	76.7 (0–338.9)	—	6.7 (0–39)	19.7 (0–77.7)	18.8 (0–57.2)	2.8 (0–28.1)	—	10	Parente et al. (2001)
Thin fuet	1.8 ± 2.3	121.8 ± 99.4	6.5 ± 14.8	56.3 ± 62.0	105.2 ± 200.7	9.2 ± 7.3	32.8 ± 19.6	—	15	Bover-Cid et al. (1999a, 1999b)
Fuet	15.2 ± 43.5	156.9 ± 85.4	10 ± 13.6	367.2 ± 55.4	64.7 ± 56.5	10.3 ± 10.9	30.6 ± 20.5	—	23	Bover-Cid et al. (1999a, 1999b)
Fuet	2.2 (0–57.7)	190.7 (31.8–742.6)	8.7 (0–67.8)	18.9 (5.4–51.3)	71.6 (2.2–222.1)	—	—	—	11	Hernandez-Jover et al. (1997)
Salsichon	19.4 ± 27.6	198.4 ± 183.6	21.3 ± 50.4	26.4 ± 46.5	138.5 ± 143.8	5.6 ± 4.1	35.2 ± 18.2	—	19	Bover-Cid et al. (1999a, 1999b)
Salsichon	7.3 (0–150.9)	280.5 (53.3–513.4)	8.5 (0–65.1)	11.7 (0–342.3)	102.7 (5.5–400.0)	—	—	—	22	Hernandez-Jover et al. (1997)
Poličane	28.5 (2.5–32)	89 (86–92)	5 (5–5)	6 (6–6)	54 (54–54)	2.5 (2–3)	2 (2–2)	—	2	Komprda et al. (2001)
Chorizo	17.5 (0–314.3)	282.3 (29.2–626.8)	15.9 (0–87.8)	20.1 (0–658.1)	60.4 (2.6–415.6)	—	—	—	20	Hernandez-Jover et al. (1997)
Sobrasada	9.0 (2.8–143.1)	332.1 (57.6–500.6)	11.5 (0–64.8)	12.6 (3.0–41.6)	65.2 (1.8–500.7)	—	—	—	7	Hernandez-Jover et al. (1997)
Salpicão de Vinhais	0.2	13.4	—	0.1	0.4	—	—	—	1	Ferreira et al. (2007)
Chouriça de Vinhais	1.6	75.9	—	1.7	11.3	—	—	—	1	Ferreira et al. (2007)
Painho de Portalegre with 3% NaCl	0.0	356.5	3.2	1,732.1	1,047.5	8.5	12.0	—	2	Roseiro et al. (2006)
Painho de Portalegre with 6% NaCl	0.0	257.4	0.0	295.6	601.4	17.5	29.6	—	2	Roseiro et al. (2006)
Hercules with starter A	2 ± 0.1	123 ± 1.5	3 ± 0.1	15 ± 0.1	247.0 ± 0.6	4 ± 0.2	30 ± 0.1	—	2	Komprda et al. (2004)
Hercules with starter B	1 ± 0.0	1 ± 0.1	3 ± 0.1	1 ± 0.0	1 ± 0.1	3 ± 0.1	22 ± 1.5	—	2	Komprda et al. (2004)

									n	Reference
Paprikáš with starter A	0 ± 0.0	9 ± 0.3	1 ± 0.3	1 ± 0.1	12.0 ± 0.1	2 ± 0.5	21 ± 1.1	—	2	Komprda et al. (2004)
Paprikáš with starter B	1 ± 0.0	1 ± 0.1	1 ± 0.1	1 ± 0.0	1 ± 0.1	3 ± 0.1	13 ± 2.2	—	2	Komprda et al. (2004)
Salamini italiani alla cacciatora	ND	64.4 ± 0.8	ND	—	—	—	—	—	4	Coisson et al. (2004)
Cooked ham	0 (ND–0.9)	1.0 (ND–78.0)	—	0 (ND–9.5)	0 (ND–12.4)	2.1 (1.4–3.5)	21.4 (6.4–35.7)	—	20	Hernandez-Jover et al. (1997)
Spanish dry-cured ham	0.8 (ND–150.0)	0.7 (ND–46.5)	—	2.1 (ND–305.5)	2.3 (ND–17.4)	5.6 (4.4–7.3)	35.7 (24.9–62.1)	—	23	Hernandez-Jover et al. (1997)
Spanish mortadella	0 (ND–4.8)	2.5 (ND–67.0)	—	1.7 (4.3–28.8)	1.3 (ND–5.7)	4.0 (1.0–8.9)	17.2 (7.6–32.2)	—	20	Hernandez-Jover et al. (1997)
Botifarra catalana	0 (ND–1.8)	5.1 (4.0–22.0)	—	9.6 (ND–40.0)	0.7 (ND–3.7)	2.9 (1.2–5.3)	17.9 (12.5–25.8)	—	8	Hernandez-Jover et al. (1997)
Italian dry-cured ham (15 mo old)	ND	40.2 (ND–110)	—	0.30 (ND–2.0)	0.13 (ND–3)	7.8 (2.0–19)	42.1 (21–67)	—	23	Virgili et al. (2007)
Italian dry-cured ham (23 mo old)	ND	72.4 (1.0–274)	—	5.5 (1.0–15)	72.3 (19.0–218)	4.4 (ND–11)		—	9	Virgili et al. (2007)
Pressurized vacuum-packaged ham (77 d, 2°C)	1.8	10.5	—	0.1	0.7	2.8	30.3	0.5	1	Ruiz-Capillas et al. (2007)
Nonpressurized vacuum-packaged ham (77 d, 2°C)	2.3	51.9	—	1.2	0.9	2.4	30.3	0.7	1	Ruiz-Capillas et al. (2007)
Pressurized vacuum-packaged ham (77 d, 12°C)	1.4	29.6	—	1.0	0.4	2.5	29.6	0.3	1	Ruiz-Capillas et al. (2007)
Nonpressurized vacuum-packaged ham (77 d, 12°C)	2.4	57.0	—	1.3	17.5	2.5	30.0	0.1	1	Ruiz-Capillas et al. (2007)
Bovine meat	—	10.7	20.4	18.5	2.1	2.2	17.2	—	1	Vinci and Antonelli (2002)
Cheeses										
Dutch cheese	52.0 (0–350.0)	138.0 (0–625.0)	—	73.0 (1.0–140.0)	19.0 (1.0–71.0)	—	—	—	8	ten Brick et al. (1990)
Castelmagno	645.8 ± 30.2	1,009.1 ± 48.8	1,048.7 ± 51.5	310.4 ± 14.2	—	0.42 ± 0.02	449.6 ± 20.5	—	1	Gosetti et al. (2007)
Raschera	452.4 ± 19.4	153.9 ± 7.6	389.9 ± 19.6	118.9 ± 5.1	—	10.6 ± 0.47	352.7 ± 16.2	—	1	Gosetti et al. (2007)
Toma piemontee	587.6 ± 29.0	282.3 ± 14.2	255.5 ± 12.4	1.3 ± 0.06	—	6.6 ± 0.23	193.9 ± 9.1	—	1	Gosetti et al. (2007)
Blue-veined cheese	13.9 (0.8–20.2)	32.0 (8.3–58.4)	3.6 (2.1–5.4)	90.2 (12.2–190.7)	52.2 (10.1–60.4)	11.8 (6.2–17.2)	0.6 (0–1.1)	—	4	Marino et al. (2000)
Unripened cheeses	ND	0.0 (ND–0.6)	ND	0.0 (ND–1.5)	0.0 (ND–3.1)	0.3 (ND–0.8)	0.0 (ND–1.1)	ND	20	Novella-Rodriguez et al. (2003)
Hard-ripened from pasteurized milk	4.0 (ND–163.6)	7.2 (ND–301.4)	0.9 (ND–45.1)	8.0 (ND–710.1)	5.0 (ND–611.7)	5.7 (ND–43.0)	1.6 (ND–18.7)	0.0 (ND–22.0)	20	Novella-Rodriguez et al. (2003)
Hard-ripened from raw milk	18.3 (ND–391.4)	125.5 (ND–609.4)	0.3 (ND–33.8)	29.5 (0.9–368.5)	8.1 (ND–62.5)	0.1 (ND–39.6)	0.0 (ND–21.5)	0.0 (ND–27.2)	20	Novella-Rodriguez et al. (2003)
Goat cheeses	1.3 (ND–88.4)	8.5 (ND–830.5)	0.6 (ND–17.4)	0.7 (ND–88.7)	4.1 (ND–191.8)	0.9 (ND–14.5)	0.0 (ND–3.6)	0.0 (ND–3.2)	20	Novella-Rodriguez et al. (2003)
Blue cheeses	6.6 (ND–376.6)	14.4 (ND–1,585.4)	3.2 (ND–128.8)	11.3 (ND–2,101.4)	18.0 (3.0–257.2)	10.1 (ND–71.6)	0.0 (ND–18.9)	1.1 (ND–28.5)	20	Novella-Rodriguez et al. (2003)

Continued on following page

Table 2. *Continued*

Products[c]	Biogenic amine[b]								No.[a]	Reference(s)
	HI	TY	TR	CA	PU	SD	SM	AG		
Other fermented foods										
Japanese natto (mg/100 g)	3.5 (ND–45.7)	0.12 (ND–4.5)	0.91 (30.1)	2.2 (0.20–4.2)	1.7 (ND–2.7)	4.5 (ND–12.4)	0.86 (ND–7.1)	ND	36	Tsai et al. (2007)
Taiwanese natto (mg/100 g)	4.5 (ND–13.7)	ND	ND	0.05 (ND–0.30)	0.16 (ND–0.6)	2.5 (ND–5.0)	ND	ND	3	Tsai et al. (2007)
Miso (supermarket) (mg/100 g)	1.6 (ND–22.1)	1.2 (ND–2.8)	2.7 (ND–43.4)	3.2 (ND–20.1)	0.12 (ND–1.2)	ND	0.80 (ND–21.6)	0.24 (ND–6.6)	27	Kung et al. (2007b)
Miso (retail market) (mg/100 g)	0.77 (ND–10.2)	1.6 (ND–4.9)	4.9 (ND–76.2)	0.9 (ND–3.0)	0.11 (ND–0.9)	ND	0.72 (ND–9.3)	0.58 (ND–7.5)	13	Kung et al. (2007b)
White sufu (mg/100 g)	0.46 (ND–2.0)	0.80 (ND–4.0)	1.6 (ND–8.1)	0.70 (ND–4.1)	1.5 (ND–4.5)	ND	1.6 (ND–8.2)	—	12	Kung et al. (2007a)
Brown sufu	2.7 (ND–15.8)	1.1 (ND–6.5)	ND	5.0 (ND–37.1)	0.24 (ND–1.0)	ND	ND	—	10	Kung et al. (2007a)
Yogurt	<1	<1	—	<1	<1	—	—	—	6	ten Brick et al. (1990)
Sauerkraut	38.0 (1.0–104.0)	75.0 (2.0–192.0)	—	73.0 (1.0–311.0)	154.0 (6.0–550.0)	—	—	—	8	ten Brick et al. (1990)
Vegetables										
Chinese cabbage (5 d, 5°C)	0.9	2.1	—	—	11.3	11.7	0.27		2	Simon-Sarkadi et al. (1994)
Endive (5 d, 5°C)	ND	2.0	—	—	9.9	7.4	0.16		2	Simon-Sarkadi et al. (1994)
Iceberg lettuce (5 d, 5°C)	ND	1.8	—	—	15.7	6.1	0.16		2	Simon-Sarkadi et al. (1994)
Radicchio (5 d, 5°C)	ND	1.8	—	—	14.4	10.1	0.27		2	Simon-Sarkadi et al. (1994)
Maize fresh	0.4	7.4	—	4.2	11.0	—	—		1	Nishino et al. (2007)
Maize (60-d ensilage)	6.7	374.0	—	217	70.0	—	—		3	Nishino et al. (2007)
Maize (60-d ensilage inoculated with *L. casei*)	1.1	165.0	—	139.0	51.0	—	—		3	Nishino et al. (2007)
Maize (60-d ensilage inoculated with *L. buchneri*)	4.6 (3.77–5.43)	437.5 (437–438)	—	226.5 (220–233)	125.5 (110–141)	—	—		3	Nishino et al. (2007)
Festulolium, fresh	0.15	1.7	—	1.3	4.6	—	—		1	Nishino et al. (2007)
Festulolium (60-d ensilage)	195	1,063	—	1,233	443	—	—		3	Nishino et al. (2007)
Festulolium (60-d ensilage inoculated with *L. casei*)	15.0	260.0	—	313.0	103.0	—	—		3	Nishino et al. (2007)
Festulolium (60-d ensilage inoculated with *L. buchneri*)	14.6 (10.3–18.9)	668.5 (620.0–717.0)	—	328.5 (313.0–44.0)	179.5 (166.0–93.0)	—	—		3	Nishino et al. (2007)
TMR, fresh	1.9	5.0	—	5.1	18.4	—	—		1	Nishino et al. (2007)
TMR (60-d ensilage)	126.0	149.0	—	15.6	102.0	—	—		3	Nishino et al. (2007)
TMR (60-d ensilage inoculated with *L. casei*)	4.25	15.9	—	14.4	36.1	—	—		3	Nishino et al. (2007)
TMR (60-d ensilage inoculated with *L. buchneri*)	37.4 (33.4–41.3)	85.7 (78.2–93.2)	—	104.7 (94.4–115.0)	14.2 (12.2–16.1)	—	—		3	Nishino et al. (2007)
Wines/beverages										
Brazilian beer (mg/liter)	0.2 (<0.2–1.5)	2.2 (0.3–36.8)	0.5 (<0.35–10.1)	0.5 (<0.2–2.6)	3.9 (0.9–9.8)	0.7 (<0.2–6.0)	—	10.9 (2.1–46.8)	91	Gloria et al. (1999)
Beer	<1	<1	—	9.5 (0–60)	4 (0.5–8)	—	—	—	13	ten Brick et al. (1990)
Sherry	3 (0–9)	7 (0.5–17.0)	—	<1	9 (3–25)	—	—	—	28	ten Brick et al. (1990)

Wine									n[a]	Reference
Wheat beer #1	1.5 (0–6)	1.3 (0–5)	—	<1	4 (1–12)	—	—	—	20	ten Brick et al. (1990)
	0.57 (0.54–0.60)	—	—	0.48 (0.28–0.67)	6.5 (6.4–6.6)	0.82 (0.45–1.20)	0.6 (0.47–0.73)	8.4 (7.7–9.1)	2	De Borba and Rohrer (2007)
Lager beer	0.72	—	—	ND	3.0	0.14	0.33	14.9	1	De Borba and Rohrer (2007)
Cabernet Sauvignon wine	3.2 (0.8–5.5)	—	—	0.66 (0.53–0.79)	13.3 (7.1–19.4)	1.7 (1.4–1.9)	0.2 (0.21–0.19)	0.3 (0.23–0.37)	2	De Borba and Rohrer (2007)
Pinot Grigio wine	ND	—	—	ND	1.7	ND	ND	ND	1	De Borba and Rohrer (2007)
White wine, Cadiz	—	ND	—	—	—	—	—	—	5	Gil-Agusti et al. (2007)
Rose wine, Castilla	—	ND	—	—	—	—	—	—	5	Gil-Agusti et al. (2007)
Red wine, Rioja	—	4.20 ± 0.08	—	—	—	—	—	—	5	Gil-Agusti et al. (2007)
Red wine, Barcelona	—	0.59 ± 0.03	—	—	—	—	—	—	5	Gil-Agusti et al. (2007)
Ciders	1.1 (0.9–6.9)	1.3 (2.0–5.0)	—	—	3.6 (3.3–12.3)	—	—	—	24	Garai et al. (2006)

[a]Number of samples.

[b]The mean value and range (in parentheses) for biogenic amine content are given, except for values given as means ± standard deviations; all values are given in mg/kg, unless otherwise indicated. ND, not detected; —, not analyzed. HI, histamine; TY, tyramine; TR, tryptamine; PU, putrescine; CA, cadaverine; SD, spermidine; SM, spermine; AG, agmatine.

[c]d, days; TMR, total mixed ration.

normal dietary intakes of biogenic amines in the intestinal tract. The end metabolites are readily excreted in the urine (Rice et al., 1976; Bjedanes et al., 1978). The toxic dose of biogenic amines is strongly dependent on the efficiency of this detoxification process. In cases in which individuals have heightened sensitivity, exogenous factors inhibit the activity of amine oxidases, or very high levels of biogenic amines are consumed, the enzymic detoxification system may not be adequate, allowing amines to accumulate in the bloodstream. In general, an amine "overdose" may present a risk to all individuals. However, people with gastrointestinal problems are at higher risk because the activity of oxidases in their intestines is usually lower than that in healthy individuals. People with respiratory and coronary problems or those with hypertension or vitamin B_{12} deficiency are also sensitive to lower doses of biogenic amines (Rice et al., 1976; Bjedanes et al., 1978; ten Brink et al., 1990; Halasz et al., 1994). In addition, certain pharmaceutical agents which act as MAO- and DAO-inhibiting drugs have an inhibitory effect on the detoxification system, and thus, people on such medication are more sensitive to biogenic amines (ten Brink et al., 1990; Stratton et al., 1991).

Histamine toxicity

The most notorious food-borne intoxications caused by biogenic amines are related to histamine. Histamine exerts its toxicity by binding to three types of receptors, H_1, H_2, and H_3, on cellular membranes in the respiratory, cardiovascular, gastrointestinal, and hematological/immunological systems (Joosten, 1988; Cavanah and Casale, 1993) and thus can affect a number of physiological functions regulated by these receptors. Histamine causes contraction of the smooth muscle of the uterus, the intestine, and the respiratory tract (regulated by the H_1 receptor); controls gastric acid secretion (regulated by the H_2 receptor located on the parietal cells); affects liberation of adrenaline and noradrenaline from the suprarenal gland; stimulates both sensory and motor neurons; and causes dilatation of the peripheral blood vessels (Soll and Wollin, 1977; Taylor et al., 1984; Taylor, 1986; Joosten, 1988; Stratton et al., 1991). As a result of these various physiological effects, histamine poisoning is often characterized by a variety of symptoms which can be categorized as cutaneous (rash, urticaria, oedema, and localized inflammation), gastrointestinal (nausea, vomiting, diarrhea, and abdominal cramps), hemodynamic (hypotension), and neurological (headache, palpitations, tingling, burning, and itching) (Gilbert et al., 1980; Taylor, 1985, 1986; Stratton et al., 1991; Wu et al., 1997). In severe cases of histamine poisoning, bronchospasm and

respiratory distress have also been reported (Franzen and Eysell, 1969; Shalaby, 1996). However, the most common symptoms of histamine poisoning are limited to rash, diarrhea, flushing, sweating, and headache (Bartholomew et al., 1987).

Histamine poisoning has been associated with the consumption of various foods, including fish, several types of cheeses, and beer (Doeglas et al., 1967; Sumner et al., 1985; Kahana and Todd, 1981; Taylor, 1985, 1986; Uragoda and Lodha, 1979; Izquierdo-Pulido et al., 1994). Histamine (or scombroid) fish poisoning (HFP) is the most significant food-borne disease associated with histamine (Lehane and Olley, 2000). The first case of HFP was reported in 1828 (Henderson, 1830). Since then, HFP has been described in many countries and in some of them is now one of the most prevalent forms of seafood-borne disease (Lipp and Rose, 1997). HFP is caused by eating fish of the families *Scombridae* and *Scomberesocideae*, including tuna, saury, bonito, seerfish, and butterfly kingfish. Nonscombroid fish, such as sardine, pilchards, anchovies, herring, and marline, have also been implicated in cases of HFP (Taylor, 1985). The main characteristic of the species listed above is the relatively high concentration of histidine in their flesh, which leads to a favorable environment for histamine formation by bacterial decarboxylases.

Although HFP is associated generally with the consumption of fish with high levels of histamine, the exact role of histamine in the pathogenesis of the disease has not been clearly elucidated. The main evidence supporting the involvement of histamine in HFP is the identical symptoms of HFP and intravenous histamine administration and the efficacy of antihistamines in HFP therapy. Symptoms of HFP occur from several minutes to several hours after consumption of the toxic fish. The illness typically lasts a few hours but may continue for several days. In the report of Gilbert et al. (1980) on the symptoms of 150 patients affected in 30 separate outbreaks of HFP in Britain, diarrhea was the most common clinical sign followed by hot flush and rash (Figure 1). Similar symptoms were reported by Müller et al. (1992) for 10 incidents of HFP involving 22 patients in South Africa.

Despite the evidence described above implicating histamine as the causative agent of HFP, there is no clear relationship between the histamine dose and HFP symptoms. In fact, several studies have shown that consumption of fish containing a certain amount of histamine is more likely to cause toxic effects than taking the same amount of pure histamine by mouth. Taylor (1986) reported that doses of pure histamine required to produce mild toxic reactions were several times higher than doses consumed with fish. The latter observation has been attributed by several scientists to

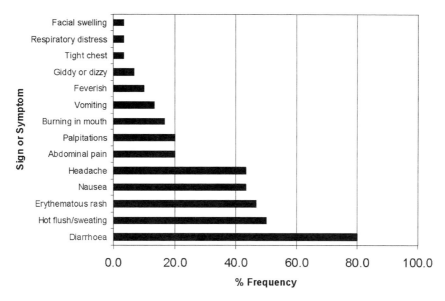

Figure 1. Frequency of symptoms of 150 patients affected in 30 separate outbreaks of HFP in Britain (data from Gilbert et al. [1980]).

the synergistic effect of histamine on other components of fish which act as potentiators (Bjeldanes et al., 1978; Taylor and Lieber, 1979; Chu and Bjeldanes, 1981; Lyons et al., 1983; Taylor 1986, 1988; Stratton et al., 1991). These potentiators can act as enzyme inhibitors of the detoxification system of histamine in humans, which is based on several routes, including oxidative diamination by DAO and methylation by HMT (histamine methyl transferase) (Rice et al., 1976; Taylor, 1986). In addition to their role as inhibitors of histamine-metabolizing enzymes, the potentiators may interfere with the possible protective action of intestinal mucin, which prevents histamine absorption by binding it (Lehane and Olley, 2000) or may cause release of endogenous (mast cell) histamine (Clifford et al., 1991; Ijomah et al., 1991, 1992; Bartholomew et al., 1987; Gessner et al., 1996). Experiments with rats and guinea pigs have indicated

that histamine potentiators include other biogenic amines normally present in fresh or spoiled fish. Mongar (1957) reported that cadaverine and putrescine inhibited DAO in guinea pig ileum, and Taylor and Lieber (1979) observed that cadaverine is a strong inhibitor of both DAO and HMT in rats. Parrot and Nicot (1966) showed that the toxicity of orally ingested histamine in guinea pigs is 10 times higher when it is administrated 40 min after oral administration of pure putrescine. Other biogenic amines that have been reported to potentiate histamine toxicity are tyramine (MAO inhibitor), tryptamine (DAO inhibitor), and β-phenylethylamine (DAO and HMT inhibitor) (Stratton et al., 1991).

Due to the great variety of individual sensitivities and the effect of potentiating factors, it is very difficult to establish acceptable levels of histamine in foods. Several studies have shown that the threshold

Table 3. Histamine poisoning dose-response data for humans

Histamine dose	Dose (mg/kg body wt)	Administration	Symptoms	Reference
180 mg		Pure/orally	No noticeable toxic effects	Weiss et al. (1932)
7 μg		Pure/intravenously	Vasodilatation and increased heart rate	Weiss et al. (1932)
67.5 mg		Pure/orally	No noticeable toxic effects	Granerus (1968)
100–180 mg	1.6–3, assuming an average wt of 60 kg	Mixed in grapefruit juice/orally	Characteristic symptoms of mild histamine poisoning (mild-to-severe headache and obvious flushing) in 1/4 volunteers	Motil and Scrimshaw (1979
100–180 mg	1.6–3, assuming an average wt of 60 kg	Mixed in 100 g high-quality tuna/ orally	Characteristic symptoms of mild histamine poisoning (mild-to-severe headache and obvious flushing) in 4/8 subjects	Motil and Scrimshaw (1979)
300 mg	5	Mixed in 50 g fresh mackerel/orally	Mild symptoms of histamine poisoning (oral tingling, headache, and flushing in some subjects)	Clifford et al. (1989)
300 mg	5	Mixed in 50 g spoiled mackerel/orally	Mild symptoms of histamine poisoning (oral tingling, headache, and flushing in some subjects)	Clifford et al. (1989)

Table 4. Safety criteria for histamine established in European Union Regulation 2073/2005 for the microbiological criteria for foodstuffs[a]

Food category[b]	Sampling plan		Limits (mg/kg body wt)		Analytical reference method[c]	Stage in which the criterion applies
	n	c	m	M		
Fishery products from fish species associated with a high amount of histidine	9	2	100	200	HPLC	Products placed on the market during their shelf-life
Fishery products which have undergone enzyme maturation treatment in brine and manufactured from fish species associated with a high amount of histidine	9	2	200	400	HPLC	Products placed on the market during their shelf-life

[a]n, number of units comprising the sample; c = number of sample units giving values over m or between m and M; m, minimum; M, maximum.
[b]Particularly fish species of the families *Scombridae, Clupeidae, Engraulidae, Coryphaenidae, Pomatomidae,* and *Scomberesocidae.*
[c]Sources: Malle et al. (1996) and Duflos et al. (1999).

toxic dose for oral consumption of histamine can vary significantly (Table 3). Shalaby (1996), after reviewing the oral toxicity of histamine to humans, considered that histamine-induced poisoning is, in general, slight at 8 to 40 mg/100 g, moderate at >40 mg/100 g, and severe at >100 mg/100 g. Based on an analysis of recent poisoning episodes, Shalaby (1996) suggested the following guideline levels for the histamine content of fish:

- <5 mg/100 g (safe for consumption)
- 5 to 20 mg/100 g (possibly toxic)
- 20 to 100 mg/100 g (probably toxic)
- >100 mg/100 g (toxic and unsafe for human consumption)

Despite the uncertainty of the histamine toxic level and because of the recurrence of histamine poisoning in many parts of the world and the importance of international trade of the fish species concerned, many countries have enacted maximal limits or guidelines on histamine levels in traded fish. The FDA (2001) has established guidelines for tuna, mahimahi, and related fish which specify 50 mg/100 g (500 ppm) as the toxicity level. In Australia and New Zealand, the level of histamine in a composite sample of fish or fish products, other than crustaceans and mollusks, must not exceed 10 mg/100 g (100 ppm). A "composite sample" is a sample, taken from each lot, consisting of five portions of equal size taken from five representative samples. This clause, which came into force in October 1994, is currently under review, with a proposal to increase the maximum allowable level of histamine in fish and fish products to 20 mg/100 g (200 ppm) (Lehane and Olley, 2000). The European Union, in the recent regulation 2073/2005 for the microbiological criteria for foodstuffs, set a food safety criterion for

the histamine level in fishery products associated with a high amount of histidine, particularly fish species of the families *Scombridae, Clupeidae, Engraulidae, Coryphafenidae, Pomatomidae,* and *Scomberesocidae* (Table 4).

Tyramine toxicity

Tyramine is included in the amine group (together with tryptamine and β-phenylethylamine) of vasoactive amines. The physiological effects of tyramine include peripheral vasoconstriction, increased cardiac output, increased respiration, elevated blood glucose, and release of norepinephrine (Shalaby, 1996). In addition to being slightly toxic itself, tyramine reacts with MAO inhibitor (MAOI) drugs, giving rise to hypertensive crisis. As has been discussed above, MAO is an enzyme in the digestive tract that catalyzes the oxidative deamination of a variety of neurotransmitters as well as the monoamines of dietary origin, i.e., dopamine, phenyletylamine, serotonin, and tyramine. MAO has two subtypes, MAO-A and MAO-B, which can be found on the short arm of the X chromosome and seem to be derived from a duplication of a common ancestral gene. MAO is found in a variety of tissues and is localized to the outer membranes of the mitochondria. Dopamine and tyramine can be metabolized by both MAO-A and MAO-B (Metcalfe, 2003). MAOI medications are an important class of drugs, which have been used since the 1950s for the treatment of a variety of psychiatric disorders, including depression with either melancholic or atypical features, dysthymia, psychotic depression, bipolar major depression, mixed anxiety and depression, and refractory depression (Gardner et al.,1996; Amsterdam, 2003; Marcason, 2005). They have also been used successfully as effective and relatively safe options in the treatment of depression in the elderly and of panic disorders and phobias (Gardner et al., 1996; Amsterdam, 2003). There is, however, concern over using these drugs, due to the serious, even fatal, interaction observed between the drugs and foods containing tyramine. The use of MAOI drugs eliminates the detoxification process, and thus, relatively high concentrations of presser amines, such as tyramine, derived from foods can be accumulated in the blood, leading to a hypertension crisis of patients, with symptoms such as severe headache, nausea, neck stiffness, palpitations, diaphoresis, and confusion, and even causing stroke and death (Gardner et al., 1996; Marcason, 2005; Blackwell, 1963; Lejonic et al., 1979; Smith and Durack, 1978).

Cheese was the food initially associated with hypertensive disturbances noted with patients undergoing treatment with MAOI (Asatoor et al., 1963; Blackwell and Mabbit, 1965), and therefore, the increase in blood pressure caused by tyramine is also known as the cheese reaction. However, several other foods have been implicated in similar hypertensive attacks, including pickled herring, meat products (Boulton et al., 1970), and chicken liver (Massey, 1976). As in the cases of other amines, there is no clear evidence to support a defined toxic dose of tyramine. In individuals taking no MAOI medication, 20 to 80 mg of tyramine injected intravenously or subcutaneously caused a marked elevation of blood pressure (Asatoor et al., 1963). On the other hand, in individuals on MAOI medications, as little as 6 mg taken orally can cause a rise in blood pressure (Asatoor et al., 1963). In the study of Til et al. (1997), the toxicity of various biogenic amines in Wistar rats was examined. The results of the study showed that the dose of tyramine required for acute oral toxicity was more than 2,000 mg/kg body weight, significantly lower than other biogenic amines. In the same study the authors reported that the no-observed-adverse-effect level of tyramine was 180 mg/kg body weight/day.

Toxicity of spermidine, spermine, putrescine, and cadaverine

The available data on the toxicity of biogenic amines other than histamine and tyramine are limited. Til et al. (1997) examined the acute and subacute toxicity of five biogenic amines, including tyramine, spermidine, spermine, putrescine, and cadaverine, in Wistar rats. The results showed that spermine and spermidine had the highest acute toxicity with 50% lethal dose values of 600 mg/kg body weight, while the 50% lethal dose values for putrescine and cadaverine were >2,000 mg/kg body weight. All amines investigated caused a dose-related decrease in blood pressure after intravenous administration, except for tyramine, for which an increase was found. In a 6-week feeding study in which biogenic amines were administered at various levels from 0 to 10,000 ppm, spermine was the most toxic, leading to emaciation, aggressiveness, convulsions, paralysis of the hind legs, slight anemia, and changes in plasma clinical chemistry. Spermine caused impaired kidney function, together with renal histopathological changes and changes in plasma electrolytes and urea. Slight increases in packed cell volume, hemoglobin concentration, and thrombocytes occurred with cadaverine. In general, adverse effects were observed in the top dose groups of all other amines. The no-observed-adverse-effect level was 2,000 ppm (180 mg/kg body weight/day) cadaverine and putrescine, 1,000 ppm (83 mg/kg body weight/day) of spermidine, and 200 ppm (19 mg/kg body weight/day) of spermine.

Biogenic Amines as Precursors of Carcinogens

Apart from the toxic effects described above, biogenic amines can play a role in other types of food-related risks for human health. Several studies have provided data on the potential role of biogenic amines as precursors of carcinogenic compounds. Various biogenic amines, such as spermidine, spermine, tyramine, putrescine, and cadaverine, when subjected to heat, can lead to the formation of secondary amines which, in the presence of nitrites, can generate nitrosamines. Nitrosamines are chemical agents considered to possess major carcinogenic properties (Patterson and Mottram, 1974; Warthesen et al., 1975). Putrescine and cadaverine on heating are converted to pyrrolidine and piperidine, respectively, from which N-nitrosopyrrolidine and N-nitrosopiperidine are formed by heating in the presence of nitrite. Joosten (1988) reported that incubation of tyramine and nitrite at 37°C and a pH of 1 to 2 for 60 min, simulating stomach conditions, led to significant amounts of 3-diazotyramine, which have been shown to induce oral cavity cancer in rats. Other nitrosatable amines which form the most known carcinogenic N-nitrosamines are secondary amines, such as agmatine and the polyamines spermine and spermidine. The formation of nitrosamines is particularly important in some food products with high biogenic amine levels and added nitrates and nitrites as preservatives, especially when these are consumed after heating. High amounts of nitrosamines of dimethylamine, diethylamine, pyrrolidine, and piperidine among others, which occur widely in fish, meat, and vegetable products (Smith, 1980), have been reported for many food products, especially fried foods (Pensabene et al., 1974; Doyle et al., 1993). Generally, there was a significant increase in urinary nitrosamine following consumption of a meal containing nitrate and amines compared to a meal without additional amines (Doyle et al., 1993).

BIOGENIC AMINES AS SPOILAGE INDICATORS OF FOODS

The fact that most biogenic amines are absent or present in low levels in fresh foods, while their concentrations increase during storage as a result of microbial metabolic activity (ten Brink et al., 1990; Halazs et al., 1994; Bardocz, 1995), led many researchers to examine their use as potential indicators of food spoilage (or freshness). Several spoilage indicators based on the concentration of individual amines or combinations of amines have been proposed mainly for fish and meat products (Table 5).

Mietz and Karmas (1977) were the first who proposed a biogenic amine-based index for monitoring canned tuna quality. This quality index comprises levels of histamine, putrescine, cadaverine, spermidine, and spermine (Table 5). The authors reported that as decomposition progressed, histamine, putrescine, and

cadaverine increased, spermidine and spermine decreased, and they set an index value of 10 as the limit of fish acceptability. The use of the Mietz and Karmas quality index was successfully validated for other seafood products, including rockfish, salmon, lobster, and shrimp (Mietz and Karmas, 1978). Veciana-Nogues et al. (1997) studied the effectiveness of the Mietz and Karmas index as a spoilage indicator of fresh tuna and reported that the limit value of 10 was not reached in samples at the end of their organoleptic acceptance, while the content of histamine in those samples exceeded the regulation limit of 100 µg/g. Veciana-Nogues et al. (1997) suggested that it would be advisable to also monitor the levels of tyramine, which is not included in the Mietz and Karmas index, and proposed a biogenic amine index (BAI) based on the sum of histamine, putrescine, cadaverine, and tyramine. Baixas-Nogueras et al. (2005), comparing the effectiveness of the BAI with the Mietz and Karmas index in monitoring the freshness of Mediterranean hake, reported that the former showed a higher correlation with the sensory results and suggested that a BAI limit of acceptability may be established in a range of 15 to 20 µg/g. Both the BAI and the Mietz and Karmas index were also found to be

Table 5. Biogenic amines proposed as spoilage indicators for fish and meat products

Product	Biogenic amine spoilage index[a]	Reference(s)
Fish		
Canned tuna	(HI+PU+CA)/(1+SD+SM)	Mietz and Karmas (1977)
Rockfish, salmon, lobster, shrimp	(HI+PU+CA)/(1+SD+SM)	Mietz and Karmas (1978)
Sardine, saury pike	CA	Yamanaka et al. (1987)
Herring, haddock	PU, CA	Fernandez-Salguero and Mackie (1987a, 1987b)
Squid	AG	Yamanaka et al. (1987b)
Rainbow trout	PU, CA	Dawood et al. (1988)
Scallop	PU	Yamanaka (1989)
Horse mackerel	PU, CA	Okozumi et al. (1990)
Tuna	HI+CA+TY+PU	Veciana-Nogues et al. (1997)
Gilt-head sea bream	PU, CA	Koutsoumanis et al. (1999)
Cold-smoked salmon	200−(31×pH)+(0.06×TY)+ (0.06×CA)+(0.04×PU)+(0.15×HI)	Jørgensen et al. (2000)
Hake	CA, AG	Ruiz-Capillas and Moral (2001)
Mediterranean hake	HI+CA+TY+PU	Baixas-Nogueras et al. (2005)
Sardine	HI+CA+TY+PU	Özogul and Özogul (2006)
Farmed rainbow trout	PU	Rezaei et al. (2007)
Meat		
Beef	PU, CA	Slemr (1981)
Beef, Pork Bologna sausage	HI+CA+TY+PU	Wortberg and Woller (1982)
Beef	PU, CA, TY	Edwards et al. (1985, 1987)
Beef	PU, TY	Sayem-El-Daher et al. (1984)
Pork, Beef, Rabbit	PU, CA	Guerrero-Legarreta and Chavez-Gallardo (1991)
Pork	HI+CA+TY+PU	Hernandez-Jover et al. (1996a)
Dry sausage	HI+CA+TY+PU	Eerola et al. (1998a, 1998b)

[a]HI, histamine; PU, putrescine; CA, cadaverine; SD, spermidine; SM, spermine; AG, agmatine; TY, tyramine.

effective in monitoring the quality of sardines stored under aerobic conditions, vacuum packaging, and modified atmosphere packaging. The BAI gave significantly higher values than the Mietz and Karmas index during storage (Ozogul and Ozogul, 2006). Yamanaka et al. (1987a) proposed cadaverine as the most effective spoilage index of sardine and saury pike and suggested a value of 15 mg/100 g as an indication of good quality, values between 15 and 20 mg/100 g as an indication of initial decomposition, and values above 20 mg/100 g as an indication of advanced decomposition. Dawood et al. (1988) reported that putrescine and cadaverine could be used to assess freshness of chilled-stored rainbow trout *(Salmo irideus)*. The use of putrescine as a quality index of rainbow trout was confirmed by Rezaei et al. (2007). Okozumi et al. (1990) showed that high levels of putrescine and cadaverine were detected at the spoilage stage of horse mackerel meat. These two diamines have also been proposed as freshness indicators for other fish species (Fernandez-Salguero and Mackie, 1987a, 1987b; Koutsoumanis et al., 1999). Yamanaka et al. (1987b) reported that in the fresh muscle of squid, agmatine was detected in small amounts, but its concentration increased with storage time and reached a very high level at the stage of advanced decomposition. The authors proposed agmatine as a useful indicator for the freshness of common squid. Agmatine has also been proposed together with cadaverine as a spoilage indicator of hake (Ruiz-Capillas and Moral, 2001). Jørgensen et al. (2000b) proposed a multiple-compound quality index for cold-smoked salmon consisting of a combination of pH and the level of four biogenic amines (tyramine, cadaverine, putrescine, and histamine). This index, which was developed using multivariate regression, was intended to correlate with the sensory score and not the storage time. In their study, Jørgensen et al. stated that the suggested multiple-compound quality index relies on measurements that may be carried out by direct electrode (pH) and biosensors for biogenic amines and is therefore of practical use for quality inspection.

Biogenic amines have also been used as quality indexes and indicators of various meat products (Table 5). Slemr (1981) and Edwards et al. (1985) reported that putrescine and cadaverine correlate well with the microbial load of fresh beef and suggested them as a quality index. Wortberg and Woller (1982) proposed a biogenic amine index consisting of the sum of putrescine, cadaverine, histamine, and tyramine for Bologna sausage, minced beef, and pork and established 500 mg/kg as the limit of quality acceptability. Hernandez-Jover et al. (1996) examined

the above index for fresh pork and suggested the following limits: <5 mg/kg for good quality; between 5 and 20 mg/kg for acceptable quality, but with initial spoilage signs; between 20 and 50 mg/kg for low meat quality; and >50 mg/kg for spoiled meat. The sum of putrescine, cadaverine, histamine, and tyramine has also been proposed by Eerola et al. (1998a, 1998b) as an indicator of quality for fermented meat products.

Despite the extensive research on the use of biogenic amines as spoilage indicators, it is not possible to identify a general biogenic amine index which would be applicable to all foods. Biogenic amine production is specific to certain organisms, and when these organisms are not present or their growth is inhibited, incorrect spoilage information is provided. Even within the same species, not all strains develop the same decarboxylating capacity, so that a low biogenic amine concentration does not always signal good microbiological quality (Santos et al., 1985; Dainty et al., 1987). Furthermore, the increase in amine concentration is the result of the utilization of a specific amino acid, but the absence of the given substrate or its presence in low quantities does not preclude spoilage. Finally, the rate of amine production and the metabolic pathways of bacteria are affected by imposed environmental conditions (e.g., pH, oxygen availability, and temperature). Thus, the applicability of a biogenic amine index as a spoilage indicator should be examined and validated to specific food products, while extrapolation to other foods may lead to erroneous results.

ANALYTICAL METHODS FOR THE DETERMINATION OF BIOGENIC AMINES IN FOODS

Several methods for the determination of biogenic amines in foods based on thin-layer chromatography (TLC), liquid chromatography, gas chromatography, biochemical assays, and capillary electrophoresis have so far been described (Onal, 2007). Because of the complexity of the food matrix and the low concentration levels of biogenic amines in foods, the extraction of amines from the samples is a crucial step for most analytical procedures. Many different solvents have been used for the extraction of biogenic amines from food samples, such as hydrochloric acid, trichloroacetic acid, perchloric acid, methanesulfonic acid, petrolum ether, and other organic solvents (Onal, 2007).

TLC is probably the simplest method for biogenic amine analysis (Shultz et al., 1976; Shalaby, 1999; Lapa-Guimaraes and Pickova, 2004). The main advantage of TLC is that it does not require special equipment. Among the disadvantages of TLC methods, the excessive time needed for analysis and the semiquantitative nature of the obtained results are the most important. High-pressure liquid chromatography (HPLC) is by far the most frequently reported technique for simultaneous separation and quantification of multiple biogenic amines in foods. A review of the published HPLC analytical procedures for the determination of biogenic amines in various foods is presented in Table 6. Although the application of gas chromatography (GC) for the analysis of biogenic amines is not very common, some methods based on GC alone or in combination with mass spectrometry have been developed for the determination of histamine and other amines (Fernandes and Ferreira, 2000; Hwang et al., 2003). Capillary electrophoresis (CE) is another method which has been used widely in recent years for the determination of biogenic amines in foods due to its high sensitivity and the short analysis time. In a comparative study, Lange et al. (2002) reported that biogenic amines were separated in less than 9 min by CE compared to the 20 min achieved by HPLC. CE methods used for the determination of biogenic amines are based on indirect UV (Arce et al., 1998; Ruiz-Jimenez and Luque de Castro, 2006), lamp-induced fluorescence (Zhang and Sun, 2004; Zhang et al., 2005), laser-induced fluorescence (Cortacero-Ramirez et al., 2005), conductometric (Kvasnicka and Voldrich, 2006), and pulsed amperometric detection (Sun et al., 2003). Nouadje et al. (1997) applied micellar electrokinetic capillary chromatography for the determination of 28 biogenic amines and amino acids in wine using a laser-induced fluorescence detector. Micellar electrokinetic capillary chromatography with laser-induced fluorescence or UV detection methods have also been applied for the quantitative analysis of biogenic amines in other foods, including salami, beer, and fish (Kovacs et al., 1999; Kalac et al., 2002).

Very promising techniques involve enzymatic methods, such as radioimmunoassays and the enzyme-linked immunosorbent assay system, which have been applied to detect histamine and other amines in foods (Guesdon et al., 1986; Stratton et al., 1991). These methods have the advantage of being rapid and not requiring expensive instrumentation (Stratton et al., 1991). Lerke et al. (1983) developed a rapid qualitative method to detect histamine in fish based on a two-step sequential enzymatic system. The first step involves diamine oxidase which catalyzes histamine breakdown with hydrogen peroxide production, while in the second step hydrogen peroxide is detected by the formation of crystal violet from the leuco base in the presence of peroxidase at 596 nm. Later, Lopez-Sabater et al. (1993) modified this method to achieve histamine quantification. Further modification of the method by Rodriguez-Jerez et al. (1994b) resulted in a rapid and reliable assay for histamine determination. Cepicka et al. (1992) developed a competitive enzyme immunoanalysis for the determination of histamine in beers. This method is very selective and not affected by the presence of other amines or their various derivatives. The proof limit of this method is 7 ng/ml. Lange and Wittmann (2002) developed an enzyme sensor array for the simultaneous determination of histamine, tyramine, and putrescine using an artificial neural which provides results within 20 min, with only one extraction and subsequent neutralization step required prior to sensor measurement. An enzymatic method for amine determination has also been developed in the form of a test strip. Gouygou et al. (1992) developed an immobilized enzyme film containing both lentil seed diamino oxidase and horse peroxidase. The film was made by coentrapping the molecules of enzyme and an inactive matrix protein (gelatine or bovine serum albumin) with a bifunctional reagent (glutaraldehyde). Another strip/dipstick test for diamines in tuna was also reported by Hall et al. (1999). This solid phase assay is based on the coupling of diamine oxide to a peroxidase/dye system. The assay is linear to 75 µM in phosphate buffer, and the minimum detectable concentration is 0.5 µM (<0.1 ppm), corresponding to 0.01 mg/100 ml in spiked extracts, while intra- and interassay precisions are <20%.

Most analytical methods for the determination of biogenic amines require a derivatization procedure. This is due to the low volatility and lack of chromophores of most biogenic amines. Different chemical reagents have been used for derivatization, such as dansyl and dabsyl chloride, benzoyl chloride, fluoresceine, 9-fluorenylmethyl chloroformate, o-phthalaldehyde (OPA), and naphthalene-2, 3-dicarboxaldehyde. OPA reacts readily with the primary amine group in the presence of a thiol group (e.g., b-mercaptoethanol) to form a fluorescent isoindole derivative. However, OPA derivatives are not very stable. Dabsyl and dansyl chloride react with both primary and secondary amino groups, with the advantages of providing stable derivatives and the possibility of measuring with both fluorescence and UV detectors. There are only a few methods in which derivatization is not required, such as enzymatic methods (Lange et al., 2002), HPLC and CE with conductometric detection (Cinquina et al., 2004), and capillary zone electrophoresis with amperometric detection (Sun et al., 2003).

Table 6. HPLC conditions for determination of biogenic amines in foods

Food sample	Sample treatment (concentration)	Column	Mobile phase	Derivatization/detection	Reference(s)
Fish	Trichloroacetic acid (5%)	RP-8 LiChrosorb	Gradient elution with methanol, acetonitrile, and 0.02 M acetic acid	Dansyl chloride/UV	Rosier and Van Peteghem (1988)
Fish	Trichloroacetic acid (5%)	Altex Ultrasphere-Si	Hexane–ethyl acetate(40:60) with addition of 0.01% aminoethanol	Dansyl chloride/fluorescence detector	Lebiedzińska et al. (1991)
Meat	Perchloric acid (0.4 M)	Spherisorb ODS 2	Ammonium acetate and acetonitrile	Dansyl chloride/UV	Buncic et al. (1993)
Sausages	Perchloric acid (0.6 M)	Nucleosil 10 7 C_{18}	A: 0.05 M hexanesulfonic acid, 0.1 M sodium dihydrogenphosphate (pH 3.5); B: eluent A and acetonitrile (3:1)	o-Phthalaldehyde/fluorescence detector	Straub et al. (1993)
Fish	Perchloric acid (0.6 M)	Nova-Pak C_{18}	A: 0.1 M sodium acetate and 10 mM octanesulfonic acid (pH 5.2); B: acetonitrile, 0.2 M sodium acetate and 10 mM octanesulfonic acid (pH 4.5)	o-Phthalaldehyde/fluorescence detector	Veciana-Nogues et al. (1995)
Cheese, red wine, sausages	Hydrochloric acid (0.1 M)	Spherisorb ODS 2	A: 4% dihydrogen phosphate, dimethylformamide, 0.18% trimethylamine (pH 6.55); B: 80% acetonitrile, 10% tertbutyl ether, 10% water	Dansyl chloride/UV	Bockhardt et al. (1996)
Meat	Perchloric acid (0.6 M)	Nova-Pak C18	Octanesulfonic acid	o-Phthalaldehyde/fluorescence detector	Hernandez-Jover et al. (1996a)
Fish, cheese, meat	Hydrochloric acid (0.1 M)	Spherisorb 3S T6	Mixture of acetonitrile and water	Dansyl chloride/UV	Moret and Conte (1996)
Fish, sauerkraut, wine	Perchloric acid (6%)	Iner+sil ODS 2	Phosphate buffer (pH 7) and acetonitrile	o-Phthalaldehyde/fluorescence detector	Beljaars et al. (1998)
Cheese	Hydrochloric acid (0.1 M)	μ Bondapak C_{18}	A: 0.2 M sodium acetate, 10 mM diethylether; B: ethanol–acetonitrile–sodium octanesulfonate (1:9:1)	o-Phthalaldehyde/fluorescence detector	Vale and Glória (1998)
Fish	Trichloroacetic acid (5%)	Phenomenex IB-Sil	0.02 M phosphate buffer (pH 7.2) and acetonitrile	Fluorescein/fluorescence detector	Gingerich et al. (1999)
Meat	Trichloroacetic acid (10%)	Spherisorb ODS 2	A: 0.1 M ammonium acetate; B: acetonitrile	Dansyl chloride/UV	Masson et al. (1999)
Cheese	Hydrochloric acid (0.1 M)	Spherisorb 3STG	Acetonitrile	Dansyl chloride/UV	Galgano et al. (2001)
Meat	Perchloric acid (0.4 M)	Supelcosil LC-18	Gradient elution with ammonium acetate (0.1 M) and acetonitrile	Dansyl chloride/variable wavelength detector	Vinci and Antonelli (2002)
Sausages	Trichloracetic acid (5%)	Zorbax Eclipse XDB C8	Gradient elution with acetate buffer (100 mM, pH 5.8) and acetonitrile	o-Phthalaldehyde/fluorescence detector	Komprda et al. (2004)
Sausages	Trichloracetic acid (5%)	Zorbax Eclipse XDB C_{18}	Gradient elution with water and acetonitrile	UV–Vis detector	Komprda et al. (2004)
Fish	Trichloracetic acid (5%)	Lichrospher 100 RP-18	Methanol–water	Benzoylchloride/UV–Vis detector	Tsai et al. (2005a, 2005b)
Vegetables	Hydrochloric acid (0.1 M)	Kromasil C_{18}	Gradient elution with phosphate buffer (0.07 M, pH 7.0) and acetonitrile	Dansyl chloride/UV–VIS detector	Moret et al. (2005)
Vegetables	Hydrochloric acid (0.1 M)	Kromasil C_{18}	Gradient elution with water and acetonitrile	o-Phthalaldehyde/ spectrofluorimetric	Moret et al. (2005)
Sausages	Perchloric acid (0.4 M)	Spherisorb ODS-2 150A	Gradient elution with ammonium acetate (0.1 M) and acetonitrile	Dansyl chloride/diode array detector	Roseiro et al. (2006)
Ciders		Nova-Pak C_{18}	Gradient elution with sodium phosphate dodecahydrate (10 mM, pH 8.3) and 1% 2-octanol in acetonitrile	o-Phthalaldehyde/fluorescence detector	Garai et al. (2006)
Sardines	Trichloracetic acid (6%)	Spherisorb ODS-2 C_{18}	Acetonitrile–water	Diode array detector	Özogul and Özogul (2006)
Trout	Trichloracetic acid (5%)	Spherisorb ODS-2 C_{18}	Methanol–water	Benzoylchloride/UV	Rezaei et al. (2007)
Miso, natto, sufu	Trichloracetic acid (6%)	LiChrospher 100 RP-18	Gradient elution with ethanol and water	Benzoylchloride/UV–VIS detector	Kung et al. (2007a, 2007b)
					Tsai et al. (2007)
Cereals	Trichloroacetic acid	Asahipak ODP-50	50 mmol/liter sodium acetate buffer (pH 9.9) and acetonitrile containing 2 mmol/liter o-phthalaldehyde and 2 mmol/liter N-acetyl-L-cysteine	o-Phthalaldehyde and N-acetyl-L-cysteine/fluorescence detector	Nishino et al. (2007)

REFERENCES

Ababouch, L., M. E. Afilal, and H. Benabdeljelil. 1991. Quantitative changes in bacteria, amino acids and biogenic amines in sardine *(Sardina pilchardus)* stored at ambient temperature (25–28°C) and in ice. *Int. J. Food Sci. Technol.* **26:**297–306.

Aerny, J. 1985. Origine de l'histamine dans les vins. Connaissances actuelles. *Bull. Off. Intern. Vin* **656–657:**1016–1019.

Amsterdam, J. D. 2003. A double blind placebo controlled trial of the safety and efficacy of selegiline transdermal system without dietary restrictions in patients with major depressive disorder. *J. Clin. Psychiatry* **64:**208–214.

Ansorena, D., M. C. Montelb, M. Rokkac, R. Talonb, S. Eerolac, A. Rizzoc, M. Raemaekersd, and D. Demeyerd. 2002. Analysis of biogenic amines in northern and southern European sausages and role of flora in amine production. *Meat Sci.* **61:**141–147.

Arce, L., A. Rios, and M. Valcarcel. 1998. Direct determination of biogenic amines in wine by integrating continuous flow clean-up and capillary electrophoresis with indirect UV detection. *J. Chromatogr. A* **803:**249–260.

Arnold, S. H., and W. D. Brown. 1978. Histamine toxicity from fish products. *Adv. Food Res.* **24:**113–154.

Arnold, S. H., R. J. Price, and W. D. Browen. 1980. Histamine formation by bacteria isolated from skipjack tuna *Katsuwonus plamis. Bull. Jpn. Sot. Sci. Fish.* **46:**991–995.

Asatoor, A. M., A. J. Levi, and M. D. Milne. 1963. Tranylcypromine and cheese. *Lancet* **ii:**733–734.

Askar, A., and H. Treptow. 1986. *Biogene Amine in Lebensmitteln. Vorkommen, Bedeutung und Bestimmung.* Eugen Ulmer GmbH and Co., Stuttgart, Germany.

Baixas-Nogueras, S., S. Bover-Cid, M. T. Veciana-Nogués, A. Mariné-Font, and M. C. Vidal-Carou. 2005. Biogenic amine index for freshness evaluation in iced Mediterranean hake *(Merluccius merluccius). J. Food Prot.* **68:**2433–2438.

Baranowski, J. D., P. A. Brust, and H. A. Frank. 1985. Growth of Klebsiella pneumoniae UH-2 and properties of its histidine decarboxylase system in resting cells. *J. Food Biochem.* **9:**349–360.

Bardocz, S. 1995. Polyamines in foods and their consequences for food quality and human health. *Trends Food Sci. Technol.* **6:**341–346.

Bartholomew, B. A., P. R. Berry, J. C. Rodhouse, and R. J. Gilbert. 1987. Scombrotoxic fish poisoning in Britain: features of over 250 suspected incidents from 1976 to 1986. *Epidemiol. Infect.* **99:**775–782.

Behling, A. R., and S. L. Taylor. 1982. Bacterial histamine production as a function of temperature and time of incubation. *J. Food Sci.* **47:**1311–1317.

Beljaars, P. R., R. Van Dijk, K. M. Jonker, L. J. Schout, G. Brans, J. Boomkamp, J. C. Bosman, and E. Wijma. 1998. Liquid chromatographic determination of histamine in fish, sauerkraut, and wine: interlaboratory study. *J. AOAC Int.* **81:**991–998.

Ben-Gigirey, B., J. M. Vieites Baptista de Sousa, T. G. Villa, and J. Barros-Velaquez. 1998. Changes in biogenic amines and microbiological analysis in albacore *(Thunnus alalunga)* muscle during frozen storage. *J. Food Prot.* **61:**608–615.

Bjedanes, L. F., D. E. Schutz, and M. M. Morris. 1978. On the aetiology of scromboid poisoning: cadaverine potentiation of histamine toxicity in the guinea pig. *Food Cosmet. Toxicol.* **167:**157–159.

Blackwell, B. 1963. Hypertensive crisis due to monoamine-oxidase inhibitors. *Lancet* **ii:**849.

Blackwell, B., and L. Mabbit. 1965. Tyramine in cheese related to hypertension crisis after monoamine-oxidase inhibition. *Lancet* **ii:**938–940.

Bockhardt, A., I. Krause, and H. Klostermeyer. 1996. Determination of biogenic amines by RP-HPLC of the dabsyl derivates. *Z. Lebensm. Unters. Forsch.* **203:**65–70.

Bover-Cid, S., J. Migué lez-Arrizado, and M. C. Vidal-Carou. 2001a. Biogenic amine accumulation in ripened sausages affected by the addition of sodium sulphite. *Meat Sci.* **59:**391–396.

Bover-Cid, S., M. Hugas, M. Izquierdo-Pulido, and M. C. Vidal-Carou. 2001b. Amino acid-decarboxylase activity of bacteria isolated from fermented pork sausages. *Int. J. Food Microbiol.* **66:**185–189.

Bover-Cid, S., and W. H. Holzapfel. 1999. Improved screening procedure for biogenic amine production by lactic acid bacteria. *Int. J. Food Microbiol.* **53:**33–41.

Bover-Cid, S., M. Izquierdo-Pulido, and M. C. Vidal-Carou. 1999a. Effect of proteolytic starter cultures of *Staphylococcus* spp. on biogenic amine formation during the ripening of dry fermented sausages. *Int. J. Food Microbiol.* **46:**95–104.

Bover-Cid, S., S. Schoppen, M. Izquierdo-Pulido, and M. C. Vidal-Carou. 1999b. Relationship between biogenic amine contents and the size of dry fermented sausages. *Meat Sci.* **51:**305–311.

Buncic, S., L. Paunovic, V. Teodorovic, D. Radisic, G. Vojinovic, D. Smiljanic, and M. Baltic. 1993. Effects of gluconodeltalactone and Lactobacillus plantarum on the production of histamine and tyramine in fermented sausages. *Int. J. Food Microbiol.* **17:**303–309.

Butturini, A., P. Aloisi, R. Tagliazucchi, and C. Cantoni. 1995. Ammine biogene prodotte da enterobatteri e batteri lattici. *Industrie Alimentari* **24:**105–107.

Cantoni, C., C. Bersani, L. Damenis, and G. Comi. 1994. Ammine biogene negli insaccati crudi stagionati. *Industrie Alimentari* **23:**1239–1243.

Cavanah, D. K., and T. B. Casale. 1993. Histamine, p. 321–342. *In* M. A. Kaliner and D. D. Metcalfe (ed.), *The Mast Cell in Health and Disease.* Marcel Dekker, New York, NY.

Cepicka, J., P. Rychetsky, I. Hochel, F. Strejcek, and P. Rauch. 1992. Immunoenzymic determination of histamine in beer. *Monatsschr. Brauwiss.* **42:**161–164.

Chander, H., V. H. Batish, S. Babu, and R. S. Singh. 1989. Factors affecting amine production by a selected strain of Lactobacillus bulgaricus. *J. Food Sci.* **54:**940–942.

Chen, C. M., L. C. Lin, and G. G. Yen. 1994. Relationship between changes in biogenic amine contents and freshness of pork during storage at different temperatures. *J. Chin. Agric. Chem. Soc.* **32:**47–60.

Chu, C. H., and L. F. Bjeldanes. 1981. Effect of diamines, polyamines and tuna fish extracts on the binding of histamine to mucin in vitro. *J. Food Sci.* **47:**79–88.

Cinquina, A. L., A. Calı, F. Longoa, L. De Santis, A. Severoni, and F. Abballe, 2004. Determination of biogenic amines in fish tissues by ion-exchange chromatography with conductivity detection. *J. Chromatogr. A* **1032:**73–77.

Clifford, M. N., R. Walker, J. Wright, R. Hardy, and C. K. Murray. 1989. Studies with volunteers on the role of histamine in suspected scombrotoxicosis. *J. Sci. Food Agric.* **47:**365–375.

Clifford, M. N., R. Walker, P. Ijomah, J. Wright, C. K. Murray, and R. Hardy. 1991. Is there a role for amines other than histamines in the aetiology of scombrotoxicosis? *Food Addit. Contam.* **8:**641–652.

Coisson, J. D., C. Cerutti, F. Travaglia, and M. Arlorio. 2004. Production of biogenic amines in "Salamini italiani alla cacciatora PDO." *Meat Sci.* **67:**343–349.

Cooper, R. A. 1997. On the amine oxidases of *Klebsiella aerogenes* strain W70. *FEMS Microbiol. Lett.* **146:**85–89. *J. Food Prot.* **49:**423–427.

Cortacero-Ramírez, S., D. Arráez-Román, A. Segura-Carretero, and A. Fernández-Gutiérrez. 2005. Determination of biogenic amines in beers and brewing-process samples by capillary electrophoresis coupled to laser-induced fluorescence detection. *Food Chem.* **100**:383–389.

Dainty, R. H., R. A. Edwards, C. M. Hibard, and S. V. Ramantanis. 1987. Amines in fresh beef of normal pH and the role of bacteria in changes in concentration observed during storage in vacuum packs at chill temperatures. *J. Appl. Bacteriol.* **63**:427–434.

Dapkevicius, M. L. N. E., M. J. R. Nout, F. M. Rombouts, J. H. Houben, and W. Wymenga. 2000. Biogenic amine formation and degradation by potential fish silage starter microorganisms. *Int. J. Food Microbiol.* **57**:107–114.

Dawood, A. A., J. Karkalas, R. N. Roy, and C. S. Williams. 1988. The occurrence of non-volatile amines in chilled-stored rainbow trout *(Salmo irideus)*. *Food Chem.* **27**:33–45.

De Borba, B. M., and J. S. Rohrer. 2007. Determination of biogenic amines in alcoholic beverages by ion chromatography with suppressed conductivity detection and integrated pulsed amperometric detection. *J. Chromatogr. A* **1155**:22–30.

de Llano, D. G., P. Cuesta, and A. Rodriguez. 1998. Biogenic amine production by wild lactococcal and leuconostoc strains. *Lett. Appl. Microbiol.* **26**:270–274.

Diaz-Cinco, M. E., C. L. Fraijo, P. Grajeda, J. Lozano-Taylor, and E. Gonzalez de Mejia. 1992. Microbial and chemical analysis of Chihuahua cheese and relationship to histamine. *J. Food Sci.* **57**:355–356, 365.

Doeglas, H. M. G., J. Huisman, and J. P. Nater. 1967. Histamine intoxication after cheese. *Lancet* ii:1361–1362.

Doyle, M. E., C. E. Steinhart, and B. A. Cochrana. 1993. *Food Safety*, p. 254–259. Marcel Dekker, New York, NY.

Duflos, G., C. Dervin, P. Malle, and S. Bouquelet. 1999. Relevance of matrix effect in determination of biogenic amines in plaice (Pleuronectes platessa) and whiting (Merlangus merlangus). *J. AOAC Int.* **82**:1097–1101.

Edwards, R. A., R. H. Daintry, and C. M. Hibard. 1983. The relationship of bacterial numbers and types to diamine concentration in fresh and aerobically stored beef, pork, and lamb. *J. Food Technol.* **18**:777–788.

Edwards, R. A., R. H. Dainty, and C. M. Hibard. 1985. Putrescine and cadaverine formation in vacuum packed beef. *J. Appl. Bacteriol.* **58**:13–19.

Edwards, R. A., R. H. Dainty, C. M. Hibard, and S. V. Ramantanis. 1987. Amines in fresh beef of normal pH and the role of bacteria in changes in concentration observed during storage in vacuum packs at chill temperatures. *J. Appl. Bacteriol.* **63**:427–434.

Edwards, S. T., and S. L. Sandine. 1981. Public health significance of amine in cheese. *J. Dairy Sci.* **64**:2431–2438.

Eerola, H. S., A. X. Roig Sagués, and T. K. Hirvi. 1998b. Biogenic amines in Finnish dry sausages. *J. Food Saf.* **18**:127–138.

Eerola, S., R. Maijala, A. X. Roig Sagués, M. Salminen, and T. Hirvi. 1996. Biogenic amines in dry sausages as affected by starter culture and contaminant amine-positive *Lactobacillus. J. Food Sci.* **61**:1243–1246.

Eerola, S., R. Maijala, A. X. Roig-Sagués, M. Salminen, and T. K. Hirvi. 1998a. Biogenic amines in dry sausages as affected by starter culture and contaminant amine-positive *Lactobacillus. J. Food Sci.* **61**:1243–1246.

Eitenmiller, R. S., and S. C. De Souza. 1984. *Ragelis EP Seafood Toxins*. ACS Symposium Series, American Chemical Society, Washington, DC.

Eitenmiller, R. S., P. S. Koehler, and J. O. Reagan. 1978. Tyramine in fermented sausages: factors affecting formation of tyramine and tyrosine decarboxylase. *J. Food Sci.* **43**:689–693.

Emborg, J., and P. Dalgaard. 2006. Formation of histamine and biogenic amines in cold-smoked tuna: an investigation of psychrotolerant bacteria from samples implicated in cases of histamine fish poisoning. *J. Food Prot.* **69**:897–906.

FDA. 2001. *Fish and Fishery Products Hazards and Controls Guide*, 3rd ed. FDA, Center for Food Safety and Applied Nutrition, Office of Seafood, Washington, DC.

Ferencik, M. 1970. Formation of histamine during bacterial decarboxylation of histidine in the flesh of some marine fishes. *J. Hyg. Epidemiol. Microbiol. Immunol.* **14**:52–60.

Fernandes, J. O., and M. A. Ferreira. 2000. Combined ion-pair extraction and gas chromatography–mass spectrometry for the simultaneous determination of diamines, polyamines and aromatic amines in Port wine and grape juice. *J. Chromatogr. A* **886**:183–195.

Fernandez-Salguero, J., and I. M. Mackie. 1987a. Comparative rates of spoilage of fillets and whole fish during storage of haddock *(Melanogrammus aeglefinus)* and herring *(Clupea harengus)* as determined by the formation of non-volatile and volatile amines. *Int. J. Food Sci. Technol.* **22**:385–390.

Fernandez-Salguero, J. and M. Mackie. 1987b. Preliminary survey of the content of histamine and other higher amines in some samples of Spanish canned fish. *Int. J. Food Sci. Technol.* **22**:409–412.

Ferreira, V., J. Barbosa, J. Silva, S. Vendeiro, A. Mota, F. Siva, M. J. Monteiro, T. Hogg, P. Gibbs, and P. Teixeira. 2007. Chemical and microbiological characterisation of "Salpicão de Vinhais" and "Chouriça de Vinhais": traditional dry sausages produced in the north of Portugal. *Food Microbiol.* **24**:618–623.

Frank, H. A., J. D. Baranowski, M. Chongsiriwatana, P. A. Brust, and R. J. Premaratue. 1985. Identification and decarboxylase activities of bacteria isolated from decomposed mahimahi after incubation at 0 and 32°C. *Int. J. Food Microbiol.* **2**:331–340.

Franzen, F., and K. Eysell. 1969. *Biologically Active Amines in Man*. Pergamon Press, Oxford, United Kingdom.

Gale, E. F. 1942. The oxidation of amines by bacteria. *Biochem. J.* **36**:54.

Galgano, F., G. Suzzi, F. Favati, M. Caruso, M. Martuscelli, F. Gardini, and G. Salzano. 2001. Biogenic amines during ripening in 'Semicotto Caprino' cheese: role of enterococci. *Int. J. Food Sci. Technol.* **36**:153–160.

Garai, G., M. T. Duenas, A. Irastorza, P. J. Martin-Alvarez, and M. V. Moreno-Arribas. 2006. Biogenic amines in natural ciders. *J. Food Prot.* **69**:3006–3012.

Gardini, F., M. Martuscelli, M. C. Caruso, F. Galgano, M. A. Crudele, F. Favati, M. E. Guerzoni, and G. Suzzi. 2001. Effects of pH, temperature and NaCl concentration on the growth kinetics, proteolytic activity and biogenic amine production of Enterococcus faecalis. *Int. J. Food Microbiol.* **64**:105–117.

Gardini, F., M. Martuscelli, M. A. Crudele, A. Paparella, and G. Suzzi. 2002. Use of Staphylococcus xylosus as a starter culture in dried sausages: effect on the biogenic amine content. *Meat Sci.* **61**:275–281.

Gardner D. M., K. I. Shulman, S. E. Walker, and S. A. Tailor. 1996. The making of a user friendly MAOI diet. *J. Clin. Psychiatry* **57**:99–104.

Genccelep, H., G. Kaban, and M. Kaya. 2007. Effects of starter cultures and nitrite levels on formation of biogenic amines in sucuk. *Meat Sci.* **77**:424–430.

Gessner, B., Y. Hokama, and S. Isto. 1996. Scombrotoxicosis-like illness following the ingestion of smoked salmon that demonstrated low histamine levels and high toxicity on mouse bioassay. *Clin. Infect. Dis.* **23**:1316–1318.

Gil-Agusti, M., S. Carda-Broch, L. Monferrer-Pons, and J. Esteve-Romero. 2007. Simultaneous determination of tyramine and tryptamine and their precursor amino acids by micellar liquid chromatography and pulsed amperometric detection in wines. *J. Chromatogr. A* **1156**:288–295.

Gilbert, R. J., G. Hobbs, C. K. Murray, J. G. Cruickshank, and S. E. J. Young. 1980. Scombrotoxic fish poisoning: features of the first 50 incidents to be reported in Britain 1976–1979. *Br. Med. J.* **281**:71–72.

Gingerich, T. M., T. Lorca, G. J. Flick, M. D. Pierson, and H. M. McNair. 1999. Biogenic amine survey and organoleptic changes in fresh, stored, and temperature-abused bluefish (*Pomatomus saltatrix*). *J. Food Prot.* **6**:1033–1037.

Gloria, M. B. A., and M. Izquierdo-Pulido. 1999. Levels and significance of biogenic amines in Brazilian beers. *J. Food Composition Anal.* **12**:129–136.

Gosetti, F., E. Mazzucco, V. Gianotti, S. Polati, and M. C. Gennaro. 2007. High performance liquid chromatography/tandem mass spectrometry determination of biogenic amines in typical Piedmont cheeses. *J. Chromatogr. A* **1149**:151–157.

Gouygou, J. P., C. Sinquin, M. Etienne, A. Landrein, and P. Durand. 1992. Quantitative and qualitative determination of biogenic amines in fish, p. 178–186. *In* J. R. Burt, R. Hardy, and K. J. Whittle (ed.), *Pelagic Fish: the Resource and Its Exploitation.* Fishing News, Oxford, United Kingdom.

Granerus, G. 1968. Urinary excretion of histamine, methylhistamine and methylimidazoleacetic acids in man under standardized dietary conditions. *Scandinavian J. Clin. Lab. Invest.* **104**(Suppl.):59–68.

Guerrero-Legarreta, I., and A. M. Chavez-Gallardo. 1991. Detection of biogenic amines as meat spoilage indicators. *J. Muscle Foods* **2**:263–278.

Guesdon, J. L., J. C. Chenrier, B. Mazie, B. David, and S. Avrameas. 1986. Monoclonal antihistamine antibody. Preparation, characterization, and application to enzyme immunoassay of histamine. *J. Immunol. Methods* **87**:69–78.

Halasz, A., A. Barath, L. Simon-Sarkadi, and W. Holzapfel. 1994. Biogenic amines and their production by microorganisms in food. *Trends Food Sci. Technol.* **5**:42–48.

Hall, M., P. A. Sykes, D. L. Fairclough, L. J. Lucchese II, P. Rogers, W. Staruszkiewicz, and R. C. Bateman, Jr. 1999. A test strip for diamines in tuna. *J. AOAC Int.* **82**:1102–1108.

Havelka, B. 1967. Role of the Hafiia bacteria in the rise of histamine in tuna fish meat. *Cesk. Hyg.* **12**:343–352.

Henderson, P. B. 1830. Case of poisoning from the bonito (Scomber pelamis). *Edinb. Med. J.* **34**:317–318.

Henry Chin, K. D., and P. E. Koehler. 1986. Effects of salt concentration and incubation temperature on formation of histamine, phenethylamine tryptamine and tyramine during miso fermentation. *J. Food Prot.* **49**:423–427.

Hernandez-Jover, T., M. Izquierdo-Pulido, M. T. Veciana-Nogues, and M. C. Vidal-Carou. 1996a. Biogenic amine sources in cooked cured shoulder pork. *J. Agric. Food Chem.* **44**:3097–3101.

Hernandez-Jover, T., M. Izquierdo-Pulido, M. T. Veciana-Nogues, and M. C. Vidal-Carou. 1996b. Ion-pair high performance liquid chromatographic determination of biogenic amines in meat and meat products. *J. Agric. Food Chem.* **44**:2710–2715.

Hernandez-Jover, T., M. Izquierdo-Pulido, M. T. Veciana-Nogues, A. Marine-Font, and M. C. Vidal-Carou. 1997. Biogenic amine and polyamine contents in meat and meat products. *J. Agric. Food Chem.* **45**:2098–2102.

Huis in't Veld, J. H. J., H. Hose, G. J. Schaafsma, H. Silla, and J. E. Smith. 1990. Health aspects of food biotechnology, p. 2.73–2.97. *In* P. Zeuthen, J. C. Cheftel, C. Ericksson, T. R. Gormley, P. Link, and K. Paulus (ed.), *Processing and Quality of Food. vol. 2. Food Biotechnology: Avenues to Healthy and Nutritious Products.* Elsevier Applied Science, London, United Kingdom.

Hwang, B. S., J. T. Wang, and Y. M. Choong. 2003. A rapid gas chromatographic method for the determination of histamine in fish and fish products. *Food Chem.* **82**:329–334.

Ibe, A., T. Nishima, and N. Kasai. 1992a. Bacteriological properties of and amine-production conditions for tyramine- and histamine-producing bacterial strains isolated from soybean paste (miso) starting materials. *Jpn. J. Toxicol. Environ. Health* **38**:403–409.

Ibe, A., T. Nishima, and N. Kasai. 1992b. Formation of tyramine and histamine during soybean paste (miso) fermentation. *Jpn. J. Toxicol. Environ. Health* **38**:181–187.

Ienistea, C. 1973. Significance and detection of histamine in food. *In* B. C. Hobbs and J. H. B. Christian (ed.), *The Microbiological Safety of Foods*, Academic Press, New York, NY.

Ijomah, P. L., M. N. Clifford, R. Walker, J. Wright, R. Hardy, and C. K. Murray. 1992. Further volunteer studies on scombrotoxicosis, p. 194–199. *In* J. R. Burt, R. Hardy, and K. J. Whittle (ed.), *Pelagic Fish: the Resource and Its Exploitation.* Fishing News Books, Oxford, United Kingdom.

Ijomah, P., M. N. Clifford, R. Walker, J. Wright, R. Hardy, and C. K. Murray. 1991. The importance of endogenous histamine relative to dietary histamine in the aetiology of scombrotoxicosis. *Food Addit. Contam.* **8**:531–542.

Izquierdo-Pulido, M., T. Hemandez-Jover, A. Marine-Font, and M. C. Vidal-Carou. 1994. Biogenic amine contents in European beers, p. 65–71. *In Proceedings of the International European Food Toxicology IVth conference.* Polish Academy of Science, Olsztyn, Poland.

Izquierdo-Pulido, M., J. Font-Fabrégas, J. P. Carceller-Rosa, A. Mariné-Font, and M. C. Vidal-Carou. 1996. Biogenic amine changes related to lactic acid bacteria during brewing. *J. Food Prot.* **59**:175–180.

Joosten, H. M. L. G. 1988. The biogenic amine contents of Dutch cheese and their toxicological significance. *Neth. Milk Dairy J.* **42**:2542.

Joosten, H. M. L. J., and M. D. Northold. 1989. Detection, growth and amine-producing capacity of lactobacilli in cheese. *Appl. Environ. Microbiol.* **55**:2356–2359.

Jørgensen, L. V., H. H. Huss, and P. Dalgaard. 2000a. The effect of biogenic amine production by single bacterial cultures and metabiosis on cold-smoked salmon. *J. Appl. Microbiol.* **89**:920–934.

Jørgensen, L.V., P. Dalgaard, and H. H. Huss. 2000b. Multiple compound quality index for cold-smoked salmon (Salmo salar) developed by multivariate regression of biogenic amines and pH. *J. Agric. Food Chem.* **48**:2448–2453.

Kahana, L. M., and E. Todd. 1981. Histamine intoxication in a tuberculosis patient on isoniazid. *Can. Dis. Wkly. Rep.* **7**:79–80.

Kalac, P., J. Savel, M. Krizek, T. Pelikánová, and M. Prokopová. 2002. Biogenic amine formation in bottled beer. *Food Chem.* **79**:431–434.

Kalac, P., and M. Krizek. 2003. A review of biogenic amines and polyamines in beer. *J. Inst. Brew.* **109**:123–128.

Kimata, M. 1961. The histamine problem. *In* G. Borgstrom (ed.), *Fish as Food.* Academic Press, New York, NY.

Klausen, N. K., and E. Lund. 1986. Formation of biogenic amines in herring and mackerel. *Z. Lebensm. Unters. Forsch.* **182**:459–463.

Klausen, N. K., and H. H. Huss. 1987. Growth and histamine production by Morganella morganii under various temperature conditions. *Int. J. Food Microbiol.* **5**:147–156.

Koessler, K. K., M. T. Hanke, and M. S. Sheppard. 1928. Production of histamine, tyramine, brochospastic and arteriospastic substance in blood broth by pure cultures of microorganisms. *J. Infect. Dis.* **3**:363–377.

Komprda, T., J. Neznalova., S. Standara, and S. Bover-Cid. 2001. Effect of starter culture and storage temperature on the content of biogenic amines in dry fermented sausages Poličane. *Meat Sci.* **59**:267–276.

Komprda, T., D. Smela, P. Pechova, L. Kalhotka, J. Stencl, and B. Klejdus. 2004. Effect of starter culture, spice mix and storage time and temperature on biogenic amine content of dry fermented sausages. *Meat Sci.* **67**:607–616.

Koutsoumanis, K., K. Lambropoulou, and G. J. E. Nychas. 1999. Biogenic and sensory changes associated with the microbial flora of Mediterranean gilt-head seabream (Sparus aurata) stored aerobically at 0, 8, and 15°C. *J. Food Prot.* **62**:392–402.

Kovacs, A., L. Simon-Sarkadi, and K. Ganzler. 1999. Determination of biogenic amines by capillary electrophoresis. *J. Chromatogr.* **836**:305–313.

Kung, H. F., Y. H. Lee, S. C. Chang, C. I. Wei, and Y. H. Tsai. 2007a. Histamine contents and histamine-forming bacteria in sufu products in Taiwan. *Food Control* **18**:381–386.

Kung, H. F., Y. H. Tsai, and C. I. Wei. 2007b. Histamine and other biogenic amines and histamine-forming bacteria in miso products. *Food Chem.* **101**:351–356.

Kurihara, K., Y. Wagatuma, T. Jujii, and M. Okuzumi. 1993. Effect of reaction conditions on L-histidine decarboxylation activity of halophilic histamine-forming bacteria. *Bull. Jpn. Soc. Sci. Fisher.* **59**:1745–1748.

Kvasnicka, F., and M. Voldrich. 2006. Determination of biogenic amines by capillary zone electrophoresis with conductometric detection. *J. Chromatogr. A* **1103**:145–149.

Lange, J., K. Thomas, and C. Wittmann. 2002. Comparison of a capillary electrophoresis method with high-performance liquid chromatography for the determination of biogenic amines in various food samples. *J. Chromatogr. B* **779**:229–239.

Lange, J., and C. Wittmann. 2002. Enzyme sensor array for the determination of biogenic amines in food samples. *Anal. Bioanal. Chem.* **372**:276–283.

Lapa-Guimaraes, J., and J. Pickova. 2004. New solvent systems for thin layer chromatographic determination of nine biogenic amines in fish and squid. *J. Chromatogr. A* **1045**:223–232.

Lebiedzińska, A., H. Lamparczyk, Z. Ganowiak, and K. I. Eller. 1991. Differences in biogenic amine patterns in fish obtained from commercial. *Z. Lebensm. Unters. Forsch.* **192**:240–243.

Lehane, L., and J. Olley. 2000. Histamine fish poisoning revisited. *Int. J. Food Microbiol.* **58**:1–37.

Lejonic, J. L., D. Gusmini, and P. Brochard. 1979. Isoniazid and reaction to cheese. *Ann. Int. Med.* **91**:793.

Lepez-Sabater, E. I., J. L. Rodriguez-Jerez, M. Hemandez-Herrero, and M. T. Mora-Ventura. 1994. Evaluation of histamine decarboxylase activity of bacteria isolated from sardine (Sadina pilchardus) by an enzymic method. *Lett. Appl. Microbiol.* **19**:70–75.

Lerke, P. A., S. B. Werner, S. L. Taylor, and L. S. Guthertz. 1978. Scombroid poisoning. Report of an outbreak. *West. J. Med.* **129**:381.

Lerke, P. A., M. N. Porcuna, and H. B. Chin. 1983. Screening test for histamine in fish. *J. Food Sci.* **48**:155–157.

Leroi, F., J. J. Joffraud, and F. Chevalier. 2000. Effect of salt and smoke on the microbiological quality of cold-smoked salmon during storage at 5 degrees C as estimated by the factorial design method. *J. Food Prot.* **63**:502–508.

Leuschner, R. G. K., M. Heidel, and W. P. Hammes. 1998. Histamine and tyramine degradation by food fermenting microorganisms. *Int. J. Food Microbiol.* **39**:1–10.

Lipp, E. K., and J. B. Rose. 1997 The role of seafood in foodborne diseases in the United States of America. *Rev. Sci. Tech.* **16**:620–640.

Lonvaud-Funel, A., and A. Joyeux. 1994. Histamine production by wine lactic acid bacteria: isolation of a histamine-producing strain of Leuconostoc oenos. *J. Appl. Bacteriol.* **77**:401–407.

Lonvaud-Funel, A. 2001. Biogenic amines in wines: role of lactic acid bacteria. *FEMS Microbiol. Lett.* **199**:9–13.

Lopez-Sabater, E. I., J. J. Rodriguez-Jerez, A. X. Roig-Sagues, and M. T. Mora-Ventura. 1993. Determination of histamine in fish using an enzymic method. *Food Addit. Contam.* **10**:593–602.

Lovenberg, W. 1973. Some vaso- and psychoactive substances in food: amines, stimulates, depressants, and hallucinogens, p. 170–188. *In* Committee on Food Protection, NRC (ed.), *Toxicants Occurring Naturally in Foods*. National Academy of Science, Washington, DC.

Luten, J. B., W. Bouquet, L. A. J. Seuren, M. M. Burggraaf, G. Riekwel-Booy, P. Durand, M. Etienne, J. P. Gouyou, A. Landrein, A. Ritchie, M. Leclerq, and R. Guinet. 1992. Biogenic amines in fishery products: standardization methods within E.C., p. 427–439. *In* H. H. Huss, M. Jakobsen, and J. Liston (ed.), *Quality Assurance in the Fish Industry*, Elsevier Science Publishers B. V., Amsterdam, The Netherlands.

Lyons, D. E., P. Beery, S. A. Lyons, and S. L. Taylor. 1983. Cadaverine and aminoguanidine potentiate the uptake of histamine in vitro in perfused intestinal segments of rats. *Toxicol. Appl. Pharmacol.* **70**:445–458.

Maijala, R., S. Eerola, M. Aho, and J. Hirn. 1993. The effect of GDL-induced pH decrease on the formation of biogenic amines in meat. *J. Food Prot.* **56**:125–129.

Maijala, R., F. Nurmi, and A. Fischer. 1995. Influence of processing temperature on the formation of biogenic amines in dry sausages. *Meat Sci.* **39**:19–22.

Malle, P., M. Valle, and S. Bouquelet. 1996. Assay of biogenic amines involved in fish decomposition. *J. AOAC Int.* **79**:43–49.

Marcason, W. 2005. What is the bottom line for dietary guidelines when taking monoamine oxidase inhibitors? *J. Am. Diet. Assoc.* **105**:163.

Marino, M., M. Maifreni, S. Moret, and G. Rondinini. 2000. The capacity of Enterobacteriaceae species to produce biogenic amines in cheese. *Lett. Appl. Microbiol.* **31**:169–173.

Martuscelli, M., M. A. Crudele, F. Gardini, and G. Suzzi. 2000. Biogenic amine formation and oxidation by Staphylococcus xylosus strains from artisanal fermented sausages. *Lett. Appl. Microbiol.* **31**:228–232.

Massey, S. M. 1976. MAOIs and food—fact and fiction. *Int. J. Food Sci. Nutr.* **30**:415–419.

Masson, F., R. Talon, and M. C. Montel. 1996. Histamine and tyramine production by bacteria from meat products. *Int. J. Food Microbiol.* **32**:199–207.

Masson, F., G. Johansson, and M. C. Montel. 1999. Tyramine production by a strain of Carnobacterium divergens inoculated in meat-fat mixture. *Meat Sci.* **52**:65–69.

Medina, M. A., J. L. Urdiales, C. Rodriguez-Caso, F. J. Ramirez, and F. Sanchez-Jimenez. 2003. Biogenic amines and polyamines: similar biochemistry for different physiological missions and biomedical applications. *Crit. Rev. Biochem. Mol. Biol.* **38**:23–59.

Metcalfe, D. D. 2003. Food allergy: adverse reactions to foods and food additives. *In* H. A. Sampson and R. A. Simon (ed.), *Pharmacologic Food Reactions*, 3rd ed. Blackwell Publishing, Oxford, United Kingdom.

Middlebrooks, B. L., P. M. Toom, W. L. Douglas, R. E. Harrison, and S. McDowell. 1988. Effects of storage time and temperature on the microflora and amine development in Spanish mackerel (Scomberomorus macularus). *J. Food Sci.* **53**:1024–1029.

Mietz, J. L., and E. Karmas. 1977. Chemical quality index of canned tuna as determined by high pressure liquid chromatography. *J. Food Sci.* **42**:155–158.

Mietz, J. L., and E. Karmas. 1978. Polyamines and histamine content of rockfish, salmon, lobster, and shrimp as an indicator of decomposition. *J. Assoc. Off. Anal. Chem.* **61**:139–145.

Mongar, J. L. 1957. Effect of chain length of aliphatic amines on histamine potentiation and release. *Br. J. Pharmacol.* **12:**140–148.

Montel, M. C., F. Masson, and R. Talon. 1999. Comparison of biogenic amine content in traditional and industrial French dry sausages. *Sci. Aliments* **19:**247–254.

Moret, S., D. Smela, T. Populin, and L. S. Conte. 2005. A survey on free biogenic amine content of fresh and preserved vegetables. *Food Chem.* **89:**355–361.

Moret, S., and L. S. Conte. 1996. High-performance liquid chromatographic evaluation of biogenic amines in foods and analysis of different methods of sample preparation in relation to food characteristics. *J. Chromatogr. A* **729:**363–369.

Morii, H., D. C. Cann, and L. Y. Tayor. 1988. Histamine formation by luminous bacteria in mackarel stored at low temperature. *Nipp. Suis. Gakk.* **54:**299–305.

Morii, H., D. C. Cann, L. Y. Taylor, and C. K. Murray. 1986. Formation of histamine by luminous bacteria isolated from scombroid fish. *Bull. Jpn. Soc. Sci. Fisher.* **52:**2135–2141.

Motil, K. J., and N. S. Scrimshaw. 1979. The role of exogenous histamine in scombroid poisoning. *Toxicol. Lett.* **3:**219–223.

Müller, G. J., J. H. Lamprecht, J. M. Barnes, R. V. P. de Villiers, B. R. Honeth, and B. A. Hoffman. 1992. Scombroid poisoning. Case series of 10 incidents involving 22 patients. *S. Afr. Med. J.* **81:**427–430.

Nishino, N., H. Hattori., H. Wada, and E. Touno. 2007. Biogenic amine production in grass, maize and total mixed ration silages inoculated with *Lactobacillus casei* or *Lactobacillus buchneri. J. Appl. Microbiol.*, in press.

Nouadje, G., N. Siméon, F. Dedieu, M. Nertz, P. Puig, and F. Couderc. 1997. Determination of twenty eight biogenic amines and amino acids during wine aging by micellar electrokinetic chromatography and laser-induced fluorescence detection. *J. Chromatogr. A* **765:**337–343.

Novella-Rodriguez, S., M. T. Veciana-Nogues, M. Izquierdo-Pulido, and M. C. Vidal-Carou. 2003. Distribution of biogenic amines and polyamines in cheese. *J. Food Sci.* **68:**750–755.

Okuzumi, M., S. Okuda, and M. Awano. 1981. Isolation of psychrophilic and halophilic histamine-forming bacteria from Scomber japonicus. *Nipp. Suis. Gakk.* **47:**1591–1598.

Okozumi, M., I. Fukumoto, and T. Fujii. 1990. Changes in bacterial flora and polyamines contents during storage of horse mackerel meat. *Nipp. Suis. Gakk.* **56:**1307–1312.

Olley, J., and T. A. McMeekin. 1985. Use of temperature function integrators, p. 24–29. *In Histamine Formation in Marine Products: Production by Bacteria, Measurement and Prediction of Formation,* FAO Fisheries Technical Paper No. 252. Food and Agriculture Organization of the United Nations, Rome, Italy.

Omure, Y., R. J. Price, and H. S. Olcott. 1978. Histamine forming bacteria isolated from spoiled skipjack tuna and jack mackerel. *J. Food Sci.* **43:**1779–1781.

Onal, A. 2007. A review: current analytical methods for the determination of biogenic amines in foods. *Food Chem.* **103:**1475–1486.

Özogul, F., and Y. Özogul. 2006. Biogenic amine content and biogenic amine quality indices of sardines (Sardina pilchardus) stored in modified atmosphere packaging and vacuum packaging. *Food Chem.* **99:**574–578.

Pan, B. S. 1985. Use of Arrhenius plot to estimate histamine formation in mackerel and bonito, p. 21–23. *In Histamine Formation in Marine Products: Production by Bacteria, Measurement and Prediction of Formation,* FAO Fisheries Technical Paper No. 252. Food and Agriculture Organization of the United Nations, Rome, Italy.

Parente, E., M. Martuscelli, F. Gardini, S. Grieco, M. A. Crudele, and G. Suzzi. 2001. Evolution of microbial populations and biogenic amine production in dry sausages produced in Southern Italy. *J. Appl. Microbiol.* **90:**882–891.

Parrot, J., and G. Nicot. 1966. Pharmacology of histamine, p. 148–161. *In Handbuch der Experimeutellin Pharmakologie,* vol. 18. Springer-Verlag, New York, NY.

Patterson, R. L., and D. S. Mottram. 1974. The occurrence of volatile amines in uncured and cured pork: their possible role in nitrosamine formation in bacon. *J. Sci. Food Agric.* **25:**1419–1425.

Pensabene, J. W., W. Fiddler, R. A. Gates, J. C. Fagen, and A. E. Wasserman. 1974. Effect of frying and other cooking conditions on nitrosopyrrolidine formation in bacon. *J. Food Sci.* **39:**314–316.

Pereira, C. I., M. T. Barreto Crespo, and M. V. San Romão. 2001. Evidence for proteolytic activity and biogenic amines production in Lactobacillus curvatus and Lactobacillus homohiochii. *Int. J. Food Microbiol.* **68:**211–216.

Pons-Sanchez-Cascado, S., M. T. Veciana-Nogues, S. Bover-Cid, A. Marine-Font, and M. C. Vidal-Carou. 2005. Volatile and biogenic amines, microbiological counts, and bacterial amino acid decarboxylase activity throughout the salt-ripening process of anchovies (Engraulis encrasicholus). *J. Food Prot.* **68:**1683–1689.

Recsie, P. A., and E. E. Snell. 1982. Histidine decarboxylase of Lactobacillus 30a. Comparative properties of wild type and mutant proenzymes and their derived enzymes. *J. Biol. Chem.* **257:**7196–7202.

Rezaei, M., N. Montazeri, H. E. Langrudi, B. Mokhayer, M. Parviz, and A. Nazarinia. 2007. The biogenic amines and bacterial changes of farmed rainbow trout (Oncorhynchus mykiss) stored in ice. *Food Chem.* **103:**150–154.

Rice, S. L., R. R. Eitenmiller, and P. E. Koehler. 1976. Biologically active amines in foods. A review. *J. Milk Food Technol.* **39:**353–358.

Ritchie, A. H., and I. M. Mackie. 1979. The formation of diamines and polyamines during storage of mackerel (Scomber scombrus), p. 489–494. *In* J. J. Connell (ed.), *Advances in Fish Science and Technology.* Fishing News Books, Surrey, United Kingdom.

Rodrigues-Jerez, J. J., G. Colavita, V. Giaccone, and E. Parisi. 1994a. *Bacillus macerans,* a new potential histamine producing bacteria isolated from Italian cheese. *Food Microbiol.* **11:**409–415.

Rodriguez-Jerez, J. J., M. A. Grassi, and T. Civera. 1994b. A modification of Lerke enzymic test for histamine quantification. *J. Food Prot.* **57:**1019–1021.

Rodriguez-Jerez, J. J., M. T. Mora-Ventura, E. I. Lepez-Sabater, and M. Hemandez-Herrero. 1994c. Histamine, lysine and ornithine decarboxylase bacteria in Spanish salted semi-preserved anchovies. *J. Food Prot.* **57:**784–791.

Roig-Sagues, A. X., M. Hernandez-Herrero, E. I. Lopez-Sabater, J. J. Rodriguez-Jerez, and M. T. Mora-Ventura. 1996. Histidine decarboxylase activity of bacteria isolated from raw and ripened Salsichon, a Spanish cured sausage. *J. Food Prot.* **59:**516–520.

Roseiro, C., C. Santos, M. Sol, L. Silva, and I. Fernandes. 2006. Prevalence of biogenic amines during ripening of a traditional dry fermented pork sausage and its relation to the amount of sodium chloride added. *Meat Sci.* **74:**557–563.

Rosier, J., and C. Van Peteghem. 1988. A screening method for the simultaneous determination of putrescine, cadaverine, histamine, spermidine and spermine in fish by means of high pressure liquid chromatography of their 5-dimethylaminonaphthalene-l-sulphonyl. *Z. Lebensm. Unters. Forsch.* **186:**25–28.

Ruiz-Capillas, C., and F. Jiménez-Colmenero. 2004. Biogenic amines in meat and meat products. *Crit. Rev. Food Sci. Nutr.* **44:**489–499.

Ruiz-Capillas, C., and A. Moral. 2001. Production of biogenic amines and their potential use as quality control indices for hake (Merluccius merluccius, L.) stored in ice. *J. Food Sci.* **66:**1030–1032.

Ruiz-Capillas, C., J. Carballo, and F. Jimenez Colmenero. 2007. Biogenic amines in pressurized vacuum-packaged cooked sliced ham under different chilled storage conditions. *Meat Sci.* **75:**397–405.

Ruiz-Jiménez, J., and M. D. Luque de Castro. 2006. Pervaporation as interface between solid samples and capillary electrophoresis: determination of biogenic amines in food. *J. Chromatogr. A* **1110:**245–253.

Santos, C., M. Jalon, and A. Marine. 1985. Contenido de tiramina en alimentoa de origen animal. I. came derivados carnicos y productos relacionados. *Rev. Agroquim. Tecnol. Aliment.* **25:**362–368.

Santos-Buelga, C., M. J. Pena-Egido, and J. C. Rivas-Gonzalo. 1986. Changes in tyramine during Chorizo-sausage ripening. *J. Food Sci.* **51:**518–527.

Satake, K., S. Ando, and H. Fujita. 1953. Bacterial oxidation of some primary amines. *J. Biochem.* **40:**299–315.

Sayem-El-Daher, N., R. E. Simard, and J. Fillion. 1984. Changes in the amine content of ground beef during storage and processing. *Lebensm. Wiss. Technol.* **17:**319–323.

Schomburg, D., and D. Stephan. 1993. *Enzyme Handbook.* Springer-Verlag, Berlin, Germany.

Shalaby, A. R. 1993. Survey on biogenic amines in Egyptian foods: sausages. *J. Sci. Food Agric.* **62:**219–224.

Shalaby, A. R. 1996. Significance of biogenic amines to food safety and human health. *Food Res. Int.* **29:**675–690.

Shalaby, A. R. 1999. Simple, rapid and valid thin layer chromatographic method for determining biogenic amines in foods. *Food Chem.* **65:**117–121.

Shultz, D. E., G. W. Chang, and L. F. Bjeldanes. 1976. Rapid thin layer chromatographic method for the determination of histamine in fish products. *J. AOAC Int.* **59:**1224–1225.

Silla-Santos, M. H. 1996. Biogenic amines: their importance in foods. *Int. J. Food Microbiol.* **29:**213–231.

Silla Santos, M. H. 1998. Amino acid decarboxylase capability of microorganisms isolated in Spanish fermented meat products. *Int. J. Food Microbiol.* **39:**227–230.

Simon-Sarkadi, L., W. H. Holzapfel, and A. Halasz. 1994. Biogenic amine content and microbial contamination of leafy vegetables during storage at 5°C. *J. Food Biochem.* **17:**407–418.

Sinell. H. J. 1978. Biogenic amine als risikofaktoren in der Fischhygiene. *Arch. Lebensm.* **29:**206–210.

Slemr, T. 1981. Biogenic amine als potentioller chemescher qualitatsindikator fur fleish. *Fleischwirtsschaft* **61:**921–924.

Smith, C. K., and D. T. Durack. 1978. lsoniazid and reaction to cheese. *Ann. Intern. Med.* **88:**520–521.

Smith, T. A. 1980. Amines in food. *Food Chem.* **6:**169–200.

Snell, E. E., P. A. Rescei, and H. Misono. 1975. Histidine decarboxylase from *Lactobacillus* 30a: nature of conversion of proenzyme to active enzyme, p. 213–219. *In* S. Shaltiel (ed.), *Metabolic Interconversion of Enzymes*, Springer-Verlag, Berlin, Germany.

Soll, A. H., and A. Wollin. 1977. The effects of histamine, prostaglandin Es, and secretin on cyclic AMP in separated canine fimdic mucosal cells. *Gastroenterology* **72:**1166.

Stratton, J. E., R. W. Hutkins, and S. L. Taylor. 1991. Biogenic amines in cheese and other fermented foods. A review. *J. Food Prot.* **54:**460–470.

Stratton, J. E., and S. L. Taylor. 1991. Scombroid poisoning, p. 331–351. *In* D. Ward and C. Hackney (ed.), *Microbiology of Marine Food Products*. Spectrum, New York, NY.

Straub, B., M. Schollenberger, M. Kicherer, B. Luckas, and W. P. Hammes. 1993. Extraction and determination of biogenic amines in fermented sausages and other meat products using reversed-phase-HPLC. *Z. Lebensm. Unters. Forsch.* **197:**230–232.

Straub, B. W., P. S. Tichaczek, M. Kicherer, and W. P. Hammes. 1994. Formation of tyramine by Lactobacillus curvatus LTH 972. *Z. Lebensm. Unters. Forsch.* **1:**9–12.

Straub, B. W., M. Kicherer, S. M. Schilcher, and W. P. Hammes. 1995. The formation of biogenic amines by fermentation organisms. *Z. Lebensm. Unters. Forsch.* **201:**79–82.

Sumner, S. S., M. W. Speckhard, E. B. Somers, and S. L. Taylor. 1985. Isolation of histamine-producing *Lactobacillus buchneri* from Swiss cheese implicated in a food poisoning outbreak. *Appl. Environ. Microbiol.* **50:**1094–1096.

Sumner, S. S., F. Roche, and S. L. Taylor. 1990. Factors controlling histamine production in Swiss cheese inoculated with Lactobacillus buchneri. *J. Dairy Sci.* **73:**3050–3058.

Sun, X., X. Yang, and E. Wang. 2003. Determination of biogenic amines by capillary electrophoresis with pulsed amperometric detection. *J. Chromatogr. A* **1005:**189–195.

Suzzi, G., and F. Gardini. 2003. Biogenic amines and dry fermented sausages: a review. *Int. J. Food Microbiol.* **88:**41–54.

Taylor, S. L., L. S. Guthertz, M. Leatherwood, F. Tillman, and E. R. Lieber. 1978. Histamine production by food-borne bacterial species. *J. Food Saf.* **1:**173–187.

Taylor, S. L., and E. R. Lieber. 1979. In vivo inhibition of rat intestinal histamine-metabolizing enzymes. *Food Cosmet. Toxicol.* **17:**237–240.

Taylor, S. L., L. S. Guthertz, M. Leatherwood, and E. R. Lieber. 1979. Histamine production by *Klebsiella pneumonia* and an incident of scombroid fish poisoning. *Appl. Environ. Microbiol.* **37:**274–278.

Taylor, S. L., and M. W. Speckhard. 1984. Inhibition of bacteria histamine production by sorbate and other antimicrobial agents. *J. Food Prot.* **47:**508–511.

Taylor, S. L., J. Y. Hui, and D. E. Lyous. 1984. Toxicology of scombroid poisoning, p. 417. *In* E. P. Ragils (ed.), *Seafood Toxins*, ACS Symposium Series, No. 262. Washington, DC.

Taylor, S. L. 1985. *Histamine Poisoning Associated with Fish, Cheese, and Other Foods*. VPH/FOS/ 85.1. World Health Organization, Geneva, Switzerland.

Taylor, S. L. 1986. Histamine food poisoning: toxicology and clinical aspects. *Crit. Rev. Toxicol.* **17:**91–128.

Taylor, S. L. 1988. Marine toxins of microbial origin. *Food Technol.* **42:**94–98.

ten Brink, B., C. Damink, H. M. L. J. Joosten, and J. H. J. Huis in 't Veld. 1990. Occurrence and formation of biologically active amines in foods. *Int. J. Food Microbiol.* **11:**73–84.

Teodorovic, V., S. Buncic, and D. Smiljanić. 1994. A study of factors influencing histamine production in meat. *Fleischwirtsch* **74:**170–172.

Tiecco, G., G. Tantillo, E. Francioso, A. Paparella, and G. De Natale. 1986. Ricerca quali-quantitativa di alcune amine biogene in insaccati nel corso della stagionatura. *Industrie Alimentari* **5:**209–213.

Til, H. P., H. E. Falke, M. K. Privisen, and M. I. Willems. 1997. Acute TNF-alpha: a possible and subacute toxicity of tyramine, spermidine, spermine, putrescine and cadaverine in rats. *Food Chem. Toxicol.* **35:**337–348.

Tsai, Y. H., S. C. Chang, and H. F. Kung. 2007. Histamine contents and histamine-forming bacteria in natto products in Taiwan. *Food Control* **18:**1026–1030.

Tsai, Y. H., H. F. Kung, T. M. Lee, H. C. Chen, S. S. Chou, C. I. Wei, and D. F. Hwang. 2005a. Determination of histamine in canned mackerel implicated in a food borne poisoning. *Food Control* **16:**579–585.

Tsai, Y. H., C. Y. Lin, S. C. Chang, H. C. Chenc, H. F. Kung, C. I. Weid, and D. F. Hwange. 2005b. Occurrence of histamine and histamine-forming bacteria in salted mackerel in Taiwan. *Food Microbiol.* **22**:461–467.

Uragoda, G. G., and S. C. Lodha. 1979. Histamine intoxication in a tuberculosis patient after ingestion of cheese. *Tubercle* **60**: 56–61.

Vale, S., and M. B. A. Glória. 1998. Biogenic amines in Brazilian. *Food Chem.* **63**:343–348.

Vandekerckhove, P. 1977. Amines in dry fermented sausages. *J. Food Sci.* **42**:283–285.

Veciana-Nogues, M. T., T. Hernandez-Jover, A. Marine-Font, and M. C. Vidal-Carou. 1995. Liquid chromatographic method for determination of biogenic amines in fish and fish products. *J. AOAC Int.* **78**:1045–1050.

Veciana-Nogues, M. T., A. Marine-Font, and M. C. Vidal-Carou. 1997. Biogenic amines as hygienic quality indicators of tuna. Relationship with microbial counts, ATP-related compounds, volatile amines, and organoleptic changes. *J. Agric. Food Chem.* **45**:2036–2041.

Vidal, M. C., and A. Marine. 1984. Histamina en pescados y derivados. Formation y posible papel coma indicador de1 estado de 10s mismos. *Aliment.* **15**:93–97, 99–102.

Vinci, G., and M. L. Antonelli. 2002. Biogenic amines: quality index of freshness in red and white meat. *Food Control* **13**:519–524.

Virgili, R., G. Saccani, L. Gabba, E. Tanzi, and C. Soresi Bordini. 2007. Changes of free amino acids and biogenic amines during extended ageing of Italian dry-cured ham. *Lebensm. Wiss. Technol.* **40**:871–878.

von Beutling, D. 1993. Studies on the formation of tyramine by microbes with food hygienic relevance. *Arch. Lebensmittelhyg.* **44**:83–87.

Warthesen, J. J., R. A. Scanlan, D. D. Bills, and L. M. Libbely. 1975. Formation of heterocyclic N-nitrosamines from the reaction of nitrite and selected primary diamines and amino acids. *J. Agric. Food Chem.* **23**:898–902.

Watts, D. A., and W. D. Brown. 1982. Histamine formation in abusively stored Pacific mackerel—effect of CO2-modified atmosphere. *J. Food. Sci.* **47**:1386–1387.

Weiss, S., G. P. Robb, and L. B. Ellis. 1932. The systemic effects of histamine in man. *Arch. Int. Med.* **49**:360.

Wendakoon, C. N., and M. Sakaguchi. 1993. Combined effect of sodium chloride and clove on growth and biogenic amine formation of Enterobacter aerogenes in Mackerel muscle extract. *J. Food Prot.* **56**:410–413.

Wortberg, B., and R. Woller. 1982. Quality and freshness of meat and meat products as related to their content of biogenic amines. *Fleischwirtsch* **62**:1457–1463.

Wu, M. L., C. C. Yang, G. Y. Yang, J. Ger, and J. F. Deng. 1997. Scombroid fish poisoning: an overlooked marine food poisoning. *Vet. Hum. Toxicol.* **39**:236–241.

Yamanaka, H. 1989. Changes in polyamines and amino acids in scallop aductor muscle during storage. *J. Food Sci.* **54**:1113–1115.

Yamanaka, H., K. Shiomi, and T. Kikuchi. 1987a. Agmatine as a potential index for freshness of common squid (Tobarodes pacificus). *J. Food Sci.* **52**:936–938.

Yamanaka, H., K. Shimakura, K. Shiomi, and T. Kikuchi. 1987b. Changes in nonvolatile amine contents of the meat of sardine and saury pike during storage. *Bull. Jpn. Soc. Sci. Fisher.* **52**:127.

Yatsunami, K., and T. Echigo. 1991. Isolation of salt tolerant histamine-forming bacteria from commercial rice-brane pickles of sardine. *Nipp. Suis. Gakk.* **57**:1723–l728.

Yatsunami, K., and T. Echigo. 1992. Non-volatile amine contents of commercial rice-bran pickles of sardine. *J. Food Hyg. Soc. Jpn.* **33**:310–313.

Yoshinaga, D. H., and H. A. Frank. 1982. Histamine-producing bacteria in decomposing skipjack tuna *(Katsuwonus pelamis)*. *Appl. Environ. Microbiol.* **44**:447–452.

Zhang, L. Y., and M. X. Sun. 2004. Determination of histamine and histidine by capillary zone electrophoresis with pre-column naphthalene-2,3-dicarboxaldehyde derivatization and fluorescence detection. *J. Chromatogr. A* **1040**:133–140.

Zhang, L. Y., X. C. Tang, and M. X. Sun. 2005. Simultaneous determination of histamine and polyamines by capillary zone electrophoresis with 4-fluor-7-nitro-2,1,3-benzoxadiazole derivatization and fluorescence detection. *J. Chromatogr. B* **820**:211–219.

Pathogens and Toxins in Foods: Challenges and Interventions
Edited by V. K. Juneja and J. N. Sofos
© 2010 ASM Press, Washington, DC

Chapter 17

Fungal and Mushroom Toxins

CHARLENE WOLF-HALL

The biological kingdom of the fungi contains many toxigenic species. Unlike many of the microorganisms of food safety concern, the fungi can be macroscopic, colorful, and, in this author's opinion, beautiful and endlessly fascinating to observe. The world of toxigenic fungi is a diverse and interesting one, and information abounds. This chapter is a mere snapshot of that information, and the reader is encouraged to explore the literature and reliable websites for more detail; the book by Hudler (1998) is a particularly good resource as a place to start learning more about the fungi.

This chapter discusses species that pose the most significant food safety risks. The species include those of the filamentous fungi, also known as molds, which produce mycotoxins. The species discussed also include members of the higher fungi responsible for the most commonly consumed toxic mushrooms. Types of toxins, toxicoses, the conditions under which toxins are produced, and control methods are also discussed. Keep in mind that although there is much information published about mycotoxins and toxic mushrooms, there are still many unanswered questions about the risks posed by these toxins. This chapter attempts to point out areas where more information is needed.

MYCOTOXINS

As mold propagules are ubiquitous in most environments, there is much potential for them to become established, grow, and possibly produce mycotoxins in foods and food ingredients. Humans are likely exposed to some level of mold or mold metabolic byproducts by almost any food they consume, and zero tolerances for any mold metabolite could have dramatic economic impacts and lead to food shortages.

The challenge is to keep the risk of harm due to these metabolites at an acceptable level. The major mycotoxigenic mold genera of agronomical concern include *Aspergillus* (Hocking, 2007), *Pencillium* (Pitt, 2007), and *Fusarium* (Bullerman, 2007). However, several other genera are of concern as well, and some examples of these are discussed below.

Molds may enter the food chain at almost any point. Many are of preharvest concern in plant materials harvested for human food and are commonly referred to as field fungi. Molds that grow in or on developing plant tissue include plant pathogens, endophytes, epiphytes, and commensals. Saprophytic fungi are common on senescent plant material. Several species within the *Fusarium* genus fall into the category of plant pathogens and are known to produce mycotoxins in developing cereal grains and other food commodities. Although *Aspergillus* species can be preharvest plant pathogens in crops such as corn and peanuts, they are also of concern in stored or postharvest points of the food chain. *Penicillium* species are mainly a postharvest concern and are major spoilage problems in a wide variety of foods and ingredients. For an overview of grain fungi, including color images, see the reference guide published by the Grain Inspection, Packers and Stockyards Administration (GIPSA) of the United States Department of Agriculture (http://archive.gipsa.usda.gov/pubs/mycobook.pdf).

Storage molds become food safety problems when intrinsic and extrinsic factors allow them to become metabolically active and produce mycotoxins. This tends to occur in food systems in which the rapid growth of bacteria is suppressed by low pH, high osmotic pressure, or low water activity, giving the molds a competitive advantage. Examples of intrinsic conditions that favor mold growth in stored

Charlene Wolf-Hall • Great Plains Institute of Food Safety, 1523 Centennial Blvd., 114A Van Es Hall, North Dakota State University, Fargo, ND 58105.

products include adequate water activity and oxygen, and examples of extrinsic conditions that favor growth are storage temperature and time. The more xerophylic species, such as some within the *Aspergillus* genus, will grow at a minimum water activity of 0.70 (Jay et al., 2005). However, the toxigenic *Aspergillus* species are thought to require a minimum water activity of 0.80 (Hocking, 2007). Murphy et al. (2006) summarize studies done to determine optimal temperature and water activity conditions for production of the major food mycotoxins.

Food mycotoxins can be defined as secondary metabolites produced by filamentous fungi that are toxic to animals and humans. Mycotoxicoses are the manifestations of intoxication by mycotoxins. The major food safety concerns with mycotoxins are their stability to food processing techniques and the fact that many can be transmitted through animal tissue, milk, and eggs from animals fed mycotoxin-contaminated feed. Mycotoxins have likely always had an impact on humankind. There are studies published from the 1800s and early 1900s about mold intoxications (Barger, 1931; Pitt, 2007); however, mycotoxins were not thoroughly scientifically evaluated until the 1960s. Since then, thousands of mycotoxins have been discovered, but not all are considered significant food safety hazards. The most significant mycotoxins of food safety concern and the most studied include ergot alkaloids, aflatoxins, trichothecenes, fumonisins, zearalenone, ochratoxins, and patulin (Richard and Payne, 2003; Murphy et al., 2006). The chemical structures for the major mycotoxins are given by Murphy et al. (2006) and can be freely viewed on the Internet. For a review of mycotoxins of concern in the pet food industry, see Leung et al. (2006). The following sections introduce control methods and provide details about the major mycotoxin groups.

Control of Mycotoxins in Foods

The management practices for controlling risks due to mycotoxins include physical, chemical, and/or biological interventions. The mycotoxins are typically quite heat stable and survive most food processing methods. Physical interventions can include exclusion practices, in which contaminated material is excluded from the food production chain. In some cases, as with the trichothecene deoxynivalenol, it is allowable to dilute mycotoxins to "safe" concentrations by blending them with "clean" material. Exclusion and dilution are the most common forms of mycotoxin control. Exclusion can result in dramatic economic impacts, shifting food safety risks to other areas of concern. Dilution may be a practical solution, but a better understanding of interactions with other

toxins would be prudent; the topic of cocontamination is discussed more below.

The United States Economic Research Service stated, "To minimize the initial risk of mycotoxin contamination and consequently lessen the likelihood that tolerance levels will be exceeded, private sector actors or public agencies can consider implementing process standards based on [good agricultural practices] GAPs, [good manufacturing practices] GMPs, and [hazard analysis critical control point] HACCP principles. Developing countries are likely to require technical assistance and economic support to implement these strategies" (Dohlman, 2003). For a review of control methods for mycotoxins, the reader is referred to the *Manual on the Application of the HACCP System in Mycotoxin Prevention and Control* (FAO, 2001) and the Codex Alimentarius Commission code of practice (CAC, 2003). For a summary of guidelines and regulations for mycotoxins in food and feed, including those for the United States, the European Union, and the Codex Alimentarius Commission, see Murphy et al. (2006). Murphy et al. (2006) also summarize data from surveys of mycotoxin content in foods and feeds done in several countries around the world, as well as chemical and biological approaches of control. For a more comprehensive overview, including risk assessment, modeling of exposure, analysis, and controls, see Magan and Olsen (2004). For a review of the costs and societal impacts of mycotoxin control, see http://www.apsnet.org/online/feature/mycotoxin/. For up-to-date information about appropriate and rapid testing methods, see the GIPSA information at http://www.gipsa.usda.gov/GIPSA/webapp?area=home&subject=grpi&topic=rd-my and the European Mycotoxin Awareness Network at http://www.mycotoxins.org/.

The aflatoxins are the only mycotoxins for which the United States Food and Drug Administration has set an "action level." There are FDA guidelines set for patulin, fumonisins, and deoxynivalenol and an interest in monitoring ochratoxin A levels. The FDA has posted information related to mycotoxins in domestic as well as imported foods (see food compliance programs 7307.001 and 7309.006, respectively, at http://www.fda.gov/Food/GuidanceComplianceRegulatoryInformation/ComplianceEnforcement/ucm071496.htm).

Ergot Alkaloids

Ergot alkaloids are produced by species within several mold genera, including *Acremonium, Balansia, Aspergillus,* and *Penicillium,* but the most well known is *Claviceps purpurea* (Boichenko et al., 2001). Images of the sclerotia of *Claviceps purpurea* and some of the symptoms of cattle intoxication can be viewed from the North Dakota State University Extension website at http://www.ag.ndsu.edu/pubs/plantsci/crops/pp551w.htm.

The ergot fungus, *Claviceps purpurea,* has a long history of human medicinal uses (Bove, 1970). Two types of ergot intoxication, including neurological and gangrenous ergotism, are known to occur (Weidenborner, 2001). The neurological effects are due to lysergic acid derivatives, which are produced in high concentrations in the sclerotia of the fungus and cause symptoms in humans, including sustained spasms, muscle cramps, twitching, numbness of the extremities, tingling of the skin, constriction of blood vessels, and hallucinations. The gangrenous form of ergotism starts with a feeling of lassitude and is followed by a prickling sensation in the extremities and sensations of hot and cold. As the blood flow becomes constricted to the extremities, gangrene can set in. This form often leads to spontaneous amputation of the affected extremities. The intense sensations of burning led to the use of the term "St. Anthony's fire" to describe the syndrome in the Middle Ages.

Ergotism has probably affected humans since grasses were first cultivated as food sources. The first recorded outbreak occurred in Germany in 857 AD (Barger, 1931). Many other outbreaks occurred during the Middle Ages in Europe. The picture on the cover of the CAST report (Richard and Payne, 2003) (see the following site for a picture of the cover and an interpretive summary: http://www.cast-science.org/website uploads/pdfs/mycotoxins_is.pdf) is from the painting known as "The Beggars" by the Flemish artist Pieter Bruegel the Elder (ca. 1525–1569) and depicts some likely victims of gangrenous ergotism. The following poem by Leah Esterianna and Richard the Poor of Ely (see http://www.pbm.com/~lindahl/lod/vol3/dancing_mania.html) is thought to describe manic dancing from the age due to the neurotoxic effects of ergotism:

Amidst our people here is come
The madness of the dance.
In every town there now are some
Who fall upon a trance.
It drives them ever night and day,
They scarcely stop for breath,
Till some have dropped along the way
And some are met by death.

There are stories of minstrels following the victims playing music to sooth their tortured souls, as these afflicted victims would dance until they died from exhaustion. Caporael (1976) famously linked ergotism to the unusual behaviors and afflictions involved in the Salem, Massachusetts' witch trials of the 1600s, demonstrating environmental conditions of the period, which would have been favorable for ergot development in rye, a staple food source for the population.

Modern grain handling includes physical methods for removing the larger, dark ergot sclerotia from the grain or for grading and diverting ergoty grain, so the occurrence of ergot in the human food supply is very rare. Economic losses occur due to price discounts and unacceptability, as well as through reduction of yield (see http://www.gipsa.usda.gov/GIPSA/documents/GIPSA_Documents/wheatinspection.pdf for examples of U.S. standards for wheat or http://www1.agric.gov.ab.ca/$department/deptdocs.nsf/all/prm2402 for examples of Canadian standards for other grains). Yields are occasionally reduced by as much as 5% in rye and 10% in wheat. Pasture poisoning remains a concern in livestock. The presence of ergot in barley has recently been a concern in the upper Midwest of the United States (P. B. Schwarz, M. S. Hill, and G. E. Rottinghaus, presented at the Annual Meeting of the American Society of Brewing Chemists, Savannah, GA, 2005; Schwarz et al., 2006). Outbreaks of ergotism have occurred in human populations during modern times due to economic upheaval and wars (Murphy et al., 2006).

Aflatoxins

There are six major chemical forms of the aflatoxins, including aflatoxins B_1, B_2, G_1, G_2, M_1, and M_2 (Murphy et al., 2006; Weidenborner, 2001). The B and G forms are produced mainly by toxigenic strains of *Aspergillus flavus* and *Aspergillus parasiticus.* Aflatoxins B_1 and B_2 fluoresce blue under shortwave UV, while aflatoxins G_1 and G_2 fluoresce green. Aflatoxins M_1 and M_2 are the forms that B forms are converted to through metabolic processes in animals and are excreted in milk. Aflatoxin B_1 is the most common form found in foods. Other forms of aflatoxin that are of less agronomical and acute toxicological concern are aflatoxicol (occurs in breast milk and pistachio nuts, also known as aflatoxin R_0), aflatoxicol H1 (the hydroxylated oxidative metabolite of aflatoxicol), aflatoxin B_{2a} (water adduct of aflatoxin B_1, also known as aflatoxin W), aflatoxin B_3 (either a precursor or a breakdown product), aflatoxin DB_1 (major product of ammoniation of B_1), aflatoxin G_{2a} (a water adduct that retains toxicity), aflatoxin GM_1 (4-hydroxylated derivative of G_1), aflatoxin M_4 (found in cow's milk), aflatoxin P_1 (urinary metabolite of B_1), and aflatoxin Q_1 (microsomal metabolite of B_1) (Weidenborner, 2001).

Aflatoxins are among the most potently toxic mycotoxins, and aflatoxin B_1 is the most carcinogenic natural toxin known to man. The International

Agency for Research on Cancer has designated aflatoxin B$_1$ and mixtures of the other forms as group I carcinogens (IARC, 1993a). The mechanisms of aflatoxin B$_1$ genotoxicity have been well studied (Wang and Groopman, 1999). The aflatoxins are known to produce acute necrosis, cirrhosis, and carcinoma of the liver in several animal species, and all species tested are susceptible to acute poisoning of aflatoxins with 50% lethal dose (LD$_{50}$) values ranging from 0.5 to 10 mg/kg body weight (FDA, 2006b). The United States Food and Drug Administration's action level for aflatoxin in foods is 20 ppb, and for aflatoxin M$_1$ in milk, it is 0.5 ppb (http://www.fda.gov/Food/Guid anceComplianceRegulatoryInformation/GuidanceDo cuments/ChemicalContaminantsandPesticides/ ucm077969.htm#afla). See Murphy et al. (2006) for a review of the toxicological mechanisms of aflatoxins.

Recently, 317 cases of acute human aflatoxicosis occurred in Kenya in 2004, which resulted in a fatality rate of 39% (Azziz-Baumgartner et al., 2004; Nyikal et al., 2004). The strain of *A. flavus* isolated from the contaminated maize was identified as a member of the S strain, a strain implicated in two previous epidemics in Kenya (Probst et al., 2007). This S strain was a particularly proficient aflatoxin-producing strain (Coty, 1989). This type of aflatoxicosis seems to occur regularly in Kenya, with outbreaks in 2004 and 2005 resulting in over 150 deaths, maize samples containing up to 46,400 ppb of total aflatoxins, and over 55% of maize samples exceeding regulatory limits in Kenya (Lewis et al., 2005; Ngugi and Wilson, 2006). In the 2004 outbreak, contaminated maize was purchased at the local level and entered the commercial distribution system, causing increased exposure to the highly contaminated grain and leading to the recommendation to consider the market system when attempting to control such an outbreak (Lewis et al., 2005).

In regions of the world where safe storage of dried commodities can be a challenge and food security in the sense of an adequate food supply can be a problem, the concern over aflatoxin contamination is quite serious. As an example, Koirala et al. (2005) highlighted a survey done in Nepal, where 18% of the samples tested were found to be contaminated with aflatoxins at concentrations above the 30-ppb threshold set for Nepal and with concentrations as high as 1,806 ppm. This type of result was also found with samples of peanut products in Bangladesh, where 36% of samples contained aflatoxins at concentrations above the recommended level (Dawlatana et al., 2002). There are many other such surveys being reported from less-developed regions of the world.

Wu (2006) demonstrates the risks and benefits of utilization of genetically modified corn in relation to concerns with aflatoxin and other mycotoxin contamination (see also several other relevant publications listed at http://www.pitt.edu/~few8/Publications.htm). Wu (2006) points out that with the implementation of lower aflatoxin concentration regulations in richer countries, poorer nations are more at risk of acute aflatoxicosis outbreaks. The poorer nations tend to sell the good-quality grain and end up consuming the lower-quality material. As the food supply becomes increasingly globalized, it will be important to take economic and societal impacts into consideration with the development of control strategies for food safety hazards due to mycotoxins such as aflatoxin.

Trichothecenes

The trichothecenes are a group of toxins which share similar chemical structures. There are two groups, including type A and type B trichothecenes (Weidenborner, 2001). Type A trichothecenes include some of the most acutely toxic forms, including T-2 toxin, HT-2 toxin, diacetoxyscirpenol, monoacetoxyscirpenol, and neosolaniol. Type B trichothecenes contain a carbonyl group at the C-8 position and include nivalenol, deoxynivalenol, fusarenon X, and diacetylnivalenol. Deoxynivalenol (also known as DON or vomitoxin) is the most commonly detected and most abundant of the trichothecenes found in foods; however, it is also thought to be one of the least toxic forms within this family of over 180 trichothecenes. Human mycotoxicoses due to trichothecenes have included alimentary toxic aleukia, which was endemic in Russia during World War II, with mortality rates as high as 80%, and akakabi byo or red mold disease, which has occurred in Japan, Korea, India, and China (Bullerman, 2007; Weidenborner, 2001). Trichothecenes have also been implicated as biological terror weapons (Desjardins, 2003).

Pestka and Smolinski (2005) reviewed the topic of deoxynivalenol toxicity. They indicated that at the molecular level, deoxynivalenol inhibits protein synthesis, leading to alterations of cell signaling involved with proliferation, differentiation, and apoptosis. Only 0.05 mg/kg body weight was sufficient to induce vomiting in pigs and dogs, two of the most sensitive animal models. Vomiting is the most common acute effect of deoxynivalenol mycotoxicosis. Much less is clearly understood about the chronic effects of deoxynivalenol exposure over long periods of time. Chronic effects observed with animal models include anorexia, decreased growth, and immune function changes (both enhancement and suppression). Bony et al. (2006) demonstrated the genotoxic potential of deoxynivalenol at concentrations compatible with human exposure levels and recommended additional studies for thorough risk assessment.

The United States Food and Drug Administration has set a guideline of 1 ppm for deoxynivalenol in finished wheat products intended for human consumption (http://www.fda.gov/Food/GuidanceCompliceRegulatoryInformation/GuidanceDocuments/NaturalToxins/ucm120184.htm). Estimates of $637 million in direct economic impact of deoxynivalenol in United States-produced human food crops have been reported (Richard and Payne, 2003), and there are no currently known control practices to significantly reduce these costs.

Although of less agronomical concern, many other species of molds are known to produce trichothecene mycotoxins. An example would be *Trichoderma* species, which are known producers of trichothecenes, including trichodermin and harzianum A (Blumenthal, 2004; Corley et al., 1994; Godtfredsen and Vangedal, 1964, 1965; Harman et al., 2004; Nielsen et al., 2005). The term "trichothecene" was actually coined from the *Trichoderma* genus name to describe the chemically related sesquiterpenoids within this mycotoxin family (Weidenborner, 2001). *Trichoderma* trichothecene production may be of more significance than realized, particularly with the application of some of these species for industrial enzyme production and as biocontrol agents of plant pathogens (Nielsen et al., 2005), for which there might be a potential for introduction of these mycotoxins into the food chain. This example highlights the need to screen molds used for industrial and agricultural applications for mycotoxigenicity for risk assessment.

Zearalenone

Zearalenone is produced by species of *Fusarium* and most commonly in grains by toxigenic strains of *F. graminearum* and *F. culmorum*. Zearalenone is thermostable, and control practices are similar to those for deoxynivalenol, a common cocontaminant.

The biological activity of zearalenone mimics that of estrogenic compounds. Mycotoxicoses in animals from zearalenone are known to lead to fertility problems, and there is concern for human populations regarding disruption of normal hormone activities. In controlled doses, derivatives of zearalenone have medicinal value as growth promoters in cattle and have been developed into commercially available feed additives for beef cattle (Murphy et al., 2006). Zearalenone belongs to a group of over a million compounds of similar structure and pharmacological properties (Alldrick et al., 2004).

The Joint FAO/WHO Expert Committee on Food Additives (JECFA) reported on the incidence of zearalenone and found much variability related to food type, climate, harvest, and storage conditions, with maize from Africa containing the highest concentrations (JECFA, 2001). There is contradiction in the literature regarding the bioaccumulation of zearalenone in tissues of animals fed contaminated feed (Alldrick et al., 2004). Pigs appear to be a sensitive animal species to the effects of zearalenone, and a recommendation for concentrations of zearalenone in pig feed is not to exceed 200 ppb (Gaumy et al., 2001).

Fumonisins

Fumonisins are among the most recently discovered mycotoxins of food safety concern (Gelderblom et al., 1988). Fumonisins are produced primarily by *Fusarium proliferatum* and *Fusarium verticillioides* (former species name *moniliforme*). The United States Food and Drug Administration's guideline for the combined concentrations of fumonisin B_1, B_2, and B_3 in degermed dry-milled corn products is 2 ppm (http://www.fda.gov/Food/GuidanceComplianceRegulatoryInformation/GuidanceDocuments/ChemicalContaminantsandPesticides/ucm109231.htm).

Fumonisin B1 is known to cause equine leukoencephalomalacia in horses and porcine pulmonary edema in pigs, two calamities documented in South Africa leading to the discovery of the fumonisins (Jackson and Jablonski, 2004). The fumonisins are considered probable human carcinogens (IARC, 1993b) and are thought to cause birth defects, such as neural tube malformation (Marasas et al., 2004). The incidence of fumonisins in maize-based products (Bullerman, 2007) is of significant concern, as maize is a dietary staple for many populations. These products may also concurrently contain several other mycotoxins, for which little is understood about their interactions.

Afolabi et al. (2006) were able to demonstrate that physical sorting of grain by quality attributes could significantly lower the concentrations of fumonisins in maize grown in Nigeria. The concern remains with what happens to the highly contaminated material in areas where food and feed supplies are limited or where there is economic incentive to misuse the contaminated grain.

Ochratoxins

Species of molds that produce ochratoxins include *Aspergillus ochraceus*, *Aspergillus carbonarius*, and *Penicillium verrucosum*. There are three types of ochratoxins, A, B, and C, of which the A and C forms contain a chlorine molecule (Weidenborner, 2001). Ochratoxin A is the most common, toxic, and abundantly produced form in foods.

The IARC has classified ochratoxin A as a class 2B carcinogen, which indicates possible carcinogenicity to humans (http://monographs.iarc.fr/ENG/Classification/ListagentsCASnos.pdf). Ochratoxin A is also described as nephrotoxic, teratogenic, and immunotoxic (Murphy et al., 2006). The European Commission Scientific Committee on Food (SCF) and the Joint FAO/WHO Expert Committee on Food Additives (JECFA) have established tolerable intake levels of ochratoxin A for humans (SCF, 1998; JECFA, 2001). Ochratoxin A is of most concern in portions of Europe and Africa (Murphy et al., 2006) and is thought to play a major role in endemic human Balkan nephropathy (Castegnaro et al., 2006). The European Union has set maximum levels for ochratoxin A in cereal-based products at 3 ppb, and at 10 ppb for dried vine fruits (http://www.efsa.europa.eu/cs/BlobServer/Scientific_Opinion/contam_op_ej365_ochratoxin_a_food_en.pdf?ssbinary=true). The United States currently has no guidelines set for ochratoxin A but is considering the data available.

Ochratoxin A has been found in grains, beer, coffee, chocolate, dried fruits, wine, herbs, cheeses, and processed meats and can also be transmitted through tainted feed into animal tissue (Aish et al., 2004). Because of the diversity of food systems that ochratoxin A can be found in, there is a lack of standardized analytical methodology (Murphy et al., 2006). Almela et al. (2007) observed differences in concentrations of ochratoxin A in peppers sampled from Peru, Brazil, Zimbabwe, and Spain and suggested relationships between climatic conditions and cultural practices. They also mentioned implications for emerging risks due to global trade. Currently, the United States lacks survey data to understand the extent of ochratoxin A content in domestic foods, much less imported foods. There is limited understanding of the effects of processing techniques on the stability of ochratoxin A (Murphy et al., 2006); however, it is thought to be relatively heat stable.

Patulin

Patulin is a mycotoxin of concern with several types of fruit, including apricots, grapes, peaches, pears, apples, and olives, as well as fruit juices and some cereals (Murphy et al., 2006; Sperjers, 2004). The *Penicillium* species that produce patulin invade damaged tissue; therefore, a physical control method is necessary to keep harvested fruit from being damaged. Other species known to produce patulin include *Aspergillus clavatus, Aspergillus terreus, Byssochlamys nivea,* and *Byssochlamys fulva* (Weidenborner, 2001). The acidic pH of fruit juices is thought to contribute to the heat stability of patulin during pasteurization (Murphy et al., 2006). The United States Food and Drug Administration's guideline for patulin in apple juice, apple juice concentrate, and apple juice products is 50 ppb (http://www.fda.gov/ora/compliance_ref/cpg/cpgfod/cpg510-150.htm).

Cocontamination

The presence of multiple mycotoxins in a food system is known to occur. Schwarz et al. (2006) demonstrated multiple mycotoxins, including ergot alkaloids, deoxynivalenol, zearalenone, and other trichothecenes present in barley. Murphy et al. (2006) mention this concern for food mycotoxins. Little is known about the interactions of multiple toxins and the effects on animal and human health. Leung et al. (2006) discuss studies evaluating synergistic interactions among mycotoxins in domestic animals. Of particular concern are studies demonstrating increased nephrotoxic and carcinogenic effects of ochratoxin A in combinations with citrinin and penicillic acid (Kanisawa, 1984; Stoev et al., 2002). Masked mycotoxins are also a concern. Some mycotoxins, like the trichothecenes, may be covalently bound to food components such as proteins and carbohydrates but are later metabolized in the human/animal body to release the parent toxin. The low levels of the parent toxins reported for foods may actually be misleading (Berthiller et al., 2005). Further research is warranted for a thorough risk assessment of synergisms and other interactions that might occur when two or more mycotoxins are present in a diet.

MUSHROOM TOXINS

Mushrooms are the reproductive structures of higher basidiomycetes or ascomycetes. These structures produce copious amounts of microscopic spores but are themselves macroscopic. Many mushrooms are edible and have become highly prized culinary delicacies. However, many species are also toxigenic and known causes of human intoxication and, in some cases, death. A misnomer for toxic mushrooms is the term toadstool, which is derived from the German todesstuhl, for death's stool (FDA, 2006a). Mycetism is the term for intoxications due to toxic mushrooms and is distinct from mycotoxicoses, described in the previous section. Poisonous mushrooms have been employed throughout human history for nefarious purposes. The book by Benjamin (1995) includes an interesting chapter on the "History of mushroom eating and mushroom poisoning," covering prehistoric humans through modern times.

Mushrooms are becoming increasingly more prominent as cultivated foods. This includes the common *Agaricus bisporus,* known as the button mushroom, as well as several "gourmet" mushrooms of other species. Globally, mushroom production, consumption, and trade are on the rise (Mayett et al., 2006).

Harvesting mushrooms in the wild is also becoming a more popular hobby. The term "shroomer," a colloquialism for wild-mushroom gatherer, was added to the online *Oxford English Dictionary* in 2004. A possible result of the increasing popularity of becoming a shroomer, has been an increase in reported cases of mushroom poisoning. An old saying among mushroom hunters is, "There are old mushroom hunters and there are bold mushroom hunters, but there are no old bold mushroom hunters." As gourmet mushrooms gain popularity, particularly those that cannot yet be successfully cultivated, there is more incentive to harvest these in the wild, leading to a higher risk of mistakes in mushroom identification.

The United States Centers for Disease Control and Prevention reported that for the span of 1998 to 2002, there were only two outbreaks of mushroom intoxications, which included six cases and no deaths (Lynch et al., 2006). The American Association of Poison Control Centers reported that in 2004, 8,601 cases of mushroom poisonings with five fatalities were reported (Watson et al., 2005). About 80% of mushrooms involved in these cases were unidentified. It is thought that 60% of toxic mushroom exposures occur in children younger than 6 years.

Statistics on mushroom poisonings are incomplete or nonexistent, especially for countries other than the United States. Reports on the ProMED-mail listserve indicate that the Ukrainian Ministry of Health (see http://www.korrespondent.net/main/162630/) reported 82 cases of mushroom poisoning due to wild mushrooms in 2006, as of September, of which 19 cases were children, and three of the adults died.

The populations most at risk for mushroom intoxication include young children, foragers who misidentify wild mushrooms, foragers seeking psychoactive little brown mushrooms, and victims of attempted homicides. Immigrants who are mushroom foragers are also a high-risk group. In December of 1981, seven cases of mushroom poisoning among Laotian refugees in a Sonoma County, CA, hospital were reported, for which the mushrooms were not identified (CDC, 1982). In September of 2006, seven cases of mushroom poisoning occurred among an immigrant population of Hmong in St. Paul, MN (http://www.startribune.com/462/story/678165.html). Immigrants may have been collecting specimens that resemble edible species in their homelands, but even the same species of mushrooms grown in different locations can be toxic. The intrinsic and extrinsic conditions of the production area can lead to induction of or increased toxin production for some strains of toxigenic mushrooms. There is much unknown about variations in toxin production in wild and cultivated mushrooms.

The major classes of mushroom toxins are discussed below in this section. The chemical structures for the major mushroom toxins are available on the Internet (FDA, 2006a). For pictures of the mushrooms produced by toxigenic species, the reader is referred to Lincoff (1981), with the warning that more than this field guide may be needed for accurate identification. For a more detailed review of other adverse reactions to mushrooms, including panic reactions, bacterial food poisoning, pesticide contamination, heavy metal contamination, radioisotope contamination, and intestinal obstruction, see Benjamin (1995). The handbook by Spoerke and Rumack (1994) is also a resource for learning more details about toxigenic mushrooms. The most effective control measure for food safety is to accurately identify edible mushrooms and to process and store them properly.

Protoplasmic Toxins

Protoplasmic toxins, produced by some highly poisonous mushrooms, are molecules that cause catastrophic damage to cells, which can lead to organ failure and death. Amatoxins, hydrazines, and orellanine fall into this category (FDA, 2006a). Cooking does not inactivate these toxins.

Amatoxins

Mushroom amatoxins are a family of cyclic octapeptides and are produced by mushroom species including *Amanita phalloides*, *Amanita virosa*, and *Galerina autumnalis*. The conventional names for the mushrooms of the *Amanita* species include death cap, destroying angel, and fool's mushroom, while the *G. autumnalis* mushroom is commonly referred to as the autumn skullcap (FDA, 2006a). Outbreaks of amatoxin poisoning in the United States seem to cluster in California and to involve immigrant populations (Benjamin, 1995).

Symptoms of amatoxin intoxication have three phases: (i) the gastrointestinal phase which involves abdominal pain, vomiting, and watery diarrhea and has an onset period of 6 to 36 hours, depending on the dosage consumed, (ii) the honeymoon phase, which lasts about 48 to 72 hours, in which symptoms seem to dissipate, and (iii) the terminal phase, which lasts about 72 to 96 hours, in which abdominal symptoms recur along with jaundice, liver failure, possible kidney, cardiac, and skeletal muscle damage, and death (FDA, 2006a; Benjamin, 1995). Death can occur within 72 hours or up to 10 days after exposure, depending on how rapidly the severe organ damage occurs. Benjamin (1995) indicates a 10 to 15% mortality rate, while the FDA (2006a) indicates a 50 to 90% mortality rate. The estimated LD_{50} for alpha-amatoxin in humans is 0.1 mg/kg body weight (Weiland, 1968), and one mushroom cap could easily contain an amount leading to this dosage (Benjamin, 1995).

Hydrazine derivatives

Gyromitrin, a volatile hydrazine derivative, leads to acute intoxication, with symptoms similar to those of the amatoxins, but less severe. The mortality rate is much lower than that of amatoxins, at only 2 to 4%, as reported by the FDA (2006a).

Orellanine

Orellanine is produced by *Cortinarius orellanus*, *Cortinarius speciosissimus*, *Cortinarius orellanoides*, and possibly *Cortinarius splendens* and causes a syndrome known as delayed-onset renal failure (Benjamin, 1995). This syndrome is characterized by a long asymptomatic latent period ranging from 3 to 14 days, followed by polydipsia and polyuria, nausea, headache, muscular pains, chills, spasms, and loss of consciousness (FDA, 2006a). The mortality rate is approximately 15%, with death typically due to renal failure several weeks later.

Neurotoxins

The mushroom toxins that cause neurological effects can cause a range of symptoms including sweating, coma, convulsions, hallucinations, excitement, depression, and spastic colon. These include some of the most infamous forms of mushroom intoxication, and cooking does not inactivate these toxins.

Muscarine poisoning

Several species of the *Inocybe*, *Clitocybe*, and *Omphalotus* genera, as well as *Boletus calopus*, *B. luridus*, and *Entoloma rhodopolium*, produce significant amounts (3 to 5% of mushroom mass) of muscarine toxins (Benjamin, 1995; FDA 2006a). The onset time for symptoms is quite short, ranging from 15 minutes to 1 hour. Characteristic symptoms are profuse sweating, salivation, and lacrimation. Larger doses lead to abdominal pain, severe nausea, diarrhea, blurred vision, bradycardia, and an urge to urinate. Estimates for LDs of muscarine range from 40 to 180 mg per exposure. Death is rare and usually due to cardiac or respiratory failure.

Psilocybin poisoning

Psilocybin intoxication puts the victim into a state similar to that of alcohol intoxication but is sometimes accompanied by hallucinations. The symptoms appear very rapidly and usually subside in a few hours. Symptoms include uncontrollable laughter, hallucinations, euphoria, dilated pupils, confusion, vertigo, muscular weakness, and increased deep-tendon reflexes (Benjamin, 1995). Cases of intentional intoxication may be cause for concern of drug overdose, especially if other drugs are taken in combination (FDA, 2006a). Species purposely sought for the intoxicating effects include *Psilocybe cubensis*, *Psilocybe mexicana,* and *Conocybe cyanopus*, sometimes referred to as "little brown mushrooms" or "magic mushrooms." Death is extremely rare, and usually these cases are children and involve seizures. It is illegal to possess or sell these mushrooms in the United States, as they are considered an illegal substance. Illegally purchased mushrooms vary in content of active psilocybin and may contain added substances like LSD (Benjamin, 1995).

Gastrointestinal Irritants

Gastrointestinal irritants produced by many types of mushrooms can cause a range of symptoms, including rapid nausea, vomiting, abdominal pain, and diarrhea. Symptoms are similar to the early stages of protoplasmic poisonings, but the onset is much more rapid. Symptoms can last for several days. Death is extremely rare and due to severe loss of fluids. The toxins involved are not characterized, and much more information is needed (FDA, 2006a).

Benjamin (1995) warned against eating raw mushrooms and recommended that all edible mushrooms be thoroughly cooked. He rested his argument on the documented cases of adverse reactions, including severe gastrointestinal upset after eating edible species of mushrooms raw, and indicated that there are several heat-labile toxins in many types of edible mushrooms, including the morels (*Morchella* spp.) and the honey mushrooms (*Armillaria mellea* complex). Stamets (2000) also warned of gastrointestinal upset due to improper storage of cultivated edible mushrooms, such as the oyster mushroom *(Pleurotus pulmonarius)*.

Disulfiram-Like Toxins

Antabuse syndrome

The inky cap mushroom *(Coprinus atramentarius)* is an edible mushroom which produces varying amounts of the amino acid coprine. Coprine is converted through human metabolism to cyclopropanone hydrate, which interferes with alcohol metabolism. When this type of mushroom is consumed with alcohol, concurrently or within 70 hours after eating the mushrooms, symptoms of headache, nausea, vomiting, flushing, and cardiovascular disturbances occur and last for 2 to 3 hours (FDA, 2006a). The effect of the symptoms is similar to the effect of treating alcoholics with disulfiram to discourage drinking. The symptoms, although unpleasant, are usually self-limited. The most severe cases are those involving alcoholics already on disulfiram and who have other complicating factors (Benjamin, 1995).

Carcinogens

Hydrazines

Mushrooms can contain a variety of hydrazines that have been shown to be highly mutagenic (Benjamin, 1995). Even the common button mushroom, *Agaricus bisporus,* produces hydrazines, but at concentrations considered not to be of significant risk. The risk of cancer from wine or beer is higher than that from consuming most edible mushrooms. Some mushrooms have been shown to have anticancer compounds as well.

Idiosyncratic Reactions

Adverse reactions to edible mushrooms are individualistic in many cases and are quite unpredictable. An inability to metabolize some component of the mushroom or an allergic response is thought to trigger such reactions, factors that do not seem to be based on age, sex, or general health (FDA, 2006a).

CONCLUDING REMARKS

There is much that is not clearly understood about the biological activities and health consequences of fungal toxins. The toxins of highest risk for acute toxicoses have been studied the most. There is a clear need to have a better understanding of the acute and chronic effects of fungal toxins. There is also a need to determine fungal toxin effects in combination with other toxic chemicals in food, as well as interactions with other complicating intrinsic factors.

With the recent impact of increasing global fuel prices resulting in vastly increased production of biofuels, there is much concern being expressed regarding the concentration of mycotoxins in the byproducts of the biofuel manufacturers. Great quantities of these materials will go into animal feed. There is a tremendous interest in control measures for mycotoxins in these materials to prevent introduction of increased levels of mycotoxins into the food supply chain.

Fungi as food-borne allergens is an area of research for which more information is needed. Children who have respiratory allergies to molds are more prone to food allergies. A link between mold aeroallergens and allergens from food has been demonstrated, with positive skin prick reactions to mold proteins from *Alternaria alternata, Cladosporium herbarum,* and/or *Aspergillus fumigatus* by individuals who also tested positive for reactions to extracts of mushrooms *(Agaricus bisporus)* (Herrara-Mozo et al., 2006). Little is known of the allergenic potential of many specialty mushrooms, which may account

for many idiosyncratic reactions to edible mushrooms. Also, as novel foods are developed from molds, such as mycoprotein (Miller and Dwyer, 2001), a better understanding of the potential for allergenicity would be helpful.

As more is learned about the fungal toxins and their roles in human and animal health, there needs to be consideration of appropriate, practical, and economically feasible control strategies to minimize the health risks. Governments and private industry should be more proactive in facilitating research efforts to understand and deal with both the acute and chronic issues of all fungal toxins. This will become increasingly clear as the world's food supply becomes steadily more globalized.

REFERENCES

Afolabi, C. G., R. Bandyopadhyay, J. F. Leslie, and E. J. A. Ekpo. 2006. Effect of sorting on incidence and occurrence of fumonisins and *Fusarium verticillioides* on maize from Nigeria. *J. Food Prot.* 69:2019–2023.

Aish, D., E. H. Rippon, T. Barlow, and S. J. Hattersley. 2004. Ochratoxin A, p. 307–338. *In* N. Magan and M. Olsen (ed.), *Mycotoxins in Food: Detection and Control.* CRC Press, New York, NY.

Alldrick, A. J., Campden and Chorleywood Food Research Association UK, and M. Hajselova. 2004. Zearalenone, p. 353–366. *In* N. Magan and M. Olsen (ed.), *Mycotoxins in Food: Detection and Control.* CRC Press, New York, NY.

Almela, L., V. Rabe, B. Sanchez, F. Torrilla, J. P. Lopez-Perez, J. A. Gabaldon, and L. Guardiola. 2007. Ochratoxin A in red paprika: relationship with the origin of the raw material. *J. Food Microbiol.* 24:319–327.

Azziz-Baumgartner, E., K. Lindblade, K. Gieseker, H. Schurz Rogers, S. Kieszak, H. Njapau, R. Schleicher, L. F. McCoy, A. Misore, K. DeCock, C. Rubin, L. Slutsker, and the Aflatoxin Investigative Group. 2004. Case-control study of an acute aflatoxicosis outbreak, Kenya, 2004. *Environ. Health Perspect.* 113:1779–1783.

Barger, G. 1931. *Ergot and Ergotism.* Guerney and Jackson, London, United Kingdom.

Benjamin, D. R. 1995. *Mushrooms Poisons and Panaceas: a Handbook for Naturalists, Mycologists, and Physicians.* W. H. Freeman and Company, New York, NY.

Berthiller, F., C. Dall'Asta, R. Schuhmacher, M. Lemmens, G. Adam, and R. Krska. 2005. Masked mycotoxins: determination of a deoxynivalenol glucoside in artificially and naturally contaminated wheat by liquid chromatography-tandem mass spectrometry. *J. Agric. Food Chem.* 53:3421–3425.

Blumenthal, C. Z. 2004. Production of toxic metabolites in *Aspergillus niger, Aspergillus oryzae,* and *Trichoderma reesei:* justification of mycotoxin testing in food grade enzyme preparations derived from the three fungi. *Regul. Toxicol. Pharmacol.* 39:214–228.

Boichenko, L. V., D. M. Boichenko, N. G. Vinokurova, T. A. Reshetilova, and M. U. Arinbasarov. 2001. Screening for ergot alkaloid producers among microscopic fungi by means of the polymerase chain reaction. *Microbiology* 70:306–310.

Bony, S., M. Carcelen, L. Olivier, and A. Devaux. 2006. Genotoxicity assessment of deoxynivalenol in the Caco-2 cell line model using the Comet assay. *Toxicol. Lett.* 166:67–76.

Bove, F. J. 1970. *The Story of Ergot*. S. Krager, Basel, Switzerland.

Bullerman, L. B. 2007. Fusaria and toxigenic molds other than aspergilli and penicillia, p. 563–578. *In* M. P. Doyle and L. R. Beuchat (ed.), *Food Microbiology: Fundamentals and Frontiers*, 3rd ed. ASM Press, Washington, DC.

CAC. 2003. Code of practice for the prevention and reduction of mycotoxin contamination in cereals, including annexes on ochratoxin A, zearalenone, fumonisins and trichothecenes. CAC/RCP 51-2003. Codex Alimentarius Commission, Rome, Italy. http://www.codexalimentarius.net/download/standards/406/CXC_051e.pdf. Accessed 30 May 2007.

Caporael, L. R. 1976. Ergotism: the satan loosed in Salem? *Science* 192:21–26.

Castegnaro, M., D. Danadas, T. Vrabcheva, T. Petkova-Bocharova, I. N. Chernozemsky, and A. Pfohl-Leszkowicz. 2006. Balkan endemic nephropathy: role of ochratoxins A through biomarkers. *Mol. Nutr. Food Res.* 50:519–529.

CDC. 1982. Mushroom poisoning among Laotian refugees—1981. *MMWR Morb. Mortal. Wkly. Rep.* 31:287–288.

Corley, D. G., M. Miller-Wideman, and R. C. Durley. 1994. Isolation and structure of harzianum A: a new trichothecene from *Trichoderma harzianum*. *J. Nat. Prod.* 57:422–425.

Coty, P. J. 1989. Virulence and cultural characteristics of two *Aspergillus flavus* strains pathogenic to cotton. *Phytopathology* 79:808–814.

Dawlatana, M., R. D. Coker, M. J. Nagler, C. P. Wild, M. S. Hassan, and G. Blunden. 2002. The occurrence of mycotoxins in key commodities in Bangladesh: surveillance results from 1993–1995. *J. Nat. Toxins* 11:379–386.

Desjardins, A. E. 2003. Trichothecenes: from yellow rain to green wheat. *ASM News* 69:182–185.

Dohlman, E. 2003. Mycotoxin hazards and regulations, p. 98–108. *In* J. Buzby (ed.), *International Trade and Food Safety: Economic Theory and Case Studies*. Agricultural Economic Report no. AER828. Economic Research Service, U.S. Department of Agriculture, Washington, DC. http://www.ers.usda.gov/publications/aer828/. Accessed 27 May 2007.

FAO. 2001. *Manual on the Application of the HACCP System in Mycotoxin Prevention and Control*. FAO/IAEA Training and Reference Centre for Food and Pesticide Control, Vienna, Austria. http://www.fao.org/docrep/005/y1390e/y1390e00.htm. Accessed 20 May 2007.

FDA. 2006a. Mushroom toxins. *In Bad Bug Book*. Center for Food Safety and Applied Nutrition, U.S. Food and Drug Administration, Washington, DC. http://www.cfsan.fda.gov/~mow/chap40.html. Accessed 10 May 2007.

FDA. 2006b. Aflatoxins. *In Bad Bug Book*. Center for Food Safety and Applied Nutrition, U.S. Food and Drug Administration, Washington, DC. http://www.cfsan.fda.gov/~mow/chap41.html. Accessed 25 May 2007.

Gaumy, J. L., J. D. Bailey, G. Benard, and P. Guerre. 2001. Zearalenone: origin and effects on farm animals. *Rev. Med. Vet.* 152:123–136.

Gelderblom, W. C., K. Jaskiewicz, and W. F. O. Marassas. 1988. Fumonisins—novel mycotoxins with cancer-promoting activity produced by *Fusarium moniliforme*. *Appl. Environ. Microbiol.* 54:1806–1811.

Godtfredsen, W. O., and S. Vangedal. 1964. Trichodermin, a new antibiotic, related to trichothecin. *Proc. Chem. Soc.* June:188–189.

Godtfredsen, W. O., and S. Vangedal. 1965. Trichodermin, a new sesquiterpene antibiotic. *Acta Chem. Scand.* 19:1088–1102.

Harman, G. E., C. R. Howell, A. Viterbo, I. Chet, and M. Lorito. 2004. *Trichoderma* species—opportunistic, avirulent plant symbionts. *Nat. Rev. Microbiol.* 2:43–56.

Herrera-Mozo, I., B. Ferrer, J. L. Rodríguez-Sanchez, and C. Juarez. 2006. Description of a novel panallergen of cross-reactivity between moulds and foods. *Immunol. Invest.* 35:181–197.

Hocking, A. 2007. Toxigenic *Aspergillus* species, p. 537–550. *In* M. P. Doyle and L. R. Beuchat (ed.), *Food Microbiology: Fundamentals and Frontiers*, 3rd ed. ASM Press, Washington, DC.

Hudler, G. W. 1998. *Magical Mushrooms and Mischievous Molds*. Princeton University Press, Princeton, NJ.

IARC. 1993a. Aflatoxins: naturally occurring aflatoxins (Group 1), aflatoxins M1 (Group 2B). *Int. Agency Res. Cancer* 56:245.

IARC. 1993b. Toxins derived from *Fusarium moniliforme*: fumonisins B1, Bx and fusarin C: monograph on the evaluation of carcinogenic risk to humans. *Int. Agency Res. Cancer* 56:445–466.

Jackson, L., and J. Jablonski. 2004. Fumonisins, p. 367-405. *In* N. Magan and M. Olsen (ed.), *Mycotoxins in Food: Detection and Control*. CRC Press, New York, NY.

Jay, J. M., M. J. Loessner, and D. A. Golden. 2005. Intrinsic and extrinsic parameters of foods that affect microbial growth, p. 39–59. *In Modern Food Microbiology*, 7th ed. Springer Science, New York, NY.

JECFA. 2001. Safety evaluation of certain mycotoxins in food. Food and Agriculture Organisation of the United Nations (FAO) food and nutrition paper 74, World Health Organization (WHO) food additive series, 47. http://www.inchem.org/documents/jecfa/jecmono/v47je01.htm. Accessed 29 May 2007.

Kanisawa, M. 1984. Synergistic effect of citrinin on hapatorenal carcinogenesis of ochratoxin A in mice. *Dev. Food Sci.* 7:245–254.

Koirala, P., S. Kumar, B. K. Yadav, and K. C. Premarajan. 2005. Occurrence of aflatoxin in some of the food and feed in Nepal. *Indian J. Med. Sci.* 59:331–336.

Leung, M. C. K., G. Diaz-Llang, and T. K. Smith. 2006. Mycotoxins in pet food: a review on worldwide prevalence and preventative strategies. *J. Agric. Food Chem.* 54:9623–9635.

Lewis, L., M. Onsongo, H. Njapau, H. Schurz-Rogers, G. Luber, S. Kieszak, J. Nyamongo, L. Backer, A. M. Dahiye, A. Misore, K. DeCock, C. Rubin, and the Kenya Aflatoxicosis Investigation Group. 2005. Aflatoxin contamination of commercial maize products during an outbreak of acute aflatoxicosis in eastern and central Kenya. *Environ. Health Perspect.* 113:1763–1767.

Lincoff, G. H. 1981. *National Audubon Society Field Guide to North American Mushrooms*. Knopf, New York, NY.

Lynch, M., J. Painter, R. Woodruff, and C. Braden. 2006. Surveillance for foodborne-disease outbreaks—United States, 1998–2002. *MMWR Morb. Mortal. Wkly. Rep.* 55:1–34.

Magan, N., and M. Olsen (ed.). 2004. *Mycotoxins in Food: Detection and Control*. CRC Press, New York, NY.

Marasas, W. F. O., T. R. Riley, K. A. Hendricks, V. L. Stevens, T. W. Sadler, J. Glineau-van Waes, S. A. Missmer, J. Cabrera, O. Torres, W. C. Gelderblom, J. Allegood, C. Martinez, J. Maddoz, J. D. Miller, L. Starr, M. C. Sullands, A. V. Roman, K. A. Voss, E. Wang, and A. H. Merril. 2004. Fumonisins disrupt sphingolipids metabolism, folate transport, and neural tube development in embryo culture and *in vivo*: a potential risk factor for human neural tube defects among populations consuming fumonisin-contaminated maize. *J. Nutr.* 134:711–716.

Mayett, Y., D. Martinez-Carrera, M. Sainchez, A. Macaas, S. Mora, and A. Estrada-Torres. 2006. Consumption trends of edible mushrooms in developing countries: the case of Mexico. *J. Int. Food Agribus. Market* 18:151–176.

Miller, S. A, and J. T. Dwyer. 2001. Evaluating the safety and nutritional value of mycoprotein. *Food Technol.* 55:42–47.

Murphy, P., S. Hendrich, C. Landgren, and C. B. Bryant. 2006. Food mycotoxins: an update. *J. Food Sci.* 71:51–65.

Ngugi, H. K., and D. M. Wilson. 2006. An overview of aflatoxin exposure in Kenya and challenges for monitoring and management, p. S152. Abstr. Spec. Session Presentations APS-CPS-MSA Joint Meet., Quebec City, Quebec, Canada, 29 July to 2 August. http://www.apsnet.org/meetings/2006/abstracts/ss06ma134. htm. Accessed 12 May 2007.

Nielson, K. F., T. Grafenhan, D. Zafari, and U. Thrane. 2005. Trichothecene production by Trichoderma brevicompactum. *J. Agric. Food Chem.* **53:**8190–8196.

Nyikal, J., A. Misore, C. Nzioka, C. Njuguna, E. Muchiri, J. Njau, S. Maingi, J. Njoroge, J. Mutiso, J. Onteri, A. Langat, I. K. Kilei, J. Nyamongo, G. Ogana, B. Muture, P. Tukei, C. Onyango, W. Ochieng, C. Tettch, S. Likimani, P. Nguku, T. Galgalo, S. Kibet, A. Manya, A. Dahiye, J. Mwihia, L. Mugoya, J. Onsongo, A. Ngindu, K. M. DeCock, K. Lindblade, L. Slutsker, P. Amornkul, D. Rosen, D. Feiken, T. Thomas, P. Mensah, N. Eseko, A. Nejjar, M. Onsongo, F. Kesell, H. Njapau, D. L. Park, L. Lewis, G. Luber, H. Rogers, L. Baker, C. Rubin, K. E. Gieseker, E. Azzia-Baumgartner, W. Chege, and A. Bowman. 2004. Outbreak of aflatoxin poisoning—Eastern and Central Provinces, Kenya, January–July 2004. *MMWR Morb. Mortal. Wkly. Rep.* **53:**790–793.

Pestka, J. J., and A. T. Smolinski. 2005. Deoxynivalenol: toxicology and potential effects on humans. *J. Toxicol. Environ. Health B* **8:**39–69.

Pitt, J. I. 2007. Toxigenic *Penicillium* species, p. 551–562. *In* M. P. Doyle and L. R. Beuchat (ed.), *Food Microbiology: Fundamentals and Frontiers*, 3rd ed. ASM Press, Washington, DC.

Probst, C., H. Njapau, and P. J. Coty. 2007. Outbreak of an acute aflatoxicosis in Kenya in 2004: identification of the causal agent. *Appl. Environ. Microbiol.* **73:**2762–2764.

Richard, J. L., and G. A. Payne (ed.). 2003. *Mycotoxins: Risks in Plant, Animal, and Human Systems.* Council for Agricultural Science and Technology, Ames, IA.

SCF. 1998. Opinion of the Scientific Committee on Food on ochratoxin A. Scientific Committee on Food, European Commission, http://ec.europa.eu/food/fs/sc/scf/out14_en.html. Accessed 25 May 2007.

Schwarz, P. B., S. M. Neate, and G. E. Rottinghaus. 2006. Widespread occurrence of ergot in upper midwestern U.S. barley, 2005. *Plant Dis.* **90:**527.

Sperjers, G. J. A. 2004. Patulin, p. 339–352. *In* N. Magan and M. Olsen (ed.), *Mycotoxins in Foods: Detection and Control,* CRC Press, Boca Raton, FL.

Spoerke, D. G., and B. H. Rumack (ed.). 1994. *Handbook of Mushroom Poisoning: Diagnosis and Treatment.* CRC Press, Boca Raton, FL.

Stamets, P. 2000. *Growing Gourmet and Medicinal Mushrooms.* Ten Speed Press, Berkeley, CA.

Stoev, S. D., H. Daskalov, B. Radie, A. M. Domijan, and M. Peraica. 2002. Spontaneous mycotoxin nephropathy in Bulgarian chickens with unclarified mycotoxin aetiology. *Vet. Res.* **33:**83–93.

Wang, J. S., and J. D. Groopman. 1999. DNA damage by mycotoxins. *DNA Res.* **424:**167–181.

Watson, W. A., T. L. Litovitz, G. C. Rodgers, W. Klein-Schwartz, N. Reid, J. Youniss, A. Flannigan, and K. M. Wruk. 2005. 2004 Annual report of the American Association of Poison Control Centers Toxic Exposure Surveillance System. *Am. J. Emerg. Med.* **23:**589–666.

Weidenborner, M. 2001. *Encyclopedia of Food Mycotoxins.* Springer-Verlag, New York, NY.

Weiland, T. 1968. Poisonous principles of mushrooms of the genus *Aminita. Science* **159:**946–952.

Wu, F. 2006. An analysis of Bt corn's benefits and risks for national and regional policymakers considering Bt corn adoption. *Int. J. Technol. Global.* **2:**115–136.

Pathogens and Toxins in Foods: Challenges and Interventions
Edited by V. K. Juneja and J. N. Sofos
© 2010 ASM Press, Washington, DC

Chapter 18

Critical Evaluation of Uncertainties of Gluten Testing: Issues and Solutions for Food Allergen Detection

Carmen Diaz-Amigo and Jupiter M. Yeung

Food allergy has emerged as an important public health concern based on its increasing prevalence, its persistence throughout life for those who are sensitized to the foods that can trigger severe reactions, the potential for fatal outcomes, and the lack of preventive treatment other than food avoidance. Food allergy is defined as an adverse immune response to proteins present in food that is caused by immunoglobulin E (IgE)-mediated or non-IgE-mediated (cellular) mechanisms (Sicherer and Sampson, 2006). While most food allergies, such as peanut allergy, are IgE-mediated diseases, celiac disease is a cell-mediated disease of food intolerance. Food-induced allergic reactions are responsible for a variety of symptoms ranging from hives or gastrointestinal discomfort to anaphylaxis. Anaphylaxis is a severe and life-threatening systemic allergic reaction characterized by hives, fall of blood pressure, upper airway obstruction, and severe wheezing. In severe cases, unmanaged anaphylaxis can lead to death (Sampson et al., 2006). Food allergy is frequently accompanied by asthma, and asthma is an important risk factor for severe allergic reactions to food.

Food allergy accounts for about 35 to 50 percent of emergency room visits for anaphylaxis and causes about 30,000 episodes of anaphylaxis and 100 to 200 deaths per year in the United States (Burks and Ballmer-Weber, 2006; Sampson, 2003). Severe, life-threatening reactions occur mostly in adolescents and young adults, and peanuts and tree nuts are the most common causes of such reactions. Currently, the only treatments for food allergy are allergen avoidance and management of reactions caused by allergen exposure. In addition to the psychological effects of the risk of death and the stigma of avoiding common foods, food allergy has nutritional impacts on the health, development, and lifestyle of children. Hence, food allergy has emerged as an important public health concern with a considerable social and economic impact on industrialized society.

Although any food can provoke an allergic reaction, relatively few foods are responsible for the vast majority of significant food-induced allergic reactions. More than 90 percent of food allergies are caused by eight major food allergens: peanuts, tree nuts, milk, eggs, soybeans, finfish, shellfish, and wheat. Food allergens are naturally occurring proteins that are resistant to heat, digestion, proteolysis, and acid (Hefle et al., 1996). Even with diligent efforts to avoid the food allergens, each year approximately one of every four food-allergic individuals will have an accidental exposure that leads to a food-induced reaction (Keet and Wood, 2007).

Undeclared allergens in foods pose a major risk for allergic individuals. Consumers with food allergies and food intolerances rely on accurate food labeling to enable them to make informed choices about the foods they eat. In the United States, the Food Allergen and Consumer Protection Act of 2004 (Public Law 108-282), which took effect on 1 January 2006, mandates that foods containing the major food allergens (milk, eggs, fish, crustacean shellfish, peanuts, tree nuts, wheat, and soy) must be declared in plain language on the ingredient list. Therefore, reliable detection and quantification methods for food allergens are necessary to ensure compliance with food labeling and to ensure consumer protection. This review focuses on a specific wheat allergen and the appropriate methods for its detection in food products to demonstrate the complexity of analytical challenges that the scientific community is facing. Such challenges apply to food

Carmen Diaz-Amigo • Center for Food Safety and Applied Nutrition, U.S. Food and Drug Administration, Laurel, MD 20708. **Jupiter M. Yeung** • Grocery Manufacturers Association, Washington, DC 20005.

allergen detection in general (Westphal, 2006; Yeung, 2006).

PROPERTIES OF GLUTEN: HEALTHY OR TOXIC?

Cereal grains and their products can be present in foods for nutritional and health reasons. They are sources of dietary protein, fiber, vitamins, and minerals or serve as replacers, such as wheat starch, replacing fat or wheat gluten as a meat substitute in vegetarian foods. Because of their technological properties, cereal fractions, like wheat starch and wheat gluten, are commonly used as dough strengtheners, formulation aids, processing aids, stabilizers, thickeners, surface finishing agents, and texturizing agents (Day et al., 2006; Lynn et al., 1997). A component of some cereals, called gluten, can cause celiac disease in sensitive individuals. Celiac disease is a food intolerance characterized by a T-cell-mediated intestinal inflammation following exposure to gluten proteins (Periolo and Chernavsky, 2006). Although celiac disease may be asymptomatic, the classical form affects the villous structure of the mucosa of the small intestine and is characterized by diarrhea, malabsorption, and malnutrition. Celiac disease is responsible for difficulty to thrive in some children. Nonclassical presentation of the disease, also called "silent celiac disease," may include anemia and dermatitis herpetiformis, among others (McGough and Cummings, 2005). Celiac disease has become a major health concern since the prevalence of the disease is higher than previously thought (Hischenhuber et al., 2006). In the United States, it has been estimated that the prevalence ranges from 1 in 250 (Not et al., 1998) to 1 in 133 (Fasano et al., 2003). Currently, there is neither a cure nor treatment for the disorder. However, the strict avoidance of the offending cereals is effective to attenuate and mitigate the symptoms of celiac disease.

There are three key elements of improving the quality of life of celiac individuals (Cranney et al., 2007; Zarkadas et al., 2006): (i) early diagnosis of the disease, (ii) education of physicians, patients, their families and friends, and care-takers, and (iii) availability and diversification of gluten-free products. A gluten-free diet does not necessarily mean "zero gluten," since low levels of this protein complex seem to be tolerated by patients. It has been estimated that a 10- to 100-mg daily intake may be a safe amount of gluten (Hischenhuber et al., 2006). In a double-blind, placebo-controlled food-challenged trial, Catassi et al. (2007) determined that the ingestion of gluten should be kept below 50 mg gluten per day. As required by the Food Allergen and Consumer Protection Act, the U.S. Food and Drug Administration (FDA) has proposed the definition of "gluten-free" claims (21 CFR Part 101) for voluntary use in food labels. Products containing ingredients derived from wheat, rye, and barley that have been processed to remove gluten (like starch) and grains other than the three mentioned can be labeled "gluten-free" if the gluten content is 20 ppm or less.

The availability of validated quantitative analytical techniques will be a key element in ensuring the accuracy of food labels. Antibody-based immunoassays have been developed since the mid-1980s to target gluten. But, what is gluten? Gluten is a heterogeneous group of proteins which are considered one of the most complex groups of proteins in nature. They are abundant in cereals. While gluten from wheat, rye, and barley has been shown to be toxic to people suffering from celiac disease (Marsh, 1992), there is open debate regarding the safety of oats to a subpopulation of celiac patients. Due to the complexity of wheat proteins, several classifications have been proposed (Table 1). According to Osborne (1924), gluten proteins can be classified according to their solubility in alcohol solutions (prolamins) or acid/alkaline (glutelins) solutions. Prolamins are called gliadins in wheat, secalins in rye, and hordeins in barley. In wheat, the glutelin fraction comprises high- and low-molecular-weight (HMW and LMW) glutenins. However, solubility is not as clear-cut between these two groups since the reduced form of glutelin, which is present mostly in cereals as

Table 1. Classification of gluten proteins

Based on protein solubility[a]

   Glutelins
- Wheat: HMW and LMW glutenins

   Prolamins
- Wheat: ω-gliadins, α-gliadins, γ-gliadins
- Rye: secalins
- Barley: hordeins

Based on structural and evolutionary relationship[b]

   HMW prolamins
- Wheat: HMW glutenins
- Rye: HMW secalins
- Barley: D hordein

   S-rich prolamins
- Wheat: α-gliadin, γ-gliadin, B- and C-type LMW glutenins
- Rye: B-hordeins
- Barley: γ-hordeins

   S-poor prolamins
- Wheat: ω-gliadin, D-type LMW glutenins
- Rye: ω-secalins
- Barley: C-hordeins

[a]Osborne (1924).
[b]Shewry and Tatham (1990).

polymers, is also soluble in aqueous alcohol. On the other hand, Shewry and Tatham (1990) established a different classification of gluten proteins according to their structural relationship. In this chapter, we use the Osborne nomenclature because it is most cited among immunoassay developers.

THE INTRICACY OF DETECTION METHODS: THE RESULT OF TARGET COMPLEXITY AND FOOD PROCESSING

There is a general agreement that immunoassays have to target the toxic gluten proteins from wheat, rye, and barley. Moreover, they have to be able to detect these proteins in processed foods, which poses a major challenge for developing analytical techniques. Other important challenges include design of the antibody (Ab) and sample preparation. Aqueous ethanol has been traditionally used for the extraction of gluten. However, a novel extraction procedure using a cocktail protocol has been shown to be more suitable for the extraction of gluten components from heated foods (Garcia et al., 2005). Different approaches have been used to develop assays for gluten. Some assays use polyclonal Abs (Aubrecht and Toth, 1995; Chirdo et al., 1995; Nicolas et al., 2000), a combination of polyclonal and monoclonal Abs (mAbs) (Bermudo Redondo et al., 2005; Ellis et al., 1998) or a combination of mAbs (Sanchez et al., 2007; Sorell et al., 1998). Another approach is based on the use of a single mAb. Skerritt and Hill (1990) used a single mAb specific for ω-gliadin, which shows reduced sensitivity for barley prolamins. Valdés et al. (2003) reported an enzyme-linked immunosorbent assay (ELISA) system using the single R5 mAb, raised against rye secalins and specific for celiac toxic peptides (Kahlenberg et al., 2006), which can quantify wheat, rye, and barley in a similar manner. Some of these Abs are being used in some commercial kits. Tepnel uses Skerritt's Ab, and R-Biopharm employs R5 Ab in their ELISA. Assays using these Abs have undergone interlaboratory collaborative trials (Gabrovska et al., 2006; Mendez et al., 2005; Skerrit and Hill, 1991). In the U.S. market, other gluten commercial kit providers include Neogen Corp., whose ELISA is based on two mAbs, for rye and wheat, and Morinaga, whose ELISA uses polyclonal antibodies for wheat proteins. There is no published information about the antibodies used by these two companies.

To protect public health, risk assessors and managers rely on the accuracy of data provided by commercial kits. However, it is not clear how the complexity of gluten and the impact of food processing on these proteins affect the performance of the kits. Given the lack of reference material, the wide variety of choices of test kits, and various options offered for the detection methods, comparing results is a very difficult task. Here we use the gluten test to demonstrate the complexity of detection methods for food allergens in processed samples and the uncertainties and implications of results.

Gluten content in foods can be evaluated by either commercial ELISA kits or lateral flow devices (LFD). We evaluated the ability of four commercial ELISA kits from Morinaga, Neogen, R-Biopharm, and Tepnel to detect gluten proteins from wheat, and we determined the impact of different sample extraction solutions on the final results. Both the Neogen and R-Biopharm kits offer two options for sample preparation. Samples can be extracted by traditional aqueous ethanol or by a two-step extraction using cocktail solution and then aqueous ethanol. The cocktail contains a reducing agent, mercaptoethanol, and a disaggregating agent, guanidine hydrochloride. The extraction of samples analyzed by the Tepnel system is carried out exclusively by aqueous ethanol, while the Morinaga kit uses buffer containing a detergent, sodium dodecyl sulfate (SDS), and mercaptoethanol. To further evaluate assay specificity, we performed immunoblot analyses using the peroxidase-conjugated antibodies (signal or detector antibody) included in the kits. The assay specifications of each kit are summarized in Table 2.

Evaluation of Gluten-Spiked Food Matrices by ELISA and LFD

Four gluten levels were evaluated in this study (50, 25, 10, and 5 ppm gluten) in two different matrices, gluten-free high-fiber bread mix and matrix-free 60% ethanol. Guar gum was also used at a single level (50 ppm). The gluten reference material SRM 8418, obtained from the National Institute of Standards and Technology (NIST), was used to spike the samples since it is commercially available, well characterized, and without lot-to-lot variability. To ensure sample homogeneity, four subsamples of 50-ppm ($P = 0.23$) and 25-ppm ($P = 0.27$) gluten spikes were extracted with aqueous ethanol and analyzed by the Neogen ELISA. Samples were homogeneous when P values were greater than 0.05 by an analysis of variance F test.

Blank samples

All samples were extracted and analyzed according to the manufacturers' instructions. Table 3 summarizes the gluten content at all spiked levels tested by the different commercial assays. Gluten-free matrices (0 ppm gluten) confirmed the absence of gluten residue. Divergent results were obtained, depending

Table 2. Commercial gluten test kits[d]

Company	Product	Antibody	Target antigen	LOD (ppm)	LOQ (ppm)	Extraction procedure
Neogen Corp.	Veratox	2 mAb	Gliadin	NA	10	40% Ethanol or cocktail + 55% alcohol
R-Biopharm AG	Ridascreen	R5 mAb	α γ ω-Gliadin	3	5	60% Ethanol or cocktail + 80% alcohol
	Ridascreen Fast	R5 mAb	α γ ω-Gliadin	10	10	60% Ethanol or cocktail + 80% alcohol
	RidaQuick[a]	R5 mAb	α γ ω-Gliadin	5[c]	NA	40% Ethanol
Tepnel Biosystem	Gluten assay	Skerritt mAb	ω-Gliadin	1	10	40% ethanol
Morinaga Inc.	Wheat protein	Wheat pAb	Wheat proteins	0.3[b]	3.12[b]	Denaturing and reducing buffer, overnight

[a]Lateral flow device.
[b]Concentrations are ppm of wheat protein.
[c]Cutoff value for a qualitative assay.
[d]LOD, limit of detection; NA, data not available; pAb, polyclonal antibody.

on the kit used. Gluten was not detected by the Neogen, R-Biopharm, and Tepnel kits in either matrix. However, concentrations slightly above the limit of quantitation (LOQ) of the Morinaga assay were observed for guar gum and bread mix. This type of positive result, being so close to the LOQ for blank samples, needs to be interpreted with caution since it is difficult to know whether this result is a "true positive" due to sample contamination or because its sensitivity is lower than the other kits, or whether it is a "false positive" due to matrix effect. Issues like this can be answered by using confirmatory methods like

PCR, quantitative competitive PCR (Dahinden et al., 2001; Sandberg et al., 2003), and mass spectrometry (Camafeita and Mendez, 1998; Camafeita et al., 1998; Mamone et al., 2000; Salplachta et al., 2005). The Morinaga kit differs in several aspects from the others. It has the lowest LOQ and different reporting units, and it is the only one that uses polyclonal antibodies. In addition, the samples are extracted under reducing and denaturing conditions over an extended period of time (overnight for >12 h). It is unknown if such conditions might increase extraction efficiency. Nevertheless, it is likely that the positives are the

Table 3. Effects of food matrices, extraction methods, and ELISA types in determination of gluten levels in spiked samples[a]

Spiked gluten (ppm)	Matrix	R-Biopharm Ridascreen Fast gliadin (ppm gluten)[b]		R-Biopharm Ridascreen gliadin (ppm gluten)[c]		R-Biopharm RidaQuick (ppm gluten)[d]	Neogen Veratox gliadin (ppm gluten)[e]		Tepnel BioKit gluten (ppm gluten)[f]	Morinaga wheat protein (ppm wheat protein)[g]
		Ethanol	Cocktail	Ethanol	Cocktail	Ethanol	Ethanol	Cocktail	Ethanol	Morinaga buffer
0	Guar gum	ND	ND	ND	ND	–	ND	ND	ND	3.3
50	Guar gum	ND	26.3	16.4	30.4	+	41.8	50.2	55	54.7
0	Bread mix	ND	ND	ND	ND	–	ND	ND	ND	3.8
5	Bread mix	ND	ND	ND	ND	+	ND	ND	ND	9.5
10	Bread mix	ND	ND	ND	ND	+	13.2	12.8	ND	17.8
25	Bread mix	ND	ND	ND	ND	+	26.0	22.2	21.5	34.2
50	Bread mix	16.2	32.6	15.5	39.3	+	46.0	50.8	42.5	58.7
0	60% EtOH	ND	ND	ND	ND	–	ND	ND	ND	ND
5	60% EtOH	ND	ND	ND	ND	+	ND	ND	ND	5.7
10	60% EtOH	ND	ND	ND	7.2	+	11.0	12.4	11.2	14.7
25	60% EtOH	ND	12.2	ND	13.1	+	23.6	23.2	21.0	25.9
50	60% EtOH	20.8	30.4	17.4	36.3	+	51.0	45.0	39.4	52.6

[a]ND, not detected.
[b]LOQ of 10 ppm gluten.
[c]LOQ of 5 ppm gluten.
[d]LOQ with a cutoff value of 5 ppm.
[e]LOQ of 10 ppm gluten.
[f]LOQ of 10 ppm gluten.
[g]LOQ of 3.12 ppm wheat protein.

result of matrix effect since no gluten was detected in the 60% ethanol blank sample.

Spiked samples

It would be ideal if results provided by the different kits were equivalent, regardless of the sample preparation procedures and the type of antibodies used. However, discrepancy can be observed after the analysis of the spiked samples, as shown in Table 3.

For nonprocessed gluten-spiked flours, the Neogen ELISA kit provided a high degree of consistency at all levels and between the two extraction preparations. No false negatives or false positives were observed. The recoveries ranged between 80 and slightly over 100%, which are excellent values for this type of assay. It is possible that Neogen uses the same NIST reference material that we used for spiking. Similar recoveries were provided by the Tepnel kit, which only used a single extraction procedure. However, their kit failed to detect a 10-ppm gluten-spiked sample at its LOQ. This false negative is not unexpected since recoveries are rarely 100%.

Values obtained by the Morinaga ELISA did not differ substantially from those of Neogen and Tepnel, even though they use a different extraction solution. However, we did not just compare the numbers, we judged the results by the numbers with the reporting unit. It is important to note that all test kits except Morinaga report results in ppm gliadin or gluten. Morinaga expresses the data in ppm wheat protein. In the absence of a conversion factor for wheat protein to wheat gluten, it is impossible to compare their results. For the same reason, the recovery data for the Morinaga kit are somehow meaningless since we spiked the samples with ppm gluten, and the results using the Morinaga kit are given as ppm wheat protein. Reporting units should be standardized and harmonized by test kit manufacturers, and they should be in agreement with the units defined by the scientific community, if regulatory action levels are to be based on sound science.

The three assays from R-Biopharm use the same mAb R5; therefore, equivalent results were expected. However, they provided divergent results, depending on the extraction buffer and assay used. RidaQuick, a qualitative LFD with a cutoff value of 5 ppm gluten, provided the most reliable results of the three R-Biopharm assays (Table 3). The extraction protocol for this test was carried out with aqueous ethanol only, as recommended by the manufacturer. Results obtained with LFD met the manufacturer's claim expectations. Blank samples were negatives, while the rest of the samples containing 5 ppm gluten or more were positive. Surprisingly, although both R-Biopharm assays, Ridascreen Fast gliadin and

Ridascreen gliadin, provided similar values for the same sample, they struggled to detect gluten proteins below the highest gluten level of 50 ppm, with recovery percentages in the low 30s for aqueous ethanol extracts and between 60 and 80% for the cocktails. Only cocktail extracts from 25-ppm gluten-spiked 60% blank ethanol matrix resulted in positive results, with recoveries around 50%.

Before drawing any conclusion, it is necessary to evaluate potential answers to the questions raised by these results. In a situation like this, it is very likely that the analyst, who may not be familiar with assay design and development, may conclude problems are due to poor extraction efficiency or matrix effect. However, there are other issues that may remain unnoticed. The R-Biopharm kit results can be used to illustrate the problems associated with the selection and use of reference materials. In this study we have used NIST gluten SRM 8418 to evaluate the kits. Kit manufacturers may have used a different reference material to prepare the control standards and to calibrate the assay. We know, in fact, that R-Biopharm uses the European gliadin standard ERM-DA-480 (formerly IRMM-480), which is prepared from a mixture of 28 European wheat cultivars (Mendez et al., 2005; van Ecker et al., 2006). The content of gliadin varies with cultivar, ranging between 33 and 45% (Table 4). As a consequence, it cannot be concluded that the R-Biopharm kit has problems of protein recovery or matrix effect. Results are affected by an external factor, such as the lack of a common reference material, and this is the principal reason why allergen kits cannot really be evaluated or compared.

There are also internal factors affecting the performance of R-Biopharm ELISA kits. The differences in concentrations observed between cocktail and aqueous alcohol analyzed by R-Biopharm ELISAs are related to how the manufacturer prepares and uses the control standards versus how the sample is extracted, and on the specificity of the antibody. The control standards used by R-Biopharm kits are prepared by extracting the reference material ERM-DA-480 with 60% ethanol after removing the albumin/globulin fraction (Mendez et al., 2005; van Ecker et al., 2006). It is known that most of the HMW and LMW glutenins are insoluble in alcohol preparations (DuPont et al., 2005; Osborne, 1924). R5 Ab binds the repetitive unit QQPFP (Valdés et al., 2003), which is also present in the LMW glutenin. As a consequence, the cocktail extracts contain additional targets (not present, or present at very low levels, in the control standard and alcohol extracts), and this may account for the higher concentrations recovered by the cocktail. This can explain the previously reported

Table 4. Detection of wheat protein fractions by commercial ELISAs

| Wheat fraction (ppm) | Extraction solution (ppm gluten)[a] | | | | | |
| | Aqueous ethanol | | | Cocktail | | |
	R-Biopharm	Neogen	Tepnel	R-Biopharm	Neogen	Tepnel
Albumins/Globulins						
1,000	>80	>100	49.2	>80	>100	37.4
100	76.6	36.2	ND	53.5	42.6	ND
10	2.1	ND	ND	ND	ND	ND
1	ND	ND	ND	ND	ND	ND
Gliadins						
1,000	>80	>100	>200	>80	>100	121.1
100	>80	>100	23.1	>80	>100	15.9
10	32.1	45.2	ND	35.5	52.4	ND
1	ND	ND	ND	ND	ND	ND
Glutenins						
1,000	>80	>100	>200	>80	>100	>200
100	39.4	23.6	>200	34.2	28.2	>200
10	ND	ND	14.0	15.5	ND	10.3
1	ND	ND	ND	ND	ND	ND

[a]ND, nondetected or <10 ppm gluten.

observations of up to a threefold difference in gluten content between the two extracts of starch and unheated products (Garcia et al., 2005; Hasselberg et al., 2004) and can be prevented by preparing the standard material and the samples in a similar way. Establishing a reference material and making it available are critically important for validation by both manufacturers and kit users. It should help prevent misinterpretations of results obtained from kits of different manufacturers.

The Neogen kit, which also offers two extraction procedures, uses a different approach. Instead of using one single mAb, it uses different mAbs as capture and detector antibodies. The capture Ab is specific for rye prolamins, and the detector Ab is specific for wheat gliadins (M. Abouzied, personal communication). Neogen assays gave the same results with both extracts, but this may apply only to wheat gluten in unheated samples, which were evaluated in this study. It is unknown if the same observation can be extended to the analysis of rye and barley from unheated samples and processed foods containing any of the three cereals. Contrary to those of the R5 kit, the characterization and performance of the Neogen assay have not been made public, which makes it difficult to provide a critical evaluation of the kit.

The fact that a manufacturer offers two different options for sample preparation that yield divergent results may cause confusion and create uncertainty in the making of management decisions and potential liability issues. For example, it would be of particular concern if the values obtained with the two different

extraction buffers fell above and below the tolerance limit set by the regulatory agency.

Antibody Specificity to the Different Wheat Protein Fractions

According to the kit manufacturers, the antibodies in their ELISAs target several gluten, gliadin, or wheat proteins. It is important to recognize that targeting multiple proteins is a more complex approach than targeting a single protein, and it is even more complicated when the sample is a heat-processed food. Since ELISA does not differentiate the targeted proteins but only the overall signals, we cannot truly quantify, if the assay is missing targets in different extracts from different samples. In assay development and assay evaluation, it is very important to monitor the presence of the targets in different extracts from different food samples (processed and nonprocessed) and antibody specificity/cross-reactivity to ensure the suitability of the extraction solution and the suitability of the antibody to be utilized. Immunoblotting is a powerful tool to delineate this, since the proteins to be detected have been separated previously.

The peroxidase-conjugated antibodies (detector antibodies) from three commercial kits, Neogen, R-Biopharm, and Tepnel, were evaluated for specificity to wheat protein fractions in immunoblots. The three wheat protein fractions were obtained from the wheat variety *Triticum aestivum* L. "Butte86." The albumin/globulin fraction was prepared according to a procedure recently described (Hurkman and Tanaka, 2004), and both the gliadin and glutenin fractions were obtained by the method developed by DuPont et al. (2005). The purity of each fraction is estimated to be at least 90% for gliadin and 95% for the glutenin and albumin/globulin fractions. The three fractions were separated by SDS-polyacrylamide gel electrophoresis (PAGE) (Figure 1A). Figure 1B shows

analysis of the gel image and the estimated molecular weight of the proteins present in the each fraction.

As expected, all of the kits were able to detect wheat proteins. However, the antibodies showed different specificities, as observed with the diverse

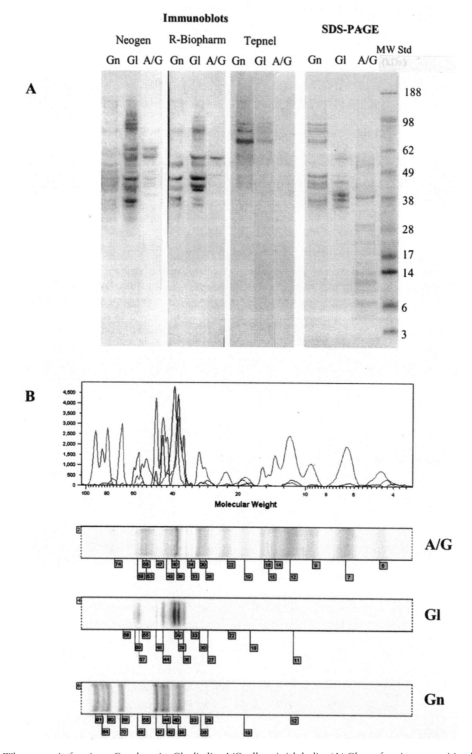

Figure 1. Wheat protein fractions. Gn, glutenin; Gl, gliadin; A/G, albumin/globulin. (A) Gluten fraction recognition by detector antibody from commercial kits. MW Std, molecular weight standards (in thousands). (B) Image analysis of gluten fractions separated by SDS-PAGE.

protein pattern recognition. Both Neogen and R-Biopharm antibodies are specific for the total gliadin fraction. In addition, they also bind some LMW glutenins, which have a molecular weight similar to those of α/β- and γ-gliadins (Figure 1B). Both antibodies also recognize some proteins in both the glutenin and albumin/globulin fractions, which are gliadin residues remaining in those fractions after purification. Those bands are not clearly visible with SDS-PAGE, but they are very distinct in the blots due to the high affinity of both antibodies to the gliadins.

Even though both Neogen and R-Biopharm mAbs recognize the gliadin fraction, the protein patterns are not exactly the same. We know very little about the Neogen mAb antibody used in the kit. However, the mAb used by R-Biopharm has been very well characterized. R5 was produced after immunizing mice with ethanolic rye extracts (Sorell et al., 1998). The greatest reactivity is directed to the toxic repetitive unit QQPFP present in all gliadins, but it also recognizes other homologous repeats, including QQTFP (Kahlenberg et al., 2006; Osman et al., 2001). Because the sequence QQPFP is more abundant in ω-gliadin than in α/β- and γ-gliadins, the antibody has been shown to be 14 times more reactive to ω-gliadin. Moreover, the two sequences were blasted in Expasy Blast (www.expasy.org/tools/blast/). The software retrieved matches to LMW glutenins, which may explain the binding to the glutenin fraction observed with the immunoblot and the higher results from the cocktail extracts from samples spiked with gluten compared to the aqueous alcohol extracts (Table 3). A couple of assays have been developed using a mAb PN3 raised against the toxic peptide LGQQQPFPPQQPYPQPQPF (Bermudo Redondo et al., 2005; Ellis et al., 1998). Similar to R5, PN3 recognizes the sequence QQQPFP, and it was not only reactive to the gliadin proteins but also to the LMW glutenins. However, it did not show reactivity to the HMW glutenins.

The Tepnel kit uses a mAb developed by Skerritt and Hill (1990) (Hill and Skerrit, 1989, 1990). The immunoblot shows how the antibody distinctively binds HMW glutenins as well as ω-gliadin with a lower degree of reactivity. Because of the strong binding to HMW glutenins, it is evident that traces of HMW glutenins are present in the gliadin fraction. While Hill and Skerrit (1990) reported that the antibody is specific to ω-gliadin, results of our cross-reactivity study between the different wheat protein fractions clearly showed that cross-reactivity for HMW glutenins was significantly higher than those for ω-gliadin. The Skerritt and Hill (1991) assay underwent a validation trial to obtain the official first action by AOAC. The method became AOAC Official Method 991.19 in 1995 to detect levels of gluten

>160 ppm (80 ppm gliadin). This level is below the values permitted by the Codex Alimentarius standard for the use of the gluten-free claim. However, the Codex Alimentarius is in the process of reviewing the definition of gluten free and has released a draft for the new standard for the use of a gluten-free statement, with lower levels of gluten allowed. In the United States, the proposed gluten-free claim is 20 ppm gluten or lower. To be applicable to the new standard, Tepnel modified their protocol by increasing incubation times to achieve lower sensitivity. Even though the components of the kit are the same for the high- and low-sensitivity protocols, the "Official Method" claim can apply only to the original, less sensitive procedure.

In the immunoblots, color intensity of bands due to antibody binding can be misleading if it is not interpreted correctly. The color intensity shown in membranes incubated with Neogen antibodies is stronger than those incubated with R-Biopharm and Tepnel Abs. In this study, color intensity is not a measure of the concentrations of analyte, which are the same for all three cases, but a combination effect of the analyte concentration and the titer of the antibody conjugate. Thus, color intensity of the blot cannot be used to predict concentration of the analyte in the ELISA. ELISA kits have other components, like coating antibody, dilution factors, and a calibrated standard curve, that contribute to quantification.

While immunoblot proteins are detected by a single antibody, the signal in a direct sandwich ELISA requires that the protein be bound by two antibodies (capture and detector). In order to determine the protein specificity of the Ab, three wheat protein fractions were analyzed by ELISA. We determined the lowest dilution (sensitivity) of each protein fraction being detected by the assays, and we used this value to evaluate the relative sensitivity and specificity of the Ab. Results of the ELISA fully correlate with the binding/specificity observed with the immunoblots. The lowest gliadin concentration detected was given by Neogen and R-Biopharm, whose values were lower than those from the other fractions. This confirms the specificities of the antibodies to gliadin. The Tepnel mAb appeared to be more specific to the glutenin fraction than to gliadins. As shown in the immunoblots, that signal was mostly due to glutenin residue in the gliadin fraction. The three mAbs showed binding in the higher concentrations of the albumin/globulin fraction by the ELISA, in which the signal is probably due to gliadin/glutenin traces in the albumin/globulin fraction. There are no significant differences between the proteins solubilized in cocktail and in 60% ethanol solutions.

Processed Foods: Protein Recovery and Antibody Binding to Wheat Proteins in Bread

Wheat flour is rarely consumed without being heat processed, and it is known that wheat proteins undergo physical-chemical modifications, depending on the extent of the thermal treatment. Such modification may have a detrimental impact caused by aggregation of proteins and, consequently, reduced protein recovery by the sample extraction buffer. To visualize the impact of baking on the detection of wheat proteins, we used all-purpose flour bread as a food model. The crust and the bread crumb were analyzed separately for comparison purposes. Four sample extraction buffers and procedures were followed: cocktail, cocktail followed by aqueous alcohol, aqueous alcohol, and extraction with Morinaga buffer. Protein profiles were obtained by separation by SDS-PAGE, and antibody binding was evaluated by immunoblotting (Figure 2). The SDS-PAGE protein profiles from different extracts and their relative color intensities revealed that both aqueous alcohol and Morinaga buffer are not as efficient as the cocktail solution offered by R-Biopharm and Neogen (Figure 2). In a similar study, only 17% of gliadin spiked in flour was recovered from bread crust by 40% ethanol (Ellis et al., 1998). The more intense bands in Morinaga extracts correspond to proteins present in the extraction solution supplied by the kit manufacturer.

It has been reported previously that wheat proteins aggregate and cross-link with sugars in a Maillard reaction when subjected to heat treatments (Hansen et al., 1975; Hansen and Millington, 1979).

Some of the covalent modifications negatively affect the solubility of wheat proteins. It is thought that S-rich gliadins and glutenins are covalently linked, as a result of heating, which might involve the formation of disulfide bonds between the proteins of the two fractions (Schofield et al., 1983; Wieser, 1998). The extent of the effects of treatment on protein solubility depends on the treatment conditions and the food system (Cuq et al., 2000; Li and Lee, 1997, 1998; Rumbo et al., 2001; Weegels et al., 1994). Solubility may be improved to some degree using disaggregating additives, detergents, and/or reducing agents. This may explain why cocktail extraction is more efficient than alcohol in extracting insoluble components from heated foods, particularly the S-rich gliadins α-/β- and γ-gliadins (Wieser, 1998). However, the combination of SDS and mercaptoethanol, components of Morinaga buffer, does not work as well as the cocktail. Both cocktail and Morinaga buffers use mercaptoethanol as reducing agent, but they differ in the other component. SDS (Morinaga) and guanidine HCl (cocktail) are denaturing agents involved in the disruption of hydrophobic interactions. However, guanidine HCl seems to be more efficient than SDS in resolubilizing gluten proteins after heating. Ellis et al. (1998) used a reducing buffer containing 1% β-mercaptoethanol and 2 M urea for extracting gluten proteins from bread crumb and crust. Samples and standards were diluted 100-fold in PBS/Tween prior to analysis by ELISA. However, the use of this buffer resulted in dropping signals of the control standards possibly due to the denaturing effect of the buffer on the capture antibodies.

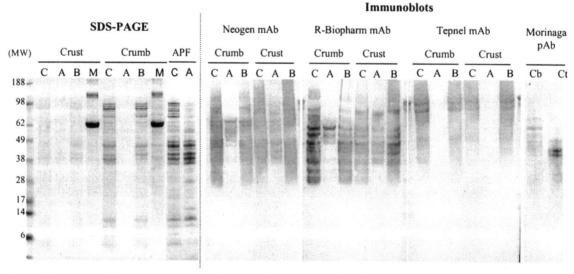

Figure 2. Detection of gluten proteins in bread crumb and crust extracted with cocktail (C), 60% ethanol (A), cocktail and aqueous ethanol (B), and Morinaga (M) buffers by detector antibody from commercial ELISAs. Protein profiles from SDS-PAGE of bread samples are compared to those of wheat flour (APF). MW, molecular weight (in thousands).

Bermudo Redondo et al. (2005) used 0.1% dithio-threitol as a reducing agent in their extraction buffer, and they also reported a drop of sensitivity, but the limit of detection was not affected in 60% ethanol.

Those assays that rely only on the aqueous ethanol extraction will be limited primarily to the detection of ω-gliadins in processed samples. This gliadin, which is a minor component of the wheat prolamin fraction, is the gliadin less affected by heat treatments (Hill and Skerrit, 1990), probably because most of the ω-gliadins lack cysteines and therefore do not cross-link to the other gluten proteins at high temperatures.

Even though the cocktail buffer is more efficient than Morinaga buffer and aqueous ethanol in extracting proteins from heated samples, its efficiency can be limited in highly heated samples, like bread crust (SDS-PAGE in Figure 2). By using cocktail buffer, the protein profile from the crumb does not differ much from the profile of wheat flour. However, much less protein was observed in the bread crust samples. Because the crust is exposed to higher temperatures than the bread crumb, the impact of heat on protein solubility differs considerably depending on the location of the sampling. Simonato et al. (2001) reported that the impact of temperature is different in the bread crumb (that reaches temperatures lower than 100°C), with respect to bread crust for which temperatures can be higher than 200°C. The treatment with SDS and reducing agent can reverse protein aggregation in bread crumb (Simonato et al., 2001). However, this combination is not enough to resolubilize proteins from the bread crust, which may be the result of stronger cross-links, like those resulting from the Maillard reaction (Ames, 1992). Therefore, a limited protein recovery is expected from samples exposed to high temperature, regardless of the sample extraction buffer used. The selection of the extraction buffer has to be optimized for the selected target proteins to provide the best recovery in cooked food samples. Previously, Morinaga successfully showed the buffer used in their egg kit was very efficient in extracting egg proteins from processed samples (Watanabe et al., 2005); hence, they include the same buffer for their wheat protein kit. Our study clearly demonstrates one size does not fit all, and each application should be validated separately.

The signal provided by antibody depends on the concentration of target present in the sample (Figure 2). This is why the signals are stronger in cocktail extracts than in aqueous ethanol extracts in bread crumb and crust. The Neogen and R-Biopharm mAbs bind only to ω-gliadins in aqueous ethanol extracts. The Tepnel mAb provides a very strong binding to HMW

glutenins from cocktail extracts, which is not the buffer provided by the kit. It is possible that large protein aggregates were present in the extracts from bread crust that could not enter the gel; therefore, they could not be visualized either in the gels or in the immunoblots. Moreover, the presence of aggregates in the sample may also affect the antibody binding since many of the target epitopes may be inaccessible in the interior of such protein aggregates.

Cocktail sample extraction solution is the most efficient buffer tested in this study. Consistent with our finding, Garcia et al. (2005) found that gliadin recoveries in spiked maize bread ranged from 89 to 105% for cocktail extractions, which were double those recovered in aqueous ethanol (31 to 60%).

Sample Extraction Procedures and Antibody Cross-Reactivity to Wheat, Barley, and Rye

The ability of each buffer to extract gluten proteins from wheat, barley, and rye was examined. Protein profiles were analyzed by SDS-PAGE. In the case of the procedure using cocktail followed by aqueous ethanol extraction, we analyzed only the cocktail extract to see its contribution to the final protein solubilization, and we contrasted this information with the protein profile of the aqueous ethanol extract. The samples were prepared according to the manufacturer's ELISA procedures. All the samples were extracted using the same proportion of sample to buffer, and the same volume of each extract was loaded in the gel. It is apparent that HMW glutelins are more abundant in the cocktail extracts of wheat, barley, and rye and that they are also present in low concentrations in the aqueous ethanol extract. It has been reported that a small proportion of HMW and LMW glutenins are soluble in alcohol solutions but only as monomers, dimers, or small polymers (DuPont et al., 2005). The two cocktail and aqueous ethanol extracts from the same cereal (wheat, rye, or barley) showed a similar protein profile of gliadins, secalins, and hordeins, respectively, in SDS-PAGE.

The immunoblots reveal that the conjugated antibodies from the four assays bind gluten proteins from wheat, barley, and rye differently (Figure 3). The binding confirms similar prolamin contents in both the cocktail and ethanol extracts of wheat, barley, and rye and the higher abundances of HMW glutenins in the cocktail extracts. It is expected that those antibodies binding HMW glutenins provide a higher signal for the cocktail extract in ELISA than for its alcoholic counterpart. The two antibodies binding HMW glutenins are those from the Morinaga and Tepnel assays. However, Tepnel, Neogen, and R-Biopharm antibodies bind very strongly not only to the prolamins but also to HMW glutelins of rye, but

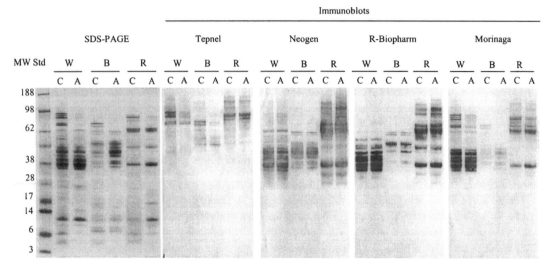

Figure 3. SDS-PAGE of wheat (W), barley (B), and rye (R) extracted with cocktail (C) and 60% ethanol (A). Specificity and cross-reactivity of detector antibodies from commercial ELISA evaluated by immunoblot. MW Std, molecular weight standards (in thousands).

not so to those of barley. The profiles shown by Neogen and R-Biopharm Abs for the two extracts of the wheat and rye are almost identical, but again, this is not the case for barley. Garcia et al. (2005) reported no difference between the two extracts when they analyzed wheat flour, rye-contaminated oats, barley-spiked corn flour, and barley-contaminated oats in ELISA with R5 mAb.

Is the Gliadin Control Standard in ELISA Adequate To Measure Barley and Rye Gluten Content?

It is a general belief that commercial or published ELISAs that are developed for wheat gluten are also suitable for the detection and quantification of gluten from rye and barley. Wheat gluten or gliadin is traditionally the preferred reference material used by the kits. While antibodies used in different assays are able to detect gluten from cereals involved in the pathogenesis of celiac disease, some assumptions have been made when the same reference material is used interchangeably for the quantification of gluten from wheat, rye, and barley. These assumptions, which follow, are not necessarily valid:

1. The gluten/prolamin fractions are similar in the three cereals; therefore, gliadin can be used as a control standard for quantification of gluten, regardless of the source.

2. The epitopes or amino acid sequences targeted by antibodies are present in an equivalent number in wheat, rye, and barley.

3. The stability levels of gluten from the three cereals are the same.

4. Buffers used for extraction are equally efficient for the extraction of proteins from the three grains.

5. The ratio of prolamin/glutenin is 1:1 for all cereals; hence, a factor of 2 can be used to convert ppm gliadin to ppm gluten.

Gliadin or wheat gluten is used as a control standard in all the immunoassays evaluated. Gluten fractions and gluten proteins from wheat have been better characterized than their rye and barley counterparts. It has been reported that the main target of the R5 mAb is the repetitive amino acid sequence QQPFP from gliadin, which is also present in rye and barley. However, the distribution and number of this repetitive unit in gluten from rye and barley cereals are not so well known. Valdés et al. (2003) estimated that the QQPFP motif is present around 14, 2, and 11 times in ω-, α-, and γ-gliadin, respectively. If the repetitive unit is more (or less) frequent in different prolamin fractions from rye and barley than it is in those from wheat, the signal obtained in the kit for these cereals (at a known concentration of gluten) will be higher (or lower). Consequently, differences in the frequencies of the repetitive motifs may not interfere with the qualitative detection of gluten, but they will definitely impact the quantitative result of the assay.

It has been reported that barley may contain about twice as much ω-type hordein than ω-gliadin (Shewry and Tatham, 1999). Valdés et al. (2003)

reported that the R5-based ELISA is equally efficient in the detection of wheat gliadins, rye secalins, and barley hordeins. However, they did not calculate the gluten content in rye and barley by a method other than the ELISA. The method, which used gliadin as a control standard, will provide a value that may or may not be close to the real concentration of secalins or hordeins. In order to evaluate the suitability of the assay for the quantification of rye and barley, secalin and hordein control proteins should be used. In addition, the slopes of the curves for wheat, rye, and barley appeared different (Valdés et al., 2003). This means that the recognition of gluten protein by R5 is different, depending on the grain being analyzed. This observation is also supported by others (Kanerva et al., 2006) who reported that R5 overestimated barley content. However, the results were "corrected" by the use of hordein as a reference material.

The ratio of 1:1, based on the equal proportion between gliadin and glutenin fractions, is used to report the results as ppm gluten instead of ppm gliadin. This ratio may change due to several factors:

- *Cereal type and cultivar.* The ratio of 1:1 is based on wheat. However, it is not well known how constant this ratio is among the commercial cultivars of wheat, rye, and barley. Table 5 shows the variability of prolamin and glutelin content in the three grains. The wheat prolamin, ω-gliadin, is the most variable component in wheat, and it may range between 6 and 20% of the total gliadin fraction, depending on the wheat variety (Wieser et al., 1994). Therefore, using methods based on the detection of gliadin may create a large uncertainty factor for rye and barley.
- *Extraction buffer and specificity of antibody.* In the case of wheat, most of the protein extracted by the aqueous ethanol solution is prolamin. However, the cocktail also extracts glutenins, and we have shown that R5 also binds LMW glutenins. Therefore, the ratio of 1:1 does not really apply to cocktail extraction

since the ratio of 1:1 corresponds to gliadin/glutenin, and in the case of cocktail extraction the ratio is actually gliadin + LMW glutenin/HMW glutenin. On the other hand, aqueous ethanol extracts most of the gluten proteins from rye and barley (Kanerva et al., 2006).
- *Food processing.* While there is abundant information regarding the technological properties of gluten, information regarding the impact of food processing on detection of the different cereals is very limited. If the prolamin fraction is not completely extracted from heated samples, as happens when aqueous ethanol is used, the application of a factor of 2 for gluten conversion may underestimate the real content of gluten in the sample. As previously discussed, ω-gliadin is the gliadin most resistant to heat. Barley C-hordein, the ω-gliadin counterpart, has also been shown to be more resistant to heat than B-hordein (Nakai et al., 2002). However, their solubilities are improved by the addition of reducing agent to the extraction buffer.
- *Cereal fraction used as ingredient.* In ingredient preparation, the content of the different gluten protein fractions may be modified. For example, depending on how starches are prepared, it is possible that γ-gliadin remains more abundant than α- and ω-gliadins (Ellis et al., 1998). Consequently, the quantification of this ingredient may also be compromised.

R5 and Skerritt mAbs can detect partially hydrolyzed gliadins if the epitopes are intact (Skerrit and Hill, 1990; Valdés et al., 2003). The peptide resulting from hydrolysis needs to be large enough and contain at least two epitopes to allow the binding of the capture and detector Abs. Smaller peptides may be bound only by the capture antibody, but because of steric hindrance there is no room for the detector antibody to reach its target (Bermudo Redondo et al., 2005). Therefore, results should be interpreted with caution.

CONCLUDING REMARKS

We have provided a detailed discussion on the characteristics of R5 (R-Biopharm) and Skerritt (Tepnel) mAbs based on our findings and published information. We are optimistic that more information on commercial kits is starting to emerge in peer-reviewed literature. This allows the analytical community to comment on it, which may offer an opportunity for improvement in test kits to yield more accurate and

Table 5. Protein composition of cereals[b]

| Cereal | Crude protein[a] | Composition (%): | | | |
		Prolamin	Glutelin	Albumin	Globulin
Wheat	10.6	33–45	40–46	9–15	6–7
Rye	8.7	21–42	25–40	10–44	10–19
Barley	11.0	25–52	52–55	12–14	8–12

[a]Dry matter.
[b]Data derived from Alais and Linden (1991) and Eliasson and Larsson (1993).

reproducible results. Such detection systems will provide appropriate tools to accurately determine risk and exposure for risk assessment and risk management. This will ultimately benefit consumers.

The detection of gluten proteins is quite difficult due to the complexity of the targets and specificities of the Ab. Moreover, the detection of these proteins in processed foods makes the issue even more complicated. The kits are really developed to quantify wheat gluten since the signal is compared to the signal from control standard gliadin. However, the accuracy of the assays for the quantification of barley and rye remains questionable. It is even more complicated if the other commingling grains are inadvertently present in variable amounts in an agricultural commodity, e.g., the potential presence of wheat, barley, and/or rye in oat.

We have identified four major areas for improvement:

- Improvement of knowledge of characteristics of the targets not only in raw materials but also in processed foods.
- Standardization of reference material and reporting units to make results from the different kits comparable.
- Further evaluation of the suitability of a single assay for the quantitative detection of wheat, rye, and barley. Alternative grain-specific assays should be considered. The benefits of having a specific assay for the three grains would allow for a more accurate determination of the source of cereal grain contamination, which is usually not obvious. Such specific kits are needed to comply with regulations that require labeling of the source of gluten.
- Fulfillment of the need for commercial confirmatory methods, like PCR, to verify the presumptive positives and source of gluten in the sample.

From our data and others', we noticed that there are some inaccurate declarations in the kit labels and package inserts that can cause confusion regarding their intended applications. These inaccuracies are related to Ab specificity. The "gliadin," "ω –gliadin," and "wheat protein" claims refer only to wheat. Since the kits also react to rye and barley, it would be more suitable to use the terms "prolamin" or "gluten." Moreover, by using the term "wheat" or "wheat proteins" in the label, the kit could be misused for the detection of wheat allergens, unless the kit inserts contain an "unintended use" statement.

While the current gluten test is not perfect, excellent progress has been made since Skerritt first reported the gluten test. The issues identified and solutions offered for the gluten test here also apply to the detection of food allergen. It is hoped that this review serves as a model for critically reviewing other food allergen detection systems. Commercial test kits provide a very valuable tool for ensuring safe foods, to the extent possible, for celiac consumers. Regulatory agencies, kit manufacturers, the food industry, and academics are dedicated to protecting people with celiac disease and to ensuring that food labels are truthful and unambiguous.

REFERENCES

Alais, C., and G. Linden (ed.). 1991. *Food Biochemistry*. Ellis Horwood Ltd., New York, NY.

Ames, J. M. 1992. The Maillard reaction, p. 99–153. *In* B. J. F. Hudson (ed.), *Biochemistry of Food Proteins*. Blakie Academic & Professional, London, United Kingdom.

Aubrecht, E., and A. Toth. 1995. Investigation of gliadin content of wheat flour by ELISA method. *Acta Aliment.* 24:23–29.

Bermudo Redondo, M. C., P. B. Griffin, M. Garzon Ransanz, H. J. Ellis, P. J. Ciclitira, and C. K. O'Sullivan. 2005. Monoclonal antibody-based competitive assay for the sensitive detection of coeliac disease toxic prolamins. *Anal. Chim. Acta* 551:105–114.

Burks, W., and B. K. Ballmer-Weber. 2006. Food allergy. *Mol. Nutr. Food Res.* 50:595–603.

Camafeita, E., and E. Mendez. 1998. Screening of gluten avenins in foods by matrix-assisted laser desorption/ionization time-of-flight mass spectrometry. *J. Mass Spectrom.* 33:1023–1028.

Camafeita, E., J. Solis, P. Alfonso, J. A. Lopez, L. Sorell, and E. Mendez. 1998. Selective identification by matrix-assisted laser desorption/ionization time-of-flight mass spectrometry of different types of gluten in foods made with cereal mixtures. *J. Chromatogr. A* 823:299–306.

Catassi, C., E. Fabiani, G. Iacono, C. D'Agate, R. Francavilla, F. Biagi, U. Volta, S. Accomando, A. Picarelli, I. De Vitis, G. Pianelli, R. Gesuita, F. Carle, A. Mandolesi, I. Bearzi, and A. Fasano. 2007. A prospective, double-blind, placebo-controlled trial to establish a safe gluten threshold for patients with celiac disease. *Am. J. Clin. Nutr.* 85:160–166.

Chirdo, F. G., C. A. Fossati, and M. C. Añón. 1995. Optimization of a competitive ELISA with polyclonal antibodies for quantitation of prolamins in foods. *Food Agric. Immunol.* 7:333–343.

Cranney, A., M. Zarkadas, I. D. Graham, J. D. Butzner, M. Rashid, R. Warren, M. Molloy, S. Case, V. Burrows, and C. Switzer. 2007. The Canadian celiac health survey. *Dig. Dis. Sci.* 52:1087–1095.

Cuq, B., F. Bourtot, A. Redl, and V. Lullien-Pellerin. 2000. Study of the temperature effect on the formation of wheat gluten network: influence on mechanical properties and protein solubility. *J. Agric. Food Chem.* 10:2954–2959.

Dahinden, I., M. von Büren, and J. Lüthy. 2001. A quantitative competitive PCR system to detect contamination of wheat, barley or rye in gluten-free food for coeliac patients. *Eur. Food Res. Technol.* 212:228–233.

Day, L., M. A. Augustin, I. L. Batey, and C. W. Wrigley. 2006. Wheat-gluten uses and industry needs. *Trends Food Sci. Technol.* **17:**82–90.

DuPont, F. M., R. Chan, R. Lopez, and W. H. Vensel. 2005. Sequential extraction and quantitative recovery of gliadins, glutenins, and other proteins from small samples of wheat flour. *J. Agric. Food Chem.* **53:**1575–1584.

Eliasson, A. C., and K. Larsson (ed.). 1993. *Cereals in Breadmaking: a Molecular Colloidal Approach.* Marcel Dekker, Inc., New York, NY.

Ellis, H. J., S. Rosen-Bronson, N. O'Reilly, and P. J. Ciclitira. 1998. Measurement of gluten using a monoclonal antibody to a coeliac toxic peptide of A-gliadin. *Gut* **43:**190–195.

Fasano, A., I. Berti, T. Gerarduzzi, T. Not, R. B. Colletti, S. Drago, Y. Elitsur, P. H. Green, S. Guandalini, I. D. Hill, M. Pietzak, A. Ventura, M. Thorpe, D. Kryszak, F. Fornaroli, S. S. Wasserman, J. A. Murray, and K. Horvath. 2003. Prevalence of celiac disease in at-risk and not-at-risk groups in the United States: a large multicenter study. *Arch. Intern. Med.* **163:**286–292.

Gabrovska, D., J. Rysova, V. Filova, J. Plicka, P. Cuhra, M. Kubik, and S. Barsova. 2006. Gluten determination by gliadin enzyme-linked immunosorbent assay kit: interlaboratory study. *J. AOAC Int.* **89:**154–160.

Garcia, E., M. Llorente, A. Hernando, R. Kieffer, H. Wieser, and E. Mendez. 2005. Development of a general procedure for complete extraction of gliadins for heat processed and unheated foods. *Eur. J. Gastroenterol. Hepat.* **17:**529–539.

Hansen, L. P., P. H. Johnston, and R. E. Ferrel. 1975. Heat-moisture effects on wheat flour. I. Physical and chemical changes of flour proteins resulting from thermal processing. *Cereal Chem.* **52:**459–472.

Hansen, L. P., and R. J. Millington. 1979. Blockage of protein enzymatic digestion (carboxypeptidase-B) by heat-induced sugar-lysine reactions. *J. Food Sci.* **44:**1173–1177.

Hasselberg, A., B. Kruse, and I. M. Yman. 2004. Quantification of gluten in food samples—comparison of two assays, p. 109–118. *In* M. Stern (ed.), *Proceedings of the 18th Meeting of Working Group on Prolamin Analysis and Toxicity.* Verlag Wissenschaftliche Scripten, Zwichau, Germany.

Hefle, S. L., J. A. Nordlee, and S. L. Taylor. 1996. Allergenic foods. *Crit. Rev. Food Sci. Nutr.* **36**(Suppl.):S69–S89.

Hill, A., and J. H. Skerrit. 1990. Determination of gluten in foods using a monoclonal antibody-based competition enzyme immunoassay. *Food Agric. Immunol.* **2:**21–35.

Hill, A. S., and J. H. Skerrit. 1989. Monoclonal antibody-based two-site enzyme-immunoassays for wheat gluten proteins. 1. Kinetic characteristics and comparison with other ELISA formats. *Food Agric. Immunol.* **1:**147–160.

Hischenhuber, C., R. Crevel, B. Jarry, M. Maki, D. A. Moneret-Vautrin, A. Romano, R. Troncone, and R. Ward. 2006. Safe amounts of gluten for patients with wheat allergy or coeliac disease. *Aliment. Pharmacol. Ther.* **23:**559–575.

Hurkman, W. J., and C. K. Tanaka. 2004. Improved methods for separation of wheat endosperm proteins and analysis by two-dimensional gel electrophoresis. *J. Cereal Sci.* **40:**295–299.

Kahlenberg, F., D. Sanchez, I. Lachmann, L. Tuckova, H. Tlaskalova, E. Mendez, and T. Mothes. 2006. Monoclonal antibody R5 for detection of putatively coeliac-toxic gliadin peptides. *Eur. Food Res. Technol.* **222:**78–82.

Kanerva, P. M., T. S. Sontag-Strohm, P. H. Ryöppy, P. Alho-Lehto, and H. O. Salovaara. 2006. Analysis of barley contamination in oats using R5 and omega-gliadin antibodies. *J. Cereal Sci.* **44:**347–352.

Keet, C. A., and R. A. Wood. 2007. Food allergy and anaphylaxis. *Immunol. Allergy Clin. North Am.* **27:**193–212.

Li, M., and T. C. Lee. 1998. Effect of cysteine in the molecular weight distribution and disulfide cross-link of wheat flour proteins in extrudates. *J. Agric. Food Chem.* **46:**846–853.

Li, M., and T. C. Lee. 1997. Relationship of the extrusion temperature and the solubility and disulfide bond distribution of wheat proteins. *J. Agric. Food Chem.* **45:**2711–2717.

Lynn, A., R. D. M. Prentice, M. P. Cochrane, A. M. Cooper, F. Dale, C. M. Duffus, R. P. Ellis, I. M. Morrison, L. Paterson, J. S. Swanston, and S. A. Tiller. 1997. Cereal starches. Properties in relation to industrial uses, p. 69–106. *In* G. M. Campbell, C. Webb, and S. L. McKee (ed.), *Cereals: Novel Uses and Processes.* Plenum Press, New York, NY.

Mamone, G., P. Ferranti, L. Chianese, L. Scafuri, and F. Addeo. 2000. Qualitative and quantitative analysis of wheat gluten proteins by liquid chromatography and electrospray mass spectrometry. *Rapid Commun. Mass Spectrom.* **14:**897–904.

Marsh, M. N. 1992. Gluten, major histocompatibility complex, and the small intestine. A molecular and immunobiologic approach to the spectrum of gluten sensitivity ('celiac sprue'). *Gastroenterology* **102:**330–354.

McGough, N., and J. H. Cummings. 2005. Coeliac disease: a diverse clinical syndrome caused by intolerance of wheat, barley and rye. *Proc. Nutr. Soc.* **64:**434–450.

Mendez, E., C. Vela, U. Immer, and F. W. Janssen. 2005. Report of a collaborative trial to investigate the performance of the R5 enzyme linked immunoassay to determine gliadin in gluten-free food. *Eur. J. Gastroenterol. Hepatol.* **17:**1053–1063.

Nakai, R., H. Ashida, and G. Danno. 2002. Effect of different heating conditions on the extractability of barley hordeins. *J. Nutr. Sci. Vitaminol.* (Tokyo) **48:**149–154.

Nicolas, Y., S. Denery-Papini, J. P. Martinant, and Y. Popineau. 2000. Suitability of a competitive ELISA using anti-peptide antibodies for determination of the gliadin content of wheat flour: comparison with biochemical methods. *Food Agric. Immunol.* **12:**53–65.

Not, T., K. Horvath, I. D. Hill, J. Partanen, A. Hammed, G. Magazzu, and A. Fasano. 1998. Celiac disease risk in the USA: high prevalence of antiendomysium antibodies in healthy blood donors. *Scand. J. Gastroenterol.* **33:**494–498.

Osborne, T. B. 1924. *The Vegetable Proteins,* 2nd ed. Longmans and Co., London, United Kingdom.

Osman, A. A., H. H. Uhlig, I. Valdes, M. Amin, E. Mendez, and T. Mothes. 2001. A monoclonal antibody that recognizes a potential coeliac-toxic repetitive pentapeptide epitope in gliadins. *Eur. J. Gastroenterol. Hepatol.* **13:**1189–1193.

Periolo, N., and A. C. Chernavsky. 2006. Coeliac disease. *Autoimmun. Rev.* **5:**202–208.

Rumbo, M., F. G. Chirdo, C. A. Fossati, and M. C. Anon. 2001. Analysis of the effects of heat treatment on gliadin immunochemical quantification using a panel of anti-prolamin antibodies. *J. Agric. Food Chem.* **49:**5719–5726.

Salplachta, J., M. Marchetti, J. Chmelik, and G. Allmaier. 2005. A new approach in proteomics of wheat gluten: combining chymotrypsin cleavage and matrix-assisted laser desorption/ionization quadrupole ion trap reflectron tandem mass spectrometry. *Rapid Commun. Mass Spectrom.* **19:**2725–2728.

Sampson, H. A. 2003. Anaphylaxis and emergency treatment. *Pediatrics* **111:**1601–1608.

Sampson, H. A., A. Munoz-Furlong, R. L. Campbell, N. F. Adkinson, Jr., S. A. Bock, A. Branum, S. G. Brown, C. A. Camargo, Jr., R. Cydulka, S. J. Galli, J. Gidudu, R. S. Gruchalla, A. D. Harlor, Jr., D. L. Hepner, L. M. Lewis, P. L. Lieberman, D. D. Metcalfe, R. O'Connor, A. Muraro, A. Rudman, C. Schmitt, D. Scherrer, F. E. Simons, S. Thomas, J. P. Wood, and W. W. Decker. 2006. Second symposium on the definition and management of anaphylaxis: summary report—Second National

Institute of Allergy and Infectious Disease/Food Allergy and Anaphylaxis Network Symposium. *J. Allergy Clin. Immunol.* **117:**391–397.

Sanchez, D., L. Tuckova, M. Burkhard, J. Plicka, T. Mothes, I. Hoffmanova, and H. Tlaskalova-Hogenova. 2007. Specificity analysis of anti-gliadin mouse monoclonal antibodies used for detection of gliadin in food for gluten-free diet. *J. Agric. Food Chem.* **55:**2627–2632.

Sandberg, M., L. Lundberg, M. Ferm, and I. M. Yman. 2003. Real time PCR for the detection and discrimination of cereal contamination in gluten free foods. *Eur. Food Res. Technol.* **217:**344–349.

Schofield, J. D., R. C. Bottomley, M. F. Timms, and M. R. Booth. 1983. The effect of heat on wheat gluten and the involvement of sulphydryl-disulphide interchange reactions. *J. Cereal Sci.* **1:**241–253.

Shewry, P. R., and A. S. Tatham. 1999. The characteristics, structures and evolutionary relationships of prolamins, p. 11–33. *In* P. R. Shewry and R. Casey (ed.), *Seed Proteins.* Kluwer Academic Publishers, Dordrecht, The Netherlands.

Shewry, P. R., and A. S. Tatham. 1990. The prolamin storage proteins of cereal seeds: structure and evolution. *Biochem. J.* **267:**1–12.

Sicherer, S. H., and H. A. Sampson. 2006. 9. Food allergy. *J. Allergy Clin. Immunol.* **117:**S470–S475.

Simonato, B., G. Pasini, M. Giannattasio, A. D. Peruffo, F. De Lazzari, and A. Curioni. 2001. Food allergy to wheat products: the effect of bread baking and in vitro digestion on wheat allergenic proteins. A study with bread dough, crumb, and crust. *J. Agric. Food Chem.* **49:**5668–5673.

Skerrit, J. H., and A. Hill. 1990. Monoclonal antibody sandwich enzyme immunoassay for determination of gluten in foods. *J. Agric. Food Chem.* **38:**1771–1778.

Skerrit, J. H., and A. S. Hill. 1991. Enzyme immunoassay for determination of gluten in foods: collaborative study. *J. AOAC Int.* **74:**257–264.

Sorell, L., J. A. López, I. Valdés, P. Alfonso, E. Camafeita, B. Acevedo, F. Chirdo, J. Gavilondo, and E. Méndez. 1998. An innovative sandwich ELISA system based on an antibody cocktail for gluten analysis. *FEBS Lett.* **439:**46–50.

Valdés, I., E. Garcia, M. Llorente, and E. Mendez. 2003. Innovative approach to low-level gluten determination in foods using a novel sandwich enzyme-linked immunosorbent assay protocol. *Eur. J. Gastroenterol. Hepatol.* **15:**465–474.

van Ecker, R., E. Berghofer, P. J. Ciclitira, F. Chirdo, S. Denery-Papini, H. J. Ellis, P. Ferranti, P. R. Goodwin, U. Immer, G. Mamone, E. Mendez, T. Mothes, S. Novalin, A. Osman, M. Rumbo, M. Stern, L. Thorrell, A. Whim, and H. Wieser. 2006. Towards a new gliadin reference material-isolation and characterisation. *J. Cereal Sci.* **43:**331–341.

Watanabe, Y., K. Aburatani, T. Mizumura, M. Sakai, S. Muraoka, S. Mamegosi, and T. Honjoh. 2005. Novel ELISA for the detection of raw and processed egg using extraction buffer containing a surfactant and a reducing agent. *J. Immunol. Methods* **300:**115–123.

Weegels, P. L., A. M. G. de Groot, J. A. Verhoek, and R. J. Hamer. 1994. Effects on gluten heating at different moisture contents. II. Changes in physicochemical properties and secondary structure. *J. Cereal Sci.* **19:**39–47.

Westphal, C. D. 2006. Approaches to the detection of food allergens, from a food science perspective, p. 189–218. *In* S. J. Maleki, A. W. Burks, and R. M. Helm (ed.), *Food Allergy.* ASM Press, Washington, DC.

Wieser, H. 1998. Investigations on the extractability of gluten proteins from wheat bread comparison with flour. *Z. Lebensm. Unters. Forsch.* **207:**128–132.

Wieser, H., W. Seilmeier, and H. D. Belitz. 1994. Quantitative determination of gliadin subgroups from different wheat varieties. *J. Cereal Sci.* **19:**149–155.

Yeung, J. M. 2006. Enzyme-linked immunosorbent assays (ELISAs) for detecting allergens in food, p. 109–124. *In* S. J. Koppelman and S. L. Hefle (ed.), *Detecting Allergens in Food.* CRC Press, Boca Raton, FL.

Zarkadas, M., A. Cranney, S. Case, M. Molloy, C. Switzer, I. D. Graham, J. D. Butzner, M. Rashid, R. E. Warren, and V. Burrows. 2006. The impact of a gluten-free diet on adults with coeliac disease: results of a national survey. *J. Hum. Nutr. Diet.* **19:**41–49.

Pathogens and Toxins in Foods: Challenges and Interventions
Edited by V. K. Juneja and J. N. Sofos
© 2010 ASM Press, Washington, DC

Chapter 19

Naturally Occurring Toxins in Plants

ANDREA R. OTTESEN AND BERNADENE A. MAGNUSON

This chapter discusses naturally occurring toxins that have been associated with the consumption of foods and that are not included in other chapters in this volume. There is an exceedingly large number of additional compounds that could have been included; however, a detailed discussion of a few compounds should be of greater value to the reader than a long list of compounds with little information on each. The compounds selected for inclusion (Table 1) are those that are either common or of particular interest to the authors.

ACKEE

Characteristics and Occurrence

The toxicity associated with ackee fruit *(Blighia sapida)*, the national fruit of Jamaica, is due to the extremely high amounts of hypoglycin A that are present in the unripe arils (the extra seed covering)

Table 1. Selected compounds

Compound(s)	Topics covered
Ackee	Jamaican vomiting sickness caused by ackee *(Blighia sapida)*
Antinutrients	Proteinase inhibitors and tannins
Cyanogenic glycosides	Cyanophoric foodstuffs, including cassava *(Manihot esculenta)*, lima beans *(Phaseolus* spp.), and others, and associated diseases
Favic agents	Fava beans *(Vicia faba)* and the hemolytic effect on people with G6PD enzyme deficiency
Grayanotoxins	Toxins in honey of *Rhododendron* spp.
Lathyrogens	Toxicity and disease presentation from exposure to chickling pea *(Lathyrus sativus)* and AMPA-mediated disease presentation associated with cycads *(Cycas* spp.)
Sesquiterpene lactones	Star anise *(Illicium* spp.)

of ackee fruit. Hypoglycin A and B are both toxic compounds associated with the ackee fruit, but hypoglycin B is thought to exist only in the inedible seed—not the aril. Hypoglycin A was conclusively linked to "Jamaican vomiting sickness" by Tanaka et al. (1976).

The mechanism of the toxicity of hypoglycin A has yet to be fully characterized; however, it is known to interrupt metabolic pathways, resulting in a state of hypoglycemia in humans. Hypoglycin A and its metabolite, methylenecyclopropyl acetic acid, interfere with fatty acid metabolism and gluconeogenesis. Methylenecyclopropyl acetic acid competitively inhibits enzymes involved in the breakdown of acyl coenzyme A (CoA) compounds by binding irreversibly to coenzyme A and carnitine, reducing their bioavailability to the CoA transferase system that is responsible for the transport of long-chain fatty acids into mitochondria. This reaction inhibits the beta-oxidation of fatty acids and subsequently interrupts the cell's supply of ATP, NADH, and acetyl-CoA. The body then increases its use of glucose, which depletes glycogen stores and leads to hypoglycemia. With mitochondrial beta-oxidation severely impaired, the long-chain fatty acids are believed to undergo omega-oxidation by oxidases in the liver, resulting in accumulation of short- and medium-chain dicarboxylic acids, such as glutaric acid.

The plant part of *B. sapida* that is consumed is known botanically as an aril. An aril is an outgrowth of the seed's outer covering. Other commonly consumed foods that are classified as arils are pomegranates, lychees, mangosteens, and the "mace" of nutmeg.

The ackee aril is a fleshy, whitish-yellow substance with a texture that has been described as "brain-like." When ripe, there is less than 0.1 ppm of hypoglycin A in ackee; however, before it ripens,

Andrea R. Ottesen • Plant Sciences and Landscape Architecture, University of Maryland, College Park, MD 20742. **Bernadene A. Magnuson** • Cantox Health Sciences International, 2233 Argentia Road, Suite 308, Mississauga, Ontario, Canada L5N 2X7.

there is approximately 1,000 ppm of hypoglycin in the fruit (Brown et al., 1992). There are also extremely high concentrations of hypoglycin A and B in the seeds and hulls, but these plant parts are not consumed.

The observation was made by Henry et al. (1998) that malnourished individuals and children appear to be more susceptible to ackee poisoning. Joskow et al. (2006) suggest that children may be more susceptible due to a greater vulnerability to malnutrition and less robust ability to maintain glucose homeostasis, in addition to a less sophisticated ability to distinguish subtle differences in fruit ripeness. A 1998 study of seven deaths of children aged 2 to 5 years in Burkina Faso in West Africa found that the only risk factor consistently present in all deaths was the proximity of an ackee tree (Meda et al., 1999).

Sources

Ackee, *Blighia sapida,* was introduced to the Americas from its native West Africa in 1778 (Joskow et al., 2006). The ackee fruit seems to be most famous for its cherished place in Jamaican cuisine; however, many cultures enjoy the fruit, and the tree has been naturalized in Central America, Florida, and California. There is a high demand for ackee in the United States.

Ackee was banned by the U.S. Food and Drug Administration in 1973 (Ware, 2002), and the ban remained in effect until several companies were able to demonstrate the safety of their canned products. Personal consumption of the fruit was never restricted, and travelers were allowed to return to the United States with personal quantities of canned ackee.

Food Processing That Influences Stability

Even when unripe ackee is cooked, it retains its toxicity. The only way for the consumer to avoid ackee toxicity is to be familiar with the morphology of the fully ripened fruit. When the ackee is ripening, it changes from greenish to yellow and finally to a bright pinkish-red color. Fully ripened ackee fruit will split open into three sections revealing three large, black seeds, each with its fleshy, yellowish-white aril. The fruit should never be consumed before the hull has split open of its own accord.

Interventions

Treatment of poisoning with ackee involves management of hypoglycemia. Administration of a 50% dextrose bolus (large, single dose) is sometimes followed by carefully monitored intravenous dextrose infusion. Other potentially therapeutic agents for treatment of hypoglycin-induced hypoglycemia, such as L-carnitine, glycine, and clofibrate (ethyl

p-chlorophenoxyisobutrate), have been examined with animals (Mehta et al., 2000).

Preventative Education

As children are often poisoned by ackee, educational programs targeting young children might be assisted by the development of an illustrated version of the maturity "ripeness" scale presented by Brown et al. (1992), accompanied by graphic depictions of symptoms of illness associated with consumption of unripe ackee.

Detection

Acidemia and aciduria have been reported to be diagnostic indications of ackee poisoning (Joskow et al., 2006). Indications of hypoglycin poisoning are elevated concentrations of dicarboxylic acids, such as glutaric acid as high as 4- to 200-fold greater than normal levels, and elevated concentrations of carnitine, such as octanoylcarnitine in blood and urine (Joskow et al., 2006; Meda et al., 1999). Measurement of hypoglycin A concentrations can be achieved by ion exchange chromatographic determination, high-pressure liquid chromatography, and reverse-phase liquid chromatography (Ware, 2002).

ANTINUTRIENTS: PROTEINASE INHIBITORS AND TANNINS

Proteinase Inhibitors

Proteinase inhibitors are plant, bacterial, and animal proteins that inhibit the activities of proteinases, resulting in impaired protein digestion and nutritional deficiencies in animals. Proteinase inhibitors are classified by the type of proteinases that they inhibit such as serine-, sulfhydryl-, acid-, and metalloproteinases (Xavier-Filho and Campos, 1989). Consumption by some animals of raw foods, such as soybeans, leads to growth depression, hypersecretion of pancreatic enzymes into the gut lumen, and eventual enlargement of the pancreas (Liener, 1979).

In the human diet, the Bowman-Birk serine proteinase inhibitor family is among the more commonly encountered inhibitors, being found in *Leguminosae* (one of the largest families of flowering plants), including soybeans. Soybeans also contain a trypsin proteinase inhibitor known as the Kunitz soybean inhibitor (Liener, 1979). Proteinase inhibitors are usually detected using specific proteinase assays wherein a reduction in hydrolysis of the target protein or peptide is indicative of the presence of a proteinase inhibitor (Xavier-Filho and Campos, 1989).

Proteinase inhibitors are proteins and are, therefore, denatured and inactivated by heat during cooking

(Liener, 1979). Because foods that contain proteinase inhibitors are cooked to improve human acceptability and palatability, an adverse effect of proteinase inhibitors has not been observed with humans. The presence of proteinase inhibitors in plants, however, has had an impact on food production, as it limits the sources of available plants for animal feeds due to the need for expensive heat treatments.

Tannins

Tannins are present in a number of foods, including dry beans, green peas, cereal products, leafy and green vegetables, tea, coffee, and wine. Tannins are antinutritional because they interfere with protein digestion through interactions with proteins, either by inactivating the digestive enzymes or by reducing the susceptibility of the substrate proteins to cleavage by proteases. The outcome for animals is reduced feed efficiency and growth impairment. The tannin content of food legumes ranges from 0 to 2,000 mg/100 g of beans, depending on the bean species (Reddy et al., 1985). Light and dark red kidney beans (*Phaseolus vulgaris* L.) contain 100 to 150 mg tannins/100 g, and faba beans (*Vicia faba* L.) have been reported to contain the highest concentrations, with up to 2,000 mg/100 g.

The amount of tannins needed to result in toxicity and growth depression in humans is not known. With experimental animals, the amount in the diet needed to depress growth appears to vary with species and ranges from 0.5 to 5% of the diet (Reddy et al., 1985). Processing methods, including dehulling, soaking, cooking, and germination, reduce the antinutritional activity of tannins.

CYANOGENIC GLYCOSIDES

Characteristics and Occurrence

Cyanogenic glycosides are labile compounds comprised of a sugar component—most commonly a mono- or disaccharide—and a cyanide precursor compound (cyanogen). The term "glycoside" describes an inclusive range of sugars, and the term "glucoside" is used when the sugar moiety is specifically glucose. Cyanogenesis refers to the process of generating hydrogen cyanide (HCN) or hydrocyanic acid by liberating the cyanide from the sugar group and other aglycones. This usually occurs in two steps: first, a hydrolytic enzyme, β-glycosidase, acts on the cyanogenic glycoside to produce a sugar and a nitrile (α-hydroxynitrile or cyanohydrin), and then a second enzyme, hydroxynitrile lyase, further dissociates the cyanogen into hydrogen cyanide and the corresponding aldehyde, ketone, or, in some rare instances, acid

(Concon, 1988; Selmar et al., 1989; Tewe and Iyayi, 1973). Most cyanogenic glycosides are derived from the five hydrophobic protein amino acids, tyrosine, phenylalanine, valine, leucine, and isoleucine (Poulton, 1990).

Cyanogens are found in a wide range of organisms: higher plants, bacteria, fungi, *Myriapoda*, and *Insecta* (Tewe and Iyayi, 1973). In plants, injury from food preparation and natural causes allows the hydrolytic enzymes to come in contact with the cyanogenic glycosides. Enzymes and cyanogens are usually separated physically from each other, either compartmentalized within the same cells or between different cells and tissues (Poulton, 1990). When damage occurs, hydrolytic enzymes come in contact with cyanophoric compounds, and hydrolysis reactions begin to liberate hydrogen cyanide. Cyanogenesis appears to be generated exclusively by the plant's endogenous enzymes or by microbial microflora present in the mammalian digestive tracts; mammals appear not to produce the necessary hydrolytic enzymes themselves (Jones, 1998).

There are several types of poisonings associated with cyanogenic foodstuffs. One, described as "acute cyanide poisoning," is characterized by cytotoxic anoxia (cellular oxygen deprivation) (Concon, 1988). Cyanide binds reversibly to the iron moiety of cytochrome a_3, the terminal oxidase of the respiratory chain in the cytochrome *c* oxidase complex, rendering the enzyme unable to participate in the electron transport chain, thus interrupting the ability of cells to accept oxygen from the blood and ultimately resulting in respiratory failure. Symptoms of poisoning are hyperventilation, headache, nausea, abdominal pain, and vomiting; severity of poisoning increases to include paralysis, convulsions, and eventual respiratory arrest. Death may occur anywhere from 3 to 20 min and up to a matter of hours in incidences where smaller amounts of cyanophoric compounds are consumed. Poisoning characteristics are influenced by numerous other factors, such as age and size of the victim, previous exposure to cyanogens, type of cyanogen, and amount of foodstuff consumed. Lethal doses of cyanide in humans range from 0.5 to 3.5 mg/kg body weight (Dahler et al., 1995).

A second form of poisoning associated with consumption of cyanogenic glycosides results from a chronic exposure that manifests in the form of neurological degenerative disease that can be characterized by disintegration of the myelin lamellae and the presence of myelin bodies in macrophages, indicating digestion of myelin (Concon, 1988). Myelin is a protective component of neurons and its degeneration is associated with numerous neurological disorders. The body's failure to dispose of cyanide from diets of primarily cassava origin, coupled with low protein intake, is believed to be correlated with tropical ataxic neuropathy, konzo, and amblyopia (lazy eye). Tropical ataxic neuropathy has been reported in Nigeria (Jones, 1998), Senegal (Concon, 1988), Tanzania, and many other African countries. It has also been reported in the West Indies and Asia (http:www.anu.edu.au/BoZo/CCDN/). Tropical ataxic neuropathy has a slow onset and affects adults in their 50s and 60s whose diet is predominantly cassava. Symptoms include loss of sensation in the hands and a tingling, prickling, or numbness (paresthesia), blurring vision, and weakening and thinning of the legs. Konzo is an upper motor neuron disease characterized by irreversible spastic paraparesis that has been reported in Mozambique, Tanzania, the Democratic Republic of the Congo, Central African Republic, and Cameroon (Ernesto et al., 2002). In the Democratic Republic of the Congo, 100,000 cases have been estimated (Bradbury, 2005). In all reports of konzo, there has been a correlation with high cyanogenic intake due to a predominantly cassava diet. Cassava (Manihot esculenta) is eaten by 500 million people worldwide, including 300 million in sub-Saharan Africa. This root crop is drought resistant and nutritionally very poor and contains (like many plants) cyanogenic glycosides that release hydrogen cyanide. Outbreaks have increased after periods of drought because the cyanide potential of cassava can become elevated when the plant is stressed by drought and also because there is a shortage of anything but the "bitter" varieties of cassava and the staple derivative, cassava flour (Ernesto et al., 2002).

A third form of poisoning associated with chronic exposure to cyanogens in the diet is characterized by goitrogeny (Concon, 1988). In animals, the cyanide ion is detoxified by glucosidases and sulphur transferases (rhodanese) to thiocyanate, which is an active goitrogenic agent. Thiocyanate and similar compounds affect hormone synthesis either by inhibiting iodine uptake or by competitively interfering with the activity of thyroid peroxidase, which inhibits the organification of iodide or iodination of tyrosine in the thyroglobulin and the coupling reaction (Chandra et al., 2004). Severe iodine deficiencies can result in goiter formation and numerous other symptoms associated with hypothyroidism.

One of the earliest descriptions of a fatal cyanide poisoning from plant foodstuffs, in this instance, lima beans (Phaseolus lunatus), was recorded in 1884 by Davidson and Stevenson. Numerous Phaseolus lunatus poisonings were reported in Europe following a large-scale importation of Phaseolus beans to relieve food pressures caused by World War I. Following these incidents,

legislation was introduced in the United States and several European countries to limit the permissible cyanide content of edible beans to between 10 and 20 mg HCN/100 g (Montgomery, 1965).

Sources

Approximately 60 to 75 different cyanogenic glycosides have been reported from an estimated 2,500 to 3,000 species spanning 130 angiosperm families (Aikman et al., 1996; Poulton, 1990). Studies have demonstrated that cyanogenic glycosides play a role in plant defense from herbivory (Jones, 1998). Higher amounts of cyanogens are present in young plant tissue (such as shoots and new leaves) than in tissues of advanced maturity.

Many food plants contain cyanogenic glycosides in varying concentrations throughout different parts of the same plant. Even within genera and species there is variation of cyanogenic potential (Aikman et al., 1996). Table 2 presents a few well-characterized cyanogens and the foods in which they are found.

Table 2. Cyanogens in foods[a]

Cyanogenic glycoside	Sugar + other hydrolysis products	Plant/food source
Amygdalin	Gentiobiose, hydrogen cyanide, benzaldehyde	Almond, cherry, apricot, peach, pear, and plum (Prunus spp.) (Pyrus); apple (Malus); quince (Cydonia)
Dhurrin	D-Glucose, hydrogen cyanide, p-hydroxybenzaldehyde	Wheat, (Tricticum spp.), sorghums, rye (Secale cereale), macadamia nut (Macadamia spp.)
Linamarin	D-Glucose, hydrogen cyanide, acetone	Lima bean (Phaseolus lunatus), cassava (Manihot esculenta), oats, passion fruit (Passiflora spp.)
Lotaustralin	D-Glucose, hydrogen cyanide, acetone	Lima bean (Phaseolus lunatus), cassava (Manihot esculenta), wheat (Tricticum), kidney bean (Phaseolus)
Prunasin	D-Glucose, hydrogen cyanide, benzaldehyde	Prunus spp., apple (Malus), papaya (Carica), macadamia (Macadamia)

[a]Data derived from Concon (1988), Conn (1981), and Jones (1998).

The most important cyanogenic-linked toxicities are not surprisingly associated with plant parts that are high in cyanogenic glycosides in the part of the plant that is directly consumed. Important food plants that fall into this category are the root tubers of cassava, lima beans *(Phaseolus lunatus)*, red kidney beans *(Phaseolus* sp.), bitter almonds *(Prunus dulcis)*, bamboo shoots *(Bambusa* sp.), bean sprouts *(Vigna* sp. and others), macadamia nuts *(Macadamia* sp.), passion fruit *(Passiflora* sp.), and pepperwort *(Lepidium* sp.) seedlings (Jones, 1998). Taro *(Colocasia esculenta)* and giant taro *(Alocasia macrorrhiza)* are other important food plants that have been reported to have varying contents of cyanogens. Loquot *(Eriobotrya* sp.), elderberry *(Sambucus* spp.), quince *(Cydonia* sp.), and sapote *(Pouteria caimito)* have also been reported to contain cyanogenic glycosides (Jones, 1998).

The seeds of apples *(Malus domestica)*, cherries, peaches, apricots, and plums *(Prunus* spp.) contain amounts of cyanogenic glycosides that can prove poisonous, in rare cases, when they are consumed; however, these plant parts are usually discarded in food preparation. Apple cider, however, is made using entire apples. People with the potential for Leber's disease (hereditary optic neuropathy) can go blind because of the effects of cyanide in cider (Jones, 1998).

The lethal human dose of cyanide is 0.5 to 3.5 mg/kg of body weight. This is equivalent to an oral dose somewhere between 30 and 210 mg for a 60-kg adult (Tewe and Iyayi, 1973); any amount of cyanide between 50 and 100 mg/100 g of food is considered to be moderately toxic and above 100 mg/100 g is considered highly toxic. Chronic exposure does seem to acclimate some organisms' ability to consume cyanophoric materials. The golden bamboo lemur *Hapalemur aureus* may eat 50 mg/kg of body weight of cyanide from bamboo species *(Cephalostachyum cf. viguieri)* a day without adverse effect (Jones, 1998). For many concentrations of cyanogens in plants, a human would have to eat almost his or her own weight in cyanophoric foodstuff before succumbing to toxic effects. There are, however, situations wherein concentrations of cyanogens are high enough to cause poisoning in a single meal. There is a bitter to sweet range of cassava cultivars, and often the more bitter the cultivar, the higher the cyanide potential. There are certainly exceptions to this rule, but generally, bitterness is correlated with higher concentrations of cyanogenic compounds, and in many instances, the bitter cultivars are preferred for the quality of the resulting flour and also for crop vigor (Chiwona-Karltun et al., 2004). Table 3 presents data on the concentrations of cyanide in a number of foods.

Table 3. Cyanide as HCN in foods[a]

Food plant	HCN (mg/100 g)	Reference
Bitter almond	205–411.5	Dicenta et al. (2002)
Sweet almond	0–10.5	Dicenta et al. (2002)
Cassava dried root cortex	245	Collens (1915)
Cassava whole root	53	deBruijn (1973)
Cassava leaves	104	deBruijn (1973)
Lima bean *(Phaseolus lunatus)*, black cultivar	300	Viehoever (1940)
Lima bean *(Phaseolus lunatus)*, white cultivar	10	Montgomery (1964)
Bamboo shoots (tips) *(Bambusa* spp.)	800	Baggchi and Ganguli (1943)
Bamboo immature stem *(Bambusa* spp.)	300	Baggchi and Ganguli (1943)

[a]Adapted from Montgomery (1980).

Food Processing That Influences Stability

The cyanide content of foods can be greatly reduced with various methods of preparation. Cassava is almost always peeled, as some of the highest concentrations of cyanogens are located in the "skin" of the root. In some geographical areas, it is subsequently boiled. When boiling high cyanide varieties, it is recommended that the lid be removed and that the water be changed as cyanide will remain in the water. African regions use a method of "sun drying," in which the tubers are peeled, cut into pieces, and then baked in the sun. In times of drought, "heap fermentation" is preferred over simple sun drying, as the tubers are higher in cyanogens to begin with during drought (concentrations of HCN are approximately 2 or 3 times higher than during nondrought times), and "heap fermentation" generally removes twice as much cyanide as do "sun drying" methods (Cumbana et al., 2007). Heap fermentation generally takes much longer, as the cut tubers must be placed in a heap in the sun for 3 to 5 days to allow the fermentation to release the cyanide. Approximately 12 to 16% of the cyanide still remains after "heap fermentation." The preparations for "gari," or flour, appear to be the most thorough methods to remove cyanide. Tubers are peeled and grated and then moistened and put into a bag. After several days, the water is squeezed out, and the cassava is dried and roasted. Mixing cassava flour with water and spreading it in a thin layer to bake for around 5 hours has been shown to reduce cyanide content by about 84% (Cumbana et al., 2007). It has been suggested that if this method were taken up by rural women in East, Southern, and Central Africa, these areas could see the end of konzo in those areas (J. H. Bradbury, personal communication). In general, the more cells destroyed in the process, the better the enzymes are at liberating the cyanide preingestion. The same is true for bitter almonds. If they are ground in a mortar and pestle and left for a few hours, the bitter flavor is lost and the food becomes sweet because cyanogenesis releases the volatile HCN and the sugar is left behind, making the food "sweeter"(Jones, 1998).

Environmental conditions, such as nitrogen content of soil and drought, can influence cyanogen content in plants, and a wilted plant may become more cyanogenic. Wheat or grasses used for feeds become toxic after drought and wilting episodes as a result of their increased cyanogenic content (Wertheim, 1974).

Lima beans are traditionally boiled or soaked in water to remove cyanide. The 1884 poisoning incident in Europe reported by Davidson and Stevenson (1884) reportedly took place after consumption of twice-boiled beans, with water that was changed between boiling. This incident occurred prior to legislative regulations and breeding programs that subsequently lowered acceptable cyanogen content in most beans that are grown in or imported into the United States.

Detection Methods

Because cyanogens are released by enzymes, acids, alkalis, and heat, the most common method for analysis is to liberate hydrogen cyanide by acid or enzyme hydrolysis and then to quantify released cyanide with a variety of methods ranging from spectrophotometry to chromatography. A picrate kit, developed by Howard Bradbury's group, called "Kit A," was designed to measure cyanide concentration in cassava roots or flour. It uses linamarase to hydrolyze cyanogens from foodstuffs and picrate paper to react with released cyanide, which can then be quantified colorimetrically as it changes from yellow to brown or spectrophotometrically at an absorbance of 510 nm (Bradbury et al., 1999; Egan et al., 1998). The complete protocol for "Kit A" is available from the website of the Cassava Cyanide Disease Network (CCDN). The kits are available free of charge to health workers and agriculturalists in developing countries and can be used in the field or in the laboratory with little training.

Cyanide can also be measured colorimetrically with the Konig reaction, which involves the addition of oxidizing and coupling reagents (Tsuge et al., 2001). Chloramine-T oxidizes the hydrogen cyanide, which is then coupled to another reagent that yields a stable purple complex that can be measured at 580 to 583 nm with a spectrophotometer. Absorption of light at this wavelength is an estimate of cyanide concentration. Other methods include thin-layer chromatography, nuclear magnetic resonance, gas-liquid chromatography of trimethylsilyl ethers of cynogenic

glycosides, medium-pressure liquid chromatography and vacuum chromatography (Brinker and Seigler, 1989; Seigler et al., 2002). Studies have demonstrated with statistical significance that the ability to identify highly cyanophoric varieties simply by tasting the root can also be quite efficient (Chiwona-Karltun et al., 2004).

FAVISM

Characteristics and Occurrence

Faba or fava beans *(Vicia faba)* contain vicine and convicine, which are reported to induce hemolytic anemia in patients with a specific hereditary predisposition, namely, a deficient version of glucose-6-phosphate dehydrogenase (G6PD). The identification of the genetically linked trait of G6PD deficiency occurred due to the observation that in certain people, the antimalarial drug primaquine induced severe hemolytic anemia (Carson et al., 1956). This was the first discovery to shed light on the genetic basis for the enzyme deficiency that can result in acute hemolytic anemia in response to certain drugs or favic agents. "Favism" is named after the fava bean because fava metabolites can serve as exogenous agents capable of inducing hemolysis in erythrocytes. It is primarily the metabolites of the pyrimidine glycosides vicine and convicine that are hydrolyzed by intestinal flora to the highly oxidative pyrimidine aglycones divicine and isouramil.

G6PD deficiency is one of the most common genetically determined enzymatic abnormalities in humans (Yoshida, 1973). There are over 80 variants of the G6PD enzyme, with diverse kinetic activities (biochemical variants are numbered at 387) (Mehta et al., 2000). Some of the variants are not associated with any clinical manifestations, and some are associated with chronic nonspherocytic hemolytic anemia.

G6PD catalyzes the first step in the hexose and pentose monophosphate oxidation pathways of carbohydrate metabolism causing reduction of $NADP^+$ to NADPH (Yoshida, 1973). The PPP (pentose phosphate pathway) is involved in the conversion of glucose into pentose sugars used for the biosynthesis of important molecules such as ATP, CoA, NAD, flavin adenine dinucleotide, RNA, and DNA. Because red blood cells lack mitochondria, the PPP is their only source of NADPH (Mehta et al., 2000). Inability to reduce $NADP^+$ to NADPH at the normal rate makes cells more susceptible to oxidative damage. In non-G6PD-deficient cells, exposure to oxidative agents will result in increased G6PD expression that maintains normal levels of NADPH and glutathione (GSH). Under normal conditions, the ability of the deficient G6PD to use NAD as an alternate hydrogen source allows the enzyme-deficient erythrocyte to maintain adequate levels of GSH. However, when G6PD-deficient cells are exposed to the oxidative stresses that divicine and isouramil induce, the metabolic equilibrium is severely upset. The PPP of G6PD-deficient cells is already functioning at full capacity and cannot "upregulate," and subsequently levels of NADPH and GSH fall (Mehta et al., 2000). GSH is irreversibly oxidized, and the oxidized disulfide form of GSH is catabolized, leading to the eventual destruction of the erythrocyte (Mager et al., 1980). Consumption of fresh, frozen, cooked, raw, or dried fava beans by susceptible individuals, therefore, results in serious anemia, jaundice, and hemoglobinuria.

More than 400 million people, of whom 90% are male, are affected by G6PD deficiency. The ratio of male-to-female manifestations of the disease is about 21:1, with the highest frequency occurring in children under 10 years of age, specifically in males between the ages of 2 and 4. The mutation for the deficient G6PD enzyme is on the X chromosome; thus, women are carriers twice as often as are men, but their normal allele is able to counteract the mutated allele. Homozygous (XY) men have no normal allele to provide a normally functioning enzyme (Mager et al., 1980; Yoshida, 1973). The statistically rare occurrence of homozygous mutant alleles in females is the reason that the incidence among females is so low.

There is evidence of mother-to-womb transmission of "favism" (Corchia et al., 1995). Although both mother and daughter discussed in the report were heterozygous for G6PD deficiency, because of the much lower body weight of the child, it is likely that only the child was affected by the meal of fava beans 5 days prior to her stress-induced birth. This seems consistent with the predominance of manifestations of favism in children (71%) compared with that in adults (29%) (Corchia et al., 1995). The amount of fava beans consumed in relation to body weight may play a significant role in disease presentation.

Interestingly, the geographical occurrence of G6PD-deficient populations overlaps areas that were once or are presently afflicted with malaria, suggesting that the G6PD-deficient trait was advantageous in malaria-afflicted areas. There is evidence that the deficient version of the G6PD enzyme may actually create a "more hostile" environment for the malaria parasite *Plasmodium falciparum* (Wajcman and Galacteros, 2004), possibly due to the reduced levels of NADPH in the cell that seem to be needed for *P. falciparum* infection (Mehta et al., 2000).

A study by Luzzatto et al. (1969) used "G6PD A" heterozygous females to examine parasitization preference. The heterozygous females have roughly half of their red blood cells with deficient G6PD levels and the other half with normal levels. The two types of cells could be distinguished under the microscope, and parasitization of normal blood cells was observed 2 to 80 times more frequently than that of G6PD-deficient cells.

In the 1900s, a fatality rate of 6 to 8% was associated with poisoning by favic agents, but current medical expertise, specifically the availability of blood transfusions, has made favism-associated fatalities very rare. A seasonal peak that correlates with faba bean harvest has been reported (Mager et al., 1980).

Symptoms

The symptoms of favism are fatigue, nausea, abdominal or back pain, fever and chills, hemoglobinuria, jaundice (including neonatal jaundice), and even acute renal failure (Mager et al., 1980).

Sources

Fava, an important source of protein, is one the oldest crops and is grown all over the world.

Food Processing That Influences Stability

The favism-inducing compounds are thermostable (Gutierrez et al., 2006). The only method that has been reported to completely remove vicine or convicine from beans is the continuous-flow soaking in tap water for 72 hours at 50°C, 60 hours at 55°C, or 48 hours at 60°C, with a continuous flow rate of 0.5 ml/min for all procedures (Jamalian and Ghorbani, 2005). Researchers have utilized various other methods—dehulling, soaking in different solutions, preparing protein isolates, autoclaving, germinating, fermenting, cooking, drying, pureeing, and hydrolyzing with enzymes—with little or moderate success (Jamalian and Ghorbani, 2005). Attempts to breed or engineer a plant that produces seed with little or no vicine content have yet to be successful; however, the discovery of a spontaneously occurring mutant allele that yields a 10- to 20-fold decrease in vicine and convicine content may pave the way to a vicine- and convicine-free faba plant (Gutierrez et al., 2006).

Prevention and Control

The only safe measure for sensitive individuals to avoid the toxic effects of fava beans remains to avoid "favic" agents. Folic acid (5 mg) daily is a treatment to be given for 2 to 3 weeks following an acute hemolytic event and daily for long-term patients with chronic hemolysis (Mehta et al., 2000).

GRAYANOTOXINS

Characteristics and Occurrence

Grayanotoxins (GTX), formerly known as andromedotoxins, rhodotoxins, and acetylandromedol, occur in multiple parts of ericaceous plants (Furbee and Wermuth, 1997). The most significant human food poisonings associated with these compounds to date have been related to the consumption of honey made from the nectar of species of mountain laurel (*Kalmia latifolia*) and sheep laurel (*Kalmia angustifolia*), *Rhododendron* spp. (including azaleas), and andromeda (*Pieris* spp.) (Holstege et al., 2001).

GTXs are heterocyclic diterpenes (diterpene polyalcohols) with a tetracyclic A-nor-B homo ent-kaurane (andromedane) skeleton (Holstege et al., 2001). GTX I, II, and III are considered the most toxic of over 30 characterized GTX. GTX II and III appear to be derivatives of GTX I (Furbee and Wermuth, 1997).

The GTX molecule's mechanism of toxicity is associated with its tendency to bind to closed sodium channels of cell membranes, creating a modified, continuously open channel that generates a slow, steady current, increasing the resting permeability (shown in squid axon membranes) and causing a large depolarization of the membrane (Bull et al., 2004; Seyama and Narahashi, 1981).

Poisoning symptoms of GTX ingestion, such as respiratory depression, bradycardia, hypotension, and seizure, are all consistent with symptoms associated with shifts in the electrical potential of neuron transmission (Furbee and Wermuth, 1997; Holstege et al., 2001). Modulations in the rate of ion transport through sodium/potassium pumps can effectively depolarize membrane potentials leading to a depleted pool of releasable neurotransmitters, which in turn leads to the inability to properly relay messages from one neuron to the next (Bull et al., 2004).

One of the earliest food poisonings ever recorded is believed to be due to GTX in honey made by bees that had fed on ericaceous plants. The incident was described in *Anabasis*, the written work of Xenophon, a member of Cyrus's army in approximately 400 BC. Xenophon described Cyrus's troops en route to battle, after the troops had consumed honey encountered along the way: "… all of the soldiers that ate of the honey combs lost their senses, vomited, and were affected with purging, and none of them was able to stand upright; such as had eaten only a little were like men greatly intoxicated, and such as had eaten much were like mad men and some like persons at the point of death…. The next day, no one of them was found dead; and they recovered their senses about the same hour they had lost them on the preceding day."

The men are believed to have consumed honey from bees that had been feeding on *Rhododendron ponticum* (Gossinger et al., 1983). To this day, poisoning by GTX in honey is known as "mad honey poisoning." The geographical site of this ancient poisoning is the northeastern coast of the Black Sea in present-day Turkey, a region still renowned for its honey and, interestingly, with a few exceptions, the region from which most of the recent reports of honey poisoning originate (Gossinger et al., 1983; Gunduz et al., 2006; Kerkvliet, 1981; Ozhan et al., 2004; Sutlupinar et al., 1993; Yilmaz et al., 2006).

Sutlupinar et al. (1993) describe instances of self-medication with honey. Among 11 hospitalizations of adult males, many of the patients were reported to have used a traditional folk medicine, known in the Black Sea region of Turkey as "deli-bal" or "tutan bal" (toxic honey), to address hypertension and gastrointestinal disorders. This honey is also believed to be a sexual stimulant. Also known as "mad honey" or "bitter honey," it reportedly causes a sharp burning sensation in the throat (Gunduz et al., 2006).

Food Processing That Influences Stability

For safe consumption of honey, the recommendation is to avoid honey made in the spring when members of the *Ericaceae* plant family are blooming. One should avoid honey from regions with high frequency of the genera *Rhododendron*, *Pieris*, and *Kalmia*. Summer harvest can be preferable as the ericaceous species are finished blooming by the end of spring. Processing plants use honey from multiple sources to ensure that no one batch will contain high levels of GTX.

Interventions

The appropriate interventions depend on the symptoms exhibited by the victim of toxicity. Activated charcoal can be administered to a victim of GTX toxicity if ingestion has occurred within 2 hours (Furbee and Wermuth, 1997). If bradycardia occurs, it can be treated with intravenous atropine, or in severe cases, a cardiac pacemaker may be appropriate (Furbee and Wermuth, 1997; Gunduz et al., 2006; Ozhan et al., 2004; Yilmaz et al., 2006). Lidocaine or other sodium channel-blocking antiarrhythmics can be administered for ventricular arrhythmias (Furbee and Wermuth, 1997). If hypotension is a presenting symptom, antihypotensive agents, such as aramine (metaraminol and *n*-hydroxynorephedrine) and Novadral (norfenefrine HCL), can be administered. With development of hemorrhagic urticaria, treatment with steroids may be applicable

(Sutlupinar et al., 1993). For symptoms of gastric acidity, antacids in suspension are appropriate, and to increase dieresis, furosemide (40 mg) can be added to an antacid infusion (Sutlupinar et al., 1993).

Detection

Methods include detection of ericaceous pollens in honey by scanning electron microscopy or regular microscopy (Kerkvliet, 1981) or detection of toxins by liquid chromatography–mass spectrometry (Holstege et al., 2001) and thin-layer chromatography (Furbee and Wermuth, 1997; Scott et al., 1971; Sutlupinar et al., 1993).

Prevention and Control

Honey from several different sources is often mixed before it is sold to prevent possible poisoning associated with high amounts of toxins in a particular batch. Honey from early spring is much more likely to cause problems because most *Rhododendron*, *Pieris,* and *Kalmia* species flower early in the spring, and the blossoms are often completely gone by summer.

LATHYROGENS

Characteristics and Occurrence

Lathyrogens include a number of chemical components of the *Lathyrus* species that have toxic properties. Lathyrism, which is a neurological disorder characterized by spastic paraparesis, occurs in both humans and animals; however, the characteristics of the disease differ with species. In humans, the disease (neurolathyrism) involves the central nervous systems and the causative factor is thought to be the neurotoxic β-*N*-oxaloylamino-L-alanine (L-BOAA). Victims display nervous paralysis of the lower limbs and are forced to walk on the toes in short, jerky motions. Symptoms may manifest suddenly when brought on by exposure to a wet environment and/or hard work and exhaustion. The presence of uncommon amino acids in the plants may also contribute to the disease. The study of human lathyrism has been difficult because oral L-BOAA administration to animals does not reproduce the human disease. Experimental lathyrism in animals (osteolathyrism) involves bone and connective tissue and is induced with administration of β-aminopropionitrile (Spencer, 1999).

The disease manifests only when consumption of seeds of certain *Lathyrus* spp., such as *L. sativus* (chickling vetch or chickling pea), becomes a major source of food intake (up to one-third of food consumption), as may occur in times of drought or flood. Reports of the toxicity of peas to humans date back to the times of Hippocrates (460 BC), and outbreaks occurred in Ethiopia as recently as 1977 in over 2,500 individuals (Spencer and Palmer, 2003).

The cultivation of *L. sativus* was banned in many parts of the world after discovery of its toxicity; however, because it is of economic value and is easily grown under adverse conditions, the ban is sometimes ignored. *Lathyrus sativus* is known as the grass pea or chickling pea (not the chick pea). The plant has extraordinary environmental tolerance to both drought and flood; it fixes nitrogen in soil and requires no inputs or pesticides. Grass pea has served throughout history as an insurance crop as well as a basic food item in certain parts of the world. Its agricultural assets, combined with its high protein content and tasty flavor, have resulted in the nickname, "manna from heaven"—except for the single non-protein amino acid excitotoxin (L-BOAA) that acts as a potent α-amino-3-hydroxyl-5-methyl-4-isoxazole-propionate (AMPA) agonist, causing excessive neuronal excitation and nerve cell degeneration (Tshala-Katumbay and Spencer, 2007).

Continued consumption results in a motor system disease (lathyrism) in which degeneration of the principal motor pathway from brain to spinal cord results in a clinical picture of spastic para- or tetraparesis. The precise pathogenic mechanisms of the neural degeneration associated with lathyrism are yet to be fully understood; however, it is postulated that excessive glutamatergic stimulation of nerve cells causes an excessive influx of sodium and chloride ions that results in membrane depolarization. ATP is subsequently depleted as the sodium/potassium ATPase pump attempts without success to restore the transmembrane ionic gradient. Osmotic pressure results in a massive influx of water to the cytoplasm, leading to postsynaptic vacuolation of dendrites and perikarya that culminates in cell death (Tshala-Katumbay and Spencer, 2007).

Lathyrism once plagued the European continent but was last seen there in association with the Spanish Civil War and Second World War. Today, the disease is extant in Ethiopia, India, and Bangladesh, with some additional cases in China and perhaps parts of South America. In Ethiopia, grass pea production increased by 20% between 1981 and 1987, in response to food shortages (Spencer and Palmer, 2003).

Populations subsistent on cassava, who show evidence of sulfur protein deficiency, develop a disease almost indistinguishable from lathyrism. Cyanide is detoxified to cyanate and thiocyanate; the latter (and perhaps the former) acts as a chaotropic agent that in vitro selectively increases the binding of glutamate to the AMPA receptor. It is therefore possible, perhaps likely, that it is AMPA-mediated mechanisms underlying the neurotoxic properties of cassava and grass pea. Even more concerning than the large number of irreversibly crippled people in cassava-dependent villages of Africa is the possibility that thiocyanate (a thyroid toxin) is impairing the physical and brain development of children.

Another source of a plant toxin that acts in a manner similar to that of L-BOAA from the grass pea is beta-*N*-methylamino-L-alanine (L-BMAA) from cycads (*Cycas* spp.). L-BMAA is a cyanobacterial amino acid that resides in highly neurotoxic and carcinogenic cycad plants that have been used for food and medicine throughout history. Whereas L-BOAA acts as a potent glutamate analog on the surface of neurons, L-BMAA is a weak excitotoxin but is also taken up and apparently incorporated into brain tissue, possibly into protein. Uses of cycad in food and medicine are strongly associated with a remarkable and terrible neurodegenerative disease, amyotrophic lateral sclerosis/parkinsonism-dementia complex, which affects native people of Guam, West Papua, and Kii Peninsula, Honshu Island, Japan. Amyotrophic lateral sclerosis/parkinsonism-dementia complex in these three genetically distinct groups has declined in all three populations with the advent of modernity and the decline/discontinuation of cycad use. While we cannot state that cycad is the cause of this disease, use of the plant for food and medicine is the leading environmental candidate. Cycad clearly causes neuromuscular disease in cattle, and a second cycad toxin is an established developmental rodent neurotoxin capable of blocking the formation of cerebellum or cortex, depending on when the agent (cycasin [methylazoxymethanol-beta-D-glucoside]) is administered. Cracking the etiology of this disorder is of great importance not only for the affected population but also to open up understanding of look-a-like diseases (Lou Gehrig's disease, Parkinson's disease, and Alzheimer's disease) worldwide.

Food Processing That Influences Stability

Soaking grass peas in water can be an effective method to leach out up to 80% to 90% of the neurotoxic content of the pea (Liener, 1979; Spencer and Palmer, 2003). It is recommended to cook in water and then discard the water, or to soak the peas overnight and then roast, steam, or sun dry them (Liener, 1979). Ghotu, an Indian staple, is a porridge created by mixing grass pea with rice and water; this food type is thought to precipitate lathyrism more quickly than the Indian bread chapati, also prepared with grass pea (Spencer and Palmer, 2003). The Ethiopian bread kitta is thought to contain even higher levels of neurotoxin than the starting material that was used to prepare it (Teklehaimanot et al., 1993).

Prevention and Control

The observation was made by Spencer and Palmer (2003) that few studies have addressed the potential neurotoxicity of different food preparations of grass pea. This is clearly a research gap that, when addressed, will assist with prevention and control of grass pea-associated outbreaks of lathyrism. Another important research initiative, yet to be funded, would be to test a low L-BOAA strain of grass pea to determine if it is safe for human consumption. The use of breeding or genetic manipulation to develop nontoxic strains of *Lathyrus* would provide a source of safe, nutritious food for humans and animals. Another excellent effort would be the development of a lathyrism disease network whose mission would be to create awareness of the methods of preparation that can be used to reduce toxic levels in foods, as well as to coordinate an awareness and response on the part of the international community to provide adequate quantities of cereals (or other staples) to regions that for one reason or another (drought or war) have been forced into consumption of dangerous levels of grass pea simply to subsist.

SESQUITERPENE LACTONES

Characteristics and Occurrence

The convulsant activity, as well as the culinary appeal, of the seeds of *Illicium anisatum* has been recognized for several centuries (Yamada et al., 1968). Sesquiterpene lactones are a large and diverse group of biologically active molecules that have been characterized from *Illicium* species as well as many members of the *Compositae/Asteraceae* family. The highly oxygenated sesquiterpene lactones play a protective role in plant biochemistry, functioning as antifeedants in numerous species (Harbourne, 2001).

The first research group to characterize anisatin was Lane et al. (1952). In 1952, they reported the isolation of a pure toxic compound, anisatin. This compound and its sister compound neoanisatin were fully characterized by Yamada et al. (1968). Anisatin is a known potent noncompetitive antagonist of gamma-amino butyric acid (GABA)-dependent neurons in mammals (Schmidt et al., 1998). The GABA receptor is one of two ligand-gated ion channels responsible for mediating the effects of GABA, the major inhibitory neurotransmitter in the brain. A neurotoxic sesquiterpene lactone, repin, has been implicated in causing a condition similar to Parkinson's disease in horses (Cornell Animal Sciences, 2005).

Seedpods of *Illicium verum*, commonly known as Chinese star anise, are used to prepare teas and other herbal and culinary dishes. Batches of *Illicium verum* seedpods are occasionally adulterated with seedpods from sister taxa, *Illicium anisatum* and *Illicium lanceolatum*, which contain potent neurotoxic sesquiterpene lactones, such as anisatin, neoanisatin, and pseudoanisatin.

Caribbean, Latino, and Asian cultures traditionally use an infusion of the seedpods as a carminative and sedative for the treatment of infant colic. This particular practice has resulted in cases of infant hospitalizations in instances in which batches of Chinese star anise are believed to have been adulterated with Japanese star anise (synonyms: *Illicium anisatum*, *Illicium japonicum*, and *Illicium religiosum*) (Fernandez et al., 2002; Ize-Ludlow et al., 2004a, 2004b; Minodier et al., 2003; Yamada et al., 1968). Several cases of poisoning symptoms among European adults have also been reported following consumption of "mulled wine" prepared with star anise pods (wine is warmed and steeped with popular spices). Poisoning symptoms range from seizures and vomiting to jitteriness and irritability (Ize-Ludlow et al., 2004b).

Sources

Star anise pods are most commonly grown in Asia. When the seedpod is still associated with the plant, it is not difficult to tell the species apart; however, if anise pods from Japanese star anise have mistakenly been added to Chinese star anise pods, then it becomes virtually impossible to tell one species from the other with morphological or chemical methods.

Detection

There is currently no efficient method to detect whether batches of *I. verum* have been adulterated with *I. anisatum* or *I. lanceolatum*. Star anise seedpods are sold in bags or bulk. In this state, morphological taxonomic methods, such as carpel number and size, are not sufficient to distinguish among seedpods of *I. anisatum*, *I. verum*, or *I. lanceolatum*. Anisatin, neoanisatin, and pseudoanisatin are all potently toxic sesquiterpene lactones found in *I. anisatum* and *I. lanceolatum*.

All *Illicium* species, however, contain sesquiterpene lactones and the secondary metabolites of these compounds. *Illicium verum* contains sesquiterpene veranisatins A, B, and C that when chromatographed by high-pressure liquid chromatography resemble peaks formed by the more toxic sesquiterpene lactones anisatin, neoanisatin, and pseudoanisatin, making it extremely difficult to reliably characterize either species.

All chemical methods to distinguish among *Illicium* species have been confounded by similar chemistries, and FDA scientists are currently reviewing the efficiency of molecular methods involving the use of barcoding regions of DNA to reliably distinguish species in adulterated batches of *I. verum* (Ottesen and Ziobro, 2006).

SUMMARY

In this chapter, we have covered very briefly a few naturally occurring toxicants primarily from food plant sources. A clear delineation of food safety cannot be established without a comprehensive understanding of the nature and mechanism of action of each potential toxicant and the biological distinctiveness of the individual consuming it. This is obviously an endeavor of enormous magnitude, and although we have made astounding progress in the past 20 years, we have many miles to go before research and education provide all people with the tools to best manage food consumption for their own unique health. Some of the biggest research gaps in the science of food safety still center on natural components in foods and long-term health hazards that may be associated with their chronic consumption.

REFERENCES

Aikman, K., D. Bergman, J. Ebinger, and D. Siegler. 1996. Variation of cyanogenesis in some plant species of the midwestern United States. *Biochem. Syst. Ecol.* **24:**637–645.

Baggchi, K. N., and H. D. Ganguli. 1943. Toxicology of young shoots of common bamboos (*Bambusa arundinacdea* Willd.). *Indian Med. Gaz.* **78:**40–42.

Bradbury, J. H. 2005. *Working Together To Eliminate Cyanide Poisoning, Konzo and Tropical Ataxic Neuropathy (TAN).* Australian National University, School of Botany and Zoology, Cassava Cyanide Disease Network (CCDN) News, Caberra, Australia, No. 6. Available at: http://www.anu.edu.au/BoZo/CCDN/newsletters/CCDNNEWS6.pdf.

Bradbury, M. E., S. V. Egan, and J. H. Bradbury. 1999. Picrate paper kits for determination of total cyanogens in cassava roots and all forms of cyanogens in cassava products. *J. Sci. Food Agric.* **79:**593–601.

Brinker, A. M., and D. S. Seigler. 1989. Methods for the detection and quantitative determination of cyanide in plant materials. *Phytochem. Bull.* **21:**24–31.

Brown, M., R. P. Bates, C. Mcgowan, and J. A. Cornell. 1992. Influence of fruit maturity on the hypoglycin-A level in ackee (*Blighia sapida*). *J. Food Saf.* **12:**167–177.

Bull, M., J. Ailts, C. Arndt, M. Ahrens, B. Sager, and B. Wilson. 2004. Neuron depolarization. *In NeuroScience Technical Bulletin*, No. 6. Neuroscience, Osceola, WI. https://www.neurorelief.com/newsletterarchive.php?issue=236.

Carson, P. E., C. L. Flanagan, C. E. Ickes, and A. S. Alving. 1956. Enzymatic deficiency in primaquine-sensitive erythrocytes. *Science* **124:**484–485.

Chandra, A. K., S. Mukhopadhyay, D. Lahari, and S. Tripathy. 2004. Goitrogenic content of Indian cyanogenic plant foods and their in vitro anti-thyroidal activity. *Indian J. Med. Res.* **119**:180–185.

Chiwona-Karltun, L., L. Brimer, J. D. Saka, A. R. Mhone, J. Mkumbira, L. Johansson, M. Bokanga, N. Z. Mahungu, and H. Rosling. 2004. Bitter taste in cassava roots correlates with cyanogenic glucoside levels. *J. Sci. Food Agric.* **84**:581–590.

Collens, A. E. 1915. Alcohol from cassava. *Bull. Dept. Agric. Trinidad Tobago* **14**:54.

Concon, J. M. 1988. Endogenous toxicants in foods derived from higher plants, p. 281–404. *In Food Toxicology. Part A: Principles and Concepts.* Marcel Dekker, Inc., New York, NY.

Conn, E. E. 1981. Secondary plant products, p. 479–500. *In* P. K. Stumpf and E. E. Conn (ed.), *The Biochemistry of Plants: a Comprehensive Treatise.* Academic Press, New York, NY.

Corchia, C., A. Balata, G. F. Meloni, and T. Meloni. 1995. Favism in a female newborn infant whose mother ingested fava beans before delivery. *J. Pediatr.* **127**:807–808.

Cornell Animal Sciences. 2005. Sesquiterpene lactones and their toxicity to livestock. *In Plants Poisonous to Livestock.* Department of Animal Science, Cornell University, Ithaca, NY. http://www.ansci.cornell.edu/plants/toxicagents/sesqlactone/sesqlactone.html.

Cumbana, A., E. Mirione, J. Cliff, and J. H. Bradbury. 2007. Reduction of cyanide content of cassava flour in Mozambique by the wetting method. *Food Chem.* **101**:894–897.

Dahler, J. M., C. A. McConchie, and C. G. N. Turnbull. 1995. Quantification of cyanogenic glycosides in the seedlings of three macadamia species. *Aus. J. Bot.* **43**:619–928.

Davidson, A., and T. Stevenson. 1884. Poisoning by pois d'Achery (*Phaseolus lunatus* Lr.). *Practitioner* **XXXII**:35.

deBruijn, G. H. 1973. The cyanogen character of cassava (Manihot esculenta), p. 43–48. *In* B. Nestel and R. MacLentyre (ed.), *Chronic Cassava Toxicity; Proceedings of an Interdisciplinary Workshop,* International Development Research Centre, Ottawa, Canada.

Dicenta, F., P. Martinez-Gomez, N. Grane, M. L. Martin, A. Leon, J. A. Canovas, and V. Berenguer. 2002. Relationship between cyanogenic compounds in kernels, leaves, and roots of sweet and bitter kernelled almonds. *J. Agric. Food Chem.* **50**:2149–2152.

Egan, S. V., H. H. Yeoh, and J. H. Bradbury. 1998. Simple picrate paper kit for determination of the cyanogenic potential of cassava flour. *J. Sci. Food Agric.* **76**:39–48.

Ernesto, M., A. P. Cardoso, D. Nicala, E. Mirione, F. Massaza, J. Cliff, M. R. Haque, and J. H. Bradbury. 2002. Persistent Konzo and cyanogen toxicity from cassava in northern Mozambique. *Acta Trop.* **82**:357–362.

Fernandez, G. C., G. P. Pintado, B. A. Blanco, M. R. Arrieta, R. R. Fernandez, and R. F. Rosa. 2002. Cases of neurological symptoms associated with star anise consumption used as a carminative. *An. Esp. Pediatr.* **57**:290–294.

Furbee, B., and M. Wermuth. 1997. Life-threatening plant poisoning. *Med. Toxicol.* **13**:849–888.

Gossinger, H., K. Hurby, A. Haubenstock, A. Pohl, and S. Davogg. 1983. Cardiac arrhythmias in a patient with grayanotoxin honey poisoning. *Vet. Hum. Toxicol.* **25**:328–329.

Gunduz, A., T. Suleyman, H. Uzan, and M. Topbas. 2006. Mad honey poisoning. *Am. J. Emerg. Med.* **24**:595–598.

Gutierrez, N., C. M. Avila, G. Duc, P. Marget, M. J. Suso, M. T. Moreno, and A. M. Torres. 2006. CAPs markers to assist selection for low vicine and convicine contents in faba bean (Viciea faba L.). *Theor. Appl. Genet.* **114**:59–66.

Harbourne, J. 2001. Twenty five years of chemical ecology. *Nat. Prod. Rep.* **18**:361–379.

Henry, S. H., S. W. Page, and P. M. Bolger. 1998. Hazard assessment of ackee fruit (Blighia sapida). *Hum. Ecol. Risk Assess.* **4**:1175–1187.

Holstege, D. M., B. Puschner, and T. Le. 2001. Determination of grayanotoxins in biological samples by LC-MS/MS. *J. Agric. Food Chem.* **49**:1648–1651.

Ize-Ludlow, D., S. Ragone, I. S. Bruck, J. N. Bernstein, M. Duchowny, and B. M. Garcia-Pena. 2004a. Neurotoxicities in infants seen with the consumption of atar anise tea. *Pediatrics* **114**:653–656.

Ize-Ludlow, D., S. Ragone, I. S. Bruck, M. Duchowny, and B. M. Garcia-Pena. 2004b. Chemical composition of Chinese star anise *(Illicium verum)* and neurotoxicity in infants. *JAMA* **291**:562–563

Jamalian, J., and M. Ghorbani. 2005. Extraction of favism-inducing agents from whole seeds of faba bean (Vicia faba L var major). *J. Sci. Food Agric.* **85**:1055–1060.

Jones, D. A. 1998. Why are so many food plants cyanogenic? *Phytochemistry* **47**:155–162.

Joskow, R., M. Belson, H. Vesper, L. Backer, and C. Rubin. 2006. Ackee fruit poisoning: an outbreak investigation in Haiti 2000–2001, and review of the literature. *Clin. Toxicol.* **44**:267–273.

Kerkvliet, J. D. 1981. Analysis of a toxic *Rhododendron* honey. *J. Apicult. Res.* **20**:249–253.

Lane, J. F., W. T. Koch, N. S. Leeds, and G. Gorin. 1952. On the toxin of Illicium Anisatum. I. the isolation and characterization of a convulsant principle: anisatin. *J. Am. Chem. Soc.* **74**:3211–3215.

Liener, I. 1979. Significance for humans of biologically active factors in soybeans and other food legumes. *J. Am. Oil Chem. Soc.* **56**:121–129.

Luzzatto, L., E. A. Usanga, and S. Reddy. 1969. Glucose 6-phosphate dehydrogenase deficient red cells resistance to infection by malarial parasites. *Science* **164**:839–842.

Mager, J., M. Chevion, and G. Glaser. 1980. Favism, p. 265–294. *In* I. E. Liener (ed.), *Toxic Constituents of Plant Foodstuffs,* 2nd ed. Academic Press, Inc., New York, NY.

Meda, H. A., B. Diallo, J. P. Buchet, D. Lison, H. Barennes, A. Ouangre, M. Sanou, S. Cousens, F. Tall, and P. Van de Perre. 1999. Epidemic of fatal encephalopathy in preschool children in Burkina Faso and consumption of unripe ackee (Blighia sapida) fruit. *Lancet* **353**:536–540.

Mehta, A., P. J. Mason, and T. J. Vulliamy. 2000. Glucose-6-phosphate dehydrogenase deficiency. *Baillieres Clin. Haematol.* **13**:21–38.

Minodier, P., P. Pommier, E. Moulene, K. Retournaz, N. Prost, and L. Deharo. 2003. Star anise poisoning in infants. *Arch. Pediatr.* **10**:619–621.

Montgomery, R. D. 1964. Observations on the cyanide content and toxicity of tropical pulses. *West Indian Med. J.* **13**:1–11.

Montgomery, R. D. 1965. The medical significance of cyanogen in plant foodstuffs. *Am. J. Clin. Nutr.* **17**:103–113.

Montgomery, R. D. 1980. Cyanogens, p. 143–160. *In* I. E. Liener (ed.), *Toxic Constituents of Plant Foodstuffs,* 2nd ed. Academic Press, Inc., New York, NY.

Ottesen, A., and G. C. Ziobro. 2006. Use of "DNA barcoding" to detect adulteration of Star Anise *(Illicium verum)* by other *Illicium* species, abstr. C-21. *2006 FDA Sci. Forum Poster Abstr.* Center for Food Safety and Applied Nutrition, U.S. Food and Drug Administration, College Park, MD. http://www.cfsan.fda.gov/~frf/forum06/C-21.htm.

Ozhan, H., R. Akdemir, M. Yazici, H. Gunduz, S. Duran, and C. Uyan. 2004. Cardiac emergencies caused by honey ingestion: a single centre experience. *Emerg. Med. J.* **21**:742–744.

Poulton, J. E. 1990. Cyanogenesis in plants. *Plant Physiol.* **94**:401–405.

Reddy, N. R., M. D. Pierson, S. K. Sathes, and D. K. Salunkhe. 1985. Dry bean tannins: a review of nutritional implications. *J. Am. Oil Chem. Soc.* **62**:541–549.

Schmidt, T. J., H. M. Schmidt, E. Muller, W. Peters, F. R. Fronczek, A. Truesdale, and N. H. Fischer. 1998. New sesquit-erpene lactones from *Illicium floridanum. J. Nat. Prod.* **61:**230–236.

Scott, P. M., B. B. Coldwell, and G. S. Wiberg. 1971. Grayanotoxins. Occurrence and analysis in honey and a comparison of toxicities in mice. *Food Cosmet. Toxicol.* **9:**179–184.

Seigler, D. S., G. F. Pauli, A. Nahrstedt, and R. Leen. 2002. Cyanogenic allosides and glucosides from *Passiflora edulis* and *Carica papaya. Phytochemistry* **60:**873–882.

Selmar, D., R. Lieberei, B. Biehl, and E. E. Conn. 1989. α-Hydroxynitrile lyase in Hevea brasiliensis and its significance for rapid cyanogenesis. *Physiol. Plant.* **75:**97–101.

Seyama, I., and T. Narahashi. 1981. Modulation of sodium chan-nels of squid nerve membranes by grayanotoxin I. *J. Pharmacol. Exp. Ther.* **219:**614–624.

Spencer, P. S. 1999. Food toxins, AMPA receptors, and motor neu-ron diseases. *Drug Metab. Rev.* **31:**561–587.

Spencer, P. S., and V. S. Palmer. 2003. Lathyrism: aqueous leaching reduces grass-pea neurotoxicity. *Lancet* **362:**1775–1776.

Sutlupinar, N., A. Mat, and Y. Satganoglu. 1993. Poisoning by toxic honey in Turkey. *Arch. Toxicol.* **67:**148–150.

Tanaka, K., E. A. Kean, and B. Johnson. 1976. Jamaican vomiting sickness: biochemical investigations of two cases. *N. Engl. J. Med.* **295:**461–467.

Teklehaimanot, R., B. M. Abegaz, E. Wuhib, A. Kassina, Y. Kidane, N. Kebede, T. Alemu, and P. S. Spencer. 1993. Pattern of *Lathyrus sativus* (grass pea) consumption and beta-N-oxalyl-alpha-beta-diaminoproprionic acid (beta-Odap) content of food samples in the lathyrism endemic region of northwest Ethiopia. *Nutr. Res.* **13:**1113–1126.

Tewe, O. O., and E. A. Iyayi. 1973. Cyanogenic glycosides, p. 43–96. *In* P. R. Cheeke (ed.), *Toxicants of Plant Origin, Volume II. Glycosides.* CRC Press, Inc., Boca Raton, FL.

Tshala-Katumbay, D. D., and P. S. Spencer. 2007. Toxic disorders of the upper motor neuron system, p. 353–372. *In* A. Eisen and P. Shaw. *Handbook of Clinical Neurology. Motor Neuron Disorders and Related Diseases,* vol. 82, 3rd series. Elsevier, Amsterdam, New York, NY.

Tsuge, K., M. Kataoka, and Y. Seto. 2001. Rapid determination of cyanide and azide in beverages by microdiffusion spectrophoto-metric method. *J. Anal. Toxicol.* **25:**228–236.

Viehoever, A. 1940. Edible and poisonous beans of the lima type (Phaseolus lunatus L.): a comparative study, including other similar beans. *Thai. Sci. Bull.* **2:**1–99.

Wajcman, H., and F. Galacteros. 2004. Glucose 6-phosphate dehydrogenase deficiency: a protection against malaria and a risk for hemolytic accidents. *Comptes Rendus Biol.* **327:**711–720.

Ware, G. M. 2002. Method validation study of hypoglycin A determination in ackee fruit. *J. AOAC Int.* **85:**933–938.

Wertheim, A. H. 1974. Raw beans, anyone?, p. 32–38. *In The Natural Poisons in Natural Foods.* Lyle Stuart, Inc., Secaucus, NJ.

Xavier-Filho, J., and F. A. P. Campos. 1989. Proteinase inhibitors, p. 1–27. *In* P. R. Cheeke (ed.), *Toxicants of Plant Origin, Vol. III: Proteins and Amino Acids.* CRC Press, Boca Raton, FL.

Yamada, K., S. Takada, S. Nakamura, and Y. Hirata. 1968. The structures of anisatin and neoanisatin. *Tetrahedron* **24:**199–229.

Yilmaz, O., M. Eser, A. Sahiner, L. Altintop, and O. Yesildag. 2006. Hypotension, bradycardia and syncope caused by honey poisoning. *Resuscitation* **68:**405–408.

Yoshida, A. 1973. Hemolytic anemia and G6PD deficiency. *Science* **179:**532–537.

Pathogens and Toxins in Foods: Challenges and Interventions
Edited by V. K. Juneja and J. N. Sofos
© 2010 ASM Press, Washington, DC

Chapter 20

Chemical Residues: Incidence in the United States

Stanley E. Katz and Paula Marie L. Ward

Chemical residues have occurred in foods since the beginning of civilization and perhaps considerably before any written records existed. Since *Homo sapiens* emerged, at least, foods, whether grains or animal parts, have been cooked over fires. Such food preparation resulted in thermal conversion, degradation, and/or synthetic products of unknown safety. Safety was never a consideration, since life was unfortunately short, difficult, and brutal. As the world has become aware of dangers in foods from many of its cultivation or production practices, whether for plant or animal food products, there has been a greater emphasis on the safety aspects of foods.

Residues, more often than not, are related to pesticide, antibiotic, drug, and/or whatever synthetic chemicals are used in the commercial aspects of food production, including the leachates from packaging materials. It should be remembered that other residues abound in the food supply. Excess heavy metals, salts, and products of fungal and bacterial growth as well as pathogenic and nonpathogenic microorganisms are commonly found on or in foods. Substances from air and or water, especially in areas with higher pollution loads than those found in more rural areas, are not uncommon. Foods, in general, whether of plant, animal, aquatic, or avian origin, are not produced under either sterile or especially controlled circumstances. Often ignored is the water necessary for food production. All too often irrigation waters are far less than pristine. The residues of human handling add to the mix, often with implications more serious than those of the chemicals routinely used. Residues are many and varied, and all should be aware that there is a dearth of knowledge related to the effects of the many combinations and permutations present.

From a practical point of view for this chapter, the use of the term residue relates primarily to the

trace amounts of materials found in foods of various origins resulting from the application of chemicals during production. Wherever foods are produced using some form of chemicals for plant, animal protection, and/or increased growth or yield enhancement, residues will occur. The reasons are manyfold, and starting with the obvious include (i) ignorance, (ii) misuse or use for purposes not intended, (iii) deliberate nonadherence to label uses and/or withdrawal times, (iv) unintended mistakes, (v) climatic conditions, (vi) correct uses which can lead to residues, and (vii) no understandable reason. Many of these situations will result in the presence of residues, most below the tolerance level as well as others above the established limits.

The definition of a residue remains, as always, subject to some interpretation. The most logical definition is the trace amount of chemical remaining in a product after a time period defined by the kinetics of disappearance. The actual level of an acceptable residue concentration is defined by inputs correlating toxicity, intake levels, and suitable safety factors related to the vagaries of animal testing. However, one must always realize that when a chemical is used on a product and the analytical result indicates that the chemical cannot be measured (not detected), the result does not necessarily indicate a zero concentration. The result indicates only that if any amount is present, that amount is below the level of detectability. The kinetics of disappearance are usually first order in nature, which in turn indicates that the disappearance becomes asymptotic but may never be zero. This analytical factoid is usually lost on interpreters of residue data, who unfortunately look for the presence, levels above tolerance or violative levels, and absence of measurable levels. One should be cognizant of the analytical realities. A lack of measurable

Stanley E. Katz and Paula Marie L. Ward • Department of Biochemistry and Microbiology, Rutgers, The State University of New Jersey, School of Environmental & Biological Sciences, 76 Lipman Drive, New Brunswick, NJ 08901-8525.

residues is an extremely desirable situation and something that should be a goal associated with all chemicals used in the production of foods.

This discourse will focus primarily on residues that occur from the use of chemicals in the production of foodstuffs but will include chemical conversion products and food processing products, as well as some of the environmental problems that are present or can occur if care and/or understanding is lacking.

SAMPLING

The question remains of how to sample, or more realistically, sample the various commodities in a fashion that is representative of the scale of the commodity being examined. There is no way that the food supply can be sampled to ensure that all foods are in total compliance or have residues below tolerance. The best approach and perhaps the only realistic approach is to use a statistical approach not dissimilar to those used in political polling. Otherwise, how does one obtain a representative sample of billions of pounds of commodities that will reflect the realities of production and consumption? Couple this with the vast array of agricultural chemicals that are possible, reflecting both legal and illegal usages.

In order to focus on the type of sampling and the underlying statistical basis, the approach used for estimating the incidence of residues in animal products is presented. The incidence of residues in animals produced for food (antibiotics, drugs, and pesticides), as reported by the Food Safety Inspection Service (FSIS), is a combination of four different sampling components. The most common component is the monitoring program, which takes samples from healthy animals in a random fashion. This component yields the greatest amount of residue frequency data. The surveillance component encompasses samples taken for which residue problems are known to exist and is used to reduce any abnormal level of residues. The enforcement segment is exactly what the term implies, the monitoring of producers with a record of violations. The most difficult to define are the special projects for which there are limited data on the volume of animals, any history of violations, and the need to develop residue data where information is limited or nonexistent.

There are four levels of frequency sampling, 460, 300, 230, and 90 samples per year, and all are based upon the assumption that normal-appearing, healthy, inspected, and passed carcasses will be reasonably homogeneous and will yield a sampling that is indicative of the animal food supply. The sampling frequency of 300 samples per year is consistent with a 95% confidence level, which indicates that a violative

sample occurring in 1% of the samples could be detected 95% of the time. Obviously, the greater the number of samples, the greater is the probability of detection of a violative sample. Statistically, this makes a certain amount of sense, considering the number of assays performed on each sample and the cost in time and personnel to perform the assays. For a complete discussion of the sampling, the reader should make reference to the yearly reports of the FSIS (USDA/FSIS, 2004, 2005). However, considering both the number of animals and the tonnage of edible product, this is a very small number of samples. Similarly, the import residues program is based upon the assumption that there is an equivalence between the exporter's inspection program and that of the FSIS inspection. Equivalence is a necessary aspect of export eligibility. Unfortunately, with the recent attempts of China to export cooked chicken to the United States (Henderson, 2007), the concept of equivalence in inspection practices might fall by the wayside, as the decision to allow such imports might be more in the political than in the legal or scientific arena. This scenario remains to be played to completion.

However, one has to question whether the inspection system has worked well with other potential imports from China, from where an estimated 200 shipments per month have been rejected because of residues of pesticides, antibiotics, and other potentially harmful chemicals and a lack of required labeling and/or manufacturers' information (Bodeen, 2007). At the same time, the melamine problem, in which residues have appeared in some hogs and chickens, is a matter of concern. Since both species in some areas of the United States contained low levels of melamine residues, there has been some "tap dancing" concerning safety issues and perhaps allowing the products to be sold for either human or animal consumption. Although there are no reported illnesses at this point in time and some consider the products to have a low potential for causing health problems (Anonymous, 2007), the animals are adulterated and should not enter the human or animal food chains. Melamine has never been cleared for human consumption, and there are no tolerances for such residues. This tabloid will indicate whether the producers' or regulatory (the consumer) interests are predominant. It is hoped that Michael Doyle (director of the Center for Food Safety at the University of Georgia) may be mistaken when quoted in a weekly news magazine (Volland, 2007) as saying, "Our system is out of date, and it's broken." Although this comment is editorial in nature, there seems to be more than a modicum of validity to it. For example, during the time period of 1994 to 2006, the value of food imports (related to both dollar value and quantity) increased from approximately \$34 billion to \$65 billion. The FDA inspection incidence, as a percent of imports, decreased from approximately 3% to 0.5% (Bottari, 2007). It would seem that as the numbers of imports increase and, more important, as the tonnage and/or dollar value increases, that the degree of inspection would at least remain constant and not fall significantly. It would be logical to have a statistical sampling program based upon the parameters of volume and worth.

Recently, HHS Secretary M. Leavitt stated at the Import Safety Summit on 9 July 2008 (IWGIS, 2007) that "increasing the current inspection strategy won't work." Instead, the thought is to utilize respected inspection systems in Australia and the European Union to inspect drug manufacturing facilities for pharmaceuticals/medicinals and foods. Historically, this has been a procedure used by the FSIS which has been effective in countries with similar regulatory approaches. It is a wide-open question whether this approach would be effective throughout the rest of the world. A second aspect of Secretary Leavitt's report concerns farm-raised shrimp. The establishment of "third-party" programs to certify compliance with FDA standards would be the central core of the program. Unfortunately, memoranda of understanding and/or memoranda of agreements will not ensure product compliance with chemical residue requirements. At best, a good number of years will be required to establish adequate working arrangements between international regulatory entities as well as projected self-assessment programs. All one has to do is study the economic problems related to the U.S. housing and/or investment institutions to understand that self-regulation usually will take a backseat to self-centered economic "opportunities." The effectiveness of these initiatives will be evaluated some years from now and remains a question to be answered.

SURVEY DATA

Residues in Animals

In this segment, data from the last 2 years that such data were available will be presented (USDA/FSIS, 2004, 2005). Past history can teach a great deal concerning past mistakes and/or experience, but the most current data are necessary to give one the greatest understanding of any existing problems. One of the great problems in monitoring an animal food supply is the relationship of the assay data to the total number of animals produced/slaughtered. Table 1 shows the estimated production/slaughter numbers for the years 2004 and 2005 (NMDRDB, 2002, 2003). A quick perusal of the numbers of animals

Table 1. Estimated livestock and poultry production[a]

Production class	No. slaughtered	
	2004	2005
Bovine	31,984,400	31,930,250
Veal	830,501	726,580
Sheep	2,660,592	2,545,347
Porcine	102,063,729	104,151,297
Chickens	8,876,798,073	9,141,543,766
Turkeys	252,198,225	250,556,291
Goats	558,703	541,109
Ducks	26,003,644	27,974,170
Geese	268,297	252,462
Other fowl	1,381,173	1,299,089
Rabbits	340,096	384,863
Horses	62,200	93,768
Bison	26,213	35,763

[a]Data derived from NMDRDB (2003).

makes one question the validity of the sampling system, since the numbers of animals in some production categories are extremely large in comparison to the number of samples taken for analysis. At the same time, how does one collect reasonable data? Whatever the inadequacies of the current statistical approach to sample taking, there is no way to do otherwise currently.

The data presented in Table 1 are a modification of the FSIS (Food Safety Inspection Service) system of production classes. Bovines include all cattle, except calves and veal calves, which are classified under veal. The porcine category includes all types of hogs, boars, sows, roaster pigs, etc. Chickens and turkeys include both young and mature birds. Sheep include both lambs and mature animals. Similarly, the residue data were combined into larger categories. The antibiotic/antimicrobial categories include all antibiotics, including chloramphenicol, florfenicol, and the sulfonamides; the CHC/COP/PCB/phenylbutazone grouping includes the chlorinated hydrocarbons, the chlorinated organophosphate, the polychlorinated biphenyls, and phenylbutazone; the β-agonists include clenbuterol, cimaterol, salbutamol, and some ractopamine; hormones include melengestrol, trenbolone, and zeranol; nonsteroidal anti-inflammatory drugs include thyreostats and a number of thiouracils; the miscellaneous grouping combines drugs such as carbadox, nitrofurans, and nitroimidizoles.

The residue profiles for the various categories of drugs and chemicals for 2004, as shown in Table 2, do not show any drastic trend of residue problems for the years 2000 to 2003. For 2004, it was not unexpected that the veal category would be somewhat higher than the other production categories, considering the propensity for drug usage in that group. The presence of arsenic residues in chickens,

Table 2. Residue profiles for production classes for 2004

Production class	No. of samples	No. nonviolative	No. violative	% Violative
Antibiotic/ antimicrobial				
Bovine	2,725	6	4	0.15
Veal	1,886	19	31	1.64
Sheep	452	2	0	0.00
Porcine	3,065	62	6	0.20
Chickens	1,029	1	1	0.10
Turkeys	319	0	1	0.31
Egg products	299	0	0	0.00
Heavy metal (arsenic)				
Goats	68	0	0	0.00
Chickens	547	248	0	0.00
Turkeys	377	14	0	0.00
Egg products	301	0	0	0.00
CHC/COP/PCB/ phenylbutazone				
Bovine	1,753	53	0	0.00
Veal	608	10	1	0.16
Sheep	400	17	0	0.00
Porcine	944	13	2	0.21
Goats	222	1	0	0.00
Chickens	587	0	0	0.00
Turkeys	466	1	0	0.00
Egg products	288	0	0	0.00
Antiparasitics				
Bovine	562	11	2	0.36
Veal	285	1	0	0.00
Sheep	74	1	1	1.35
Goats	232	0	12	5.17
B-agonists				
Bovine	254	0	0	0.00
Veal	248	0	0	0.00
Porcine	274	0	0	0.00
Hormones				
Bovine	238	20	0	0.00
Miscellaneous				
Bovine	401	0	5	1.25

albeit at nonviolative levels, occurred in 45% of the samples. This is not surprising, considering that arsenicals have been reported to be used in 70% of chickens (IWGIS, 2007). Organic arsenic residues are not a terribly toxic mammalian residue problem, but the same organic arsenicals can be broken down in soil to inorganic arsenic of varying valence states (considering soil disposal of chicken wastes). Continual soil disposal of chicken wastes can result in a soil buildup of arsenic and even potential leach into surface and ground waters.

The pesticide residue profiles in the animal tissue are unremarkable, with very few samples containing violative levels. Only in the bovine production class did the level reach approximately 3%, twice that of the other production classes. Somewhat surprising is the presence of violative levels of antiparasitic drugs

in goats, over 5% of the samples. The β-agonist residue picture showed no trends. Hormone residues were not detected. The miscellaneous drug category showed no patterns but did have a higher-than-expected violative incidence of 1.25%.

Table 3 shows the sampling distribution for 2004 between production classes. The monitoring samples follow the same patterns that were reported in the years 2000 to 2003 and the 1990s (USDA/FSIS, 1998; 1999). In contrast, the enforcement sampling focused primarily on the bovine and veal production classes, with the porcines accounting for a relatively modest 2.9%. The primary focus of the enforcement analytical work was related to antibiotic/antimicrobial assays and accounted for >99% of the samples. Although the number of hormone enforcement samples was low, proportionally, it accounted for 29% of the monitoring samples.

Referring again to Table 1, the numbers of animals in the production classes remained fairly constant in 2004 and 2005. There have been both modest increases and decreases in some classes, with the greatest number gain for chickens, which is not surprising since chicken production has been increasing continually for years and is related to increased national usage and exports. There were no substantial increases in samples taken for analysis. Table 4 presents the scheduled or monitoring samples taken for analysis for the various production classes. As was shown with the 2004 data, the veal category in 2005 had the highest percentage of violative antibiotic/antimicrobial samples, 1.74%, but was essentially at the same level as that found in 2004. The frequency of violative samples in the pesticide group was low, with only goats having a 1% incidence. As in past years, pesticide residues were minimal in most animal production classes.

Surprisingly, for 2005, the year after a 45% nonviolative incidence of arsenic residues in chickens, no arsenic assays were reported for chicken samples; instead, arsenic residues were looked at with egg products. There is a dearth of information on arsenical levels in chickens, in spite of the fact that 70% of chickens were reported to be fed arsenical growth promotants (Hileman, 2007). Increased information should be gathered on arsenical residue levels in chickens as a priority. Antiparasitic drugs, overall, have a higher frequency of violative residues in the production classes in which they are used. In bovines, the incidence is low, 0.14%, but in veal and goats the incidence is 1.2% and 2.2%, respectively. In sheep the incidence is 2.2%, approximately 2.5 times that reported in 2004, while in goats the 2005 frequency is 43% of that of 2004. The β-agonists, hormones, and miscellaneous drugs have a very low incidence of both nonviolative and violative residues in the production classes sampled.

Table 3. Distribution of samples and analytical assays for 2004

Class	No. of samples			
	Monitoring	Enforcement	Surveillance	Exploratory
Production				
Bovine	6,362	100,427	16	1,485
Veal	2,893	48,050	0	1
Sheep	926	283	12	0
Porcine	4,065	4,523	173	1,593
Chickens	2,163	9	0	1,200
Turkeys	1,162	0	0	0
Goats	522	31	0	0
Miscellaneous	0	117	0	692
Egg products	888	2	0	0
Total	19,001	153,425	201	4,348
Compound				
Antibiotics/antimicrobials	9,388	153,462	148	1,711
Heavy metal (As, Pd, Cd)	1,293	5	0	2,456
CHC/COP/PCB/phenylbutazone	5,974	2	0	63
Hormones	238	70	0	0
Antiparasitics	931	0	0	0
β-Agonists	776	0	44	0
Miscellaneous				
Carbadox, ractopamine, flunixin	401	0	0	0
Total	19,001	153,542	201	4,348

Table 5 presents the distribution of exposure or monitoring samples, as well as samples from suspect animals (enforcement). The frequency of exposure samples is essentially the same as in 2004, 21,479 versus 19,001, with the greatest number of assays related to antibiotics/antimicrobials. Some half of the exposure samples were taken from the bovine production class; the veal class exposure samples were approximately 25% of the total; the combined porcine category was some 11%. Again, antibiotic/antimicrobial analyses were predominant.

As was seen in 2004, the reported data indicated that the incidences of violative residues occurred in a relatively minor portion of the samples assayed. The same was the case for the nonviolative results. The published data indicated animal products were reasonably free of antibiotic/antimicrobial residues, which is good news because of the increasing problems with antibiotic-resistant bacterial infections, both from hospital and community origins. This factoid is especially important in trying to assess the underlying causes of the resistant bacterial infections. If there is any criticism of the residue sampling program, it is that the assays for nonantibiotic residues should be a greater part of the program and the program should not be so narrowly focused on the antibiotic/antimicrobial area. A greater amount of information is needed in other areas.

Residues in Milk

Milk is one product that holds a special place in the thoughts of the American public, since it is so intimately related to raising babies and children. Thus, residues occurring in the milk supply primarily from the treatment of mastitis can be very hurtful to the dairy industry. The saga of the evolution of the current milk program can be found in *Antimicrobials in Food*, 3rd edition (Katz and Ward, 2005) and need not be repeated herein. Again, the most important aspects are the residue data in recent available surveys and their significance.

Antibiotic and antimicrobial residues occur in milk primarily from mastitis treatments and nonadherence to withdrawal times, although overdosage and the use of drugs not intended for mastitis treatments remain part of the overall portrait. The concept at one time was to cut short the withdrawal times and incorporate the milk into larger volumes, a purification by dilution approach. Fortunately, in recent years, the overwhelming majority of producers adhere closely to label use directions and dairies have much stronger quality assurance systems to keep antibiotic, antimicrobial, and drug residues out of the milk supply.

Table 6 summarizes the results of 3 years of monitoring the milk supply (NMDRDB, 2003, 2004, 2005). The data report basically four sources of samples, namely, (i) bulk milk, which is raw milk from the dairy farm wherein samples are taken daily on receipt of every tanker at the dairy; (ii) pasteurized fluid milk and milk products, which are finished products, such as milks, cream, condensed and dried milk, and whey products, with a minimum of four samples tested for each product every 6 months; (iii) producer milk, which

Table 4. Residue profiles for production classes for 2005

Production class or residues	No. of samples	No. nonviolative	No. violative	% Violative
Antibiotic/ antimicrobial				
Bovine	2,882	5	1	0.035
Veal	2,011	25	35	1.74
Sheep	159	0	0	0.00
Porcine	1,019	25	7	0.69
Chickens	297	0	0	0.00
Turkeys	258	0	0	0.00
Egg products	189	0	0	0.00
CHC/COP/PCB/ phenylbutazone				
Bovine	1,975	55	2	0.10
Veal	636	14	1	0.16
Sheep	346	15	0	0.00
Porcine	997	4	0	0.00
Goats	199	4	2	1.00
Chickens	503	2	0	0.00
Turkeys	360	0	0	0.00
Horses	78	3	0	0.00
Egg products	178	0	0	0.00
Heavy metal (arsenic)				
Egg products	25	0	0	0.00
Antiparasitics				
Bovine	1,392	19	2	0.14
Veal	269	10	3	1.16
Goats	180	6	4	0.22
Sheep	211	1	1	0.47
Horses	76	1	0	0.00
β-agonists				
Bovine	562	0	0	0.00
Hormones				
Bovine	350	0	0	0.00
Veal	2,182	0	0	0.00
Miscellaneous drugs[a]				
Bovine	1,159	0	1	0.09
Porcine	317	1	0	0.00
Turkeys	251	0	0	0.00
Veal	109	0	0	0.00

[a]Includes carbadox, nitrofurans, nitroimidazoles, and ractopamine.

is raw milk obtained from bulk tanks on the farm, for which samples are reported by the permitting state (with each producer being tested at least four times every 6 months); and (iv) an other category that includes samples from milk plant silos and tanker trucks, with sampling conducted on a random basis. A positive residue is a result obtained for a drug using an assay procedure acceptable for taking regulatory action.

The results over the 3-year period covering approximately 4 million samples per year are essentially superimposable. In all cases, the results indicate that there were few positive (antibiotic residue) results in every category, with results, for the most part, under 0.2% of the samples. Positive industrial samples ranged from 0.065 to 0.079%, a very good

record. Regulatory samples, much fewer in number than the industrial samples, usually had a considerably higher percentage of positives, ranging from 0.00 to 0.226%. Why this disparity exists is somewhat unknown, leaving one to muse upon the reasons. It may simply be a reflection of some disconnect in the quality assurance program to keep antibiotic/ antimicrobial residues out of milk products sold to the public. Regardless, in the pasteurized milk products category, the products that the public uses, the incidence is very low by both sampling results. However, there should be no residues in market milk and related products. The industrial range for the 3 years was 0.00 to 0.025% versus the regulatory range of 0.00 to 0.014%. Translated, the public has little to be concerned about with regard to unwanted residues in milk products. Although there were a number of antibiotics in the assay protocols, the overwhelming number of assays was for the β-lactam (penicillin) family simply because of the potential allergic reactions from this family of antibiotics. The results overall are a far cry from the results found in 1989–1990 academic and federal milk surveys (Katz and Ward, 2005). The improvement in the antibiotic/antimicrobial residue picture for milk has been dramatic, and the industry should be complimented on its efforts.

ACRYLAMIDE RESIDUES

As was implied in the introduction, there can be residues formed through some of the normal food preparation processes, such as frying, baking, grilling, roasting, toasting, or cooking at temperatures greater than 120°C or 250°F. A common acrylamide-forming reaction occurs when naturally occurring sugars react with asparagine at elevated temperatures. It also occurs in the browning reactions between carbonyl compounds, such as sugars, carbohydrates, aldehydes, ketones, lipids and proteinaceous materials, phospholipids, and amino acids (USFDA-CFSAN, 2006; Yasuhara et al., 2003). There is a question as to whether or not acrylamide residues have cancer-related properties. This discourse will not attempt to offer an opinion on this matter but instead will focus on the residues in different types of foods and compare the levels with established tolerances, or on total diet studies. The survey data upon which Table 7 is derived have the following quantitation limits. The limit of quantitation is 10 ppb, and values below the limit of quantitation but above 0 are reported as <10 ppb.

Table 7 presents the acrylamide content of various food groupings, as reported by the FDA (2003 data updated in 2006). Unlike the FDA presentation, Table 7 summarizes the results without using brand names and places products in categories. For a more

Table 5. Distribution of samples and analytical assays for 2005

Class	No. of samples tested			
	Exposure	Exploratory	Suspect animals	Suspect population
Production				
Bovine	10,268	571	108,436	15
Veal	5,340		17,920	16,643
Sheep	716		337	13
Porcine	2,485	631	2,530	15
Chickens	800		1	
Turkeys	869		6	
Other poultry (ducks, geese, and ostrich)			10	
Goats	379		11	2
Horses	230	34	117	
Other			846	
Egg products	392			
Total	21,479	1,236	130,214	16,686
Compounds				
Antibiotics/antimicrobials	7,043	647	129,824	16,648
Heavy metal (As, Pd, and Cd)	25	340		
CHC/COP/PCB/phenylbutuzone	6,146	9	34	
Hormones	2,532			
Antiparasitics	2,098	7	2	0
β-agonists	1,020			40
Miscellaneous drugs (carbodox, nitrofurans, nitroimidizoles)	1,546	72	5	
NSAIDS[a]		661		

[a]NSAIDS, nonsteroidal anti-inflammatory drugs.

complete listing, the readers should refer to the FDA report on acrylamide in food (USFDA-CFSAN, 2006).

Perusal of the abbreviated and product-categorized data (November 2002) shows the relation between higher-temperature-prepared foods and the upper-range level of acrylamide. Fast food french fries, baked fries, potato chips, and nonpotato chip snack foods were among the foods with the highest levels of acrylamide, which in turn were a function of the temperature of preparation. Dried foods, to some extent, indicate the same temperature relationship. Baby products were overall low, as were fruits and vegetables. Basically, where higher temperatures were not a significant component of preparation, the acrylamide levels were relatively modest. In the 2003–2004 data, the temperature of the food product preparation appears to result in higher acrylamide levels, as represented by the upper-range values for the products. French fries again had the highest upper-range level, along with cookies, potato chips, and nonpotato chip snack foods, while tacos, tostadas, and tortillas had moderately high upper-range values. Restaurant and restaurant takeout food had relatively modest upper levels. Baby formula again had very low or nonmeasurable levels.

What significance are these levels to the average consumer? There is no consensus as to what the health-related problems are or might be from acrylamide ingestion from foods. Health Canada estimates that the average Canadian adult is exposed to approximately 0.4 µg/kg body weight/day, a level that appears to be consistent with estimated exposures in other nations. The most important point to be made is that acrylamide is considered a potential carcinogen and that the levels found in foods can be reduced during food preparation. At the same time, one should never conclude that acrylamide is a major driving force in causing cancers in humans, considering the multitude of chemicals in the diet, as well as the environment to which the average individual is exposed. Perhaps, a labeling approach might be the way that the industry can be induced into lowering acrylamide levels. Just as trans fat concentrations have become part of the label on many products, there is no reason why the acrylamide levels cannot be placed on a label. A well-informed consumer could be the most powerful driving force to reduce acrylamide levels in baked and fried foods and foods heated above 120°C for extended periods of time.

RESIDUES IN WATER

It may be a bit disconcerting to realize that our waters, freshwater streams and rivers, lakes, and ground waters might be considered somewhat of a chemical soup. The fact that all medications taken by

Table 6. Results of sampling and analyses from 1 October to 30 September of the periods 2000–2001, 2001–2002, and 2002–2003[b]

Yr or source[a]	Industrial			Regulatory			Total		
	No. of samples	No. positive	% Positive	No. of samples	No. positive	% Positive	No. of samples	No. positive	% Positive
2000–2001									
Grade A milk									
Bulk	2,913,675	1,801	0.062	24,244	50	0.198	2,988,919	1,851	0.063
Past	7,958	2	0.025	55,128	0	0.000	63,086	2	0.003
Prod	550,132	855	0.155	153,568	248	0.161	703,700	1,097	0.150
Other	79,206	18	0.022	6,720	8	0.119	84,936	26	0.030
Grade B milk									
Bulk	339,346	336	0.099	1,771	4	0.226	340,117	346	0.100
Past	258	0	0.00	682	0	0.00	910	0	0.00
Prod	52,440	117	0.22	2,240	1	0.045	54,680	118	0.216
Other	14,302	6	0.042	976	1	0.102	15,278	7	0.046
Total	3,987,287	8,135	0.079	246,329	312	0.084	4,203,616	8,401	0.081
2001–2002									
Grade A milk									
Bulk	3,266,029	1,925	0.059	25,151	42	0.167	3,291,180	1,967	0.056
Past	5,522	0	0.00	50,153	3	0.006	55,675	3	0.006
Prod	516,557	857	0.166	184,356	263	0.148	700,915	1,120	0.160
Other	75,004	9	0.012	6,144	11	0.179	81,148	20	0.025
Grade B milk									
Bulk	332,242	245	0.075	280	0	0.00	832,522	245	0.074
Past	24	0	0.00	897	0	0.00	921	0	0.00
Prod	47,558	140	6.29	3,747	1	0.027	51,305	141	0.275
Other	12,025	4	0.033	725	0	0.00	12,750	4	0.031
Total	4,254,961	3,180	0.075	271,453	320	0.118	4,526,414	3,500	0.079
2002–2003									
Grade A milk									
Bulk	3,243,092	1,658	0.051	22,402	42	0.187	3,265,494	1,700	0.052
Past	4,933	1	0.020	49,494	7	0.014	54,427	8	0.015
Prod	471,976	714	0.157	148,047	204	0.138	620,017	918	0.148
Other	72,215	15	0.021	6,848	12	0.175	79,023	27	0.034
Grade B milk									
Bulk	305,040	198	0.065	1,300	1	0.077	306,340	199	0.065
Past	159	0	0.00	346	0	0.00	505	0	0.00
Prod	42,851	91	0.212	2,759	0	0.00	45,610	91	0.199
Other	10,659	2	0.019	859	0	0.00	11,518	2	0.017
Total	4,150,919	2,679	0.065	232,055	266	0.115	4,382,971	2,945	0.067

[a]Prod, producer; Past, pasteurized.
[b]Data derived from NMDRDB (2003, 2004).

the general public have the potential of being present at low levels in sewage and depending upon the treatment processes, appearing in receiving surface waters. Add to this antibiotics, antimicrobials, growth promotants, and a host of other drugs fed to animals leaching into surface waters. Pesticides from the treatment of animals and crops as well as personal care products, plasticizers, and cleaning chemicals are all known to be present in wastewaters (Kolpin et al., 2002).

As early as 1977, there was evidence of the presence of pharmaceuticals in the effluent coming from the Kansas City treatment plant (Hignite and Azarnoff, 1977). Richardson and Bowron (1985) reported pharmaceuticals in surface waters; their report was followed, after at least a decade, by a number of reports that indicated the almost ubiquitous presence of pharmaceuticals in surface waters in Europe and Brazil (Richardson and Bowron, 1985; Herberer and Stan, 1997; Stumpf et al., 1999; Hally-Sorenson et al., 1998). Kolpin et al. (2002) of the USGS (U.S. Geological Survey) in a national "reconnaissance" looked at the occurrence and concentrations of a wide spectrum of pharmaceuticals and personal care products and found a significant number of the target compounds in the streams surveyed. Underlying all the investigations concerning the presence of all the compounds in waters, although the concentrations were in the low-ppb range, was the question of biological significance. Early insight into

Table 7. Acrylamide levels in foods[b]

Data to 15 November 2002			Data from 20 November 2003 – 7 October 2004		
Product type	No. of brands	No. or range ppb[a]	Product type	No. of brands	No. or range ppb
Baby fruits	2	ND	Infant formula	4	ND
Baby vegetables	2	ND–121	Breads/bakery products	33	ND–59
Baby cereals	3	ND–<10	Cereals	20	ND–534
Baby dinners	2	30–75	Cookies	34	ND–1,540
Baby biscuits	5	ND–130	Fruit/vegetable products	9	ND–239
Infant formula	4	4	Olives	16	ND–798
French fries			Potato chips	13	412–1,570
Fast foods	8	117–1,030			
Baked	5	119–1,325			
Not baked	5	20–212			
Potato chips	6	117–2,712	Snack foods (not potato chips)	23	462–1,340
Breads/bakery products	9	ND–343	Taco, tostada, and tortilla	6	29–794
Cereals	2	47–266	French fries	24	ND–408
Snack food (not potato chips)	9	12–1,168	Restaurant takeout		
Protein foods	10	ND–116	Asian	22	ND–84
Gravies/seasonings	10	ND–115	Hispanic	33	ND–154
Nuts/nut butters	8	ND–457	Southern/Cajun	19	ND–202
Crackers	7	26–505			
Cookies	4	36–199			
Chocolate products	9	ND–909			
Canned fruits/vegetables	7	ND–83			
Coffee (not brewed)	5	175–351			
Frozen vegetables	7	<10			
Dried fruits	7	<10–1,184			
Dairy	3	ND–43			

[a]ND, not detected.
[b]Data derived from USFDA-CFSAN (2006).

some interactions between antibiotics and pesticides was provided by Bordas et al. (1997), who found an enhancement in antibiotic resistance in a sensitive strain of *Staphylococcus aureus* after exposure to both types of compounds. Kleiner et al. (2007) determined the ability of different combinations of multiple agrochemical compounds, at the 10-ppb level, to influence the resistance profile of the sensitive *S. aureus* strain, as measured by changes in the baseline MIC. Kleiner et al. (2007) used 17 different agrochemical compounds and found that single-compound exposure increased the MIC values in 21.8% of the exposures. Two-compound exposures resulted in increases in 94.9% of the exposures. Most notable was the extent of the increases: 34% were four to eight times and 60.6% were greater than eight times the baseline MIC. For six-compound combinations, the results showed that 94.1% of the MIC values increased, with 78.5% of the increases greater than eightfold and 15.6% having increases between 4 and 8 times the baseline.

Care must be taken in interpreting these data since the concentrations of each compound were 10 ppb, but the accumulated levels were approximately 20 ppb for the two-compound mixtures and approximately 60 ppb for the six-compound

exposure. These values were many times the levels found by Kolpin et al. (2002). What is important are the facts that combinations of compounds did result in a biological effect, in this case increases in the MIC (antibiotic resistance) and that the scientific community has little experience handling many-compound exposures.

PESTICIDE RESIDUES

The administration of the pesticide regulatory system is quite complex. The responsibility for registration and for pesticide residue tolerances in foods is divided among three agencies. The EPA (Environmental Protection Agency) registers the use of pesticides and sets tolerances for residues in foods; the USDA's AMS (Agricultural Marketing Service) is responsible for raw agricultural product and processed food residue testing, while the FSIS monitors meat, poultry, and some egg products. The FDA requires incidence/level data on commodity/pesticide combinations, carries out market basket surveys, and enforces tolerances in both domestic and imported foods, including animal feeds shipped via interstate commerce.

The FDA uses both single-residue methodology in limited cases for foods and the less resource-intense multiresidue methodology, which can determine about half the 400 pesticides with and without EPA-set tolerances, including metabolites and impurities. The multiresidue methodology is used to detect the organochlorides, organophosphates, organosulphates, organonitrogens, N-methyl carbamates, pyrethroids, triazines, and conozoles/triazoles, as well as nearly 200 different insecticides, herbicides, fungicides, plant growth regulators, their metabolite degradation products, and isomers. All residues tentatively identified must be validated by mass spectrometry or another secondary system (Punzi et al., 2005).

The USDA Pesticide Data Program (Punzi et al., 2005) collected pesticide residue data since 1991. The data collected before 1994 are considered somewhat unreliable. In 1994, 38.5% of food samples analyzed showed no detectable residues. In 2002, the percentage of samples with nondetectable levels rose to 57.9%. Also, in 1994, 9.8% of the samples contained four or more residues per sample; in 2002, 8.2% of the samples analyzed had such multiple residues. Of the more than 100,000 samples collected throughout the Pesticide Data Program period, some 65% were fruits and vegetables and 20% were processed commodities (Table 8). Processed commodities tended to have violative detections lower than those of fresh fruits and vegetables. Between 1993 and 2002, violative detections with fruits and vegetables were narrowly ranged. In 2003, 2,344 domestic samples were collected from 45 states and Puerto Rico, and 4,890 samples were collected from 99 countries. Pesticide residues were not detected in 62.7% of domestic and 71.8% of imported samples. Of those samples, 2.4% of domestic and 6.1% of imported samples had violative levels. Table 9 shows the breakdown of incidences of violative samples in broad categories of foods. Nine domestic and 26 imported samples contained residue levels which exceeded tolerances; samples containing pesticide residues not registered for use in foods in the United States included 48 domestic and 270 import samples.

Table 8. Incidence of pesticide residue in domestic and imported foods, 1993–2003[a]

Sample type	% Detectable
All samples	58
Grains and grain products	47
Processed commodities	34
Fruits and vegetables	65
Milk	15
Beef and chicken	10
Water	61

[a]Summarized from Punzi et al. (2005).

Table 9. Violative pesticide residues in domestic and imported foods, 2003[a]

Sample type	% Violative	
	Domestic	Imported
All samples	2.4	6.1
Grains and grain products	0.0	1.4
Milk, dairy products, eggs, fish	0.0	0.0
Fruits	2.2	5.3
Vegetables	1.9	6.7
Other	16.7	14.1

[a]Summarized from Punzi et al. (2005).

Table 10 lists the common fruits and vegetables in descending ranking of pesticide usage (1993 to 2002) and the sample analyses for 2003. With the exceptions of cherries, strawberries, peaches, cucumbers, and green beans, the violative residues on individual commodities were low.

Animal feed residues pose a dual problem, not only in the feed itself but possibly for the animal being fed, not from toxicity but as a potential residue problem. In the animal feeds sampled in 2003, malathion, methoxychlor, diazinon, and chlorpyrofos-methyl accounted for 76.7% of all residues. Of the 438 domestic samples assayed, 69.2% contained no violative residues, while 1.8% showed residues

Table 10. Listing of common fruits and vegetables, in descending rank of pesticide usage[a]

Commodity	2003		
	Total no. of samples	Without residues (%)	Violative residues (%)
Peaches	95	32.6	2.1
Apples	183	48.6	1.1
Sweet bell peppers	35	74.3	0.0
Celery	23	8.7	0.0
Strawberries	79	44.3	3.8
Cherries	19	27.8	16.7
Pears	43	46.5	0.0
Grapes (imported)	23	67.9	0.0
Spinach	22	36.4	4.5
Lettuce (leaf)	49	40.8	6.1
Potatoes	105	64.8	0.0
Carrots	56	69.6	0.0
Green beans	61	67.2	3.3
Hot peppers	5	100.0	0.0
Cucumbers	44	70.5	6.8
Grapes (domestic)	28	67.9	0.0
Oranges	120	30.8	2.5
Grapefruit	12	25.0	0.0
Mushrooms	3	66.7	0.0
Tomatoes	89	76.4	0.0
Broccoli	27	81.5	0.0
Cabbage	73	80.8	0.0
Bananas	12	100.0	0.0
Onions	64	92.2	0.0

[a]From Punzi et al. (2005).

exceeding tolerances. Interestingly, the 1.8% of the samples (eight samples) that were in violation contained 11 residues. Of the 60 imported feed samples taken, 83.3% (50 samples) contained no violative pesticide residues; 5% (3 samples) contained violative residues.

OVERVIEW

During the discourse in this chapter, there have been several comments made concerning the adequacy of the sampling system. No doubt it is easier to criticize than to offer solutions. But, to be quiet about what is a problem is tantamount to acceptance. Hence, any comments concerning the perceived inadequacies of the sampling system are made in the spirit of constructive suggestions.

How does one sample a food supply as large and diverse as that found in the United States? Obviously, one cannot analyze every animal, food batch, bulk tank, produce field, etc. A statistical approach must be taken, but will the few thousand monitoring samples of what appear to be healthy animals adequately represent the approximately 9 billion chickens, 32 million bovines, and 100 million porcine produced annually in the nation? Albeit there are many more enforcement or suspect samples taken, is there a better way to give better statistical representation of the overall harvest of animals?

Analytically, in the vast majority of animal samples, the analyte is an antibiotic or antimicrobial. This is very understandable with the milk supply because of mastitis treatments and the problems with allergic reactions, especially with the β-lactams and chloramphenicol. Is the preponderance of analyses for the antibiotics/antimicrobials the best approach for many of the production classes, especially among the combined bovine production class? Would increased surveillance for hormones in beef cattle be more in line with the public's concerns? Would increased surveillance for arsenical residues in chickens be an important aspect, since it is reported (Hileman, 2007) that 70% of chickens produced are fed arsenicals for growth enhancement? These are but a few of the questions that have evolved from the review of residue levels in animal production.

On the positive side, industry has done a very solid and responsible job in lowering antibiotic/antimicrobial residues in milk. The public should feel quite good about the overall quality of the market milk and milk products available. The animal product supply is also of high quality, with very few samples exhibiting any nonviolative and/or violative levels of drug residues. Similarly, pesticide residues in fruits and vegetables remain at a low violative

incidence, also a positive for the regulatory services. A similar question must be asked concerning the sampling protocols, namely, how can one handle, statistically, the very extensive list of pesticides approved for use to ensure a screening for all registered as well as many unregistered and therefore illegal compounds?

Overall, the data indicate that the U.S. food supply has relatively low levels of chemical residues, and the public should be assured concerning its safety. The problems with the U.S. food supplies are not the overall-occurring chemical residues but more likely the wide variety of bacteriological residues occurring from processing and distribution systems.

REFERENCES

Anonymous. 2007. Transcript of USDA-FDA update on adulterated animal feed. Release no. 0158.07. United States Department of Agriculture, Washington, DC. http://news.tradingcharts.com/futures/6/9/94022496.html.

Bodeen, C. 2007. China's food safety woes expand overseas. *Yahoo News*, 12 April.

Bordas, A. A., M. S. Brady, M. Siewierski, and S. E. Katz. 1997. In vitro enhancement of antibiotic resistance development: interaction of residue levels of pesticides and antibiotics. *J. Food Prot.* 60:532–536.

Bottari, M. 2007. Trade deficit in food safety. *Public Citizen's Global Trade Watch*, July.

Hally-Sorenson, B., S. Nors-Nielson, P. F. Lanzky, F. Ingersleu, H. C. Holten Lutzhoff, and S. F. Jorgensen. 1998. Occurrence, fate and effects of pharmaceutical substances in the environment. *Chemosphere* 36:357–392.

Henderson, D. 2007. U.S. proposal to allow chicken imports from China raises health concern. *International Herald Tribune*, 9 May.

Herberer, T., and H. J. Stan. 1997. Determination of clofibric acid and N-(phenylsulfonyl)-sarcosine in sewage, river and drinking water. *Int. J. Environ. Anal. Chem.* 67:113–124.

Hignite, C., and D. L. Azarnoff. 1977. Drugs and drug metabolites as environmental contaminants: chlorophenoxyisobutyrate and salicyclic acid in sewage water effluent. *Life Sci.* 20:337–341.

Hileman, B. 2007. Arsenic in chicken production. *Chemical Engineering News*, 9 April.

IWGIS. 2007. Action Plan for Import Safety: a Roadmap for Continual Improvement. A Report to the President. Interagency Working Group on Import Safety, Washington, DC. http://www.importsafety.gov/report/actionplan.pdf.

Katz, S. E., and P. M. Ward. 2005. Antibiotic residues in food and their significance, p. 599–618. In P. M. Davidson, J. N. Sofos, and A. L. Branen (ed.), *Antimicrobials in Food*, 3rd ed. Taylor and Frances, New York, NY.

Kleiner, D. K., S. E. Katz, and P. M. L. Ward. 2007. Development of in vitro antimicrobial resistance in bacteria exposed to residue level exposures of antimicrobial drugs, pesticides and veterinary drugs. *Chemotherapy* 53:132–136.

Kolpin, D. W., E. T. Furlong, M. T. Meyer, E. M. Thurman, S. D. Zaugg, H. B. Barber, and H. T. Buxton. 2002. *Environ. Sci. Technol.* 36:1202–1211.

NMDRDB. 2003. National Milk Drug Residue Data Base. Fiscal Year 2002 Annual Report. October 1, 2001–September 2002. GLH Incorporated, Lighthouse Point, FL.

NMDRDB. 2004. National Milk Drug Residue Data Base. Fiscal Year 2003 Annual Report. October 1, 2002–September 2003. GLH Incorporated, Lighthouse Point, FL.

NMDRDB. 2005. National Milk Drug Residue Data Base. Fiscal Year 2004 Annual Report. October 1, 2003–September 2004. GLH Incorporated, Lighthouse Point, FL.

Punzi, J. S., M. Lamont, D. Haynes, and R. L. Epstein. 2005. USDA pesticide data program: pesticide residues on fresh and processed fruit and vegetables, grains, meats, milk, and drinking water. *Outlooks Pest Manag.* **16:**131–137.

Richardson, M. L., and J. M. Bowron. 1985. The fate of pharmaceutical chemicals in the aquatic environment. *J. Pharm. Pharmacol.* **37:**1–12.

Stumpf, M., T. A. Ternes, R.-D. Wilkens, S. V. Rodrequez, and W. Bauman. 1999. *Total Environ.* **225:**135–141.

USDA/FSIS. 1998. *Food Safety and Inspection Service 1998 National Residue Program Domestic Residue Data Book.* Food Safety Inspection Service, U.S. Department of Agriculture, Washington, DC.

USDA/FSIS. 1999. *Food Safety and Inspection Service 1999 National Residue Program Domestic Residue Data.* Food Safety Inspection Service, U.S. Department of Agriculture, Washington, DC.

USDA/FSIS. 2004. *2004 FSIS National Residue Program Data.* Food Safety Inspection Service, U.S. Department of Agriculture, Washington, DC.

USDA/FSIS. 2005. *2005 National Residue Program Data.* Food Safety Inspection Service, U.S. Department of Agriculture, Washington, DC.

USFDA-CFSAN. 2006. Survey Data on Acrylamide in Food: Industrial Food Products. Center for Food Safety and Applied Nutrition, United States Food and Drug Administration, Washington, DC. http://www.cfsan.fda.gov/~dms/acrydata.html.

Volland, A. 2007. A human connection? *U.S. News and World Report,* 14 May.

Yasuhara, A., Y. Tanaka, M. Hengel, and T. Shibamota. 2003. Gas chromatographic investigation of acrylamide formation in browning model systems. *J. Agric. Food Chem.* **51:**3999–4003.

Pathogens and Toxins in Foods: Challenges and Interventions
Edited by V. K. Juneja and J. N. Sofos
© 2010 ASM Press, Washington, DC

Chapter 21

A European Food Safety Perspective on Residues of Veterinary Drugs and Growth-Promoting Agents

MARTIN DANAHER AND DEIRDRE M. PRENDERGAST

Veterinary drugs are essential in modern intensive agriculture to maintain the health and yields of food-producing animals. Residues of veterinary drugs can occur in food and may give rise to human health concerns through the direct consumption of meat products containing residues or through indirect exposure through the selection of resistance in pathogens. In recent years, steady progress has been made in the control of veterinary drug and growth promoter residues in food in the European Union. This has been achieved through research activities (between and within member states), improvements in residue control, and developments in analytical technology. These developments have identified a number of high-profile residue findings, such as nitrofurans, chloramphenicol (CAP), and streptomycin (Anonymous, 2002a; O'Keeffe et al., 2004). It is likely that the rate of violations will increase in the future through the development of improved chemical assays that test for a wider range of residues or apply lower detection levels using more sophisticated techniques, such as liquid chromatography coupled to tandem mass spectrometry (LC-MS/MS) or ultra-performance liquid chromatography coupled to time-of-flight mass spectrometry (UPLC-TOF) (Kaufmann et al., 2008; Kinsella et al., 2007; Radeck and Gowik, 2005; Stolker et al., 2008). Such technologies have been applied in surveys and to different and sometimes more exotic food matrices (Ding et al., 2006).

In this chapter, European contributions to food safety aspects of veterinary drugs and growth-promoting agents (GPAs) are described. This includes a rounded discussion, including licensing of veterinary drugs, setting of maximum residue limits (MRLs), and the impact of legislation on the availability of veterinary medicines. This is followed by a description of the current European Union strategy for residue control, the design of national monitoring plans, the most recently published results for European Union residue testing, and concerns over indirect exposure. The chapter concludes with recent developments in the area of LC-MS/MS, which has become the technique of choice for surveillance of residues in food; validation of the performance of methods; and future perspectives, including the development of a risk-based surveillance program.

VETERINARY DRUGS, FEED ADDITIVES, GPAs, AND THEIR USE IN FOOD-PRODUCING ANIMALS IN THE EUROPEAN UNION

There are several groups of substances approved for treating food-producing animals in the European Union, including antibacterials, anthelmintics, anticoccidials (or antiprotozoan drugs), sedatives, pesticides (applied as veterinary drugs), and anti-inflammatory drugs (steroidal and nonsteroidal). GPAs are banned, but a number of products are allowed under certain animal health grounds, e.g., clenbuterol is allowed for delaying labor and treatment of respiratory infection in animals (Anonymous, 1990).

In recent years, there has been a reduction in the number of effective veterinary drugs and feed additives due to the development of drug resistance, removal of many feed additives from the market, and increasing hurdles that the pharmaceutical industry has to overcome to gain market authorization (White et al., 2002a; McEvoy, 2002; Anonymous, 2003). The most controversial has been the precautionary bans on antibiotic feed additives because of the

Martin Danaher and Deirdre M. Prendergast • Food Safety Department, Teagasc, Ashtown Food Research Centre, Ashtown, Dublin 15, Ireland.

potential risk for bacterial resistance to humans (McEvoy, 2002; Pugh, 2002). Feed additives are added to feed to enhance production efficiency or animal health. Concern was raised over this practice in the 1960s, because it was suggested that the non-essential use of antibiotics to enhance animal production would devalue their usefulness in the treatment of zoonoses. Further concern over the development of drug resistance in bacteria and human health implications led to questions about the use of antibiotics for therapeutic or growth-promoting purposes. In the United Kingdom, the debate led to the main recommendation that no antibiotic used for therapeutic purposes in animals or humans should be used for enhancing animal production (Swann, 1969). As a result, a number of new antibiotics were developed solely as feed additives for application as enhancers in animal production. With this approach, it was presumed that any resultant resistance would be of no consequence to animal or human therapy.

However, a number of antibiotic feed additives were banned in the European Union in the late 1990s because of concerns over the possible links between veterinary drug residues in food and antimicrobial resistance. There was a perception of widespread use of antibiotic feed additives and the transfer of antibiotic-resistant organisms and resistance genes to humans as a result of veterinary and zootechnical use in food animals (McEvoy, 2002). Feed additives include a range of substances, including performance-enhancing antibiotic growth promoters, many but not all anticoccidials, binding agents, and enzymes (McEvoy, 2002). Avoparcin and ardacin were banned in 1997 (Anonymous, 1997), and bans on carbadox and olaquindox (Anonymous, 1998a); and bacitracin, spiramycin, tylosin, and virginiamycin (Anonymous, 1998b) followed.

The availability of veterinary medicines has become a serious problem in recent years, owing mainly to the fact that few new products are coming onto the market. A further factor is the restriction on obtaining certain types of medicines. In an attempt to reduce the frequent and the supposed unnecessary usage of antibiotics, they have been made available as prescription-only medicines. More recently, it has been suggested that anthelmintics should no longer be distributed as over-the-counter medicines and should receive prescription-only status. It has been proposed that there are now few effective anthelmintic agents and that their lifetimes could be extended through restriction of their usage and through alternative control approaches. However, the mechanism for distributing anthelmintics has been left in the charge of the competent authority in individual member states.

EUROPEAN FOOD SAFETY LEGISLATION FOR THE CONTROL OF AGROCHEMICAL RESIDUES IN FOOD

Veterinary Drugs and GPAs in the European Union

Legislation can be categorized into the following areas: (i) evaluation of safety, (ii) procedures for the establishment of MRLs, (iii) measures to control residues in food produced within the European Union and imported from developing countries, and (iv) guidelines for validation of analytical methods.

Evaluation of safety and establishment of European Union MRLs

The legal treatment of food-producing animals with approved veterinary medicinal products may result in residues in edible tissues that pose a risk to human health. In order to reduce this risk, MRLs have to be established, where necessary, in food matrices, such as meat, fish, milk, eggs, and honey. MRLs are established at a European Union level as an addition to consumer protection; they also facilitate free movement of foodstuffs and veterinary medicinal products between member states. The procedure to establish MRLs for veterinary medicinal products in foodstuffs of animal origin is described in Council Regulation No. 2377/90 (Anonymous, 1990). An MRL is defined as the concentration of residue legally permitted or recognized as acceptable in or on a food that occurs in edible tissues after treatment with a veterinary medicinal product (expressed in mg kg^{-1} or µg kg^{-1} on a fresh weight basis). A list of approved and banned pharmacologically active substances used in veterinary medicinal products is included in the Annexes of Council Regulation 2377/90 (Anonymous, 1990). Annex I contains substances for which MRLs have been established. Substances that have been considered safe and do not require MRLs are included in Annex II. Annex III contains substances that have provisional MRLs listed because submitted studies or documentation was incomplete. Annex IV contains veterinary medicinal products for which MRLs cannot be established. The administration of the substances listed in Annex IV to food-producing animals is prohibited in the European Union.

Safety assessments of veterinary medicinal products are carried out by the Committee for Veterinary Medicinal Products (CVMP), which is part of the European Medicinal Evaluation Agency (EMEA). Safety assessments take into account assessments by international organizations, in particular the Codex Alimentarius Commission, or by other scientific committees established within the European Union. MRLs are usually established using the acceptable daily intake (ADI) concept, which is based on multiple-dose toxicological studies, usually of long duration, that are supposed to represent chronic exposure to residues. MRLs for antibiotics are increasingly set on the basis of establishing MIC values. MIC values are the concentration of antibiotic which has no inhibitory effect on the bacteria in the human gut. ADI is established by applying a safety factor to a MIC value or a no-observed-adverse-effect level (NOAEL) value that has been identified in the most sensitive species. However, in the event that metabolic and pharmacokinetic data identify an alternative species that is more suitable for extrapolation to humans, then the NOAEL for this species can be used. The NOAEL is divided by a safety factor to establish an ADI. A safety factor of 100 is usually applied, which is based on the assumption that the human is 10 times more sensitive to the substance than experimental animals and that there is a 10-fold range in sensitivity within the human population (10×10). However, alternative safety factors can be applied depending on (i) the quality of the toxicological data, (ii) the toxicokinetic and/or toxicodynamic data, (iii) the nature of toxicity, (iv) the presence of highly sensitive groups in the population, (v) whether the source of ADI is based on human studies, and (vi) the presence of the substance in common foodstuffs as a contaminant. In the European Union, safety factors applied to veterinary drugs are generally in the range of 100 to 1,000.

Control of Veterinary Drug Residues in Food of Animal Origin

European reference laboratory structure

A complex laboratory structure comprising national reference laboratories (NRLs) and community reference laboratories (CRLs) has been established to provide seamless implementation of residue control within the European Union. NRL(s) are established at a member state level, and their role is to provide expert monitoring of residues. The role of NRLs is to provide support for residue control at a member state level, including provision of expert laboratory analysis, input to annual national monitoring plans, and acting as a contact point with the CRLs. CRLs are established at a European Union level to provide overarching expertise for different substance groups and/or foods. In this capacity, they are funded by the European Commission for (i) management activities, (ii) method development and validation, (iii) organization of proficiency tests and provision of reference materials, and (iv) provision of technical support to member states and the commission. The current CRLs and NRLs are listed in Council Regulations 776/2006/EC and 130/2006/EC, respectively.

Residue control plans

The control of veterinary drug residues in live animals and their food is described in Council Directive 96/23/EC (Anonymous, 1996a). The substances or residues to be monitored in food are listed in Annex 1 of this document and are divided into two groups, A and B. Group A substances are listed in Table 1 and include substances having anabolic effect and unauthorized substances. Group A also includes substances listed in Annex IV to Council Regulation (EEC) No. 2377/90 (namely, CAP, nitrofurans, and nitromidazoles). Group B substances are listed in Table 2 and include veterinary drugs and contaminants. Substances listed in Annex I are monitored in live animals, their excrement, body fluids,

Table 1. Group A residue or substance to be detected by type of animal, their feeding stuffs, including drinking water, and primary animal products

Group	Substances having anabolic effect and unauthorized substances	Bovine, ovine, caprine, porcine, equine animals	Poultry	Aquaculture animals	Milk	Eggs	Rabbit meat and the meat of wild game and farmed game[a]	Honey
A1	Stilbenes, stilbene derivatives, and their salts and esters	×	×	×			×	
A2	Antithyroid agents	×	×				×	
A3	Steroids	×	×	×			×	
A4	Resorcylic acid lactones, including zeranol	×	×				×	
A5	Beta-agonists	×	×				×	
A6	Compounds included in Annex IV of Council Regulation (EEC) no. 2377/90 of 26 June 1990	×	×	×	×	×	×	

[a]Only chemical elements are relevant where wild game is concerned.

Table 2. Group B residue or substance to be detected by type of animal, their feeding stuffs, including drinking water, and primary animal products

Group	Veterinary drugs and contaminants[a]	Bovine, ovine, caprine, porcine, equine animals	Poultry	Aquaculture animals	Milk	Eggs	Rabbit meat and the meat of wild game and farmed game[b]	Honey
B1	Antibacterial substances, including sulphonomides, quinolones	×	×	×	×	×	×	×
B2a	Anthelmintics	×	×	×	×		×	
B2b	Anticoccidials, including nitroimidazoles	×	×			×	×	
B2c	Carbamates and pyrethroids	×	×				×	×
B2d	Sedatives	×						
B2e	NSAIDs	×	×		×		×	
B2f	Other pharmacologically active substances							
B3	Other substances and environmental contaminants							
B3a	Organochlorine compounds, including PCBs	×	×	×	×	×	×	×
B3b	Organophosphorus compounds	×			×			×
B3c	Chemical elements	×	×	×	×		×	×
B3d	Mycotoxins	×	×	×	×			
B3e	Dyes			×				
B3f	Others							

[a] Veterinary drugs include unlicensed substances which could be used for veterinary purposes. PCBs, polychlorinated biphenyls.
[b] Only chemical elements are relevant where wild game is concerned.

tissue, animal products, animal feed, and drinking water.

Monitoring plans are drawn up by the competent authority in each member state and submitted by the end of March each year. A list of animal species and the relevant substances to be monitored in them are listed in Annex II of 96/23/EC (also described in Tables 1 and 2). Sampling for group A and B substances is described in Annexes III (sampling strategy) and IV (sampling levels and frequency). The monitoring plan is designed to identify the cause of residues in food. Samples are collected such that the element of surprise is constantly maintained. A targeted sampling approach is adopted for the control of residues in food of animal origin in the European Union. It is stated that a randomized sampling protocol may be adopted only if justified. The targeted approach for the control of residues is described in Table 3. The number of samples to be taken each year is prepared based on the total production number for each member state for the previous year. The minimum number of samples to be collected for each species is defined for group A and B substances. The legislation allows the scope for the individual member states to increase sample numbers for the different subgroups based on potential abuse or misuse of products. Samples may also be further divided to allow collection of residues from farms, slaughter houses, processing plants, or wholesalers.

Control of prohibited substances

A number of substances have been banned or prohibited from use in the European Union, resulting in their categorization as group A substances. These include GPAs such as hormones and β-agonists, described in Council Directive 96/22/EC (Anonymous, 1996b). A number of these agents are applied widely in the treatment of food-producing species in various non-European Union countries. However, they are banned in the European Union because there is concern that increased exposure to hormones can be associated with an increased risk of cancer and detrimental effects in development. Most of these effects have been demonstrated following high-dose exposure of experimental animals, as well as human beings, at different stages of development (including adulthood). However, particularly with regard to the subject of estrogenic effects during development, there is no compelling evidence suggesting that these effects do not also occur at low doses. More important, if endogenous levels of hormones are associated with lifetime risk of disease, for example, human breast cancer, then continuous additional exposure, even at low doses, is likely to add to this risk but, as yet, cannot be quantified. The

Table 3. Sampling

Sampling	Species description
Bovine	
Numbers	0.25% and 0.15% from groups A and B, based on annual production from previous year
Site	Samples to be taken on farm (50%) and at slaughter house (50%)
Minimum samples per group (%)	5% from each subgroup of A. 30, 30, and 15% for B1, B2, and B3, respectively
Porcine	
Numbers	0.02 and 0.03% from groups A and B, based on annual production from previous year
Site	2% of group A sampling must be from on-farm visits.
Minimum samples per group (%)	5% from each subgroup of A. 30, 30, and 15% for B1, B2, and B3, respectively
Sheep and goats	
Numbers	0.01 and 0.04% from groups A and B, based on annual production from previous year
Site	Samples to be taken where appropriate
Minimum samples per group (%)	5% from each subgroup of A. 30, 30, and 15% for B1, B2, and B3, respectively
Equine	
Numbers	To be determined by each member state in relation to the problems identified
Poultry	
Numbers	Minimum total (groups A and B) of 1 sample per 200 tonnes of annual production (dead weight)
Site	One-fifth of group A samples must be taken on-farm.
Minimum no. of samples per group	A minimum of 100 samples for each subgroup of A and B (if production is >5,000 tonnes) or 5% from each subgroup of A and 30, 30, and 15% for B1, B2, and B3, respectively
Aquaculture: finfish	
Numbers	Minimum total (groups A and B) of 1 sample per 100 tonnes of production
Site	All samples must be taken on-farm for group A substances. Sampling for group B should be carried out preferably on-farm with fish ready for the marketplace or at processing or wholesale.
Minimum no. of samples per group	One-third of total samples must be for group A. Two-thirds of total samples must be for group B.
Aquaculture: other	When member states have reason to believe that veterinary medicine or chemicals are being applied to the other aquaculture products or when environmental contamination is suspected, then these species must be included in the sampling plan in proportion to their production as additional samples to those taken for finfish farming products.

Sub-Committee on Veterinary Public Health (SCVPH) report of April 1999 identified several important considerations linked to the use of hormones for growth-promoting purposes in food-producing animals (Anonymous, 2002b). These include

- the possible consequences of continuous daily exposure—even to low levels of hormones—of all segments of the human population, including at the most susceptible periods (in utero and prepubertal);
- the risk to the consumer posed by the use of synthetic hormones in implants, although the database for a comprehensive evaluation of these synthetic compounds is obviously incomplete;
- the risk posed by higher hormone doses that might be present in meat, when hormonal implants are inappropriately used; these implant residues may end up in the diet of either the population at large or an individual consumer.

Substances in Annex IV of Council Regulation 2377/90/EC, for which no MRL could be established

because residues of these substances, at whatever limit, in foodstuffs of animal origin constitute a hazard to the health of the consumer include CAP, nitrofurans, nitroimidazoles, and malachite green.

Minimum required performance limits (MRPLs) were established for analytical methods for some substances for which no permitted limits were established, in order to ensure the same level of protection of consumers throughout the European Union. These limits are set on the basis of technological capabilities rather than a toxicological basis. MRPLs have been established for the compounds listed in Table 4. Recently, MRPLs are being replaced by reference points for action. The reason for this change is the growing concern that MRPLs were being considered pseudo MRLs. Limits for these substances may be lowered in the future on the advice of the CRL to the commission.

Analytical methods and performance criteria

Analytical methods that are applied in national residue surveillance in the European Union come from the following sources:

- Methods published in peer-reviewed literature or conference proceedings,

Table 4. Listing of MRPLs

Substance and/or metabolite	Matrix	MRPL
Chloramphenicol	Meat Eggs Milk Urine Aquaculture products Honey	0.3 µg kg^{-1}
Medroxyprogesterone acetate	Pig kidney fat	1 µg kg^{-1}
Nitrofuran metabolites (furazolidone furaltadone, nitrofurantoin, nitrofurazone)	Poultry meat Aquaculture products	1 µg kg^{-1} for all
Sum of malachite green and leucomalachite green	Meat of aquaculture products	2 µg kg^{-1}

- Methods developed in-house,
- Method developed by CRLs, and
- Monograph methods developed to support the registration of a veterinary drug.

The normal approach adopted is to source multiresidue methods from literature or the CRL and adapt these methods in-house. A number of groups develop their own methods and publish literature. The adoption of monograph methods is becoming less frequent because such methods are generally unsuitable for routine residue surveillance because they are generally single-residue methods. Until the late 1990s, veterinary residues were routinely detected by chemical assays (high-performance liquid chromatography [HPLC] thin-layer chromatography, and gas chromatography [GC]) or screening assays (inhibitory tests and immunoassays). However, in recent years much research has focused on the development of methods that can detect multiple residues through the application of LC-MS/MS, UPLC-TOF, and biosensor technologies. Since the year 2000, LC-MS/MS has become a routine tool in European Union reference laboratories. It is proposed that this occurred with the identification of nitrofurans and CAP residues in foods imported into the European Union from developing countries (Anonymous, 2002a; O'Keeffe et al., 2004). This crisis resulted in the implementation of MRPLs for these substances, limits which could be best reached through the application of LC-MS/MS technology.

OUTCOME AND FUTURE TRENDS IN RESIDUE TESTING IN THE EUROPEAN UNION

The results from residue testing from the European Union member states are published through the European Union website. However, the results can be regarded as historical in nature because only results from the 2005 residue testing are currently available. More up-to-date residue testing results can be viewed through consultation of the results from individual member states (namely, Ireland, Finland, and the United Kingdom). In this section, an overview of the 2005 European Union results will be discussed (Anonymous, 2005a).

Group A Substances

In 2005, stilbenes (group A1) were not detected in any samples, and they have not been detected since 2000. However, thyrostats (group A2) were detected in 8 out of 657 samples taken in France for the first time in the European Union since 2000. The noncompliant samples (all thiouracil) were attributed to an improvement in the sensitivity of the assay from 50 to 1 µg kg^{-1}. It was proposed that the positives could be explained through natural levels of thiouracil in feeding constituents. Hormones (steroids and zeranol derivatives) were detected in 0.13% of samples taken from bovines (similar to 2004) and in 0.44% of samples taken from pigs (compared to 0.3% in 2004). The majority of noncompliant bovine samples were due to nandrolone, estradiol, and progesterone. Zearalenone, a fungal metabolite, accounted for a large number of the positives in pigs. The majority of noncompliant samples for A3 substances were due to corticosteroids, namely, dexamethasone and betamethasone ($n = 187$). A low incidence of positives were observed for prednisolone ($n = 3$). Steroids were also found in aquaculture samples ($n = 5$). β-agonists (in the form of clenbuterol) were detected in a total of 43 samples.

Banned veterinary drugs belonging to group A6 (CAP, malachite green, nitrofurans, and nitroimidazoles) were detected in a number of different foodstuffs. Banned antibacterials (CAP and nitrofurans) were detected in bovine (0.11%) and porcine samples (0.05%). CAP, nitrofurans, and nitroimidazoles ($n = 5$) were also found in aquaculture. Noncompliant nitrofuran results ($n = 2$) were found with samples of rabbit. The banned dye malachite green continued to be a problem residue for aquaculture products and was found in 45 samples. Its major metabolite, leucomalachite green, was observed with samples. The results suggest that the number of positives might be inflated due to double counting of noncompliant samples. It was noted that the number of positives had decreased from the previous year for this substance.

Group B: Licensed Veterinary Drugs

Animal tissue

In 2005, the noncompliance rate for targeted samples was approximately 0.2% (721 samples out of 395,032 taken). Bovine and porcine had positive rates of 0.29 and 0.14%, respectively. This was similar to previous results in 2003 and 2004. A total of 1,299 noncompliant samples were observed for B1 substances from approximately 95,000 suspect samples (1.4% positive). B1 substances accounted for 50 and 82% of noncompliant results across all substances. In the area of B2 substances, the rate of positives was generally less than 0.14% of cattle, sheep, and goat targeted samples. In the case of bovine animals, there was a particularly high rate of positives for suspect animals (4.9%). However, these data can be regarded as biased because 113 out of the 138 samples were dexamethasone positives from Belgium. A higher proportion of positives was observed with horses and poultry at 0.44 and 0.75%, respectively. A total of 3 of the 697 samples taken were found to be positive for phenylbutazone. In the case of poultry, the high incidence of positives can be largely attributed to anticoccidials. It is expected that the rate of positives for anticoccidials in poultry will increase in the future with the application of LC-MS/MS, which will result in a lowering of the detection limits by a factor of 10. In addition, there might be some ambiguity over the definition of a noncompliant sample because MRLs are not set for many compounds. This is highlighted in the case of nicarbazin, which has a JECFA MRL of 200 $\mu g\ kg^{-1}$ in edible tissue (muscle and liver). In Ireland (and the United Kingdom), liver is used as the matrix to investigate the presence of anticoccidials. Danaher et al. (2008a) published the results of a survey of nicarbazin in poultry (n = 736 liver samples; n = 342 muscle samples). The results of the survey highlighted that the level of nicarbazin was found to be greater than the JECFA MRL in 12% of liver samples. However, no muscle sample contained nicarbazin at a level greater than the JECFA MRL. It can be concluded that, depending on the test matrix and the limit of reporting or MRL applied, the incidence of reported positives can vary greatly.

Other foodstuffs

In milk, 59 samples were found to be noncompliant for antibacterials. Additional positives were comprised of levamisole (n = 1), nonsteroidal anti-inflammatory drugs (NSAIDs) (n = 5), and salicyclic acid (n = 1). In honey, antibacterials made up the majority of the noncompliant samples (n = 66). With eggs, antibacterials (n = 27) and anticoccidials (n = 55) were the only noncompliant residues found. A total of seven aquaculture samples were found to be noncompliant due to the presence of anthelmintic residues (n = 7). Antibacterials (n = 3), anticoccidials (n = 9), and NSAIDs were found in rabbit.

DIRECT EXPOSURE TO RESIDUES: RECENT RESULTS FROM RESIDUE SURVEILLANCE IN THE EUROPEAN UNION

Since 2002, there have been several high-profile scandals relating to residues of banned veterinary drugs in food imported into the European Union from developing countries. The residues at the center of the scandal were mainly antibiotics, such as CAP and nitrofurans. As a result, there was a ban or a restriction of imports from these countries over the last few years. These residues continue to be detected in imports into the European Union (European Commission, 2006). In 2006, nitrofuran residues were detected in aquaculture products from several countries, but mainly in shrimp from Bangladesh (27 RASSF notifications), India (20), Vietnam (3), China (1), Indonesia (1), Thailand (1), and Venezuela (1). Noncompliant residues were due mainly to nitrofurazone metabolite (40 notifications), furazolidone metabolite (15), and furaltadone metabolite (2). In 2006, nitrofuran residues were detected in honey (one sample from Argentina) and two samples of meat. CAP residues have been found in imported honey or royal jelly from several countries, including China, Russia, Switzerland, and the United States. CAP has also been found in imported meat (two samples) and poultry (one sample). The image of honey has been much tainted after being associated with a number of other residues, including streptomycin, sulphonamides, tylosin, and tetracyclines. In 2006, residues of sulphonamides, tetracycline, trimethroprim, and tylosin were found in honey samples. Other banned substances that have been detected in imports include malachite green (18 samples) and crystal violet (5 samples) in aquaculture.

INDIRECT EXPOSURE TO VETERINARY DRUG RESIDUES THROUGH ENVIRONMENTAL CONTAMINATION

The most widely administered veterinary drugs to treat disease and protect the health of food-producing animals are antimicrobials (Sarmah et al., 2006), and routine use of these has become a serious concern for public health, particularly where the same classes of antimicrobials are being used to treat humans. There is increasing evidence that antimicrobial use in animals selects for resistant food-borne pathogens that may be transmitted to humans as food contaminants (White et al., 2002a). Transmission of resistant bacteria may occur through direct contact with animals or to consumers via the food chain (Kemper, 2008). Researchers have characterized several mobile genetic elements, such as plasmids, transposons, insertion sequences (also known as an IS, an insertion sequence element, or an IS element), and integrons, used by bacteria in the rapid spread of antibiotic resistance genes among bacterial populations (Aarestrup, 1999; Levy, 1987; Ochman et al., 2000). Antimicrobial resistance phenotypes have been recognized among a number of zoonotic pathogens, including *Salmonella enterica* serovar Typhimurium (Perron et al., 2007), *Escherichia coli* (White et al., 2002b), *Campylobacter coli* (Thakur and Gebreyes, 2005), *Listeria monocytogenes* (Srinivasan et al., 2005), and *Yersinia enterocolitica* (Stock and Wiedemann, 1999).

A 7-year study in Spain reported a significant increase in resistance to antimicrobial agents used for therapy (Prats et al., 2000). This study reported that between 1985 and 1987 and 1995 and 1998 a greater increase in the rate of resistance was observed with *Campylobacter jejuni* for quinolones (from 1 to 82%) and tetracycline (from 23 to 72%) and with salmonellae for ampicillin (from 8 to 44%), CAP (from 1.7 to 26%), and trimethoprim-sulfamethoxazole and nalidixic acid (from <0.5 to 11%). In addition, multidrug resistance was detected with several *Salmonella* serotypes (Prats et al., 2000). A multidrug resistance gene cluster, which is a complex class 1 integron, is located within *Salmonella* genomic island 1 (SGI1). A distinctive feature associated with most *S.* Typhimurium DT104 isolates is a multiple antibiotic resistance phenotype to ampicillin, chloramphenicol/florfenicol, streptomycin, sulfonamides, and tetracycline (ACSSuT type) (Herikstad et al., 1997; Glynn et al., 1998; White et al., 2002a). Some strains exhibit resistance to trimethoprim and fluoroquinolones (CDC, 1997; Threnlfal et al., 1996). In the United Kingdom, from 1993 to 1995, trimethoprim-resistant DT104 (R-type ACSSuTTm) increased from 1% to 27% of isolates, and ciprofloxacin-resistant DT104 (R-type ACSSuTCp) increased from 0 to 6% of isolates (Threnlfal et al., 1996). In 1995, a total of 30 *S.* Typhimurium R-type ACSSuT isolates were obtained from 10 states and were sent to the United Kingdom for phage typing; of these, 25 (83%) were DT104 (CDC, 1997).

In order to elucidate the sources and causes of antimicrobial resistance, more information is needed about the reservoirs of bacterial resistance. It is only in recent years that concerns over antibiotics in the environment and the potential impact on human and animal health have emerged (Boxall et al., 2003, 2006; Boxall, 2004), and concerns have been raised over possible toxic effects of antibiotics on aquatic and edaphic organisms (Davis et al., 2006). While there is a lot of information in regard to the effect of other classes of substance (e.g., pesticides and nutrients) that are applied to soil, there is limited information on the concentration, transport, fate, and behavior of these compounds in the soil and water environment (Boxall

et al., 2002; Sarmah et al., 2006) or the antibiotic runoff from agricultural fields (Davis et al., 2006). Antibiotic resistance rates for enteric bacteria seem to be much higher than those for pathogens found in a variety of environments and closely associated to the host environment (Mayrhofer et al., 2004). Following administration of antibiotics to food-producing animals, they are disseminated into the environment through a variety of different sources (Figure 1) either directly in feces or indirectly through the application of slurry and manure as a fertilizer (Boxall et al., 2002), and once in the environment, their effectiveness is dependent on a number of factors, such as physicochemical properties, prevailing climatic conditions, and soil types (Kemper, 2008; Sarmah et al., 2006). Some antibiotics, for example, oxytetracycline, may persist in the environment for up to approximately 1 year after application (Kay et al., 2004), and if they are not efficiently degraded, it is possible that these residues may assist in maintaining or in the development of antibiotic-resistant microbial populations (Witte, 1998). Tetracyclines enter the environment in significant concentrations via repeated fertilizations with liquid manure, build up persistent residues, and accumulate in soil (Hamscher et al., 2002).

A number of workers have detected antibiotics in manure, water resources, and agricultural fields, but detailed information on transport mechanisms is limited (Davis et al., 2006). Schlusener et al. (2003) reported that tiamulin and salinomycin were detected at concentrations of 43 and 11 $\mu g\ kg^{-1}$, respectively, in manure samples. Haller et al. (2002) detected sulfonamides, including sulfamethazine and sulfathiazole, in liquid manure at concentrations of up to 20 mg kg^{-1}. Multiple classes of antimicrobial compounds

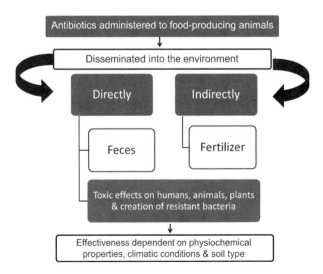

Figure 1. Overview of direct and indirect effects of veterinary drug residues on food safety.

commonly at concentrations of >100 µg liter^{-1} have been detected with swine waste storage lagoons, and the highest concentration in lagoons was chlortetracycline (1,000 µg liter^{-1}), followed by sulfamethazine (400 µg liter^{-1}) (Campagnolo et al., 2002). These researchers also tested water samples, i.e., surface water, tile drainage, and wells, and reported that 31% and 67% of antibiotics tested were detected near swine and poultry farms, respectively. Hamscher et al. (2002) investigated the distribution and persistence of tetracyclines in a field fertilized with liquid manure and detected 4.0 mg kg^{-1} tetracycline and 0.1 mg kg^{-1} chlortetracycline in the liquid manure. In addition, three sublayers of soil were tested; average tetracycline concentrations of 86.2 (0 to 10 cm), 198.7 (10 to 20 cm), and 171.7 (20 to 30 cm) µg kg^{-1} were detected; and 4.6 to 7.3 µg kg^{-1} chlortetracycline was detected in all three sublayers. These researchers concluded that tetracyclines may have a potential risk, and investigations of the environmental effects of these antibiotics are necessary.

Davis et al. (2006) sprayed an antibiotic solution including tetracycline, chlortetracycline, sulfathiazole, sulfamethazine, erythromycin, tylosin, and monensin on the soil surface 1 h before rainfall simulation (average intensity = 60 mm h^{-1} for 1 h). Runoff samples were collected continuously and analyzed for aqueous and sediment antibiotic concentrations. They found that monensin had the highest concentration in runoff (1.20 µg liter^{-1}), and the lowest concentrations were for tetracycline (0.03 µg liter^{-1}), chlortetracycline (0.04 µg liter^{-1}), and tylosin (0.09 µg liter^{-1}). On the other hand, sediment concentrations were highest for erythromycin (17.0 µg kg^{-1}), followed by monensin (10.5 µg kg^{-1}) and tylosin (8.0 µg kg^{-1}).

Boxall et al. (2006) investigated the potential for a representative range of veterinary medicines, including antibiotics that have a high potential to be released into the environment (sulfonamides, tetracyclines, fluoroquinolones, macrolides, and β-lactams) and enter the food chain via uptake from soil into food plants. They found that following spiking of soil, amoxicillin was rapidly dissipated and that, with the exception of enrofloxacin, all of the other test substances had 50% dissipation values of <15 weeks. However, florfenicol, enrofloxacin, and oxytetracycline had 90% dissipation values of >150 days, indicating that for these compounds residues could persist in the soil environment for >6 months following application.

STATE-OF-THE-ART LC-MS/MS DETECTION OF VETERINARY DRUGS AND GPAs

In the last 5 years there have been several advances in the area of chemical instrumentation. In the area of MS, a number of vendors are marketing instruments that allow rapid switching between positive and negative modes. Vendors have also developed fast scanning instruments that are compatible with fast chromatography (or UPLC). The combination of these developments allows the sensitive analysis of several different substance groups in a single injection cycle. More recently, groups have started to investigate the application of high-resolution MS technology, such as TOF, Fourier transform, and Orbitrap in veterinary drug residue analysis (Hernando et al., 2006; Nielen et al., 2007). Such applications, when combined with newer high-resolution chromatographic techniques (UPLC and rapid resolution LC), have been demonstrated to allow the determination of 100 residues in a single injection (Kaufmann et al., 2007, 2008; Stolker et al., 2008). It is the objective of this section to review the current state-of-the-art methods in residue analysis using LC-MS/MS technology with different veterinary drug groups.

Antibiotics

Antibiotics in food are normally screened through the application of inhibitory plate tests or receptor or immunochemical assays. These methods offer several advantages in terms of cost, sample throughput, and in some cases the potential for application in nonspecialist laboratories. The main disadvantages of these methods are the potential for generating false-negative results and nonspecificity. The development of a multiresidue method (screening or quantitative) for the analysis of total antibiotics has been a solution sought by several groups for many years. A number of groups have demonstrated the value of the application of LC-MS/MS for the analysis of multiple antibiotic residues in food.

Becker et al. (2004) developed a method for the quantitative determination of 15 penicillin and cephalosporin residues in bovine muscle, kidney, and milk. Samples were extracted using acetonitrile and purified by Oasis HLB SPE prior to determination by LC-MS/MS operating in positive and negative electrospray. Sample extracts were injected onto the LC-MS/MS and separated in 35 min; a further 25 min was required to wash the column prior to the next injection. The method was validated according to 2002/657/EC criteria and showed good performance during the validation studies. The method was shown to be very precise with coefficients of variations of <9, <15, and 10% for muscle, kidney, and milk, respectively. This resulted in suitably low values for the decision limits (CCβ) and the detection capabilities (CCα) of the method, which were close to the MRL. The method was applied to samples that screened positive for antibiotic residues (bovine kidney [n = 8], bovine muscle [n = 1], sheep muscle [n = 1], and milk [n = 3]). Amoxicillin, cloxacillin, cefquinome, and penicillin G were found to be present in three, one, one, and eight samples, respectively.

Granelli and Branzell (2007) developed an LC-MS/MS screening method for detecting 19 residues from five different classes of antibiotics (tetracyclines, sulfonamides, quinolones, β-lactams, and macrolides) in the muscle and kidney of pig, cattle, sheep, deer, horse, and reindeer. Antibiotic residues were isolated from samples using a single extraction using 70% methanol diluted with water and injected in the LC–MS/MS. Two injections were required to analyze the 19 residues (total run time of 15 min); this was due to the LC-MS/MS technology used. It was proposed that the method could replace existing inhibitory tests for the detection of antibiotic residues. The method allowed the analysis of 60 samples in 24 h, but this sample throughput could be improved through the pooling of samples.

Hammel et al. (2008) developed an LC-MS/MS method for the detection of 42 antibiotic residues in honey samples (including tetracyclines, macrolides, aminoglycosides, β-lactams, amphenicols, and sulfonamides). Honey samples were sequentially extracted with different solvent systems to prevent degradation of certain residues and maintain high recovery. Extracts were reconstituted and injected sequentially onto the LC-MS/MS system using a stacking system. The limit of quantitation (LOQ) of the method for 37 of the 42 residues studied was 20 μg kg^{-1}. A total of five compounds did not perform favorably in the method, namely, amoxicillin, ampicillin, desmycosin, penethamate, and sulfanilamide.

In recent years, a number of groups have investigated the application of UPLC-TOF for the analysis of veterinary drug residues. Kaufmann et al. (2007) investigated the application of this technology for the determination of more than 100 veterinary drug residues in urine samples. Urine samples were simply diluted and injected unfiltered onto the TOF instrument. The sensitivity of the assay was typically 5 μg liter^{-1} or better in urine samples. Injection run times were typically <7 min. Although no MRL is listed for antibiotics in urine, this work demonstrated the power of UPLC-TOF technology for the detection of multiple veterinary drugs (several classes of antibiotics, anthelmintics, nitromidazoles, tranquilizers, and dyes).

Anticoccidials

Anticoccidials are a diverse range of chemicals that are used in the treatment of coccidiosis, which is a disease that can occur in intensively reared species. Analysis of anticoccidials is further complicated due to the ban of a number of drugs in the European

Union (mainly the mutagenic nitroimidazoles and certain ionophores) and the absence of or changing situation of MRLs for other substances. Nitromidazoles are generally analyzed separately from the other anticoccidials using HPLC, GC-MS, or LC-MS/MS. However, LC-MS/MS has now become the method of choice due to the need to satisfy low detection levels and confirmatory criteria. LC-MS/MS offers an advantage over GC-MS/MS because it allows the analysis of nitromidazole residues without the need for derivatization and thus allows discrimination between ronidazole and HMMNI (2-hydroxymethyl-1-methyl-5-nitroimidazole). Fraselle et al. (2007) developed a method for the analysis of seven nitromidazole residues in pig plasma by LC-MS/MS. Samples were purified using the sample preparation procedure previously developed and applied in the GC-MS/MS analysis of nitroimidazole residues (Polzer et al., 2004). The method employed the use of six deuterated internal standards to allow the reliable detection of low levels of residues. CCα and CCβ values were determined to be <1.25 μg liter^{-1} based on identification point (IP) criteria.

There have been considerable developments in the methodology to detect other anticoccidials residues in food in the European Union in the last 5 years. In early work, LC-MS/MS was applied to determine residues of nicarbazin, halofuginone, or ionophores (monensin, salinomycin, narasin, and lasalocid). In recent years, a number of groups have developed generic sample preparation procedures based on LC-MS/MS detection. Mortier et al. (2003) developed a method to determine five chemical-type anticoccidials in eggs, including dimetridazole. Dubois et al. (2004) developed an LC-MS/MS method to detect nine coccidiostat (monensin, salinomycin, narasin, maduramicin, lasalocid, halofuginone, robenidine, diclazuril, and dintrocarbanilide) residues in egg and muscle at levels ≤1 μg kg^{-1}. Samples were mixed with anhydrous sodium sulphate, extracted with acetonitrile, and purified by passing through a silica SPE column. The method was shown to be suitable for the quantitative and confirmatory analysis of residues in muscle and egg. CCα and CCβ values varied from 0.07 to 0.6 and 0.1 to 1.0 μg kg^{-1} for salinomycin and halofuginone, respectively. Mulder et al. (2005) developed and applied an LC-MS/MS method to investigate the metabolism of halofuginone and toltrazuril residues in eggs. The limit of detection of the method was 2, 10, and 30 μg kg^{-1} for halofuginone, toltrazuril, and toltrazuril sulphone residues, respectively.

Radeck and Gowik (2005) developed an LC-MS/MS method to determine 25 anticoccidials in egg, muscle, and liver. The method allows the detection of certain nitroimidazoles, ionophores (laidlomycin, lasalocid, maduramicin, monensin, narasin, salinomycin, and semduramicin), and chemical anticoccidials (amprolium, aprinocid, clopidol, diclazuril, ethopabate, halofuginone, nicarbazin, robenidine, toltrazuril, and toltrazuril sulphone). Samples (5 g) are simply extracted with acetonitrile (15 ml), evaporated to dryness, and further purified using Oasis HLB cartridges. The LOQ of the method was typically around 1 µg kg^{-1}, but was higher for toltrazuril, at 50 µg kg^{-1}.

Antiparasitic Drugs

It can be concluded from consultation with scientific literature and conference papers that the methodology for the determination of antiparasitic drug residues in food needs to be greatly improved (Danaher et al., 2006, 2007). However, a number of groups have developed effective HPLC fluorescence methods for macrocyclic lactones. Zuidema et al. (2006) reported false-positive results using HPLC fluorescence for emamectin in milk. However, this may depend on the analytical conditions (cleanup or chromatographic separation) or reporting levels. It has been found in the author's laboratory that matrix interference peaks rarely present a problem for emamectin in animal tissue and milk. The major deficiencies in the area of antiparasitic drug residues are that few multiresidue methods are available to detect flukicide residues and that coverage of levamisole and/or benzimidazole metabolites is often poor. The benzimidazoles pose a difficult challenge to residue scientists. In particular, triclabendazole residues have proved difficult to include in analytical methods due to poor extractability because of tight binding to proteins, while other benzimidazoles, namely, 5-hydroxy thiabendazole, albendazole, and fenbendazole are also difficult residues to include in methods because of their susceptibility to oxidation. In this section, it is shown that LC-MS/MS is the best solution available at present.

Several groups have recently developed extensive methods for the detection of multiple benzimidazole and levamisole residues in food through the application of LC-MS/MS (Edder et al., 2004; Radeck and Gowik, 2008). Radeck and Gowik (2008) reported an assay capable of detecting 21 benzimidazoles and levamisole residues in milk. This group found that triclabendazole metabolites became tightly bound to proteins after fortification and required acid hydrolysis to release residues. Liberated metabolites were extracted with solvent (acetonitrile and ethyl acetate) and purified by liquid-liquid partitioning prior to determination by LC-MS/MS. The method allowed the determination of residues to less than 10 µg kg^{-1}

but typically to as low as 1 µg kg^{-1}. The most significant development in antiparasitic drug residue analysis was the development of a method to determine some 39 antiparasitic drugs and pesticide residues in liver, muscle, and milk (Kinsella et al., 2007). The method included benzimidazoles, levamisole, macrocyclic lactones, flukicide agents, and certain pesticide residues (coumaphos, dichlorvos, and haloxon). This development was achieved through the application of the QuEChERS sample preparation approach, a technique that has been applied successfully to the analysis of multiple-pesticide residues in food. The method offers several advantages over existing methods, including the number of residues covered, the ability to process 20 samples in a 2-hour period, and low limits of quantitation (10 µg kg^{-1} for dichlorvos and 5 µg kg^{-1} for other residues).

Some effective screening methods have been developed for the determination of antiparasitic drugs, namely, the enzyme-linked immunosorbent assay (Brandon et al., 1998) and biosensor (Crooks et al., 2003; Keegan et al., 2008; Samsanova et al., 2002). However, the major difficulty faced at present is the availability of antibodies for flukicides, the cross-reactivity profiles of antibodies, and the ability to harness different antibodies in a single multichannel detection system. Multichannel systems are desirable for surveillance purposes but are costly or require complex technology for microspotting, a technique which is generally unavailable in routine laboratories. Single-channel systems are still suitable for application in investigative work in which a specific residue or residue group is being investigated.

NSAIDs

A total of four NSAIDs (carprofen, flunixin, meloxicam, and tolfenamic acid) are approved for treatment of food-producing animals in the European Union. An additional 17 NSAIDs have the potential to be misused in food-producing animals; phenylbutazone is probably one of the most infamous (Gentili, 2007). A number of these drugs are available as over-the-counter human medicines. A review of the literature indicates that few methods have been developed to determine the presence of NSAIDs in food, but they have been investigated widely with wastewater. Gentili (2007) reviewed the current methodology in this area and indicated that the most extensive multiresidue assay was an LC-MSn method to determine six NSAIDs in bovine muscle, with CCβ values ranging between 2 and 68 µg kg^{-1} (Van Hoof et al., 2004). This paper indicates that significant improvement needs to be made in methodology to detect the presence of NSAIDs in food.

Corticosteroids

The administration of corticosteroids for fattening purposes is banned, according to 96/22/EC. Corticosteroids are strictly regulated in the European Union because of the potential abuse due to their pharmacological and physiological effects and collateral effects linked with suppression of the immune response and interference with other hormones. However, four corticosteroids are approved for therapeutic purposes in the European Union, namely, dexamethasone, betamethasone, methylprednisolone, and prednisolone. MRLs have been set for the residues of these drugs in edible tissues of food-producing animals. The corticosteroids are a challenging group of substances to determine. An obstacle encountered by researchers is the separation of dexamethasone and its isomer, bethamethasone. It has been reported that this is difficult to achieve using methanol as a modifier but is achievable using acetonitrile. The fragmentation of corticosteroids in LC-MS/MS has been described by a number of groups (Antignac et al., 2001, 2004; Greig et al., 2003; Sangiorgi et al., 2003). Researchers have investigated the application of a number of different MS/MS scanning modes, including product ion scan, precursor scan, and neutral loss scan. Antignac et al. (2001) developed a method to determine and confirm the presence of 12 corticosteroids by LC-MS/MS using negative electrospray ionization. It was concluded that the diagnostic ions chosen (precursor ion $[M + acetate]^-$ gave greater selectivity than the two fragment ions $[M–H]^-$ and $[M–H–CH_2O]^-$).

The detection limits of the developed methods were between 2.9 and 9.3 µg kg^{-1} for hair samples and in the range of 40 to 70 ng kg^{-1} for the urine or meat samples. Draisci et al. (2001) reported a method to determine three fluorinated corticosteroids in liver samples. Samples were extracted using accelerated solvent extraction and purified extracts were determined on an LC-MS/MS equipped with an atmospheric pressure chemical ionization (CI) interface operating in negative mode.

GPAs

The use of β-agonist and hormones for growth-promoting purposes in the European Union is prohibited under Council Regulation 96/22/EC. The residue analysis of GPAs has been well described recently in a paper by Stolker et al. (2007). The authors will not try to replicate or summarize this review but will describe some of the current trends in this area. The ban of GPAs has led to the circulation of a range of black-market products in the European Union. This has led to intense research of new methods

to detect the presence of banned residues using MS and immunochemical detection systems. The analysis of these substances is very complex due to similarity in structures and most important the ambiguity over their source (endogenous or exogenous). There is growing concern over possible moves by black marketers toward the application of cocktails of multiple GPAs that are more difficult to detect in animals. This has led groups to develop alternative approaches for residue detection. TOF MS is becoming widely applied in residue analyses because it enables retrospective analysis of residues. For example, if a previously unidentified banned substance was discovered on a premises, its abuse could be investigated through consultation of previous analytical runs. Ambiguity over marker residues for some natural hormones has led researchers to investigate alternative markers and matrices. Steroid esters in hair have been investigated by some groups as a method to differentiate between endogenous and exogenous hormones. Some groups are investigating the possible use of biomarkers as indicators of potential abuse of anabolic agents.

PERFORMANCE OF ANALYTICAL METHODS

The performance criteria for analytical methods for detecting residues are described in Commission Decision 2002/657/EC criteria. The document describes the necessary characteristics of methods required for screening and quantitative and confirmatory analysis of residues of both group A and B substances in food. Screening assays are required to have low-rate false-negative results at the level of interest (i.e., a β-error of ≤5%; β-error means the rate of false compliant results). Confirmatory methods are required to provide unambiguous confirmation of the identity of a molecule. In the case of group B substances, residues can be confirmed by using two independent techniques (e.g., immunochemical detection followed by HPLC-UV is acceptable). Alternatively, HPLC fluorescence is a suitable technique for confirming the presence of residues alone. The validity of this approach has been called into question by some groups, who state that LC-MS/MS is the only technique that is suitable for the confirmation of residues. It is the author's experience that when noncompliant results are observed for MRL substances, they are generally at high levels and there is little doubt. In addition, immunobiosensor in conjunction with HPLC-UV has shown to be a very suitable technique to screen and confirm the presence of nicarbazin residues in noncompliant samples (Danaher et al., 2003).

In the cases of group A substances, residues can be screened using a suitably sensitive technique, as long as a β-error of ≤5% can be achieved. However,

Table 5. Relationship between the different MS techniques and identification points earned

MS technique[a]	No. of identification points earned per ion
LR-MS	1.0
LR-MSn precursor ion	1.0
LR-MSn transition products	1.5
HRMS	2.0
HR-MSn precursor ion	2.0
HR-MSn transition products	2.5

[a]LR, low resolution; HR, high resolution.

to confirm the presence of residues, IP criteria have to be met, a concept that is associated mostly with MS. A total of three and four IPs are required to confirm the presence of group B and A substances, respectively. The relationship between the MS techniques and IPs is shown in Table 5. One IP can be achieved for a precursor ion, while an additional 1.5 IPs can be achieved for each product ion produced by collision-induced dissociation, to give a total of four IPs. This concept also applies to GC-MS/MS and high resolution MS (HRMS). However, higher points are awarded for HRMS to take into account high-mass accuracy. However, in the case of GC-MS, electron impact and CI techniques are regarded as different techniques and the ions generated by each technique are counted as separate. In addition, if different derivatization reaction chemistries are employed, the ions generated by each derivative are counted as separate ions. It can be concluded that group A and B substances are most commonly confirmed by LC-MS/MS using a precursor and two product ions. In routine practice, if ions are found to be present in samples, their relative ion intensities have to fall within the limits described in Table 6.

A number of different approaches have been set out in the 2002/657/EC criteria for validation of screening and confirmatory methods. In all cases, selectivity or specificity of a method needs to be determined through analysis of blank samples from at least 20 animals representative of different storage times and feeding regimes. In addition, the

Table 6. Maximum permitted tolerances for relative ion intensities using a range of MS techniques

Relative intensity (% of base peak)	EI-GC-MS (%)[a]	CI-GC-MS, GC-MSn, LC-MSn, LC-MSn (relative %)
>50	± 10	± 20
>20–50	± 15	± 25
>10–20	± 20	± 30
≤10	± 50	± 50

[a]EI, electron impact.

ruggedness of the method needs to be determined through investigating different factors, such as different operators and small changes to the method. Experiments are also required to investigate the stability of residues (in matrix and sample extracts) and standard materials during various storage conditions. Repeatability involves the assay of at least 18 identical fortified blank samples on three separate occasions by a single analyst. The within-laboratory reproducibility involves the analysis of at least 18 different or fortified identical blank samples on separate occasions by three different analysts. The within-laboratory reproducibility study is most important because it is used in the calculation of robust CCα and CCβ values. CCβ has to be determined for all assay types (screening and confirmatory), while CCα needs only to be determined for confirmatory methods. There has been considerable discussion of CCα and CCβ since their implementation, which is mainly due to the different approaches that are adopted for their calculation. In the case of group A substances, CCα can be determined by analyzing 20 blank samples and calculating the signal-to-noise ratio at the retention time window in which the analyte is expected. CCα is calculated as three times the signal-to-noise ratio. CCβ is calculated by fortifying 20 samples at CCα and by adding 1.64 times the standard deviation of the within-laboratory reproducibility to the mean value measured. An alternative approach is to spike samples at and around the level of interest (possibly the LOQ) and determine CCα and CCβ using extrapolation. The problem with this approach is that below the lowest calibration level, CCα or CCβ can give very low values. However, disregarding the approach adopted, CCα and CCβ should be verified through spiking studies. In addition, confirmatory criteria should be satisfied using the relative ion intensities. Similar approaches can be adopted to determine CCα and CCβ for MRL substances. In the case of MRL substances, it is important to highlight the importance of the precision of the method. The lower the within-laboratory reproducibility, the closer CCα and CCβ are to the MRL, which is most desirable.

The determination of CCα and CCβ has proven to be a difficult concept for many scientists to grasp due to the several approaches that can be adopted. At present, the validation of multiresidue screening assays is a particular problem faced by residue scientists. This is highlighted for assays that can detect several different residue types at a time but cannot discriminate between them. The result is that present guidelines recommend that assays should be validated for all drug, matrix, species, and concentration combinations. This often leads to an impractical number of experiments in laboratories. One group has adopted a reduced factorial design approach to validate methods, resulting in fewer experiments being needed (Gowik et al., 2008). An alternative approach that might be adopted is that key residues or those responsible for frequent noncompliant results should be selected. Another practical problem is the validation of methods for the determination of residues for which a sum of MRLs is listed. Radeck and Gowik (2008) proposed an alternative approach for the validation of methods when a summed MRL is applied.

DEVELOPMENT OF IMPROVED STRATEGIES FOR MONITORING AND SURVEILLANCE

In the European Union, a targeted approach has been applied for surveillance of residues in food of animal origin, as laid down in 96/23/EC. The approach is targeted on the basis that residues are tested in food that they are likely to occur in (e.g., malachite green in aquaculture products). There has been a call in the recent *Reflection Paper on Residues in Foodstuffs of Animal Origin* for the adoption of a risk-based approach for monitoring in surveillance (Anonymous, 2003). In this approach, residues would be targeted based on risk through consideration of several factors: nature of toxicity, potency of substance, proportion of animals that might be treated in a particular year, groups that might be subject to high exposure, and incidence of positives. Several other factors could be factored into this approach, including data on sales of products and metabolism/persistence studies on food-producing animals. At present, information on the sales of veterinary medicinal products is not widely available and is in many cases incomplete. However, if these data could be accessed, they would be a very powerful tool for risk-based surveillance. Logically, it can be concluded that a different approach would have to be adopted for the control of banned substances because information on usage is unavailable.

At present, two such approaches are being applied for prioritizing residues in food. In the European Union, the Veterinary Residues Committee adopted a matrix-ranking approach for prioritizing substances as part of their nonstatutory surveillance scheme in the United Kingdom (VRC, 2006). In the United Kingdom, CAP and nitrofurans are ranked as the highest priority. It has to be highlighted that the system was designed to prioritize substances for testing in imported foods. However, this approach is not applied for the control of residues in statutory surveillance in the United Kingdom. The second approach is the one that is applied by the U.S. Department of

Agriculture Food Safety and Inspection Service (FSIS) for "selecting, scoring and ranking candidate veterinary drugs" (Anonymous, 2005b). In the FSIS system, antibiotics are given the highest priority. The major difference between the two systems is that the approaches adopted to rank the toxicity of the substance differ markedly. At present, no standardized approach is available to determine the weighting to be applied to each substance. It is expected that if a risk-based approach is adopted in the European Union to rank residues that this obstacle will have to be overcome.

It can be concluded that the Food Safety and Inspection Service (FSIS) ranking system is systematic but is very complex due to the several equations that have to be applied. In particular, the categorization of species according to animal age is of particular benefit when prioritizing residues. Animals in their first season are more susceptible to infection due to their developing immune system. Similarly, older animals, particularly dairy cows, are more susceptible to infection. Taking these factors into consideration, these categories of animals are more likely to require therapeutic treatment, which indicates that residues are more likely to be detected in cow beef, lamb, and veal. In many cases, it is suggested that a seasonal approach should be adopted to target the monitoring of residues in food, while taking epidemiological factors and zoonoses into consideration. It is proposed that anthelmintic drug residues could be best monitored using such an approach. Anthelmintic residues are more likely to be detected in milk during spring due to treatments prior to calving during the dry cow period. In particular, residues of macrocyclic lactones have been shown to be highly persistent and can be detected up to 150 days posttreatment with dairy products (Danaher et al., 2008b). It can be concluded that there is a high potential of detecting noncompliant anthelmintic residues in milk because many products are approved for treatment of animals during the dry cow period but have no MRLs listed. Similarly, residues of benzimidazoles are likely to be detected in lamb meat due to intensive prophylactics administered from April to June. There is a call by many groups in Europe to adopt a risk-based approach to surveillance of residues in food of animal origin (Anonymous, 2003). It is expected that this approach will place a higher emphasis on certain drugs and result in reduced testing for others, such as contaminants. This new approach will introduce a bias into sampling (that may present a problem for the risk communicators) but should result in greater success in detecting noncompliant residues, which will result in greater consumer protection.

CONCLUSIONS

There is a growing trend in the European Union to detect veterinary drugs and GPAs using LC-MS/MS technologies. In the past, LC-MS/MS was regarded as expensive and was considered as a specialist tool that should be applied only for analyzing banned substances. It is expected that LC-MS/MS will replace HPLC in the next 5 years in the majority of laboratories. A number of laboratories have developed applications using HRMS that allow the detection of several classes of veterinary drugs. Some methods include as many as 100 veterinary drug residues. However, it is difficult to envisage these methods being applied widely due to the complex testing structure in many countries, in which more than one laboratory is responsible for surveillance activities.

In 2003, a review was carried out on residue legislation and the impact that it has on the availability of veterinary medicines, trade, and food safety in the European Union. As part of this review, it was proposed that a risk-based approach should be adopted for the control of residues that should run in parallel with the existing targeted surveillance program. It is expected that if this legislation is enacted, there will be a greater emphasis placed on ranking the different residues using information on toxicology, food consumption data, and interrogation of residue testing plan results. This legislation may support the inclusion of multiclass and/or commodity-based methods that allow the detection of multiple veterinary drugs. The impact of such developments on methodologies and legislation will provide a clearer knowledge of exposure to veterinary drug residues through improvements in the scope of residue surveillance programs.

REFERENCES

Aarestrup, F. 1999. Association between the consumption of antimicrobial agents in animal husbandry and the occurrence of resistant bacteria among food animals. *Int. J. Antimicrob. Agents.* **12**:279–285.

Anonymous. 1990. Council Regulation No. 90/2377/EC of 26 June 1990 laying down a community procedure for the establishment of maximum residue limits of veterinary medicinal products in foodstuffs of animal origin. *Off. J. Eur. Comm.* **L224**:1–136.

Anonymous. 1996a. Council Directive No. 96/23/EC of 29 April 1996 on measures to monitor certain substances and residues thereof in live animals and animal products and repealing Directives 85/358/EEC and 86/469/EEC and Decisions 89/187/EEC and 91/664/EEC. *Off. J. Eur. Comm.* **L125**:1–28.

Anonymous. 1996b. Council Directive No. 96/22/EC of 29 April 1996 concerning the prohibition on the use in stockfarming of certain substances having a hormonal or thyrostatic action and of beta-agonists, and repealing Directives 81/602/EEC, 88/146/EEC and 88/299/EEC. *Off. J. Eur. Comm.* **L125**:1–14.

Anonymous. 1997. Commission Directive No. 97/6/EC of 30 January 1997 amending Council Directive 70/524/EEC concerning additives in feedingstuffs. *Off. J. Eur. Comm.* **L035**:11–13.

Anonymous. 1998a. Commission Regulation (EC) No. 2788/98 of 22 December 1998 amending Council Directive 70/524/EEC concerning additives in feedingstuffs as regards the withdrawal of authorisation for certain growth promoters. *Off. J. Eur. Comm.* **L347**:3.

Anonymous. 1998b. Council Regulation No. 2821/98 of 17 December 1998 amending, as regards withdrawal of the authorisation of certain antibiotics, Directive 70/524/EEC concerning additives in feedingstuffs. *Off. J. Eur. Comm.* **L351**:4–8.

Anonymous. 2002a. Chloramphenicol and streptomycin in Chinese honey and in blended honey containing Chinese honey, Food alert notification 2002.058. Food Safety Authority of Ireland, Dublin, Ireland.

Anonymous. 2002b. Opinion of the scientific committee on veterinary measures relating to public health on review of previous SCVPH opinions of 30 April 1999 and 3 May 2000 on the potential risks to human health from hormone residues in bovine meat and meat products. European Commission, Brussels, Belgium.

Anonymous. 2003. *Reflection Paper on Residues in Foodstuffs of Animal Origin.* European Commission, Brussels, Belgium.

Anonymous. 2005a. Commission staff working document on the implementation of national monitoring plans in the member states in 2005 (Council Directive 96/23/EC), Brussels, 8.2.2007 SEC (2007). **196**:1–231.

Anonymous. 2005b. *2005 FSIS National Residue Program Scheduled Sampling Plans, Blue Book.* Food Safety and Inspection Service, U.S. Department of Agriculture, Washington, DC.

Antignac, J.-P., B. Le Bizec, F. Monteau, F. Poulain, and F. Andre. 2001. Multi-residue extraction–purification procedure for corticosteroids in biological samples for efficient control of their misuse in livestock production. *J. Chromatogr. B* **757**:11–19.

Antignac, J.-P., F. Monteau, J. Negriolli, F. Andre, and B. Le Bizec. 2004. Application of hyphenated mass spectrometric techniques to the determination of corticosteroid residues in biological matrices. *Chromatography* **59**:1–10.

Becker, M., E. Zittlau, and M. Petz. 2004. Residue analysis of 15 penicillins and cephalosporins in bovine muscle, kidney and milk by liquid chromatography–tandem mass spectrometry. *Anal. Chim. Acta* **520**:19–32.

Boxall, A. B. A., P. Blackwell, R. Cavallo. P. Kay, and J. Tolls. 2002. The sorption and transport of a sulphonamide antibiotic in soil systems. *Toxicol. Lett.* **131**:19–28.

Boxall, A. B. A., D. W. Kolpin, B. Halling Sørensen, and J. Toll. 2003. Are veterinary medicines causing environmental risks? *Environ. Sci. Technol.* **37**:286A–294A.

Boxall, A. B. A. 2004. The environmental side effects of medication. *EMBO Rep.* **5**:1110–1116.

Boxall, A. B. A., P. Johnson, E. J. Smith, C. J. Sinclair, E. Stutt, and L. S. Levy. 2006. Uptake of veterinary medicines from soils into plants. *J. Agric. Food Chem.* **54**:2288–2297.

Brandon, D. L., K. P. Holland, J. S. Dreas, and A. C. Henry. 1998. Rapid screening for benzimidazole residues in bovine liver. *J. Agric. Food Chem.* **46**:3653–3656.

Campagnolo, E. R., K. R. Johnson, A. Karpati, C. S. Rubin, D. W. Kolpin, M. T. Meyer, and J. E. Esteban. 2002. Antimicrobial residuals in animal waste and water resources proximal to large-scale swine and poultry feeding operations. *Sci. Total Environ.* **299**:89–95.

CDC. 1997. Multidrug-resistant *Salmonella* serotype Typhimurium in the United States, 1996. *MMWR Morb. Mortal Wkly. Rep.* **46**:308–310.

Crooks, S. R. H., B. McCarney, I. M. Traynor, C. S. Thompson, S. Floyd, and C. T. Elliott. 2003. Detection of levamisole residues in bovine liver and milk by immunobiosensor. *Anal. Chim. Acta* 483:181–186.

Danaher, M., S. Yakkundi, E. Capurro, G. Kennedy, M. O'Keeffe, C. Elliott, A. Anastasio, and M. L. Cortesi. 2003. Comparison between three techniques—Biacore biosensor, HPLC-UV and LC-MS/MS—for determination of Nicarbazin residues in poultry, p. 403–407. *In Proceedings of Euro Food Chem XII: Strategies for Safe Food*, Brugge, Belgium.

Danaher, M., L. C. Howells, S. R. H. Crooks, V. Cerkvenik-Flajs, and M. O'Keeffe. 2006. Review of methodology for the determination of macrocyclic lactone residues in biological matrices. *J. Chromatogr. B* 845:175–203.

Danaher, M., H. De Ruyck, S. R. H. Crooks, G. Dowling, and M. O'Keeffe. 2007. Review of methodology for the determination of benzimidazole residues in biological matrices. *J. Chromatogr. B* 845:1–37.

Danaher, M., K. Campbell, M. O'Keeffe, E. Capurro, G. Kennedy, and C. Elliott. 2008a. Survey of the anticoccidial feed additive nicarbazin (as dinitrocarbanilide residues) in poultry and eggs. *Food Addit. Contam.* 25:1, 32–40.

Danaher, M. F. Fabbrocile, H. Cantwell, P. Byrne, A. Callaghan, and L. Clarke. 2008b. Investigation of anti-parasitic drug residues in farmhouse cheese. *In Proceedings of the Farmhouse Cheese Workshop*, Teagasc, Moorepark, County. Cork, Ireland.

Davis, J. G., C. C. Truman, S. C. Kim, J. C. Ascough, and K. Carlson. 2006. Antibiotic transport via runoff and soil loss. *J. Environ. Qual.* 35:2250–2260.

Ding, T., J. Xu, C. Shen, Y. Jiang, H. Chen, B. Wu, Z. Zhao, G. Li, J. Zhang, and F. Liu. 2006. Determination of three nitroimidazole residues in royal jelly by high performance liquid chromatography-tandem mass spectrometry. *Chin. J. Chrom.* 24: 331–334.

Draisci, R., C. Marchiafava, L. Palleschi, P. Cammarata, and S. Cavalli. 2001. Accelerated solvent extraction and liquid chromatography-tandem mass spectrometry quantitation of corticosteroid residues in bovine liver. *J. Chromatogr. B* 753: 217–223.

Dubois, M., G. Pierret, and P. Delahaut. 2004. Efficient and sensitive detection of residues of nine coccidiostats in egg and muscle by liquid chromatography–electrospray tandem mass spectrometry. *J. Chromatogr. B* 813:181–189.

Edder, P., D. Ortelli, P. Jan, and C. Corvi. 2004. Analysis of benzimidazoles and levamisole residues by liquid chromatography tandem mass spectrometry detection, p. 566. *In* L.A. van Ginkel and A. A. Bergwerff (ed.), *Proceedings of EuroResidue V Conference on Residues of Veterinary Drugs in Food, Noordwijkerhout, The Netherlands*. Faculty of Veterinary Medicine, University of Utrecht, Utrecht, The Netherlands.

European Commission. 2006. The Rapid Alert System for Food and Feed (RASFF) Annual Report. European Commission, Brussels, Belgium.

Fraselle, S., V. Derop, J.-M. Degroodt, and J. Van Loco. 2007. Validation of a method for the detection and confirmation of nitroimidazoles and the corresponding hydroxy metabolites in pig plasma by high performance liquid chromatography–tandem mass spectrometry. *Anal. Chim. Acta* 586:383–393.

Gentili, A., 2007. LC-MS methods for analyzing anti-inflammatory drugs in animal-food products. *Trends Anal. Chem.* 26:595–608.

Glynn, M. K., C. Bopp, W. Dewitt, P. Dabney, M. Mokhtar, and F. J. Angulo. 1998. Emergence of multidrug-resistant *Salmonella enterica* serotype Typhimurium DT104 infections in the United States. *N. Engl. J. Med.* 338:1333–1338.

Gowik, P., J. Polzer, and S. Uhlig. 2008. Determination of MU in residue analysis by means of statistical factorial design, p. 1205–1209. *In* L. A. van Ginkel and A. A. Bergwerff (ed.), *Proceedings of EuroResidue VI Conference on Residues of Veterinary Drugs in Food*, Egmond aan Zee, The Netherlands.

Granelli, K., and C. Branzell. 2007. Rapid multi-residue screening of antibiotics in muscle and kidney by liquid chromatography-electrospray ionization—tandem mass spectrometry. *Anal. Chim. Acta* 586:289–295.

Greig, M. J., B. Bolanos, T. Quenzer, and J. M. R. Bylund. 2003. Fourier transform ion cyclotron resonance mass spectrometry using atmospheric pressure photoionization for high-resolution analyses of corticosteroids. *Rapid Commun. Mass Spectrom.* 17:2763–2768.

Haller, M. Y., S. R. Muller, C. S. McArdell, A. C Alder, and M. J. F. Suter. 2002. Quantification of veterinary antibiotics (sulfonamides and trimethoprim) in animal manure by liquid chromatography-mass spectrometry. *J. Chromatogr. A.* 952:111–120.

Hammel, Y.-A., R. Mohamed, E. Gremaud, M.-H. LeBreton, and P. A. Guy. 2008. Multi-screening approach to monitor and quantify 42 antibiotic residues in honey by liquid chromatography–tandem mass spectrometry. *J. Chromatogr. A* 1177:58–76.

Hamscher, G., S., Sczesny, H. Hoper, and H. Nau. 2002. Determination of persistent tetracycline residues in soil fertilized with liquid manure by high-performance liquid chromatography with electrospray ionization tandem mass spectrometry. *Anal. Chem.* 74:1509–1518.

Herikstad, H., P. Hayes, M. Mokhtar, M. L. Fracaro, E. J. Threlfall, and F. J. Angulo. 1997. Emerging quiolone-resistant *Salmonella* in the United States. *Emerg. Infect. Dis.* 3:371–372.

Hernando, M. D., M. Mezcuaa, J. M. Suarez-Barcenab, and A. R. Fernandez-Alba. 2006. Liquid chromatography with time-of-flight mass spectrometry for simultaneous determination of chemotherapeutant residues in salmon. *Anal. Chim. Acta* 562:176–184.

Kaufmann, A., P. Butcher, K. Maden, and M. Widmer. 2007. Ultra-performance liquid chromatography coupled to time of flight mass spectrometry (UPLC–TOF): a novel tool for multi-residue screening of veterinary drugs in urine. *Anal. Chim. Acta* 586:13–21.

Kaufmann, A., P. Butcher, K. Maden, and M. Widmer. 2008. Quantitative multiresidue method for about 100 veterinary drugs in different meat matrices by sub 2-µm particulate high-performance liquid chromatography coupled to time of flight mass spectrometry. *J. Chromatogr. A* 1194:66–79.

Kay, P., P. A. Blackwell, and A. B. A. Boxall. 2004. Fate and transport of veterinary antibiotics in drained clay soils. *Environ. Toxicol. Chem.* 23:1136–1144.

Keegan, J., M. Danaher, S. R. H. Crroks, C. T. Elliott, and R. O'Kennedy. 2008. Detection of benzimidazole carbamate residues in liver tissue using a surface plasmon resonance biosensor, p. 1153. *In* L. A. van Ginkel and A. A. Bergwerff (ed.), *Proceedings of EuroResidue VI Conference on Residues of Veterinary Drugs in Food*. Egmond aan Zee, The Netherlands.

Kemper, N. 2008. Veterinary antibiotics in the aquatic and terrestrial environment. *Ecol. Indic.* 8:1–13.

Kinsella, B., S. J. Lehotay, K. Matsakova, A. R. Lightfield, A. Furey, and M. Danaher. 2007. New method for the analysis of anthelmintic residues in animal tissues. *In Proceedings of 121st AOAC International Annual Meeting and Exposition*. Anaheim, CA.

Levy, S. 1987. Antibiotic use for growth promotion in animals: ecological and public health consequences. *J. Food Prot.* 50:616–620.

Mayrhofer, S., P. Peter Paulsen, F. J. M. Smulders, and F. Hilbert. 2004. Antimicrobial resistance profile of five major food-borne pathogens isolated from beef, pork and poultry. *Int. J. Food Microbiol.* **97**:23–29.

McEvoy, J. D. G. 2002. Contamination of animal feeding stuffs as a cause of residues in food: a review of regulatory aspects, incidence and control. *Anal. Chim. Acta* **473**:3–26.

Mortier, L., E. Daeseleire, and P. Delahaut. 2003. Simultaneous detection of five coccidiostats in eggs by liquid chromatography–tandem mass spectrometry. *Anal. Chim. Acta* **483**:27–37.

Mulder, P. P. J., P. Balzer-Rutgers, E. M. te Brinke, Y. J. C. Bolck, B. J. A. Berendsen, H. Gercek, B. Schat, and J. A. van Rhijn. 2005. Deposition and depletion of the coccidiostats toltrazuril and halofuginone in eggs. *Anal. Chim. Acta* **529**:331–337.

Nielen, M. W. F., M. C. van Engelen, R. Zuiderent, and R. Ramaker. 2007. Screening and confirmation criteria for hormone residue analysis using liquid chromatography accurate mass time-of-flight, Fourier transform ion cyclotron resonance and orbitrap mass spectrometry techniques. *Anal. Chim. Acta* **586**:122–129.

Ochman, H., J. Lawrence, and E. Groisman. 2000. Lateral gene transfer and the nature of bacterial innovation. *Nature* **405**:299–304.

O'Keeffe, M., A. Conneely, K. M. Cooper, D. G. Kennedy, L. Kovacsics, A. Fodor, P. P. J. Mulder, J. A. van Rhijn, and G. Trigueros. 2004. Nitrofuran antibiotic residues in pork: the FoodBRAND retail survey. *Anal. Chim. Acta* **520**:125–131.

Perron, G., S. Quessy, and G. Bell. 2007. Genotypic diversity and antimicrobial resistance in asymptomatic *Salmonella enterica* serotype Typhimurium DT104. *Infect. Genet. Evol.* **7**:223–228.

Polzer, J., C. Stachel, and P. Gowik. 2004. Treatment of turkeys with nitroimidazoles; impact of the selection of target analytes and matrices on an effective residue control. *Anal. Chim. Acta* **521**:189–200.

Prats, G., B. Mirelis, T. Llovet, C. Munoz, E. Miro, and F. Navarro. 2000. Antibiotic resistance trends in enteropathogenic bacteria isolated in 1985–1987 and 1995–1998 in Barcelona. *Antimicrob. Agents Chemother.* **44**:1140–1145.

Pugh, D. M. 2002. The EU precautionary bans on animal feed additive antibiotics. *Tox. Lett.* **128**:35–44.

Radeck, W., and P. Gowik. 2005. Validation of a multiresidue method for confirmation and quantification of 25 anticoccidial drugs in egg, muscle and liver. *In Proceedings of the 5th International Symposium on Hormones and Veterinary Drug Residue Analysis.* Antwerp, Belgium.

Radeck, W., and P. Gowik. 2008. Validation of a multiresidue method for the determination and quantitation of anthelmintics in milk, p. 1181–1186. *In* L. A. van Ginkel and A. A. Bergwerff (ed.), *Proceedings of EuroResidue VI Conference on Residues of Veterinary Drugs in Food.* Egmond aan Zee, The Netherlands.

Samsonova, J. V., G. A. Baxter, S. R. H. Crooks, A. E. Small, and C. T. Elliott. 2002. Determination of ivermectin in bovine liver by optical immunobiosensor. *Biosenso Bioelectron.* **17**:523–529.

Sangiorgi, E., M. Curatolo, W. Assini, and E. Bozzoni. 2003. Application of neutral loss mode in liquid chromatography-mass spectrometry for the determination of corticosteroids in bovine urine. *Anal. Chim. Acta* **483**:259–267.

Sarmah, A. K., M. T. Meyer, and A. B. A. Boxall. 2006. A global perspective on the use, sales, exposure pathways, occurrence, fate and effects of veterinary antibiotics (Vas) in the environment. *Chemosphere* **65**:725–759.

Schlusener, M. P., K. Bester, and M. Spiteller. 2003. Determination of antibiotics such as macrolides, ionophores, and tiamulin in liquid manure by HPLC-MS/MS. *Anal. Bioanal. Chem.* **375**:942–947.

Srinivasan, V., H. M. Nam, L. T. Nguyen, B. Tamilselvam, S. E. Murinda, and S. P. Oliver. 2005. Prevalence of antimicrobial resistance genes in Listeria monocytogenes isolated from dairy farms. *Foodborne Pathog. Dis.* **2**:201–211.

Stock, I., and B. Wiedemann. 1999. An in-vitro study of the antimicrobial susceptibilities of *Yersinia enterocolitica* and the definition of a database. *J. Antimicrob. Chemother.* **43**:37–45.

Stolker, A. A. M., T. Zuidema, and M. W. F. Nielen. 2007. Residue analysis of veterinary drugs and growth-promoting agents. *Trends Anal. Chem.* **26**:969–979.

Stolker, A. A. M., P. Rutgers, E. Oosterlink, J. J. P. Lasaroms, R. J. B. Peters, J. A. van Rhihn, and M. W. F. Neilen. 2008. Comprehensive screening and quantification of veterinary drugs in milk using UPLC-ToF-MS. *Anal. Bioanal. Chem.* **391**:2309–2322.

Swann, M. M. 1969. *Swann Report,* Cmnd 4190. Joint Committee on the Use of Antibiotics in Animal Husbandry and Veterinary Medicine, HMSO, London, England.

Thakur, S., and W. A. Gebreyes. 2005. *Campylobacter coli* in swine production: antimicrobial resistance mechanisms and molecular epidemiology. *J. Clin. Microbiol.* **43**:5705–5714.

Threnlfal E. J., J. A. Frost, L. R. Ward, and B. Rowe. 1996. Increasing spectrum of resistance in multiresistant *Salmonella typhimurium*. *Lancet* **347**:1053–1054.

Van Hoof, N., K. De Wasch, S. Poelmans, H. Noppe, and H. De Brabander. 2004. Multi-residue liquid chromatography/tandem mass spectrometry method for the detection of non-steroidal anti-inflammatory drugs in bovine muscle: optimisation of ion trap parameters. *Rapid Commun. Mass Spectrom.* **18**:2823–2829.

VRC. 2006. *Annual Report on Surveillance for Veterinary Residues in Food in the UK, 2005.* The Veterinary Residues Committee, Surrey, United Kingdom.

White, D. G., S. Zhao, S. Simjee, D. D. Wagner, and P. F. McDermott. 2002a. Antimicrobial resistance of foodborne pathogens. *Microbes Infect.* **4**:405–412.

White, D. G., S. Zhao, P. F. McDermott, S. Ayers, S. Gaines, S. Friedman, D. D. Wagner, J. Meng, D. Needle, M. Davis, and C. Debroy. 2002b. Characterization of antimicrobial resistance among *Escherichia coli* O111 isolates of animal and human origin *Microb. Drug Resist.* **8**:139–146.

Witte, W. 1998. Medical consequences of antibiotic use in agriculture. *Science* **279**:996–997.

Zuidema, T., P. P. Mulder, J. J. Lasaroms, S. Stappers, and J. A. Van Rhijn. 2006. Can high-performance liquid chromatography coupled with fluorescence detection under all conditions be regarded as a sufficiently conclusive confirmatory method for B-group substances? *Food Addit. Contam.* **23**:1149–1156.

Pathogens and Toxins in Foods: Challenges and Interventions
Edited by V. K. Juneja and J. N. Sofos
© 2010 ASM Press, Washington, DC

Chapter 22

Prions and Prion Diseases

DRAGAN MOMCILOVIC

The term "prions" was coined from the terms "proteinaceous" and "infectious" and initially defined as proteinaceous, infectious particles that resist inactivation by procedures that modify nucleic acids (Prusiner, 1982). Subsequently, the definition was modified to characterize prions as proteinaceous, infectious particles that lack nucleic acid (Prusiner, 1997). Prions are generally regarded as the most reliable markers of so-called prion diseases. With the advent of the protein-only concept of the transmission of prion diseases, which is predicated on the idea that prions are the infectious agents causing prion disease (Prusiner, 1982), and with the 1997 Nobel Prize for Medicine awarded for the prion protein concept (Nobel Assembly, 1997), prions have firmly asserted their inclusion in the club of infectious agents, along with bacteria, viruses, fungi, and parasites, the club's traditional members. The ultimate proof of the protein-only concept is lacking, meaning that the question of whether prions are only the product of prion diseases or are also the transmissible agent causing them, continues to be incompletely answered. Prions are regarded as a permanent source of concern with respect to animal and human health because of their associations with prion diseases.

PRION DISEASES

Prion diseases, also known as transmissible spongiform encephalopathies (TSEs), are a group of invariably fatal, progressively degenerative neurological diseases of animals and humans characterized by the accumulation of an abnormal form of prion protein in tissues. The diseases are *transmissible* because they can be transmitted to other individuals, *spongiform* because they induce sponge-like changes visible microscopically in the central nervous system (CNS),

and *encephalopathies* because they affect the brain. TSEs in animals include scrapie of sheep and goats, chronic wasting disease (CWD) of deer and elk, feline spongiform encephalopathy of cats, transmissible mink encephalopathy (TME) of mink, and bovine spongiform encephalopathy (BSE) of cattle. TSEs in humans include kuru, Creutzfeldt-Jakob Disease (CJD), variant Creutzfeldt-Jakob Disease (vCJD), Gerstmann-Straussler-Scheinker syndrome, and fatal familial insomnia.

Under experimental conditions, TSEs can be transmitted to a wide range of mammalian species, but under natural conditions, interspecies transmission is limited. Some TSEs spread horizontally by direct or indirect contact (e.g., scrapie and CWD), while others do not (e.g., BSE and vCJD). TSEs are transmitted via oral exposure to a variety of sources of infectivity, including the placenta and fetal fluids, in the case of scrapie (Race et al., 1998); products such as meat and bone meal containing the infectious agent, in the case of BSE (Wilesmith et al., 1991); contaminated tissues, in the case of vCJD (Will et al., 1996) and TME (Hartsough and Burger, 1965); or oral contact with saliva from an infected animal, in the case of CWD (Mathiason et al., 2006). BSE is the only TSE known to transmit to humans through consumption of food. Other animal TSEs, such as scrapie, TME, and CWD, are not recognized as being zoonotic.

History

Human experience with TSEs spans more than two-and-a-half centuries. First occurrences of scrapie were described for sheep in 1732, and first occurrences of vCJD in humans, the last identified TSE, were described in 1996 (Will et al., 1996). Regardless of the length of time that has elapsed since humankind

---

Dragan Momcilovic • Center for Veterinary Medicine, Food and Drug Administration, Rockville, MD 20855.

learned about individual members of the group, TSEs continue to be considered a novel group of diseases primarily because they have been elucidated relatively recently, in the second half of the 20th century (Prusiner, 1982). Before the prion protein theory gained wider acceptance, TSEs were considered to be caused by genetic mutations, slow viruses, viroid-like nucleic acids, or replicating polysaccharides (Prusiner, 1998). Two Nobel Prizes have been awarded for work on TSEs. In 1976, Daniel Gajdusek was awarded the Nobel Prize in Medicine for his research on kuru (Nobel Assembly, 1976). He elucidated a disease that was affecting the South Fore people of New Guinea in the 1950s and 1960s, by connecting the occurrence of kuru with the practice of cannibalistic eating of brains of deceased relatives (Nobel Assembly, 1976). In 1997, Stanley Prusiner was awarded the Nobel Prize in Medicine for "his pioneering discovery of an entirely new genre of disease-causing agents and the elucidation of the underlying principles of their mode of action" (Nobel Assembly, 1997).

NORMAL AND ABNORMAL PRIONS

The term "prions" is used to denote both the normal and abnormal forms of the prion protein. Normal prions, also known as "cellular," are associated with the cell membrane, are sensitive to degradation with proteinase K, and are denoted in literature as PrP^C or PrP^{Sen}. The superscripts "C" and "Sen" stand for "cellular" and "sensitive" (sensitive to digestion with proteinase K), respectively. Abnormal prions are much more resistant to proteinase K than normal prions and are denoted in literature as PrP^{Sc} or PrP^{Res}; "Sc" stands for "scrapie," and "Res" for "resistant." Other superscripts sometimes used in the literature are "BSE," "CJD," and "CWD" for individual TSEs and "DIS" or "d" for "diseased."

Only the normal form of the prion protein is encoded in the DNA. The full version of the normal protein consists of approximately 209 amino acid residues (Oesch et al., 1985), while the abnormal prion partially digested with proteinase K consists of 142 amino acid residues (James et al., 1997). Normal prions are synthesized just like any other cellular protein predestined for positioning in the cell membrane. Thus, the prion gene is transcribed in the cell nucleus, the mRNA translated, and protein synthesized in the endoplasmic reticulum before it matures in the Golgi and *trans*-Golgi network; the process finishes its course with the protein anchored with the glycosylphosphatidylinositol anchor in the cell membrane (Baron and Caughey, 2003). The process by which normal prions are converted to the abnormal form has yet to be fully explained.

PROTEIN-ONLY THEORY

According to the protein-only theory, PrP^{Sc} is an isoform of PrP^C. This was supported by a finding of identical amino acid compositions, but of different proportions of α-helix and β-sheet structures (42%:3%) in PrP^C and (30%:43%) in PrP^{Sc} (Pan et al., 1993). A molecule that changes its functional/physiochemical properties through changing its conformation rather than through its primary composition, i.e., its amino acid sequence, could be compared conceptually to an accordion. The stretched and collapsed accordions have the same compositions, but different shapes and different properties. Somewhat stretched (PrP^C) and somewhat compressed (PrP^{Sc}) prions have the same composition but remarkably different properties, as summarized in Table 1.

The essence of the protein-only theory is that the abnormal prion enters the organism and converts the host's normal prions into abnormal prions. Although widely accepted as the theory that best matches the observations, this novel concept, in which an infectious agent lacks genetic material and in which abnormal prions are the result and the cause of TSE diseases, remains controversial. One discrepancy is the existence of multiple scrapie strains, or abnormal prion protein conformations, when only a single gene codes for the normal prion protein (Basler et al., 1986). A single coding exon (Basler et al., 1986) means that the cell can synthesize only one type of prion protein. The existence of various strains of abnormal prion protein, therefore, cannot be explained by alternative splicing of the prion protein mRNA. This means that the existence of abnormal prion protein strains must be the result of posttranslational changes.

PRION PROTEIN CONVERSION

The protein-only theory does not exactly explain how the protein conversion occurs. The process

Table 1. Overview of prion protein properties

PrP^C	PrP^{Sc}
Normal protein	Abnormal protein
Physiologic functions incompletely known	Associated with prion diseases
Synthesized within the cell	Converted from the normal prion
Anchored in the cell membrane	Not bound to membrane
Secondary structure α-helix rich	Secondary structure β-sheet rich
Soluble and sensitive to proteases	Insoluble and resistant to proteases
Sensitive to heat	Extremely heat resistant

assumes recruiting of a nascent normal prion protein and configuring it according to the template provided by the abnormal prion protein (Prusiner, 1998). The efficiency of the conversion is influenced by the degree of amino acid sequence homology between the two PrPs involved, so that the greater the degree of homology, the greater the rate of conversion (Prusiner, 1998). Several lines of studies have provided support for this theory. In one study, genetically modified animals lacking the ability to synthesize normal prion proteins could not contract scrapie because no new infectious particles were generated (Bueler et al., 1993). In another study, prion strains were determined by the shape of the molecule, and distinct shapes could be created solely by varying the polymerization temperature (Tanaka et al., 2004). In a third study, once a stable prion strain had emerged, it would efficiently transmit to daughter cells during cell division in yeast (King and Diaz-Avalos, 2004).

It was postulated that prion protein conversion is catalyzed by a so-called protein X (Telling et al., 1995), the existence of which has not been confirmed. One finding suggested that protein X may be RNA, not a protein (Deleault et al., 2003). The authors observed no conversion of normal to abnormal prion protein in a brain homogenate with all RNA removed. The addition of normal brain homogenate, however, accelerated the conversion sixfold, and the addition of excess RNA spiked the rate of conversion 24-fold (Deleault et al., 2003).

SPONTANEOUS TSEs

Proteins sometimes misfold after synthesis (Chesebro, 2004). Such misfolding can result in the de novo synthesis of abnormal prions in the cell (Ma and Lindquist, 2002). Facing the reality of dealing with faulty synthesized proteins, the cell has developed two basic mechanisms for dealing with the problem (Chesebro, 2004). One of the two mechanisms is employing molecular chaperone proteins to prevent premature and incorrect protein folding, and the other is employing proteasomes to degrade faulty synthesized and misfolded proteins. Inhibition of proteasome function resulted in the synthesis of faulty PrP proteins resembling PrP^{Sc} (Ma et al., 2002), which in turn could produce TSE (Ma and Lindquist, 2002). The faulty synthesized prion protein observations have laid a scientific foundation for an expectation of the existence of spontaneous (sporadic) occurrences of TSEs.

PRIONS ON THE MOVE

It is well established that the ultimate target of abnormal prions is the CNS and that in most cases abnormal prions enter the organism via the alimentary tract. It is unknown exactly how abnormal prions get moved from one point to another, i.e., from the alimentary tract to the CNS, from guts to brain. The exact point of prion entry in the organism from the alimentary tract has not been determined precisely. Studies have shown that abnormal prions enter through either the M cells in Peyer's patches in the small intestine (Heppner et al., 2001) or across villous enterocytes in the immediate vicinity of the M cells (Jeffrey et al., 2006). Such observations are in line with the finding that human enterocytes express normal prion protein (Morel et al., 2004). Following de novo replication of abnormal prions in Peyer's patches 30 days after inoculation (Jeffrey et al., 2006), abnormal prions move into the brain, either directly via the nervus vagus or indirectly via the spinal cord (Hoffmann et al., 2007). Another alimentary route of infection is the tongue. Bartz et al. (2003) reported TSE infection being 100,000-fold more efficient by inoculation of abnormal prions into the tongues of hamsters than by oral ingestion.

Although the overall infectivity spread in one animal could be deduced from findings of the infectivity in various tissues, it is not known exactly how the infectivity spreads at the local tissue level. It is believed that abnormal prions can spread locally on the adjacent cells with exosomes that exit and enter the cell (Fevrier et al., 2004) or possibly via tunneling nanotubes, which, according to Rustom et al. (2004), connect adjacent cells temporarily with bridges that support exchange of molecules and even small organelles.

A simple explanation for mechanisms that transport prions throughout one organism seems to be lacking, and questions, such as how prions get transported from the alimentary tract to the nerve endings, how they are transported along the nerves, and how they bridge the synapses, remain incompletely answered (Glatzel at al., 2004). The exact mode of the abnormal prion circulation in the organism may be even more complicated than previously thought. It was shown that PrP^{Sc} accumulated in the spleens of transgenic mice following an oral (Andreoletti et al., 2006) or intracerebral injection (Crozet et al., 2007). The latter observation was rather unexpected because the mice were lacking the ability to produce PrP^{C} in peripheral tissues (Crozet et al., 2007). It was proposed that nerve fibers may promote the circulation of PrP^{Sc} between the CNS and the periphery within the host in a centripetal and/or centrifugal manner (Crozet et al., 2007).

Several studies suggest that white blood cells may be involved in the process. It was observed that gut-associated macrophages initially remove TSE infectivity, but once their ability to remove the infectivity completely is overcome, macrophages may facilitate the spread of infectivity (Maignien et al., 2005). Similarly, an inoculum containing abnormal prions was moved in the intestine in a fashion attributable to dendritic cells or macrophages (Jeffrey et al., 2006). In line with these findings in which white blood cells possibly transport abnormal prions, another study reported that mice infected with scrapie accumulated abnormal prions in their parenchymal organs, including the liver, kidneys, and pancreas, only when these organs were acutely inflamed (Heikenwalder et al., 2005), suggesting that white blood cells migrating into the inflamed organs brought along abnormal prions. Similarly, TSE infectivity was detected in urine samples of scrapie-infected mice suffering from acute nephritis, but not in scrapie-infected mice not suffering from acute nephritis (Seeger et al., 2005). Finally, abnormal prions were detected in the mammary glands of ewes infected with scrapie and suffering from acute mastitis, but not in ewes infected with scrapie but unaffected by mastitis (Ligios et al., 2005). Both the mice and the sheep studies seem to support the idea of white blood cells being involved in abnormal prion protein mobility. Similar studies on cattle have not been performed. Given the differences in lympho-reticular involvement, it is unclear whether abnormal prions would be found in inflamed bovine tissues.

DISTRIBUTION OF INFECTIVITY

The amount of infectivity in individual tissues varies with the type of TSE, animal species, concurrent inflammatory disease, the initially received dose, and the stage of incubation. TSE infectivity has been demonstrated in the brain, spinal cord, dorsal root ganglion, eyes, tonsils, thymus, trigeminal ganglia, and distal ileum. The exact location of the infectivity depends on animal species, type of TSE, and the stage of incubation. As confirmed with bioassays with highly sensitive transgenic mice, BSE infectivity in a terminally ill cow was remarkably well constrained in the cow's CNS, unlike scrapie infectivity in sheep, mice, and hamsters which was present in the lymphatic system (Buschmann and Groschup, 2005). Besides the organ predilection, there appears to be a relatively high degree of predilection for a local distribution within an organ. Thus, the pattern of BSE distribution was very similar across the nine brains of cattle that died of BSE (Vidal et al., 2005). The observed differences in the staining intensities were attributed to various stages of incubation that the cattle were in at the times of death (Vidal et al., 2005). Extremely small amounts of the infectivity were found in inflamed parenchymal organs, such as the liver, pancreas, kidneys (Heikenwalder et al., 2005; Seeger et al., 2005; Siso et al., 2006), and mammary gland (Ligios et al., 2005), and in muscle (Buschmann and Groschup, 2005; Angers et al., 2006).

SOURCES

Those tissues (e.g., CNS and lymphatic) in which TSE infectivity has been detected are potential sources of infectivity. Considering that various TSEs have various ways of spreading, one's consideration of TSE infectivity spread through the system when starting with an animal infected with TSE as the definitive source of infection will depend on a particular TSE. Muscle and milk are the two primary products derived from ruminant animals and used in human nutrition. In numerous studies based on biological assays with mice and experimental and practical feedings of milk derived from BSE-positive animals to calves, no BSE was found in milk. There was no evidence in two recently reported studies of the presence of BSE infectivity in the somatic cell-rich fraction of milk collected from cows incubating experimentally induced BSE (Everest et al., 2006), or in the colostrum collected from an advanced clinical BSE case and tested in highly sensitive transgenic mice (Buschmann and Groschup, 2005).

Also, there is very little evidence for the presence of TSE infectivity in muscle. Research evidence for the presence of TSE infectivity in ruminant muscle is limited to three reports of small amounts of infectivity in skeletal muscle of sheep experimentally infected with scrapie (Andreoletti et al., 2004), deer infected with CWD (Angers et al., 2006), and in a single muscle collected from a cow infected with BSE (Buschman and Groschup, 2005). In the latter case, only 1 of 10 transgenic Tgbov mice expressing high levels of normal bovine prion protein produced a small amount of abnormal prions in the brain following an injection of the extract prepared from one of the two examined muscles (Buschmann and Groschup, 2005). It should be noted that the amounts of TSE infectivity reported in these three studies were extremely small and that two of the three TSEs in these reports (scrapie and CWD) are not zoonotic.

INACTIVATION

Abnormal prions are remarkably resistant to inactivation by heat, irradiation, and chemical reagents. They are particularly resistant to heat degradation, as boiling water and autoclaving at 121°C for 15 minutes are ineffective in reducing TSE infectivity (Taylor et al., 1999). Exposure of scrapie-infected brain tissue to dry heat at 150 to 300°C for 5 minutes produced only negligible infectivity reduction, while 15-minute exposures reduced the infectivity for 2 to 3 logs but did not eliminate it (Brown et al., 2000). Moreover, 5 out of 18 tissue samples exposed to 600°C for 15 minutes transmitted the disease (Brown et al., 2000). In a repeated study, 2 out of 21 1-gram samples that burned for 15 minutes transmitted scrapie (Brown et al., 2004). These two studies suggest that even the ash can be infectious. Considering that organic substances get carbonized at such high temperatures, it was proposed that inorganic "fossil templates" form and propagate infectivity long after the prion itself is gone (Brown et al., 2000).

Hyperbaric rendering at temperatures not less than 133°C, at pressure not less than 3 bars, and for no less than 20 minutes was found capable of reducing the starting level of TSE infectivity by at least 3 logs (Taylor et al., 1997). The European Union required, before instituting the complete ban on feeding animal proteins to nonruminant animals in the year 2000, that all animal materials be rendered at the "133°C, 3 bar, 20 minute" conditions (Taylor and Woodgate, 2003). Rendering under atmospheric pressure alone is not considered as effective in reducing TSE infectivity as the 3-bar pressure rendering, but it is thought that a sufficient level of risk reduction using atmospheric rendering can be achieved in conjunction with control measures such as a ruminant-to-ruminant feed ban (Taylor and Woodgate, 2003). An ultra-high-pressure approach was proposed for reduction of prion infectivity in meat (Brown et al., 2003).

Abnormal prions are resistant to ionizing irradiation (Safar et al., 2005) and are particularly resistant to conventional sterilization techniques, as they are capable of binding to surfaces of metal and plastic without losing infectivity (Weissmann et al., 2002).

The resistance to degradation with protease K and several other enzymes is not absolute, and prolonged exposure to protease K will result in the protein's complete degradation (Jackson et al., 2005).

Treatment of an instrument's surfaces and tissues with chemical reagents, such as 1M sodium hydroxide, was effective in making abnormal prion protein sensitive to proteinase K degradation (Kasermann and Kempf, 2003), and so was copper hydrogen peroxide (Solassol et al., 2006). Alkaline hydrolysis using 1M sodium or potassium hydroxide for 3 hours at 150°C and under 4-bar pressure digests large carcasses, producing a protein-free concoction of amino acids, peptides, sugars, and soaps (Taylor and Woodgate, 2003). Elimination of TSE infectivity on surgical instruments and other surfaces appears to be very difficult and cumbersome but achievable by use of certain chemical reagents.

INTERVENTIONS

No procedure is currently available for complete removal of TSE infectivity from food and/or animal feed. Therefore, the best strategy has been to exclude animals affected by TSE from the population so that no product derived from such animals enters the food and/or the animal feed supply. The problem, however, is that affected animals may not show clinical symptoms until the last stages of the disease incubation. To prevent potentially undetected infectivity from entering the human food supply, governments of many countries designated tissues known to carry the greatest risk of harboring the infectivity specified risk materials (SRM). Such designated materials are required to be removed from food channels at slaughter. In the United States, the Departments of Agriculture and Health and Human Services, reacting to the first case of BSE in the country, designated brain, skull, eyes, trigeminal ganglia, spinal cord, vertebral column (excluding the vertebrae of the tail, the transverse processes of the thoracic and lumbar vertebrae, and the wings of the sacrum), dorsal root ganglia in cattle 30 months of age and older, and the tonsils and distal ileum of the small intestine in cattle of any age SRM (USDA, 2004).

In addition, the World Health Organization recommends that materials derived from TSE-positive animals not be used in human or animal nutrition (WHO, 2000).

DETECTION

The exact minimum amount of material that can spread BSE or other TSEs is not known. It has been shown that the minimum oral infectious dose in cattle may be less than 1 milligram of the brain stem (Wells et al., 2007). Detecting small levels of infectivity in small amounts of material is difficult.

Detection of TSEs is based on detection of histological changes or detection of a TSE marker, typically abnormal prion protein (see Figure 1).

Bioassay involves taking a sample from the tested animal and injecting it into the recipient animal.

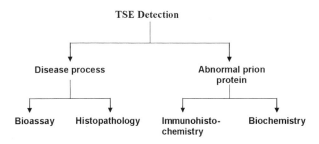

Figure 1. Common approaches to detecting TSEs.

Following the injection, the recipient animal is observed for clinical signs of TSE or euthanized at a certain point postinjection and examined histologically for changes consistent with TSE. Bioassay is slow and expensive; it is slow because it takes a long time to obtain the results, and it is expensive because of the cost associated with maintaining the recipient animals.

Histopathology involves preparation of a histological slide from a tissue (typically brain) of the suspect animal and examination for typical signs of TSE, such as gliosis, spongiosis, and neuronal loss. The test is visual and fast relative to bioassay, as it takes several days for its completion. The test, however, can be performed only on nonautolysed samples, can add some artificial changes to the sample during the tissue preparation, and requires a trained pathologist to read the slide.

Immunohistochemistry (IHC) combines histopathology with the use of antibodies specific for prion proteins to stain the histological slide. This test improves the sensitivity and specificity of testing relative to histopathology alone and is considered to be the gold standard for TSE testing (O'Rourke et al., 2003).

Biochemistry-based tests are very commonly used today and include the use of monoclonal antibodies in an enzyme-linked immunosorbent assay (ELISA) or Western blot format. The advantage of the two formats is high speed (in hours) and throughput combined with high sensitivity and specificity. Both ELISA and Western blot use digestion with proteinase K as the preparatory step.

All these methods are capable of detecting TSE in tissues collected from live or dead animals. Whether the tested animal will be dead or alive at testing will depend on whether the sample could be taken without sacrificing the animal. In TSEs that are characterized by peripheral distribution of abnormal prions (e.g., scrapie and CWD in tonsils of sheep and deer), the sampling could be achieved without sacrificing the animals. In TSEs that are characterized by central distribution of abnormal prions (e.g., BSE in cattle brain), the sampling could not be achieved without sacrificing the animal. This is to say that in the absence of a practical way of sampling CNS tissue from a live animal, there is no test for diagnostics of BSE in live animals.

In anticipation of development of live animal tests for BSE, the European Commission published criteria that such a test would have to meet in order to be considered acceptable (EFSA, 2004). The Commission recently reported that one test subjected to the validation process failed to meet the set criteria and was therefore found to be unacceptable (EFSA, 2006a).

TWO TSEs CONSIDERED IN MORE DETAIL

The remainder of this chapter focuses on two TSEs: BSE, because of its public health importance, and CWD, because of its threat to the U.S. deer and elk populations.

Bovine Spongiform Encephalopathy

Introduction

Like other TSEs, BSE is characterized by a long incubation period that averages about 5 years, an absence of clinical signs during the incubation phase, an accumulation of abnormal prions in the CNS, and the death of the infected animal. BSE was recognized in 1986 as a novel disease affecting cattle (Wells et al., 1987). Since then, over 190,000 BSE cases have been officially diagnosed worldwide, with about 185,000 cases in the United Kingdom alone (OIE, 2007) (Figure 2). The true number of BSE cases was likely much greater. In France, where as of June 2000 only 103 BSE cases were officially confirmed, it was estimated that there were more than

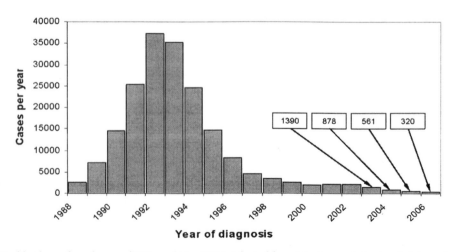

Figure 2. Worldwide confirmed cases of BSE as of May 2007. Adapted from http:/www.oie.int./eng/info/en_esbmonde.htm.

300,000 undetected cases of BSE in cattle, meaning that the French epidemic of the late 1980s went completely undetected (Supervie and Costagliola, 2004).

Origin

Several theories have been proposed to explain the origin of the first BSE case. The leading theory on BSE origin is the scrapie theory. According to the theory, cattle were fed rendered products derived from sheep affected by one of the several scrapie strains, which adapted to its new cattle host. Once the cattle became infected and their remnants rendered and fed to more cattle, more cattle got ill. The process was running its course uninterrupted until BSE was recognized and epidemiologically associated with feed. Several events provide support for this theory. BSE started in the United Kingdom because (i) there was a relatively high sheep-to-cattle ratio; (ii) rendered sheep and cattle materials were used extensively as protein supplements in cattle feed (Brown, 2001), particularly in feed of young calves, which are considered to be most susceptible to BSE at the age of about 6 months (Arnold and Wilesmith, 2004); and (iii) the rendering industry switched to a lower temperature of fat extraction, which could help BSE to survive and get established (Brown, 2001).

According to other theories, BSE could have developed from feeding cattle the remains of either wild animals (Schreuder, 1993) or humans (Colchester and Colchester, 2005). This theory is based on three hypotheses: (i) BSE was acquired from a human TSE (prion disease); (ii) the route of infection was oral, through animal feed containing imported mammalian raw materials contaminated with human remains; and (iii) the origin was the Indian subcontinent. It was recently suggested that BSE could have originated from an atypical form of BSE called bovine amyloidotic spongiform encephalopathy (BASE) (Capobianco et al., 2007). The suggestion is based on a previously reported similarity between sporadic CJD in humans and BASE (Casalone et al., 2004) and an observation of conversion of BASE into BSE (Capobianco et al., 2007).

Epidemiological evidence linked BSE with the practice of feeding cattle meat and bone materials derived from cattle carcasses containing the BSE agent. The disease could spread easily in cattle because temperatures used at rendering were not sufficiently high to inactivate fully the BSE agent and because of the absence of a species barrier when cattle material was fed back to cattle. Perhaps the strongest evidence that BSE spreads by animal feed was the effect of prohibiting ruminant protein in feed for cattle in the United Kingdom in 1988 (Statutory Instrument, 1988). The prohibition was followed by a precipitous drop in the numbers of new BSE cases. Considering that BSE infectivity does not appear to leave infected cattle via any other route (e.g., milk, saliva, or feces), and considering that clear evidence for vertical transmission of BSE (from the cow to the fetus in utero) has not been provided, feeding materials contaminated with the BSE agent may be the only practical way in which BSE spreads.

BSE in other species

BSE has also been reported in species other than cattle in which it either occurred naturally or was induced experimentally. Epidemiological and strain type analyses have suggested that naturally transmitted BSE caused feline spongiform encephalopathy in cats (Pearson et al., 1992), vCJD in humans (Will et al., 1996), and BSE in a French goat (Eloit et al., 2005). It was postulated that the sheep believed to be the most resistant genetically to prion diseases may be silent carriers of the BSE agent (Bencsik and Baron, 2007). A comprehensive retrospective study found no evidence for the presence of BSE in sheep (Stack et al., 2006).

Experimentally, BSE was induced by feeding brain tissue of a BSE-positive cow to sheep and goats (Foster et al., 1993) and monkeys (Lasmezas et al., 2005). Though BSE has been induced in sheep by experimental feeding, it is not known whether and to what extent it is transmitted to sheep under normal feeding conditions. There was a concern that if BSE entered the sheep population early in the course of the United Kingdom BSE epidemic, when sheep were fed the same materials that were fed to cattle, it could be confused by scrapie. Considering that BSE is zoonotic and scrapie is not, it became essential to know whether BSE in sheep would behave as scrapie or as BSE. A study showed that when experimentally induced and vertically transmitted from two ewes to their two lambs, BSE behaved in the brains of the lambs as BSE in cattle, not as scrapie in sheep (Bellworthy et al., 2005). This was also the first case of "natural" transmission of BSE within species other than bovine.

Similar to the concern about BSE in sheep is the concern about the possible presence of BSE in the United Kingdom deer population because early in the BSE epidemic, United Kingdom deer were likely fed materials infected with BSE (SEAC, 2006). No CWD or other TSE has been found in United Kingdom deer.

Atypical BSE

For some time, analytical and other evidence supported the idea that only a single strain of abnormal prions that originated in the United Kingdom causes all cases of BSE. Then, the uncharacteristic behavior of abnormal prions recovered from the brains of a 23-month-old Japanese steer (Yamakawa et al., 2003) and three 8-, 10-, and 15-year-old French cattle (Biacabe et al., 2004) was reported. Western blot analyses in both reports showed remarkable changes in the electrophoretic patterns of migration of abnormal prions on sodium dodecyl sulfate-polyacrylamide gel electrophoresis.

Partial digestion of abnormal prions with proteinase K yields three fractions differing in their masses and the number of carbohydrate molecules attached to each of the fractions (Figure 3). The heaviest fraction carried two attached carbohydrate molecules (di-glycosylated) and migrated the shortest distance, i.e., stayed at the top. The lightest fraction had no carbohydrate attached (nonglycosylated) and migrated the farthest, i.e., appeared at the bottom. The monoglycosylated fraction fell in between. The two studies produced two new patterns of distribution of abnormal prion fractions that differed from the pattern characteristic for a case of typical BSE. The primary difference was in the molecular weight of the nonglycosylated fraction, which was less in the Japanese case and greater in the French cases (Figure 3) than the same fraction in a typical BSE. Later the two major patterns were named L (for lower molecular weight than that of typical BSE) and H (for higher molecular weight than that of typical BSE) (Brown et al., 2006). Interestingly, the French atypical cases (the H pattern) showed the migration pattern of abnormal prions on the gel being identical to the migration pattern of scrapie prions collected from a cow that was intracerebrally injected with sheep scrapie (Biacabe et al., 2004).

Subsequent to the Japanese and French reported cases, atypical cases of BSE were reported in Italy (Casalone et al., 2004), Belgium (De Boscherre et al., 2004), Poland (Polak et al., 2004), Germany (Buschmann et al., 2006), the United Kingdom (DEFRA, 2007), and the United States (Richt et al., 2007). As summarized in Figure 3, either L or H atypical patterns were reported. In addition to

exhibiting the L pattern, the Italian atypical BSE case revealed several other characteristics in which it differed from all other cases of BSE. One difference was the finding of amyloid plaques for the first time in a BSE case in cattle, which led the authors to propose the name "bovine amyloidotic spongiform encephalopathy" (BASE) for this type of BSE case (Casalone et al., 2004). Another difference was the molecular signature of prions in this BASE case that was remarkably similar to a case of sporadic CJD in humans (Casalone et al., 2004).

Subsequent transmissibility studies have shown that atypical cases of BSE preserve their atypical properties after successful transmission to new generations, i.e., upon transmission atypical BSE shows characteristics distinct from typical BSE (Baron et al., 2006; Buschmann et al., 2006). Interestingly, studies on transgenic mice showed that the incubation time of 185 days for L-type abnormal prions was substantially shorter than the 230-day incubation time for the typical BSE, or 322 days for the H type (Buschmann et al., 2006).

Preventing BSE

Preventing BSE in the U.S. cattle population is based on several complementary measures, which are designed to (i) prevent introduction of BSE in native cattle via restricting importation of cattle from BSE-positive countries, (ii) prohibit in feed for cattle and other ruminants animal materials that could contain the BSE agent, and (iii) monitor the domestic cattle population for the presence of the disease via active surveillance. To minimize the potential for importing apparently healthy animals incubating undetected BSE, the United States prohibited importation of cattle from the United Kingdom after BSE was detected there in 1986.

Because consumption of feed contaminated with the BSE agent is the only documented route of natural transmission of BSE, prohibiting animal proteins that could contain undetected BSE infectivity from use in animal feed is the single most important measure in preventing transmission of BSE to cattle. Therefore, to prevent amplification and spread of BSE in animal feed, many countries have introduced national feed rules, which regulate the use of animal proteins in animal feed. In the United States, the Food and Drug Administration (FDA) prohibited in 1997 the use of most mammalian proteins in feed for cattle and other ruminants (FDA, 1997). In the European Union, the European Council, which initially prohibited in 1994 certain processed animal protein in ruminant feed, restricted feeding processed animal proteins back to farm animals in 2001 (EC, 2000).

Reliable national feed rules are a component of an effective feed ban, which would also include credible

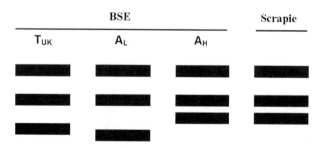

Figure 3. Schematic of migration patterns of abnormal prion protein reported in cases of typical BSE (T_{UK}), atypical BSE (A_L or light and A_H or heavy) and scrapie. A_L has been reported in Japan, Italy, and Belgium, and A_H in France, the United Kingdom, and the United States.

structure of manufacturing; distribution chain; and convincing surveillance, monitoring, and inspection systems (EC, 2001). An effective feed ban would include feed tests, capable of detecting prohibited ingredients, as an essential tool for assessing compliance with feed regulations (EC, 2001). Feed testing is not designed to detect BSE, but instead to detect animal materials that could be contaminated with the BSE agent. Current tests for prohibited animal proteins in general are based on detection of animal tissues by use of microscopy, genetic material by use of PCR, and proteins by use of immunological tests (Momcilovic and Rasooly, 2000; Gizzi et al., 2003).

The ideal test for prohibited animal ingredients in feed would differentiate between permitted and prohibited animal proteins rendered under various conditions and would be quantitative. In spite of considerable research, no such test or test combination is presently available. Development of a test for prohibited animal material in feed is plagued by limitations of a technical and regulatory nature. Two primary technical limitations are the very highly variable composition of products containing prohibited animal materials, i.e., they could include complete carcasses or only certain tissues of a single or multiple animal species, and the heat treatments used in the rendering process, which modify chemical and physical properties of individual components of rendered feeds (Momcilovic and Rasooly, 2000). The variable composition greatly complicates the selection of a marker, which is a component of feed that is specific and representative of the feed. Typically, a marker is a protein, DNA, or some other component in prohibited material. Heat treatments, which are applied at various levels and durations at rendering, denature proteins and DNA and thereby affect the performance of tests that are based on detection of proteins and DNA (Momcilovic and Rasooly, 2000). The regulatory aspect of limitations is reflected in the fact that mammalian materials, such as milk, blood, gelatin, and pure porcine and equine proteins, are not prohibited from use in ruminant feed in the United States (FDA, 1997). The presence of such materials in feed complicates their identification and separation from the prohibited materials. Quantitative determination of the amount of prohibited animal material in feed is important because knowing the level of inclusion would help further characterize BSE risk associated with that feed and distinguish intentional inclusion from cross-contamination. Presently available feed tests are inadequate for achieving precise quantitative determination of prohibited material in feed.

In the United States, the Department of Agriculture (USDA) is in charge of active surveillance. USDA inspects cattle at slaughter and samples them for

testing for BSE. Since 1 June 2004 the USDA (2007a) analyzed over 780,000 bovine samples for BSE using a rapid screening test and then performed confirmatory testing for any samples that tested inconclusive with the screening test. In addition, to ensure that samples inappropriate for the rapid screen test were still tested, the USDA (2007a) tested in the same period over 9,000 samples by using IHC tests. BSE diagnostics has been performed by using one of several approved tests. Twelve validated rapid tests are approved in the European Union (EFSA, 2006b).

The two most important controls to prevent human exposure to the BSE agent are (i) preventing BSE in cattle and (ii) removing SRM from the food chain. The first measure is accomplished by preventing spread of BSE through animal feed. The second measure, which is perhaps the single most important public health protective measure, is designed to remove from the food chain the tissues known to contain BSE infectivity, should BSE be present. Removal of such tissues ensures that undetected BSE does not enter the food chain.

The list of SRM in cattle was first based on the results of experiments on sheep affected by scrapie (EFSA, 2005). Later the list was modified to include tissues that tested positive in pathogenesis studies with cattle experimentally infected with BSE. It was shown that BSE infectivity builds up in an infected animal over time, so that the infective load in a particular animal will depend on the length of time since the animal was infected with BSE (SSC, 1999).

In addition to the total infective load, the distribution of the BSE infectivity in the animal's body also was found to change over time, i.e., at early stages of the incubation the intestines are infective, while at later stages the CNS carries significantly higher infective loads (Wells et al., 1998). It was estimated that approximately 90% of total BSE infectivity resides in the brain and spinal cord (SSC, 1999) and 10% in the distal ileum (Comer and Huntly, 2004).

Based on data from sheep scrapie studies that showed detectable infectivity in the CNS approximately at 54% of the incubation period (Kimberlin and Walker, 1989), it was assumed that a similar value of 50% might apply to BSE in cattle (EFSA, 2005). This assumption led to selecting the 30-month cut-off point for removal of SRM in cattle. In more recent studies, BSE infectivity in cattle may appear in the CNS at about three-fourths of the incubation time (EFSA, 2005), or about 45 months postinfection, which suggests that the 30-month cut-off point provides a considerable margin of safety.

Reacting to the first detected case of BSE in the United States, the USDA (2004) introduced legislation that designated SRM the distal ileum and tonsils from all cattle and the brain, spinal cord, dorsal root ganglia, trigeminal ganglia, and eyes from cattle 30 months of age and older. In addition, the USDA (2004) prohibited food materials derived from non-ambulatory disabled cattle and mechanically separated beef and the use of air-injection stunning at slaughter. The primary purpose of the latter measure is to prevent dissemination of the brain tissue into other tissues at slaughter, which occurs when air-injection stunning is used.

CWD

History

CWD is the only known TSE that occurs naturally in cervids, such as mule deer, whitetail deer, Rocky Mountain elk (Williams, 2005), and wild moose (Miller, 2006). CWD was observed for the first time in 1967 with captive deer in Colorado and recognized as a spontaneously occurring form of spongiform encephalopathy in 1980 (Williams and Young, 1980). The disease subsequently spread into several other states and Canada presumably by wild animal migration and/or marketing of infected animals. The only reports of CWD outside the North American continent have come from South Korea, where the disease was diagnosed in animals imported from Canada—a single elk imported in 1997 (Sohn et al., 2002) and deer imported in 2001 (Kim et al., 2005). No case of CWD has been identified with the deer populations in the European Union (Doherr, 2006). While the origin of CWD is unknown, it has been proposed that CWD originated from scrapie, occurred spontaneously, or developed from another unidentified source of infection (Williams, 2005).

Transmission

The exact mode of CWD transmission is not entirely known. It is known, however, that CWD spreads horizontally, with efficiency that can be as high as 89% (Miller and Williams, 2003). The accumulation of the abnormal prion protein in the peripheral lymphatic tissue (Miller and Williams, 2003) and saliva (Mathiason et al., 2006) is crucial for the horizontal transmissibility of the disease. A meticulously executed study showed that in addition to the direct contact, the environment could play an equally important role in transmission of the disease (Miller et al., 2004). In this study, healthy deer developed CWD after being placed in paddocks that were contaminated in two ways: remnants of a CWD-positive deer were left to decay for 1.8 years before the commencement of the study, or CWD-positive deer were allowed to stay in the paddock for some time and then the paddock was emptied and kept empty for 2.2 years before the study. These results showed that CWD eradication may be difficult to achieve because the CWD agent may remain infective in the environment long after infected animals are gone. Vertical transmission is not considered to play a role in CWD transmission (Miller and Williams, 2003).

Susceptibility of other species to CWD

Data relative to spread to species other than deer and elk are scarce. Approximately a year following a study which showed that CWD could be transmitted orally to moose (Kreeger et al., 2006), CWD was detected in wild moose (Miller, 2006). The prevalence of CWD in the moose populations is unknown (Miller, 2006).

Epidemiological and experimental evidence suggests strongly that CWD does not affect humans (Belay et al., 2004). Even though CWD prions converted human prions in an in vitro cell-free environment, the rate of conversion was very low, thereby indicating the existence of a species barrier at the molecular level (Raymond et al., 2000). An intracerebral inoculation of the CWD-positive extract in transgenic mice expressing human prion protein produced no disease in the mice (Tamguney et al., 2006). The disease has been transmitted to squirrel monkeys via an intracerebral inoculation (Marsh et al., 2005).

CWD does not appear to spread to cattle because cattle fed high doses of brain from mule deer with CWD remained TSE-free for more than 7 years post-feeding (Williams, 2005). Also, cattle that commingled with CWD-positive deer in CWD-endemic areas remained clinically healthy for more than 7 years postexposure (Williams, 2005). Furthermore, CWD prions were inefficient in converting bovine prions in vitro (Raymond et al., 2000) and produced no CWD in mice genetically modified to express bovine prion protein (Tamguney et al., 2006).

Preventing CWD

Preventing CWD is complicated by an incomplete understanding of the disease and the role that the environment seems to play. In the United States, state agencies have the primary responsibility to contain and eradicate CWD. State agencies perform many activities, including monitoring farmed and wild cervid populations for CWD, providing specimen collection trainings and testing cervids for CWD, regulating importation of cervids into a state, and regulating movement of captive cervids within state borders.

The USDA (2007b) supports the surveillance of farmed and wild animals and, with the federal government providing supporting services, provides assistance to state agencies for quarantine of infected animals and premises, humane euthanasia, and testing affected and exposed animals. The USDA also provides indemnity to animal owners for the depopulation of animals identified as CWD positive. The Centers for Disease Control and Prevention (CDC) monitors epidemiological data for any evidence of CWD affecting humans. The CDC has advised hunters to avoid eating meat from deer and elk that appear sick or that test positive for CWD (http://www.cdc.gov/ncidod/dvrd/cwd/). Also, the FDA has recommended that no material derived from CWD-positive animals be used in any animal feed or feed ingredients (FDA, 2003).

Diagnostics of CWD

The principles of CWD diagnostics parallel the principles of diagnostic testing of other TSEs. IHC on the obex tissue of the brain is the "gold standard" diagnostic test for CWD (USDA, 2007b), and several ELISA-based tests have been approved for use in the United States (USDA, 2007b). An important difference exists between testing cervids for CWD and testing cattle for BSE. Because the CWD agent accumulates in the peripheral lymph nodes (Spraker et al., 2002), tissues such as the tonsils (Wild et al., 2002) or rectal mucosa (Spraker et al., 2006) could be collected from a cervid for testing without sacrificing the animal.

CONCLUDING REMARKS

Abnormal prions have found a way to multiply and propagate via a process that apparently requires no involvement of genetic material. Further scientific characterization is necessary to explain what made the prion protein, which is only one of approximately 100,000 proteins in the human organism, so different. In spite of ongoing scientific efforts that consistently provide new information elucidating the nature and biological properties of TSEs, many gaps in our understanding of TSEs continue to exist. One gap is determining the origin of a TSE. This is important because knowing the origin would help identify the cause of a TSE and shed light on mechanisms that produced the disease. Finding the first case is not easy, however, as it resembles astronomers trying to witness the big bang. While they could observe events that occurred immediately following the big bang, they cannot see the big bang itself. We may extrapolate so much but may never know exactly how TSEs started.

Control measures are reducing the incidence of BSE cases worldwide. Whether BSE will ever be eradicated may depend in large part on further characterization of atypical BSE, i.e., should science show that

atypical BSE occurs as a purely spontaneous event, then the spectrum of our intervention options to eradicate BSE will be significantly reduced. It is possible that one or more tests will be developed for testing live animals for BSE. While testing of species other than bovine (e.g., sheep, goats, and deer) may detect occasional cases of BSE, it is unlikely that such testing will show that BSE is deeply entrenched in any of those species. For example, the ongoing EU surveillance program detected no BSE in over 400,000 sheep tested between 2005 and 2006 (EFSA, 2007).

The environmental component will continue to complicate the efforts to eradicate CWD. More research is necessary to explain exactly how it spreads and whether it could affect species other than cervids. It is also possible that more definitive tests for detection of prohibited animal protein in feed will be developed.

Regulatory measures introduced to prevent the spread of TSEs to humans and other animal species are likely to stay. It is possible, however, that more research and better tests will more adequately characterize risk, which in turn may lead to a gradual loosening of some of the currently existing restrictions following a carefully performed risk assessment.

REFERENCES

Andreoletti, O., N. Morel, C. Lacroux, V. Rouillon, C. Barc, G. Tabouret, P. Sarradin, P. Berthon, P. Bernardet, J. Mathey, S. Lugan, P. Costes, F. Corbiere, J. C. Espinosa, J. M. Torres, J. Grassi, F. Schelcher, and F. Lantier. 2006. Bovine spongiform encephalopathy agent in spleen from an ARR/ARR orally exposed sheep. *J. Gen. Virol.* 87:1043–1046.

Andreoletti, O., S. Simon, C. Lacroux, N. Morel, G. Tabouret, A. Chabert, S. Lugan, F. Corbiere, P. Ferre, G. Foucras, H. Laude, F. Eychenne, J. Grassi, and F. Schelcher. 2004. PrP^{Sc} accumulation in myocytes from sheep incubating natural scrapie. *Nat. Med.* 10:591–593.

Angers, R. C., S. R. Browning, T. S. Seward, C. J. Sigurdson, M. W. Miller, E. A. Hoover, and G. C. Telling. 2006. Prions in skeletal muscles of deer with chronic wasting disease. *Science* 311:1117.

Arnold, M. E., and J. W. Wilesmith. 2004. Estimation of the age-dependent risk of infection to BSE of dairy cattle in Great Britain. *Prev. Vet. Med.* 66:35–47.

Baron, G. S., and B. Caughey. 2003. Effect of glycosylphosphatidylinositol anchor-dependent and -independent prion protein association with model raft membranes on conversion to the protease-resistant isoform. *J. Biol. Chem.* 278:14883–14892.

Baron, T. G. M., A. G. Biacabe, A. Bencsik, and J. P. M. Langeveld. 2006. Transmission of new bovine prion to mice. *Emerg. Infect. Dis.* 12:1125–1128.

Bartz, J. C., A. E. Kincaid, and R. A. Bessen. 2003. Rapid prion neuroinvasion following tongue infection. *J. Virol.* 77:583–591.

Basler, K., B. Oesch, M. Scott, D. Westaway, M. Walchli, D. F. Groth, M. P. McKinley, S. B. Prusiner, and C. Weissmann. 1986. Scrapie and cellular PrP isoforms are encoded by the same chromosomal gene. *Cell* 46:417–428.

Belay, E. D., R. A. Maddox, E. S. Williams, M. W. Miller, P. Gambetti, and L. B. Schonberger. 2004. Chronic wasting disease and potential transmission to humans. *Emerg. Infect. Dis.* 10:977–984.

Bellworthy, S. J., G. Dexter, M. Stack, M. Chaplin, S. A. C. Hawkins, M. M. Simmons, M. Jeffrey, S. Martin, L. Gonzales, and P. Hill. 2005. Natural transmission of BSE between sheep within an experimental flock. *Vet. Rec.* 157:206.

Benczik, A., and T. Baron. 2007. Bovine spongiform encephalopathy agent in a prion protein (PrP)^ARR/ARR genotype sheep after peripheral challenge: complete immunohistochemical analysis of disease-associated PrP and transmission studies to ovine-transgenic mice. *J. Infect. Dis.* 195:989–996.

Biacabe, A. G., J. L. Laplanche, S. Ryder, and T. Baron. 2004. Distinct molecular phenotypes in bovine prion diseases. *EMBO Rep.* 5:110–114.

Brown, P. 2001. Bovine spongiform encephalopathy and variant Creutzfeldt-Jakob disease. *Br. Med. J.* 322:841–844.

Brown, P., E. H. Rau, B. K. Johnson, A. E. Bacote, C. J. Gibbs, and D. C. Gajdusek. 2000. New studies on the heat resistance of hamster adapted scrapie agent: threshold survival after ashing at 600°C suggests an inorganic template of replication. *Proc. Nat. Acad. Sci. USA* 97:3418–3421.

Brown, P., E. H. Rau, P. Lemieux, B. K. Johnson, A. E. Bacote, and D. C. Gajdusek. 2004. Infectivity studies of both ash and air emissions from simulated incineration of scrapie-contaminated tissues. *Environ. Sci. Technol.* 38:6155–6160.

Brown, P., L. M. McShane, G. Zanusso, and L. Detwiler. 2006. On the question of sporadic or atypical bovine spongiform encephalopathy and Creuzfeldt-Jakob disease. *Emerg. Infect. Dis.* 12:1816–1821.

Brown, P., R. Meyer, F. Cardone, and M. Pocchiari. 2003. Ultra-high-pressure inactivation of prion infectivity in processed meat: a practical method to prevent human infection. *Proc. Natl. Acad. Sci. USA* 100:6093–6097.

Bueler, H., A. Aguzzi, A. Sailer, R. A. Greiner, P. Autenried, M. Aguet, and C. Weissmann. 1993. Mice devoid of PrP are resistant to scrapie. *Cell* 73:1339–1347.

Buschmann, A., and M. H. Groschup. 2005. Highly bovine spongiform encephalopathy-sensitive transgenic mice confirm the essential restriction of infectivity to the nervous system in clinically diseased cattle. *J. Infect. Dis.* 192:934–942.

Buschmann, A., A. Gretzschel, A. G. Biacabe, K. Schiebel, C. Corona, C. Hoffmann, M. Eiden, T. Baron, C. Casalone, and M. H. Groschup. 2006. Atypical BSE in Germany—proof of transmissibility and biochemical characterization. *Vet. Microbiol.* 117:103–116.

Capobianco, R., C. Casalone, S. Suardi, M. Mangieri, C. Miccolo, L. Limido, M. Catania, G. Rossi, G. D. Fede, G. Giaccone, M. G. Bruzzone, L. Minati, C. Corona, P. Acutis, D. Gelmetti, G. Lombardi, M. H. Groschup, A. Buschmann, G. Zanusso, S. Monaco, M. Caramelli, and F. Tagliavini. 2007. Conversion of the BASE prion strain into the BSE strain: the origin of BSE? *PLoS Pathog.* 3:e31. [Epub ahead of print.]

Casalone, C., G. Zanusso, P. Acutis, S. Ferrari, L. Capucci, F. Tagliavini, S. Monaco, and M. Caramelli. 2004. Identification of a second bovine amyloidotic spongiform encephalopathy: molecular similarities with sporadic Creutzfeldt-Jakob disease. *Proc. Natl. Acad. Sci. USA* 101:3065–3070.

Chesebro, B. 2004. A fresh look at BSE. *Science* 305:1918–1921.

Colchester, A. C., and N. T. H. Colchester. 2005. The origin of bovine spongiform encephalopathy the human prion disease hypothesis. *Lancet* 366:856–861.

Comer, P. J., and P. J. Huntly. 2004. Exposure of the human population to BSE infectivity over the course of the BSE epidemic in Great Britain and the impact of changes to the Over Thirty Month Rule. *J. Risk Res.* 7:523–543.

Crozet, C., S. Lezmi, F. Flamant, J. Samarut, T. Baron, and A. Benczik. 2007. Peripheral circulation of the prion infectious agent in transgenic mice expressing the ovine prion protein gene in neurons only. *J. Infect. Dis.* 195:997–1006.

De Bosschere, H., S. Roels, and E. Vanopdenbosch. 2004. Atypical case of bovine spongiform encephalopathy in an East-Flemish cow in Belgium. *Int. J. Appl. Res. Vet. Med.* 2:52–54.

DEFRA. 2007. 2005 case of H-type BSE in a cow, Information Bulletin, 71/07. Department for Environment, Food and Rural Affairs, London, United Kingdom. http://www.defra.gov.uk/news/2007/070309b.htm.

Deleault, N. R., R. W. Lucassen, and S. Supattapone. 2003. RNA molecules stimulate prion protein conversion. *Nature* 425:717–720.

Doherr, M. G. 2006. Brief review on the epidemiology of transmissible spongiform encephalopathies. *Vaccine* [Epub ahead of print.] doi:10.1016/j.vaccine.2006.10.059.

EC. 2000. Council decision of 4 December 2000 concerning certain protection measures with regard to transmissible spongiform encephalopathies and the feeding of animal protein. *Off. J. Euro. Comm.* 306:32–33.

EC. 2001. Effective feed ban; guidance note for the third countries. European Commission, Brussels, Belgium. http://ec.europa.eu/food/fs/bse/bse30_en.pdf.

EFSA. 2006a. Evaluation of a rapid *ante mortem* BSE test. *EFSA J.* 95:1–14.

EFSA. 2007. Quantitative risk assessment on the residual BSE risk in sheep meat and meat products. *EFSA J.* 442:1–44.

EFSA. 2004. *Scientific Report of the European Food Safety Authority on the Design of a Field Trial Protocol for the Evaluation of BSE Tests for Live Cattle.* European Food Safety Authority, Parma, Italy. http://www.efsa.europa.eu/en/science/tse_assessments/bse_tse/612.html.

EFSA. 2005. Annex to the Opinion. Report of the Working Group on the assessment of the age limit in cattle for the removal of certain specified risk materials (SRM). *EFSA J.* 220:1–21.

EFSA. 2006b. Scientific Report of the European Food Safety Authority on Transmissible Spongiform Encephalopathy (TSE) on a request from the European Commission on the evaluation of a rapid ante mortem BSE test. *EFSA J.* 95:1–14.

Eloit, M., K. Adjou, M. Coulpier, J. J. Fontaine, R. Hamel, T. Lilin, S. Messiaen, O. Andreoletti, T. Baron, A. Bencsik, A. G. Biacabe, V. Beringue, H. Laude, A. Le Dur, J. L. Vilotte, E. Comoy, J. P. Deslys, J. Grassi, S. Simon, F. Lantier, and P. Sarradin. 2005. BSE agent signatures in a goat. *Vet. Rec.* 156:523–524.

Everest, S. J., L. T. Thorne, J. A. Hawthorn, R. Jenkins, C. Hammersley, A. M. Ramsay, S. A. Hawkins, L. Venables, L. Flynn, R. Sayers, J. Kilpatrick, A. Sach, J. Hope, and R. Jackman. 2006. No abnormal prion protein detected in the milk of cattle infected with the bovine spongiform encephalopathy agent. *J. Gen. Virol.* 87:2433–2441.

FDA. 1997. Substances prohibited from use in animal food or feed: animal proteins prohibited in ruminant feed. *Fed. Reg.* 62:30936–30978.

FDA. 2003. Use of material from deer and elk in animal feed, guidance for industry no. 158. Food and Drug Administration, Rockville, MD. http://www.fda.gov/cvm/Documents/guide 158.pdf.

Fevrier, B., D. Vilette, F. Archer, D. Loew, W. Faigle, M. Vidal, H. Laude, and G. Raposo. 2004. Cells release prions in association with exosomes. *Proc. Natl. Acad. Sci. USA* 101: 9683–9688.

Foster, J. D., J. Hope, and H. Fraser. 1993. Transmission of bovine spongiform encephalopathy to sheep and goats. *Vet. Rec.* 133:339–341.

Gizzi, G., L. W. D. van Raamsdonk, V. Beaten, I. Murray, G. Berben, G. Brambilla, and C. von Holst. 2003. An overview of tests for animal tissues in feeds applied in response to public health concerns regarding bovine spongiform encephalopathy. *Rev. Sci. Techn. Off. Int. Epiz.* 22:311–331.

Glatzel, M., O. Giger, N. Braun, and A. Aguzzi. 2004. The peripheral nervous system and the pathogenesis of prion diseases. *Curr. Mol. Med.* 4:355–359.

Hartsough, G. R., and D. Burger. 1965. Encephalopathy of mink. I. Epizootiologic and clinical observations. *J. Infect. Dis.* 115:387–392.

Heikenwalder, M., N. Zeller, H. Seeger, M. Prinz, P. C. Klohn, P. Schwarts, N. H. Ruddle, C. Weissmann, and A. Aguzzi. 2005. Chronic lymphocytic inflammation specifies the organ tropism of prions. *Science* 307:1107–1110.

Heppner, F. L., A. D. Christ, M. A. Klein, M. Prinz, M. Fried, J. P. Kraehenbuhl, and A. Aguzzi. 2001. Transepithelial prion transport by M cells. *Nat. Med.* 7:976–977.

Hoffmann, C., U. Ziegler, A. Buschmann, A. Weber, L. Kupfer, A. Oelschlegel, B. Hammerschmidt, and M. H. Groschup. 2007. Prions spread via the autonomic nervous system from the gut to the central nervous system in cattle incubating bovine spongiform encephalopathy. *J. Gen. Virol.* 88:1048–1055.

Jackson, G. S., E. McKintosh, E. Flechsig, K. Prodromidou, P. Hirsch, J. Linehan, S. Brandner, A. R. Clarke, C. Weissmann, and J. Collinge. 2005. An enzyme-detergent method for effective prion decontamination of surgical steel. *J. Gen. Virol.* 86:869–878.

James, T. L., H. Liu, N. B. Ulyanov, S. Farr-Jones, H. Zhang, D. G. Donne, K. Kaneko, D. Groth, I. Mehlhorn, S. B. Prusiner, and F. E. Cohen. 1997. Solution structure of a 142-residue recombinant prion protein corresponding to the infectious fragment of the scrapie isoform. *Proc. Natl. Acad. Sci. USA* 94:10086–10091.

Jeffrey, M., L. Gonzales, A. Espenes, C. M. Press, S. Martin, M. Chaplin, L. Davis, T. Landsverk, C. MacAldowie, S. Eaton, and G. McGovern. 2006. Transportation of prion protein across the intestinal mucosa of scrapie-susceptible and scrapie-resistant sheep. *J. Pathol.* 209:4–14.

Kasermann, F., and C. Kempf. 2003. Sodium hydroxide renders the prion protein PrP^{Sc} sensitive to proteinase K. *J. Gen. Virol.* 84:3173–3176.

Kim, T. Y., H. J. Shon, Y. S. Joo, U. K. Mun, K. S. Kang, and Y. S. Lee. 2005. Additional cases of chronic wasting disease in imported deer in Korea. *J. Vet. Med. Sci.* 67:753–759.

Kimberlin, R. H., and C. A. Walker. 1989. Pathogenesis of scrapie in mice after intragastric infection. *Virus Res.* 12:213–220

King, C. Y., and R. Diaz-Avalos. 2004. Protein-only transmission of three yeast prion strains. *Nature* 428:319–323.

Kreeger, T. J., D. L. Montgomery, J. E. Jewell, W. Schultz, and E. S. Williams. 2006. Oral transmission of chronic wasting disease in captive Shira's moose. *J. Wildl. Dis.* 42:640–645.

Lasmezas, C. I., E. Comay, S. Hawkins, C. Herzog, F. Mouthon, T. Konold, F. Auvre, E. Correia, N. Lescoutra-Etchegaray, N. Sales, G. Wells, P. Brown, and J.-P. Deslys. 2005. Risk of oral infection with bovine spongiform encephalopathy agent in primates. *Lancet* 365:781–783.

Ligios, C., C. J. Sigurdson, C. Santucciu, G. Carcassola, G. Manco, M. Basagni, C. Maestrale, M. G. Cancedda, L. Madau, and A. Aguzzi. 2005. PrP^{Sc} in mammary glands of sheep affected by scrapie and mastitis. *Nat. Med.* 11:1137–1138.

Ma, J., and S. Lindquist. 2002. Conversion of PrP to a self-perpetuating PrP^{Sc}-like conformation in the cytosol. *Science* 298:1785–1788.

Ma, J., R. Wollmann, and S. Lindquist. 2002. Neurotoxicity and neurodegeneration when PrP accumulates in the cytosol. *Science* 298:1781–1785.

Maignien, T., M. Shakweh, P. Calvo, D. Marce, N. Sales, E. Fattal, J. P. Deslys, P. Couvreur, and C. I. Lasmezas. 2005. Role of gut macrophages in mice orally contaminated with scrapie or BSE. *Int. J. Pharm.* 298:293–304.

Marsh, R. F., A. E. Kincaid, R. A. Bessen, and J. C. Bartz. 2005. Interspecies transmission of chronic wasting disease prions to squirrel monkeys *(Saimiri sciureus)*. *J. Virol.* 79:13794–13796.

Mathiason, C. K., J. G. Powers, S. J. Dahmes, D. A. Osborn, K. V. Miller, R. J. Warren, G. L. Mason, S. A. Hays, J. Hayes-Klug, D. M. Seelig, M. A. Wild, L. L. Wolfe, T. R. Spraker, M. W. Miller, C. J. Sigurdson, G. C. Telling, and E. A. Hoover. 2006. Infectious prions in the saliva and blood of deer with chronic wasting disease. *Science* 314:131–136.

Miller, M. W., and E. S. Williams. 2003. Horizontal prion transmission in mule deer. *Nature* 425:35–36.

Miller, M. W. 2006. Chronic wasting disease in Colorado, 2005–2007. Colorado Division of Wildlife, Denver, CO. http://wildlife.state.co.us/NR/rdonlyres/763F5731-F895-4D52-9F27-2-B8D5BE91175/0/CWDReport2005_2006.pdf.

Miller, M. W., E. S. Williams, N. Thompson-Hobbs, and L. L. Wolfe. 2004. Environmental sources of prion transmission in mule deer. *Emerg. Infect. Dis.* 10:1003–1006.

Momcilovic, D., and A. Rasooly. 2000. Detection and analysis of animal materials in food and feed. *J. Food Prot.* 63:1602–1609.

Morel, E., S. Fouquet, D. Chateau, L. Yvernault, Y. Frobert, M. Pincon-Raymond, J. Chambaz, T. Pillot, and M. Rousset. 2004. The cellular prion protein PrPc is expressed in human enterocytes in cell-cell junctional domains. *J. Biol. Chem.* 279:1499–1505.

Nobel Assembly. 1976. Press release of 14 October 1976. Nobel Foundation, Stockholm, Sweden. http://nobelprize.org/nobel_prizes/medicine/laureates/1976/press.html.

Nobel Assembly. 1997. Press release of 6 October 1997. Karolinska, Institutet, Stockholm, Sweden. http://www.csb.ki.se/kisv/nobel.97.

Oesch, B., D. Westaway, M. Walchli, M. P. McKinley, S. B. H. Kent, R. Aebersold, R. A. Barry, P. Tempst, D. B. Teplow, L. E. Hood, S. B. Prusiner, and C. Weissmann. 1985. A cellular gene encodes scrapie PrP 27–30 protein. *Cell* 40:735–746.

OIE. 2007. World animal health situation. World Organization for Animal Health, Paris, France. http://www.oie.int/eng/info/en_esbmonde.htm.

O'Rourke, K. I., D. Zhuang, A. Lyda, G. Gomez, E. S. Williams, W. Tuo, and M. W. Miller. 2003. Abundant PrP^{CWD} in tonsil from mule deer with preclinical chronic wasting disease. *J. Vet. Diagn. Invest.* 15:320–323.

Pan, K.-M., M. Baldwin, J. Nguyen, M. Gasset, A. Serban, D. Groth, I. Mehlhorn, Z. Huang, R. J. Fletterick, F. E. Cohen, and S. B. Prusiner. 1993. Conversion of alpha-helices into beta-sheets features in the formation of the scrapie prion protein. *Proc. Natl. Acad. Sci. USA* 90:10962–10966.

Pearson, G. R., J. M. Wyatt, T. J. Gruffydd-Jones, J. Hope, A. Chong, R. J. Higgins, A. C. Scott, and G. A. Wells. 1992. Feline spongiform encephalopathy: fibril and PrP studies. *Vet. Rec.* 131:307–310.

Polak, M. P., W. Rozek, J. Rola, and J. F. Zmudzinski. 2004. Prion protein glycoforms from BSE cases in Poland. *Bull. Vet. Inst. Pulawy* 48:201–205.

Prusiner, S. B. 1982. Novel proteinaceous infectious particles cause scrapie. *Science* 216:136–144.

Prusiner, S. B. 1997. Prion diseases and the BSE crisis. *Science* 278:245–251.

Prusiner, S. B. 1998. Prions. *Proc. Natl. Acad. Sci. USA* 95:13363–13383.

Race, R., A. Jenny, and D. Sutton. 1998. Scrapie infectivity and proteinase K-resistant prion protein in sheep placenta, brain, spleen, and lymph node: implications for transmission and antemortem diagnosis. *J. Infect. Dis.* 178:949–953.

Raymond, G. J., A. Bossers, L. D. Raymond, K. I. O'Rourke, L. E. McHolland, P. K. Bryant III, M. W. Miller, E. S. Williams, M. Smits, and B. Caughey. 2000. Evidence of a molecular barrier limiting susceptibility of humans, cattle and sheep to chronic wasting disease. *EMBO J.* 19:4425–4430.

Richt, J. A., R. A. Kunkle, D. Alt, E. M. Nicholson, A. N. Hamir, S. Czub, J. Kluge, A. J. Davis, and S. Mark Hall. 2007.

356 MOMCILOVIC

Identification and characterization of two bovine spongiform encephalopathy cases diagnosed in the United States. *J. Vet. Diagn. Invest.* 19:142–154.

Rustom, A., R. Saffrich, I. Markovic, P. Walther, and H. H. Gerdes. 2004. Nanotubular highways for intercellular organelle transport. *Science* 303:1007–1010.

Safar, J. G., K. Kellings, A. Serban, D. Groth, J. E. Cleaver, S. B. Prusiner, and D. Riesner. 2005. Search for a prion-specific nucleic acid. *J. Virol.* 79:10796–10806.

Schreuder, B. E. C. 1993. General aspects of transmissible spongiform encephalopathies and hypotheses about the agents. *Vet. Q.* 15:167–174.

SEAC. 2006. Chronic wasting disease in UK deer update. Spongiform Encephalopathy Advisory Committee, London, United Kingdom. http://www.seac.gov.uk/papers/93-2.pdf.

Seeger, H., M. Heikenwalder, N. Zeller, J. Kranich, P. Schwarz, A. Gaspert, B. Seifert, G. Miele, and A. Aguzzi. 2005. Coincident scrapie infection and nephritis lead to urinary prion excretion. *Science* 310:324–326.

Siso, S., L. Gonzales, M. Jeffrey, S. Martin, F. Chianini, and P. Steele. 2006. Prion protein in kidneys of scrapie-infected sheep. *Vet. Rec.* 159:327–328.

Sohn, H. J., J. H. Kim, K. S. Choi, J. J. Nah, Y. S. Joo, J. H. Jean, S. W. Ahn, O. K. Kim, D. Y. Kim, and A. Balachadran. 2002. A case of chronic wasting disease in an elk imported to Korea from Canada. *J. Vet. Med. Sci.* 64:855–858.

Solassol, J., M. Pastore, C. Crozet, V. Perrier, and S. Lehmann. 2006. A novel copper-hydrogen peroxide formulation for prion decontamination. *J. Infect. Dis.* 194:865–869.

Spraker, T. R., R. R. Zink, B. A. Cummings, C. J. Sigurdson, M. W. Miller, and K. I. O'Rourke. 2002. Distribution of protease-resistant prion protein and spongiform encephalopathy in free-ranging mule deer (Odocoileus hemionus) with chronic wasting disease. *Vet. Pathol.* 39:546–556.

Spraker, T. R., T. L. Gidlewski, A. Balachadran, K. C. VerCauteren, L. Creekmore, and R. D. Munger. 2006. Detection of PrP(CWD) in postmortem rectal lymphoid tissues in Rocky Mountain elk (Cervus alaphus nelsoni) infected with chronic wasting disease. *J. Vet. Diagn. Invest.* 18:553–557.

SSC. 1999. *Opinion on the Human Exposure Risk (HER) via Food with Respect to BSE.* Scientific Steering Committee, European Commission, Brussels, Belgium. http://ec.europa.eu/food/fs/sc/ssc/out67_en.pdf.

Stack, M., M. Jeffrey, S. Gubbins, S. Grimmer, L. Gonzalez, S. Martin, M. Chaplin, P. Webb, M. Simmons, Y. Spencer, P. Bellerby, J. Hope, J. Wilesmith, and D. Matthews. 2006. Monitoring for bovine spongiform encephalopathy in sheep in Great Britain, 1998–2004. *J. Gen. Virol.* 87:2099–2107.

Statutory Instrument. 1988. The bovine spongiform encephalopathy order 1988 (SI 1988/1039). HMSO, London, United Kingdom.

Supervie, V., and D. Costagliola. 2004. The unrecognized French BSE epidemic. *Vet. Res.* 35:349–362.

Tamguney, G., K. Giles, E. Bouzamondo-Bernstein, P. J. Bosque, M. W. Miller, J. Safar, S. J. DeArmond, and S. B. Prusiner. 2006. Transmission of elk and deer prions to transgenic mice. *J. Virol.* 80:9104–9114.

Tanaka, M., P. Chien, N. Naber, R. Cooke, and J. S. Weissman. 2004. Conformational variations in an infectious protein determine prion strain differences. *Nature* 428:323–328.

Taylor, D. M., K. Fernie, I. McConnell, and P. J. Steele. 1999. Survival of scrapie agent after exposure to sodium dodecyl sulphate and heat. *Vet. Microbiol.* 67:13–16.

Taylor, D. M., and S. L. Woodgate. 2003. Rendering practices and inactivation of transmissible spongiform encephalopathy agents. *Rev. Sci. Tech. Off. Int. Epiz.* 22:297–310.

Taylor, D. M., S. L. Woodgate, A. J. Fleetwood, and R. J. G. Cawthorne. 1997. The effect of rendering procedures on scrapie agent. *Vet. Rec.* 141:643–649.

Telling, G. C., M. Scott, J. Mastrianni, R. Gabizon, M. Torchia, F. E. Cohen, S. J. DeArmond, and S. B. Prusiner. 1995. Prion propagation in mice expressing human and chimeric PrP transgenes implicates the interaction of cellular PrP with another protein. *Cell* 83:79–90.

USDA. 2007a. BSE test results. United States Department of Agriculture, Washington, DC. http://www.aphis.usda.gov/lpa/issues/bse_testing/test_results.html.

USDA. 2007b. Chronic wasting disease. United States Department of Agriculture, Washington, DC. http://www.aphis.usda.gov/vs/nahps/cwd/. Accessed 15 March 2007.

USDA. 2004. Prohibition of the use of specified risk materials for human food and requirements for the disposition of non-ambulatory disabled cattle. *Fed. Reg.* 69:1861–1874.

Vidal, E., M. Marquez, M. Ordonez, A. J. Raeber, T. Struckmeyer, B. Oesch, S. Siso, and M. Pumarola. 2005. Comparative study of the PrPBSE distribution in brains from BSE field cases using rapid tests. *J. Virol. Methods* 127:24–32.

Weissmann, C., M. Enari, P. C. Klohn, D. Rossi, and E. Flechsig. 2002. Transmission of prions. *Proc. Nat. Acad. Sci. USA* 99(Suppl. 4):16378–16383.

Wells, G. A. H., A. C. Scott, C. T. Johnson, R. F. Gunning, R. D. Hancock, M. Jeffrey, M. Dawson, and M. Bradley. 1987. A novel progressive spongiform encephalopathy in cattle. *Vet. Rec.* 121:419–420.

Wells, G. A. H., S. A. C. Hawkins, R. B. Green, A. R. Austin, I. Dexter, Y. I. Spencer, M. J. Chaplin, M. J. Syack, and M. Dawson. 1998. Preliminary observations on the pathogenesis of experimental bovine spongiform encephalopathy (BSE): an update. *Vet. Rec.* 142:103–106.

Wells, G. A. H., T. Konold, M. E. Arnold, A. R. Austin, S. A. C. Hawkins, M. Stack, M. M. Simmons, Y. H. Lee, D. Gavier-Widen, M. Dawson, and J. W. Wilesmith. 2007. Bovine spongiform encephalopathy: the effect of oral exposure dose on attack rate and incubation period in cattle. *J. Gen. Virol.* 88:1363–1373.

Wild, M. A., T. R. Spraker, C. J. Sigurdson, K. I. O'Rourke, and M. W. Miller. 2002. Preclinical diagnosis of chronic wasting disease in captive mule deer (Odocoileus hemionus) and white-tailed deer (Odocoileus virginianus) using tonsillar biopsy. *J. Gen. Virol.* 83:2629–2634.

Wilesmith, J. W., J. B. M. Ryan, and M. J. Atkinson. 1991. Bovine spongiform encephalopathy—epidemiologic studies on the origin. *Vet. Rec.* 128:199–203.

Will, R. G., J. W. Ironside, M. Zeidler, S. N. Cousens, K. Estibeiro, A. Alperovitch, S. Poser, M. Pocchiari, A. Hofman, and P. G. Smith. 1996. A new variant of Creutzfeldt-Jakob disease in the UK. *Lancet* 347:921–925.

Williams, E. S. 2005. Chronic wasting disease. *Vet. Pathol.* 42:530–549.

Williams, E. S., and S. Young. 1980. Chronic wasting disease of captive mule deer: a spongiform encephalopathy. *J. Wildl. Dis.* 16:89–98.

World Health Organization. 2000. *WHO Consultation on Public Health and Animal Transmissible Spongiform Encephalopathies: Epidemiology, Risk and Research Requirements*, p. 52. World Health Organization, Geneva, Switzerland.

Yamakawa, Y., K. Hagiwara, K. Nohtomi, Y. Nakamura, M. Nishijima, Y. Higuchi, Y. Sato, T. Sata, and the Expert Committee for BSE Diagnosis, Ministry of Health, Labor and Welfare of Japan. 2003. Atypical proteinase K-resistant prion protein (PrPres) observed in an apparently healthy 23-month-old Holstein steer. *Jpn. J. Infect. Dis.* 56:221–222.

Pathogens and Toxins in Foods: Challenges and Interventions
Edited by V. K. Juneja and J. N. Sofos
© 2010 ASM Press, Washington, DC

Chapter 23

Interventions for Hazard Control in Foods Preharvest

Jarret D. Stopforth, Balasubrahmanyam Kottapalli, and John N. Sofos

Food safety has become one of the most important issues of public concern in recent times. The aftermath of food-borne illness outbreaks has involved many losing their lives, companies going bankrupt, employees losing their jobs, and the public becoming sensitized and sceptical toward the safety of the associated food product. For purposes of this chapter, food-borne illness is defined as disease contracted by ingesting microbiologically contaminated food. The term "preharvest" as described in this chapter refers to the period of time when a food commodity is growing, prior to the harvest of the crop or livestock, and since most food is grown on farms (i.e., meat animals and crops), with the exception of certain fish and seafood, the term "farm" will be used to describe the confined production system used to grow food and the associated environment. Food-borne hazards may include physical, chemical, or biological agents. The focus of this chapter is biological hazards, such as viruses, parasites, bacteria, fungi, and any of their associated toxins. It should be noted that biological hazards pose the greatest public health threat at the preharvest stage since chemical and physical hazards do not reproduce and increase in the food and in the environment with time, as do microbiological hazards. It is important to note that there is a fundamental difference between biological hazards and risks. A biological hazard is an agent in food with the potential to cause human illness, while risk is an estimate of the probability and severity for exposed populations of the adverse effect(s) resulting from a biological hazard in food.

Food safety involves complex interactions among environments, insects, plants, animals, and humans. A number of microorganisms are pathogenic to animals and plants, while others are commensals; however, food safety in this chapter refers to the impact of food-borne illness on public health. The human pathogens of concern in preharvest environments include a wide array of viruses, parasites, fungi, and bacteria and their immediate effects, and the severity thereof on humans is largely dependent on pathogen- and host-specific properties. Such pathogens may be found in all kinds of different foods and typically are acquired at the farm level before the food actually reaches processing facilities and undergoes further handling and distribution. With many food products, it can be difficult to prevent transmission of pathogens to consumers once the product leaves the farm since postharvest mitigations, if any, may not always be sufficient or completely effective, since many food products, such as fruits, vegetables, nuts, and some fish/seafood, are actually eaten raw.

Consumption of food, especially fresh foods that are not sterile, carries inherent risks of food-borne disease. There are many steps in the process of conveying food from the farm to the "table," and each step provides the opportunity for contamination and, ultimately, risk of food-borne illness. While it is recognized that the preharvest sector of food production is not the most suitable, or in some cases, the most effective/critical, stage at which control can be exerted, cost-effective interventions with the potential for controlling pathogens are essential, even at this stage of the continuum of hazard reduction for public health. Thus, preharvest food safety needs to encompass a range of strategies for preventing, reducing, or eliminating the presence or proliferation of pathogens in food prior to harvest. However, eliminating pathogens from the preharvest environment is practically impossible, and as such, the focus should be on reducing the level of pathogens in food and minimizing the frequency, extent, and distribution of pathogens at this stage.

Jarret D. Stopforth • PURAC, Arkelsedijk 46, 4206 AC, Gorinchem, The Netherlands. **Balasubrahmanyam Kottapalli** • Kraft Foods, 200 Deforest Ave., East Hanover, NJ 07936. **John N. Sofos** • Center for Meat Safety & Quality, Food Safety Cluster of Infectious Diseases Supercluster, Department of Animal Sciences, Colorado State University, 1171 Campus Delivery, Fort Collins, CO 80523-1171.

The following sections address the hazards associated with animal products, plant products, and aquatic products and the interventions that may be applied to meet the challenges faced at the preharvest (farm) sector.

ANIMAL PRODUCTS

Pathogens of Concern

Animal production food safety involves hazards that are associated with meat, poultry, or other animal products, such as milk and eggs, that form part of the human diet. The safety of foods of animal origin is determined by events that may happen in the chain including environmental inputs on the farm; continuing with transportation, the slaughterhouse, further processing steps, handling, and distribution; and finally ending with the consumer (Ahl and Buntain, 1997). Pathogens entering the chain at any stage can be transferred to the final product and result in food-borne illness. There are two types of biological hazards: those that reproduce and those that do not reproduce. Biological hazards that do not reproduce in animal products include viruses and parasites. Pathogens that do reproduce in animal products are associated primarily with the intestinal flora of the animals and are spread via fecal material; such organisms are predominantly bacterial. Animal production food safety is an issue of cross-contamination since the edible commodities before harvest (muscle tissue, milk, and egg contents) are essentially sterile. The pathogens that are encountered in the gastrointestinal tract and on the outside of the animal are invariably introduced on the farm (water, air, soil, pasture, etc.) and spread to the food during harvesting/processing.

The types and levels of pathogens on/in the farm environment are determined largely by a complex combination of production and environmental factors. Factors that have been associated with the increased occurrence of food-borne pathogens in animals include (i) high animal density, (ii) stress, (iii) warm humid environment, (iv) contaminated water supplies, (v) contaminated bedding, (vi) unsanitary handling of waste materials, (vii) commingling of infected and uninfected animals or carriers, (viii) recycling of farm waste materials, (ix) poorly nourished animals, (x) high insect populations, (xi) contaminated feeds and wild animals, and (xii) rodents or vermin in production areas (Draughon, 2006). A number of risk factors which have been responsible for the emergence of new pathogens and the spread of zoonoses (defined as a disease-causing agent that can be transmitted between animal species including humans) have been identified: (i) increased demand for animal protein, leading to changes in farming practices; (ii) increased animal density; (iii) coproduction of animals and food on the same land; and (iv) encroachment of animals on the habitat of other animals. Table 1 presents a list of zoonotic pathogens that are commonly associated with domestic farm animals.

Pathogens on Meat

Red meat is derived from a number of animal species (cattle, swine, sheep, goats, and horses) and has the potential to carry many food-borne pathogens to consumers; the organisms that may constitute a hazard in red meat products are *Salmonella* sp., pathogenic *Escherichia coli*, *Yersinia enterocolitica*, *Campylobacter jejuni*, *Campylobacter coli*, *Staphylococcus aureus*, *Listeria monocytogenes*, *Clostridium perfringens*, *Clostridium botulinum*, *Bacillus cereus*, *Toxoplasma gondii*, *Sarcocystis* sp., *Trichinella spiralis*, *Taenia saginata*, and *Taenia solium*. The organisms that constitute a hazard in poultry products are *Salmonella* sp., *Campylobacter* spp., *C. perfringens*, *C. botulinum*, *S. aureus*, and *L. monocytogenes* (ICMSF, 2005).

Pathogens in Milk

Before the enforcement of pasteurization, raw milk was a source of contamination by numerous pathogens; however, since the implementation of pasteurization pathogens associated with pasteurized milk represent accidental postprocess contamination. Common pathogens isolated from milk products are *Coxiella burnetii*, *Brucella* spp., *Salmonella* sp., *C. jejuni*, *Y. enterocolitica*, *L. monocytogenes*, pathogenic *E. coli*, and *S. aureus*. Furthermore, milk may contain mycotoxins, including aflatoxin B_1, aflatoxin G_1, sterigmatocystin, ochratoxin, and especially aflatoxin M_1 (ICMSF, 2005).

Pathogens in/on Eggs

Salmonella sp. is the most important pathogen associated with eggs and can in fact also enter the egg via transovarian transmission (ICMSF, 2005).

Preharvest Hazard Control for Animal Products

The major problem with preharvest control of food-borne pathogens is the wide variety of food products, the range of pathogens, and the complexity of environmental conditions. This complicates mitigation efforts to target specific organisms, and, thus, there is a need to give priority to those pathogens with the highest health impact (Isaacson et al., 2005). Priority may be ranked according to the risks posed

Table 1. Common zoonotic pathogens transmitted by domestic farm animals to humans[a]

Pathogen agent	Primary host(s)	Disease in humans
Viruses		
Vaccinia (cowpox)	Cattle, horses	Papulovesiculopustular skin lesions
Paravacciia	Cattle	Papulovesiculogranulomatous skin lesions
Vesicular stomatitis	Cattle, swine, horses	Fever, chills, headache, myalgia
Protozoans		
Cryptosporidium	Cattle, horses, swine	Diarrhea, enteritis, dysentery
Cyclospora	Cattle, horses, swine	Diarrhea, enteritis, dysentery
Giardia	Cattle, horses, swine	Diarrhea, enteritis, dysentery
Toxoplasma gondii	Cattle, swine	Stillbirth, congenital defects
Parasites		
Taenia saginata	Cattle	Abdominal pain, diarrhea, weight loss
Trichinella	Cattle, swine	Myositis, conjunctivitis, weight loss
Bacteria		
Bacillus anthracis	Cattle, sheep, horses, goats, swine	Gastroenteritis, pneumonitis, skin lesions
Brucella	Cattle, sheep, horses, goats, swine	Fever, malaise, bacteremia, osteomyelitis
Campylobacter	Cattle, sheep, swine, poultry	Gastroenteritis, placentitis, endocarditis, meningitis
Listeria monocytogenes	Cattle, sheep, horses, goats, poultry	Meningitis, fetal listeriosis, septicemia, gastroenteritis
Mycobacterium bovis	Cattle, horses, swine	Pulmonary tuberculosis, gastroenteritis
Mycobacterium avium subsp. *paratuberculosis*	Cattle, sheep, horses, swine	Chronic inflammatory bowel disease
Pathogenic *Escherichia coli*	Cattle, sheep, swine, poultry	Diarrhea, gastroenteritis, septicemia, reactive arthritis
Salmonella	Cattle, sheep, swine, poultry	Diarrhea, gastroenteritis, septicemia, reactive arthritis
Yersinia	Cattle, sheep, swine, poultry	Diarrhea, gastroenteritis, septicemia, reactive arthritis

[a]Adapted from Draughon (2006).

to susceptible individuals, risks posed to the greatest number of individuals, risks of the most severe health consequences, risks posed by various products, or risk based on monetary value (Isaacson et al., 2005). Risk mitigation interventions should be selected based on the maximum risk reduction for the minimum cost. Control measures for reducing pathogens in food animals at the farm level should focus on minimizing initial contamination of the animals/products, reduction of contamination, and inhibition of growth following contamination. The next sections address strategies that should be considered at the farm level in the continuum of the food safety chain (Stopforth and Sofos, 2006).

Minimizing Initial Contamination

The interaction of animals with their environment is the primary factor affecting the health and occurrence of food-borne pathogens in/on the animal; that is, a clean environment promotes a healthy animal. Thus, environmental control is critical in minimizing initial contamination of animals by pathogens. Management practices at the farm level to control pathogens in the immediate environment form the basis of "good agricultural practices" (GAP). The farm environment encompasses four primary components: air, water, soil, and fomites (Dowd et al., 2004). The occurrence, survival, and fate of pathogens in the farm environment are dependent on a number of biotic and abiotic factors, such as temperature, competition, moisture, sunlight, predation, oxygen availability, and desiccation.

Air

Carriage of pathogens via aerosols is a process that can rapidly contaminate an entire farm, and thus, airborne transmission is the primary mode of pathogen dispersal among livestock (Proux et al., 2001). Control of bioaerosols is relatively impossible in open-air environments, such as open range or feedlots (cattle, sheep, etc.); however, on farms with housing (swine, poultry, etc.), there are options for minimizing the effect of bioaerosols. Such controls may include ventilation, filtration, UV treatment, biocidal agents, and physical isolation (Dowd and Maier, 1999). Ventilation creates a dilution-type effect by disturbing settlement in areas where pathogen accumulation may occur, and successful implementation is affected by the quality of the input air and the turnover rate within the occupied area in relation to the amount of contamination. Ventilation is further enhanced by the use of filters (filtration) that can remove infectious particles from outdoor air. Biocidal control can be combined with filtration to reduce the microorganisms in the air via superheating, superdehydration, ozonization, and UV irradiation and can be applied to both input air and exhaust air from other indoor areas. Physical control of bioaerosols can be achieved by the use of positive or negative pressurized air pressure gradients and airtight seals.

Water

On-farm water environments include incoming sources, such as rainwater, runoff, groundwater, surface water (rivers, streams, lakes, etc.), wells, and wastewater (Asandhi et al., 2006). Control of waterborne pathogens assumes a barrier approach that involves source water protection and physical (filtration and sedimentation) and chemical (chemical and physical disinfection) treatment (Dowd et al., 2004; Rangarajan et al., 2000). Typically, source water is pumped from drinking water wells and stored in a central container before being distributed throughout the farm. It is important to ensure that the drinking water wells be located upstream of the farm to prevent groundwater that passes under the farm from carrying pathogens to the well. Surface water contamination should be minimized by limiting human and animal access and deposition and accumulation of organic material from the environment (Asandhi et al., 2006; Dowd et al., 2004). Since this is not always practical, the best way to protect source waters is geographical, hydrological, and physical isolation of the source and the use of watering and pump systems that can carry the water to central disinfection and storage areas. Although it is critical to ensure that contamination of source water is controlled, it is arguably as important that wastewater disposal and treatment, protection of distribution systems, and flood and runoff water be controlled to prevent pathogen transport and dissemination (Redman et al., 2001). Storage and distribution systems should be maintained to prevent accumulation of biofilms, decay, or accidental breach and contamination by wastewater and regularly treated using a good chlorination program. Additional control can be exerted at watering reservoirs or points of access, since horizontal transfer of pathogens and contamination/cross-contamination of animals is highly likely. Strategies to minimize contamination of animals at this stage of the process are interventions to reduce pathogen survival in the water. This may be achieved via constant replenishing of water, periodic cleaning of reservoirs, chlorination of water, ozonation, and screens to reduce organic solids (LeJeune et al., 2004).

Soil

Soil is the most complex environment on the farm since it contains solid, liquid, and gas components that interact to ultimately determine the fate of pathogen contaminants (Rangarajan et al., 2000). Once soil is contaminated, it is more difficult to disinfect than air and water because of the difficulties faced with penetration of disinfectants, organic material rendering disinfectants less effective, and cross-contamination and spread of organisms via underground water passages. Horizontal transmission and propagation of pathogens in the soil is a problem and can be minimized only by application of localized surface disinfectants.

Fomites

Any inanimate (nonsoil or water) surface in/on the farm environment is considered a fomite and has the ability to harbor and transfer pathogen contamination (Weese and Rousseau, 2006). Common surfaces include floors, walls, gates, and fences. The ability of a fomite to sustain microorganisms is determined by the nature of the material (e.g., it is more difficult for organisms to colonize hard metal surfaces versus wood, rope, etc.) and availability of nutrients and water. Most fomites are poor harborage sites for pathogens and are easily controlled through desiccation, temperature, sunlight, oxygen-related stresses, and disinfection.

It is not only the farm environment where introduction of contamination should be minimized, but also the interface between the preharvest and harvest environments, which is also critical. Implementation of GAP extends to the period of time when animals or their products leave the farm until they reach further processing environments; the following are control measures that must be maintained to minimize internal and external contamination of animals before harvest: (i) feed should be withheld for the immediate time frame before transport to reduce fecal shedding; (ii) equipment and vehicles associated in transport should be designed to minimize cross-contamination from animal-to-animal where possible; (iii) transport equipment and vehicles should be cleaned and sanitized between loads; (iv) stress and starvation should be reduced during holding prior to harvest to reduce shedding; (v) time of transit and holding should be limited to reduce the potential to cross-contaminate other animals; (vi) holding facilities should be cleaned and sanitized to reduce carryover of pathogens; and (vii) intestinal fill should be avoided to reduce breakage of intestinal contents during slaughter (Stopforth and Sofos, 2006).

Reduction/Inhibition of Contamination

The implementation of "Hazard Analysis Critical Control Points" (HACCP) principles at the postharvest stage (in food processing and food service environments) has been successful in management of interventions that control food-borne hazards. It has been proposed that the preharvest stage of food production adopt the principles and concepts of this system; however, it should be realized that unlike the postharvest environments, the farm environment has multiple sources for pathogen entry and no proven effective interventions to act as critical control points (CCP). The main approach at the preharvest level is to minimize pathogen sources and the risk of exposure or contamination transfer rather than prevention or elimination of pathogenic agents. While mandatory HACCP is appropriate for postharvest processing of animal and seafood products at present, the application of the same system at the farm level is not, since there is insufficient information regarding the efficacy of interventions at this stage (Troutt et al., 1999). A HACCP-like approach that considers the identification of new interventions focused on CCP to assist in risk reduction is a far more achievable goal. The CCP for reducing the prevalence and level of pathogens on the farm would pave the way for designing new interventions; however, these CCP need to target specific pathogens or groups of pathogens in specific environments. The following are potential intervention strategies targeting food-borne pathogens in animals at the farm level.

Dietary changes/feed supplements

The majority of research on the effect of dietary changes as a means of reducing pathogens in animals has been conducted with cattle and sheep. With cattle, it has been suggested that increased amounts of fiber or roughage assist in reducing food-borne pathogens; however, the effect appears to be somewhat inconsistent (Callaway et al., 2003a; Hancock et al., 1999). The mechanism is thought to be related to the increased ruminal pH and a decrease in volatile fatty acid concentration, while the opposite is true with high-concentrate (grain) diets (Magnuson et al., 2000). Furthermore, hay-fed cattle shed pathogens more effectively than those fed a concentrate diet (Hovde, 1999). Although increased hay or roughage in the diet may result in reduction of pathogens in cattle, other factors that need to be considered are cost, productivity, practicality, and marketing issues. While it is recognized that grass-fed cattle may have a niche market, the switching of the industry as a whole to these practices is infeasible and cattle would require additional feeding time, making the option economically challenging (Callaway et al., 2003a). It was also shown that feeding sheep a high-fiber diet reduced enteric pathogen carriage compared to feeding them a high-concentrate diet (Kudva et al., 1997). Thus,

factors to be considered before including forage as an integral part of dietary management to control food-borne pathogens are (i) type and quality of the forage, (ii) cost, (iii) practicality, and (iv) effect on meat quality. Whole cottonseed has been investigated as a supplement to high-concentrate finishing diets and has shown efficacy in reducing enteric pathogen carriage (S. Younts-Dahl, M. Brashears, M. Galyean, and G. Loneragan, presented at the Annual Meeting of the International Association of Food Protection, New Orleans, LA, 2003). Tasco 14 is an extract from the seaweed *Ascophyllum nodosum*, which has been reported to be effective in reducing enteric pathogens in cattle when used as a feed supplement during the final 2 weeks of high-concentrate feeding (Braden et al., 2004).

Direct-fed microbials/competitive exclusion

Use of native or introduced microorganisms to reduce pathogenic organisms in the gut has been termed a "competitive enhancement strategy" (Anderson et al., 2006). The objective of such a strategy is to feed groups of beneficial bacteria that are competitive, or antagonistic, to pathogens (Ouwenhand et al., 1999). There are essentially three competitive enhancement strategies: (i) competitive exclusion, (ii) probiotics, and (iii) prebiotics. Competitive exclusion involves the addition of a nonpathogenic bacterial culture to the intestinal tract of food animals in order to reduce colonization or decrease populations of pathogenic bacteria in the gastrointestinal tract (usually through fierce competition for available nutrients) (Steer et al., 2000). Competitive exclusion cultures may be composed of one or more strains or species of bacteria; however, they should be derived from the animal of interest and, as such, take advantage of synergies acquired during coevolution of host and microorganism (Callaway et al., 2004). The objective of competitive exclusion may be the exclusion of pathogens from the gut of a neonatal animal or the displacement of an established pathogenic bacterial population, and for this reason the mixture of bacteria chosen for the culture must be designed for the specific animal and production scenario (Nurmi et al., 1992). The majority of research has focused on controlling *Salmonella* colonization in newly hatched poultry chicks with gastrointestinal tracts (Callaway et al., 2004). Competitive exclusion cultures have also been developed for other species, including swine (Anderson et al., 1999; Harvey et al., 2003) and cattle (Brashears et al., 2003; Zhao et al., 2003).

Probiotics are preparations or products containing viable, defined microorganisms in sufficient numbers that alter the microflora by implantation or colonization in a compartment of the host to exert a beneficial effect on the same (Schrezenmeir and De Vrese, 2001). Unlike competitive exclusion products, probiotics typically consist of single or mixed species of lactic acid bacteria or yeasts and are not necessarily of animal origin (Wiemann, 2003). Probiotics usually fall into the following categories: (i) live cultures of yeasts or bacteria, (ii) heat-treated (or otherwise inactivated) cultures of yeasts or bacteria, or (iii) fermentation end products from incubations of yeasts or bacteria (Wiemann, 2003). The most common probiotic bacteria are *Bifidobacterium* and *Lactobacillus* and have been reported to reduce food-borne pathogens in animals (Tkalcic et al., 2003). Research has demonstrated the effect of *Lactobacillus casei* to reduce postweaning diarrhea in swine (Kyriakis et al., 2001) and the effect of *Lactobacillus acidophilus* in sheep (Lema et al., 2001) and cattle (Brashears et al., 2003; Ransom et al., 2003).

Prebiotics are organic compounds that are unavailable to, or indigestible by, the host animal but digestible by specific microbiological populations (Schrezenmeir and De Vrese, 2001). A well-known prebiotic is fructo-oligosaccharides, sugars that are not degraded by intestinal enzymes and can pass through to the cecum and colon to become "colonic food" (Willard et al., 2000). Prebiotics can function by providing energy or limiting nutrients to the intestinal mucosa, as well as substrates for intestinal bacterial fermentation, resulting in enhanced production of vitamins and antioxidants that further directly benefit the animal (Collins and Gibson, 1999). To date, prebiotics have really been used successfully only in swine (Mosenthin and Bauer, 2000). The paired use of competitive exclusion cultures and prebiotics is known as synbiotics, which may find application in reducing food-borne pathogens in animals prior to slaughter (Bomba et al., 2002).

Vaccination

Use of the animal's native immune system to reduce the pathogen load is a strategy that has also been explored. Vaccination of poultry against *Salmonella enterica* serovar Enteritidis has been effective in reducing the pathogen within flocks and eggs and indeed the incidence of salmonellosis in humans (Adak et al., 2002). Vaccines against *Salmonella* have also been developed for use in swine and dairy cattle (House et al., 2001). Vaccination has also been successful in combating postweaning *E. coli* edema in young pigs (Gyles, 1998). Vaccines have been developed for use in cattle to combat *E. coli* O157:H7 by reducing fecal shedding (Moxley et al., 2003). Recently, a vaccine containing type III secreted proteins against *E. coli* O157:H7 demonstrated an average efficacy of 58.7% less risk of recovering

E. coli O157:H7 in vaccinated versus control cattle (Potter et al., 2004). Vaccines may have good potential for the cattle industry for numerous reasons: (i) cattle producers are familiar with vaccine administration, (ii) incorporation into management practices would be relatively simple, and (iii) vaccines could be used in all segments of the industry (Loneragan and Brashears, 2005).

Antibiotics

Antibiotics may alter the microflora of the intestinal tract and increase the animal's growth rate and efficiency. Antibiotics generally target one of the following cellular processes: (i) protein synthesis, (ii) cell wall synthesis, (iii) nucleic acid replication and synthesis, or (iv) folate metabolism (Hooper, 2001). Bacteria have many mechanisms to resist antibiotics, and the widespread use of these drugs in animals and humans has led to the dissemination of antibiotic resistance genes (Busz et al., 2002). In all food animals there is some degree of therapeutic and subtherapeutic use of antibiotics to combat animal diseases; however, there is not very wide use of antibiotics to control food-borne pathogens in animals for food safety purposes. Most attention has been focused on the use of neomycin sulfate, a broad-spectrum antibiotic, to target enteric pathogens in cattle, and the drug is approved for in-feed and in-water administration for the control or treatment of colibacillosis (Loneragan and Brashears, 2005). Research has shown that neomycin sulphate as an in-water administration resulted in substantial reductions in fecal shedding and hide contamination in cattle (Loneragan and Brashears, 2005), and similar results were observed when it was added to feed (Ransom et al., 2003). Ionophores, particularly monensin, have traditionally been used as coccidiostats in poultry, and it was hypothesized that its use in other animals could reduce food-borne pathogens; however, no effect was observed with sheep (Edrington et al., 2003a), swine, or cattle (Dealy and Moeller, 1977).

Bacteriophages

Bacteriophages are obligate parasites, bacterial viruses that can infect and lyse (kill) bacterial host cells (Greer, 2005). These bacterial viruses target the host cell biosynthetic machinery to replicate and eventually deplete all intracellular nutrients, causing the host cell to burst and release daughter phages which subsequently infect more host cells (Callaway et al., 2004). Their ability to kill a specific target species has been explored in an effort to design a natural antimicrobial intervention. Bacteriophages are natural, nontoxic alternatives to the use of antibiotics in reducing food-borne pathogens. The benefits of phage antimicrobial therapy are (i) increased specificity, (ii) no toxicity, (iii) reduced cost, (iv) a single dose (self-perpetuating; a few animals can spread the virus through the entire population at a production stage), (v) the inability of pathogens to develop phage resistance, and (vi) reduced virulence of resistant mutants (Greer, 2005). Bacteriophages have been reported as effective in reducing enteric pathogens in in vivo experimental trials using cattle, sheep, swine, and poultry (Huff et al., 2002; Kudva et al., 1999; Smith and Huggins, 1982); however, there has been limited success in real-world settings. The application of the nontherapeutic use of phages to control bacterial pathogens at the farm level has been reported by Goodridge and Abedon (2003), who suggested the terminology "bacteriophage control." Bacteriophages have limited approval for use in human food (i.e., against *Listeria monocytogenes* in ready-to-eat foods), and the issues that need to be resolved before they are widely applied are mechanism of delivery to reduce inactivation in the intestinal tract, time of application, host range of phage mixtures, multiplicity of infection, phage-resistant mutants, and physiology of host bacteria in vivo (Joerger, 2003).

Antimicrobial proteins

Certain bacteria have the ability to produce and release extracellular proteins with antimicrobial properties, termed bacteriocins (Lakey and Slatin, 2001). The use of bacteriocins, specifically colicins (to target specific *E. coli* populations), has only recently been considered for application in reducing food-borne pathogens in food animals (Jordi et al., 2001; Schamberger and Diez-Gonzalez, 2002). Substantially more research is needed to determine if this control strategy is a viable option and the biggest challenges faced are optimizing efficacy and scaling up production so that it can be an economically feasible addition to food rations (Callaway et al., 2004).

Metabolic inhibitors

The antimicrobial effect of sodium chlorate against enteric pathogens is well known. Enteric pathogens respire by reduction of nitrate to nitrite using nitrate reductase; however, the enzyme does not differentiate between the nitrate and its analog, chlorate, and thus supplemented chlorate will be reduced in the intestine of animals to chlorite, which when accumulated is toxic to bacteria (Stewart, 1988). The effect of chlorate treatment has been demonstrated in vitro (Anderson et al., 2000) as well as in vivo with swine and sheep intestinal tracts (Anderson et al., 2001; Edrington et al., 2003b) and with poultry intestinal contents (Byrd et al., 2003; Jung et al., 2003). Studies demonstrated that soluble chlorate

administered in drinking water was effective against populations of *E. coli* and *E. coli* O157:H7 in cattle (Anderson et al., 2002; Callaway et al., 2003b). Sodium chlorate is toxic to animals in large amounts and is not currently approved for use in food animals; however, further research is needed to determine the quantity required to be effective and the risk of inherent toxicity to animals.

PLANT PRODUCTS

Pathogens of Concern

Several major outbreaks of food-borne illness linked to the consumption of fresh fruits and vegetables have highlighted the importance of these commodities as a serious threat to public health (Table 2). Raw produce can be contaminated at any point in the chain from the farm to consumption and may as such harbor a diverse range of human pathogens, including *Aeromonas* spp., *Bacillus cereus*, *Campylobacter* spp., *Clostridium botulinum*, *Cryptosporidium* sp., pathogenic *E. coli*, *Giardia lamblia*, hepatitis viruses, *Listeria monocytogenes*, Norwalk-like viruses, *Salmonella* sp., *Shigella* spp., *Staphylococcus aureus*, *Vibrio cholerae*, and *Yersinia enterocolitica* (Castillo and Rodriguez-Garcia, 2004; Warriner, 2005; Zhao, 2005). Based on the incidence of food-borne illness in produce historically, the pathogens of primary concern (up to 80% of produce-related outbreaks) appear to be *Salmonella*, *E. coli* O157:H7, *Shigella*, Norwalk-like viruses *(Norovirus)*, and pathogenic protozoa, and their incidence is increasing (Castillo and Rodriguez-Garcia, 2004). It is difficult to establish which types of produce carry a greater risk of contamination with food-borne pathogens and its public health impact. Table 3 places fresh produce into microbiological risk categories based on the growing characteristics and final use by consumers. Generally, root vegetables (i.e., carrots, etc.) have a carriage rate of human pathogens higher than that of leafy vegetables (i.e., lettuce, etc.) (Prazak et al., 2002); however, leafy vegetables (lettuce in particular) have been implicated in a greater number of food-borne outbreaks than root vegetables. Between 1998 and 2006, five commodity groups were responsible for 76% of produce-related outbreaks (Buchanan, 2006), namely, (i) lettuce/leafy greens (30%), (ii) tomatoes (17%), (iii) cantaloupe (13%), (iv) herbs (basil and parsley) (11%), and (v) green onions (5%).

Sources of Contamination for Raw Produce

Preharvest microbial contamination of raw produce is associated primarily with soil, irrigation water, manure/compost, aerosols, and contact with animals (domestic and wild) (Beuchat, 2006). Most

Table 2. Major food-borne outbreaks in the United States caused by fresh/fresh-cut produce in the last 10 years[a]

Known/suspect vehicle	Yr	Recall/FDA alert	Pathogen	Location	Venue	No. of cases/deaths
Green onions	1997		Cryptosporidium parvum	Washington	Restaurant	54/0
Baby lettuce	1997	No	Cyclospora cayetanensis	Florida	Restaurants/cruise ship	1,465/0
Raspberries	1997	Growers voluntarily suspended shipment	Cyclospora cayetanensis	United States and Canada	Various	1,012/0
Basil	1997		Cyclospora cayetanensis	Multistate, United States	Retail/catered events	>308/0
Lettuce	1997		E. coli O157:H7	Montana	Retail	70/0
Cantaloupe	1997	Voluntary	Salmonella	California	Home, stores	24/0
Raspberries	1998	Import alert—FDA	Cyclospora cayetanensis	Ontario, Canada	Various	315/0
Salad	1998		E. coli O157:H7	California	Restaurant	2/0
Fruit salad	1998		E. coli O157:H7	Wisconsin	Catered event	47/0
Coleslaw	1998	No	E. coli O157:H7	Indiana	Restaurant	33/0
Lettuce	1998		Shigella sonnei	Minnesota	Various	160/0
Parsley	1998	Market recall	Shigella sonnei	Minnesota, Massachusetts, California	Restaurants	400/0
Parsley	1998	Import alert—FDA	Shigella sonnei	Canada	Food fair	35/0
Tomatoes	1998–1999		Salmonella	Florida	Various	85/3
Blackberries	1999	Import alert—FDA	Cyclospora cayetanensis	Ontario, Canada	Banquet hall	104/0
Romaine lettuce	1999		E. coli O157:H7	Oregon	Community	3/0
Romaine lettuce	1999		E. coli O157:H7	Pennsylvania	Retirement community	41/0
Romaine lettuce	1999		E. coli O157:H7	California	Community	8/0
Coleslaw (cabbage)	1999	No	E. coli O157:H7	Indiana		27/0
Coleslaw (cabbage)	1999	No	E. coli O157:H7	Ohio		19/0
Romaine lettuce	1999		E. coli O157:H7	Washington	Community	6/0
Cantaloupe	2000	Voluntary recall	Salmonella	Multistate, United States	Various	43/2
Cantaloupe	2001	No	Salmonella	Eight states	Various	30/0
Lettuce	2002	No	E. coli O157:H7	Washington	Camp	28/0
Lettuce	2003	No	E. coli O157:H7	California	Restaurant	50/0
Green onions	2003	Market recall	Norovirus	Pennsylvania	Restaurant	601/4
Spinach	2003		E. coli O157:H7	California	Nursing home	16/2
Spring mix salad	2004	No	Cyclospora	Texas	Restaurant	38/0
Tomatoes	2004	Stopped production	Salmonella	United States and Canada	Various	561/0
Mesculin/spring mix salad	2004	No	Cyclospora	Illinois	Restaurant	57/0
Iceberg lettuce	2004		Salmonella	Maryland	Restaurant/deli	97/0
Parsley	2005		E. coli O157:H7	Washington, Oregon	Restaurant	60/0
Lettuce	2005	Yes	E. coli O157:H7	Wisconsin, Oregon, Minnesota	Various	23/0
Lettuce	2005		Norovirus	Michigan	Restaurant	55/0
Spinach	2006	Yes	E. coli O157:H7	Multistate, United States	Various	204/3

[a]Summarized from CDC (2007).

of these are essential elements needed to grow crops and are called "inputs" in the production sector. Inputs may also include containers, equipment, tools, chemicals, hardware, and humans, and while these inputs are no less important, their overall contribution and risk of pathogen hazards are substantially less than those of soil, irrigation water, manure/compost, and animal contact, since these serve as harborage for pathogens. The following section describes the risk of contamination with

Table 3. Microbiological risk categories of fresh produce[a]

Class 1 (grown in or close to the ground and can be eaten uncooked)	Class 2 (grown off the ground or protected by skin and can be eaten uncooked)	Class 3 (generally eaten cooked)	Class 4 (grown over 3 feet above ground, with an edible skin)	Class 5 (grown over 3 feet above ground, with the skin not normally eaten)
Cabbages	Beans	Artichokes	Apples	Avocado
Carrot	Berry fruits	Asparagus	Apricots	Banana
Celery	Broccolis	Beetroots	Cherries	Custard apples
Herbs	Capsicum	Broccolis	Grapes	Grapefruit
Leeks	Cauliflowers	Brussel sprouts	Nectarines	Kiwifruit
Lettuces	Chillis	Cabbages	Olives	Papaya
Mushrooms	Cucumbers	Corn	Peaches	Passion fruit
Onions	Garlic	Eggplant	Pears	Persimmon
Radishes	Melons	Ginger	Plums	Pineapples
Spinach	Peas	Parsnip		Lemons
Strawberries	Tomatoes	Peas		Limes
		Peanuts		Lychees
		Potatoes		Mandarins
		Pumpkin		Mangoes
		Rhubarb		Nuts
		Squash		Oranges
		Turnips		
		Zucchini		

[a]Class 1, highest risk; class 5, lowest risk. Adapted from Australian Government (2004). Available at http://www.horticulture.com.au/docs/publications/Guidelines_for_On_Farm_Food_Safety_for_Fresh_Produce_Second_Edition.pdf.

pathogens for soil, water, manure, and animal contact.

Soil

Soil is a natural habitat for several food-borne pathogens, including *Aeromonas* spp., *B. cereus*, *C. botulinum*, *C. perfringens*, and *L. monocytogenes* and, to a lesser extent, pathogenic *E. coli*, *Salmonella*, enteric viruses, and protozoa (*Giardia lamblia* and *Cryptosporidium* spp.) (Warriner, 2005). Enteric pathogens are able to persist longer in moist soils at cool temperatures (Guo et al., 2002). Enteric bacteria are thought to struggle for survival in soil due to the inherent stresses associated with nutrient limitation, water activity, and temperature extremes. In addition to these stresses, they have to compete (for nutrients, against antimicrobials produced by the resident flora, etc.) with the endogenous microflora to become established in the soil; however, enteric bacteria have been successful surviving in soil by modifying their immediate microenvironment (Ibekwe and Grieve, 2004). Association with root rhizospheres also enhances survival of enteric pathogenic bacteria (Gagliardi and Karns, 2002). Although enteric viruses are easily inactivated on the surface of the soil, they have been shown to survive for up to 4 months within soil and are effectively dispersed in the soil during times of heavy rainfall (Santamaria and Toranzos, 2003). Pathogens can also be transferred through the air attached to dry manure, dry soil, or dust, and

in fact, research has found significant levels of airborne pathogenic bacteria (Lee et al., 2006).

Water

Although crop contamination may be affected by soil, manure, aerosols, and animals, water is arguably the most likely vehicle of contamination and may have the most expansive impact. Sewage spills, runoff from concentrated animal production facilities, storm-related contamination of surface waters, illicit discharge of waste, and other sources can introduce pathogens and affect the quality of surface and groundwater used for produce irrigation (Suslow et al., 2003). Use of recycled municipal wastewater for irrigation may increase the risk of introducing pathogens (Exall, 2004). The method of irrigation can also affect the risk of pathogen contamination of raw produce (NACMCF, 1999). The various methods used for irrigation include gravity (flood) irrigation, spray (overhead) irrigation, drip/trickle irrigation, and subirrigation (FDA-CFSAN, 2001). Flood and spray irrigation are associated with the greatest risk of contamination, as any pathogens in the water may be deposited directly onto the edible parts of the crops (FDA-CFSAN, 2001). While the majority of contamination most likely occurs through direct contact between edible portions of crops and contaminated water, researchers (Solomon et al., 2002; Wachtel et al., 2002) have illustrated that pathogens may also enter the plant tissue through the root system. The ability of contamination to occur through the root system is highly dose dependent (high concentrations of pathogens need to enter the plant tissue); however, the specific thresholds have not been established, and its importance to food safety is unproven.

Manure

Animal manure is a source of macro- and micronutrients and is an effective fertilizer, often used preferentially to synthetic fertilizers; however, manure may be the source of pathogens and subsequently contaminate produce crops (Natvig et al., 2002). Despite the significant number of produce-related outbreaks, their epidemiological association with the use of manure/compost is not established; however, the knowledge that improper use of manure can transfer pathogens onto crops and result in human illness is undeniable. Sewage effluents (of animals and humans) represent the most significant source of pathogens recovered in water, soil, and produce (Cieslak et al., 1993), and application of untreated or improperly treated manure or sewage is a direct route by which produce may be contaminated. Manure may be used predominantly in organic cultivation systems but is also applied by conventional growers (USDA, 2001). Because organic growers rely primarily on animal manure for fertilization of soil, it is thought that organic produce may represent a greater risk, with regard to contamination by enteric pathogens, compared to their conventional counterparts (Johannessen et al., 2004). Research (Loncarevic et al., 2005; Mukherjee et al., 2004) has found that organic produce may have a higher incidence of enteric microorganisms compared to conventional produce. The potential for pathogenic contamination of produce through manure depends on four primary factors (Suslow et al., 2003): (i) pathogen populations in the animal feces (type and level), (ii) treatment/storage/processing methods for manure, (iii) biological activity and structural stability of the soil to which it is applied, and (iv) relative timing and location of manure in crop rotation. The standards for biosolids derived from manure treatment and destined for general fertilizer use are <1,000 fecal coliform MPN/gram, <4 *Salmonella* MPN/gram, and <4 enteric virus plaque-forming units/gram biosolids (Mechie et al., 1997). The survival of pathogens in manure is dependent on temperature, solid content, level of contamination, aeration, and holding time (Ajariyakhajorn et al., 1997). In general, persistence of enteric pathogens in manure is favored under low-temperature and high-moisture conditions, and they can indeed survive for up to 4 months under such conditions (Enriquez et al., 2003).

Animal contact

Domestic livestock (especially cattle) are a primary source of food-borne pathogens, while wild animals may also be a source. The enteric pathogens most often associated with such animals are *Salmonella* sp. and pathogenic *E. coli*. The risk of produce contamination with pathogens of domestic livestock origin is affected by (i) the location of the animal operation (distance from farm, topography [on higher elevation than farm], water bodies in the immediate vicinity [especially upstream of the farm], and typical wind direction); (ii) time of year (prevalence of enteric pathogens in cattle is highest in spring and summer); and (iii) preventative barriers on the animal operation to restrict animal movement. Wild mammals are an increased source of pathogen contamination since their access to crops is somewhat unrestricted. Research has indicated that rodents are also a potential source of pathogenic bacteria, especially when living in close proximity to livestock and fecal material (Nielsen et al., 2004). Commensal rodents (those living in closer contact with human and livestock waste) have a much higher prevalence of enteric pathogens than field rodents and as such pose a higher risk of contamination to produce (Meerburg et al., 2004). Enteric pathogens have been isolated from deer sharing a rangeland with cattle; however, evidence suggests that they are not a major pathogen reservoir (Fischer et al., 2001; Sargeant et al., 1999). Feral pigs carrying protozoa such as *Cryptosporidium* and *Giardia lamblia* have been implicated as potential sources of contamination in a large outbreak involving spinach (Atwill et al., 1997; Robertson and Gjerde, 2000). In a separate and unrelated outbreak involving spinach, feral pigs carrying *E. coli* O157:H7 from a cattle operation were implicated as the source of contamination (Jay et al., 2007). Other potential vectors of food-borne pathogens include invertebrates, such as beetles, mealworms, houseflies, fruit flies, and slugs; however, research has indicated that the prevalence of pathogens in invertebrates is very low (Sproston et al., 2006). Similarly, birds, especially those associated with pastoral environments (woodpeckers, chickadees, nuthatches, etc.), are unlikely to carry food-borne pathogens. Although the overall prevalence of food-borne pathogens is low among wild birds, it should be noted that the prevalence of pathogens may be higher in birds associated with human waste (e.g., gulls, etc.) (Hancock et al., 1998).

Preharvest Hazard Control for Produce

There have been no government-mandated food safety guidelines regarding fresh produce; however, increased concern of consumers and regulators has resulted in development and implementation of voluntary commodity-specific food safety guidelines and GAP programs throughout the industry. The intent of these GAP and food safety programs is to prevent food-borne disease and illness by control through identification and minimization of microbiological hazards at all stages of the chain, including growing, harvesting, packing, handling, processing, food preparation, and food service. The postharvest processing and handling operations may be well managed by the application of a HACCP systems approach. The approach of such a system is one that identifies hazards and employs controls to prevent, minimize, or inhibit the same. The implementation of HACCP in its full sense cannot be implemented in fresh produce operations since currently there is no definite "kill" step available for produce eaten raw. The real application of HACCP principles at the farm level is to identify and reduce the potential for produce to become contaminated with food-borne pathogens. Much of the control that is exercised by fresh produce growers is achieved through application of GAP programs that manage all aspects of crop production. Thus, control of microbiological hazards at the farm level is achieved through on-farm food safety (OFFS) programs that incorporate GAP to manage the sources of hazard and reduce contamination of raw produce (Chapman and Powell, 2005). The following section introduces the sources/inputs of microbial hazards in fresh produce and the control strategies to reduce the risk to public health.

Minimizing Initial Contamination

In many ways, GAP are the agricultural industry's equivalent to good manufacturing practices used to support the HACCP systems at the food processing stage. The elements of GAP applied to fresh produce should include (i) site (land) history and management, (ii) waste and pollution management, (iii) seed program, (iv) fertilizer usage, (v) manure and biosolids, (vi) soil and substrate management, (vii) agriculture and washing water, (viii) worker hygiene and sanitary work practices, (ix) containers, equipment, and tools, (x) vehicle management, (xi) pest control program, and (xii) wildlife management program (Early, 2005).

The following discussion of the elements comprising a GAP program is adapted and summarized from *On-Farm Food Safety Guidelines for Fresh Fruit and Vegetables in Canada* (Canadian Horticultural Council, 2004), *Guide to Minimize Microbial Food Safety Hazards for Fresh Fruits and Vegetables* (FDA-CFSAN, 1998), *Food Safety Begins on the Farm: a Grower's Guide* (Rangarajan et al., 2000), "Good agricultural practice and HACCP in fruit and vegetable

cultivation" (Early, 2005), and "Implementing on-farm food safety programs in fruit and vegetable cultivation" (Chapman and Powell, 2005). These resources should be consulted for an in-depth description of GAP programs that should be implemented in an effort to minimize contamination of fresh produce with food-borne pathogens.

Site (land) history and management

Fruit and vegetables can acquire pathogens from the soil, and it is imperative that the production site be assessed for (i) previous uses (previous crop production, nonagricultural uses, and past livestock production), (ii) animal access to the production site (access assessment and preharvest inspection), (iii) adjacent field usage (adjacent crop production, livestock production, manure storage sites, or human-related activity), and (iv) location relative to surface water and runoff/drift from animal operations, to ensure that it does not represent a risk for microbial hazards.

Waste and pollution management

Waste must be handled and stored in a manner that does not contaminate fresh produce. This element requires establishment of plans and procedures for controlling waste and pollutants, identification of waste and pollutants, and disposal of wastes in approved containers and ways.

Seed program

A critical element in minimizing pathogen contamination of produce is selection and use of seeds that are not contaminated with pathogens. Application of other GAP and control strategies is rendered ineffective when the basic raw material (seeds) itself is contaminated, because not only is there a high likelihood that produce from such seed will be contaminated, but the seed is able to contaminate the soil and thereby cross-contaminate other plants. Seed should be purchased from certified suppliers that have a seed-testing program.

Fertilizer usage

Commercial fertilizer should be purchased from certified suppliers, applied at appropriate times and according to recommended rates, and stored in designated areas and under proper conditions.

Manure and biosolids

Manure must be obtained from an approved supplier with certification that it is processed adequately to kill pathogens. Manure must be stored in a designated area separate from where produce and other inputs are stored and not in a pile near the field when harvesting. Application of manure should be timed appropriately. In fall, manure should be applied when soils are warm and nonsaturated, and have a cover crop (annual, biennial, or perennial grown as a monoculture to manage and improve soil quality in the off-season for the major crop); in spring/summer, it should be incorporated 2 weeks prior to planting major crops. Produce should not be harvested within 120 days after manure application. Raw (untreated) sewage sludge must not be applied to fields; growers must receive a certificate of fitness for agricultural use from the supplier before using sewage sludge on fields.

Soil/substrate management

An important part of minimizing contamination of fresh produce is to maintain the integrity of the soil or substrate used to grow the crop. Sites should be mapped by soil type and appropriate methods of cultivation should be applied. Soil erosion should be avoided to prevent disruption of the structure and ability to "hold" dust. Soil fumigation should be controlled and ideally applied prior to planting.

Agriculture and washing water

Agricultural water is used for chemical application, fertilizer application, and irrigation and does not need to be potable. Washing water is used for cleaning and rinsing produce, fluming, and product cooling. It is critical to identify sources of agricultural water (surface water, well or groundwater, or tertiary/gray water [water receiving treatment from a municipal or government sewage treatment plant]) and adopt appropriate preventative measures to reduce the risk of source water contamination. The risk of microbiological contamination is higher if the water is applied to the edible parts of produce. This may include overhead (spray) irrigation, washing water, water in postharvest dips/sprays, water in unloading tanks and troughs, and water in hydrocoolers and the top-icing of packages. Although agricultural water does not need to be potable, it is suggested that such water, especially irrigation water, contain no more than 1,000 total coliforms/100 ml of water (and ideally no more than 100 fecal coliforms/100 ml of water and 20 CFU of $E.\ coli$/100 ml of water). Surface water (pond, river, stream, etc.) may be protected from contamination using the following measures: (i) spread manure during dry weather, when the probability of rain is low for 48 hours after spreading; (ii) leave a manure-free protective strip at least 30-feet wide around ponds; (iii) control runoff by planting sod around ponds, grass waterways, and vegetative buffers; (iv) prevent livestock and wild animal access to ponds via a protective barrier (fence, etc.); (v) identify

and divert contamination risks upstream from the water intake; and (vi) ensure that the water intake is not situated in an area of stagnant water. Well or groundwater may be protected from contamination by the following measures: (i) verify seals and well casings for tightness, (ii) maintain wells in good condition, (iii) prevent livestock and wild animal access from the area immediately surrounding the water source, and (iv) level the ground around the water source to prevent runoff. Tertiary/gray water (nonindustrial wastewater generated from domestic processes, such as dishwashing, laundry, bathing, etc.) should be received from a certified supplier (e.g., municipality), and producers must verify that such water can be used under state agricultural and environmental regulations. Washing water (used for washing produce; hand washing; cleaning equipment, containers, and tools) must be potable and contain less than 10 total coliforms/100 ml water. Washing water must be tested on a routine basis to ensure adequate microbiological quality, and if using municipal water, producers must ensure that they are notified immediately if the water becomes contaminated.

Worker hygiene and sanitary work practices

Workers who come into contact, either directly or indirectly, with fresh produce pose a risk for product contamination (contamination is often transmitted by the fecal-oral route) and should follow good personal hygiene and sanitary practices. The basic elements of a guide to "personal hygiene and food handling practices" should address (i) worker illnesses and diseases, (ii) worker injuries, (iii) worker garments and accessories, (iv) jewelry and personal effects, (v) worker cleanliness and hand washing, (vi) worker personal habits, and (vii) good hand-washing technique. Sanitary hygienic facilities are an important risk factor for produce contamination and, when properly controlled, help in reducing fecal-oral route transmission of contamination. Field workers should have access to portable toilets, and if these are not available, the producer must provide breaks and transportation to sanitary facilities. During field operations workers should also have access to alcohol-based hand washes. The setup of sanitary hygienic facilities should meet the following requirements: (i) not be located near a water source; (ii) not open directly into rooms housing containers, equipment, and tools used to contact produce or rooms used to handle produce; (iii) have adequate sanitary paper, water, pump soap, disposable paper towels, and waste containers; (iv) have hand-wash stations with soap and preferably hot and cold water; and (v) have a maintenance and cleaning program that is performed on a routine basis. Personal hygiene and sanitary practices can be only as effective as the training programs supporting them; thus, training must be provided that includes (i) good personal hygiene, (ii) good food handling practices, (iii) food safety hazards, (iv) instruction on use of chemicals to aid in cleaning, (v) application of the sanitation program, and (vi) instruction on sanitary facility location and operation. Furthermore, written instructions, visual aids, and checklists should be developed to aid in the delivery and success of such programs.

Containers, equipment, and tools

Containers, equipment, and tools that may contact raw produce should be sanitized and maintained in order to reduce the risk of cross-contamination. Containers used for carrying harvested produce should be used only for this purpose and should be constructed of nontoxic material, easy to clean, and in good repair. Produce contact implements should be cleaned using adequate washing, sanitizing, and rinsing protocols, and the frequency should be determined and a schedule maintained. Cleaning of implements should be performed in a separate area and at appropriate times to prevent contamination of growing produce. Storage of these implements should be in a clean area separate from that of manure/compost and other material that may contaminate them.

Vehicle management

A critical element that is often overlooked as a risk of cross-contamination to raw produce during growing is farm vehicles. The access to vehicles should be minimized, and only those designated for crop operations should be permitted to access the crops. Vehicles should be well maintained and sanitized to remove risk of cross-contamination from contamination on vehicle exteriors, especially tires. Vehicles should be washed down before entering a separate field to reduce contamination that may have been acquired en route. Minimize physical contact of the vehicle with crops where possible; in the event that direct contact is required, ensure adequate sanitation of parts in physical contact with crops.

Pest control program

All animals and pests constitute sources of contamination for raw produce. In order to minimize the impact pests have on the microbiological safety of raw produce, a pest prevention/control program should be implemented to (i) keep grounds clear of waste that attract pests; (ii) remove unnecessary articles that could serve as habitats for rodents and insects (i.e., old equipment, boxes, etc.); (iii) remove dropped produce where possible; (iv) maintain grass and vegetation to minimize breeding, harboring, and feeding activities of pests; (v) minimize entry of pests into buildings (use of mesh and screens where appropriate); and (vi) carry out a pest-monitoring program, including removal and handling of dead or trapped pests.

Wildlife management program

Wildlife is an important source of contamination for produce, and their access to growing crops should be minimized. Where possible, the use of barriers (fences) should be considered to prevent access of wildlife animals to crops. Direct contact with birds may be controlled in ground crops (i.e., strawberries, etc.) that permit the use of mesh screens. Barriers to prevent animal traffic in crops should be supported by a monitoring and maintenance program that ensures the integrity and effectiveness of the barrier.

Reducing Contamination

The primary control for food-borne pathogen contamination of produce at the farm level is achieved through implementation of an OFFS with well-monitored GAP that actually minimize introduction or exposure to contamination rather than reduce contamination. For produce, where there are very few direct antimicrobial interventions that are applied at the farm level to reduce pathogen contamination of the actual product, the approach is better defined as "risk-reducing interventions." The next section describes how contamination may be reduced in the sources that carry risk of pathogen contamination. The four primary sources of contamination that were identified—soil, water, manure, and animal contact—should be managed to reduce pathogen contamination and thus risk of cross-contamination to crops.

Soil

Commensal microflora in the soil may reduce pathogen contamination (Suslow et al., 2003) through antagonistic functions (nutrient competition, production of antimicrobials, attachment, etc.) (Johannessen et al., 2005). To date there has been no suggestion to selectively promote proliferation or indeed inoculate soil with high levels of nonpathogenic commensal organisms as a means of excluding pathogen persistence (Adams et al., 2002). A similar approach has been adopted in livestock by means of competitive exclusion and probiotic strategies. The soil is a complex natural system, and thus, there are limited control strategies to reduce pathogen contamination in the same due to its ever-changing dynamic. The potential for introducing microflora (as a biomass) into the soil environment to "treat" the soil will be affected by (i) the ability to produce large quantities economically, (ii) the uniformity with which it can be distributed throughout the soil, (iii) the ability of the organism/ mixture of organisms to survive in adverse conditions associated with the soil environment, and (iv) the antagonistic capacity of the organism(s) for a wide array of enteric pathogens. The sustainability of the "inoculum" in the soil will depend on the soil structure (aeration, pH, nutrient content, etc.), the underlying water table and channels, and climatic conditions (especially rain). The concern with good distribution may be addressed by "inoculating" manure with the competitive exclusion cultures rather than the soil itself, as this may ensure adequate spreading (Darby et al., 2006). It should be noted that this is only conceptual, and there are no data supporting the application of this strategy to reduce pathogen contamination of soil. A potential physical barrier to reduce soil contamination is reducing the effect of airborne depositions containing pathogens. This may be achieved by installation of structures such as windbreaks or vegetation. Furthermore, sites may be covered with plastic covers during rotation to prevent wind deposition of pathogens.

Water

Monitoring and treating pathogens in agricultural source water as a means to reduce pathogen contamination in water that will contact crops are relatively costly, and there are no data to indicate the benefit of monitoring and treating water compared to implementation and monitoring of GAP to minimize contamination of source water. The value of monitoring source waters is recognized though, since there are effective "shock" treatments that effectively disinfect water. Where there is a significant risk of contaminating produce from contaminated water, reduction in such risks may be achieved by treatment of the water source with chemical and nonchemical sanitizing methods (Blatchley et al., 2007; Hambidge, 2001; Hijnen et al., 2006; Schijven et al., 2003). Such options include (i) chlorine, (ii) chlorine dioxide, (iii) chloro-bromine compounds, (iv) hydrogen peroxide, (v) peracetic acid, (vi) peroxy compounds (hydrogen peroxides and peracetic acid), (vii) ozone, and (viii) UV light. There are many factors that control the efficacy of a sanitizer to reduce pathogen contamination in water: (i) the type and number of microorganisms present, (ii) chemical properties of the water (hardness, pH, etc.), (iii) quality of the water (level of organic material, soil, etc.), and (iv) concentration of sanitizer applied. These treatments may be effective and economical in specific situations, in particular, in water intended for drip or micro-sprinkler irrigation, since these methods use a limited volume of water. Filtration (through perennial forage and/or grasses) of overland flow may be used to reduce pathogen contamination in water (Tate et al., 2006). The efficiency of filtration depends on water flow, soil type, slope, and vegetated buffers used. Vegetated treatment systems, planted areas through which water is directed (grassed waterways, vegetated ponds or basins, and constructed wetlands), with a settling basin can achieve significant reduction of pathogen contamination in water (Koelsch et al., 2006). Constructed wetlands are highly effective in that they reduce pathogen contamination in water by removal with filtration through dense vegetation, sedimentation of particles, microbial competition and predation, high temperatures, and UV disinfection (Greenway, 2005).

Manure

The primary control strategies to reduce pathogen contamination in manure and cross-contamination of crops are through adequate processing, storage, and application to crops. The Environmental Protection Agency (EPA) advised that pathogen reduction in manure is achieved by composting, although "aging" if done thoroughly (55 to 65°C [preferably 60 to 65°C] for at least 3 days) may be as effective (EPA, 1993). Composting uses the same directed aerobic fermentation approach with the same target temperatures as "aging"; however, it requires continuous aeration, moisture addition, and storage for at least 3 months (Suslow et al., 2003). Composting is more effective than "aging" in reducing pathogen contamination in manure. Even so, the periods of time suggested above may need to be revised based on factors such as load of pathogen in manure, size of manure pile, aeration practice (blower control systems or tilling), and heat transfer coefficients. It has been suggested that aging for a minimum of 6 months and composting for a minimum of 6 weeks are necessary to effectively reduce pathogen contamination (Australian Government, 2004). Raw animal manure for organic cultivation must be composted, applied to land for crops not intended for human consumption, or incorporated into the soil at least 120 days before harvesting of crops intended for human consumption. Composting of manure as mentioned above should occur between 55 and 76°C from 3 to 15 days, depending on the composting system (Suslow et al., 2003). It is important to develop manure treatment protocols that more effectively reduce pathogen contamination to a level that significantly reduces the risk for cross-contamination of crops. This may require development of longer and more thorough composting programs, resulting in delays in production which would have a negative impact on the supply chain.

Animal contact

Reducing the risk of pathogen contamination of crops from animal contact is best accomplished by reducing access of animals to crops. This is discussed in Pest control program and Wildlife management program in Plant Products: Minimizing initial contamination, earlier in this chapter.

Application of these GAP and risk-reducing interventions as part of an OFFS program is the best way to control pathogen contamination of produce at the farm level. Table 4 summarizes and highlights the sources (inputs) of contamination and the approaches to minimize/reduce contamination of produce at various stages of on-farm operations.

Table 4. Sources and control points for microbial hazards in fresh produce during preharvest farm operations

Farm operation	Inputs (sources)	Control strategy	
		GAPs	Risk-reducing interventions
Select and prepare growing site	Soil, fertilizers, implements, human	(i) Site history and management; (ii) fertilizer usage; (iii) containers, equipment, and tools; (iv) worker hygiene and sanitary work practices; (v) soil and substrate management	Soil, animal contact, water (overflow and runoff from adjacent farms)
Planting	Seeds, implements, human	(i) Seed program; (ii) containers, equipment, and tools; (iii) vehicle management; (iv) worker hygiene and sanitary work practices; (v) soil and substrate management	Soil, animal contact
Preplant fertilization	Water, implements, human	(i) Agriculture and washing water; (ii) containers, equipment, and tools; (iii) vehicle management; (iv) worker hygiene and sanitary work practices	
Irrigating	Water	(i) Agriculture and washing water	Water
Crop nutrition	Manure, water, implements, human	(i) Manure and biosolids; (ii) agriculture and washing water; (iii) containers, equipment, and tools; (iv) vehicle management; (v) worker hygiene and sanitary work practices; (vi) soil and substrate management	Manure, water
Pest/disease management	Water, implements, animals, human	(i) Agriculture and washing water; (ii) containers, equipment, and tools; (iii) pest control program; (iv) wildlife management; (v) worker hygiene and sanitary work practices; (vi) waste and pollution management	Animal contact
Weed control	Implements	(i) Containers, equipment, and tools; (ii) worker hygiene and sanitary work practices; (iii) soil and substrate management	Soil
Pruning/training	Implements, human	(i) Containers, equipment, and tools; (ii) worker hygiene and sanitary work practices; (iii) waste and pollution management; (iv) vehicle management	
Crop/growth regulation	Water, implements	(i) Containers, equipment, and tools; (ii) worker hygiene and sanitary work practices; (iii) agriculture and washing water	Water
Weather/vermin protection	Implements, human	(i) Containers, equipment, and tools; (ii) pest control program; (iii) wildlife management; (iv) worker hygiene and sanitary work practices; (v) waste and pollution management	
Harvest	Implements, vehicle, human	(i) Containers, equipment, and tools; (ii) worker hygiene and sanitary work practices; (iii) waste and pollution management; (iv) vehicle management	
Field-packed or transported to processing/packing shed			

FISH, MOLLUSKS, CRUSTACEANS, AND AQUATIC PLANT PRODUCTS

Aquaculture is one of the fastest growing food production sectors in the world and is defined as the farming of aquatic organisms, including fish, mollusks, crustaceans, and aquatic plants (WHO, 1999). The process, much like livestock raising and crop growing, involves some sort of intervention in the rearing process to enhance production and safety. Such products are generally regarded as safe when harvested from their natural environment; however, products from aquaculture have inherent food safety issues arising from risk of contamination by pathogenic organisms (Chinabut et al., 2006; Hastein et al., 2006). There are different methods of "farming" fish, including intensive (commercial operation), semi-intensive, and subsistence (small-scale) systems, and the associated risks vary according to the type of farming, management practices, and immediate aquatic environment. The physical farming environments include rice fields and swamps, ponds, tanks and raceways, cages, and pens and can be located inland (typically freshwater) and in coastal areas. The shift of aquaculture to a major food-producing sector highlights the need for increased efforts in assessment and control of food safety risks (Hastein et al., 2006).

Pathogens of Concern

Parasitic hazards are associated generally with fish (both marine and freshwater), which are the intermediate host of the parasite, while humans become the definitive host following ingestion. The following are important parasites affecting fish: (i) trematodes, (ii) cestodes, and (iii) nematodes. Trematodes of importance in fish that cause food-borne illness are *Clonorchis sinensis*, *Opisthorchis viverrini*, *Opisthorchis felineus*, *Paragonimus westermani*, *Heterophyes heterophyes*, *Heterophys nocens*, *Heterophyes continua*, and *Metagonimus yokogawai* (WHO, 1999). Cestodes of importance in fish that cause food-borne illness are *Diphyllobothrium latum*, *Diphyllobothrium yonagoense*, *Diphyllobothrium pacificum*, *Diphyllobothrium cameroni*, *Diphyllobothrium scoticum*, *Diphyllobothrium hians*, and *Diphyllobothrium ditremum* (WHO, 1999). Fishborne nematodiases in humans can be considered incidental since the natural definitive hosts for nematodes are marine mammals, birds, and swine. Nematodes of importance in fish that cause food-borne illness are *Capillaria philippinensis*, *Gnathostoma spinigerum*, and *Anisakis simplex* (WHO, 1999). Bacterial hazards are associated primarily with finfish and crustaceans and are either indigenous to the

aquatic environment or introduced via contamination by human or animal feces. Bacteria that result in food-borne illness due to consumption of aquatic products include *Salmonella* sp., pathogenic *E. coli*, *Campylobacter* spp., *Vibrio* spp. (*V. cholerae*, *V. mimicus*, *V. vulnificus*, and *V. parahaemolyticus*), *Aeromonas* spp., *Plesiomonas* spp., *C. botulinum*, *L. monocytogenes*, and *Mycobacterium marinum* (WHO, 1999). Viruses, especially adenoviruses, are associated with food-borne disease resulting from consumption of aquatic products (since the animal-specific adenoviruses infect a wide range of species, including fish) (Russell and Benk, 1999). Further biological hazards are associated with the consumption of fish that have acquired cyanobacterial toxins or of mollusks that have paralytic-shellfish-poisoning toxins.

Preharvest Hazard Control for Aquatic Products

The approach of OFFS programs to minimize and reduce biological hazards associated with aquaculture products is very recent, and development of GAP is in its infancy. The underlying principles for controlling the environment and the inputs (sources) to control any pathogenic contamination associated with them are similar to those of raising livestock or growing crops (Khamboonruang et al., 1997).

Minimizing Initial Contamination

The application of GAP to aquaculture should consider the following elements: (i) site selection/preparation; (ii) water quality; (iii) breeding stock; (iv) feed supply; (v) worker hygiene and sanitary work practices; (vi) containers, equipment, and tools; (vii) vehicle management; (viii) pest control program; and, (ix) wildlife management program (Ababouch, 2000; Khamboonruang et al., 1997; Roth and Rosenthal, 2006).

Site selection/preparation

Site selection, design, and construction of aquaculture farms need to include evaluation of the immediate environment of the bottom soil, proximity to land, and soil/water interface designated for rearing aquatic products (Khamboonruang et al., 1997). Soil properties such as pH not only affect the ability to rear fish effectively but also affect the microorganisms present. The proximity to land affects the water and soil deposits since runoff from nearby or adjacent agricultural land may change the pH of the soil and carry enteric pathogenic organisms. Aquaculture facilities should be located in areas where the risk of contamination from soil, land, and water is minimized (Linton, 2007; Lupin, 1999).

Water quality

The quality of water in the area immediately surrounding the facility poses the most important risk of contaminating aquatic products with pathogenic microorganisms (Linton, 2007). The use of wastewater or animal manure to fertilize "farms" may result in an increased risk of contamination from pathogenic bacteria and parasites. Growers should use only wastewater from certified suppliers and should follow international guidelines proposed by the World Health Organization (WHO) (Mara and Cairncross, 1989). The increased presence of aquatic birds in the vicinity of the facility is an important source for introducing enteric pathogens into the water. Parasites and their intermediate snail hosts are risk factors that should be monitored and eliminated from the water supply (Hastein et al., 2006; Khamboonruang et al., 1997).

Breed stock

It is important that "farms" be stocked with breeding stock free of pathogenic organisms. The introduction of breeding stock with food-borne pathogens renders all other control measures ineffective, as the environment will inherently be contaminated and thus the offspring will invariably be contaminated with the same organisms.

Feed supply

Biological hazards may persist in manufactured feed or feed derived from organic waste. A control point to minimize introduction of contamination into "farms" is to purchase feed from a certified supplier that can guarantee the microbiological quality of their products.

Worker hygiene and sanitary work practices

Workers who come into contact either directly or indirectly with the aquatic environment pose a risk for product contamination and should follow good personal hygiene and sanitary practices (Hastein et al., 2006; Linton, 2007; Lupin, 1999).

Containers, equipment, and tools

Containers, equipment, and tools that may contact the aquatic environment should be sanitized and maintained in order to reduce the risk of cross-contamination (Hastein et al., 2006; Linton, 2007; Lupin, 1999).

Vehicle management

The access of vehicles (aquatic craft) should be minimized, and only those designated for farming operations should be permitted. Vehicles should be well maintained and sanitized to remove the risk of cross-contamination from contamination on vehicle exteriors (Hastein et al., 2006; Linton, 2007; Lupin, 1999).

Pest control program

All animals and pests are risks of contamination for aquatic produce because they harbor or serve as vectors for pathogens. In order to minimize the impact pests have on the microbiological safety of raw produce, a pest prevention/control program should be implemented.

Wildlife management program

Wildlife may be an important source for contamination of water and their access to immediate water bodies should be minimized. Where possible, the use of barriers (fences) should be considered to prevent access of wildlife animals to water sources.

Reducing Contamination

There are three primary interventions that may be applied to aquaculture farms to reduce the risk of contamination by pathogenic organisms: (i) water treatment compounds, (ii) pesticides, and (iii) disinfectants (Hastein et al., 2006; Linton, 2007; Lupin, 1999; WHO, 1999).

Water treatment compounds

Lime-based compounds, such as agricultural limestone (pulverized calcium carbonate or dolomite), lime (calcium/magnesium oxide), and hydrated lime (calcium/magnesium hydroxide), are sometimes applied to ponds and soils to regulate pH and also to "sanitize" pond soils between production cycles. Other water treatment compounds used are oxidizing agents and flocculants/coagulants. Oxidizing agents, such as potassium permanganate, hydrogen peroxide and calcium peroxide, calcium hypochlorite, and sodium nitrate, are used for controlling phytoplankton, killing pathogenic organisms, or oxidizing bottom soils (Boyd, 1995). Flocculants, such as aluminium sulfate, ferric chloride, calcium sulfate, and zeolite, are applied to pond waters to cause suspended particles to precipitate and may aid in precipitating and removing suspended pathogenic organisms.

Pesticides

Algicides and herbicides are occasionally applied to aquaculture ponds to control blooms of algae in an effort to reduce dissolved oxygen demand and combat growth and possible toxin production by cyanobacteria (Schrader et al., 2003).

Disinfectants

Disinfectants (benzalkonium chloride, polyvidone iodine, glutaraldehyde, formalin, and hypochlorites) in aquaculture are used typically between production cycles for disinfecting portable equipment and holding units.

CONCLUDING REMARKS

Preharvest control of biological hazards in foods is a relatively new concept, although there have always been "quality assurance" programs in the production sector. With the increased demand for safer foods, there is a need for all industries in the production sector to implement OFFS programs that are focused on development of GAP that minimize initial food-borne pathogen contamination and interventions strategies that reduce these pathogens in/on raw agricultural products. The implementation of such programs will strengthen the food safety chain in the "farm-to-fork" continuum. The approach is to minimize initial pathogen contamination of products and to reduce the level of contamination, since it is practically impossible to prevent contamination in products exposed to the environment and to multiple handling operations. The OFFS programs should not be compared to HACCP programs applied in the processing, retail, and food service industries, since there are numerous, uncontrolled contamination points and cannot be managed the same way as HACCP or HACCP-based programs. The implementation of OFFS programs includes GAP, which are the equivalent of good manufacturing practices at the processing/retail/food service level, and direct interventions, where appropriate. Since these programs are still in their infancy for many raw agricultural products, there is a need for further research to narrow the knowledge gaps regarding hazards, associated risks, and control strategies.

REFERENCES

Ababouch, L. 2000. Potential of *Listeria* hazard in African fishery products and possible control measures. *Int. J. Food Microbiol.* 62:211–215.

Adak, G. K., S. M. Long, and S. J. O'Brien. 2002. Trends in indigenous foodborne disease and deaths. *Gut* 51:832–841.

Adams, T. T., M. A. Eiteman, and B. M. Hanel. 2002. Solid state fermentation of broiler litter for production of biocontrol agents. *Bioresour. Technol.* 82:33–41.

Ahl, A. S., and B. Buntain. 1997. Risk and the food safety chain: animal health, public health, and the environment. *Rev. Sci. Tech.* 16:322–330.

Ajariyakhajorn, C., S. M. Goyal, R. A. Robinson, L. J. Johnston, and C. A. Clanton. 1997. The survival of *Salmonella anatum*, pseudorabies virus and porcine reproductive and respiratory syndrome virus in swine slurry. *New Microbiol.* 20:365–369.

Anderson, R. C., L. H. Stanker, C. R. Young, S. A. Buckley, K. J. Genovese, R. B. Harvey, J. R. DeLoach, N. K. Keith, and D. J. Nisbet. 1999. Effect of competitive exclusion treatment on colonization of early-weaned pigs by *Salmonella* serovar choleraesuis. *Swine Health Prod.* 12:155–160.

Anderson, R. C., S. A. Buckley, L. F. Kubena, L. H. Stanker, R. B. Harvey, and D. J. Nisbet. 2000. Bactericidal effect of sodium chlorate on *Escherichia coli* O157:H7 and *Salmonella typhimurium* DT104 in rumen contents in vitro. *J. Food Prot.* 63:1038–1042.

Anderson, R. C., S. A. Buckley, T. R. Callaway, K. J. Genovese, L. F. Kubena, R. B. Harvey, and D. J. Nisbet. 2001. Effect of sodium chlorate on *Salmonella typhimurium* concentrations in the pig gut. *J. Food Prot.* 64:255–259.

Anderson, R. C., T. R. Callaway, T. J. Anderson, L. F. Kubena, N. K. Keith, and D. J. Nisbet. 2002. Bactericidal effect of sodium chlorate on *Escherichia coli* concentrations in bovine ruminal and fecal contents in vivo. *Microbiol. Ecol. Health Dis.* 14:24–29.

Anderson, R. C., K. J. Genovese, R. B. Harvey, T. R. Callaway, and D. J. Nisbet. 2006. Preharvest food safety applications of competitive exclusion cultures and probiotics, p. 273–284. *In* I. Goktepe, V. K. Juneja, and M. Ahmedna (ed.), *Probiotics in Food Safety and Human Health*. CRC Press, Boca Raton, FL.

Asandhi, A. A., H. Schoorlemmer, W. Adiyoga, L. Dibyantoro, M. van der Voort, N. Sulastrini, and I. Sulastrini. 2006. Development of a Good Agricultural Practice to improve food safety and product quality in Indonesian vegetable production. Research Report 03 2006 of Horticultural Research Cooperation between Indonesia and the Netherlands (HORTIN). Applied Plant Research, Lelystad, The Netherlands. http://documents.plant.wur.nl/ppo/agv/hortin.pdf. Accessed 19 November 2007.

Atwill, E. R., R. A. Sweitzer, M. das Gracas C. Pereira, I. A. Gardner, D. van Vuren, and W. M. Boyce. 1997. Prevalence of and associated risk factors for shedding *Cryptosporidium parvum* oocysts and *Giardia* cysts within feral pig populations in California. *Appl. Environ. Microbiol.* 63:3946–3949.

Australian Government. 2004. Guidelines for On-farm Food Safety for Fresh Produce, 2nd ed. Department of Agriculture, Fisheries and Forestry, Australian Government, Canberra, Australia. http://www.horticulture.com.au/docs/publications/Guidelines_for_On_Farm_Food_Safety_for_Fresh_Produce_Second_Edition.pdf. Accessed 1 April 2007.

Beuchat, L. R. 2006. Vectors and conditions of preharvest contamination of fruits and vegetables with pathogens capable of causing enteric diseases. *Brit. Food J.* 108:38–53.

Blatchley, E. R., III, W. L. Gong, J. E. Alleman, J. B. Rose, D. E. Huffman, M. Otaki, and J. T. Lisle. 2007. Effects of wastewater disinfection on waterborne bacteria and viruses. *Water Environ. Res.* 79:81–92.

Bomba, A., R. Nemcova, D. Mudronova, and P. Guba. 2002. The possibilities of potentiating the efficacy of probiotics. *Trends Food Sci. Technol.* 13:121–126.

Boyd, C. E. 1995. Potential of sodium nitrate to improve environmental conditions in aquaculture ponds. *World Aquaculture* 26:38–40.

Braden, K. W., J. R. lanton, V. G. Allen, K. R. Pond, and M. F. Miller. 2004. Ascopyllum nodosum supplementation: a preharvest intervention for reducing *Escherichia coli* O157:H7 and *Salmonella* spp. in feedlot steers. *J. Food Prot.* 67:1824–1828.

Brashears, M. M., D. Jaroni, and J. Trimble. 2003. Isolation, selection and characterization of lactic acid bacteria for a competitive exclusion product to reduce shedding of *Escherichia coli* O157:H7 in cattle. *J. Food Prot.* 66:355–363.

Buchanan, R. 2006. Spinach outbreak as part of broader concerns about produce safety—A FDA perspective. Center for Food Safety and Applied Nutrition, United States Food and Drug

Administration, Washington, DC. http://www.foodprotection. org/meetingsEducation/Rapid%20Response%20Presentations/ Buchanan,%20Robert.pdf. Accessed 15 May 2007.

Busz, H. W., T. A. McAllister, L. J. Yanke, M. E. Olson, D. W. Morck, and R. R. Read. 2002. Development of antibiotic resistance among *Escherichia coli* in feedlot cattle. *J. Anim. Sci.* 80:102.

Byrd, J. A., R. C. Anderson, T. R. Callaway, R. W. Moore, K. Knape, L. F. Kubena, R. L. Ziprin, and D. J. Nisbet. 2003. Effect of experimental chlorate product administration in the drinking water on *Salmonella typhimurium* contamination of broilers. *J. Poult. Sci.* 82:1403–1406.

Callaway, T. R., R. O. Elder, J. E. Keen, R. C. Anderson, and D. J. Nisbet. 2003a. Forage feeding to reduce pre-harvest *E. coli* populations in cattle: a review. *J. Dairy Sci.* 86:852–860.

Callaway, T. R., T. S. Edrington, R. C. Anderson, K. J. Genovese, T. L. Poole, R. O. Elder, J. A. Byrd, K. M. Bischoff, and D. J. Nisbet. 2003b. *Escherichia coli* O157:H7 populations in sheep can be reduced by chlorate supplementation. *J. Food Prot.* 66:194–199.

Callaway, T. R., R. C. Anderson, T. S. Edrington, K. J. Genovese, R. B. Harvey, T. L. Poole, and D. J. Nisbet. 2004. Recent preharvest supplementation strategies to reduce carriage and shedding of zoonotic enteric bacteria in food animals. *Anim. Health Res. Rev.* 5:35–47.

Canadian Horticultural Council. 2004. *On-Farm Food Safety Guidelines for Fresh Fruit and Vegetables in Canada.* Agriculture and Agri-food Canada, Canadian Horticultural Council, Ontario, Canada. http://www.bcveg.com/2004%20%20CHC%20Guide lines%20for%20Fresh%20Fruit%20&%20Vegetable %20Producers.pdf. Accessed 14 April 2007.

Castillo, A., and M. O. Rodriguez-Garcia. 2004. Bacterial hazards in fresh and fresh-cut produce: sources and control, p. 43–57. *In* R. C. Beier, S. D. Pillai, T. D. Phillips, and R. L. Ziprin (ed.), *Preharvest and Postharvest Food Safety.* Blackwell Publishing Limited, Oxford, United Kingdom.

CDC. 2007. Bacterial foodborne and diarrheal disease national case surveillance annual reports. http://www.cdc.gov/foodbo. rneoutbreaks/outbreak_data.htm. Accessed March 2008.

Chapman, B. J., and D. A. Powell. 2005. Implementing on-farm food safety programs in fruit and vegetable cultivation, p. 268–292. *In* W. Jongen (ed.), *Improving the Safety of Fresh Fruit and Vegetables.* Woodhead Publishing Ltd., Cambridge, United Kingdom.

Chinabut, S., T. Somsiri, C. Limsuwan, and S. Lewis. 2006. Problems associated with shellfish farming. *Rev. Sci. Tech.* 25:627–635.

Cieslak, P. R., T. J. Barrett, P. M. Griffin, K. F. Gensheimer, G. Beckett, J. Buffington, and M. G. Smith. 1993. *Escherichia coli* O157:H7 infection from a manured garden. *Lancet* 342:8867.

Collins, D. M., and G. R. Gibson. 1999. Probiotics, prebiotics, and synbiotics: approaches for modulating the microbial ecology of the gut. *Amer. J. Clin. Nutr.* 69:1052S–1057S

Darby, H. M., A. G. Stone, and R. P. Dick. 2006. Compost and manure mediated impacts on soilborne pathogens and soil quality. *Soil Sci. Soc. Am. J.* 70:347–358.

Dealy, J., and M. W. Moeller. 1977. Influence of bambermycins on *Salmonella* infection and antibiotic resistance in calves. *J. Anim. Sci.* 44:734–738.

Dowd, S. E., and R. M. Maier. 1999. Aeromicrobiology, p. 91–122. *In* R.M. Maier, I. L. Pepper, and C.P. Gerba (ed.), *Environmental Microbiology.* Academic Press, San Diego, CA.

Dowd, S. E., J. A. Thurston-Enriquez, and M. Brashears. 2004. Environmental reservoirs and transmission of foodborne pathogens, p. 161–200. *In* R. C. Beier, S. D. Pillai, T. D. Phillips, and R. L. Ziprin (ed.), *Preharvest and Postharvest Food Safety.* Blackwell Publishing Limited, Oxford, United Kingdom.

Draughon, A. 2006. Human health impacts of animal agriculture. Paper presented at the John M. Airy Symposium: Visions for Animal Agriculture and the Environment, Kansas City, MO, Kansas City, MO. http://www.iowabeefcenter.org/content/Airy/ DRAUGHON%20Abstract.pdf. Accessed 1 April 2007.

Early, R. 2005. Good agricultural practice and HACCP in fruit and vegetable cultivation, p. 228–268. *In* W. Jongen (ed.), *Improving the Safety of Fresh Fruit and Vegetables.* Woodhead Publishing Ltd., Cambridge, United Kingdom.

Edrington, T. S., T. R. Callaway, K. M. Bischoff, K. J. Genovese, R. O. Elder, R. C. Anderson, and D. J. Nisbet. 2003a. Effect of feeding the ionophores monensin and laidlomycin propionate and the antimicrobial bambermycin to sheep experimentally infected with *E. coli* O157:H7 and *Salmonella* Typhimurium. *J. Anim. Sci.* 81:553–560.

Edrington, T. S., T. R. Callaway, R. C. Anderson, K. J. Genovese, Y. S. Jung, R. O. Elder, K. M. Bischoff, and D. J. Nisbet. 2003b. Reduction of *E. coli* O157:H7 populations in sheep by supplementation of an experimental sodium chlorate product. *Small Ruminant Res.* 49:173–181.

Enriquez, C., A. Alum, E. M. Suarez-Rey, C. Y. Choi, G. Oron, and C. P. Gerba. 2003. Bacteriophages MS2 and PRD1 in turfgrass by subsurface drip irrigation. *J. Environ. Eng.* 129:852–857.

EPA. 1993. A guide to the Federal EPA Rule for land application of domestic septage to non-public contact sites. Office of Wastewater Enforcement and Compliance, Environmental Protection Agency, Washington, DC. http://www.epa.gov/OW-OWM.html/pdfs/ septage_guide.pdf. Accessed 11 April 2007.

Exall, K. 2004. A review of water reuse and recycling, with reference to Canadian practice and potential: 2. Applications. *Water Qual. Res. J. Can.* 39:13–28.

FDA-CFSAN. 1998. *Guide to Minimize Microbial Food Safety Hazards for Fresh Fruits and Vegetables.* Center for Food Safety and Applied Nutrition, United States Food and Drug Administration, Washington, DC. http://www.cfsan.fda. gov/~acrobat/prodguid.pdf. Accessed 20 April 2007.

FDA-CFSAN. 2001. Analysis and evaluation of preventative control measures for the control/elimination of microbial hazards on fresh and fresh-cut produce. Center for Food Safety and Applied Nutrition, United States Food and Drug Administration, Washington, DC. http://vm.cfsan.fda.gov.html. Accessed 14 May 2007.

Fischer, J. R., T. Zhao, M. P. Doyle, M. R. Goldberg, C. A. Brown, C. T. Sewell, D. M. Kavanaugh, and C. D. Bauman. 2001. Experimental and field studies of *Escherichia coli* O157:H7 in white-tailed deer. *Appl. Environ. Microbiol.* 67:1218–1224.

Gagliardi, J. V., and J. S. Karns. 2002. Persistence of *Escherichia coli* O157:H7 in soil and on plant roots. *Environ. Microbiol.* 4:89–96.

Goodridge, L., and S. T. Abedon. 2003. Bacteriophage biocontrol and bioprocessing: application of phage therapy to industry. *Soc. Ind. Microbiol. News* 53:254–262.

Greenway, M. 2005. The role of constructed wetlands in secondary effluent treatment and water reuse in subtropical Australia. *Ecol. Eng.* 25:501–509.

Greer, G. G. 2005. Bacteriophage control of foodborne bacteria. *J. Food Prot.* 68:1102–1111.

Guo, X., J. Chen, R. E. Brackett, and L. R. Beuchat. 2002. Survival of *Salmonella* on tomatoes stored at high relative humidity, in soil, and on tomatoes in contact with soil. *J. Food Prot.* 65:274–279.

Gyles, C. L. 1998. Vaccines and shiga toxin-producing *Escherichia coli* in animals, p. 434–444. *In* J. B. Kaper and A. D. O'Brien (ed.), Escherichia coli O157:H7 and Other Shiga Toxin-Producing

E. coli *Strains*. American Society for Microbiology, Washington, DC.

Hambidge, A. 2001. Reviewing efficacy of alternative water treatment techniques. *Health Estate* **55**:23–25.

Hancock, D. D., T. E. Besser, D. H. Rice, E. D. Ebel, D. E. Erriott, and L. V. Carpenter. 1998. Multiple sources of *Escherichia coli* O157 in feedlots and dairy farms in the northwestern USA. *Prev. Vet. Med.* **35**:11–19.

Hancock, D. D., T. Besser, C. Gill, and C. J. Howde-Bohach. 1999. Cattle, hay, and *E. coli*. *Science* **284**:49.

Harvey, R. B., R. C. Ebert, C. S. Scmitt, K. Andrews, K. J. Genovese, R. C. Anderson, H. M. Scott, T. R. Callaway, and D. J. Nisbet. 2003. Use of a porcine-derived, defined culture of commensal bacteria as an alternative to antibiotics used to control *E. coli* disease in weaned pigs, p. 72–74. *In* R. Ball (ed.), *Proceedings of the 9th International Symposium on Digestive Physiology in Pigs*. Department of Agriculture, Food and Nutritional Science, University of Alberta, Edmonton, Alberta, Canada.

Hastein, T., B. Hjeltnes, A. Lillehaug, J. Utne Skare, M. Berntssen, and A. K. Lundebye. 2006. Food safety hazards that occur during the production stage: challenges for fish farming and the fishing industry. *Rev. Sci. Tech.* **25**:607–625.

Hijnen, W. A., E. F. Beerendonk, and G. J. Medema. 2006. Inactivation credit of UV radiation for viruses, bacteria and protozoan (oo)cysts in water: a review. *Water Res.* **40**:3–22.

Hooper, D. C. 2001. Mechanisms of action of antimicrobials: focus on fluoroquinolones. *Clin. Infect. Dis.* **32**:9–15.

House, J. K., M. M. Ontiveros, N. M. Blackmer, E. L. Dueger, J. B. Fitchhorn, G. R. McArthur, and B. P. Smith. 2001. Evaluation of an autogenous *Salmonella* bacterin and a modified live *Salmonella* serotype Cholaraesuis vaccine on a commercial dairy farm. *Am. J. Vet. Res.* **62**:1897–1902.

Hovde, C. J., P. R. Austin, K. A. Cloud, C. J. Williams, and C. W. Hunt. 1999. Effect of cattle diet on *Escherichia coli* O157:H7. *J. Food Prot.* **61**:802–807.

Huff, W. E., G. R. Huff, N. C. Rath, J. M. Balog, H. Xie, P. A. Moore, Jr., and A. M. Donoghue. 2002. Prevention of *Escherichia coli* respiratory infection in broiler chickens with bacteriophage (SPR02). *Poult. Sci.* **81**:437–441.

Ibekwe, A. M., and C. M. Grieve. 2004. Changes in developing plant microbial community structure as affected by contaminated water. *FEMS Microbiol. Ecol.* **48**:239–248.

ICMSF. 2005. *Microorganisms in Foods 6. Microbial Ecology of Food Commodities*, 2nd ed. Aspen Publishers, Gaithersburg, MD.

Isaacson, R. E., M. Torrence, and M. R. Buckley. 2005. Preharvest food safety and security. Report from the American Academy of Microbiology, Washington, DC. http://www.asm.org/ASM/files/ccLibraryFiles/FILENAME/000000001318/ASM-Preharvest%20Food.pdf. 28 March 2007.

Jay, M. T., M. Cooley, D. Carychao, G. W. Wiscomb, R. A. Sweitzer, L. Crawford-Miksza, J. A. Farrar, D. K. Lau, J. O'Connell, A. Millington, R. V. Asmundson, E. R. Atwill, and R. E. Mandrell. 2007. *Escherichia coli* O157:H7 in feral swine near spinach fields and cattle, Central California Coast. *Emerg. Infect. Dis.* **13**:1908–1911.

Joerger, R. D. 2003. Alternatives to antibiotics: bacteriocins, antimicrobial peptides and bacteriophages. *Poult. Sci.* **82**:640–647.

Johannessen, G. S., R. B. Froseth, L. Solemdal, J. Jarp, Y. Wasteson, and L. M. Rorvik. 2004. Influence of bovine manure as fertilizer on the bacteriological quality of organic Iceberg lettuce. *J. Appl. Microbiol.* **96**:787–794.

Johannessen, G. S., G. B. Bengtsson, B. T. Heier, S. Bredholt, Y. Wasteson, and L. M. Rorvik. 2005. Potential uptake of *Escherichia coli* O157:H7 from organic manure into crisphead lettuce. *Appl. Environ. Microbiol.* **71**:2221–2225.

Jordi, B. J. A. M., K. Boutaga, C. M. E. van Heeswijk, F. van Knapen, and L. J. A. Lipman. 2001. Sensitivity of shiga toxin-producing *Escherichia coli* (STEC) strains for colicins under different experimental conditions. *FEMS Microbiol. Lett.* **204**:329–344.

Jung, Y. S., R. C. Anderson, J. A. Byrd, T. S. Edrington, R. W. Moore, T. R. Callaway, J. L. McReynolds, and D. J. Nisbet. 2003. Reduction of *Salmonella typhimurium* in experimentally challenged broilers by nitrate adaptation and chlorate supplementation in drinking water. *J. Food Prot.* **66**:660–663.

Khamboonruang, C., R. Keawvichit, K. Wongworapat, S. Suwanrangsi, M. Hongpromyart, K. Sukhawat, K. Tonguthai, and C. A. Lima dos Santos. 1997. Application of hazard analysis critical control point (HACCP) as a possible control measure for *Opisthorchis viverrini* infection in cultured carp *(Puntius gonionotus)*. *Southeast Asian J. Trop. Med. Public Health* **28**:65–72.

Koelsch, R. K., J. C. Lorimor, and K. R. Mankin. 2006. Vegetative treatment systems for management of open lot runoff: review of literature. *Appl. Eng. Agric.* **22**:141–154.

Kudva, I. T., C. W. Hunt, C. J. Williams, U. M. Nance, and C. J. Hovde. 1997. Evaluation of dietary influences on *Escherichia coli* O157:H7 shedding by sheep. *Appl. Environ. Microbiol.* **63**:3878–3886.

Kudva, I. T., S. Jelacic, P. I. Tarr, P. Youderian, and C. J. Hovde. 1999. Biocontrol of *Escherichia coli* O157 with O157-specific bacteriophages. *Appl. Environ. Microbiol.* **65**:3767–3773.

Kyriakis, S. C., V. K. Tsiloyiannis, J. Vlemmas, K. Sarris, A. C. Tsinas, C. Alexopoulos, and L. Jansegers. 2001. The effect of probiotic LSP 122 on the control of post-weaning diarrhea syndrome of piglets. *Res. Vet. Sci.* **67**:223–238.

Lakey, J. H., and S. L. Slatin. 2001. Pore-forming colicins and their relatives, p. 131–161. *In* F.G. Van Der Goot (ed.), *Current Topics in Microbiology and Immunology. Pore-Forming Toxins*, vol. 257. Springer-Verlag, Berlin, Germany.

Lee, S. A., A. Adhikari, S. A. Grinshpun, R. McKay, R. Shukla, and T. Reponen. 2006. Personal exposure to airborne dust and microorganisms in agricultural environments. *J. Occup. Environ. Hygiene* **3**:118–130.

LeJeune, J. T., T. E. Besser, D. H. Rice, J. L. Berg, R. P. Stillborn, and D. D. Hancock. 2004. Longitudinal study of fecal shedding of *Escherichia coli* O157:H7 in feedlot cattle: predominance and persistence of specific clonal types despite massive cattle population turnover. *Appl. Environ. Microbiol.* **70**:377–385.

Lema, M., L. Williams, and D. R. Rao. 2001. Reduction of fecal shedding of enterohemorrhagic *Escherichia coli* O157:H7 in lambs by feeding microbial feed supplement. *Small Ruminant Res.* **39**:31–39.

Linton, R. H. 2007. A HACCP approach for aquaculture products. Bulletin, Department of Food Science, Purdue University, West Lafayette, IN. http://aquagenic.org/publicat/state/il-in/ces/linton.pdf. Accessed January 2008.

Loncarevic, S., G. S. Johannessen, and L. M. Rorvik. 2005. Bacteriological quality of organically grown lettuce in Norway. *Lett. Appl. Micrtobiol.* **41**:186–189.

Loneragan, G. H., and M. M. Brashears. 2005. Pre-harvest interventions to reduce carriage of *E. coli* O157 by harvest-ready feedlot cattle. *Meat Sci.* **71**:72–78.

Lupin, H. M. 1999. Producing to achieve HACCP compliance of fishery and aquaculture products for export. *Food Control* **10**:267–275.

Magnuson, B. A., M. Davis, S. Hubele, P. R. Austin, I. T. Kudva, C. J. Williams, C. W. Hunt, and C. J. Hovde. 2000. Ruminant gastrointestinal cell proliferation and clearance of *Escherichia coli* O157:H7. *Infect. Immun.* **68**:3808–3814.

Mara, D., and S. Cairncross. 1989. Guidelines for the safe use of wastewater and excreta in agriculture and aquaculture:

measures for public health protection. World Health Organization, Geneva, Switzerland. http://www.who.int/water_sanitation_health/wastewater/wastreusexecsum.pdf. Accessed 25 May 2007.

Mechie, S. C., P. A. Chapman, and C. A. Siddons. 1997. A fifteen month study of *Escherichia coli* O157:H7 in a dairy herd. *Epidemiol. Infect.* 118:17–25.

Meerburg, B. G., M. Bonde, F. W. A. Brom, S. Endepols, A. N. Jensen, H. Leirs, J. Lodal, G. R. Singleton, H. J. Pelz, T. B. Rodenburg, and A. Kijlstra. 2004. Towards sustainable management of rodents in organic animal husbandry. *NJAS Wageningen J. Life Sci.* 52:195–206.

Mosenthin, R., and E. Bauer. 2000. The potential use of prebiotics in pig nutrition. *Asian Australas. J. Anim. Sci.* 13:315–325.

Moxley, R. A., D. Smith, T. J. Klopfenstein, G. Erickson, J. Folmer, C. Macken, S. Hinkley, A. Potter, and B. Finlay. 2003. Vaccination and feeding a competitive exclusion product as intervention strategies to reduce the prevalence of *Escherichia coli* O157:H7 in feedlot cattle, p. 23. *In Proceedings of the 5th International Symposium on Shiga Toxin-producing* Escherichia coli *Infections.* Kyoto, Japan.

Mukherjee, A., D. Speh, E. Dyck, and F. Diez-Gonzales. 2004. Preharvest evaluation of coliforms, *Escherichia coli, Salmonella,* and *Escherichia coli* O157:H7 in organic and conventional produce grown by Minnesota farmers. *J. Food Prot.* 67: 894–900.

NACMCF. 1999. Microbiological safety evaluations and recommendations on sprouted seeds. *Int. J. Food Microbiol.* 52:123–153.

Natvig, E. E., S. C. Ingham, B. H. Ingham, L. R. Cooperband, and T. R. Roper. 2002. *Salmonella enterica* serovar Typhimurium and *Escherichia coli* contamination of root and leaf vegetables grown in soils with incorporated bovine manure. *Appl. Environ. Microbiol.* 68:2737–2744.

Nielsen, E. M., M. N. Skov, J. J. Madsen, J. Lodal, J. B. Jespersen, and D. L. Baggesen. 2004. Verocytotoxin-producing *Escherichia coli* in wild birds and rodents in close proximity to farms. *Appl. Environ. Microbiol.* 70:6944–6947.

Nurmi, E., L. Nuotio, and H. Ito. 1992. The competitive exclusion concept: development and future. *Int. J. Food Microbiol.* 15:237–240.

Ouwenhand, A. C., P. V. Kirjavainen, C. Shortt, and S. Salminen. 1999. Probiotics: mechanisms and established effects. *Int. Dairy J.* 9:43–52.

Potter, A. A., S. Klashinsky, Y. Li, E. Frey, H. Townsend, D. Rogan, G. Erickson, S. Hinkley, T. Klopfenstein, R. A. Moxley, D. R. Smith, and B. B. Finlay. 2004. Decreased shedding of *Escherichia coli* O157:H7 by cattle following vaccination with type III secreted proteins. *Vaccine* 22:362–369.

Prazak, A. M., E. A. Murano, I. Mercado, and G. R. Acuff. 2002. Prevalence of *Listeria monocytogenes* during production of and post-harvest processing of cabbage. *J. Food Prot.* 65: 1728–1734.

Proux, K., R. Cariolet, P. Fravalo, C. Houdayer, A. Keranflech, and F. Madec. 2001. Contamination of pigs by nose-to-nose contact or airborne transmission of *Salmonella* Typhimurium. *Vet. Res.* 32:591–600.

Rangarajan, A., E. A. Bihn, R. B. Gravani, D. L. Scott, and M. P. Pritts. 2000. *Food Safety Begins on the Farm: a Grower's Guide.* Good Agricultural Practices (GAPs) Publications, College of Agricultural and Life Sciences, Cornell University, Ithaca, NY. http://www.sfc.ucdavis.edu/pubs/articles/foodsafetybeginsonthefarm.pdf. Accessed 13 November 2007.

Ransom, J. R., K. E. Belk, J. N. Sofos, J. A. Scanga, M. L. Rossman, G. C. Smith, and J. D. Tatum. 2003. Investigation of on-farm management practices as pre-harvest beef microbiological interventions. Final Report, National Cattlemen's Beef Association,

Centennial, CO. http://www.beefusa.org/uDocs/ACF3A9B.pdf. Accessed 27 March 2007.

Redman, J. A., S. B. Grant, T. M. Olson, and M. K. Estes. 2001. Pathogen filtration, heterogeneity, and the potable reuse of wastewater. *Environ. Sci. Technol.* 35:1798–1805.

Robertson, L. J., and B. Gjerde. 2000. Isolation and enumeration of *Giardia* cysts, *Cryptosporidium* oocysts, and *Ascaris* eggs from fruit and vegetables. *J. Food Prot.* 63:775–778.

Roth, E., and H. Rosenthal. 2006. Fisheries and aquaculture industries involvement to control product health and quality safety to satisfy consumer-driven objectives on retail markets in Europe. *Mar. Pollut. Bull.* 53:599–605.

Russell, W. C., and M. Benk. 1999. Animal viruses, p. 14–21. *In* A. Granoff and R.G. Webster (ed.), *Encyclopedia of Virology,* 2nd ed. Academic Press, London, United Kingdom.

Santamaria, J., and G. A. Toranzos. 2003. Enteric pathogens and soil: a short review. *Int. Microbiol.* 6:5–9.

Sargeant, J. M., D. J. Hafer, J. R. Gillespie, R. D. Oberst, and S. J. Flood. 1999. Prevalence of *Escherichia coli* O157:H7 in whitetailed deer sharing rangeland with cattle. *J. Amer. Vet. Med. Assoc.* 215:792–794.

Schamberger, G. P., and F. Diez-Gonzalez. 2002. Selection of recently isolated colicingenic *Escherichia coli* strains inhibitory to *Escherichia coli* O157:H7. *J. Food Prot.* 65:1381–1387.

Schijven, J., P. Berger, and I. Miettinen. 2003. Removal of pathogens, surrogates, indicators, and toxins using riverbank filtration, p. 73–116. *In* C. Ray, G. Melin, and R. B. Linsky (ed.), *Riverbank Filtration: Improving Source-Water Quality.* Kluwer Academic Publisher, Norwell, MA.

Schrader, K. K., N. P. Nanayakkara, C. S. Tucker, A. M. Rimando, M. Ganzera, and B. T. Schaneberg. 2003. Novel derivatives of 9,10-anthraquinone are selective algicides against the mustyodor cyanobacterium *Oscillatoria perornata. Appl. Environ. Microbiol.* 69:5319–5327.

Schrezenmeir, J., and M. De Vrese. 2001. Probiotics, prebiotics, and synbiotics—approaching a definition. *Am. J. Clin. Nutr.* 73:354–361.

Smith, H. W., and M. B. Huggins. 1982. Successful treatment of experimental *Escherichia coli* infection in mice using phage: its general superiority over antibiotics. *J. Gen. Microbiol.* 128: 307–318.

Solomon, E. B., C. J. Potenski, and K. R. Matthews. 2002. Effect of irrigation method on transmission to and persistence of *Escherichia coli* O157:H7 on lettuce. *J. Food Prot.* 65:673–676.

Sproston, E. L., M. Macrae, I. D. Ogden, M. J. Wilson, and N. J. Strachan. 2006. Slugs: potential novel vectors of *Escherichia coli* O157. *Appl. Environ. Microbiol.* 72:144–149.

Steer, T., H. Carpenter, K. Tuohy, and G. R. Gibson. 2000. Perspectives on the role of the human gut microbiota and its modulation by pro and prebiotics. *Nutr. Res. Rev.* 13:229–254.

Stewart, V. J. 1988. Nitrate respiration in relation to facultative metabolism in enterobacteria. *Microbiol. Rev.* 52:190–232.

Stopforth, J. D., and J. N. Sofos. 2006. Recent advances in pre- and postslaughter intervention strategies for control of meat contamination, p. 66–86. *In* V. K. Juneja, J. P. Cherry, and M. H. Tunick (ed.), *Advances in Microbial Food Safety,* American Chemical Society, Washington, DC.

Suslow, T. V., M. P. Oria, L. R. Beuchat, E. H. Garrett, M. E. Parish, L. J. Harris, J. N. Farber, and F. F. Busta. 2003. Production practices as risk factors in microbial food safety of fresh and fresh-cut produce. *Comp. Rev. Food Sci. Food Saf.* 2S:38–77.

Tate, K. W., J. W. Bartolome, and G. Nader. 2006. Significant *Escherichia coli* attenuation by vegetative buffers on annual grasslands. *J. Environ. Qual.* 35:795–805.

Tkalcic, S., T. Zhao, B. G. Harmon, M. P. Doyle, C. A. Brown, and P. Zhao. 2003. Fecal shedding of enterohemorrhagic

Escherichia coli in weaned calves following treatment with probiotic *Escherichia coli*. *J. Food Prot.* **66:**1184–1189.

Troutt, H. F., J. Gillespie, and B. I. Osburn. 1999. Implementation of HACCP program on farms and ranches, p. 36–51. *In* A. M. Pearson and T. R. Dutson (ed.), *HACCP in Meat, Poultry, and Fish Processing*, vol. 10. Springer, New York, NY.

USDA. 2001. Fruit and vegetable agricultural practices—1999. National Agricultural Statistics Service, U.S. Department of Agriculture, Washington, DC. http://usda.nass/pubs/rpts106.htm. Accessed 15 May 2007.

Wachtel, M. R., L. C. Whitehand, and R. E. Mandrell. 2002. Association of *Escherichia coli* O157:H7 with preharvest leaf lettuce upon exposure to contaminated irrigation water. *J. Food Prot.* **65:**18–25.

Warriner, K. 2005. Pathogens in vegetables, p. 3–43. *In* W. Jongen (ed.), *Improving the Safety of Fresh Fruit and Vegetables.* Woodhead Publishing Ltd., Cambridge, United Kingdom.

Weese, J. S., and J. Rousseau. 2006. Survival of *Salmonella* Copenhagen in food bowls following contamination with experimentally inoculated raw meat: effects of time, cleaning, and disinfection. *Can. Vet. J.* **47:**887–889.

WHO. 1999. Food safety issues associated with products from aquaculture. WHO Technical Report Series (883), Report of a joint FAO/NACA/WHO Study Group, World Health Organization, Geneva, Switzerland. http://www.who.int/foodsafety/publications/fs_management/en/aquaculture.pdf. Accessed 4 April 2007.

Wiemann, M. 2003. How do probiotic feed additives work? *Int. Poultry Prod.* **11:**7–9.

Willard, M. D., R. B. Simpson, N. D. Cohen, and J. S. Clancy. 2000. Effects of dietary fructooligosaccharide on selected bacterial populations in feces of dogs. *Am. J. Vet. Res.* **61:**820–825.

Zhao, T., S. Tkalcic, M. P. Doyle, B. G. Harmon, C. A. Brown, and P. Zhao. 2003. Pathogenicity of enterohemorrhagic *Escherichia coli* in neonatal calves and evaluation of fecal shedding by treatment with probiotic *Escherichia coli*. *J. Food Prot.* **66:**924–930.

Zhao, Y. 2005. Pathogens in fruit, p. 44–88. *In* W. Jongen (ed.), *Improving the Safety of Fresh Fruit and Vegetables.* Woodhead Publishing Ltd., Cambridge, United Kingdom.

Chapter 24

Interventions for Hazard Control in Foods during Harvesting

MAYRA MÁRQUEZ-GONZÁLEZ, KERRI B. HARRIS, AND ALEJANDRO CASTILLO

A large number of cases and outbreaks of disease associated with contaminated food are reported every year worldwide. An increased occurrence of food-borne disease seems to be associated with changes in the food supply, lifestyles, and dietary habits (Collins, 1997). Today, greater amounts of food are produced, and the number of people potentially exposed if the product is contaminated with a food safety hazard may result in a large outbreak. Many foods are manufactured as ready-to-eat products in order to satisfy the needs of busy consumers. Uncooked ready-to-eat products, such as fresh-cut fruits and vegetables, commonly have an increased surface area as well as receive more handling; therefore, the chances for contamination are greater. An increase in the number of outbreaks associated with the consumption of fresh fruits and vegetables has been reported in association with an increase in the consumption of fresh fruits and vegetables (Sivapalasingam et al., 2004).

According to the Centers for Disease Control and Prevention (CDC), raw meat and poultry, raw eggs, unpasteurized milk, and raw shellfish are most likely to be contaminated. The CDC also states that there is a concern with fruits and vegetables consumed raw (CDC, 2005). Bacterial pathogens were responsible for 55% of the food-borne outbreaks which occurred between 1998 and 2002 for which the etiology was determined, whereas virus, chemical agents, and parasites caused 33, 10, and 1% of these outbreaks, respectively (CDC, 2000, 2006). The causal agent associated with each of the food categories is listed in Table 1.

In order to reduce the incidence of food-borne illness, regulatory agencies in the United States have mandated food safety regulations. Many of these regulations are designed to improve the safety of foods

by reducing the occurrence and number of pathogenic microorganisms. Currently, the United States has regulations that mandate hazard analysis and critical control point programs for fish and seafood (U.S. FDA, 1995a), meat and poultry products (USDA/FSIS, 1996b), and juices (U.S. FDA, 2001a). There are other guidance documents, such as for fresh fruits and vegetables (U.S. FDA, 1998a) and for eggs (U.S. FDA, 2000a), that also are related to the safety of food products.

Product contamination may occur at different stages during production; therefore, several intervention steps have been developed to minimize contamination. The interventions for hazard control in foods during preharvest are discussed in chapter 23. Different methods are currently being used for hazard control in foods during harvesting. These methods include physical and chemical processes intended to reduce pathogen populations in or on foods. Most of the methods discussed here have been evaluated under laboratory conditions, measuring in some cases the reduction in total bacterial counts as a prediction of pathogen reduction. However, it should be noted that this approach may not be accurate when developing a treatment for further use as a measure for pathogen control. This chapter will discuss the interventions applied during harvest of different food commodities.

FRESH MEATS

The National Advisory Committee on Microbiological Criteria for Foods (NACMCF) and the U.S. Department of Agriculture (USDA) identified the receiving and holding of cattle and dehiding and evisceration as the major sites of contamination during the harvest of beef cattle. Additional contamination

Mayra Márquez-González, Kerri B. Harris, and Alejandro Castillo • Department of Animal Science, Texas A&M University, College Station, TX 77843-2471.

Table 1. Principal food categories associated with causal agents of food-borne outbreaks that occurred in the United States from 1998 through 2002[a]

Food category	Causal agent			
	Bacterial	Chemical	Parasitic	Viral
Seafood	*Bacillus cereus*	Ciguatoxin		HAV
	Campylobacter	Scombrotoxin		Norovirus
	Clostridium botulinum	Shellfish toxin		
	Salmonella, Shigella			
	Vibrio			
Beef and pork	*Clostridium perfringens*			Norovirus
	Escherichia coli			
	Listeria monocytogenes			
	Salmonella, Shigella			
	Staphylococcus aureus			
	Yersinia enterocolitica			
Poultry	*Bacillus cereus*			Norovirus
	Campylobacter			
	Clostridium perfringens			
	Escherichia coli			
	Listeria monocytogenes			
	Salmonella, Shigella			
	Staphylococcus aureus			
	Streptococcus, Vibrio parahaemolyticus			
Fruits and vegetables	*Bacillus cereus*	Heavy metals	*Cyclospora cayetanensis*	Norovirus
	Campylobacter	Mushroom toxin		HAV
	Clostridium botulinum			
	Clostridium perfringens			
	Escherichia coli			
	Salmonella, Shigella			
	Staphylococcus aureus			
Dairy	*Brucella*			HAV
	Campylobacter, E. coli			Norovirus
	Listeria monocytogenes			
	Salmonella			
Eggs	*Clostridium perfringens*			
	Salmonella, Shigella			

[a]Data summarized from CDC (2006).

may occur during splitting of the carcasses and in further meat fabrication, such as trimming and grinding (USDA-NACMCF, 1993). Current interventions applied in order to reduce hazards during the harvesting of beef cattle include physical and chemical treatments. These treatments are applied during harvesting and pre- and postevisceration and at the end of the process.

Dehairing

Chemical dehairing is a decontamination process that involves the removal of hair from animals by using a sodium sulfide solution followed by a rinse with a hydrogen peroxide solution and water prior to dehiding (Bowling and Clayton, 1992). Schnell et al. (1995) reported that dehairing with a 10% sodium sulfide solution after exsanguinating enhanced the visual cleanliness of the carcasses, but

the efficacy in reducing aerobic plate counts (APCs) or coliform counts was limited. Castillo et al. (1998a) reported reductions of 3.4, 3.9, and >4.3 log CFU/cm^2 of APCs, coliforms, and *Escherichia coli*, respectively, on inoculated bovine skin pieces by using a 10% sodium sulfide solution. Counts of *E. coli* O157:H7 and *Salmonella enterica* serovar Typhimurium artificially inoculated on the hides were also reduced by >4.6 log CFU/cm^2 after chemical dehairing. With an in-plant test of cattle dehairing before slaughter, Nou et al. (2003) reported the APCs and *Enterobacteriaceae* counts on carcasses obtained from dehaired cattle to be 1.8 or 2 log CFU/100 cm^2 lower than those obtained from the untreated cattle. Also, the prevalence of *E. coli* O157:H7 was significantly lower on hides and carcasses obtained from treated animals than that in control hides and carcasses.

Carcass contamination may occur during the evisceration process. Removal of intact viscera reduces spread of contamination (Gill, 2005). Regardless of where the point of contamination occurs, further decontamination steps may include cleaning methods, such as trimming, water washing, and steam vacuuming, as well as sanitizing methods by organic acids rinses, hot water rinse, steam pasteurization, trisodium phosphate treatments, or other treatments. The effect on bacterial reduction of these treatments has been tested individually and as combined interventions (Castillo et al., 2002; USDA-NACMCF, 1993).

Knife Trimming

The USDA's Food Safety and Inspection Service (FSIS) has a long-standing policy of zero tolerance, prohibiting visible feces, ingesta, or milk on carcasses. The approved methods to meet this requirement are knife trimming and vacuuming beef carcasses with hot water or steam, when such contamination is less than 1 square inch (USDA-FSIS, 1996a). Removal of physical contamination by trimming has been reported to reduce bacterial counts in beef carcasses (Gorman et al., 1995a; Hardin et al., 1995). By using knife trimming alone, Gorman et al. (1995a) obtained reductions in APCs and E. coli counts of 2.0 and 2.2 log CFU/cm^2 on inoculated beef brisket. In another study, populations of E. coli O157:H7 and S. Typhimurium inoculated on beef carcass surfaces were reduced by 4.4 and 3.9 log CFU/cm^2, respectively, by knife trimming under laboratory conditions (Hardin et al., 1995). Under similar laboratory conditions, Castillo et al. (1998b) obtained reductions of 3.3 and 3.6 log CFU/cm^2 in S. Typhimurium and E. coli O157:H7 inoculated on different beef carcass regions. In contrast, in a study conducted at a beef packing plant, Gill et al. (1996) found no differences in total bacterial and E. coli counts from carcasses before or after trimming. The reductions obtained by trimming were enhanced by subsequent application of treatments, such as washing with automated spray washers (Delmore et al., 1997; Reagan et al., 1996) and hot water wash or organic acid sprays (Castillo et al., 1998b; Hardin et al., 1995). Reagan et al. (1996) also reported a significant reduction of the incidence of Listeria spp. from 43.7% to 12.6% and Salmonella spp. from 30.3% to 1.4% on beef carcasses after knife trimming of visible fecal contamination followed by spray washing (28 to 42°C at 410 to 2,758 kPa and washing times of 18 to 39 s).

Water Wash

Water wash is a common practice during beef slaughter to remove bone dust, blood, hair, and other physical compounds to improve the visual quality of carcasses. The efficacy of water wash in reducing physical contamination depends upon factors such as water pressure, water flow rate, and speed of movement of meat through spray (Anderson et al., 1975). The impact of carcass washing in regard to microbial load has been studied extensively, and various results for the efficacy of such treatments have been reported. Some authors reported that increasing water wash pressure enhanced the reduction of bacterial populations (Anderson et al., 1981; Hardin et al., 1995; Kotula et al., 1974), while others reported that no significant differences were obtained by increasing the pressure of water wash treatments (Crouse et al., 1988; DeZuniga et al., 1991).

Kotula et al. (1974) reported the effect of washing beef forequarters by using two different pressures and different temperatures on the water wash. These authors obtained reductions of APCs when applying a high pressure wash (2.4 MPa) that were 1.2 log CFU/cm^2 larger than the reductions with a low pressure wash (0.5 MPa). These authors did not find any effect of the water temperature (51.7°C versus 12.8°C) at 2.4 MPa. At 0.5 MPa, larger reductions (1.4 log CFU/cm^2) were reported for washing the forequarters with water at 51.7°C than for those at 12.8°C. Anderson et al. (1981) compared the reduction of APCs on beef carcass halves after hand washing with tap water at 15°C and the reduction of APCs obtained with an experimental beef carcass washing unit. Overall, reductions of 1.0 and 1.1 log CFU/cm^2 were obtained by hand washing by applying water at 1.2 MPa for 1.7 min at a rate of 31 liters/min and by using an automated washing unit at 2.1 MPa for 15 s at a rate of 290 liter/min, respectively (Anderson et al., 1981). Hardin et al. (1995) inoculated different carcass surface regions with a fecal slurry containing E. coli O157:H7 and S. Typhimurium. After removing the gross fecal matter using a hand sprayer, an automated pressure wash (5 liters, 1.7 MPa for 5 s at 35°C followed by 2.8 MPa for 4 s) was applied to the inoculated carcass regions. This type of wash resulted in reductions between 2.0 and 3.5 log CFU/cm^2.

Crouse et al. (1988) evaluated the reduction of APCs and Enterobacteriaceae counts in beef carcasses after washing with water at 13°C using different pressures and different chain speeds. There was no significant difference in reduction numbers of Enterobacteriaceae on beef carcasses, as affected by the spray pressure or the chain speed. DeZuniga et al. (1991) also reported no significant differences due to pressure for the removal of Enterobacteriaceae from inoculated meat surfaces. De Zuniga et al. (1991) addressed the effect of a high-pressure water wash on the penetration of bacteria on meat surfaces by using an insoluble dye (blue lake). This study determined that pressures above 4.1 MPa had a significant effect on penetration of blue lake and concluded that bacteria might penetrate into the meat as a result of such high pressure.

Dorsa et al. (1996b) studied the effect of using hot water to wash the surface of beef carcasses artificially inoculated with feces. A reduction of APCs near 3.0 log CFU/cm^2 was obtained when 82.2°C water was delivered to the carcass surface. Gorman et al. (1995b) reported APCs and *E. coli* reductions of 3.0 and 3.3 log CFU/cm^2, respectively, on inoculated beef brisket adipose tissue after spray washing with hot water (74°C) for 12 s at 2,068 kPa. Gill et al. (1995) obtained *E. coli* reductions of 2.5 log CFU/cm^2 on whole pig carcasses treated with water at 85°C for 20 s. Castillo et al. (1998c) reported reductions for *E. coli* O157:H7 and *S.* Typhimurium of 3.7 and 3.8 log CFU/cm^2, respectively, when a hot-water treatment (97°C at 165 kPa for 5 s) was applied.

Steam Vacuum

Vacuuming with hot water or steam is a method accepted by FSIS that can be used to remove visible fecal, ingesta, and milk contamination from beef carcasses, when such contamination is less than 1 square inch (USDA/FSIS, 1996a). The steam vacuum system delivers hot water (>82.2°C) and steam directly onto the carcass surface while removing physical contamination with a vacuum. Kochevar et al. (1997) reported that the use of steam vacuum systems was at least as effective as knife trimming for decontaminating beef carcasses with visible fecal contamination of a dimension no greater than 1 square inch. These authors reported reductions of APCs and total coliform counts up to 2.0 and 2.1 log CFU/cm^2, respectively, on surface carcasses with visible fecal contamination after steam vacuuming. Dorsa et al. (1996b) inoculated beef carcasses with fecal slurry, and after double washing them with water at 72 and 30°C, the carcasses were treated with the steam vacuum. They did not indicate differences in APCs between the water wash alone and the water wash followed by steam vacuum. In a different study, the steam vacuum system reduced *E. coli* O157:H7 by 5.5 log CFU/cm^2 when inoculated on beef plates (Dorsa et al., 1996a). In a separate study, Dorsa et al. (1997) inoculated beef carcasses with fecal slurry, and after steam vacuuming, they observed reductions of 1.6, 2.0, 2.0, and 2.1 log CFU/cm^2 for APCs, *Listeria innocua*, lactic acid bacteria, and *E. coli* counts, respectively.

Steam Pasteurization

Steam pasteurization treatments have been used in meat processing. Phebus et al. (1997) treated cutaneous trunci steer muscles inoculated with *E. coli* O157:H7, *S.* Typhimurium, and *Listeria monocytogenes* by applying steam at 99 to 101°C for 15 s, and then the surface of the meat was cooled with chilled water spray at 6 to 8°C. This treatment resulted in a 3.4- to 3.7-log-CFU/cm^2 reduction of these pathogens. Nutsch et al. (1997) conducted commercial evaluations of the steam pasteurization process in a beef processing plant. The carcasses were subjected to a preliminary water wash and then passed through potent air blowers to reduce the chances for condensation during steam application. The carcasses then were passed through a steam chamber and then to another section of the cabinet where cold water was applied. By applying this treatment, APCs were reduced from 2.1 to 2.2 to 0.6 to 0.8 log CFU/cm^2. No differences were observed for APCs between 6 and 8 s of the steam treatment. Counts of *E. coli* were also reduced from original counts of 0.6 to 1.5 log CFU/cm^2 to <0.6 log CFU/cm^2 after 6 or 8 s of steam treatment.

Organic Acids

Organic acids have been used extensively for carcass decontamination, predominantly L-lactic and acetic acid. Extensive research has reported the effectiveness of these two acids in reducing naturally occurring microflora (Anderson et al., 1977; Biemuller et al., 1973; Delmore et al., 2000; Hardin et al., 1995; Ockerman et al., 1974). Anderson et al. (1977) reported reductions of 2.6 logs for counts of viable microorganisms on meat by spraying 3% acetic acid; the reductions were significantly greater than the 0.2-log reduction obtained by spraying hypochlorite solution at 200 to 250 mg/liter. Reductions of pathogenic bacteria by spraying organic acids have also been reported. Total reductions in counts of various pathogens on beef have been reported and vary between 2 and 4.3 log cycles after spraying 2% acetic acid (Dickson, 1991; Dickson and Anderson, 1991; Hardin et al., 1995; Tinney et al., 1997). Variations in reductions between studies may be due to factors such as the temperature (25°C to 55°C), volume of the acid solution applied (0.2 to 0.5 ml/cm^2), and the reduction of the meat surface pH. Brackett et al. (1994) reported acetic, citric, and lactic acid solutions at different concentrations to be unable to reduce *E. coli* O157:H7 on beef sirloin pieces, regardless of the concentration and temperature of the acid solution. These authors reported meat surface pH values ranging from 5.0 to 5.5 after treatment with 0.5, 1.0, and 1.5% acid solutions at 20 and 55°C. In contrast, Anderson and Marshall (1990) reduced the surface pH of beef dipped in 3% lactic acid solution, from 5.6 (untreated meat) to 3.9, and Hardin et al. (1995)

obtained pH values for beef carcass surfaces of 2.6 to 2.8 after spraying with 2% L-lactic acid and of 3.1 to 3.5 after spraying with 2% acetic acid. Even though *E. coli* O157:H7 has been reported to be resistant to low-pH environments, recent studies indicate that L-lactic or acetic acid sprays, when applied at 55°C, effectively reduced levels of *Salmonella* and *E. coli* O157:H7 from ca. 5 log CFU/cm^2 to levels near or below the detection limit of 0.5 log CFU/cm^2 (Castillo et al., 1998b; Hardin et al., 1995).

Lactic acid has been demonstrated to have a strong antibacterial capacity. Hardin et al. (1995) reported that L-lactic acid was more effective than acetic acid in reducing *E. coli* O157:H7 and as effective as acetic acid in reducing *S.* Typhimurium on beef carcass surfaces. Woolthuis et al. (1984) found immersing porcine livers for 5 min in a 0.2% lactic acid solution to be significantly more effective in reducing total bacterial counts and lactic acid bacteria than immersing them in hot water (65°C) for 15 s, whereas *Enterobacteriaceae* counts were reduced at the same rate after both treatments were applied. Prasai et al. (1991) reduced the APCs of beef carcasses at two slaughter plants by ca. 2 log CFU/cm^2 by spraying 1% L-lactic acid at 55°C after dehiding and eviscerating. In addition to lactic and acetic acids, other organic acids have been tested for the ability to reduce bacterial populations on beef (Anderson et al., 1992; Cutter and Siragusa, 1994b; Garcia-Zepeda et al., 1994; Podolak et al., 1996; Reynolds and Carpenter, 1974).

By using a combination of decontamination treatments, the reductions of microbial populations obtained are higher compared to those of individual treatments. Bacon et al. (2000) reported reductions ranging from 2.2 to 5.4, 2.5 to 5.9, and 4.3 to 6.0 log CFU/100 cm^2 for total plate counts, total coliform counts, and *E. coli* counts, respectively, on beef carcasses after applying multiple sequential interventions at eight commercial slaughter plants. Reductions in the prevalence of *Salmonella* from 15.4% to 1.3% have also been reported by Bacon et al. (2002), after applying combined treatments at commercial slaughter plants. The interventions at some or all the plants included (i) steam vacuuming (104 to 110°C, 138 to 345 kPa steam, -7 to 12 mm of Hg vacuum); (ii) pre-evisceration carcass washing (29 to 38°C water at 193 to 331 kPa, 6 to 8 s); (iii) pre-evisceration application of organic acid solution rinsing (1.6 to 2.6% lactic or acetic acid solution, 43 to 60°C, 317 to 324 kPa, 2 to 4 s); (iv) thermal pasteurizing (71 to 77°C water, 69 to 228 kPa, 10 to 14 s); (v) final carcass washing (16 to 32°C water, 483 to 897 kPa, 10 to 14 s); and (vi) postevisceration application of organic acid solution rinsing (1.6 to 2.6% lactic or acetic acid solution, 43 to 60°C, 317 to 324 kPa, 2 to 4 s) (Bacon et al., 2000, 2002).

Other Antimicrobials

Several chemicals with antimicrobial activity have been tested for usefulness as carcass sanitizers. Chorine and chlorine-based compounds have been approved and are used for decontamination of meat surfaces. Although meats treated with chlorine have lower bacterial counts (Cutter and Siragusa, 1995a; Emswiler et al., 1976), other chemicals have been shown to be more effective against pathogenic bacteria than chlorine. Acidified sodium chlorite (ASC) was reported to reduce *E. coli* O157:H7 and *S.* Typhimurium on inoculated beef carcasses by 3.8- to 4.6 log CFU/cm^2 (Castillo et al., 1999). The highest reductions were obtained by acidifying ASC with citric acid (4.5- to 4.6-log-CFU/cm^2 reductions) rather than by acidifying with phosphoric acid (3.8- to 3.9-log-CFU/cm^2 reductions). The U.S. Food and Drug Administration (FDA) approved the use of ASC for meat processing (U.S. FDA, 1998c, 2000c). Cetylpyridinium chloride (CPC) is another compound that has been reported to reduce pathogenic bacteria on beef. Cutter et al. (2000) reported reductions of 5 to 6 log CFU/cm^2 of *E. coli* O157:H7 and *S.* Typhimurium on lean beef surfaces sprayed with 1% CPC for 15 s. Although CPC has been approved to treat the surface of raw poultry carcasses prior to immersion in a chiller, CPC has not yet been approved for use as an antimicrobial in beef production. Chlorine dioxide at concentrations up to 20 mg/liter has been reported to be no more effective than regular water in reducing bacteria of fecal origin on beef carcass tissue (Cutter and Dorsa, 1995).

Trisodium phosphate (TSP) has been tested by dipping and spraying methods on inoculated beef surfaces with pathogens such as *S.* Typhimurium, *Listeria monocytogenes*, *E. coli* O157:H7, and non-O157:H7 enterohemorrhagic *E. coli*. Reductions between 0.8 and 3 log CFU/cm^2 have been reported (Cutter and Rivera-Betancourt, 2000; Dickson et al., 1994; Kim and Slavik, 1994). Delmore et al. (2000) reported reductions of naturally occurring bacteria on beef variety meats by dipping or spraying with TSP (12%, pH 12.5, and temperature of 40 to 50°C for 10 s). These authors reported 0.7- to 1.2-log-CFU/g reductions on APCs, 0.1- to 2.6-log-CFU/g reductions on total coliform counts, and 0.2- to 2.2-log-CFU/g reductions on *E. coli* counts.

Nisin has been shown to reduce bacteria from meat tissues. Cutter and Siragusa (1994a) reported reductions of 2 to 3.6 log CFU/cm^2 for *Brochothrix thermosphacta* counts and 2.8 log CFU/cm^2 or greater for *Carnobacterium divergens* counts after spraying lean and adipose beef carcass tissue with nisin solution (5,000 activity units/ml). In a separate study, Cutter and Siragusa (1995b) reported reductions up to 0.4 log CFU/cm^2 in populations of *S.* Typhimurium and *E. coli* O157:H7 on inoculated beef tissue after applying nisin combined with lactate or nisin combined with EDTA. Neither the use of TSP nor the use of nisin has yet been approved for decontamination of meat carcasses. Activated lactoferrin has been reported to achieve higher reductions (99.9%) of *E. coli* O157:H7 counts when beef carcasses were treated for 10 s with 1% activated lactoferrin spray after washing with cold and hot water (82.2°C) and applying 2% lactic acid spray than the 72% reduction obtained when the final lactoferrin spray was excluded (Naidu, 2002). The use of activated lactoferrin for decontamination of beef carcasses is approved by the USDA/FSIS (2002).

FSIS Directive 7120.1 provides a current list of all Safe and Suitable Ingredients Used in the Production of Meat and Poultry Products. The directive provides the approved antimicrobials, the products that they can be applied to the amounts allowed, and any labeling requirements. The directive is updated as new compounds are approved or as the approval for application changes.

Hide Wash

Hide washes with and without sanitizers have been tested for reduction of pathogens and indicator organisms in several studies. Bosilevac et al. (2005a) observed reductions from 44% to 16% in the isolation rate of *E. coli* O157:H7 on cattle hide after stunning after the application of a hide wash using a pilot cabinet in a slaughter establishment. The frequency of this pathogen on carcasses obtained from cattle subjected to hide wash was 2%, significantly lower than the 17% observed on carcasses to which no hide wash was applied. These investigators later improved the hide wash cabinet, tested it at a fed-beef processing plant, and reported the prevalence of *E. coli* O157:H7 and *Salmonella* to have been reduced from 97.6% to 89.6% and from 94.8% to 68.8%, respectively (Arthur et al., 2007). The addition of antimicrobials, such as CPC, ozone, or acidic electrolyzed oxidizing water, to the hide wash to enhance the reduction of pathogens has also been tested (Bosilevac et al., 2004, 2005b). However, in another study no differences in bacterial reductions were observed between water alone, chlorine, or L-lactic acid wash for cattle hides at a commercial establishment (Mies et al., 2004). Reductions of ca. 3 log cycles were reported by Baird et al. (2006) by spraying the hides with CPC, L-lactic acid, and hydrogen peroxide.

Irradiation

The FDA approved the irradiation of refrigerated uncooked meat products at doses not to exceed 4.5 kGy and up to 7 kGy for frozen uncooked meat products for the control of food-borne pathogens (U.S. FDA, 1997). These doses are sufficient to eliminate pathogenic microorganisms in meat products. Thayer (1995) reported that doses <3 kGy can significantly reduce or eliminate pathogens such as *Campylobacter jejuni*, *E. coli* O157:H7, *Staphylococcus aureus*, *Salmonella* spp., *L. monocytogenes*, and *Aeromonas hydrophila*. Arthur et al. (2005) reported reductions ≥4 log CFU/cm^2 in *E. coli* O157:H7 counts after irradiating inoculated chilled beef carcass surfaces (400 cm^2) at a 1-kGy dose. Collins et al. (1996) reported that a dose of 1.5 kGy was sufficient to reduce 5 and 7 log CFU/g of *Arcobacter butzleri* and *C. jejuni*, respectively, on ground pork.

The sensitivity of pathogenic bacteria to irradiation may vary by type of microorganism, the type of meat that is irradiated, and the conditions under which irradiation is carried out (López-González et al., 1999; Thayer et al., 1995). Thayer et al. (1995) reported no significant differences in the thermal resistance levels (*D* values) for *E. coli* O157:H7 and *L. monocytogenes* on five different types of meat tested (beef, lamb, pork, turkey breast, and turkey leg) and no significant differences in the *D* values for *Salmonella* on beef, lamb, and turkey meats. A significantly lower *D* value for *Salmonella* was reported for pork. López-González et al. (1999) observed greater *D* values for *E. coli* O157:H7 in inoculated ground beef patties when irradiation was applied at –15°C compared to 5°C, when the patties were packed in high-oxygen barrier film (0.49 cm^3/100 in^2/24 h/atm), or when the energy source was gamma rays versus electron beams.

POULTRY

The steps in poultry slaughter for which contamination is most likely to occur are the scalding and defeathering points (Abu-Ruwaida et al., 1994). According to Thomas and McMeekin (1984), submerging poultry carcasses in aqueous fluids changes the tissue microtopography, and contamination may occur easily. These authors hypothesized that prolonged immersion times might result in an increased uptake of contaminants by the carcass. Contamination during evisceration has also been reported by other authors (Notermans et al., 1980).

One of the most utilized treatments to reduce bacterial populations in poultry carcasses is the application of cold (20°C) or hot (55 to 60°C) water wash either by aspersion or by immersion. The efficacy of these washes on bacterial reduction is generally related to the level of initial contamination, the temperature used, and the time of exposure. Decontamination procedures include the use of heat and chemical interventions (Hwang and Beuchat, 1995; May, 1974; Morrison and Fleet, 1985; Notermans et al., 1980).

Water Wash, Hot Water, and Steam

The effectiveness of reducing bacterial populations by the spray washing of poultry carcasses with cold water has been evaluated by May (1974) and Notermans et al. (1980). Notermans et al. (1980) evaluated the effect of the application of several spray washes during evisceration of poultry carcasses on APCs and *Enterobacteriaceae* counts. In this study a spray cleaning apparatus consisting of a frame with four adjustable nozzles at a rate of 15 to 20 liters/min was used. An increase in *Enterobacteriaceae* counts up to 1.0 log CFU/g of skin after evisceration was seen when no additional spray cleaners were used, compared to the counts obtained after defeathering. When additional spray cleaners were used during the evisceration process, no significant differences were observed between the *Enterobacteriaceae* counts after defeathering and after the evisceration process.

Increasing the temperature of the scalding water has been determined to have a positive effect on reduction of bacteria on broilers. Avens et al. (2002) studied the effect on APCs caused by immersion of broiler carcasses in boiling water (95°C) at different time periods (0 to 8 min) and exposure to flowing steam (96–98°C) on skin. These investigators reported reductions in APCs from levels of 10^3 and 10^4 CFU/cm^2 to <10^1 CFU/cm^2 both by immersion in boiling water and exposure to flowing steam for 3 and >3 min. Such treatments may cause visible changes on the skin of the broiler carcasses. Therefore, the authors recommend the use of those treatments for skinless processed poultry products. Göksoy et al. (2001) evaluated the effect of different heat treatments and different exposure times on the appearance of vacuum-packaged chicken. They reported that treatment at 60 and 90°C for 60 s and 2 s, respectively, did not result in a significant change in appearance. In the same study, Göksoy et al. (2001) inoculated skin-on chicken pieces with *E. coli* serotype O80 at levels of 6 log CFU/cm^2. These authors reported no significant reductions in counts of *E. coli* serotype O80. Morrison and Fleet (1985) immersed inoculated *Salmonella* chicken carcasses in water at 18 or 60°C for 10 min. They reported reductions up to 2.5 log CFU/carcass, as estimated from counts conducted in a 300-ml rinse, after immersion in water at 60°C.

Decontamination of poultry carcasses by using steam has been reported by James et al. (2000). Using steam at atmospheric pressure (100°C for 10 s), these authors were able to reduce *E. coli* serotype O80 by 1.9 log CFU/cm^2 on inoculated skin-on chicken breast portions. Avens et al. (2002) reduced APCs from 4 log CFU/cm^2 of chicken skin to <1 log CFU/cm^2 after treating chicken carcasses with steam (96 to 98°C) for 3 min.

Other Antimicrobials

In 1995, the FDA approved the use of chlorine dioxide to control the microbial population in poultry process water (U.S. FDA, 1995b). Chlorine dioxide has been reported to be four to seven times more effective at reducing bacterial concentrations on poultry chilling water than chlorine gas at the same concentration (Lillard, 1979). Lillard (1980) evaluated the bactericidal effect of chlorine dioxide in commercial broiler processing plants. Total APCs on chicken breast skin were reduced 0.9 log/g after using chilling water with 5 m/liter of chlorine dioxide.

Tamblyn and Conner (1997) compared the effectiveness of different organic acids in reducing *S.* Typhimurium counts of inoculated chicken skin at different processing conditions (chilling, dipping, and scalding). Scalder applications of 2% and 4% organic acid solutions (50°C, 2 min) achieved *S.* Typhimurium reductions between 0.7 and 2.4 log per skin sample (10-cm diameter).

The effectiveness of ASC in reducing microorganisms naturally occurring on broiler carcasses was evaluated by Kemp et al. (2000). These authors reported reductions of 1.0, 2.3, and 1.9 log CFU/ml on APCs, *E. coli*, and total coliforms, respectively, as estimated from counts conducted in a 400-ml rinse, after total immersion of five carcasses into a container holding 18.9 liters of ASC solution (1,200 mg/liter) for 5 s. The use of ASC is approved in poultry processing by the FDA (U.S. FDA, 1996, 2000d). Spraying 0.1% CPC solution for 1 min on chicken surfaces reduced *S.* Typhimurium by 0.9 and 1.7 log cycles at 15 and 50°C, respectively (Kim and Slavik, 1996). Similar *S.* Typhimurium reductions of 1.0 log CFU/skin square (2.5 by 2.5 cm) were obtained when poultry skin squares were immersed in 2 mg/liter CPC solution for 1 min (Breen et al., 1997). Higher reductions of *S.* Typhimurium up to 4.6 log CFU/skin square (2.5 by 2.5 cm) were obtained by increasing the concentration of CPC to 8 mg/liter and the immersion time to 10 min (Breen et al., 1997). The use of CPC is approved to treat the surface of raw poultry carcasses prior to immersion in a chiller. The FSIS directive 7120.1 (USDA-FSIS, 2005) states that CPC can be used to treat the surface of raw poultry carcasses prior to immersion in a chiller and that it is applied as a fine mist spray of an ambient-temperature aqueous solution. It does not mention a maximum level; the FDA approves its use at a level not to exceed 0.3 g of CPC per pound of raw poultry carcass (U.S. FDA, 2004).

Irradiation

Several studies on the irradiation of poultry carcasses have shown its effectiveness in reducing pathogenic bacteria. A dose of 2.5 kGy has been reported to be enough to reduce naturally occurring *Salmonella* spp., *Campylobacter* spp., *Yersinia* spp., and coliforms on chicken carcasses to undetectable levels and also to reduce total bacterial counts by ca. 2.1 to 3.1 log CFU/g (Abu-Tarboush et al., 1997). Lewis et al. (2002) reported reductions in the numbers of boneless, skinless chicken breast samples testing positive for *Salmonella* and *Campylobacter* spp., of 40% and 13%, respectively, to none, after irradiation at 1 kGy. Sarjeant et al. (2005) reported reductions of 4 to 5 log CFU/g in populations of inoculated *Salmonella* spp. on chicken breast fillets after irradiation at 2.0 and 3.0 kGy. The FDA approves irradiation of poultry products at doses not to exceed 3 kGy for control of food-borne pathogens in fresh or frozen, uncooked poultry products (U.S. FDA, 1990).

FRUITS AND VEGETABLES

Different food-borne pathogens have been associated with fresh and fresh-cut produce. Outbreaks that occurred in Canada during 1981 to 1999 revealed a list of products responsible for the 18 outbreaks, including alfalfa sprouts, raspberries, cantaloupes, coleslaw, lettuce, parsley, and blackberries as well as potato salad and vegetable salad (Sewell and Farber, 2001). The presence of pathogens in produce may be the result of exposure to fecal sources, either directly or indirectly through contaminated water, utensils, or field workers (Beuchat, 1996; Geldreich and Bordner, 1971; Knabel et al., 2003), or other nonfecal sources, such as the environment or improperly sanitized equipment (DeRoever, 1998; Knabel et al., 2003).

Washing produce by spraying or dipping in potable water can effectively remove physical contamination and partially remove microbial populations if done properly (Brackett, 1992). Garg et al. (1990) surveyed microflora populations in commercial processing lines used to prepare fresh-cut vegetables.

These authors reported reductions of 0.4 log CFU/g on spinach passed through a water bath and then conveyed under sprays to the centrifuges used for the removal of the excess water. Beuchat et al. (1998) reported 1.6-log-CFU/cm^2 *Salmonella* reductions on inoculated apples that were soaked and rinsed 1 min in water. In the same study, Beuchat et al. (1998) did not find significant differences on *E. coli* O157:H7 counts on inoculated apples at 1.5 log CFU/cm^2 by applying the same treatment. The differences in the efficacies of the same treatments in reducing bacteria populations on produce have been addressed by some authors. Singh et al. (2002) suggested that such differences may be caused by the inoculation method used, the incubation time after inoculation, population size, and multiple washing. The effect of the sanitizer application method has also been reported by Ukuku and Fett (2004).

The surface of the produce is another factor that plays an important role in the effectiveness of the washing treatments. Annous et al. (2004) reported no significant differences in APCs and *E. coli* counts on inoculated cantaloupes after washing at room temperature for 3 min. These authors suggest the structure of the rind netting as the cause of the failure to dislodge microorganisms attached to the surface.

Surface Pasteurization

Annous et al. (2004) reported reductions of 3.6 to 4 log CFU/cm^2 in *E. coli* counts of inoculated cantaloupe surfaces pasteurized at 76°C for 2 to 3 min by dipping in hot water and then cooled on ice bath for 5 min. In the same study, these authors compared the effectiveness of pasteurizing at 76°C for 3 min in commercial-scale conditions cantaloupes inoculated with *Salmonella enterica* serovar Poona at 3.6 log CFU/cm^2. *S.* Poona loads were significantly reduced to 0.1 log CFU/cm^2. Pao and Davis (1999) reported a 5-log-CFU/cm^2 reduction on oranges inoculated with *E. coli* and pasteurized by immersing the inoculated fruit at 80 and 70°C for 2 and 4 min, respectively. Mild heat treatments may be used only for pasteurization of products with thick rinds. Damage of the flesh under the rind of cantaloupes has been reported to occur at temperatures of 96°C for 4 min (Annous et al., 2004). Solomon et al. (2006) developed a hot-water dip for cantaloupes, capable of reducing *Salmonella* by 4.5 log CFU/cm^2 while not affecting the firmness of the rind or the flesh under the rind.

Organic Acids

Although organic acids are widely used for meat decontamination (Castillo et al., 2002), their use with fresh produce is not common. Zhang and Farber

(1996) studied the effect of lactic and acetic acid to reduce *L. monocytogenes* counts on inoculated shredded lettuce. Reductions of 0.5 and 0.2 log CFU/g were obtained after dipping inoculated shredded lettuce for 10 min at 22°C in 1% lactic and 1% acetic acid, respectively. These authors (Zhang and Farber, 1996) reported enhanced reductions for combinations of organic acids with 100 mg/liter chlorine solution. Reductions up to 1.1 log CFU/g of *L. monocytogenes* counts were obtained with solutions of lactic acid plus 100 mg/liter chlorine. The combined effect of organic acids and heat has been addressed by Alvarado-Casillas et al. (2007). These authors reported reductions of almost 3 log CFU/cm^2 on cantaloupes and 3.6 log CFU/cm^2 on bell peppers inoculated with *S.* Typhimurium and *E. coli* O157:H7 for the spraying of 2% lactic acid at 55°C. Delaquis et al. (1999) inoculated mung bean seeds with *S.* Typhimurium, *E. coli* O157:H7, and *L. monocytogenes* for testing the effectiveness of gaseous acetic acid. Counts of *E. coli* O157:H7 at 3 log CFU/g and *S.* Typhimurium at 5 log CFU/g were reduced to nondetectable levels after exposure of bean seed to 242 µl of acetic acid per liter of air for 12 h at 22°C. *L. monocytogenes* was recovered by enrichment from 2 out of 10 samples treated. Rodgers et al. (2004) reported *L. monocytogenes* and *E. coli* O157:H7 reductions of >4.9 log CFU/g on whole lettuce leaves, apples, strawberries, and cantaloupes after dipping for 5 min with 80 mg/liter peroxyacetic acid at 21 to 23°C. Reductions of *L. monocytogenes* and *E. coli* O157:H7 of ca. 4.4 log CFU/g were obtained for shredded lettuce and sliced apples by applying the same treatment of peroxyacetic acid. Lower reductions of *L. monocytogenes* with peroxyacetic acid sanitizer were reported in a different study by Beuchat et al. (2004). These authors simulated the conditions used by commercial fresh-cut produce processors and treated inoculated lettuce pieces with a peroxyacetic acid sanitizer (80 µg/ml at 4°C for 30 s and a lettuce/sanitizer ratio of 1:100 [wt/vol]). Reductions of 1.0 to 1.8 log CFU/g were significantly different than the reductions obtained with washing in water when lettuce pieces and shredded iceberg lettuce were treated. There were no significant differences between the reductions obtained with romaine lettuce pieces by applying the same peroxyacetic acid treatment and the reductions obtained by water wash.

Other Antimicrobials

The use of chlorinated water will further reduce microbial populations on produce. Garg et al. (1990) found that the use of chlorinated ice water reduced 2.7 log APC/g on lettuce under commercial processing operations. Beuchat et al. (1998) sprayed chlorinated water at 200 and 2,000 mg/liter of free chlorine on apples, lettuce, and tomatoes inoculated with pathogenic bacteria. These authors reported significant reductions of *Salmonella*, *E. coli* O157:H7, and *L. monocytogenes* after treatment by washing and spraying chlorinated water at 200 mg/liter. The effectiveness of chlorine in inactivating *Salmonella enterica* serovar Montevideo on fresh tomatoes was studied by Zhuang et al. (1995). They reported that surface reductions of 1.5 log CFU/cm^2 of *S.* Montevideo counts were obtained by dipping inoculated batches of six tomatoes in 1 liter of chlorinated water (320 mg/liter) for 2 min and applying constant agitation. Zhang and Farber (1996) reported that *L. monocytogenes* reductions of 1.3 log CFU/g on fresh-cut lettuce were obtained by dipping it in 200 mg/liter chlorine for 10 min. Alvarado-Casillas et al. (2007) reported reductions of 2.1 and 1.5 log CFU/cm^2 for whole cantaloupes inoculated with *S.* Typhimurium and *E. coli* O157:H7 after spraying for 15 s with 200 mg/liter sodium hypochlorite solution at pH 6.0.

Chlorine destabilization by the presence of organic matter, metals and its reaction with light, reduces its sanitizing efficacy. The contact of chlorine with open wounds, crevices, or other tissues of the fresh produce may decrease the effectiveness of this compound in reducing pathogens in produce by increased amounts of organic matter (Castillo and Rodríguez-García, 2004). While the produce industry commonly uses hypochlorite in produce washes at concentrations >50 mg/ml as recommended by the FDA (U.S. FDA, 2001b), this treatment is intended to keep the wash water disinfected and so prevent cross-contamination between product units, rather than to sanitize the product.

Chlorine dioxide is a secondary food additive and was approved to be used in the reduction of microbial populations of fresh produce by the FDA (U.S. FDA, 1998b). The efficacy of chlorine dioxide in reducing bacteria populations on different produce surfaces has been studied. Singh et al. (2002) reported *E. coli* O157:H7 reductions of 1.5 log CFU/g on inoculated shredded lettuce at 7.8 log CFU/g after immersion in aqueous solution of chlorine dioxide (10 mg/liter) for 10 min. In the same study, a significantly higher reduction of 1.9 log CFU/g was obtained with a lower inoculum level of 3.7 log CFU/g. The effect of chorine dioxide on fresh-cut lettuce inoculated with *L. monocytogenes* was studied by Zhang and Farber (1996). These authors reported 0.8-log-CFU/g reductions by dipping fresh-cut lettuce on 5 mg/liter chlorine dioxide solution at 22°C for 10 min. Reductions of 1.1 log CFU/g of *L. monocytogenes* counts were obtained by treatments with 5 mg/liter chlorine dioxide at 4°C for 10 min.

Chlorine dioxide gas has been used as an alternative to aqueous sanitizers. Du et al. (2002) reported significant reductions of *L. monocytogenes* counts on inoculated apples after treatment with chlorine dioxide gas at 4 mg/liter for 10 min. Reductions up to 3.2 and 3.5 log per inoculation spotted site were reported for the calyx and stem cavity, whereas reductions of 5.5 log per inoculation spotted site were reported for the pulp surface. Han et al. (2000) reported 6.4-log reductions on *E. coli* O157:H7, counts on inoculated surface-injured green peppers after treatment with 1.24 mg/liter of chlorine dioxide gas for 30 min. Sy et al. (2005) studied the reductions of *Salmonella*, *E. coli* O157:H7, and *L. monocytogenes* on different fresh produce surfaces by using chlorine dioxide gas. These authors reported reductions for each of the pathogen microorganisms greater than 3, 5, and 1.5 log CFU/g on fresh-cut cabbage, carrots, and lettuce, respectively, after exposure to 4.1 mg/liter of chlorine dioxide for 30 min.

Ozone is a compound that is generally recognized as safe (U.S. FDA, 2001c), and its effectiveness in reducing microbial populations has been tested. Kim et al. (1999) reported a decrease of natural microbiota of shredded lettuce of 1.6 log CFU/g by bubbling ozone gas (4.9% [vol/vol], 0.5 liter/min) in lettuce water mixture (1:20, wt/wt). Singh et al. (2002) reported *E. coli* O157:H7 reductions of 1.5 log CFU/g on inoculated shredded lettuce after immersion for 10 min in ozonated water (10 mg/liter). Rodgers et al. (2004) reported effective reductions higher than 5 log CFU/g on *E. coli* O157:H7 and *L. monocytogenes* counts by treating with ozone (3 mg/liter for 5 min) artificially inoculated whole lettuce leaves, apples, strawberries, and cantaloupes as well as sliced apples and shredded lettuce.

Irradiation

Studies on irradiation of fruits and vegetables have been conducted on cilantro leaves (Fan et al., 2003a), lettuce (Fan and Sokorai, 2002), green onion leaves (Fan et al., 2003b), and sliced cantaloupe (Palekar et al., 2004). However, these reports deal with quality issues rather than hazard control. Niemira et al. (2003) reported reductions of 2.6 and 4.0 log CFU/g on *L. monocytogenes* counts of inoculated leaf pieces of endive after irradiating at 0.42 and 0.84 kGy, respectively. Lee et al. (2006) irradiated ready-to-use vegetables inoculated with *S.* Typhimurium, *E. coli*, *S. aureus*, and *Listeria ivanovii*. A dose of 2 kGy was necessary to eliminate *S.* Typhimurium from levels of 7.3 log CFU/g to nondetectable levels from seasoned spinach and seasoned burdock, whereas a dose of 3 kGy was necessary to yield the same reduction on

cucumber. Lower doses (1 kGy) reduced *S.* Typhimurium counts about 2.5 to 3.1 log CFU/g. A 3-kGy dose was necessary to reduce *E. coli*, *S. aureus*, and *L. ivanovii* counts to nondetectable levels in all the vegetables. Lower doses (2 kGy) were able to reduce pathogen counts to nondetectable levels; however, the microorganisms were able to grow after 24 h of storage at 10, 20, and 30°C. Schmidt et al. (2006) irradiated cut tomatoes inoculated with *Salmonella enterica* serovar Agona and *S.* Montevideo. These authors reported reductions up to 2.2 log CFU/g after irradiating at 0.95 kGy. Effectiveness of irradiation on seeds has also been studied. Bari et al. (2003) reported that irradiation doses of 2 kGy completely eliminated *E. coli* O157:H7 from alfalfa and mung bean seeds, whereas a dose of 2.5 kGy was necessary to completely remove the pathogen from radish seeds.

The FDA approved the irradiation of produce at doses <1 kGy for growth and maturation inhibition of fresh foods and for disinfestation of arthropod pests in foods (U.S. FDA, 1986) and <8 kGy for control of microbial pathogens of seeds for sprouting (U.S. FDA, 2000b). To date, there is no approved irradiation dose for the control of pathogens on fresh and fresh-cut produce.

Internalization of Pathogens

One of the causes of the reductions in the effectiveness of decontamination procedures for fresh produce is the internalization of pathogens in the commodities through the core, stem scar region, or injuries present in the surface of the fruits and vegetables (Annous et al., 2004; Buchanan et al., 1999; Han et al., 2000; Ibarra-Sanchez et al., 2004). Also, rough surfaces of some produce make it difficult to remove the attached microorganisms (Annous et al., 2004). The effect of temperature differentials among produce and washing solution has been addressed by Buchanan et al. (1999). These authors reported that by immersing warm apples in cold water with dye (red food dye no. 40), approximately 16% of the immersed apples had substantial accumulation of the dye in the inner core region. The absorption of the dye occurred through an open channel in the core region or through the peeling, particularly in bruised or punctured regions.

SEAFOOD

Contamination of fish and seafood comes from the environmental water (Kelly and Dinuzzo, 1985) and also may be caused by human contamination (Fleet, 1978). Physical and chemical processes have been reported to reduce the bacterial load on shellfish.

Cold Temperatures

The use of cold temperatures for the reduction or growth inhibition of pathogens in seafood has been evaluated by several authors. Thompson et al. (1976) reported that populations of *Vibrio parahaemolyticus* in shucked oysters held at 3°C for 7 days decreased from >1.1×10^4 to 3.6×10^{-1} most probable numbers (MPN)/g. Oliver (1981) reported the reduction of *V. vulnificus* on oysters artificially inoculated and held at 0.5°C on ice. Muntada-Garriga et al. (1995) reported the inactivation of *V. parahaemolyticus* in oyster meat homogenates at low temperatures. According to these authors (Muntada-Garriga et al., 1995), initial counts of *V. parahaemolyticus* (5 log CFU/g) may be reduced to nondetectable levels by freezing at –18 and –24°C for 12 weeks.

Depuration

Depuration is one of the most common processes used to reduce bacterial contaminations on shellfish. Depuration is a controlled process that allows shellfish to purge sand and grit from the gut to clean seawater. This process reduces bacterial counts and increases the shelf life of refrigerated products (Su and Liu, 2007). The efficacy of bacteria reductions by this method is limited. Kelly and Dinuzzo (1985) reported that oysters artificially inoculated with *V. vulnificus* may require 16 days to reduce the counts from 25 CFU/g to <0.1 CFU/g by using sterile seawater. Croci et al. (2002) studied the effect of depuration by using seawater with an ozone treatment system in blue mussels inoculated with *E. coli*, *V. cholerae* O1, and *V. parahaemolyticus*. These authors (Croci et al., 2002) used a pilot unit of 250 liters of seawater and exchanged the water from the aquarium and the ozone treatment tank at a flow rate of 6 liters/min. Initial *E. coli* counts of 1.1×10^5 MPN/g were reduced to 4.6×10^3 MPN/g after 5 h of depuration, whereas *V. cholerae* and *V. parahaemolyticus* counts of 9.3×10^3 and 7.4×10^3 MPN/g were reduced to 2.9×10^3 MPN/g and 2.4×10^3 MPN/g, respectively. After 44 h of treatment, the counts on these two pathogens were 1.1×10^3 MPN *V. parahaemolyticus*/g and 9.3×10^2 MPN *V. cholerae*/g.

Heat

Andrews et al. (2000) studied the effect of using low pasteurization (50°C water treatment for 0, 5, 10, and 15 min followed by cooling to 2 to 4°C using a clean ice water bath) to reduce the levels of *V. parahaemolyticus and V. vulnificus* on inoculated oysters. These authors reported reductions (≥99.9%) of *V. vulnificus* and *V. parahaemolyticus* from 10^5 to 9.3×10^1 MPN/g using an exposure time of 5 min

and to <3 MPN/g in 10 min of exposure. In a similar study, Andrews et al. (2003b) tested the efficacy of the pasteurization treatment previously described for highly virulent and process-resistant strains of *V. parahaemolyticus* O3:K6 inoculated in oysters. These authors recommend a total processing time of at least 22 min in water at 52°C to ensure a reduction of *V. parahaemolyticus* from a 5-log concentration to nondetectable levels (<3 MPN/g). These processing conditions are required for oysters to reach 50°C and keep the pasteurization temperature for 10 min.

Kozempel et al. (2001) utilized the steam vacuum technology to clean the surface of catfish. These authors reported reductions of 1.82 log CFU/ml of *L. innocua* counts for rinses of inoculated catfish after 4 cycles at 143°C for 0.10 s per cycle.

High Pressure

High-pressure processing is an alternative treatment used to eliminate pathogens from foods with no thermal pasteurization and extend the shelf life of foods (Smelt, 1998). Calik et al. (2002) reported effective reductions of *V. parahaemolyticus* on inoculated oysters by applying treatments of 345 MPa for 90 s from levels of 10^5 to 10^7 CFU/g to nondetectable levels (<10 CFU/g). In a different study, Cook (2003) reported that a pressure treatment of 241 MPa for 120 s was enough to reduce *V. vulnificus* in shell oysters from 1.2×10^5 MPN/g to <1.8 MPN/g. In the same study, Cook (2003) reported that a more severe treatment of 300 MPa for 120 s was needed to achieve the same 5-log reduction of *V. parahaemolyticus* serotype O3:K6 on inoculated oysters.

Inactivation of viruses by high-pressure treatments has also been studied (Calci et al., 2005; Grove et al., 2006). Calci et al. (2005) inoculated live oysters with hepatitis A virus (HAV) at levels of >5 log PFU/oyster. These authors reported HAV reductions of 3 log PFU/oyster by treating oysters at 400 MPa for 1 min. In the same study, Calci et al. (2005) treated oysters inoculated with lower concentrations of HAV (3 to 4 log PFU/oyster). The authors reported no infectious virus after treatment of these oysters at 400 MPa for 1 min, with a detection limit of 1.5 log PFU/oyster.

Other Antimicrobials

The decontaminant agent most used in the seafood industry is chlorine. According to Chaiyakosa et al. (2007), reductions of more than 90% of *V. parahaemolyticus* counts were achieved by immersing inoculated shrimp in a 50 mg/liter chlorine solution at 5 to 7°C for 30 min.

Shin et al. (2004) studied the effectiveness of combining cold and chemical treatments by using an antimicrobial ice formulation capable of releasing gaseous chlorine dioxide. In their study, they tested the antimicrobial ice against *E. coli* O157:H7, *S.* Typhimurium, and *L. monocytogenes* on selective media and on fish skin. *E. coli* O157:H7, *S.* Typhimurium, and *L. monocytogenes* counts were reduced by 5 log, 4 log, and 1 log, respectively, on selective media after 10 min of treatment with chlorine dioxide ice (25, 50, and 100 mg/liter). Application of chlorine dioxide ice on mackerel skin reduced *E. coli* O157:H7 counts from 5.3 log CFU/cm^2 to 3.1 log CFU/cm^2 after 60 min at 25 mg/liter, 2 log CFU/cm^2 after 120 min at 50 mg/liter, and 0.3 log CFU/cm^2 after 120 min at 100 mg/liter. The final count for *S.* Typhimurium was reduced from 5.2 to 2.9 log CFU/cm^2 after 120 min at 50 mg/liter and 2.1 log CFU/cm^2 after 120 min at 100 mg/liter. *L. monocytogenes* counts were also significantly reduced after treatment with chlorine dioxide ice from 5.9 log CFU/cm^2 to 3.6 and 2.0 log CFU/cm^2 after 60 min at 50 and 100 mg/liter.

Irradiation

Some studies developed to test the effectiveness of radiation in decontamination of seafood have demonstrated that doses of 3 kGy or below are enough to eliminate pathogenic bacteria. Andrews et al. (2003a) reported that a dose of 0.75 to 1 kGy reduced 10^6 CFU/g of *V. vulnificus* on oyster meat to nondetectable levels (<3 MPN/g). According to these authors, higher doses (1.0 to 1.5 kGy) are required to reduce *V. parahaemolyticus* O3:K6 to nondetectable levels. Jakabi et al. (2003) reduced *Salmonella enterica* serovar Enteritidis and *Salmonella enterica* serovar Infantis by 5 to 6 log MPN/g from inoculated oysters by irradiating at 3.0 kGy. In the same study, Jakabi et al. (2003) reported that a dose of 1 kGy was sufficient to reduce *V. parahaemolyticus* inoculated at a level of 10^6 MPN/g to 0.3 MPN/g. Jaczynski and Park (2003) reported reductions of 6.1 log of *S. aureus* counts on surimi paste after irradiation at 2 kGy. These authors reported sterilization of *S. aureus* in surimi at irradiation doses of 4 kGy. The FDA has approved the use of irradiation at levels of <5.5 kGy in or on fresh or frozen molluscan shellfish to control *Vibrio* bacteria and other food-borne microorganisms (U.S. FDA, 2005).

CONCLUSIONS

As discussed above, there are multiple interventions for hazard control during harvest that are capable of reducing pathogenic bacteria in foods of different origin. The effectiveness of such interventions may vary among commodities and the type of contamination. Likewise, there is not a unique treatment to be used in a specific commodity. Every producer should adopt the appropriate interventions according to their products/processes and capabilities of application. Then the producer must validate these interventions within their process. It is important to note that the application of interventions/decontamination methods should never be used as an alternative for proper good manufacturing practices or good agriculture practices during production and/or harvest of various food commodities.

Also, each producer must be aware of the regulatory requirements for using interventions to control hazards. As noted in the discussion above, USDA's FSIS and the FDA have specific regulations and guidelines for the approved use of various interventions. Producers must ensure that they are complying with these requirements.

Overall, the proper use of approved interventions/decontamination methods will continue to enhance the safety of the food supply and allow producers to meet the demands and expectations of the consumers.

REFERENCES

Abu-Ruwaida, A. S., W. N. Sawaya, B. H. Dashti, M. Murad, and H. A. Al-Othman. 1994. Microbiological quality of broilers during processing in a modern commercial slaughterhouse in Kuwait. *J. Food Prot.* 57:887–892.

Abu-Tarboush, H. M., H. A. Al-Kahtani, M. Atia, A. A. Abou-Arab, A. S. Bajaber, and M. A. El-Mojaddidi. 1997. Sensory and microbial quality of chicken as affected by irradiation and postirradiation storage at 4.0 °C. *J. Food Prot.* 60:761–770.

Alvarado-Casillas, S., S. Ibarra-Sánchez, O. Rodríguez-García, N. Martínez-Gonzáles, and A. Castillo. 2007. Comparison of rinsing and sanitizing procedures for reducing bacterial pathogens on fresh cantaloupes and bell peppers. *J. Food Prot.* 70:655–660.

Anderson, M. E., and R. T. Marshall. 1990. Reducing microbial populations on beef tissues: concentration and temperature of lactic acid. *J. Food Saf.* 10:181–190.

Anderson, M. E., R. T. Marshall, and J. S. Dickson. 1992. Efficacies of acetic, lactic and two mixed acids in reducing numbers of bacteria on surfaces of lean meat. *J. Food Saf.* 12:139–147.

Anderson, M. E., R. T. Marshall, H. D. Naumann, and W. C. Stringer. 1975. Physical factors that affect removal of yeasts from meat surfaces with water sprays. *J. Food Sci.* 40:1232–1235.

Anderson, M. E., R. T. Marshall, W. C. Stringer, and H. D. Naumann. 1977. Combined and individual effects of washing and sanitizing on bacterial counts of meat —a model system. *J. Food Prot.* 40:668–670.

Anderson, M. E., R. T. Marshall, W. C. Stringer, and H. D. Naumann. 1981. Evaluation of a prototype beef carcass washer in a commercial plant. *J. Food Prot.* 44:35–38.

Andrews, L., M. Jahncke, and K. Mallikarjunan. 2003a. Low dose gamma irradiation to reduce pathogenic *Vibrios* in live oysters (*Crassostrea virginica*). *J. Aquat. Food Prod. Technol.* 12:71–82.

Andrews, L. S., S. DeBlanc, C. D. Veal, and D. L. Park. 2003b. Response of *Vibrio parahaemolyticus* 03:K6 to a hot water/cold shock pasteurization process. *Food Addit. Contam.* **20:**331–334.

Andrews, L. S., D. L. Park, and Y. P. Chen. 2000. Low temperature pasteurization to reduce the risk of *Vibrio* infections from raw shell-stock oysters. *Food Addit. Contam.* **19:**787–791.

Annous, B. A., A. Burke, and J. E. Sites. 2004. Surface pasteurization of whole fresh cantaloupes inoculated with *Salmonella* Poona or *Escherichia coli*. *J. Food Prot.* **67:**1876–1885.

Arthur, T. M., T. L. Wheeler, S. D. Shackelford, J. M. Bosilevac, X. Nou, and M. Koohmaraie. 2005. Effects of low-dose, low-penetration electron beam irradiation of chilled beef carcass surface cuts on *Escherichia coli* O157:H7 and meat quality. *J. Food Prot.* **68:**666–672.

Arthur, T. M., J. M. Bosilevac, D. M. Brichta-Harhay, N. Kalchayanand, S. D. Shackelford, T. L. Wheeler, and M. Koohmaraie. 2007. Effects of a minimal hide wash cabinet on the levels and prevalence of *Escherichia coli* O157:H7 and *Salmonella* on the hides of beef cattle at slaughter. *J. Food Prot.* **70:**1076–1079.

Avens, J. S., S. N. Albright, A. S. Morton, B. E. Prewitt, P. A. Kendall, and J. N. Sofos. 2002. Destruction of microorganisms on chicken carcasses by steam and boiling water immersion. *Food Control* **13:**445–450.

Bacon, R. T., K. E. Belk, J. N. Sofos, R. P. Clayton, J. O. Reagan, and G.C. Smith. 2000. Microbial populations on animal hides and beef carcasses at different stages of slaughter in plants employing muliple-sequential interventions for decontamination. *J. Food Prot.* **63:**1080–1086.

Bacon, R. T., J. N. Sofos, K. E. Belk, D. R. Hyatt, and G. C. Smith. 2002. Prevalence and antibiotic susceptibility of *Salmonella* isolated from beef animal hides and carcasses. *J. Food Prot.* **65:**284–290.

Baird, B. E., L. M. Lucia, G. R. Acuff, K. B. Harris, and J. W. Savell. 2006. Beef hide antimicrobial interventions as a means of reducing bacterial contamination. *Meat Sci.* **73**(2):245–248.

Bari, M. L., E. Nazuka, Y. Sabina, S. Todoriki, and K. Isshiki. 2003. Chemical and irradiation treatments for killing *Escherichia coli* O157:H7 on alfalfa, radish and mung bean seeds. *J. Food Prot.* **66:**767–774.

Beuchat, L. R. 1996. Pathogenic microorganisms associated with fresh produce. *J. Food Prot.* **59:**204–216.

Beuchat, L. R., B. B. Adler, and M. M. Lang. 2004. Efficacy of chlorine and a peroxyacetic acid sanitizer in killing *Listeria monocytogenes* on iceberg and romaine lettuce using simulated commercial processing conditions. *J. Food Prot.* **67:**1238–1242.

Beuchat, L. R., B. V. Nail, B. B. Adler, and M. R. S. Clavero. 1998. Efficacy of spray application of chlorinated water in killing pathogenic bacteria on raw apples, tomatoes, and lettuce. *J. Food Prot.* **61:**1305–1311.

Biemuller, G. W., J. A. Carpenter, and A. E. Reynolds. 1973. Reduction of bacteria on pork carcasses. *J. Food Sci.* **38:**261–263.

Bosilevac, J. M., T. M. Arthur, T. L. Wheeler, S. D. Shackelford, M. Rossman, J. O. Reagan, and M. Koohmaraie. 2004. Prevalence of Escherichia coli O157 and levels of aerobic bacteria and Enterobacteriaceae are reduced when hides are washed and treated with cetylpyridinium chloride at a commercial beef processing plant. *J. Food Prot.* **67:**646–650.

Bosilevac, J. M., X. Nou, M. S. Osborn, D. M. Allen, and M. Koohmaraie. 2005a. Development and evaluation of an on-line hide decontamination procedure for use in a commercial beef processing plant. *J. Food Prot.* **68:**265–272.

Bosilevac, J. M., S. D. Shackelford, D. M. Brichta, and M. Koohmaraie. 2005b. Efficacy of ozonated and electrolyzed oxidative waters to decontaminated hides of cattle before slaughter. *J. Food Prot.* **68:**1393–1398.

Bowling, R. A., and R. P. Clayton. September 1992. Method for dehairing animals. U. S. patent 5,149,295.

Brackett, R. E. 1992. Shelf stability and safety of fresh produce as influenced by sanitation and disinfection. *J. Food Prot.* **55:**808–814.

Brackett, R. E., Y. Y. Hao, and M. P. Doyle. 1994. Ineffectiveness of hot acid sprays to decontaminate *Escherichia coli* O157:H7 on beef. *J. Food Prot.* **57:**198–203.

Breen, P. J., H. Salari, and C. M. Compadre. 1997. Elimination of *Salmonella* contamination from poultry tissues by cetylpyridinium chloride solutions. *J. Food Prot.* **60:**1019–1021.

Buchanan, R. L., S.G. Edelson, R. L. Miller, and G. M. Sapers. 1999. Contamination of intact apples after immersion in an aqueous environment containing *Escherichia coli* O157:H7. *J. Food Prot.* **62:**444–450.

Calci, K. R., G. K. Meade, R. C. Tezloff, and D.H. Kingsley. 2005. High-pressure inactivation of hepatitis A virus within oysters. *Appl. Environ. Microbiol.* **71:**339–343.

Calik, H., M. T. Morrissey, P. W. Reno, and H. An. 2002. Effect of high-pressure processing on *Vibrio parahaemolyticus* strains in pure culture and Pacific oysters. *J. Food Sci.* **67:**1506–1510.

Castillo, A., J. S. Dickson, R. P. Clayton, L. M. Lucia, and G. R. Acuff. 1998a. Chemical dehairing of bovine skin to reduce pathogenic bacteria and bacteria of fecal origin. *J. Food Prot.* **61:**623–625.

Castillo, A., M. D. Hardin, G. R. Acuff, and J. S. Dickson. 2002. Reduction of microbial contamination on carcasses, p. 351–381. *In* V. K. Juneja and J. N. Sofos (ed.), *Control of Foodborne Microorganisms*. Marcel Dekker, New York, NY.

Castillo, A., L. M. Lucia, K. J. Goodson, J. W. Savell, and G. R. Acuff. 1998b. Comparison of water wash, trimming, and combined hot water and lactic acid treatments for reducing bacteria of fecal origin on beef carcasses. *J. Food Prot.* **61:**823–828.

Castillo, A., L. M. Lucia, K. J. Goodson, J. W. Savell, and G. R. Acuff. 1998c. Use of hot water for beef carcass decontamination. *J. Food Prot.* **61:**19–25.

Castillo, A., L. M. Lucia, G. K. Kemp, and G. R. Acuff. 1999. Reduction of *Escherichia coli* O157:H7 and *Salmonella* Typhimurium on beef carcass surfaces using acidified sodium chlorite. *J. Food Prot.* **62:**580–584.

Castillo, A., and M. O. Rodríguez-García. 2004. Bacterial hazards in fresh and fresh-cut produce: sources and control, p. 43–57. *In* R. C. Beier, S. D. Pillai, and T. D. Phillips (ed.), *Preharvest and Postharvest Food Safety. Contemporary Issues and Future Directions.* Blackwell, Ames, IA.

CDC. 2000. Surveillance for foodborne disease outbreaks —United States, 1993–1997. *MMWR Morbid. Mortal. Wkly. Rep.* **49:**1–51.

CDC. 2006. Surveillance for foodborne-disease outbreaks —United States, 1998–2002. *MMWR A Morbid. Mortal. Wkly. Rep.* **55**(SS10):1–34.

CDC. 2005. Foodborne Illness. Centers for Disease Control and Prevention, Atlanta, GA. http://www.cdc.gov/ncidod/dbmd/diseaseinfo/ foodborneinfections_g.htm#riskiestfoods.

Chaiyakosa, S., W. Charernjiratragul, K. Umsakul, and V. Vuddhakul. 2007. Comparing the efficiency of chitosan with chlorine for reducing *Vibrio parahaemolyticus* in shrimp. *Food Control* **18:**1031–1035.

Collins, C. I., E. A. Murano, and I. V. Wesley. 1996. Survival of *Arcobacter butzleri* and *Campylobacter jejuni* after irradiation treatment in vacuum-packaged ground pork. *J. Food Prot.* **59:**1164–1166.

Collins, J. E. 1997. Impact of changing consumer lifestyles on the emergence/reemergence of foodborne pathogens. *Emerg. Infect. Dis.* **3**:471–479.

Cook, D. W. 2003. Sensitivity of *Vibrio* species in phosphate-buffered saline and in oysters to high-pressure processing. *J. Food Prot.* **66**:2276–2282.

Croci, L., E. Suffredini, L. Cozzi, and L. Toti. 2002. Effects of depuration of molluscs experimentally contaminated with *Escherichia coli*, *Vibrio cholerae* O1 and *Vibrio parahaemolyticus*. *J. Appl. Microbiol.* **92**:460–465.

Crouse, J. D., M. E. Anderson, and H. D. Naumann. 1988. Microbial decontamination and weight of carcass beef as affected by automated washing pressure and length of time of spray. *J. Food Prot.* **51**:471–474.

Cutter, C. N., and W. J. Dorsa. 1995. Chlorine dioxide spray washes for reducing fecal contamination on beef. *J. Food Prot.* **58**:1294–1296.

Cutter, C. N., W. J. Dorsa, A. Handie, S. Rodriguez-Morales, X. Zhou, P. J. Breen, and C. M. Compadre. 2000. Antimicrobial activity of cetylpyridinium chloride washes against pathogenic bacteria on beef surfaces. *J. Food Prot.* **63**:593–600.

Cutter, C. N., and M. Rivera-Betancourt. 2000. Interventions for the reduction of *Salmonella* Typhimurium DT 104 and non-O157:H7 enterohemorrhagic *Escherichia coli* on beef surfaces. *J. Food Prot.* **63**:1326–1332.

Cutter, C. N., and G. R. Siragusa. 1994a. Decontamination of beef carcass tissue with nisin using a pilot scale model carcass washer. *Food Microbiol.* **11**:481–489.

Cutter, C. N., and G. R. Siragusa. 1994b. Efficacy of organic acids against *Escherichia coli* O157:H7 attached to beef carcass tissue using a pilot scale model carcass washer. *J. Food Prot.* **57**:97–103.

Cutter, C. N., and G. R. Siragusa. 1995a. Application of chlorine to reduce populations of *Escherichia coli* on beef. *J. Food Saf.* **15**:67–75.

Cutter, C. N., and G. R. Siragusa. 1995b. Treatments with nisin and chelators to reduce *Salmonella* and *Escherichia coli* on beef. *J. Food Prot.* **58**:1028–1030.

Delaquis, P. J., P. L. Sholberg, and K. Stanich. 1999. Desinfection of mung bean seed with gaseous acetic acid. *J. Food Prot.* **62**:953–957.

Delmore, L. R. G., J. N. Sofos, J. O. Reagan, and G. C. Smith. 1997. Hot-water rinsing and trimming/washing of beef carcasses to reduce physical and microbiological contamination. *J. Food Sci.* **62**:373–378.

Delmore, R. J., Jr, J. N. Sofos, G. R. Schmidt, K. E. Belk, W. R. Lloyd and G. C. Smith. 2000. Interventions to reduce microbiological contamination of beef variety meats. *J. Food Prot.* **63**:44–50.

DeRoever, C. 1998. Microbiological safety evaluations and recommendations on fresh produce. *Food Control* **9**:321–347.

DeZuniga, A. G., M. E. Anderson, R. T. Marshall, and E. L. Iannotti. 1991. A model system for studying the penetration of microorganisms into meat. *J. Food Prot.* **54**:256–258.

Dickson, J. S. 1991. Control of *Salmonella typhimurium*, *Listeria monocytogenes*, and *Escherichia coli* O157:H7 on beef in a model spray chilling system. *J. Food Sci.* **56**:191–193.

Dickson, J. S., and M. E. Anderson. 1991. Control of *Salmonella* on beef tissue surfaces in a model system by pre- and post-evisceration washing and sanitizing, with or without spray chilling. *J. Food Prot.* **54**:514–518.

Dickson, J. S., C. N. Cutter, and G. R. Siragusa. 1994. Antimicrobial effects of trisodium phosphate against bacteria attached to beef tissue. *J. Food Protect.* **57**:952–955.

Dorsa, W. J., C. N. Cutter, and G. R. Siragusa. 1996a. Effectiveness of a steam-vacuum sanitizer for reducing *Escherichia coli* O157:H7 inoculated to beef carcass surface tissue. *Lett. in Appl. Microbiol.* **23**:61–63.

Dorsa, W. J., C. N. Cutter, and G. R. Siragusa. 1997. Effects of steam-vacuuming and hot water spray wash on the microflora of refrigerated beef carcass surface tissue inoculated with *Escherichia coli* O157:H7, *Listeria innocua*, and *Clostridium sporogenes*. *J. Food Prot.* **60**:114–119.

Dorsa, W. J., C. N. Cutter, G. R. Siragusa, and M. Koohmaraie. 1996b. Microbial decontamination of beef and sheep carcasses by steam, hot water spray washes, and a steam-vacuum sanitizer. *J. Food Prot.* **59**:127–135.

Du, J., Y. Han, and R. H. Linton. 2002. Inactivation by chlorine dioxide gas (ClO$_2$) of *Listeria monocytogenes* spotted onto different apple surfaces. *Food Microbiol.* **19**:481–490.

Emswiler, B. S., A. W. Kotula, and D. K. Rough. 1976. Bactericidal effectiveness of three chlorine sources used in beef carcass washing. *J. Anim Sci.* **42**:1445–1450.

Fan, X., B. A. Niemira, and K. J. B. Sokorai. 2003a. Sensorial, nutritional and microbiological quality of fresh cilantro leaves as influenced by ionizing radiation and storage. *Food Res. Int.* **36**:713–719.

Fan, X., B. A. Niemira, and K. J. B. Sokorai. 2003b. Use of ionizing radiation to improve sensory and microbial quality of fresh-cut green onion leaves. *J. Food Sci.* **68**:1478–1483.

Fan, X., and K. J. B. Sokorai. 2002. Sensorial and chemical quality of gamma-irradiated fresh-cut iceberg lettuce in modified atmosphere packages. *J. Food Prot.* **65**:1760–1765.

Fleet, G. H. 1978. Oyster depuration—a review. *Food Technol. Autr.* **30**:444–454.

Garcia-Zepeda, C. M., C. L. Kastner, B. L. Willard, R. K. Phebus, J. R. Schwenke, B. A. Fijal, and R. K. Prasai. 1994. Gluconic acid as a fresh beef decontaminant. *J. Food Prot.* **57**:956–962.

Garg, N., J. J. Churey, and D. F. Splittstoesser. 1990. Effect of processing conditions on the microflora of fresh cut vegetables. *J. Food Prot.* **53**:701–703.

Geldreich, E. E., and R. H. Bordner. 1971. Fecal contamination of fruits and vegetables during cultivation and processing for market. A review. *J. Milk Food Technol.* **34**:184–195.

Gill, C. O. 2005. Sources of microbial contamination at slaughtering plants, p. 231–243. *In* J. N. Sofos (ed.), *Improving the Safety of Fresh Meat*. CRC Press, Boca Raton, FL.

Gill, C. O., M. Badoni, and T. Jones. 1996. Hygienic effects of trimming and washing operations in a beef-carcass-dressing process. *J. Food Prot.* **59**:666–669.

Gill, C. O., D. S. McGinnis, J. Bryant, and B. Chabot. 1995. Decontamination of commercial, polished pig carcasses with hot water. *Food Microbiol.* **12**:143–149.

Göksoy, E. O., C. James, J. E. L. Corry, and S. J. James. 2001. The effect of hot-water immersions on the appearance and microbiological quality of skin-on chicken-breast pieces. *Int. J. Food Sci. Technol.* **36**:61–69.

Gorman, B. M., J. B. Morgan, J. N. Sofos, and G. C. Smith. 1995a. Microbiological and visual effects of trimming and/or spray washing for removal of fecal material from beef. *J. Food Prot.* **58**:984–989.

Gorman, B. M., J. N. Sofos, J. B. Morgan, G. R. Schmidt, and G. C. Smith. 1995b. Evaluation of hand-trimming, various sanitizing agents, and hot water spray-washing as decontamination interventions for beef brisket adipose tissue. *J. Food Prot.* **58**:899–907.

Grove, S. F., A. Lee, T. Lewis, C. M. Stewart, H. Chen, and D. G. Hoover. 2006. Inactivation of foodborne viruses of significance by high pressure and other processes. *J. Food Prot.* **69**:957–968.

Han, Y., D. M. Sherman, R. H. Linton, S. S. Nielsen, and P. E. Nelson. 2000. The effects of washing and chlorine dioxide gas on survival and attachment of *Escherichia coli* O157:H7 to green pepper surfaces. *Food Microbiol.* 17:521–533.

Hardin, M. D., G. R. Acuff, L. M. Lucia, J. S. Oman, and J. W. Savell. 1995. Comparison of methods for decontamination from beef carcass surfaces. *J. Food Prot.* 58:368–374.

Hwang, C. A., and L. R. Beuchat. 1995. Efficacy of selected chemicals for killing pathogenic and spoilage microorganisms on chicken skin. *J. Food Prot.* 58:19–23.

Ibarra-Sanchez, L. S., S. Alvarado-Casillas, M. O. Rodriguez-Garcia, N. E. Martinez-Gonzales, and A. Castillo. 2004. Internalization of bacterial pathogens in tomatoes and their control by selected chemicals. *J. Food Prot.* 67:1353–1358.

Jaczynski, J., and J. W. Park. 2003. Microbial inactivation and electron penetration in surimi seafood during electron beam processing. *J. Food Sci.* 68:1788–1792.

Jakabi, M., D. S. Gelli, J. C. M. D. Torre, M. A. B. Rodas, B. D. G. M. Franco, M. T. Destro, and M. Landgraf. 2003. Inactivation by ionizing radiation of *Salmonella* Enteritidis, *Salmonella* Infantis, and *Vibrio parahaemolyticus* in oysters *(Crassostrea brasiliana)*. *J. Food Prot.* 66:1025–1029.

James, C., E. O. Göksoy, J. E. L. Corry, and S. J. James. 2000. Surface pasteurisation of poultry meat using steam at atmospheric pressure. *J. Food Eng.* 45:111–117.

Kelly, M. T., and A. Dinuzzo. 1985. Uptake and clearance of *Vibrio vulnificus* from Gulf Coast oysters *(Crassostrea virginica)*. *Appl. Environ. Microbiol.* 50:1548–1549.

Kemp, G. K., M. L. Aldrich, and A. L. Waldroup. 2000. Acidified sodium chlorite antimicrobial treatment of broiler carcasses. *J. Food Prot.* 63:1087–1092.

Kim, J. G., A. E. Yousef, and G. W. Chism. 1999. Use of ozone to inactivate microorganisms on lettuce. *J. Food Saf.* 19:17–34.

Kim, J. W., and M. F. Slavik. 1994. Trisodium phosphate (TSP) treatment of beef surfaces to reduce *Escherichia coli* O157:H7 and *Salmonella typhimurium*. *J. Food Sci.* 59:20–22.

Kim, J. W., and M. F. Slavik. 1996. Cetylpyridinium chloride (CPC) treatment on poultry skin to reduce attached *Salmonella*. *J. Food Prot.* 59:322–326.

Knabel, S. J., P. Fatemi, J. Patton, L. F. Laborde, B. Annous, and G. M. Sapers. 2003. On farm contamination of horticultural products in the USA and strategies for decontamination. *Food Austr.* 55:580–582.

Kochevar, S. L., J. N. Sofos, R. R. Bolin, J. O. Reagan, and G. C. Smith. 1997. Steam vacuuming as a pre-evisceration intervention to decontaminate beef carcasses. *J. Food Prot.* 60:107–113.

Kotula, A. W., W. R. Lusby, J. D. Crouse, and B. de Vries. 1974. Beef carcass washing to reduce bacterial contamination. *J. Anim. Sci.* 39:674–679.

Kozempel, M. F., D. L. Marshall, E. R. Radewonuk, O. J. Scullen, N. Goldberg, and M. F. A. Bal'a. 2001. A rapid surface intervention process to kill *Listeria innocua* on catfish using cycles of vacuum and steam. *J. Food Sci.* 66:1012–1016.

Lee, N. Y., C. Jo, D. H. Shin, W. G. Kim, and M. W. Byun. 2006. Effect of gamma-irradiation on pathogens inoculated into ready-to-use vegetables. *Food Microbiol.* 23:649–656.

Lewis, S. J., A. Velásquez, S. L. Cuppett, and S. R. McKee. 2002. Effect of electron beam irradiation on poultry meat safety and quality. *Poult. Sci.* 81:896–903.

Lillard, H. S. 1979. Levels of chlorine and chlorine dioxide of equivalent bactericidal effect in poultry processing water. *J. Food Sci.* 44:1594–1597.

Lillard, H. S. 1980. Effect on broiler carcasses and water of treating chiller water with chlorine or chlorine dioxide. *Poult. Sci.* 59:1761–1766.

López-González, V., P. S. Murano, R. E. Brennan, and E. A. Murano. 1999. Influence of various commercial packaging conditions on survival of *Escherichia coli* O157:H7 to irradiation by electron beam versus gamma rays. *J. Food Prot.* 62:10–15.

May, K. N. 1974. Changes in microbial numbers during final washing and chilling of commercially slaughtered broilers. *Poult. Sci.* 53:1282–1285.

Mies, P. D., B. R. Covington, K. B. Harris, L. M. Lucia, G. R. Acuff, and J. W. Savell. 2004. Decontamination of cattle hides prior to slaughter using washes with and without antimicrobial agents *J. Food Prot.* 67:579–582.

Morrison, G. J., and G. H. Fleet. 1985. Reduction of *Salmonella* on chicken carcasses by immersion treatments. *J. Food Prot.* 48:939–943.

Muntada-Garriga, J. M., J. J. Rodriguez-Jerez, E. I. Lopez-Sabater, and M. T. Mora-Ventura. 1995. Effect of chill and freezing temperatures on survival of *Vibrio parahaemolyticus* inoculated in homogenates of oyster meat. *Lett. Appl. Microbiol.* 20:225–227.

Naidu, A. S. 2002. Activated lactoferrin —a new approach to meat safety. *Food Technol.* 56:40–45.

Niemira, B. A., X. Fan, K. J. B. Sokorai, and C. H. Sommers. 2003. Ionizing radiation sensitivity of *Listeria monocytogenes* ATCC 49594 and *Listeria innocua* ATCC 51742 inoculated on endive *(Cichorium endiva)*. *J. Food Prot.* 66:993–998.

Notermans, S., R. J. Terbijhe, and M. van Schothorst. 1980. Removing faecal contamination of broilers by spray-cleaning during evisceration. *Br. Poult. Sci.* 21:115–121.

Nou, X., M. Rivera-Betancourt, J. M. Bosilevac, T. L. Wheeler, S. D. Shackelford, B. L. Gwartney, J. O. Reagan, and M. Koohmaraie. 2003. Effect of chemical dehairing on the prevalence of *Escherichia coli* O157:H7 and the levels of aerobic bacteria and *Enterobacteriaceae* on carcasses in a commercial beef processing plant. *J. Food Prot.* 66:2005–2009.

Nutsch, A. L., R. K. Phebus, M. J. Riemann, D. E. Schafer, J. E. Boyer, Jr., R. C. Wilson, J. D. Leising, and C. L. Kastner. 1997. Evaluation of a steam pasteurization process in a commercial beef processing facility. *J. Food Prot.* 60:485–492.

Ockerman, H. W., R. J. Borton, V. R. Cahill, N. A. Parrett, and H. D. Hoffman. 1974. Use of acetic and lactic acid to control the quantity of microorganisms on lamb carcasses. *J. Milk Food Technol.* 37:203–204.

Oliver, J. D. 1981. Lethal cold stress of *Vibrio vulnificus* in oysters. *Appl. Environ Microbiol.* 41:710–717.

Palekar, M. P., E. Cabrera-Diaz, A. Kalbasi-Ashtari, J. E. Maxim, R. K. Miller, L. Cisneros-Zevallos, and A. Castillo. 2004. Effect of electron beam irradiation on the bacterial load and sensorial quality of sliced cantaloupe. *J. Food Sci.* 69:M267–M273.

Pao, S., and C. L. Davis. 1999. Enhancing microbiological safety of fresh orange juice by fruit immersion in hot water and chemical sanitizers. *J. Food Prot.* 62:756–760.

Phebus, R. K., A. L. Nutsch, D. E. Schafer, R. C. Wilson, M. J. Riemann, J. D. Leising, C. L. Kastner, J. R. Wolf, and R. K. Prasai. 1997. Comparison of steam pasteurization and other methods for reduction of pathogens on surfaces of freshly slaughtered beef. *J. Food Prot.* 60:476–484.

Podolak, R. K., J. F. Zayas, C. L. Kastner, and D. Y. C. Fung. 1996. Reduction of bacterial populations on vacuum-packaged ground beef patties with fumaric and lactic acids. *J. Food Prot.* 59:1037–1040.

Prasai, R. K., G. R. Acuff, L. M. Lucia, D. S. Hale, J. W. Savell, and J. B. Morgan. 1991. Microbiological effects of acid decontamination of beef carcasses at various locations in processing. *J. Food Prot.* 54:868–872.

Reagan, J. O., G. R. Acuff, D. R. Buege, M. J. Buyck, J. S. Dickson, C. L. Kastner, J. L. Marsden, J. B. Morgan, R. Nickelson II,

G. C. Smith, and J. N. Sofos. 1996. Trimming and washing of beef carcasses as a method of improving the microbiological quality of meat. *J. Food Prot.* **59**:751–756.

Reynolds, A. E., and J. A. Carpenter. 1974. Bactericidal properties of acetic and propionic acids on pork carcasses. *J. Anim. Sci.* **38**:515–519.

Rodgers, S. L., J. N. Cash, M. Siddiq, and E. T. Ryser. 2004. A comparison of different chemical sanitizers for inactivating *Escherichia coli* O157:H7 and *Listeria monocytogenes* in solution and on apples, lettuce, strawberries, and cantaloupe. *J. Food Prot.* **67**:721–731.

Sarjeant, K. C., S. K. Williams, and A. Hinton, Jr. 2005. The effect of electron beam irradiation on the survival of *Salmonella enterica* serovar Typhimurium and psychrotrophic bacteria on raw chicken breasts stored at four degrees Celsius for fourteen days. *Poult. Sci.* **84**:955–958.

Schmidt, H. M., M. P. Palekar, J. E. Maxim, and A. Castillo. 2006. Improving the microbiological quality and safety of fresh-cut tomatoes by low-dose electron beam irradiation. *J Food Prot.* **69**:575–581.

Schnell, T. D., J. N. Sofos, V. G. Littlefield, J. B. Morgan, B. M. Gorman, R. P. Clayton, and G. C. Smith. 1995. Effects of postexsanguination dehairing on the microbial load and visual cleanliness of beef carcasses. *J. Food Prot.* **58**:1297–1302.

Sewell, A. M., and J. M. Farber. 2001. Foodborne outbreaks in Canada linked to produce. *J. Food Prot.* **64**:1863–1877.

Shin, J. H., S. Chang, and D. H. Kang. 2004. Application of antimicrobial ice for reduction of foodborne pathogens (*Escherichia coli* O157:H7, *Salmonella* Typhimurium, *Listeria monocytogenes*) on the surface of fish. *J. Appl. Microbiol.* **97**:916–922.

Singh, N., R. K. Singh, A. K. Bhunia, and R. L. Stroshine. 2002. Effect of inoculation and washing methods on the efficacy of different sanitizers against *Escherichia coli* O157:H7 on lettuce. *Food Microbiol.* **19**:183–193.

Sivapalasingam, S., C. R. Friedman, L. Cohen, and R. V. Tauxe. 2004. Fresh produce: a growing cause of outbreaks of foodborne illness in the United States, 1973 through 1997 *J. Food Prot.* **67**:2342–2353.

Smelt, J. P. P. M. 1998. Recent advances in the microbiology of high pressure processing. *Trends Food Sci. Technol.* **9**:152–158.

Solomon, E. B., L. Huang, J. E. Sites, and B. A. Annous. 2006. Thermal inactivation of *Salmonella* on cantaloupes using hot water. *J. Food Sci.* **71**:M25–M30.

Su, Y. C., and C. Liu. 2007. *Vibrio parahaemolyticus*: concern of seafood safety. *Food Microbiol.* **24**:549–558.

Sy, K. V., M. B. Murray, M. D. Harrison, and L. R. Beuchat. 2005. Evaluation of gaseous chlorine dioxide as a sanitizer for killing *Salmonella*, *Escherichia coli* O157:H7, *Listeria monocytogenes*, and yeasts and molds on fresh and fresh-cut produce. *J. Food Prot.* **68**:1176–1187.

Tamblyn, K. C., and D. E. Conner. 1997. Bactericidal activity of organic acids against *Salmonella typhimurium* attached to broiler chicken skin. *J. Food Prot.* **60**:629–633.

Thayer, D. W. 1995. Use of irradiation to kill enteric pathogens on meat and poultry. *J. Food Saf.* **15**:181–192.

Thayer, D. W., G. Boyd, J. B. Fox, Jr., L. Lakritz, and J. W. Hampson. 1995. Variations in radiation sensitivity of foodborne pathogens associated with the suspending meat. *J. Food Sci.* **60**:63–67.

Thomas, C. J., and T. A. McMeekin. 1984. Effect of water uptake by poultry tissues on contamination by bacteria during immersion in bacterial suspensions. *J. Food Prot.* **47**:398–402.

Thompson, C. A., Jr, C. Vanderzant, and S. M. Ray. 1976. Effect of processing, distribution and storage on *Vibrio parahaemolyticus* and bacterial counts of oysters (*Crassostrea virginica*). *J. Food Sci.* **41**:123–127.

Tinney, K. S., M. F. Miller, C. B. Ramsey, L. D. Thompson, and M. A. Carr. 1997. Reduction of microorganisms on beef surfaces with electricity and acetic acid. *J. Food Prot.* **60**:625–628.

Ukuku, D. O., and W. F. Fett. 2004. Method of applying sanitizers and sample preparation affects recovery of native microflora and *Salmonella* on whole cantaloupe surfaces. *J. Food Prot.* **67**:999–1004.

U.S. FDA. 1986. Irradiation in the production, processing, and handling of food, final rule. *Fed. Reg.* **51**:13376–13399.

U.S. FDA. 1990. Irradiation in the production, processing and handling of food. final rule. *Fed. Reg.* **55**:18538–18544.

U.S. FDA. 1995a. Procedures for the safe and sanitary processing and importing of fish and fishery products, final rule. *Fed. Reg.* **61**:65095–65202.

U.S. FDA. 1995b. Secondary direct food additives permitted in food for human consumption, final rule. *Fed. Reg.* **60**:11899–11900.

U.S. FDA. 1996. Secondary direct food additives permitted in food for human consumption, final rule. *Fed. Reg.* **61**:17828–17829.

U.S. FDA. 1997. Irradiation in the production, processing, and handling of food, final rule. *Fed. Reg.* **62**:64107–64121.

U.S. FDA 1998a. Guidance for industry. Guide to minimize microbial food safety hazards for fresh fruits and vegetables.U.S. Food and Drug Administration, Washington, DC. http://www.foodsafety.gov/~dms/prodguid.html.

U.S. FDA. 1998b. Secondary direct food additives permitted in food for human consumption, final rule. *Fed. Reg.* **63**:38746–38747.

U.S. FDA. 1998c. Secondary direct food additives permitted in food for human consumption, final rule. *Fed. Reg.* **63**:11118–11119.

U.S. FDA. 2000a. Food labeling, safe handling statements, labeling of shell eggs; refrigeration of shell eggs held for retail distribution, final rule. *Fed. Reg.* **65**:76091–76114.

U.S. FDA. 2000b. Irradiation in the production, processing, and handling of food, final rule. *Fed. Reg.* **65**:64605–64607.

U.S. FDA. 2000c. Secondary direct food additives permitted in food for human consumption, final rule. *Fed. Reg.* **65**:1776.

U.S. FDA. 2000d. Secondary direct food additives permitted in food for human consumption, final rule. *Fed. Reg.* **65**:16312.

U.S. FDA. 2001a. Hazard Analysis and Critical Control Point (HAACP); procedures for the safe and sanitary processing and importing of juice, final rule. *Fed. Reg.* **66**:6137–6202.

U.S. FDA. 2001b. Methods to reduce/eliminate pathogens from fresh and fresh-cut produce. *In Analysis and Evaluation of Preventive Control Measures for the Control and Reduction/Elimination of Microbial Hazards on Fresh and Fresh-Cut Produce*. http://www.cfsan.fda.gov/~comm/ift3-5.html.

U.S. FDA. 2001c. Secondary direct food additives permitted in food for human consumption, final rule. *Fed. Reg.* **66**:33829–33830.

U.S. FDA. 2004. Secondary direct food additives permitted in food for human consumption, final rule. *Fed. Reg.* **69**:17297–17298.

U.S. FDA. 2005. Irradiation in the production, processing, and handling of food, final rule. *Fed. Reg.* **70**:48057–48073.

USDA/FSIS. 1996a. Notice of policy change; achieving the zero tolerance performance standard for beef carcasses by knife trimming and vacuuming with hot water or steam; use of acceptable carcass interventions for reducing carcass contamination without prior agency approval. *Fed. Reg.* **61**:15024–15027.

USDA/FSIS. 1996b. Pathogen reduction: Hazard Analysis and Critical Control Point (HACCP) systems, final rule. *Fed. Reg.* **61**:38805–38989.

USDA/FSIS. 2002. Safe and suitable ingredients used in the production of meat and poultry products. FSIS Directive 7120.1. USDA-FSIS, Washington, DC.

USDA/FSIS. 2005. Safe and suitable ingredients used in the production of meat and poultry products. FSIS Directive 7120.1 Amend 3. April 7, USDA-FSIS, Washington, DC.

USDA-NACMCF. 1993. Generic HACCP for raw beef. *Food Microbiol.* **10**:449–488.

Woolthuis, C. H. J., D. A. A. Mossel, J. G. Van Logtestijn, J. M. De Kruijf, and F. J. M. Smulders. 1984. Microbial decontamination of porcine liver with lactic acid and hot water. *J. Food Prot.* **47**:220–226.

Zhang, S., and J. M. Farber. 1996. The effects of various disinfectants against *Listeria monocytogenes* on fresh-cut vegetables. *Food Microbiol.* **13**:311–321.

Zhuang, R. Y., L. R. Beuchat, and F. J. Angulo. 1995. Fate of *Salmonella montevideo* on and in raw tomatoes as affected by temperature and treatment with chlorine. *Appl. Environ. Microbiol.* **61**:2127–2131.

Pathogens and Toxins in Foods: Challenges and Interventions
Edited by V. K. Juneja and J. N. Sofos
© 2010 ASM Press, Washington, DC

Chapter 25

Interventions for Hazard Control during Food Processing

IFIGENIA GEORNARAS AND JOHN N. SOFOS

Microbiological changes that occur in foods from production through consumption may result in human food-borne illness due to growth of pathogenic microorganisms and/or microbial toxin production, as well as economic losses due to proliferation of spoilage microorganisms. Food preservation technologies, some of which have been used for centuries to minimize deterioration of foods and to enhance their safety, include salting, heating, chilling, freezing, drying, acidifying (either directly or by fermentation), modifying packaging atmosphere, and using chemical antimicrobial compounds. Over the years, the emergence of food safety issues and changes in consumer preferences have led to modifications in how existing methods are applied, especially with the advent of hurdle technology (Leistner and Gould, 2002). In addition, a number of novel preservation methods have been developed with the dual purpose of microbial control and maintenance of the sensorial, nutritional, and functional properties of foods. Advances in food preservation may have also been necessary due to the potential adaptation of microorganisms to food and/or environmental stresses, the emergence or reemergence of pathogenic microorganisms that are resistant to traditional preservation processes or that can cause illness with low infectious doses, and increases in human populations at risk for food-borne illness (Juneja and Sofos, 2002). Growing consumer demand for "healthy," nutritious, and convenient foods that are safe, minimally processed, and additive free and with an extended shelf life has also been a driving force for reevaluation of food preservation methods. Food processors are thus challenged with meeting the demands of the consumer, but at the same time with producing a food product that is microbiologically safe for human consumption and microbiologically stable over its shelf life.

Traditional methods of food preservation have been reviewed extensively and excellently by numerous authors (Doyle and Beuchat, 2007; Juneja and Sofos, 2002; Koutsoumanis et al., 2006; Rahman, 1999; Sofos, 1993, 1994; Zeuthen and Bøgh-Sørensen, 2003) and, as such, are only briefly described in this chapter. Emphasis is placed on recent advances and novel technologies or tools for inhibition or inactivation of microbial contamination in foods.

APPROACHES TO FOOD PRESERVATION

The major approaches used to enhance the microbiological quality and safety of food include the application of procedures that (i) prevent or minimize access of microorganisms to the product, (ii) reduce contamination that has gained access to the product, (iii) inactivate microorganisms on the product without cross-contamination, and (iv) prevent or inhibit growth of microorganisms which have gained access and have not been inactivated, during product storage (Gould, 1999; Juneja and Sofos, 2002; Koutsoumanis et al., 2006; Sofos, 1994, 2008; Stopforth and Sofos, 2006). The foundation of success of any preservation method is the restriction or minimization of the access of microorganisms to products; this includes mainly proper and adequate cleaning, sanitation and hygienic procedures, and effective packaging of processed products. Procedures that reduce contamination that has gained access include washing and decontamination or sanitization interventions. Intervention technologies utilized by U.S. beef slaughtering facilities to reduce contamination include carcass cleaning processes, such as knife trimming and/or steam vacuuming for removal of visible fecal contamination and carcass washing and decontamination interventions, such as rinsing or spraying

Ifigenia Geornaras and John N. Sofos • Center for Meat Safety & Quality, Food Safety Cluster of Infectious Diseases Supercluster, Department of Animal Sciences, Colorado State University, Fort Collins, CO 80523-1171.

with cold and/or hot (>74°C) water and/or chemical solutions (e.g., chlorine and warm [55°C] lactic acid) and/or steam pasteurization (Geornaras and Sofos, 2005; Koutsoumanis et al., 2006; Stopforth and Sofos, 2006). Decontamination intervention strategies are also applied to produce (Beuchat, 1998; Castillo and Rodríguez-García, 2004; Sapers, 2003) and have been reviewed excellently in other chapters of this book; this chapter focuses on procedures that inactivate or prevent/suppress the growth of microorganisms remaining on products. Preservation processes utilized by food processors can be of a physical, physicochemical, and/or biological nature. Physical methods include freezing, refrigeration, packaging (e.g., vacuum, modified atmospheres), heating, and irradiation. Physicochemical processes include acidification (low pH), reduced water activity or drying, smoking, and addition of antimicrobial agents or preservatives. Processes of a biological nature, also known as biopreservation, include use of microorganisms (e.g., lactic acid bacteria), their metabolites (bacteriocins, such as nisin), or both, as well as fermentates (Gould, 1999, 2001; Juneja and Sofos, 2002; Koutsoumanis et al., 2006). Regardless of the type of preservation method(s) used, however, the goal is to enhance product safety, maintain product quality, and extend shelf life by elimination or inhibition of the growth of pathogenic and spoilage microorganisms (Sofos, 1994).

The use of an intelligent combination of two or more processes (hurdles), each applied at individually sublethal levels to improve the microbial stability and safety of foods, known as hurdle technology, has gained widespread popularity and application in industrialized and developing countries (Leistner and Gould, 2002). Effective application of hurdle technology involves knowing microbial cell function, modes of antimicrobial action of various hurdles, and metabolic exhaustion and stress reactions of microorganisms in order to select effective hurdles, intensities, and sequences of application (Samelis and Sofos, 2003). In multitarget preservation, different hurdles are applied at the same time, or sequentially, with the purpose of attacking different cellular targets (e.g., cell membrane, enzyme systems, and DNA). As a result, homeostasis of the cell is disrupted, forcing it to activate repair mechanisms and synthesize stress shock proteins and other compounds. This process, in turn, may lead to metabolic exhaustion and death (Geornaras and Sofos, 2005; Leistner, 2000; Samelis and Sofos, 2003). A major advantage of using the multihurdle approach is that it involves the use of hurdles of lower intensity and, in so doing, has less of an adverse effect on the sensory, nutritive, and economic properties of

foods (Geornaras and Sofos, 2005; Leistner and Gould, 2002).

PROCESSING FOR MICROBIAL INHIBITION

Preservation technologies that delay or prevent growth of microbial contamination in/on foods are based on the manipulation of factors affecting microbial survival and growth. These factors include storage temperature (refrigeration and freezing), relative humidity, product composition, water activity, pH, redox potential (which is associated with the gaseous atmosphere surrounding the product), and naturally present or added antimicrobials (Gould, 2001; Juneja and Sofos, 2002). These factors have been discussed extensively in various reviews, chapters, and books (Doyle and Beuchat, 2007; Juneja and Sofos, 2002; Rahman, 1999; Sofos, 1993; Zeuthen and Bøgh-Sørensen, 2003); therefore, only some recent and interesting advances are presented in this chapter, including the use of natural antimicrobials, specifically plant essential oils, active packaging, and bacteriophages to inhibit or delay growth of microbial contamination in foods.

Natural Antimicrobials and Plant Essential Oils

Antimicrobials are used widely by the food industry to control growth of mainly spoilage, but also pathogenic, microorganisms on processed products during storage (Davidson et al., 2005). Moreover, they are used to reduce initial contamination (Beuchat, 1998; Castillo and Rodríguez-García, 2004; Koutsoumanis et al., 2006; Sapers, 2003; Sofos, 2005; Stopforth and Sofos, 2006) levels at the pre- and postharvest stages of processing (these aspects are also covered in other chapters in this book). Traditional antimicrobials approved by regulatory agencies for use in foods include lactic, acetic, benzoic, propionic, and sorbic acids and their salts, as well as nitrites, sulfites, phosphates, alkyl esters of p-hydroxybenzoic acid, and dimethyl dicarbonate (Davidson and Taylor, 2007; Davidson et al., 2005). In recent years, however, there has been an increasing demand by consumers for foods that are additive (including chemical or synthetic antimicrobial)-free and more "natural." This has led scientists and the food industry to look for and evaluate alternative antimicrobial agents, including naturally occurring antimicrobial compounds from microbial, animal, and plant sources (Davidson et al., 2005; Gould, 2001). Examples of natural antimicrobials include the microbially derived (natamycin, bacteriocins [nisin and pediocin], and ε-polylysine), the animal derived (the lactoperoxidase system, lactoferrin, lysozyme, ovotransferrin, avidin, and chitosan), and the plant derived (organic acids, phytoalexins, phenolic compounds, hops beta acids, and essential oils) (Davidson and Zivanovic, 2003; López-Malo Vigil et al., 2005; Smid and Gorris, 1999; Sofos et al., 1998). Some of these are currently approved for use in foods in the United States, including natamycin, nisin, ε-polylysine, lactoferrin, lysozyme, and hops beta acids (http://www.cfsan.fda.gov/~rdb/opa-gras.html; Davidson and Zivanovic, 2003).

It is well established that extracts obtained from spices, herbs, and other aromatic plants have antimicrobial properties (Sofos et al., 1998). The active compounds are contained commonly within the essential (volatile) oil fraction, and it is suggested that it is the structure and concentration of these active compounds that determine the antimicrobial activity of essential oils (Dorman and Deans, 2000; López-Malo Vigil et al., 2005). The antimicrobial properties of essential oils from spices are attributed mainly (Burt, 2004; Cosentino et al., 1999; Dorman and Deans, 2000) to compounds with phenolic groups (e.g., thymol, carvacrol, and eugenol). Numerous published studies conducted in vitro with microbiological media have reported on the antimicrobial effects of essential oils against a wide variety of food spoilage and pathogenic bacteria (Burt, 2004; Cermelli et al., 2008; Dorman and Deans, 2000; Hammer et al., 1999; Ouattara et al., 1997; Oussalah et al., 2007b; Preuss et al., 2005), yeasts (Giordani et al., 2004), molds (Bluma et al., 2008), parasites (de Almeida et al., 2007; Santoro et al., 2007), and viruses (Benencia and Courrèges, 2000; Cermelli et al., 2008). From these studies, it is apparent that the oils of cloves, cinnamon, oregano, rosemary, sage, thyme, and vanillin are consistently effective against various microorganisms (Davidson and Taylor, 2007; Holley and Patel, 2005). Oftentimes, however, direct comparison between quantitative data of published studies is difficult due to the use of different methods to extract essential oils from plant materials; different geographical origins, agronomic practices, and harvesting times of plant materials; use of different methods to evaluate antimicrobial efficacy; and differences in strains, culture media (type and pH), growth phases of test organisms, inoculation levels, and incubation temperatures and times (Burt, 2004; Janssen et al., 1987). Therefore, development and validation of standardized methods for evaluating the antimicrobial activity of essential oils are needed.

Compared to the number of studies conducted in vitro, there are fewer published reports on the antimicrobial effects of essential oils in food systems. Burt (2004) and Holley and Patel (2005) provide excellent reviews of such reports. Many of these studies concluded that the antimicrobial effects of essential oils

in vitro are significantly decreased when tested in vivo. Therefore, higher concentrations of the oils are needed to achieve the same effect, which may potentially have detrimental effects on the organoleptic properties of the food (Smid and Gorris, 1999). Gill et al. (2002) found that 6% cilantro oil applied as a surface treatment was ineffective in controlling growth of *Listeria monocytogenes* on vacuum-packaged ham, even though in vitro results showed that minimum inhibitory concentrations (MICs) (brain heart infusion broth, 24°C) ranged from 0.018 to 0.074%. In another study (Veldhuizen et al., 2007), carvacrol exhibited strong antilisterial effects (MIC of 1.6 mM) in tryptic soy broth incubated at 30°C, but when added to steak tartare at 5 mM/g, growth of *L. monocytogenes* was not different than that of the untreated control during storage at 10°C. Additional studies have shown that high concentrations of essential oils are needed to inhibit growth of various microorganisms on foods such as chicken frankfurters (Mytle et al., 2006), sausage (Busatta et al., 2008), cheese and ground meat (Vrinda Menon and Garg, 2001), Greek salads and pâté (Tassou et al., 1995), and pasteurized milk (Cava et al., 2007). It is generally believed that the potency of essential oils in foods is affected mainly by the fat and protein levels, which immobilize and inactivate the essential oil components, but also by the pH, water and salt contents, and storage temperature (Burt, 2004; Gutierrez et al., 2008; Singh et al., 2003; Skandamis and Nychas, 2000; Smith-Palmer et al., 2001; Tassou et al., 1995; Veldhuizen et al., 2007). Tassou et al. (1995) reported an increase in the antimicrobial effects of mint oil against *L. monocytogenes* and *Salmonella enterica* serovar Enteritidis in the order of cucumber yogurt salad (low fat, no protein) > fish roe salad > pâté (high fat, high protein), and synergistic effects were noted at low pH (4.3 [cucumber yogurt salad] versus 6.8 [pâté]). In addition, studies (Moon et al., 2006; Raybaudi-Massilia et al., 2006) have found promising antimicrobial effects of essential oils or essential oil components in fruit juices.

In order to overcome or minimize the undesirable sensory effects on foods associated with the addition of essential oils or their components and to potentially bring about synergistic effects, use of essential oils at subinhibitory levels in combination with other preservation technologies has been investigated. Synergistic effects against *L. monocytogenes* and *Bacillus subtilis* were obtained in vitro when subinhibitory concentrations of nisin Z (40 and 75 IU/ml, respectively) and thymol (0.02 and 0.03%, respectively) were used in combination (Ettayebi et al., 2000). The authors proposed that the synergistic effects were due to destabilization of the cell membrane by thymol, which resulted in increased permeability for nisin. Synergistic effects of nisin (1,000 IU/g) and thymol (0.6%) have also been reported for ground beef, and sensory evaluation of cooked samples containing 0.6% thymol was found to be acceptable by a group of trained panelists (Solomakos et al., 2008). Combining essential oils with vacuum or modified-atmosphere packaging (MAP) has also been shown to potentiate antibacterial effects. Oregano essential oil (0.8%) applied to fresh beef inoculated with 3.5 log units of *L. monocytogenes* reduced initial counts by 1 log (Tsigaridia et al., 2000). Subsequent storage (5°C) under aerobic conditions permitted pathogen growth to 5 log units by day 14, while storage under vacuum or MAP (40% CO_2/30% O_2/30% N_2) conditions resulted in listericidal effects. The antimicrobial activity of essential oils or their components has also been enhanced when combined with nonthermal processing technologies, such as high-pressure processing (HPP) (Karatzas et al., 2001; Mohácsi-Farkas et al., 2002; Palhano et al., 2004), high-intensity pulsed electric fields (PEF) (Mosqueda-Melgar et al., 2008a, 2008b), and ultrasound (Ferrante et al., 2007).

As indicated, extensive research has been conducted on the antimicrobial activity of essential oils; however, their future use and regulatory approval as food antimicrobials will require a better understanding of their activity spectrum, especially in foods, and their mechanisms of action, targets, and interactions with environmental conditions and food components, as well as toxicity studies (Davidson and Taylor, 2007). Also, further investigation of synergistic effects with other preservation methods will potentially minimize the concentrations needed to achieve antimicrobial effects and in this way reduce potential negative sensory effects and cost (Burt, 2004).

Active Packaging

Conventional packaging systems range from overwrapping of foods with oxygen permeable films for short-term chilled storage and/or retail display, to packaging systems with modified gaseous atmospheres containing high levels of oxygen and carbon dioxide (i.e., MAP) for longer-term chilled storage and/or display, to vacuum packaging or MAP systems utilizing 100% carbon dioxide (i.e., controlled-atmosphere packaging) for long-term chilled storage. The function of these traditional packaging systems is to provide a barrier between the food and the external environment and, as such, prevent microbial contamination. An additional goal in the case of vacuum packaging, MAP, and controlled-atmosphere packaging is to inhibit or delay the onset of microbial spoilage and

growth of pathogens (Jeremiah, 2001; Kerry et al., 2006; Koutsoumanis et al., 2006). Active packaging is a relatively innovative technology and encompasses those systems in which the package, product, and environment interact to extend the shelf life and/or enhance microbial safety and/or improve the sensory properties of the food, without affecting product quality (Quintavalla and Vicini, 2002; Suppakul et al., 2003; Vermeiren et al., 2002). These functions may be achieved by incorporation of additives into packaging systems, be it loosely within the package or incorporated within or attached to the inside of packaging materials (Kerry et al., 2006). The purpose of these additives is to absorb (scavenge) oxygen, moisture, ethylene, carbon dioxide, or flavors/odors or to release (emit) carbon dioxide, ethanol, antioxidants, flavors, or antimicrobial agents (López-Rubio et al., 2004; Ozdemir and Floros, 2004). Thus, any packaging technology that interacts with the internal gaseous atmosphere of the package and/or directly with the food product and has a beneficial outcome is referred to as active packaging (López-Rubio et al., 2004). Active packaging systems are being utilized in the United States, Japan, and Australia; however, little development or application has occurred in Europe (Coma, 2008; López-Rubio et al., 2004).

Antimicrobial packaging is a form of active packaging that has shown promising results with intact meat products, mainly since microbial contamination of these products occurs primarily on the surface (Koutsoumanis et al., 2006; Quintavalla and Vicini, 2002). In recent years, strategies to improve the microbial safety of intact meat products in the United States have included addition of antimicrobials into product formulations (e.g., lactate-diacetate in processed meats) and/or application of surface antimicrobial spraying or immersion treatments (e.g., lactic acid onto beef carcass surfaces during slaughter) (Koutsoumanis et al., 2006). Possible limitations of these types of applications include neutralization or partial inactivation of active compounds when coming into contact with food components and, in the case of surface applications, rapid diffusion of active compounds from the surface into the food mass (Quintavalla and Vicini, 2002). Antimicrobial-containing packaging films could be more efficient than the direct application methods described above, due to the slow release of the active compounds from the packaging material to the product surface, thus helping to maintain high concentrations where needed. In addition, a long migration period of the antimicrobial may extend its activity into all stages of the food chain, including transport, distribution, and storage (Koutsoumanis et al., 2006; Quintavalla and Vicini, 2002).

Overall, nonedible antimicrobial films can be classified as those that contain additives that (i) migrate to the surface of the food or (ii) are bound to the surface layer of the film and are effective against surface contamination without migration. The antimicrobial additives can be coated, incorporated, immobilized, or surface modified onto the packaging material (Kerry et al., 2006; Suppakul et al., 2003). Any type of food-grade antimicrobial agent or compound may be used, for example, organic acid salts, such as lactate, diacetate, benzoate, and sorbate (Ye et al., 2008); bacteriocins, such as nisin, enterocin, and sakacin (Jofré et al., 2007; Mauriello et al., 2005; Neetoo et al., 2008); antibiotics, such as natamycin (de Oliveira et al., 2007); and various plant extracts or essential oils with antimicrobial properties (Ha et al., 2001; Suppakul et al., 2008). Coating of a low-density polyethylene film with grapefruit seed extract (1%) inhibited growth of aerobic and coliform bacteria on ground beef during storage (3°C) and extended the time to reach spoilage levels (i.e., 7 log CFU/g) by approximately 5 days (Ha et al., 2001). Neetoo et al. (2008) evaluated nisin-coated films against surface contamination of cold-smoked salmon with *L. monocytogenes* (initial levels of 2.7 or 5.7 log CFU/cm^2) and found immediate reductions of the pathogen, followed by inhibition of growth during storage at 4 and 10°C. For samples inoculated with 2.7 log CFU/cm^2 of *L. monocytogenes* and wrapped in low-density polyethylene films coated with 2,000 IU/cm^2 nisin, final pathogen counts following 56 and 49 days of storage at 4 and 10°C, respectively, were 3.9 log CFU/cm^2 lower than those of the control (film with no nisin) (Neetoo et al., 2008). The benefit of using combinations of antimicrobials in active packaging systems has also been demonstrated (Jofré et al., 2007). Interleavers containing nisin (200 AU/g) combined with potassium lactate (1.8%) were more effective in inhibiting growth of *L. monocytogenes* on ham slices than interleavers containing each of the antimicrobials alone; following 30 days of storage at 6°C, *L. monocytogenes* counts for the combination treatment were 4 to 5 log CFU/g lower than the individual treatments (Jofré et al., 2007).

Edible coatings and films supplemented with antimicrobial agents are another type of antimicrobial packaging system; however, in this case, selection of active agents is limited to edible compounds. Edible films are typically comprised of three major components: polysaccharides (e.g., alginate, cellulose, chitosan, starch, carageenan, pectin, and dextrin), proteins (e.g., casein, corn zein, gelatin/collagen, wheat gluten, and soy/whey protein), and lipids (e.g., waxes, acylglycerols, and fatty acids), used alone or in combination. There are numerous published reports (Cagri et al., 2004; Joerger, 2007; Koutsoumanis et al., 2006; Lin and Zhao, 2007; Quintavalla and Vicini, 2002) related to the testing of edible coatings and films on various muscle foods and fresh and minimally processed fruits and vegetables. In the case of muscle foods, use of edible coatings and films can enhance the quality of the product by reducing moisture loss, reducing drip, and minimizing lipid oxidation and discoloration, and with the addition of antimicrobials, there is enhancement of microbial safety and shelf life extension (Cutter, 2006; Gennadios et al., 1997). Antimicrobials that have been evaluated with edible films or coatings for controlling growth of spoilage microorganisms and food-borne pathogens include organic acids (Ouattara et al., 2000), essential oils (Oussalah et al., 2006, 2007a; Zivanovic et al., 2005), nisin and other bacteriocins (Janes et al., 2002; Marcos et al., 2007), and lysozyme (Duan et al., 2007; Kim et al., 2008). Enhanced antimicrobial effects obtained by using antimicrobial-containing edible coatings in combination with other preservation technologies, such as HPP (Marcos et al., 2008a) and irradiation (Ouattara et al., 2001, 2002), have been reported.

Bacteriophages

Reports on the use of bacteriophages (phages) to treat human bacterial infections date back to more than 90 years ago. Research in this area dwindled following the discovery of antibiotics, but in recent years there has been renewed interest in the field of phage therapy due to concerns regarding the emergence of antibiotic-resistant bacteria (Hagens and Loessner, 2007; IFT Expert Panel, 2006). The possibility of using virulent (lytic) phages to control food-borne pathogens has also sparked interest among researchers (Carlton et al., 2005; Hagens and Loessner, 2007; Strauch et al., 2007). The cycle of a lytic phage includes binding of the virus to specific receptors on the bacterial cell surface, injection of the phage genome into the cell, replication of the phage genetic material, and release of the phage progeny following host cell lysis. In contrast, temperate phages have the option of entering the lysogenic cycle, which involves integration of the phage genome into the bacterial chromosome (or other genetic elements), where it replicates as part of the host genome (Hudson et al., 2005; Strauch et al., 2007). This can lead to an altered phenotype of the host and, in some cases, an increase in bacterial pathogenicity (Brüssow et al., 2004; Hagens and Loessner, 2007). Lytic phages do not possess the genes for integration into the host genome, which means that host cell infection is always followed by host cell lysis to release phage progeny, which then go on to infect other target bacterial cells

(Strauch et al., 2007). For this reason, lytic phages are better suited for use to control pathogen contamination in foods.

Attributes of lytic phages that make them attractive for use as biocontrol agents of food-borne pathogens include the following: (i) they are bacterial viruses and as such, are natural enemies of bacteria; (ii) they are ubiquitous in nature (e.g., soil, water, and effluent), and their presence in foods (Baross et al., 1978; Kennedy et al., 1984; Tsuei et al., 2007; Whitman and Marshall, 1971) suggests that they are normal inhabitants of food ecosystems; and (iii) they are host specific and generally do not cross species or genus barriers and hence will not affect starter cultures in fermentation processes and commensal bacteria in the gastrointestinal tract (Carlton et al., 2005; Hudson et al., 2005). The potential for phage-mediated control of bacterial pathogens in foods is reflected by recent studies with *L. monocytogenes* (Carlton et al., 2005; Leverentz et al., 2003, 2004), *Staphylococcus aureus* (García et al., 2007b), *Campylobacter jejuni* (Atterbury et al., 2003; Bigwood et al., 2008; Goode et al., 2003), *Escherichia coli* O157:H7 (O'Flynn et al., 2004), *Salmonella* (Bigwood et al., 2008; Goode et al., 2003; Leverentz et al., 2001), and *Enterobacter sakazakii* (Kim et al., 2007). Leverentz et al. (2001) evaluated a phage cocktail for its ability to control growth of *Salmonella* serovar Enteritidis on fresh-cut honeydew melons and apple slices. Pathogen numbers decreased by up to 3.5 log units on melon stored at 5 or 10°C; however, similar reductions were not obtained on apples. The authors reported rapid inactivation of the phage on apple slices and hypothesized that it was, at least partially, due to the low pH of this fruit (pH 4.2 versus pH 5.8 of melon). Similar findings were observed when the same fruits were contaminated with *L. monocytogenes* and treated with a mixture of lytic phages specific for *L. monocytogenes*, including serotypes 1/2a, 1/2b, and 4b (Leverentz et al., 2003). Antilisterial effects during storage of phage-treated melons were enhanced when high concentrations (8 log PLU/ml) of the phage mixture were used and by treating the melon either at the time of contamination or up to 1 h before contamination (Leverentz et al., 2004). In another study (García et al., 2007b), addition of two lytic phages to milk prior to acid and enzymatic curd manufacture reduced counts of *S. aureus* to nondetectable levels (reduction of >6 log units) within 4 h and 1 h, respectively.

The U.S. Food and Drug Administration (FDA) and U.S. Department of Agriculture Food Safety and Inspection Service (USDA/FSIS) have approved use of a preparation consisting of six lytic phages as an antimicrobial agent against *L. monocytogenes* on ready-to-eat meat and poultry products (http://www.accessdata.fda.gov/scripts/cdrh/cfdocs/cfCFR/CFRSearch.cfm?fr=172.785; http://www.fsis.usda.gov/OPPDE/rdad/FSISDirectives/7120.1Amend16.pdf). Another phage preparation (called "bacteriophage P100 preparation from *Listeria innocua*") has received generally recognized-as-safe approval for control of *L. monocytogenes* in cheese (i.e., brie, cheddar, Swiss, and other cheeses that are normally aged and ripened), meat and poultry products, and foods in general (http://www.cfsan.fda.gov/~rdb/opa-gras.html; http://www.fsis.usda.gov/OPPDE/rdad/FSISDirectives/7120.1Amend16.pdf). Preparation P100 has a broad host range within the *Listeria* genus and is characterized as being strictly lytic; the phage infected and killed more than 95% out of approximately 250 food-borne *L. monocytogenes* strains belonging to serotypes 1/2 and 4, and 5 strains of *Listeria ivanovii*. Its ability to control *L. monocytogenes* in food was evaluated on a surface-ripened Munster-type soft cheese (Carlton et al., 2005). The cheese was contaminated (approximately 1 log CFU/cm^2) with the pathogen at the beginning of the ripening period and then repeatedly treated with a low (6.3 log PLU/cm^2) or high (7.8 log PLU/cm^2) concentration of P100 during the rind-washing process. At the low concentration of P100, growth of *L. monocytogenes* was inhibited and counts were 3.5 log units lower than those of the control at the end of the 21-day ripening/storage period. At the high concentration of P100, the pathogen was inactivated and remained undetectable (confirmed by enrichment) during the entire ripening/storage period. Complete inactivation of *L. monocytogenes* was also obtained when P100 was applied as a single-dose treatment at the high concentration (7.8 log PLU/cm^2). Preparation P100 has been found to be safe based on a repeated dose oral toxicity study performed with rats (Carlton et al., 2005).

The possibility of using phages to extend the shelf life of foods has also been considered. Homologous phages and bacteriophage pools have been evaluated for controlling the growth of *Pseudomonas* spp., *Brochothrix thermosphacta*, and *Leuconostoc gelidum* on meat (Greer, 1986; Greer and Dilts, 1990, 2002; Greer et al., 2007). Also, use of phage therapy at the preharvest stage of food production to reduce fecal carriage of human bacterial pathogens in food-producing animals has been reported (Callaway et al., 2008; Fiorentin et al., 2005; Sheng et al., 2006; Wagenaar et al., 2005). Reduction of fecal carriage of pathogens would reduce their levels on the outside (hide, fleece, and feathers) of animals and birds prior to slaughter, thus reducing contamination levels on carcasses during slaughter and further down the food production chain.

Phage-mediated control of bacterial pathogens in foods is still at the experimental stage (Strauch et al., 2007). Some disadvantages that may affect their successful use for enhancing the safety of foods include their limited host range, the potential for development of phage-resistant mutants, the possible presence of virulence genes on the phage genome, the potential for lytic phages to mutate into temperate variants and lysogenize the host, and the consumer's perception of having viruses added to foods (Greer, 2005; Hudson et al., 2005; Strauch et al., 2007). Also, there is evidence (Bigwood et al., 2008; Greer, 2005; Hudson et al., 2005; Wiggins and Alexander, 1985) that a host cell threshold density is needed for phage replication to occur, which would be of critical relevance in controlling growth of pathogens in foods when the host cells are present at low levels (Greer, 2005).

PROCESSING FOR MICROBIAL INACTIVATION

Application of heat, or thermal processing, is the most widely used method for destruction of microbial contamination in food, with time and temperature of heating determining the number of microorganisms destroyed in each food type (Awuah et al., 2007; Doyle and Beuchat, 2007; Juneja and Sofos, 2002; Koutsoumanis et al., 2006; Mañas and Pagán, 2005). Although thermal processing is effective, economical, and readily available, it cannot be applied to all foods since exposure to heat for an extended period of time can trigger destructive chemical changes and can lead to a loss of nutrients and sensory properties (Awuah et al., 2007; Torres and Velazquez, 2005). For this reason, and also due to the increasing demand by the consumer for "fresh" and "natural" foods, alternative preservation technologies for microbial destruction which do not rely on heat have recently received increased attention. Nonthermal technologies include irradiation, HPP, PEF, radio frequency electric fields, UV light, pulsed light, ultrasound, and others (http://www.fda.gov/Food/ScienceResearch/ResearchAreas/SafePracticesforFoodProcesses/ucm100158.htm). Irradiation is not a new concept (Doyle and Beuchat, 2007; Gould, 1999; Juneja and Sofos, 2002; O'Bryan et al., 2008). More than 40 countries have approved its use in a variety of foods, most recently (August 2008) for fresh spinach and iceberg lettuce in the United States (http://www.fda.gov/ForConsumers/ConsumerUpdates/ucm093651.htm; Mahapatra et al., 2005). Its commercial application has been hampered by consumer concerns regarding its safety; however, social perception of irradiation may be slowly changing. Irradiated foods on a commercial scale have been introduced successfully in countries such as Belgium, China, France, Japan, South Africa, Thailand, and the United States (Loaharanu, 2003).

HPP

HPP, also referred to as high hydrostatic pressure or ultra-high pressure processing, is one of the most researched new and emerging nonthermal inactivation technologies (Devlieghere et al., 2004). Its use as a food preservation method was first reported more than 100 years ago (Hite, 1899), but it was only in the early 1990s that the first commercial application of HPP to foods (jams and jellies) was reported (Aymerich et al., 2008; Patterson, 2005). The process involves subjecting packaged or unpackaged liquid and solid foods to pressures in the range of 100 to 1,000 MPa (Aymerich et al., 2008; Cheftel, 1995; Doyle and Beuchat, 2007). What makes HPP a more desirable preservation technology over thermal processing and other preservation methods is that HPP effects are uniform and instantaneous throughout the food and therefore are independent of the food shape and size (Torres and Velazquez, 2005). Moreover, in contrast to thermal processing, HPP does not disrupt covalent bonds; thus, foods retain, to a large extent, their sensorial attributes and nutritional value (Patterson, 2005; San Martín et al., 2002). Commercial applications of HPP have been reported in the United States, Australia, Asia, and Europe (Aymerich et al., 2008; Considine et al., 2008), and it has been applied to solid and liquid foods, including juices, fruit jams, jellies, sauces, rice products, cake, fish and shellfish, poultry products, sliced ready-to-eat meats, and meal kits (containing pressure-treated cooked meats, acidified vegetables, salsa, and guacamole) (Devlieghere et al., 2004; Goh et al., 2007; Murchie et al., 2005).

Depending on the pressure applied, microbial inactivation is most likely due to a combination of factors affecting the cell membrane, cell morphology, biochemical reactions, and genetic mechanisms (Aymerich et al., 2008; Cheftel, 1995; Hoover et al., 1989). It is generally acknowledged, however, that the primary target is the cell membrane (Cheftel, 1995; Mañas and Pagán, 2005; Smelt, 1998). Effects of HPP vary among microorganisms and bacterial species and strains (Alpas et al., 1999; Patterson 2005; San Martín et al., 2002). With regard to vegetative cells, gram-positive bacteria are generally more resistant to inactivation than gram-negative bacteria (Patterson, 2005; Smelt, 1998). HPP treatments of 400 to 600 MPa applied at room temperature reduce counts of most vegetative cells by >4 log units (Devlieghere et al., 2004). Tolerance of microorganisms to HPP is influenced by the growth phase of cells; in general, cells in stationary phase are more resistant to pressure than those in exponential phase (Hayman et al., 2007). HPP (400 MPa, 23 to 24°C, 10 min) of *L. monocytogenes* cells in stationary and exponential phases resulted in reductions of 1.3 and >7 log units, respectively (Mackey et al., 1995). As with other preservation methods, bacterial spores are very resistant to HPP, with spores of *Clostridium botulinum* being the most pressure tolerant. *C. botulinum* spores can survive treatment with more than 1,000 MPa, unless the process is carried out at temperatures close to 100°C (Cheftel, 1995; Patterson, 2005; Smelt, 1998). Spores can be triggered to germinate by treating them with relatively low pressures (<200 MPa); thus, a cyclical two-stage process can be used, in which the first pressure treatment results in spore germination and the second process, applied at a higher pressure, inactivates the now germinated spores (Black et al., 2007; Patterson, 2005). Inactivation of viruses, yeasts, and molds with HPP has also been reported, although these studies are limited (Goh et al., 2007; Grove et al., 2006; Murchie et al., 2005; Patterson, 2005). Recent evidence shows that HPP may also have an effect on prions (Heindl et al., 2008; Patterson, 2005). Heindl et al. (2008) found a decrease of infectivity, by up to 6 to 8 log units, of hamster-adapted scrapie strain 263K following pressure treatment at 800 MPa at 60 to 80°C for 5 (80°C) or 30 (60 and 80°C) min.

Apart from microbial characteristics, the efficacy of HPP on microbial inactivation is influenced by process and product/substrate parameters (Mañas and Pagán, 2005). Process parameters include pressure magnitude, temperature, and time (Considine et al., 2008; Patterson, 2005; Simpson and Gilmour, 1997a). HPP treatments applied at temperatures below or above the optimum growth temperature of the microorganism are generally more effective (Hugas et al., 2002; Raso and Barbosa-Cánovas, 2003). Product/substrate parameters include its chemical composition, pH, and water activity (Considine et al., 2008; Patterson, 2005). Certain food constituents such as proteins, carbohydrates, and lipids can have a protective effect, as does low water activity (Simpson and Gilmour, 1997b); thus, it is important not to extrapolate data obtained in buffer systems to real food systems (Smelt, 1998).

To further enhance the safety and shelf life of HPP-treated products and to explore the possibility of synergistic effects, the process has been evaluated in combination with other preservation (physical and chemical) technologies. In this regard, the combined effect of HPP and heat is the most extensively investigated (Devlieghere et al., 2004; Raso and Barbosa-Cánovas, 2003) for enhanced inactivation of vegetative cells (Kalchayanand et al., 1998; Patterson and Kilpatrick, 1998; Simpson and Gilmour, 1997a) and

spores (Black et al., 2007; Gao and Ju, 2008; Reddy et al., 2006). HPP has also been tested together with antimicrobials, including potassium lactate, lactate-diacetate, bacteriocins (nisin, enterocin, pediocin, and sakacin), and lysozyme in laboratory media (Kalchayanand et al., 1998), as formulation ingredients of meat models/products (Aymerich et al., 2005; Garriga et al., 2002; Jofré et al., 2008b; Marcos et al., 2008b), additives to milk and fruit juices (Nakimbugwe et al., 2006), or active packaging ingredients (Jofré et al., 2008a; Marcos et al., 2008a). By applying an additional hurdle to the former combination treatments, such as storage of foods at refrigeration temperatures, additional antimicrobial effects may be obtained. For example, application of an HPP (400 MPa, 17°C, 10 min) treatment to inoculated cooked ham slices reduced initial *L. monocytogenes* levels by 3.4 log CFU/cm^2, while storage of the product at 6°C between alginate films containing enterocins (2,000 AU/cm^2) inhibited growth of the pathogen during vacuum-packaged storage for 60 days (Marcos et al., 2008a). In comparison, HPP treatment without subsequent exposure to enterocin-containing packaging films inhibited growth of *L. monocytogenes* for only 8 days (Marcos et al., 2008a). Discovery of combination treatments with synergistic effects is important, as they would potentially allow use of less-severe high-pressure treatments, which would not only minimize any possible detrimental effects on nutritional and sensory aspects of foods, but would also potentially reduce operation costs of this process (http://www.fda.gov/Food/ScienceResearch/ResearchAreas/SafePracticesforFoodProcesses/ucm100158.htm).

PEF

Another nonthermal processing technology is high-intensity PEF, also referred to as PEF or high electric field pulses. It was introduced in the 1960s but has only recently stimulated interest in the food industry due to technological developments that offer the possibility of scaling up the technology (Wouters et al., 2001). The PEF process involves application of short-duration (less than 1 s) high-intensity electric field pulses (typically 20 to 80 kV/cm) to fluid foods placed between two electrodes. The process is applied at subambient to just-above-ambient temperatures and overall does not affect the nutritional or sensory aspects of the food (http://www.cfsan.fda.gov/~comm/ift-toc.html; Cserhalmi et al., 2006). Various PEF systems have been evaluated for inactivation of pathogenic and spoilage microorganisms in buffers, various fruit juices (e.g., apple, grapefruit, orange, pear, tomato, melon, watermelon, etc.), milk, liquid egg whites, and liquid whole eggs (Barsotti and Cheftel, 1999; Craven et al., 2008; Min et al., 2007;

Mosqueda-Melgar et al., 2008a, 2008b; Wouters et al., 2001). Depending on a number of factors, described below, reductions ranging from 2 to 9 log units have been reported for vegetative bacteria and yeasts in these substrates (Barsotti and Cheftel, 1999; Min et al., 2007; Wouters et al., 2001).

Microbial inactivation by PEF is a result of irreversible damage to the cell membrane (García et al., 2007a; Min et al., 2007; Wouters et al., 2001). The magnitude of microbial inactivation depends on process parameters (i.e., electric field strength, pulse length and shape, number of pulses, and treatment time and temperature), product parameters (i.e., composition, conductivity, ionic strength, pH, and water activity), and microbial characteristics (type and size of microorganism, initial microbial load, and growth phase of the organism) (Toepfl et al., 2007; Wouters et al., 2001). In general, the order of sensitivity to PEF is as follows: yeasts > gram-negative vegetative cells > gram-positive vegetative cells > bacterial spores and mold ascospores (Wouters et al., 2001). For example, Aronsson et al. (2001) reported 6.3-, 5.4-, 1.6-, and 1.4-log-unit reductions of *Saccharomyces cerevisiae*, *E. coli*, *Leuconostoc mesenteroides*, and *L. innocua*, respectively, following PEF treatment (30 kV/cm, 20 pulses, 4-µs duration). As with other preservation technologies, bacterial cells in the exponential phase of growth are more sensitive to PEF than cells in stationary phase (Cebrián et al., 2007). Also, inactivation due to PEF treatment is reportedly enhanced in an acidic environment and may be due to the inability of sublethally injured cells to recover in a low pH environment. PEF treatment (28 kV/cm, 400 µs) of *L. monocytogenes* in a buffer medium adjusted to pH 7.0, 6.5, 5.0, and 3.5 reduced counts by 1.5, 2.3, 3.0, and 6.0 log cycles, respectively (Gómez et al., 2005).

There are several reports on enhanced microbial inactivation when PEF treatment is combined with other methods, such as moderate heat (<60°C) (Craven et al., 2008; Nguyen and Mittal, 2007; Toepfl et al., 2007), bacteriocins (Nguyen and Mittal, 2007; Smith et al., 2002; Sobrino-López and Martín-Belloso, 2008), essential oils (Mosqueda-Melgar et al., 2008a, 2008b), organic acids (Mosqueda-Melgar et al., 2008a, 2008b; Somolinos et al., 2007), and enzymes (Smith et al., 2002; Sobrino-López and Martín-Belloso, 2008). PEF treatment (35 kV/cm, 4-µs pulse duration) of melon and watermelon juices treated with 2 and 1.5% citric acid, respectively, or 0.2% cinnamon bark oil, reduced populations of *E. coli* O157:H7, *Salmonella* Enteritidis, and *L. monocytogenes* by 5 to 7 log CFU/ml, whereas reductions obtained without the addition of antimicrobials were 3 to 4 log CFU/ml (Mosqueda-Melgar et al., 2008b).

Sensory analysis of the treated juices, however, revealed negative effects on the odor and taste of the products due to the addition of citric acid and cinnamon bark oil. In another study (Smith et al., 2002), a 7-log reduction of natural microbial contamination was obtained when raw skim milk was treated with PEF (80 kV/cm, 50 pulses) in combination with mild heat (52°C), nisin (38 IU/ml), and lysozyme (1638 IU/ml); the PEF treatment in combination with mild heat (i.e., without nisin and lysozyme) only reduced counts by 1.3 logs.

Although there have been reports of one company in the United States that produced PEF-treated juices and blends on a commercial scale (Clark, 2006; Min et al., 2007), more studies are needed to verify the microbiological and chemical safety of PEF-treated foods before the process can be applied on a larger scale (Min et al., 2007).

Other Technologies

Several other food processing technologies for microbial inactivation have been described in scientific publications. These include radio frequency heating, ohmic and conductive heating, UV light, pulsed light, ultrasound, magnetic fields, high-voltage arc discharge, and dense-phase carbon dioxide. Many of these are still in the experimental stage of development and therefore will not be discussed in detail. For additional information, we refer the reader to reviews by the FDA's Center for Food Safety and Applied Nutrition (http://www.fda.gov/Food/ScienceResearch/ResearchAreas/SafePracticesforFoodProcesses/ucm100158.htm) and Morris et al. (2007), as well as studies cited below.

Radio frequency heating (capacitive dielectric heating) uses electromagnetic energy to heat food uniformly and rapidly and has large penetration depth and lower energy consumption than conventional cooking methods (Piyasena et al., 2003a; Rowley, 2001; Tang et al., 2006). Microbial inactivation using radio frequency heating has been evaluated in milk (Awuah et al., 2005) and caviar (Al-Holy et al., 2004). Radio frequency electric fields processing, on the other hand, is a nonthermal process, similar to PEF. The difference between the two is that in PEF, the field is applied in pulses using a pulse generator, whereas for the radio frequency electric fields process, the field is applied continuously using an AC generator (Geveke and Brunkhort, 2008; Geveke et al., 2007). The pulsed-light system is another nonthermal process. This technology involves application of intense, short-duration pulses of broad spectrum (UV to near-infrared) light, and has been evaluated as a surface decontamination method of foods and packaging materials (Elmnasser et al., 2007; Gómez-López et al., 2007). When applied to various minimally processed vegetables, aerobic mesophilic populations were reduced by 0.6 to 2.0 log CFU/g (Gómez-López et al., 2005). Ultrasound (sound waves with a frequency of >20 kHz) has been reported to have weak bactericidal effects on its own, but enhanced lethal effects can be obtained by combining it with mild heat, pressure (manosonication), manosonication combined with temperature (manothermosonication), UV irradiation, and antimicrobials (Ferrante et al., 2007; Pagán et al., 1999; Piyasena et al., 2003b).

The above technologies all aim to enhance the microbiological safety of foods in a way that does not affect the sensory and nutritional properties of the product and in a manner that is energy efficient, cost effective, and environmentally friendly. However, before these technologies can be safely applied to foods, a better understanding of their mechanisms of inactivation is needed, as well as evaluation of possible synergistic effects when combined with other preservation methods or processes.

CONCLUSIONS AND FUTURE OUTLOOK

The goal of any food preservation method is to enhance product safety, maintain product quality, and extend product shelf life by inactivating or retarding the growth of pathogenic and spoilage microorganisms. Traditionally, the most popular food preservation processes used to accomplish this were, and still are, heating, manipulation of the product's pH and/or water activity, use of chemical antimicrobial compounds, and controlling the product's storage temperature and atmosphere. Of late, increasing consumer demands for foods that are convenient, fresh, minimally processed without the addition of synthetic chemical preservatives, and "healthy" have led to significant research efforts by the scientific community, as well as devotion of considerable resources by the food industry, to produce foods and food products that are wholesome and microbiologically safe and have an extended shelf life. Successful implementation of novel preservation methods, such as the use of essential oils and other natural antimicrobials, antimicrobial packaging, bacteriophages, and nonthermal processes for microbial inactivation, among others, will depend on a better understanding of the mechanisms of microbial inactivation in foods, the behavior of microbial cells during and after application of the treatment, toxicity studies where necessary, and evaluation of synergistic effects with one or more of the existing or novel preservation methods or processes to minimize product sensory effects and possible development of stress-hardened cells.

REFERENCES

Al-Holy, M., J. Ruiter, M. Lin, D.-H. Kang, and B. Rasco. 2004. Inactivation of *Listeria innocua* in nisin-treated salmon *(Oncorhynchus keta)* and sturgeon *(Acipenser transmontanus)* caviar heated by radio frequency. *J. Food Prot.* 67:1848–1854.

Alpas, H., N. Kalchayanand, F. Bozoglu, A. Sikes, C. P. Dunne, and B. Ray. 1999. Variation in resistance to hydrostatic pressure among strains of food-borne pathogens. *Appl. Environ. Microbiol.* 65:4248–4251.

Aronsson, K., M. Lindgren, B. R. Johansson, and U. Rönner. 2001. Inactivation of microorganisms using pulsed electric fields: the influence of process parameters on *Escherichia coli, Listeria innocua, Leuconostoc mesenteroides* and *Saccharomyces cerevisiae. Innov. Food Sci. Emerg. Technol.* 2:41–54.

Atterbury, R. J., P. L. Connerton, C. E. R. Dodd, C. E. D. Rees, and I. F. Connerton. 2003. Application of host-specific bacteriophages to the surface of chicken skin leads to a reduction in recovery of *Campylobacter jejuni. Appl. Environ. Microbiol.* 69:6302–6306.

Awuah, G. B., H. S. Ramaswamy, and A. Economides. 2007. Thermal processing and quality: principles and overview. *Chem. Eng. Process.* 46:584–602.

Awuah, G. B., H. S. Ramaswamy, A. Economides, and K. Mallikarjunan. 2005. Inactivation of *Escherichia coli* K-12 and *Listeria innocua* in milk using radio frequency (RF) heating. *Innov. Food Sci. Emerg. Technol.* 6:396–402.

Aymerich, T., A. Jofré, M. Garriga, and M. Hugas. 2005. Inhibition of *Listeria monocytogenes* and *Salmonella* by natural antimicrobials and high hydrostatic pressure in sliced cooked ham. *J. Food Prot.* 68:173–177.

Aymerich, T., P. A. Picouet, and J. M. Monfort. 2008. Decontamination technologies for meat products. *Meat Sci.* 78:114–129.

Baross, J. A., J. Liston, and R. Y. Morita. 1978. Incidence of *Vibrio parahaemolyticus* bacteriophages and other *Vibrio* bacteriophages in marine samples. *Appl. Environ. Microbiol.* 36:492–499.

Barsotti, L., and J. C. Cheftel. 1999. Food processing by pulsed electric fields. II. Biological aspects. *Food Rev. Int.* 15:181–213.

Benencia, F., and M. C. Courrèges. 2000. *In vitro* and *in vivo* activity of eugenol on human herpesvirus. *Phytother. Res.* 14:495–500.

Beuchat, L. R. 1998. Surface decontamination of fruits and vegetables eaten raw: a review. WHO/FSF/FOS/98.2. Food Safety Unit, World Health Organization, Geneva, Switzerland.

Bigwood, T., J. A. Hudson, C. Billington, G. V. Carey-Smith, and J. A. Heinemann. 2008. Phage inactivation of foodborne pathogens on cooked and raw meat. *Food Microbiol.* 25:400–406.

Black, E. P., P. Setlow, A. D. Hocking, C. M. Stewart, A. L. Kelly, and D. G. Hoover. 2007. Response of spores to high-pressure processing. *Compr. Rev. Food Sci. Food Saf.* 6:103–119.

Bluma, R., M. R. Amaiden, J. Daghero, and M. Etcheverry. 2008. Control of *Aspergillus* section *Flavi* growth and aflatoxin accumulation by plant essential oils. *J. Appl. Microbiol.* 105:203–214.

Brüssow, H., C. Canchaya, and W.-D. Hardt. 2004. Phages and the evolution of bacterial pathogens: from genomic rearrangements to lysogenic conversion. *Microbiol. Mol. Biol. Rev.* 68:560–602.

Burt, S. 2004. Essential oils: their antibacterial properties and potential applications in foods—a review. *Int. J. Food Microbiol.* 94:223–253.

Busatta, C., R. S. Vidal, A. S. Popiolski, A. J. Mossi, C. Dariva, M. R. A. Rodrigues, F. C. Corazza, M. L. Corazza, J. Vladimir Oliveira, and R. L. Cansian. 2008. Application of *Origanum majorana* L. essential oil as an antimicrobial agent in sausage. *Food Microbiol.* **25**:207–211.

Cagri, A., Z. Ustunol, and E. T. Ryser. 2004. Antimicrobial edible films and coatings. *J. Food Prot.* **67**:833–848.

Callaway, T. R., T. S. Edrington, A. D. Brabban, R. C. Anderson, M. L. Rossman, M. J. Engler, M. A. Carr, K. J. Genovese, J. E. Keen, M. L. Looper, E. M. Kutter, and D. J. Nisbet. 2008. Bacteriophage isolated from feedlot cattle can reduce *Escherichia coli* O157:H7 populations in ruminant gastrointestinal tracts. *Foodborne Pathog. Dis.* **5**:183–191.

Carlton, R. M., W. H. Noordman, B. Biswas, E. D. de Meester, and M. J. Loessner. 2005. Bacteriophage P100 for control of *Listeria monocytogenes* in foods: genome sequence, bioinformatic analyses, oral toxicity study, and application. *Regul. Toxicol. Pharmacol.* **43**:301–312.

Castillo, A., and M. O. Rodríguez-García. 2004. Bacterial hazards in fresh and fresh-cut produce: sources and control, p. 43–57. *In* R. C. Beier, S. D. Pillai, T. D. Phillips, and R. L. Ziprin (ed.), *Preharvest and Postharvest Food Safety: Contemporary Issues and Future Directions*. Blackwell Publishing Professional, Ames, IA.

Cava, R., E. Nowak, A. Taboada, and F. Marin-Iniesta. 2007. Antimicrobial activity of clove and cinnamon essential oils against *Listeria monocytogenes* in pasteurized milk. *J. Food Prot.* **70**:2757–2763.

Cebrián, G., N. Sagarzazu, R. Pagán, S. Condón, and P. Mañas. 2007. Heat and pulsed electric field resistance of pigmented and non-pigmented enterotoxigenic strains of *Staphylococcus aureus* in exponential and stationary phase of growth. *Int. J. Food Microbiol.* **118**:304–311.

Cermelli, C., A. Fabio, G. Fabio, and P. Quaglio. 2008. Effect of eucalyptus essential oil on respiratory bacteria and viruses. *Curr. Microbiol.* **56**:89–92.

Cheftel, J. C. 1995. Review: High-pressure, microbial inactivation and food preservation. *Food Sci. Technol. Int.* **1**:75–90.

Clark, J. P. 2006. Pulsed electric field processing. *Food Technol.* **60**:66–67.

Coma, V. 2008. Bioactive packaging technologies for extended shelf life of meat-based products. *Meat Sci.* **78**:90–103.

Considine, K. M., A. L. Kelly, G. F. Fitzgerald, C. Hill, and R. D. Sleator. 2008. High-pressure processing—effects on microbial food safety and food quality. *FEMS Microbiol. Lett.* **281**:1–9.

Cosentino, S., C. I. G. Tuberoso, B. Pisano, M. Satta, V. Mascia, E. Arzedi, and F. Palmas. 1999. *In-vitro* antimicrobial activity and chemical composition of Sardinian *Thymus* essential oils. *Lett. Appl. Microbiol.* **29**:130–135.

Craven, H. M., P. Swiergon, S. Ng, J. Midgely, C. Versteeg, M. J. Coventry, and J. Wan. 2008. Evaluation of pulsed electric field and minimal heat treatments for inactivation of pseudomonads and enhancement of milk shelf-life. *Innov. Food Sci. Emerg. Technol.* **9**:211–216.

Cserhalmi, Z., Á. Sass-Kiss, M. Tóth-Markus, and N. Lechner. 2006. Study of pulsed electric field treated citrus juices. *Innov. Food Sci. Emerg. Technol.* **7**:49–54.

Cutter, C. N. 2006. Opportunities for bio-based packaging technologies to improve the quality and safety of fresh and further processed muscle foods. *Meat Sci.* **74**:131–142.

Davidson, P. M., and T. M. Taylor. 2007. Chemical preservatives and natural antimicrobial compounds, p. 713–745. *In* M. P. Doyle and L. R. Beuchat (ed.), *Food Microbiology: Fundamentals and Frontiers*, 3rd ed. ASM Press, Washington, DC.

Davidson, P. M., and S. Zivanovic. 2003. The use of natural antimicrobials, p. 5–30. *In* P. Zeuthen, and L. Bøgh-Sørensen (ed.), *Food Preservation Techniques*. CRC Press, Boca Raton, FL.

Davidson, P. M., J. N. Sofos, and A. L. Branen. 2005. *Antimicrobials in Food*, 3rd ed. CRC Press/Taylor & Francis, Boca Raton, FL.

de Almeida, I., D. S. Alviano, D. P. Vieira, P. B. Alves, A. F. Blank, A. H. C. S. Lopes, C. S. Alviano, and M. S. Rosa. 2007. Antigiardial activity of *Ocimum basilicum* essential oil. *Parasitol. Res.* **101**:443–452.

de Oliveira, T. M., N. F. F. Soares, R. M. Pereira, and K. F. Fraga. 2007. Development and evaluation of antimicrobial natamycin-incorporated film in Gorgonzola cheese conservation. *Packag. Technol. Sci.* **20**:147–153.

Devlieghere, F., L. Vermeiren, and J. Debevere. 2004. New preservation technologies: possibilities and limitations. *Int. Dairy J.* **14**:273–285.

Dorman, H. J. D., and S. G. Deans. 2000. Antimicrobial agents from plants: antibacterial activity of plant volatile oils. *J. Appl. Microbiol.* **88**:308–316.

Doyle, M. P., and L. R. Beuchat. 2007. *Food Microbiology: Fundamentals and Frontiers*, 3rd ed. ASM Press, Washington, DC.

Duan, J., S.-I. Park, M. A. Daeschel, and Y. Zhao. 2007. Antimicrobial chitosan-lysozyme (CL) films and coatings for enhancing microbial safety of Mozzarella cheese. *J. Food Sci.* **72**:M355–M362.

Elmnasser, N., S. Guillou, F. Leroi, N. Orange, A. Bakhrouf, and M. Federighi. 2007. Pulsed-light system as a novel food decontamination technology: a review. *Can. J. Microbiol.* **53**:813–821.

Ettayebi, K., J. E. Yamani, and B.-D. Rossi-Hassani. 2000. Synergistic effects of nisin and thymol on antimicrobial activities in *Listeria monocytogenes* and *Bacillus subtilis*. *FEMS Microbiol. Lett.* **183**:191–195.

Ferrante, S., S. Guerrero, and S. M. Alzamora. 2007. Combined use of ultrasound and natural antimicrobials to inactivate *Listeria monocytogenes* in orange juice. *J. Food Prot.* **70**:1850–1856.

Fiorentin, L., N. D. Vieira, and W. Barioni, Jr. 2005. Oral treatment with bacteriophages reduces the concentration of *Salmonella* Enteritidis PT4 in caecal contents of broilers. *Avian Pathol.* **34**:258–263.

Gao, Y.-L., and X.-R. Ju. 2008. Exploiting the combined effects of high pressure and moderate heat with nisin on inactivation of *Clostridium botulinum* spores. *J. Microbiol. Methods* **72**:20–28.

García, D., N. Gómez, P. Mañas, J. Raso, and R. Pagán. 2007a. Pulsed electric fields cause bacterial envelopes permeabilization depending on the treatment intensity, the treatment medium pH and the microorganism investigated. *Int. J. Food Microbiol.* **113**:219–227.

García, P., C. Madera, B. Martínez, and A. Rodríguez. 2007b. Biocontrol of *Staphylococcus aureus* in curd manufacturing processes using bacteriophages. *Int. Dairy J.* **17**:1232–1239.

Garriga, M., M. T. Aymerich, S. Costa, J. M. Monfort, and M. Hugas. 2002. Bactericidal synergism through bacteriocins and high pressure in a meat model system during storage. *Food Microbiol.* **19**:509–518.

Gennadios, A., M. A. Hanna, and L. B. Kurth. 1997. Application of edible coatings on meats, poultry and seafoods: a review. *Lebensm. Wiss. Technol.* **30**:337–350.

Geornaras, I., and J. N. Sofos. 2005. Combining physical and chemical decontamination interventions for meat, p. 433–460. *In* J. N. Sofos (ed.), *Improving the Safety of Fresh Meat*. CRC Press, Boca Raton, FL.

Geveke, D. J., and C. Brunkhorst. 2008. Radio frequency electric fields inactivation of *Escherichia coli* in apple cider. *J. Food Eng.* **85**:215–221.

Geveke, D. J., C. Brunkhorst, and X. Fan. 2007. Radio frequency electric fields processing of orange juice. *Innov. Food Sci. Emerg. Technol.* 8:549–554.

Gill, A. O., P. Delaquis, P. Russo, and R. A. Holley. 2002. Evaluation of antilisterial action of cilantro oil on vacuum packed ham. *Int. J. Food Microbiol.* 73:83–92.

Giordani, R., P. Regli, J. Kaloustian, C. Mikaïl, L. Abou, and H. Portugal. 2004. Antifungal effect of various essential oils against *Candida albicans*. Potentiation of antifungal action of amphotericin B by essential oil from *Thymus vulgaris*. *Phytother. Res.* 18:990–995.

Goh, E. L. C., A. D. Hocking, C. M. Stewart, K. A. Buckle, and G. H. Fleet. 2007. Baroprotective effect of increased solute concentrations on yeast and moulds during high pressure processing. *Innov. Food Sci. Emerg. Technol.* 8:535–542.

Gómez, N., D. García, I. Álvarez, S. Condón, and J. Raso. 2005. Modelling inactivation of *Listeria monocytogenes* by pulsed electric fields in media of different pH. *Int. J. Food Microbiol.* 103:199–206.

Gómez-López, V. M., F. Devlieghere, V. Bonduelle, and J. Debevere. 2005. Intense light pulses decontamination of minimally processed vegetables and their shelf-life. *Int. J. Food Microbiol.* 103:79–89.

Gómez-López, V. M., P. Ragaert, J. Debevere, and F. Devlieghere. 2007. Pulsed light for food decontamination: a review. *Trends Food Sci. Technol.* 18:464–473.

Goode, D., V. M. Allen, and P. A. Barrow. 2003. Reduction of experimental *Salmonella* and *Campylobacter* contamination of chicken skin by application of lytic bacteriophages. *Appl. Environ. Microbiol.* 69:5032–5036.

Gould, G. W. 1999. *New Methods of Food Preservation*. Aspen Publishers, Inc., Gaithersburg, MD.

Gould, G. W. 2001. New processing technologies: an overview. *Proc. Nutr. Soc.* 60:463–474.

Greer, G. G. 1986. Homologous bacteriophage control of *Pseudomonas* growth and beef spoilage. *J. Food Prot.* 49:104–109.

Greer, G. G. 2005. Bacteriophage control of foodborne bacteria. *J. Food Prot.* 68:1102–1111.

Greer, G. G., and B. D. Dilts. 1990. Inability of a bacteriophage pool to control beef spoilage. *Int. J. Food Microbiol.* 10:331–342.

Greer, G. G., and B. D. Dilts. 2002. Control of *Brochothrix thermosphacta* spoilage of pork adipose tissue using bacteriophages. *J. Food Prot.* 65:861–863.

Greer, G. G., B. D. Dilts, and H.-W. Ackermann. 2007. Characterization of a *Leuconostoc gelidum* bacteriophage from pork. *Int. J. Food Microbiol.* 114:370–375.

Grove, S. F., A. Lee, T. Lewis, C. M. Stewart, H. Chen, and D. G. Hoover. 2006. Inactivation of foodborne viruses of significance by high pressure and other processes. *J. Food Prot.* 69:957–968.

Gutierrez, J., C. Barry-Ryan, and P. Bourke. 2008. The antimicrobial efficacy of plant essential oil combinations and interactions with food ingredients. *Int. J. Food Microbiol.* 124:91–97.

Ha, J.-U., Y.-M. Kim, and D.-S. Lee. 2001. Multilayered antimicrobial polyethylene films applied to the packaging of ground beef. *Packag. Technol. Sci.* 15:55–62.

Hagens, S., and M. J. Loessner. 2007. Application of bacteriophages for detection and control of foodborne pathogens. *Appl. Microbiol. Biotechnol.* 76:513–519.

Hammer, K. A., C. F. Carson, and T. V. Riley. 1999. Antimicrobial activity of essential oils and other plant extracts. *J. Appl. Microbiol.* 86:985–990.

Hayman, M. M., R. C. Anantheswaran, and S. J. Knabel. 2007. The effects of growth temperature and growth phase on the inactivation of *Listeria monocytogenes* in whole milk subject to high pressure processing. *Int. J. Food Microbiol.* 115:220–226.

Heindl, P., A. F. Garcia, P. Butz, B. Trierweiler, H. Voigt, E. Pfaff, and B. Tauscher. 2008. High pressure/temperature treatments to inactivate highly infectious prion subpopulations. *Innov. Food Sci. Emerg. Technol.* 9:290–297.

Hite, B. H. 1899. The effect of pressure in the preservation of milk. *Bull. West Virginia Univ. Agric. Exp. Station* 58:15–35.

Holley, R. A., and D. Patel. 2005. Improvement in shelf-life and safety of perishable foods by plant essential oils and smoke antimicrobials. *Food Microbiol.* 22:273–292.

Hoover, D. G., C. Metrick, A. M. Papineau, D. F. Farkas, and D. Knorr. 1989. Biological effects of high hydrostatic pressure on food microorganisms. *Food Technol.* 43:99–107.

Hudson, J. A., C. Billington, G. Carey-Smith, and G. Greening. 2005. Bacteriophages as biocontrol agents in food. *J. Food Prot.* 68:426–437.

Hugas, M., M. Garriga, and J. M. Monfort. 2002. New mild technologies in meat processing: high pressure as a model technology. *Meat Sci.* 62:359–371.

IFT Expert Panel. 2006. Antimicrobial resistance: implications for the food system. An expert report funded by the Institute of Food Technologists Foundation. *Compr. Rev. Food Sci. Food Saf.* 5:71–137.

Janes, M. E., S. Kooshesh, and M. G. Johnson. 2002. Control of *Listeria monocytogenes* on the surface of refrigerated, ready-to-eat chicken coated with edible zein film coatings containing nisin and/or calcium propionate. *J. Food Sci.* 67:2754–2757.

Janssen, A. M., J. J. C. Scheffer, and A. Baerheim Svendsen. 1987. Antimicrobial activity of essential oils: a 1976-1986 literature review. Aspects of the test methods. *Planta Med.* 53:395–398.

Jeremiah, L. E. 2001. Packaging alternatives to deliver fresh meats using short- or long-term distribution. *Food Res. Int.* 34:749–772.

Joerger, R. D. 2007. Antimicrobial films for food applications: a quantitative analysis of their effectiveness. *Packag. Technol. Sci.* 20:231–273.

Jofré, A., M. Garriga, and T. Aymerich. 2007. Inhibition of *Listeria monocytogenes* in cooked ham through active packaging with natural antimicrobials and high-pressure processing. *J. Food Prot.* 70:2498–2502.

Jofré, A., T. Aymerich, and M. Garriga. 2008a. Assessment of the effectiveness of antimicrobial packaging combined with high pressure to control *Salmonella* sp. in cooked ham. *Food Control* 19:634–638.

Jofré, A., M. Garriga, and T. Aymerich. 2008b. Inhibition of *Salmonella* sp., *Listeria monocytogenes* and *Staphylococcus aureus* in cooked ham by combining antimicrobials, high hydrostatic pressure and refrigeration. *Meat Sci.* 78:53–59.

Juneja, V. K., and J. N. Sofos. 2002. *Control of Foodborne Microorganisms*. Marcel Dekker, Inc., New York, NY.

Kalchayanand, N., A. Sikes, C. P. Dunne, and B. Ray. 1998. Interaction of hydrostatic pressure, time and temperature of pressurization and pediocin AcH on inactivation of foodborne bacteria. *J. Food Prot.* 61:425–431.

Karatzas, A. K., E. P. W. Kets, E. J. Smid, and M. H. J. Bennik. 2001. The combined action of carvacrol and high hydrostatic pressure on *Listeria monocytogenes* Scott A. *J. Appl. Microbiol.* 90:463–469.

Kennedy, J. E., Jr., J. L. Oblinger, and G. Bitton. 1984. Recovery of coliphages from chicken, pork sausage and delicatessen meats. *J. Food Prot.* 47:623–626.

Kerry, J. P., M. N. O'Grady, and S. A. Hogan. 2006. Past, current and potential utilisation of active and intelligent packaging systems for meat and muscle-based products: a review. *Meat Sci.* **74**:113–130.

Kim, K.-P., J. Klumpp, and M. J. Loessner. 2007. *Enterobacter sakazakii* bacteriophages can prevent bacterial growth in reconstituted infant formula. *Int. J. Food Microbiol.* **115**:195–203.

Kim, K. W., M. Daeschel, and Y. Zhao. 2008. Edible coatings for enhancing microbial safety and extending shelf life of hardboiled eggs. *J. Food Sci.* **73**:M227–M235.

Koutsoumanis, K. P., I. Geornaras, and J. N. Sofos. 2006. Microbiology of land muscle foods, p. 52.1–52.43. *In* Y. H. Hui (ed.), *Handbook of Food Science, Technology and Engineering*, vol. 1. CRC Press, Taylor & Francis Group, Boca Raton, FL.

Leistner, L. 2000. Basic aspects of food preservation by hurdle technology. *Int. J. Food Microbiol.* **55**:181–186.

Leistner, L., and G. W. Gould. 2002. *Hurdle Technologies: Combination Treatments for Food Stability, Safety and Quality*. Kluwer Academic/Plenum Publishers, New York, NY.

Leverentz, B., W. S. Conway, Z. Alavidze, W. J. Janisiewicz, Y. Fuchs, M. J. Camp, E. Chighladze, and A. Sulakvelidze. 2001. Examination of bacteriophage as a biocontrol method for *Salmonella* on fresh-cut fruit: a model study. *J. Food Prot.* **64**:1116–1121.

Leverentz, B., W. S. Conway, M. J. Camp, W. J. Janisiewicz, T. Abuladze, M. Yang, R. Saftner, and A. Sulakvelidze. 2003. Biocontrol of *Listeria monocytogenes* on fresh-cut produce by treatment with lytic bacteriophages and a bacteriocin. *Appl. Environ. Microbiol.* **69**:4519–4526.

Leverentz, B., W. S. Conway, W. Janisiewicz, and M. J. Camp. 2004. Optimizing concentration and timing of a phage spray application to reduce *Listeria monocytogenes* on honeydew melon tissue. *J. Food Prot.* **67**:1682–1686.

Lin, D., and Y. Zhao. 2007. Innovations in the development and application of edible coatings for fresh and minimally processed fruits and vegetables. *Compr. Rev. Food Sci. Food Saf.* **6**:60–75.

Loaharanu, P. 2003. *Irradiated Foods*, 5th ed. American Council on Science and Health, New York, NY.

López-Malo Vigil, A., E. Palou, and S. M. Alzamora. 2005. Naturally occurring compounds—plant sources, p. 429–451. *In* P. M. Davidson, J. N. Sofos, and A. L. Branen (ed.), *Antimicrobials in Food*, 3rd ed. CRC Press/Taylor & Francis, Boca Raton, FL.

López-Rubio, A., E. Almenar, P. Hernandez-Muñoz, J. M. Lagarón, R. Catalá, and R. Gavara. 2004. Overview of active polymer-based packaging technologies for food applications. *Food Rev. Int.* **20**:357–387.

Mackey, B. M., K. Forestière, and N. Isaacs. 1995. Factors affecting the resistance of *Listeria monocytogenes* to high hydrostatic pressure. *Food Biotechnol.* **9**:1–11.

Mahapatra, A. K., K. Muthukumarappan, and J. L. Julson. 2005. Applications of ozone, bacteriocins and irradiation in food processing: a review. *Crit. Rev. Food Sci. Nutr.* **45**:447–461.

Mañas, P., and R. Pagán. 2005. Microbial inactivation by new technologies of food preservation. *J. Appl. Microbiol.* **98**:1387–1399.

Marcos, B., T. Aymerich, J. M. Monfort, and M. Garriga. 2007. Use of antimicrobial biodegradable packaging to control *Listeria monocytogenes* during storage of cooked ham. *Int. J. Food Microbiol.* **120**:152–158.

Marcos, B., T. Aymerich, J. M. Monfort, and M. Garriga. 2008a. High-pressure processing and antimicrobial biodegradable packaging to control *Listeria monocytogenes* during storage of cooked ham. *Food Microbiol.* **25**:177–182.

Marcos, B., A. Jofré, T. Aymerich, J. M. Monfort, and M. Garriga. 2008b. Combined effect of natural antimicrobials and high pressure processing to prevent *Listeria monocytogenes* growth after

a cold chain break during storage of cooked ham. *Food Control* **19**:76–81.

Mauriello, G., E. De Luca, A. La Storia, F. Villani, and D. Ercolini. 2005. Antimicrobial activity of a nisin-activated plastic film for food packaging. *Lett. Appl. Microbiol.* **41**:464–469.

Min, S., G. A. Evrendilek, and H. Q. Zhang. 2007. Pulsed electric fields: processing system, microbial and enzyme inhibition, and shelf life extension of foods. *IEEE Trans. Plasma Sci.* **35**:59–73.

Mohácsi-Farkas, C., G. Kiskó, L. Mészáros, and J. Farkas. 2002. Pasteurisation of tomato juice by high hydrostatic pressure treatment or by its combination with essential oils. *Acta Alimentaria* **31**:243–252.

Moon, K. D., P. Delaquis, P. Toivonen, S. Bach, K. Stanich, and L. Harris. 2006. Destruction of *Escherichia coli* O157:H7 by vanillic acid in unpasteurized juice from six apple cultivars. *J. Food Prot.* **69**:542–547.

Morris, C., A. L. Brody, and L. Wicker. 2007. Non-thermal food processing/preservation technologies: a review with packaging implications. *Packag. Technol. Sci.* **20**:275–286.

Mosqueda-Melgar, J., R. M. Raybaudi-Massilia, and O. Martín-Belloso. 2008a. Non-thermal pasteurization of fruit juices by combining high-intensity pulsed electric fields with natural antimicrobials. *Innov. Food Sci. Emerg. Technol.* **9**:328–340.

Mosqueda-Melgar, J., R. M. Raybaudi-Massilia, and O. Martín-Belloso. 2008b. Combination of high-intensity pulsed electric fields with natural antimicrobials to inactivate pathogenic microorganisms and extend the shelf-life of melon and watermelon juices. *Food Microbiol.* **25**:479–491.

Murchie, L. W., M. Cruz-Romero, J. P. Kerry, M. Linton, M. F. Patterson, M. Smiddy, and A. L. Kelly. 2005. High pressure processing of shellfish: a review of microbiological and other quality aspects. *Innov. Food Sci. Emerg. Technol.* **6**:257–270.

Mytle, N., G. L. Anderson, M. P. Doyle, and M. A. Smith. 2006. Antimicrobial activity of clove *(Syzygium aromaticum)* oil in inhibiting *Listeria monocytogenes* on chicken frankfurters. *Food Control* **17**:102–107.

Nakimbugwe, D., B. Masschalck, G. Anim, and C. W. Michiels. 2006. Inactivation of gram-negative bacteria in milk and banana juice by hen egg white and lambda lysozyme under high hydrostatic pressure. *Int. J. Food Microbiol.* **112**:19–25.

Neetoo, H., M. Ye, H. Chen, R. D. Joerger, D. T. Hicks, and D. G. Hoover. 2008. Use of nisin-coated plastic films to control *Listeria monocytogenes* on vacuum-packaged cold-smoked salmon. *Int. J. Food Microbiol.* **122**:8–15.

Nguyen, P., and G. S. Mittal. 2007. Inactivation of naturally occurring microorganisms in tomato juice using pulsed electric field (PEF) with and without antimicrobials. *Chem. Eng. Process.* **46**:360–365.

O'Bryan, C. A., P. G. Crandall, S. C. Ricke, and D. G. Olson. 2008. Impact of irradiation on the safety and quality of poultry and meat products: a review. *Crit. Rev. Food Sci. Nutr.* **48**:442–457.

O'Flynn, G., R. P. Ross, G. F. Fitzgerald, and A. Coffey. 2004. Evaluation of a cocktail of three bacteriophages for biocontrol of *Escherichia coli* O157:H7. *Appl. Environ. Microbiol.* **70**:3417–3424.

Ouattara, B., S. F. Sabato, and M. Lacroix. 2001. Combined effect of antimicrobial coating and gamma irradiation on shelf life extension of pre-cooked shrimp *(Penaeus* spp.). *Int. J. Food Microbiol.* **68**:1–9.

Ouattara, B., S. F. Sabato, and M. Lacroix. 2002. Use of gamma-irradiation technology in combination with edible coating to produce shelf-stable foods. *Radiat. Phys. Chem.* **63**:305–310.

Ouattara, B., R. E. Simard, R. A. Holley, G. J.-P. Piette, and A. Bégin. 1997. Antibacterial activity of selected fatty acids and essential oils against six meat spoilage organisms. *Int. J. Food Microbiol.* **37:**155–162.

Ouattara, B., R. E. Simard, G. Piette, A. Bégin, and R. A. Holley. 2000. Inhibition of surface spoilage bacteria in processed meats by application of antimicrobial films prepared with chitosan. *Int. J. Food Microbiol.* **62:**139–148.

Oussalah, M., S. Caillet, S. Salmiéri, L. Saucier, and M. Lacroix. 2006. Antimicrobial effects of alginate-based film containing essential oils for the preservation of whole beef muscle. *J. Food Prot.* **69:**2364–2369.

Oussalah, M., S. Caillet, S. Salmiéri, L. Saucier, and M. Lacroix. 2007a. Antimicrobial effects of alginate-based films containing essential oils on *Listeria monocytogenes* and *Salmonella* Typhimurium present in bologna and ham. *J. Food Prot.* **70:**901–908.

Oussalah, M., S. Caillet, L. Saucier, and M. Lacroix. 2007b. Inhibitory effects of selected plant essential oils on the growth of four pathogenic bacteria: *E. coli* O157:H7, *Salmonella* Typhimurium, *Staphylococcus aureus* and *Listeria monocytogenes*. *Food Control* **18:**414–420.

Ozdemir, M., and J. D. Floros. 2004. Active food packaging technologies. *Crit. Rev. Food Sci. Nutr.* **44:**185–193.

Pagán, R., P. Mañas, I. Alvarez, and S. Condón. 1999. Resistance of *Listeria monocytogenes* to ultrasonic waves under pressure at sublethal (manosonication) and lethal (manothermosonication) temperatures. *Food Microbiol.* **16:**139–148.

Palhano, F. L., T. T. B. Vilches, R. B. Santos, M. T. D. Orlando, J. A. Ventura, and P. M. B. Fernandes. 2004. Inactivation of *Colletotrichum gloeosporioides* spores by high hydrostatic pressure combined with citral or lemongrass essential oil. *Int. J. Food Microbiol.* **95:**61–66.

Patterson, M. F. 2005. Microbiology of pressure-treated foods. *J. Appl. Microbiol.* **98:**1400–1409.

Patterson, M. F., and D. J. Kilpatrick. 1998. The combined effect of high hydrostatic pressure and mild heat on inactivation of pathogens in milk and poultry. *J. Food Prot.* **61:**432–436.

Piyasena, P., C. Dussault, T. Koutchma, H. S. Ramaswamy, and G. B. Awuah. 2003a. Radio frequency heating of foods: principles, applications and related properties—a review. *Crit. Rev. Food Sci. Nutr.* **43:**587–606.

Piyasena, P., E. Mohareb, and R. C. McKellar. 2003b. Inactivation of microbes using ultrasound: a review. *Int. J. Food Microbiol.* **87:**207–216.

Preuss, H. G., B. Echard, M. Enig, I. Brook, and T. B. Elliott. 2005. Minimum inhibitory concentrations of herbal essential oils and monolaurin for gram-positive and gram-negative bacteria. *Mol. Cell Biochem.* **272:**29–34.

Quintavalla, S., and L. Vicini. 2002. Antimicrobial food packaging in meat industry. *Meat Sci.* **62:**373–380.

Rahman, M. S. 1999. *Handbook of Food Preservation.* Marcel Dekker, Inc., New York, NY.

Raso, J., and G. V. Barbosa-Cánovas. 2003. Nonthermal preservation of foods using combined processing techniques. *Crit. Rev. Food Sci. Nutr.* **43:**265–285.

Raybaudi-Massilia, R. M., J. Mosqueda-Melgar, and O. Martín-Belloso. 2006. Antimicrobial activity of essential oils on *Salmonella* Enteritidis, *Escherichia coli*, and *Listeria innocua* in fruit juices. *J. Food Prot.* **69:**1579–1586.

Reddy, N. R., R. C. Tetzloff, H. M. Solomon, and J. W. Larkin. 2006. Inactivation of *Clostridium botulinum* nonproteolytic type B spores by high pressure processing at moderate to elevated high temperatures. *Innov. Food Sci. Emerg. Technol.* **7:**169–175.

Rowley, A. T. 2001. Radio frequency heating, p. 163–177. *In* P. Richardson (ed.), *Thermal Technologies in Food Processing.* Woodhead Publishing, Cambridge, England.

Samelis, J., and J. N. Sofos. 2003. Strategies to control stress-adapted pathogens, p. 303–351. *In* A. E. Yousef and V. K. Juneja (ed.), *Microbial Stress Adaptation and Food Safety.* CRC Press, Boca Raton, FL.

San Martín, M. F., G. V. Barbosa-Cánovas, and B. G. Swanson. 2002. Food processing by high hydrostatic pressure. *Crit. Rev. Food Sci. Nutr.* **42:**627–645.

Santoro, G. F., M. das Graças Cardoso, L. G. L. Guimarães, A. P. S. P. Salgado, R. F. S. Menna-Barreto, and M. J. Soares. 2007. Effect of oregano (*Origanum vulgare* L.) and thyme (*Thymus vulgaris* L.) essential oils on *Trypanosoma cruzi* (Protozoa: Kinetoplastida) growth and ultrastructure. *Parasitol. Res.* **100:**783–790.

Sapers, G. M. 2003. Washing and sanitizing raw materials for minimally processed fruit and vegetable products, p. 221–253. *In* J. S. Novak, G. M. Sapers, and V. K. Juneja (ed.), *Microbial Safety of Minimally Processed Foods.* CRC Press, Boca Raton, FL.

Sheng, H., H. J. Knecht, I. T. Kudva, and C. J. Hovde. 2006. Application of bacteriophages to control intestinal *Escherichia coli* O157:H7 levels in ruminants. *Appl. Environ. Microbiol.* **72:**5359–5366.

Simpson, R. K., and A. Gilmour. 1997a. The resistance of *Listeria monocytogenes* to high hydrostatic pressure in foods. *Food Microbiol.* **14:**567–573.

Simpson, R. K., and A. Gilmour. 1997b. The effect of high hydrostatic pressure on *Listeria monocytogenes* in phosphate-buffered saline and model food systems. *J. Appl. Microbiol.* **83:**181–188.

Singh, A., R. K. Singh, A. K. Bhunia, and N. Singh. 2003. Efficacy of plant essential oils as antimicrobial agents against *Listeria monocytogenes* in hotdogs. *Lebensm. Wiss. Technol.* **36:**787–794.

Skandamis, P. N., and G.-J. E. Nychas. 2000. Development and evaluation of a model predicting the survival of *Escherichia coli* O157:H7 NCTC 12900 in homemade eggplant salad at various temperatures, pHs, and oregano essential oil concentrations. *Appl. Environ. Microbiol.* **66:**1646–1653.

Smelt, J. P. P. M. 1998. Recent advances in the microbiology of high pressure processing. *Trends Food Sci. Technol.* **9:**152–158.

Smid, E. J., and L. G. M. Gorris. 1999. Natural antimicrobials for food preservation, p. 285–308. *In* M. S. Rahman (ed.), *Handbook of Food Preservation.* Marcel Dekker, Inc., New York, NY.

Smith, K., G. S. Mittal, and M. W. Griffiths. 2002. Pasteurization of milk using pulsed electrical field and antimicrobials. *J. Food Sci.* **67:**2304–2308.

Smith-Palmer, A., J. Stewart, and L. Fyfe. 2001. The potential application of plant essential oils as natural food preservatives in soft cheese. *Food Microbiol.* **18:**463–470.

Sobrino-López, A., and O. Martín-Belloso. 2008. Enhancing the lethal effect of high-intensity pulsed electric field in milk by antimicrobial compounds as combined hurdles. *J. Dairy Sci.* **91:**1759–1768.

Sofos, J. N. 1993. Current microbiological considerations in food preservation. *Int. J. Food Microbiol.* **19:**87–108.

Sofos, J. N. 1994. Microbial growth and its control in meat, poultry and fish, p. 359–403. *In* A. M. Pearson and T. R. Dutson (ed.), *Quality Attributes and Their Measurement in Meat, Poultry and Fish Products.* Blackie Academic and Professional, Glasgow, United Kingdom.

Sofos, J. N. 2005. *Improving the Safety of Fresh Meat.* CRC Press, Boca Raton, FL.

Sofos, J. N. 2008. Challenges to meat safety in the 21st century. *Meat Sci.* **78:**3–13.

Sofos, J. N., L. R. Beuchat, P. M. Davidson, and E. A. Johnson. 1998. Naturally occurring antimicrobials in food. *Regul. Toxicol. Pharmacol.* **28:**71–72.

Solomakos, N., A. Govaris, P. Koidis, and N. Botsoglou. 2008. The antimicrobial effect of thyme essential oil, nisin, and their combination against *Listeria monocytogenes* in minced beef during refrigerated storage. *Food Microbiol.* 25:120–127.

Somolinos, M., D. García, S. Condón, P. Mañas, and R. Pagán. 2007. Relationship between sublethal injury and inactivation of yeast cells by the combination of sorbic acid and pulsed electric fields. *Appl. Environ. Microbiol.* 73:3814–3821.

Stopforth, J. D., and J. N. Sofos. 2006. Recent advances in pre- and postslaughter intervention strategies for control of meat contamination, p. 66–86. *In* V. K. Juneja, J. P. Cherry, and M. H. Tunick (ed.), *Advances in Microbial Food Safety.* American Chemical Society, Washington, DC.

Strauch, E., J. A. Hammerl, and S. Hertwig. 2007. Bacteriophages: new tools for safer food? *J. Verbr. Lebensm.* 2:138–143.

Suppakul, P., J. Miltz, K. Sonneveld, and S. W. Bigger. 2003. Active packaging technologies with an emphasis on antimicrobial packaging and its applications. *J. Food Sci.* 68:408–420.

Suppakul, P., K. Sonneveld, S. W. Bigger, and J. Miltz. 2008. Efficacy of polyethylene-based antimicrobial films containing principal constituents of basil. *LWT Food Sci. Technol.* 41:779–788.

Tang, X., J. G. Lyng, D. A. Cronin, and C. Durand. 2006. Radio frequency heating of beef rolls from *biceps femoris* muscle. *Meat Sci.* 72:467–474.

Tassou, C. C., E. H. Drosinos, and G. J. E. Nychas. 1995. Effects of essential oil from mint *(Mentha piperita)* on *Salmonella enteritidis* and *Listeria monocytogenes* in model food systems at 4°C and 10°C. *J. Appl. Bacteriol.* 78:593–600.

Toepfl, S., V. Heinz, and D. Knorr. 2007. High intensity pulsed electric fields applied for food preservation. *Chem. Eng. Process.* 46:537–546.

Torres, J. A., and G. Velazquez. 2005. Commercial opportunities and research challenges in the high pressure processing of foods. *J. Food Eng.* 67:95–112.

Tsigarida, E., P. Skandamis, and G.-J. E. Nychas. 2000. Behaviour of *Listeria monocytogenes* and autochthonous flora on meat stored under aerobic, vacuum and modified atmosphere packaging conditions with or without the presence of oregano essential oil at 5°C. *J. Appl. Microbiol.* 89:901–909.

Tsuei, A.-C., G. V. Carey-Smith, J. A. Hudson, C. Billington, and J. A. Heinemann. 2007. Prevalence and numbers of coliphages and *Campylobacter jejuni* bacteriophages in New Zealand foods. *Int. J. Food Microbiol.* 116:121–125.

Veldhuizen, E. J. A., T. O. Creutzberg, S. A. Burt, and H. P. Haagsman. 2007. Low temperature and binding to food components inhibit the antibacterial activity of carvacrol against *Listeria monocytogenes* in steak tartare. *J. Food Prot.* 70:2127–2132.

Vermeiren, L., F. Devlieghere, and J. Debevere. 2002. Effectiveness of some recent antimicrobial packaging concepts. *Food Addit. Contam.* 19:163–171.

Vrinda Menon, K., and S. R. Garg. 2001. Inhibitory effect of clove oil on *Listeria monocytogenes* in meat and cheese. *Food Microbiol.* 18:647–650.

Wagenaar, J. A., M. A. P. Van Bergen, M. A. Mueller, T. M. Wassenaar, and R. M. Carlton. 2005. Phage therapy reduces *Campylobacter jejuni* colonization in broilers. *Vet. Microbiol.* 109:275–283.

Whitman, P. A., and R. T. Marshall. 1971. Isolation of psychrophilic bacteriophage-host systems from refrigerated food products. *Appl. Microbiol.* 22:220–223.

Wiggins, B. A., and M. Alexander. 1985. Minimum bacterial density for bacteriophage replication: implications for significance of bacteriophages in natural ecosystems. *Appl. Environ. Microbiol.* 49:19–23.

Wouters, P. C., I. Alvarez, and J. Raso. 2001. Critical factors determining inactivation kinetics by pulsed electric field food processing. *Trends Food Sci. Technol.* 12:112–121.

Ye, M., H. Neetoo, and H. Chen. 2008. Control of *Listeria monocytogenes* on ham steaks by antimicrobials incorporated into chitosan-coated plastic films. *Food Microbiol.* 25:260–268.

Zeuthen, P., and L. Bøgh-Sørensen. 2003. *Food Preservation Techniques.* CRC Press, Boca Raton, FL.

Zivanovic, S., S. Chi, and A. F. Draughon. 2005. Antimicrobial activity of chitosan films enriched with essential oils. *J. Food Sci.* 70:M45–M51.

Pathogens and Toxins in Foods: Challenges and Interventions
Edited by V. K. Juneja and J. N. Sofos
© 2010 ASM Press, Washington, DC

Chapter 26

Interventions for Hazard Control in Retail-Handled Ready-To-Eat Foods

ALEXANDRA LIANOU AND JOHN N. SOFOS

In response to changes in consumer eating preferences, driven by corresponding societal and dietetic changes, food retailers have modified their practices to increase the role of the retail industry in the food processing, preparation, and marketing chain (Collins, 1997; Zink, 1997; Codron et al., 2005). Efforts to meet consumer demands for quality, safety, and convenience have resulted in a marked increase in ready-to-eat (RTE) or ready-to-cook products available, including in-store (i.e., bakeries, convenience stores, and supermarkets) prepared (i.e., in delicatessen sections) foods (Codron et al., 2005; USDHHS-FDA-CFSAN, 2006a).

The intensive efforts that have been made during the past decade to comply with new regulations designed to enhance food safety (USDHHS-FDA-CFSAN, 1995; USDA-FSIS, 1996; USDHHS-FDA, 2001; USDA-FSIS, 2003; USDHHS-FDA-CFSAN, 2006a, 2006b) may have rendered food processing facilities less likely sources of food-borne hazards than in the past. On the other hand, the lack of strict federal regulations, less-frequent inspections, and the relatively open and publicly accessible character of retail operations, compared to manufacturing establishments, may raise concern for potential introduction of hazards in foods (Martin et al., 1999). Hazards that may be present in retail foods and cause injury, illness, or death include biological agents (i.e., bacteria and their toxins, parasites, and viruses), physical objects (e.g., jewelry, stones, and glass), and chemical contaminants (e.g., natural toxins, unlabelled allergens, cleaning compounds, non-food-grade lubricants, and insecticides); these hazards may be either inherent in the products or acquired at retail premises (USDHHS-FDA-CFSAN, 2006a). The purpose of this chapter is to consider only biological hazards and, more specifically, viral and bacterial pathogens.

Although the incidence and transmission dynamics of food-borne pathogens in food processing environments have been studied extensively, data regarding the role of retail operations in pathogen transmission and contamination of foods are limited. Nevertheless, increased consumption of foods handled by food workers in commercial processing and retail/food service establishments has been recognized to contribute to changes in the epidemiology of food-borne diseases, resulting in greater consumer exposure to food-borne pathogens (Hedberg et al., 1994; Collins, 1997). Food products that have strongly been associated with food-borne illness outbreaks include foods that are frequently prepared and sold at retail premises, such as sandwiches, fresh produce items, and salads (NACMCF, 1999; USDHHS-FDA-CFSAN, 1999; Long et al., 2002). Along with consumption trends, initial levels of contamination of retail foods with pathogenic organisms have been shown to be strongly related to the risk of food-borne illness (Bahk et al., 2006). Potential sources of food-borne pathogens in retail environments include foods, equipment, environment, employees, and customers or vendors (PSU, 2006). Foods from unsafe sources, improper holding temperatures, contaminated equipment, and poor personal hygiene have been identified as the most significant food-borne illness risk factors that may be encountered in retail establishments in the United States (CDC, 2000a; USDHHS-FDA-CFSAN, 2000). Therefore, according to the U.S. Food and Drug Administration (FDA), effective control of the above risk factors should constitute the primary target of food safety systems in retail operations (USDHHS-FDA-CFSAN, 2000).

Alexandra Lianou • Laboratory of Food Microbiology and Hygiene, Department of Food Science and Technology, Faculty of Agriculture, Aristotle University of Thessaloniki, Thessaloniki 54124, Greece. **John N. Sofos** • Center for Meat Safety & Quality, Food Safety Cluster of Infectious Diseases Supercluster, Department of Animal Sciences, Colorado State University, Fort Collins, Colorado 80523-1171.

Not all approaches that have been shown to be effective and are commonly applied at food processing facilities for the control of food-borne pathogens (e.g., positive air pressure, foot baths, and sanitizer misters) may be applicable at retail settings (ILSI Research Foundation/Risk Science Institute, 2005; PSU, 2006). The development and implementation of appropriate control measures may be hindered by certain operational attributes of retail environments, including the complexity, variety, and highly dynamic character of operations; the openness to the public; the need to display, handle, and repackage certain products; their high dependency on part-time or temporary food handlers; and the associated high employee turnover (Reimers, 1994; Mortlock et al., 1999; McSwane and Linton, 2000; Mortlock et al., 2000). The U.S. Food Code, developed by the FDA, is a reference document for local, state, and federal jurisdictions responsible for food safety in retail food stores and food service establishments, and its implementation, although voluntary, is strongly encouraged as a means of developing or upgrading food safety rules and promoting the establishment of uniform national food safety standards (USDHHS-FDA-CFSAN, 2005).

This chapter provides an overview of the hazards and risks associated with the presence and transmission of viral and bacterial pathogens in retail-handled RTE foods and discusses potential control interventions. The term "retail-handled RTE foods" refers to RTE products that fall in the "Receive-Store-Prepare-Hold-Serve" food preparation process (USDHHS-FDA-CFSAN, 2006a), meaning that they do not undergo a cooking step at retail premises but are handled (e.g., sliced, cut, and repackaged) or used in the preparation of other RTE foods (e.g., salads or sandwiches). Products included in this category can be foods cooked at the manufacturing level (e.g., delicatessen meats, cheese, and other pasteurized products) and exposed to the environment at retail, or raw or minimally processed RTE products, such as seafood or produce items (USDHHS-FDA-CFSAN, 2006a).

HAZARDS AND RISKS

Viral Pathogens

Viral infections have been recognized worldwide as an important category of food-borne illness. Information derived from formal surveillance systems and population-based studies in the United States demonstrates that, despite the fact that viruses contribute considerably less to food-borne illness deaths than bacteria, they account for a significant proportion of illnesses and hospitalizations (Mead et al., 1999). Approximately 67% of illnesses and 35% of hospitalizations attributed to food-borne transmission annually in the United States are caused by viruses (Mead et al., 1999). Norwalk-like viruses or noroviruses (NoV) and hepatitis A virus (HAV) are highly infectious and therefore likely to cause widespread outbreaks. Although NoV and HAV are normally transmitted directly from person to person, they also constitute the two most important food-borne viral infections (Cliver, 1997; Appleton, 2000).

NoV, formerly known as small round-structured viruses, have emerged as the most common causative agent of nonbacterial gastroenteritis, if not of food-borne illness in general, among adults in several parts of the world (Vinje et al., 1997; Fankhauser et al., 1998; Maguire et al., 1999; Mead et al., 1999; Appleton, 2000; Deneen et al., 2000; Hedlund et al., 2000; Svensson, 2000; Greening et al., 2001; Parashar and Monroe, 2001; Fankhauser et al., 2002; CDC, 2003a; Van Duynhoven et al., 2005; Blanton et al., 2006), causing both epidemic and sporadic disease (Cliver, 1994; Appleton, 2000; McCarthy et al., 2000; Bresee et al., 2002; Buesa et al., 2002). NoV caused 49% of food-borne disease outbreaks of known etiology that were reported by sites of the Foodborne Diseases Active Surveillance Network (FoodNet) of the United States Centers for Disease Control and Prevention (CDC) in 2005 (CDC, 2006a). Characteristics of NoV that favor their involvement in food-borne epidemic disease and may play an important role in their transmission within retail environments include their low infectious dose (i.e., <100 viral particles), the prolonged viral shedding that may occur among asymptomatic individuals, their environmental stability, and the wide strain diversity resulting in repeated infections (CDC, 2001).

Regarding the primary mode of NoV transmission (i.e., food borne or person to person), findings of outbreak investigations are not consistent (Fankhauser et al., 1998, 2002; Blanton et al., 2006). Observations made during some investigations have indicated that the predominant transmission route may vary with the setting at which the outbreak occurred or with viral strain (CDC, 2003a; Van Duynhoven et al., 2005). Although food-borne transmission of NoV has been associated primarily with consumption of shellfish (Hedberg and Osterholm, 1993; Shieh et al., 2000; Svensson, 2000; CDC, 2001; Hamano et al., 2005), human infections have also been associated with various other food items, including sandwiches, cakes, frostings, fruits (e.g., raspberries and sliced melon), and vegetables (e.g., green salad and celery), as well as with drinking water and ice (Lo et al., 1994; NACMCF, 1999; Anderson et al., 2001; CDC, 2001;

Rutjes et al., 2006). Similarly, foods implicated in HAV-associated outbreaks include shellfish, dairy products, baked products, sandwiches, salads, and fruits and vegetables, such as lettuce, raspberries, frozen strawberries, and sliced tomatoes (Feinstone, 1996; Cliver, 1997; NACMCF, 1999).

Outbreaks of NoV gastroenteritis have occurred in multiple environments, including nursing homes, hospitals, restaurants, events with catered food, schools and universities, day care centers, retirement centers, and other settings (Vinje et al., 1997; Fankhauser et al., 1998; CDC, 2001; Fankhauser et al., 2002; Blanton et al., 2006; Rutjes et al., 2006). The limited number of reports of outbreaks of viral infections originating from retail-handled foods should not be considered as evidence that viral pathogens are not of importance at retail. Due to difficulties associated with development and application of simple and sensitive viral diagnostic methods and the fact that the modes of transmission of viruses are not limited to food, verification and assessment of their food-borne transmission may be more complex and underreporting is more likely than in the case of bacterial pathogens (Mead et al., 1999; Wheeler et al., 1999; Svensson, 2000; CDC, 2001; Bresee et al., 2002; Koopmans et al., 2003; Blanton et al., 2006; Rutjes et al., 2006). Nevertheless, the findings of numerous outbreak investigations illustrate the critical role of food handlers in the transmission of viruses in foods (Hedberg and Osterholm, 1993; USDHHS-FDA-CFSAN, 1999; Van Duynhoven et al., 2005). In a review of published scientific articles for the period 1975 to 1998, undertaken by the FDA, it was demonstrated that NoV and HAV accounted for 60% of the food-borne disease outbreaks that were identified to have resulted from contamination of food by food workers (USDHHS-FDA-CFSAN, 1999). Subsequent investigations of food-borne outbreaks of viral infections in the United States, in which food handlers were implicated as the source of contamination of foods, are listed in Table 1. Such epidemiological findings, coupled with the characteristics of the food-borne viruses mentioned above and the operational requirements of retail establishments, indicate that viral pathogens, especially NoV, may constitute important human health hazards transmitted through consumption of RTE foods handled at retail settings.

Bacterial Pathogens

Bacteria contribute considerably to the burden of food-borne illness. It has been estimated that 30% of illnesses, 60% of hospitalizations, and 72% of deaths associated with food-borne transmission in the United States are caused by bacterial agents (Mead

Table 1. Examples of investigations of food-borne outbreaks of NoV gastroenteritis and HAV infection in the United States implicated to be associated with food handlers

Pathogen	Implicated foods	Food preparation setting	Reference
HAV	Ice-slush beverage	Convenience market	CDC (1990)
NoV	Crumb cake, pie, cinnamon rolls, ice cream	Military base dining facility	CDC (1999)
HAV	Cold foods[a]	Catering facility	Massoudi et al. (1999)
NoV	Potato salad	Restaurant	CDC (2000b)
NoV	Deli ham	Cafeteria deli bar	Daniels et al. (2000)
NoV	Salads and lunch-meat sandwiches	Country club	Eddy et al. (2000)
NoV	Side salads	Catering facility	Anderson et al. (2001)
NoV	Tossed salad	Catering facility	Kassa (2001)
HAV	Ready-to-eat items (primarily sandwiches)	Restaurant	CDC (2003b)
NoV	Cake	Bakery	Friedman et al. (2005)
NoV	Sandwiches	Restaurant	CDC (2006b)

[a]Specific food item was not identified.

et al., 1999). Specifically, *Salmonella* (nontyphoidal) and *Listeria monocytogenes* have been associated with the highest number of deaths attributable to food-borne transmission and caused by known pathogens, accounting for 31 and 28% of estimated food-related deaths, respectively (Mead et al., 1999). Similar are the observations made with regard to the epidemiology of food-borne disease outbreaks in Australia, with salmonellosis as well as listeriosis being responsible for 40% of food-borne illness deaths (Dalton et al., 2004). *Salmonella* was the most common bacterial etiologic agent of food-borne disease outbreaks reported in the United States by FoodNet sites in 2005, being associated with 18% of outbreaks of known etiology and coming second only to NoV (CDC, 2006a). According to preliminary surveillance FoodNet data, the overall incidence of salmonellosis in the United States in 2006 was 148.1 cases per million people (CDC, 2007).

Considering that a wide variety of food products, including both raw and RTE foods, may be introduced and handled in retail operations, virtually any known bacterial pathogen might constitute a hazard for retail-handled foods. It has been suggested that establishment of microbiological criteria for retail foods should be based on knowledge of retail systems and on conditions associated with handling and use of such products (PFMG-IFST, 1995). Depending on the natural habitat and the epidemiology of a bacterial agent, certain events or practices within retail operations may have more important implications than others, with respect to its transmission in these environments and potential contamination of RTE foods. For instance, pathogens such as enterohemorrhagic *Escherichia coli* and *Salmonella,* which commonly originate from raw foods of animal origin (Jay, 2000), may be present in RTE foods if conditions in retail environments allow for cross-contamination. Given their potentially low infectious dose (Jay, 2000), these pathogens may cause disease, even under proper temperature control. However, given its ubiquitous nature (Schuchat et al., 1991) and its high epidemiological association with foods that are either consumed raw or RTE and susceptible to postprocessing contamination (Bell and Kyriakides, 2005), *L. monocytogenes* may be considered a bacterial pathogen with particularly serious public health implications for retail-handled foods. *L. monocytogenes* has been found to be present in many retail RTE foods, at various prevalence rates and occasionally at high concentrations (Lianou and Sofos, 2007). Among products handled and thus subjected to potential recontamination or cross-contamination in retail operations, RTE meat and poultry products, particularly delicatessen meats, are those involving the highest risk of listeriosis (USDHHS-FDA-CFSAN/USDA-FSIS, 2003).

The association of retail premises with bacterial food-borne disease was made epidemiologically as early as the 1960s (Howie, 1968). Since then, the findings of several epidemiological investigations have demonstrated that bacterial transmission and contamination of RTE foods at retail operations are likely (Table 2). Cross-contamination of foods in

delicatessen counters was thought to have contributed to the magnitude of a large listeriosis outbreak in France in 1992, in which contaminated pork tongue-in-jelly was identified as the major vehicle for human infections; the epidemic strain was isolated from various delicatessen products handled in the same retail stands where pork tongue-in-jelly was sold, as well as from utensils used in these environments (Jacquet et al., 1995; Bell and Kyriakides, 2005). A break in the cold chain at the retail level was investigated as an event likely to have contributed to a *Clostridium botulinum* outbreak in Italy in 1996 that was associated with consumption of "mascarpone," a commercial cream cheese (Aureli et al., 2000). Conditions of temperature abuse at retail could have favored the germination of *C. botulinum* spores that were present in the product and the production of botulinum toxin (Franciosa et al., 1999; Aureli et al., 2000). During the investigation of a suspected salmonellosis outbreak associated with a retail premises in the United Kingdom in 2003, it was revealed that the premises was extensively contaminated with *Salmonella enterica* serotype Enteritidis; the organism was recovered from various food samples, including raw or ready-to-use ingredients and RTE foods, and from numerous environmental swabs (Duncanson et al., 2003).

Retail environments may also play an important role in sporadic bacterial food-borne illness. A case-control study, undertaken to evaluate the role of food in sporadic listeriosis in the United States, demonstrated that 17% of cases occurring between 1988 and 1990 could be linked to consumption of food from a store delicatessen counter and that listeriosis patients were more likely to have consumed foods from delicatessen counters than controls (Schuchat et al., 1992). More recent epidemiological findings suggest that interventions specifically directed at retail

Table 2. Characteristic outbreaks of bacterial food-borne disease associated with retail RTE foods

Yr	Country	Pathogen	Food[a]	Retail system failure	Reference
1964	England	*Salmonella enterica* serovar Typhi	Corned beef	Cross-contamination,[b] temp abuse	Howie (1968)
1992	France	*Listeria monocytogenes*	Pork tongue-in-jelly	Cross-contamination[b]	Jacquet et al. (1995)
1995	England	*Escherichia coli* O157	Cold meats in sandwiches	Supplier credibility	McDonnell et al. (1997)
1996	Italy	*Clostridium botulinum* (type A)	Cream cheese (mascarpone)	Cold-chain break	Aureli et al. (2000)
1997	Australia	*Salmonella enterica* serovar Bredeney	Meat/chicken product	Cross-contamination[b]	Baker et al. (1998)
1999	England	*Listeria monocytogenes*	Sandwiches	—[c]	Anonymous (1999)
2003	United Kingdom	*Listeria monocytogenes*	Sandwiches	—[c]	Dawson et al. (2006)

[a]Food product that was implicated as the major outbreak vehicle.
[b]Event resulting in secondary contamination of additional food products handled in the same retail premises.
[c]—, prepackaged products contaminated at the production level (manufacturer or caterer).

environments and not limited to foods widely recognized as high risk may be required in order to reduce the incidence of sporadic food-borne listeriosis (Varma et al., 2007).

The transmission potential of bacterial pathogens within retail environments and the likelihood of contamination of retail-handled RTE foods have also been demonstrated by microbiological food survey data. Seven out of 10 samples of soft cheeses that were found positive for *L. monocytogenes* during a retail survey in Norway were cut in the same store and were contaminated with low concentrations of strains of the same serotype; thus, the retail store appeared to be the most likely source of the pathogen (Rørvic and Yndestad, 1991). In another survey undertaken in the United States, the presence of *E. coli* and, to a lesser extent, *L. monocytogenes* in vegetable salads served at restaurants and supermarkets was attributed to contamination during handling; all contaminated samples were obtained from only 5 of 31 establishments surveyed, and 4 of these establishments served contaminated salads at two different times (Lin et al., 1996). RTE muscle foods (i.e., meat, poultry, and seafood), sampled from delicatessens in retail markets in New Zealand, were more frequently contaminated with *Aeromonas*, *Yersinia enterocolitica*, and *Listeria innocua* compared to similar prepackaged products (Hudson et al., 1992). Nichols et al. (1998) reported significantly higher *L. monocytogenes* prevalences for pâtés that were open than those for prepackaged products sampled from retail outlets in the United Kingdom. Recontamination of RTE meat products at retail was indicated by the findings of a survey undertaken in supermarkets in Belgium, according to which higher incidence rates of *L. monocytogenes* were obtained for whole cooked meat products after slicing than before (Uyttendaele et al., 1999). According to data collected for RTE foods from retail markets in the United States, higher *L. monocytogenes* prevalence rates were found with in-store-packaged products (i.e., deli salads, luncheon meats, and seafood salads) than in manufacturer-packaged products (Gombas et al., 2003). Potential reasons for these findings include additional handling of foods at retail stores or inappropriate refrigerated storage of in-store-packaged products in retail delicatessen cases (Gombas et al., 2003).

Contamination Routes

The risk of food-borne illness is associated with both the presence of a source of contamination and events that contribute to its transmission (Hedberg et al., 2006). Knowledge of potential sources of pathogenic organisms and likely routes of contamination

of RTE foods handled in retail establishments would be beneficial in the identification of appropriate control measures. Potential sources of viral or bacterial pathogens in retail environments include the source of foods entering the premises, environmental surfaces, and food handlers (USDHHS-FDA-CFSAN, 1999). Viruses, requiring a viable host cell for growth, do not multiply or produce toxins in foods, and, therefore, the latter cannot be considered to be contributing to the active transfer of viral pathogens (Cliver, 1997; Jaykus, 1997; Appleton, 2000; Eddy et al., 2000). Nevertheless, once viruses are introduced in foods, they may remain viable for long periods of time under various storage and processing conditions (Jaykus, 1997; Eddy et al., 2000; Koopmans and Dulxer, 2004). With the transfer dynamics of food-borne bacteria being studied more extensively than those of viruses, it has been illustrated that, regardless of the original source of pathogenic bacteria, contamination of retail-handled RTE foods may be the result of multiple, direct or indirect, routes of bacterial transmission.

At-source contamination of foods entering retail establishments may refer either to contamination of foods at their growing/harvesting areas or to postprocessing contamination at the manufacturing center (Appleton, 2000; Anderson et al., 2001). Manufacturer-packaged RTE products are frequently opened, handled (e.g., sliced, weighed), and repackaged in retail operations. Thus, if such products are contaminated postprocessing at the production level, they may serve as vehicles for cross-contamination of other RTE foods handled in retail environments. Pathogen and particularly viral contamination of food vehicles, primarily bivalve mollusks and fresh produce, at their growing/harvesting areas has been attributed mainly to sewage polluted water (Cliver, 1994; Appleton, 2000). Molluscan shellfish and produce items are likely to be consumed uncooked, undercooked, or after minimal processing that may not ensure viral inactivation (Nguyen-the and Carlin, 1994; Appleton, 2000), and thus, handling of such foods in retail premises may constitute an important means of pathogen transmission within these environments. Since human enteric viruses have been shown to retain their infectivity under certain processing and preservation conditions (e.g., temperatures of up to 60°C, acidic environments, and freezing), residual viral activity may be present after minimal processing (Kirkland et al., 1996; Appleton, 2000; CDC, 2001; Koopmans and Dulxer, 2004; Hewitt and Greening, 2006).

With regard to environmental contamination, food contamination routes that may be encountered in retail operations include use of contaminated equipment and utensils, water, or ice. Handling (e.g., slicing, squeezing, shredding etc.) of RTE foods using contaminated utensils may result in cross-contamination events in the absence of adequate cleaning and sanitation (Tauxe, 1997). Due to limitations associated with currently applied detection methods as well as the special methodologies needed for NoV culturing, available information regarding the environmental stability of these viruses is limited (Appleton, 2000). However, observations made during epidemiological investigations and studies of infectivity in volunteers indicate that viral agents are able to survive on inanimate surfaces and that NoV and HAV are the most resistant to inactivation enteric viruses (Appleton, 2000). Similarly, pathogenic bacteria, such as *Salmonella*, *Staphylococcus aureus*, and *Campylobacter jejuni*, may remain viable on food contact surfaces and therefore constitute a food contamination hazard for long periods of time, even in the absence of moisture (Kusumaningrum et al., 2003; Moore et al., 2003; Cools et al., 2005; Cliver, 2006).

Similar to food processing environments, harborage sites or niches of pathogenic bacteria established in retail premises may have serious public health implications (Tompkin, 2002). Although available data on potential bacterial pathogen harborage sites in retail environments are limited, they should be similar to those found in manufacturing environments and may include any difficult-to-clean place that can trap food and water and can therefore support microbial residence and multiplication (PSU, 2006; Grinstead and Cutter, 2007). Such places can be found both in food contact (e.g., knives, slicers, cutting boards, preparation tables, serving utensils, food containers, and trays in display cases and refrigerators) and nonfood contact (e.g., drains, walls, floors, sinks, grease traps, air/cooling ventilation systems, shopping carts, door/equipment handles, and cleaning and maintenance tools) surfaces, and serve as vehicles for cross-contamination of RTE foods (Sauders et al., 2004; CFP, 2006; PSU, 2006; Grinstead and Cutter, 2007).

Contaminated refrigerator surfaces may serve as means of transmission of psychrotrophic pathogens in retail-handled RTE foods. *L. monocytogenes* was isolated from 1.7% of refrigerators sampled in retail food stores in Greece, with the positive samples corresponding to surfaces in contact with cheese, sausages, and miscellaneous products (Sergelidis et al., 1997). Although the presence of pathogenic bacteria on food contact surfaces is obviously of higher concern, important vehicles for pathogen transmission in foods handled in retail operations may also be sponges, cloths or towels, and aprons, particularly

when they are not cleaned and sanitized on a systematic basis (Little and de Louvois, 1998; Montville et al., 2001; Sattar et al., 2001; Cogan et al., 2002; Michaels et al., 2002; Sagoo et al., 2003a; Lues and Van Tonder, 2007). Environmental sampling in catering and retail premises in the United Kingdom revealed that cleaning cloths or towels were more heavily contaminated with bacteria, including *E. coli*, than food preparation surfaces, while pathogenic organisms (i.e., *Campylobacter* spp., *S. aureus*, and *Salmonella* spp.) also were detected (Sagoo et al., 2003a). Dry fabrics and all-cotton materials have been associated with lower bacterial transfer rates compared to moist fabrics and materials made from both cotton and polyester (Sattar et al., 2001).

Bacterial transfer between surfaces used in food preparation and finally to food has been shown to be highly variable and depends on several factors, including surface type and moisture content, contact pressure and time, friction, initial levels of contamination, and type of food (Bloomfield and Scott, 1997; Chen et al., 2001; Sattar et al., 2001; Kusumaningrum et al., 2003; Montville and Schaffner, 2003; Rodríguez and McLandsborough, 2007). Wachtel et al. (2003) reported a prolonged transfer of *E. coli* O157:H7 from a contaminated cutting board to cut lettuce pieces. Such findings, coupled with the low infectious dose of this organism, dictate the need for adequate segregation during handling of potentially contaminated foods (e.g., raw meat) from RTE foods or foods intended to be consumed without cooking. Nevertheless, even in the case of pathogens for which high illness risk is associated with high contamination levels, such as *L. monocytogenes* (Chen et al., 2003), transfer of low pathogen concentrations in retail-handled RTE foods may also be hazardous if these foods support pathogen growth during extended storage in retail or domestic settings (Burnett et al., 2005; Lianou et al., 2007a, 2007b). The risk of cross-contamination of RTE foods during handling becomes even more apparent when taking into account the ability of certain pathogens (e.g., *Salmonella* and *L. monocytogenes*) to adhere and form biofilms on various food contact surfaces (Blackman and Frank, 1996; Norwood and Gilmour, 1999; Beresford et al., 2001; Lundén et al., 2002; Kim and Wei, 2007).

Slicing also may contribute considerably to pathogen transmission in retail-handled RTE products, as indicated by food survey data (Uyttendaele et al., 1999) and environmental sampling of retail premises (Humphrey and Worthington, 1990; Hudson and Mott, 1993; Little and de Louvois, 1998). Slicing machines may become contaminated with pathogenic bacteria, such as *L. monocytogenes*, from the outside of the packages of whole pieces of luncheon meats when the packages are not removed during slicing and subsequently cross-contaminate delicatessen products (Hudson and Mott, 1993). The extent of bacterial transfer and contamination of delicatessen meats during slicing or cutting may be affected, in addition to contamination level and product type, by blade composition and wear and scoring of the blade, as well as operational characteristics of the equipment, such as slicer design and rotation speed of the slicer blade (Lin et al., 2006; Vorst et al., 2006a, 2006b). Worn and scored blades of slicing machines or knives may provide harborages for bacteria, rendering cleaning and sanitation ineffective (Vorst et al., 2006a, 2006b).

Contact of produce items with contaminated water (e.g., during washing of fresh produce items used in salads or sandwiches) also may result in cross-contamination events (Tauxe, 1997; Ackers et al., 1998; Brackett, 1999; NACMCF, 1999; Wachtel and Charkowski, 2002). The practice of "crisping" (i.e., soaking in tepid water followed by refrigeration), commonly applied to improve the appearance of lettuce, was identified during the investigation of an outbreak of *E. coli* O157:H7 in the United States as a likely means of cross-contamination among several batches of leaf lettuce (Ackers et al., 1998). Ice (e.g., used to cool RTE foods or drinks) also may be a cause of concern for cross-contamination of RTE foods in retail operations, when its microbiological quality is low, as a result of improper ice production (e.g., improper cleaning of ice machines) or storage (Khan et al., 1994; Nichols et al., 2000). Introduction of pathogens to foods, such as produce items, via ice may occur either by directly contacting water from melting ice contaminated with pathogenic organisms or by pathogen transfer from contaminated to uncontaminated food items via melted ice (Kim and Harrison, 2008).

Manually handled foods are considered of higher risk than commercially processed foods, with regard to pathogen incidence and transmission (Koopmans and Dulxer, 2004). The critical role of food handlers in the spread of food-borne illness has been illustrated by epidemiological data (USDHHS-FDA-CFSAN, 1999; Van Duynhoven et al., 2005; Todd et al., 2007a), with infected food handlers constituting the most important vehicle for viral transmission in foods other than shellfish (Dalton et al., 1996; Cliver, 1997; CDC, 1999; NACMCF, 1999; Bidawid et al., 2000; Eddy et al., 2000). Foods that require handling but no cooking prior to consumption are those of the highest risk, and multi-ingredient cold items, such as sandwiches, salads, and bakery and confectionary products, have been implicated frequently in food-

borne outbreaks, primarily viral infections, linked to food handlers (Appleton, 2000; CDC, 2001; Greig et al., 2007).

Food handlers constitute a critical factor, with respect to the safety of retail-handled RTE foods, as they may carry pathogenic viruses or bacteria on their hands (Kerr et al., 1993; USDHHS-FDA-CFSAN, 1999; Lues and Van Tonder, 2007). A review of data of 816 food worker-associated outbreaks, which occurred from 1927 until the first quarter of 2006 and were linked to various segments of the food industry, showed that viruses, mainly NoV and HAV, caused the majority of the outbreaks, while outbreaks associated with bacterial agents were primarily attributed to *Salmonella*, *S. aureus*, *Shigella*, and *Streptococcus* (Greig et al., 2007). Only 12 of the 816 outbreaks reviewed were linked to retail food outlets, with 11 of these outbreaks being associated with supermarkets and one with a butcher shop (Todd et al., 2007a). According to the researchers, such a relatively low outbreak incidence may be due to the fact that many retail foods have protective packaging or are purchased raw and are subsequently rendered safe via cooking (Todd et al., 2007a). An additional factor that may contribute to this low incidence is underreporting of food worker-associated outbreaks in retail environments. Factors that may hinder the investigation of food-borne outbreaks in retail settings and result in underreporting include difficulties in identifying and observing symptoms of food workers, given that many of them work in shifts or on a part-time basis or are assigned to specific jobs on a periodic basis, and the uncertainty regarding the role of food workers (victims or the cause of infections) in food-borne outbreaks (Greig et al., 2007; Todd et al., 2007b).

Food workers may contaminate foods during handling either by serving as the primary source of pathogenic organisms, in the case of actively infected food handlers, or by mediating their transfer from contaminated foods, objects (e.g., money, soiled clothing, and door knobs), or surfaces (e.g., food contact surfaces and counters) (Cruickshank, 1990; Paulson, 2000, 2005; Greig et al., 2007; Todd et al., 2007b). In both cases, hand contact appears to be the most important food contamination route and has been identified or implied as contributing to pathogen transmission in numerous food handler-associated outbreaks (USDHHS-FDA-CFSAN, 1999, 2003; Todd et al., 2007b). Poor personal hygiene, open sores on hands or arms, and snacking on foods have been included among the findings of outbreak investigations in several instances (USDHHS-FDA-CFSAN, 1999; Todd et al., 2007b). With a considerable proportion of NoV infections being asymptomatic, food workers that are asymptomatic carriers of viral particles or have sick family members also may contribute to the spread of viruses within retail environments (Graham et al., 1994; Lo et al., 1994; NACMCF, 1999; Daniels et al., 2000; CDC, 2001; Koopmans and Dulxer, 2004; Greig et al., 2007; Todd et al., 2007b).

Gastroenteritis and hepatitis viruses are able to survive on hands, and therefore, fecally contaminated fingers of infected individuals may contaminate food and work surfaces, if frequent and appropriate hand washing is not practiced (Appleton, 2000). Sick food handlers have been identified as playing a very important role in produce-associated NoV outbreaks, primarily due to the fact that fresh produce items may receive extensive direct hand contact during preparation in retail operations (NACMCF, 1999). Quantitative data relative to transfer of human enteric viruses from hands to foods are limited. In a study undertaken to assess the transfer dynamics of HAV from artificially contaminated finger pads of volunteers to pieces of fresh lettuce, a transfer rate of approximately 9% was reported; such a transfer rate could potentially result in a viral concentration capable of causing infection to susceptible individuals (Bidawid et al., 2000). However, adsorption and persistence of viral pathogens appear to vary among different produce items, with lettuce demonstrating higher HAV adsorption capacity than other fresh vegetables (Croci et al., 2002). Furthermore, food workers are more likely than other workers (e.g., clerical workers) to carry bacterial contaminants on their hands (Kerr et al., 1993). *L. monocytogenes* has been shown to be able to survive and persist on experimentally inoculated fingertips, particularly in the presence of food residues, and hand washing may not be adequate to eliminate this pathogen when present at high levels (Snelling et al., 1991; Kerr et al., 1993). According to Kerr et al. (1995), a predominant molecular subtype of *L. monocytogenes* was isolated from the hands of three workers in the same delicatessen, indicating either the predominance of certain strains of the organism in retail establishments or that certain strains are better adapted than others to survival on hands.

Poor food handling practices have been identified as one of the most important contributing factors of food-borne illnesses in the United States (CDC, 2000a). Identification of practices that may compromise food safety is expected to provide the basis for constructive changes and improvements in retail establishments. Thus, observational studies recording food handling practices in retail operations as well as findings derived from outbreak investigations provide information that, apart from enhancing our understanding with respect to pathogen transmission routes, also may facilitate the determination of problem areas in these operations. Food handling malpractices that, when encountered in food establishments, may result in contamination or cross-contamination of RTE foods with viral or bacterial pathogens include intermingling of contaminated and uncontaminated food items (CDC, 2003c), mixing liquid ingredients or scooping ice with bare hands (Kuritsky et al., 1984; Khan et al., 1994; Friedman et al., 2005), using a single common sink for food preparation and employee hand washing (CDC, 2006b), handling uncovered raw food with bare hands (Angelillo et al., 2000), using common areas or equipment for serving or handling RTE foods and other items (Little and de Louvois, 1998), handling RTE meats after handling raw meats using the same gloves or without washing hands (Little and de Louvois, 1998; Pérez-Rodriguez et al., 2006), soaking or storing fresh produce items in water, particularly if the latter is not changed frequently and is used for processing of numerous product batches (Ackers et al., 1998; Wachtel and Charkowski, 2002), and inadequate hand washing (Green and Selman, 2005; Green et al., 2005, 2006). Occasions when hand washing may not be practiced by food workers include prior to putting on gloves; when changing gloves; after handling money; after sneezing, coughing, eating, or drinking; after taking a break; or after touching their face, hair, or clothes (Green and Selman, 2005).

CONTROL OF HAZARDS AND RISKS

As mentioned above, human enteric viruses are environmentally inert, and therefore, viral multiplication in food or environmental settings is not a concern (Cliver, 1997; Jaykus, 1997). Hence, control of pathogenic viruses in RTE foods handled or used in the preparation of other foods at the retail level can be based on measures aiming at either prevention or destruction of contamination. Improved detection of food-borne viruses in food and environmental samples, as provided by simple and sensitive detection methods, is also expected to contribute to their effective control in retail operations. Prevention and destruction of food contamination are equally important for the control of both viral and bacterial pathogens. Interventions aiming at prevention of contamination should address all potential sources and routes of contamination that may be encountered in retail operations, as outlined above. In addition to prevention of contamination, prevention or inhibition of growth of pathogenic bacteria is of critical importance for their control in retail-handled RTE foods.

Prevention of Contamination

Efforts to prevent contamination of retail-handled RTE foods with pathogenic organisms should be based on a series of good retail practices (GRP) and appropriate sanitation standard operating procedures (SSOP). The complexity of retail operations may render determination of the original source of pathogens (food, environment, or food handlers) in these environments difficult (Medus et al., 2006). Nevertheless, minimizing the likelihood of contamination at the manufacturing level and preventing cross-contamination at the retail level are expected to provide adequate pathogen control (ILSI Research Foundation/Risk Science Institute, 2005).

At-Source Contamination

The risk of pathogen transmission by at-source contaminated foods entering retail premises may be reduced by selecting credible and reliable suppliers. Assurance for hazard control on incoming products can be derived from purchase specifications and supplier audits (Reimers, 1994). Prevention of cross-contamination of retail-handled RTE foods (e.g., delicatessen products) with pathogenic organisms originating from raw foods (meat, poultry, seafood, and produce) can be achieved via GRP. Such GRP include development of appropriate flow plans for foods from receiving to sale, segregation of RTE foods and raw ingredients during storage and display (display in separate case or physically separated by cleanable dividers within cases), minimization of traffic between raw and RTE areas, handling of RTE and raw foods by designated personnel using designated utensils (e.g., utilizing color coding), and cleaning and sanitation of utensils and equipment according to proper SSOP (Anonymous, 1991; Linton, 1996; ILSI Research Foundation/Risk Science Institute, 2005; PSU, 2006).

Specifically with respect to viral pathogens, particular attention should be given to shellfish and produce items, given the high association of these products with human enteric viruses as a result of contamination at their growing/harvesting locations. Retailers can reduce the risk of contaminated fresh produce entering their premises by selecting growers and processors capable of assuring GAP and GMP, respectively, such as use of properly treated manure and good quality water (USDHHS-FDA-CFSAN, 1998). Similarly, information regarding shellfish harvesters and processors should be taken into account by retailers. The waters where shellfish are harvested in the United States are monitored and regulated under the National Shellfish Sanitation Program, and approved listings for harvesters and dealers are

available from the FDA (USDHHS-FDA-CFSAN, 2000). Harvesters or dealers that depurate, ship, or reship shellstock shall affix source identification tags or labels to shellstock containers. These tags or labels should indicate the harvest location and date, the dealer's name and address, and the certification number assigned by the shellfish control authority (USDHHS-FDA-CFSAN, 2005, 2007). Facilities that sell or serve raw molluscan shellfish shall maintain the identity of the source of shellstock by retaining shellstock tags or labels for 90 days from the date that is recorded on the tag or label (USDHHS-FDA-CFSAN, 2005, 2007). According to the FDA Retail Food Program Database of Foodborne Illness Risk Factors, maintenance of harvest records through retention of shellstock tags in seafood departments of retail food stores is of critical importance for the control of foodborne illness and constitutes an area of concern in the retail segment of the food industry (USDHHS-FDA-CFSAN, 2000). On the occasion that shellstock are removed from their original tagged or labeled containers, their source identification shall be maintained by using a record keeping system and by ensuring that shellstock from one container are not commingled with shellstock from other containers with different certification numbers or harvest or growing areas (USDHHS-FDA-CFSAN, 2007).

Concerning bacterial contamination, it has been shown that the extent of cross-contamination of foods depends on inoculum concentration (Lin et al., 2006). Inventory control and appropriate product rotation at retail operations may reduce the likelihood of cross-contamination of surfaces or RTE foods from contaminated products, by limiting bacterial growth via minimization of storage time (ILSI Research Foundation/Risk Science Institute, 2005). For this purpose, the "first in, first out" concept or novel alternative approaches may be used (ILSI Research Foundation/Risk Science Institute, 2005; Koutsoumanis et al., 2005).

Environmental Contamination

Potential sources of environmental pathogen contamination of RTE foods in retail premises include contaminated water or ice and inadequately cleaned and sanitized food contact surfaces, equipment, and utensils (Mbithi et al., 1991; Tauxe, 1997; Eddy et al., 2000; Moore et al., 2003; ILSI Research Foundation/Risk Science Institute, 2005). GRP and an environmental sanitation program comprised of a series of appropriate SSOP are vital for the interruption of this potential route of pathogen transmission.

In order to avoid cross-contamination of retail-handled foods with water-associated pathogens, high-quality water needs to be used in food preparation. Particular attention should be paid to the microbiological quality of water used for washing, spraying, and maintaining the appearance of fresh produce items in grocery stores (Tauxe, 1997). Prevention of cross-contamination of successive loads of fruits and vegetables during washing through immersion water may be achieved via the application of compounds demonstrating sufficient virucidal and bactericidal activity in water, such as chlorine-based disinfectants (Chaidez et al., 2003; Pao et al., 2007). In general, in addition to heating, viral inactivation in water may be achieved with ultraviolet light or other strong oxidizing agents (Cliver, 1997). Furthermore, in order for cross-contamination events to be avoided and the efficacy of applied disinfectants to be maintained, water used to rinse fresh produce should be discarded and renewed frequently. Research findings suggest that the turbidity created by the presence of organic and inorganic material in water may reduce the efficacy of disinfectants (Chaidez et al., 2003). In addition to water, ice used in retail operations should also be of high microbiological quality and for this purpose should be stored in well-insulated containers that are regularly cleaned, and dispensed using clean utensils (Nichols et al., 2000).

The use of multiple color-coded food contact surfaces or utensils (e.g., cutting boards) may prevent cross-contamination of RTE foods with pathogens originating from other foods such as raw meat or poultry products (Green and Selman, 2005). Additional measures to prevent environmental transmission of pathogenic organisms and to limit their presence in foods include appropriate disposal of waste and soiled materials and adequate cleaning and sanitation (CDC, 2003a; Jean et al., 2003). Plastic surfaces may be more appropriate than wooden surfaces for use in food operations with respect to cleanability (Welker et al., 1997). Nevertheless, research undertaken to model cross-contamination events associated with wooden and plastic surfaces has generated conflicting results, and the potential for cross-contamination associated with each surface type appears to be difficult to predict (Welker et al., 1997; Cliver, 2006). Therefore, it has been acknowledged that, irrespective of composition, proper cleaning and sanitation of food contact surfaces are of vital importance if the risk of cross-contamination of RTE foods is to be reduced (Deza et al., 2007). The fact that a good sanitation program is an efficient and consistent means of preventing contamination of RTE foods with pathogens should motivate retail managers toward routine implementation and monitoring (Robbins and McSwane, 1994).

Retail display cases should be kept clean at all times, while serving and cleaning utensils should be cleaned thoroughly after use and stored under dry and sanitary conditions (Brackett, 1999; ILSI Research Foundation/Risk Science Institute, 2005; Grinstead and Cutter, 2007). For equipment, such as slicers, that is used frequently at ambient temperatures and has harborage points, it is important that the equipment be disassembled into cleanable sections and cleaned at regular intervals during operations (Grinstead and Cutter, 2007). When cleaning environmental surfaces (both food contact and nonfood contact), in addition to proper use of cleaning utensils, care must also be taken not to use high-pressure sprays which could potentially create aerosols and splashing and result in spreading of contamination to food contact surfaces or foods (Grinstead and Cutter, 2007). Cleaning high-risk (RTE food) and low-risk (raw food) areas with different and separately stored implements and the presence of well-documented cleaning schedules and records also are important with respect to the microbiological quality of environmental surfaces used in the preparation of RTE foods in retail premises (Sagoo et al., 2003a).

Variables that may affect cleaning performance and should therefore be taken into account when developing cleaning programs include efficacy of the cleaning agent, appropriate concentration and application, contact time with soiled surfaces, mechanical action required to assist in removal of soils, and temperature, since reasonably higher temperatures often improve the performance of cleaners and melt fat (PSU, 2006; Grinstead and Cutter, 2007). Thorough rinsing, as part of detergent-based cleaning, also needs to be ensured in order to achieve a hygienic state of food contact surfaces or cloths (Cogan et al., 2002). However, since residual contamination may be present even after thorough cleaning, use of antimicrobial agents may also be advisable (Cogan et al., 2002). Numerous disinfectants have been available for use in food operations and include quaternary ammonium compounds, amphoteric products, biguanides, iodophors, peroxy acids, and chlorine-containing compounds, with the last being one of the most common disinfection methods (Deza et al., 2007). Parameters that need to be taken into account when assessing the efficacy and the application potential of disinfectants are, in addition to concentration of the active ingredient, the type of surface to which they will be applied, the contact time, the ambient temperature, and the presence of organic residues (Taylor et al., 1999; Jean et al., 2003). Hence, it is important to evaluate disinfectant efficacy under conditions likely to be encountered in food-related environments, including retail operations; final validation and verification steps in the form of field trials undertaken at food operations by the end users may also be valuable (Taylor et al., 1999).

Most of the current knowledge on the virucidal efficacy of disinfectants has been associated with health care settings, and only limited information is available relative to their efficacy under conditions likely to be encountered in the food industry. Research findings have demonstrated that HAV is relatively resistant to commercial disinfectants, particularly when these disinfectants are applied on contaminated surfaces rather than in solution (Mbithi et al., 1990; Jean et al., 2003). Disinfectants containing glutaraldehyde or chlorine-based formulations, such as sodium hypochlorite, have been found to be effective against HAV attached to food contact surfaces, particularly at high concentrations of the active ingredients (Mbithi et al., 1990; Jean et al., 2003). Moreover, a quaternary ammonium compound containing 23% HCl demonstrated major virucidal activity against HAV attached to stainless steel, while iodine-based products, alcohols, and acetic, peracetic, citric, and phosphoric acids were not effective (Mbithi et al., 1990). Disinfection using at least a 1:50 solution of bleach has been recommended as a measure to prevent environmental transmission of NoV (CDC, 2003a). Disinfection agents that have been shown to be effective against feline calicivirus, used as a surrogate model for NoV, on food contact surfaces include formulations containing quaternary ammonium compounds, as well as sodium bicarbonate (Jimenez and Chiang, 2006; Malik and Goyal, 2006). Phenolic compounds also may be effective, but due to their toxicity, their use as surface disinfectants should be considered only for the purpose of outbreak control (Gulati et al., 2001).

With respect to bacterial contaminants, acidic electrolyzed water and neutral electrolyzed water (NEW) appear to be promising alternatives for disinfection of cleaned food contact surfaces (Venkitanarayanan et al., 1999; Park et al., 2002; Deza et al., 2005, 2007). NEW has been shown to be effective against bacterial pathogens such as *S. aureus* and *L. monocytogenes* on various food contact surfaces (i.e., plastic and wooden cutting boards, stainless steel, and glass surfaces) and at the same time capable of preventing cross-contamination originating from the washing solutions generated after treatment of surfaces (Deza et al., 2005, 2007). Furthermore, NEW appears to be advantageous over other disinfectants, including commonly used agents such as sodium hypochlorite, due to its safety, noncorrosive nature, ease of handling, and long storage life (Deza et al., 2007).

As mentioned previously, bacterial attachment and biofilm formation on food contact surfaces constitute a serious food safety concern. Thorough cleaning combined with correct application of appropriate sanitizers may be effective in inactivation or removal of bacterial biofilms from surfaces (Kim et al., 2001; Somers and Lee Wong, 2004; Ayebah et al., 2005; Kreske et al., 2006; Midelet et al., 2006). Nevertheless, normal cleaning and sanitizing procedures may not be effective in areas where niches of pathogenic bacteria are established (Tompkin, 2002). Therefore, environmental sampling programs need to be in place in retail operations in order to detect areas of persistent bacterial contamination that may serve as reservoirs of pathogens and cross-contaminate RTE foods (Tompkin, 2002; ILSI Research Foundation/Risk Science Institute, 2005). Food contact surfaces may vary in their susceptibilities to bacterial attachment, depending on their type and physical or chemical treatments applied (Arnold and Bailey, 2000; Arnold and Silvers, 2000). Thus, materials and surface finishes selected for use in the manufacture of retail food contact surfaces (e.g., knife or delicatessen slicer blades) should be less favorable for bacterial adherence and biofilm formation in order to potentially reduce the risk of bacterial persistence (Arnold and Bailey, 2000; Midelet et al., 2006).

Given the unique operational requirements of retail establishments, the development of effective and practical antimicrobial interventions, applicable at specified intervals during operations when thorough cleaning and sanitation cannot be performed, would be of great value for the control of viral and bacterial pathogens in these environments. Such interventions could be application (e.g., short-term immersion) of hot water or sanitizers on equipment and utensils (e.g., knives) and could provide an efficient means of interruption of pathogen transmission during handling and prevention of cross-contamination of retail-handled RTE foods (Taormina and Dorsa, 2007).

Food Handlers

Food handlers may contaminate RTE foods with pathogens either directly, during the course of an infection, or indirectly when contributing to passive pathogen transmission from a contaminated source to foods through inappropriate food handling practices (Cruickshank, 1990; Paulson, 2000, 2005). Therefore, control approaches aiming at prevention of contamination of retail-handled RTE foods should address both of these routes of pathogen transmission.

The vast majority of food handler-associated outbreaks of food-borne disease have been linked to food workers who were ill either prior to or at the time of the outbreaks (USDHHS-FDA-CFSAN, 1999). Findings from outbreak investigations clearly demonstrate the need for exclusion of symptomatically ill food handlers from work if transmission of pathogenic organisms, particularly viral pathogens, is to be prevented (Cruickshank, 1990; CDC, 2003b; USDHHS-FDA-CFSAN, 2005). The period of maximum viral shedding and infectivity during the course of infections depends on the organism; shedding of NoV is thought to occur primarily at the time of illness, while HAV appears to have maximum infectivity during the latter half of the incubation period, which varies from 15 to 50 days (USDHHS-FDA-CFSAN, 1999). Early epidemiological data indicated that exclusion of ill food handlers for as long as 48 to 72 h after resolution of illness would be an efficient means of prevention of food-borne NoV gastroenteritis (Thornhill et al., 1975; CDC, 2001). Findings of subsequent epidemiological investigations (Curry et al., 1987; Iversen et al., 1987; Lo et al., 1994; Parashar et al., 1998; CDC, 1999; USDHHS-FDA-CFSAN, 1999) and studies with volunteers (Graham et al., 1994), utilizing more sensitive diagnostic methods, suggested that NoV shedding may occur for longer periods than previously believed, even in the absence of clinical symptoms. Nevertheless, additional research is required in order for the validity and epidemiological significance of such findings to be ascertained (CDC, 2001).

The existing uncertainty relative to the duration of NoV shedding may complicate the development of an ill worker exclusion policy in retail operations. The 2005 Food Code encourages the adoption of NoV and HAV containment recommendations by state health departments and provides detailed information regarding the application of employee exclusion or restriction policies. The code also describes conditions that should be met in order for removal, adjustment, or retention of employee exclusion/restriction policies to be considered (USDHHS-FDA-CFSAN, 2005). Assessment of the general health of food handling personnel by a qualified health care worker via proper questionnaires also has been suggested as a means of prevention of pathogen transmission in food establishments (Cruickshank, 1990). Regarding prevention of transmission of HAV to retail-handled foods, vaccination of people employed as food handlers might also be useful (CDC, 2003b). Overall, with respect to contamination through infected food handlers, prompt reporting of illness by food handlers coupled with a paid sick leave policy, when exclusion from work is required, is expected to contribute the most to prevention of retail food contamination (Cruickshank, 1990; Bresee et al., 2002; CDC, 2003b).

Relative to contamination transmission by healthy retail food workers to RTE foods from other foods or environmental surfaces, proper food handling practices, such as minimization of the number of times a food is handled or repackaged, are very important (PSU, 2006). Furthermore, good hygiene practices (GHP) including regular and thorough hand washing and use of barriers to bare-hand contact with RTE foods, such as gloves, deli wraps, and utensils, can contribute considerably to the interruption of both of the food handler-associated routes of pathogen transmission described above (USDHHS-FDA-CFSAN, 1999; Paulson, 2000; CDC, 2003b; PSU, 2006). Indeed, GHP appear to be the most important hazard control intervention, with regard to food handlers that may be asymptomatic carriers of pathogenic organisms. Inadequate and inappropriate hand washing has been identified as one of the most important food-borne illness risk factors in the retail segment of the food industry (USDHHS-FDA-CFSAN, 2000, 2003). Factors that may have a negative impact on proper hand washing include inadequate or inconvenient sinks, inadequate resources (e.g., soap), and time pressure as a result of either high work volume or inadequate staffing (Green and Selman, 2005; Todd et al., 2007b). In order for hand hygiene practices to be improved, retail operations should ensure that functional and fully equipped hand washing facilities are available and accessible to food workers (Eddy et al., 2000; Allwood et al., 2004a; Green et al., 2007). Furthermore, appropriate hand washing and sanitization procedures should be outlined in detailed standard operating procedures. Parameters that need to be considered when developing a hand washing technique include duration, water temperature, hand drying method, frequency, and hand washing agents, such as detergents, soaps, and sanitizers (USDHHS-FDA-CFSAN, 1999).

Available knowledge regarding the efficacy of hand washing measures against pathogenic viruses is limited. Bidawid et al. (2000) found that water and topical agents (i.e., nonmedicated soap and an antimicrobial agent) were effective in reducing the amount of HAV remaining on finger pads and thus the likelihood of viral transfer to produce during handling. With regard to the efficacy of fingernail sanitation practices, it has been shown that liquid soap plus a nailbrush may be more effective than alcohol-based hand sanitizers (Lin et al., 2003). Hand sanitizers have been shown to be effective against bacterial pathogens and can therefore constitute a potential intervention for prevention of contamination of RTE foods (McCarthy, 1996; Schaffner and Schaffner, 2007). Although alcohol-based hand sanitizers have been recommended by the CDC as a suitable alternative to hand washing in health care settings, substitution of regular and conventional hand washing for hand sanitizing kits may not be prudent in retail operations; use of hand sanitizing kits following hand washing may be recommended (USDHHS-FDA-CFSAN, 2003, 2005). In general, the efficacy of hand or any sanitizers may be reduced by organic residues (e.g., proteinaceous material or fatty substances) present on hands of food handlers or other surfaces (Paulson, 1994; McCarthy, 1996; Charbonneau et al., 2000; USDHHS-FDA-CFSAN, 2003). Results are conflicting concerning the efficacy of antimicrobial soaps compared to that of regular soaps (Mahl, 1989; Paulson, 1994; Lin et al., 2003).

The benefits of hand washing may be negated if food is handled with bare hands and subsequent food storage conditions favor pathogen proliferation (Pérez-Rodriguez et al., 2006). Thus, in addition to good hand washing practices, use of gloves is recommended for food workers handling RTE foods, as they can reduce considerably pathogen transfer from hands to food and between foods if kept clean and changed frequently (USDHHS-FDA-CFSAN, 1999; Montville et al., 2001; Gill and Jones, 2002). Parameters that need to be taken into account with respect to glove use include glove material and permeability, duration of wearing, and hand washing techniques prior to wearing and after removal of gloves (USD-HHS-FDA-CFSAN, 1999). Nevertheless, use of gloves by food handlers involves the risk of creating a false sense of hygienic security (Eddy et al., 2000), which may result in behaviors that might compromise food safety. Such behaviors include wearing the same gloves for extended periods of time and less frequent hand washing than when handling foods with bare hands (Green and Selman, 2005; Lynch et al., 2005; Green et al., 2006, 2007). Since there is a considerable potential for secondary pathogen transmission to foods by food handlers, use of sanitized serving utensils (e.g., tongs) should be preferred, when feasible, over direct contact of gloved or bare hands with RTE foods (Paulson, 1997, 2000). Procedures such as double hand washing and use of nailbrushes may constitute effective pathogen control measures alternative to gloving and allow for bare-hand contact of food handlers with RTE food (Snyder, 2007).

Glove use as well as hand washing, in addition to food preparation, should also occur during other activities taking place in retail operations, such as handling dirty equipment (Green et al., 2007). A simulation study, conducted to determine the amount of surrogate NoV (feline calicivirus strain) transferred from contaminated surfaces to gloved hands, demonstrated that a significant viral load was transferable (Paulson, 2005). The best approach from a food safety perspective is responsibility for specific tasks within retail operations to be clearly identified among members of the workforce (Paulson, 2000) and for employees who prepare, handle, or serve RTE foods not to engage in other activities (e.g., equipment maintenance or cleaning and handling money). However, in some cases, and particularly in small-size operations, restriction of tasks among food workers may not be feasible. On occasions where movement in and out of RTE food areas is required or when food handlers need to perform additional tasks, inadequate hand washing or gloving (i.e., failure to use clean gloves and change them between different tasks) may result in contamination of food or equipment with pathogenic organisms (CFP, 2006; PSU, 2006).

Unlike manufacturing environments, access to retail operations is not limited to employees. Customers, vendors, or visitors also may constitute a source of contamination of RTE foods with viral or bacterial pathogens and therefore should not be allowed into RTE food preparation areas (CFP, 2006). Moreover, in premises providing self-serve RTE foods (e.g., salad bars), customers may contaminate foods through their hands or serving utensils (Doser, 1999). In this case, along with the GRP and SSOP discussed above, protection of foods and serving utensils via properly designed covers (i.e., sneeze-guards, lids, or doors), as well as use of equipment and utensils that allow for easy cleaning and inspection may also be valuable (Doser, 1999).

Destruction of Contamination in Foods

In contrast to those for food processing facilities, measures designed to control (i.e., eliminate or reduce) pathogens in retail-handled RTE foods are limited. Adequate cooking appears to be the sole effective and reliable method for inactivation of viral and bacterial pathogens in foods (Cliver, 1997). Considering that most retail-handled RTE foods do not undergo any cooking step after handling by food workers and prior to consumption, the potential for development and application of alternative inactivation methods needs to be assessed. Moreover, water rinsing may not ensure pathogen removal from fresh produce (Croci et al., 2002). As is common practice for the control of pathogenic and spoilage microorganisms in fresh fruits or vegetables (Beuchat and Ryu, 1997), thorough washing with appropriate disinfectants/sanitizers can constitute an effective decontamination intervention in retail operations that use these commodities to prepare RTE products (e.g., salads and sandwiches).

Only limited information is available with respect to the effectiveness of currently available and used produce sanitizers in inactivating human enteric viruses (USDHHS-FDA-CFSAN, 1998; Gulati et al., 2001; USDHHS-FDA-CFSAN, 2001; Allwood et al., 2004b). Sanitizer efficacy testing, which would provide useful information relative to removal of NoV from contaminated produce, has been problematic; limitations include the lack of an in vitro NoV growth system, as well as the inability of currently applied detection methods to differentiate viable from nonviable viral particles (Allwood et al., 2004b). Findings of studies utilizing feline calicivirus as a surrogate model for NoV suggest that viruses may be resistant to commercial produce sanitizers (Gulati et al., 2001; Allwood et al., 2004b). The produce sanitizers sodium hypochlorite, a quaternary ammonium compound, and peroxyacetic acid-hydrogen peroxide, tested at the manufacturers' recommended concentration, were found ineffective against feline calicivirus in strawberries and lettuce; only the product containing peroxyacetic acid and hydrogen peroxide used at a concentration fourfold higher than the recommended concentration was effective (Gulati et al., 2001). Thus, it appears that until the current stage of knowledge regarding viral inactivation in foods is extended, prevention of contamination constitutes the most reliable and promising intervention for the control of pathogenic viruses in non-reheated RTE foods handled at retail.

Various agents have been evaluated and shown to be effective in reducing bacterial populations, including pathogenic organisms, on fresh produce (Izumi, 1999; Koseki et al., 2001; Deza et al., 2003; Kim et al., 2003; Hellström et al., 2006; Kondo et al., 2006; Alvarado-Casillas et al., 2007). Despite the fact that produce bacterial contaminants may be reduced considerably through washing with chlorine solution, more efficient sanitizers may be required for pathogens with low infectious doses, such as enterohemorrhagic E. coli (Wachtel and Charkowski, 2002). Research findings encourage the application of acidic electrolyzed water and particularly NEW for decontamination of fresh produce (Izumi, 1999; Deza et al., 2003; Kim et al., 2003). In addition to its strong bactericidal activity, NEW does not appear to affect the surface color, general appearance, or organoleptic characteristics of fruits and vegetables (Izumi, 1999; Deza et al., 2003). According to Alvarado-Casillas et al. (2007), spraying of cantaloupes and bell peppers with 2% l-lactic acid resulted in important reductions in populations of Salmonella enterica serotype Typhimurium and E. coli O157:H7, with none of the pathogens being able to grow during subsequent refrigerated storage. The antibacterial efficacy of chemical compounds and thus their potential utilization in decontamination of RTE foods have

also been demonstrated with foods other than fresh produce (Dupard et al., 2006).

Virus-Specific Control Interventions

The difficulties associated with prevention of viral food-borne disease, particularly viral gastroenteritis, have been attributed partially to the lack of simple and sensitive detection methods (Bresee et al., 2002). Utilization of appropriate methods for the detection of food-borne viruses, allowing for rapid results of food and environmental sample testing, is required in order for effective control interventions to be developed and implemented (Duncanson et al., 2003; Koopmans and Dulxer, 2004; Blanton et al., 2006). Availability of sensitive, robust, and rapid detection methods is expected to provide better insight into the modes of viral transmission and food contamination and thus to facilitate the determination, validation, and verification of critical control points (CCP) in retail operations (Jaykus, 1997; Schwab et al., 2000).

Detection of human enteric viruses in foods other than shellfish has been problematic (Rutjes et al., 2006). During the last decade, traditional detection methods, based on mammalian cell culture infectivity assays, have been replaced by immunological (enzyme-linked immunosorbent assay) or molecular (nucleic acid hybridization and nucleic acid amplification methods) techniques; however, such methods have been successful in detecting viruses primarily with shellfish and are not, in general, available for routine testing of foods (Cliver, 1997; Jaykus, 1997; Appleton, 2000; Bresee et al., 2002; Sair et al., 2002; Koopmans and Dulxer, 2004). Problems that have been encountered with regard to detection of viruses in foods, and which novel detection methods need to overcome, include low contamination levels and recovery efficacy, inhibition exerted by food components that interfere with the enzymatic amplification of viral RNA (PCR inhibitors), and the lack of animal models or cell culture systems, as well as the great antigenic and genetic diversity of NoV (Jaykus, 1997; Shieh et al., 2000; Dentinger et al., 2001; Rutjes et al., 2006). Furthermore, it has been demonstrated that the efficacy of virus concentration and extraction procedures may vary with food product, rendering optimal detection methods product specific (Rutjes et al., 2006). Recent advances in detection methods, such as enhanced processing of viruses and food samples, and application of nested reverse transcriptase PCR (RT-PCR), coupled with the increasing knowledge of NoV sequence diversity, are expected to overcome the above limitations (Green et al., 1998; Schwab et al., 2000). The application and evaluation of an RT-PCR method with deli products (i.e., ham, turkey, and roast beef) associated with a NoV outbreak at a university cafeteria in the United States (Daniels et al., 2000) demonstrated that molecular-based methods may also be applicable for the detection of human enteric viruses in nonshellfish food products (Schwab et al., 2000).

Another area of scientific interest relative to improved detection of human enteric viruses in the environment and RTE foods is the identification of organisms that would reliably indicate their presence (Jaykus, 1997). Such an intervention would be of great value in the determination of the prevalence of pathogenic viruses in different types of retail RTE foods as well as in the assessment of potential contamination routes in retail environments (Allwood et al., 2004c). Observations that have been made with respect to correlation between the presence of indicator organisms (i.e., E. coli and bacteriophages) and human enteric viruses in foods have not been consistent (Doré et al., 2000; Allwood et al., 2004c; Koopmans and Dulxer, 2004). Although the association between presence of bacteriophages and presence of NoV in foods requires further clarification, evidence indicates that the absence of bacteriophages may be a reliable indicator of the absence of enteric viruses (Doré et al., 2000). Assessment of the occurrence of reliable viral indicators in various retail RTE products should constitute the objective of future research (Allwood et al., 2004c).

Bacteria-Specific Control Interventions

Prevention or inhibition of growth of pathogenic bacteria is very important for their control in retail-handled RTE foods, particularly in the case of pathogens with relatively high infectious doses (ILSI Research Foundation/Risk Science Institute, 2005) and can be based primarily on antimicrobial interventions applied at the manufacturing level and appropriate control of storage temperatures and times in retail operations.

Several interventions have been shown to be effective against pathogenic bacteria in foods and include postpackaging decontamination technologies, chemical compounds, biopreservation approaches, and plant extracts with antibacterial activity (Leverentz et al., 2003; Ross et al., 2003; Burt, 2004; Zhu et al., 2005). Given the distinct attributes of retail operations compared to food processing facilities, implementation of the above interventions in retail environments is difficult. However, it has been demonstrated that antimicrobial agents applied at the production level may also inhibit pathogen growth under handling, contamination, and storage conditions

encountered subsequently in the food chain, in retail and domestic settings (Cagri et al., 2004; Lianou et al., 2007a, 2007b).

Retail storage or display of foods at abusive temperatures may result in high pathogen contamination levels, if present. In addition to causing directly food-borne illness, high contamination levels may lead to transfer of higher pathogen levels during handling to food contact surfaces, utensils, and other foods. A strong association between the bacteriological quality of retail foods and storage temperature has been demonstrated in several instances (Richardson and Stevens, 2003; Elson et al., 2004; Dallaire et al., 2006; Lewis et al., 2006). Loss of temperature control at the retail level can be due to mishandling of foods after distribution from the supplier, delivery of warm product coupled with subsequent ineffective cooling, or operation of retail refrigerators at high temperatures (Gill et al., 1995). Abusive temperatures have been documented both in retail refrigerators and RTE products (Sergelidis et al., 1997; Lyhs et al., 1998; Nichols et al., 1998; Audits International/FDA, 1999). According to the findings of a nationwide food temperature evaluation study undertaken in the United States, deli-counter meat of temperature exceeding 5°C was found in 71% of retail refrigerators (Audits International/FDA, 1999).

Temperature control is particularly important in the case of psychrotrophic pathogens, such as L. monocytogenes, Y. enterocolitica, and nonproteolytic C. botulinum strains (Jay, 2000). Specifically, good temperature control is critical for the control of L. monocytogenes in RTE foods (USDHHS-FDA-CFSAN/USDA-FSIS, 2003), since even slightly abusive storage temperatures may result in prolific growth of this pathogen in certain foods (McCarthy, 1997; Warke et al., 2000; Burnett et al., 2005). Additional food-borne pathogens that may grow at mildly abusive temperatures include strains of certain Salmonella serotypes, Shigella spp., enterohemorrhagic E. coli, S. aureus, and Bacillus cereus (Jay, 2000). Bahk et al. (2006) utilized a mathematical modeling approach for assessment of the risk associated with S. aureus levels in RTE kimbab at the time of consumption; temperature control under 10°C was identified as a CCP in retail establishments in order to prevent growth of toxigenic strains of this pathogen to levels sufficient for enterotoxin production to occur. Hence, retail operations need to ensure good temperature control in refrigerators and display cases, with RTE products that support growth of psychrotrophic pathogens being always stored at temperatures at or below 4°C (Brackett, 1999; Doser, 1999; Grinstead and Cutter, 2007). Since certain practices within retail operations (e.g., placement of foods close to lights during display in retail cabinets or irregular defrost cycles in refrigeration equipment) may result in abusive product temperatures, maintenance of temperature close to the freezing point (–2 to –1°C) in retail distribution of refrigerated foods may be prudent (Snyder, 1991).

Strict temperature control appears to be even more important for minimally processed RTE foods, such as produce, and needs to be in place throughout the food production chain, including retail storage and display, in order for the safety of such products to be ensured (Beuchat, 1996; Little and Mitchell, 2004). Particularly increased attention should be paid to foods such as vegetables in salad bars, due to the fact that such products may be exposed to multiple sources of contamination (Doser, 1999). Temperatures of salad bar items obtained from deli operations of supermarkets or grocery stores in the United States ranged from 5.1 to 18.9°C; conditions that may support growth of pathogenic bacteria if present or are introduced during handling by food workers or customers and intrinsic properties of foods are favorable to growth (Albrecht et al., 1995). Since refrigerators in self-serve operations may be subject to frequent openings, while salad bar foods can be easily exposed to ambient temperatures, the cooling capacities of such systems, when present in retail operations, need to be evaluated and corrected accordingly (Doser, 1999).

Determination of appropriate storage times for RTE products at retail or at home would be facilitated by concise label-dating of products by manufacturers, as well as by in-store label-dating of products that are handled within retail establishments (ILSI Research Foundation/Risk Science Institute, 2005). The importance of the length of storage of RTE foods as a risk factor for food-borne listeriosis, particularly under abusive temperature conditions and in the absence of antilisterial interventions, has been illustrated (USDHHS-FDA-CFSAN/USDA-FSIS, 2003; FAO/WHO, 2004; Yang et al., 2006; Lianou et al., 2007a, 2007b). Therefore, regardless of where in the food chain (i.e., processing or retail level) label-dating is applied, it may be necessary to be safety oriented if growth of pathogens to hazardous levels is to be prevented (NACMCF, 2005). The 2005 Food Code, via "date marking" requirements developed to not allow more than 1 log of *L. monocytogenes* growth, recommends that retail storage of potentially hazardous RTE foods (foods prepared in retail premises or opened packages of processed foods) should not exceed 7 days at 5°C or 4 days at 7°C (USDHHS-FDA-CFSAN, 2005). However, more stringent than the above requirements, dictating lower storage

temperatures or shorter storage times may be needed, with regard to refrigerated storage of certain RTE products, such as poultry products, particularly when no antimicrobial ingredients are incorporated in their formulation (Lianou et al., 2007b). Storage times of retail RTE products need to be operation and product specific. The length of refrigerated storage of such products needs to be determined based on the type of product and its potential to support growth of psychrotrophic pathogens and handling and storage conditions expected to be encountered in retail establishments, as well as on the application and anticipated effectiveness of other antimicrobial hurdles. Under all circumstances, however, rotation of RTE foods using the "First in, First out" rule would be useful (PSU, 2006).

Certain interventions that have been shown to be promising in controlling bacterial pathogens in foods may be regarded as interventions to be considered for potential application at the retail level. Natural occurrence of a wide variety of bacteriocin-producing lactic acid bacteria in retail food samples suggests a potentially important role of bacteriocins as biopreservatives in retail-handled foods (Garver and Muriana, 1993). The need to evaluate biopreservation approaches, such as bacteriophage biocontrol strategies, under conditions encountered in commercial establishments has been acknowledged (Greer, 2005). The novel concept of "antimicrobial packaging," which has been identified as a promising antimicrobial intervention (Vermeiren et al., 2002; Mauriello et al., 2004; Ercolini et al., 2006), also may have application at the retail level. Packaging materials made active against pathogenic bacteria via the incorporation of antimicrobial substances may be used at retail operations for repackaging of foods that, due to on-premise handling, are subject to surface contamination (e.g., delicatessen meats or cheeses cut at retail premises). Nevertheless, the intimate contact between food and packaging material, required for this technology to be effective (Vermeiren et al., 2002), may not be ensured with retail-handled foods and particularly in products where contamination may be located in areas that cannot come in contact with the package (e.g., within the product or between slices). Thus, the applicability and antibacterial efficacy of this and other interventions need to be assessed under the operational conditions existing in retail premises.

Food Safety Programs and Training

The microbiological quality of RTE foods from retail and food service environments has been associated strongly with the existence of a documented hazard analysis system and with food hygiene training of managers and food handlers (Gillespie et al., 2001;

Little et al., 2003; Richardson and Stevens, 2003). Implementation of hazard analysis and critical control point (HACCP) principles in certain food processing establishments in the United States is mandated by regulations developed by the FDA and the United States Department of Agriculture. This is not the case for retail operations where, although the HACCP philosophy is embraced in the Food Code, its implementation is voluntary (USDHHS-FDA-CFSAN, 2005). Nevertheless, in addition to food manufacturing establishments, HACCP concepts can be useful in the control of food-borne pathogens in retail environments, assisting retail managers in ensuring food safety and quality as part of a total-quality management approach (Reimers, 1994).

The inherent characteristics of retail operations render implementation of HACCP systems more challenging than in food processing operations (Mortlock et al., 1999). However, similar to manufacturing environments, top management commitment is vital for successful implementation of HACCP concepts in retail operations. Any HACCP-based process control systems intended to be applied in retail establishments need to be simple, easy-to-use, and flexible (Reimers, 1994; Mortlock et al., 1999; McSwane and Linton, 2000). It has been suggested that HACCP systems should be tailored to the needs of retail operations through modifications in the practical application or interpretation of the system, with its methodology and definitions maintained as consistent as possible throughout the food industry (McSwane and Linton, 2000).

Both viral and bacterial pathogens should be considered when developing food safety control and management systems, such as GHP, GRP, and HACCP (Koopmans and Dulxer, 2004). Since many different types of foods may be handled simultaneously in retail operations, a unit operations "process approach" may be more appropriate when conducting a hazard analysis. This is a method of categorizing food operations into broad categories, namely, food preparation processes, to each of which hazard identification and control measures are subsequently ascribed (USDHHS-FDA-CFSAN, 2006a). Moreover, in order for retail HACCP systems to be simple and easy to use, CCP should be as few as possible and able to be monitored on an ongoing basis by in-house personnel and under active supervision (Bryan, 1990; Reimers, 1994). Timely application of corrective actions can be facilitated via the use of decision charts and flow diagrams at identified CCP locations (Reimers, 1994). Verification procedures may be performed by the manager, supervisor, a food safety professional, or by health department staff or other regulatory personnel (Bryan, 1990; USDHHS-FDA-CFSAN,

2006a). Record-keeping also may be challenging in a constantly changing retail environment. Record-keeping systems documenting process rather than product information, and utilization of efficient management information systems may simplify record-keeping in retail operations (Reimers, 1994; USDHHS-FDA-CFSAN, 2006a). Prerequisites for effective implementation of HACCP or HACCP-based systems in retail operations include clarification of the role of regulatory personnel and of the criteria by which compliance is determined and open and consistent communication between the retail industry and regulatory authorities (Linton et al., 1998; McSwane and Linton, 2000).

Various resources are available and may assist retailers in the development and successful implementation of food safety programs, such as GRP and HACCP, including professional societies, trade associations, universities, and government agencies (Linton, 1996; AFDO, 2004; USDHHS-FDA-CFSAN, 2005, 2006a; PSU, 2006). Universities that have extension programs may constitute an inexpensive and valuable source of assistance for retail operations, providing voluntary guidelines pertinent to application, verification, and maintenance of food safety programs and control measures (Linton, 1996; PSU, 2006). The FDA has recently issued a manual which provides information that can help retailers in the development and implementation of food safety management systems using HACCP principles (USDHHS-FDA-CFSAN, 2006a). Internal or third-party audits, routine microbial testing, development of a recall plan, and regular reviews of these procedures are parameters that can be useful in maintenance of pathogen control programs (PSU, 2006). Routine inspections have limitations, with regard to the identification of problem areas within retail operations, primarily because they cannot evaluate the overall performance of their food safety programs (Hedberg et al., 2006; Phillips et al., 2006). Nevertheless, HACCP principles may also be used by regulatory food safety professionals in conducting risk-based inspections which may assist them in the evaluation and improvement of food safety management systems applied by retail operations (USDHHS-FDA-CFSAN, 2006b).

Effective risk communication, education, and training of food workers, supervisors, and managers are considered prerequisites for the successful implementation of hazard control measures at retail (Mortlock et al., 1999; Bryan, 2002). According to Eddy et al. (2000), "An informed food handler will become a responsible food handler." However, retail premises appear to be more likely than other sectors of the food industry to employ food handlers with limited food hygiene qualifications (Mortlock et al., 2000).

Food safety training of food handlers on GRP and HACCP and food safety knowledge and certification of managers constitute important measures for the control of food-borne pathogens in retail operations (Hedberg et al., 2006). The Food Code states that the person in charge in each establishment shall ensure that employees receive proper food safety training according to their duties, while managers should demonstrate food safety knowledge by passing a test that is part of an accredited program (USDHHS-FDA-CFSAN, 2005).

Managing food handler illnesses and enhancing food handlers' awareness of the presence and transmission of viral and bacterial pathogens in retail environments are areas that need to be underscored in food safety training programs (Eddy et al., 2000; Koopmans and Dulxer, 2004; Hedberg et al., 2006). The positive impact of education and training programs on food safety knowledge and behaviors of retail food handlers has been demonstrated (Angelillo et al., 2000). Adequate training on food hygiene along with HACCP-specific training of food handlers and managers is required for effective implementation of HACCP-based control measures (McSwane and Linton, 2000; Little et al., 2003; Sagoo et al., 2003b). HACCP-specific training among production, retail, and catering businesses in the United Kingdom was significantly associated with adoption of all seven HACCP principles (Mortlock et al., 1999). However, given the high employee turnover that is commonly encountered in retail operations, it is important that HACCP training is performed on an ongoing basis (McSwane and Linton, 2000), while refresher training also may be required in order for a long-term positive effect to be achieved (Vaz et al., 2005). Education and training approaches, such as videos, computer-based training, online training programs, food safety icons, self-audit checklists, and verification of specific learning tasks, can be useful in retail operations (ILSI Research Foundation/Risk Science Institute, 2005; CDC, 2006b). Parameters that may be associated with food handling behavior and should therefore be taken into account when developing food safety education and training programs include age, risk perception, work responsibilities, and type of premises (Clayton et al., 2002; Green et al., 2005).

Awareness of food safety risks by food handlers may not translate into safe practices. A survey of grocery store seafood employees showed that only 40% of those aware of the risks associated with molluscan shellfish and the purpose of shellfish tags also kept the tags for 90 days (Stivers and Gates, 2000). Food safety programs need to address all factors that may affect food preparation behaviors, including the structural environment of retail operations, the availability and functionality of equipment and resources supportive of food safety, worker characteristics, and food safety education and training (Green and Selman, 2005). Finally, with human intentional behavior being considerably complex and depending on both cultural and personal values, management leadership and commitment are essential if changes in food handlers' behavior are to be achieved (Paulson, 2000).

CONCLUDING REMARKS

Viral and bacterial pathogens constitute important food-borne hazards in retail premises and may contaminate retail-handled RTE foods via multiple routes. Effective control of pathogenic organisms at retail is of vital importance as part of a comprehensive "farm-to-fork" food safety approach and can be attainable only if based on a good understanding of pathogen transmission dynamics in retail operations. With the incidence, modes of transmission, and association with food-borne illness of pathogenic bacteria being studied more extensively than those of pathogenic viruses, factors that warrant further clarification are those associated with virus survival, persistence, removal, and inactivation in retail foods and environments. The operational requirements of retail establishments render control of food-borne pathogens in these environments challenging. Nevertheless, a food safety management system that integrates the concepts of HACCP and dictates adherence to good facility sanitation and GRP is expected to provide adequate hazard control in retail operations.

REFERENCES

Ackers, M. L., B. E. Mahon, E. Leahy, B. Goode, T. Damrow, P. S. Hayes, W. F. Bibb, D. H. Rice, T. J. Barrett, L. Hutwagner, P. M. Griffin, and L. Slutsker. 1998. An outbreak of *Escherichia coli* O157:H7 infections associated with leaf lettuce consumption. *J. Infect. Dis.* 177:1588–1593.

AFDO. 2004. Guidance for processing fresh-cut produce in retail operations. Association of Food and Drug Officials, York, PA. http://www.afdo.org/afdo/upload/Fresh-cutProduce.pdf.

Albrecht, J. A., F. L. Hamouz, S. S. Sumner, and V. Melch. 1995. Microbial evaluation of vegetable ingredients in salad bars. *J. Food Prot.* 58:683–685.

Allwood, P. B., T. Jenkins, C. Paulus, L. Johnson, and C. W. Hedberg. 2004a. Hand washing compliance among retail food establishment workers in Minnesota. *J. Food Prot.* 67:2825–2828.

Allwood, P. B., Y. S. Malik, C. W. Hedberg, and S. M. Goyal. 2004b. Effect of temperature and sanitizers on the survival of feline calicivirus, *Escherichia coli*, and F-specific coliphage MS2 on leafy salad vegetables. *J. Food Prot.* 67:1451–1456.

Allwood, P. B., Y. S. Malik, S. Maherchandani, K. Vought, L.-A. Johnson, C. Braymen, C. W. Hedberg, and S. M. Goyal. 2004c. Occurrence of *Escherichia coli*, noroviruses, and F-specific coliphages in fresh market-ready produce. *J. Food Prot.* 67:2387–2390.

Alvarado-Casillas, S., S. Ibarra-Sánchez, O. Rodríguez-García, N. Martínez-Gonzáles, and A. Castillo. 2007. Comparison of rinsing and sanitizing procedures for reducing bacterial pathogens on fresh cantaloupes and bell peppers. *J. Food Prot.* 70:655–660.

Anderson, A. D., V. D. Garrett, J. Sobel, S. S. Monroe, R. L. Fankhauser, K. J. Schwab, J. S. Bresee, P. S. Mead, C. Higgins, J. Campana, R. I. Glass, and the Outbreak Investigation Team. 2001. Multistate outbreak of Norwalk-like virus gastroenteritis associated with a common caterer. *Am. J. Epidemiol.* 154:1013–1019.

Angelillo, I. F., N. M. A. Viggiani, L. Rizzo, and A. Bianco. 2000. Food handlers and foodborne diseases: knowledge, attitudes, and reported behavior in Italy. *J. Food Prot.* 63:381–385.

Anonymous. 1991. *Listeria monocytogenes* recommendations by the National Advisory Committee on Microbiological Criteria for Foods; fact sheet for retail and institutional food service employees. *Int. J. Food Microbiol.* 14:238–241.

Anonymous. 1999. Listeriosis linked to retail outlets in hospitals—caution needed. *Commun. Dis. Rep. Wkly.* 9:219, 222.

Appleton, H. 2000. Control of foodborne viruses. *Brit. Med. Bull.* 56:172–183.

Arnold, J. W., and G. W. Bailey. 2000. Surface finishes on stainless steel reduce bacterial attachment and early biofilm formation: scanning electron and atomic force microscopy study. *Poult. Sci.* 79:1839–1845.

Arnold, J. W., and S. Silvers. 2000. Comparison of poultry processing equipment surfaces for susceptibility to bacterial attachment and biofilm formation. *Poult. Sci.* 79:1215–1221.

Audits International/FDA. 1999. U.S. food temperature evaluation. Audits International, Northbrook, IL, and Food and Drug Administration, Washington, DC. http://www.foodriskclearinghouse.umd.edu/Audits-FDA_temp_study.htm.

Aureli, P., M. Di Cunto, A. Maffei, G. De Chiara, G. Franciosa, L. Accorinti, A. M. Gambardella, and D. Greco. 2000. An outbreak in Italy of botulism associated with a dessert made with mascarpone cream cheese. *Eur. J. Epidemiol.* 16:913–918.

Ayebah, B., Y.-C. Hung, and J. F. Frank. 2005. Enhancing the bactericidal effect of electrolyzed water on *Listeria monocytogenes* biofilms formed on stainless steel. *J. Food Prot.* 68:1375–1380.

Bahk, G. J., C. H. Hong, D. H. Oh, S. D. Ha, K. H. Park, and E. C. Todd. 2006. Modeling the level of contamination of *Staphylococcus aureus* in ready-to-eat kimbab in Korea. *J. Food Prot.* 69:1340–1346.

Baker, D. F., E. Kraa, and S. J. Corbett. 1998. A multi-state outbreak of *Salmonella enterica* serotype Bredeney food poisoning: a case control study. *Aust. N. Z. J. Public Health* 22:552–555.

Bell, C., and A. Kyriakides. 2005. *Listeria*: a practical approach to the organism and its control in foods. Blackwell Publishing, Oxford, United Kingdom.

Beresford, M. R., P. W. Andrew, and G. Shama. 2001. *Listeria monocytogenes* adheres to many materials found in food processing environments. *J. Appl. Microbiol.* 90:1000–1005.

Beuchat, L. R. 1996. *Listeria monocytogenes*: incidence on vegetables. *Food Control* 7:223–228.

Beuchat, L. R., and J.-H. Ryu. 1997. Produce handling and processing practices. *Emerg. Infect. Dis.* 3:459–465.

Bidawid, S., J. M. Farber, and S. A. Sattar. 2000. Contamination of foods by food handlers: experiments on hepatitis A virus transfer to food and its interruption. *Appl. Environ. Microbiol.* 66:2759–2763.

Blackman, I. C., and J. F. Frank. 1996. Growth of *Listeria monocytogenes* as a biofilm on various food processing surfaces. *J. Food Prot.* 59:827–831.

Blanton, L. H., S. M. Adams, R. S. Beard, G. Wei, S. N. Bulens, M. A. Widdowson, R. I. Glass, and S. S. Monroe. 2006. Molecular and epidemiologic trends of caliciviruses associated with outbreaks of acute gastroenteritis in the United States, 2000–2004. *J. Infect. Dis.* 193:413–421.

Bloomfield, S. F., and E. Scott. 1997. Cross-contamination and infection in the domestic environment and the role of chemical disinfectants. *J. Appl. Microbiol.* 83:1–9.

Brackett, R. E. 1999. Incidence, contributing factors, and control of bacterial pathogens in produce. *Postharvest Biol. Technol.* 15:305–311.

Bresee, J. S., M.-A. Widdowson, S. S. Monroe, and R. I. Glass. 2002. Foodborne viral gastroenteritis: challenges and opportunities. *Clin. Infect. Dis.* 35:748–753.

Bryan, F. L. 1990. Hazard Analysis Critical Control Point (HACCP) systems for retail food and restaurant operations. *J. Food Prot.* 53:978–983.

Bryan, F. L. 2002. Where we are in retail food safety, how we got to where we are, and how do we get there? *J. Environ. Health* 65:29–36.

Buesa, J., B. Collado, P. López-Andújar, R. Abu-Mallouh, J. Rodríguez Díaz, A. García Díaz, J. Prat, S. Guix, T. Llovet, G. Prats, and A. Bosch. 2002. Molecular epidemiology of caliciviruses causing outbreaks and sporadic cases of acute gastroenteritis in Spain. *J. Clin. Microbiol.* 40:2854–2859.

Burnett, S. L., E. L. Mertz, B. Bennie, T. Ford, and A. Starobin. 2005. Growth or survival of *Listeria monocytogenes* in ready-to-eat meat products and combination deli salads during refrigerated storage. *J. Food Sci.* 70:M301–M304.

Burt, S. 2004. Essential oils: their antimicrobial properties and potential applications in foods—a review. *Int. J. Food Microbiol.* 94:223–253.

Cagri, A., Z. Ustunol, and E. T. Ryser. 2004. Antimicrobial edible films and coatings. *J. Food Prot.* 67:833–848.

CDC. 1990. Epidemiologic notes and reports. Foodborne hepatitis A—Alaska, Florida, North Carolina, Washington. *MMWR Morb. Mortal. Wkly. Rep.* 39:228–232.

CDC. 1999. Norwalk-like viral gastroenteritis in U.S. Army trainees—Texas, 1998. *MMWR Morb. Mortal. Wkly. Rep.* 48:225–227.

CDC. 2000a. Surveillance for foodborne-disease outbreaks—United States, 1993–1997. CDC Surveillance Summaries. *MMWR Morb. Mortal. Wkly. Rep.* 49(SS-1):1–64.

CDC. 2000b. Outbreaks of Norwalk-like viral gastroenteritis—Alaska and Wisconsin, 1999. *MMWR Morb. Mortal. Wkly. Rep.* 49:207–211.

CDC. 2001. "Norwalk-like viruses": public health consequences and outbreak management. *MMWR Morb. Mortal. Wkly. Rep.* 50(RR-9):1–17.

CDC. 2003a. Norovirus activity—United States, 2002. *MMWR Morb. Mortal. Wkly. Rep.* 52:41–45.

CDC. 2003b. Foodborne transmission of hepatitis A—Massachusetts, 2001. *MMWR Morb. Mortal. Wkly. Rep.* 52:565–567.

CDC. 2003c. Hepatitis A outbreak associated with green onions at a restaurant—Monaca, Pennsylvania, 2003. *MMWR Morb. Mortal. Wkly. Rep.* 52:1155–1157.

CDC. 2006a. Preliminary FoodNet data on the incidence of infection with pathogens transmitted commonly through food—10 states, United States, 2005. *MMWR Morb. Mortal. Wkly. Rep.* 55:392–395.

CDC. 2006b. Multisite outbreak of Norovirus associated a franchise restaurant—Kent County, Michigan, May 2005. *MMWR Morb. Mortal. Wkly. Rep.* 55:395–397.

CDC. 2007. Preliminary FoodNet data on the incidence of infection with pathogens transmitted commonly through food—10 states, 2006. *MMWR Morb. Mortal. Wkly. Rep.* **56**:336–339.

CFP. 2006. Voluntary guidelines of sanitation practices, standard operating procedures and good retail practices to minimize contamination and growth of *Listeria monocytogenes* within food establishments. Conference for Food Protection, Lincoln, CA. http://www.foodprotect.org/pdf/2006CFPLmInterventionvoluntaryguidelines.pdf.

Chaidez, C., M. Moreno, W. Rubio, M. Angulo, and B. Valdez. 2003. Comparison of the disinfection efficacy of chlorine-based products for inactivation of viral indicators and pathogenic bacteria in produce wash water. *Int. J. Environ. Health Res.* **13**:295–302.

Charbonneau, D. L., J. M. Ponte, and B. A. Kochanowski. 2000. A method of assessing the efficacy of hand sanitizers: use of real soil encountered in the food service industry. *J. Food Prot.* **63**:495–501.

Chen, Y., K. M. Jackson, F. P. Chea, and D. W. Schaffner. 2001. Quantification and variability analysis of bacterial cross-contamination rates in common food service tasks. *J. Food Prot.* **64**:72–80.

Chen, Y., W. H. Ross, V. N. Scott, and D. E. Gombas. 2003. *Listeria monocytogenes:* low levels equal low risk. *J. Food Prot.* **66**:570–577.

Clayton, D. A., C. J. Griffith, P. Price, and A. C. Peters. 2002. Food handlers' beliefs and self-reported practices. *Int. J. Environ. Health Res.* **12**:25–39.

Cliver, D. O. 1994. Epidemiology of viral foodborne disease. *J. Food Prot.* **57**:263–266.

Cliver, D. O. 1997. Virus transmission via food. *World Health Stat. Q.* **50**:90–101.

Cliver, D. O. 2006. Cutting boards in *Salmonella* cross-contamination. *J. AOAC Int.* **89**:538–542.

Codron, J.-M., K. Grunert, E. Giraud-Heraud, L.-G. Soler, and A. Regmi. 2005. Retail sector responses to changing consumer preferences, p. 32–46. *In* A. Regmi and M. Gehlhar (ed.), *New Directions in Global Food Markets*. Economic Research Service, United States Department of Agriculture. http://www.ers.usda.gov/publications/aib794/aib794.pdf.

Cogan, T. A., J. Slader, S. F. Bloomfield, and T. J. Humphrey. 2002. Achieving hygiene in the domestic kitchen: the effectiveness of commonly used cleaning procedures. *J. Appl. Microbiol.* **92**:885–892.

Collins, J. E. 1997. Impact of changing consumer lifestyles on the emergence/reemergence of foodborne pathogens. *Emerg. Infect. Dis.* **3**:471–479.

Cools, I., M. Uyttendaele, J. Cerpentier, E. D'Haese, H. J. Nelis, and J. Debevere. 2005. Persistence of *Campylobacter jejuni* on surfaces in a processing environment and on cutting boards. *Lett. Appl. Microbiol.* **40**:418–423.

Croci, L., D. De Medici, C. Scalfaro, A. Fiore, and L. Toti. 2002. The survival of hepatitis A virus in fresh produce. *Int. J. Food Microbiol.* **73**:29–34.

Cruickshank, J. G. 1990. Food handlers and food poisoning. *Brit. Med. J.* **300**:207–208.

Curry, A., T. Riordan, J. Craske, and E. O. Caul. 1987. Small round structured viruses and persistence of infectivity in food handlers. *Lancet* **ii**:864–865.

Dallaire, R., D. I. LeBlank, C. C. Tranchant, L. Vasseur, P. Delaquis, and C. Beaulieu. 2006. Monitoring the microbial populations and temperatures of fresh broccoli from harvest to retail display. *J. Food Prot.* **69**:1118–1125.

Dalton, C. B., J. Gregory, M. D. Kirk, R. J. Stafford, R. Ginvey, E. Kraa, and D. Gould. 2004. Foodborne disease outbreaks in Australia, 1995 to 2000. *Commun. Dis. Intell.* **28**:211–224.

Dalton, C. B., A. Haddix, R. E. Hoffman, and E. E. Mast. 1996. The cost of a food-borne outbreak of hepatitis A in Denver, Colo. *Arch. Intern. Med.* **156**:1013–1016.

Daniels, N. A., D. A. Bergmire-Sweat, K. J. Schwab, K. A. Hendricks, S. Reddy, S. M. Rowe, R. L. Fankhauser, S. S. Monroe, R. L. Atmar, R. I. Glass, and P. Mead. 2000. A food-borne ourbreak of gastroenteritis associated with Norwalk-like viruses: first molecular traceback to deli sandwiches contaminated during preparation. *J. Infect. Dis.* **181**:1467–1470.

Dawson, S. J., M. R. Evans, D. Willby, J. Bardwell, N. Chamberlain, and D. A. Lewis. 2006. *Listeria* outbreak associated with sandwich consumption from a hospital retail shop, United Kingdom. *Euro. Surveill.* **11**:89–91.

Deneen, V. C., J. M. Hunt, C. R. Paule, R. I. James, R. G. Johnson, M. J. Raymond, and C. W. Hedberg. 2000. The impact of food-borne calicivirus disease: the Minnesota experience. *J. Infect. Dis.* **181**:S281–S283.

Dentinger, C. M., W. A. Bower, O. V. Nainan, S. M. Cotter, G. Myers, L. M. Dubusky, S. Fowler, E. D. Salehi, and B. P. Bell. 2001. An outbreak of hepatitis A associated with green onions. *J. Infect. Dis.* **183**:1273–1276.

Deza, M. A., M. Araujo, and M. J. Garrido. 2003. Inactivation of *Escherichia coli* O157:H7, *Salmonella enteritidis* and *Listeria monocytogenes* on the surface of tomatoes by neutral electrolyzed water. *Lett. Appl. Microbiol.* **37**:482–487.

Deza, M. A., M. Araujo, and M. J. Garrido. 2005. Inactivation of *Escherichia coli*, *Listeria monocytogenes*, *Pseudomonas aeruginosa* and *Staphylococcus aureus* on stainless steel and glass surfaces by neutral electrolysed water. *Lett. Appl. Microbiol.* **40**:341–346.

Deza, M. A., M. Araujo, and M. J. Garrido. 2007. Efficacy of neutral electrolyzed water to inactivate *Escherichia coli*, *Listeria monocytogenes*, *Pseudomonas aeruginosa*, and *Staphylococcus aureus* on plastic and wooden kitchen cutting boards. *J. Food Prot.* **70**:102–108.

Doré, W. J., K. Henshilwood, and D. N. Lees. 2000. Evaluation of F-specific RNA bacteriophage as a candidate human enteric virus indicator for bivalve molluscan shellfish. *Appl. Environ. Microbiol.* **66**:1280–1285.

Doser, J. G. 1999. How safe are self-serve unpackaged foods? *J. Environ. Health* **61**:29–32.

Duncanson, P., D. R. A. Wareing, and O. Jones. 2003. Application of an automated immunomagnetic separation-enzyme immunoassay for the detection of *Salmonella* spp. during an outbreak associated with a retail premises. *Lett. Appl. Microbiol.* **37**:144–148.

Dupard, T., M. E. Janes, R. L. Beverly, and J. W. Bell. 2006. Antimicrobial effect of cetylpyridinium chloride on *Listeria monocytogenes* V7 growth on the surface of raw and cooked retail shrimp. *J. Food Sci.* **71**:M241–M244.

Eddy, C., T. Ingram, and J. Leever. 2000. The Norwalk virus: a guidance template for local environmental health professionals. *J. Environ. Health* **63**:8–11.

Elson, R., F. Burgess, C. L. Little, and R. T. Mitchell, on behalf of the Local Authorities Co-ordinators of Regulatory Services and the Health Protection Agency. 2004. Microbiological examination of ready-to-eat cold sliced meats and pâté from catering and retail premises in the UK. *J. Appl. Microbiol.* **96**:499–509.

Ercolini, D., A. Storia, F. Villani, and G. Mauriello. 2006. Effect of a bacteriocin-activated polythene film on *Listeria monocytogenes* as evaluated by viable staining and epifluorescence microscopy. *J. Appl. Microbiol.* **100**:765–772.

Fankhauser, R. L., S. S. Monroe, J. S. Noel, C. D. Humphrey, J. S. Bresee, U. D. Parashar, T. Ando, and R. I. Glass. 2002. Epidemiologic and molecular trends of "Norwalk-like viruses" associated with outbreaks of gastroenteritis in the United States. *J. Infect. Dis.* **186:**1–7.

Fankhauser, R. L., J. S. Noel, S. S. Monroe, T. Ando, and R. I. Glass. 1998. Molecular epidemiology of "Norwalk-like viruses" in outbreaks of gastroenteritis in the United States. *J. Infect. Dis.* **178:**1571–1578.

FAO/WHO. 2004. Risk assessment of *Listeria monocytogenes* in ready-to-eat foods. Food and Agriculture Organization, Rome, Italy, and World Health Organization, Geneva Switzerland. http://www.fao.org/ag/agn/jemra/listeria_en.stm.

Feinstone, S. M. 1996. Hepatitis A: epidemiology and prevention. *Eur. J. Gastroenterol. Hepatol.* **8:**300–305.

Franciosa, G., M. Pourshaban, M. Gianfranceschi, A. Gattuso, L. Fenicia, A. M. Ferrini, V. Mannoni, G. De Luca, and P. Aureli. 1999. *Clostridium botulinum* spores and toxin in mascarpone cheese and other milk products. *J. Food Prot.* **62:**867–871.

Friedman, D. S., D. Heisey-Grove, F. Argyros, E. Berl, J. Nsubuga, T. Stiles, J. Fontana, R. S. Beard, S. Monroe, M. E. McGrath, H. Sutherby, R. C. Dicker, A. DeMaria, Jr., and B. T. Matyas. 2005. An outbreak of norovirus gastroenteritis associated with wedding cakes. *Epidemiol. Infect.* **133:**1057–1063.

Garver, K. I., and P. M. Muriana. 1993. Detection, identification and characterization of bacteriocin-producing lactic acid bacteria from retail food products. *Int. J. Food Microbiol.* **19:**241–258.

Gill, C. O., M. Friske, A. K. W. Tong, and J. C. McGinnis. 1995. Assessment of the hygienic characteristics of a process for the distribution of processed meats, and of storage conditions at retail outlets. *Food Res. Int.* **28:**131–138.

Gill, C. O., and T. Jones. 2002. Effects of wearing knitted or rubber gloves on the transfer of *Escherichia coli* between hands and meat. *J. Food Prot.* **65:**1045–1048.

Gillespie, I. A., C. L. Little, and R. T. Mitchell. 2001. Microbiological examination of ready-to-eat quiche from retail establishments in the United Kingdom. *Commun. Dis. Public Health* **4:**53–59.

Gombas, D. E., Y. Chen, R. S. Clavero, and V. N. Scott. 2003. Survey of *Listeria monocytogenes* in ready-to-eat foods. *J. Food Prot.* **66:**559–569.

Graham, D. Y., X. Jiang, T. Tanaka, A. R. Opekun, H. P. Madore, and M. K. Estes. 1994. Norwalk virus infection of volunteers: new insights based on improved assays. *J. Infect. Dis.* **170:**34–43.

Green, J., K. Hensilwood, C. I. Gallimore, D. W. G. Brown, and D. N. Lees. 1998. A nested reverse transcriptase PCR assay for detection of small round-structured viruses in environmentally contaminated molluscan shellfish. *Appl. Environ. Microbiol.* **64:**858–863.

Green, L. R., V. Radke, R. Mason, L. Bushnell, D. W. Reimann, J. C. Mack, M. D. Motsinger, T. Stigger, and C. A. Selman. 2007. Factors related to food worker hand hygiene practices. *J. Food Prot.* **70:**661–666.

Green, L. R., and C. Selman. 2005. Factors impacting food workers' and managers' safe food preparation practices: a qualitative study. *Food Prot. Trends* **25:**981–990.

Green, L., C. Selman, A. Banerjee, R. Marcus, C. Medus, F. J. Angulo, V. Radke, S. Buchanan, and the EHS-Net Working Group. 2005. Food service workers' self-reported food preparation practices: an EHS-Net study. *Int. J. Hyg. Environ. Health* **208:**27–35.

Green, L. R., C. A. Selman, V. Radke, D. Ripley, J. C. Mack, D. W. Reimann, T. Stigger, M. Motsinger, and L. Bushnell. 2006. Food worker hand washing practices: an observation study. *J. Food Prot.* **69:**2417–2423.

Greening, G. E., M. Mirams, and T. Berke. 2001. Molecular epidemiology of "Norwalk-like viruses" associated with gastroenteritis outbreaks in New Zealand. *J. Med. Virol.* **64:**58–66.

Greer, G. G. 2005. Bacteriophage control of foodborne bacteria. *J. Food Prot.* **68:**1102–1111.

Greig, J. D., E. C. D. Todd, C. A. Bartleson, and B. S. Michaels. 2007. Outbreaks where food workers have been implicated in the spread of foodborne disease. Part 1. Description of the problem, methods, and agents involved. *J. Food Prot.* **70:**1752–1761.

Grinstead, D. A., and C. N. Cutter. 2007. Controlling *Listeria monocytogenes* in a retail setting. *Food Prot. Trends* **27:**22–28.

Gulati, B. R., P. B. Allwood, C. W. Hedberg, and S. M. Goyal. 2001. Efficacy of commonly used disinfectants for the inactivation of calicivirus on strawberry, lettuce, and a food-contact surface. *J. Food Prot.* **64:**1430–1434.

Hamano, M., M. Kuzuya, R. Fujii, H. Ogura, and M. Yamada. 2005. Epidemiology of acute gastroenteritis outbreaks caused by noroviruses in Okayama, Japan. *J. Med. Virol.* **77:**282–289.

Hedberg, C. W., K. L. MacDonald, and M. T. Osterholm. 1994. Changing epidemiology of food-borne disease: a Minnesota perspective. *Clin. Infect. Dis.* **18:**671–682.

Hedberg, C. W., and M. T. Osterholm. 1993. Outbreaks of foodborne and waterborne viral gastroenteritis. *Clin. Microbiol. Rev.* **6:**199–210.

Hedberg, C. W., S. J. Smith, E. Kirkland, V. Radke, T. F. Jones, C. A. Selman, and the EHS-Net Working Group. 2006. Systematic environmental evaluations to identify food safety differences between outbreak and nonoutbreak restaurants. *J. Food Prot.* **69:**2697–2702.

Hedlund, K. O., E. Rubilar-Abreu, and L. Svensson. 2000. Epidemiology of calicivirus infections in Sweden, 1994–1998. *J. Infect. Dis.* **181:**S275–S280.

Hellström, S., R. Kervinen, M. Lyly, R. Ahvenainen-Rantala, and H. Korkeala. 2006. Efficacy of disinfectants to reduce *Listeria monocytogenes* on precut iceberg lettuce. *J. Food Prot.* **69:**1565–1570.

Hewitt, J., and G. E. Greening. 2006. Effect of heat treatment on hepatitis A virus and norovirus in New Zealand greenshell mussels *(Perna canaliculus)* by quantitative real-time reverse transcription PCR and cell culture. *J. Food Prot.* **69:**2217–2223.

Howie, J. W. 1968. Typhoid in Aberdeen, 1964. *J. Appl. Bacteriol.* **31:**171–178.

Hudson, J. A., and S. J. Mott. 1993. Presence of *Listeria monocytogenes*, motile aeromonads and *Yersinia enterocolitica* in environmental samples taken from a supermarket delicatessen. *Int. J. Food Microbiol.* **18:**333–337.

Hudson, J. A., S. J. Mott, K. M. Delacy, and A. L. Edridge. 1992. Incidence and coincidence of *Listeria* spp., motile aeromonads and *Yersinia enterocolitica* on ready-to-eat fleshfoods. *Int. J. Food Microbiol.* **16:**99–108.

Humphrey, T. J., and D. M. Worthington. 1990. *Listeria* contamination of retail meat slicers. *PHLS Microbiol. Digest* **7:**57.

ILSI Research Foundation/Risk Science Institute. 2005. Achieving continuous improvement in reductions in foodborne listeriosis—a risk-based approach. *J. Food Prot.* **68:**1932–1994.

Iversen, A. M., M. Gill, C. L. Bartlett, W. D. Cubitt, and D. A. McSwiggan. 1987. Two outbreaks of foodborne gastroenteritis caused by a small round structured virus: evidence of prolonged infectivity in a food handler. *Lancet* **ii:**556–558.

Izumi, H. 1999. Electrolyzed water as a disinfectant for fresh-cut vegetables. *J. Food Sci.* **64:**536–539.

Jacquet, C., B. Catimel, R. Brosch, C. Buchrieser, P. Dehaumont, V. Goulet, A. Lepoutre, P. Veit, and J. Rocourt. 1995. Investigations related to the epidemic strain involved in the French listeriosis outbreak in 1992. *Appl. Environ. Microbiol.* **61:**2242–2246.

Jay, J. M. (ed.). 2000. *Modern Food Microbiology*, 6th ed. Aspen Publishers, Gaithersburg, MD.

Jaykus, L.-A. 1997. Epidemiology and detection as options for control of viral and parasitic foodborne disease. *Emerg. Infect. Dis.* 3:529–539.

Jean, J., J. -F. Vachon, O. Moroni, A. Darveau, I. Kukavica-Ibrulj, and I. Fliss. 2003. Effectiveness of commercial disinfectants for inactivating hepatitis A virus on agri-food surfaces. *J. Food Prot.* 66:115–119.

Jimenez, L., and M. Chiang. 2006. Virucidal activity of a quaternary ammonium compound disinfectant against feline calicivirus: a surrogate for norovirus. *Am. J. Infect. Control* 34:269–273.

Kassa, H. 2001. An outbreak of Norwalk-like viral gastroenteritis in a frequently penalized food service operation: a case for mandatory training of food handlers in safety and hygiene. *J. Environ. Health* 64:9–12, 33.

Kerr, K. G., D. Birkenhead, K. Seale, J. Major, and P. M. Hawkey. 1993. Prevalence of *Listeria* spp. on the hands of food workers. *J. Food Prot.* 56:525–527.

Kerr, K. G., P. Kite, J. Heritage, and P. M. Hawkey. 1995. Typing of epidemiologically associated environmental and clinical strains of *Listeria monocytogenes* by random amplification of polymorphic DNA. *J. Food Prot.* 58:609–613.

Khan, A. S., C. L. Moe, R. I. Glass, S. S. Monroe, M. K. Estes, L. E. Chapman, X. Jiang, C. Humphrey, E. Pon, J. K. Iskander, and L. B. Schonberger. 1994. Norwalk virus-associated gastroenteritis traced to ice consumption aboard a cruise ship in Hawaii: comparison and application of molecular method-based assays. *J. Clin. Microbiol.* 32:318–322.

Kim, C., Y.-C. Hung, R. E. Brackett, and J. F. Frank. 2001. Inactivation of *Listeria monocytogenes* biofilms by electrolyzed oxidizing water. *J. Food Process Preserv.* 25:91–100.

Kim, C., Y.-C. Hung, R. E. Brackett, and C.-S. Lin. 2003. Efficacy of electrolyzed oxidizing water in inactivating *Salmonella* on alfalfa seeds and sprouts. *J. Food Prot.* 66:208–214.

Kim, J. K., and M. A. Harrison. 2008. Transfer of *Escherichia coli* O157:H7 to romaine lettuce due to contact water from melting ice. *J. Food Prot.* 71:252–256.

Kim, S.-H., and C.-I. Wei. 2007. Biofilm formation by multidrug-resistant *Salmonella enterica* serotype Typhimurium phage type DT104 and other pathogens. *J. Food Prot.* 70:22–29.

Kirkland, K. B., R. A. Meriwether, J. K. Leiss, and W. R. Mac Kenzie. 1996. Steaming oysters does not prevent Norwalk-like gastroenteritis. *Public Health Rep.* 111:527–530.

Kondo, N., M. Murata, and K. Isshiki. 2006. Efficiency of sodium hypochlorite, fumaric acid, and mild heat in killing native microflora and *Escherichia coli* O157:H7, *Salmonella* Typhimurium DT104, and *Staphylococcus aureus* attached to fresh-cut lettuce. *J. Food Prot.* 69:323–329.

Koopmans, M., and E. Dulxer. 2004. Foodborne viruses: an emerging problem. *Int. J. Food Microbiol.* 90:23–41.

Koopmans, M., H. Vennema, H. Heersma, E. van Strien, Y. van Duynhoven, D. Brown, M. Reacher, and B. Lopman, for the European Consortium on Foodborne Viruses. 2003. Early identification of common-source foodborne virus outbreaks in Europe. *Emerg. Infect. Dis.* 9:1136–1142.

Koseki, S., K. Yoshida, S. Isobe, and K. Itoh. 2001. Decontamination of lettuce using acidic electrolyzed water. *J. Food Prot.* 64:652–658.

Koutsoumanis, K., P. S. Taoukis, and G. J. E. Nychas. 2005. Development of a safety monitoring and assurance system for chilled food products. *Int. J. Food Microbiol.* 100:253–260.

Kreske, A. C., J.-H. Ryu, C. A. Pettigrew, and L. R. Beuchat. 2006. Lethality of chlorine, chlorine dioxide, and a commercial produce sanitizer to *Bacillus cereus* and *Pseudomonas* in a liquid detergent, on stainless steel, and in biofilm. *J. Food Prot.* 69:2621–2634.

Kuritsky, J. N., M. T. Osterholm, H. B. Greenberg, J. A. Korlath, J. R. Godes, C. W. Hedberg, J. C. Forfang, A. Z. Kapikian, J. C. McCullough, and K. E. White. 1984. Norwalk gastroenteritis: a community outbreak associated with bakery product consumption. *Ann. Intern. Med.* 100:519–521.

Kusumaningrum, H. D., G. Riboldi, W. C. Hazeleger, and R. R. Beumer. 2003. Survival of foodborne pathogens on stainless steel surfaces and cross-contamination to foods. *Int. J. Food Microbiol.* 85:227–236.

Leverentz, B., W. S. Conway, M. J. Camp, W. J. Janisiewicz, T. Abuladze, M. Yang, R. Saftner, and A. Sulakvelidze. 2003. Biocontrol of *Listeria monocytogenes* on fresh-cut produce by treatment with lytic bacteriophages and a bacteriocin. *Appl. Environ. Microbiol.* 69:4519–4526.

Lewis, H. C., C. L. Little, R. Elson, M. Greenwood, K. A. Grant, and J. McLauchlin. 2006. Prevalence of *Listeria monocytogenes* and other *Listeria* species in butter from United Kingdom production, retail, and catering premises. *J. Food Prot.* 69:1518–1526.

Lianou, A., I. Geornaras, P. A. Kendall, K. E. Belk, J. A. Scanga, G. C. Smith, and J. N. Sofos. 2007a. Fate of *Listeria monocytogenes* in commercial ham, formulated with or without antimicrobials, under conditions simulating contamination in the processing or retail environment and during home storage. *J. Food Prot.* 70:378–385.

Lianou, A., I. Geornaras, P. A. Kendall, J. A. Scanga, and J. N. Sofos. 2007b. Behavior of *Listeria monocytogenes* at 7°C in commercial turkey breast, with or without antimicrobials, after simulated contamination for manufacturing, retail and consumer settings. *Food Microbiol.* 24:433–443.

Lianou, A., and J. N. Sofos. 2007. A review of the incidence and transmission of *Listeria monocytogenes* in ready-to-eat products in retail and food service environments. *J. Food Prot.* 70:2172–2198.

Lin, C.-M., S. Y. Fernando, and C. Wei. 1996. Occurrence of *Listeria monocytogenes*, *Salmonella* spp., *Escherichia coli* and *E. coli* O157:H7 in vegetable salads. *Food Control* 7:135–140.

Lin, C. M., K. Takeuchi, L. Zhang, C. B. Dohm, J. D. Meyer, P. A. Hall, and M. P. Doyle. 2006. Cross-contamination between processing equipment and deli meats by *Listeria monocytogenes*. *J. Food Prot.* 69:71–79.

Lin, C.-M., F.-M. Wu, H. -K. Kim, M. P. Doyle, B. S. Michaels, and L. K. Williams. 2003. A comparison of hand washing techniques to remove *Escherichia coli* and caliciviruses under natural or artificial fingernails. *J. Food Prot.* 66:2296–2301.

Linton, R. H., D. Z. McSwane, and C. D. Woodley. 1998. A comparison of perspectives about the critical areas of knowledge for safe food handling in food establishments. *J. Environ. Health* 60:8–15.

Linton, R. 1996. Keeping food safe in foodservice and food retail establishments. Department of Food Science, Purdue Cooperative Extension Service, West Lafayette, IN. http://www.ces.purdue.edu/extmedia/FS/FS-3.pdf.

Little, C. L., and J. de Louvois. 1998. The microbiological examination of butchery products and butchers' premises in the United Kingdom. *J. Appl. Microbiol.* 85:177–186.

Little, C. L., D. Lock, J. Barnes, and R. T. Mitchell. 2003. Microbiological quality of food in relation to hazard analysis systems and food hygiene training in UK catering and retail premises. *Commun. Dis. Public Health* 6:250–258.

Little, C. L., and R. T. Mitchell. 2004. Microbiological quality of pre-cut fruit, sprouted seeds, and unpasteurised fruit and vegetable juices from retail and production premises in the UK, and the application of HACCP. *Commun. Dis. Public Health* 7:184–190.

Lo, S. V., A. M. Connolly, S. R. Palmer, D. Wright, P. D. Thomas, and D. Joynson. 1994. The role of the pre-symptomatic food handler in a common source outbreak of food-borne SRSV gastroenteritis in a group of hospitals. *Epidemiol. Infect.* **113:**513–521.

Long, S. M., G. K. Adak, S. J. O'Brien, and I. A. Gillespie. 2002. General outbreaks of infectious intestinal disease linked with salad vegetables and fruit, England and Wales, 1992–2000. *Commun. Dis. Public Health* **5:**101–105.

Lues, J. F. R., and I. Van Tonder. 2007. The occurrence of indicator bacteria on hands and aprons of food handlers in the delicatessen sections of a retail group. *Food Control* **18:**326–332.

Lundén, J. M., T. J. Autio, and H. J. Korkeala. 2002. Transfer of persistent *Listeria monocytogenes* contamination between food-processing plants associated with a dicing machine. *J. Food Prot.* **65:**1129–1133.

Lyhs, U., M. Hatakka, N. Mäki-Petäys, E. Hyytiä, and H. Korkeala. 1998. Microbiological quality of Finnish vacuum-packaged fishery products at retail level. *Arch. Lebensmittelhyg.* **49:**146–150.

Lynch, R. A., M. L. Phillips, B. L. Elledge, S. Hanumanthaiah, and D. T. Boatright. 2005. A preliminary evaluation of the effect of glove use by food handlers in fast food restaurants. *J. Food Prot.* **68:**187–190.

Maguire, A. J., J. Green, D. W. G. Brown, U. Desselberger, and J. J. Gray. 1999. Molecular epidemiology of outbreaks of gastroenteritis associated with small round-structured viruses in East Anglia, United Kingdom, during the 1996–1997 season. *J. Clin. Microbiol.* **37:**81–89.

Mahl, M. C. 1989. New method for determination of efficacy of health care personnel hand wash products. *J. Clin. Microbiol.* **27:**2295–2299.

Malik, Y. S., and S. M. Goyal. 2006. Virucidal efficacy of sodium bicarbonate on a food contact surface against feline calicivirus, a norovirus surrogate. *Int. J. Food Microbiol.* **109:**160–163.

Martin, K. E., S. Knabel, and V. Mendenhall. 1999. A model train-the-trainer program for HACCP-based food safety training in the retail/food service industry: an evaluation. *J. Extension* **37:** 3FEA1. http://www.joe.org/joe/1999june/a1.html.

Massoudi, M. S., B. P. Bell, V. Paredes, J. Insko, K. Evans, and C. N. Shapiro. 1999. An outbreak of hepatitis A associated with an infected foodhandler. *Public Health Rep.* **114:**157–164.

Mauriello, G., D. Ercolini, A. La Storia, A. Casaburi, and F. Villani. 2004. Development of polythene films for food packaging activated with an antilisterial bacteriocin from *Lactobacillus curvatus* 32Y. *J. Appl. Microbiol.* **97:**314–322.

Mbithi, J. N., S. Springthorpe, and S. A. Sattar. 1990. Chemical disinfection of hepatitis A on environmental surfaces. *Appl. Environ. Microbiol.* **56:**3601–3604.

Mbithi, J. N., S. Springthorpe, and S. A. Sattar. 1991. Effect of relative humidity and air temperature on survival of hepatitis A virus on environmental surfaces. *Appl. Environ. Microbiol.* **57:**1394–1399.

McCarthy, S. A. 1996. Effect of sanitizers on *Listeria monocytogenes* attached to latex gloves. *J. Food Saf.* **16:**231–237.

McCarthy, S. A. 1997. Incidence and survival of *Listeria monocytogenes* in ready-to-eat seafood products. *J. Food Prot.* **60:** 372–376.

McCarthy, M., M. K. Estes, and K. C. Hyams. 2000. Norwalk-like virus infection in military forces: epidemic potential, sporadic disease, and the future direction of prevention and control efforts. *J. Infect. Dis.* **181:**S387–S391.

McDonnell, R. J., A. Rampling, S. Crook, P. M. Cockroft, G. A. Wilshaw, T. Cheasty, and J. Stuart. 1997. An outbreak of Vero cytotoxin producing *Escherichia coli* O157 infection associated with takeaway sandwiches. *Commun. Dis. Rep. CDR Rev.* **7:**R201–R205.

McSwane, D., and R. Linton. 2000. Issues and concerns in HACCP development and implementation for retail food operations. *J. Environ. Health* **62:**15–18.

Mead, P. S., L. Slutsker, V. Dietz, L. F. McCaig, J. S. Bresee, C. Shapiro, P. M. Griffin, and R. V. Tauxe. 1999. Food-related illness and death in the United States. *Emerg. Infect. Dis.* **5:**607–625.

Medus, C., K. E. Smith, J. B. Bender, J. M. Besser, and C. W. Hedberg. 2006. *Salmonella* outbreaks in restaurants in Minnesota, 1995 through 2003: evaluation of the role of infected foodworkers. *J. Food Prot.* **69:**1870–1878.

Michaels, B., T. Ayers, and W. Birbari. 2002. Hygiene issues associated with food service potholders and oven mitts. *Food Serv. Technol.* **2:**81–86.

Midelet, G., A. Kobilinsky, and B. Carpentier. 2006. Construction and analysis of fractional multifactorial designs to study attachment strength and transfer of *Listeria monocytogenes* from pure or mixed biofilms after contact with a solid model food. *Appl. Environ. Microbiol.* **72:**2313–2321.

Montville, R., Y. Chen, and D. W. Schaffner. 2001. Glove barriers to bacterial cross-contamination between hands to food. *J. Food Prot.* **64:**845–849.

Montville, R., and D. W. Schaffner. 2003. Inoculum size influences bacterial cross contamination between surfaces. *Appl. Environ. Microbiol.* **69:**7188–7193.

Moore, C. M., B. W. Sheldon, and L.-A. Jaykus. 2003. Transfer of *Salmonella* and *Campylobacter* from stainless steel to romaine lettuce. *J. Food Prot.* **66:**2231–2236.

Mortlock, M. P., A. C. Peters, and C. J. Griffith. 1999. Food hygiene and Hazard Analysis Critical Control Point in the United Kingdom food industry: practices, perceptions, and attitudes. *J. Food Prot.* **62:**786–792.

Mortlock, M. P., A. C. Peters, and C. J. Griffith. 2000. A national survey of the food hygiene training and qualification levels in the UK food industry. *Int. J. Environ. Health Res.* **10:**111–123.

NACMCF. 1999. Microbiological safety evaluations and recommendations on fresh produce. *Food Control* **10:**117–143.

NACMCF. 2005. Considerations for establishing safety-based consume-by date labels for refrigerated ready-to-eat foods. *J. Food Prot.* **68:**1761–1775.

Nguyen-the, C., and F. Carlin. 1994. The microbiology of minimally processed fresh fruits and vegetables. *Crit. Rev. Food Sci. Nutr.* **34:**371–401.

Nichols, G., I. Gillespie, and J. de Louvois. 2000. The microbiological quality of ice used to cool drinks and ready-to-eat food from retail and catering premises in the United Kingdom. *J. Food Prot.* **63:**78–82.

Nichols, G., J. McLauchlin, and J. de Louvois. 1998. The contamination of pâté with *Listeria monocytogenes*—results from the 1994 European Community-coordinated food control program for England and Wales. *J. Food Prot.* **61:**1299–1304.

Norwood, D. E., and A. Gilmour. 1999. Adherence of *Listeria monocytogenes* strains to stainless steel coupons. *J. Appl. Microbiol.* **86:**576–582.

Pao, S., D. F. Kelsey, M. F. Khalid, and M. R. Ettinger. 2007. Using aqueous chlorine dioxide to prevent contamination of tomatoes with *Salmonella enterica* and *Erwinia carotovora* during fruit washing. *J. Food Prot.* **70:**629–634.

Parashar, U. D., L. Dow, R. L. Fankhauser, C. D. Humphrey, J. Miller, T. Ando, K. S. Williams, C. R. Eddy, J. S. Noel, T. Ingram, J. S. Bresee, S. S. Monroe, and R. I. Glass. 1998. An outbreak of viral gastroenteritis associated with consumption of sandwiches: implications for the control of transmission by food handlers. *Epidemiol. Infect.* **121:**615–621.

Parashar, U. D., and S. S. Monroe. 2001. "Norwalk-like viruses" as a cause of foodborne disease outbreaks. *Rev. Med. Virol.* **11:**243–252.

Park, H., Y.-C. Hung, and C. Kim. 2002. Effectiveness of electrolyzed water as a sanitizer for treating different surfaces. *J. Food Prot.* **65:**1276–1280.

Paulson, D. S. 1994. A comparative evaluation of different hand cleansers. *Dairy Food Environ. Sanit.* **14:**524–528.

Paulson, D. S. 1997. Foodborne disease: controlling the problem. *J. Environ. Health* **59:**15–19.

Paulson, D. S. 2000. Handwashing, gloving and disease transmission by the food preparer. *Dairy Food Environ. Sanit.* **20:**838–845.

Paulson, D. S. 2005. The transmission of surrogate Norwalk-like virus from inanimate surfaces to gloved hands: Is it a threat? *Food Prot. Trends* **25:**450–454.

Pérez-Rodríguez, F., E. C. D. Todd, A. Valero, E. Carrasco, R. M. García, and G. Zurera. 2006. Linking quantitative exposure assessment and risk management using the food safety objective concept: an example with *Listeria monocytogenes* in different cross-contamination scenarios. *J. Food Prot.* **69:**2384–2394.

PFMG-IFST. 1995. Microbiological criteria for retail foods. *Lett. Appl. Microbiol.* **20:**331–332.

Phillips, M. L., B. L. Elledge, H. G. Basara, R. A. Lynch, and D. T. Boatright. 2006. Recurrent critical violations of the Food Code in retail food service establishments. *J. Environ. Health* **68:**24–30.

PSU (Pennsylvania State University). 2006. Control of *Listeria monocytogenes* in retail establishments. College of Agricultural Sciences, Pennsylvania State University, University Park, PA. http://pubs.cas.psu.edu/FreePubs/pdfs/UK137.pdf.

Reimers, F. 1994. HACCP in retail food stores. *Food Control* **5:**176–180.

Richardson, I. R., and A. M. Stevens. 2003. Microbiological examination of ready-to-eat stuffing from retail premises in the north-east of England. The "Get Stuffed" survey. *J. Appl. Microbiol.* **94:**733–737.

Robbins, M., and D. McSwane. 1994. Sanitation doesn't cost, it pays: Is it true and can we prove it? *J. Environ. Health* **57:**14–20.

Rodríguez, A., and L. A. McLandsborough. 2007. Evaluation of the transfer of *Listeria monocytogenes* from stainless steel and high-density polyethylene to bologna and American cheese. *J. Food Prot.* **70:**600–606.

Rørvic, L. M., and M. Yndestad. 1991. *Listeria monocytogenes* in foods in Norway. *Int. J. Food Microbiol.* **13:**97–104.

Ross, A. I. V., M. W. Griffiths, G. S. Mittal, and H. C. Deeth. 2003. Combining nonthermal technologies to control foodborne microorganisms. *Int. J. Food Microbiol.* **89:**125–138.

Rutjes, S. A., F. Lodder-Verschoor, W. H. M. van der Poel, Y. T. H. P. van Duijnhoven, and A. M. de Roda Husman. 2006. Detection of noroviruses in foods: a study on virus extraction procedures in foods implicated in outbreaks of human gastroenteritis. *J. Food Prot.* **69:**1949–1956.

Sagoo, S. K., C. L. Little, C. J. Griffith, and R. T. Mitchell. 2003a. Study of cleaning standards and practices in food premises in the United Kingdom. *Commun. Dis. Public Health* **6:**6–17.

Sagoo, S. K., C. L. Little, and R. T. Mitchell. 2003b. Microbiological quality of open ready-to-eat salad vegetables: effectiveness of food hygiene training of management. *J. Food Prot.* **66:**1581–1586.

Sair, A. I., D. H. D'Souza, and L. A. Jaykus. 2002. Human enteric viruses as causes of foodborne disease. *Compr. Rev. Food Sci. Saf.* **1:**73–89.

Sattar, S. A., S. Springthorpe, S. Mani, M. Gallant, R. C. Nair, E. Scott, and J. Kain. 2001. Transfer of bacteria from fabrics to hands and other fabrics: development and application of a quantitative method using *Staphylococcus aureus* as a model. *J. Appl. Microbiol.* **90:**962–970.

Sauders, B. D., K. Mangione, C. Vincent, J. Schermerhorn, C. M. Farchione, N. B. Dumas, D. Bopp, L. Kornstein, E. D. Fortes, K. Windham, and M. Wiedmann. 2004. Distribution of *Listeria monocytogenes* molecular subtypes among human and food isolates from New York State shows persistence of human disease-associated *Listeria monocytogenes* strains in retail environments. *J. Food Prot.* **67:**1417–1428.

Schaffner, D. W., and K. M. Schaffner. 2007. Management of risk of microbial cross-contamination from uncooked frozen hamburgers by alcohol-based hand sanitizer. *J. Food Prot.* **70:**109–113.

Schuchat, A., K. A. Deaver, J. D. Wenger, B. D. Plikaytis, L. Mascola, R. W. Pinner, A. L. Reingold, C. V. Broome, and the *Listeria* Study Group. 1992. Role of foods in sporadic listeriosis I. Case-control study of dietary risk factors. *JAMA* **267:**2041–2045.

Schuchat, A., B. Swaminathan, and C. V. Broome. 1991. Epidemiology of human listeriosis. *Clin. Microbiol. Rev.* **4:**169–183.

Schwab, K. J., F. H. Neill, R. L. Fankhauser, N. A. Daniels, S. S. Monroe, D. A. Bergmire-Sweat, M. K. Estes, and R. L. Atmar. 2000. Development of methods to detect "Norwalk-like viruses" (NLVs) and hepatitis A virus in delicatessen foods: application to a food-borne NLV outbreak. *Appl. Environ. Microbiol.* **66:**213–218.

Sergelidis, D., A. Abrahim, A. Sarimvei, C. Panoulis, P. Karaioannoglou, and C. Genigeorgis. 1997. Temperature distribution and prevalence of *Listeria* spp. in domestic, retail and industrial refrigerators in Greece. *Int. J. Food Microbiol.* **34:**171–177.

Shieh, Y., S. S. Monroe, R. L. Fankhauser, G. W. Langlois, W. Burkhardt III, and R. S. Baric. 2000. Detection of Norwalk-like virus in shellfish implicated in illness. *J. Infect. Dis.* **181:**S360–S366.

Snelling, A. M., K. G. Kerr, and J. Heritage. 1991. The survival of *Listeria monocytogenes* on fingertips and factors affecting elimination of the organism by hand washing and disinfection. *J. Food Prot.* **54:**343–348.

Snyder, O. P. 1991. HACCP in the retail food industry. *Dairy Food Environ. Sanit.* **11:**73–81.

Snyder, O. P. 2007. Removal of bacteria from fingertips and the residual amount remaining on the hand washing nailbrush. *Food Prot. Trends* **27:**597–602.

Somers, E. B., and A. C. Lee Wong. 2004. Efficacy of two cleaning and sanitizing combinations on *Listeria monocytogenes* biofilms formed at low temperature on a variety of materials in the presence of ready-to-eat meat residue. *J. Food Prot.* **67:**2218–2229.

Stivers, T. L., and K. W. Gates. 2000. Survey of grocery store seafood employees. *Dairy Food Environ. Sanit.* **20:**746–752.

Svensson, L. 2000. Diagnosis of foodborne viral infections in patients. *Int. J. Food Microbiol.* **59:**117–126.

Taormina, P. J., and W. J. Dorsa. 2007. Evaluation of hot-water and sanitizer dip treatments of knives contaminated with bacteria and meat residue. *J. Food Prot.* **70:**648–654.

Tauxe, R. V. 1997. Emerging foodborne diseases: an evolving public health challenge. *Emerg. Infect. Dis.* **3:**425–434.

Taylor, J. H., S. J. Rogers, and J. T. Holah. 1999. A comparison of the bactericidal efficacy of 18 disinfectants used in the food industry against *Escherichia coli* O157:H7 and *Pseudomonas aeruginosa* at 10 and 20°C. *J. Appl. Microbiol.* **87:**718–725.

Thornhill, T. S., A. R. Kalica, R. G. Wyatt, A. Z. Kapikian, and R. M. Chanock. 1975. Pattern of shedding of the Norwalk particle in stools during experimentally induced gastroenteritis in volunteers as determined by immune electron microscopy. *J. Infect. Dis.* **132:**28–34.

Todd, E. C. D., J. D. Greig, C. A. Bartleson, and B. S. Michaels. 2007a. Outbreaks where food workers have been implicated in the spread of foodborne disease. Part 2. Description of outbreaks by size, severity, and settings. *J. Food Prot.* **70:**1975–1993.

Todd, E. C. D., J. D. Greig, C. A. Bartleson, and B. S. Michaels. 2007b. Outbreaks where food workers have been implicated in the spread of foodborne disease. Part 3. Factors contributing to outbreaks and description of outbreak categories. *J. Food Prot.* **70:**2199–2217.

Tompkin, R. B. 2002. Control of *Listeria monocytogenes* in the food processing environment. *J. Food Prot.* **65:**709–725.

USDA-FSIS. 1996. Pathogen reduction; hazard analysis and critical control point (HACCP) systems; final rule. *Fed. Regist.* **61:**38805–38989.

USDA-FSIS. 2003. Control of *Listeria monocytogenes* in ready-to-eat meat and poultry products; final rule. *Fed. Regist.* **68:**34208–34254.

USDHHS-FDA. 2001. Hazard analysis and critical control point (HACCP); procedures for the safe and sanitary processing and importing of juice, final rule. *Fed. Regist.* **66:**6137–6202.

USDHHS-FDA-CFSAN. 1995. Procedures for the safe and sanitary processing and importing of fish and fishery products, final rule. *Fed. Regist.* **60:**65095–65202.

USDHHS-FDA-CFSAN. 1998. Guide to minimize microbial food safety hazards for fresh fruits and vegetables. Center for Food Safety and Applied Nutrition, U.S. Food and Drug Administration. http://www.foodsafety.gov/~acrobat/prodguid.pdf.

USDHHS-FDA-CFSAN. 1999. Evaluation of risks related to microbiological contamination of ready-to-eat food by food preparation workers and the effectiveness of interventions to minimize those risks. Center for Food Safety and Applied Nutrition, U.S. Food and Drug Administration. http://www.cfsan.fda.gov/~ear/rterisk.html.

USDHHS-FDA-CFSAN. 2000. Report of the FDA retail food program database of foodborne illness risk factors. Center for Food Safety and Applied Nutrition, U.S. Food and Drug Administration. http://www.cfsan.fda.gov/~acrobat/retrsk.pdf.

USDHHS-FDA-CFSAN. 2001. Analysis and evaluation of preventive control measures for the control and reduction/elimination of microbial hazards on fresh and fresh-cut produce. Center for Food Safety and Applied Nutrition, U.S. Food and Drug Administration. http://www.cfsan.fda.gov/~comm/ift3-toc.html.

USDHHS-FDA-CFSAN. 2003. Hand hygiene in retail and food service establishments. Center for Food Safety and Applied Nutrition, U.S. Food and Drug Administration. http://www.cfsan.fda.gov/~comm/handhyg.html.

USDHHS-FDA-CFSAN. 2005. 2005 Food Code. Center for Food Safety and Applied Nutrition, U.S. Food and Drug Administration. http://www.cfsan.fda.gov/~dms/fc05-toc.html.

USDHHS-FDA-CFSAN. 2006a. Managing food safety: a manual for the voluntary use of HACCP principles for operators of food service and retail establishments. Center for Food Safety and Applied Nutrition, U.S. Food and Drug Administration. http://www.cfsan.fda.gov/~acrobat/hret2.pdf.

USDHHS-FDA-CFSAN. 2006b. Managing food safety: a regulator's manual for applying HACCP principles to risk-based retail and food service inspections and evaluating voluntary food safety management systems. Center for Food Safety and Applied Nutrition, U.S. Food and Drug Administration. http://www.cfsan.fda.gov/~acrobat/hret3.pdf.

USDHHS-FDA-CFSAN. 2007. Supplement to the 2005 FDA Food Code. Center for Food Safety and Applied Nutrition, U.S. Food and Drug Administration. http://www.cfsan.fda.gov/~dms/fc05-sup.html.

USDHHS-FDA-CFSAN/USDA-FSIS. 2003. Quantitative assessment of the relative risk to public health from food-borne *Listeria monocytogenes* among selected categories of ready-to-eat foods. Center for Food Safety and Applied Nutrition, U.S. Food and Drug Administration. http://www.foodsafety.gov/~dms/lmr2-toc.html.

Uyttendaele, M., P. De Troy, and J. Debevere. 1999. Incidence of *Listeria monocytogenes* in different types of meat products on the Belgian retail market. *Int. J. Food Microbiol.* **53:**75–80.

Van Duynhoven, Y. T. H. P., C. M. de Jager, L. M. Kortbeek, H. Vennema, M. P. G. Koopmans, F. van Leusden, W. H. M. van der Poel, and J. M. van den Broek. 2005. A one-year intensified study of outbreaks of gastroenteritis in The Netherlands. *Epidemiol. Infect.* **133:**9–21.

Varma, J. K., M. C. Samuel, R. Marcus, R. M. Hoekstra, C. Medus, S. Segler, B. J. Anderson, T. F. Jones, B. Shiferaw, N. Haubert, M. Megginson, P. V. McCarthy, L. Graves, T. Van Gilder, and F. J. Angulo. 2007. *Listeria monocytogenes* infection from foods prepared in a commercial establishment: a case-control study of potential sources of sporadic illness in the United States. *Clin. Infect. Dis.* **44:**521–528.

Vaz, M. L. S., N. F. Novo, D. M. Sigulem, and T. B. Morais. 2005. A training course on food hygiene for butchers: measuring its effectiveness through microbiological analysis and the use of an inspection checklist. *J. Food Prot.* **68:**2439–2442.

Venkitanarayanan, K. S., G. O. I. Ezeike, Y.-C. Hung, and M. P. Doyle. 1999. Inactivation of *Escherichia coli* O157:H7 and *Listeria monocytogenes* on plastic kitchen cutting boards by electrolyzed oxidizing water. *J. Food Prot.* **62:**857–860.

Vermeiren, L., F. Devlieghere, and J. Debevere. 2002. Effectiveness of some recent antimicrobial packaging concepts. *Food Addit. Contam.* **19:**163S–171S.

Vinje, J., S. A. Altena, and M. P. Koopmans. 1997. The incidence and genetic variability of small round-structured viruses in outbreaks of gastroenteritis in The Netherlands. *J. Infect. Dis.* **176:**1374–1378.

Vorst, K. L., E. C. D. Todd, and E. T. Ryser. 2006a. Transfer of *Listeria monocytogenes* during mechanical slicing of turkey breast, bologna, and salami. *J. Food Prot.* **69:**619–626.

Vorst, K. L., E. C. D. Todd, and E. T. Ryser. 2006b. Transfer of *Listeria monocytogenes* during slicing of turkey breast, bologna, and salami with simulated kitchen knives. *J. Food Prot.* **69:**2939–2946.

Wachtel, M. R., and A. O. Charkowski. 2002. Cross-contamination of lettuce with *Escherichia coli* O157:H7. *J. Food Prot.* **65:**465–470.

Wachtel, M. R., J. L. McEvoy, Y. Luo, A. M. Williams-Campbell, and M. B. Solomon. 2003. Cross-contamination of lettuce (*Lactuca sativa* L.) with *Escherichia coli* O157:H7 via contaminated ground beef. *J. Food Prot.* **66:**1176–1183.

Warke, R., A. Kamat, M. Kamat, and P. Thomas. 2000. Incidence of pathogenic psychrotrophs in ice creams sold in some retail outlets in Mumbai, India. *Food Control* **11:**77–83.

Welker, C., N. Faiola, S. Davis, I. Maffatore, and C. A. Batt. 1997. Bacterial retention and cleanability of plastic and wood cutting boards with commercial food service maintenance practices. *J. Food Prot.* **60:**407–413.

Wheeler, J. G., D. Sethi, J. M. Cowden, P. G. Wall, L. C. Rodrigues, D. S. Tompkins, M. J. Hudson, and P. J. Roderick on behalf of the Infectious Intestinal Disease Study Executive. 1999. Study of infectious intestinal disease in England: rates in the community, presenting to general practice, and reported to national surveillance. *Brit. Med. J.* **318:**1046–1050.

Yang, H., A. Mokhtari, L.-A. Jaykus, R. A. Morales, S. C. Cates, and P. Cowen. 2006. Consumer phase risk assessment for *Listeria monocytogenes* in deli meats. *Risk Anal.* **26:**89–103.

Zhu, M., M. Du, J. Cordray, and D. U. Ahn. 2005. Control of *Listeria monocytogenes* contamination in ready-to-eat meat products. *Compr. Rev. Food Sci. Saf.* **4:**34–42.

Zink, D. L. 1997. The impact of consumer demands and trends on food processing. *Emerg. Infect. Dis.* **3:**467–469.

Pathogens and Toxins in Foods: Challenges and Interventions
Edited by V. K. Juneja and J. N. Sofos
© 2010 ASM Press, Washington, DC

Chapter 27

Interventions for Hazard Control at Food Service

O. Peter Snyder, Jr.

Previous chapters in this book have described how food from the farm is contaminated with a variety of biological, chemical, and physical hazards, as identified in research reports and Centers for Disease Control and Prevention publications. It is the legal responsibility of the food service operator to protect consumers' health by controlling the hazards in the food prepared for customers. This chapter will describe the science-based intervention strategies for the manager to use in a food service operation. A food establishment is defined by the Food and Drug Administration (FDA) retail Food Code (FDA, 2005) as an establishment that prepares and sells food, directly to the consumer, to be eaten in a facility, from a vehicle, or from a satellite operation, etc.

THE BIOLOGICAL, CHEMICAL, AND PHYSICAL HAZARDS AT FOOD SERVICE

A list of common biological, chemical, and physical hazards is shown in Table 1.

CHEMICAL AND PHYSICAL HAZARD CONTROL INTERVENTIONS

Chemical Hazard Control

Most of the chemical hazards shown in Table 1 cannot be eliminated or reduced by intervention strategies available to the cook, such as washing, peeling, or cooking food. Chemicals are very stable at retail food cooking temperatures of 150 to 212°F (65.6 to 100°C), with a few exceptions, such as boiling red kidney beans for 10 minutes to reduce phytohemagglutinins to a safe level. Basically, however, the food grower and supplier must provide assurance that insecticides, pesticides, and natural poisons in food are at acceptable, safe levels by preventing them from getting into the food at an unsafe level when the food is grown or produced. Foods must be labeled accurately so that the cook knows that food components causing allergies and intolerance are present. In addition, managers must train employees to use cleaning and sanitizing chemicals properly. Food chemical additives, such as nitrate used for making sausage, must be measured accurately to ensure that limits are not exceeded. If acid foods are prepared, people must be trained to avoid using copper containers if the food is more acidic, with a pH less than 6.0.

Physical Hazard Control

Requiring suppliers to use a control strategy to detect and remove physical objects in food can prevent injury. Strategies include sifting hard foreign objects (e.g., rocks) out of the food or X-ray and metal detector screening of the product and rejection if there is contamination. If this is not possible because, for example, the food establishment is buying produce directly from a local farmer, then the cook must take responsibility to reduce the hazard to a tolerable level of risk by sorting and removing rocks, sticks, nut shells, and other physical hazards from the product. People who work in the kitchen should not wear excess jewelry that might fall into food. They must be sure that pieces of equipment do not break and contaminate the product. Food servers must be sure that food prepared for children and senior citizens is cut into small enough pieces that there is no danger of a choking hazard. Hot beverages and soup must be served carefully so that there is no danger of burns to customers if spilled.

O. Peter Snyder, Jr. • Hospitality Institute of Technology and Management, 670 Transfer Road, Suite 21A, St. Paul, MN 55114.

Table 1. Common biological, chemical, and physical hazards[a]

Hazards	
Biological	
Vegetative cells	Parasites
Campylobacter jejuni, other campylobacters	*Giardia lamblia*
Diarrheagenic *Escherichia coli*	*Entamoeba histolytica*
Listeria monocytogenes	*Cryptosporidium parvum*
Salmonella	*Toxoplasma gondii*
Staphylococcus aureus	*Trichinella*
Shigella	*Cyclospora cayetanensis*
Vibrio	*Anisakis* and related worms
Yersinia enterocolitica and *Y. pseudotuberculosis*	*Diphyllobothrium* spp.
Other bacterial pathogens	*Nanophyetus* spp.
Aeromonas	*Acanthamoeba* and other free-living amoebae
Arcobacter	*Ascaris lumbricoides*
Nitrobacteria	*Trichuris trichiura* and others
Helicobacter	Viruses
Mycobacterium	Hepatitis A and E viruses
Plesiomonas	Norovirus
Streptococcus and others	Rotaviruses and other viruses
Spores	Biological toxins
Bacillus cereus and other *Bacillus* spp.	*Staphylococcus aureus*
Clostridium botulinum	*Bacillus cereus*
Clostridium perfringens	*Clostridium botulinum*
Clostridium difficile	Aflatoxin and mycotoxins
Chemical	
Biogenic amines and poisons	Residue poisons and heavy metals
Histamine, scombroid	Pesticides, herbicides, hormones, antibiotics
Ciguatera	
Paralytic shellfish poison	Lead, mercury, copper
Neurotoxic shellfish poison	Radionucleotides
Amnesic shellfish poison	Cleaning and sanitizing agents
Tyramine	
Serotonin	Prions
Catecholamine neurotransmitters	nCJD
Tryptamine	Allergens and Intolerance
Other natural poisons	Milk, eggs, wheat, peanuts, soy, tree nuts, crustacean allergen
Pyrrolizidine alkaloids	
Phytohaemagglutinin (red kidney bean poisoning)	Gluten, fat, milk intolerance
Grayanotoxin (honey intoxication)	
Mushroom poison	
Physical	
Broken teeth	Choking	Cuts in mouth
Bones	Plastic wrap	Glass
Rocks	Large, insufficiently chewed pieces of food	Burns
Metal		Temperature of foods above 170°F (77°C)
Twist ties/staples		
Pits/seeds		

[a]Source: Snyder (2007a).

Other Hazards

There are other hazards not listed in Table 1, and new hazards will be identified in the future with the use of improved analytical procedures to recover, culture, and identify hazards.

RAW FOOD CONTAMINATION PROBLEM

The retail food establishment operator would like to receive food from farms with hazards at safe levels. While hazard-free/zero-defect food may be the desired goal, it is unrealistic. It is not feasible (except under special circumstances, such as growing fresh mushrooms where the soil is pasteurized) to make the land farm or water farm environments free of biological, chemical, and physical hazards. Figure 1 illustrates the raw food contamination problem "from farm to fork." It shows the flow of food from the land farm and water farm, where there is wild animal and bird contamination, through the distribution system and processing system to the restaurant and food market and eventually to the consumer. The questions are, what is the level of the hazards on the crops, animals, fowl, and fish that are delivered to the kitchen, and what is a safe level/size for consumption?

International trade allows food to come from farms all over the world. Each country has its own food safety laws. Farmers are responsible for growing and harvesting meat, poultry, fish, fruits and vegetables, grains, dairy products, and other foods that are sold on the retail market. However, there is the risk that the level of biological hazards on the food from a farm may be at an unsafe level due to contamination from the wild animal, soil, fish, etc. pathogen reservoir. Chemical hazards can be controlled by the farmer, but the chance of a biological or physical hazard contaminating the food is significant. However, farmers can grow food for human consumption, when the risk is known, so that a food safety objective (FSO) that provides an appropriate level of protection (ALOP), as established by a country to protect human health (WTO, 1995), can be achieved. The ALOP is also called tolerable level of risk (TLR). The FSO is the level, in log value, of a hazard that most people can consume and meet the standard for a TLR or expectation of illness from a hazard in the specified food delivery system. The FSO contains three elements: (i) the hazard, (ii) the food, and (iii) the frequency or concentration of the hazard that is considered tolerable for consumer protection. The purpose of the FSO is to translate a public health goal—the desired level of consumer protection—to measurable standards that allow the operator to set control measures for processes such as cooking, cooling, and reheating that ensure a TLR to the consumer. The International Commission on Microbiological Specifications for Foods (ICMSF, 2002) identifies TLR as the effective absence of disease, which it defines as less than 1.0 case per 100,000 people per year.

Figure 1. Flow chart of the raw food contamination problem "from farm to fork." Source: Snyder (1994a).

Information from active and passive surveillance systems, outbreak investigations, and sentinel studies has been compiled by the Centers for Disease Control and Prevention (Mead et al., 1999). The ISCMF goal (level of illness risk/100,000 people) can be used to evaluate these data and determine those biological hazards that exceed the desired TLR. It is also possible to determine which pathogens currently meet a TLR of one illness or less per 100,000 people (Table 2).

This tabulation shows that *Campylobacter* spp., *Salmonella* (nontyphoidal), *Clostridium perfringens*, *Shigella* spp., and noroviruses greatly

Table 2. Level of risk for some pathogens[a]

Microorganism	No. of estimated food-borne illnesses[b]	No. of illnesses/100,000	No. of estimated deaths caused by food-borne illness[b]	No. of deaths/1,000,000
Vegetative bacteria				
Brucella spp.	777	0.3	6	0.022
Campylobacter spp.	1,963,141	719.9	99	0.363
Escherichia coli O157:H7	62,458	22.9	52	0.191
E. coli, non-O157	31,229	11.5	26	0.095
E. coli, enterotoxigenic	55,594	20.4	0	0.000
E. coli, other diarrheogenic	23,826	8.7	0	0.000
Listeria monocytogenes	2,493	0.9	499	1.830
Salmonella enterica serovar Typhi	659	0.2	3	0.011
Salmonella (nontyphoidal)	1,341,873	492.1	553	2.028
Shigella spp.	89,648	32.9	14	0.051
Staphylococcus food poisoning	185,060	67.9	2	0.007
Streptococcus (food-borne)	50,920	18.7	0	0.000
Vibrio cholerae, toxigenic	49	0.018	0	0.000
Vibrio vulnificus	47	0.017	18	0.066
Vibrio, other	5,122	1.9	13	0.048
Yersinia enterocolitica	86,731	31.8	2	0.007
Parasites				
Cryptosporidium parvum	30,000	11.0	7	0.026
Cyclospora cayetanensis	14,638	5.4	0	0.000
Giardia lamblia	200,000	73.3	1	0.004
Toxoplasma gondii	112,500	41.3	375	1.375
Trichinella spiralis	52	0.019	0	0.000

Continued on following page

Table 2. *Continued*

Microorganism	No. of estimated food-borne illnesses[b]	No. of illnesses/100,000	No. of estimated deaths caused by food-borne illness[b]	No. of deaths/1,000,000
Viral				
Noroviruses	9,200,000	3,373.8	124	0.455
Rotavirus	39,000	14.3	0	0.000
Astrovirus	39,000	14.3	0	0.000
Hepatitis A	4,170	1.5	4	0.015
Spore-forming bacteria				
Bacillus cereus	27,360	10.0	0	0.000
Clostridium perfringens	248,520	91.1	7	0.026
Clostridium botulinum (adults)	58	0.021	4	0.015

[a]Based on U.S. population of 272,690,813.
[b]Source: Mead et al. (1999).

exceed the ICMSF goal of one illness per 100,000 people per year. As will be shown, these bacteria are appropriate pathogens to choose as target organisms for intervention strategies.

When the farmer does not control the hazards at a level that meets an FSO, the food processor or cook must execute a validated intervention/hazard control at a critical control point in the kitchen to reduce the hazard to an ALOP. There are currently no national requirements for food safety management programs, such as hazard analysis critical control points (HACCP), on farms. There are only desired goals and guidelines. In spite of this, most food from farms is safe. The hazards are at a minimal risk level. Considering that there are 300,000,000 people in the United States eating three meals a day and that only 70,000,000 estimated people get ill per year, this is about one meal in 4,700 causing illness. These illness statistics include foods prepared in the home, and home food preparers receive no required education in safe food preparation, as do most personnel in retail food establishments. Occasionally, food-borne illness outbreaks are caused by a chain of multiple events that occur when contaminated food from a farm is harvested and placed in distribution to be purchased by both homemakers and retail food operators. If food processors, food distributors, and food preparers do not use adequate intervention controls, such as adequately pasteurizing food, shipping and storing food at controlled temperatures, washing fruits and vegetables, and cooking raw products adequately before the food is consumed, a food-borne illness or outbreak will occur.

SCIENTIFIC BASIS FOR RETAIL FOOD ESTABLISHMENT INTERVENTION STRATEGIES/PROCESS CONTROLS

The ICMSF (2002) also provides a scientific basis for developing the process controls and process performance standards that can be used for the intervention strategies in retail food operations: $H_0 + \Sigma I - \Sigma R \leq FSO$. H_0, in log value, is the estimated level of the hazard on the food from the farm as it arrives at the kitchen. ΣI, in log value, is the increase in the hazard in preparation for the consumer. Based on $H_0 + \Sigma I$, ΣR is the intervention reduction, in log value, appropriate to reducing the hazard to a safe FSO by the consumer. Figure 2 illustrates this equation.

The process steps are adjusted by the process developer/cook until the H_0 plus total increase ΣI minus reduction ΣR in a hazard results in a level of hazard in the food eaten by the consumer that is less than or equal to the FSO. As an example, if the H_0 for *Salmonella* in hamburger is 3 log; if there is no increase in the kitchen because the temperature of the food is 41°F (5.0°C); and if the FSO is –2 log, then the ΣR must be 5 log.

RISK MANAGEMENT AND HAZARD CONTROL OF THE FARM-TO-FORK RETAIL PROCESS

Some farmers do have excellent food safety management systems (FSMSs). They know the hazards on their farms and have selected and applied validated controls to the hazards on the farm. For example, the farmer begins testing irrigation water for *Escherichia coli* and/or implements a warm sanitizer water wash on tomatoes that are to be sold fresh. Another intervention strategy is the composting of manure to reduce pathogenic bacteria to a TLR.

Controls are monitored to ensure that the processes are in control and the hazards are at a TLR when the food leaves the farm. The risk that still exists is dependent on the completeness of the food safety knowledge that farmers have gained through education, years of practical experience in raising

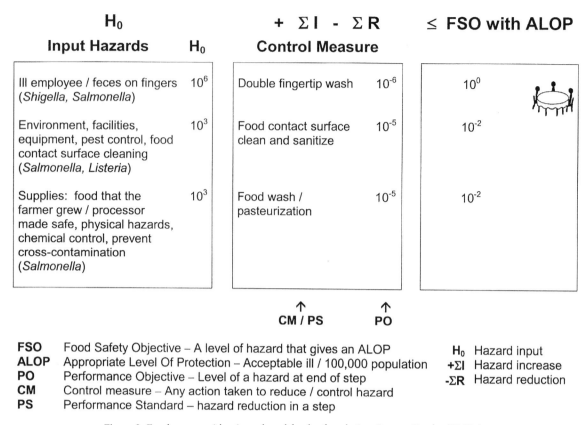

$$\text{H}_0 \qquad\qquad + \Sigma\text{I} - \Sigma\text{R} \qquad \leq \text{FSO with ALOP}$$

Input Hazards $\quad$ H₀ $\qquad$ **Control Measure**

Ill employee / feces on fingers (*Shigella, Salmonella*)	10^6
Environment, facilities, equipment, pest control, food contact surface cleaning (*Salmonella, Listeria*)	10^3
Supplies: food that the farmer grew / processor made safe, physical hazards, chemical control, prevent cross-contamination (*Salmonella*)	10^3

Double fingertip wash	10^{-6}
Food contact surface clean and sanitize	10^{-5}
Food wash / pasteurization	10^{-5}

10^0

10^{-2}

10^{-2}

↑
CM / PS

↑
PO

FSO   Food Safety Objective – A level of hazard that gives an ALOP
ALOP   Appropriate Level Of Protection – Acceptable ill / 100,000 population
PO   Performance Objective – Level of a hazard at end of step
CM   Control measure – Any action taken to reduce / control hazard
PS   Performance Standard – hazard reduction in a step

H₀   Hazard input
+ΣI   Hazard increase
-ΣR   Hazard reduction

Figure 2. Food system with science-based food safety design. Source: Snyder (2007a).

product that has not made anyone ill, and diligence to make sure that controls are always exactly applied. It also depends on the competency of people such as state agricultural inspectors, auditors, and agricultural extension agents used by the farmers to assist them in identifying hazards and specifying controls.

Under most conditions, the level of hazards on the food coming from the farm is sufficiently low that, except for immune-compromised individuals, the immune systems of typical consumers are adequate to protect their health, and they can eat the food "raw." We call this food "safe," and even though there are pathogens at low levels, our immune systems protect us. Table 3 provides guidelines for the target hazards.

People throughout the world eat raw vegetables and grains, drink raw milk, and eat dairy products prepared from unpasteurized milk. They do so without becoming ill, because their immune systems are adequate to protect them from the normally low levels of pathogens coming from carefully managed water farms or land farms. Some Americans, for instance, enjoy rare hamburgers and salads prepared with untreated produce. Many Europeans like raw pork sausage. People from the Middle East consume raw beef and lamb. Because of the low risk

for immune-complete people, the FDA Food Code (FDA, 2005) allows individual consumers to have underprocessed/raw food, such as raw eggs, providing there is a consumer advisory notice in the retail establishment reminding the consumer of the increased risk.

Immune-compromised babies, young children, and senior citizens should not consume raw or minimally processed foods, because their immune systems are incomplete, and they have a low hazard threshold for illness. The best intervention strategy for immune-compromised individuals (or parents of children/caregivers) is to choose products that are pasteurized and/or sterilized (canned).

Occasionally, there is a loss of control/deviation of the hazard control processes on the farm. For example, irrigation water becomes contaminated for a few hours or days with untreated sewage or manure runoff water, a wild animal wanders onto the farm and contaminates the fruits or vegetables, or human fecal material contaminates a fresh oyster bed. If the safety of the food is not documented, the cook must treat all unprocessed food as being contaminated. Sometimes there are human errors. These incidents are rare occurrences, but when hazardous portions of a product are produced and consumed, some people will become ill.

Table 3. Food hazards (H_0) and FSOs[a]

Hazard	Raw product contamination (H_0)	Process performance criteria (ΣR)	FSO
Microbiological infective			
Vegetative pathogens, infection			
Salmonella spp.	10^3 CFU/g	10^{-5} CFU/g; reduce	10^{-2} CFU/g
Shigella spp.	10^3 CFU/g	10^{-5} CFU/g; reduce	10^{-2} CFU/g
Escherichia coli O157:H7	10^3 CFU/g	10^{-5} CFU/g; reduce	10^{-2} CFU/g
Parasites			
Cryptosporidium parvum	1 cyst	Prevent/reduce	Undetectable
Toxoplasma gondii	1 cyst	Prevent/reduce	Undetectable
Trichinella spiralis	1–500 larvae	Prevent/reduce	Undetectable
Viruses			
Hepatitis A	>10 viruses/g	Prevent/reduce	Undetectable
Noroviruses	>100 viruses/g	Prevent/reduce	Undetectable
Toxin producing			
Staphylococcus aureus (exotoxin)	10^3 CFU/g	$<10^3$-CFU/g increase	$<10^6$ CFU/g (toxin dose of <1 microgram)
Spores			
Clostridium botulinum (exotoxin)	10^0 spores/g	$<10^3$-CFU/g increase	$<10^3$ CFU/g (toxin dose of ≤2 nanograms)
Bacillus cereus (exotoxin, enterotoxin)	10^2 spores/g	$<10^3$-CFU/g increase	$<10^5$ CFU/g (toxin dose unknown)
Clostridium perfringens (enterotoxin)	10^2 spores/g	$<10^3$-CFU/g increase	$<10^5$ CFU/g (toxin dose unknown)
Chemical			
Sulfites	Variable	None added	<10 ppm
Nitrates	Variable	<500 ppm added	<500 ppm
Nitrites	Variable	<200 ppm added	<200 ppm
Monosodium glutamate	Variable	≤0.5 g/serving added	<3.0 g/meal
Aflatoxins (from mold)	<20 ppb	No increase	<20 ppb
Histamine (from fish, cheese)	<20 ppm	No increase	<20 ppm
Physical			
Hard foreign objects (broken tooth)	>1/16-inch diameter	Prevent/remove	Undetectable
Choking	>1/4-inch diameter	Cut ≤1/4 inch	<1/4 inch

[a]Source: Snyder (1994b).

The retail food establishment must be ready to deal with these levels of hazards.

FIVE RETAIL FOOD SERVICE USDA PROCESS GROUPS

How are intervention strategies applied to the design of biological process interventions in the kitchen that achieve FSOs? The retail food establishment produces products that can be grouped into five HACCP process groups, shown in Table 4 (USDA/FSIS, 2007), for hazard control intervention. Note that the term "heat treated," according to the USDA/FSIS (2001), means a 6.5-log reduction of *Salmonella* spp. in meat and a 7-log reduction in poultry. The FDA uses the alternate term "pasteurized," which varies somewhat with the food, but ground meat and fish are to receive a 5-log *Salmonella* reduction, roast cuts of meat a 6.5-log reduction (same as the USDA), and poultry a 7-log reduction (FDA, 2005) (same as the USDA). FDA retail control measures are used in Table 4. All meat *Salmonella* process performance

standards originated from the research study by the American Bacteriological and Chemical Research Corporation (Goodfellow and Brown, 1978). Note that group V is commercially sterile food. Most people do not think that a kitchen is capable of producing sterile food; however, pressure cookers at 15 pounds of pressure can be used to shorten cooking times for less-tender cuts of meat in soups and stews containing vegetables. At 15 pounds of pressure, the cooking temperature is 250°F (121.1°C). When products reach and are held at 250°F (121.1°C) for 3 minutes, there is a 12-log reduction of proteolytic *Clostridium botulinum* spores; thus, a commercially sterile product is produced (Stumbo, 1973).

The current FDA retail code provides "safe harbors" intervention strategies for two of these processes—I, not heat treated and not shelf stable and III, fully cooked and not shelf stable. The other three processes require the operator to do a process validation, write a HACCP plan with intervention strategies, and get the plan approved by the local regulatory authority.

Table 4. Food groups for food safety process analysis[a]

Group no.	HACCP process groups prerequisite/GMPs working[b]	Control	Shelf life
I	Not heat treated, not shelf stable (raw). Not PHF; raw meat, fish; sushi, sashimi; eggs, raw fruits, and vegetables.	Grown or supplier made safe, with $H_0 \leq$ FSO. May require temperature control for quality.	<14 days (bacterial spoilage)
II	Not heat treated, with inhibitors to make shelf stable. a_w control: baked bread and muffins, pastry, nuts, chocolate candy, spices and herbs, oil, salted, dried fish, fresh pasta Fermentation control: pepperoni, salami; olives; dairy (cheese, yogurt, sour cream/milk/crème fraîche); sauerkraut; kimchee; beer, wine Acid control: salad dressing; coleslaw; salsa; condiments; pickled eggs; tomato sauce	Grown safe, made safe by supplier, with H_0 that, with $+\Sigma I - \Sigma R$ (5-log *Salmonella* reduction), meets FSO. Does not require TCS because of product a_w, pH, or additive stabilizers.	>2 years, 70°F (21.1°C) (chemical spoilage)
III	Fully cooked, not shelf stable. Hot or cooled, refrigerated ready-to-eat food; meat, fish, poultry; fruits, vegetables, dairy, pastry filling, pudding	Pasteurized (5 log to 7 log *Salmonella*) so that $H_0 + \Sigma I - \Sigma R$ meets or exceeds FSO. Requires TCS.	>130°F (54.4°C) safe <30°F (1.1°C) safe ≤41°F (5.0°C), 7 days; 130 to 41°F (54.4 to 5.0°C), 4 h
IV	Fully cooked, with inhibitors to make shelf stable. Marinara sauce; fruit pie fillings; cake icing, bread and pastry, dry cereals, dry pasta, smoked fish; packaged, "pickled" low-pH fruits and vegetables	Pasteurized (5 log to 7 log *Salmonella*) so that $H_0 + \Sigma I - \Sigma R$ meets FSO. Does not require TCS because of product a_w, pH, or additive stabilizers.	>5 yr
V	Commercially sterile, shelf stable. "Packaged" meat, fish, poultry, fruits, vegetables, dairy/UHT milk	Sterilized, *Clostridium botulinum* spores reduced 9 to 12 log. Does not require TCS.	>5 yr

[a]PHF, potentially hazardous food; UHT, ultra-high temperature; H_0, starting hazard; Σ, summary; I, increase; R, reduction; TCS, temperature control for safety. Source: Snyder (2007a).
[b]USDA/FSIS (2007).

THE RETAIL KITCHEN HAZARD CONTROL SYSTEM

Figure 3 illustrates an FSMS with the five food process control groups. Current FDA Retail Food Code food establishment "safe harbors" intervention strategies are as follows. (They are not HACCP, science-based rules, because no H_0 or FSO is specified, and there is no reference to critical limit or process control validation.)

- Wash hands for 20 seconds for control of bacterial cross-contamination.
- Wash and sanitize equipment and food contact surfaces.
- Keep facility free of insects and pests that can contaminate ready-to-eat food.
- Refrigerated processed food from the wholesale system: hold at 41°F (5.0°C) with no time limit, if not opened.
- Raw (uncut) fruits and vegetables (except for sprouts): no temperature requirement and no time limit.
- Shelf-stable products (e.g., canned fruits and vegetables and sauces): no temperature requirement and no time limit.

- Cooking time from refrigeration to greater than 130°F (54.4°C): no time limit. (This acknowledges that when food reaches 130°F [54.4°C], pathogenic vegetative bacteria, such as vegetative cells of *Clostridium perfringens* or *Salmonella,* that grew during cooking are reduced to a TLR.)
- Cooking roasts to 130°F (54.4°C) and holding for 112 minutes for a 6.5-log *Salmonella* reduction.
- Cooking solid steaks and chops to 145°F (62.8°C) for 15 seconds or freezing to −4°F (20.0°C) for 7 days to reduce parasites to a TLR.
- Pasteurization of ground meat and fish at 145°F (62.8°C) for 3 minutes/150°F (65.6°C) for 1 minute/155°F (68.3°C) for 15 seconds for a 5-log reduction of *Salmonella*.
- Cooking poultry to 165°F (73.9°C) for 15 seconds for a safe reduction of *Salmonella*.
- Safe hot holding of food at 130 to 135°F (54.4 to 57.2°C) to prevent the outgrowth of the spores of *C. perfringens, Clostridium botulinum,* and *Bacillus cereus.*
- Cooling food from 135 to 41°F (57.2 to 5.0°C) in 6 hours; adequate to prevent germination

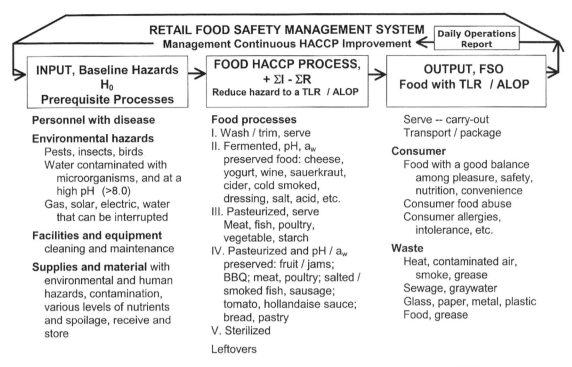

Figure 3. Food safety management system with five process groups. Source: Snyder (2007a).

and multiplication of *C. perfringens* by greater than a 1-log increase.

- Cold holding of cooked food at 41°F (5.0°C) for 7 days; a performance criterion to ensure a less than 3-log increase of *Listeria monocytogenes* in kitchen-prepared ready-to-eat food.

THE FLOW CHART FOR THE FSMS

The flow chart for the FSMS is shown in Figure 4. The first requirement/intervention for safe food is that there is good management of the FSMS. This includes commitment, self-inspection, training, and quality control. The FDA Food Code, Annex 4 (FDA, 2005) uses "Active Managerial Control" (AMC) as the foundation of an FSMS to prevent, eliminate, or reduce the occurrence of food-borne illness risk factors. Note: it is not possible to ensure the control of hazards in the food to be served if there is no FSMS with AMC. Using AMC by the manager/cook to establish an FSMS HACCP involves the following steps:

1. Identify hazards, the H_0 and FSO for the hazards, and which are significant risks (frequent and severe).

2. Select, validate, and implement food safety intervention strategies/policies, procedures, and standards that will reduce expected H_0s to ALOPs.

3. Train employees to perform and monitor the intervention strategies.

4. Do self-inspections of operations to look for new hazards and opportunities to improve the intervention strategies.

DEVELOPING SCIENCE-BASED INTERVENTION STRATEGIES

In order to develop HACCP-based intervention strategies/control measures and critical limits based on science so that necessary reductions in biological

Figure 4. Flow chart for hazard control at food service for the food safety management system. Source: Snyder (2007a).

hazards can be achieved and an FSO met, it is necessary to begin by understanding the process control characteristics of the biological hazards. (Note that chemical and physical hazards are controlled by prerequisite processes, as explained earlier.)

Biological hazards can be controlled in various ways. Tables 5 and 6 (ICMSF, 1996; Meng, et al., 2007; Nachamkin, 2007; Setlow and Johnson, 2007; Snyder, 1994b 2007a) show summaries of the hazard control critical limits (e.g., time and temperature for growth and death, pH of growth and death, and water activity [a_w]) for control of the biological hazards.

RETAIL FOOD HACCP OPERATIONS INTERVENTION PLAN

Using the information in Tables 5 and 6, intervention strategies can be developed for the prerequisite and food process hazards and controls. Table 7 is a retail food HACCP operations intervention plan for the prerequisite processes and food HACCP processes in retail establishments.

For each component of the FSMS, the operational hazard is identified in the first column. In the second column, the target organism/hazard is identified, along with the H_0 and the FSO. The third column describes the science-based intervention/control strategy.

PREREQUISITES INTERVENTION STRATEGY

Personal Hygiene Hazard Controls

When employees come to work or use the toilet, it should be assumed that they get a high level of fecal pathogens on their fingertips by ineffective use of toilet paper. As a result, 10^5 CFU per gram *Salmonella* spp., *Shigella* spp., or noroviruses can be transferred to the fingers. The intervention strategy must reduce this level to undetectable in a 10-ml rinse of the fingertips and under the fingernails. An effective intervention strategy is use of the double wash for hands and fingertips (Snyder, 2007b) when coming from the toilet, which utilizes a nail brush, soap, friction, and flowing water over the hands and fingertips on the first wash to get a 3-log reduction in 10 seconds, which is followed by a second wash for about 10 seconds, but without a nail brush, for another 2-log reduction. Note that the basic mechanism for removal of transient microbial pathogens is friction and dilution. Therefore, the wash must be done under flowing water. Hands and fingertips are dried with a paper towel, giving a final 1-log reduction. The double wash for hands and fingertips after

Table 5. Food pathogen control data summary: infective microorganisms (inactivated by pasteurization)[a]

Microorganism and source	Temp range for growth and atmosphere	pH range and minimal a_w for growth	G, D, and z[b]
Yersinia enterocolitica; feces/food	29.3–111°F (−1.5–43.9°C); aerobic but may be facultative	4.6–9.0 pH, 0.975 a_w	G (32°F [0°C]) = 2 days; G (41°F [5.0°C]) = 17 h D (145°F [62.8°C]) = 0.24–0.96 min z = 9.2–10.4°F (5.1–5.8°C)
Listeria monocytogenes; environment	29.3–112°F (−1.5–44.4°C); facultative, grows in 5% salt	4.5–9.5 pH, 0.90–0.93 a_w	G (32°F [0°C]) = 7.5 days; G (40°F [4.4°C]) = 1 day D (140°F [60.0°C]) = 2.85 min z = 10.4–11.3°F (5.8–6.3°C)
Vibrio parahaemolyticus; seafood/water	41–109.4°F (5.0–43.0°C); facultative anaerobe	4.5–11.0 pH, 0.937 a_w	G (68°F [20.0°C]) = 27 min D (120°F [48.9°C]) = 0.35–0.72 min z = 10.0–22.0°F (5.6–12.2°C)
Salmonella spp.; feces/food	41.5–114°F (5.3–45.6°C); facultative	4.1–9.0 pH, 0.92–0.95 a_w	G (96.8°F [36.0°C]) = 50 min D (150°F [65.6°C]) = 0.17 min z = 10°F (5.6°C)
Escherichia coli O157:H7; feces/food	44.6–114°F (7.0–45.6°C); facultative	4.0–9.0 pH, 0.95 a_w	G (98.6°F [37.0°C]) = 22 min D (140°F [60.0°C]) = 1.7 min
Campylobacter jejuni; feces/food	90–113°F (32.2–45.0°C); obligate microaerophile	4.9–8.0 pH, >0.98 a_w	G (95°F [35.0°C]) = 2.8–3.3 h D (137°F [58.3°C]) = 12–21 s z = 10.6–11.4°F (5.9–6.3°C)
Staphylococcus aureus; human skin	43.8–122°F (6.6–50.0°C); facultative, grows in 10% salt and produces toxin	4.5–9.3 pH, 0.83 a_w	Vegetative cells: G (96.8°F [36.0°C]) = 56 min D (140°F [60.0°C]) = 5.2–7.8 min z = 9.7–10.4°F (5.4–5.8°C)
	Toxin production: 50–114.8°F (10.0–46.0°C)	5.15–9.0 pH, 0.86 a_w	Toxin destruction: D (210°F [98.9°C]) = >2 h z = about 50°F (27.8°C)

[a]Source: Snyder (2007a).
[b]G, growth or doubling time; D, death rate for 10:1 reduction time; z, temperature increase for 10× faster kill.

Table 6. Food pathogen control data summary: toxin producers and/or spore-formers (not inactivated by pasteurization)[a]

Microorganism and source	Temp range for growth and atmosphere	pH range and minimal a_w for growth	G, D, and/or z[b]
Clostridium botulinum (type E and other nonproteolytic strains); food	38–113°F (3.3–45.0°C); anaerobic, grows in up to 5% salt	5.0–9.0 pH; 0.97 a_w	Spores: D (180°F [82.2°C]) = 0.49–0.74 min z = 9.9–19.3°F (5.5–10.7°C) Toxin destruction (any botulinal toxin): D (185°F [85.0°C]) = 5 min z = 7.2–11.2°F (4.0–6.2°C)
Bacillus cereus: dirt and food	39.2–122°F (4.0–50.0°C); aerobic but can be facultative	4.35–9.0 pH; 0.912 – 0.95 a_w	Vegetative cells: D (140°F [60.0°C]) = 1 min z = 12.4°F (6.9°C) Spores: D (212°F [100.0°C]) = 2.7–3.1 min z = 11°F (6.1°C) D (212°F [100.0°C]) = 5 min z = 18°F (10.0°C) Toxin destruction: Diarrheal: D (133°F [56.1°C]) = 5 min Emetic: stable at 249.8°F (121.0°C)
Clostridium botulinum (type A and proteolytic B strains); dirt and food	50–118°F (10.0–47.8°C); anaerobic, grows in up to 5% salt	4.6–9.0 pH; 0.95 a_w	Spores: D (250°F [121.1°C]) = 0.2 min z = 18°F (10.0°C) D (212°F [100.0°C]) = 50 min Toxin destruction: D (185°F [85.0°C]) = 5 min z = 7.2–11.2°F (4.0–6.2°C)
Clostridium perfringens; dirt and food	59–126.1°F (15.0–52.3°C); facultative anaerobe	5.0–9.0 pH; 0.93 – 0.95 a_w	Vegetative cells: G (105.8°F [41.0°C]) = 7.2 min D (138°F [58.9°C]) = 7.2 min z = 6.8°F (3.8°C) Spores: D (210°F [98.9°C]) = 26–31 min z = 13°F (7.2°C) D (212°F [100.0°C]) = 0.3–20 min z = 18–54°F (10.0–30°C)

[a]Source: If (2007a).
[b]G, growth or doubling time; D, death rate for 10:1 reduction time; z, temperature increase for 10× faster kill.

using the toilet is a very important intervention strategy in the prevention of food-borne illness, since about half of employee-caused outbreaks are due to asymptomatic employees (Todd et al., 2007). It is equally important when sick employees, who should stay home, come to work and handle food.

Environment/Facility Hazard Controls

The environment surrounding the facility can be contaminated with pests, birds, insects, standing water, and possible sewage backup. There can be *Salmonella* spp. and *L. monocytogenes* at up to an estimated 10^3 CFU per gram or ml. The intervention strategy is cleaning, sanitizing, and maintaining the environment outside the facility to ensure that these pathogens are kept outside the facility at a low H_0, because, if the outside is clean and the walls, doors, and windows are tight, hazards can be prevented inside the facility. Inside the facility, the hazards must be kept at an established low level, for example, on the floor, so that they are undetectable in 25 grams of food produced in the facility. The standard cleaning and sanitizing procedure is to sweep and remove gross soil, scrub and wash with an appropriate chemical solution, and finally, sanitize and air dry. This friction, dilution, and sanitizing procedure can be expected to reduce pathogens on a facility surface so that pathogens in the facility can be kept at such a low level that they will not contribute detectable pathogens to the food being processed.

Equipment Hazard Controls

Campylobacter spp. are a common bacterial contaminant of raw poultry products (Stern and Pretanik, 2006). Some raw poultry can be expected to have more than 10^5 CFU per carcass. Thus, when raw chicken touches a food contact surface, such as a cutting board or knife, it can be expected to leave 10^3 CFU per 50 cm^2 on the surfaces (Schaffner et al., 2004). The food contact surface intervention strategy (cleaning and sanitizing) must reduce the transfer of pathogenic microorganisms to ready-to-eat food

Table 7. Retail food HACCP operation intervention plan[a]

Operational hazard	Target hazard; H_0; and FSO	Intervention (prevention, elimination, and reduction)
Prerequisite process		
Personal hygiene		
Fecal contamination on fingers by and after toilet and failure to wash fingertips	B: *Salmonella/Shigella/*norovirus H_0: 10^7 in feces; 10^5 gets transferred to fingers because of failure in toilet paper use. FSO: Undetectable in 10-ml rinse of fingers and under the fingertips	10^6 reduction by double handwash with nail brush and paper towel dry.
Environment/facility		
Pests, birds, insects Water, sewage back-up	B: *Salmonella, L. monocytogenes* H_0: SSOP/GMP ensure a safe level FSO: Absent in 25-gram portion of food	Through effective use of SSOP/GMPs. Prevent/eliminate pests, birds, insects, water, and sewage backup with appropriate cleaning, sanitizing, and maintenance.
Equipment		
Contact with raw, pathogen-contaminated food (chicken)	B: *Campylobacter* spp. H_0: $\leq 10^4$ from chicken transferred to cutting board and 10^3 transfer from food contact surface to food that contacts the surface FSO: Undetectable in 25 grams of food	Rinse and brush food contact surfaces, and then wash, rinse, and sanitize for a 10^5 reduction.
Leaching	C: Copper and other heavy metal surfaces H_0: Use of copper kettle for cooking FSO: ≤ 1.3 ppm (drinking water standard)[b]	Do not prepare food with pH of <6 in a copper or other heavy metal cooking utensil.
Supplies, ingredients		
Biological hazards: supplier makes safe	B: *Salmonella, L. monocytogenes*, parasites H_0: Supplier processed; no evidence of a significant risk FSO: Undetectable in 25 grams of food	Supplier provides proof of hazard control for the product it supplies.
Physical hazards: supplier and cook make safe	P: Bones, rocks, nut shells H_0: >0.0625 inch (1.6 mm) FSO: <0.0625 inch (1.6 mm)	Sort and remove if necessary.
Chemical hazards: supplier and cook make safe	C: Intentional additives (e.g., nitrate, MSG); accidental additives (e.g., cleaning chemicals) H_0: Depends on additive FSO: Depends on additive	Segregate chemicals; supplier controls; labeling; accurate measurement.
Food process		
Prepreparation: raw meat, fish, poultry, fruits and vegetables	B: Vegetative cells (*Salmonella; L. monocytogenes*); parasites H_0: *Salmonella* 10^3/gram; *L. monocytogenes* 10^2/gram; parasites 1/gram FSO: Undetectable in 25 grams of food	Hold food at $\leq 41°F$ ($\leq 5.0°C$) to reduce spoilage. Wash fruits and vegetables eaten uncooked for 2- to 4-log reduction. Fermentation, 5-log reduction. Acidify, 5-log reduction. Freeze for parasite reduction and to limit bacterial growth.
Preparation: cook/pasteurize	B: Vegetative cells (*Salmonella*) H_0: 10^3/gram FSO: Undetectable in 25 grams of food	Pasteurize, 5-log reduction. Bakery items, a_w of <0.95 to control *C. botulinum* spore germination.
Hot hold, take out, cool	B: Spores – *C. perfringens* H_0: 10^3/gram FSO: Vegetative cells – *C. perfringens*, 10^5/gram	Hold at >130°F (54.4°C) or hold at any temp <4 h. Cool 120 to 55°F (48.9 to 12.8°C) <6 h, <1-log increase.
Cold holding	B: *Bacillus cereus* H_0: 10^2 FSO: 10^5/gram	Hold at $\leq 41°F$ (5.0°C), <3-log increase.
Consumer risk communication	C: Allergens H_0: >10 ppm allergen protein FSO: ≤ 10 ppm allergen protein P: Choking on food H_0: >0.25 inch (6.4 mm) FSO: <0.25 inch (6.4 mm)	The menu must list ingredients that cause allergies and intolerances, or the server must do so. Server offers to cut food into small pieces for children.

[a]Source: Snyder (2007a). B, biological; C, chemical; P, physical; SSOP, sanitation standard operating procedure; GMP, good manufacturing practices.
[b]EPA (2006).

prepared next on the food contact surface so that it is undetectable in 25 grams of the ready-to-eat food. The standard surface rinse, scrub with detergent, rinse, and sanitizing give a sufficient reduction so that pathogen levels are undetectable. This can be accomplished by taking the cutting board/equipment to the sink and wiping off scraps under flowing water and using a brush with a little detergent to scrub the surface until soil is removed, for 10 to 20 seconds. This gives a 3-log bacterial reduction (Snyder, 1997). The cutting board/equipment is then put into the wash sink and scrubbed again with the brush to get another 2-log reduction on the surface. Finally, the item is rinsed, sanitized, and allowed to air dry.

Leaching of Heavy Metals from Food Preparation Equipment

Some foods, especially salads, pie fillings, and fruits, have a pH of less than 6.0. The FDA Food Code (FDA, 2005) says that if food has a pH of less than 6, it may not be prepared in a food container made of heavy metal such as copper, because the acid will dissolve the metal. Copper cooking containers are common in candy shops and bakeries because of the uniform heat transfer. An effective intervention strategy requires that the cook knows the pH of the food being prepared. If the pH is greater than 6, as is the case for most candy and egg custards, copper can be used. If not, an acid-safe container such as stainless steel or food-grade plastic must be used to prepare and store the food.

Supplies and Ingredients: Hazards and Controls

Two kinds of ingredients are prepared in retail food service operations. First are the ingredients that suppliers have made safe and can be consumed without a food preparer's intervention. These are foods with a pH below 4.2 for *Salmonella* spp. control, those with an a_w below 0.92 for *B. cereus* control, and those that are sterilized for a 9- to 12-log *C. botulinum* reduction. Chemical hazards are controlled by excluding toxins and poisons and by indicating all allergen and intolerance ingredients on the label. All physical hazards, such as rocks, are reduced to a TLR. The FSO for the vegetative pathogen hazards in the food from the supplier is that specified vegetative pathogens are undetectable in 25 grams of food. The chemical hazards must be identified or, if poisonous, at a safe level. The physical hazards must be at a TLR.

The second kind of food/ingredient is one that the cook makes safe when he/she prepares the food. For physical hazards, the cook must inspect the ingredient and remove all hard objects that are greater than 0.0625 inch (1.6 mm) or reduce them to be less than

this size. If the hazard is a chemical additive, either intentional or unintentional, the intervention strategy is to add intentional additives according to written recipe procedures and to prevent/eliminate the accidental addition of chemicals, such as cleaning chemicals, by training employees and putting the chemicals in a location where they cannot be mistaken for food.

FOOD PROCESS HACCP INTERVENTION

Again, using the information in Tables 5 and 6, intervention strategies can be developed for the food process steps.

Prepreparation Hazards and Controls

Raw meat, fish, poultry, and fruits and vegetables can be expected to have *Salmonella* at 10^3 CFU per gram, *Listeria* spp. at 10^2 CFU per gram, and parasites at 1 cyst per gram (H_0). Food service refrigerators are designed to operate at 38 ± 2°F (3.3 ± 1°C) (NSF International, 2007). Therefore, refrigeration temperatures of 40 to 41°F (4.4 to 5.0°C) are reasonable for receiving storage. This temperature range, plus the high levels of spoilage microorganisms on raw food, will limit growth so that the ΣI, or increase in population, can be assumed to be negligible, even though the *Listeria* spp. and *Yersinia* spp. begin to multiply at 29.3°F (−1.5°C). A 5-log pasteurization is more than adequate to achieve an FSO of undetectable in 25 grams of food. Fruits and vegetables that are to be eaten uncooked should be brushed and double washed to give a 2-log reduction. If desired, a strong sanitizer can be used in a third wash to give, perhaps, a total 4-log reduction (FDA, 2007). If the ingredients are fermented (e.g., sauerkraut and buttermilk), the production of acid in the first 48 hours is important to ensure a 10^4 to 10^5 reduction of target organisms. Parasites must be controlled by freezing at −4°F (−20.0°C) for 7 days (or equivalent) or by specifying that the supplier certify that it has performed another intervention strategy for a 3-log reduction. If a product is cold smoked or cured, the sodium nitrite (limited to 200 ppm), in combination with the disinfectant properties of smoke and heat destruction of microbial cells, can combine to give a reduction of 10^5 CFU per gram, thus enabling the product to meet an FSO of "undetectable in 25 grams of food."

Preparation Hazards and Controls

The goal of food preparation is not, first, to make the food taste delicious; rather, it is to make the food safe to eat. As discussed previously, food from the farm has always had the risk of being randomly contaminated with biological, chemical, and physical hazards, and the primary job of the cook is to reduce the biological hazards to a TLR. The FDA Food Code (FDA, 2005) has specified this as a 5-log reduction of *Salmonella* and 4-log reduction of *L. monocytogenes*. This is based on USDA baseline studies (Federal Register, 2001), which have shown that the H_0 for contaminated meat can be set at 10^3 CFU *Salmonella* per gram. *Salmonella* is chosen as the target organism because it is frequently found in many foods, and the severity of the illness is high, with over two deaths per 1,000,000 people (Table 2). The FSO for *Salmonella* is to reduce it to 1 cell per 25 grams of food, or basically, a 5-log reduction. The *D* value for *Salmonella* at 150°F (65.6°C) is 0.17 minute, and the *z* value is 10°F (5.6°C). Every vegetative pathogen of concern has a different *D* and *z* value, based on factors such as the a_w, pH, and fat level of the food, as well as other factors. The H_0 of *Salmonella* will rarely be 10^3 CFU per gram, and consumers with adequate immune systems will not be made ill by a few *Salmonella* bacteria. A 5-log reduction of *Salmonella*, as described by the FDA Food Code (FDA, 2005), has proven to be an adequate intervention pasteurization strategy. Occasionally, there are foods, such as phytohaemagglutin in red kidney beans, that need to be boiled to make them safe (Table 1). Cookbooks describe these procedures, and cooks inform one another, so the risk is tolerable. Another important change to consider is moisture loss that occurs when items are baked. Bread, cookies, and muffins become shelf stable and do not require temperature control for safety. If the a_w of the baked products is reduced to less than 0.95 (Table 6), *C. botulinum* is controlled, and the food does not need refrigeration.

Hot Hold, Take Out, and Cool

After the food is pasteurized, vegetative pathogens have been reduced to a TLR, but spores of *C. perfringens*, *B. cereus*, and *C. botulinum* have been activated and must be prevented from germinating and multiplying, and any increase in vegetative cells must be limited. Table 6 shows that *C. perfringens* is the spore-forming pathogen that will germinate and begin to multiply at the highest temperature of 126.1°F (52.3°C). The FDA Food Code (FDA, 2005) allows roasts to be hot held at 130°F (54.4°C), and this has proven to be safe. It also allows the cook to meet the customer's desire for rare roast beef. There is no time limit if the temperature is 130°F (54.4°C) or above. If the food temperature is not controlled, however, the FDA Food Code (FDA, 2005) provides a "safe harbors" provision of 4 hours, as long as the food is discarded at the end of 4 hours. This can be used as an effective intervention strategy for take-out food, for which temperature is not normally controlled.

The target pathogen for cooling food is *C. perfringens* because of the rapid multiplication of vegetative cells once there is germination (Table 6). While the FDA Food Code (FDA, 2005) specifies "safe harbors" of 2 hours from 135 to 70°F (57.2 to 21.1°C) and 4 hours from 70 to 41°F (21.1 to 5.0°C), the USDA's *Draft Compliance Guidelines for Ready-to-Eat Meat and Poultry Products* (USDA FSIS, 2001) allows cooling from 120 to 55°F (48.9 to 12.8°C) in 6 hours, with continued cooling to 40°F (4.4°C) before boxing, or about 14.2 hours from 120 to 40°F (48.9 to 4.4°C). This ensures that the current "safe harbors" performance standard of a less than 1-log increase of *C. perfringens* is met. These USDA guidelines are widely used in process plants and have been proven to be safe. In practical terms, this cooling time is equivalent to the time it takes to cool a pan of food 2 inches (5.1 cm) deep from 130 to 41°F (54.4 to 5.0°C) in a standard retail refrigerator.

Cold Holding

The target pathogen for cold holding is nonproteolytic *C. botulinum*, and the temperature is 38°F (8.8°C) or less if the food is cooked fish (Table 6). For all other food, the target organism is *B. cereus*, and the temperature is 40 to 41°F (4.4 to 5.0°C) or less to prevent multiplication (Table 6). Since the expected level of germinated cells from *B. cereus* spores in food is 10^2 per gram, and the FSO is 10^5 per gram, a 3-log increase is allowed. This has proven to be safe, as judged by the safety history of chilled commercial foods, whereby processors set shelf lives that often range from 45 to 90 days, and there are no problems, in spite of the fact that grocery store and home refrigerators can be, occasionally, at 45 to 50°F (7.2 to 10.0°C).

Consumer Risk Communication

An important intervention strategy for food served in food service operations and as takeout food is communicating risk to consumers. Some customers have allergic reactions to some ingredients and also may have intolerances. The cause of food allergen and intolerance cannot be controlled by cooking; it must be controlled by prevention. The consumer should not eat the food. The hazard can be controlled when the customer tells the server that he/she has a sensitivity, and the server asks the cook if the food contains that ingredient and informs the customer. If a food with an allergen is being prepared in the kitchen, food contact surfaces can become contaminated, and the surface must be cleaned. If surfaces are cleaned to be visually clean and have less than 10 ppm protein residual, as measured by a protein swab, the cross-contamination hazard is controlled to an ALOP.

SUMMARY

This chapter has described how to establish, using AMC, a science-based HACCP FSMS with policies, procedures, and standards. The food service operator must first identify the hazards in the food coming into the food establishment. This is accomplished by reviewing food-borne illness outbreaks and their causes and then determining the risk factors in foods. The operator groups the food prepared in his/her retail food service operation according to one of the five processes that make the food safe. Some foods (e.g., canned foods, fully cooked meat items, and prepared frozen entrees) are made safe by the supplier, while other foods are unprocessed and must be made safe by the cook/food workers in the operation. For those foods that the cook makes safe, processes are designed based on hazards that must be prevented, reduced, or eliminated to an ALOP/TLR (commonly, a 5-log *Salmonella* reduction). Employees are trained to handle and prepare food according to the operation's food safety rules. This includes correct hand washing and personal hygiene, equipment and facility cleaning and maintenance, pest control, food pasteurization and washing, service procedures, and policies regarding leftovers/carry-out. These correct practices are the intervention strategies for which trained employees are responsible. Employees are also trained to monitor what they do, to include taking food temperatures, monitoring refrigeration temperatures, and taking corrective actions when critical limits are not met. A regular self-inspection (daily, weekly, and monthly) must be completed to verify that intervention strategies with policies, procedures, and standards are being carried out and to determine if there are any process deviations that need corrective action. In this way, a food establishment will have a very high probability of never causing a customer illness, avoiding allegations of involvement in food-borne illness outbreak and, at the same time, continuously improving processes and the consistency of the food served and consumer satisfaction.

REFERENCES

FDA. 2005. Food Code. U.S. Public Health Service, U.S. Department of Health and Human Services, Washington, D.C. http://www.cfsan.fda.gov/~dms/fc05-toc.html.

FDA. 2007. *Guide to Minimize Microbial Food Safety Hazards of Fresh-Cut Fruits and Vegetables.* Center for Food Safety and Applied Nutrition, U.S. Food and Drug Administration, College Park, MD. http://www.cfsan.fda.gov/~dms/prodgui3.html.

Federal Register. 2001. Performance standards for the production of processed meat and poultry products, proposed rule. 66:12589–12636.

Goodfellow, S. J., and W. L. Brown. 1978. Fate of *Salmonella* inoculated into beef for cooking. *J. Food Prot.* 41:598–605.

ICMSF. 1996. *Microorganisms in Foods 5: Microbiological Specifications of Food Pathogens.* Blackie Academic & Professional, New York, NY.

ICMSF. 2002. *Microorganisms in Foods 7: Microbiological Testing and Food Safety Management.* Kluwer Academic Plenum Publishers, New York, NY.

Mead, P. S., L. Slutaker, V. Dietz, L. F. McCaig, J. S. Bresee, C. Shapiro, P. M. Griffin, and R. V. Tauxe. 1999. Food-related illness and death in the United States. *Emerg. Infect. Dis.* 5:606–625.

Meng, J, M. P. Doyle, T. Zhoa, and S. Zhoa. 2007. Enterohemorrhagic *Escherichia coli*, p. 249–270. *In* M. P. Doyle and L. R. Beuchat (ed.), *Food Microbiology: Fundamentals and Frontiers.* ASM Press, Washington, DC.

Nachamkin, I. 2007. *Campylobacter jejuni*, p. 237–248. *In* M. P. Doyle and L. R. Beuchat (ed.), *Food Microbiology: Fundamentals and Frontiers,* ASM Press, Washington, DC.

NSF International. 2007. *American National Standard/NSF International Standard for Food Equipment—Commercial Refrigerators and Freezers.* NSF International, Ann Arbor, MI.

Schaffner, D. W., S. Sithole, and R. Montville. 2004. Use of microbial modeling and Monte Carlo simulation to determine performance criteria for bacterial populations on plastic cutting boards in use in foodservice kitchens. *Food Prot. Trends* 24:14–19.

Setlow, P., and E. A. Johnson. 2007. Spores and their significance, p. 69–86. *In* M. P. Doyle and L. R. Beuchat (ed.), *Food Microbiology: Fundamentals and Frontiers.* ASM Press, Washington, DC.

Snyder, O. P. 1994a. *Developing and Implementing HACCP-Based Retail Food Operations,* 2007 ed. Hospitality Institute of Technology and Management, St. Paul, MN.

Snyder, O. P. 1994b. *Technology of HACCP-Based, Chilled Food Production Systems,* 2007 ed. Hospitality Institute of Technology and Management, St. Paul, MN.

Snyder, O. P. 1997. The microbiology of cleaning and sanitizing a cutting board. Hospitality Institute of Technology and Management, St. Paul, MN. http://www.hi-tm.com/Documents/Cutboard.html.

Snyder, O. P. 2007a. *HITM Archives.* Hospitality Institute of Technology and Management, St. Paul, MN.

Snyder, O. P. 2007b. Removal of bacteria from fingertips and the residual amount remaining on the hand washing nailbrush. *Food Prot. Trends* 27:597–602.

Stern, N., and S. Pretanik. 2006. Counts of *Campylobacter* spp. on U.S. broiler carcasses. *J. Food Prot.* 69:1034–1039.

Stumbo, C. R. 1973. *Thermobacteriology in Food Processing,* 2nd ed. Academic Press, New York, NY.

Todd, E. C. D., J. D. Greig, C. A. Bartleson, and B. S. Michaels. 2007. Outbreaks where food workers have been implicated in the spread of foodborne disease. Part 2. Description of outbreaks by size, severity, and settings. *J. Food Prot.* 70:1975–1993.

USDA/FSIS. 2001. *Draft Compliance Guidelines for Ready-to-Eat Meat and Poultry Products.* Food Safety and Inspection Service, U.S. Department of Agriculture, Washington, DC. http://www.fsis.usda.gov/OPPDE/rdad/FRPubs/97–013P/RTEGuide.pdf.

USDA/FSIS. 2007. Code of Federal Regulations (CFR). Title 9, Animal and Animal Products. Part 417, Hazard analysis and critical control point (HACCP) systems. U.S. Government Printing Office, Washington, DC. http://www.access.gpo.gov/nara/cfr/waisidx_07/9cfr417_07.html.

WTO. 1995. The WTO agreement on the application of sanitary and phytosanitary measures (SPS agreement). World Trade Organization, Geneva, Switzerland. http://www.wto.org/IAB english/tratop_e/sps_e/spsagr_e.htm.

Pathogens and Toxins in Foods: Challenges and Interventions
Edited by V. K. Juneja and J. N. Sofos
© 2010 ASM Press, Washington, DC

Chapter 28

Recent Developments in Rapid Detection Methods

Lawrence D. Goodridge and Mansel W. Griffiths

The continued presence of pathogenic microorganisms and their toxins in food and drinking water has necessitated the ongoing need for newer, more sensitive and robust analytical systems capable of rapid detection of these contaminants in complex samples. There are significant challenges to the detection of food-borne pathogens. The ideal detection method should be capable of rapidly detecting and confirming the presence of food-borne pathogenic microorganisms directly from complex samples with no false-positive or false-negative results. Furthermore, these assays should be user-friendly, cost-effective, and capable of testing for multiple pathogens simultaneously. Finally, as food-borne pathogen testing increasingly moves from the laboratory to the food processing environment, such methods should be portable and amenable to testing with minimal equipment.

The infectious dose of food-borne pathogens, such as *Escherichia coli* O157:H7, is low (Teunis et al., 2004). Therefore, detection assays that identify these microorganisms must be capable of detecting low concentrations of target microorganisms without interference from background flora or from the food matrix. Specificity is as important as sensitivity in the detection of food-borne pathogens. High specificity is important to minimize background signals and false-positive results from foods that often contain uncharacterized components. In most cases, assays that are currently utilized to detect food-borne pathogens are incapable of detecting these agents directly from foods at levels equal to or below the infective dose because of their lack of sensitivity. Consequently, all food-borne detection methods require an initial enrichment step, in which the numbers of the target microorganism are increased to a detectable level. The disadvantages to enrichment are obvious: in addition to increasing the total detection time by as much as 24 to 72 hours, many agents of food-borne disease, such as toxins and viruses, cannot be enriched. Viruses (including noroviruses and hepatitis A virus [HAV]) account for as much as 70% of food-borne illnesses of known etiology (Mead et al., 1999).

Generally, nucleic acid-based detection systems are more sensitive than affinity-based detection systems. For example, the real-time reverse transcription-PCR (RT-PCR) can detect 10 or fewer microorganisms in less than 6 hours, depending on the sample matrix (Tsai et al., 2006). However, RT-PCR requires a clean sample and is unable to detect protein toxins. Furthermore, cultures of the target organism are often not available for additional tests following RT-PCR analysis. In contrast, affinity-based assays, such as immunoassays, are capable of detecting toxins but are not as specific as nucleic acid-based methods, and the high fat and protein content present in foods such as ground beef and the high numbers of background flora in food samples cause interference during the assay. There have been many reviews written regarding the use of nucleic acid-based or affinity-based assays to effect rapid detection of food-borne pathogens. Therefore, these technologies are considered to be "emerged" and will not be discussed in detail. The reader is referred to the following for additional information on these methodologies: Mozola (2006), Malorny et al. (2003), Fung (2002), de Boer and Beumer (1999), Feng (1997), and Hill (1996).

Conventional culture techniques continue to be the gold standard for the isolation, detection, and identification of target pathogens. As discussed, these methods increase detection times by hours to days, causing preliminary test results to be delayed. The recombinant DNA era has led to the development of rapid methods that attempt to detect pathogenic microorganisms and toxins without the need for

Lawrence D. Goodridge • Department of Animal Sciences, Colorado State University, Fort Collins, CO 80523-1171. **Mansel W. Griffiths** • Canadian Research Institute for Food Safety and Department of Food Science, University of Guelph, Guelph, Ontario, Canada N1G 2W1.

enrichment. These methods have a similar theme, in that there is a focus on enhancing the detection signal instead of increasing the concentration of the target pathogen. Furthermore, these methods combine a biological or chemical recognition system with a physical process to increase the sensitivity and specificity of the test method. As such, many of these developing technologies must be paired with a transducer that can transform the test response into an analyzable signal. Several test approaches are emerging based on these principles. For the purpose of this review, these assays are defined as affinity, cell/tissue, and nucleic acid technologies. While these technologies still require the use of enrichment procedures, the enrichment process is usually rapid and allows for detection of the target within hours as opposed to days.

Immunological Detection

Antibody-based and other affinity probes are being employed in many of the emerging technologies for rapid detection of food-borne pathogens. Detection methods based on this principle utilize the specificity of a ligand to capture the target(s) of interest. These "shape recognition" technologies are capable of detecting a wide range of potential food-borne pathogenic agents, including bacteria (vegetative cells and spores), viruses, prions, and toxins. Currently, a substantial amount of research is concentrated on improving the stability, specificity, and sensitivity of antibody-based assays through the generation of recombinant antibodies, antibody fragments, and bacteriophage (phage) display antibody probes (Lim et al., 2005). Other recognition elements are currently under investigation, including the use of ganglioside receptors for capture of toxins (Newsome, 2003), aptamers, and peptide ligands (Goldschmidt, 2006). A major advantage of the use of affinity probes is their potential to isolate live organisms for culture and further analysis (Kramer et al., 2002b). Several reviews have described the extensive research regarding the use of affinity probes in detection of pathogens (Hock et al., 2002; Brody and Gold, 2000; Iqbal et al., 2000).

Antibodies and Fragments

While conventional antibodies have traditionally been used in affinity-based detection assays, various components of whole antibodies, including mono- and divalent antibody fragments and single-chain variable regions, have been utilized in an attempt to improve the stability, specificity, and sensitivity of these affinity probes, as compared to whole antibodies (Petrenko and Vodyanoy, 2003; Kramer et al., 2002a;

Emanuel et al., 2000). Filpula (2007) has recently contributed an excellent review in this discipline. Using recombinant technology, Petrenko and Vodyanoy (2003) have demonstrated that antibody fragments can be modified to increase their binding kinetics, improving their usefulness as affinity probes. These authors also demonstrated the ability of phage display in combination with the use of antibodies and their fragments to effect detection of pathogenic microorganisms. Phage display libraries have led to the development of improved antibody-based probes by facilitating the biopanning of libraries containing thousands of possible peptides to select the ones that bind to a specific antigen with the highest specificity. The identified antibodies or fragments can then be chemically synthesized or produced in large quantities in a recombinant host; alternatively, the phage expressing the antibody or fragment may also be used as the probe (Petrenko and Sorokulova, 2004). Emanuel et al. (2000) used phage display to develop an assay for detection of *Clostridium botulinum* toxin. In this work the authors generated a recombinant antibody fragment (Fab) directed against botulinum neurotoxin A/B and demonstrated that the Fab was sensitive in a variety of assay formats, including surface plasmon resonance, flow cytometry, enzyme-linked immunosorbent assay, and a hand-held immunochromatographic assay.

Aptamers

Antibody-based detection systems are still considered to be the gold standard of affinity-based testing methods. However, the use of antibodies or antibody fragments in multianalyte diagnostics can be problematic, due in part to their physical structure and methods used to synthesize the antibodies (Tombelli et al., 2007). The discovery that nucleic acids (RNA in particular) can assume stable secondary structures and can be easily synthesized and functionalized led to the creation of nucleic acid ligands called aptamers (Ellington and Szostak, 1990; Tuerk and Gold, 1990). Aptamers are small DNA or RNA ligands that recognize a target by shape, as opposed to sequence, and they are generated by a combinatorial selection process known as systematic evolution of ligands by exponential enrichment (Tuerk and Gold, 1990). During this enrichment process, a large library of oligonucleotides with different sequences (generally 10^{15} different sequences) is subjected to iterative cycles of selection and amplification. Following incubation with the specific target and the portioning of the binding from the nonbinding molecules, the oligonucleotides that are selected are amplified to create a new mixture, enriched only in those nucleic acid sequences that possess a higher affinity for the target. After several iterations of the selection process, the pool is enriched in the high-affinity sequences at the expense of the low-affinity binding sequences.

Aptamers offer several advantages over the use of antibodies in the identification of food-borne microorganisms and toxins. For example, while the generation of antibodies against nonimmunogenic molecules is often difficult, aptamers can be generated against virtually any target. Antibodies are typically identified under physiological conditions that limit the extension to which the antibodies can be functionalized (Tombelli et al., 2007). In contrast, the aptamer selection process can be modified to allow for aptamers to be identified that bind selectively under different binding (pH and temperature) conditions. Furthermore, upon selection, aptamers are produced by chemical synthesis and purified to a very high degree by eliminating the batch to batch variation typically observed with antibodies. Another advantage is the higher temperature stability of aptamers compared to antibodies. Antibodies are large proteins that can undergo irreversible denaturation when exposed to high temperatures. By comparison, aptamers are very stable at higher temperatures, and they can recover their native structure following denaturation (Tombelli et al., 2007).

The primary limitation of the use of aptamers in detection assays has been related to their nuclease sensitivity, which affects their use as affinity probes (Famulok et al., 2000). Two approaches have been developed to solve this problem. Pieken et al. (1991) have shown that chemical modification of the ribose ring at the 2′ position improves the stability of these ligands. Alternatively, Klussman et al. (1996) developed a stabilization method based on the selection of aptamers that recognize the stereoisomer of the intended target molecule and on chemical synthesis of the mirror image of the selected sequences. Due to the molecular symmetry of the two molecules, the mirror-image aptamer (L-ribose) binds to the natural target molecule. The substitution of D-ribose with L-ribose makes the mirror-image aptamer completely stable (Klussman et al., 1996).

Several recent publications have highlighted the use of aptamers to detect food-borne agents of disease. Aptamers have been selected that recognize food-borne toxins (staphylococcal enterotoxin B) and bacteria (Mycobacterium avium subsp. Paratuberculosis and Campylobacter jejuni) (Bannantine et al., 2007; McMasters and Stratis-Cullum, 2006). McMasters and Stratis-Cullum (2006) developed a fluorescently labeled DNA aptamer which was used

to detect C. jejuni by capillary electrophoresis and laser-induced fluorescence detection. The aptamer exhibited strong binding affinity toward C. jejuni, with minimal cross reactivity against other food-borne pathogens, including E. coli O157:H7 and Salmonella enterica serovar Typhimurium. Recently, an aptamer-linked immobilized sorbent assay was developed by Vivekananda and Keil (2006) who used a set of 25 unique single-stranded DNA sequences to identify Francisella tularensis in a sandwich format. The researchers showed that the assay had good sensitivity and selectivity for F. tularensis and could detect as few as 25 ng of F. tularensis subspecies japonica and tularensis antigen.

Affinity Probe-Based Biosensors

Biosensors have been defined as the offspring of the combination of biology and electronics (DeYoung, 1983). Modern biosensors have effectively combined both disciplines, with electronics/information technology exemplified by microcircuits and optical fibers and with biology exemplified in the form of enzymes or affinity probes (Richter, 1993). The basic principle of a biosensor employs electronic or optical transduction technology to monitor a parameter of the reaction between an affinity probe and an analyte and to display the parameter as a quantifiable electrical or optical signal (Griffiths and Hall, 1993). The signal can be related to concentration. Analytes that are not recognized by the respective affinity probe will not produce a signal. Various biosensors are currently being developed that make use of affinity probes. One area of biosensor research that has the potential to revolutionize microbial diagnostics is the development of cantilever-based assays. Cantilevers are often employed as mechanical sensors, and they were first utilized in atomic force microscopy (Binning et al., 1986). A cantilever is a diving board-shaped, single-clamped, suspended beam. Cantilever sensors function on the principle that any physical, chemical, or biological stimulus can affect the mechanical characteristics of the micromechanical transducers in such a way that the resulting change can be measured using electronic, optical, or other methods (Sarid, 1991). There are two main sensing methods in microcantilever-based systems. In stress detection mode, a biochemical reaction is performed selectively on one side of the cantilever, resulting in a measurable bending of the cantilever and label-free detection of the target analyte (Bhattacharya et al., 2007). The stress detection mode has not been demonstrated in the detection of large analytes such as cells. The other sensing method is known as mass detection mode. In this mode, the cantilever vibrates at its resonant frequency by an external disturbance. The resonance frequency shift is measured to find added masses, or analytes, bound on the cantilevers.

A large number of cantilever-based biosensors have been described in the literature. With respect to detection of microorganisms, Illic et al. (2001) have demonstrated very sensitive detection of bacteria by using cantilevers in the resonance mode. The mass sensitivity of the designed cantilever was such that the sensor could detect the mass changes corresponding to the attachment of a single *E. coli* cell on its surface. Other studies have demonstrated the use of cantilevers to detect and monitor the growth of *E. coli* cells and *Aspergillus niger* spores (Gfeller et al., 2005; Nugaeva et al., 2005). Gupta et al. (2004a, 2004b) have applied the use of resonant cantilever biosensors to detection of bacteria and viruses. These researchers were able to detect a single *Listeria innocua* cell and a single vaccinia virion using this technique. Davila et al. (2007) were able to detect as few as 50 to 100 *Bacillus anthracis* spores with the use of a planar rectangular geometry cantilever. Other researchers have applied cantilever technology to detection of food-borne pathogens. For example, Campbell and Mutharasan (2005) constructed a cantilever-based biosensor and used it to detect *E. coli* O157:H7. A composite self-excited millimeter-sized lead zirconate titanate glass cantilever was fabricated for the detection of *E. coli* O157:H7. The researchers immobilized affinity-purified monoclonal antibody (anti-*E. coli* O157:H7) at the cantilever glass tip and then immersed the cantilever in solutions containing the pathogen at several concentrations (70 to 7×10^7 cells/ml). Results indicated that the cantilever sensor could detect as few as 700 cells/ml.

Cell/Tissue-Based Assays

Tissue or cell-based detection systems use the intrinsic response of a specific cell type to a potentially toxic or infectious foreign substance to identify a pathogenic agent (Lim et al., 2005). In these assays, the cells constituting the sensors produce an action potential signal that can be measured by an electrode or optical detector (Kovacs, 2003). The cells that comprise the detector may originate from a specific unicellular organism or tissue type and may be primary or immortalized. Ghosh et al. (2007) developed a whole cell biosensor for detection of *Staphylococcus aureus* alpha-toxin. Alpha-toxin is also known to induce permeability to endothelial cell monolayers in vitro due to the formation of interendothelial gaps. The researchers utilized this principle in the development of the whole-cell-based biosensor. The biosensor, consisting of a confluent monolayer of human umbilical vein endothelial cells on a potassium ion-selective

electrode, takes advantage of cell permeability dysfunction to detect the presence of small quantities of alpha-toxin. When a confluent monolayer of cells was formed on the membrane surface, the response of the electrode toward the marker ion, potassium, was inhibited. Upon exposing this sensor to various concentrations of alpha-toxin for 20 min, an increase in sensor response to potassium was observed. The response obtained was indirectly related to the concentration of alpha-toxin. The detection limit of this sensor for alpha-toxin was found to be 0.1 ng/ml. Cell monolayers were stained with silver nitrate to quantify the formation of intercellular gaps as well as to study the effect of this toxin on human umbilical vein endothelial cell morphology. A strong positive correlation was observed between the response obtained from the biosensor and the area of the intercellular gaps (Ghosh et al., 2007).

Rider and colleagues (2003) have developed a B lymphocyte-based assay, termed cellular analysis and notification of antigen risks and yields (CANARY), and have demonstrated the sensitive detection of various microorganisms. The researchers chose the B lymphocyte as the basis for their biosensor for several reasons. First, B cells are a component of the immune system, and they display antibodies on their surface as pathogen receptors. An immense diversity of antibodies with various specificities can be selected and expressed in these cells with the use of recombinant methods. In addition, B lymphocytes link the surface immunoglobulin–receptor complex to a rapid-response system. The cross-linking of surface antibodies by cognate polyvalent antigen initiates a cascade of events that leads to the mobilization of intracellular calcium stores within seconds and, ultimately, to activation and proliferation of the cells. To develop the biosensor, Rider et al. (2003) engineered the B lymphocytes to express aequorin, a protein found in jellyfish that emits light in response to calcium flux, and they also modified these same cells so that they expressed surface antibodies with specificity for bacterial or viral agents of interest (including *Yersinia pestis*, *Bacillus anthracis*, and vaccinia virus). The responses of the engineered B lymphocytes to as few as 50 bacteria were detected in a luminometer in less than 5 minutes, including the time needed to prepare the sample.

CANARY was also shown to be capable of sensitive detection of food-borne pathogens in food matrices. B cells specific for *E. coli* O157:H7 detected as little as 500 CFU/g in lettuce in less than 5 min, which included the initial sample preparation time. Nevertheless, while the CANARY exhibits rapid and sensitive detection of microorganisms, the fact that it is based on affinity (antibody) recognition of the target agent means that this assay is subject to the same cross-reactivity problems observed with other antibody-based assays. Another issue to consider is the stability (i.e., storage and maintenance) of the assay, which is a problem observed with other cell-based sensor systems (Lim et al., 2005).

Nucleic Acid-Based Assays

The specificity of base pair matching is at the heart of all nucleic acid detection assays. Any microorganism that contains DNA or RNA can be detected using nucleic acid-based assays, but a limitation of these diagnostics is their inability to detect protein-based agents of disease, such as toxins or prions. Two of the emerging methods for microbial detection within this class of assays include nucleic acid sequence-based amplification (NASBA) and microarray technology.

NASBA

NASBA is a sensitive, isothermal, transcription-based amplification system specifically designed for the detection of RNA targets. In some NASBA systems, DNA is also amplified, though very inefficiently and only in the absence of the corresponding RNA target or in case of an excess (>1,000-fold) of target DNA over RNA (Lim et al., 2005). During NASBA, a primer binds to the target RNA sequence and a reverse transcriptase step produces a cDNA strand. RNase is used to digest the template RNA and a second primer binds to the cDNA, which the reverse transcriptase uses to form a double-stranded cDNA molecule. The addition of bacteriophage T7 RNA polymerase facilitates the production of RNA transcripts via the amplification process. Deiman and colleagues have described the applications, advantages, and disadvantages of NASBA (Deiman et al., 2002). NASBA-based assays have been developed and evaluated for many pathogenic microorganisms, including viruses (Casper et al., 2005), bacteria (Rodriguez-Lazaro et al., 2004), fungi (Yoo et al., 2005), and protozoa (Schneider et al., 2005). NASBA-based detection of microorganisms in clinical (Rodriguez-Lazaro et al., 2004) and environmental (Baeumner et al., 2003) samples has also been reported in the literature.

Several NASBA-based methods have been described for food-borne pathogens, including *Campylobacter* spp., *L. monocytogenes,* and *Salmonella enterica* serovar Enteritidis in various foods and for *Cryptosporidium parvum* in water, as reviewed by Cook (2003). Both 16S rRNA and various mRNAs have been used as target molecules for detection. Most of the methods to detect pathogens in foods

have employed enrichment in nutrient medium prior to NASBA, as this can ensure sensitivity of detection and encourage the detection of only viable target cells. Although a relatively recent method, NASBA has the potential for adoption as a diagnostic tool for food-borne pathogens.

Nadal et al. (2007) recently described a molecular beacon-based real-time NASBA (QNASBA) assay for detection and identification of *L. monocytogenes.* The assay targeted a sequence from the mRNA transcript of the *hly* gene and included an internal amplification control to detect failure of the reaction. The assay consistently detected as few as 100 target molecules and 40 exponentially growing *L. monocytogenes* cells per reaction. The researchers also demonstrated the accurate quantification of target RNA molecules in the presence of DNA in the sample. In combination with a short RNase treatment prior to nucleic acid extraction, the QNASBA assay specifically detected viable *L. monocytogenes* cells. The authors demonstrated the successful application of the assay to rapid detection of *L. monocytogenes* in meat and salmon products, and the authors concluded that the assay could be a useful tool for the study of the growth of *L. monocytogenes* in food samples (Nadal et al., 2007).

D'Souza and Jaykus (2003) developed a NASBA assay for the detection of *Salmonella* serovar Enteritidis in several foods. A previously reported primer and probe set based on mRNA sequences of the *dnaK* gene were used in this study. To evaluate the assay, 25-g samples of representative food samples, including fresh meats, poultry, fish, ready-to-eat salads, and bakery products, were preenriched with and without *Salmonella* serovar Enteritidis inoculation. RNA isolation was followed by NASBA amplification and electrochemiluminescent (ECL) detection. The results indicated that the end-point detection limit of the NASBA-ECL assay was equivalent to 10^1 CFU of *S.* Enteritidis per amplification reaction. The noninoculated samples did not produce any false-positive results. However, some of the food matrices inhibited the NASBA-ECL reaction, and this problem was negated by diluting the associated RNA 10-fold prior to amplification. For all food items tested, positive ECL signals were achieved after 18 h of preenrichment and subsequent NASBA at initial inoculum levels of 10^1 and 10^2 CFU per 25 g food sample.

Other researchers have applied NASBA to the detection of food-borne viral pathogens. For example, Kou and coworkers (2006) rapidly detected noroviruses in fecal samples and shellfish with NASBA. During this work the authors utilized 58 fecal samples that had previously tested positive for the presence of noroviruses to develop a NASBA assay for these viruses. Oligonucleotide primers targeting the polymerase coding region were used to amplify the viral RNA, and the amplicons were detected by hybridization with digoxigenin-labeled oligonucleotide probes that were highly specific for norovirus genogroups I and II. The expected band (327 bp) was readily visualized using denaturing agarose gel electrophoresis in the absence of any nonspecific bands. All fecal samples were confirmed as positive for noroviruses by electron microscopy and by NASBA. The sensitivity of the norovirus NASBA method was high. Target RNA concentrations as low as 5 pg/ml were detected in fecal specimens, and when the assay was applied to artificially contaminated shellfish, the assay was capable of detecting 100 pg of nucleic acid per 1.5 g shellfish tissue. The authors concluded that the norovirus NASBA assay provided a rapid and efficient way of detecting these viruses in fecal samples, and the study demonstrated its potential for detecting noroviruses in food and environmental samples with high specificity and sensitivity (Kou et al., 2006).

Abd el-Galil et al. (2005) used NASBA in combination with a molecular beacon for the real-time detection and quantification of hepatitis A virus (HAV). A 202-bp, highly conserved 5′ noncoding region of HAV was the target of the NASBA assay. The sensitivity of the real-time NASBA assay was tested with 10-fold serial dilutions of viral RNA, and a detection limit of 1 PFU/ml was obtained. The specificity of the assay was evaluated by testing for the presence of HAV against background microflora, including environmental pathogens and indicator microorganisms, with only HAV positively identified. When combined with immunomagnetic separation, the NASBA assay successfully detected as few as 10 PFU/ml of HAV from seeded lake water samples. The authors concluded that because of its isothermal nature, its speed, and its similar sensitivity compared to the real-time RT-PCR assay, the real-time NASBA method should have broad applications for the rapid detection of HAV in contaminated food or water (Abd el-Galil et al., 2005).

Taken collectively, these studies indicate that NASBA is a potentially useful method for the sensitive, specific, and rapid detection of microbial pathogens in food and environmental samples. The method may also be applicable to the detection and analysis of viable organisms if mRNA is used as the template (Cook, 2003).

DNA Microarrays

DNA microarrays represent one of the newer advances in molecular detection technology. This technology utilizes DNA-DNA hybridization for

detection of unknown DNA sequences, which takes advantage of the fact that DNA molecules are an ideal and naturally optimized reagent with which to identify other DNA molecules. This is due to the strong sequence-specific base pairing between complimentary DNA strands. In microarray (and other hybridization) methods, one strand of DNA is considered the target (and is located in the sample to be analyzed) and the other strand is considered the probe. The analysis of a sample consists of evaluating whether the DNA in the sample can hybridize to the probe. If this occurs, a double-stranded DNA complex is formed. Since the sequence of the probe is known, information regarding the DNA sequences in the sample can be ascertained (Bhattacharya et al., 2007). Some of the earliest microarray hybridization experiments were conducted on homogeneous solutions (Cantor and Smith, 1999). Typically, the hybridization was allowed to proceed for a fixed amount of time and was followed by column-based separation of the double-stranded DNA molecules. The amount of double-stranded DNA was then quantified using a radioisotopic label on the probe or target.

Modern microarray hybridization protocols involve immobilization of the probe on a solid support using chemical methodologies. In a typical array, single strands of known sequences (the probe) are placed at specifically known sites on the solid support. This is accomplished using optical or electrical methods. The optical approach involves selective deprotection of sites where known sequences of single strands can be built base by base (Bhattacharya et al., 2007). The electrical method takes advantage of the net negative charge of a DNA molecule, which can be electrophoretically transported to specified locations on chip surfaces, as described by Bashir (2004). To analyze a sample of unknown DNA target sequences, the DNA in the sample is first labeled with a fluorophore and added to the microarray chip, where hybridization will occur if complimentarity exists between the target and probe. The nonhybridized DNA is washed off the array, and the surface of the array is analyzed using a microarray scanner to detect the fluorescent signal.

The main advantage of the microarray technique is the ability to rapidly analyze many samples in a miniaturized format. In combination with bioinformatics, microarrays provide unparalleled opportunities for simultaneous detection of thousands of genes or target DNA sequences and offer tremendous potential for the detection and study of food-borne and other microorganisms. For example, in addition to the relatively straightforward application of detection of food-borne microorganisms, microarray technology can be further utilized to characterize microorganisms by providing information for specific identification of isolates, as well as data related to pathogenesis based on the presence of virulence genes, leading to an indication of the mechanisms by which new pathogenic strains evolve epidemiologically and phylogenetically (molecular subtyping).

Due to the low numbers of pathogenic microorganisms in food samples, the target organisms must be enriched in order to obtain enough target DNA for the microarray. This is typically accomplished through nonspecific or specific enrichment of the target microorganism(s). As previously discussed, this limits the speed of detection. An alternative strategy is to amplify the target DNA by PCR (which is particularly useful for viruses, since these pathogens can rarely be enriched) and then perform microarray-based detection of the target DNA. Several reports have described the use of multiplex PCR to amplify a number of pathogen-specific genetic markers that are subsequently detected using a DNA microarray. This strategy takes advantage of the inherent sensitivity of PCR, and the sample analysis capabilities of the microarray. This strategy has been used to develop assays for food-borne bacteria (E. coli O157:H7 [Call et al., 2001]; Campylobacter spp. [Keramas et al., 2004]; Vibrio spp. [Panicker et al., 2004]; Listeria spp. [Volokhov et al., 2002]; and a mixture of pathogens [Wilson et al., 2002]). Salazar and Caetano-Anollés (1996) were the first to introduce this strategy when they used arbitrarily primed PCR to amplify DNA fragments from different strains of E. coli O157:H7. These fragments were then hybridized to a membrane-bound array composed of 11-mer oligonucleotide probes. Sixty-four strains were analyzed using the assay, and hybridization patterns were identified that classified the strains into 14 unique subtypes. Beattie (1997) demonstrated that the subtyping approach could be configured in a microarray format, and Willse et al. (2004) followed this work by developing a microarray-based genome-independent microbial fingerprinting method that provided high-resolution differentiation between various Salmonella enterica isolates.

It is often necessary to validate newer detection methodologies against the gold standard of bacterial culture before these methods become widely employed. Keramas et al. (2004) compared the use of culture, PCR analysis, and DNA microarrays for detection of C. jejuni and Campylobacter coli in chicken feces. Sixty-five pooled chicken cloacal swab samples from 650 individual broiler chickens were included in the study. The results of the microarray-based detection were compared to those obtained by conventional culture and gel electrophoresis. By conventional culture, 60% of the samples were positive

for either C. jejuni or C. coli. PCR and capillary electrophoresis detected Campylobacter spp. in 95% of the samples, while DNA microarrays detected Campylobacter spp. in 100% of the samples. By application of DNA microarray analysis, the isolates in 4 samples (6%) could not be identified to the species level, whereas by PCR-capillary electrophoresis, the isolates in 12 samples (19%) were unidentified. PCR-capillary electrophoresis analysis revealed that two (3%) of the samples were positive for both C. jejuni and C. coli, while DNA microarray analysis revealed that nine (14%) of the samples were positive for both species. Of 65 samples, 2 samples were identified to contain C. coli by conventional culture but were positive for C. jejuni by both PCR-capillary electrophoresis and DNA microarray analysis.

The discrepancy between the different methods in the study by Keramas et al. (2004) highlights an important issue with all nucleic acid-based detection methods. While these researchers used Campylobacter spp. 16S rRNA and C. jejuni- and C. coli-specific genes as the targets in their analysis, the detection of genes involved in virulence (which is a typical approach in nucleic acid detection assays) as a way to specifically identify a given pathogen may be problematic, since virulence genes are known to be exchanged between bacterial species by horizontal and bacteriophage-based gene transfer (Arnold et al., 2007). If these genes are used as the detection targets, there is the real possibility of false-positive detection of the target organism. Therefore, all nucleic acid detection assays should include a species-specific target in addition to other target sequences to alleviate the possibility of false detection.

Another disadvantage of microarray-based detection methods is the possibility of false-negative signal generation caused by the photo-bleaching or wash-off of the fluorophore used to label the target sequence(s). Obviously, the possibility of a false-negative signal may have dramatic consequences since a food product contaminated with a pathogen could be erroneously determined as pathogen free. A molecular beacon DNA microarray system for rapid detection of E. coli O157:H7 that eliminates the risk of false-negative signals was recently developed by Kim et al. (2007). The assay employed a color changing molecular beacon as the probe for the optical detection of the target sequence. A computer-controlled detection platform was utilized to measure the target hybridization-induced change of fluorescence color due to the fluorescent resonance energy transfer between a pair of spectrally shifted fluorophores conjugated to the opposite ends of the beacon (oligonucleotide probe). The presence of two fluorescence molecules allowed for active visualization of both hybridized and unhybridized states of the beacon. This aspect of the assay was used to eliminate false-negative signal detection characteristic of traditional microarray assays in which bleaching or washout of the fluorophore is indistinguishable from the absence of the target DNA sequence. The two-color design also allowed for the concentration of the target DNA in a sample to be quantified at concentrations as low as ≤ 1 ng/μl. The authors concluded that the fluorescent resonance energy transfer-based microarray method was suitable for simultaneous and reliable detection of hundreds of DNA target sequences in one test run using a series of beacons immobilized on a single substrate (Kim et al., 2007).

In addition to analysis of numerous numbers of samples, one of the main advantages of microarray hybridization technology is the ability to specifically detect the DNA of interest in a background of competing DNA from background microflora. Kostic et al. (2007) presented an elegant approach in which competitive oligonucleotide probes were used to improve the specificity of a multiplexed microarray assay and demonstrated the ability to detect a broad range of pathogenic bacteria. The approach was tested with a set of 35 oligonucleotide probes targeting *E. coli*, *Shigella* spp., *Salmonella* spp., *Aeromonas hydrophila*, *Vibrio cholerae*, *Mycobacterium avium*, *Mycobacterium tuberculosis*, *Helicobacter pylori*, *Proteus mirabilis*, *Yersinia enterocolitica*, and *C. jejuni*. The introduction of competitive oligonucleotides in the labeling reaction successfully suppressed cross-reaction by closely related sequences, significantly improving the performance of the assay. Environmental applicability was tested using animal specimens and soil samples that harbored complex microbial communities. Detection sensitivity in the range of 0.1% was demonstrated during the study, which compares favorably with the 5% detection limit of traditional microbial diagnostic microarrays (Kostic et al., 2007).

Nanoarrays represent the next evolutionary step in the miniaturization of microarray-based diagnostics. BioForce Nanosciences (Ames, IA) (http://www.bioforcenano.com) has produced an ultraminiaturized version of the traditional microarray that can measure interactions between molecules down to resolutions of as little as 1 nm. The miniaturized array contains approximately 400 nanoarray spots in the same area as a single spot on a traditional microarray. This allows for as many as 1,500 different samples to be tested using the same area now needed for just one domain on a traditional microarray (Jain, 2003).

CONCLUSION

The advent of molecular biology and the combination of biology and electronics have resulted in the development of emerging novel, sensitive, and rapid methods for the detection of food-borne pathogens. A discussion of these new food-borne pathogen detection methodologies cannot be complete without the mention of sample preparation. While the emerging rapid detection techniques discussed above have progressed to the point where single-cell detection may be possible, the methods are limited by the inherent problems (time and labor intensiveness, well-trained staff, and expensive laboratory equipment) of sample processing. The major challenge in preparing an appropriate sample comes from the high probability of components of the food matrix, such as fat and proteins, interfering with the ability of the detection method to reliably identify the target organism. Therefore, the standard approach in traditional sample preparation methods has centered on extraction of the target microorganism from a given food matrix, using an upfront protocol. The results of some of the current and emerging research in rapid microbial diagnostics offer the prospect that target microorganisms may not always have to be extracted from the food matrix in the future. Nevertheless, sample issues will have to be solved before the sensitive and real-time detection of food-borne pathogens can be realized. In the meantime, new and emerging detection strategies will continue to be measured against traditional cultural techniques to assess the robustness of the method and the efficiency of detection.

REFERENCES

Abd el-Galil, K. H, M. A. El-Sokkary, S. M. Kheira, A. M. Salazar, M. V. Yates, W. Chen, and A. Mulchandani. 2005. Real-time nucleic acid sequence-based amplification assay for detection of hepatitis A virus. *Appl. Environ. Microbiol.* 71:7113–7116.

Arnold, D. L., R. W. Jackson, N. R. Waterfield, and J. W. Mansfield. 2007. Evolution of microbial virulence: the benefits of stress. *Trends Genet.* 23:293–300.

Baeumner, A. J., R. N. Cohen, V. Miksic, and J. Min. 2003. RNA biosensor for the rapid detection of viable *Escherichia coli* in drinking water. *Biosens. Bioelectron.* 18:405–413.

Bannantine, J. P., T. J. Radosevich, J. R. Stabel, S. Sreevatsan, V. Kapur, and M. L. Paustian. 2007. Development and characterization of monoclonal antibodies and aptamers against major antigens of *Mycobacterium avium* subsp. *paratuberculosis*. *Clin. Vaccine Immunol.* 14:518–526.

Bashir, R. 2004. BioMEMS: state-of-the-art in detection, opportunities and prospects. *Adv. Drug Deliv. Rev.* 56:1–22.

Beattie, K. L. 1997. Genomic fingerprinting using oligonucleotide arrays. *In* G. Caetano-Anollés and P. M. Gresshoff (ed.), *Protocols, Applications, and Overviews.* Wiley-Liss, New York, NY.

Bhattacharya, S., J. Jang, L. Yang, D. Akin, and R. Bashir. 2007. Biomems and nanotechnology-based approaches for rapid detection of biological entities. *J. Rapid Methods Automation Microbiol.* **15**:1–32.

Binning, G., C. F. Quate, and C. Gerber. 1986. Atomic force microscope. *Phys. Rev. Lett.* **56**:930–933.

Brody, E. N., and L. Gold. 2000. Aptamers as therapeutic and diagnostic agents. *J. Biotechnol.* **4**:5–13.

Call, D. N. R., F. J. Brockman, and D. P. Chandler. 2001. Detecting and genotyping *Escherichia coli* O157:H7 using multiplexed PCR and nucleic acid microarrays. *Int. J. Food Microbiol.* **67**:71–80.

Campbell, G. A., and R. Mutharasan. 2005. *Escherichia coli* O157:H7 detection limit of millimeter-sized PZT cantilever sensors is 700 cells/mL. *Anal Sci.* **21**:355–357.

Cantor, C. R., and C. L. Smith. 1999. Analysis of DNA sequences by hybridization, p. 120–125. *In Genomics: the Science and Technology behind the Human Genome Project.* Wiley Inter-Science Publication, Hoboken, NJ.

Casper, E. T., S. S. Patterson, M. C. Smith, and J. H. Paul. 2005. Development and evaluation of a method to detect and quantify enteroviruses using NASBA and internal control RNA (IC-NASBA). *J. Virol. Methods* **124**:149–155.

Cook, N. 2003. The use of NASBA for the detection of microbial pathogens in food and environmental samples. *J. Microbiol. Methods* **53**:165–174.

Davila, A., J. Jang, A. Gupta, T. Walter, A. Aronson, and R. Bashir. 2007. Micro-resonator mass sensors for detection of *Bacillus anthracis* sterne spores in air and water. *Biosens. Bioelectron.* [Epub ahead of print.] doi 10.1016/j.bios.2007.01.012.

de Boer, E., and R. R. Beumer. 1999. Methodology for detection and typing of foodborne microorganisms. *Int. J. Food Microbiol.* **50**:119–130.

Deiman, B., P. van Aarle, and P. Sillekens. 2002. Characteristics and applications of nucleic acid sequence-based amplification (NASBA). *Mol. Biotechnol.* **20**:163–180.

DeYoung, H. G. 1983. Biosensors, the mating of biology and electronics. *High Technol.* **11**:41–49.

D'Souza, D. H., and L. A. Jaykus. 2003. Nucleic acid sequence based amplification for the rapid and sensitive detection of *Salmonella enterica* from foods. *J. Appl. Microbiol.* **95**:1343–1350.

Ellington, A. D., and J. W. Szostak. 1990. In vitro selection of RNA molecules that bind specific ligands. *Nature* **346**:818–822.

Emanuel, P. A., J. Dang, J. S. Gebhardt, J. Aldrich, E. A. Garber, H. Kulaga, P. Stopa, J. J. Valdes, and A. Dion-Schultz. 2000. Recombinant antibodies: a new reagent for biological agent detection. *Biosens. Bioelectron.* **14**:751–759.

Famulok, M., G. Mayer, and M. Blind. 2000. Nucleic acid aptamers—from selection in vitro to application in vivo. *Acc. Chem. Res.* **33**:591–599.

Feng, P. 1997. Impact of molecular biology on the detection of foodborne pathogens. *Mol. Biotechnol.* **7**:267–278.

Filpula, D. 2007. Antibody engineering and modification technologies. *Biomol. Eng.* **24**:201–215.

Fung, D. Y. 2002. Predictions for rapid methods and automation in food microbiology. *J. AOAC Int.* **85**:1000–1002.

Gfeller, K., N. Nugaeva, and M. Hegner. 2005. Micromechanical oscillators as rapid biosensor for the detection of active growth of *Escherichia coli*. *Biosens. Bioelectron.* **21**:528–533.

Ghosh, G., L. G. Bachas, and K. W. Anderson. 2007. Biosensor incorporating cell barrier architectures for detecting *Staphylococcus aureus* alpha toxin. *Anal. Bioanal. Chem.* **387**:567–574.

Goldschmidt, M. C. 2006. The use of biosensor and microarray techniques in the rapid detection and identification of salmonellae. *J. AOAC Int.* **89**:530–537.

Griffiths, D., and G. Hall. 1993. Biosensors—what real progress is being made? *Trends Biotechnol.* **11**:122–130.

Gupta, A., D. Akin, and R. Bashir. 2004a. Detection of bacterial cells and antibodies using surface micromachined thin silicon cantilever resonators. *J. Vac. Sci. Technol. B* **22**:2785–2791.

Gupta, A., D. Akin, and R. Bashir. 2004b. Single virus particle detection using microresonators with nanoscale thickness. *Appl. Phys. Lett.* **84**:1976–1978.

Hill, W. E. 1996. The polymerase chain reaction: applications for the detection of foodborne pathogens. *Crit. Rev. Food Sci. Nutr.* **36**:123–173.

Hock, B., M. Seifert, and K. Kramer. 2002. Engineering receptors and antibodies for biosensors. *Biosensor. Bioelectron.* **17**:239–249.

Illic, B., D. Czaplewski, M. Zalalutdinov, H. G. Craighead, P. Neuzil, C. Campaglono, and C. Batt. 2001. Single cell detection with micromechanical oscillators. *J. Vac. Sci. Technol. B* **19**:2825–2840.

Iqbal, S. S., M. W. Mayo, J. G. Bruno, B. V. Bronk, C. A. Batt, and J. P. Chambers. 2000. A review of molecular recognition technologies for detection of biological threat agents. *Biosensor. Bioelect.* **15**:549–578.

Jain, K. K. 2003. Nanodiagnostics: application of nanotechnology in molecular diagnostics. *Expert Rev. Mol. Diagn.* **3**:153–161.

Keramas, G., D. D. Bang, M. Lund, M. Madsen, S. E. Rasmussen, H. Bunkenborg, P. Telleman, and C. B. Christensen. 2004. Development of a sensitive DNA microarray suitable for rapid detection of *Campylobacter* spp. *Mol. Cell. Probes* **17**:187–196.

Kim, H., M. D. Kaneb, S. Kimc, W. Dominguezc, B. M. Applegate, and S. Savikhin. 2007. A molecular beacon DNA microarray system for rapid detection of *E. coli* O157:H7 that eliminates the risk of a false negative signal. *Biosens. Bioelectron.* **22**:1041–1047.

Klussmann, S., A. Nolte, R. Bald, V. A. Erdmann, and J. P. Furst. 1996. Mirror image RNA that binds D-adenosine. *Nat. Biotechnol.* **14**:1112–1115.

Kostic, T., A. Weilharter, S. Rubino, G. Delogu, S. Uzzau, K. Rudic, A. Sessitsch, and L. Bodrossy. 2007. A microbial diagnostic microarray technique for the sensitive detection and identification of pathogenic bacteria in a background of nonpathogens. *Anal. Biochem.* **360**:244–254.

Kou, X., Q. Wu, J. Zhang, and H. Fan. 2006. Rapid detection of noroviruses in fecal samples and shellfish by nucleic acid sequence-based amplification. *J. Microbiol.* **44**:403–408.

Kovacs, G. T. A. 2003. Electronic sensors with living cellular components. *Proc. IEEE* **91**:915–929.

Kramer, K., M. Fiedler, A. Skerra, and B. Hock. 2002a. A generic strategy for subcloning antibody variable regions from the scFv phage display vector pCANTAB 5 E into pASK85 permits the economical production of Fab fragments and leads to improved recombinant immunoglobulin stability. *Biosensor. Bioelectron.* **17**:305–313.

Kramer, M. F., T. B. Tims, and D. V. Lim. 2002b. Recovery of *Escherichia coli* O157:H7 from optical waveguides used for rapid biosensor detection. *J. Rapid Methods Automat. Microbiol.* **10**:93–106.

Lim, D. V., J. M. Simpson, E. A. Kearns, and M. F. Kramer. 2005. Current and developing technologies for monitoring agents of bioterrorism and biowarfare. *Clin. Microbiol. Rev.* **18**:583–607.

Malorny, B., P. T. Tassios, P. Radstrom, N. Cook, M. Wagner, and J. Hoorfar. 2003. Standardization of diagnostic PCR for the detection of foodborne pathogens. *Int. J. Food Microbiol.* **83**:39–48.

McMasters, S., and D. N. Stratis-Cullum. 2006. Evaluation of aptamers as molecular recognition elements for pathogens using capillary electrophoretic analysis. *Proc. SPIE* **6380,** 63800B. doi:10.1117/12.686357.

Mead, P. S., L. Slutsker, V. Dietz, L. F. McCaig, J. S. Bresee, C. Shapiro, P. M. Griffin, and R. V. Tauxe. 1999. Food-related illness and death in the United States. *Emerg. Infect. Dis.* **5:** 607–625.

Mozola, M. A. 2006. Genetics-based methods for detection of *Salmonella* spp. in foods. *J. AOAC Int.* **89:**517–529.

Nadal, A., A. Coll, N. Cook, and M. Pla. 2007. A molecular beacon-based real time NASBA assay for detection of *Listeria monocytogenes* in food products: role of target mRNA secondary structure on NASBA. *J. Microbiol. Methods* **68:**623–632.

Newsome, R. 2003. Dormant microbes: research needs. 2003. *Food Technol.* **57:**38–42.

Nugaeva, N., K. Gfeller, N. Backmann, H. Lang, M. Düggelin, and M. Hegner. 2005. Micromechanical cantilever array sensors for selective fungal immobilization and fast growth detection. *Biosens. Bioelectron.* **21:**849–856.

Panicker, G., D. R. Call, M. J. Krug, and A. K. Bej. 2004. Detection of pathogenic *Vibrio* spp. in shellfish by using multiplex PCR and DNA microarrays. *Appl. Environ. Microbiol.* **70:**7436–7444.

Petrenko, V. A., and V. J. Vodyanoy. 2003. Phage display for detection of biological threat agents. *J. Microbiol. Methods* **53:**253–262.

Petrenko, V. A., and I. B. Sorokulova. 2004. Detection of biological threats. A challenge for directed molecular evolution. *J. Microbiol. Methods* **58:**147–168.

Pieken, W., D. B. Olsen, F. Benseler, H. Aurup, and F. Eckstein. 1991. Kinetic characterization of ribonuclease-resistant 20-modified hammerhead ribozymes. *Science* **253:**314–317.

Richter, E. R. 1993. Biosensors: applications for dairy food industry. *J. Dairy Sci.* **76:**3114–3117.

Rider, T. H., M. S. Petrovick, F. E. Nargi, J. D. Harper, E. D. Schwoebel, R. H. Mathews, D. J. Blanchard, L. T. Bortolin, A. M. Young, J. Chen, and M. A. Hollis. 2003. A B cell-based sensor for rapid identification of pathogens. *Science* **301:**213–215.

Rodriguez-Lazaro, D., J. Lloyd, A. Herrewegh, J. Ikonomopoulos, M. D'Agostino, M. Pla, and N. Cook. 2004. A molecular beacon-based realtime NASBA assay for detection of *Mycobacterium avium* subsp. *Paratuberculosis* in water and milk. *FEMS Microbiol. Lett.* **237:**119–126.

Salazar, N. M., and G. Caetano-Anollés. 1996. Nucleic acid scanning by hybridization of enterohemorrhagic *Escherichia coli*

isolates using oligodeoxynucleotide arrays. *Nucleic Acids Res.* **24:**5056–5057.

Sarid, D. 1991. *Scanning Force Microscopy.* Oxford University Press, New York, NY.

Schneider, P., L. Wolters, G. Schoone, H. Schallig, P. Sillekens, R. Hermsen, and R. Sauerwein. 2005. Real-time nucleic acid sequence-based amplification is more convenient than real-time PCR for quantification of *Plasmodium falciparum*. *J. Clin. Microbiol.* **43:**402–405.

Teunis, P., K. Takumi, and K. Shinagawa. 2004. Dose response for infection by *Escherichia coli* O157:H7 from outbreak data. *Risk Anal.* **24:**401–407.

Tombelli, S., M. Minunni, and M. Mascini. 2007. Aptamers-based assays for diagnostics, environmental and food analysis. *Bio. Eng.* **24:**191–200.

Tsai, T. Y., W. J. Lee, Y. J. Huang, K. L. Chen, and T. M. Pan. 2006. Detection of viable enterohemorrhagic *Escherichia coli* O157 using the combination of immunomagnetic separation with the reverse transcription multiplex TaqMan PCR system in food and stool samples. *J. Food Prot.* **69:** 2320–2328.

Tuerk, C., and L. Gold. 1990. Systematic evolution of ligands by exponential enrichment: RNA ligands to bacteriophage T4 DNA polymerase. *Science* **249:**505–510.

Vivekananda, J., and J. L. Kiel. 2006. Anti-*Francisella tularensis* DNA aptamers detect tularemia antigen from different subspecies by aptamer-linked immobilized sorbent assay. *Lab Invest.* **86:**610–618.

Volokhov, D., A. Rasooly, K. Chumakov, and V. Chizhikov. 2002. Identification of *Listeria* species by microarray-based assay. *J. Clin. Microbiol.* **40:**4720–4728.

Willse, A., T. M. Straub, S. Wunschell, J. A. Small, D. R. Call, D. Daly, and D. P. Chandler. 2004. Quantitative oligonucleotide microarray fingerprinting of closely related *Salmonella enterica* isolates. *Nucleic Acids Res.* **32:**1848–1856.

Wilson, W. J., C. L. Strout, T. Z. DeSantis, J. L. Stilwell, A. V. Carrano, and G. L. Andersen. 2002. Sequence-specific identification of 18 pathogenic microorganisms using microarray technology. *Mol. Cell. Probes* **16:**119–127.

Yoo, J. H., J. H. Choi, S. M. Choi, D. Lee, W. S. Shin, W. Min, and C. C. Kim. 2005. Application of nucleic acid sequence-based amplification for diagnosis of and monitoring the clinical course of invasive aspergillosis in patients with hematologic diseases. *Clin. Infect. Dis.* **40:**392–398.

Pathogens and Toxins in Foods: Challenges and Interventions
Edited by V. K. Juneja and J. N. Sofos
© 2010 ASM Press, Washington, DC

Chapter 29

Molecular Subtyping and Tracking of Food-Borne Bacterial Pathogens

BRANDON A. CARLSON AND KENDRA K. NIGHTINGALE

While microorganisms that cause human food-borne illness include bacteria, viruses, protozoa, and fungi, the majority of previous studies and thus current available knowledge on molecular subtyping of food-borne pathogens have been focused on bacteria. As a result, we will focus our discussions in this chapter on bacterial food-borne pathogens and particularly those food-borne bacterial pathogens that have been associated with significant mortality. This chapter provides an overview on molecular subtyping methods, including conventional banding-based methods and novel DNA sequence-based methods, to molecularly confirm that isolates belong to a given food-borne bacterial pathogen and to discriminate among isolates belonging to a given food-borne bacterial pathogen. In light of recent highly publicized multistate outbreaks of human food-borne illness (e.g., CDC, 2006, 2007), there is a clear need to develop and implement rapid, highly discriminatory molecular subtyping methods that generate portable data in order to identify outbreaks of human food-borne illness and the food vehicle responsible for disease in real time.

Both phenotypic and molecular subtyping methods have been implemented widely to differentiate isolates belonging to a given food-borne bacterial pathogen beyond the species or subspecies level (Wiedmann, 2002). When coupled with epidemiological investigations, subtyping of bacterial isolates has routinely been used to (i) detect clusters of epidemiologically related cases of human food-borne illness and to identify an outbreak of food-borne illness (Swaminathan et al., 2006), (ii) control outbreaks of human food-borne illness by identifying the food vehicle responsible for cases (Graves et al., 2005), (iii) define and detect emerging clonal groups within a known food-borne pathogen (Matsumoto et al.,

2000) or newly emerged food-borne pathogens (Mohan Nair and Venkitanarayanan, 2006), (iv) gain insight into the evolution and population structure of a food-borne pathogen (Nightingale et al., 2005a), and (v) elucidate underlying genetic differences among subtypes within a food-borne pathogen that contribute to subtype-specific virulence and tissue specificity during an infection (Nightingale et al., 2005b; Pohl et al., 2006). An acceptable subtyping method should be highly discriminatory, reproducible, rapid, capable of high-throughput analysis, and practical with respect to cost and the amount of labor required to perform subtyping. In addition, an ideal subtyping method should be able to type all isolates representing a food-borne pathogen species and should be easy to standardize across laboratories as well as generate data that are simple to interpret and are portable between laboratories.

Phenotypic subtyping methods include serotyping, phage typing, biotyping, antimicrobial resistance profiling, and multilocus enzyme electrophoresis. Molecular subtyping methods rely on nucleic acids to differentiate bacterial isolates belonging to a given species or subspecies (Wiedmann, 2002). While molecular subtyping often represents a more discriminatory, reliable, and reproducible approach to characterizing food-borne pathogens, phenotypic subtyping, particularly serotyping, continues to provide clinically relevant subtype information for food-borne pathogens, including *Salmonella*, *Escherichia coli*, and *Listeria monocytogenes*. For the purposes of this chapter, we will limit our discussion to molecular subtyping methods and these methods will be grouped into two categories: (i) conventional banding-based or DNA fingerprint-based methods and (ii) DNA sequence-based methods.

Brandon A. Carlson and Kendra K. Nightingale • Department of Animal Sciences, Colorado State University, Fort Collins, CO 80523.

Application of the "species" concept as a taxonomic unit remains a fundamental debate in bacterial systematics due to the fact that the definition of a species by successful interbreeding, which is used for higher organisms, is not directly applicable to bacteria (Ochman et al., 2005). Bacteria reproduce by binary fission, an asexual method of reproduction, in which the chromosome is replicated and segregates and a single cell splits off into two daughter cells (Maiden, 2006). Each daughter cell, in theory, represents an exact clone of the parent cell; however, due to the errors in the DNA replication process (e.g., polymerase slippage and DNA mismatch repair), silent and beneficial mutations become accumulated in the population after several generations. As a result, unlike human DNA fingerprinting, each bacterial cell characterized by molecular subtyping will not be classified as a unique subtype. Molecular subtyping of bacterial isolates belonging to a given food-borne bacterial pathogen species will permit identification of isolates that are genetically distinct as well as clonal groups containing multiple isolates that are indistinguishable by a particular molecular subtyping approach (Wiedmann, 2002).

Initial theories of bacterial population structure were centered on the known asexual means of bacterial reproduction and the assumption that genetic variation would be solely propagated by vertical transmission (i.e., from parent to daughter cell) of mutational events that arose from DNA replication errors within a single parent cell (Maiden, 2006). Within recent years, it has become widely accepted that both closely related and unrelated bacterial cells can exchange genetic material via sexual means, a process referred to as horizontal or lateral gene transfer (Spratt and Maiden, 1999). While some bacterial pathogens appear to have evolved slowly by accumulation and vertical transmission of point mutations, others appear to have evolved more rapidly as a result of sexual promiscuity, allowing for more frequent horizontal gene transfer events (Feil et al., 2001). It is difficult to accurately reconstruct the evolutionary history of a group of taxa that has been highly affected by horizontal gene transfer, as gene sequences are mosaic in nature and phylogenetic trees inferred from different genes do not show congruent topologies (Feil et al., 2001).

The evolution of a pathogenic lifestyle from ancestors with a nonpathogenic or an environmental lifestyle can be attributed to the acquisition of virulence traits, like the ability to adhere to or invade host cells, spread between host cells intracellularly, sequester iron, and produce toxins (Kirsch et al., 2004; Schmidt and Hensel, 2004). Virulence genes that permit emergence of a pathogenic lifestyle are often acquired as clusters through horizontal gene transfer

by the uptake of mobile genetic elements, such as plasmids and transposons, or from bacteriophage-mediated insertions (Hacker et al., 1997; Maurelli, 2007). Locations in the genome containing clusters of virulence genes acquired by horizontal gene transfer are termed pathogenicity islands (PAIs), and PAIs are typically unique to pathogenic bacteria and are thus rarely found in nonpathogenic strains of the same or related bacterial species (Groisman and Ochman, 1996).

The locus of enterocyte effacement represents a hallmark PAI found among food-borne pathogens, and this PAI is present in human and animal enteropathogenic *E. coli* and in most enterohemorrhagic *E. coli*. The LEE encodes a type III secretion system and intimin, which facilitates intimate contact between the bacterium and host epithelial cells. The virulence factors in this PAI are also responsible for the formation of the characteristic attaching and effacing lesion (Kirsch et al., 2004). The presence of PAIs is also important in the pathogenesis of other food-borne diseases, such as listeriosis. Genes encoded on the 9-kb *Listeria* PAI 1 (i.e., *hly*, *plcA*, *plcB*, *actA*, *mpl*, and *prfA*) as well as the internalin genes (i.e., *inlA*, *inlB*, *inlC*, *inlC2*, *inlD*, *inlE*, *inlF*, *inlG*, and *inlH*) play an important role in the ability of *L. monocytogenes* to transverse host cell barriers, survive and multiply intercellularly, and spread intercellularly between hosts cells, allowing this pathogen to cause severe systemic infections (Vazquez-Boland et al., 2001). Interestingly, a modified version of this PAI, which enables pathogenicity in *L. monocytogenes* and *Listeria ivanovii*, is present in apathogenic *Listeria seeligeri*. Results from a recent study suggested that the PAI may have been present in the common ancestor of *L. monocytogenes*, *L. ivanovii*, *L. seeligeri*, *L. welshimeri*, and *Listeria innocua* and was lost in *L. innocua* and *L. welshimeri*, demonstrating that gene loss can also be a mechanism for bacterial lifestyle change (Schmid et al., 2005).

MECHANISMS OF GENETIC CHANGE

Mutations

A mutation can be defined as any heritable alteration in DNA, and fitness can be defined as the ability of a new genotype to survive and proliferate in a given ecological niche. While an exact copy of the chromosome is generated during the DNA replication process most of the time, occasionally mutations are introduced due to errors (e.g., polymerase slippage and mismatch repair) in the replication process (Graur and Li, 2000). Due to the random nature of mutation occurrence and the degeneracy of the genetic code, most mutations are deleterious or harmful to the cell because they confer reduced fitness in the preferred niche. Nondeleterious mutations and mutations that confer a selective advantage leading to enhanced fitness in a particular ecological niche become accumulated in a given population and function as the unit for measuring genetic change over time. Mutations can be generally categorized by the amount of DNA that is affected. Specifically, mutations can affect (i) a single nucleotide (and these mutations are referred to as point mutations or nucleotide substitutions) and (ii) several adjacent nucleotides or large segments of DNA (and these mutations include insertions or deletions, inversions, and horizontal gene transfer events) (Graur and Li, 2000).

There are two types of nucleotide substitutions—transitions and transversions. A substitution between two purines (e.g., A to G) or two pyrimidines (e.g., C to T) is known as a transition, while a transversion occurs when a purine is substituted for a pyrimidine (e.g., A to T) or a pyrimidine is substituted for a purine (Windham et al., 2005). Nucleotide substitutions can be further classified as synonymous or nonsynonymous, where synonymous mutations do not result in an amino acid change and nonsynonymous mutations (also referred to as missense mutations) cause a different amino acid to be encoded. Because synonymous mutations do not result in amino acid replacement, they are often referred to as silent mutations, and nonsynonymous mutations that do not confer a new phenotype are also considered to be silent. It has been estimated that bacteria accumulate between 0.0001 to 0.0002 synonymous mutations along the genome per generation (estimated through sequence comparisons) (Lawrence and Ochman, 1998; Ochman et al., 1999). Since nucleotide substitutions are accumulated slowly over time, they provide the most reliable information for classifying isolates into different clonal groups (Windham et al., 2005). A nucleotide substitution that changes a codon that originally encoded an amino acid to a stop codon and leads to the production of a truncated gene product is referred to as a nonsense mutation (Windham et al., 2005). Nonsense mutations have been hypothesized to lead to a lifestyle change from a pathogenic strain to an environmentally adapted strain for *L. monocytogenes* strains that produce a truncated and secreted form of InlA, a key virulence factor (Nightingale et al., 2005a, 2005b).

Mutations resulting in the addition or removal of one or more nucleotide base(s) are known as insertions and deletions, respectively. Although the evolution of bacterial genomes is thought to be biased toward deletion events (Ochman and Davalos, 2006), it is not possible to definitively determine whether sequence differences observed in multiple sequence alignments arose from an insertion or deletion event, and as a

result, these mutations are often referred to as "indels." Indels can involve a single nucleotide base, several adjacent nucleotide bases, or large segments of DNA.

Both the acquisition of new genes or gene clusters (e.g., PAIs and antibiotic resistance cassettes) and the loss of genes play an important role in the emergence of pathogenic traits (Ochman and Moran, 2001). For example, a large chromosomal deletion flanking the *cadA* region in *Shigella*, which encodes the lysine decarboxylase (LCD) system, provides an example of a pathoadaptive mutation, as the product of LCD (cadaverine) blocks the action of *Shigella* enterotoxins, and loss of cadaverine production thus leads to virulence (Day et al., 2001; Maurelli et al., 1998). Recent studies reported that naturally occurring LCD-negative Shiga toxin-producing *E. coli* are characterized by rearrangements and missing genes in the *cad* region and that complementation of these strains with an intact *cad* operon restored LCD activity and resulted in reduced efficiency of adhesion to tissue culture cells, suggesting LCD silencing is a pathoadaptive mutation in Shiga toxin-producing *E. coli* (Jores et al., 2006; Torres et al., 2005).

Instances where a base pair or a few base pairs are inserted or deleted, not in multiples of three, cause a shift from the normal reading frame and are referred to as frameshift mutations (Windham et al., 2005). Frameshift mutations disrupt the correct reading frame and lead to a premature stop codon 3′ of the frameshift mutation. Reversible frameshift mutations that occur in homopolymetric runs may provide a means for selective phase variation in bacterial pathogens (Kearns et al., 2004; Park et al., 2000; Segura et al., 2004; Theiss and Wise et al., 1997). For example, high-frequency reversible insertion and deletion mutations in a poly-T run within *flhA* result in populations of motile and nonmotile *Campylobacter coli* cells. Loss of flagella may facilitate evasion of the host immune system during a *C. coli* infection and energy conservation in environments where motility is not required (Park et al., 2000). Inversion mutations typically arise through site-directed intrachromosomal recombination events caused by a double-strand break and rejoining or crossover between strands in opposite orientation, and as a result, an entire segment of chromosomal DNA is reversed (Windham et al., 2005). Inversion mutations result in a reversal of an entire segment of chromosomal DNA and are not particularly common among bacteria but have been observed with *Salmonella enterica* serotype Typhi (Liu and Sanderson, 1996).

Horizontal Gene Transfer

Horizontal gene transfer compels more rapid evolutionary change. For instance, extensive horizontal gene transfer has been linked to the emergence of new pathogenic serogroups, such as *E. coli* O157:H7 (Perna et al., 2001), and pandemic clones, such as *Vibrio parahaemolyticus* serotype O3:K6 (Hurley et al., 2006). Complete genome sequencing for the nonpathogenic laboratory strain *E. coli* K-12 and pathogenic *E. coli* O157:H7 illustrated a significant amount of divergence between two strains within the same species. Specifically, comparison of *E. coli* K-12 and *E. coli* O157:H7 genome sequences revealed that while 4.1 million base pairs are conserved between these two genomes, the pathogenic strain carries an additional 1.34 million base pairs encoding >1,387 new genes (Perna et al., 2001; Schaechter, 2001). Horizontal gene transfer involves the exchange of chromosomal DNA segments ranging from 500 to upwards of 10,000 kb between different genomes. Horizontal gene transfer is unidirectional and may occur via three mechanisms, including conjugation (transfer of genetic information usually through a vector from a donor to a recipient cell [referred to as bacterial sex]), transduction (phage-mediated transfer of genetic information to bacterial cells), and transformation (acquisition of exogenous or naked DNA from the environment by a bacterial cell) (Windham et al., 2005).

Horizontal gene transfer may introduce genetic material from closely related (homologous horizontal gene transfer) or distantly related (nonhomologous horizontal gene transfer) donor organisms into a cell. DNA segments that have been introduced recently by horizontal gene transfer from a distantly related organism can be differentiated from ancestral DNA due to inconsistencies in nucleotide composition (GC content) and codon usage bias (Ochman et al., 2005). Since its divergence from the *Salmonella* lineage approximately 100 million years ago, the *E. coli* genome has been altered considerably through at least 234 horizontal gene transfer events (Lawrence and Ochman, 1998). DNA segments introduced by nonhomologous horizontal gene transfer will eventually become indistinguishable from ancestral DNA over time due to mutational biases of the recipient organism (Lawrence and Ochman, 1998). Plasmids that have become integrated into the bacterial chromosome through site-specific recombination are called prophage, and these bacteriophage-mediated insertions as well as deletion and duplication events appear to be largely responsible for genetic diversity among *E. coli* O157:H7 isolates, including the acquisition of genes encoding Shiga-like toxins I and II (Kudva et al., 2002; Shaikh and Tarr, 2003). Different bacterial pathogens can demonstrate considerable diversity of population structure, ranging from *Neisseria*, a species that shows almost free and rapid recombination, to *Salmonella*, a species that appears to be highly clonal and largely unaffected by recombination (Lan and Reeves, 2001).

Natural Selection

While genetic variation is introduced into a population by the occurrence of mutations as described above, natural selection is responsible for driving a new genotype to fixation in a population or removal of that genotype from the population (Windham et al., 2005). Most mutations are either deleterious, leading to reduced fitness in nature (and the new genotype will be eliminated by negative selection), or silent and thus do not change the fitness of a genotype in nature (and silent mutations may be maintained in the population by neutral selection or lost by genetic drift). Some mutations, however, confer enhanced fitness to a new genotype in a given ecological niche (i.e., host or tissue specificity), and these mutations will become fixed in the population by positive selection (Windham et al., 2005). Estimating the rate of synonymous (silent) and nonsynonymous (amino acid-changing) substitutions is an important tool for understanding molecular evolution of DNA sequences. While synonymous mutations do not influence natural selection, nonsynonymous mutations may indicate positive selection or adaptive evolution. Estimating the ratio of the number of nonsynonymous substitutions/nonsynonymous site to the number of synonymous substitutions/synonymous site (ω) among lineages within a phylogeny (genetically related organisms that share a recent common ancestor) may indicate adaptive evolution in specific lineages. In addition, studying ω ratios among codons may allow identification of specific amino acids that evolved by positive selection and lead to inferences about function limitations at certain amino acid sites (Yang et al., 2000).

Food-borne pathogens experience several selective pressures in order to survive and proliferate in a variety of ecological niches during their transmission along the food chain and during infection of mammalian hosts. As a result, genes belonging to a food-borne pathogen that diversified by positive selection are likely to play an important role in lifestyle changes, including survival upon exposure to environmental stresses and evasion of host immune responses. Specifically, results from a previous study showed that positive selection contributed to the diversification of key *L. monocytogenes* virulence genes (i.e., *actA* and *inlA*), and the fixation of specific amino acid residues through positive selection was consistent with epidemiological associations between *L. monocytogenes* clonal groups and their isolation from either human clinical cases or food (Nightingale et al., 2005a). A previous study also indicated that extracellular domains of intimin in *E. coli* O111 diversified by positive selection of amino acid residues (Tarr and Whittam, 2002).

DNA BAND-BASED MOLECULAR SUBTYPING METHODS

DNA band-based molecular subtyping methods are the most routinely employed molecular subtyping methods used to track and control human food-borne pathogens throughout the food continuum. Specifically, pulsed-field gel electrophoresis (PFGE) is considered to be the "gold standard" molecular subtyping approach for most food-borne pathogens, and this subtyping method has proven to be instrumental in the detection and control of food-borne illness outbreaks (Gerner-Schmidt et al., 2006). Generally, band-based subtyping methods involve a combination of PCR amplification of one or more DNA fragments and a restriction digest of amplicons or macro-restriction of bacterial DNA. Resultant DNA fragments are then separated by size via gel electrophoresis. Electrophoresed amplicons yield a banding pattern, which is often referred to as a "DNA fingerprint." Banding-pattern data can be interpreted similarly to the interpretation of bar codes found on a supermarket item, with isolates that have an identical banding pattern considered to be closely related and those with different banding patterns considered more distantly related. Clustering algorithms are available to assess the similarity of banding patterns by constructing a dendrogram. The dendrogram groups banding patterns according to percent similarity; however, additional manual interpretation is usually required to discriminate banding patterns that differ by a single or a few bands and shift in band position. While a dendrogram looks similar to a phylogenetic tree, it should not be used to infer the evolutionary history of a group of isolates, since nucleotide sequence data are required for these analyses.

Multiplex PCR

Simultaneous amplification of multiple genes or gene fragments is termed multiplex PCR, and this technique was first utilized to screen for the presence of several deletion-prone exons from the Duchenne muscular dystrophy locus (Chamberlain et al., 1988). Multiplex PCR is a rapid, convenient, and repeatable screening and subtyping assay that may be utilized to determine the presence or absence of multiple genes or gene fragments, which indicates the occurrence of mutations (e.g., indels, horizontal gene transfer events, or nucleotide substitutions) that were accumulated among different bacterial isolates. Given that multiplex PCR simultaneously amplifies several

loci in a single reaction, parameters such as (i) relative concentrations of the primer pairs designed to amplify each target; (ii) concentration of PCR buffer, magnesium chloride, and deoxynucleotides; and (iii) thermal cycling conditions must be optimized to ensure reproducible amplification of all targets (Henegariu et al., 1997). In addition, separation of multiplex PCR products may require the use of different agarose gel concentrations. Specifically, when PCR products differ from each other by at least 30 bp in length, separation on a 2 to 3% agarose gel will be sufficient; however, if PCR products differ by only a few base pairs in length, then the use of a 6 to 10% polyacrylamide gel may be necessary for adequate separation (Henegariu et al., 1997). Multiplex PCR has been used to molecularly confirm that bacterial isolates belong to a given species or serotype as well as to screen for the presence of specific virulence genes. For example, a six-gene multiplex PCR assay was previously utilized to distinguish genotypes among *E. coli* O157 isolates by screening for the presence of stx_1, stx_2, *eaeA*, $fliC_{h7}$, $rfbE_{O157}$, and *hlyA* (Childs et al., 2006). In addition, a recent study (Doumith et al., 2005) developed a multiplex PCR assay to group *L. monocytogenes* isolates into one of the major human disease-associated serotypes (i.e., 1/2a, 1/2b, 1/2c, and lineage I 4b).

RAPD

Random amplified polymorphic DNA (RAPD), also referred to as arbitrarily primed PCR, uses low-stringency PCR amplification with arbitrarily designed oligonucleotide primers to amplify random DNA sequences along the bacterial chromosome (Wang et al., 1993). Arbitrary primers used for RAPD analysis are comprised of short (typically 10- to 12-bp) random nucleotide sequences, and low-stringency amplification conditions are used to allow for some mismatches (Welsh and McClelland, 1991; Williams et al., 1990). Following PCR amplification, resultant PCR products are separated by agarose gel electrophoresis and visualized by staining, and computer software is then used to analyze the resulting banding patterns. RAPD is a relatively simple molecular subtyping method that requires equipment present typically in most laboratories (i.e., thermal cycler and gel electrophoresis equipment), and RAPD typing is relatively inexpensive to perform. However, the main disadvantage of RAPD is the inability to reproduce the banding patterns (Lukinmaa et al., 2004b). Primer-template interaction, secondary structure of the template DNA, annealing temperatures, and the DNA polymerase were identified as important factors that can affect the reproducibility of banding patterns (Tyler et al., 1997; Meunier and Grimont, 1993).

RAPD was utilized during a large outbreak of food-borne illness attributed to *E. coli* O157:H7 in Japan during the late 1990s to identify radish sprouts as the probable food vehicle responsible for the outbreak (Watanabe et al., 1999).

rep-PCR

Repetitive element sequence-based PCR (rep-PCR) is a molecular subtyping method that utilizes amplification and separation of interspersed repetitive DNA sequences to differentiate bacterial isolates (Versalovic et al., 1991). Interspersed repetitive sequence elements are a group of repeated sequences that are relatively short in length (<200 bp), found in noncoding regions, and typically widely distributed throughout prokaryotic genomes (Lupski and Weinstock, 1992). Two common examples of interspersed repetitive DNA sequences are the repetitive extragenic palindrome sequence, which ranges from 33 to 40 bp in size, and the larger enterobacterial repetitive intergenic consensus sequence, which is comprised of 124 to 127 bp, both of which were identified in *Salmonella enterica* serovar Typhimurium and *E. coli* (Hulton et al., 1991; Stern et al., 1984). The gram-positive version of the repetitive extragenic palindrome and enterobacterial repetitive intergenic consensus is the 154-bp BOX element that was ultimately discovered in the genome of *Streptococcus pneumonia* (Martin et al., 1992). The common presence of these repetitive DNA elements in the genomes of microorganisms enables rapid identification of a given bacterial species and intraspecific discrimination among strains (Versalovic et al., 1991). Briefly, DNA is isolated from the organism of interest, and PCR is then performed under high-stringency conditions utilizing a single or multiple primer pairs. PCR products are separated in an agarose gel, visualized, and analyzed using computer algorithms (Hiett, 2005). The throughput of rep-PCR was improved when the method was determined to be discriminatory when whole-cell methods were employed, eliminating the time required to isolate DNA (Woods et al., 1993).

rep-PCR is very practical, as it requires basic laboratory equipment, including a thermal cycler and gel electrophoresis equipment; however, prior knowledge of sequence data for a particular organism of interest is required. The benefits of a rapid, convenient, and discriminatory method such as the rep-PCR method prove useful in nosocomial settings as well as outbreak investigations (Tyler et al., 1997). During an evaluation of subtyping methods, rep-PCR was identified as the simplest and most rapid method that could be performed at a relatively low cost to distinguish *E. coli* O157:H7 from other *E. coli*

serotypes (Hahm et al., 2003). It is, however, imperative to employ optimal PCR and thermal cycling conditions to ensure reproducibility of the banding-pattern data (Tyler et al., 1997). To reduce interlaboratory variation, automated rep-PCR systems have been developed and marketed to further improve convenience, consistency, reproducibility, and user-friendliness (Healy et al., 2005). During an outbreak of *Salmonella enterica* serovar Saintpaul, rep-PCR differentiated outbreak cases from unrelated sporadic infections and assisted in the identification of the food vehicle responsible for the outbreak (Beyer et al., 1998).

AFLP

Amplified fragment length polymorphism (AFLP) is a molecular subtyping method that relies on selective PCR amplification of digested fragments from total genomic DNA, and this method can be utilized to obtain DNA fingerprinting data from samples of any origin or complexity (Blears et al., 1998). AFLP begins with restriction enzyme digestion of a small amount of purified genomic DNA with two restriction enzymes, including one enzyme with an average cutting frequency (e.g., EcoRI) followed by a second with a higher cutting frequency (e.g., MseI or TaqI). Double-stranded oligonucleotide adapters are then ligated to the digested DNA fragments. Ligated fragments are then subjected to two rounds of PCR amplification with adapter-specific primers that contain 3′ nucleotide extensions that permit amplification of DNA fragments only with nucleotides flanking the restriction sites complementary to primers. Fragments are then separated by polyacrylamide gel electrophoresis, and different banding patterns indicate that mutations are present within restriction sites (or sites adjacent to restriction sites) or complementary 3′ nucleotide extensions in adapter-specific primers or that indels are present within digested fragments (Blears et al., 1998; Savelkoul et al., 1999). A fluorescent variation of AFLP includes primers labeled with a fluorescent dye, and labeled PCR amplicons can be visualized by capillary electrophoresis using a DNA sequence analyzer or real-time PCR methods (Lukinmaa et al., 2004b).

AFLP typing is a broadly applicable and highly discriminatory molecular typing method that provides a whole genome analysis of an organism without the need for prior DNA template sequencing. Additionally, AFLP has the ability to detect more point mutations compared to other PCR-based typing methods (Blears et al., 1998). However, the software required to interpret the complex banding patterns produced by AFLP typing and the equipment needed for fluorescent AFLP applications are costly initial investments. As with other molecular typing methods, there is need for a standardization of laboratory reagents and methods to allow for comparison of AFLP patterns across laboratories and compilation of accurate genetic databases. AFLP was employed as the molecular subtyping method during two different outbreaks of *Salmonella* Typhimurium DT 126 in Australia and effectively discriminated between isolates from the two different outbreaks. However, AFLP was not able to differentiate all epidemiologically unrelated DT 126 isolates from sporadic cases and outbreak isolates (Ross and Heuzenroeder, 2005).

PCR-RFLP

PCR-restriction fragment length polymorphism (PCR-RFLP) is similar to AFLP, but PCR-RFLP is conducted in reverse order compared to AFLP. Specifically, in PCR-RFLP analysis, genes or gene fragments are amplified by PCR, and resulting PCR amplicons are then subjected to digestion by one or more restriction enzymes (Blears et al., 1998). Digested DNA fragments are electrophoresed, and resultant banding patterns are compared (Blears et al., 1998). The discriminatory ability of PCR-RFLP can be increased with the addition of multiple restriction enzymes (Hiett, 2005). There are two critical disadvantages to PCR-RFLP: (i) prior knowledge of sequence data for the targeted gene or genes is required to design primers and to identify discriminatory restriction enzymes and (ii) PCR-RFLP analyzes only a small portion of the genome (Blears et al., 1998; Hiett, 2005). PCR-RFLP analysis of highly polymorphic virulence or virulence-associated genes has been implemented to characterize the allelic variation in several food-borne pathogens. For example, PCR-RFLP analysis of *L. monocytogenes* virulence genes that play important roles in the intracellular lifecycle of this pathogen (i.e., *hly*, *inlA*, and *actA*) has been used to group isolates into major genetic lineages, and virulence gene allelic profiles ascertained through PCR-RFLP may be associated with strain-specific pathogenic potential (Wiedmann et al., 1997). Specifically, *L. monocytogenes* isolates carrying a 105-bp deletion of a praline-rich repeat in *actA* (*actA* type 3) were associated with a more invasive clinical manifestation of listeriosis (encephalitis) in cattle compared to *actA* allelic variants carrying four proline-rich repeats (*actA* type 4), which were more commonly linked to septicemia and fetal infections (Pohl et al., 2006). In addition, PCR-RFLP has been used to characterize allelic diversity within *eae*, which encodes intimin, for enteropathogenic and Shiga

toxin-producing *E. coli* (Ramachandran et al., 2003) as well as *stx* genes among Shiga toxin-producing *E. coli* (Eklund et al., 2002).

PFGE

PFGE is a molecular typing method that allows for the analysis and comparison of total bacterial DNA (including the bacterial chromosome and plasmid DNA). PFGE employs one or more restriction enzymes, referred to as "rare-cutters," which recognize specific DNA sequences and cut the DNA infrequently, yielding a relatively small number of DNA fragments that are large in size (>50 kb). These large DNA fragments are susceptible to mechanical shearing by pipetting action and progression through agarose gels during electrophoresis. In order to prevent DNA breakage, whole bacterial cells are embedded into an agarose plug prior to cell lysis (Lukinmaa et al., 2004b). Once cells are embedded into the agarose plug, they are subjected to lysis detergents and enzymes and extensively washed to remove remaining cellular or chemical contamination. After washing, DNA within the plug is digested with a restriction enzyme. Commonly used restriction enzymes utilized during PFGE include AscI, ApaI, AvrII, SpeI, XbaI, SmaI, SfiI, and NotI (Swaminathan et al., 2001). Following digestion, a portion of the digested plug is loaded into an agarose gel and separated using a unique electrophoresis apparatus.

The first method that successfully separated these large DNA fragments was the orthogonal-field alternating gel electrophoresis (OFAGE) system. This system relied on four electrodes oriented at 90° angles relative to each other and designed to separate DNA fragments ranging in size from 50 kb to >750 kb (Carle and Olson, 1984). The major disadvantage with the OFAGE system was lack of homogeneity of the electric field, resulting in distorted bands at the bottom of the lanes, preventing adequate comparisons (Bustamante et al., 1993). This dilemma was solved with the development of the contour-clamped homogeneous electric field electrophoresis system, which uses 24 electrodes evenly spaced in a hexagonal arrangement. This electrophoresis system produces uniform and homogeneous electric fields across the gel by constantly alternating current at 120° angles, eliminating distorted bands at the bottom of lanes. The contour-clamped homogeneous electric field apparatus can sufficiently separate a greater range of DNA fragments compared to the OFAGE system (< 50 kb to up to 2 Mb) (Chu et al., 1986).

The banding patterns produced from PFGE typing with one or more restriction enzymes can be complex, analysis and interpretation require sophisticated computer software, and manual evaluation is often required to distinguish highly similar banding patterns (e.g., banding patterns resulting from the shift of a single band). The discriminatory power of PFGE can be improved through the reanalysis of isolates within a cluster with the use of additional restriction enzymes. A generally accepted rule for the interpretation of PFGE data is that isolates are considered to be closely related when they differ only by up to three DNA fragments, possibly related when the difference is between four and six DNA fragments, and unrelated when the difference is greater than seven DNA fragments (Tenover et al., 1995).

PFGE is considered to be one of the most discriminatory molecular subtyping methods available due to its ability to probe an organism's whole genome without prior knowledge of genomic sequences (Hiett, 2005; Hahm et al., 2003; Lukinmaa et al., 2004a; Martin et al., 1996). The initial investment for the specialized PFGE equipment and analysis software is costly, but once equipment is obtained, it can be used to analyze several organisms. PFGE is a lengthy and labor-intensive method that can take 2 to 4 days before analysis is complete, increasing response times during an outbreak. As previously mentioned, it is imperative that all laboratories use standardized protocols to allow for interlaboratory comparison and database formation.

PFGE is considered the "gold standard" technique for typing of most food-borne bacterial pathogens (Silbert et al., 2003). The U.S. Centers for Disease Control and Prevention (CDC) has chosen PFGE as the standard method of molecular typing for the PulseNet program (Swaminathan et al., 2001). PulseNet is a national surveillance program, which was created by the CDC in collaboration with the Association of Public Health Laboratories (APHL), to provide scientists at public health laboratories with access to a searchable electronic database of PFGE patterns to facilitate real-time comparison of bacterial PFGE patterns isolated from cases of food-borne illness throughout the entire country (Swaminathan et al., 2001). The CDC, along with state public health laboratories, employed PFGE to identify epidemiologically and genetically related isolates associated with clinical cases during recent outbreaks of food-borne illness attributed to E. coli O157:H7 and Salmonella in fresh spinach and peanut butter, respectively (CDC, 2006, 2007). Numerous studies have concluded that PFGE is the most discriminatory technique available to type strains of E. coli O157:H7, L. monocytogenes, and Campylobacter jejuni, and after serotyping, PFGE typing is also very discriminatory for Salmonella isolates representing the same serotype (Fitzgerald et al., 2001; Foley et al., 2006;

Hahm et al., 2003; Hänninen et al., 1998; Heir et al., 2000; Krause et al., 1996; Liebana et al., 2001; Louie et al., 1996; Lukinmaa et al., 2004a, 2004b).

Ribotyping

Ribotyping is a molecular subtyping method that relies on the analysis of the genetic components that code for rRNA (i.e., 5S, 16S, and 23S rRNA). Generally, bacteria contain multiple rRNA operons at different positions along the chromosome, permitting both species-level identification and intraspecies discrimination by probing these genes (Stull et al., 1988). Because rRNA genes are located throughout the chromosome, ribotyping represents an effective approach to probe genetic diversity along the entire genome. rRNA genes are housekeeping genes that make up the ribosomes, which are involved in protein synthesis and are thus required for cell survival. Analysis of rRNA genes thus provides a stable target for assigning molecular subtypes, since rRNA genes are likely to diversify slowly. Ribotyping is completed by releasing total bacterial DNA (i.e., chromosomal and plasmid) by cell lysis, which is followed by digestion with a "frequent cutter" restriction enzyme. Digested DNA is electrophoresed to resolve DNA fragments by size, and DNA fragments are captured and immobilized on a nylon membrane. Once on the membrane, the DNA is hybridized with a chemically labeled rRNA operon probe. After hybridization, the membrane is extensively washed and treated with blocking buffer and conjugate. Unbound conjugate is removed, and a chemiluminscent substrate is added, allowing bands containing rRNA to be visualized (Lefresne et al., 2004; Stull et al., 1988). Unlike PFGE, none of the visible bands resulting from ribotyping will be from plasmid DNA because plasmids do not contain rRNA genes. Fragment pattern data are then normalized for band intensity and positioned against a molecular marker, and similarity analysis is used to classify isolates into a ribotype based on an existing database.

Although several restriction enzymes can be used for ribotyping, EcoRI has been reported to yield the most homogeneous distribution of DNA fragments, regardless of species (Bruce, 1996). With the exception of cell lysis, ribotyping may be fully automated if performed using the Riboprinter system (DuPont Qualicon) and thus requires minimal labor. The development of fully automated ribotyping systems improved the repeatability and reproducibility of interlaboratory ribotyping (Bruce, 1996). The Riboprinter can process eight samples in an 8-hour run, and up to 32 samples can be analyzed in a single day (Bruce, 1996). Disadvantages of automated ribotyping include costly initial investments for this highly specialized piece of equipment and the expensive reagents required for ribotyping. Since ribotyping relies on analyzing an evolutionarily stable set of genes, this approach has been used to gain insight into the population structure of food-borne pathogens, particularly *L. monocytogenes*. Specifically, ribotyping showed that *L. monocytogenes* isolates can be grouped into two major genetic lineages and a third minor genetic lineage (e.g., Wiedmann et al., 1997), and these genetic lineages demonstrate different epidemiological associations with distinct sources of isolation (Gray et al., 2004; Jeffers et al., 2001). In addition, large-scale ribotyping databases have been shown to be useful for identifying virulence-attenuated clonal groups within *L. monocytogenes* (Nightingale et al., 2007). Ribotyping has also proved to be a useful molecular subtyping technique to identify niches in a food processing plant where a particular *L. monocytogenes* strain persists over time (Lappi et al., 2004).

DNA SEQUENCE-BASED MOLECULAR SUBTYPING METHODS

DNA sequencing of one or more genes or the whole genome, probing the presence of specific repeat sequences, sequence-specific hybridizations, and differentiating allelic types from single-nucleotide polymorphisms represent DNA sequence-based molecular subtyping methods used to differentiate food-borne pathogens. DNA sequence-based molecular subtyping methods take advantage of the wealth of available full-genome sequence data for microorganisms and generally provide a greater level of discriminatory power to differentiate isolates compared to DNA band-based molecular subtyping methods. DNA sequence data are easily interpreted and can be readily transferred between laboratories. DNA sequence data can also be used to re-create the evolutionary history of a group of isolates and for population genetics studies. Specifically, DNA sequence data can be used to construct phylogenetic trees using a variety of approaches (i.e., distance, parsimony, Bayesian, and maximum likelihood methods), and these data can be used to determine to what degree horizontal gene transfer and positive selection contributed to the diversity of a given food-borne pathogen or virulence traits (Windham et al., 2005). In addition, statistical approaches have been combined with phylogenetic trees to identify clades (a group of closely related isolates) within a phylogenetic tree that are significantly associated with different sources of isolation. Statistically significant same-source isolate clades within a phylogenetic tree indicate that isolates obtained from

the same source may be more genetically related to each other compared to isolates obtained from different sources, suggesting niche adaptation of clonal groups (Nightingale et al., 2007).

MLST

The DNA sequence-based molecular subtyping approach selected to characterize a specific food-borne bacterial pathogen is dependent on the level of clonality and mechanisms of genetic change that contribute to the evolution of the particular organism of interest. Specifically, *E. coli* O157:H7 represents a recently emerged and thus highly clonal serogroup within pathogenic *E. coli*, and multilocus sequence typing (MLST) does not have sufficient discriminatory power to differentiate *E. coli* O157:H7 isolates (Noller et al., 2003a). On the other hand, MLST has been widely implemented as an effective molecular subtyping tool for more genetically diverse food-borne pathogens, such as *L. monocytogenes* (e.g., Nightingale et al., 2005a) and *Campylobacter jejuni* (Dingle et al., 2001). One of the main advantages of DNA sequence-based subtyping over band-based methods is that interpretation of sequence data is much less subjective and arbitrary than deciphering banding patterns, and sequence data are universally transferable between laboratories. In addition, data generated from MLST analysis may also be used to study the evolution and population structure of a group of taxa (Urwin and Maiden, 2003). Selection of loci to be targeted in development of a MLST scheme is not standardized, and as a result multiple MLST schemes have been developed for the same bacterial food-borne pathogen in several instances. For example, several MLST schemes have been developed to characterize *L. monocytogenes* isolates, based on different combinations of housekeeping and virulence genes (i.e., Cai et al., 2002; Meinersmann et al., 2004; Nightingale et al., 2005a; Salcedo et al., 2003; Ward et al., 2004).

MLST was conceptualized in the late 1990s as a universal method for characterizing human pathogens, and the first MLST scheme was developed for *Neisseria meningitis* (Maiden, 2006). The conceptual framework behind MLST was derived from MLEE, but MLST assigns alleles at each targeted loci based on DNA sequence data instead of the electrophoretic mobility of cellular enzymes (Maiden, 2006; Enright and Spratt, 1999; Urwin and Maiden, 2003). MLST schemes typically include partial or full nucleotide sequences, ranging from 450 to 600 nucleotide bases in length, from multiple housekeeping genes distributed around the chromosome. Unique combinations of polymorphisms at each locus are used to assign alleles, and unique combinations of alleles are used to assign MLST types to each bacterial isolate characterized (Maiden, 2006; Enright and Spratt, 1999; Urwin and Maiden, 2003). Housekeeping genes have been targeted primarily for MLST schemes because they are more likely to have evolved primarily by accumulation of point mutations and are thus less likely to have been influenced by positive selection and horizontal gene transfer, making these genes more suitable to re-create the evolutionary history of a given group of isolates (Maiden, 2006). World Wide Web-based databases have been designed to promote the exchange of MLST data (e.g., www.mlst.net and www.pathogentracker.net), and these databases promote the global exchange of MLST data for several food-borne pathogens (e.g., *L. monocytogenes*, *Salmonella*, *E. coli*, *Campylobacter jejuni*, *Bacillus cereus*, and *Staphylococcus aureus*). Additionally, www.mlst.net provides a Web-based platform for analysis of MLST data, including allele assignment, identification of clonal complexes, measurement of linkage disequilibrium, and generation of simple phylogenetic trees.

Studies have demonstrated that incorporating virulence genes in a MLST scheme allowed improved discriminatory power to differentiate closely related isolates and provided important insight into the evolution of virulence (e.g., Nightingale et al., 2005a; Ward et al., 2004). In addition, inclusion of virulence genes in MLST typing schemes has revealed insight into strain-specific virulence differences, host specificity, and tissue specificity within an infected host. For example, sequencing the 3′ portion of *inlA*, a key *L. monocytogenes* virulence gene that plays an important role in crossing the mucosal and possibly placental borders (Lecuit et al., 1997; 2004), revealed multiple distinct point and frameshift mutations, leading to premature stop codons and thus the production of a truncated and secreted gene product (InlA) in *L. monocytogenes* (Nightingale et al., 2005a, 2005b; Orsi et al., in press). Interestingly, *L. monocytogenes* isolates carrying these premature stop codon mutations in *inlA* demonstrated attenuated invasion efficiency in human intestinal epithelial cells and were overrepresented among food isolates but rarely linked to human disease (Nightingale et al., 2005a, 2005b). MLST studies for *Campylobacter jejuni* showed that some clonal complexes within this species are associated exclusively or predominantly with specific host populations, including clonal complexes that were associated with humans and specific animal host species, such as sheep or poultry (Colles et al., 2003; Dingle et al., 2002). Similarly, Luan et al. (2005) used MLST analysis to describe a specific clonal complex among invasive group B *Streptococcus* isolates

associated with neonatal infections but which rarely caused invasive disease in adults.

MLVA

While MLST has been shown to be a powerful molecular subtyping tool to differentiate isolates belonging to certain food-borne pathogens (e.g., *L. monocytogenes*), MLST fails to discriminate isolates belonging to other food-borne pathogens, such as *E. coli* O157:H7 isolates (Noller et al., 2003a). Bacterial genome sequences are known to contain a large number of repetitive DNA elements, and they appear to be rich in tandem repeat (TR) sequences. TRs may be clustered in one area of the genome or dispersed throughout the whole genome, and loci that include TRs are typically among the most variable regions within a bacterial genome. Specifically, TRs are targets for genetic events such as DNA polymerase slippage or dissociation, mismatch repair errors, and recombination, which results in TR size changes and are referred to as variable-number TRs (VNTRs). Frequent occurrences of these genetic events result in a high degree of polymorphism within loci containing VNTRs, and probing VNTRs across multiple loci may differentiate isolates belonging to highly monomorphic bacterial pathogens (Lindstedt, 2005). Multiple-locus variable-number tandem repeat analysis (MLVA) is a PCR-based approach that relies on the detection of TR sequences in multiple loci to classify bacterial isolates into subtypes (Keim et al., 2000). A full-genome sequence is usually required to identify VNTRs, and an MLVA typing scheme typically includes the use of several PCR primer sets designed to flank VNTR sequences located either in open reading frames or noncoding regions. The size of PCR amplicons is then determined by either conventional or capillary gel electrophoresis. MLVA assays may be further optimized by amplification of VNTR loci with fluorescent-labeled primers and then performing multiplexed separation of labeled VNTRs via capillary electrophoresis, where software recognizes peaks of different sizes and colors to assign subtypes (Lindstedt, 2005).

Most of the early work using VNTRs as genetic markers was performed for potential bioterrorism agents, such as *Bacillus anthracis* (Keim et al., 2000) and *Yersinia pestis* (Klevytska et al., 2001), that emerged relatively recently and are thus highly monomorphic. More recently, MLVA typing schemes have also been developed and implemented to characterize food-borne bacterial pathogens, such as *E. coli* O157:H7 (Lindstedt et al., 2004; Noller et al., 2003a, 2003b; 2006), *Clostridium perfringens* (Sawires and Songer, 2005), *Salmonella enterica* serotype Enteritidis (Boxrud et al., 2007), *Salmonella enterica* serotype Typhimurium (Lindstedt et al., 2004), and *L. monocytogenes* (Murphy et al., 2006). A standardized protocol for MLVA characterization of Shiga toxin-producing *E. coli* O157 based on nine VNTRs was recently proposed for public health laboratories that participate in PulseNet (Hyytia-Trees et al., 2006). MLVA appears to be a particularly useful subtyping tool to study the short-term evolution of food-borne pathogens due to the rapid evolution of VNTRs. For example, a MLVA typing scheme developed for *E. coli* O157:H7 correctly identified isolates from outbreaks, while discriminating those from sporadic cases (Noller et al., 2003b). In addition, MLVA analysis of *E. coli* O157:H7 isolates was able to differentiate isolates that grouped into the same PFGE type (Noller et al., 2003b). Similar to MLST typing schemes, selection of loci containing VNTRs is subjective, and inclusion of highly unstable loci may fail to accurately distinguish bacterial genotypes. The increased number of available genome sequences belonging to a pathogenic bacterial species will facilitate identification of informative and stable VNTRs and design of primer sequences. While MLVA shows great promise to be useful for detecting outbreaks of food-borne illness and identifying a food vehicle responsible for a given outbreak, this method has limited value for long-term molecular epidemiology or evolution studies. A TR database (http://minisatellites.u-psud.fr) and software (Benson, 1999) are available to facilitate identification of stable VNTRs and placement of primers to flank these sequences.

Genome Sequencing

The Sanger sequencing method has functioned as the workhorse for genome sequencing projects for the last 30 years, including the human genome-sequencing project. Currently, more than 500 microbial genome sequences have been completed, and many more microbial sequencing projects are in process (http://cmr.tigr.org). The massive expansion of available genome sequence data has provided the foundation for the "omics" era, leading to a surge of interest in projects directed at the probing transcriptome and proteome. Advances in genome sequencing technologies within the next 5 years are expected to reduce the cost of genome sequencing by at least 100-fold, and sequencing technological advances are projected to move toward a $1,000 human genome sequence within the next 10 years (Chan, 2005).

Several novel sequencing technologies have recently been introduced, and these methods are significantly less costly, time consuming, and labor intensive compared to the current gold standard Sanger method (Metzker, 2005). New sequencing technologies can be grouped into two major categories, including combined amplification and sequencing approaches (e.g., Bennett, 2004; Margulies et al., 2005; Shendure et al., 2004) and single-molecule approaches (e.g., Braslavsky et al., 2003). The 454 Life Sciences (Margulies et al., 2005), Polony (Shendure et al., 2004), and Solexa (Bennett, 2004) sequencing methods rely on high-throughput in vitro DNA amplification methods, rather than the labor-intensive in vivo cloning approach used for shotgun sequencing by the Sanger method, and involve massive parallel sequencing on an immobilized platform (Hall, 2007). The first described single-molecule sequencing approach (Braslavsky et al., 2003) involves hybridization of DNA to complementary primers bound to a silica surface followed by primer extension of Cy3- and Cy5-labeled nucleotides, and a camera mounted on a microscope captures each base as it is added (Hall, 2007). These new sequencing methods, however, have certain limitations, including short read lengths, lack of paired-end reads and sequencing artifacts in homopolymetric runs that may complicate assembly of contigs, and authentication of point and frameshift mutations resulting in premature stop codons (Goldberg et al., 2006; Hall, 2007).

At least one full-genome sequence is available for most food-borne pathogens (http://cmr.tigr.org), and initial studies comparing genome sequences from pathogenic and nonpathogenic strains belonging to the same species or genus have been conducted, including comparisons of *E. coli* K-12 and *E. coli* O157:H7 genomes (Perna et al., 2001) as well as *L. innocua* and *L. monocytogenes* genomes (Glaser et al., 2001). Results from these initial comparative genomics studies permitted identification of virulence or virulence-associated genes that play a role in the pathogenesis of human food-borne diseases and provided important insight into the evolution of a pathogenic lifestyle (Glaser et al., 2001; Perna et al., 2001). Recent and future projected advances in genome sequencing technologies, including reduced cost and higher throughput, will facilitate genome sequencing for multiple isolates representing a given food-borne pathogen for large-scale comparative genomics studies (Hall, 2007). At the time this book chapter was prepared, *L. monocytogenes* represents one of the most highly sequenced human pathogens, and currently, two finished (with gap closure) and two unfinished (without gap closure) *L. monocytogenes* genome sequences were completed using the Sanger method (Nelson et al., 2004; Glaser et al., 2001). In addition, draft sequences were recently released for 18 *L. monocytogenes* isolates, which were sequenced using the 454 Life Sciences technology (http://www.broad.mit.edu).

The explosion of microbial genome sequencing projected for the near future using the newest sequencing technologies will provide an abundance of genome sequence data for comparative genomics analyses and functional genomics studies. The availability of a significant number of genome sequences belonging to a given food-borne pathogen will not only allow the identification of new genes involved in various processes such as responses to stresses encountered in host and nonhost environments but will allow us to gain fundamental biological knowledge of how these genes evolved and how they function. For example, positive selection and horizontal gene transfer are likely to play a key role in the adaptation of strains within a given food-borne pathogen to be better suited to either infect mammalian hosts or persist in a nonhost environment such as food (Nightingale et al., 2005a). The availability of multiple genome sequences belonging to a given food-borne pathogen will thus allow alignment of orthologs and subsequent genome-level analysis to identify allelic types and amino acid residues that may confer enhanced fitness of certain strains or clonal groups in a particular ecological niche.

Microarrays

Full-genome sequences permit the design of microarrays, which can be used to analyze basically every open reading frame in the reference genome (for which a genome sequence is available) in a single hybridization experiment. Microarrays represent a high-throughput miniaturized dot blot assay, in which fluorescently labeled RNA, cDNA, or DNA probes are hybridized to a membrane or coated glass slide spotted with thousands of long DNA oligonucleotides or PCR products, which have been designed to represent each gene in the reference strain genome. Microarrays are imaged following posthybridization washing treatments using a high-resolution scanner (Call et al., 2003; Schoolnik, 2002; Ye et al., 2001). Microarrays have been utilized predominantly to assay the transcriptome (through RNA or cDNA probe experiments) of strains grown under specific conditions to identify genes that are differentially expressed under various conditions and to define the regulon of transcriptional regulator genes (Lockhart and Winzeler, 2000). Microarrays also show utility as a whole-genome-level molecular subtyping tool by comparing the gene profiles across isolates to identify genes that are present or absent in a group of isolates belonging to the same species or closely related species through DNA probe experiments. DNA probe hybridization experiments are limited, however, such that they can recognize only genes shared between the reference strain (used to design

the microarray) and test strains and genes that are unique to the reference strain and thus cannot identify genes that are unique to the test strain (Ochman and Santos, 2005).

However, microarrays provide a practical, rapid, and high-throughput approach to inventory genes that comprise the genomes for bacterial strains belonging to the same species and bacterial strains representing closely related pathogenic and nonpathogenic species. Specifically, DNA microarrays have been used to identify genes that are present or absent across genomes of isolates belonging to the same food-borne pathogen and to identify single-nucleotide polymorphisms (SNPs) at the genomic level for highly monomorphic organisms. DNA probe microarray experiments have been used to identify genes that are conserved among isolates belonging to a bacterial food-borne pathogen, and results from these studies show that the percentage of shared genes varies greatly, ranging from 54% for *Salmonella enterica* isolates representing the six described subspecies (Chan et al., 2003) to 99% for *Vibrio cholerae* isolates (Dziejman et al., 2002). Whole-genome microarray analyses revealed considerable differences in genome contents for *L. monocytogenes* isolates representing the two major genetic lineages (i.e., lineages I and II) within this pathogen (Doumith et al., 2004; Zhang et al., 2003), consistent with the hypothesis that these two divergent lineages may represent distinct subspecies (Nightingale et al., 2005a). In a recent study, Zhang et al. (2006) developed a comparative genome sequencing microarray to probe genetic diversity at the nucleotide level by screening for the presence of SNPs in 1,199 *E. coli* O157 chromosomal genes and in the large virulence plasmid pO157. A total of 906 SNPs among 11 *E. coli* O157 isolates selected to represent *E. coli* O157 isolates from diverse geographical origins were identified. Among these 906 SNPs identified, 248 SNPs appeared to be informative (able to differentiate two closely related strains) for genotyping *E. coli* O157 at the population level (Zhang et al., 2006). Additionally, many of these 248 informative SNPs were found in genes encoding known *E. coli* O157 virulence factors, including genes that encode flagella and fimbriae, adhesion molecules, and the type III secretion system (Zhang et al., 2006).

SNP Typing

Completion of the human genome sequencing project provided the impetus behind the current drive to develop and implement high-throughput and cost-effective assays to detect SNPs as genetic markers associated with a given biological trait (Gut, 2004).

SNP typing assays have also been used to detect food-borne pathogens (Best et al., 2006), provide information on the microevolution of food-borne pathogens (Hommais et al., 2005), and differentiate strains associated with the majority of human food-borne illness from virulence-attenuated strains with limited ability to cause disease (Ducey et al., 2007). Comparisons of sequence data (i.e., MLST alignments, whole-genome sequence alignments, and comparative genome sequencing microarray data) can be used to identify SNPs that differentiate a group of bacterial isolates. Computer algorithms have been developed to identify highly informative SNPs for bacterial pathogens that have MLST data deposited in MLST databases (Robertson et al., 2004). DNA sequencing methods are limited as a molecular subtyping tool by cost and the amount of time required to determine the nucleotide sequence of several gene fragments (i.e., for MLST typing) or complete genomes, and these methods cannot be multiplexed. In addition, the majority of nucleotides sequenced will be static or invariant and are thus noninformative for subtyping purposes. SNP typing assays may be developed to directly target nucleotides that have been shown to discriminate allelic types. Another major advantage of SNP typing over conventional banding-based or other DNA sequence-based molecular subtyping techniques is that SNP genotyping assays can be designed to target slowly accumulated genetic variations in protein-coding genes (e.g., synonymous mutations), providing the most reliable inference of genetic relatedness.

A large number of SNP typing methodologies have been developed over recent years, which incorporate different molecular mechanisms (e.g., allele specific hybridization, primer extension or minisequencing, oligonucleotide ligation, and invasive cleavage), readouts (e.g., fluorescence, luminescence, and mass determination), and assay formats (e.g., homogeneous solution or solid support) to differentiate allelic types (Sorbrino et al., 2005). For an in-depth review on SNP typing methodologies, please refer to Sorbrino et al. (2005). SNP typing technologies that have multiplexing capabilities, such as primer extension or minisequencing assays have the greatest potential utility for high-throughput and cost-effective genotyping. In addition, SNP typing methods that rely on equipment and software that is commonplace in research, clinical, and diagnostic laboratories (i.e., instruments to perform real-time PCR, DNA sequencing, and microarray slide scanning) are practical for routine SNP typing of bacterial food-borne pathogens.

CONCLUDING REMARKS

DNA band-based molecular subtyping approaches continue to represent the most common molecular subtyping techniques employed to differentiate isolates belonging to a food-borne pathogen beyond the species level. In combination with epidemiological investigations, PFGE typing and the PulseNet network have provided critical information to recognize and diffuse several multistate outbreaks of food-borne illness (e.g., CDC, 2006, 2007). PFGE typing currently represents the most highly standardized band-based molecular subtyping approach to differentiate food-borne pathogens. More recently, DNA sequence-based approaches have, however, provided new perspectives on the evolution and population structure of human food-borne pathogens. In addition, DNA sequence-based subtyping techniques (particularly MLST and MLVA typing schemes) are becoming more standardized across laboratories and databases that contain DNA sequenced-based molecular subtyping data continue to rapidly expand. Multiplexed SNP typing assays that differentiate allelic types based on SNPs present within housekeeping genes known to diversify on an evolutionary time scale show great promise to characterize food-borne pathogens.

REFERENCES

Bennett, S. 2004. Solexa Ltd. *Pharmacogenomics* 5:433–438.

Benson, G. 1999. Tandem repeats finder: a program to analyze DNA sequences. *Nucleic Acids Res.* 27:573–580.

Best, E. L., A. J. Fox, R. J. Owen, J. Cheesbrough, and F. J. Bolton. 2006. Specific detection of *Campylobacter jejuni* from faeces using single nucleotide polymorphisms. *Epidemiol. Infect.* 17:1–8.

Beyer, W., F. M. Mukendi, P. Kimmig, and R. Böhm. 1998. Suitability of repeititive-DNA-sequence-based PCR fingerprinting for characterizing epidemic isolates of *Salmonella enterica* serovar Saintpaul. *J. Clin. Microbiol. Biotechnol.* 36:1549–1554.

Blears, M. J., S. A. De Grandis, H. Lee, and J. T. Trevors. 1998. Amplified fragment length polymorphism (AFLP): a review of the procedure and its applications. *J. Ind. Microbiol. Biotechnol.* 21:99–114.

Boxrud, D., K. Pederson-Gulrud, J. Wotton, C. Medus, E. Lyszkowicz, J. Besser, and J. M. Bartkus. 2007. Comparison of multiple-locus variable-number tandem repeat analysis, pulsed-field gel electrophoresis, and phage typing for subtype analysis of *Salmonella enterica* serotype Enteritidis. *J. Clin. Microbiol.* 45:536–543.

Braslavsky, I., B. Hebert, E. Kartalov, and S. R. Quake. 2003. Sequence information can be obtained from single DNA molecules. *Proc. Natl. Acad. Sci. USA* 100:3960–3964.

Bruce, J. 1996. Automated system rapidly identifies and characterizes microorganisms in food. *Food Technol.* 50:77–81.

Bustamante, C., S. Gurrieri, and S. B. Smith. 1993. Towards a molecular description of pulsed-field gel electrophoresis. *Trends Biotechnol.* 11:23–30.

Cai, S., D. Y. Kabuki, A. Y. Kuaye, T. G. Cargioli, M. S. Chung, R. Nielsen, and M. Wiedmann. 2002. Rational design of DNA sequence-based strategies for subtyping *Listeria monocytogenes*. *J. Clin. Microbiol.* **40:**3319–3325.

Call, D. R., M. K. Borucki, and F. J. Loge. 2003. Detection of bacterial pathogens in environmental samples using DNA microarrays. *J. Microbiol. Methods* **53:**235–243.

Carle, G. F., and M. V. Olson. 1984. Separation of chromosomal DNA molecules from yeast by orthogonal-field-alternation gel electrophoresis. *Nucleic Acids Res.* **12:**5647–5664.

CDC. 2007. Multistate outbreak of *Salmonella* serotype Tennessee infection associated with peanut butter—United States, 2006–2007. *MMWR Morbid. Mortal. Wkly. Rep.* **56:**521–524.

CDC. 2006. Ongoing multistate outbreak of *Escherichia coli* serotype O157:H7 infections associated with consumption of fresh spinach—United States, September 2006. *MMWR Morbid. Mortal. Wkly. Rpt.* **55:**1045–1046.

Chamberlain, J. S., R. A. Gibbs, J. E. Ranier, P. N. Nguyen, and C. T. Caskey. 1988. Deletion screening of the Duchenne muscular dystrophy locus via multiplex DNA amplification. *Nucleic Acids Res.* **16:**11141–11156.

Chan, E. Y. 2005. Advances in sequencing technology. *Mutat. Res.* **573:**13–40.

Chan, K., S. Baker, C. C. Kim, C. S. Detweiler, G. Dougan, and S. Falkow. 2003. Genomic comparison of *Salmonella enterica* serovars and *Salmonella bongori* by use of an *S. enterica* serovar Typhimurium DNA microarray. *J. Bacteriol.* **185:**553–563.

Childs, K. D., C. A. Simpson, W. Warren-Serna, G. Bellenger, B. Centrella, R. A. Bowling, J. Ruby, J. Stefanek, D. J. Vote, T. Choat, J. A. Scanga, J. N. Sofos, G. C. Smith, and K. E. Belk. 2006. Molecular characterization of *Escherichia coli* O157:H7 hide contamination routes: feedlot to harvest. *J. Food Prot.* **69:**1240–1247.

Chu, G., D. Vollrath, and R. W. Davis. 1986. Separation of large DNA molecules by contour clamped homogeneous electric fields. *Science* **234:**1582–1585.

Colles, F. M., K. Jones, R. M. Harding, and M. C. J. Maiden. 2003. Genetic diversity of *Campylobacter jejuni* isolates from farm animals and the farm environment. *Appl. Environ. Microbiol.* **69:**7409–7413.

Day, W. A., R. E. Fernandez, and A. T. Maurelli. 2001. Pathoadaptive mutations that enhance virulence: genetic organization of the *cadA* regions of *Shigella* spp. *Infect. Immun.* **69:**7471–7480.

Dingle, K. E., F. M. Colles, D. R. Wareing, R. Ure, A. J. Fox, F. E. Bolton, and H. J. Bootsma. 2001. Multilocus sequencing typing system for *Campylobacter jejuni*. *J. Clin. Microbiol.* **39:**14–23.

Dingle, K. E., F. M. Colles, R. Ure, J. A. Wagenaar, B. Duim, F. J. Bolton, A. J. Fox, D. R. A. Wareing, and M. C. J. Maiden. 2002. Molecular characterization of *Campylobactor jejuni* clones: a basis for epidemiologic investigation. *Emerg. Infect. Dis.* **8:**949–955.

Doumith, M., C. Jacquet, P. Gerner-Smidt, L. M. Graves, S. Loncarevic, T. Mathisen, A. Morvan, C. Salcedo, M. Torpdahl, J. A. Vazquez, and P. Martin. 2005. Multicenter validation of a multiplex PCR assay for differentiating the major *Listeria monocytogenes* serovars 1/2a, 1/2b, 1/2c, and 4b: toward an international standard. *J. Food Prot.* **68:**2648–2650.

Doumith, M., C. Cazalet, N. Simones, L. Frangeul, C. Jacquet, F. Kunst, P. Martin, P. Cossart, P. Glaser, and C. Buchrieser. 2004. New aspects regarding evolution and virulence of *Listeria monocytogenes* revealed by comparative genomics and DNA arrays. *Infect. Immun.* **72:**1072–1083.

Ducey, T. F., B. Page, T. Usgaard, M. K. Borucki, K. Pupedis, and T. J. Ward. 2007. A single-nucleotide-polymorphism-based

multilocus genotyping assay for subtyping lineage I isolates of *Listeria monocytogenes*. *Appl. Environ. Micrbiol.* **73:**133–147.

Dziejman, M., E. Balon, D. Boyd, C. M. Fraser, J. F. Heidelberg, and J. J. Mekalanos. 2002. Comparative genomics of *Vibrio cholerae*: genes that correlate with cholera endemic and pandemic disease. *Proc. Natl. Acad. Sci. USA* **99:**1556–1561.

Eklund, M., K. Leino, and A. Siitonen. 2002. Clinical *Escherichia coli* strains carrying *stx* genes: *stx* variants and *stx*-positive virulence profiles. *J. Clin. Microbiol.* **40:**4585–4593.

Enright, M. C., and B. G. Spratt. 1999. Multilocus sequence typing. *Trends Microbiol.* **7:**482–487.

Feil, E. J., E. C. Holmes, D. E. Bessen, M. Chan, N. P. J. Day, M. C. Enright, R. Goldstein, D. W. Hood, A. Kalia, C. E. Moore, J. Zhou, and B. G. Spratt. 2001. Recombination within natural populations of pathogenic bacteria: short-term empirical estimates and long-term phylogenetic consequences. *Proc. Natl. Acad. Sci. USA* **98:**182–187.

Fitzgerald, C., L. O. Helsel, M. A. Nicholson, S. J. Olsen, D. L. Swedlow, R. Flahart, J. Sexton, and P. I. Fields. 2001. Evaluation of methods for subtyping *Campylobacter jejuni* during an outbreak involving a food handler. *J. Clin. Microbiol.* **39:**2386–2390.

Foley, S. L., D. G. White, R. F. McDermott, R. D. Walker, B. Rhodes, P. J. Fedorka-Cray, S. Simjee, and S. Zhao. 2006. Comparison of subtyping methods for differentiating *Salmonella enterica* serovar Typhimurium isolates obtained from animal sources. *J. Clin. Microbiol.* **44:**3569–3577.

Gerner-Schmidt, P., K. Hise, J. Kincaid, S. Hunter, S. Rolando, E. Hyytia-Trees, E. M. Ribot, B. Swaminathan, and Pulsenet Taskforce. 2006. PulseNet USA: a five-year update. *Foodborne Pathog. Dis.* **3:**9–19.

Glaser, P., L. Frangeul, C. Buchrieser, C. Rusnoik, A. Amend, F. Baquero, P. Berche, H. Bloecker, P. Brandt, T. Chakraborty, A. Charbit, F. Chetouani, E. Couvelin, A. de Daruvar, P. Dehoux, E. Domann, G. Dominguez-Bernal, E. Duchaud, L. Durant, O. Dussurget, K. D. Entian, H. Fsihi, F. G. Portillo, P. Garrido, L. Gautier, W. Goebel, N. Gomez-Lopez, T. Hain, J. Hauf, D. Jackson, L. M. Jones, U. Kaerst, J. Kreft, M. Kuhn, F. Kunst, G. Kurapkat, E. Madueno, A. Maitournam, J. M. Vicente, E. Ng, H. Nedjari, G. Nordsiek, S. Novella, B. de Pablos, J. C. Perez-Diaz, R. Purcell, B. Remmel, M. Rose, T. Schlueter, N. Simones, A. Tierrez, J. A. Vaszquez-Boland, H. Voss, J. Wehland, and P. Cossart. 2001. Comparative genomics of *Listeria* species. *Science* **294:**849–852.

Goldberg, S. M., J. Johnson, D. Busam, T. Feldblyum, S. Ferriera, R. Friedman, A. Halpern, H. Khouri, S. A. Kravitz, F. M. Lauro, K. Li, Y. Rogers, R. Strausberg, G. Sutton, L. Tallon, T. Thomas, E. Venter, M. Frazier, and J. C. Venter. 2006. A Sanger/pyrosequencing hybrid approach for the generation of high-quality draft assemblies of marine microbial genomes. *Proc. Natl. Acad. Sci. USA* **103:**11240–11245.

Graur, D., and W. Li. 2000. Genes, genetic codes, and mutation, p. 5–38. *In* D. Graur and W. Li (ed.), *Fundamentals of Molecular Evolution*, 2nd ed. Sinauer Associates, Sunderland, MA.

Graves, L. M., S. B. Hunter, A. R. Ong, D. Schoonmaker-Bopp, K. Hise, L. Kornstein, W. E. DeWitt, P. S. Hayes, E. Dunne, P. Mead, and B. Swaminathan. 2005. Microbiological aspects of the investigation that traced the 1998 outbreak of listeriosis in the United States to contaminated hot dogs and establishment of molecular subtyping-based surveillance for *Listeria monocytogenes* in the PulseNet network. *J. Clin. Microbiol.* **43:**2350–2355.

Gray, M. J., R. N. Zadoks, E. D. Fortes, B. Dogan, S. Cai, Y. Chen, V. N. Scott, D. E. Gombas, K. J. Boor, and M. Wiedmann. 2004. Food and human isolates of *Listeria monocytogenes* form distinct but overlapping populations. *Appl. Environ. Microbiol.* **70:**5833–5841.

Groisman, E. A., and H. Ochman. 1996. Pathogenicity islands: bacterial evolution in quantum leaps. *Cell* **87:**791–794.

Gut, I. G. 2004. DNA analysis by MALDI-TOF mass spectrometry. *Hum. Mutat.* **23:**437–441.

Hacker, J., G. Blum-Oehler, I. Mühldorfer, and H. Tschäpe. 1997. Pathogenicity islands of virulent bacteria: structure, function and impact on microbial evolution. *Mol. Microbiol.* **23:**1089–1097.

Hahm, B.-K., Y. Maldonado, E. Schreiber, A. K. Bhunia, and C. H. Nakatsu. 2003. Subtyping of foodborne and environmental isolates of *Escherichia coli* by multiplex-PCR, rep-PCR, PFGE, ribotyping and AFLP. *J. Microbiol. Methods* **53:**387–399.

Hall, N. 2007. Advanced sequencing technologies and their wider impact in microbiology. *J. Exp. Biol.* **209:**1518–1525.

Hänninen, M.-L., S. Pajarre, M.-L. Klossner, and H. Rautelin. 1998. Typing of human *Campylobacter jejuni* isolates in Finland by pulsed-field gel electrophoresis. *J. Clin. Microbiol.* **36:**1787–1789.

Healy, M., J. Huong, T. Bittner, M. Lising, S. Frye, S. Raza, R. Schrock, J. Manry, A. Renwick, R. Nieto, C. Woods, J. Versalovic, and J. R. Lupski. 2005. Microbial DNA by automated repetitive-sequence-based PCR. *J. Clin. Microbiol.* **43:**199–207.

Heir, E., B.-A. Lindstedt, T. Vardund, Y. Wasteson, and G. Kapperud. 2000. Genomic fingerprinting of shigatoxin-producing *Escherichia coli* (STEC) strains: comparison of pulsed-field gel electrophoresis (PFGE) and fluorescent amplified-fragment-length polymorphism. *Epidemiol. Infect.* **125:**537–548.

Henegariu, O., N. A. Heerema, S. R. Dlouhy, G. H. Vance, and P. H. Vogt. 1997. Multiplex PCR: critical parameters and step-by-step protocol. *BioTechniques* **23:**504–511.

Hiett, K. L. 2005. Molecular typing methods for tracking pathogens, p. 592–606. *In* J. N. Sofos, *Improving the Safety of Fresh Meat.* CRC Press, Boca Raton, FL.

Hommais, F., S. Pereira, C. Acquaviva, P. Escobar-Paramo, and E. Denamur. 2005. Single-nucleotide polymorphism phylotyping of *Escherichia coli. Appl. Environ. Microbiol.* **71:**4784–4792.

Hulton, C. S., C. F. Higgins, and P. M. Sharp. 1991. ERIC sequences: a novel family of repetitive elements in the genomes of *Escherichia coli, Salmonella typhimurium* and other enterobacteria. *Mol. Microbiol.* **5:**825–834.

Hurley, C. C., A. Quirke, F. J. Reen, and E. F. Boyd. 2006. Four genomic islands that mark post-1995 pandemic *Vibrio parahaemolyticus* isolates. *BMC Genomics* **7:**104.

Hyytia-Trees, E., S. C. Smole, P. A. Fields, B. Swaminathan, and E. M. Ribot. 2006. Second generation subtyping: a proposed PulseNet protocol for multiple-locus variable-number tandem repeat analysis of shiga toxin-producing *Escherichia coli* O157 (STEC O157). *Foodborne Path. Dis.* **3:**118–131.

Jeffers, G. T., J. L. Bruce, P. L. McDonough, J. Scarlet, K. J. Boor, and M. Wiedmann. 2001. Comparative genetic characterization of *Listeria monocytogenes* isolates from human and animal listeriosis cases. *Microbiology* **147:**1095–1104.

Jores, J., A. G. Torres, S. Wagner, C. B. Tutt, J. B. Kaper, and L. H. Wieler. 2006. Identification and characterization of "pathoadaptive mutations" of the *cadBA* operon in several intestinal *Escherichia coli. Int. J. Med. Microbiol.* **296:**547–552.

Kearns, D. B., F. Chu, R. Rudner, and R. Losick. 2004. Genes governing swarming in *Bacillus subtilis* and evidence for a phase variation mechanism controlling surface motility. *Mol. Microbiol.* **52:**357–369.

Keim, P., L. B. Price, A. M. Klevytska, K. L. Smith, J. M. Schupp, R. Okinaka, P. J. Jackson, and M. E. Hugh-Jones. 2000. Multilocus-variable number-tandem repeat analysis reveals relationships within *Bacillus anthracis. J. Bacteriol.* **182:**2928–2936.

Kirsch, P., J. Jores, and L. H. Wieler. 2004. Plasticity of bacterial genomes: pathogenicity islands and the locus of enterocyte effacement (LEE). *Berl. Munch. Tierarztl. Wochenschr.* **117:**116–129.

Klevytska, A. M., L. B. Price, J. M. Schupp, P. L. Worsham, J. Wong, and P. Kiem. 2001. Identification and characterization of variable-number tandem repeats in the *Yersinia pestis* genome. *J. Clin. Microbiol.* **39:**3179–3185.

Krause, U., F. M. Thomson-Carter, and T. H. Pennington. 1996. Molecular epidemiology of *Escherichia coli* O157:H7 by pulsed-field gel electrophoresis and comparison with that by bacteriophage typing. *J. Clin. Microbiol.* **34:**959–961.

Kudva, I. T., P. S. Evans, N. T. Perna, T. J. Barrett, F. M. Ausubel, F. R. Blattner, and S. B. Calderwood. 2002. Strains of *Escherichia coli* O157:H7 differ primarily by insertions or deletions, not single-nucleotide polymorphisms. *J. Bacteriol.* **184:**1873–1879.

Lan, R., and P. R. Reeves. 2001. When does a clone deserve a name? A perspective on bacterial species based on population genetics. *Trends Microbiol.* **9:**419–424.

Lappi, V. R., J. Thimothe, K. K. Nightingale, K. Gall, V. N. Scott, and M. Wiedmann. 2004. Longitudinal studies on *Listeria* in smoked fish plants: impact of intervention strategies on contamination patterns. *J. Food Prot.* **67:**2500–2514.

Lawrence, J. G., and H. Ochman. 1998. Molecular archaeology of the Escherichia coli genome. *Proc. Natl. Acad. Sci. USA* **95:**9413–9417.

Lecuit, M., D. M. Nelson, S. D. Smith, H. Khun, M. Huerre, M. C. Vacher-Lavenu, J. I. Gordon, and P. Cossart. 2004. Targeting and crossing of the human maternofetal barrier by *Listeria monocytogenes:* role of internalin interaction with trophoblast E-cadherin. *Proc. Natl. Acad. Sci. USA* **101:**6152–6157.

Lecuit, M., H. Ohayon, L. Braun, J. Menguad, and P. Cossart. 1997. Internalin of *Listeria monocytogenes* with an intact leucine-rich repeat region is sufficient to promote internalization. *Infect. Immun.* **65:**5309–5319.

Lefresne, G., E. Latrille, F. Irlinger, and P. A. D. Grimont. 2004. Repeatability and reproducibility of ribotyping and its computer interpretation. *Res. Microbiol.* **155:**154–161.

Liebana, E., D. Guns, L. Garcia-Migura, M. J. Woodward, F. A. Clifton-Hardley, and R. H. Davies. 2001. Molecular typing of *Salmonella* serotypes prevalent in animals in England: assessment of methodology. *J. Clin. Microbiol.* **39:**3609–3616.

Lindstedt, B. A. 2005. Multiple-locus variable number tandem repeats analysis for genetic fingerprinting of pathogenic bacteria. *Electrophoresis* **26:**2567–2582.

Lindstedt, B. A., T. Vardund, and G. Kapperud. 2004. Multiple-locus variable-number tandem-repeats analysis of *Escherichia coli* O157 using PCR multiplexing and multi-colored capillary electrophoresis. *J. Microbiol. Methods* **58:**213–222.

Liu, S., and K. E. Sanderson. 1996. Highly plastic chromosomal organization in *Salmonella typhi. Proc. Natl. Acad. Sci. USA* **93:**10303–10308.

Lockhart, D. J., and E. A. Winzeler. 2000. Genomics, gene expression and DNA arrays. *Nature* **405:**827–836.

Louie, M., P. Jayarante, I. Luchsinger, J. Devenish, J. Yao, W. Schlech, and A. Simor. 1996. Comparison of ribotyping, arbitrarily primed PCR, and pulsed-field gel electrophoresis for molecular typing of *Listeria monocytogenes. J. Clin. Microbiol.* **34:**15–19.

Luan, S., M. Granlund, M. Sellin, T. Lagergard, B. G. Spratt, and M. Norgren. 2005. Multilocus sequence typing of Swedish invasive group B *Streptococcus* isolates indicates a neonatally associated genetic lineage and capsule switching. *J. Clin. Microbiol.* **43:**3727–3733.

Lukinmaa, S., K. Aarnisalo, M.-L. Suihko, and A. Siitonen. 2004a. Diversity of *Listeria monocytogenes* isolates of human and food origin studied by serotyping, automated ribotyping and pulsed-field gel electrophoresis. *Clin. Microbiol. Infect.* **10:**562–568.

Lukinmaa, S., U.-M. Nakari, M. Eklund, and A. Siitonen. 2004b. Application of molecular genetic methods in diagnostics and epidemiology of food-borne bacterial pathogens. *APMIS* **112:**908–929.

Lupski, J. R., and G. M. Weinstock. 1992. Short, interspersed repetitive DNA sequences in prokaryotic genomes. *J. Bacteriol.* **174:**4525–4529.

Maiden, M. C. 2006. Multilocus sequence typing of bacteria. *Annu. Rev. Microbiol.* **60:**561–588.

Margulies, M., M. Egholm, W. E. Attiya, J. S. Bader, L. A. Bemben, J. Berka, M. S. Braverman, Y. J. Chen, Z. Chen, S. B. Dewell, L. Du, J. M. Fierro, X. V. Gomes, B. C. Godwin, W. He, S. Helgesen, C. H. Ho, G. P. Irzyk, S. C. Jando, M. L. Alenquer, T. P. Jarvie, K. B. Jirage, J. B. Kim, J. R. Knight, J. R. Lanza, J. H. Leamon, S. M. Lefkowitz, M. Lei, J. Li, K. L. Lohman, H. Lu, V. B. Makhijani, K. E. McDade, M. P. McKenna, E. W. Myers, E. Nickerson, J. R. Nobile, R. Plant, B. P. Puc, M. T. Ronan, G. T. Roth, G. J. Sarkis, J. F. Simons, J. W. Simpson, M. Srinivasan, K. R. Tartaro, A. Tomasz, K. A. Vogt, G. A. Volkmer, S. H. Wang, Y. Wang, M. P. Weiner, P. Yu, R. F. Begley, and J. M. Rothberg. 2005. Genome sequencing in microfabricated high-density picolitre reactors. *Nature* **437:**376–380.

Martin, B., O. Humbert, M. Camara, E. Guenzi, J. Walker, T. Mitchell, P. Andrew, M. Prudhomme, G. Alloing, R. Hakenbeck, D. A. Morrison, G. J. Boulnois, and J. Claverys. 1992. A highly conserved repeated DNA element located in the chromosome of *Streptococcus pneumonia. Nucleic Acids Res.* **20:**3479–3483.

Martin, I. E., S. D. Tyler, K. D. Tyler, R. Khakhria, and W. M. Johnson. 1996. Evaluation of ribotyping as epidemiologic tool for typing *Escherichia coli* serogroup O157 isolates. *J. Clin. Microbiol.* **34:**720–723.

Matsumoto, C., J. Okuda, M. Ishibashi, M. Iwanaga, P. Garg, T. Rammamurthy, H. C. Wong, A. Depaola, Y. B. Kim, M. J. Albert, and M. Nishibuchi. 2000. Pandemic spread of an O3:K6 clone of *Vibrio parahaemolyticus* and emergence of related strains evidenced by arbitrarily primed PCR and *toxRS* sequence analysis. *J. Clin. Microbiol.* **38:**578–585.

Maurelli, A. T. 2007. Black holes, antivirulence genes, and gene inactivation in the evolution of bacterial pathogens. *FEMS Microbiol. Lett.* **267:**1–8.

Maurelli, A. T., R. E. Fernandez, C. A. Block, C. K. Rode, and A. Fasano. 1998. "Black holes" and bacterial pathogenesis: a large genomic deletion that enhances the virulence of *Shigella* spp. and enteroinvasive *Escherichia coli. Proc. Natl. Acad. Sci. USA* **95:**3943–3948.

Meinersmann, R. J., R. W. Phillips, M. Wiedmann, and M. E. Berrang. 2004. Multilocus sequence typing of *Listeria monocytogenes* by use of hypervariable genes reveals clonal and recombination histories of three lineages. *Appl. Environ. Microbiol.* **70:**2193–2203.

Metzker, M. L. 2005. Emerging technologies in DNA sequencing. *Genome Res.* **15:**1767–1776.

Meunier, J. R., and P. A. D. Grimont. 1993. Factors affecting reproducibility of random amplified polymorphic DNA fingerprinting. *Res. Microbiol.* **144:**373–379.

Mohan Nair, M. K., and K. S. Venkitanarayanan. 2006. Cloning and sequencing of the *ompA* gene of *Enterobacter sakazakii* and

development of an *ompA*-targeted PCR for rapid detection of *Enterobacter sakazakki* in infant formula. *Appl. Environ. Microbiol.* **72:**2539–2546.

Murphy, M., D. Corocoran, J. F. Buckley, M. O'Mahony, P. Whyte, and S. Fanning. 2006. Development and application of multiple-locus variable number of tandem repeat analysis (MLVA) to subtype a collection of *Listeria monocytogenes. Int. J. Food Microbiol.* **115:**187–194.

Nelson, K. E., D. E. Fouts, E. F. Mongodin, J. Ravel, R. T. DeBoy, J. F. Kolonay, D. A. Rasko, S. V. Angiuoli, S. R. Gill, I. T. Paulsen, J. Peterson, O. White, W. C. Nelson, W. Nierman, M. J. Beanan, L. M. Brinkac, S. C. Daugherty, R. J. Dodson, A. S. Durkin, R. Madupu, D. H. Haft, J. Selengut, S. Van Aken, H. Khouri, N. Fedorova, H. Forberger, B. Tran, S. Kathariou, L. D. Wonderling, G. A. Uhlich, D. O. Bayles, J. B. Luchansky, and C. M. Fraser. 2004. Whole genome comparisons of serotype 4b and 1/2a strains of the food-borne pathogen *Listeria monocytogenes* reveal new insights into the core genome components of this species. *Nucleic Acids Res.* **32:**2386–2395.

Nightingale, K. K., K. Windham, P. Jalan, R. Nielsen, and M. Wiedmann. 2007. *L. monocytogenes* contains clonal groups within the major genetic lineages that show distinct ecological preferences in host and non-host environments. *J. Clin. Microbiol.* **44:**3742–3751.

Nightingale, K. K., K. Windham, and M. Wiedmann. 2005a. Evolution and molecular phylogeny of *Listeria monocytogenes* from human and animal cases and foods. *J. Bacteriol.* **187:**5537–5551.

Nightingale, K. K., K. Windham, K. E. Martin, M. Yeung, and M. Wiedmann. 2005b. Selected *Listeria monocytogenes* stubypes commonly found in food show reduced invasion in human intestinal cells due to distinct nonsense mutations in *inlA* leading to expression of truncated and secreted internalin A. *Appl. Environ. Microbiol.* **12:**8764–8772.

Noller, A. C., M. C. McEllistrem, K. A. Shutt, and L. H. Harrison. 2006. Locus-specific mutational events in a multilocus variable-number tandem repeat analysis of *Escherichia coli* O157:H7. *J. Clin. Microbiol.* **44:**374–377.

Noller, A. C., M. C. McEllistrem, O. C. Stine, J. G. Morris, D. J. Boxrud, B. Dixon, and L. H. Harrison. 2003a. Multilocus sequence typing reveals a lack of diversity among *Escherichia coli* O157:H7 isolates that are distinct by pulsed-field gel electrophoresis. *J. Clin. Microbiol.* **41:**675–679.

Noller, A. C., M. C. McEllistrem, A. G. Pacheco, D. J. Boxrud, and L. H. Harrison. 2003b. Multilocus variable-number tandem repeat analysis distinguishes outbreak and sporadic *Escherichia coli* O157:H7 isolates. *J. Clin. Microbiol.* **41:**5389–5397.

Ochman, H., and L. M. Davalos. 2006. The nature and dynamics of bacterial genomes. *Science* **311:**1730–1733.

Ochman, H., and S. R. Santos. 2005. Exploring microbial microevolution with microarrays. *Infect. Genet. Evol.* **5:**103–108.

Ochman, H., E. Lerat, and V. Daublin. 2005. Examining bacterial species under the specter of gene transfer and exchange. *Proc. Natl. Acad. Sci. USA* **102:**6595–6599.

Ochman, H., and N. A. Moran. 2001. Genes lost and genes found: evolution of bacterial pathogenesis and symbiosis. *Science* **292:**1096–1099.

Ochman, H., S. Elwyn, and N. A. Moran. 1999. Calibrating bacterial evolution. *Proc. Natl. Acad. Sci. USA* **96:**12638–12643.

Orsi, R. H., D. Ripoll, K. K. Nightingale, and M. Wiedmann. Evolution of *inlA* in lineage I and lineage II *Listeria monocytogenes* isolates from humans, animals, foods, and pristine environments. *Microbiology*, in press.

Park, S. F., D. Purdy, and S. Leach. 2000. Localized reversible frameshift mutation in the *flhA* gene confers phase variability to flagellin gene expression in *Campylobacter coli*. *J. Bacteriol.* 182:207–210.

Perna, N.T., G. Plunkett III, V. Burland, B. Mau, J. S. Glasner, D. J. Rose, G. F. Mayhew, P. S. Evans, J. Gregor, J. A. Kirkpatrick, G. Posfai, J. Hackett, S. Klink, A. Boutin, Y. Shao, L. Miller, E. J. Grothbeck, N. W. Davis, A. Lim, E. T. Dimalanta, K. D. Potamousis, J. Apodaca, T. S. Anantharaman, J. Lin, G. Yen, D. C. Schwartz, R. A. Welch, and F. R. Blattner. 2001. Genome sequence of enterohaemorrhagic *Escherichia coli* O157:H7. *Nature* 409:529–533.

Pohl, M. A., M. Wiedmann, and K. K. Nightingale. 2006. Associations among *Listeria monocytogenes* genotypes and distinct clinical manifestations of listeriosis in cattle. *Am. J. Vet. Res.* 67:616–626.

Ramachandran, V., K. Brett, M. A. Hornitzky, M. Dowton, M. J. Walker, and S. P. Djordjevic. 2003. Distribution of intimin subtypes among *Escherichia coli* isolates from ruminant and human sources. *J. Clin. Microbiol.* 41:5022–5032.

Robertson, G. A., V. Thiruvenkataswamy, H. Shilling, E. P. Price, F. Huygens, F. A. Henskens, and P. M. Giffard. 2004. Identification and interrogation of highly informative single nucleotide polymorphism sets defined by bacterial multilocus sequence typing databases. *J. Med. Microbiol.* 53:35–45.

Ross, I. L., and M. W. Heuzenroeder. 2005. Use of AFLP and PFGE to discriminate between *Salmonella enterica* serovar Typhimurium DT126 isolates from separate food-related outbreaks in Australia. *Epidemiol. Infect.* 133:635–644.

Salcedo, C., L. Arreaza, B. Alcala, L. de la Fuente, and J. A. Vazquez. 2003. Development of a multilocus sequence typing method for analysis of *Listeria monocytogenes* clones. *J. Clin. Microbiol.* 41:757–762.

Savelkoul, P. H. M., H. J. M. Aarts, J. De Haas, L. Dijkshoorn, B. Duim, M. Otsen, J. L. W. Rademaker, L. Schouls, and J. A. Lenstra. 1999. Amplified-fragment length polymorphism analysis: the state of an art. *J. Clin. Microbiol.* 37:3083–3091.

Sawires, Y. S., and J. G. Songer. 2005. Multiple-locus variable-number tandem repeat analysis for strain typing of *Clostridium perfringens*. *Anaerobe* 11:262–272.

Schaechter, M., and the View From Here Group. 2001. *Escherichia coli* and *Salmonella* 2000: the view from here. *Microbiol. Mol. Biol. Rev.* 65:119–130.

Schmid, M. W., E. Y. Ng, R. Lampidis, M. Emmerth, M. Walcher, J. Kreft, W. Goebel, M. Wagner, and K. H. Schleifer. 2005. Evolutionary history of the genus *Listeria* and its virulence genes. *Syst. Appl. Microbiol.* 28:1–18.

Schmidt, H., and M. Hensel. 2004. Pathogenicity islands in bacterial pathogenesis. *Clin. Microbiol. Rev.* 17:14–56.

Schoolnik, G. K. 2002. Functional and comparative genomics of pathogenic bacteria. *Curr. Opin. Microbiol.* 5:20–26.

Segura, A., A. Hurtado, E. Duque, and J. L. Ramos. 2004. Transcriptional phase variation at the *flhB* gene of *Pseudomonas putida* DOT-T1E is involved in response to environmental changes and suggests the participation of the flagellar export system in solvent tolerance. *J. Bacteriol.* 186:1905–1909.

Shaikh, N., and P. I. Tarr. 2003. *Escherichia coli* O157:H7 Shiga toxin-encoding bacteriophages: integrations, excisions, truncations, and evolutionary implications. *J. Bacteriol.* 185:3596–3605.

Shendure, J., R. D. Mitra, C. Varma, and G. M. Church. 2004. Advanced sequencing technologies: methods and goals. *Nat. Rev. Genet.* 5:335–344.

Silbert, S., L. Boyken, R. J. Hollis, and M. A. Pfaller. 2003. Improving typeability of multiple bacterial species using pulsed-field gel electrophoresis and thiourea. *Diagn. Microbiol. Infect. Dis.* 47:619–621.

Sobrino, B., M. Brion, and A. Carracedo. 2005. SNPs in forensic genetics: a review on SNP typing methodologies. *Forensic Sci. Int.* 154:181–194.

Spratt, B. G., and M. C. Maiden. 1999. Bacterial population genetics, evolution and epidemiology. *Philos. Trans. R. Soc. Lond. B* 354:701–710.

Stern, M. J., G. F.-L. Ames, N. H. Smith, E. C. Robinson, and C. F. Higgins. 1984. Repetitive extragenic palindromic sequences: a major component of the bacterial genome. *Cell* 37:1015–1026.

Stull, T. L., J. J. LiPuma, and T. D. Edlind. 1988. A broad-spectrum probe for molecular epidemiology of bacteria: ribosomal RNA. *J. Infect. Dis.* 157:280–285.

Swaminathan, B., T. J. Barret, and P. Fields. 2006. Surveillance for human *Salmonella* infections in the United States. *J. AOAC Int.* 89:553–550.

Swaminathan, B., T. J. Barrett, S. B. Hunter, and R. V. Tauxe. 2001. Pulsenet: the molecular subtyping network for foodborne bacterial disease surveillance, United States. *Emerg. Infect. Dis.* 7:382–389.

Tarr, C. L., and T. S. Whittam. 2002. Molecular evolution of the intimin gene in O111 clones of pathogenic *Escherichia coli*. *J. Bacteriol.* 184:479–487.

Tenover, F. C., R. D. Arbeit, R. V. Goering, P. A. Mickelsen, B. A. Murray, D. H. Persing, and B. Swaminathan. 1995. Interpreting chromosomal DNA restriction patterns produced by pulsed-field gel electrophoresis: criteria for bacterial strain typing. *J. Clin. Microbiol.* 33:2233–2239.

Theiss, P., and K. S. Wise. 1997. Localized frameshift mutation generates selective, high-frequency phase variation of a surface lipoprotein encoded by a *Mycoplasma* ABC transporter operon. *J. Bacteriol.* 179:4013–4022.

Torres, A. G., R. C. Vazquez-Jaurez, C. B. Tutt, and J. G. Garcia-Gallegos. 2005. Pathoadaptive mutation that mediates adherence of Shiga toxin-producing *Escherichia coli* O111. *Infect. Immun.* 73:4766–4776.

Tyler, K. D., G. Wang, S. D. Tyler, and W. M. Johnson. 1997. Factors affecting reliability and reproducibility of amplification-based DNA fingerprinting of representative bacterial pathogens. *J. Clin. Microbiol.* 35:339–346.

Urwin, R., and M. C. J. Maiden. 2003. Multi-locus sequence typing: a tool for global epidemiology. *Trends Microbiol.* 11:479–487.

Vazquez-Boland, J., G. Dominguez, B. Gonzalaz-Zorn, J. Kreft, and W. Goebel. 2001. Pathogenicity islands and virulence evolution in *Listeria*. *Microbes Infect.* 3:571–584.

Versalovic, J., T. Koeuth, and J. R. Lupski. 1991. Distribution of repetitive DNA sequences in eubacteria and application to fingerprinting of bacterial genomes. *Nucleic Acids Res.* 19:6823–6831.

Wang, G., T. S. Whittam, C. M. Berg, and D. E. Berg. 1993. RAPD (arbitrarily primed) PCR is more sensitive that multilocus enzyme electrophoresis for distinguishing related bacterial strains. *Nucleic Acids Res.* 21:5930–5933.

Ward, T. J., L. Gorski, M. K. Borucki, R. E. Mandrell, J. Hutchins, and K. Pupedis. 2004. Intraspecific phylogeny and lineage group identification based on the *prfA* virulence gene cluster of *Listeria monocytogenes*. *J. Bacteriol.* 15:4994–5002.

Watanabe, Y., K. Ozasa, J. H. Mermin, P. M. Griffin, K. Masuda, S. Imashuku, and T. Sawada. 1999. Factory outbreak of *Escherichia coli* O157:H7 infection in Japan. *Emerg. Infect. Dis.* 5:424–428.

Welsh, J., and M. McClelland. 1991. Genomic fingerprinting using arbitrarily primed PCR and a matrix of pairwise combinations of primers. *Nucleic Acids Res.* 19:5275–5279.

Wiedmann, M. 2002. Subtyping of bacterial foodborne pathogens. *Nutr. Rev.* 60:201–208.

Wiedmann, M., J. L. Bruce, C. Keating, A. E. Johnson, P. L. McDonough, and C. A. Batt. 1997. Ribotypes and virulence gene polymorphisms suggest three distinct *Listeria monocytogenes* lineages with differences in pathogenic potential. *Infect. Immun.* 65:2707–2716.

Williams, J. G., A. R. Kubelik, K. J. Livak, J. A. Rafalski, and S. V. Tingey. 1990. DNA polymorphisms amplified by arbitrary primers are useful as genetic markers. *Nucleic Acids Res.* 18:6531–6535.

Windham, K., K. K. Nightingale, and M. Wiedmann. 2005. Molecular evolution and diversity of foodborne pathogens. p. 1259–1291. *In* K. Shetty, G. Paliyath, A. Pometto, and R. E. Levin (ed.), *Food Biotechnology*, 2nd ed. CRC Press, Boca Raton, FL.

Woods, C. R., J. Versalovic, T. Koeuth, and J. R. Lupski. 1993. Whole-cell repetitive element sequence-based polymerase chain reaction allows rapid assessment of clonal relationships of bacterial isolates. *J. Clin. Microbiol.* 31:1927–1931.

Yang, Z., R. Nielsen, N. Goldman, and A. M. Pedersen. 2000. Codon-substitution models for heterogeneous selection pressure at amino acid sites. *Genetics* 155:431–449.

Ye, R. W., T. Wang, L. Bedzyk, and K. M. Croker. 2001. Applications of DNA microarrays in microbial systems. *J. Microbiol. Methods* 47:257–272.

Zhang, C., M. Zhang, J. Ju, J. Nietfeldt, J. Wise, P. M. Terry, M. Olson, S. D. Kachman, M. Wiedmann, M. Samadpour, and A. K. Bensen. 2003. Genome diversification in phylogenetic lineages I and II of *Listeria monocytogenes*: identification of segments unique to lineage II populations. *J. Bacteriol.* 185:5573–5584.

Zhang, W., W. Qi, T. J. Albert, A. S. Motiwala, D. Alland, E. K. Hyytia-Trees, E. M. Ribot, P. I. Fields, T. S. Whittam, and B. Swaminathan. 2006. Probing genomic diversity and evolution of *Escherichia coli* O157 by single nucleotide polymorphisms. *Genome Res.* 16:757–767.

Pathogens and Toxins in Foods: Challenges and Interventions
Edited by V. K. Juneja and J. N. Sofos
© 2010 ASM Press, Washington, DC

Chapter 30

Food Safety Management Systems

VIRGINIA N. SCOTT AND YUHUAN CHEN

TYPES OF FOOD SAFETY MANAGEMENT SYSTEMS

A food safety management system can be defined as a group of interacting or interdependent elements forming a network to ensure that food presents a minimal risk to consumers. Thus, depending on the food chain sector, a food safety management system will encompass different elements. FAO/WHO has defined a food safety management system for industry as "a holistic system of controls that manage food safety in a food business. Includes [good hygiene practices] GHPs, the [hazard analysis and critical control point] HACCP system, management policies and traceability/recall systems" (FAO/WHO, 2007).

Food safety management can be viewed from different levels: broadly from the perspective of government food safety management systems or more narrowly from that of industry-wide food safety management systems or that of food safety management within an individual establishment. It is conceivable that there could be a "horizontal" food safety system that extends from farm to fork, but practically, food safety efforts are segmented into multiple systems. Industry systems may be tailored to sectors, such as farmers, slaughterers, manufacturers, or restaurants. Within these sectors, there will be further delineation of programs, with respect to specific types of operations (e.g., manufacturing may include fresh, frozen, and canned products, for which food safety management systems will differ); the complete food safety system for a specific operation will be based on the types of products, processes, and sources of ingredients.

Ultimately, management of food safety on all levels must be achieved to ensure that foods are safe with respect to pathogens and toxins. The most effective systems will focus on prevention. However, since we will never be able to completely prevent contamination of food with pathogens, our systems must have effective, rapid means of detecting contamination that could result in food-borne illness, and then the systems should include appropriate responses to minimize the potential for illness to occur. Effective food safety systems, in addition to focusing on prevention, will provide for a variety of interventions—interventions to prevent contamination and interventions that can address contamination that has been detected. For a country to have an effective food safety system, there must be integration of various elements within all sectors, from production through manufacturing to the consumer, who has an important role in ensuring that the food safety system is effective. Communication across the entire food chain is also necessary to ensure an effective food safety management system within a country.

The role of the food industry is to provide safe foods; industry is responsible for establishing food safety management systems that ensure that foods present a minimal risk to the consumer. One of the primary roles of government is to verify that industry is implementing appropriate food safety management systems. The government has additional roles in defining what is safe, in establishing appropriate targets or metrics, in providing guidance to industry on ways to achieve safety, and in informing consumers when breakdowns in the systems have resulted in unsafe foods.

Elements of a governmental food safety management system are shown in Table 1. For many of these elements there is also an industry role. The system begins with the laws and regulations that define requirements for food safety, which require inspection and enforcement capabilities. However, "inspection," in the form of audits to ensure compliance with an establishment's food safety system, is also a function

Virginia N. Scott and Yuhuan Chen • Grocery Manufacturers Association, 1350 I St. N.W., Suite 300, Washington, DC 20005.

Table 1. Elements of a government food safety management system and elements for which industry has a responsibility[a]

Element	Responsible sector
Food laws, regulations, standards, and policies	Government
Inspection of food establishments	Government, industry
Enforcement capabilities	Government
Food-borne illness surveillance	Government
Monitoring for food-borne hazards	Government, industry
Scientific capabilities	Government, industry
Analytical laboratories	Government, industry
Emergency response capabilities	Government, industry
Training	Government, industry
Public information, education, and communication	Government, industry
Risk analysis capabilities	Government, industry

[a]Elements of government food safety management system are based on FAO/WHO (2006).

of industry. Food-borne illness surveillance, along with monitoring for food-borne hazards using analytical laboratories, provides information on hazards that need to be managed and in which commodities these hazards are a concern. While food-borne illness surveillance will be a government responsibility, monitoring for food-borne hazards will be the responsibility of both government and industry. Thus, both government and industry require scientific capabilities to establish appropriate food safety management systems. Since all problems cannot be prevented, it is critical that both government and industry have emergency response capabilities to address recalls and illnesses from specific foods, and training is required throughout the food safety management system so that everyone understands their roles and responsibilities in ensuring safe food. Public information, education, and communication are primarily the responsibility of government, but industry will also play a role, with respect to specific products, especially when problems occur with a product.

The final element of the food safety management system involves risk analysis capabilities. The recent application of risk analysis in the development of food safety policies within regulatory food safety management systems provides a means to link hazards with risks to public health and a scientific basis for various components of a food safety management system (FAO/WHO, 2006). Risk analysis allows the development of risk-based metrics, such as food safety objectives (FSOs), performance objectives (POs), and performance criteria (PCs), which are addressed at the end of this chapter. The government has a responsibility to use transparent processes when conducting risk analyses, such that there is open communication and sharing of information with stake-holders, resulting in broader acceptance of food safety management options that become part of the system. While risk analysis will be primarily the role of government, there are aspects of it that can be applied by industry. For example, many large companies have expertise in risk assessment, which they use in developing risk management options for the products they produce.

While there are other hazards, we will confine our approach to the focus of this book, which is pathogens and toxins. These are usually considered to be the most significant hazards to be addressed by food safety management systems.

INFORMATION NEEDED TO DEVELOP EFFECTIVE FOOD SAFETY MANAGEMENT SYSTEMS

Clearly, in order to develop a food safety management system for pathogens and toxins, it is essential to understand which of these are resulting in food-borne illnesses. This requires a food-borne disease surveillance system that can identify illnesses and appropriately attribute them to a food. Many countries have passive food-borne illness reporting systems in which food-borne illnesses identified at the local level are reported to a central location where data are compiled. This provides information on outbreaks. However, most cases of food-borne illness are sporadic, and many outbreaks are too small for investigation leading to association with a specific food. Moreover, simply attributing illness to a specific food does not provide all the information needed to develop an effective food safety management system. It is important to also identify the source of the problem—the root cause—because this will help determine where to focus resources to manage the safety of the food. For example, if illness results from growth of *Clostridium perfringens* in a roasted turkey due to improper cooling at a restaurant, this should be addressed within the food safety management system applied at food service, not at eliminating *C. perfringens* from turkeys. Thus, full investigation of food-borne illness to the extent possible is important in identification of points in the food chain where control can be effected. Information on food-borne sources of illness and factors contributing to illness must come from government activities.

In addition to understanding the sources of food-borne illness, it is important to develop information on interventions that can eliminate pathogens or toxins or reduce them to acceptable levels. It is also important to know the impact of interventions on the pathogen or toxin of concern. (This is discussed further in the section below on Practical Considerations in HACCP Verification and Validation.) Some interventions have a much greater impact than others, e.g., a pathogen kill step, and given limited resources, it may be prudent to select high-impact interventions for the food safety management system. However, it would be unwise to ignore smaller-impact interventions that, when taken together, can have a measurable impact. This concept may be particularly important in the production of food products for which there is no pathogen kill step, such as fresh-cut produce, where on-farm good agricultural practices (GAPs), field worker GHPs, and plant good manufacturing practices (GMPs) are combined to manage the safety of the product. Information on interventions can come from multiple sources: industry and industry trade associations, academia, and governmental research groups, such as the USDA Agricultural Research Service. While in some instances interventions developed by industry are proprietary, it has become more common to share information on practices that are effective in reducing the risk from pathogens, making food safety a noncompetitive issue. Discussions on specific interventions to reduce or control pathogens at various points in the food chain are covered in other chapters.

HACCP: A BRIEF OVERVIEW

The HACCP system is the approach of choice for industry to manage the safety of foods. Although it is frequently stated that HACCP is applicable throughout the food chain ("farm to fork"), in practice the classic HACCP system is applied most frequently in the processing sector, while in many other food sectors HACCP-based approaches are applied.

Origin of HACCP

HACCP originated in the United States in the early 1960s in connection with manned space flights. The Pillsbury Company, the National Aeronautics and Space Administration, and the U.S. Army Natick Laboratories developed the first HACCP system to meet the demand of producing safe foods for astronauts in the space program (Stevenson, 2006). HACCP, based on the concept of prevention, was developed in light of the recognition that sampling and testing of finished products, the quality assurance paradigm at the time, in and of itself was not adequate to ensure product safety. With low levels of contamination, chances are that testing would likely miss contaminated product units in a lot. After extensive evaluation of various concepts, the "Modes of Failure" concept developed by the Natick Laboratories was adapted to the production of foods, and the concept of prevention underlying HACCP was born,

with a basic premise that food safety would be ensured by establishing control over the entire food manufacturing process, including the control of raw materials, the processing environment, and the people involved (Bernard, 1998; Stevenson, 2006). This entails gathering information on the ingredients, product, and process to predict what hazards (e.g., *Salmonella*) might occur and where they might occur and put in place control measures to prevent the hazards from occurring in the finished product.

The HACCP approach was first unveiled to the public at the National Conference on Food Protection in 1971 (Bauman, 1971). Howard Bauman from the Pillsbury Company presented three principles that were in the original HACCP system, i.e., identification and assessment of hazards associated with the product from growing/harvesting to marketing/preparation, determination of critical control points (CCPs) to control the hazards identified, and establishment of monitoring procedures for the CCPs (Bauman, 1971; Bernard, 1998). While there was considerable interest in HACCP upon its introduction and the FDA applied the HACCP approach in the promulgation of the low-acid and acidified canned food regulations to control *Clostridium botulinum* in the 1970s, early uses of HACCP by the food industry as well as the government were limited and unsuccessful in many cases, partly due to a lack of understanding of HACCP (Bernard, 1998). The inadequate elaboration of the concept of prevention and a lack of guidance on how to apply HACCP principles hindered widespread application of HACCP in the food industry. A milestone occurred in 1985 when a subcommittee of the Food Protection Committee of the National Academy of Sciences issued a report (NAS, 1985) that strongly endorsed the HACCP approach to food safety. The subcommittee also recommended the formation of an expert panel to provide guidance on HACCP application.

The National Advisory Committee on Microbiological Criteria for Foods (NACMCF) was formed in 1988 as an advisory panel to the U.S. government agencies with food safety jurisdiction. The NACMCF has since encouraged the adoption of the HACCP approach to food safety through further elaboration of the HACCP principles and development of application guidelines. The latest revision of the NACMCF guidance document was published in 1998. Available guidance from the NACMCF (1998), from industry sources (e.g., Scott and Stevenson, 2006; Sperber et al., 1998), and from government sources (e.g., FDA, 2001b and 2004; FSIS, 1999–2006; FSIS, 2005; CFIA, 2007) has enabled and greatly promoted HACCP implementation in a wide range of industry segments. The NACMCF guidance was also used as the scientific basis for HACCP regulations in the United States.

HACCP Application in Food Processing and Other Industry Sectors

In the United States, the meat, poultry, seafood, and juice industry segments are required by law to implement HACCP in their operations. The U.S. Department of Agriculture's Food Safety and Inspection Service (USDA FSIS) published a regulation that mandates HACCP for meat and poultry operations (FSIS, 1996), and the U.S. Food and Drug Administration (FDA) published regulations that mandate HACCP for seafood (FDA, 1995) and for juice operations (FDA, 2001a). The mandatory application of HACCP principles applies to all seafood, juice, and meat and poultry products that are produced domestically or that are imported from a foreign country. Processors of many different products, such as cheese, salad dressing, bread, flour, frozen dough, and breakfast cereal, participated in an FDA voluntary pilot program that was designed to gather information on the feasibility of mandatory HACCP for all industry sectors in the United States (Hontz and Scott, 2006). Many industry sectors in the United States have voluntarily adopted the HACCP approach to food safety, including those participating in the FDA pilot program, the egg industry (Bernard and Scott, 2007), parts of the dairy industry (Anonymous, 2007; FDA, 2003), and the retail and food service sector (FDA, 2006).

HACCP is also used widely in countries outside of the United States, including but not limited to member states in the European Union, Canada, Australia, New Zealand, and Japan (Hontz and Scott, 2006). In developing countries, e.g., China (Bai et al., 2007), HACCP implementation in the food industry has primarily been driven by demands of the export markets. Application of HACCP principles to ensure food safety is mandatory for all food business operators in the European Union (European Union, 2004); the mandatory requirement for HACCP applies to food imports into the European Union. In Canada, development and implementation of HACCP are mandatory for meat and poultry establishments and encouraged for all food establishments, including dairy, honey, maple syrup, processed fruit and vegetable, shell egg, processed egg establishments, and poultry hatchery sectors (CFIA, 2007; CDJ, 2007).

HACCP AS A FRAMEWORK TO DETERMINE SIGNIFICANT HAZARDS AND CONTROL RISK

Seven HACCP principles form the framework of a systematic approach to ensure the determina-

tion and control of significant food safety hazards associated with a product and process. The HACCP principles, as defined by the NACMCF (1998), are the following:

1. Conduct a hazard analysis.
2. Determine the CCPs.
3. Establish critical limits.
4. Establish monitoring procedures.
5. Establish corrective actions.
6. Establish verification procedures.
7. Establish record-keeping and documentation procedures.

In order to effectively apply these principles and to sustain HACCP implementation in an establishment, it is important that the establishment has a solid foundation of prerequisite programs in place (see discussion on prerequisite programs below). The NACMCF (1998) also recommended that five preliminary steps be taken prior to conducting a hazard analysis: (i) assemble the HACCP team, (ii) describe the food and its distribution, (iii) describe the intended use and consumers of the food, (iv) develop a flow diagram that describes the process, and (v) verify the flow diagram. Most experts also agree that another requisite preliminary step is to gain management support prior to HACCP development and implementation (NACMCF, 1998; Bernard and Scott, 2007).

The Codex Alimentarius Commission, in an effort to promote the use of HACCP to ensure food safety globally, adopted guidelines detailed in its document *Hazard Analysis and Critical Control Point (HACCP) System and Guidelines for Its Application.* The HACCP guidelines are an annex to the international code of practice for food hygiene. The current Codex HACCP document (CAC, 2003) provides additional context for the principles on monitoring, corrective action, verification, and record-keeping and describes a sequence of 12 steps for HACCP application, i.e., 5 preliminary steps plus 7 steps for the individual HACCP principles. Despite some differences, the basic framework adopted by Codex and many other technical bodies around the world is very similar to that adopted by the NACMCF, in that the same seven principles form the core of a HACCP food safety management system.

A short description of the HACCP principles is presented below to illustrate the systematic nature of the HACCP framework. More in-depth discussion of the principles and more examples on how to apply these principles have been elaborated in the literature, e.g., a book chapter with emphasis on the use of HACCP to control microbiological hazards (Bernard and Scott, 2007) and a comprehensive manual for developing and implementing a HACCP plan (Scott and Stevenson, 2006).

Principles 1 and 2: Conduct a Hazard Analysis and Determine CCPs

After completing the preliminary steps, which include developing a flow diagram of the process, the HACCP team conducts a hazard analysis to determine significant food safety hazards associated with the product and process under consideration. The hazard analysis also includes the identification of appropriate control measures for the significant hazards and an assessment of which of these are applied at CCPs in the process, in which control is essential to prevent, eliminate, or reduce a hazard to an acceptable level.

A food safety hazard, according to the NACMCF (1998), is "a biological, chemical, or physical agent that is reasonably likely to cause illness or injury in the absence of its control." The process of hazard analysis involves collecting and evaluating information on various potential hazards associated with a food and the process used to make the food, in order to decide which hazards are significant, i.e., reasonably likely to occur in the absence of control, and thus must be addressed in the HACCP plan (NACMCF, 1998). Food safety hazards under consideration include biological, physical, and chemical hazards that may be introduced, enhanced, or controlled in the manufacturing process.

Biological hazards are microorganisms that are capable of causing food-borne illnesses, including pathogenic bacteria, viruses, and parasites. Examples of potential biological hazards are commonly known pathogens, such as *Salmonella* spp., *Escherichia coli* O157:H7, *Listeria monocytogenes*, *Staphylococcus aureus*, *Clostridium botulinum*, *Campylobacter jejuni*, *Campylobacter coli*, hepatitis A, *Cyclospora cayetanensis*, and *Cryptosporidium parvum*. Potential chemical hazards include the many chemicals used at various points of food production that, when excessive levels are present, may cause acute or chronic illnesses. Examples of such chemicals are pesticides used in growing crops and in-plant pest control, antibiotics and hormones used in raising livestock, sanitizers used in food plant sanitation, and processing aids used in production. Other examples of potential chemical hazards are natural toxins, such as mycotoxins (e.g., aflatoxin), fish toxins (e.g., ciguatoxin and scombrotoxin), and shellfish toxins (e.g., saxitoxin). Although food allergens, such as certain proteins in peanuts and milk, are a part of the food supply and are not harmful for most consumers, undeclared allergens are conventionally considered a

chemical hazard because they may trigger allergic reactions in a small fraction of the population. Potential physical hazards are hard or sharp foreign objects that may cause injury in the consumer, e.g., glass fragments, pieces of metal, or hard plastic.

The NACMCF (1998) recommended the use of a two-stage process to conduct a hazard analysis. Stage 1, hazard identification, is a brainstorming session, in which the HACCP team develops a list of potential biological, chemical, and physical hazards that may be associated with each ingredient, each processing step under the direct control of the establishment, and the finished product. Gathering thorough information about the raw materials and/or ingredients used in the product, the processing method and equipment used, the activities conducted at each processing step, the type of packaging, the product distribution system, and the intended use by the consumer facilitates the determination of where and how food safety hazards may be introduced. Knowledge of illnesses or injuries associated with the product or ingredients also facilitates hazard identification. When identifying potential hazards, it is important to be specific so that, should the potential hazard be determined to be significant, appropriate control measures may be identified. For example, rather than listing "pathogens," the HACCP team should identify the specific pathogen and/or the specific concern related to the pathogen, e.g., *Salmonella*, *S. aureus* growth and enterotoxin production, etc. Such specifics are needed because the control measure may be different, e.g., heat treatment to kill *Salmonella* versus refrigeration to control *S. aureus* growth and prevent toxin production (Bernard and Scott, 2007).

Stage 2 of the hazard analysis is hazard evaluation. After a list of potential hazards is assembled, the HACCP team evaluates each of the potential hazards to determine whether the hazard, if left uncontrolled, is reasonably likely to cause illness or injury to the consumer. Hazard evaluation is based on two criteria: severity of the hazard when it does occur (the magnitude and duration of illness or injury) and its likelihood of occurrence. The HACCP team relies upon epidemiological data, information in the scientific literature, as well as their experience, to conduct the hazard evaluation. In essence, hazard analysis is analogous to a qualitative risk assessment because the determination of the likelihood of occurrence and severity of the consequence of a hazard, subsequent to its identification, are two main inputs in a risk assessment (Bernard and Scott, 2007). Table 2 shows a tool adapted from one developed by Bernard et al. (2006a) and NACMCF (1998) to illustrate how to evaluate a potential hazard according to a qualitative risk assessment thought process. The hazard analysis

Table 2. Application of the two-stage approach to conduct hazard analysis[a]

Example of ingredient/ processing step	Stage 1: hazard identification	Stage 2: hazard evaluation		
	Step 1: determine potential hazards associated with the ingredient[b]	Step 1: assess severity of health consequences of potential hazard when it is present	Step 2: determine likelihood of occurrence of potential hazard in the absence of control	Step 3: using information from steps 1 and 2, determine if the potential hazard must be addressed in the HACCP plan
Raw beef trimmings used to manufacture frozen cooked beef patties	Enteric pathogens, such as *Salmonella* and *E. coli* O157:H7, from the trimmings	Epidemiological data indicate that these pathogens cause moderate or severe health effects, including death, especially among children and the elderly. Undercooked beef patties have been linked to disease caused by these pathogens.	The likelihood of occurrence is moderate for *Salmonella* and low for *E. coli* O157:H7 in raw beef trimmings.	The HACCP team decides that enteric pathogens are significant hazards for the finished product, if not properly controlled. The pathogens must be addressed in the HACCP plan.
Deboning step in manufacturing frozen precooked chicken for further processing	*S. aureus* growth and toxin production	Enterotoxin produced by certain strains of *S. aureus* causes a moderately severe food-borne illness.	The product may be contaminated with *S. aureus* due to human handling during deboning of the cooked chicken. Enterotoxin capable of causing illness will occur only after the organism grows to 10^5–10^6 CFU/g. Good employee practices minimize contamination. Operating procedures (relatively short time and cool ambient temperatures) during deboning minimize growth of *S. aureus*. The potential for enterotoxin formation is very low.	The HACCP team concludes that enterotoxin formation is not reasonably likely to occur due to prerequisite programs in place. The potential hazard does not need to be addressed in the HACCP plan.

[a]Adapted from a similar table developed by Bernard et al. (2006a) and NACMCF (1998).
[b]Only biological hazards associated with the ingredient or processing step are considered for the purpose of illustrating the two-stage analytical thought process.

of frozen cooked beef patties resulted in the conclusion that enteric pathogens such as *Salmonella* organisms are significant hazards and must be controlled in the HACCP plan, while *S. aureus* growth and toxin production in a frozen precooked chicken operation is not likely to occur and the potential hazard does not need to be addressed in the HACCP plan (Table 2).

Hazard evaluation is one of the most challenging aspects of HACCP plan development because of the complexity of potential hazards and the uncertainty associated with likelihood of occurrence, given, among other factors, diverse ingredients and their sources and varying operation conditions, including the functioning of prerequisite programs that have been implemented. Ultimately, the HACCP team must make a decision based on its collective knowledge and experience and document their decision-making process to help future reassessment of the hazard analysis.

The outcome of hazard evaluation is the determination of, usually from a long list of potential hazards,

a much shorter list of food safety hazards that are truly significant and must be controlled in the operation to make the product safe for consumption. The HACCP team then identifies appropriate control measure(s) for each of the significant hazards. The step at which the control measure can be applied and where control is essential to prevent or eliminate a food safety hazard or reduce it to an acceptable level becomes a CCP in the process. Each significant hazard must be controlled by at least one CCP. Moreover, more than one CCP may be employed to control a hazard, and multiple hazards may be controlled by one CCP.

Principles 3 to 5: Establish Critical Limits, Monitoring Procedures, and Corrective Actions

The three principles that follow hazard analysis and CCP determination are the points at which common sense application of science and technology to control the identified food safety hazards takes place.

At each CCP, the HACCP team needs to establish parameters that can be used to define whether the

CCP is "in" or "out" of control. These parameters are referred to as "critical limits." The NACMCF (1998) defined a critical limit as a maximum or minimum value to which a biological, chemical, or physical parameter must be controlled to reduce to an acceptable level, prevent, or eliminate a food safety hazard at a CCP. As HACCP has evolved, it has become acceptable to express a critical limit as an attribute, e.g., presence of an intact filter, a metal detector is on and functioning, etc. Quantitative critical limits usually are physical or chemical parameters that lend themselves to real-time monitoring, e.g., temperature, time, flow rate, pH, water activity, and sanitizer concentration. Critical limits are set based on scientific principles and data, which may be found in sources such as papers published in peer-reviewed scientific journals, government regulations or guidelines, trade association guidelines, and university extension publications. Sometimes it may be necessary to consult experts in areas such as food microbiology and processing technologies in order to establish science-based critical limits that will meet appropriate food safety targets, such as applicable performance standards. New research may be necessary when sufficient data are not available to establish critical limits. In-plant studies may be conducted when the HACCP team needs to ensure critical limits are suitable to and practical for the capacities of the establishment. For example, a laboratory study (Mazzotta, 2001) showed that to achieve a 5-log reduction of vegetative pathogens of public health concern in fruit juices, pasteurization must achieve a minimum of 71.1°C (160°F) for at least 3 seconds. In a continuous pasteurizer it is more practical to monitor flow rate and diameter and length of the hold tube than to monitor residence time. In this case, the appropriate flow rate for a specific hold tube can be established through engineering calculations. An in-plant study is used to determine that the critical limits for flow rate and temperature can be met. Critical limits should be established based on safety; other limits set for nonsafety reasons, such as those for product quality or customer specifications, should not be designated critical limits.

After appropriate critical limits are set, the HACCP team establishes procedures to monitor each CCP in order to determine whether a specific product/process operation is conducted in a manner that will control the identified hazard(s). Another important function of monitoring is to produce records that document the control of the CCP. The NACMCF (1998) describes monitoring as a planned sequence of observations or measurements to assess whether a CCP is under control and to produce an accurate record for future use in verification. Monitoring

procedures are established for what will be monitored, how it will be monitored, how often it will be monitored, and who will be responsible for the activities. In outlining how the monitoring will be conducted, a monitoring device (appropriately calibrated) is usually specified for critical limits that are numerical values. If the monitoring procedure is to monitor an attribute, the operator responsible for the activity must be trained to provide objective and accurate observations. Monitoring can be conducted through continuous or periodic measurements or observations. While continuous monitoring is preferred, it is not always feasible or practical. When necessary, discontinuous monitoring can be used and it must be designed in a way that measurement or observation of a parameter is conducted with sufficient frequency to detect any deviation from the critical limit. This usually requires knowledge about the variability of the operation regarding the parameter being measured and other factors.

When monitoring indicates a deviation from the critical limit or when monitoring does not occur at the frequency specified in the HACCP plan, the control of the CCP is in question and the safety of the product is unknown. Corrective actions must be taken to address the problem.

Corrective action procedures for deviations at CCPs are designed to address the product in question as well as the cause of the deviation. Although specific procedures to follow may depend on the process parameters and the type of product, the objective of corrective action is to prevent an unsafe product from leaving the establishment and to identify and correct the cause of the deviation. The product affected by the deviation is segregated and held, and a review is conducted by a trained individual, which may include consultation with an expert, to determine the safety and appropriate disposition of the product in question. Corrective action procedures are also established to correct the cause of the deviation and bring the CCP back under control, and, where appropriate, identify measures to prevent recurrence of the problem. Corrective actions are documented to produce a record that appropriate actions were taken.

In a HACCP system, multiple layers of control build upon each other. Monitoring is the first line of defense in the preventive HACCP system. In a well-designed and well-implemented HACCP system, deviations to critical limits occur infrequently. However, recognizing that HACCP is not a "zero defects" system, monitoring combined with corrective action serves as a means to catch and correct a food safety problem before the product leaves the establishment. Another layer of control lies in principle 6, verification, which is fundamentally a check on the monitoring and other components in the HACCP system.

Principles 6 and 7: Establish Verification Procedures and Record-Keeping and Documentation Procedures

Verification is defined by the NACMCF as those activities, other than monitoring, that determine the validity of the HACCP plan and that the HACCP system is operating according to the plan (NACMCF, 1998). The HACCP system is the outcome of implementing the HACCP plan (the written document) in the establishment's operation. The verification principle has two components: (i) to determine if the HACCP system is operating in accordance with the plan, i.e., whether the procedures outlined in the plan are being followed as specified; and (ii) to determine if the HACCP plan is adequate to control food safety hazards, including whether all food safety hazards associated with the product and process have been identified appropriately in the first place. Since a wide range of activities may be conducted in either validation of the plan or verification of compliance with the plan, there has been some confusion about which activities belong to compliance verification and which ones belong to efficacy validation (Scott et al., 2006). Many food safety experts now believe that the verification principle could be clarified by establishing validation as a principle separate from verification.

In order to ensure that the HACCP system is operating according to the HACCP plan, various types of verification activities are conducted, including routine verification at the CCPs and HACCP system compliance audits. CCP verification focuses on activities conducted at individual CCPs to make sure that the CCPs are in control. Typical CCP verification activities include review of CCP records (i.e., monitoring, corrective action, and CCP verification records), calibration of instruments and devices used for monitoring and for verification, and an independent check on the adequacy of the CCP to control the identified hazard (e.g., sampling and testing, observation of monitoring activity by a second person, etc.). A HACCP system compliance audit is broader in scope than CCP verification. It may include review of the HACCP plan, review of CCP records, verification of the flow diagram, verification of compliance with prerequisite programs that support the HACCP system, on-site observations, and sampling and testing. Validation centers on the determination of whether the HACCP plan is adequate to control hazards and whether the HACCP plan has been designed properly to facilitate implementation. Validation activities include initial validation when the HACCP plan is first implemented and periodic HACCP plan reassessment, which may be conducted on an annual basis. In addition to verification conducted by the establishment, regulatory inspections and third-party audits may also serve as HACCP verification; these usually involve HACCP system compliance verification and/or an evaluation of whether the HACCP plan has been appropriately validated.

Record-keeping is an integral part of a working HACCP system. Activities conducted for the verification principle, as well as those for the previous five HACCP principles, generate records. After a written HACCP plan is developed, record-keeping during implementation provides written evidence to show whether the procedures outlined in the plan are being followed and whether requirements in the HACCP plan are being met. HACCP records facilitate verification and validation because record-keeping provides the written information for review. Timely and careful review of HACCP records is essential to identify potential problems and allow corrective actions to be taken before the finished product leaves the establishment or a public health problem occurs (Weddig and Stevenson, 2006). HACCP records provide written evidence that can be verified (e.g., through records review) to ascertain that the product has been produced under conditions that ensure safety. If questions arise after the product enters commerce, the records may be the most crucial means, if not the only mechanism, to determine whether the product has been produced under conditions that ensure safety according to the establishment's HACCP plan (Weddig and Stevenson, 2006).

The NACMCF (1998) recommended that four types of records be kept as part of a HACCP record-keeping system: (i) the summary of the hazard analysis, (ii) the HACCP plan, (iii) supporting documentation (the rationale for the HACCP plan and its components), and (iv) daily operational records (i.e., monitoring, corrective action, and verification records). The HACCP plan should include a summary table (Table 3) and a list of the HACCP team and their responsibilities, a description of the product, its distribution, intended use and consumer, and a flow diagram of the manufacturing process that has been verified. HACCP records should be kept for a specified period of time, usually 1 or 2 years, depending on the product.

The strength of HACCP as a preventive food safety management system lies in all of its seven principles, which form the framework of a system that is designed to not only control food safety hazards associated with a product and process but also to document the safety of the product being produced.

Table 3. HACCP plan summary—frozen breaded fish sticks (example)[a]

CCP	Hazard(s) to be addressed in HACCP plan	Critical limits for each control measure	Monitoring				Corrective action	Verification activities	Record-keeping procedures
			What	How	Frequency	Who			
CCP1(P): cut saw	Metal fragments	No broken or missing metal parts from saw blade	Presence of broken or missing metal parts from saw blade	Visually check saw blade for broken or missing parts	Prior to start-up; end of operations; after saw blade malfunction	Saw operator	Stop production; adjust or modify equipment to reduce risk of recurrence; hold product from last acceptable check; run product through calibrated operable metal detector	QA[b] to inspect saw blade once per personnel shift; QA supervisor or designated employee to review monitoring and corrective action records daily; review of metal detector calibration records	Saw blade inspection log; corrective action log; QA verification log; metal detector calibration log
CCP2(B): batter	*Staphylococcus aureus* toxin formation	Hydrated batter temp should not exceed 50°F for more than 12 hours and should not exceed 70°F for more than 3 hours[c]	Temp of hydrated batter (exposure time will be monitored by frequency of checks)	Manually check temp in hold tank with digital indicating thermometer	Approx every hour	Batter operator	Dump batter and clean batter storage tank if temp is over 50°F for more than 12 hours or over 70°F for more than 3 hours; make repairs to and/or adjust batter refrigeration equipment; hold product involved since last good check to evaluate the total time/temp exposure	QA personnel verify batter temp once per personnel shift; QA supervisor or designated employee to review monitoring and corrective action records daily; calibrate digital indicating thermometer daily	Batter/breading inspection log; corrective action log; thermometer calibration log; QA verification log
CCP3(P): batter/bread	Metal fragments	No broken or missing metal parts from mesh conveyor belt	Presence of broken or missing metal parts from conveyor belt	Visually check conveyor belt for broken or missing parts	Prior to start-up; end of operations; after conveyor belt malfunction	Batter/bread operator	Stop production; adjust or modify equipment to reduce risk of recurrence; hold product from last acceptable check; run product through operable metal detector	QA to inspect conveyor belt once per personnel shift; QA supervisor or designated employee to review monitoring and corrective action records daily; review of metal detector calibration records	Batter/breading inspection log; corrective action log; QA verification log

[a] Scott and Stevenson (2006).
[b] QA, quality assurance.
[c] The SOP requires operator to cool batter if temperature exceeds 50°F and determine exposure time should batter temperature exceed 70°F.

ROLE OF PREREQUISITE PROGRAMS IN FOOD SAFETY MANAGEMENT

For HACCP to be effective, it must be integrated with prerequisite programs that form the foundation on which HACCP plans are based and provide ongoing support for the HACCP system (CAC, 2003; NACMCF, 1998). The specific prerequisite programs will depend on the operation. Broadly speaking, these prerequisite programs include GAPs, GHPs, and GMPs. While there are some who make distinctions between GHPs and GMPs, in the United States GHPs are part of GMPs. In the United States, as in many other countries, many of the prerequisite programs are part of existing regulations for sanitary control of operations.

In addition to those related to GMP or sanitation regulations, prerequisite programs can include other programs, such as ingredient specifications, consumer complaint management, glass control programs, allergen management programs, microbiological monitoring of the plant environment, ingredient-to-product traceability programs, and supplier approval programs (Bernard et al., 2006b). Without these programs in place and performing effectively, HACCP may be ineffective in ensuring the production of safe foods. Some prerequisite programs are more important than others, depending on the operation. For example, environmental monitoring can play a critical role in processing operations such as those producing ready-to-eat foods exposed to the environment, whereby the products may be contaminated with *L. monocytogenes*, or in operations producing low-moisture products (roasted nuts, peanut butter, chocolate, etc.) that may be contaminated with *Salmonella* after a process designed to kill the pathogen of concern.

One of the biggest challenges for the food safety team is the determination of what is managed within a prerequisite program and what is included in a HACCP program (Bernard et al., 2006b; Bernard and Scott, 2007). The basic differences between issues addressed through prerequisite programs and those covered in a HACCP plan include the following (Sperber et al., 1998):

- Prerequisite programs usually deal only indirectly with food safety issues, while HACCP plans deal solely and directly with food safety issues.
- Prerequisite programs are more general and may be applicable throughout the plant, crossing multiple product lines, while HACCP plans are more specific.
- Failures to meet a prerequisite program requirement seldom result in a food safety concern, while deviations from a HACCP plan critical

limit typically indicate the potential for a food safety hazard to exist.

Because many prerequisite programs are plant-wide, the adherence to (or failure to adhere to) the program often cannot be associated with a specific lot of product, e.g., pest control or employee training (Sperber et al., 1998). Thus, these programs are usually managed outside the HACCP system, which is an appropriate approach, as long as a failure to adhere to the prerequisite program does not have a direct impact on the safety of the product (Bernard et al., 2006b; Bernard and Scott, 2007; Sperber et al., 1998).

The concept of "reasonably likely to occur" was addressed in describing HACCP principle 1, conduct a hazard analysis. In many cases it is a functioning prerequisite program that makes a potential hazard "not reasonably likely to occur." For example, a plant making chicken salad may purchase cooked chicken from another establishment. The cooked chicken may have been exposed to the environment and possible contamination with *L. monocytogenes* during packaging. However, if the supplier has a *Listeria* control program that includes verification by environmental monitoring and if audits of the facility indicate appropriate adherence to the *Listeria* control program, this may provide the purchaser with sufficient assurance that *L. monocytogenes* is not reasonably likely to occur in this product. Thus, the hazard analysis relies on several prerequisite programs—the purchaser's own supplier control program and the supplier's *Listeria* control program, as well as environmental monitoring and auditing programs—to address this potential hazard. Thus, it is essential to review existing prerequisite programs and evidence of adherence to the programs when conducting a hazard analysis. The HACCP team needs to consider the degree of control necessary to feel comfortable that the potential hazard is adequately managed through the prerequisite program (Bernard et al., 2006a).

The above example touched on two very important programs in a food safety system, supplier control programs and audits. In implementing a food safety system, there will always be a need to rely on food safety controls that are outside the control of the establishment. Managing suppliers is a key element and is a prerequisite program that can have a direct impact on the safety of a product. Knowing the supplier, the supplier's food safety programs, and how the supplier manages these programs is critical to ensuring that that element of an establishment's food safety management system is functioning to minimize the potential for hazards in ingredients and ultimately the risk to the consumer from the finished product.

Having a good auditing program for suppliers is a component in knowing the supplier.

Often plants will refer to their food safety management system as the combination of the HACCP system and the prerequisite programs. Verification elements such as audits are also part of the food safety management system. These concepts are captured in ISO 22000, *Food Safety Management Systems—Requirements for Any Organization in the Food Chain* (ISO, 2005), which specifies requirements for a food safety management system that combines the key elements of interactive communication, system management, prerequisite programs, and HACCP principles.

PRACTICAL CONSIDERATIONS IN HACCP VERIFICATION AND VALIDATION

Once a HACCP plan is developed and implemented, verification plays an important role in sustaining and maintaining the HACCP program. Validation is defined by the NACMCF as the element of verification that focuses on collecting and evaluating scientific and technical information to determine if the HACCP plan, when properly implemented, will effectively control food safety hazards (NACMCF, 1998). An initial validation is recommended when the HACCP plan is first implemented to ensure that the plan is adequate to control food safety hazards and to verify that the plan can be implemented as written, i.e., the HACCP plan has been adequately designed (Scott et al., 2006). Verification and validation are, in addition to the hazard analysis, the more complex aspects of HACCP to understand and apply. As the industry gains more experience in HACCP application, more and more attention is being given to verification and validation in order to achieve an effective as well as efficient HACCP system.

Practical considerations in HACCP verification, i.e., in determining compliance with the HACCP plan, include the development of standard operating procedures (SOPs) for HACCP activities (e.g., monitoring activities), training, and management review and oversight. A well-developed SOP will outline the objective for an activity, stepwise procedures, frequency of conducting the activity, measurable acceptance criteria, actions to be taken when the criteria are not met, records to be kept, who (job position) is responsible for carrying out the task, who is responsible for verification, and the approval of the SOP by the HACCP team or the plant management. Included in the development of SOPs is the design of task assignments and record-keeping forms that will streamline the performance of the task (Stevenson and Barach, 2006). SOPs can be used in training for

operation personnel and can serve as a reference for verification of performance. Deficiencies uncovered during ongoing CCP verification and regularly scheduled audits of the HACCP system should be corrected in a timely fashion through, for example, retraining or reviewing/revising SOPs to facilitate implementation. Periodic reviews of performance (e.g., records review and on-site observations) are also essential to ensure that follow-up activities for corrective actions are performed and that root causes for deviations are determined and eliminated where feasible. It is well recognized that the success of a HACCP program depends on management's support and commitment to verification and continuous improvement of the program. Members of the HACCP team responsible for verification should have received appropriate training in HACCP principles and application of the principles in their products and processes.

The industry approach to HACCP validation involves primarily validating that parameters of critical limits are effective in controlling the identified food safety hazards (Scott, 2005) and that the plant operation is capable of delivering the target critical limits during production (Scott et al., 2006). Broadly speaking, however, validation encompasses other elements of the HACCP plan, e.g., the justification in the hazard analysis for why a hazard is or is not reasonably likely to occur, the justification for a monitoring frequency that is adequate to detect deviations, the determination of product disposition when taking corrective action in response to a deviation, etc. Sources of scientific and technical information useful for validation include publications in scientific journals, historical knowledge, regulations and regulatory guidance documents, survey data, and predictive microbiology models and databases (Scott, 2005). In the absence of data and information (e.g., for a new hazard or newly recognized hazard), inoculated pack/microbiological challenge studies or other scientific experiments may be necessary for validation, e.g., to identify critical limits capable of controlling the identified hazards.

Generally, data and information obtained from multiple sources are combined and applied to validate elements in the HACCP plan. For example, in 2001 FDA started to require juice processors to comply with the juice HACCP regulation, along with a performance standard for a 5-log reduction of microbial pathogens. The adequacy of the 5-log reduction performance standard was established by FDA, which relied on recommendations of the NACMCF (1997). NACMCF determined that a 5-log reduction of pathogens for juice would result in a "tolerable level of risk." In supporting the validity of a 5-log reduction in protecting public health, the NACMCF took into consideration, among other factors, a probability of yearly illness of less than 10^{-5} based on the consumption of a serving of juice daily (NAC-MCF, 1997). The mean log concentration of pathogens found in citrus juice, based on the NACMCF's analysis of enumeration data from approximately 15,000 survey samples submitted by the state of Florida, was also considered (R. Buchanan, FDA, personal communication). On the other hand, there were no data available in the scientific literature for juice processors to use to set critical limits for heat treatments that are adequate to inactivate the pathogens. The National Food Processors Association conducted a study on the thermal inactivation of three pathogens (*E. coli* O157:H7, *Salmonella*, and *L. monocytogenes*) in three juices (apple, orange, and white grape). The study generated D and z values for the individual pathogen/juice combinations and established that a general process of 71.1°C (160°F) for 3 seconds is adequate to deliver a 5-log reduction of the pathogens in single-strength juices with pH 4.0 or below (Mazzotta, 2001). For a juice processor producing a single-strength juice (e.g., pasteurized orange juice), validation of the heat process would include citing the published study to justify the heating time and temperature used to inactivate the pathogens of concern and in-plant data to confirm that the pasteurization system used in the facility will deliver the required time and temperature to the juice.

Predictive microbiology models and databases are a scientific resource that has been used and will be used to a greater extent in the future as a HACCP validation and support tool. Well-known modeling software tools include the USDA pathogen modeling program (http://ars.usda.gov/Services/docs.htm?docid=6788); ComBase and its modeling toolbox, ComBase Predictor and the Perfringens Predictor (http://www.combase.cc); and the American Meat Institute Process Lethality Determination Spreadsheet (AMI, 2007). In September, 2007, USDA FSIS unveiled a new Internet resource, the Predictive Microbiology Information Portal (http://www.ars.usda.gov/naa/errc/mfsru/portal). The portal is one of the most comprehensive resources available for food processors to use in making science-based decisions on controlling microbial hazards. Among many benefits the portal can offer, predictive models, data, and information accessible at the portal may be valuable for the development and validation of HACCP plans.

CHALLENGES IN APPLICATION OF HACCP

It is often claimed that HACCP can be applied "farm to fork," suggesting that HACCP can be applied at all levels of the food chain from food production (farming, fishing, etc.) to consumer use. In practice, hazards at many levels of the food chain are managed by "HACCP-type" systems in which some but not all seven HACCP principles are applied. At the farm level, we can conduct a hazard analysis and determine some controls, but we may not be able to fully determine appropriate CCPs that will reduce the hazards to acceptable levels at this point, in part because we do not know all the points at which the hazards are introduced into the system nor what acceptable levels are for certain products. And even when a CCP can be determined, e.g., irrigation, the establishment of critical limits at CCPs may be difficult, and validation that these critical limits are adequate to control the hazards may be impossible.

For example, at the farm level we can do a hazard analysis and determine that enteric pathogens need to be addressed in the production of animals or plants for food. However, the control measures for lettuce, for example, are not as clear-cut as in a processing operation for which there is a pathogen kill step. When the source of the hazard is not clearly identified (Contaminated water? Fecal contamination from wild animals? Runoff from nearby cattle pastures?), the CCPs may be difficult to determine and the controls usually only reduce the potential for contamination rather than eliminate the pathogen. When the controls are limited to GAPs, GHPs, GAPs, etc. to minimize the potential for contamination with pathogens, validation may not be possible, or even necessary; e.g., how does one establish a valid distance between a lettuce field and a cattle pasture? How does one control the hazards associated with the presence of wild birds in apple orchards? Ultimately this must rely on expert consensus on "best practices" that evolve as we learn more about where enteric pathogens enter the production chain. Verification that these practices are adhered to becomes more important than validation, at least until there is information that a control is no longer adequate (valid), with respect to controlling a hazard. For example, we currently do not have sufficient information to establish procedures for testing irrigation water that can be validated as adequate to ensure water does not serve as a source of enteric pathogens. If procedures for ensuring the safety of irrigation water involve testing water for specific organisms with a certain frequency, and contamination of produce is traced to contaminated irrigation water that was not detected by the testing procedures used, the procedures would no longer be considered valid. New procedures would need to be established.

Another operation for which the application of HACCP is a challenge is animal slaughter. Again, enteric pathogens are the concern. To date we do not have preslaughter or slaughter interventions that will completely eliminate enteric pathogens from carcasses (Scott and Elliott, 2006). Moreover, while the importance of reducing pathogens on raw animal carcasses cannot be understated, it is not clear what level of prevalence or how may pathogens are appropriate for protection of public health, especially in a raw product that will be cooked prior to consumption. This makes the application of all elements of HACCP difficult, especially the setting of valid critical limits.

A third area that presents challenges with respect to HACCP is with small businesses (small- and medium-size enterprises or small and/or less-developed businesses) (FAO/WHO, 2007). The challenges include lack of resources (time, labor, and financial) and lack of technical expertise. Similarly, restaurants and catering businesses present challenges in the application of the HACCP principles, with the numerous food items and changing recipes. In these operations the concept of "active managerial control" focusing on control of food-borne illness risk factors using a "process approach" has been described (FDA, 2006). The process approach divides food preparation in an establishment into broad categories based on operational steps in the flow of food through an establishment (e.g., receiving, storing, preparing, cooking, cooling, reheating, holding, assembling, packaging, serving, etc.), analyzing the hazards and placing managerial controls on each grouping. Active managerial control of risk factors can be achieved either by designating certain operational steps CCPs or by implementing prerequisite programs (FDA, 2006). Record-keeping may be much less formal than with processing establishments and will usually be specific to processes rather than to products.

FOOD SAFETY METRICS—COMPONENTS OF A FOOD SAFETY SYSTEM THAT LINK TO PUBLIC HEALTH GOALS

Codex Alimentarius has articulated a number of general principles for managing microbiological risks, including that control measures in a food safety system should be science and risk based; the stringency of food safety control systems should be commensurate with the risk to public health; and the performance of a food safety control system should be verifiable (CAC, 2007). Even before these principles were articulated, they have often been an underlying basis for traditional control measures, which included product criteria (e.g., pH <4.6 to control *C. botulinum*), process criteria (e.g., the times and temperatures for milk pasteurization), and microbiological

criteria (e.g., absence of detectable enteric pathogens in ready-to-eat products). While these metrics have been, and continue to be, useful, they are not directly linked to protection of public health but have been developed based on what is practical and achievable. With the development of quantitative microbiological risk assessment techniques, it has become possible to predict the effects of specific control measures and criteria on public health. This has led to the development of new food safety metrics for managing microbiological risks: FSOs, POs, and PCs. These are defined in Table 4. The Codex Committee on Food Hygiene has developed guidance on the use of microbiological risk management metrics, both traditional and emerging risk analysis-based metrics, to more objectively and transparently relate the level of stringency of control measures or entire food safety control systems to the required level of public health protection (CAC, 2007).

The establishment of FSOs is the responsibility of national governments because they are linked to the "appropriate level of protection" (ALOP), which is a clear function of the government. An FSO serves as a means of expressing the ALOP, as well as serving as the "performance target" for the establishment of other metrics, such as the PO. The PO will differ from the FSO based on whether or not the hazard can increase (or decrease) between the point at which the PO is applied and the time of consumption. If the frequency and/or concentration of the hazard is not

Table 4. Food safety metrics[a]

ALOP[b]	Level of protection deemed appropriate to protect human, animal, or plant life
FSO	The maximum frequency and/or concentration of a hazard in a food at the time of consumption that provides or contributes to the ALOP
PO	The maximum frequency and/or concentration of a hazard in a food at a specified step in the chain before the time of consumption that provides or contributes to an FSO or ALOP, as applicable
PC	The effect in frequency and/or concentration of a hazard in a food that must be achieved by the application of one or more control measures to provide or contribute to a PO or an FSO
PdC	A chemical or physical characteristic of a food (e.g., pH or water activity) that, if met, contributes to food safety
PcC	The conditions of treatment that a food must undergo at a specific step in its manufacture to achieve a desired level of control of a microbiological hazard
MC	A criterion defining the acceptability of a product or a food lot, based on the absence or presence, or number of microorganisms, including parasites, and/or quantity of their toxins/metabolites, per unit(s) of mass, volume, area, or lot

[a]CAC (1997, 2007, 2008a).
[b]Anonymous (1994). PdC, product criterion; PcC, process criterion; MC, microbiological criterion.

likely to change, then the PO and the FSO can be the same. A quantitative microbiological risk assessment can be used to determine the relationship between the PO and the FSO (CAC, 2007). POs may be set by governments but may also be established by industry when the government has set an FSO.

PCs are usually applied to microbiocidal or microbiostatic processes to define the amount of inactivation or limitations in growth. As with POs, PCs may also be set by government or by industry; when established by law or regulation, they are called performance standards. For example, recent U.S. regulations have set a requirement for a 5-log reduction of the pathogen of concern in juice (FDA, 2001a); a 7-log reduction of *Salmonella* in poultry products (FSIS, 1999a); a 6.5-log reduction for *Salmonella* in certain beef products, such as roast beef (FSIS, 1999a); and a requirement for an increase of less than 1 log in *C. perfringens* during cooling of certain meat products (FSIS, 1999a). In some cases PCs have been set by precedent many years ago, e.g., the 12-log reduction of *C. botulinum* in low-acid canned foods. However, the number of foods for which PCs have been set are few, so in many cases industry establishes criteria to ensure foods are safe and unadulterated.

PCs can easily be translated into product or process criteria by industry or by government. For example, FSIS has established time and temperature combinations (FSIS, 1999b) to meet either a 6.5-log or 7-log reduction of *Salmonella* in cooked beef, roast beef, and cooked corned beef, which comply with the required performance standard. Product criteria such as pH or water activity can be established to limit the growth of pathogens in many food products. These are familiar metrics to any food microbiologist or food technologist.

FSOs, POs, and PCs can also lead to the establishment of microbiological criteria, which are not a new concept, but the linking of them to public health outcomes is. For example, the Codex Committee on Food Hygiene has developed microbiological criteria for *L. monocytogenes,* based on risk assessments that show that low numbers present little risk, even to the most susceptible population (CAC, 2008b).

In integrating a farm-to-fork food safety system, there will be a variety of metrics at points along the food chain (CAC, 2007). These metrics should relate to each other such that a metric established at one point in the food chain can be related to the outcome at another point, and ultimately these metrics will relate to a public health outcome (CAC, 2007). Because of the interrelationship of FSOs, POs, PCs, and microbiological criteria, establishing one can be used to specify the others and ultimately can be used to infer the ALOP of the country establishing the metrics.

The application of these concepts is growing as many countries begin establishing transparent, science- and risk-based food safety systems.

REFERENCES

AMI. 2007. Process lethality determination spreadsheet. American Meat Institute Foundation, Washington, DC. http://www.amif.org/ht/d/sp/i/1870/pid/1870.

Anonymous. 1994. Agreement on the application of sanitary and phytosanitary measures. World Trade Organization, Geneva, Switzerland. http://www.wto.org/english/docs_e/legal_e/15sps_01_e.htm.

Anonymous. 2007. Keeping up dairy's grade A food safety initiatives: an interview with Allen R. Sayler, International Dairy Foods Association. *Food Saf. Mag.* July:48–55.

Bai, L., C.-L. Ma, Y.-S. Yang, S.-K. Zhao, and S.-L. Gong. 2007. Implementation of HACCP system in China: a survey of food enterprises involved. *Food Control* 18:1108–1112.

Bauman, H. 1971. *Proceedings of the National Conference on Food Protection.* Department of Health, Education, and Welfare, Public Health Service, Washington, DC.

Bernard, D. 1998. Developing and implementing HACCP in the USA. *Food Control* 9:91–95.

Bernard, D. T., and V. N. Scott. 2007. Hazard analysis and critical control point system: use in controlling microbiological hazards, p. 971–985. *In* M. P. Doyle and L. R. Beuchat (ed.), *Food Microbiology: Fundamentals and Frontiers,* 3rd ed. ASM Press, Washington, DC.

Bernard, D. T., K. E. Stevenson, and V. N. Scott. 2006a. Hazard analysis, p. 57–68. *In* V. N. Scott and K. E. Stevenson (ed.), *HACCP: a Systematic Approach to Food Safety,* 4th ed. Food Products Association, Washington, DC.

Bernard, D. T., N. G. Parkinson, and Y. Chen. 2006b. Prerequisites to HACCP, p. 5–12. *In* V. N. Scott and K. E. Stevenson (ed.), *HACCP: a Systematic Approach to Food Safety,* 4th ed. Food Products Association, Washington, DC.

CAC. 1997. *Principles for the Establishment and Application of Microbiological Criteria for Foods.* CAC/GL 21-1997. Codex Alimentarius Commission, World Health Organization-Food and Agriculture Organization of the United Nations, Rome, Italy.

CAC. 2003. *Recommended International Code of Practice General Principles of Food Hygiene.* CAC/RCP 1-1969, Revision 4 (2003). Codex Alimentarius Commission, World Health Organization-Food and Agriculture Organization of the United Nations, Rome, Italy.

CAC. 2007. Principles and guidelines for the conduct of microbiological risk management. Annex II. Guidance on microbiological risk management metrics. *In Report of the 39th Session of the Codex Committee on Food Hygiene.* ALINORM 08/31/13 (appendix IV). Codex Alimentarius Commission, World Health Organization-Food and Agriculture Organization of the United Nations, Rome, Italy.

CAC. 2008a. *Procedural Manual,* 18th ed. Codex Alimentarius Commission, World Health Organization-Food and Agriculture Organization of the United Nations, Rome, Italy.

CAC. 2008b. Microbiological criteria for *Listeria monocytogenes* in ready-to-eat foods. *In Report of the 40th Session of the Codex Committee on Food Hygiene.* ALINORM 09/32/13 (appendix II). Codex Alimentarius Commission, World Health Organization-Food and Agriculture Organization of the United Nations, Rome, Italy.

CDJ. 2007. *Meat Inspection Regulations, 1990 (SOR/90-288),* amended 2004. Canada Department of Justice, Ontario, Canada. http://laws.justice.gc.ca/en/ShowTdm/cr/SOR-90-288///en.

CFIA. 2007. *Food Safety Enhancement Program Manual.* Canadian Food Inspection Agency, Ontario, Canada. http://www.inspection.gc.ca/english/fssa/polstrat/haccp/manue/tablee.shtml.

European Union. 2004. Corrigendum to Regulation (EC) No 852/2004 of the European Parliament and of the Council of 29 April 2004 on the hygiene of foodstuffs (OJ L 139, 30.4.2004. (Corrected version in OJ L 226, 25.6.2004.) *Off. J. Eur. Union* L 226:3–21.

FAO/WHO. 2006. *Food Safety Risk Analysis: a Guide for National Food Safety Authorities.* Food and Agriculture Organization/World Health Organization, Rome, Italy. ftp://ftp.fao.org/docrep/fao/009/a0822e/a0822e00.pdf.

FAO/WHO. 2007. *FAO/WHO Guidance to Governments on the Application of HACCP in Small and/or Less-Developed Food Businesses—FAO Food and Nutrition Paper 86.* Food and Agriculture Organization/World Health Organization, Rome, Italy. ftp://ftp.fao.org/docrep/fao/009/a0799e/a0799e00.pdf.

FDA. 1995. Procedures for the safe and sanitary processing and importing of fish and fishery products, final rule. *Fed. Regist.* 60:65096–65202.

FDA. 2001a. Hazard analysis and critical control point (HAACP) [sic]: procedures for the safe and sanitary processing and importing of juice, final rule. *Fed. Regist.* 66:6138–6202.

FDA. 2001b. *Fish and Fishery Products Hazards and Controls Guidance,* 3rd ed. Food and Drug Administration, Washington, DC. http://www.fda.gov/Food/GuidanceComplianceRegulatoryInformation/GuidanceDocumentsSeafood/FishandFisheriesProductsHazardsandControlsGuide/default.htm.

FDA. 2003. *National Conference on Interstate Milk Shipments (NCIMS): Evaluation of the NCIMS HACCP Pilot Program Phase II Expansion. Dairy Grade A Voluntary HACCP.* Food and Drug Administration, Washington, DC. http://www.fda.gov/Food/FoodSafety/HazardAnalysisCriticalControlPointsHACCP/DairyGradeAVoluntaryHACCP/ucm115433.htm.

FDA. 2004. *Guidance for the Industry: Juice HACCP Hazards and Controls Guidance,* 1st ed. Food and Drug Administration, Washington, DC. http://www.fda.gov/Food/GuidanceComplianceRegulatoryInformation/GuidanceDocuments/Juice/ucm072557.htm.

FDA. 2006. *Managing Food Safety: a Manual for the Voluntary Use of HACCP Principles for Operators of Food Service and Retail Establishments.* Food and Drug Administration, Washington, DC. http://www.fda.gov/Food/FoodSafety/RetailFoodProtection/ManagingFoodSafetyHACCPPrinciples/Operators/default.htm.

FSIS. 1996. Pathogen reduction: hazard analysis and critical control point (HACCP) systems, final rule. *Fed. Regist.* 61:38806–38989.

FSIS. 1999–2006. *Guidebook for the Preparation of HACCP Plans and Generic HACCP Models.* USDA Food Safety and Inspection Service, Washington, DC. http://www.fsis.usda.gov/Science/Generic_HACCP_Models/index.asp.

FSIS. 1999a. Performance standards for the production of certain meat and poultry products. *Fed. Regist.* 64:732–749.

FSIS. 1999b. Appendix A: compliance guidelines for meeting lethality performance standards for certain meat and poultry products. USDA Food Safety and Inspection Service, Washington, DC. http://www.fsis.usda.gov/Frame/FrameRedirect.asp?main=http://www.fsis.usda.gov/OPPDE/rdad/FRPubs/95-033F/95-033F_Appendix_A.htm.

FSIS. 2005. *Meat and Poultry Hazards and Controls Guide.* USDA Food Safety and Inspection Service, Washington, DC. http://www.fsis.usda.gov/OPPDE/rdad/FSISDirectives/5100.2/Meat_and_Poultry_Hazards_Controls_Guide_10042005.pdf.

Hontz, L. R., and V. N. Scott. 2006. HACCP and the regulatory agencies, p. 113–120. *In* V. N. Scott and K. E. Stevenson (ed.), *HACCP: a Systematic Approach to Food Safety,* 4th ed. Food Products Association, Washington, DC. http://www.codexalimentarius.net/download/report/686/al31_13e.pdf.

ISO. 2005. *Food Safety Management Systems—Requirements for Any Organization in the Food Chain,* ISO 22000:2005, 1st ed. International Organization for Standardization, Geneva, Switzerland. http://www.iso.org/iso/iso_catalogue/catalogue_tc/catalogue_detail.htm?csnumber=35466.

Mazzotta, A. S. 2001. Thermal inactivation of stationary-phase and acid-adapted *Escherichia coli* O157:H7, *Salmonella,* and *Listeria monocytogenes* in fruit juices. *J. Food Prot.* 64:315–320.

NACMCF. 1997. Recommendations on fresh juice. National Advisory Committee on Microbiological Criteria for Foods, FSIS, USDA, Washington, DC. http://www.cfsan.fda.gov/~mow/nacmcf.html.

NACMCF. 1998. Hazard analysis and critical control point principles and applications guidelines. *J. Food Prot.* 61:762–775.

NAS. 1985. *An Evaluation of the Role of Microbiological Criteria for Foods and Food Ingredients.* National Academy Press, Washington, DC.

Scott, V. N. 2005. How does industry validate elements of HACCP plans? *Food Control* 16:497–503.

Scott, V. N., and P. H. Elliott. 2006. Influence of food processing practices and technologies on consumer-pathogen interactions. *In* M. Potter (ed.), *Food Consumption and Disease Risk.* CRC Press, Boca Raton, FL.

Scott, V. N., and K. E. Stevenson (ed.). 2006. *HACCP: a Systematic Approach to Food Safety,* 4th ed. Food Products Association, Washington, DC.

Scott, V. N., K. E. Stevenson, and D. E. Gombas. 2006. Verification procedures, p. 91–98. *In* V. N. Scott and K. E. Stevenson (ed.), *HACCP: a Systematic Approach to Food Safety,* 4th ed. Food Products Association, Washington, DC.

Sperber, W. H., K. E. Stevenson, D. T. Bernard, K. E. Deibel, L. J. Moberg, L. R. Hontz, and V. N. Scott. 1998. The role of prerequisite programs in managing a HACCP system. *Dairy Food Environ. Sanit.* 18:418–423.

Stevenson, K. E. 2006. Introduction to hazard analysis and critical control point systems, p. 1–4. *In* V. N. Scott and K. E. Stevenson (ed.), *HACCP: a Systematic Approach to Food Safety,* 4th ed. Food Products Association, Washington, DC.

Stevenson, K. E., and J. T. Barach. 2006. Organizing and managing HACCP programs. p. 107–112. *In* V. N. Scott and K. E. Stevenson (ed.), *HACCP: a Systematic Approach to Food Safety,* 4th ed. Food Products Association, Washington, DC.

Weddig, L. M., and K. E. Stevenson. 2006. Record-keeping, p. 99–106. *In* V. N. Scott and K. E. Stevenson (ed.), *HACCP: a Systematic Approach to Food Safety,* 4th ed. Food Products Association, Washington, DC.

INDEX

Acetic acid
  hazard control in meat, 382–383
  hazard control in plant products, 387
Acidic electrolyzed water, 421, 424
Acinetobacter lwoffi, 250
Ackee
  characteristics, 301–302
  food processing that influences stability, 302
  indications of poisoning, 302
  occurrence, 301–302
  plant sources, 302
  preventive education, 302
  treatment of poisoning, 302
Acremonium, mycotoxins, 275–277
Acrylamide residues, 319–320, 322
Active packaging, 399–400
Acute cyanide poisoning, 303
Adaptive evolution, 463
Adenovirus, enteric, 218–220
Aerolysins, 182
Aeromonas
  aerolysins, 182
  antimicrobial resistance, 182
  biofilms, 182–183
  characteristics of organism, 181–182
  clinical disease, 181–182
  control in animal products, 384
  control in plant products, 363, 365
  control in RTE foods, 415
  control in seafood, 372
  detection and identification, 183
  in environment, 182
  food processing operations that influence, 183
  food processing operations to guard against, 183
  outbreaks, 182–183
  rapid detection methods, 457
  sources and incidence in foods, 182
  survival and growth in foods, 182–183
  temperature tolerance, 182
  trace-back programs, 183
  virulence factors, 182
Aeromonas caviae, 181
Aeromonas hydrophila, 181, 250, 384, 457
Aeromonas jandaei, 181
Aeromonas salmonicida, 181
Aeromonas schubertii, 181
Aeromonas trota, 181
Aeromonas veronii, 181

Affinity probe(s), 451
Affinity probe-based biosensors, 452–453
Aflatoxins, 276–278, 358
  food safety objectives, 441
AFLP (amplified fragment length polymorphism) typing, 465–466
Agaricus bisporus, 283
Agmatine, 248–249, 264–265
Agricultural water, 368
Aichi virus, 219–220, 226
Airborne pathogens, preharvest control, 359–360
Akakabi byo, 278
Albendazole residues, 336
Aleukia, 278
Alexandrium, 234–235
Allergen management program, 448, 487
Allergy, gluten testing, 286–300
ALOP, *see* "Appropriate level of protection"
Alpha-toxin
  C. perfringens, 54, 65
  S. aureus, 453
Amanita phalloides, 281
Amanita virosa, 281
Amatoxins, 281
American Meat Institute Process Lethality Determination Spreadsheet, 489
Amines, biogenic, *see* Biogenic amine(s)
Aminoglycoside residues, 335
Amnesic shellfish poisoning, 233, 235–236
Amoxicillin residues, 334–335
Amphenicol residues, 335
Amphoteric products, hazard control in RTE foods, 420
Ampicillin residues, 335
Amprolium residues, 336
Amygdalin, 304
Amyotrophic lateral sclerosis/parkinsonism-dementia complex, 310
Angiostrongylus cantonensis, 208–210
  life cycle, 210
Animal feed, 323–324
  additives, 326–342
  preharvest hazard control, 361
Animal products, *see also* Poultry
  hazard control during harvesting, 379–386
    antimicrobials, 383–384
    dehairing, 380–381
    hide wash, 384
    irradiation, 384

knife trimming, 381
   organic acids, 382–383
   steam pasteurization, 382–383
   steam vacuuming, 382–383
   water wash, 381–382
pathogens in, 358
preharvest hazard control, 358–363
   airborne pathogens, 359–360
   antibiotic administration, 362
   antimicrobial proteins, 363
   bacteriophage therapy, 362–363
   dietary changes/feed supplements, 361
   direct-fed microbials/competitive exclusion, 361–362
   fomite-borne pathogens, 360
   metabolic inhibitors, 363
   minimizing initial contamination, 359–360
   reduction/inhibition of contamination, 360–363
   soilborne pathogens, 360
   vaccination, 362
   waterborne pathogens, 360
prion diseases, 343–356
Anisakis simplex, 207–208, 372
   life cycle, 207
Anisatin, 310–311
Antabuse syndrome, 282
Anthelmintic drug residues, 335, 339
Anthrax, 13
Antibody-based detection methods, 451
Anticoccidial residues, 329, 332, 335–336
Antimicrobial(s)
   hazard control during food processing, 397–398
   hazard control in meat, 383–384
   hazard control in plant products, 387–388
   hazard control in poultry, 385–386
   hazard control in seafood, 389–390
   inhibition of *E. coli*, 81
   preharvest hazard control in animals, 362
   Salmonella control during poultry processing, 114–115
Antimicrobial packaging, 399–400, 426
Antimicrobial residues, 316–320, 322, 327, 329, 331–335
Antimicrobial resistance, 327, 332–333, 362
   Aeromonas, 182
   C. coli, 333
   C. jejuni, 333
   E. coli, 333
   L. monocytogenes, 333
   S. aureus, 322
   Salmonella, 111, 116, 333
   Shigella, 136, 142
   Y. enterocolitica, 333
Antinutrients, in plants, 301–303
Antiparasitic drug residues, 317–320, 336
Antithyroid agent residues, 328
"Appropriate level of protection" (ALOP), 437, 439, 490–491
Aprinocid residues, 336
Aptamers, 451–452
Aquaculture, 372–374
Aquatic craft, 373
Arcobacter

characteristics of organism, 183–184
clinical disease, 183–184
detection and identification, 184, 186
in environment, 184–185
food processing operations that influence, 185
food processing operations to guard against, 185–186
outbreaks, 185
sources and incidence in foods, 184–185
survival and growth in foods, 185
trace-back programs, 186
Arcobacter butzleri, 183–185, 384
Arcobacter cibarius, 183
Arcobacter cryaerophilus, 183–185
Arcobacter halophilus, 183–184
Arcobacter nitrofigilis, 183–184
Arcobacter skirrowii, 183–185
Arcobacter sulfidicus, 184
Ardacin residues, 327
Armillaria mellea, 282
Arsenic residues, 316–320
Aspergillus, mycotoxins, 275–280
Aspergillus carbonarius, 279
Aspergillus clavatus, 280
Aspergillus flavus, 277–278
Aspergillus niger, 453
Aspergillus ochraceus, 279
Aspergillus parasiticus, 277
Aspergillus terreus, 280
Astrovirus, 219–220, 439
Avian influenza virus H5N1, 218, 221–222, 226
Avidin, 397
Avoparcin residues, 327
Azaspiracids, 237

Bacillary dysentery, 131–145
Bacillus, 1–19
   biogenic amine production, 250
   taxonomy, 1
Bacillus anthracis, 1, 6, 13
   control in animal products, 359
   molecular subtyping, 469
   rapid detection methods, 453
Bacillus cereus
   BcET toxin, 4, 10
   characteristics of organism, 2
   clinical disease, 2–5
   control in animal products, 358, 380
   control in plant products, 363, 365, 380
   control in RTE foods, 425
   control in seafood, 380
   CytK protein, 4, 8, 10
   detection and identification
      antibodies against spores and vegetative cells, 10
      cultural methods, 9
      molecular-based assays, 9–10
      noncultural methods, 9–11
   diarrheal syndrome, 3–6
   emetic syndrome, 3, 5–7
   EntFM toxin, 4
   food safety objectives, 441

HBL toxin, 3–5, 8, 10–12
heat tolerance, 8–9
level of risk, 439
molecular subtyping, 469
Nhe toxin, 3–4, 8, 10–11
psychrotrophic strains, 7–9
sources and incidence in food, 6–8
spores, 2, 7, 13
    germination and outgrowth, 2, 8–9
    heat resistance, 8–9
survival and growth in foods, 7–9
survival of pasteurization, 445
toxin detection, 11–12
treatment and prevention, 13
typing, 11
vegetative growth, 2
Bacillus circulans, 12
Bacillus firmus, 12
Bacillus fusiformis, 12
Bacillus licheniformis, 1, 12
Bacillus macerans, 251
Bacillus megaterium, 12
Bacillus mojavensis, 12
Bacillus mycoides, 1, 11
Bacillus pumilus, 1, 12
Bacillus simplex, 12
Bacillus sphaericus, 13
Bacillus subtilis, 1, 12, 298
Bacillus thuringiensis, 1, 6, 11–13
Bacillus weihenstephanensis, 1–2, 8, 12
Bacitracin residues, 327
Bacteriocins, 156
    control of *L. monocytogenes*, 101
    hazard control during food processing, 397, 400
    hazard control in RTE foods, 426
    preharvest hazard control in animals, 363
Bacteriophage assay, *E. coli*, 85
Bacteriophage therapy
    control of *L. monocytogenes*, 101
    hazard control during food processing, 400–402
    preharvest hazard control in animals, 362–363
Balansia, mycotoxins, 276–277
Barbecue spice aquaresin, 185
BcET toxin, *B. cereus*, 4, 10
Benzimidazole residues, 336, 339
Benzoic acid/benzoate, hazard control during food
    processing, 397, 400
Beta-agonist residues, 316–320, 328–329, 331, 337
Beta-lactam residues, 334–335
Betamethasone residues, 331, 337
Beta-*N*-methylamino-ʟ-alanine (ʟ-BMAA), 309
Beta-*N*-oxaloylamino-ʟ-alanine (ʟ-BOAA),
    309–310
Beta-toxin, *C. perfringens*, 54, 65
Bifidobacterium, probiotics, 362
Biguanides, hazard control in RTE foods, 420
Biofilms
    Aeromonas, 182–183
    E. coli, 76
    food contact surfaces, 421

Biogenic amine(s), 248–274
    analytical methods, 265–267
    classification, 249
    degradation, 254
    derivatization, 266–267
    food spoilage indicators, 264–265
    formation, 248–254
        availability of precursor amino acids, 249–250
        cadaverine, 251–252
        histamine, 250–251
        microorganisms responsible for, 250–252
        packaging atmosphere and, 253
        pH effect, 253
        preservative effects, 253–254
        putrescine, 261–262
        salt effect, 253
        temperature effect, 252–253
        tyramine, 251
        water activity effect, 253
    occurrence in foods, 254–259
    precursors of carcinogens, 263–264
    structure, 248–249
    toxicity, 254–263
        cadaverine, 263
        histamine, 260–262
        putrescine, 263
        spermidine, 263
        spermine, 263
        tyramine, 262–263
Biogenic amine index, 264–265
Biological hazards, food service, 436–437, 482
Biopreservation, 426
Biosecurity, "on-farm" *Salmonella* control in
    poultry, 113
Biosensors
    affinity probe-based, 452–453
    B. cereus, 10
    E. coli, 85
    Shigella, 142
    Vibrio, 158
    whole cell, 453
Bioterrorism
    B. anthracis, 13
    botulinum neurotoxins, 33–34
Black pepper, 25
Bleach, hazard control in RTE foods, 421
Boletus calopus, 282
Boletus luridus, 282
Boticin, 31
"Botulinum cook," 40
Botulinum neurotoxins, 31–52
    bioterrorism, 33–34
    characteristics, 32
    detection and quantification, 43–45
        endopeptidase assays, 43–45
        immunochemical methods, 43–44
    therapeutic and cosmetic use, 34
Botulism, 31–52, *see also Clostridium botulinum*
    clinical disease, 32–34
    food-borne

epidemiology, 34–38, 46
    outbreaks associated with nonproteolytic
        C. botulinum, 36–38
    outbreaks associated with proteolytic C. botulinum,
        35–36
    infant, 33
    wound, 33
Bovine amyloidotic spongiform encephalopathy, 349–350
Bovine spongiform encephalopathy, 343, 346–352
    atypical, 349–350
    origin, 349
    prevention, 350–352
    in species other than cattle, 349
    worldwide distribution, 348–349
Brevetoxins, 233, 236, 240–241
Brevibacterium linens, 254
Brochothrix thermosphacta, 383, 401
Brucella
    control in animal products, 359, 380
    level of risk, 438
Byssochlamys fulva, 280
Byssochlamys nivea, 280

Cadaverine, 248–249, 251–252, 263–265
Caffeic acid, 185
CaliciNet, 227
Calicivirus
    feline, 421, 424
    swine enteric, 222
Campylobacter
    acid tolerance, 23
    clinical disease, 20–21
    control in animal products, 359, 380
    control in plant products, 363, 380
    control in poultry, 386
    control in seafood, 372, 380
    detection and identification, 25–26
    in environment, 21–23
    food processing interventions to guard against, 24–25
    food processing operations that influence, 23–24
    level of risk, 438
    rapid detection methods, 454, 456
    salt tolerance, 23, 25
    sources and incidence in foods, 21–23
    survival and growth in foods, 23
    taxonomy, 20
    temperature tolerance, 20–21, 23–25
    trace-back programs, 25–26
    typing, 25–26
    viable, but nonculturable state, 23
Campylobacter coli, 21–23, 26, 333, 456, 462
Campylobacter jejuni
    antimicrobial resistance, 333
    Arcobacter versus, 183–185
    characteristics of organism, 20–21
    clinical disease, 20–21
    control during food processing, 401
    control in animal products, 358, 384
    control in RTE foods, 416
    inactivation by pasteurization, 444

molecular subtyping, 467–469
    rapid detection methods, 452, 456–457
    toxins, 21
Campylobacter lari, 21
Campylobacter upsaliensis, 21
Campylobacteriosis, 20–21
CANARY (cellular analysis and notification of antigen
        risks and yields), 453–454
Candida, biogenic amine production, 251
Cantilever sensors, 452–453
Capillaria philippinensis, 208, 372
    eggs, 209
    life cycle, 209
Capillary electrophoresis, biogenic amines, 266, 268
Carbadox residues, 316–317, 319–320, 327
Carbamate residues, 323, 329
Carcass decontamination, 382–384
Carcinogens, from biogenic amines, 263–264
Carnobacterium, biogenic amine production,
        250–251
Carnobacterium divergens, 383–384
Carprofen residues, 336
Carvacrol, 59, 185, 398
CCP, see Critical control point
Celiac disease
    clinical manifestations, 287
    gluten testing, 286–300
    treatment, 287
Cell-based detection systems, 453–454
Cephalosporin residues, 334–335
Cereals, gluten testing, 286–300
Cereulide, 5–8, 12
Cestodes, 200–203
    control in seafood, 372
    eggs, 198
Cetylpyridinium chloride
    hazard control in meat, 383
    hazard control in poultry, 114, 385–386
Cheese reaction, 263
Chemical hazards, food service, 436–437, 482
Chemical residues, see also specific substances
    in animals, 316–320, 326–342
    definition, 314
    European Union, 326–342
    in milk, 318–319, 321
    sampling, 315–316, 324
    survey data, 316–319
    United States, 314–325
    in water, 320–322
Chitosan, 59, 397
Chloramphenicol residues, 316–317, 326, 330–332
Chlorinated hydrocarbon residues, 316–320
Chlorine dioxide
    control of Salmonella, 114
    hazard control in plant products, 387–388
    hazard control in poultry, 385
    hazard control in seafood, 389
Chlorine-containing compounds
    carcass decontamination, 383
    hazard control in plant products, 387

hazard control in RTE foods, 420–421, 424
hazard control in seafood, 389
Chlorine-sodium hypochlorite, control of *Salmonella*, 114
Chlortetracycline residues, 334
Choking hazard, 436–437, 441
Cholera, 146–148, 150–151
Cholera toxin, 147–148, 151
Chronic wasting disease of deer and elk, 343, 346
  diagnostics, 353
  history, 352
  prevention, 352–353
  spread to species other than deer and elk, 352
  transmission, 352
Ciguatera, 236, 238
Ciguatoxins, 236, 238, 241, 380
Cilantro oil, 398
Cimaterol residues, 316–317
Cinnamaldehyde, 59, 185
Cinnamon, 185, 398
Citrobacter, biogenic amine production, 250
Claviceps purpurea, 276–277
Clenbuterol residues, 316–317, 326, 331
Clitocybe, 282
Clonorchis sinensis, 195, 372
  clinical disease, 197
  eggs, 198
  life cycle, 196–197
Clopidol residues, 336
Clostridium argentinense, 31–32
Clostridium baratii, 31–34
Clostridium botulinum, 31–52, *see also* Botulism
  acid resistance, 39–42
  characteristics of organism, 31–32
  clinical disease, 32–34
  control during food processing, 403
  control in animal products, 358
  control in plant products, 363, 365
  control in RTE foods, 414, 425
  control in seafood, 372, 380
  detection and identification, 42–46
    botulinum neurotoxins, 43–45
    molecular methods, 45–46
  in environment, 38–39
  food safety objectives, 441
  heat resistance, 39–42
  level of risk, 439
  nonproteolytic, 31–32, 46
    culture, 42–43
    in environment, 38
    food processing operations to guard against, 40–42
    outbreaks, 36–38
  proteolytic, 31–33
    culture, 42–43
    in environment, 38
    food processing operations to guard against, 39–40
    outbreaks, 35–36
  rapid detection methods, 451
  salt tolerance, 39–42
  sources and incidence in foods, 39
  spores, 38–41, 403

survival of pasteurization, 445
temperature tolerance, 39–42
typing, 45–46
Clostridium butyricum, 31–34
Clostridium perfringens
  characteristics of organism, 53–54
  clinical disease, 53–54
  control in animal products, 358, 380
  control in plant products, 365, 380
  cpe-positive, 55
  detection and identification in food, 62–66
  enterotoxin, 53–54, 63–64
  in environment, 54–55
  food processing operations that influence, 57–58
  food processing operations to guard against, 58–66
  food safety objectives, 441
  heat resistance, 55–56, 59–60
  histamine production, 250
  level of risk, 439
  molecular subtyping, 469
  pH tolerance, 56
  regulatory requirements, 62
  salt tolerance, 56–57, 60
  sources and incidence in foods, 54–55
  spores, 53
    germination and outgrowth, 54, 57–61
  survival and growth in foods, 55–56
  survival of pasteurization, 445
  temperature tolerance, 55–56
  toxins, 52–53, 63–65
  under aerobic conditions, 56
  water activity requirement, 56
Cloves, 25, 185, 398
Clupeotoxin, 237
Coccidiostats, 362
Codex Alimentarius Commission, 481
Cold holding, 448
Cold shock response, 171
ComBase, 489
Competitive enhancement strategy, 361
Competitive exclusion culture, preventing *Salmonella* in
    poultry, 113
Conocybe cyanopus, 282
Conozole/triazole residues, 323
Consumer complaint management, 487
Consumer risk communication, 448
Controlled-atmosphere packaging, 399
Convicine, 306–307
Coprine, 282
Coprinus atramentarius, 282
Corrective action, for deviations at CCPs, 483–486
Corticosteroid residues, 331, 337
Cortinarius orellanoides, 282
Cortinarius orellanus, 282
Cortinarius speciosissimus, 282
Cortinarius splendens, 282
Coumaphos residues, 336
Coxsackievirus, 226
Creutzfeldt-Jakob disease, 343, 349–350
  variant, 343, 349

"Crisping," 417
Critical control point (CCP), 480, *see also* HACCP system
    determination, 482–483, 489
        monitoring, 484
        verification, 485
Critical limits, establishment, 483–485
Crohn's disease, 188–189
Cryptosporidium
    control in animal products, 359
    control in plant products, 363–366
    food safety objectives, 441
    level of risk, 438
    rapid detection methods, 454
Crystal violet residues, 332
Culture
    B. cereus, 9
    C. botulinum, 42–43
    E. coli, 82–84
    Salmonella, 115
    Shigella, 140
    Vibrio, 157
    Yersinia, 173–174
Cutting boards, 416, 420, 447
Cyanogenesis, 303
Cyanogenic glycosides, 301, 303–306
    characteristics, 303–304
    detection of cyanide, 306
    food processing that influences stability, 305–306
    occurrence, 303–304
    plant sources, 304–305
Cyclospora cayetanensis
    control in animal products, 359
    control in plant products, 364, 380
    level of risk, 438
CytK protein, *B. cereus*, 4, 8, 10
Cytotoxins, *see specific toxins*

Debaryomyces, biogenic amine production, 251
Decontamination fluid, inactivation of *E. coli*, 80
Dehairing, 380–381
Deletions, 462–463
Delicatessen products, 411–434
Delta-toxin, *C. perfringens*, 54
Dendrogram, 464
Deoxynivalenol, 276, 278–279
Depuration, shellfish, 154, 224, 389
Desmycosin residues, 335
Dexamethasone residues, 331–332, 337
Dhurrin, 304
Diacetate, hazard control during food processing, 400
Diacetoxyscirpenol, 278–279
Diacetylnivalenol, 278–279
Diarrhetic shellfish poisoning, 233, 236–237, 241
Diatoms, shellfish toxins, 233
Dichlorvos residues, 336
Diclazuril residues, 335–336
Dimethyl dicarbonate, hazard control during food processing, 397
Dimetridazole residues, 335
Dinitrocarbanilide residues, 335

Dinoflagellates, seafood toxins, 233
Dinophysis, 237
Dinophysistoxins, 237
Diphyllobothrium ditremum, 372
Diphyllobothrium hians, 372
Diphyllobothrium latum, 202–203, 372
    eggs and strobila, 198, 204
    holdfast, 204
    life cycle, 203
    proglottid and scolex, 204
Diphyllobothrium pacificum, 372
Diphyllobothrium yonagoense, 372
Disinfectants
    for aquaculture, 374
    control of viruses, 225
    hazard control in RTE foods, 420–421
Display case, retail, 420
Disulfiram-like toxins, 282
DIVINE-Net, 227
DNA band-based molecular subtyping, 464–468
DNA fingerprint, 464
DNA microarray, 455–457
    C. perfringens, 65–66
    molecular subtyping, 471
    Vibrio, 158
Documentation, HACCP system, 485
Domoic acid, 233–236, 238–239, 241

Echinostoma, 195
    clinical disease, 197
    life cycle, 196–197
Edible coatings/films, 400
Eggs, pathogens in, 358
ELISA, gluten testing, 288–293, 295–297
Emamectin residues, 336
Emesis, staphylococcal enterotoxin-induced, 123–124
Endotoxins, *see specific toxins*
Enrofloxacin residues, 334
Enterobacter aerogenes, 250
Enterobacter cloacae, 250, 253
Enterobacter sakazakii, 401
Enterocins, 403
Enterococcus, biogenic amine production, 250
Enterococcus faecalis, 251
Enterococcus faecium, 251
Enterotoxins
    C. perfringens, 53–54, 63–64
    staphylococcal, *see* Staphylococcal enterotoxin(s)
Enterotropic viruses, 218–220
Enterovirus, 218–219, 221, 224
EntFM toxin, *B. cereus*, 4
Entoloma rhodopolium, 282
Environmental hazard control, 445
Environmental monitoring, 487
Enzymatic methods, biogenic amine analysis, 266
Epsilon-toxin, *C. perfringens*, 54, 64–65
Equipment hazard control, 445–447
Ergot alkaloids, 276–277
Ergotism, 277
Erythromycin residues, 334

Escherichia coli O157, *see Escherichia coli,*
    diarrheagenic
Escherichia coli
    antimicrobial resistance, 333
    biofilms, 76
    biogenic amine production, 250
    characteristics of organism, 71–72
    classification, 75
    clinical disease, 72
    cold sensitivity, 82
    colonization of host cells, 73–74
    control during food processing, 401, 404
    control in animal products, 358–359, 380–385
    control in plant products, 363–366, 380, 388
    control in RTE foods, 414–417, 424–425
    control in seafood, 372, 390
    detection and identification, 82–85, 140–141
        phenotypic indicators, 84
        rapid methods, 83, 450, 453, 456–457
        recent advances, 84–85
    diarrheagenic, 71–94
    diffuse adherent (DAEC), 72–73, 75, 83
    enrichment protocols, 82–84
    enteroaggregative (EAEC), 72–74, 82–83
    enterohemorrhagic (EHEC), 72–74, 83–85, 461
    enteroinvasive (EIEC), 72–74, 82–84, 132, 141
    enteropathogenic (EPEC), 72–73, 82–83, 461
    enterotoxigenic (ETEC), 72–73, 78–79, 82–84
    in environment, 75
    food processing operations that influence, 76–79
    food processing operations to guard against
        inactivation, 80–81
        inhibition, 81–82
        minimizing contamination, 79–80
        safer supply chains, 79
    food safety objectives, 441
    genetic change, 463
    heat-labile toxin (LT), 74
    heat-stable toxin (ST), 74
    hemolysins, 74
    inactivation by pasteurization, 444
    in juice, 489
    level of risk, 438
    molecular subtyping, 464–471
    outbreaks and food recalls, 76–79
    phenotypic indicators, 75
    relatedness to *Shigella*, 132
    serine protease inhibitors, 74
    Shiga-toxin producing (STEC), 74–75, 77–78, 80,
        83–85, 462, 466
    sources and incidence in foods, 75–76
        meat products, 77, 80
        produce, 78–80
        unpasteurized dairy products, 77, 81
        unpasteurized juices, 77–78, 81
    surveillance, 85–86
    toxins, 74
    trace-back programs, 85–86
    vaccine, 362
    virulence factors, 72–75

Essential oils, plant, hazard control during food process-
    ing, 397–398, 400
Estradiol residues, 331
Eta-toxin, *C. perfringens*, 54
Ethopabate residues, 336
Eugenol, 185, 398
European Union, safety perspective on chemical residues
    in food, 326–342

Facility hazard control, 445
Farm vehicles, 369
Fasciola gigantica, 195
    clinical disease, 197
    life cycle, 196–197
Fasciola hepatica, 195
    clinical disease, 197
    eggs, 198
    life cycle, 196–197, 199
Fasciolopsis buski, 195
    clinical disease, 197
    eggs, 198
    life cycle, 196–197
Fatal familial insomnia, 343
Favic agents, 301, 306–307
    characteristics, 306–307
    clinical disease, 307
    control, 307
    food processing that influences stability, 307
    food sources, 307
    occurrence, 306–307
Favism, 306–307
Feed additives, 326–342
Feline spongiform encephalopathy, 343
Fenbendazole residues, 336
Fertilizer, 367, 371
Fingernail sanitation, 422–423, 444
Florfenicol residues, 316–317, 334
Flukes, 195–200, *see also* Trematodes
Flukicide residues, 336
Flunixin residues, 336
Fluoroquinolone residues, 334–335
Fomites, preharvest control of pathogens, 360
Food additives, control of *Vibrio* in seafood, 156
Food allergen, 448
Food allergy, gluten testing, 286–300
Food Code, 412, 422, 436
Food contact surfaces, 416, 419–421, 445, 447–448
Food handlers, 415, 417, 421–423
Food preservation, approaches, 396–397
Food processing
    control of seafood toxins, 238
    effect on plant toxins
        ackee, 302
        cyanogenic glycosides, 305–306
        favic agents, 307
        grayanotoxins, 308
        lathyrogens, 310
    gluten testing of processed foods, 294–295, 297
    hazard control during, 396–410
        active packaging, 399–400

antimicrobials, 397–398
bacteriophage, 400–402
high-pressure processing, 402–403
irradiation, 402
manothermosonication, 405
microbial inactivation, 396, 402–405
microbial inhibition, 396–402
plant essential oils, 397–398
pulsed electric field process, 403–404
pulsed-light system, 404
radio frequency electric fields processing, 404
radio frequency heating, 404
UV irradiation, 405
operations that influence pathogen content
    Aeromonas, 183
    Arcobacter, 185
    C. perfringens, 57–58
    Campylobacter, 23–24
    E. coli, 76–79
    H. pylori, 187
    L. monocytogenes, 97–100
    P. shigelloides, 190
    Salmonella, 111–112
    Shigella, 138–139
    Streptococcus, 191
    viruses, 224–226
    Yersinia, 172
operations to guard against pathogens
    Aeromonas, 183
    Arcobacter, 185–186
    C. botulinum, 39–42
    C. perfringens, 58–66
    Campylobacter, 24–25
    E. coli, 79–82
    H. pylori, 187
    L. monocytogenes, 97–100
    Mycobacterium, 188
    P. shigelloides, 190
    Salmonella, 112–115
    Shigella, 139–142
    Streptococcus, 191
    Vibrio, 154–156
    viruses, 226–229
preservation processes, 396–410
Food safety management system, 478–492
developing intervention strategies, 442–445
elements, 478–479
farm-to-fork retail process, 439–441
flow chart, 443
food process control groups, 442–443
HACCP system, *see* HACCP system
information needed to develop, 479–480
risk analysis capabilities, 479
Food safety metrics, 479, 490–491
Food safety objectives (FSO), 437, 439, 441, 479, 490–491
Food safety programs, 426–428
Food service
biological hazards, 436–437, 482
chemical hazards, 436–437, 482

hazard control, 436–449
    chemical hazards, 436–437
    cold holding, 448
    consumer risk communication, 448
    environmental hazards, 445
    equipment hazards, 445–447
    facility hazards, 445
    heavy metals leaching into food from equipment, 447
    hot hold, take out, and cool, 448
    personal hygiene, 444–445
    physical hazards, 436–437
    preparation hazards, 447–448
    prepreparation hazards, 447
    supplies and ingredients, 447
physical hazards, 436–437, 482
raw food contamination problem, 437–439, 441
Food spoilage indicators, biogenic amines, 264–265
Food utensils, 416, 419–420
FoodNet, 86, 138, 141
Fourier transform infrared spectroscopy, *B. cereus*, 10
Frameshift mutations, 462
Francisella tularensis, 452
Fructo-oligosaccharides, 362
FSO, *see* Food safety objectives
Fumonisins, 276, 279
Fungal toxins, 275–280
Fusarenon X, 278–279
Fusarium, mycotoxins, 275–280
Fusarium culmorum, 279
Fusarium graminearum, 279
Fusarium proliferatum, 279
Fusarium verticillioides, 279

Galerina autumnalis, 281
Gambierdiscus toxicus, 236
Gamma-toxin, *C. perfringens*, 54
GAP, *see* Good agricultural practices
Garlic, 25
Gas chromatography, biogenic amines, 266
Genetic change
    horizontal gene transfer, 461–463
    mutations, 461–462
    natural selection, 463–464
Genome sequencing, 470–471
Gerstmann-Straussler-Scheinker syndrome, 343
GHP, *see* Good hygiene practices
Giardia lamblia
    control in animal products, 359
    control in plant products, 363, 365–366
    level of risk, 438
Glass control program, 487
Glove use, by food handlers, 423
Glucose-6-phosphate dehydrogenase deficiency, 306–307
Glutaraldehyde
    control of viruses, 225
    hazard control in RTE foods, 421
Glutelins, 287
Gluten
    classification of gluten proteins, 287–288
    healthy food or toxic, 287–288

properties, 287–288
Gluten testing, 286–300
   antibody cross-reactivity to wheat, barley, and rye,
      295–296
   antibody specificity to different wheat protein fractions,
      291–293
   gliadin control standard in ELISA, 296–297
   processed foods, 294–295, 297
   sample extraction procedures, 295–296
   testing gluten-spiked food matrices
      blank samples, 288–290
      ELISA, 288–293, 295–297
      lateral flow devices, 288–291
      spiked samples, 290–291
Gluten-free diet, 287
Glycosides, cyanogenic, *see* Cyanogenic glycosides
GMP, *see* Good manufacturing practices
Gnathostoma, 210–211
   life cycle, 211
Gnathostoma hispidum, 211
Gnathostoma spinigerum, 209–211, 372
Goitrogens, 304
Gonyaulax catenella, 234
Good agricultural practices (GAP), 359, 367, 371, 480,
   487
Good hygiene practices (GHP), 422, 478, 487
Good manufacturing practices (GMP), 480, 487
Good retail practices, 419
Grayanotoxins, 301, 308–309
   characteristics, 308
   clinical manifestations, 308
   detection in honey, 309
   food processing that influences stability, 308
   occurrence, 308
   prevention and control of poisoning, 309
   treatment of poisoning, 308–309
Green tea catechins, 59
Growth-promoting agent residues, *see also* Veterinary
      drug residues
   European Union, 326–342
Guillain-Barré syndrome, 21
Gymnodimines, 237
Gymnodinium catenatum, 234
Gyromitrin, 281–282

HACCP pathogen reduction plan
   Salmonella in poultry, 114
   Shigella, 138–139
HACCP (hazard analysis, critical control point) system,
   439, 478–492
   application in food processing, 481
   CCP determination, 482–483
   challenges in application, 489–490
   corrective actions, 483–486
   determining significant hazards and control risk,
      481–486
   documentation, 485
   establishing critical limits, 483–485
   farm level application, 489
   food process control groups, 441–443

food process HACCP intervention, 447–448
   hazard analysis, 482–483, 486–487
   monitoring procedures, 483–486
   origin, 480–481
   prerequisite programs, 444–447, 487–488
   principles, 481–486
   in retail environments, 427–428
   record keeping, 485–486
   retail food HACCP operations intervention plan, 444,
      446
   for slaughter operations, 489–490
   for small businesses, 490
   standard operating procedures, 488
   validation, 488–489
   verification procedures, 485–489
   worldwide application, 481
Hafnia alvei, 250
Halofuginone residues, 335–336
Haloxon residues, 336
Hand sanitizers, 226, 422–423
Hand washing, 226, 418, 422–423, 444–445
Harvesting, hazard control during
   animal products, 379–386
   plant products, 380, 386–388
   poultry, 380, 384–386
   seafood, 388–390
Harzianum A, 279
HAV, *see* Hepatitis A virus
Hazard analysis
   biological hazards, 482
   chemical hazards, 482
   HACCP system, 482–483, 486–487
   hazard evaluation, 482–483
   hazard identification, 482–483
   physical hazards, 482
Hazard control
   farm-to-fork retail process, 439–441
   during food processing, 396–410
   food service operation, 436–449
   during harvesting, 379–395
   preharvest, 357–378
   ready-to-eat foods, 411–434
HBL toxin, *B. cereus*, 3–5, 8, 10–12
Heat shock response, 58–59, 96
Heat treatment
   control of *Vibrio* in seafood, 155
   control of viruses, 224–226
   hazard control in seafood, 389
Heat-labile toxin (LT), *E. coli*, 74
Heat-stable toxin (ST), *E. coli*, 74
Heavy metals, leaching into food from equipment, 447
Helicobacter pullorum, 187
Helicobacter pylori
   characteristics of organism, 186
   clinical disease, 186
   detection and identification, 187
   in environment, 186–187
   food processing operations that influence, 187
   food processing operations to guard against, 187
   outbreaks, 187

rapid detection methods, 457
sources and incidence in foods, 186–187
survival and growth in foods, 187
trace-back programs, 187
Hemolysin, *E. coli*, 74
Hemolytic uremic syndrome, 72, 74–75, 77–78, 133
Hemorrhagic colitis, 72, 74–75
Hepatitis A virus (HAV), 218–220, 222–227, 229, 380
control in plant products, 363
control in RTE foods, 412–413, 416–418, 421–422
control in seafood, 389
food safety objectives, 441
level of risk, 439
rapid detection methods, 455
vaccines, 227
Hepatitis E virus (HEV), 218–219, 221–222
Hepatotropic viruses, 220–221
Herbs, 398
Heterophyes continua, 372
Heterophyes heterophyes, 195, 372
clinical disease, 197
eggs, 198
life cycle, 196–197
HEV, *see* Hepatitis E virus
Hide contamination, 79–80
Hide wash, 384
High-pressure liquid chromatography
biogenic amines, 266–268
seafood toxin assay, 239
High-pressure processing
control of *L. monocytogenes*, 101
control of prions, 347
control of *Vibrio* in seafood, 155–156
control of viruses, 224, 226–227
hazard control during food processing, 402–403
hazard control in seafood, 389
Histamine, 248–253, 260–262, 264–266
food safety objectives, 441
Histamine fish poisoning, 260–262
HMMN1 residues, 335
Hops beta acids, 398
Horizontal gene transfer, 461–463
Hormone residues, 316–320, 329, 331, 337
Hot holding, 448
Hot water wash, poultry, 385
HT-2 toxin, 278–279
Hurdle technology, 396, 403–404
Hydrazines, 282–283
Hydrogen cyanide, 303
5-Hydroxy thiabendazole residues, 336
Hydroxybenzoic acid esters, hazard control during food processing, 397
Hypoglycin A, 301–302

Ice, contaminated, 417–420
Imported food, 315, 332, 437
Indels, 462
Infant botulism, 33
Ingredients, hazard control, 447
Ingredient-to-product traceability program, 487

Inocybe, 282
Insertions, 462–463
International Commission on Microbiological Specifications for Foods (ICMSF), 437, 439
International trade, 437
Inversions, 462
Iodophors, hazard control in RTE foods, 420
Iota-toxin, *C. perfringens*, 54, 65
Irradiation
control of *E. coli*, 81
control of *L. monocytogenes*, 101–102
control of prions, 347
control of *Salmonella*, 114
control of *Shigella*, 139–142
control of *Vibrio* in seafood, 156
hazard control during food processing, 402
hazard control in meat, 384
hazard control in plant products, 388
hazard control in poultry, 386
hazard control in seafood, 390
Isoeugenol, control of *L. monocytogenes*, 103

Jamaican vomiting sickness, 301–302

Kappa-toxin, *C. perfringens*, 54
Karenia brevis, 236
Klebsiella
biogenic amine degradation, 254
biogenic amine production, 250
Klebsiella oxytoca, 250
Klebsiella pneumoniae, 250, 253
Knife trimming, 381
Kocuria, biogenic amine production, 250–251
Konzo, 304
Kunitz soybean inhibitor, 302
Kuru, 343–344

Label-dating, hazard control in RTE foods, 426
Lactate-diacetate, 399
Lactic acid
hazard control during food processing, 397, 399–400
hazard control in meat, 382–383
hazard control in plant products, 387
hazard control in RTE foods, 424
Lactobacillus
biogenic amine degradation, 254
biogenic amine production, 250–251
probiotics, 362
Lactobacillus acidophilus, 250–251, 362
Lactobacillus alimentarius, 251
Lactobacillus arabinose, 251
Lactobacillus bavaricus, 251
Lactobacillus bulgaricus, 250–251, 253
Lactobacillus carnis, 251
Lactobacillus casei, 251
Lactobacillus curvatus, 251
Lactobacillus delbreuckii, 253
Lactobacillus divergens, 251
Lactobacillus farciminis, 251
Lactobacillus helveticus, 251

Lactobacillus homohiochii, 251
Lactobacillus plantarum, 251
Lactobacillus reuteri, 251
Lactobacillus sakei, 250–251
Lactococcus, biogenic amine production, 250–251
Lactoferrin, 384, 397
Lactoperoxidase system, 397
Laidlomycin residues, 336
Lambda-toxin, *C. perfringens*, 54
Lasalocid residues, 335–336
Lateral flow devices, gluten testing, 288–291
Lathyrism, 309–310
Lathyrogens, 301, 309–310
    characteristics, 309
    clinical manifestations, 309
    food processing that influences stability, 310
    occurrence, 309–310
    prevention and control of poisoning, 310
Lauric arginate, control of *L. monocytogenes*, 103
Leucomalachite green residues, 331
Leuconostoc, biogenic amine production, 250
Leuconostoc gelidum, 401
Leuconostoc mesenteroides, 404
Levamisole residues, 332, 336
Lichenysin, 12
Linamarin, 304
Lipophilic toxins, in seafood, 237
Liquid chromatography-mass spectroscopy, detection of
        veterinary drugs, 334–337
Liquid smoke, control of *L. monocytogenes*, 103
Listeria, rapid detection methods, 456
Listeria innocua, 382, 404, 415, 453, 461, 470
Listeria ivanovii, 388, 401, 461
Listeria monocytogenes
    acid tolerance, 97
    antimicrobial resistance, 333
    characteristics of organism, 95
    clinical disease, 95–96
    control during food processing, 398–401, 403
    control in animal products, 358–359, 380–381, 384
    control in plant products, 363, 365, 387–388
    control in RTE foods, 413–416, 418, 425–426
    control in seafood, 372, 390
    detection and identification, 103–104
    in environment, 96
    food processing operations that influence, 97–100
    food processing operations to guard against, 97–100
        biological interventions, 101
        chemical interventions, 102–103
        cold chain management, 100
        design of equipment and facilities, 99
        environmental monitoring, 100
        kill step, 98–99
        physical interventions, 101–102
        recent advances, 100–103
        sanitation, 99–100
        segregation of raw and treated areas, 98
        sourcing and storage of ingredients, 98
    inactivation by pasteurization, 444
    in juice, 489

    level of risk, 438
    molecular subtyping, 464, 466–471
    osmotic adaptation, 97
    outbreaks, 96–97
    rapid detection methods, 454
    sources and incidence in foods, 96
    survival and growth in foods, 96–97
    temperature tolerance, 96–97
    trace-back programs, 103–104
    typing, 103
    virulence, 461, 463
Listeria seeligeri, 461
Listeria welshimeri, 461
Listeriosis, 95–96
Little brown mushrooms, 282
Livestock, preharvest contamination of crop plants, 366
Locus of enterocyte effacement, 461
Lotaustralin, 304
Lysergic acid derivatives, 277
Lysozyme, 397, 400, 404

Macrocyclic lactone residues, 336, 339
Macrolide residues, 335
Maduramicin residues, 335–336
Magic mushrooms, 282
Malachite green residues, 330–332
Manosonication, 405
Manothermosonication, 405
Manure, on crop plants, 366–368, 370–371
MAO inhibitor drugs, 262–263
Matrix-ranking approach, monitoring veterinary drug
        residues, 339
Maximum residue limits, 326–328
    definition, 327
Meat, *see* Animal products
Medroxyprogesterone acetate residues, 331
Melamine, 315
Melengestrol residues, 316–317
Meloxicam residues, 336
Metabolic inhibitors, preharvest hazard control in
        animals, 363
Metagonimus yokogawai, 195
    clinical disease, 197
    eggs, 198
    life cycle, 196–197, 199
    in seafood, 372
Methylprednisolone residues, 337
Microarray, *see* DNA microarray
Micrococcus
    biogenic amine degradation, 254
    biogenic amine production, 250–251
Mietz and Karmas index, 264–265
Milk, pathogens in, 358
Miller Fisher syndrome, 21
Mint oil, 398
MLST, *see* Multilocus sequence typing
MLVA (multiple-locus variable-number tandem repeat
        analysis), 469–470
"Modes of failure" concept, 480
Modified atmosphere packaging, 82, 399

Molecular beacon, 455–456
Molecular subtyping, 460–477
    AFLP, 465–466
    applications, 460
    DNA band-based methods, 464–468
    DNA microarrays, 471
    DNA sequence-based methods, 468–472
    genome sequencing, 470–471
    ideal method, 460
    MLST, 468–469, 472
    MLVA, 469–470
    multiplex PCR, 464
    PCR-RFLP, 466
    PFGE, 464
    RAPD, 464–465
    rep-PCR, 465
    ribotyping, 467–468
    SNP typing, 471–472
Monensin residues, 334–336, 362
Monitoring procedures, HACCP system, 483–486
Monoacetoxyscirpenol, 278–279
Monosodium glutamate, 441
Morchella, 282
Morganella morganii, 250, 253–254
Multilocus sequence typing (MLST), 468–469, 472
    L. monocytogenes, 104
    Salmonella, 116
Multiplex PCR, molecular subtyping, 464
Muscarine poisoning, 282
Mushroom toxins, 280–283, 380
    disulfiram-like, 282
    gastrointestinal irritants, 282
    hydrazines, 283
    idiosyncratic reactions, 283
    neurotoxins, 282
    protoplasmic, 281–282
Mutations, 461–462
Mu-toxin, *C. perfringens*, 54
Mycetism, 280
Mycobacterium, 187–189
    characteristics of organism, 187–188
    clinical disease, 187–188
    control in animal products, 359
    control in seafood, 372
    detection and identification, 189
    in environment, 188
    food processing operations to guard against, 188
    outbreaks, 188
    rapid detection methods, 452, 457
    salt tolerance, 188
    sources and incidence in foods, 188
    survival and growth in foods, 188
    temperature tolerance, 188
    trace-back programs, 189
Mycobacterium avium, 188–189, 452, 457
Mycobacterium leprae, 188
Mycobacterium marinum, 372
Mycobacterium tuberculosis, 187, 457
Mycotoxins, 275–280
    in animal products, 329

cocontamination, 280
control in foods, 276

Nandrolone residues, 331
Nanoarray, 457
Narasin residues, 335–336
NASBA (nucleic acid sequence-based amplification) assay, 454–455
    NASBA-ECL, 454
    QNASBA, 454
    real-time, 455
Natamycin
    antimicrobial films, 400
    hazard control during food processing, 397
National Advisory Committee on Microbiological Criteria for Foods (NACMCF), 480
Natural selection, 463–464
Neisseria meningitidis, 468
Nematodes, 203–210
    control in seafood, 372
Neoanisatin, 310–311
Neomycin sulfate, 362
Neosolaniol, 278–279
Neurolathyrism, 309
Neurotoxic shellfish poisoning, 233, 236
Neurotoxins
    botulinum, *see* Botulinum neurotoxins
    mushroom, 282
Neurotropic viruses, 221
Neutral electrolyzed water, 421, 424
Nhe toxin, *B. cereus*, 3–4, 8, 10–11
Nicarbazin residues, 332, 335–336
Nipah virus, 218, 221–222
Nisin, 156, 185, 397, 399–400, 404
    hazard control in meat, 383–384
Nitrates, 441
Nitrites
    food safety objectives, 441
    hazard control during food processing, 397
    inhibition of *C. perfringens*, 57
Nitrofuran residues, 316–317, 319–320, 326, 330–332
Nitroimidazole residues, 316–317, 319–320, 329–331, 335–336
Nitrosamines, formation, 248, 263–264
Nivalenol, 278–279
Nonsense mutations, 462
Nonsynonymous mutations, 462–463
Norovirus, 218–227, 229, 380
    control in plant products, 363
    control in RTE foods, 412–413, 416–418, 421–423, 425
    food safety objectives, 441
    level of risk, 439
    rapid detection methods, 454–455
Norwalk virus, 219
Norwalk-like agent, 363
NSAID residues, 316–317, 320, 329, 332, 336
Nucleotide substitutions, 462
Nu-toxin, *C. perfringens*, 54

Ochratoxins, 276, 279–280, 358
Oenococcus oeni, 251
OFAGE (orthogonal-field alternating gel electrophoresis), 466
Okadaic acid, 233, 236–237, 239, 241
Olaquindox residues, 327
Omphalotus, 282
Opisthorchis felineus, 195, 372
   clinical disease, 197
   eggs, 198
   life cycle, 196–197
Opisthorchis viverrini, 195, 372
   clinical disease, 197
   life cycle, 196–197
Oregano, 25, 59, 398–399
Orellanine, 282
Organic acids
   control of *Arcobacter*, 185
   control of *C. perfringens*, 58–59
   control of *L. monocytogenes*, 102
   hazard control during food processing, 397–398, 400
   hazard control in meat, 382–383
   hazard control in plant products, 386–387
Organic cultivation systems, 366
Organochloride residues, 323
Organochlorine residues, 329
Organonitrogen residues, 323
Organophosphate residues, 316–320, 323, 329
Organosulfate residues, 323
Osteolathyrism, 309
Ostreopsis, 237
Ovotransferrin, 397
Oxidizing solutions, control of *L. monocytogenes*, 102
Oxytetracycline residues, 333–334
Ozone
   control of *L. monocytogenes*, 102
   control of *Salmonella*, 114
   hazard control in plant products, 388

Palytoxins, 237, 241
Paragonimus
   clinical disease, 197
   life cycle, 196–197
Paragonimus westermani, 195, 372
   clinical disease, 197
   eggs, 198
   life cycle, 196–197, 200
Paralytic shellfish poisoning, 233–235, 237–238, 240–241
Parasites, 195–217
   control in animal products, 358–359
   control in seafood, 372
Paravaccinia virus, 359
Parechovirus, 219
Pasteurization, microorganisms inactivated by, 444–445
Pathoadaptive mutations, 462
Pathogenicity island, 461
Patulin, 276, 280
PC, *see* Performance criteria
PCR-RFLP typing, 466
Pectenotoxins, 237

Pediocin, 397
Pediococcus
   biogenic amine degradation, 254
   biogenic amine production, 250–251
Pediococcus damnosus, 251
Penethamate residues, 335
Penicillin residues, 334–335
Penicillium, mycotoxins, 275–280
Penicillium verrucosum, 279
Performance criteria (PC), 479, 490–491
Performance objectives (PO), 479, 490–491
Peroxy acids, hazard control in RTE foods, 420
Peroxyacetic acid
   control of *Salmonella*, 114
   hazard control in plant products, 387
Peroxyacetic acid-hydrogen peroxide, hazard control in RTE foods, 424
Personal hygiene hazard control, 444–445
Pest control
   aquaculture, 373
   crop plants, 369
Pesticide residues, 317, 322–324, 336, 373
PFGE, *see* Pulsed-field gel electrophoresis
Phage display, 451
Phenolic compounds
   hazard control during food processing, 398
   hazard control in RTE foods, 421
Phenotypic subtyping, 460
Phenylbutazone residues, 316–320, 336
Phenylethylamine, 248–249
Phosphates, hazard control during food processing, 397
Photobacterium, biogenic amine production, 250
Physical hazards, food service, 436–437, 482
Phytoalexins, 398
Phytoplankton, seafood toxins, 233–247
Pimento leaf essential oil, 185
Pinnatoxins, 237
Plant products
   hazard control during harvesting, 380, 386–388
      antimicrobials, 387–388
      irradiation, 388
      organic acids, 386–387
      surface pasteurization, 386
   internalization of pathogens, 388
   preharvest hazard control, 363–371
      agriculture and washing water, 368, 371
      animal/wildlife contact, 366, 369
      containers, equipment, and tools, 369, 371
      contamination of raw produce, 363–366
      fertilizer usage, 367, 371
      land history and management, 367, 371
      manure on crop plants, 366–368, 370–371
      minimizing contamination, 367–369
      pest control program, 369
      reducing contamination, 369–371
      seed program, 367, 371
      soil/substrate management, 368–370
      soilborne pathogens, 364
      vehicle management, 369
      waste and pollution management, 367

water monitoring and water treatment, 370
waterborne pathogens, 365–366
worker hygiene and sanitary work practices, 368–369, 371
Plant toxins, 301–313
Plasmid pYV, 166
Platyhelminths, 195–200
Plesiomonas shigelloides
  characteristics of organism, 189
  clinical disease, 189
  control in seafood, 372
  in environment, 189
  food processing operations that influence, 190
  food processing operations to guard against, 190
  histamine production, 250
  outbreaks, 189–190
  sources and incidence in foods, 189
  survival and growth in foods, 189–190
Pleurotus pulmonarius, 282
Pneumotropic viruses, 221
PO, *see* Performance objectives
Poliovirus, 219, 221
Polychlorinated biphenyl residues, 316–320, 329
ε-Polylysine, 397
Potassium sorbate, control of amine formation, 254
Poultry, *see also* Animal products
  hazard control during harvesting, 380, 384–386
    antimicrobials, 385–386
    irradiation, 386
    water wash, hot water, and steam, 385
Prebiotics, 361
Predictive Microbiology Information Portal, 489
Predictive microbiology models, 489
Prednisolone residues, 331, 337
Pregnancy, listeriosis, 95
Preharvest hazard control, 357–378
  in animals, 358–363
  in plants, 363–371
  in seafood, 372–374
Prerequisite programs, HACCP system, 444–447, 487
Prion(s), 343–356
  definition, 343
Prion diseases, 343–356
Prion proteins
  detection in tissues, 347–348
  inactivation, 346–347
  movement within animal body, 345–346
  normal and abnormal, 344
  protein conversion, 344–345
  protein-only theory, 344
Proaerolysin, 182
Probiotics, 361–362
Process criterion, 490
Produce, *see* Plant products
Produce sanitizers, 423–424
Product criterion, 490
Progesterone residues, 331
Prolamins, 287
Prophage, 463

Propionic acid, hazard control during food processing, 397
Prorocentrolides, 237
Prorocentrum, 237
Protein X, 345
Proteinase inhibitors, 302–303
Proteinase K, 347
Protein-only theory, transmissible spongiform encephalopathy, 344
Proteome, 470
Proteus, biogenic amine production, 250–251
Proteus mirabilis, 457
Protoperidinium, 237
Protozoans, 210–214
  control in plant products, 363, 365
Prunasin, 304
Pseudoanisatin, 311
Pseudomonas
  biogenic amine production, 250–252
  control during food processing, 401
Pseudomonas putrefaciens, 250
Pseudonitzschia, 236
Psilocybe cubensis, 282
Psilocybe mexicana, 282
Psilocybin, 282
Puffer fish poisoning, 235, 238
Pulsed electric field process, hazard control during food processing, 403–404
Pulsed-field gel electrophoresis (PFGE)
  C. perfringens, 65
  E. coli, 85–86
  L. monocytogenes, 104
  molecular subtyping, 464
  Salmonella, 116
  Shigella, 141
Pulsed-light system, 404
PulseNet, 86, 116, 141, 467, 470, 472
Pumilacidins, 12
Putrescine, 248–249, 251–252, 263–266
Pyrethroid residues, 323, 329
Pyrodinium bahamense, 233–234

Quaternary ammonium compounds
  control of viruses, 225
  hazard control in RTE foods, 420–421, 424
Quinolone residues, 329

Ractopamine residues, 316–317, 319
Radiation, *see* Irradiation
Radio frequency electric fields processing, 404
Radio frequency heating, 404
RAPD (random amplified polymorphic DNA) typing, 464–465
Rapid detection methods, 450–459
  affinity probe-based biosensors, 452–453
  antibodies and fragments, 451
  aptamers, 451–452
  cell/tissue-based, 453–454
  DNA microarrays, 455–457
  immunological detection, 451

NASBA assay, 454–455
nucleic acid-based, 450, 454–457
Raw food contamination problem, 437–439, 441
Ready-to-eat (RTE) foods
    bacterial pathogens, 413–415
    contamination routes, 415–418
    hazard control, 411–434
        at-source contamination, 415, 419
        bacteria-specific treatments, 425–426
        destruction of contamination in food, 433–434
        environmental contamination, 415–416, 419–421
        food contact surfaces, 416
        food handlers, 415, 417, 421–423
        food safety programs and training, 426–428
        HACCP concepts, 427–428
        label-dating, 426
        prevention of contamination, 419
        storage times, 426
        virus-specific treatments, 424–425
    hazards and risks, 412–418
    viral pathogens, 412–413, 415
Record keeping, HACCP system, 485–486
Red mold disease, 278
Red tide, 233–234
Refrigeration
    control of *Vibrio* in seafood, 155
    hazard control in RTE foods, 425–426
Refrigerator surface, contaminated, 416
Relaying
    control of *Vibrio* in seafood, 156
    control of viruses, 224
Renal failure, delayed-onset, 282
Rep-PCR (repetitive element sequence based-PCR), 465
Resorcylic acid lactone residues, 328
Retail operations, hazard control
    food service, 436–449
    RTE foods, 411–434
Ribotyping, 467–468
    C. perfringens, 65
    L. monocytogenes, 104
    Salmonella, 116
Risk management, 439–441
Robenidine residues, 335–336
Ronidazole residues, 335
Rosemary, 398
Rotavirus, 219–220, 227, 439
    vaccine, 227
RTE foods, *see* Ready-to-eat foods

Saccharomyces cerevisiae, 404
"Safe harbors" intervention strategies, 441–442, 448
Sage, 25, 398
Sakacin, 400
Salad bars, 411–434
Salbutamol residues, 316–317
Salicylic acid residues, 332
Salinomycin residues, 333, 335–336
Salmonella
    acid tolerance, 110
    antimicrobial resistance, 111, 116, 333

biogenic amine production, 250
characteristics of organism, 108–109
clinical disease, 108–109
competitive exclusion cultures, 113
control during food processing, 398, 401
control in animal products, 358–359, 380–381,
    383–384
control in plant products, 363–366, 380, 387–388
control in poultry, 385–386
control in RTE foods, 413–414, 416–417, 424–425
control in seafood, 380, 390
culture, 115
detection and identification, 115–116
in environment, 109–110
food processing operations that influence, 111–112
food processing operations to guard against, 112–115
    antimicrobials, 114
food safety objectives, 441
heat resistance, 110–111
inactivation by pasteurization, 444
interventions to protect poultry, 113–114
in juice, 489
level of risk, 438
molecular subtyping, 465, 469–471
nomenclature, 109
outbreaks, 110–111
psychotrophic, 111
rapid detection methods, 115, 454, 456–457
salt tolerance, 110
sources and incidence in foods, 109–110
survival and growth in foods, 110–111
trace-back programs, 115–116
typing, 115–116
vaccine, 112–113, 362
Salmonella bongori, 109
Salmonella enterica, 109
Salmonellosis, 108–109
Sampling, for chemical residues, 315–316, 324
Sanger sequencing method, 470
Sanitary work practices, 368–369, 371, 373
Sanitation, 99–100
Sapovirus, 218
Sapporo virus, 219
Sarcocystis, control in animal products, 358
SARS coronavirus, 218, 221–222
Saxiphillins, 241
Saxitoxins, 233–235, 239–241
Scalding water wash, hazard control in poultry, 385
Scombroid fish poisoning, 260–262
Scombrotoxin, 380
Scrapie, 343, 346, 350
Seafood
    aquaculture, 372–374
    hazard control during harvesting, 388–390
        antimicrobials, 389–390
        cold temperatures, 389
        depuration, 154, 224, 389
        heat treatment, 389
        high-pressure processing, 389
        irradiation, 390

preharvest hazard control, 372–374
   breeding stock, 373
   containers, equipment, and tools, 373
   disinfectants, 373
   feed supply, 373
   minimizing initial contamination, 372–373
   pathogens of concern, 372
   pest control program, 373
   pesticide application, 373
   reducing contamination, 373–374
   site selection/preparation, 372–373
   vehicle management, 373
   water quality, 373
   water treatment, 373
   wildlife management, 373
   worker hygiene and sanitary practices, 373
  RTE foods, 419
  vibrios in, 146–163
  viruses in, 221
Seafood toxins, 233–247
  analyses, 239–240
  assays, 239–241
  characteristics, 234
  clinical disease, 234–237
  control in food processing, 238
  detection and identification, 239
  families of toxins, 234–237
  management, 237–239
Sedative residues, 329
Seed decontamination, 79, 367, 371
Semduramicin residues, 336
Serine proteinase inhibitors, 74, 302–303
Sesquiterpene lactones, 301, 310–311
  characteristics, 310–311
  detection, 311
  occurrence, 310–311
  plant sources, 311
Shelf life, 396
Shellfish, *see* Seafood
Shiga toxin, 74, 133
Shiga-like toxin, 133
Shigella
  acid tolerance, 136–137
  antimicrobial resistance, 136, 142
  asymptomatic carriers, 133
  biogenic amine production, 250
  characteristics of organism, 132–134
  clinical disease, 132–134
  control in animal products, 380
  control in plant products, 363–364, 380
  control in RTE foods, 417, 425
  culture, 140
  detection and identification, 140–141
  in environment, 134–136
  food processing operations that influence, 138–139
  food processing operations to guard against, 139–142
  food safety objectives, 441
  geographic distribution of species, 131–132
  level of risk, 438
  outbreaks, 134–138

  pathoadaptive mutations, 462
  rapid detection methods, 457
  relatedness to *E. coli*, 132
  salt tolerance, 137
  Shiga toxin, 133
  sources and incidence in foods, 134–135
  surveillance, 135, 141
  survival and growth in foods, 137–138
  survival and growth in laboratory, 136
  taxonomy, 132
  temperature tolerance, 136–137
  trace-back programs, 140
  typing, 141
  waterborne, 135–136
Shigella boydii, 131–145
Shigella dysenteriae, 131–145
Shigella flexneri, 131–145
Shigella sonnei, 131–145, 364
Shigellosis, 131–145
Shroomer, 280–281
Silent mutations, 462
Slicing machines, 416–417, 420
Smoke-derived flavoring/extracts, control of *L. monocytogenes*, 103
SNP (single-nucleotide polymorphism) typing, 471–472
Sodium bicarbonate, hazard control in RTE foods, 421
Sodium chlorate, preharvest hazard control in animals, 363
Sodium chlorite, acidified
  carcass decontamination, 383
  control of *Salmonella*, 114–115
  hazard control in poultry, 385
Sodium citrate, inhibition of *C. perfringens*, 58
Sodium hypochlorite
  control of *Salmonella*, 114
  control of viruses, 225
  hazard control in RTE foods, 421, 424
  reducing *E. coli* counts, 79
Sodium lactate, inhibition of *C. perfringens*, 60
Sodium nitrite, control of amine formation, 253
Sodium pyrophosphate, inhibition of *C. perfringens*, 59–60
Sodium sulfide, control of amine formation, 253
Soilborne pathogens, preharvest control, 360, 365
Sorbic acid/sorbate, hazard control during food processing, 397, 400
Species concept, 461
Spermidine, 248–249, 252, 263–265
Spermine, 248–249, 252, 263–265
Spices, 398
Spiramycin residues, 327
Spirolides, 237
Spore
  B. cereus, 2, 7, 13
    germination and outgrowth, 2, 8–9
    heat resistance, 8–9
  C. botulinum, 38–41, 403
  C. perfringens, 53
    germination and outgrowth, 54, 57–61
Standard operating procedures, HACCP activities, 488

Staphylococcal enterotoxin(s), 119
  biophysical characteristics, 122–123
  classification, 122
  detection, 126
  distribution among staphylococci, 125–126
  environmental regulation, 122–123
  induction of emesis, 123–124
  potency in vivo, 124
  rapid detection methods, 452
  structural characteristics, 123–124
Staphylococcal enterotoxin-like proteins, 122–123
Staphylococcal food poisoning, 119–130, see also
        Staphylococcus aureus
  clinical disease, 124–125
  incidence, 125
  outbreaks, 124–126
Staphylococcus
  biogenic amine production, 250–251
  level of risk, 438
Staphylococcus aureus, see also Staphylococcal food
        poisoning
  acid tolerance, 120
  antibiotic resistance, 322
  characteristics of organism, 119–122
  control during food processing, 401
  control in animal products, 358, 380, 384
  control in plant products, 363, 380, 388
  control in RTE foods, 416–417, 425
  detection and identification, 119–120
  enterotoxins, see Staphylococcal enterotoxin(s)
  in environment, 121–122
  food safety objectives, 441
  inactivation by pasteurization, 444
  molecular subtyping, 469
  osmotolerance, 120–121
  prevalence in foods, 121–122
  reservoirs, 121–122
  resistance, 120–121
  salt tolerance, 120–121
  superantigen activity, 124
  survival and growth in foods, 120–121
  taxonomy, 119–120
  temperature tolerance, 120–121
  toxin detection, 453
  water activity requirement, 120–121
Staphylococcus capitis, 126
Staphylococcus carnosus, 120, 251–252, 254
Staphylococcus chromogenes, 122
Staphylococcus cohnii, 120–121, 126
Staphylococcus delphini, 119
Staphylococcus epidermidis, 122
Staphylococcus equorum, 121, 126
Staphylococcus gallinarum, 126
Staphylococcus haemolyticus, 122, 126
Staphylococcus hominis, 120, 126
Staphylococcus hyicus, 119, 126
Staphylococcus intermedius, 119, 122, 125–126
Staphylococcus lentus, 126
Staphylococcus piscifermentans, 251–252
Staphylococcus saprophyticus, 121, 126

Staphylococcus schleiferi, 119–120, 126
Staphylococcus sciuri, 121–122
Staphylococcus simulans, 122
Staphylococcus succinus, 120
Staphylococcus warneri, 126
Staphylococcus xylosus, 121–122, 126, 251
Steam pasteurization, hazard control in meat, 382–383
Steam treatment, hazard control in poultry, 385
Steam vacuum, hazard control in meat, 382–383
Sterigmatocystin, 358
Steroid residues, 328
Stilbene residues, 328, 331
Streptococcus
  characteristics of organism, 190
  clinical disease, 190
  control in animal products, 380
  control in RTE foods, 417
  in environment, 190–191
  food processing operations that influence, 191
  food processing operations to guard against, 191
  level of risk, 438
  outbreaks, 191
  sources and incidence in foods, 190–191
  survival and growth in foods, 191
Streptococcus agalactiae, 190
Streptococcus anginosus, 190
Streptococcus avium, 190
Streptococcus bovis, 190
Streptococcus canis, 190
Streptococcus constellatus, 190
Streptococcus didelphis, 190
Streptococcus durans, 190
Streptococcus dysgalactiae, 190
Streptococcus equi, 190
Streptococcus equinus, 190
Streptococcus faecalis, 190
Streptococcus faecium, 251
Streptococcus gallolyticus, 190
Streptococcus infantarius, 190
Streptococcus iniae, 190–191
Streptococcus intermedius, 190
Streptococcus lactis, 251
Streptococcus lutetiensis, 190
Streptococcus mitis, 251
Streptococcus pasteurianus, 190
Streptococcus pneumoniae, 190
Streptococcus porcinus, 190
Streptococcus pyogenes, 190–191
Streptococcus suis, 190
Streptomycin residues, 326, 332
Sulfamethazine residues, 334
Sulfanilamide residues, 335
Sulfathiazole residues, 334
Sulfites
  food safety objective, 441
  hazard control during food processing, 397
Sulfonamide residues, 316–317, 329, 332–335
Superantigen activity, S. aureus, 124
Supplier approval program, 487
Surface pasteurization

control of *L. monocytogenes*, 102
   hazard control in plant products, 386
Synbiotics, 362
Synonymous mutations, 462–463

T-2 toxin, 278–279
Taenia, 200–202
   eggs, 198
   life cycle, 201
Taenia asiatica, 200–202
Taenia saginata, 200–202
   control in animal products, 358–359
   life cycle, 201
   proglottid and scolex, 204
Taenia solium, 200–202, 358
   life cycle, 201
   proglottid and scolex, 204
   scolex and eggs, 201
Tandem repeats, 469–470
Tannic acid, 185
Tannins, 303
Tapeworms, 200–203, *see also* Cestodes
TDH-related hemolysin, *V. parahaemolyticus*, 148
Tetracycline residues, 332–335
Tetrodotoxin, 233, 235, 238, 240
Thermal stable direct hemolysin (TDH),
      V. parahaemolyticus, 148
Theta-toxin, *C. perfringens*, 54
Thin-layer chromatography, biogenic amines, 265–266
Thiouracil residues, 316–317, 331
Thyme, 398
Thymol, 59, 185, 398–399
Thyreostat residues, 316
Tiamulin residues, 333
Tissue-based detection systems, 453–454
TLR, *see* Tolerable level of risk
Tolerable level of risk (TLR), 437–438
Tolfenamic acid residues, 336
Toltrazuril residues, 335–336
Toxin(s), *see specific toxins*
Toxin coregulated pilus, *V. cholerae*, 147, 151
Toxoplasma gondii, 195, 210–214
   clinical disease, 213–214
   control in animal products, 358–359
   detection and identification, 214
   food safety objectives, 441
   level of risk, 438
   life cycle, 211–213
Toxoplasmosis, 210–214
Trace-back programs
   Aeromonas, 183
   Arcobacter, 186
   Campylobacter, 25–26
   E. coli, 85–86
   H. pylori, 187
   L. monocytogenes, 103–104
   Mycobacterium, 189
   Salmonella, 115–116
   Shigella, 140
   Vibrio, 157–158

   Y. enterocolitica, 172–173
   Y. pseudotuberculosis, 172–173
Tranquilizer residues, 335
Transcriptome, 470
Transition mutations, 462
Transmissible mink encephalopathy, 343
Transmissible spongiform encephalopathy (TSE), 343–356
   detection of prions, 347–348
   distribution of infectivity, 346
   exclusion from food supply, 347
   history, 343–344
   organ predilection, 346
   prion inactivation, 346–347
   sources of infectivity, 346
   spontaneous, 345
   transmission, 343
      protein-only theory, 344
Transversion mutations, 462
Trematodes
   clinical disease, 197
   control in seafood, 372
   digenetic, 195–200
   eggs, 198
   geographic distribution, 197
   life cycles, 196–197
Trenbolone residues, 316–317
Triazine residues, 323
Trichinella, 203–207
Trichinella britovi, 204
Trichinella murrelli, 204–205
Trichinella nativa, 204–205
Trichinella nelsoni, 204
Trichinella papuae, 204–205
Trichinella pseudospiralis, 204–205
Trichinella spiralis, 204–205
   control in animal products, 358–359
   food safety objectives, 441
   larvae in muscle cells, 206
   level of risk, 438
   life cycle, 205
Trichinella zimbabwensis, 204–205
Trichoderma, mycotoxins, 279
Trichodermin, 279
Trichothecenes, 278–279
Triclabendazole residues, 336
Trimethoprim residues, 332
Trisodium phosphate, control of *Salmonella*, 114–115
Tropical ataxic neuropathy, 304
Trypsin proteinase inhibitors, 302–303
Tryptamine, 248–249
TSE, *see* Transmissible spongiform encephalopathy
Tylosin residues, 327, 332, 334
Typing, *see also* Molecular subtyping
   B. cereus, 11
   C. botulinum, 45–46
   Campylobacter, 25–26
   L. monocytogenes, 103
   Salmonella, 115–116
   Shigella, 141
Tyramine, 248–249, 251, 253, 262–266

Ultra-low temperature treatment, control of *Vibrio* in seafood, 155
UV radiation, hazard control during food processing, 405
USDA pathogen modeling program, 489

Vaccine
   cholera, 147–148
   E. coli, 362
   hepatitis A, 227
   rotavirus, 227
   Salmonella, 112–113, 362
Vaccinia virus, 359, 453
Vanillin, 398
Verification procedures, HACCP program, 485–489
Vesicular stomatitis virus, 359
Veterinary drug residues
   in animals, 326–342
   control in animal products
      analytical methods and performance criteria, 330–331
      European reference laboratory structure, 328
      prohibited substances, 329–330
      residue control plans, 328–329
   direct exposure, 332
   European Union, 326–342
   exposure through environmental contamination, 332–334
   Group A substances, 328–331
   Group B substances, 328–332
   maximum residue limits, 327–328
   residue detection
      LC-MS/MS, 334–337
      performance of analytical methods, 337–339
   strategies for monitoring and surveillance, 339
   United States, 317–320
Viable, but nonculturable state
   Campylobacter, 23
   Vibrio, 154
Vibrio
   acid tolerance, 154
   biogenic amine degradation, 254
   control in seafood, 380
   culture, 157
   detection and identification, 157–158
   enumeration of pathogenic strains, 157–158
   in environment, 150–153
   food processing operations to guard against, 154–156
   heat resistance, 154
   rapid detection methods, 456
   salt tolerance, 154
   in seafood, 146–163, 372
   selective differential agar media, 157
   sources and incidence in foods, 150–153
   species identification, 157
   surveillance, 158
   survival and growth in foods, 153–154
   temperature tolerance, 153–155
   trace-back programs, 157–158

   utilization of chitinous exoskeleton of plankton, 146–147, 153
   viable, but nonculturable state, 154
Vibrio alginolyticus, 146
Vibrio carchariae, 146
Vibrio cholerae
   characteristics of organism, 146–148
   clinical disease, 146–148
   control in plant products, 363
   control in seafood, 372
   in environment, 150–151
   level of risk, 438
   molecular subtyping, 471
   rapid detection methods, 457
   salt tolerance, 150
   sources and incidence in foods, 150–151
   species identification, 157
   temperature tolerance, 150
   toxin, *see* Cholera toxin
   toxin coregulated pilus, 147, 151
   vaccine, 147–148
Vibrio cincinnatiensis, 146
Vibrio damsela, 146
Vibrio fluvialis, 146
Vibrio furnissii, 146
Vibrio hollisae, 146
Vibrio metschnikovii, 146
Vibrio mimicus, 146
Vibrio parahaemolyticus, 146
   characteristics of organism, 148
   clinical disease, 147–148
   control in seafood, 372, 389–390
   in environment, 151–152
   genetic change, 463
   inactivation by pasteurization, 444
   "pandemic" strain, 151–152
   salt tolerance, 150
   sources and incidence in foods, 151–152
   species identification, 157
   TDH-related hemolysin, 148
   temperature tolerance, 150, 152
   thermal stable direct hemolysin, 148
Vibrio vulnificus, 146
   characteristics of organism, 148–150
   clinical disease, 147–150
   "clinical" genotypes, 152
   control in seafood, 372, 389–390
   encapsulation, 149–150
   in environment, 152–153
   level of risk, 438
   salt tolerance, 150
   sources and incidence in foods, 152–153
   species identification, 157
   temperature tolerance, 150
Vicine, 206–207
Virginiamycin residues, 327
Viruses, 218–232
   acid tolerance, 223–224
   at-risk foods, 221–222
   characteristics of food-borne viruses, 218–221

clinical disease, 218–221
control in RTE foods, 412–413, 415, 424–425
detection and identification, 227–229, 424
enterotropic, 218–220
food processing operations that influence, 224–226
food processing operations to guard against, 226–229
hepatotropic, 220–221
neurotropic, 221
outbreaks, 223–224
pneumotropic, 221
in prepared and RTE foods, 225–226
in produce, 221, 225
in shellfish, 221
sources and incidence in foods, 221–223
surveillance, 227
survival in foods, 223–224
temperature tolerance, 223
zoonotic transmission, 222
Vomitoxin, *see* Deoxynivalenol

Washing water, for produce, 368, 371, 417, 420, 423
Water quality, for aquaculture, 373
Water wash
    hazard control in meat, 381–382
    hazard control in poultry, 385
Waterborne pathogens, preharvest control, 360, 365–366
Weed control, 371
Wheat, gluten testing, 286–300
Wheat allergen, 286–300
Whole cell biosensors, 453
Wildlife
    preharvest contamination of crop plants, 366, 369
    preharvest contamination of seafood, 373
Worker hygiene, 368–369, 371, 373
Wound botulism, 33

Yersinia
    control in animal products, 359
    control in poultry, 386
Yersinia enterocolitica
    acid tolerance, 171
    animal reservoirs, 167–168
    antimicrobial resistance, 333
    characteristics of organism, 165–166
    clinical disease, 164–165
    cold tolerance, 171–172
    control in animal products, 358, 380
    control in plant products, 363
    control in RTE foods, 415, 425
    detection and identification, 172–173
        cultural methods, 173–174
        molecular methods, 175
        using PCR, 173

ecological and geographical distribution, 166
in environment, 168–169
food processing operations that influence, 172
heat tolerance, 172
inactivation by pasteurization, 444
level of risk, 438
outbreaks, 171–172
rapid detection methods, 457
salt tolerance, 171
sources and incidence in foods, 169–170
survival and growth in foods, 171–172
trace-back programs, 172–173
transmission to humans, 169–171
virulence factors, 166–167
yersiniabactin, 166–167
Yst toxin, 166
Yersinia pestis
    characteristics of organism, 165
    molecular subtyping, 469
    rapid detection methods, 453
Yersinia pseudotuberculosis
    acid tolerance, 171
    animal reservoirs, 167–168
    characteristics of organism, 165–166
    clinical disease, 164–165
    cold tolerance, 171–172
    detection and identification, 172–173
        cultural methods, 173–174
        molecular methods, 175
        using PCR, 173
    ecological and geographical distribution, 166
    in environment, 168–169
    food processing operations that influence, 172
    heat tolerance, 172
    outbreaks, 164–165, 171–172
    salt tolerance, 171
    sources and incidence in foods, 169–170
    surveillance, 164–165
    survival and growth in foods, 171–172
    trace-back programs, 172–173
    transmission to humans, 169–171
    virulence factors, 166–167
    YPM toxin, 167
Yersiniabactin, 166–167
Yersiniosis, 164–180
Yessotoxins, 237
Yops proteins, 166
YPM toxin, *Y. pseudotuberculosis*, 167
Yst toxin, *Y. enterocolitica*, 166

Zearalenone, 279, 331
Zeranol residues, 316–317, 328
Zoonotic transmission, viruses, 222